Year	Recipient*	Prize	Area of Research	Pages in Text
1978	Werner Arber Daniel Nathans Hamilton Smith	M & P	Restriction endonuclease technology	427
	Peter Mitchell	Chemistry	Chemiosmotic mechanism of oxidative phosphorylation	195
1976	D. Carleton Gajdusek	M & P	Prion-based diseases	67
1975	David Baltimore Renato Dulbecco Howard M. Temin	M & P	Reverse transcriptase and tumor virus activity	696
1974	Albert Claude Christian de Duve George E. Palade	M & P	Structure and function of internal components of cells	282, 284
1972	Gerald Edelman Rodney R. Porter	M & P	Immunoglobulin structure	712
	Christian B. Anfinsen	Chemistry	Relationship between primary and tertiary structure of proteins	64
1971	Earl W. Sutherland	M & P	Mechanism of hormone action and cyclic AMP	632
1970	Bernard Katz Ulf S. von Euler	M & P	Nerve impulse propagation and transmission	167
	Luis F. Leloir	Chemistry	Role of sugar nucleotides in carbohydrate synthesis	295
1969	Max Delbrück Alfred D. Hershey Salvador E. Luria	M & P	Genetic structure of viruses	22, 434
1968	H. Gobind Khorana Marshall W. Nirenberg	M & P	Genetic code	471
	Robert W. Holley		Transfer RNA structure	475
1966	Peyton Rous	M & P	Tumor viruses	695
1965	Francois Jacob Andre M. Lwoff Jacques L. Monod	M & P	Bacterial operons and messenger RNA	517, 441
1964	Dorothy C. Hodgkin	Chemistry	Structure of complex organic molecules	764
1963	John C. Eccles Alan L. Hodgkin Andrew F. Huxley	M & P	Ionic basis of nerve membrane potentials	166
1962	Francis H. C. Crick James D. Watson Maurice H. F. Wilkins	M & P	Three-dimensional structure of DNA	406
	John C. Kendrew Max F. Perutz	Chemistry	Three-dimensional structure of globular proteins	58
1961	Melvin Calvin	Chemistry	Biochemistry of CO_2 assimilation during photosynthesis	234
1960	F. MacFarlane Burnet Peter B. Medawar	M & P	Clonal selection theory of antibody formation	707
1959	Arthur Kornberg Severo Ochoa	M & P	Synthesis of DNA and RNA	556, 442
1958	George W. Beadle Joshua Lederberg Edward L. Tatum	M & P	Gene expression	440
	Frederick Sanger	Chemistry	Primary structure of proteins	56

*In a few cases, corecipients whose research was in an area outside of cell and molecular biology have been omitted from this list.

**Medicine and Physiology

Cell and Molecular Biology

Third Edition

CellView CD

Students can visualize and comprehend the most fundamental concepts of cell and molecular biology using the *CellView CD*. The interactive CD packaged with this text includes stepped-out animations of text illustrations and self-tests to help students gain a clear understanding.

These animations come alive as interactive, full-color, three-dimensional animations and are keyed to the illustrations in the text, as indicated by a CD icon. Students can manipulate the figures, visualize the processes in sequence, repeat key steps within animations, and navigate through the animations at their own speed. Among the animated concepts on *CellView* are glycolysis, the cell cycle, chromatography, and the Calvin cycle.

In addition, self-tests for each chapter include multiple choice questions (with immediate feedback) and short answer questions designed to help students test their knowledge of key text concepts.

Installation Instructions for Mac Users:
Place CD in CD-ROM drive. Open CD by double clicking the *CellView* icon on the desktop screen. To run *CellView* directly from the CD, open the Cell View-Mac folder and double-click on the CellViewStart icon. To install *CellView* on your hard drive, follow the instructions in the ReadMe file on the CD by double clicking it.

Installation Instructions for Windows Users:
Place CD in CD ROM drive. Open CD: for example, double click the CD drive icon ("CellView") in Windows Explorer. To run *CellView* directly from the CD, double click on the CellViewStart icon. To install *CellView* on your hard drive, follow the instructions in the ReadMe file on the CD by double clicking it, then run Setup.

More detailed instructions can be found in the ReadMe file located on the CD.

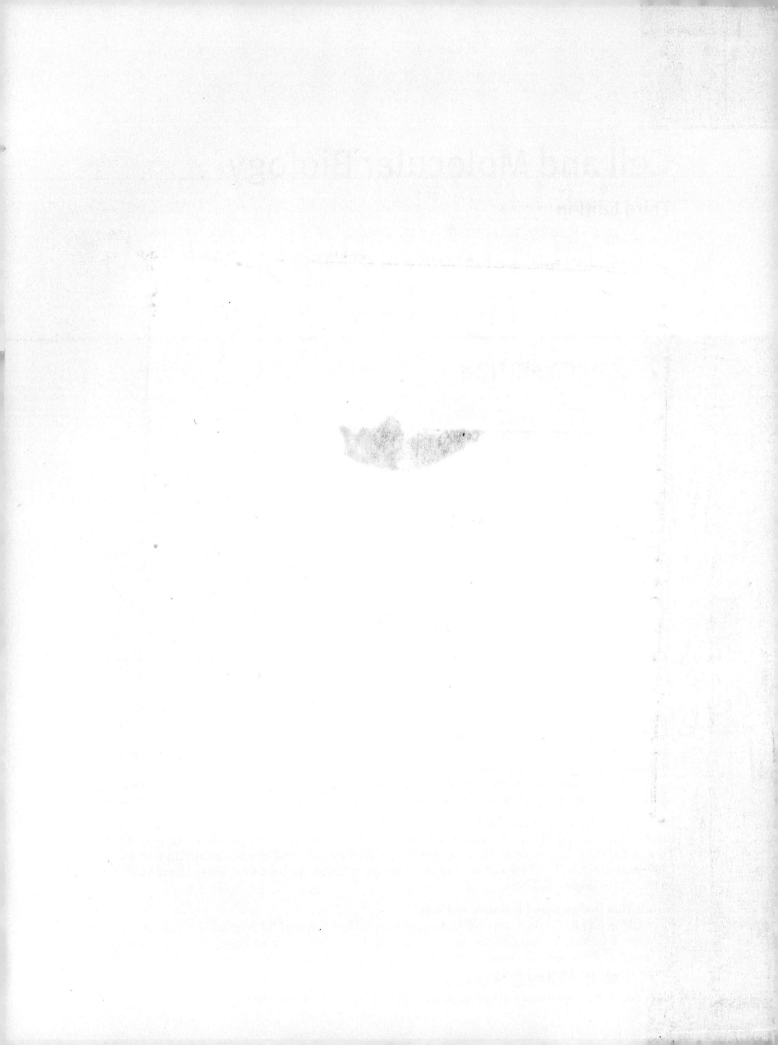

CELL AND MOLECULAR BIOLOGY

Concepts and Experiments

THIRD EDITION

Gerald Karp

John Wiley & Sons, Inc.

Acquisitions Editor	Keri Witman
Marketing Manager	Clay Stone
Production Editor	Barbara Russiello
Senior Designer	Harry Nolan
Interior Design	Lisa Delgado
Cover Design	Howard Grossman
Illustration Editor	Anna Melhorn
Illustrations	Michelle North-Klug/Imagineering
Photo Editor	Hilary Newman
Photo Researcher	Jennifer Atkins
Cover Photograph	Courtesy Christopher S. Chen, Clifford Brangwynne and Donald E. Ingber. From *Trends in Cell Biology*, Vol. 9, 1999, page 283, Fig. 2. Reproduced with permission from Elsevier Science.

This book was set in 10.5/12 Adobe Caslon by UG / GGS Information Services, Inc.,
and printed and bound by Von Hoffman Press. The cover was printed by
Von Hoffman Press.

This book is printed on acid free paper. ∞

UPDATE 2003: ISBN 0-471-26890-9

Library of Congress Cataloging-in-Publication Data:

Karp, Gerald
 Cell and molecular biology : concepts and experiments / Gerald Karp.—3rd ed.
 p. cm.
 Includes bibliographical references and index.
 ISBN 0-471-38913-7 (cloth : alk paper)
 1. Cytology. 2. Molecular biology. I. Title.
 QH581.2 .K369 2001
 571.6—dc21
 20010245558

To Patsy and Jenny

About the Author

Gerald C. Karp received a bachelor's degree from UCLA and a Ph.D. from the University of Washington. He conducted postdoctoral research at the University of Colorado Medical Center before joining the faculty at the University of Florida. Karp is the author of numerous research articles on the cell and molecular biology of early development. His interests have included the synthesis of RNA in early embryos, the movement of mesenchyme cells during gastrulation, and cell determi-nation in slime molds. For 13 years, he taught courses in molecluar, cellular, and developmental biology at the University of Florida. During this period, Gerry coau-thored a text in developmental biology with N. John Berrill and authored a text in cell and molecular biology. Finding it impossible to carry on life as both full-time professor and author, Gerry gave up his faculty position to concentrate on writing. He hopes to revise this text every three years.

About the Cover

The cover of this text displays a brightly stained endothelial cell—a type of cell that lines the inner surface of blood vessels. This cell is square because it has spread itself over a tiny square-shaped patch of an ad-hesive protein called fibronectin that was applied to a cul-ture dish. The cell appears to be mounted in a green frame because it was treated with a green fluorescent antibody that binds to the cytoplasmic protein actin, which makes up part of the cell's "skeleton." Studies with these square-shaped cells have provided valuable information about an important subject in cell biology, namely cell death. Most cells are programmed to commit suicide under certain ad-verse conditions, as might occur, for example, if a cell were struck with a high dose of DNA-damaging radia-tion. Cells that adhere to a very small island of square-shaped fibronectin are unable to spread out, which trig-gers a program leading to the death of the cell. In contrast, cells that are able to spread over a larger fibronectin sur-face, as is the case depicted on the cover, do not commit suicide. Instead, such cells undergo chromosome duplica-tion in preparation for cell division (see pages 249, 366, and 662 for discussion of these subjects). (*Reprinted from* Trends Cell Biol. *vol. 9, Christopher S. Chen, Clifford Brangwynne, and Donald E. Ingber, Squaring up to the cell-shape debate, p. 283, Copyright 1999, with permission from Elsevier Science.*)

Preface for the Third Edition

Before I began work on the *first* edition of this text, I drew up a number of basic guidelines regarding the type of book I planned to write.

- I wanted a text suited for a course that ran either a single semester or 1-2 quarters that would be taken in the sophomore or junior year. I set out to draft a text about 800 pages that would not overwhelm or discourage students at this level.

- I wanted a text that elaborated on fundamental concepts, such as the relationship between molecular structure and function, the dynamic character of cellular organelles, the use of chemical energy in running cellular activities and ensuring accurate macromolecular biosynthesis, unity and diversity at the macromolecular and cellular levels, and the mechanisms that regulate cellular activities.

- I wanted a text that was grounded in the experimental approach. Cell and molecular biology is an experimental science and, like most instructors, I believe students should gain some knowledge of how we know what we know. With this in mind, I decided to approach the experimental nature of the subject in two ways. As I wrote each chapter, I included enough experimental evidence to justify many of the conclusions that were being made. Along the way, I described the salient features of key experimental techniques and referred the reader to a more detailed discussion in the last chapter on methodologies. Chapters 8 and 9, for example, contain introductory sections on techniques that have proven most important in the analysis of cytomembranes and the cytoskeleton, respectively. I included brief discussions of selected experiments of major importance in the body of the chapters to reinforce the experimental basis of our knowledge.

 For students and instructors that wanted to explore the experimental approach in greater depth, I included the "Experimental Pathways," at the end of each chapter. Each of these narratives describes some of the key experimental findings that have led to our current understanding of a particular subject that is relevant to the chapter at hand. Because the scope of the narrative is limited, the design of the experiments can be considered in some detail. The figures and tables provided in these sections are often those that appeared in the original research article, which provides the reader an opportunity to examine original data and to realize that its analysis is not beyond their means. The Experimental Pathways also illustrate the stepwise nature of scientific discovery, showing how the result of one study raises questions that provide the basis for subsequent studies.

- I wanted a text that was interesting and readable. To make the text more relevant to undergraduate readers, particularly premedical students, I included The Human Perspective. These sections illustrate that virtually all human disorders can be traced to disruption of activities at the cellular and molecular level. Furthermore, they reveal the importance of basic research as the pathway to understanding and eventually treating most disorders. In Chapter 11, for example, The Human Perspective describes how ribozymes may prove to be an important new tool in the treatment of cancer and viral diseases, including AIDS. In this same chapter, the reader will learn how ribozymes were first discovered in studies on the processing of ribosomal RNA in a protozoan. It becomes evident that one can never predict the practical importance of basic research in cell and molecular biology. I have also tried to include relevant information about human biology and clinical applications throughout the body of the text.

- I wanted a high-quality illustration program that helped students visualize complex cellular and molecular processes. To meet this goal, many of the illustrations have been "stepped-out" so that information can be more easily broken down into manageable parts. Events occurring at each step are described in the figure legend and/or in the corresponding text. I also sought to include a large number of micrographs to enable students to see actual representations of most subjects being discussed. Included among the photographs are many fluorescence micrographs that illustrate either the dynamic properties of cells or provide a means

to localize a specific protein or nucleic acid sequence. Wherever possible, I have tried to pair line art drawings with micrographs to help students compare idealized and actual versions of a structure.

I have been gratified by the mail I have received from teachers and students containing both praise and criticism of the first two editions. These communications, together with the numerous fine reviews of the current manuscript, have guided the preparation of the third edition. The most important changes in the third edition can be delineated as follows.

■ The body of information in cell and molecular biology is continually changing, which provides much of the excitement we all feel about our selected field. Even though only three years have passed since the publication of the second edition, nearly every discussion in the text has been modified to a greater or lesser degree. This has been done without allowing the chapters to increase significantly in length. In addition to revising and updating, new topics have been added throughout the text. These include: the use of molecular techniques to discover previously unsuspected prokaryote diversity (Chapter 1); energy landscape models of protein folding (Chapter 2); determining the three-dimensional organization of membrane proteins in the absence of data from X-ray crystallography (Chapter 4); the mechanism by which energy stored in the proton gradient across the inner mitochondrial membrane can be used to drive the rotary machinery of the ATP synthase (Chapter 5); photoinhibition of the photosynthetic machinery of chloroplasts by bright light (Chapter 6); the discovery of protocadherins and their apparent role in providing specificity during the formation of synaptic junctions (Chapter 7); the use of green fluorescent protein and cell-free systems in the study of membrane trafficking (Chapter 8); recent models of the mechanism of nonmuscle motility based on actin polymerization (Chapter 9); recent insights from the sequencing of the human genome (Chapter 10); our enhanced understanding of the mechanism of translation following publication of the X-ray crystallographic structure of the ribosome (Chapter 11); the use of DNA microarrays in monitoring changes in gene expression as cells undergo basic changes in metabolism (Chapter 12); the role of a recently discovered class of DNA polymerases that function in replication at sites of DNA lesions (Chapter 13); the role of condensin and cohesin complexes in chromosome compaction and sister-chromatid separation during mitosis (Chapter 14); the role of G protein-coupled receptors in vision, taste, and smell (Chapter 15); the results of successful clinical trials for cancer therapies utilizing dendritic cell vaccines, tyrosine kinase inhibitors, and genetically modified lytic viruses (Chapter 16); proposed mechanisms for positive and negative selection of T cells in the thymus (Chapter 17). The content of the text has been updated through February of 2001.

■ Two discussions have been moved. The import of proteins into peroxisomes, mitochondria, and chloroplasts is now discussed in Chapter 8 and tied into the general subject of protein trafficking within the cell. The discussion of protein degradation by proteasomes has been moved to Chapter 12 and considered as a mechanism of posttranslational regulation.

■ Several of the Experimental Pathways from the 2nd edition have been removed from the text and placed on the World Wide Web. Of the 17 Experimental Pathways from the 2nd edition, 10 have been retained in the text (Chapters 1, 2, 3, 4, 8, 10, 11, 14, 16, and 17), whereas the other 7 can be found on the web at www.wiley.com/college/karp. The latter Experimental Pathways are indicated by a "web icon" where they are referred to in the text. Experimental Pathways have been updated where appropriate.

■ Instructors reported that their students responded very positively to the Human Perspectives. Additional Human Perspectives on stem cells and their potential use in cell transplantation and the role of protein misfolding in the development of Alzheimer's disease have been added.

■ The illustration program received high marks and I have continued with the same approach in the third edition. Now that line art illustrations are drawn by computer and stored electronically, they are much easier to render and modify than in earlier days. Every illustration in the second edition has been scrutinized and many of those that were reutilized in the third edition have been modified to some extent. A number of illustrations in the first and second editions were hand painted and unavailable for the instructor's CD; all of these illustrations have been redrawn as electronic pieces and are now present on the CD. Many of the drawings from the second edition have been deleted to make room for approximately 90 new pieces, numerous of which follow the stepped-out model. Instructors have expressed particular approval for figures that juxtapose line art and micrographs, and this style of illustration has been expanded in the third edition. Altogether, the third edition contains approximately 90 new micrographs and computer-derived images, all of which were provided by the original source.

■ **The student CD** The CD contains animations based on various illustrations presented in the text. Those subjects that have been animated are indicated in the text by a new "animation icon."

Supplements

Cell View CD-ROM

Many of the illustrations presented in the text have been animated and placed on a CD-ROM to add to the clarity and depth of student understanding. Those subjects that have been animated are indicated within the text

with a new "animation icon." These images can also be easily used for classroom presentation. In addition to the animations of the art program from the text, students can test their understanding of the concepts for each chapter by working through the quiz questions on the CD-ROM. Cell View is included in every new text purchased.

Web Site (www.wiley.com/college/karp)
This text-specific web site provides students with additional resources and extends the chapters of the text to the resources of the World Wide Web.

Instructor's Resource CD-ROM (0-471-09052-2)
Prepared by Joel Piperberg of Millersville University, this multi-platform CD-ROM includes the entire Instructor's Manual and PowerPoint Slides. The Instructor's Manual is a comprehensive resource that includes for each text chapter: Chapter Objective, Lecture Outline, Lecture Hints, Critical Thinking Questions, Experimental Pathways Questions, Human Perspectives Questions and Art Questions.

PowerPoint Slides (0-471-09052-2)
Every illustration from the text is provided in PowerPoint format for classroom presentation and can be found on the Instructor's Resource CD-ROM. They are available both as a pre-set presentation and as graphic files for easy inclusion in other presentation tools.

Full Color Transparencies (0-471-12770-1)
Full-color acetates for key illustrations are provided for classroom presentation. The images have been enlarged and fonts bolded for better projection and clarity in the classroom.

Take Note! (0-471-12775-2)
A spiral bound notebook that contains noteworthy figures from the text, allowing students to take notes directly on the page during class or lecture. This is a perfect student companion to presentations using PowerPoint slides or Transparencies.

Student Study Guide: (0-471-09029-8)
Written by Nancy L. Pruitt of Colgate University. The study guide offers students a great way to review materials from the text and test their knowledge. Each chapter in the text has a corresponding chapter in the study guide. The following six tools have been included to help students master the material: Learning Objectives, Key Terms and Phrases, Reviewing the Chapter, Key Figures, Questions for Thought, Review Problems.

Laboratory Investigations in Cell and Molecular Biology, Fourth Edition, by Allyn Bregman, State University of New York, New Paltz (0-471-20133-2)— This lab manual contains over 20 projects which cover key concepts in cell and molecular biology, such as: biochemistry and cytochemistry; organelles and their physiology; and more advanced molecular topics, including restriction mapping strategies.

Update 2003

Cell and molecular biology is in a constant state of change with new developments almost daily. At the time of this printing, a little more than one year has elapsed since publication of the Third Edition. Since publication, some key areas of interest have seen important advances. We are bringing you this 2003 Update Edition to ensure that some of these new discoveries are accessible to you, the student.

At the end of the text, a new Update section has been added to highlight these recent advances. The author has concentrated on updating the Human Perspectives as they tend to be a focus of ongoing research. Those Human Perspectives that have been updated have been highlighted in the Table of Contents and an icon has been added within the text to make it simple to find the topics that have been updated and extended in the Update section.

To the Student

At the time I began college, biology would have been at the bottom of a list of potential majors. I enrolled in a physical anthropology course to fulfill the life science requirement by the easiest possible route. During that course, I learned for the first time about chromosomes, mitosis, and genetic recombination, and I became fascinated by the intricate activities that could take place in such a small volume of cellular space. The next semester, I took Introductory Biology and began to seriously consider becoming a cell biologist. I am burdening you with this personal trivia so you will understand why I wrote this book and to warn you of possible repercussions.

Even though many years have passed, I still find cell biology the most fascinating subject to explore, and I still love spending the day reading about the latest findings by colleagues in the field. Thus, for me, writing a text in cell biology provides a reason and an opportunity to keep abreast with what is going on throughout the field. My primary goal in writing this text is to help generate an appreciation in students for the activities in which the giant molecules and minuscule structures that inhabit the cellular world of life are engaged. Another goal is to provide the reader with an insight into the types of questions that cell and molecular biologists ask and the experimental approaches they use to seek answers. As you read the text, think like a researcher; consider the evidence that is presented, think of alternate explanations, plan experiments that could lead to new hypotheses.

You might begin this approach by looking at one of the many electron micrographs that fill the pages of this text. To take this photograph, you would be sitting in a small, pitch-black room in front of a large metallic instrument whose column rises several meters above your head. You are looking through a pair of binoculars at a vivid, bright green screen. The parts of the cell you are examining appear dark and colorless against the bright green background. They are dark because they've been stained with heavy metal atoms that deflect a fraction of the electrons within a beam that is being focused on the viewing screen by large electromagnetic lenses in the wall of the column. The electrons that strike the screen are accelerated through the evacuated space of the column by a force

of tens of thousands of volts. One of your hands may be gripping a knob that controls the magnifying power of the lenses. A simple turn of this knob can switch the image in front of your eyes from that of a whole field of cells to a tiny part of a cell, such as a few ribosomes or a small portion of a single membrane. By turning other knobs, you can watch different parts of the specimen glide across the screen, giving you the sensation that you're driving around inside a cell. Once you have found a structure of interest, you can turn a handle that lifts the screen out of view, allowing the electron beam to strike a piece of film and produce a photographic image of the specimen.

Since the study of cell function generally requires the use of considerable instrumentation, the investigator is physically removed from the subject being studied. To a large degree, cells are like tiny black boxes. We have developed many ways to probe the boxes, but we are always groping in an area that cannot be fully illuminated. A discovery is made or a new technique is developed and a new thin beam of light penetrates the box. With further work, our understanding of the structure or process is broadened, but we are always left with additional questions. We generate more complete and sophisticated constructions, but we can never be sure how closely our views approach reality. In this regard, the study of cell and molecular biology can be compared to the study of an elephant as conducted by six blind men in an old Indian fable. The six travel to a nearby palace to learn about the nature of elephants. When they arrive, each approaches the elephant and begins to touch it. The first blind man touches the side of the elephant and concludes that an elephant is smooth like a wall. The second touches the trunk and decides that an elephant is round like a snake. The other members of the group touch the tusk, leg, ear, and tail of the elephant, and each forms his impression of the animal based on his own limited experiences. Cell biologists are limited in a similar manner to what they can learn by using a particular technique or experimental approach. Although each new piece of information adds to the preexisting body of knowledge to provide a better concept of the activity being studied, the total picture remains uncertain.

Before closing these introductory comments, let me take the liberty of offering the reader some advice: Don't accept everything you read as being true. There are several reasons for urging such skepticism. Undoubtedly, there are errors in this text that reflect the author's ignorance or misinterpretation of some aspect of the scientific literature. But, more importantly, we should consider the nature of biological research. Biology is an empirical science; nothing is ever proved. We compile data concerning a particular cell organelle, metabolic reaction, intracellular movement, etc., and draw some type of conclusion. Some conclusions rest on more solid evidence than others. Even if there is a consensus of agreement concerning the "facts" regarding a particular phenomenon, there are often several possible interpretations of the data. Hypotheses are put forth and generally stimulate further research, thereby leading to a reevaluation of the original proposal. A theory is constructed in terms of the concepts and prevailing perspectives of the time. As new techniques and information become available, new insights are made. Most hypotheses that remain valid undergo a sort of evolution and, when presented in the text, should not be considered wholly correct or incorrect. Remain skeptical.

Acknowledgments

Many people contributed to the development of this book. I want to express my deepest gratitude to Michelle North-Klug, who prepared all of the new illustrations for the second edition of this text and nearly all those for the present edition. Michelle is a remarkably creative artist, capable of turning a near-scribble into a striking electronic illustration. Her talents speak for themselves on most pages of the text. Michelle dedicated herself for many months to this project, driven by a feeling of personal responsibility that every piece meet with my satisfaction and be completed on time. Working with Michelle put some fun in a long, difficult project. I am also indebted to the production staff at John Wiley & Sons, who are absolutely top-notch. Barbara Russiello is a remarkably skillful production editor, who was responsible for coordinating the work of all the myriad people involved in this project. Despite a relentless storm of impossible deadlines, Barbara managed to keep the ship afloat and headed toward port, which at the time of this writing is still well beyond the horizon. If this book should happen to be published on time, Barbara deserves most of the credit. I am also indebted to Hilary Newman and Anna Melhorn, who were responsible for the photo and line-art programs, respectively. I've worked with Hilary on all three editions of this text, and am always impressed anew at her skill and perserverance. I don't think there's a photo in existence that Hilary can't track down—and have it cropped, labeled, and positioned as needed. This was the first time I've worked with Anna, and I really enjoyed our interaction. Anna is great fun to deal with and dedicated to producing the best possible illustration program. Keri Whitman, my new biology editor, joined the company in the middle of the project. Keri has given her enthusiastic support to the book and provided a wealth of useful ideas. I was fortunate, once again, to have Harry Nolan as the senior designer. Harry has directed a vibrant style for the chapters and an elegant design for the cover. Thanks in advance to Clay Stone, the marketing manager for the text, whose work lies mostly ahead. Thanks also to Laura Ierardi, who handled the page layout. Laura is extremely good at this undertaking, which is crucially important yet often goes unappreciated. I would also like to thank the artists at Imagineering for creating a number of illustrations and Jack Haley, in particular, for playing such an important role in coordinating the art program. Thanks also to Professors David Asai and Ken Robinson of Purdue University for providing a number of interesting analytic questions in Chapters 2–5. I would also like to thank Kelli Coaxum and Geraldine Osnato, who handled most of the editorial communications; Joseph Pastore, who copy edited the manuscript; Dottie Jahoda, who prepared the index; and Dr. Elizabeth Coolidge-Stolz, who wrote the glossary.

I am especially grateful to the many biologists who have contributed micrographs for use in this book; more than any other element it is these images that bring the study of cell biology to life on a printed page. Finally, I would like to apologize, in advance, for any errors that may occur in the text, and express my heartfelt embarrassment. Any comments or criticisms from readers would be greatly appreciated. They can be directed to: Biology Editor, John Wiley & Sons, 605 Third Avenue, New York, NY 10158.

In preparing the final version of the manuscript for the third edition, I sought the advice of numerous scientists whose work I have admired. I asked these individuals to review a chapter or two, and most were very gracious in agreeing to help with the project. I am grateful for the constructive criticism and sound advice from the following reviewers:

LINDA AMOS
MRC LABORATORY OF MOLECULAR BIOLOGY

GERALD T. BABCOCK
MICHIGAN STATE UNIVERSITY

JAMES BARBER
IMPERIAL COLLEGE OF SCIENCE—WOLFSON LABORATORIES

JOHN D. BELL
BRIGHAM YOUNG UNIVERSITY

DANIEL BRANTON
HARVARD UNIVERSITY

THOMAS R. BREEN
SOUTHERN ILLINOIS UNIVERSITY

K. H. ANDY CHOO
ROYAL CHILDREN'S HOSPITAL—THE MURDOCH INSTITUTE

RONALD H. COOPER
UNIVERSITY OF CALIFORNIA—LOS ANGELES

MICHAEL EDIDIN
THE JOHNS HOPKINS UNIVERSITY

ROBERT FILLINGAME
UNIVERSITY OF WISCONSIN MEDICAL SCHOOL

ARTHUR HORWICH
YALE UNIVERSITY SCHOOL OF MEDICINE

JOEL A. HUBERMAN
ROSWELL PARK CANCER INSTITUTE

WERNER KÜHLBRANDT
MAX-PLANCK-INSTITUT FÜR BIOPHYSIK

JAMES LAKE
UNIVERSITY OF CALIFORNIA—LOS ANGELES

VISHWANATH R. LINGAPPA
UNIVERSITY OF CALIFORNIA—SAN FRANCISCO

ARDYTHE A. MCCRACKEN
UNIVERSITY OF NEVADA—RENO

MIKE O'DONNELL
ROCKEFELLER UNIVERSITY

HUGH R. B. PELHAM
MRC LABORATORY OF MOLECULAR BIOLOGY

JOHATHAN PINES
WELLCOME/CRC INSTITUTE

RANDY SCHEKMAN
UNIVERSITY OF CALIFORNIA—BERKELEY

SANDRA SCHMID
THE SCRIPPS RESEARCH INSTITUTE

JENNIFER W. SHULER
WAKE FOREST UNIVERSITY

ROD SCOTT
WHEATON COLLEGE

BRUCE STILLMAN
COLD SPRINGS HARBOR LABORATORY

NIGEL UNWIN
MRC LABORATORY OF MOLECULAR BIOLOGY

CHRIS WATTERS
MIDDLEBURY COLLEGE

Mitchell Chernin
Bucknell University

Thomas C. Chiles
Boston College

Randy W. Cohen
California State University—Northridge

Dennis O. Clegg
Univeristy of California—Santa Barbara

Guy E. Farish
Adams State College

Susannah Gal
SUNY, Binghamton

Francine S. Glazer
Kean University

Margaret Johnson
University of Alabama

David Knecht
University of Connecticut—Storrs

Robert N. Leamnson
University of Massachusetts—Dartmouth

Esther M. Leise
University of North Carolina—Greensboro

Alan C. Leonard
Florida Institute of Technology

Edward J. Macarak
University of Pennsylvania

Luis A. Materon
University of Texas—Pan American

Elizabeth J. Moore
Rowan University

Dennis G. Searcy
University of Massachusetts

Diane Shakes
College of William and Mary

David H. Vickers
University of Central Florida

Anne E. K. Zayaitz
Kutztown University

Thanks are still owed to the following reviewers from the second edition:

David J. Asai
Purdue University

John D. Bell
Brigham Young University

Barbara Berg
University of Puget Sound

Niels Bols
University of Waterloo, Ontario, Canada

Thomas R. Breen
Southern Illinois University—Carbondale

David K. Bruck
San Jose State University

Reviewers of First Edition

Robert E. Bast
Cleveland State University

Catherine P. Chia
University of Nebraska—Lincoln

Sherri Clark
Eastern New Mexico University

Julia Dragolovich
University of Maryland

Karl Drlica
Public Health Research Institute

Terrence G. Frey
San Diego State University

David Fromson
California State University—Fullerton

David S. Gilmour
Pennsylvania State University—University Park

R. Jane Hanas
University of Central Oklahoma

Thomas Kistenmacher
Johns Hopkins University

Hallie M. Krider
The University of Connecticut

Mary Lee S. Ledbetter
College of the Holy Cross

Joel Piperberg
Millersville University of Pennsylvania

Nancy L. Pruitt
Colgate University

Thomas M. Roberts
Florida State University

Robert Seagull
Hofstra University

Joel Sheffield
Temple University

Sheldon Shen
Iowa State University

John Tyson
Virigina Polytechnic Institute

Fred Warner
Syracuse University

Xilin Zhao
Public Health Research Institute

Gerald Karp

Brief Table of Contents

Contents

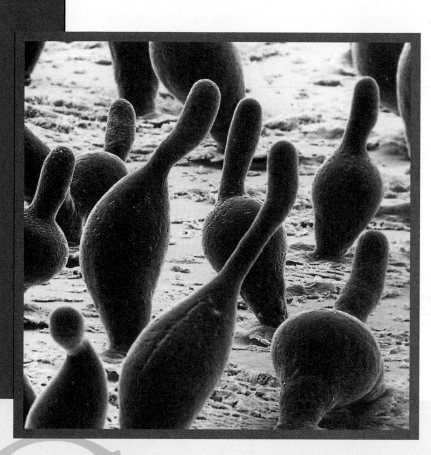

Introduction to the Study of Cell and Molecular Biology

ells, and the structures they comprise, are too small to be directly seen, heard, or touched. Yet, in spite of this tremendous handicap, cells are the subject of thousands of publications each year, with virtually every aspect of their minuscule structure coming under scrutiny. In many ways, the study of cell and molecular biology stands as a tribute to human curiosity for seeking to discover, and to human creative intelligence for devising the complex instruments and elaborate techniques by which these discoveries can be made. This is not to imply that cell biologists have a monopoly on these noble traits. At one end of the scientific spectrum, astronomers are learning of objects at the outer fringe of the universe that have properties very different from those on Earth. At the other end of the spectrum, nuclear physicists are focusing their attention on particles of subatomic dimensions that have equally inconceivable properties. Clearly, our universe consists of worlds within worlds, all aspects of which make for fascinating study.

As will be apparent throughout this book, cellular and molecular biology is *reductionist*; that is, it is based on the view that knowledge of the parts of the whole can explain the character of the whole. When viewed in this way, our feeling for the wonder and mystery of life may be replaced by the need to explain everything in terms of the workings of the "machinery" of the living

Scanning electron micrograph of cellular aggregates of a mutant strain of the slime mold Dictyostelium discoideum *in the process of forming fruiting bodies.* (COURTESY OF MARK GRIMSON, SCIENCES ELECTRON MICROSCOPY LABORATORY, TEXAS TECH. UNIVERSITY.)

system. To the degree to which this occurs, it is hoped that this loss can be replaced by an equally strong appreciation for the beauty and complexity of the mechanisms underlying cellular activity. ■

1.1 THE DISCOVERY OF CELLS

We do not know when humans first discovered the remarkable ability of curved-glass surfaces to bend light and form images. Spectacles were first made in Europe in the thirteenth century, and the first compound (double-lensed) microscopes were constructed by the end of the sixteenth century. By the mid-1600s, a handful of pioneering scientists had used their handmade microscopes to uncover a world that would never have been revealed to the naked eye. The discovery of cells (Figure 1.1) is generally credited to Robert Hooke, an English microscopist who, at age 27, was awarded the position of curator of the Royal Society, England's foremost scientific academy. One of the many questions Hooke attempted to answer was why stoppers made of cork (part of the bark of trees) were so well suited to holding air in a bottle. As he wrote in 1665: "I took a good clear piece of cork, and with a Pen-knife sharpen'd as keen as a Razor, I cut a piece of it off, and . . . then examining it with a *Microscope*, me thought I could perceive it to appear a little porous . . . much like a Honeycomb." Hooke called the pores *cells* because they reminded him of the cells inhabited by monks living in a monastery. In actual fact, Hooke had observed the empty cell walls of dead plant tissue, walls that had originally been produced by the living cells they surrounded.

Meanwhile, Anton van Leeuwenhoek, a Dutchman who earned a living selling clothes and buttons, was spending his spare time grinding lenses and constructing microscopes of remarkable quality. For 50 years, Leeuwenhoek sent letters to the Royal Society of London describing his microscopic observations—along with a rambling discourse on his daily habits and the state of his health. Leeuwenhoek was the first to examine a drop of pond water under the microscope and, to his amazement, to observe the teeming microscopic "animalcules" that darted back and forth before his eyes. He was also the first to describe various forms of bacteria, which he obtained from water in which pepper had been soaked and from scrapings of his teeth. His initial letters to the Royal Society describing this previously unseen world were met with such skepticism that the society dispatched its curator, Robert Hooke, to confirm the observations. Hooke did just that, and Leeuwenhoek was soon a worldwide celebrity, receiving visits in Holland from Peter the Great of Russia and the queen of England.

It wasn't until the 1830s that the widespread importance of cells was realized. In 1838, Matthias Schleiden, a German lawyer turned botanist, concluded that, despite differences in the structure of various tissues, plants were made of cells and that the plant embryo arose from a single

Figure 1.1 **The discovery of cells.** Microscope used by Robert Hooke. (Inset) Hooke's drawing of a thin slice of cork, showing the honeycomb-like network of "cells." (FROM THE GRANGER COLLECTION; (INSET) CORBIS-BETTMANN.)

cell. In 1839, Theodor Schwann, a German zoologist and colleague of Schleiden's, published a comprehensive report on the cellular basis of animal life. Schwann concluded that the cells of plants and animals are similar structures and proposed these two tenets of **cell theory:**

■ All organisms are composed of one or more cells.

■ The cell is the structural unit of life.

Schleiden and Schwann's ideas on the *origin* of cells proved to be less insightful; both agreed that cells could arise from noncellular materials. Given the prominence that these two scientists held in the scientific world, it took a number of years before observations by other biologists were accepted as demonstrating that cells did not arise in this manner any more than organisms arose by spontaneous generation. By 1855, Rudolf Virchow, a German pathologist, had made a convincing case for the third tenet of the cell theory:

■ Cells can arise only by division from a preexisting cell.

1.2 BASIC PROPERTIES OF CELLS

Just as plants and animals are alive, so too are cells. Life, in fact, is the most basic property of cells, and cells are the smallest units to exhibit this property. Unlike the parts of a cell, which simply deteriorate if isolated, whole cells can be removed from a plant or animal and cultured in a laboratory where they will grow and reproduce for extended periods of time. The first culture of human cells was begun by George Gey of Johns Hopkins University in 1951. The cells were obtained from a malignant tumor and named HeLa cells after the donor, Henrietta Lacks. HeLa cells—descended by cell division from this first cell sample—are still being grown in laboratories around the world today (Figure 1.2). Because they are so much simpler to study than cells situated within the body, cells grown **in vitro** (i.e., in culture, outside the body) have become an essential tool of cell and molecular biologists. In fact, much of the information that will be discussed in this book has been obtained using cells grown in laboratory cultures.

We will begin our exploration of cells by examining a few of their most fundamental properties.

Cells Are Highly Complex and Organized

Complexity is a property that is evident but difficult to describe. For the present, we can think of complexity in terms of order and consistency. The more complex a structure, the greater the number of parts that must be in their proper place, the less tolerance of errors in the nature and interactions of the parts, and the more regulation or control that must be exerted to maintain the system. During the course of this book, we will have occasion to consider the complexity of life at several different levels. We will discuss the organization of atoms into small-sized molecules; the organization of these molecules into giant polymers; and the organization of different types of polymeric molecules into complexes, which in turn are organized into subcellular organelles and finally into cells. As will be apparent, there is a great deal of consistency at every level. Each type of cell has a consistent appearance in the electron microscope; that is, its organelles have a particular shape and location, from one individual of a species to another. Similarly, each type of organelle has a consistent composition of macromolecules, which are arranged in a predictable pattern. Consider the cells lining your intestine that are responsible for removing nutrients from your digestive tract (Figure 1.3).

The epithelial cells that line the intestine are tightly connected to each other like bricks in a wall. The apical ends of these cells, which face the intestinal channel, have long processes (*microvilli*) that facilitate absorption of nutrients. The microvilli are able to project outward from the apical cell surface because they contain an internal skeleton made of filaments, which in turn are composed of two rows of protein (*actin*) molecules twisted around each other in a

15 µm

Figure 1.2 HeLa cells, such as the ones pictured here, were the first human cells to be kept in culture for long periods of time and are still in use today. Unlike normal cells, which have a finite lifetime in culture, these cancerous HeLa cells can be cultured indefinitely as long as conditions are favorable to support cell growth and division. (KEITH PORTER/PHOTO RESEARCHERS.)

characteristic array. At their basal ends, intestinal cells have large numbers of mitochondria that provide the energy required to fuel various membrane transport processes. Each mitochondrion is composed of a defined pattern of internal membranes, which in turn are composed of a consistent array of proteins, including an ATP-synthesizing enzyme that projects from the inner membrane like a ball on a stick. Each of these various levels of organization is illustrated in the series of insets provided in Figure 1.3.

Fortunately for cell and molecular biologists, evolution has moved rather slowly at the levels of biological organization with which they are concerned. Whereas a human and a cat, for example, have very different anatomical features, the cells that make up their tissues, and the organelles that make up their cells, are very similar. The actin filament portrayed in Figure 1.3, Inset 3, and the ATP-synthesizing enzyme of Inset 6 are virtually identical to similar structures found in such diverse organisms as humans, snails, yeast, and redwood trees. Information obtained by studying cells from one type of organism often has direct application to other forms of life. Many of the most basic processes, such as the synthesis of proteins, the conservation of chemical energy, or the construction of a membrane, are remarkably similar in all living organisms. Considered in evolutionary terms, many of the molecules present in our cells must be very similar to those that were present in our primitive cellular ancestors that lived more than 3 billion years ago.

Cells Possess a Genetic Program and the Means to Use It

Organisms are built according to information encoded in a collection of genes. The human genetic program contains enough information, if converted to words, to fill

Figure 1.3 **Levels of cellular and molecular organization.** The brightly colored photograph of a stained section shows the microscopic structure of a villus of the wall of the small intestine, as seen through the light microscope. Inset 1 shows an electron micrograph of the epithelial layer of cells that lines the inner intestinal wall. The apical surface of each cell, which faces the channel of the intestine, contains a large number of microvilli involved in nutrient absorption. The basal region of each cell contains large numbers of mitochondria, where energy is made available to the cell. Inset 2 shows the apical microvilli; each microvillus is seen to contain a bundle of microfilaments. Inset 3 shows the double row of actin protein molecules that make up each filament. Inset 4 shows an individual mitochondrion similar to those found in the basal region of the epithelial cells. Inset 5 shows a portion of an inner membrane of a mitochondrion including the stalked particles (upper arrow) that project from the membrane and correspond to the sites where ATP is synthesized. Insets 6 and 7 show molecular models of the ATP-synthesizing machinery, which is discussed at length in Chapter 5. (LIGHT MICROGRAPH CECIL FOX/PHOTO RESEARCHERS; INSET 1 COURTESY OF SHAKTI P. KAPUR, GEORGETOWN UNIVERSITY MEDICAL CENTER; INSET 2 COURTESY OF MARK S. MOOSEKER AND LEWIS G. TILNEY, J. CELL BIOL. 67:729, 1975, BY COPYRIGHT PERMISSION OF THE ROCKEFELLER UNIVERSITY PRESS; INSET 3 COURTESY OF KENNETH C. HOLMES; INSET 4 COURTESY OF KEITH R. PORTER/PHOTO RESEARCHERS; INSET 5 COURTESY OF HUMBERTO FERNANDEZ-MORAN; INSET 6 COURTESY OF RODERICK A. CAPALDI; INSET 7 COURTESY OF WOLFGANG JUNGE, HOLGER LILL, AND SIEGFRIED ENGELBRECHT, UNIVERSITY OF OSNABRÜCK, GERMANY.)

millions of pages of text. Remarkably, this vast amount of information is packaged into a set of chromosomes that occupy the space of a cell nucleus—thousands of times smaller than the dot on this *i*.

Genes are more than storage lockers for information: they constitute the blueprints for constructing cellular structures, the directions for running cellular activities, and the program for making more of themselves. Discovering the mechanisms by which cells use their genetic information to accomplish these functions has been one of the greatest achievements of science in recent decades.

Cells Are Capable of Producing More of Themselves

Just as individual organisms are generated by reproduction, so too are individual cells. Cells reproduce by division, a process in which the contents of a "mother" cell are distributed into two "daughter" cells. Prior to division, the genetic material is faithfully duplicated, and each daughter cell receives a complete and equal share of genetic information. In most cases, the two daughter cells produced by division have approximately equal volume. In some cases, however, as occurs when a human oocyte undergoes division, one of the cells can retain nearly all of the cytoplasm, even though it receives only half of the genetic material (Figure 1.4).

Cells Acquire and Utilize Energy

Developing and maintaining complexity requires the constant input of energy (Figure 1.5). Virtually all of the energy required by life on the Earth's surface arrives in the form of electromagnetic radiation from the sun. The energy of light is trapped by light-absorbing pigments present in the membranes of photosynthetic cells. Light energy is converted by photosynthesis into chemical energy that is stored in energy-rich carbohydrates, such as sucrose or starch. For most animal cells, energy arrives prepackaged, often in the form of the sugar glucose. In humans, glucose is released by the liver into the blood where it circulates through the body delivering chemical energy to all the cells. Once in a cell, the glucose is disassembled in such a way that its energy content can be stored in a readily available form (usually as ATP) and later put to use in running all of the cell's myriad energy-requiring activities.

Cells Carry Out a Variety of Chemical Reactions

Cells function like miniaturized chemical plants. Even the simplest bacterial cell is capable of hundreds of different chemical transformations, none of which occurs at any significant rate in the inanimate world. Virtually all chemical changes that take place in cells require enzymes—molecules that greatly increase the rate at which a chemical reaction occurs. The sum total of the chemical reactions in a cell represents that cell's **metabolism**.

20 μm

Figure 1.4 Cell reproduction. This mammalian oocyte has recently undergone a highly unequal cell division in which most of the cytoplasm has been retained within the large oocyte, which has the potential to be fertilized and develop into an embryo. The other cell is a nonfunctional remnant that consists almost totally of nuclear material (indicated by the blue-staining chromosomes, arrow). (COURTESY OF JONATHAN VAN BLERKOM.)

Cells Engage in Numerous Mechanical Activities

Cells are sites of bustling activity. Materials are transported from place to place, structures are assembled and then rapidly disassembled, and, in many cases, the entire cell moves itself from one site to another (Figure 1.6). These types of activities are based on dynamic, mechanical changes within cells, most of which are initiated by changes in the shape of certain "motor" proteins.

Cells Are Able to Respond to Stimuli

Some cells respond to stimuli in obvious ways; a single-celled protist, for example, moves away from an object in its path or moves toward a source of nutrients. Cells within a multicellular plant or animal respond to stimuli less obviously, but they respond, nonetheless. Most cells

Figure 1.5 Acquiring energy. A living cell of the filamentous alga *Spirogyra*. The ribbon-like chloroplast, which is seen to zigzag through the cell, is the site where energy from sunlight is captured and converted to chemical energy during photosynthesis. (M. I. WALKER/PHOTO RESEARCHERS.)

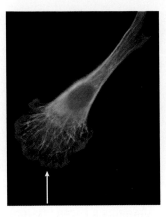

Figure 1.6 **Cell locomotion.** This fibroblast (a type of connective tissue cell that is widely used in laboratory cultures) was caught in the act of moving over the surface of a culture dish. The cell has been stained with fluorescent antibodies to reveal the distribution of actin filaments and microtubules (Chapter 9). The rounded edge of the cell is leading the way; the clusters of actin filaments at the leading edge (arrow) are sites where the force for movement is generated. (COURTESY OF J. VICTOR SMALL.)

are covered with *receptors* that interact with substances in the environment in highly specific ways. Cells possess receptors to hormones, growth factors, extracellular materials, as well as to substances on the surfaces of other cells. A cell's receptors provide doorways through which external agents can evoke specific responses in target cells. Cells may respond to specific stimuli by altering their metabolic activities, preparing for cell division, moving from one place to another, or even committing suicide.

Cells Are Capable of Self-Regulation

In addition to requiring energy, maintaining a complex, ordered state requires constant regulation. The importance of a cell's regulatory mechanisms becomes most evident when they break down. For example, failure of a cell to correct a mistake when it duplicates its DNA may result in a debilitating mutation, or a breakdown in a cell's growth control can transform the cell into a cancer cell with the capability of destroying the entire organism. We are gradually learning more and more about how the cell controls its activities, but much more is left to discover.

Consider the following experiment conducted in 1891 by Hans Driesch, a German embryologist. Driesch found that he could completely separate the first two or four cells of a sea urchin embryo and each of the isolated cells would proceed to develop into a normal embryo (Figure 1.7). How can a cell that is normally destined to form only part of an embryo regulate its own activities and form an entire embryo? How does the isolated cell recognize the absence of its neighbors, and how does this recognition redirect the course of the cell's development? How can a part of an embryo have a sense of the whole? We are not able to answer these questions much better today than we were more than a hundred years ago when the experiment was performed.

Throughout this book we will be discussing processes that require a series of ordered steps, much as the assembly-line construction of an automobile in which workers add, remove, or make specific adjustments as the car moves along. In the cell, the information for product design resides in the nucleic acids, and the construction workers are primarily proteins. It is the presence of these two types of macromolecules that, more than any other factor, sets the chemistry of the cell apart from that of the nonliving world. In the cell, the workers must act without the benefit of conscious direction. Each step of a process must occur spontaneously in such a way that the next step is automatically triggered. All the information required to direct a particular activity—whether it be the synthesis of a protein, the secretion of a hormone, or the contraction of a muscle fiber—must be present within the system itself. In many ways, cells operate in a manner analogous to the orange-squeezing contraption discovered by "The Professor" and shown in Figure 1.8. Each type of cellular activity requires a unique set of molecular tools and machines. A primary goal of cell and molecular biologists is to understand the structure and role of each component involved

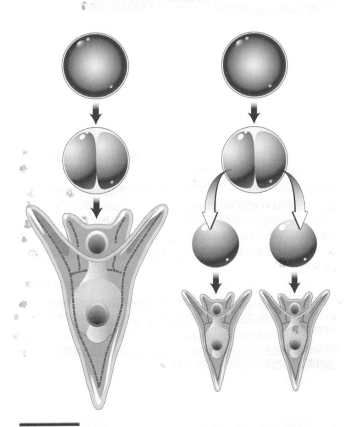

Figure 1.7 **Self-regulation.** The left panel depicts the normal development of a sea urchin in which a fertilized egg gives rise to a single embryo. The right panel depicts an experiment in which the cells of an embryo are separated from one another after the first division, and each cell is allowed to develop in isolation. Rather than developing into half of an embryo, as it would if left undisturbed, each isolated cell recognizes the absence of its neighbor, regulating its development to form a complete (although smaller) embryo.

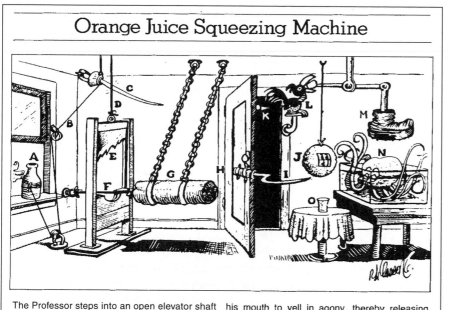

The Professor steps into an open elevator shaft and when he lands at the bottom he finds a simple orange squeezing machine. Milkman takes empty milk bottle (**A**), pulling string (**B**) which causes sword (**C**) to sever cord (**D**) and allow guillotine blade (**E**) to drop and cut rope (**F**) which releases battering ram (**G**). Ram bumps against open door (**H**), causing it to close. Grass sickle (**I**) cuts a slice off end of orange (**J**)–at the same time spike (**K**) stabs "prune hawk" (**L**) he opens his mouth to yell in agony, thereby releasing prune and allowing diver's boot (**M**) to drop and step on sleeping octopus (**N**). Octopus awakens in a rage and, seeing diver's face which is painted on orange, attacks it and crushes it with tentacles, thereby causing all the juice in the orange to run into glass (**O**).

Later on you can use the log to build a log cabin where you can raise your son to be President like Abraham Lincoln.

Figure 1.8 **Cellular activities** are often analogous to this Rube Goldberg machine in which one event "automatically" triggers the next event in a reaction sequence. Figure 15.2 provides a good example of this concept. (RUBE GOLDBERG™ AND © OF RUBE GOLDBERG, INC. DISTRIBUTED BY UNITED MEDIA.)

in a particular activity, the means by which these components interact, and the mechanism by which these interactions are regulated.

REVIEW ?

1. What are the fundamental properties that are shared by all cells? Describe the importance of each of these properties.

2. Describe some of the features of cells that suggest that all living organisms are derived from a common ancestor.

3. What is the source of energy that supports life on Earth? How is this energy passed from one organism to the next?

1.3 TWO FUNDAMENTALLY DIFFERENT CLASSES OF CELLS

Once the electron microscope became widely available, biologists were able to examine the internal structure of a wide variety of cells. It became apparent from these studies that there were two basic classes of cells—prokaryotic and eukaryotic—distinguished by their size and the types of internal structures, or **organelles**, they contain (Figure 1.9). The existence of two distinct classes of cells, without any known intermediates, represents one of the most fundamental evolutionary discontinuities in the biological world. The structurally simpler, **prokaryotic cells** are represented by all of the various forms of bacteria (discussed below). All other types of organisms—protists, fungi, plants, and animals—consist of structurally more complex **eukaryotic cells**. Prokaryotic cells living today are remarkably similar to fossilized cells found in rocks from Australia and South Africa that date back more than 3.5 billion years (Figure 1.10). In fact, prokaryotic cells are thought to have been the sole living residents of the planet for nearly 2 billion years before the appearance of the first eukaryotes.

Characteristics That Distinguish Prokaryotic and Eukaryotic Cells

The following brief comparison between prokaryotic and eukaryotic cells displays many basic differences, as well as similarities (see Figure 1.9). The similarities and differences between the two types of cells are listed in Table 1.1. The similarities reflect the fact that eukaryotic cells almost certainly evolved from prokaryotic ancestors. Because of their common ancestry, both types of cells share an identical genetic language, a common set of metabolic pathways, and many common structural features. For example, both types of cells are bounded by plasma membranes of

Figure 1.9 **The structure of cells.** Schematic diagrams of a "generalized" bacterial (*a*), plant (*b*), and animal (*c*) cell. Note: Organelles are not drawn to scale.

Bacterial flagellum

Ribosomes

DNA of nucleoid

Plasma membrane
Cell wall

Capsule

(*a*)

Nucleus
— Nuclear envelope
— Nucleoplasm
— Nucleolus

Chloroplast

Rough endoplasmic reticulum

Smooth endoplasmic reticulum

Cell wall

Plasma membrane

Peroxisome

Plasmodesma

Golgi complex

Mitochondrion

Ribosomes

Vacuole

Vesicle

Cytosol

Microtubules

(*b*)

Ribosomes

Nuclear envelope
Nucleoplasm

Mitochondrion

Nucleus

Golgi complex

Lysosome

Nucleolus

Smooth endoplasmic reticulum

Microfilaments

Rough endoplasmic reticulum

Plasma membrane

Peroxisome

Centriole

Cytosol

Microtubule

Vesicle

(*c*)

Figure 1.10 **The world's most ancient known life.** A 3.5 billion-year-old cast of a filamentous cyanobacterium from western Australia. (COURTESY OF S. M. AWRAMIK.)

Table 1.1	A Comparison of Prokaryotic and Eukaryotic Cells

Features held in common by the two types of cells:

- Plasma membrane of similar construction
- Genetic information encoded in DNA using identical genetic code
- Similar mechanisms for transcription and translation of genetic information, including similar ribosomes
- Shared metabolic pathways (e.g., glycolysis and TCA cycle)
- Similar apparatus for conservation of chemical energy as ATP (located in the plasma membrane of prokaryotes and the mitochondrial membrane of eukaryotes)
- Similar mechanism of photosynthesis (between cyanobacteria and green plants)
- Similar mechanism for synthesizing and inserting membrane proteins
- Proteasomes (protein digesting structures) of similar construction (between archaeabacteria and eukaryotes)

Features of eukaryotic cells not found in prokaryotes:

- Division of cells into nucleus and cytoplasm, separated by a nuclear envelope containing complex pore structures
- Complex chromosomes composed of DNA and associated proteins that are capable of compacting into mitotic structures
- Complex membranous cytoplasmic organelles (includes endoplasmic reticulum, Golgi complex, lysosomes, endosomes, peroxisomes, and glyoxisomes)
- Specialized cytoplasmic organelles for aerobic respiration (mitochondria) and photosynthesis (chloroplasts)
- Complex cytoskeletal system (including microfilaments, intermediate filaments, and microtubules)
- Complex flagella and cilia
- Capable of ingesting fluid and particulate material by enclosure within plasma membrane vesicles (endocytosis and phagocytosis)
- Cellulose-containing cell walls (in plants)
- Cell division using a microtubule-containing mitotic spindle that separates chromosomes
- Presence of two copies of genes per cell (diploidy), one from each parent
- Sexual reproduction requiring meiosis and fertilization

similar construction that serve as a selectively permeable barrier between the living and nonliving worlds.

Internally, eukaryotic cells are much more complex—both structurally and functionally—than prokaryotic cells (Figure 1.9). Both contain a nuclear region, which houses the cell's genetic material, surrounded by cytoplasm. The genetic material of a prokaryotic cell is present in a **nucleoid**: a poorly demarcated region of the cell that lacks a boundary membrane to separate it from the surrounding cytoplasm. In contrast, eukaryotic cells possess a **nucleus**: a region bounded by a complex membranous structure called the *nuclear envelope*. This difference in nuclear structure is the basis for the terms *prokaryotic* (*pro* = before, *karyon* = nucleus) and *eukaryotic* (*eu* = true, *karyon* = nucleus). Prokaryotic cells contain relatively small amounts of DNA; the total length of the DNA of a bacterium ranges from about 0.25 mm to about 3 mm, which is sufficient to encode between several hundred and several thousand proteins.[1] Although the simplest eukaryotic cells have only slightly more DNA (4.6 mm in yeast encoding about 6200 proteins) than the most complex prokaryotes, most eukaryotic cells contain an order of magnitude more genetic information. Both prokaryotic and eukaryotic cells have DNA-containing chromosomes. Each of the numerous chromosomes of a eukaryotic cell, however, contains a linear molecule of DNA associated tightly with protein, whereas the single circular chromosome of a prokaryotic cell consists essentially of "naked" DNA.

The cytoplasm of the two types of cells is also very different. The cytoplasm of a eukaryotic cell is filled with a great diversity of structures, as is readily apparent from the most superficial examination of an electron micrograph of nearly any cell of a plant or animal (Figure 1.11). Most notably, eukaryotic cells contain an array of membranous and membrane-bound organelles. For example, plant and animal cells typically contain mitochondria, where chemical energy is made available to fuel cellular activities; an endoplasmic reticulum, where many of a cell's proteins and lipids are manufactured; Golgi complexes, where materials are sorted, modified, and transported to specific cellular destinations; and a variety of simple membrane-bound vesicles of varying dimension. Plant cells contain additional membranous organelles, including chloroplasts, which are the sites of photosynthesis, and often a single large vacuole that can occupy most of the volume of the cell. Taken as a group, the membranes of the eukaryotic cell serve to divide the cytoplasm into compartments within which specialized activities can take place. In contrast, the cytoplasm of prokaryotic cells is essentially devoid of membranous structures. The complex photosyn-

[1]One millimeter of DNA contains approximately 3 million base pairs.

Figure 1.11 **The structure of a eukaryotic cell.** The internal structure varies greatly from one type of cell to another. This particular epithelial cell lines a portion of the male reproductive tract in the rat. A number of different organelles are indicated and depicted in schematic diagrams around the border of the figure. (DAVID PHILLIPS/VISUALS UNLIMITED.)

Lysosome

Plasma membrane

Golgi complex

Nucleus

Smooth endoplasmic reticulum

Mitochondrion

Rough endoplasmic reticulum

Chromatin

Nucleolus

thetic membranes of the cyanobacteria are a major exception to this generalization (see Figure 1.15).

The cytoplasmic membranes of eukaryotic cells form a system of interconnecting channels and vesicles that function in the direct transport of substances from one part of a cell to another, as well as between the inside of the cell and its environment. Because of their small size, directed intracytoplasmic communication is less important in prokaryotic cells, where the necessary movement of materials can be accomplished by simple diffusion.

Eukaryotic cells also contain numerous structures lacking a surrounding membrane. Included in this group are the elongated tubules and filaments of the cytoskeleton, which participate in cell contractility, movement, and support. Comparable structures are absent from prokaryotic cells. Both eukaryotic and prokaryotic cells do, however, possess ribosomes, which are nonmembranous particles that function as "workbenches" on which the proteins of the cell are manufactured. Even though ribosomes of prokaryotic and eukaryotic cells have considerably different dimensions (those of prokaryotes are smaller and contain fewer components), these structures participate in the assembly of proteins by a similar mechanism in both types of cells. Both types of cells may be surrounded by a rigid, nonliving *cell wall* that protects the delicate life form within. Although the cell walls of prokaryotes and eukaryotes may have similar functions, their chemical composition is very different.

Other major differences between eukaryotic and prokaryotic cells can be noted. Eukaryotic cells divide by a complex process of mitosis in which duplicated chromosomes condense into compact structures that are segregated by an elaborate microtubule-containing apparatus (Figure 1.12). This apparatus, which is called a *mitotic spindle*, allows each daughter cell to receive an equivalent array of genetic material. In prokaryotes, there is no compaction of the chromosome and no mitotic spindle. The DNA is duplicated, and the two copies are separated simply and accurately by the growth of an intervening cell membrane.

For the most part, prokaryotes are nonsexual organisms. They contain only one copy of their single chromosome and have no processes comparable to meiosis, gamete formation, or true fertilization. Even though true sexual reproduction is lacking among prokaryotes, some are capable of *conjugation*, in which a piece of DNA is passed from one cell to another (Figure 1.13). However, the recipient almost never receives a whole chromosome from the donor, and the condition in which the recipient cell contains both its own and its partner's DNA is fleet-

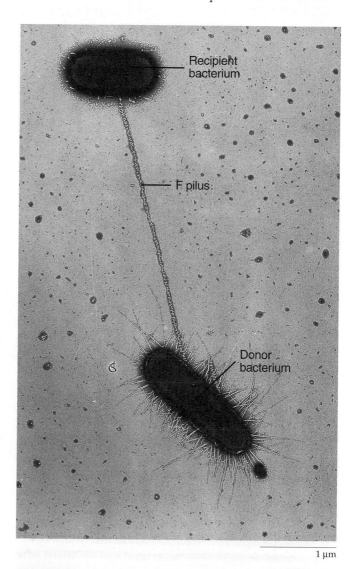

1 μm

Figure 1.13 Bacterial conjugation. Electron micrograph showing a conjugating pair of bacteria joined by a structure of the donor cell, termed the F pilus, through which DNA is thought to be passed. (COURTESY OF CHARLES C. BRINTON, JR., AND JUDITH CARNAHAN.)

4 μm

Figure 1.12 Cell division in eukaryotes requires the assembly of an elaborate chromosome-separating apparatus called the mitotic spindle, which is constructed primarily of microtubules. The microtubules in this micrograph appear green because they are specifically bound by an antibody that is linked to a green fluorescent dye. The chromosomes, which were about to be separated into two daughter cells when this cell was fixed, are stained blue. (COURTESY OF CONLY L. RIEDER.)

ing. The cell soon reverts back to possession of a single chromosome.

Eukaryotic cells possess a variety of complex locomotor mechanisms, whereas those of prokaryotes are quite simple. The movement of a prokaryotic cell may be accomplished by a thin protein filament, called a *flagellum*, which protrudes from the cell and rotates (Figure 1.14*a*). The rotations of the flagellum exert pressure against the surrounding fluid, causing the cell to be propelled through the medium. Certain eukaryotic cells, including many protists and sperm cells, also possess flagella, but the eukaryotic versions are much more complex than the simple protein filaments of bacteria, and they generate movement by a different mechanism (Figure 1.14*b*).

In the preceding paragraphs, many of the most important differences between the prokaryotic and eukaryotic levels of cellular organization were mentioned. We will elaborate on many of these points in later chapters. Before you dismiss prokaryotes as inferior, keep in mind that these organisms have remained on Earth for more than 3 billion years, and at this very moment, millions of them are clinging to the outer surface of your body or feasting on the nutrients within your digestive tract. Consider also that, metabolically, prokaryotes are very sophisticated, highly evolved organisms. For example, a bacterium, such as *Escherichia coli*, a common inhabitant of both the human digestive tract and the laboratory culture dish, has the ability to live and prosper in a medium containing one or two low-molecular-weight organic compounds and a few inorganic ions. Other bacteria are able to live on a diet consisting solely of inorganic substances. One species of bacteria has recently been found in wells more than a thousand meters below the Earth's surface living on basalt rock and molecular hydrogen (H_2) produced by inorganic reactions. In contrast, even the most metabolically talented cells in your body require a variety of organic compounds, including a number of vitamins and other essential substances they cannot make on their own. In fact, many of these essential dietary ingredients are produced by the bacteria that normally live in the large intestine.

Types of Prokaryotic Cells

Under current classification schemes, prokaryotes are divided into two major groups, or domains: the Archaea (or archaebacteria) and the Bacteria (or eubacteria). The domain Archaea includes several groups of organisms whose evolutionary ties to one another are revealed by similarities in the nucleotide sequences of their nucleic acids. The best known Archaea are species that live in extremely inhospitable environments; they are often referred to as "extremophiles." Included among the Archaea are the methanogens [prokaryotes capable of converting CO_2 and H_2 gases into methane (CH_4) gas]; the halophiles (prokaryotes that live in extremely salty environments, such as the Dead Sea or the Great Salt Lake); acidophiles (acid-loving prokaryotes that live at a pH as low as 0); and

(a) 1 µm

 30 nm

(b)

 0.12 µm

Figure 1.14 The difference between prokaryotic and eukaryotic flagella. (*a*) The bacterium *Salmonella* with its numerous flagella. Inset shows a high-magnification view of a portion of a single bacterial flagellum, which consists largely of a single protein called flagellin. (*b*) Each of these human sperm cells is powered by the undulatory movements of a single flagellum. The inset shows a cross section through a mammalian sperm flagellum near its tip. The flagella of eukaryotic cells are so similar that this cross section could just as well have been taken of a flagellum from a protist or green alga. (A: FROM BERNARD R. GERBER, LEWIS M. ROUTLEDGE, AND SHIRO TAKASHIMA, J. MOL. BIOL. 71:322, 1972. COPYRIGHT 1972 BY PERMISSION OF THE PUBLISHER ACADEMIC PRESS; INSET COURTESY OF JULIUS ADLER AND M. L. DEPAMPHILIS; B: DAVID M. PHILLIPS/VISUALS UNLIMITED; (INSET) DON W. FAWCETT/VISUALS UNLIMITED.)

thermophiles (prokaryotes that live at very high temperatures). Included in this latter group are hyperthermophiles, such as *Pyrolobus fumarii*, which lives in the hydrothermal vents of the ocean floor and is capable of reproducing in superheated water at temperatures up to 113°C. In fact, *P. fumarii* cannot grow at temperatures below about 90°C. As discussed at length in the Experimental Pathways at the end of the chapter, the domain Archaea is thought to include our closest living prokaryotic ancestors.

All other types of prokaryotes are classified in the domain Bacteria. This domain includes the smallest living cells, the mycoplasma (0.2 μm diameter), which are also the only prokaryotes lacking a cell wall. Bacteria are present in every conceivable habitat on earth, from the permanent ice shelf of the Antarctic to the driest African deserts, to the internal confines of plants and animals. Bacteria have even been found living in rock layers situated several kilometers beneath the Earth's surface. Some of these bacterial communities are thought to have been cut off from life on the surface for more than one hundred million years. The most complex prokaryotes are the cyanobacteria. Cyanobacteria contain elaborate arrays of cytoplasmic membranes, which serve as sites of photosynthesis (Figure 1.15*a*). The cytoplasmic membranes of cyanobacteria are very similar to the photosynthetic membranes present within the chloroplasts of plant cells. As in eukaryotic plants, photosynthesis in cyanobacteria is accomplished by splitting water molecules, which releases molecular oxygen.

Many cyanobacteria are capable not only of photosynthesis, but also of **nitrogen fixation**, the conversion of nitrogen (N_2) gas into reduced forms of nitrogen (such as ammonia, NH_3) that can be used by cells in the synthesis of nitrogen-containing organic compounds, including amino acids and nucleotides. Those species capable of both photosynthesis and nitrogen fixation can survive on the barest of resources—light, N_2, CO_2, and H_2O. It is not surprising, therefore, that cyanobacteria are usually the first organisms to colonize the bare rocks rendered lifeless by a scorching volcanic eruption. Another unusual habitat occupied by cyanobacteria is illustrated in Figure 1.15*b*.

Prokaryotic Diversity For the most part, microbiologists are familiar only with those microorganisms they are able to grow in a culture medium. When a patient suffering from a respiratory or urinary tract infection sees his or her physician, one of the first steps taken is to culture the pathogen. Once it has been cultured, the organism can be identified and the proper treatment prescribed. It has proven relatively easy to culture *most* disease-causing prokaryotes, but the same is not true for those living free in nature. The problem is compounded by the fact that prokaryotes are barely visible in a light microscope and their morphology is often not very distinctive. As a result, we have been largely unaware of the majority of prokaryotic organisms with which we share the planet. Our ap-

(a)

(b)

Figure 1.15 Cyanobacteria. (*a*) Electron micrograph of a cyanobacterium showing the cytoplasmic membranes that carry out photosynthesis. These concentric membranes are very similar to the thylakoid membranes present within the chloroplasts of plant cells, a feature that supports the hypothesis that chloroplasts evolved from symbiotic cyanobacteria. (*b*) Cyanobacteria living inside the hairs of these polar bears are responsible for the unusual greenish color of their coat. (A: COURTESY OF NORMA J. LANG; B: COURTESY ZOOLOGICAL SOCIETY OF SAN DIEGO.)

preciation for the diversity of prokaryotic communities has increased dramatically in recent years through the introduction of molecular techniques that do not require the isolation of a particular organism.

Suppose one wanted to learn about the diversity of prokaryotes that live in the upper layers of the Pacific Ocean off the coast of California. Rather than trying to culture such organisms, which would prove largely futile, a researcher could concentrate the cells from a sample of ocean water, extract the DNA, and analyze certain DNA sequences present in the preparation. All organisms share certain genes, such as those that code for the RNAs present in ribosomes or the enzymes of certain metabolic pathways. Even though all organisms may share such genes, the sequences of the nucleotides that make up the

genes vary considerably from one species to another. This is the basis of biological evolution. By using techniques that reveal the variety of sequences of a particular gene in a particular habitat, one immediately learns about the numbers of species that live in that habitat. By carefully analyzing the sequences that are present in the extracted DNA and comparing those sequences to their counterparts from known organisms, one can also learn about the phylogenetic types of organisms that live in that habitat. In one recent study of the prokaryotes living in a single pool in Yellowstone National Park, for example, it was found that 30 percent of the sequences were from bacteria that could not be grouped into any of the 14 previously recognized divisions within the domain Bacteria. Given the considerable differences among the unaffiliated sequences, these previously unidentified bacteria were tentatively divided into 12 new divisions. Thus as the result of a single study on a single pool of water, the known diversity within the domain Bacteria was almost doubled.

By using these sequence-based molecular techniques, biologists have found that most habitats on Earth are teeming with previously unidentified prokaryotic life. Even the Archaea, whose members were once thought to be restricted to the harshest environments (page 27), have been found to be common and abundant members of such non-extreme habitats as the ocean, lakes, and soil. One recent estimate of the sheer numbers of prokaryotes in the major habitats of the Earth is given in Table 1.2. It is noteworthy that more than 90 percent of these organisms are now thought to live in the subsurface sediments well beneath the oceans and upper soil layers. It was only a decade or so ago that these deeper sediments were thought to be only sparsely populated by living organisms. Table 1.2 also provides an estimate of the amount of carbon that is sequestered in the world's prokaryotic cells. To put this number into more familiar terms, it is roughly comparable to the total amount of carbon present in all of the world's plant life.

Types of Eukaryotic Cells: Cell Specialization

In many regards, the most complex eukaryotic cells are not found inside of plants or animals, but rather among the single-celled (*unicellular*) protists, such as those pictured in Figure 1.16. All of the machinery required for the complex activities in which this organism engages—sensing the environment, trapping food, expelling excess fluid, evading predators—is housed within the confines of a single cell.

Complex unicellular organisms represent one evolutionary pathway. An alternate pathway has led to the evolution of multicellular organisms in which different activities are conducted by different types of specialized cells. Some of the advantages provided by the division of labor among cells can be appreciated by examining the relatively simple life cycle of another protist, the cellular slime mold *Dictyostelium*.

	Table 1.2	Number and Biomass of Prokaryotes in the World

Environment	No. of prokaryotic cells, $\times 10^{28}$	Pg of C in prokaryotes*
Aquatic habitats	12	2.2
Oceanic subsurface	335	303
Soil	26	26
Terrestrial subsurface	25–250	22–215
Total	415–640	353–546

*1 Pg = 10^{15}g
Source: W. B. Whitman et al., *Proc. Nat'l. Acad. Sci. U.S.A.* 95:6581, 1998.

At most stages of their life cycle, slime mold cells exist as solitary, independent amoebas, crawling over their substratum. Each cell is a complete, self-sufficient organism (Figure 1.17*a*). When food supplies become scarce, how-

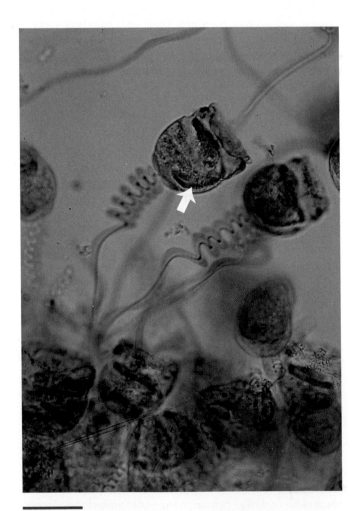

Figure 1.16 Vorticella, a complex ciliated protist. A number of these unicellular organisms are seen here; most have withdrawn their "heads" due to shortening of the blue-stained contractile ribbon in the stalk. Each cell has a single, large nucleus, called a macronucleus (arrow), which contains many copies of the genes. (CAROLINA BIOLOGICAL SUPPLY CO./PHOTOTAKE.)

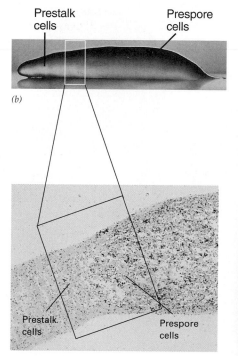

Prestalk cells

Prespore cells

(b)

Prestalk cells

Prespore cells

(a)

(c)

Spores

Stalk

(d)

Figure 1.17 The life cycle of a slime mold. (*a*) Amoebas undergoing aggregation by migrating toward a common center. (*b*) Following aggregation, the cells form a slug (or pseudoplasmodium) that migrates over the substratum leaving a trail of "slime" in its path. The cells at the front end of the slug (which will become stalk cells) can be distinguished from the cells at the rear end (which will become spore cells). As shown in the inset, the prespore cells of the rear end of the slug incorporate [³H]fucose, a radioactively labeled sugar that becomes part of the spore coat, whereas the prestalk cells lack this activity. The incorporation of the radioactive sugar is revealed by the black silver grains over the prespore cells. (*c*) The slug has ceased migration, has rounded up, and is beginning to lift off the substratum. Cells that will become part of the stalk are seen at the upper end. (*d*) The sporangium (fruiting body) consists of a slender stalk holding up a mass of spores at its upper end. Each spore will give rise to an independent amoeba, which begins the life cycle anew. (A, D: COURTESY OF JOHN TYLER BONNER; B: COURTESY OF DAVID FRANCIS, INSET BY G. KARP; C: COURTESY OF KENNETH B. RAPER.)

ever, a new type of activity is triggered among the cells. They stream toward each other to form an aggregate called a *pseudoplasmodium*, or simply a slug (Figure 1.17*b*), which migrates slowly over the substratum leaving a trail of "slime." What were previously single organisms are now small parts of a much larger, multicellular individual. Examination of the interior of the slug reveals that the cells are no longer a homogeneous population. Rather, the cells of the anterior third of the slug (called *prestalk cells*) can be distinguished from those of the posterior section (called *prespore cells*) by a number of criteria (Figure 1.17*b*, inset). After a while, a dramatic series of events occurs: the slug stops its migration, rounds up on the substratum (Figure 1.17*c*), and then extends upward into the air as an elongated sporangium (or fruiting body; Figure 1.17*d*). The sporangium is composed of a slender stalk (derived from the prestalk cells) supporting a rounded mass of dormant, encapsulated spores (derived from the prespore

cells). The stalk and spore cells have very different functions that require different types of cytoplasmic specializations. Stalk cells provide the mechanical support to hold the spore mass above the substratum, whereas spore cells are destined to scatter in the wind and develop into the next generation of amoebas.

The process whereby a relatively unspecialized cell, such as a slime mold amoeba, becomes a highly specialized cell, such as a stalk cell or spore cell, is called **differentiation**. A slime mold amoeba has two alternate paths of differentiation available to it as it enters the aggregation stage. In contrast, when the egg of a vertebrate is fertilized and progresses through its course of embryonic development, there are hundreds of possible pathways of differentiation. Some cells become part of a particular digestive gland, others part of a large skeletal muscle, others part of a bone, and so forth (Figure 1.18). The pathway of differentiation followed by each embryonic cell depends pri-

Nerve cells

Loose connective tissue
with fibroblasts

Red blood cells

Smooth
muscle cells

Bone tissue
with osteocytes

Fat (adipose) cells

Striated muscle cells

Intestinal epithelial cells

Figure 1.18 **Pathways of cell differentiation.** A few of the types of differentiated cells present in a human fetus.

marily on the signals it receives from the surrounding environment; these signals in turn depend on the position of that cell within the embryo. As discussed in the accompanying Human Perspective, researchers are learning how to control the process of differentiation in the culture dish and applying this knowledge to the treatment of complex human diseases.

As a result of differentiation, different types of cells acquire a distinctive appearance and contain unique materials. Skeletal muscle cells contain a network of precisely aligned filaments composed of unique contractile pro-

teins; cartilage cells become surrounded by a characteristic matrix containing polysaccharides and the protein collagen, which together provide mechanical support; red blood cells become disk-shaped sacks filled with a single protein, hemoglobin, which transports oxygen; and so forth. Despite their many differences, however, the various cells of a multicellular plant or animal are composed of similar organelles. Mitochondria, for example, are found in essentially all types of cells. In one type, however, they may have a rounded shape, whereas in another they may be highly elongated and thread-like. Consider the two

THE HUMAN PERSPECTIVE

Replacing Cells and Organs

UPDATE For the very latest, go to page U-1

To a person whose heart or liver is failing, an organ transplant is the best hope for survival and the return to a normal life. Organ transplantation is one of the great successes of modern medicine, but its scope is greatly limited by the availability of donor organs and the high risk of immunologic rejection. Imagine the possibilities if we could grow cells and organs in the laboratory and use them to replace damaged or nonfunctional parts within our bodies. Although this prospect remains largely in the realm of science fiction, studies in recent years have given researchers hope that one day this type of therapy will be commonplace. To better understand the concept of cell replacement therapy, we can consider a present-day procedure known as *bone marrow transplantation* in which cells are extracted from the interior of the pelvic bones of a donor and infused into the body of a recipient.

Bone marrow transplantation is used primarily to treat lymphomas and leukemias, which are cancers that affect the number of white blood cells. To carry out the procedure, the patient is exposed to a high level of radiation and/or toxic chemicals, which kills the cancerous cells, but also kills all of the cells involved in the formation of red and white blood cells. This treatment has this effect because blood-forming cells are particularly sensitive to radiation and toxic chemicals. Once a person's blood-forming cells have been destroyed, they are replaced by transplantation of bone marrow cells from a healthy donor. Bone marrow can regenerate the blood tissue of the transplant recipient because it contains a small population of cells that can proliferate and restock the patient's own blood-forming bone marrow tissue.[1,2] These "special" cells in the bone marrow are referred to as **hematopoietic stem cells**, and they are normally responsible for replacing the millions of red and white blood cells that age and die every day in our bodies. Amazingly, a *single* hematopoietic stem cell is capable of reconstituting the entire hematopoietic (blood-forming) system of an irradiated mouse.

It was thought for many years that hematopoietic stem cells were only capable of differentiating into the various types of blood cells found in the body. This belief came into question when it was shown that laboratory animals that had received bone marrow transplants showed evidence that donor cells had differentiated into unrelated types of cells, including muscle cells. It was subsequently shown that bone marrow stem cells injected into the brains of mice appeared to develop into nerve cells. In 2000 it was reported that the livers of human patients that had received bone marrow transplants several years earlier contained large numbers of donor liver

cells (hepatocytes). The donor cells could be identified in a recipient's liver because the transplant had come from a member of the opposite sex, whose cells are readily distinguished by their chromosome content. It would appear that the fate of a particular hematopoietic stem cell depends largely on the environment in which it ultimately finds itself, rather than the part of the body from which it is originally derived.

Stem cells are defined as undifferentiated cells that are capable of both (1) self-renewal, that is, production of cells like themselves, and (2) commitment, that is, production of cells that are committed to differentiate. The adult human body contains a variety of different types of stem cells that give rise to the specialized cells of the organs in which they are found.

Stem cells hold great promise in the replacement of cells that have been damaged by injury or disease. Stem cells can serve as the "raw material" for cell replacement because they can grow and divide extensively in culture, and then differentiate to produce large numbers of specialized cells. Most differentiated cells, in contrast, are not capable of growth and division so that they cannot be obtained in quantities that are useful for transplantation. Researchers are presently trying to determine how undifferentiated stem cells can be directed to differentiate in a culture dish into a particular, desired cell type, such as heart muscle cells or dopamine-producing nerve cells, which could be used for transplantation.

The greatest potential for stem cell therapy may lie in the treatment of neurodegenerative diseases such as Parkinson's disease, which is characterized by tremors and muscular rigidity, and Alzheimer's disease, which is characterized by loss of memory and cognitive function. Until recently it was believed that the human brain lacked all capacity to produce new nerve cells. Although most of our nerve cells have been with us since childhood, a couple of areas of the adult brain have been found to continually produce new nerve cells. These nerve cells are derived from a population of *neural stem cells* that can be isolated from brain tissue removed during surgery. Like hematopoietic stem cells, neural stem cells can proliferate in a laboratory culture dish. Depending on the culture conditions, neural stem cells can be kept undifferentiated or induced to differentiate into specialized cell types, typically nerve cells or glial cells (the supportive cells of the brain), but other types of cells as well. When injected into the brains of mice, neural stem cells can differentiate into nerve cells that become functionally integrated into the animal's own brain.

It was shown in the early 1990s that the symptoms of Parkinson's disease in humans could be greatly alleviated by injection of nerve cells into the affected part of the brain. The donated nerve cells were taken from the brains of aborted human fetuses. Figure 1 shows a brain scan of an individual with Parkinson's disease who had received an injection of fetal nerve cells on one side of their brain 10 years earlier. Many of the transplanted cells are still alive and functioning,

[1]Bone marrow transplantation can be contrasted to a simple blood transfusion where the recipient receives only the differentiated blood cells (especially red blood cells and platelets) present in the circulation.

[2]Bone marrow transplantation has also been used in the treatment of certain other cancers, most notably breast cancer. Patients with advanced breast cancer may be given high doses of radiation and chemotherapy to kill the cancer cells, which also has the effect of destroying the patient's blood-forming tissue, thus requiring the transplant.

Figure 1 Long-term survival of transplanted fetal neurons. *(left)* A PET scan of a normal human brain reveals the distribution of labeled dopa (shown in red/white), a marker for the dopamine-producing nerve cells that are selectively destroyed by Parkinson's disease. *(right)* Brain scan of a person with Parkinson's disease that had received a graft of fetal nerve cells on one side of their brain 10 years earlier. The transplanted cells are still alive and functional, unlike most of the patient's own dopaminergic neurons that have degenerated. (COURTESY OF PAOLA PICCINI, MRC, ICSM, HAMMERSMITH HOSPITAL, LONDON, UK.)

transplantation studies. This conclusion is based on studies in mice that have shown that *undifferentiated* ES cells have a high potential of becoming cancerous when transplanted to adult animals, forming tumors called teratomas (or teratocarcinomas). Teratomas are highly unusual tumors because they contain a strange mixture of malignant undifferentiated ES cells mixed in with various differentiated cells they have given rise to. It is not unusual, for example, to find teratomas that are covered with hair produced by skin cells that have differentiated from their malignant ES precursors.

A major question at the present time is whether adult stem cells hold as much promise as ES cells as the basis for replacement cell therapy. The discovery that adult stem cells, such as those of the bone marrow, have broader powers of differentiation than previously thought has made them more attractive prospects for use in cell therapy. Adult stem cells have the advantage over ES cells in that they can be isolated from the individual who is being treated and, thus, will not face immunological rejection when used in subsequent cell replacement. On the other hand, adult stem cells do not seem to have an unlimited potential to proliferate as is characteristic of ES cells. In addition, it may be possible to "customize" ES cells so that they possess the same genetic makeup of the individual who is being treated, and thus would not be subject to attack by the recipient's immune system. This can likely be accomplished by a roundabout procedure, shown in

even though progression of the disease has ravaged the surrounding tissue. Human fetuses, however, are not a suitable source of transplant tissue. There are major ethical considerations in this procedure, and the demand for such cells far outweighs the potential supply. Regardless, these studies have clearly demonstrated that neurodegenerative diseases can be treated by nerve cell replacement. With this in mind, clinical trials are planned in which patients with Parkinson's disease will be injected with neural stem cells harvested from their own brains.

We have restricted the discussion to this point to stem cells found in adults. The greatest excitement in this field, as well as the greatest controversy, has come not from studies on adult stem cells but on **embryonic stem (ES) cells**, which are isolated from very young mammalian embryos (see Figure 18.45). These are the cells in the early embryo that give rise to all of the various structures of the mammalian fetus. In 1998, James Thomson and his colleagues at the University of Wisconsin were the first to isolate human ES cells and to grow them for extended periods in cell culture. These cells were prepared from unused embryos that had been donated by couples to infertility clinics. ES cells have the capacity to differentiate into a diverse array of cell types in culture. The long-range goal of clinical researchers is to learn how to get ES cells to differentiate into desired cell types that can be used for cell replacement therapy. Some progress has already been made in this direction. For example, addition of retinoic acid (a derivative of vitamin A) to a culture of ES cells induces most of them to differentiate into nerve cells. When ES-derived nerve cells were injected into rats with spinal cord injuries, the animals showed marked improvements in mobility.

It will be particularly important when working with ES cells that only the differentiated cell products be used in

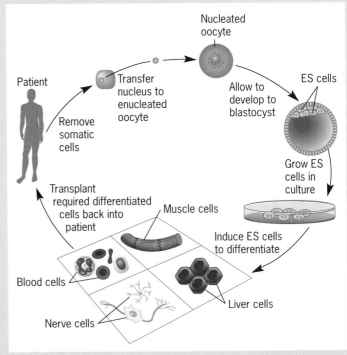

Figure 2 A procedure for obtaining differentiated cells for use in cell replacement therapy. A small piece of tissue is taken from the patient, and a nucleus from one of the cells is implanted into a donor oocyte whose own nucleus had been previously removed. The resulting oocyte (egg) is allowed to develop into an early embryo, and the ES cells are harvested and grown in culture. A population of ES cells are induced to differentiate into the required cells, which are subsequently transplanted into the patient to restore organ function.

Figure 2, that begins with an unfertilized egg—a cell that is readily obtained from the ovaries of an unrelated woman donor. In this approach, the nucleus of the unfertilized egg would be replaced by the nucleus of a cell from the patient to be treated, which would cause the egg to have the same chromosome composition as that of the patient. The egg would then be allowed to develop to an early embryonic stage, and the ES cells would be removed, cultured, and induced to differentiate into the type of cells needed by the patient. Because this procedure involves the formation of a human embryo that is used only as a source of ES cells, there are major ethical questions that must be settled before it could be routinely practiced. Alternatively, there may be ways to customize ES cells without resorting to the formation of a potentially viable human embryo.

Some researchers are looking beyond the practice of simple cell replacement therapy and envisioning a day when an entire organ might be replaced by a "new model" that had been engineered within the lab. Most organs are more than a relatively homogeneous mixture of cells. A kidney, for example, has a highly complex three-dimensional microstructure so that one part of the organ is very different from another part. When cells are grown on a culture dish, they generally form a simple layer that covers the bottom surface. To advance beyond this simple arrangement toward the construction of a large, three-dimensional structure, it is first necessary to construct some type of porous scaffold in the shape of the replacement organ to which different types of cells can adhere. The ideal scaffold is composed of an inert, branched polymer that can be digested away by the cells as they take over the interior space of the synthetic organ. Several such polymers have been developed and appear to work quite well.

The initial forays into the field of organ engineering have focused on relatively simple structures, such as the urinary bladder, blood vessels, cartilage, and skin. Let's look more closely at the bladder, which is a hollow organ whose wall consists primarily of two types of cells; a thick outer layer composed of smooth muscle cells that maintain pressure on the fluid contents, and a thin inner layer composed of epithelial cells that control the exchange of salts and fluids. In one

Figure 3 **A bioengineered bladder.** (COURTESY OF ANTHONY ATALA, REPRINTED WITH PERMISSION FROM SCIENCE 284,425,1999. COPYRIGHT © 1999 AMERICAN ASSOCIATION FOR THE ADVANCEMENT OF SCIENCE.)

series of studies, researchers at Harvard University constructed a scaffold in the shape of a urinary bladder and seeded the structure with the two major types of bladder cells. Muscle cells were seeded onto the outer portions of the scaffold, and epithelial cells were seeded onto the scaffold's inner surface. The "organ" was then placed in a culture medium where the cells proliferated to produce a structure with the cellular organization of a typical bladder (Figure 3). When this laboratory-grown bladder was placed into laboratory animals whose own bladder had been removed, the synthetic scaffold was gradually degraded and the replacement organ became infiltrated with blood vessels and innervated by nerve cells. In fact, the laboratory-engineered organ appears to meet the needs of the animal by storing urine and releasing it when appropriate. Researchers have gone on to engineer a human version of a urinary bladder. It is hoped that these organs will soon be available to persons whose own bladder has been damaged by injury or disease.

connective tissue cells shown in Figure 1.19. The cell on the left is a brown adipose cell whose primary function is to generate heat from the chemical energy stored in fat. These cells have numerous fat droplets and large numbers of mitochondria where energy conversion takes place. In contrast, the plasma cell depicted in Figure 1.19b is specialized for production of antibody (protein) molecules. Plasma cells have a relatively small number of mitochondria, but an extensive endoplasmic reticulum where protein synthesis takes place. In each case, the number, appearance, and location of the various organelles can be correlated with the activities of the particular cell type. An analogy might be made to a variety of orchestral pieces: all are composed of the same notes, but varying arrangement gives each its unique character and beauty.

Model Organisms Living organisms are highly diverse, and the results obtained from a particular experimental

analysis may depend on the particular organism being studied. As a result, cell and molecular biologists have focused their research activities on a small number of "representative" or **model organisms**, in the hope that a comprehensive body of knowledge of these systems can be obtained. Five model systems representing a wide range of eukaryotes have captured most of the attention: a budding yeast, *Saccharomyces cerevisiae*; a mustard plant, *Arabidopsis thaliana*; a nematode, *Caenorhabditis elegans*; a fruit fly, *Drosophila melanogaster*; and a mouse, *Mus musculus*. Each of these organisms has specific advantages that make it particularly useful as a research system for answering certain types of questions. Several of these organisms are pictured in Figure 1.20, and a few of their advantages as research systems are described in the accompanying legend. We will concentrate in this text on results obtained from studies on mammalian systems—mostly on the mouse and on cultured mammalian cells—because these findings

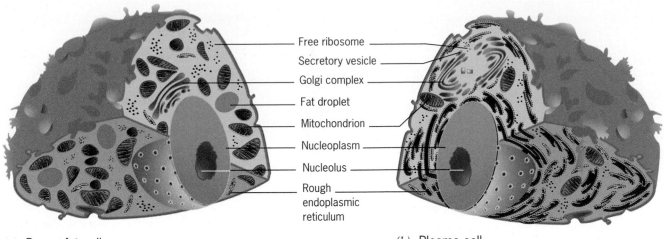

Free ribosome

Secretory vesicle

Golgi complex

Fat droplet

Mitochondrion

Nucleoplasm

Nucleolus

Rough endoplasmic reticulum

(a) Brown fat cell

(b) Plasma cell

Figure 1.19 **Different types of cells have the same organelles,** but they differ in number, shape, and distribution. The cell depicted in *a* is a brown fat cell whose function is to convert fat present in the many cytoplasmic oil droplets to thermal energy, which warms the individual. In humans, these cells are found predominantly in infants. These cells have very little endoplasmic reticulum because they are not synthesizing much protein. In contrast, the plasma cell depicted in *b* is specialized for the production and secretion of antibodies, which are proteins. The plasma cell has an extensive endoplasmic reticulum and Golgi complex, but relatively few mitochondria or fat droplets.

are most applicable to humans. Even still, we will have many occasions to describe research carried out on other systems. You may be surprised to discover how similar you are at the cell and molecular level to these much smaller and simpler organisms.

The Sizes of Cells and Their Components

Figure 1.21 shows the relative size of a number of structures of interest in cell biology. Two units of linear measure are most commonly used to describe structures within a cell: the **micrometer** (μm) and the **nanometer** (nm).

(a)

(b)

(c)

Figure 1.20 **Three model organisms.** (*a*) *Arabidopsis thaliana*, a member of a genus of mustard plants, has an unusually small genome (120 megabases, Mb) for a higher plant, a rapid generation time, and large seed production. (*b*) *Caenorhabditis elegans*, a nematode, consists of a defined number of cells (roughly 1000), each of which develops according to a precise pattern of cell divisions. The animal has a transparent body wall, a short generation time, and facility for genetic analysis. This micrograph shows the larval nervous system, which has been labeled with the green fluorescent protein (GFP). (*c*) *Mus musculus*, the common house mouse, is easily kept and bred in the laboratory. Many different genetic strains have been developed, including the "nude mouse" pictured here which develops without a thymus gland and, therefore, is able to accept human tissue grafts that are not rejected. Other model organisms—budding yeast and a fruit fly—are pictured in Figure 8.7 and 10.4, respectively. (A: JEAN CLAUDE REVY/PHOTOTAKE; B: FROM KARLA KNOBEL, KIM SCHUSKE, AND ERIK JORGENSEN, TRENDS GENETICS, VOL. 14, COVER #12, 1998; C: G. ROBERT BISHOP/STONE.)

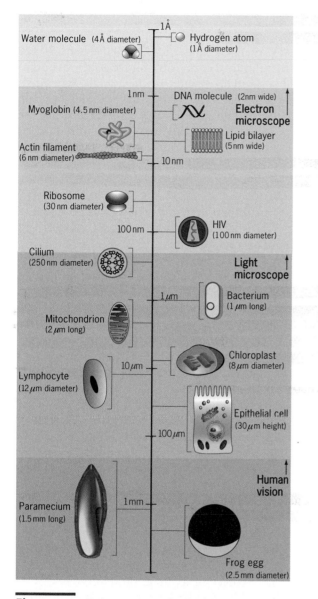

Figure 1.21 Relative sizes of cells and cell components.

number of reasons why most cells are so small. Consider the following.

- Most eukaryotic cells contain a single nucleus that contains only two copies of most genes. Because genes serve as templates for the production of information-carrying messenger RNAs, a cell can only produce a limited number of these messenger RNAs in a given amount of time. The greater a cell's cytoplasmic volume, the longer it will take to synthesize the number of messages required by that cell.

- As a cell increases in size, the surface area/volume ratio decreases.[2] The ability of a cell to exchange substances with its environment is proportional to its surface area. If a cell were to grow beyond a certain size, its surface would not be sufficient to take up the substances (e.g., oxygen, nutrients) needed to support its metabolic activities. Cells that are specialized for absorption of solutes, such as those of the intestinal epithelium, typically possess microvilli, which greatly increase the surface area available for exchange (see Figure 1.3). The interior of a large plant cell is typically filled by a large, fluid-filled vacuole rather than metabolically active cytoplasm (see Figure 8.36).

- A cell depends to a large degree on the random movement of molecules (*diffusion*). Oxygen, for example, must diffuse from the cell's surface through the cytoplasm to the interior of its mitochondria. The time required for diffusion is proportional to the square of the distance to be traversed. For example, O_2 requires only 100 microseconds to diffuse a distance of 1 micrometer, but requires 10^6 times as long to diffuse a distance of 1 millimeter. As a cell becomes larger, and the distance from the surface to the interior becomes greater, the time required for diffusion to move substances in and out of a metabolically active cell becomes prohibitively long.

One μm is equal to 10^{-6} meters, and one nm is equal to 10^{-9} meters. Although no longer formally accepted in the metric nomenclature, the **angstrom** (Å), which is equal to one-tenth of a nm, is still commonly employed by molecular biologists for atomic dimensions. One angstrom is roughly equivalent to the diameter of a hydrogen atom. Myoglobin, a typical globular protein, is approximately 4.5 nm × 3.5 nm × 2.5 nm; highly elongated proteins (such as collagen or myosin) are over 100 nm in length; and DNA is approximately 2.0 nm in width. Complexes of large molecules, such as ribosomes, microtubules, and microfilaments, are between 5 and 25 nm in diameter. Larger organelles, such as nuclei (approximately 5–10 μm in diameter) or mitochondria (approximately 2 μm in length), are more easily defined in micrometers.

Bacteria typically range in length from about 1 to 5 μm, eukaryotic cells from about 10 to 30 μm. There are a

REVIEW ?

1. Compare a prokaryotic and eukaryotic cell on the basis of structural, functional, and metabolic differences.
2. What is the importance of cell differentiation?
3. Why are cells almost always microscopic?
4. If a mitochondrion were 2 micrometers in length, how many angstroms would it be? How many nanometers? How many millimeters?

[2]You can verify this statement by calculating the surface area and volume of a cube whose sides are 1 cm in length versus a cube whose sides are 10 cm in length. The surface area/volume ratio of the smaller cube is considerably greater than that of the larger cube.

1.4 VIRUSES

By the end of the nineteenth century, the work of Louis Pasteur and others had convinced the scientific world that the infectious diseases of plants and animals were due to the presence of bacteria. But studies of tobacco mosaic disease in tobacco plants and hoof-and-mouth disease in cattle soon pointed to the existence of another type of infectious agent. It was found, for example, that sap from a diseased tobacco plant could transmit mosaic disease to a healthy plant, even when the sap showed no evidence of bacteria when examined in the light microscope. Furthermore, the sap from a diseased plant remained infective even after being forced through filters whose pores were so small that they retarded the passage of the smallest known bacteria. Investigators concluded that certain diseases were caused by pathogens that were even smaller, and pre-

sumably simpler, than the smallest bacteria. They named these pathogens **viruses**.

In 1935, Wendell Stanley of the Rockefeller Institute reported that the virus responsible for tobacco mosaic disease could be crystallized and that the crystals were infective. Substances that form crystals have a highly ordered, well-defined structure and are vastly less complex than the simplest cells. Stanley mistakenly concluded that tobacco mosaic virus (TMV) was a protein. In fact, TMV is a rod-shaped particle consisting of a single molecule of RNA surrounded by a helical shell composed of protein subunits (Figure 1.22).

Viruses are responsible for dozens of human diseases, including AIDS, polio, influenza, cold sores, measles, and a few types of cancer (see Section 16.2). Viruses occur in a wide variety of very different shapes, sizes, and constructions, but all of them share certain common properties. All viruses are obligatory intracellular parasites; that is, they cannot reproduce unless present within a host cell, which depending on the specific virus may be a plant, animal, or bacterial cell. Outside of a living cell, the virus exists as a particle, or **virion**, which is little more than a macromolecular package. The virion contains a small amount of genetic material that, depending on the virus, can be single-stranded or double-stranded, RNA or DNA. Remarkably, some viruses have as few as three or four different genes, but others may have as many as several hundred. The fewer the genes, the more the virus relies on enzymes and other proteins encoded by its host cell's genes.

The genetic material of the virion is surrounded by a protein capsule, or *capsid*, which is generally made up of a specific number of subunits. There are numerous advantages to construction by subunit, one of the most apparent being an economy of genetic information. If a viral coat is made of many copies of a single protein, as is that of TMV, or a few proteins, as are the coats of many other viruses, the virus needs only one or a few genes to code for its protein container.

Many viruses have a capsid whose subunits are organized into a polyhedron, that is, a structure having planar faces. A particularly common polyhedral shape of viruses is the 20-sided *icosahedron*. For example, adenovirus, which causes respiratory infections in mammals, has an icosahedral capsid (Figure 1.23a). In many animal viruses, including the *human immunodeficiency virus (HIV)* responsible for AIDS, the protein capsid is surrounded by a lipid-containing outer envelope that is derived from the modified plasma membrane of the host cell as the virus buds from the host cell surface (Figure 1.23b). Bacterial viruses, or *bacteriophages*, are among the most complex viruses (Figure 1.23c). The T bacteriophages (which were used in key experiments that revealed the structure and properties of the genetic material) consist of a polyhedral head containing DNA, a cylindrical stalk through which the DNA is injected into the bacterial cell, and tail fibers, which together cause the particle to resemble a landing module for the moon.

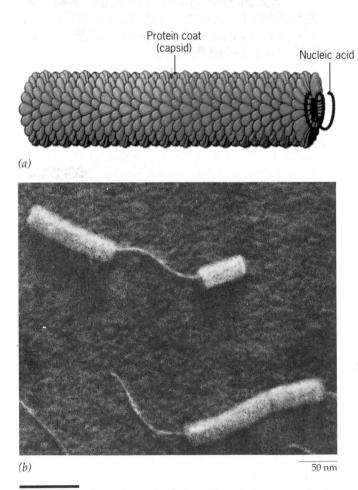

Protein coat
(capsid)

Nucleic acid

(a)

(b)

50 nm

Figure 1.22 Tobacco mosaic virus (TMV). (*a*) Diagram of a portion of a TMV particle. The protein subunits, which are identical along the entire rod-shaped particle, enclose a single helical RNA molecule. The RNA is shown protruding at the end where the protein has been stripped away. (*b*) Electron micrograph of TMV particles taken after treatment with phenol has removed the protein subunits from the middle part of the upper particle and the ends of the lower particle. Intact rods are approximately 300 nm long and 18 nm in diameter. (B: COURTESY OF M. K. CORBETT.)

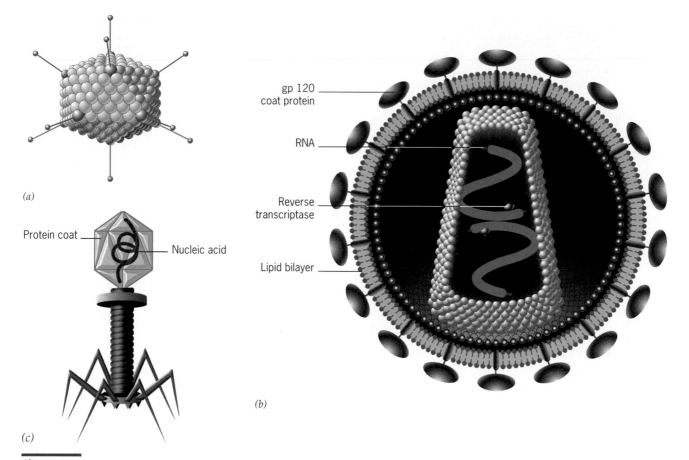

gp 120
coat protein

RNA

Reverse
transcriptase

Lipid bilayer

(a)

Protein coat

Nucleic acid

(b)

(c)

Figure 1.23 **Virus diversity.** The structures of (*a*) an adenovirus, (*b*) a human immunodeficiency virus (HIV), and (*c*) a T-even bacteriophage. (Note: These viruses are not drawn to the same scale.)

Each virus has on its surface a protein that is able to bind to a particular surface component of its host cell. For example, the protein that projects from the surface of the HIV particle (labeled gp120 in Figure 1.23*b*, which stands for glycoprotein of molecular mass 120,000 daltons[3]) interacts with a specific protein on the surface of certain white blood cells, facilitating entry of the virus into its host cell. The interaction between viral and host proteins determines the specificity of the virus, that is, the types of host cells that the virus can enter and infect. Some viruses have a wide *host range*, being able to infect cells from a variety of different organs or host species. The virus that causes rabies, for example, is able to infect many different types of mammalian hosts, including dogs, bats, and humans. Most viruses, however, have a relatively narrow host range. This is generally true, for example, of human cold and influenza viruses, which are only able to infect the respiratory epithelial cells of human hosts. A change in host cell specificity can have dramatic consequences. The worst influenza epidemic in modern history came in 1918, when a strain referred to as the Spanish influenza virus was responsible for the deaths of more than 20 million

people worldwide. Recent studies suggest that this strain may have been so virulent because it was able to infect a wide variety of cells in addition to those of the respiratory tract.

Virions are macromolecular aggregates, inanimate particles that, by themselves, are unable to reproduce, metabolize, or carry on any of the other activities associated with life. For this reason, viruses are not considered to be organisms and are not described as being alive. Nevertheless, once a virus has become attached to the outer surface of a host cell and has passed through that cell's outer membrane, the genetic material of the virus contains the necessary information to alter the activities of the host cell.

There are two basic types of viral infection. (1) In most cases, the virus arrests the normal synthetic activities of the host and redirects the cell to use its available materials to manufacture viral nucleic acids and proteins, which assemble into new virions. Viruses, in other words, do not grow like cells; they are assembled from components directly into the mature-sized virions. Ultimately, the infected cell ruptures (*lyses*) and releases a new generation of viral particles capable of infecting neighboring cells. An example of this type of *lytic* infection is shown in Figure 1.24*a*. (2) In other cases, the infecting virus does not lead

[3]One dalton is equivalent to one unit of atomic mass, the mass of a single hydrogen (^1H) atom.

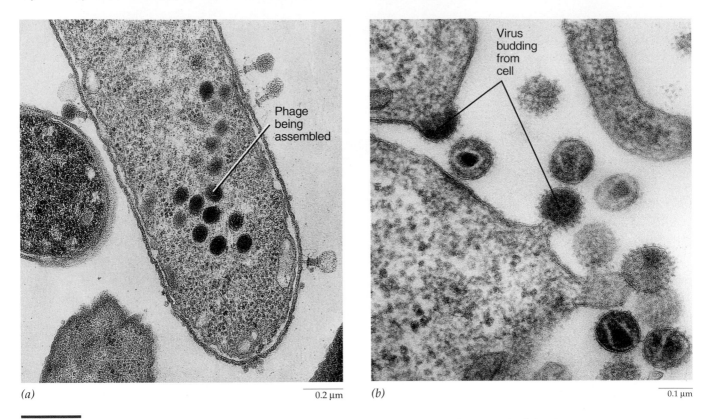

(a) 0.2 µm *(b)* 0.1 µm

Figure 1.24 **A virus infection.** (*a*) Micrograph showing a late stage in the infection of a bacterial cell by a bacteriophage. Virus particles are being assembled within the cell, and empty phage coats are still present on the cell surface. (*b*) Micrograph showing HIV particles budding from an infected human lymphocyte. (A: COURTESY OF JONATHAN KING AND ERIKA HARTWIG; B: COURTESY OF HANS GELDERBLOM.)

to the death of the host cell, but instead inserts (*integrates*) its DNA into the DNA of the host cell's chromosomes. The integrated viral DNA is called a **provirus**. An integrated provirus can have different effects depending on the type of virus and host cell. For example,

- Bacterial cells containing a provirus behave normally until exposed to a stimulus, such as ultraviolet radiation, that activates the dormant viral DNA, leading to the lysis of the cell and release of viral progeny.

- Some animal cells containing a provirus produce new viral progeny that bud at the cell surface without lysing the infected cell. Human immunodeficiency virus (HIV) acts in this way; an infected cell may remain alive for a period, acting as a factory for the production of new virions (Figure 1.24*b*).

- Some animal cells containing a provirus lose control over their own growth and division and become malignant. This phenomenon is readily studied in the laboratory by infecting cultured cells with the appropriate tumor virus.

Because of their simple construction, one might conclude that viruses represent a primitive "evolutionary stage" similar to those that existed on Earth before the evolution of prokaryotic cells. However, once you consider that viruses are totally dependent on the living cells they invade, it becomes evident that viruses could not have arrived on the scene before their hosts. Because viruses share the same genetic language with one another as well as with prokaryotic and eukaryotic cells, they could not have arisen independently as a primitive form *after* other cells had evolved. It is most reasonable to assume that viruses represent a degenerate form, that is, one derived from a cellular organism. Viruses have presumably evolved from small fragments of cellular chromosomes that were able to maintain some type of autonomous existence within the cell. Over time, these autonomous genetic elements acquired a protein wrapper and became modified into agents capable of infecting other cells. Considering the tremendous diversity among viruses, it is likely that different groups evolved independently from different cellular organisms.

Viruses are not without their virtues. Because the activities of viral genes mimic those of host genes, investigators have used viruses for decades as a research tool to study the mechanism of DNA replication and gene expression in their much more complex hosts. In addition, viruses are now being used as a means to introduce foreign genes into human cells, a technique that will serve as the

basis for the treatment of human diseases by gene therapy. Last, insect-killing viruses may play an increasing role in the future in the war against insect pests.

Viroids

It came as a surprise in 1971 to discover that viruses are not the simplest types of infectious agents. In that year, T. O. Diener of the U.S. Department of Agriculture reported that potato spindle-tuber disease, which causes potatoes to become gnarled and cracked, is caused by an infectious agent consisting of a small circular RNA molecule that totally lacks a protein coat. Diener named the pathogen a **viroid**. The RNAs of viroids range in size from about 240 to 600 nucleotides, a tenth the size of the smaller viruses. No evidence has been found that the naked viroid RNA codes for any proteins. Rather, any biochemical activities in which viroids engage take place using host cell proteins. For example, duplication of the viroid RNA within an infected cell utilizes the host's RNA polymerase II, an enzyme that normally transcribes the host's DNA into messenger RNAs. Viroids are thought to cause disease by interfering with the cell's nor-

mal path of gene expression. The effect on crops can be serious: a viroid disease called cadang-cadang has devastated the coconut palm groves of the Philippines, and another viroid has wreaked havoc on the chrysanthemum industry in the United States. The discovery of a different type of infectious agent even simpler than a viroid is described in the Human Perspective in Chapter 2.

REVIEW ?

1. What properties distinguish a virus from a bacterium?

2. Why are viruses thought to have evolved from cellular life forms rather than the reverse?

3. Compare and contrast: nucleoid and nucleus; the flagellum of a bacterium and a sperm; an archaeon and a cyanobacterium; prespore and prestalk cells of a cellular slime mold; nitrogen fixation and photosynthesis; bacteriophages and tobacco mosaic virus; a provirus and a virion.

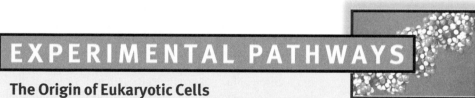

EXPERIMENTAL PATHWAYS

The Origin of Eukaryotic Cells

We have seen in this chapter that cells can be conveniently divided into two groups: prokaryotic cells and eukaryotic cells. Almost from the time this division of cellular life was first proposed, biologists have been fascinated by the question: What is the origin of the eukaryotic cell? It has been generally agreed for decades that prokaryotic cells (1) arose before eukaryotic cells and (2) gave rise to eukaryotic cells. The first point can be verified directly from the fossil record, which shows that prokaryotic cells were present in rocks approximately 3.5 billion years old (see Figure 1.10), which is 1 to 2 billion years before any evidence is seen of eukaryotes. The second point follows from the fact that the two types of cells have to be related to one another because they share many complex traits (e.g., very similar genetic codes, enzymes, metabolic pathways, and plasma membranes) that could not have evolved independently in different organisms.

Up until about 1970, it was generally believed that eukaryotic cells evolved from prokaryotic cells by a process of gradual evolution in which the organelles of the eukaryotic cell became progressively more complex. Acceptance of this concept changed dramatically about that time largely through the work of Lynn Margulis, then at Boston University. Margulis resurrected an idea that had been proposed earlier, and dismissed, that certain organelles of a eukaryotic cell—most notably the mitochondria and chloroplasts—had

evolved from smaller prokaryotic cells that had taken up residence in the cytoplasm of a larger host cell.[1,2] This hypothesis is referred to as the **endosymbiont theory** because it describes how a single "composite" cell of greater complexity can evolve from two or more separate, simpler cells living in a symbiotic relationship with one another.

Our earliest prokaryotic ancestors were presumed to have been anaerobic heterotrophic cells: *anaerobic* because they derived their energy from food matter without employing molecular oxygen (O_2), and *heterotrophic* because they were unable to synthesize organic compounds from inorganic precursors (such as CO_2 and water), but instead had to obtain preformed organic compounds from their environment. Because they gave rise to all modern organisms, these early prokaryotes must have possessed many of the characteristics that are universal in living cells today.

According to one version of the endosymbiosis theory, a large, anaerobic, heterotrophic prokaryote ingested a small, aerobic prokaryote (step 1, Figure 1). Resisting digestion within the cytoplasm, the small aerobic prokaryote took up residence as a permanent endosymbiont. As the host cell reproduced, so did the endosymbiont, so that a colony of these composite cells was soon produced. Over many generations, endosymbionts lost many of the traits that were no longer required for survival, and the once-independent oxygen-

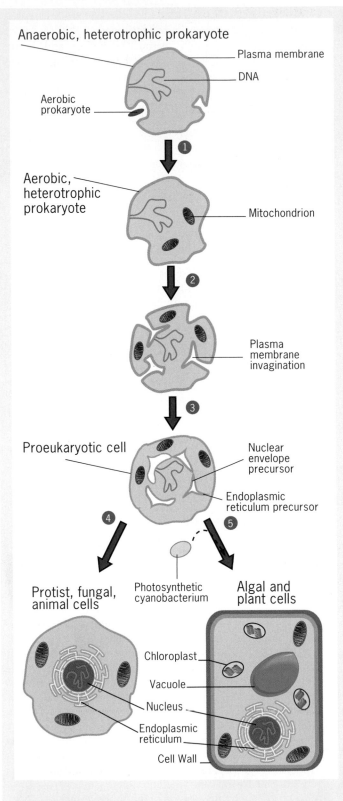

Figure 1 A model depicting possible steps in the evolution of eukaryotic cells, including the origin of mitochondria and chloroplasts by endosymbiosis. In step 1, a large anaerobic, heterotrophic prokaryote takes in a small aerobic prokaryote. Evidence strongly indicates that the engulfed prokaryote was an ancestor of modern-day rickettsia, a group of bacteria that causes typhus and other diseases. In step 2, the aerobic endosymbiont has evolved into a mitochondrion. In step 3, a portion of the plasma membrane has invaginated and is seen in the process of evolving into a nuclear envelope and associated endoplasmic reticulum. The primitive eukaryote depicted in step 3 gives rise to two major groups of eukaryotes. In one path (step 4), the primitive eukaryote evolves into nonphotosynthetic protist, fungal, and animal cells. In the other path (step 5), the primitive eukaryote takes in a photosynthetic prokaryote, which will become an endosymbiont and evolve into a chloroplast. (Note: The engulfment of the symbiont shown in step 1 could have occurred after development of some of the internal membranes, but evidence suggests it was a relatively early step in the evolution of eukaryotes.)

teristics are proposed to have arisen by a gradual process of evolution, rather than in a single step as might occur through acquisition of an endosymbiont. The endoplasmic reticulum and nuclear membranes, for example, might have been derived from a portion of the cell's outer plasma membrane that became internalized and then modified into a different type of membrane (step 3, Figure 1). A cell that possessed these various internal compartments would have been an ancestor of a heterotrophic eukaryotic cell, such as a fungal cell or a protist (step 4, Figure 1). The oldest fossils thought to be the remains of eukaryotes date back about 1.8 billion years.

It is proposed that the acquisition of another endosymbiont, specifically a cyanobacterium, could have converted an early heterotrophic eukaryote into an ancestor of photosynthetic eukaryotes: the green algae and plants (step 5, Figure 1). The acquisition of chloroplasts must have been one of the last steps in the sequence of endosymbioses because these organelles are only present in plants and algae. In contrast, all groups of eukaryotes either (1) possess mitochondria or (2) show definitive evidence they have evolved from organisms that possessed these organelles.[a]

The division of all living organisms into two categories, prokaryotes and eukaryotes, reflects a basic dichotomy in the structures of cells, but it is not necessarily an accurate phylogenetic distinction, that is, one that reflects the evolutionary relationships among living organisms. Assume for a moment that modern eukaryotes (excluding their mitochondria and chloroplasts) evolved from one particular group of ancient prokaryotes, whereas most modern prokaryotes evolved

respiring microbes evolved into precursors of modern-day mitochondria (step 2, Figure 1).

A cell whose ancestors had formed through the sequence of symbiotic events just described could have given rise to a line of cells that evolved other basic characteristics of eukaryotic cells, including a system of membranes (a nuclear membrane, endoplasmic reticulum, Golgi complex, lysosomes), a cytoskeleton, and a mitotic type of cell division. These charac-

[a]There are a number of anaerobic unicellular eukaryotes that lack mitochondria. For years, these organisms formed the basis for a proposal that mitochondrial endosymbiosis was a late event that took place after the evolution of these eukaryotic lineages. However, recent analysis of the nuclear DNA of these organisms indicates the presence of genes that were likely transferred to the nucleus from mitochondria, suggesting that the ancestors of these organisms lost their mitochondria during the course of evolution.

from a different ancient group. It is conceivable under these circumstances that a particular group of living prokaryotes—descendants of the same group that gave rise to the first eukaryote—might be more closely related to living eukaryotes than they are to other living prokaryotes.

How does one determine evolutionary relationships among organisms that have been separated in time for billions of years, such as prokaryotic and eukaryotic organisms? Most taxonomic schemes that attempt to classify organisms are based heavily on anatomic or physiologic characteristics. In 1965, Emile Zuckerkandl and Linus Pauling suggested an alternate approach based on comparisons of the structure of informational molecules (proteins and nucleic acids) of living organisms.[3] Differences between organisms in the sequence of amino acids that make up a protein or the sequence of nucleotides that make up a nucleic acid are the result of mutations in DNA that have been transmitted to offspring. Mutations can accumulate in a given gene at a relatively constant rate over long periods of time. Consequently, comparisons of amino acid or nucleotide sequences can be used to determine how closely organisms are related to one another. For example, two organisms that are closely related, that is, have diverged only recently from a common ancestor, should have fewer sequence differences in a particular gene than two organisms that are distantly related, that is, do not have a recent common ancestor. Using this type of sequence information as an "evolutionary clock," researchers can construct phylogenetic trees showing proposed pathways by which different groups of living organisms may have diverged from one another during the course of evolution.

Beginning in the mid-1970s, Carl Woese and his colleagues at the University of Illinois began a series of studies on different organisms, comparing the nucleotide sequence of the RNA molecule that resides in the small subunit of the ribosome. This RNA—which is called the 16S rRNA in prokaryotes or the 18S rRNA in eukaryotes—was chosen because it is present in large quantities in all cells, it is easy to purify, and it tends to change only slowly over long periods of evolutionary time, which means that it could be used to study relationships of very distantly related organisms. There was one major disadvantage: nucleic acid sequencing at the time required laborious, time-consuming methods. In their approach, they purified the 16S rRNA from a particular source, then subjected the preparation to an enzyme, T1 ribonuclease, that digests the molecule into short fragments called oligonucleotides. The oligonucleotides in the mixture were then separated from one another by two-dimensional electrophoresis to produce a two-dimensional "fingerprint" as shown in Figure 2. Once they were separated, the nucleotide sequence of each of the oligonucleotides could be determined and the sequences from various organisms compared. In one of their first studies, Woese and his colleagues analyzed the 16S rRNA present in the ribosomes of chloroplasts from the photosynthetic protist *Euglena*.[4] They found that the sequence of this chloroplast rRNA molecule was much more similar to that of the 16S rRNA found in ribosomes of cyanobacteria than it was to its counterpart in the ribosomes from eukaryotic cytoplasm. This finding provided strong evidence for the symbiotic origin of chloroplasts from cyanobacteria.

In 1977, Woese and George Fox published a landmark paper in the study of molecular evolution.[5] They compared the nucleotide sequences of small-subunit rRNAs that had been

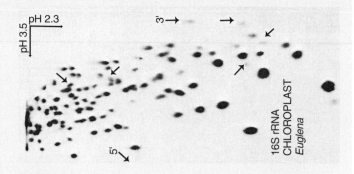

Figure 2 Two-dimensional electrophoretic fingerprint of a T1 digest of chloroplast 16S ribosomal RNA. The RNA fragments were electrophoresed in one direction at pH 3.5 and then in a second direction at pH 2.3. (FROM L. B. ZABLEN, ET AL., PROC. NAT'L ACAD. SCI. U.S.A. 72:2419, 1975.)

purified from a total of 13 different prokaryotic and eukaryotic species. The data from a comparison of all possible pairs of these organisms are shown in Table 1. The numbers along the top identify the organisms by the same numbers used along the left margin of the table. Each value in the table is an association constant that reflects the similarity in sequence between rRNAs from the two organisms being compared: the lower the number, the less similar the two sequences. They found that the sequences clustered into three distinct groups, as indicated in the table. It is evident that the rRNAs within each group (numbers 1–3, 4–9, and 10–13) are much more similar to one another than they are to rRNAs of the other two groups. The first of the groups shown in the table contain only eukaryotes; the second group contains the "typical" bacteria (gram-positive, gram-negative, and cyanobacteria); and the third group contains several species of methanogenic (methane-producing) bacteria. Woese and Fox concluded, to their surprise, that the methanogenic bacteria "appear to be no more related to typical bacteria than they are to eukaryotic cytoplasms." These results suggest that the members of these three groups represent three distinct evolutionary lines that branched apart from one another at a very early stage in the evolution of cellular organisms. Consequently, they assigned these organisms to three different kingdoms, which they named the Urkaryotes, Eubacteria, and Archaebacteria, a terminology that divided the prokaryotes into two distinct groups.

Subsequent research provided support for the concept that prokaryotes could be divided into two distantly related lineages, and it expanded the ranks of the archaebacteria to include at least two other groups, the thermophiles, which live in hot springs and ocean vents, and the halophiles, which live in very salty lakes and seas. In 1989, two published reports rooted the tree of life and suggested that the archaebacteria were actually more closely related to eukaryotes than they were to eubacteria.[6,7] Both groups of researchers compared the amino acid sequences of several proteins that were present in a wide variety of different prokaryotes, eukaryotes, mitochondria, and chloroplasts. A phylogenetic tree constructed from sequences of ribosomal RNAs, which comes to the same conclusion, is shown in Figure 3.[8] In this latter paper, Woese and colleagues proposed a revised taxonomic scheme, which has been widely accepted. In this

Table 1 Association Coefficients Between Representative Members of the Three Primary Kingdoms

	1	2	3	4	5	6	7	8	9	10	11	12	13
1. *Saccharomyces cerevisiae*, 18S	—	0.29	0.33	0.05	0.06	0.08	0.09	0.11	0.08	0.11	0.11	0.08	0.08
2. *Lemna minor*, 18S	0.29	—	0.36	0.10	0.05	0.06	0.10	0.09	0.11	0.10	0.10	0.13	0.08
3. L cell, 18S	0.33	0.36	—	0.06	0.06	0.07	0.07	0.09	0.06	0.10	0.10	0.09	0.07
4. *Escherichia coli*	0.05	0.10	0.06	—	0.24	0.25	0.28	0.26	0.21	0.11	0.12	0.07	0.12
5. *Cholorbium vibrioforme*	0.06	0.05	0.06	0.24	—	0.22	0.22	0.20	0.19	0.06	0.07	0.06	0.09
6. *Bacillus firmus*	0.08	0.06	0.07	0.25	0.22	—	0.34	0.26	0.20	0.11	0.13	0.06	0.12
7. *Corynebacterium diptheriae*	0.09	0.10	0.07	0.28	0.22	0.34	—	0.23	0.21	0.12	0.12	0.09	0.10
8. *Aphanocapsa* 6714	0.11	0.09	0.09	0.26	0.20	0.26	0.23	—	0.31	0.11	0.11	0.10	0.10
9. Chloroplast (Lemna)	0.08	0.11	0.06	0.21	0.19	0.20	0.21	0.31	—	0.14	0.12	0.10	0.12
10. *Methanebacterium thermoautotrophicum*	0.11	0.10	0.10	0.11	0.06	0.11	0.12	0.11	0.14	—	0.51	0.25	0.30
11. *M. ruminantium* strain M-1	0.11	0.10	0.10	0.12	0.07	0.13	0.12	0.11	0.12	0.51	—	0.25	0.24
12. *Methanobacterium sp.*, Cariaco isolate JR-1	0.08	0.13	0.09	0.07	0.06	0.06	0.09	0.10	0.10	0.25	0.25	—	0.32
13. *Methanosarcina barkeri*	0.08	0.07	0.07	0.12	0.09	0.12	0.10	0.10	0.12	0.30	0.24	0.32	—

Source: C. R. Woese and G. E. Fox, P.N.A.S 74:5089, 1977.

scheme, the archaebacteria, eubacteria, and eukaryotes are assigned to separate domains, which are named Archaea, Bacteria, and Eucarya, respectively.[b] Each domain can then be divided into one or more kingdoms; the Eucarya, for example, can be divided into the traditional kingdoms containing fungi, protists, plants, and animals.

According to the model in Figure 3, the first major split in the tree of life produced two separate lineages, one leading to the Bacteria and the other leading to both the Archaea and the Eucarya. If this view is correct, it was a member of the archaebacterial lineage, not the eubacterial lineage, that took in the symbionts and evolved into a eukaryotic cell. Although the host prokaryote in these symbiotic relationships was presumably an archaebacterium, the symbionts that evolved into mitochondria and chloroplasts were almost certainly eubacteria, as indicated by their close relationship with modern members of this group.

Up until 1995, phylogenetic trees of the type shown in Figure 3 were based primarily on the analysis of the gene encoding the 16S–18S rRNA. By then, phylogenetic comparisons of a number of other genes were suggesting that the scheme depicted in Figure 3 might be oversimplified. Questions about the origin of prokaryotic and eukaryotic cells came into sharp focus between 1995 and 1997 with the publication of the entire sequences of a number of prokaryotic genomes, both archaebacterial and eubacterial, and the genome of a eukaryote, the yeast *Saccharomyces cerevisiae*. Researchers could now compare the sequences of hundreds of genes simultaneously, and this analysis raised a number of puzzling questions and blurred the lines of distinction between the three domains.[9] For example, the genomes of several archaebacteria showed the presence of a significant number of eubacterial genes. For the most part, those genes in archaebacteria whose products are involved with informational processes (chromosome structure, transcription, translation, and replication) were very different from their counterparts in eubacterial cells and, in fact, resembled the corresponding genes in eukaryotic cells. This observation fit nicely with the scheme in Figure 3. In contrast, many of the genes in archaebacteria that encode the enzymes of metabolism exhibited an unmistakable eubacterial character.[10,11] The genomes of eubacterial species also showed evidence of a mixed origin, often containing a significant number of genes that bore an archaebacterial character.[12]

Most investigators who study the origin of ancient organisms have held on to the basic outline of the phylogenetic tree as demarcated in Figure 3 and argue that the presence of eubacteria-like genes in archaebacteria, and vice versa, is the result of the transfer of genes from one species to another, a phenomenon referred to as *lateral gene transfer (LGT)*.[13] According to the original premise that led to the phylogenetic tree of Figure 3, genes are inherited from one's parents, not from one's neighbors. This is the premise that allows an investigator to conclude that two species are closely related when they both possess a gene (e.g., the rRNA gene) of similar nucleotide sequence. If, however, cells can pick up genes from other species in their environment, then two species that are actually unrelated may possess genes of very similar sequence. The best measure of the extent of lateral gene transfer comes from a study that compared the genomes of two related eubacteria, *Escherichia* and *Salmonella*. It was found that 755 genes or nearly 20 percent of the *E. coli* genome is derived from "foreign" genes transferred into the *E. coli* genome over the past 100 million years, which is the time when the two species diverged. These 755 genes were acquired as the result of at least 234 separate lateral transfers from many different sources.[14] (The effect of lateral gene transfer on antibiotic resistance in pathogenic bacteria is discussed in the Human Perspective of Chapter 3.)

[b]Since this proposal, the terms archaebacteria and eubacteria have gradually faded from the literature, being replaced simply by archaea and bacteria. The earlier terms are used in this section to avoid possible confusion over use of the term "bacterial."

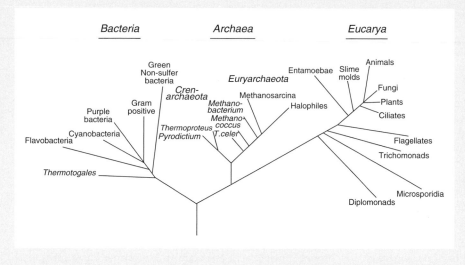

Figure 3 A phylogenetic tree based on rRNA sequence comparisons showing the three domains of life. The Archaea are divided into two subgroups as indicated. (FROM C. R. WOESE, ET AL., PROC. NAT'L ACAD. SCI. U.S.A. 87:4578, 1990.)

If genomes are a mosaic composed of genes from diverse sources, how does one choose which genes to use in determining phylogenetic relationships? According to one viewpoint, genes that are involved in informational activities (transcription, translation, replication) make the best subjects for determining phylogenetic relationships, because such genes are less likely to be transferred laterally than genes involved in metabolic reactions.[15] These authors argue that the products of informational genes (e.g., rRNAs) are parts of large complexes whose components must interact with many other molecules. It is unlikely that a foreign gene product could become integrated into the existing machinery. When "informational" genes are used as the subjects of comparison, archaebacteria and eubacteria tend to separate into distinctly different groups, whereas archaebacteria and eukaryotes tend to group together as evolutionary relatives, just as they do in Figure 3.

Analysis of eukaryotic genomes have produced similar evidence of a mixed heritage. Studies of the yeast genome show unmistakable presence of genes derived from both archaebacteria and eubacteria. The "informational genes" tend to have an archaeal character and the "metabolic genes" a eubacterial character.[16] There are several possible explanations for the mixed eukaryotic genome. Eukaryotic cells may have evolved from archaebacterial ancestors and then picked up genes from eubacteria with which they shared environments. A number of researchers have taken a more radical position and proposed that the eukaryote genome was originally derived from the fusion of an archaebacterial and a eubacterial cell followed by the integration of their two genomes.[17] Regardless of which, if any, of the current hypotheses prove correct, it is evident that no simple phylogenetic tree can account for the evolution of the entire genome of an organism. Instead, each gene or group of genes of a particular genome may have its own unique evolutionary tree, which is a thought that can be highly disconcerting to those who are attempting to determine the origin of our earliest ancestors.

References

1. SAGAN (MARGULIS), L. 1967. On the origin of mitosing cells. *J. Theor. Biol.* 14:225–274.

2. MARGULIS, L. 1970. *Origin of Eukaryotic Cells.* Yale University Press.

3. ZUCKERKANDL, E. & PAULING, L. 1965. Molecules as documents of evolutionary history. *J. Theor. Biol.* 8:357–365.

4. ZABLEN, L. B., ET AL. 1975. Phylogenetic origin of the chloroplast and prokaryotic nature of its ribosomal RNA. *Proc. Nat'l Acad. Sci. U.S.A.* 72:2418–2422.

5. WOESE, C. R. & FOX, G. E. 1977. Phylogenetic structure of the prokaryotic domain: The primary kingdoms. *Proc. Nat'l. Acad. Sci. U.S.A.* 74:5088–5090.

6. IWABE, N., ET AL. 1989. Evolutionary relationship of archaebacteria, eubacteria, and eukaryotes inferred from phylogenetic trees of duplicated genes. *Proc. Nat'l. Acad. Sci. U.S.A.* 86:9355–9359.

7. GOGARTEN, J. P., ET AL. 1989. Evolution of the vacuolar H+-ATPase: Implications for the origin of eukaryotes. *Proc. Nat'l. Acad. Sci. U.S.A.* 86:6661–6665.

8. WOESE, C., ET AL. 1990. Towards a natural system of organisms: Proposal for the domains Archaea, Bacteria, and Eucarya. *Proc. Nat'l. Acad. Sci. U.S.A.* 87:4576–4579.

9. DOOLITTLE, W. F. 1999. Lateral genomics. *Trends in Cell Biol.* 24:M5–M8 (Dec.)

10. BULT, C. J., ET AL. 1996. Complete genome sequence of the methanogenic archaeon, *Methanococcus jannaschii. Science* 273:1058–1073.

11. KOONIN, E. V., ET AL. 1997. Comparison of archaeal and bacterial genomes. *Mol. Microbiol.* 25:619–637.

12. NELSON, K. E., ET AL., 1999. Evidence for lateral gene transfer between Archaea and Bacteria from genome sequence of *Thermotoga maritima. Nature* 399:323–329.

13. OCHMAN, H., ET AL. 2000. Lateral gene transfer and the nature of bacterial innovation. *Nature* 405:299–304.

14. LAWRENCE, J. G. & OCHMAN, H. 1998. Molecular archaeology of the *Escherichia coli* genome. *Proc. Nat'l. Acad. Sci. U.S.A.* 95:9413–9417.

15. JAIN, R., ET AL. 1999. Horizontal gene transfer among genomes: The complexity hypothesis. *Proc. Nat'l. Acad. Sci. U.S.A.* 96:3801–3806.

16. RIVERA, M. C., ET AL. 1998. Genomic evidence for two functionally distinct gene classes. *Proc. Nat'l. Acad. Sci. U.S.A.* 95:6239–6244

17. MARTIN, W. & MÜLLER, M. 1998. The hydrogen hypothesis for the first eukaryote. *Nature* 392:37–41.

SYNOPSIS

The cell theory has three tenets. (1) All organisms are composed of one or more cells; (2) the cell is the basic organizational unit of life; and (3) all cells arise from preexisting cells. (p. 2)

The properties of life, as exhibited by cells, can be described by a collection of properties. Cells are very complex and their substructure is highly organized and predictable. The information to build a cell is encoded in its genes. Cells reproduce by cell division; their activities are fueled by chemical energy; they carry out enzymatically controlled chemical reactions; they engage in numerous mechanical activities; they respond to stimuli; and they are capable of a remarkable level of self-regulation. (p. 3)

Cells are either prokaryotic or eukaryotic. Prokaryotic cells are found only among archaebacteria and bacteria, whereas all other types of organisms—protists, fungi, plants, and animals—are composed of eukaryotic cells. Prokaryotic and eukaryotic cells share many common features, including a similar cellular membrane, a common system for storing and using genetic information, and similar metabolic pathways. Prokaryotic cells are the simpler type, lacking the complex membranous organelles (e.g., endoplasmic reticulum, Golgi complex, mitochondria, and chloroplasts), chromosomes, and cytoskeleton characteristic of the cells of eukaryotes. (p. 7)

Cells are almost always microscopic in size. Bacterial cells are typically 1 to 5 μm in length, whereas eukaryotic cells are typically 10 to 30 μm. Cells are microscopic in size for a number of reasons: their nuclei possess a limited number of copies of each gene; the surface area (which serves as the cell's exchange surface) becomes limiting as a cell increases in size; and the distance between the cell surface and interior becomes too great for the cell's needs to be met by simple diffusion. (p. 20)

Viruses are noncellular pathogens that can only reproduce when present within a living cell. Outside of the cell, the virus exists as a macromolecular package, or virion. Virions occur in a variety of shapes and sizes, but all of them consist of viral nucleic acid enclosed in a wrapper containing viral proteins. Viral infections may lead to either (1) the destruction of the host cell with accompanying production of viral progeny, or (2) the integration of viral nucleic acid into the DNA of the host cell, which often alters the activities of that cell. Viruses are not a primitive life form, but rather have evolved secondarily from fragments of cellular chromosomes. (p. 22)

ANALYTIC QUESTIONS

1. Consider some question about cell structure or function that you would be interested in answering. Would the data required to answer the question be easier to collect by working on an entire plant or animal or on a population of cultured cells? What might be the advantages and disadvantages of working on a whole organism versus a cell culture?

2. Figure 1.3 shows an intestinal epithelial cell with large numbers of microvilli. What is the advantage to the organism of having these microvilli? What do you expect would happen to an individual that lacked such microvilli as the result of an inherited mutation?

3. The first human cells to be successfully cultured were derived from a malignant tumor. Do you think this simply reflects the availability of cancer cells, or might such cells be better subjects for cell culture? Why?

4. The drawings of plant and animal cells in Figure 1.9*b,c* include certain structures that are present in plant cells but absent in animal cells. How do you think each of these structures affects the life of the plant?

5. It was noted that cells possess receptors on their surface that allow them to respond to specific stimuli. Many cells in the human body possess receptors that allow them to bind specific hormones that circulate in the blood. Why do you think these hormone receptors are important? What would be the effect on the physiological activities of the body if cells lacked these receptors, or if all cells had the same receptors?

6. If you were to argue that viruses are living organisms, what features of viral structure and function might you use in your argument?

7. If we presume that activities within cells do occur in a manner analogous to that shown in the Rube Goldberg cartoon of Figure 1.8, how would this differ from a human activity, such as building a car on an assembly line or shooting a free throw in a basketball game?

8. Unlike bacterial cells, the nucleus of a eukaryotic cell is bounded by a double-layered membrane studded by complex pores. How do you think this might affect traffic between the DNA and cytoplasm of a eukaryotic cell compared to that of a prokaryotic cell?

9. Examine the photograph of the ciliated protist in Figure 1.16 and consider some of the activities in which this cell engages that a muscle cell or a nerve cell in your body does not.

10. Which type of cell would you expect to achieve the largest volume: a highly flattened cell or a spherical cell? Why?

11. Suppose you were a scientist living in the 1890s and were studying a disease of tobacco crops that stunted the growth of the plants and mottled their leaves. You find that the sap from a diseased plant, when added to a healthy plant, is capable of transmitting the disease, to that plant. You examine the sap in the best light microscopes of the period and see no evidence of bacteria. You force the sap through filters whose pores are so small that they retard the passage of the smallest known bacteria, yet the fluid that passes through the filters is still able to transmit the disease. Like Dimitri Ivanovsky, a Russian biologist who conducted these experiments, you would probably conclude that the infectious agent was an unknown type of unusually small bacterium. What kinds of experiments might you perform today to test this hypothesis?

12. Most evolutionary biologists believe that all mitochondria have evolved from a single ancestral mitochondrion and all chloroplasts have evolved from a single ancestral chloroplast. In other words, the symbiotic event that gave rise to each of these organelles only occurred once. If this is the case, where on the phylogenetic tree of Figure 3, page 29, would you place the acquisition of each of these organelles?

ADDITIONAL READING

General References in Microbiology and Virology

BALOWS, A., ED. 1992. *The Prokaryotes*, 4 vols. Springer-Verlag.

BROCK, T. D., ET AL. 1993. *Biology of Microorganisms*, 7th ed. Prentice-Hall.

DULBECCO, R. & GINSBERG, H. S. 1991. *Virology*, 3d ed. Lippincott.

FIELDS, B. N., ET AL., EDS. 1995. *Fundamental Virology*, 3d ed. Raven.

LEVY, J. A., ET AL. 1994. *Virology.* 3d ed. Prentice-Hall.

STANIER, R. Y., ET AL. 1986. *The Microbial World*, 5th ed. Prentice-Hall.

TORTORA, G. J. 1999. *Microbiology: An Introduction*, 6th ed. Benjamin-Cummings.

Other Readings

BALTER, M. 2000. Evolution on life's fringes. (origin of viruses). *Science* 289:1866–1867.

BALTER, M. ET AL. 1998. Articles on AIDS. *Science* 280:1856–1900.

BONNER, J. T. 1996. *Sixty Years of Biology.* Princeton.

COFFIN, J. M., ET AL., EDS. 1998. *Retroviruses* Cold Spring Harbor Laboratory Press.

CRICK, F. H. C. 1988. *What Mad Pursuit.* Basic Books.

DAWKINS, R. 1999. *Unweaving the Rainbow: Science, Delusion and the Appetite for Wonder.* Houghton-Mifflin.

DELONG, E. F. 1998. Everything in moderation: Archaea as "non-extremophiles" *Curr. Opin. Gen. Develop.* 8:649–654.

DOOLITTLE, W. F. 1998. A paradigm gets shifty. (on the origin of eukaryotic cells). *Nature* 392:15–16.

DOOLITTLE, W. F. 2000. Uprooting the tree of life. *Sci. Am.* 282:90–95. (Feb.)

FRUTON, J. S. 1999. *Proteins, Enzymes, Genes: The Interplay of Chemistry and Biology.* Yale.

GROSS, M. 1998. *Life on the Edge: Amazing Creatures Thriving in Extreme Environments.* Plenum.

GUPTA R. S. & GOLDING, C. B. 1996. The origin of the eukaryotic cell. *Trends Biochem. Sci.* 21:166–171.

HOAGLAND, M. B. 1990. *Towards the Habits of Truth: A Life in Science.* Norton.

JUDSON, H. F. 1996. *The Eighth Day of Creation.* Cold Spring Harbor.

KERR, R. A. 1997. Life goes to extremes in the deep earth—and elsewhere. *Science* 276:703–704.

KORNBERG, A. 1989. *For the Love of Enzymes.* Harvard.

LURIA, S. 1984. *A Slot Machine, a Broken Test Tube: An Autobiography.* Harper & Row.

MADDOX, J. 1999. The unexpected science to come. *Sci. Am.* 281:62–67. (Dec.)

MADIGAN, M. T. & Marrs, B. L. 1997. Extremophiles. *Sci. Am.* 276:82–87. (April)

MAINZER, K. 1997. *Thinking in Complexity*, 3d ed. Springer-Verlag.

MALAKOFF, D. 2000. The rise of the mouse: Biomedicine's model mammal. *Science* 288:248–253.

MARGULIS, L. & SAGAN, D. 1995. *What Is Life?* Simon & Schuster.

MARGULIS, L. & SCHWARTZ, K. V. 1998. *Five Kingdoms.* Freeman.

MAZZARELLO, P. 1999. A unifying concept: The history of cell theory. *Nature Cell Biol.* 1:E13–E15.

MOORE, J. A. 1993. *Science As a Way of Knowing.* Harvard.

MORANGE, M. 1998. *A History of Molecular Biology.* Harvard.

MORELL, V. 1997. Microbiology's scarred revolutionary. *Science* 276:699–702.

ORGEL, L. E. 1994. The origin of life on the Earth. *Sci. Am.* 271:77–83. (Oct.)

PACE, N. R. 1997. A molecular view of microbial diversity and the biosphere. *Science* 276:734–740.

PENNISI, E. 2000. *Arabidopsis* comes of age. *Science* 290:32–35.

PERUTZ, M. F. 1998. *I Wish I'd Made You Angry Earlier: Essays on Science, Scientists, and Humanity.* Cold Spring Harbor Laboratory Press.

POPPER, K. 1992. *The Logic of Scientific Discovery*, 14th ed. Routledge.

PSENNER, R. & SATTLER, B. 1998. Life at the freezing point. *Science* 280:2073–2074.

RUESTOW, E. G. 1996. *The Microscope in the Dutch Republic: The Shaping of Discovery.* Cambridge.

THOMAS, L. 1974. *Lives of a Cell: Notes of a Biology Watcher.* Bantam.

WEINER, J. 1999. *Time, Love, Memory: A Great Biologist and His Quest for the Origins of Behavior.* (Biography of Seymour Benzer). Knopf.

WHITMAN, W. B., ET AL. 1998. Prokaryotes: The unseen majority. *Proc. Nat'l Acad. Sci. U.S.A.* 95:6578–6583.

WOESE, C. 1998. The universal ancestor. *Proc. Nat'l. Acad. Sci. U.S.A.* 95:6854–6858.

Human Perspective: Replacing Damaged Cells and Organs

FUCHS, E. & SEGRE, J. A. 2000. Stem cells: a new lease on life. *Cell* 100:143–155.

HINES, P. J., ET AL., 2000. Reviews on stem cells *Science* 287:1417–1446.

KELLER, G. & SNODGRASS, H. R. 1999. Human embryonic stem cells: The future is now. *Nature Med.* 5:151–152.

KEMPERMANN, G. & GAGE, F. H. 1999. New nerve cells for the adult brain. *Sci. Am.* 280:48–53. (May)

KESSLER, P. D. & BYRNE, B. J. 1999. Myoblast cell grafting into heart muscle. *Ann. Rev. Physiol.* 61:219–242.

McKAY, R. 2000. Stem cells—hype and hope. *Nature* 406:361–364.

PEDERSEN, R. A. 1999. Embryonic stem cells for medicine. *Sci. Am.* 280:69–73. (April)

SOLTER, D. & GEARHART, J. 1999. Putting stem cells to work. *Science* 283:1468–1470.

VOGEL, G. 2000. Stem cells: New excitement, persistent questions. 290:1672–1674.

The Chemical Basis of Life

e will begin this chapter with a brief examination of the atomic basis of matter, a subject that may seem out of place in a biology textbook. But the cellular level of organization is only a small step from the atomic level, as will be evident when we examine the importance of the movement of a few atoms of a molecule during such activities as muscle contraction or the transport of substances across cell membranes. The properties of cells and their organelles derive directly from the activities of the molecules of which they are composed. Consider a process such as cell division, which can be followed in considerable detail under a simple light microscope. To understand the activities that occur when a cell divides, one needs to know, for example, about the interactions between DNA and protein molecules that cause the chromosomes to condense into rod-shaped packages that can be separated into different cells; the molecular construction of protein-containing microtubules that allows them to disassemble at one moment in the cell and reassemble the next moment in an entirely different cellular location; and the properties of lipid molecules that make

A complex between two different macromolecules. A portion of a DNA molecule (shown in blue) is complexed to a protein that consists of two polypeptide subunits, one red and the other yellow. Those parts of the protein that are seen to be inserted into the grooves of the DNA have recognized and bound to a specific sequence of nucleotides in the nucleic acid molecule. (COURTESY OF A. R. FERRÉ-D'AMARÉ AND STEPHEN K. BURLEY, THE ROCKEFELLER UNIVERSITY.)

the outer cell membrane deformable so that it can be pulled into the middle of a cell, thereby pinching the cell in two. It is impossible even to begin to understand cellular function without a reasonable knowledge of the structure and properties of the major types of biological molecules. This is the goal of the present chapter: to provide the necessary information about the chemistry of life to allow the reader to understand the basis of life. We will begin by considering the types of bonds that atoms can form with one another. ■

2.1 COVALENT BONDS

The atoms that make up a molecule are joined together by **covalent bonds** in which pairs of electrons are shared between pairs of atoms. The formation of a covalent bond between two atoms is governed by the fundamental principle that an atom is most stable when its outermost electron shell is filled. Consequently, the number of bonds an atom can form depends on the number of electrons needed to fill its outer shell.

The electronic structure of a number of atoms is shown in Figure 2.1. The outer (and only) shell of a hydrogen or helium atom is filled when it contains two electrons; the outer shells of the other atoms in Figure 2.1 are filled when they contain eight electrons. Thus, an oxygen atom, with six outer-shell electrons, can fill its outer shell by combining with two hydrogen atoms, forming a molecule of water. The oxygen atom is linked to each hydrogen atom by a *single* covalent bond (denoted as H:O or H—O). The formation of a covalent bond is accompanied

ELECTRONS NEEDED FOR ATOMS IN EACH COLUMN TO ACHIEVE STABILITY

| −1 | +4 | +2 | +1 | 0 (Inert elements) |

Figure 2.1 A representation of the arrangement of electrons in a number of common atoms. Electrons are present around an atom's nucleus in "clouds" or *orbitals* that are roughly defined by their boundaries, which may have a spherical or dumbbell shape. Each orbital contains a maximum of two electrons, which is why the electrons (dark dots in the drawing) are grouped in pairs. The innermost shell contains a single orbital (thus two electrons), the second shell contains four orbitals (thus eight electrons), the third shell also contains four orbitals, and so forth. The number of outer-shell electrons is a primary determinant of the chemical properties of an element. Atoms with a similar number of outer-shell electrons have similar properties. Lithium (Li) and sodium (Na), for example, have one outer-shell electron, and both are highly reactive metals. Carbon (C) and silicon (Si) atoms can each bond with four different atoms. Because of its size, however, a carbon atom can bond to other carbon atoms, forming long-chained organic molecules, whereas silicon is unable to form comparable molecules. Neon (Ne) and argon (Ar) have filled outer shells, making these atoms highly nonreactive; they are referred to as inert gases.

Table 2.1	Electronegativities of Atoms			
+1*	+4	+5	+6	+7
H	C	N	O	F
2.2	2.5	3.0	3.5	4.0
	Si	P	S	Cl
	1.9	2.2	2.5	3.0

*+numbers correspond to the row of the atom in the periodic table.

by the release of energy, which must be reabsorbed at some later time if the bond is to be broken. The energy required to cleave C—H, C—C, or C—O covalent bonds is quite large—typically between 80 and 100 kilocalories per mole (kcal/mol)[1] of molecules—making these bonds stable under most conditions.

In many cases, two atoms can become joined by bonds in which more than one pair of electrons are shared. If two electron pairs are shared, as occurs in molecular oxygen (O_2), the covalent bond is a *double bond*, and if three pairs of electrons are shared (as in molecular nitrogen, N_2), it is a *triple bond*. Quadruple bonds are not known to occur. The type of bond between atoms has important consequences in determining the shapes of molecules. For example, atoms joined by a single bond are able to rotate relative to one another, whereas the atoms of double (and triple) bonds lack this ability.

When atoms of the same element bond to one another, as in H_2, the electron pairs of the outer shell are equally shared between the two bonded atoms. When two unlike atoms are covalently bonded, however, the positively charged nucleus of one atom exerts a greater attractive force on the outer electrons than the other. Consequently, the shared electrons tend to be located more closely to the atom with the greater attractive force, that is, the more **electronegative** atom. The electronegativity of an atom depends on two factors: (1) the number of positive charges in its nucleus (the more protons, the more electronegative) and (2) the distance of the outer electrons from the nucleus (the greater the distance, the less electronegative). Electronegativities of common atoms are rated on a scale from 0 to 4 in Table 2.1. Among the atoms most commonly present in biological molecules, nitrogen and oxygen are strongly electronegative.

Polar and Nonpolar Molecules

Let's examine a molecule of water. Water's single oxygen atom attracts electrons much more forcefully than do either of its hydrogen atoms. As a result, the O—H bonds

[1]One calorie is the amount of thermal energy required to raise the temperature of one gram of water one degree centigrade. One kilocalorie (kcal) equals 1000 calories (or one large Calorie). In addition to calories, energy can also be expressed in Joules, which is a term that was used historically to measure energy in the form of work. One kilocalorie is equivalent to 4186 Joules. A mole is equal to Avogadro's number (6×10^{23}) of molecules. A mole of a substance is its molecular weight expressed in grams.

of a water molecule are said to be *polarized*, such that one of the atoms has a partial negative charge and the other a partial positive charge. This is generally denoted in the following manner:

Molecules, such as water, that have an asymmetric distribution of charge are said to be **polar** molecules. Polar molecules of biological importance contain one or more electronegative atoms, usually O, N, S, and/or P. Molecules that lack electronegative atoms and polarized bonds, such as molecules that consist entirely of carbon and hydrogen atoms, are said to be **nonpolar**. The presence of strongly polarized bonds is of utmost importance in determining the reactivity of molecules. Large nonpolar molecules, such as waxes and fats, tend to be relatively inert. Some of the more interesting biological molecules, including proteins and phospholipids (discussed later), contain both polar and nonpolar regions, which behave very differently.

Ionization

Some atoms are so strongly electronegative that they can capture electrons from other atoms during a chemical reaction. For example, when the elements sodium (a silver-colored metal) and chlorine (a toxic gas) are mixed, the single electron in the outer shell of each sodium atom migrates to the electron-deficient chlorine atom. As a result, these two materials are transformed into charged atoms, or **ions**.

$$2Na\cdot + \overset{..}{\underset{..}{Cl}}\!\!:\!\!\overset{..}{\underset{..}{Cl}}: \rightarrow 2Na\!:\!\overset{..}{\underset{..}{Cl}}: \rightarrow 2Na^+ + 2\overset{..}{\underset{..}{Cl}}:^-$$

Because the chloride ion has an extra electron (relative to the number of protons in its nucleus), it has a negative charge (Cl^-) and is termed an **anion**. The sodium atom, which has lost an electron, has an extra positive charge (Na^+) and is termed a **cation**. When present in crystals, these two ions form sodium chloride, or common table salt.

The Na^+ and Cl^- ions depicted above are relatively stable because they possess filled outer shells. A different arrangement of electrons within an atom can produce a highly reactive species, called a *free radical*. The structure of free radicals and their importance in biology are considered in the accompanying Human Perspective.

THE HUMAN PERSPECTIVE

Free Radicals as a Cause of Aging

Why do humans have a maximum life span of approximately 100 years whereas our close relatives, the chimpanzees, live only about half this length of time? Many biologists believe that aging results from the gradual accumulation of damage to our bodily tissues. The most destructive damage probably occurs to DNA. Alterations in DNA lead to the production of faulty genetic messages that promote gradual cellular deterioration. How does cellular damage occur, and why should it occur more rapidly in a chimpanzee than a human? The answer may reside at the atomic level.

Atoms are stabilized when their shells are filled with electrons. Electron shells consist of orbitals, each of which can hold a maximum of two electrons. Atoms or molecules that have orbitals containing a single unpaired electron tend to be highly unstable—they are called **free radicals.** Free radicals may be formed when a covalent bond is broken such that each portion keeps one-half of the shared electrons, or they may be formed when an atom or molecule accepts a single electron transferred during an oxidation-reduction reaction. For example, water can be converted into free radicals when exposed to radiation from the sun:

$$H_2O \rightarrow HO\cdot + H\cdot$$
<div align="center">hydroxyl
radical
("·" indicates a free radical)</div>

Free radicals are extremely reactive and capable of chemically altering many types of molecules, including proteins, nucleic acids, and lipids. The formation of hydroxyl radicals is probably a major reason that sunlight is so damaging to skin.

In 1956, Denham Harman of the University of Nebraska proposed that aging results from tissue damage caused by free radicals. Because the subject of free radicals was not one with which biologists and physicians were familiar, the proposal didn't generate significant interest. Then, in 1969, Joe McCord and Irwin Fridovich of Duke University discovered an enzyme, superoxide dismutase (SOD), whose sole function was the destruction of the superoxide radical ($O_2\cdot^-$), a type of free radical formed when molecular oxygen picks up an extra electron. SOD catalyzes the following reaction:

$$O_2\cdot^- + O_2\cdot^- + 2H^+ \rightarrow H_2O_2 + O_2$$
<div align="center">hydrogen
peroxide</div>

Hydrogen peroxide is also a highly reactive oxidizing agent, which is why it is often used as a disinfectant and bleaching agent. If it is not rapidly destroyed, H_2O_2 can break down to form hydroxyl radicals that attack the cell's macromolecules. Hydrogen peroxide is normally destroyed in the cell by the enzyme catalase or glutathione peroxidase.

Subsequent research has revealed that superoxide radicals are formed within cells during normal oxidative metabolism and that a superoxide dismutase is present in the cells of diverse organisms, from bacteria to humans. In fact, animals possess three different versions (isoforms) of SOD: a cytosolic, mitochondrial, and extracellular isoform. It is estimated that as much as 1–3 percent of the oxygen taken into human mitochondria is converted to hydrogen peroxide rather than to water, the normal end product of respiration. The importance of SOD is most clearly revealed in studies of mutant bacteria and yeast that lack the enzyme; these cells are unable to grow in the presence of oxygen. Similarly, mice that are lacking the mitochondrial version of the enzyme (SOD2) are not able to survive more than a week or so after birth.

Although the destructive potential of free radicals, such as superoxide and hydroxyl radicals, is unquestioned, the importance of these agents as a factor in aging remains controversial. Harman's hypothesis concerning free radicals and aging makes certain predictions. For example, one would expect that animals with longer life spans might produce fewer free radicals, possess a better capability for free radical destruction, or be better able to repair the cellular damage that results from free radical reactions. Several studies have found that fruit flies genetically engineered to produce large quantities of SOD can live up to 40 percent longer than untreated controls. A study on houseflies has shown that animals that are kept caged and unable to fly live up to twice as long as those allowed to fly. In metabolic costs, flying is a very expensive activity. Consequently, flies that are unable to fly have markedly lower metabolic rates, which is thought to be the reason for the increased longevity. An animal exhibiting a lower metabolic rate requires less O_2 and consequently would be expected to produce fewer free radicals. The life span of mammals can be increased by severely restricting the calories present in the diet. As first shown in the 1930s, mice that are maintained on very strict diets typically live 30 to 40 percent longer than their littermates who are fed diets of normal caloric content. Studies of the metabolic rates of these mice have produced contradictory data, but there is general agreement that animals fed calorie-restricted diets exhibit a marked decrease in $O_2\cdot^-$ and H_2O_2 production, which could explain their increased longevity.

A related area of research concerns the study of substances called antioxidants that are able to destroy free radicals. Common antioxidants found in the body include glutathione, vitamins E and C, and beta-carotene (the orange pigment in carrots and other vegetables, and the parent compound for vitamin A). Although these substances may prove highly beneficial in the diet because of their ability to destroy free radicals, studies with rats and mice have failed to provide convincing evidence that they retard the aging process or increase maximum life span. Vitamin E has been shown to reduce the incidence of cardiovascular disease in humans. In one recent study of more than 2000 patients, dietary supplementation with vitamin E (400–800 IU per day) reduced the incidence of heart attack by 77 percent. It is not clear, however, whether this response is mediated by the vitamin's an-

tioxidant properties. A new class of synthetic antioxidant compounds has been developed that shows considerable promise. An analog of one of these compounds, phenyl tert-butyl nitrone (PBN), has been reported to reverse the effects of age-related memory loss in rodents. Whether these compounds can be safely used by humans remains to be determined.

REVIEW ?

1. Oxygen atoms have eight protons in their nucleus. How many electrons do they have? How many orbitals are in the inner electron shell? How many electrons are in the outer shell? How many more electrons can the outer shell hold before it is filled?

2. Compare and contrast: a sodium atom and a sodium ion; a double bond and a triple bond; an atom of weak and strong electronegativity; the electron distribution around an oxygen atom bound to another oxygen atom and an oxygen atom bound to two hydrogen atoms.

2.2 NONCOVALENT BONDS

Covalent bonds are strong bonds between the atoms that make up a molecule. Interactions between molecules (or between different parts of a large biological molecule) are governed by a variety of weaker linkages called noncovalent bonds. **Noncovalent bonds** do not depend on shared electrons but rather on attractive forces between atoms having an opposite charge. Individual noncovalent bonds are weak (about 1 to 5 kilocalories per mole) and are thus readily broken and reformed. As will be evident throughout this book, this feature allows noncovalent bonds to mediate the dynamic interactions among molecules in the cell.

Even though individual noncovalent bonds are weak, when large numbers of them act in concert, as between the two strands of a DNA molecule or between different parts of a large protein, their attractive forces are additive. Taken as a whole, they provide the structure with considerable stability. We will examine several types of noncovalent bonds that are important in cells.

Ionic Bonds: Attractions Between Charged Atoms

A crystal of table salt is held together by an electrostatic attraction between positively charged Na$^+$ and negatively charged Cl$^-$ ions. This type of attraction between fully charged components is called an **ionic bond** (or a *salt bridge*). Ionic bonds within a salt crystal may be quite strong. However, if a crystal of salt is dissolved in water, each of the individual ions becomes surrounded by water molecules, which inhibit oppositely charged ions from approaching one another closely enough to form ionic bonds (Figure 2.2). Because cells are composed primarily of

water, bonds between *free* ions are of little importance. In contrast, weak ionic bonds between oppositely charged groups of large biological molecules are of considerable importance. For example, when negatively charged phosphate atoms in a DNA molecule are closely associated with positively charged groups on the surface of a protein (Figure 2.3), ionic bonds between them help hold the complex together. The strength of ionic bonds in a cell is generally weak (about 3 kcal/mol) due to the presence of water, but deep within the core of a protein, where water is often excluded, such bonds can be influential.

Hydrogen Bonds

When a hydrogen atom is covalently bonded to an electronegative atom, particularly an oxygen or nitrogen atom, the single pair of shared electrons is greatly displaced toward the nucleus of the electronegative atom, leaving the hydrogen atom with a partial positive charge. As a result, the bare, positively charged nucleus of the hydrogen atom can approach near enough to an unshared pair of outer electrons of a second electronegative atom to form an attractive interaction (Figure 2.4). This weak attractive interaction is called a **hydrogen bond**.

Hydrogen bonds occur between most polar molecules and are particularly important in determining the struc-

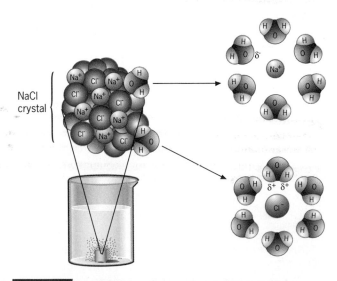

Figure 2.2 The dissolution of a salt crystal. When placed in water, the Na$^+$ and Cl$^-$ ions of a salt crystal become surrounded by water molecules, breaking the ionic bonds between the two ions. As the salt dissolves, the negatively charged oxygen atoms of the water molecules associate with the positively charged sodium ions, and the positively charged hydrogen atoms of the water molecules associate with the negatively charged chloride ions.

Figure 2.4 **Hydrogen bonds** form between a bonded electronegative atom, such as nitrogen or oxygen, which bears a partial negative charge, and a bonded hydrogen atom, which bears a partial positive charge. Hydrogen bonds (about 0.18 nm) are about twice as long as the much stronger covalent bonds.

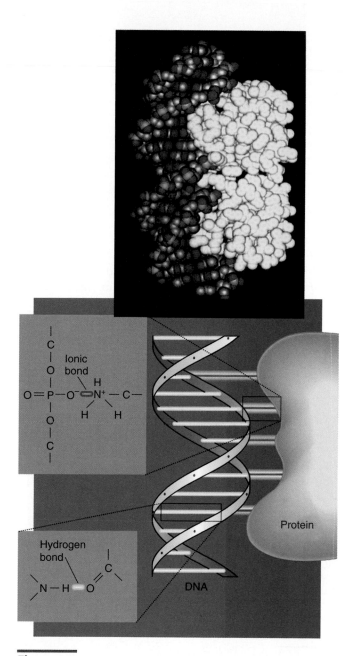

Figure 2.3 **Noncovalent ionic bonds** play an important role in holding the protein molecule on the right (yellow atoms) to the DNA molecule on the left. Ionic bonds form between positively charged nitrogen atoms in the protein and negatively charged oxygen atoms in the DNA. The DNA molecule itself consists of two separate strands held together by noncovalent hydrogen bonds (discussed in the following section). Although a single noncovalent bond is relatively weak and easily broken, large numbers of these bonds between two molecules, as between two strands of DNA, make the overall complex quite stable. (TOP IMAGE COURTESY OF STEPHEN HARRISON.)

ture and properties of water (discussed later). Hydrogen bonds also form between polar groups present in large biological molecules, as occurs between the two strands of a DNA molecule (see Figure 2.3). Because their strength is additive, the large number of hydrogen bonds between the strands makes the DNA double helix a stable structure.

However, because individual hydrogen bonds are weak (2–5 kcal/mol in aqueous solutions), the two strands can be separated to allow enzymes access to individual strands of the DNA molecule.

Hydrophobic Interactions and van der Waals Forces

Because of their ability to interact with water, polar molecules, such as sugars and amino acids (described shortly), are said to be **hydrophilic**, or "water loving." Nonpolar molecules, such as steroid or fat molecules, are essentially insoluble in water because they lack the charged regions that would attract them to the poles of water molecules. When nonpolar compounds are mixed with water, the nonpolar, **hydrophobic** ("water fearing") molecules are forced into aggregates, which minimizes their exposure to the polar surroundings (Figure 2.5). This association of nonpolar molecules is called a **hydrophobic interaction**. This is why droplets of fat molecules rapidly reappear on the surface of beef or chicken soup even after the liquid is stirred with a spoon. This is also the reason that nonpolar groups tend to localize within the interior of most soluble proteins away from the surrounding water molecules (page 55).

Hydrophobic interactions of the type just described are not classified as true bonds because they do not result from an attraction between hydrophobic molecules.[2] In addition

[2]This statement reflects an accepted hypothesis that hydrophobic interactions are driven by increased entropy (disorder). When a hydrophobic group projects into an aqueous solvent, the water molecules become ordered in a cage around the hydrophobic group. These solvent molecules become disordered when the hydrophobic group withdraws from the surrounding solvent. An alternative view holds that hydrophobic interactions are driven by the formation of weak bonds (see P. L. Privalov and S. J. Gill, *Pure Appl. Chem.* 61:1097, 1989; or G. I. Makhatadze and P. L. Privalov, *Adv. Prot. Chem.* 47:308, 1995).

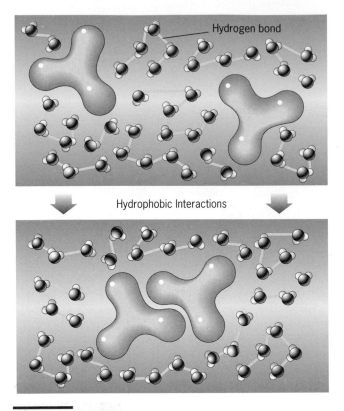

Figure 2.5 **In a hydrophobic interaction**, the nonpolar (hydrophobic) molecules are forced into aggregates, which minimizes their exposure to the surrounding water molecules.

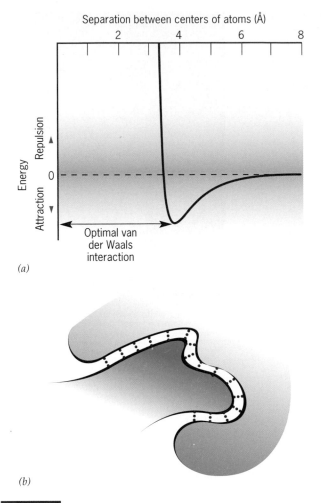

Figure 2.6 **Van der Waals forces.** (*a*) As two atoms approach each other, they experience a weak attractive force that increases up to a specific distance, typically about 4 Å. If the atoms approach more closely, their electron clouds repel one another, causing the atoms to be forced apart. (*b*) Although individual van der Waals forces are very weak, large numbers of such attractive forces can be formed if two macromolecules have a complementary surface, as is indicated schematically in this figure.

to this type of interaction, hydrophobic groups can form weak bonds with one another based on electrostatic attractions. Polar molecules associate because they contain permanent asymmetric charge distributions within their structure. Closer examination of the covalent bonds that make up a nonpolar molecule (such as H_2 or CH_4) reveals that electron distributions are not always symmetric. The distribution of electrons around an atom at any given instant is a statistical matter and, therefore, varies from one instant to the next. Consequently, at any given time, the electron density may happen to be greater on one side of an atom, even though the atom shares the electrons equally with some other atom. These transient asymmetries in electron distribution result in momentary separations of charge (*dipoles*) within the molecule. If two molecules with transitory dipoles are very close to one another and oriented in the appropriate manner, they experience a weak attractive force, called a **van der Waals force**, that bonds them together. Moreover, the formation of a temporary separation of charge in one molecule can *induce* a similar separation in an adjacent molecule. In this way, additional attractive forces can be generated between nonpolar molecules. A single van der Waals force is very weak (0.1 to 0.3 kcal/mol) and very sensitive to the distance that separates the two atoms (Figure 2.6*a*). As we will see in later chapters, however, biological molecules that interact with one another, for example, an antibody and a protein on the surface of a virus, often possess complementary shapes. As a result, many atoms of

both interactants may have the opportunity to approach each other very closely (Figure 2.6*b*), making van der Waals forces important in biological interactions.

The Life-Supporting Properties of Water

Life on Earth is totally dependent on water, and water may be essential to the existence of life anywhere in the universe. Even though it contains only three atoms, a molecule of water has a unique structure that gives the molecule extraordinary properties.[3] Most importantly,

[3]One way to appreciate the structure of water is by comparing it to H_2S. Like oxygen, sulfur has six outer-shell electrons and forms single bonds with two hydrogen atoms. But because sulfur is a larger atom, it is less electronegative than oxygen, and its ability to form hydrogen bonds is greatly reduced. At room temperature, H_2S is a gas, not a liquid. In fact, the temperature has to drop to −86°C before H_2S freezes into a solid.

Figure 2.7 Hydrogen bond formation between neighboring water molecules. Each H atom of the molecule has about four-tenths of a full positive charge, and the single O atom has about eight-tenths of a full negative charge.

1. Water is a highly asymmetric molecule with the O atom at one end and the two H atoms at the opposite end.

2. Each of the two covalent bonds in the molecule is highly polarized.

3. All three atoms in a water molecule are adept at forming hydrogen bonds.

The life-supporting attributes of water stem from these properties.

Each molecule of water can form hydrogen bonds with as many as four other water molecules, producing a highly interconnected network of molecules (Figure 2.7). Each hydrogen bond is formed when the partially positive-charged hydrogen of one water molecule becomes aligned next to a partially negative-charged oxygen atom of another water molecule. Because of their extensive hydrogen bonding, water molecules have an unusually strong tendency to adhere to one another. This feature is most evident in the thermal properties of water. For example, when water is heated, most of the thermal energy is consumed in disrupting hydrogen bonds rather than contributing to molecular motion (which is measured as an increased temperature). Similarly, evaporation from the liquid to the gaseous state requires that water molecules break the hydrogen bonds holding them to their neighbors, which is why it takes so much energy to convert water to steam. Mammals take advantage of this property when they sweat because the heat required to evaporate the water is absorbed from the body, which thus becomes cooler.

The small volume of aqueous fluid present within a cell contains a remarkably complex mixture of dissolved substances, or **solutes**. In fact, water is able to dissolve more types of substances than any other solvent. But water is more than just a solvent; it determines the structure of biological molecules and the types of interactions in which

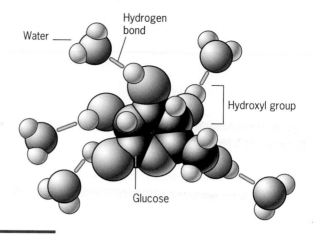

Figure 2.8 Schematic view of the types of hydrogen bonds that can form between a sugar molecule and the water it is dissolved in. The sugar molecule is shown using a space-filling model, which is a common way to depict molecular structure.

they can engage. Water is the fluid matrix around which the insoluble fabric of the cell is constructed. It is also the medium through which materials move from one compartment of the cell to another; it is a reactant or product in many cellular reactions; it protects the cell in many ways—from excessive heat, cold, or damaging radiation.

Water is such an important factor in a cell because it is able to form weak interactions with so many different types of chemical groups. Recall from page 36 how water molecules, with their strongly polarized O—H bonds, form a shell around ions, separating the ions from one another. Similarly, water molecules form hydrogen bonds with organic molecules that contain polar groups, such as amino acids and sugars (Figure 2.8), as well as the larger macromolecules of the cell. Because they are able to form weak noncovalent bonds with water, polar molecules are soluble within the cell.

REVIEW **?**

1. Describe some of the properties that distinguish covalent and noncovalent bonds.

2. Why do polar molecules, such as table sugar, dissolve so readily in water? Why do fat droplets form on the surface of an aqueous solution? Why does sweating help cool the body?

2.3 ACIDS, BASES, AND BUFFERS

Protons are not only found within atomic nuclei, they are also released into the medium, whenever a hydrogen atom loses an electron. Consider acetic acid—the distinctive ingredient of vinegar—which can undergo the following reaction, described as a **dissociation**.

Acetic Acetate Proton
acid ion (hydrogen ion)

A molecule that is capable of releasing (donating) a hydrogen ion is termed an **acid**. The proton released by the acetic acid molecule in the previous reaction does not remain in the free state; instead, it combines with another molecule. Possible reactions involving a proton include

- Combination with a water molecule to form a hydronium ion (H_3O^+).

$$H^+ + H_2O \rightarrow H_3O^+$$

- Combination with a hydroxyl ion (OH^-) to form a molecule of water.

$$H^+ + OH^- \rightarrow H_2O$$

- Combination with an amino group ($—NH_2$) in a protein to form a charged amine.

$$H^+ + —NH_2 \rightarrow —NH_3{}^+$$

Any molecule that is capable of accepting a proton is defined as a **base**. Acids and bases exist in pairs, or *couples*. When the acid loses a proton (as when acetic acid gives up a hydrogen ion), it becomes a base (in this case, acetate ion), which is termed the *conjugate base* of the acid. Similarly, when a base (such as an $—NH_2$ group) accepts a proton, it forms an acid (in this case $—NH_3{}^+$), which is termed the *conjugate acid* of that base. Thus, the acid always contains one more positive charge than its conjugate base. Water is an example of an *amphoteric* molecule, that is, one that can serve both as an acid and a base:

$$H_3O^+ \rightleftharpoons H^+ + H_2O \rightleftharpoons OH^- + H^+$$

Acid Amphoteric Base
 molecule

We will discuss another important group of amphoteric molecules, the amino acids, on page 52.

Acids vary markedly with respect to the ease with which the molecule gives up a proton. The more readily the proton is lost, that is, the less strong the attraction of a conjugate base for its proton, the stronger the acid. Hydrogen chloride is a very strong acid, one that will readily transfer its proton to water molecules. The conjugate base of a strong acid, such as HCl, is a weak base (Table 2.2). Acetic acid, in contrast, is a relatively weak acid, because for the most part it remains undissociated when dissolved in water. In a sense, one can consider the degree of dissociation of an acid in terms of the competition for protons among the components of a solution. Water is a better competitor, that is, a stronger base, than chloride ion, so HCl completely dissociates. In contrast, acetate ion is a stronger base than water, so it remains largely as undissociated acetic acid.

Table 2.2 Strengths of Acids and Bases

	Acids	Bases	
Very weak	H_2O	OH^-	Strong
Weak	NH_4^+	NH_3	Weak
	H_2S	S^{2-}	
	CH_3COOH	CH_3COO^-	
	H_2CO_3	HCO_3^-	
Strong	H_3O^+	H_2O	Very
	HCl	Cl^-	weak
	H_2SO_4	SO_4^{2-}	

The acidity of a solution is measured by the concentration of hydrogen ions[4] and is expressed in terms of **pH**.

$$pH = -\log[H^+]$$

where $[H^+]$ is the molar concentration of protons. For example, a solution having a pH of 5 contains a hydrogen ion concentration of 10^{-5} M. Because the pH scale is logarithmic, an increase of one pH unit corresponds to a tenfold decrease in H^+ concentration (or a tenfold increase in OH^- concentration). Stomach juice (pH 1.8), for example, has nearly one million times the H^+ concentration of blood (pH 7.4).

When a water molecule dissociates into a hydroxyl ion and a proton, $H_2O \rightarrow H^+ + OH^-$ (or more accurately, $2 H_2O \rightarrow H_3O^+ + OH^-$) the equilibrium constant for the reaction can be expressed as:

$$K_{eq} = \frac{[H^+][OH^-]}{[H_2O]}$$

Because the concentration of pure water is always 55.51 M, we can generate a new constant, K_W, the *ion-product constant* for water,

$$K_W = [H^+][OH^-]$$

which is equal to 10^{-14} at 25°C. In pure water, the concentration of both H^+ and OH^- is approximately 10^{-7} M. The extremely low level of dissociation of water indicates that it is a very weak acid. In the presence of an acid, the concentration of hydrogen ions rises and the concentration of hydroxyl ions drops (as a result of combination with protons to form water), so that the ion product remains at 10^{-14}.

Most biological processes are acutely sensitive to pH because changes in hydrogen ion concentration affect the ionic state of biological molecules. For example, as the hydrogen ion concentration increases, the $—NH_2$ group of the amino acid arginine becomes protonated to form

[4]In aqueous solutions, protons do not exist in the free state, but rather as hydronium ions (H_3O^+). For sake of simplicity, we will refer to them simply as protons or hydrogen ions.

—NH$_3$$^+$, which can disrupt the activity of the entire protein. Even slight changes in pH can impede biological reactions. Organisms, and the cells they comprise, are protected from pH fluctuations by **buffers**—compounds that react with free hydrogen or hydroxyl ions, thereby resisting changes in pH. Buffer solutions usually contain a weak acid together with its conjugate base. The blood, for example, is buffered by carbonic acid and bicarbonate ions, which normally hold blood pH at about 7.4.

$$HCO_3^- + H^+ \rightleftharpoons H_2CO_3$$

Bicarbonate	Hydrogen	Carbonic
ion	ion	acid

If the hydrogen ion concentration rises (as occurs during exercise), the bicarbonate ions combine with the excess protons, removing them from solution. Conversely, excess OH^- ions (which are generated during hyperventilation) are neutralized by protons derived from carbonic acid. The pH of the fluid within the cell is regulated in a similar manner by a phosphate buffer system consisting of $H_2PO_4^-$ and HPO_4^{2-}.

REVIEW ?

1. If you were to add hydrochloric acid to water, what effect would this have on the hydrogen ion concentration? on the pH? on the ionic charge of any proteins in solution?

2. What is the relationship between a base and its conjugate acid?

2.4 THE NATURE OF BIOLOGICAL MOLECULES

The bulk of an organism is water. If the water is evaporated away, most of the remaining dry weight consists of molecules containing atoms of carbon. When first discovered, it was thought that carbon-containing molecules were present only in living organisms and thus were referred to as *organic molecules* to distinguish them from *inorganic molecules* found in the inanimate world. As chemists learned to synthesize more and more of these carbon-containing molecules in the lab, the mystique associated with organic compounds disappeared. The compounds produced by living organisms are called **biochemicals**.

The chemistry of life centers around the chemistry of the carbon atom. The essential quality of carbon that has allowed it to play this role is the incredible number of molecules it can form. Having four outer-shell electrons, a carbon atom can bond with up to four other atoms (see Figure 2.1). In addition, each carbon atom is able to bond with other carbon atoms so as to construct molecules with backbones containing long chains of carbon atoms. Car-

bon-containing backbones may be linear, branched, or cyclic.

Linear	Cyclic	Branched

Cholesterol, whose structure is depicted in Figure 2.9, illustrates various arrangements of carbon atoms.

Both the size and electronic structure of carbon make it uniquely suited for generating large numbers of molecules, several hundred thousand of which are known. In contrast, silicon, which is just below carbon in the periodic table and also has four outer-shell electrons (see Figure 2.1), is too large for its positively charged nucleus to attract the outer-shell electrons of neighboring atoms with sufficient force to hold such large molecules together. We can best understand the nature of biological molecules by starting with the simplest group of organic molecules, the *hydrocarbons*, which contain only carbon and hydrogen atoms. The molecule ethane (C_2H_6) is a simple hydrocarbon

ethane

consisting of two atoms of carbon in which each carbon is bonded to the other carbon as well as to three atoms of hydrogen. As more carbons are added, the skeletons of or-

Cholesterol

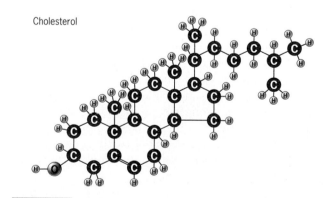

Figure 2.9 Cholesterol, whose structure illustrates how carbon atoms (represented by the black balls) are able to form covalent bonds with as many as four other carbon atoms. As a result, carbon atoms can be linked together to form the backbones of a virtually unlimited variety of organic molecules. The carbon backbone of a cholesterol molecule includes four rings, which is characteristic of steroids (e.g., estrogen, testosterone, cortisol). The cholesterol molecule shown here is drawn as a ball-and-stick model, which is another way that molecular structure is depicted.

ganic molecules increase in length and their structure becomes more complex.

Functional Groups

Hydrocarbons do not occur in significant amounts within most living cells (though they constitute the bulk of the fossil fuels formed from the remains of ancient plants and animals). Many of the organic molecules that are important in biology contain chains of carbon atoms, like hydrocarbons, but certain of the hydrogen atoms are replaced by various **functional groups**. Functional groups are particular groupings of atoms that often behave as a unit and give organic molecules their physical properties, chemical reactivity, and solubility in aqueous solution. Some of the more common functional groups are listed in Table 2.3. Two of the most common linkages between functional groups are **ester bonds**, which form between carboxylic acids and alcohols, and **amide bonds**, which form between carboxylic acids and amines.

Most of the groups in Table 2.3 contain one or more electronegative atoms (N, P, O, and/or S) and make organic molecules more polar, more water soluble, and more reactive. Many of the functional groups are capable of ionization and thus may become positively or negatively charged. The effect of the substitution of various functional groups is readily demonstrated. The hydrocarbon ethane (CH_3CH_3) depicted above is a toxic, flammable gas. Replace one of the hydrogens with a hydroxyl group (—OH) and the molecule (CH_3CH_2OH) becomes palatable—it is ethyl alcohol. Substitute a carboxyl group (—COOH) and the molecule becomes acetic acid (CH_3COOH), the strong-tasting ingredient in vinegar. Substitute a sulfhydryl group (—SH), and you have formed CH_3CH_2SH, a strong, foul-smelling agent, ethyl mercaptan, used by biochemists in studying enzyme reactions.

A Classification of Biological Molecules by Function

The organic molecules commonly found within living cells can be divided into several categories based on their role in metabolism.

1. *Macromolecules.* The molecules that form the structure and carry out the activities of cells are huge, highly organized molecules called **macromolecules**, which contain anywhere from dozens to millions of carbon atoms. Because of their size and the intricate shapes that macromolecules can assume, some of these molecular giants can perform complex tasks with great precision and efficiency. The presence of macromolecules, more than any other characteristic, endows organisms with the properties of life and sets them apart chemically from the inanimate world.

Macromolecules can be divided into four major categories: proteins, nucleic acids, polysaccharides, and certain lipids. The first three types are *polymers* composed of a large number of low-molecular-weight building blocks, or *monomers*. These macromolecules are constructed from monomers by a process that resembles coupling railroad cars onto a train (Figure 2.10). The basic structure and function of each type of macromolecule are similar in all organisms. It is not until you look very closely at the specific sequences of monomers that make up these various macromolecules that the diversity among organisms becomes apparent.

2. *The building blocks of macromolecules*. Most of the macromolecules within a cell have a short lifetime compared with the cell itself; with the exception of the cell's DNA, they are continually broken down and replaced by new macromolecules. Consequently, most cells contain a supply (or *pool*) of low-molecular-weight precursors that are ready to be incorporated into macromolecules. These include sugars, which are the precursors of polysaccharides; amino acids, which are the precursors of proteins; nucleotides, which are the precursors of nucleic acids; and fatty acids, which are incorporated into lipids.

3. *Metabolic intermediates (metabolites)*. The molecules in a cell have complex chemical structures and must be synthesized in a step-by-step sequence beginning with specific starting materials. In the cell, each series of chemical reactions is termed a **metabolic pathway**. The cell starts with compound A and converts it to compound B, then to compound C, and so on, until some end product (such as an amino acid building block of a protein) is produced that can be used in other reactions. The com-

Table 2.3	Functional Groups

Figure 2.10 Monomers and polymers. (*a*) Polysaccharides, proteins, and nucleic acids consist of monomers (subunits) linked together by covalent bonds. Free monomers do not simply react with each other to become macromolecules. Rather, each monomer is first activated by attachment to a carrier molecule that subsequently transfers the monomer to the end of the growing macromolecule. (*b*) A macromolecule is disassembled by hydrolysis of the bonds that join the monomers together. Hydrolysis is the splitting of a bond by water. All of these reactions are catalyzed by specific enzymes.

pounds formed along the pathways leading to the end products might have no function per se and are called **metabolic intermediates**.

4. ***Molecules of miscellaneous function***. This is obviously a broad category of molecules but not as large as you might expect; the vast bulk of the dry weight of a cell is made up of macromolecules and their direct precursors. The molecules of miscellaneous function include such substances as vitamins, which function primarily as adjuncts to proteins; certain steroid or amino acid hormones; molecules involved in energy storage, such as ATP or creatine phosphate; regulatory molecules such as cyclic AMP; and metabolic waste products such as urea.

REVIEW ?

1. What properties of a carbon atom are critical to life?
2. Draw the structures of four different functional groups. How would each of these groups alter the solubility of a molecule in water?

2.5 FOUR TYPES OF BIOLOGICAL MOLECULES

The molecules just described can be divided into four types or classes of organic molecules: carbohydrates, lipids, amino acids and proteins, and nucleotides and nucleic acids. The localization of these molecules in a number of cellular structures is shown in an overview in Figure 2.11.

Carbohydrates

Carbohydrates include simple sugars (or *monosaccharides*) and all larger molecules constructed of sugar building blocks. Carbohydrates function primarily as stores of chemical energy and as durable building materials for biological construction. Most sugars have the general formula $(CH_2O)_n$. The sugars of importance in cellular metabolism have values of *n* that range from 3 to 7. Sugars containing three carbons are known as *trioses*, those with four carbons as *tetroses*, those with five carbons as *pentoses*, those with six carbons as *hexoses*, and those with seven carbons as *heptoses*.

Figure 2.11 **An overview of the types of biological molecules that make up various cellular structures.**

The Structure of Simple Sugars Each sugar molecule consists of a backbone of carbon atoms linked together in a linear array by single bonds. Each of the carbon atoms of the backbone is linked to a single hydroxyl group, except for one that bears a *carbonyl* ($C=O$) group. If the carbonyl group is located at an internal position (to form a ketone group), the sugar is a *ketose*, such as fructose, which is shown in Figure 2.12*a*. If the carbonyl is located at one

(a) (b) (c) (d) (e) (f)

Figure 2.12 **The structures of sugars.** (*a*) The straight-chain formula of fructose, a ketohexose [keto indicating the carbonyl (yellow) is located internally, and hexose because it consists of six carbons]. (*b*) The straight-chain formula of glucose, an aldohexose (aldo because the carbonyl is located at the end of the molecule). (*c*) The self-reaction in which glucose is converted from an open chain to a closed ring (a pyranose ring). (*d*) Glucose is commonly depicted in the form of a flat (planar) ring with the thickened line situated closest to the reader and the H and OH groups projecting either above or below the ring. The basis for the designation α-D-glucose is discussed in the following section. (*e*) The chair conformation of glucose, which depicts its three-dimensional structure more accurately than the flattened ring of part *d*. (*f*) A ball-and-stick model of the chair conformation of glucose showing the position of the various atoms of the molecule.

end of the sugar, it forms an aldehyde group and the molecule is known as an *aldose,* as exemplified by glucose, which is shown in Figure 2.12*b–f.* Although the straight-chained formulas shown in Figure 2.12*a,b* are useful for comparing the structures of various sugars, they do not reflect the fact that sugars with five or more carbon atoms undergo a self-reaction (Figure 2.12*c*) that converts them into a closed, or ring-containing, molecule. The ring forms of sugars are usually depicted as flat (*planar*) structures (Figure 2.12*d*) lying perpendicular to the plane of the paper with the thickened line situated closest to the reader. The H and OH groups lie parallel to the plane of the paper projecting either above or below the ring of the sugar. In actual fact, the sugar ring is not a planar structure, but usually exists in a three-dimensional conformation resembling a chair (Figure 2.12*e,f*).

Stereoisomerism As noted earlier, a carbon atom can bond with four other atoms. The arrangement of the groups around a carbon atom can be depicted as in Figure 2.13*a* with the carbon placed in the center of a tetrahedron and the bonded groups projecting into its four corners. Figure 2.13*b* depicts a molecule of glyceraldehyde, which is the only aldotriose. The second carbon atom of glyceraldehyde is linked to four different groups (—H, —OH, —CHO, and —CH₂OH). If the four groups bonded to a carbon atom are all different, as in glyceraldehyde, then two possible configurations exist that cannot be superimposed on one another. These two molecules (termed *stereoisomers* or *enantiomers*) have essentially the same chemical reactivities, but their structures are mirror images of one another (not unlike a pair of right and left human hands). By convention, the molecule is called D-glyceraldehyde if the hydroxyl group of carbon 2 projects to the right, and L-glyceraldehyde if it projects to the left (Figure 2.13*c*). Because it acts as a site of stereoisomerism, carbon 2 is referred to as an *asymmetric* carbon atom.

As the backbone of sugar molecules increases in length, so too does the number of asymmetric carbon atoms and, consequently, the number of stereoisomers. Aldotetroses have two asymmetric carbons and thus can exist in four different configurations (Figure 2.14). Similarly, there are 8 different aldopentoses and 16 different aldohexoses. The designation of each of these sugars as D or L is based by convention on the arrangement of groups attached to the asymmetric carbon atom farthest from the aldehyde (the carbon associated with the aldehyde is designated C1). If the hydroxyl group of this carbon projects to the right, the aldose is a D-sugar; if it projects to the left, it is an L-sugar. The enzymes present in living cells can distinguish between the D and L forms of a sugar. Typically, only one of the stereoisomers (such as D-glucose and L-fucose) are used by cells.

The self-reaction in which a straight-chain glucose molecule is converted into a six-membered (*pyranose*) ring was shown in Figure 2.12*c*. Unlike its precursor in the open chain, the C1 of the ring bears four different groups and thus becomes a new center of asymmetry within the

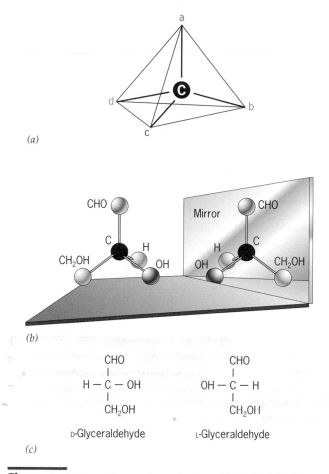

(a)

(b)

(c)

Figure 2.13 Stereoisomerism of glyceraldehyde. (*a*) The four groups bonded to a carbon atom (labeled a, b, c, and d) occupy the four corners of a tetrahedron with the carbon atom at its center. (*b*) Glyceraldehyde is the only three-carbon aldose; its second carbon atom is bonded to four different groups (—H, —OH, —CHO, and —CH₂OH). As a result, glyceraldehyde can exist in two possible configurations that are not superimposable, but instead are mirror images of each other as indicated. These two stereoisomers (or enantiomers) can be distinguished by the configuration of the four groups around the asymmetric (or *chiral*) carbon atom. Solutions of these two isomers rotate plane-polarized light in opposite directions and, thus, are said to be optically active. (*c*) Straight-chain formulas of glyceraldehyde. By convention, the D-isomer is shown with the OH group on the right.

CHO	CHO	CHO	CHO
HCOH	HOCH	HCOH	HOCH
HCOH	HCOH	HOCH	HOCH
CH₂OH	CH₂OH	CH₂OH	CH₂OH
D-Erythrose	D-Threose	L-Threose	L-Erythrose

Figure 2.14 Aldotetroses. Because they have two asymmetric carbon atoms, aldotetroses can exist in four configurations.

Figure 2.15 **Formation of an α- and β-pyranose.** When a molecule of glucose undergoes self-reaction to form a pyranose ring (i.e., a six-membered ring), two stereoisomers are generated. The two isomers are in equilibrium with each other through the open-chain form of the molecule. By convention, the molecule is an α-pyranose when the OH group of the first carbon projects below the plane of the ring, and a β-pyranose when the hydroxyl group projects upward.

sugar molecule. Because of this extra asymmetric carbon atom, each type of pyranose exists as α and β stereoisomers (Figure 2.15). By convention, the molecule is an α-pyranose when the OH group of the first carbon projects below the plane of the ring, and a β-pyranose when the hydroxyl projects upward. The difference between the two forms has important biological consequences, resulting, for example, in the compact shape of glycogen and starch molecules and the extended conformation of cellulose (discussed later).

Linking Sugars Together Sugars can be joined to one another by covalent **glycosidic bonds** to form larger molecules. Glycosidic bonds form by reaction between carbon atom C1 of one sugar and the hydroxyl group of another sugar, generating a —C—O—C— linkage between the two sugars. As discussed below (and indicated in Figures 2.16 and Figure 2.17), sugars can be joined by quite a variety of different glycosidic bonds. Molecules composed of only two sugar units are **disaccharides** (Figure 2.16). Disaccharides serve primarily as readily available energy stores. Sucrose, or table sugar, is a major component of plant sap, which carries chemical energy from one part of the plant to another. Lactose, present in the milk of most mammals, supplies newborn mammals with fuel for early growth and development. Lactose in the diet is hydrolyzed by the enzyme lactase, which is present in the plasma membranes of the cells that line the intestine. Many people lose this enzyme after childhood and find that eating lactose-containing dairy products causes digestive discomfort.

Sugars may also be linked together to form small chains called **oligosaccharides** (*oligo* = few). Most often such chains are found covalently attached to lipids and proteins, converting them into glycolipids and glycoproteins, respectively. Oligosaccharides are particularly important on the glycolipids and glycoproteins of the plasma membrane, where they project from the cell surface (see

Figure 4.4c). Because oligosaccharides may be composed of many different combinations of sugar units, these carbohydrates can play an informational role; that is, they can serve to distinguish one type of cell from another and help mediate specific interactions of a cell with its surroundings (discussed in Section 7.1).

Polysaccharides By the middle of the nineteenth century, it was known that the blood of people suffering from diabetes had a sweet taste due to an elevated level of glucose, the key sugar in energy metabolism. Claude Bernard, a prominent French physiologist of the period,

Figure 2.16 **Disaccharides.** Sucrose and lactose are two of the most common disaccharides. Sucrose is composed of glucose and fructose joined by an α(1→2) linkage, whereas lactose is composed of glucose and galactose joined by α β(1→4) linkage.

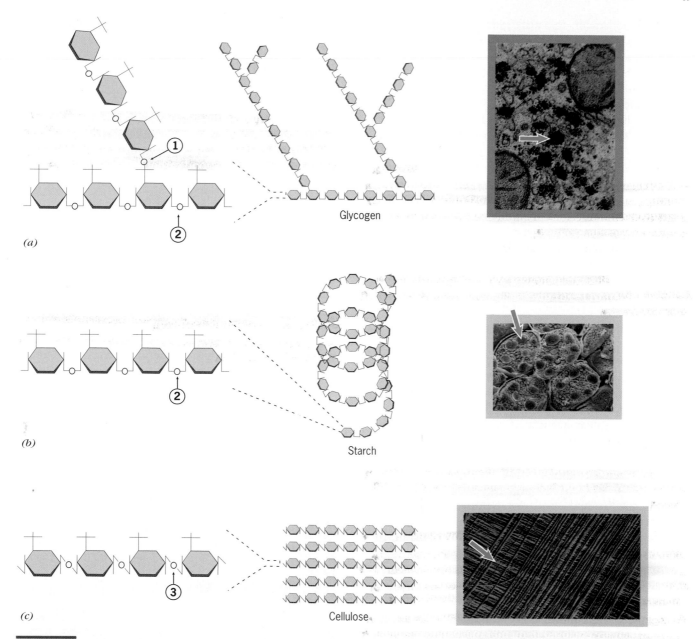

(a)

Glycogen

(b)

Starch

(c)

Cellulose

Figure 2.17 **Three polysaccharides with identical sugar monomers but dramatically different properties.**
Glycogen (*a*), starch (*b*), and cellulose (*c*) are each composed entirely of glucose subunits, yet their chemical and physical properties are very different due to the distinct ways that the monomers are linked together (three different types of linkages are indicated by the circled numbers). Glycogen molecules are the most highly branched, starch molecules assume a helical arrangement, and cellulose molecules are unbranched and highly extended. Whereas glycogen and starch are energy stores, cellulose molecules are bundled together into tough fibers that are suited for their structural role. Colorized electron micrographs show glycogen granules in a liver cell, starch grains (amyloplasts) in a plant seed, and cellulose fibers in a plant cell wall; each is indicated by an arrow. [PHOTO INSETS: (TOP) DON FAWCETT/VISUALS UNLIMITED; (CENTER) JEREMY BURGESS/PHOTO RESEARCHERS; (BOTTOM) CABISCO/VISUALS UNLIMITED.]

was looking for the cause of diabetes by investigating the source of blood sugar. It was assumed at the time that any sugar present in a human or animal had to have been previously consumed in the diet. Working with dogs, Bernard found that, even if the animals were placed on a diet totally lacking carbohydrates, their blood still contained a normal amount of glucose. Clearly, glucose could be formed in the body from other types of compounds.

After further investigation, Bernard found that glucose enters the blood from the liver. Liver tissue, he found, contains an insoluble polymer of glucose he named **glycogen**. Bernard concluded that various food materials (such as proteins) were carried to the liver where they were chemically converted to glucose and stored as glycogen. Then, as the body needed sugar for fuel, the glycogen in the liver was transformed to glucose, which was released

into the bloodstream to satisfy glucose-depleted tissues. In Bernard's hypothesis, the balance between glycogen formation and glycogen breakdown in the liver was the prime determinant in maintaining the relatively constant (homeostatic) level of glucose in the blood.

Bernard's hypothesis proved to be correct. The molecule he named glycogen is a type of **polysaccharide**—a polymer of sugar units joined together by glycosidic bonds.

Glycogen and Starch: Nutritional Polysaccharides Glycogen is a branched polymer containing only one type of monomer: glucose (Figure 2.17*a*). Most of the sugar units of a glycogen molecule are joined to one another by α(1→4) glycosidic bonds (type 2 bond in Figure 2.17*a*). Branch points contain a sugar joined to three neighboring units rather than to two, as in the unbranched segments of the polymer. The extra neighbor, which forms the branch, is linked by an α(1→6) glycosidic bond (type 1 bond in Figure 2.17*a*).

Glycogen serves as a storehouse of surplus chemical energy in most animals. Human skeletal muscles, for example, typically contain enough glycogen to fuel about 30 minutes of moderate activity. Depending on various factors, glycogen typically ranges in molecular weight from about one to four million daltons. When stored in cells, glycogen is highly concentrated in what appears as dark-staining, irregular granules in electron micrographs (Figure 2.17*a*, right).

Most plants bank their surplus chemical energy in the form of **starch**, which like glycogen is also a polymer of glucose. Potatoes and cereals, for example, consist primarily of starch. Starch is actually a mixture of two different polymers, amylose and amylopectin. Amylose is an unbranched, helical molecule whose sugars are joined by α(1→4) linkages (Figure 2.17*b*), whereas amylopectin is branched. Amylopectin differs from glycogen in being much less branched and having an irregular branching pattern. Starch is stored as densely packed granules, or *starch grains*, which are enclosed in membrane-bound organelles (*plastids*) within the plant cell (Figure 2.17*b*, right). Although animals don't synthesize starch, they possess an enzyme (*amylase*) that readily hydrolyzes starch molecules.

Cellulose, Chitin, and Glycosaminoglycans: Structural Polysaccharides Whereas some polysaccharides constitute easily digested energy stores, others form tough, durable structural materials. Cotton and linen, for example, consist largely of **cellulose**, which is the major component of plant cell walls. Cotton textiles owe their durability to the long, unbranched cellulose molecules, which are ordered into side-by-side aggregates to form molecular cables (photo, Figure 2.17*c*) that are ideally constructed to resist pulling (tensile) forces. Like glycogen and starch, cellulose consists solely of glucose monomers; its properties differ dramatically from these other polysaccharides because the glucose units are joined by β(1→4) linkages (bond 3 in Figure 2.17*c*) rather than α(1→4) linkages. Ironically, multicellular animals (with rare exception) lack the enzyme needed to

degrade cellulose, which happens to be the most abundant organic material on Earth and rich in chemical energy. Animals that "make a living" by digesting cellulose, such as termites and sheep, do so by harboring bacteria and protozoa that synthesize the necessary enzyme, cellulase.

Not all biological polysaccharides consist of glucose monomers. **Chitin** is an unbranched polymer of the sugar *N*-acetylglucosamine, which is similar in structure to glucose but has an acetyl amino group instead of a hydroxyl group bonded to the second carbon atom of the ring.

N-Acetylglucosamine

Chitin occurs widely as a structural material among invertebrates, particularly in the outer covering of insects, spiders, and crustaceans. Chitin is a tough, resilient, yet flexible material not unlike certain plastics. Insects owe much of their success to this highly adaptive polysaccharide covering (Figure 2.18).

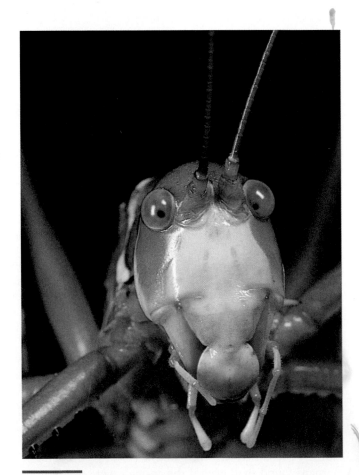

Figure 2.18 **Chitin** is the primary component of the glistening outer skeleton of this grasshopper. (FROM ROBERT AND LINDA MITCHELL.)

Another group of polysaccharides that has a more complex structure is the **glycosaminoglycans** (or **GAGs**). Unlike other polysaccharides, they have the structure —A—B—A—B—, where A and B represent two different sugars. These polysaccharides are found primarily in the spaces that surround cells, and their structure and function will be considered further in Section 7.1, where the topic of the extracellular space is discussed.

Lipids

Lipids are a diverse group of nonpolar biological molecules whose common properties are their ability to dissolve in organic solvents, such as chloroform or benzene, and their inability to dissolve in water—a property that explains many of their varied biological functions. Lipids of importance in cellular function include fats, steroids, and phospholipids.

Fats **Fats** consist of a glycerol molecule linked by ester bonds to three fatty acids; the composite molecule is termed a **triacylglycerol** (Figure 2.19a). We will begin by considering the structure of **fatty acids**. Fatty acids are long, unbranched hydrocarbon chains with a single carboxyl group at one end (Figure 2.19b). Because the two ends of a fatty acid molecule have a very different structure, they also have different properties. The hydrocarbon chain is hydrophobic, whereas the carboxyl group (—COOH), which bears a negative charge at physiological pH, is hydrophilic. Molecules having both hydrophobic and hydrophilic regions are said to be **amphipathic**; such molecules have unusual and biologically important properties. The properties of fatty acids can be appreciated by considering the use of a familiar product: soap, which consists of fatty acids. In past centuries, soaps were made by heating animal fat in strong alkali (NaOH or KOH) to break the bonds between the fatty acids and the glycerol. Today, most soaps are made synthetically. Soaps owe their grease-dissolving capability to the fact that the hydrophobic end of each fatty acid can embed itself in the grease, whereas the hydrophilic end can interact with the surrounding water. As a result, greasy materials are converted into complexes (*micelles*) that can be dispersed by water (Figure 2.20).

Fatty acids differ from one another in the length of their hydrocarbon chain and the presence or absence of double bonds. Fatty acids present in cells typically vary in length from 14 to 20 carbons. Fatty acids that lack double bonds, such as stearic acid (Figure 2.19b), are described as **saturated**; those possessing double bonds are **unsaturated**. Double bonds (of the *cis* configuration)

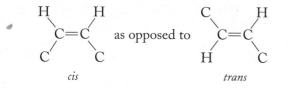

produce kinks in a fatty acid chain. Consequently, the more double bonds that fatty acid chains possess, the less

(a)

(b)

Tristearate

(c)

Linseed oil

(d)

Figure 2.19 Fats and fatty acids. (*a*) The basic structure of a triacylglycerol (also called a triglyceride or a neutral fat). The glycerol moiety, indicated in orange, is linked by three ester bonds to the carboxyl groups of three fatty acids whose tails are indicated in green. (*b*) Stearic acid, an 18-carbon saturated fatty acid that is common in animal fats. (*c*) Space-filling model of tristearate, a triacylglycerol containing three identical stearic acid chains. (*d*) Space-filling model of linseed oil, a triacylglycerol containing two different unsaturated fatty acids, derived from flax seeds. The sites of unsaturation, which produce kinks in the molecule, are indicated by the yellow-orange bars.

effectively can these long chains be packed together. This lowers the temperature at which a fatty acid-containing lipid melts. Tristearate, whose fatty acids lack double bonds (Figure 2.19c), is a common component of animal

Figure 2.20 **Soaps consist of fatty acids.** In this schematic drawing of a soap micelle, the nonpolar tails of the fatty acids are directed inward, where they interact with the greasy matter to be dissolved. The negatively charged heads are located at the surface of the micelle, where they interact with the surrounding water. Membrane proteins, which also tend to be insoluble in water, can also be solubilized in this way by extraction of membranes with detergents.

Figure 2.21 **The structure of steroids.** All steroids share the basic four-ring skeleton. The seemingly minor differences in chemical structure between cholesterol, testosterone, and estrogen generate profound biological differences.

fats and remains in a solid state well above room temperature. In contrast, the profusion of double bonds in vegetable fats accounts for their liquid state—both in the plant cell and on the grocery shelf—and for their being labeled as "polyunsaturated." Fats that are liquid at room temperature are described as **oils**. Figure 2.19*d* shows the structure of linseed oil, a highly volatile lipid extracted from flax seeds, that remains a liquid at a much lower temperature than does tristearate. Solid shortenings, such as margarine, are formed from unsaturated vegetable oils by chemically reducing the double bonds with hydrogen atoms (a process termed *hydrogenation*). A molecule of fat can contain three identical fatty acids (as in Figure 2.19*c*), or it can be a *mixed fat*, containing more than one fatty acid species (as in Figure 2.19*d*). Most natural fats, such as olive oil or butterfat, are mixtures of molecules having different fatty acid species.

Fats are very rich in chemical energy; a gram of fat contains over twice the energy content of a gram of carbohydrate (for reasons discussed in Section 3.1). Carbohydrates function primarily as a short-term, rapidly available energy source, whereas fat reserves store energy on a long-term basis. It is estimated that a person of average size contains about 0.5 kilograms (kg) of carbohydrate, primarily in the form of glycogen. This amount of carbohydrate provides approximately 2000 kcal of total energy. During the course of a strenuous day's exercise, a person can virtually deplete his or her body's entire store of carbo-

hydrate. In contrast, the average person contains approximately 16 kg of fat (equivalent to 144,000 kcal of energy), and as we all know, it can take a very long time to deplete our store of this material.

Because they lack polar groups, fats are extremely insoluble in water and are stored in cells in the form of dry lipid droplets. Since lipid droplets do not contain water as do glycogen granules, they represent an extremely concentrated storage fuel. In many animals, fats are stored in special cells (*adipocytes*) whose cytoplasm is filled with one or a few large lipid droplets. Adipocytes exhibit a remarkable ability to change their volume to accommodate varying quantities of fat.

Steroids Steroids are built around a characteristic four-ringed hydrocarbon skeleton. One of the most important steroids is *cholesterol*, a component of animal cell membranes and a precursor for the synthesis of a number of steroid hormones, such as testosterone, progesterone, and estrogen (Figure 2.21). Cholesterol is largely absent from plant cells, which is why vegetable oils are considered "cholesterol-free," but plant cells may contain large quantities of related compounds.

Phospholipids The structure of a common phospholipid is shown in Figure 2.22. The molecule resembles a fat (triacylglycerol), but has only two fatty acid chains rather than three; it is a *diacylglycerol*. The third hydroxyl of the glycerol backbone is covalently bonded to a phosphate group

Figure 2.22 The phospholipid phosphatidylcholine. The molecule consists of a glycerol backbone whose hydroxyl groups are covalently bonded to two fatty acids and a phosphate group. The negatively charged phosphate is also bonded to a small, positively charged choline group. The end of the molecule that contains the phosphorylcholine is highly water soluble, whereas the opposite end, consisting of the fatty acid tail, is insoluble in water. The structure and function of phosphatidylcholine and other phospholipids are discussed at length in Section 4.3.

(rather than a third fatty acid), which in turn is covalently bonded to a small polar group, such as choline, as shown in Figure 2.22. Thus, unlike fat molecules, phospholipids contain two ends that have very different properties: the end containing the phosphate group has a distinctly hydrophilic character; the other end composed of the two

fatty acid tails has a distinctly hydrophobic character. Because phospholipids function primarily in cell membranes, and because the properties of cell membranes depend on their phospholipid components, they will be discussed further in Section 4.3 in connection with cell membranes.

Proteins

Proteins are the macromolecules that carry out virtually all of a cell's activities; they are the molecular tools and machines that make things happen. It is estimated that the typical mammalian cell may have as many as 10,000 different proteins having a diverse array of functions. As enzymes, proteins vastly accelerate the rate of metabolic reactions; as structural cables, proteins provide mechanical support both within cells and outside their perimeters (Figure 2.23a); as hormones, growth factors, and gene activators, proteins perform a wide variety of regulatory functions; as membrane receptors and transporters, proteins determine what a cell reacts to and what types of substances enter or leave the cell; as contractile elements, proteins constitute the machinery for biological movements. Among their many other functions, proteins act as antibodies, serve as toxins, form blood clots, absorb or refract light (Figure 2.23b), and transport substances from one part of the body to another.

How can one type of molecule have so many varied functions? The explanation resides in the virtually unlimited shapes that proteins, *as a group*, can assume. In other words, proteins are capable of such a wide variety of activ-

(a) *(b)*

Figure 2.23 Two examples of the thousands of biological structures composed predominantly of protein. These include (*a*) feathers, which are adaptations in birds for thermal insulation, flight, and sex recognition; and (*b*) the lenses of eyes, as in this net-casting spider, which focus light rays. (A: FROM FRANS LANTING/ALLSTOCK, INC./STONE; B: MANTIS WILDLIFE FILMS/OXFORD SCIENTIFIC FILMS/ANIMALS ANIMALS.)

ities because they can exhibit such a great variety of structures. Each protein, however, has a unique and highly ordered structure that enables it to carry out a particular function. Most importantly, proteins have shapes that allow them to interact *selectively* with other molecules. Proteins, in other words, exhibit a high degree of **specificity**. It is possible, for example, for a particular DNA-cutting enzyme to recognize a segment of DNA containing one specific sequence of eight nucleotides, while ignoring all the other 65,535 possible sequences composed of this number of nucleotides.

The Building Blocks of Proteins Proteins are polymers made of amino acid monomers. Each protein has a unique sequence of amino acids that gives the molecule its unique properties. Many of the capabilities of a protein can be understood by examining the chemical properties of its constituent amino acids. Twenty different amino acids are commonly used in the construction of proteins, whether from a virus or a human. There are two aspects of amino acid structure to consider: that which is common to all of them, and that which is unique to each. We will begin with the shared properties.

The Structures of Amino Acids All amino acids have a carboxyl group and an amino group, which are separated from each other by a single carbon atom, the α-carbon (Figure 2.24*a,b*). In a neutral aqueous solution, the α-carboxyl group loses its proton and exists in a negatively charged state ($-COO^-$), and the α-amino group accepts a proton and exists in a positively charged state ($-NH_3^+$) (Figure 2.24*b*). We saw on page 45 that carbon atoms of sugar molecules bond to four different groups in two configurations (*stereoisomers*) that cannot be superimposed on one another. Amino acids also have asymmetric centers. With the exception of glycine, the α carbon of amino acids bonds to four different groups so that each amino acid can exist in either a D or an L form (Figure 2.25). Amino acids used in the synthesis of a protein on a ribosome are always L-amino acids. Microorganisms, however, use D-amino acids in the synthesis of certain small peptides, including those of the cell wall and several antibiotics (e.g., gramicidin A). Interestingly, D-amino acids have also been found in certain peptides produced by the skin of South American tree frogs. One of these peptides called dermorphin has been shown to act as a highly potent pain killer in rats through its interaction with opiate receptors in the brain. The D-amino acids in these frog peptides are created by enzymes acting on L-amino acids that had been initially incorporated into the peptide.

During the process of protein synthesis, each amino acid becomes joined to two other amino acids forming a long, continuous, unbranched polymer called a **polypeptide chain**. The amino acids that make up a polypeptide chain are joined by **peptide bonds** that result from the linkage of the carboxyl group of one amino acid to the amino group of its neighbor, with the elimination of a molecule of water (Figure 2.24*c*). A polypeptide chain

(a)

(b)

(c)

Figure 2.24 **Amino acid structure**. Ball-and-stick model (*a*) and chemical formula (*b*) of a generalized amino acid in which R can be any of a number of chemical groups (see Figure 2.26). (*c*) The formation of a peptide bond occurs by the condensation of two amino acids, drawn here in the uncharged state. In the cell, this reaction occurs on a ribosome as an amino acid is transferred from a carrier (a tRNA molecule) onto the end of the growing polypeptide chain (see Figure 11.50).

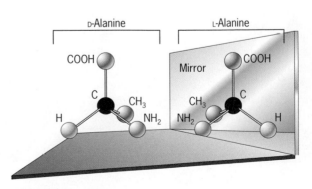

Figure 2.25 **Amino acid stereoisomerism.** Because the α carbon of all amino acids except glycine is bonded to four different groups, two stereoisomers can exist. The D and L forms of alanine are shown.

composed of a string of amino acids joined by peptide bonds has the following backbone:

Peptide bond

The "average" polypeptide chain contains about 450 amino acids. The longest known polypeptide, found in the muscle protein titin, contains nearly 27,000 amino acids.

Once incorporated into a polypeptide chain, amino acids are termed **residues**. The residue on one end of the chain, the *N-terminus*, contains an amino acid with a free (unbonded) α-amino group, whereas the residue at the opposite end, the *C-terminus*, has a free α-carboxyl group.

The Properties of the Side Chains The backbone, or main chain, of the polypeptide is composed of that part of each amino acid that is common to all of them. The **side chain** or **R group** (Figure 2.24), bonded to the α carbon, is highly variable among the 20 building blocks, and it is this variability that ultimately gives proteins their diverse structures and activities. If the various amino acid side chains are considered together, they exhibit a large variety of structural features, ranging from fully charged to hydrophobic, and they can participate in a wide variety of covalent and noncovalent bonds. As discussed in the following chapter, the side chains of the "active sites" of enzymes can facilitate (catalyze) many different organic reactions. The assorted characteristics of the side chains of the amino acids are important in both *intra*molecular interactions, which determine the structure and activity of the molecule, and *inter*molecular interactions, which determine the relationship of a polypeptide with other molecules, including other polypeptides (page 63).

Amino acids are conveniently classified on the character of their side chains. They fall roughly into four categories: polar and charged, polar and uncharged, nonpolar, and those with unique properties (Figure 2.26).

1. *Polar, charged*. Amino acids of this group include aspartic acid, glutamic acid, lysine, and arginine. These four amino acids contain side chains that become fully charged; that is, the side chains contain relatively strong organic acids and bases. The ionization reactions of glutamic acid and lysine are shown in Figure 2.27. At physiologic pH, the side chains of these amino acids are almost always present in the fully charged state. Consequently, they are able to form ionic bonds with other charged species in the cell. For example, the positively charged arginine residues of histone proteins are linked by ionic bonds to the negatively charged phosphate groups of DNA (see Figure 2.3). Histidine is also considered a polar, charged amino acid, though in most cases it is only partially charged at physiologic pH. In fact, because of its ability to gain or lose a proton in physiologic pH ranges, histidine is a particularly important residue in the active site of many proteins (see Figure 3.13 as an example).

2. *Polar, uncharged*. The side chains of these amino acids are only weakly acidic or basic. Although these groups are not fully charged at physiologic pH, they do contain atoms that have a partial negative or positive charge and thus can form hydrogen bonds with other molecules including water. These amino acids are often quite reactive. Included in this category are asparagine and glutamine (the amides of aspartic acid and glutamic acid, page 42), threonine, serine, and tyrosine.

3. *Nonpolar*. At the other extreme from those in the first category are the amino acids whose side chains are hydrophobic and are unable to form electrostatic bonds or interact with water. The amino acids of this category are alanine, valine, leucine, isoleucine, tryptophan, phenylalanine, and methionine. The side chains of the nonpolar amino acids generally lack oxygen and nitrogen. They vary primarily in size and shape, which allows one or another of them to pack tightly into a particular space within the core of a protein, associating with one another as the result of van der Waals forces and hydrophobic interactions.

4. *The other three amino acids*—glycine, proline, and cysteine—have unique properties that separate them from the others. The side chain of glycine consists of only a hydrogen atom, and glycine is a very important amino acid for just this reason. Owing to this lack of a side chain, glycine residues provide a site where the backbones of two polypeptides (or two segments of the same polypeptide) can approach one another very closely. In addition, glycine is more flexible than other amino acids and allows parts of the backbone to move or form a hinge. Proline is unique in having its α-amino group as part of a ring (making it an imino acid). Proline is a hydrophobic amino acid that does not readily fit into an ordered secondary structure (discussed later). Cysteine contains a reactive sulfhydryl (—SH) group and is often covalently linked to another cysteine residue, as a **disulfide** (—SS—) **bridge**.

Polar charged

Aspartic acid
(Asp or D)

Glutamic acid
(Glu or E)

Lysine
(Lys or K)

Arginine
(Arg or R)

Histidine
(His or H)

Properties of side chains (R groups):

Hydrophilic side chains act as acids or bases which tend to be fully charged (+ or −) under physiologic conditions. Side chains form ionic bonds and are often involved in chemical reactions.

Polar uncharged

Serine
(Ser or S)

Threonine
(Thr or T)

Glutamine
(Gln or Q)

Asparagine
(Asn or N)

Tyrosine
(Tyr or Y)

Properties of side chain:

Hydrophilic side chains tend to have partial + or − charge allowing them to participate in chemical reactions, form H-bonds, and associate with water.

Nonpolar

Alanine
(Ala or A)

Valine
(Val or V)

Leucine
(Leu or L)

Isoleucine
(Ile or I)

Methionine
(Met or M)

Phenylalanine
(Phe or F)

Tryptophan
(Trp or W)

Properties of side chain:

Hydrophobic side chain consists almost entirely of C and H atoms. These amino acids tend to form the inner core of soluble proteins, buried away from the aqueous medium. They play an important role in membranes by associating with the lipid bilayer.

Side chains with unique properties

Glycine
(Gly or G)

Cysteine
(Cys or C)

Proline
(Pro or P)

Side chain consists only of hydrogen atom and can fit into either a hydrophilic or hydrophobic environment. Glycine often resides at sites where two polypeptides come into close contact.

Though side chain has polar, uncharged character, it has the unique property of forming a covalent bond with another cysteine to form a disulfide link.

Though side chain has hydrophobic character, it has the unique property of creating kinks in polypeptide chains and disrupting ordered secondary structure.

Figure 2.26 The chemical structure of amino acids. These 20 amino acids represent those most commonly found in proteins and, more specifically, those encoded by the DNA. Other amino acids occur as the result of a modification to one of those shown here. The amino acids are arranged into four groups based on the character of their side chains, as described in the text. All molecules are depicted as free amino acids in their ionized state as they would exist at neutral pH.

(a)

(b)

Figure 2.27 The ionization of charged, polar amino acids. (*a*) The side chain of glutamic acid loses a proton when its carboxylic acid group ionizes. The degree of ionization of the carboxyl group depends on the pH of the medium: the greater the hydrogen ion concentration (the lower the pH), the smaller the percentage of carboxyl groups that are present in the ionized state. Conversely, a rise in pH leads to an increased ionization of the proton from the carboxyl group, increasing the percentage of negatively charged glutamic acid side chains. The pH at which 50 percent of the side chains are ionized and 50 percent are unionized is called the pK, which is 4.4 for the side chain of free glutamic acid. At physiologic pH, virtually all of the glutamic acid residues of a polypeptide are negatively charged. (*b*) The side chain of lysine becomes ionized when its amino group gains a proton. The greater the hydroxyl ion concentration (the higher the pH), the smaller the percentage of amino groups that are positively charged. The pH at which 50 percent of the side chains of lysine are charged and 50 percent are uncharged is 10.0, which is the pK for the side chain of free lysine. At physiologic pH, virtually all of the lysine residues of a polypeptide are positively charged. Once incorporated into a polypeptide, the pK of a charged group can be greatly influenced by the surrounding environment.

Disulfide bridges often form between two cysteines that are distant from one another in the polypeptide backbone or even in two separate polypeptides. Disulfide bridges help stabilize the intricate shapes of proteins, particularly those present outside of cells where they are subjected to added physical and chemical stress.

Not all of the amino acids described in this section are found in all proteins, nor are the various amino acids distributed in an equivalent manner. A number of other amino acids are also found in proteins, but they arise by alterations to the side chains of the 20 basic amino acids *after* their incorporation into a polypeptide chain. These amino acid alterations can generate dramatic changes in the properties and function of a protein, increasing or decreasing its solubility, for example, or modifying its interactions with other molecules.

The ionic, polar, or nonpolar character of amino acid side chains is very important in protein structure and function. Most soluble (i.e., nonmembrane) proteins are constructed so that the polar residues are situated at the surface of the molecule where they can associate with the surrounding water and contribute to the protein's solubility in aqueous solution (Figure 2.28*a*). In contrast, the nonpolar residues are situated predominantly in the core of the molecule (Figure 2.28*b*). The hydrophobic residues of the protein interior are often tightly packed together creating a type of three-dimensional jigsaw puzzle in which water molecules are generally excluded. The hydrophobic interactions among the nonpolar side chains of these residues contribute substantially to the overall stability of the protein. In many enzymes, reactive polar groups project into the nonpolar interior giving the protein its catalytic activity. For example, a nonpolar environment can enhance ionic interactions between charged groups that would be lessened by competition with water in an aqueous environment. Some reactions that might proceed at an imperceptibly slow rate in water can occur in thousandths of a second within the protein.

In addition to amino acids, many proteins contain other types of components; these are called **conjugated proteins.** Conjugated proteins include those linked covalently or noncovalently to nucleic acids, the *nucleoproteins*; to lipids, the *lipoproteins*; to carbohydrates, the *glycoproteins*; or to various low-molecular-weight materials, including metals and metal-containing groups.

The Structure of Proteins Nowhere in biology is the intimate relationship between form and function better illustrated than with proteins. Proteins are huge, complex molecules, but their structure in any given environment is completely defined and predictable. Each amino acid in a protein is located in a specific site within these giant molecules, giving the protein the structure and reactivity required for the job at hand. Protein structure can be described at several levels of organization, each emphasizing a different aspect and each dependent on different types of interactions. Customarily, four such levels are described: *primary, secondary, tertiary, and quaternary.* The first, pri-

(a)

(b)

Figure 2.28 **Disposition of hydrophilic and hydrophobic amino acid residues in the soluble protein cytochrome c.** (*a*) The hydrophilic side chains, which are shown in green, are located primarily at the surface of the protein where they contact the surrounding aqueous medium. (*b*) The hydrophobic residues, which are shown in red, are located primarily within the center of the protein, particularly in the vicinity of the central heme group. (COPYRIGHT IRVING GEIS.)

mary structure, concerns the amino acid sequence of a protein, whereas the latter three levels concern the organization of the molecule in space.

Primary Structure The primary structure of a polypeptide is the specific linear sequence of amino acids that constitute the chain. With 20 different building blocks, the number of different polypeptides that can be formed is 20^n, where n is the number of amino acids in the chain. Because most polypeptides contain over 100 amino acids (and some several thousand), the variety of possible sequences is essentially unlimited. The information for the precise order of amino acids in every protein that an organism can produce is contained within the genome of that organism.

As we will see later, the amino acid sequence provides the information required to determine the molecule's three-dimensional shape and thus its function. The sequence of amino acids, therefore, is all-important, and changes that arise in the sequence as a result of genetic mutations in the DNA may not be readily tolerated. The earliest and best-studied example of this relationship is the change in the amino acid sequence of hemoglobin that causes the disease *sickle cell anemia*. This severe, inherited anemia results solely from a single change in amino acid

sequence within the hemoglobin molecule (Figure 2.29*a,b*): a valine is present where a glutamic acid is normally located. A charged, polar amino acid has been replaced by a nonpolar one (see Figure 2.26). All the problems concerning red blood cell shape (Figure 2.29*c*) and decreased oxygen-carrying capacity in persons with sickle cell anemia result from this single change. Not all amino acid changes have such a dramatic effect, as evidenced by the differences in amino acid sequence in the same protein among related organisms. The degree to which changes in the primary sequence are tolerated depends on the degree to which the shape of the protein or the critical functional residues are disturbed.

The first amino acid sequence of a protein was determined by Frederick Sanger and coworkers at Cambridge University in the early 1950s using the protein hormone insulin. Beef insulin was chosen for the work because of its availability and its small size—two polypeptide chains of 21 and 30 amino acids each. The *sequencing* of insulin was a momentous feat in the newly emerging field of molecular biology. It revealed that proteins, the most complex molecules in cells, had a specific definable substructure that was neither regular nor repeating, as was that of polysaccharides. With the advent of techniques for rapid DNA sequencing (see Section 18.13), the primary struc-

(a) (b) (c) 1 μm

Figure 2.29 The molecular basis of sickle cell anemia. (*a,b*) Chromatograms showing the separation of peptides obtained by treatment of normal hemoglobin (*a*) or sickle cell hemoglobin (*b*) with the proteolytic enzyme trypsin. The two chromatograms are identical with the exception of one peptide (marked) that contains a single amino acid difference. (*c*) Scanning electron micrograph of a red blood cell from a person with sickle cell anemia. Compare with the micrograph of a normal red blood cell of Figure 4.32*a*. Because of their abnormal shape, these sickled blood cells can clog small blood vessels, causing pain and life-threatening crises. (A,B: FROM CORRADO BAGLIONI, BIOCHIM. BIOPHYS. ACTA 48:394, 1961; C: COURTESY OF J. T. THORNWAITE, B. F. CAMERON, AND R. C. LEIF.)

ture of a polypeptide can be deduced from the nucleotide sequence of the encoding gene. In the past few years, the complete sequences of the genomes of dozens of organisms, including humans, have been revealed. This information will eventually allow researchers to learn about every protein that an organism can manufacture. But, as we will see, translating information about primary sequence into knowledge about higher levels of protein structure remains a formidable challenge.

Secondary Structure All matter exists in space and therefore has a three-dimensional expression. Proteins are formed by linkages among vast numbers of atoms; consequently their shape is complex. The term **conformation** refers to the three-dimensional arrangement of the atoms of a molecule, that is, to their spatial organization. Secondary structure describes the conformation of portions of the polypeptide chain. The early studies on protein secondary structure were carried out by Linus Pauling and Robert Corey of the California Institute of Technology. By studying the structure of simple peptides consisting of a few amino acids linked together, Pauling and Corey concluded that polypeptide chains exist in preferred conformations that provide the maximum possible number of hydrogen bonds between neighboring amino acids.

Two conformations were proposed. In one conformation, the backbone of the polypeptide assumed the form of a cylindrical, twisting spiral called the **alpha (α) helix** (Figure 2.30*a*). The backbone lies on the inside of the helix, and the side chains project outward. The helical structure is stabilized by large numbers of hydrogen bonds

between the atoms of one peptide bond and those situated just above and below it along the spiral (Figure 2.30*b*). The X-ray diffraction patterns of actual proteins produced during the 1950s bore out the existence of the α helix, first in the protein keratin found in hair and later in various oxygen-binding proteins, such as myoglobin and hemoglobin (see Figure 2.34). Surfaces on opposite sides of an α helix may have contrasting properties. In water soluble proteins, the outer surface of an α helix often contains polar residues in contact with the solvent, whereas the surface facing inward contains nonpolar side chains.

Because it is coiled and held together by weak, noncovalent bonds, an α helix can be extended in length if subjected to pulling forces. This is illustrated by wool, whose protein fibers consist largely of α helices. When wool fibers are pulled, the hydrogen bonds are broken, and the fibers are stretched. When the tension is relieved, the bonds reform, and the fiber shortens to its original length. Human hair is less extensible than wool because the polypeptides are also stabilized by covalent disulfide bridges.

The other conformation proposed by Pauling and Corey was the **beta (β)-pleated sheet**, which consists of several segments of a polypeptide lying side by side. Unlike the coiled, cylindrical form of the α helix, the backbone of each segment of polypeptide (or **β strand**) in a β sheet assumes a folded or pleated conformation (Figure 2.31*a*). Like the α helix, the β sheet is also characterized by a large number of hydrogen bonds, but these are perpendicular to the long axis of the polypeptide chain and project across from one part of the chain to another (Figure 2.31*b*). Like the α helix, the β sheet has also been

(a)

(b)

Figure 2.34 **The three-dimensional structure of whale myoglobin.** (*a*) The tertiary structure of myoglobin. Most of the amino acids are part of α helices. The nonhelical regions occur primarily as turns, where the polypeptide chain changes direction. The position of the heme is indicated in red. (*b*) The three-dimensional structure of myoglobin (heme indicated in red). The positions of all of the molecule's atoms, other than hydrogen, are shown. (A: COPYRIGHT BY IRVING GEIS; B: KEN EWARD/PHOTO RESEARCHERS.)

centage compared with that for other proteins that have since been examined. No β-pleated sheet was found. Subsequent analyses of myoglobin using additional X-ray diffraction data provided a much more detailed picture of the molecule (Figure 2.34*a*). For example, it was shown that the heme group is situated within a pocket of hydrophobic side chains that promotes the binding of oxygen without the oxidation (loss of electrons) of the iron atom. Myoglo-

Figure 2.35 **Types of noncovalent bonds maintaining the conformation of proteins.** A molecular model showing some of these types of interactions is shown in Figure 2.42.

bin contains no disulfide bonds; the tertiary structure of the protein is held together exclusively by noncovalent interactions. All of the noncovalent bonds thought to occur between side chains within proteins—hydrogen bonds, ionic bonds, and hydrophobic interactions—were found (Figure 2.35). A three-dimensional model of the myoglobin molecule is shown in Figure 2.34*b*. Unlike myoglobin, most globular proteins contain both α helices and β sheets. Each protein has a unique tertiary structure that can be correlated with its amino acid sequence and its biological function.

Protein Domains and Motifs Many proteins, particularly the larger ones, are composed of two or more distinct modules, or **domains**, that fold independent of one another. For example, the mammalian enzyme phospholipase C, shown in the central part of Figure 2.36, consists of four distinct domains, colored differently in the drawing. The different domains of a polypeptide often represent parts that function in a semi-independent manner. For example, they might bind different factors, such as a coenzyme and a substrate or a DNA strand and another protein, or they might move relatively independent of one another. Some polypeptides containing more than one domain are thought to have arisen during evolution by the fusion of genes that coded for different ancestral proteins, with each domain representing a part that was once a separate molecule. Each domain of the mammalian phospholipase C molecule, for example, has been identified as a homologous unit in another protein (Figure 2.36). Some domains appear to have been shuffled widely about during evolution, appearing in a variety of proteins whose other

troponin C

bacterial
phospholipase C

mamalian
phospho-
lipase C

recoverin

synapto-
tagmin

Figure 2.36 **Proteins are built of structural units, or domains.** The enzyme mammalian phospholipase C is constructed of four domains, indicated in different colors. The catalytic domain of the enzyme is shown in blue. Each of the domains of this enzyme can be found independently in other proteins as indicated by the matching color. (FROM LIISA HOLM AND CHRIS SANDER, STRUCTURE 5:167, 1997.)

regions show little or no evidence of an evolutionary relationship. Shuffling of domains creates proteins with unique combinations of activities.

At first glance, the three-dimensional architecture of proteins appears almost hopelessly complex. As more tertiary structures are determined, however, biochemists have discovered recurring substructures, or **motifs**, that consist of a defined arrangement of α helices and/or β strands. Two examples are shown in Figures 2.37 and 2.38. In the *coiled coil*, which is found in various fibrous proteins, such as the muscle protein myosin (Figure 2.37*a*), two or more α helices coil around one another like entwined strands of a cable. Alpha helices that interact as coiled coils have a distinctive primary structure characterized by a regularly repeating sequence of seven amino acids. The first and fourth residues (referred to as *a* and *d*) of the repeat are hydrophobic and are situated on one side of the helix where they can form hydrophobic interactions with their counterparts on the neighboring helix (Figure 2.37*b*,*c*). It is as if one helix contained a row of "knobs" and the other helix a row of "holes" into which the knobs

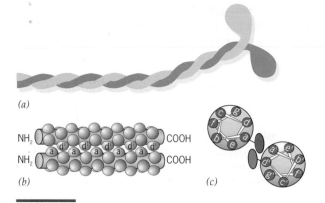

(*a*)

NH₂ ... COOH
NH₂ ... COOH

(*b*)

(*c*)

Figure 2.37 **Protein motifs: the coiled coil.** (*a*) The coiled coil, as seen in this myosin molecule, consists of two separate α-helical subunits wrapped around one another to form a rod-shaped portion of the protein. (*b*) Side view of a coiled coil showing the seven amino acid (heptad) repeat. The first and fourth amino acids of the repeat (residues a and d) are hydrophobic and interact with one another as shown. (*c*) A cross section through the two helices showing the interaction between hydrophobic residues from the two adjacent helices. Each circle of seven amino acids represents two turns of the alpha helix. (B: AFTER K. J. LUMB AND P. S. KIM, SCIENCE 268:436, 1995. COPYRIGHT © 1995 AMERICAN ASSOCIATION FOR THE ADVANCEMENT OF SCIENCE.)

Figure 2.38 **Protein motifs; the α/β barrel.** The α/β barrel of the enzyme triose phosphate isomerase as viewed from the top. The drawing on the right shows the arrangement of the β strands of the barrel as viewed from the side. (AFTER DRAWINGS BY JANE S. RICHARDSON.)

can fit. Together, the intertwined helices form a relatively rigid, rod-like structure.

One of the most complex motifs is the α/β barrel, which was originally discovered in the enzyme triose phosphate isomerase (Figure 2.38) and has since been found in up to about 10 percent of all enzymes. Each of the eight staves of the "barrel" consists of a β strand. Together, the eight parallel β strands form the interior of the barrel situated in the core of the protein. The β strands are connected to one another by sections of α helix at the barrel's surface. Common motifs may occur in proteins that are evolutionarily related and display a similar function, or they may occur in unrelated proteins having very different functions.

Dynamic Changes Within Proteins Proteins are not rigid and inflexible, but are capable of considerable internal

movements. Random, small-scale fluctuations in the arrangement of the bonds within a protein create an incessant thermal motion within the molecule. Spectroscopic techniques, such as nuclear magnetic resonance, can monitor dynamic movements within proteins, and they reveal shifts in hydrogen bonds, waving movements of external side chains, and the full rotation of the aromatic rings of tyrosine and phenylalanine residues about one of the single bonds. The extent of such random movements and the role they can play in a protein's function is illustrated in recent studies of acetylcholinesterase, the enzyme responsible for degrading acetylcholine molecules that are left behind following the transmission of an impulse from one nerve cell to another (Section 4.8). When the tertiary structure of acetylcholinesterase was first revealed by X-ray crystallography, there was no obvious pathway for acetylcholine molecules to enter the enzyme's catalytic site, which was situated at the bottom of a deep "gorge" in the molecule. In fact, the narrow entrance to the site was completely blocked by the presence of a number of bulky amino acid side chains. Using high-speed computers, researchers were able to simulate the random movements of thousands of atoms within the enzyme. These simulations indicated that dynamic movements of side chains within the protein would lead to the rapid opening and closing of a "gate" that would allow acetylcholine molecules to diffuse into the enzyme's catalytic site (Figure 2.39).

Predictable (nonrandom) movements within a protein that are triggered by the binding of a specific molecule are described as **conformational changes**. A comparison of the polypeptides depicted in Figures 3a and 3b on page 77 shows the dramatic conformational change that occurs in a bacterial protein (GroEL) when it interacts with another protein (GroES). Virtually every activity in which a protein takes part is accompanied by conformational changes within the molecule. The conformational change in the protein myosin that occurs during muscle contraction is shown in Figures 9.61 and 9.62. In this case, binding of myosin to an actin molecule leads to a small (20°) rotation of the head of the myosin, which results in a 50 to 100 Å movement of the adjacent actin filament. The importance of this dynamic event can be appreciated if you consider that the movements of your body result from the additive effect of millions of conformational changes taking place within the contractile proteins of your muscles.

Quaternary Structure Whereas many proteins such as myoglobin are composed of only one polypeptide chain, most are made up of more than one chain, or **subunit**. The subunits may be linked by covalent disulfide bonds, but more often they are held together by noncovalent bonds as occur most often between hydrophobic "patches" on the complementary surfaces of neighboring polypeptides. Proteins composed of subunits are said to have **quaternary structure**. Depending on the protein, the polypeptide chains may be identical or nonidentical. A protein composed of two identical subunits is described as

Figure 2.39 Dynamic movements within the enzyme acetylcholinesterase. A portion of the enzyme is depicted here in two different conformations: (1) a closed conformation (left) in which the entrance to the catalytic site is blocked by the presence of aromatic rings that are part of the side chains of tyrosine and phenylalanine residues (shown in red) and (2) an open conformation (right) in which the aromatic rings of these side chains have swung out of the way, opening the "gate" to allow acetylcholine molecules to enter the catalytic site. These images are constructed using computer programs that take into account a host of information about the atoms that make up the molecule, including bond lengths, bond angles, electrostatic attraction and repulsion, van der Waals forces, etc. Using this information, researchers are able to simulate the movements of the various atoms being considered, which provides images of the conformations that the protein can assume. (FROM J. ANDREW MCCAMMON, SCIENCE 281:1107, 1998)

a *homodimer*, whereas a protein composed of two nonidentical subunits is a *heterodimer*. A ribbon drawing of a homodimeric protein is depicted in Figure 2.40a. The two subunits of the protein are drawn in different colors, and the hydrophobic residues that form the complementary sites of contact are indicated. The best-studied multisubunit protein is hemoglobin, the O_2-carrying protein of red blood cells. A molecule of human hemoglobin consists of two α globin and two β globin polypeptides (Figure 2.40b), each of which binds a single molecule of oxygen. Elucidation of the three-dimensional structure of hemoglobin by Max Perutz of Cambridge University in 1959 was one of the early landmarks in the study of molecular biology. Perutz demonstrated that each of the four globin polypeptides of a hemoglobin molecule has a tertiary structure similar to that of myoglobin, a fact that strongly suggested the two proteins had evolved from a common ancestral polypeptide with a common O_2-binding function. Perutz also compared the structure of the oxygenated and deoxygenated versions of hemoglobin. In doing so, he discovered that the binding of oxygen was accompanied by the movement of the two β chains closer together within the molecule. This finding revealed for the first time that the complex functions of proteins may be carried out by means of small changes in their conformation.

(a)

(a) 20 nm *(b)*

Figure 2.41 **Pyruvate dehydrogenase: a multiprotein complex.** (*a*) Electron micrograph of a negatively stained pyruvate dehydrogenase complex isolated from *E. coli*. Each complex contains 60 polypeptide chains constituting three different enzymes. Its molecular mass approaches 5 million daltons. (*b*) A model of the pyruvate dehydrogenase complex. The core of the complex consists of a cubelike cluster of dihydrolipoyl transacetylase molecules. Pyruvate dehydrogenase dimers (black spheres) are distributed symmetrically along the edges of the cube, and dihydrolipoyl dehydrogenase dimers (small gray spheres) are positioned in the faces of the cube. (COURTESY OF LESTER J. REED.)

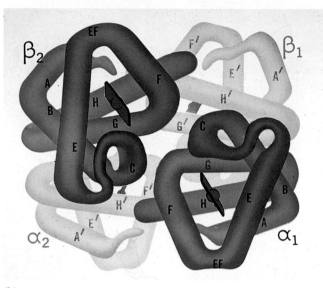

(b)

Figure 2.40 **Proteins with quaternary structure.** (*a*) Drawing of transforming growth factor-β2 (TGF-β2), a protein that is a dimer composed of two identical subunits. The two subunits are colored yellow and blue. Shown in white are the cysteine side chains and disulfide bonds. The spheres shown in yellow and blue are the hydrophobic residues that form the interface between the two subunits. (*b*) Drawing of a hemoglobin molecule, which consists of two α-globin chains and two β-globin chains joined by noncovalent bonds. When the four globin polypeptides are assembled into a complete hemoglobin molecule, the kinetics of O_2 binding and release are quite different from those exhibited by isolated polypeptides. This is because the binding of O_2 to one polypeptide causes a conformational change in the other polypeptides that alters their affinity for O_2 molecules. (A: REPRINTED WITH PERMISSION FROM S. DAOPIN, ET AL., SCIENCE 257:372, 1992, COURTESY OF DAVID R. DAVIES. COPYRIGHT © 1992 AMERICAN ASSOCIATION FOR THE ADVANCEMENT OF SCIENCE; B: COPYRIGHT BY IRVING GEIS.)

Multiprotein Complexes Even though hemoglobin consists of four subunits, it is still considered a single protein with a single function. Many examples are known in which different proteins, each with a specific function, become physically associated to form a much larger **multi-**

protein complex. One of the first multiprotein complexes to be discovered and studied was the pyruvate dehydrogenase complex of the bacterium *Escherichia coli*, which consists of 60 polypeptide chains constituting three different enzymes (Figure 2.41). The enzymes that make up this complex catalyze a series of reactions connecting two metabolic pathways, glycolysis and the TCA cycle (see Figure 5.7). Because the enzymes are physically associated, the product of one enzyme can be channeled directly to the next enzyme in the sequence without becoming diluted in the aqueous medium of the cell.

Multiprotein complexes that form within the cell are not necessarily stable assemblies, such as the pyruvate dehydrogenase complex. In fact, most proteins interact with other proteins in highly dynamic patterns, associating and dissociating depending on conditions within the cell at any given time. Interacting proteins tend to have complementary surfaces. Often a projecting portion of one molecule fits into a pocket within its partner. Once the two molecules have come into close contact, their interaction is stabilized by noncovalent bonds. These principles of protein–protein interaction are illustrated in the molecular models pictured in Figure 2.42. The reddish-colored object in Figure 2.42*a* is called an SH3 domain, and it is found as part of a number of different proteins involved in molecular signaling. SH3 domains function primarily as "knobs" that allow binding to proteins that contain a complementary "handle," in this case one that is rich in proline (a polyproline motif), shown in green. A number of different structural domains have been identified, in addition to SH3, that function as adaptors to mediate interactions be-

(a)

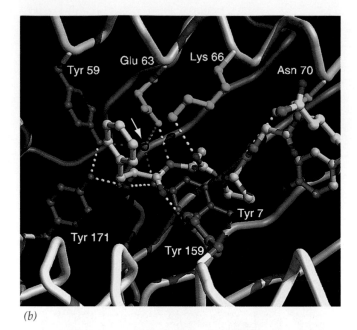

(b)

Figure 2.42 Protein-protein interactions. (*a*) A model illustrating the complementary molecular surfaces of portions of two interacting proteins. The reddish-colored molecule is a domain of the enzyme phosphatidylinositol 3-OH kinase, whose function is discussed in Chapter 15. This domain binds specifically to a variety of proline-containing peptides, such as the one shown in the space-filling model at the top of the figure. The proline residues in the peptide, which fit into pockets on the surface of the enzyme, are indicated. The polypeptide backbone of the peptide is colored yellow, and the side chains are colored green. (*b*) A closer view of the types of noncovalent interactions between two peptides having complementary surfaces. The dotted lines indicate hydrogen bonds that form (1) between amino acid residues in the two peptides and (2) between amino acid residues and a water molecule (magenta sphere shown by white arrow) that becomes positioned between the two peptides. (FROM HONGTAO YU AND STUART SCHREIBER, CELL 76:940, 1994, BY PERMISSION OF CELL PRESS; B: REPRINTED WITH PERMISSION FROM MASAZUMI MATSUMURA, ET AL., SCIENCE 257:930, 1992, COURTESY OF IAN A. WILSON. COPYRIGHT © 1992 AMERICAN ASSOCIATION FOR THE ADVANCEMENT OF SCIENCE.)

tween proteins (see Figure 15.24 for an example). In many cases, interactions between proteins are regulated by modifications, such as the addition of a phosphate group to a key amino acid, which may greatly increase or decrease its ability to bind a protein partner. Figure 2.42*b* provides a closer view of the intricate nature of the noncovalent bonding pattern that can occur between the surfaces of interacting proteins. As more and more complex molecular activities have been discovered, the importance of transient interactions among proteins is becoming increasingly apparent. For example, such diverse processes as DNA synthesis, ATP formation, and RNA processing are all accomplished by "molecular machines" that consist of a large number of interacting proteins.

Protein Folding The elucidation of the tertiary structure of myoglobin in the late 1950s led to an appreciation for the complexity of protein architecture. An important question immediately arose: How does such a complex, folded, asymmetric organization arise in the cell? The first insight into this problem was gained through a serendipitous observation in 1956 by Christian Anfinsen at the National Institutes of Health. Anfinsen was studying the properties of ribonuclease A, a small enzyme that consists of a single polypeptide chain of 124 amino acids with 4 disulfide bonds linking various parts of the chain. The

disulfide bonds of a protein are typically broken (reduced) by adding a reducing agent, such as mercaptoethanol (CH_3CH_2SH), which converts each disulfide bridge to a pair of sulfhydryl (—SH) groups (see drawing, page 53). To make all of the disulfide bonds accessible to the reducing agent, Anfinsen found that the molecule had to first be unfolded. The unfolding or disorganization of a protein is termed **denaturation**, and it can be brought about by a variety of agents, including detergents, organic solvents, radiation, heat, and compounds such as urea and guanidine chloride, all of which interfere with the various interactions that stabilize a protein's tertiary structure.

When Anfinsen treated ribonuclease molecules with mercaptoethanol and concentrated urea, he found that the preparation lost all of its enzymatic activity, which would be expected if the protein molecules had become totally unfolded. When he removed the urea and mercaptoethanol from the preparation, he found, to his surprise, that the molecules regained their normal enzymatic activity. The active molecules that had reformed from the unfolded protein were indistinguishable both structurally and functionally from the correctly folded (i.e., **native**) molecules present at the beginning of the experiment (Figure 2.43). After extensive study of these events, Anfinsen concluded that the amino acid sequence of a polypeptide contained all the information required for the

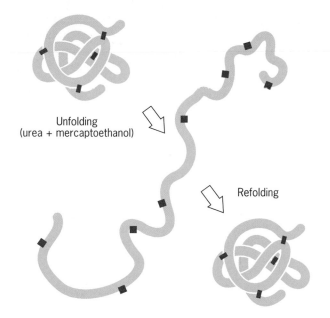

Unfolding
(urea + mercaptoethanol)

Refolding

Figure 2.43 Denaturation and refolding of ribonuclease. A native ribonuclease molecule (with intramolecular disulfide bonds indicated) is reduced and unfolded with β-mercaptoethanol and 8 M urea. After removal of these reagents, the protein undergoes spontaneous refolding. (FROM C. J. EPSTEIN, R. F. GOLDBERGER, AND C. B. ANFINSEN, COLD SPRING HARBOR SYMP. QUANT. BIOL. 28:439, 1963.)

assembly of the protein's three-dimensional conformation and, thus, that the folding process occurred by **self-assembly**. As discussed in the following chapter, events tend to progress toward states of lower energy. According to this concept, the tertiary structure that a polypeptide chain assumes after folding is the accessible structure with the lowest energy, which makes it the most thermodynamically stable structure that can be formed by that chain.

Given a polypeptide of over 100 amino acids containing over 1000 rotatable bonds, calculations indicate that, if folding occurred in a totally random manner, it could take a period of time greater than the age of the universe for that polypeptide to search through all of the possible conformations and arrive at the native state. Clearly, there must be molecular shortcuts that allow a protein to fold in a reasonable time frame (small fractions of a second to a minute or so) within the cell. Over the past couple of decades, two opposing portraits of the folding process have been proposed. According to the earlier viewpoint, folding occurs in orderly steps along a defined pathway through the formation of specific, partially folded intermediates. Once one folding intermediate is attained, clues are present that direct the process to the next intermediate stage, thereby vastly reducing the number of unstable conformations that are visited by that polypeptide. Proponents of this viewpoint cite the results of numerous studies that find evidence of discrete intermediates that appear and then disappear as certain proteins fold in the test tube.

According to the opposing viewpoint, which is often referred to as the "new view," there is no single, defined pathway by which a given protein folds. Consider what a population of unfolded polypeptides—whether denatured in a test tube or newly synthesized in a cell—might look like. A population of this type would contain molecules that are highly disordered, possessing many *different* conformations. Although different polypeptides might be expected to fold by a very different course, according to thermodynamic principles, all would progress through stages that possessed progressively lower energy (i.e., molecules having greater order and stability, Section 3.1). Because the native conformation is considered to be the state of lowest energy, all of these polypeptides will invariably arrive at the native state, regardless of the specific pathway they follow en route. The analogy has been made to the movement of water down a mountainside; even though the water may take many different routes, it ultimately flows into the same lake at the bottom. Proponents of the "new view" often describe the course of protein folding in terms of an "energy landscape," which depicts the number of molecules having various energy content at various times after folding has begun. An energy landscape for protein folding is typically depicted as a funnel, such as those shown in Figure 2.44. The top of the funnel represents the large variety of unfolded polypeptides present at the beginning of the folding process, and the bottom of the funnel represents the molecules that have reached the unique native conformation at the end of the process. The diameter of the funnel decreases during folding as the variety of conformations is reduced and their energy content decreases. Proponents of the energy landscape model of protein folding point out that many proteins, particularly smaller ones, fold very rapidly from the denatured to the native state without any evidence of discrete folding intermediates. Proteins of this type can be represented by funnels having a smooth surface, such as that of Figure 2.44*a*. Proteins that show evidence of folding intermediates are represented by funnels that exhibit hills and valleys (Figure 2.44*b*) where a polypeptide may get trapped for a period of time before continuing on its path to the native state. Proponents of energy landscapes suggest that these "kinetic traps," which may contain polypeptides that are temporarily misfolded, produce the illusion that the folding process proceeds through the formation of specific intermediates. These two views of the folding process continue to be hotly debated.

Another aspect of interest in the study of protein folding concerns the types of events that occur at various stages during the folding process. In the course depicted in Figure 2.45, the first events during protein folding lead to the formation of much of the secondary structure of the molecule. Once the α helices and β sheets are formed, subsequent folding is driven by hydrophobic interactions that force nonpolar residues together in the central core of the protein. Such hydrophobic interactions lead to the collapse of the polypeptide into a compact state, called a "molten globule," that resembles the native protein in overall shape

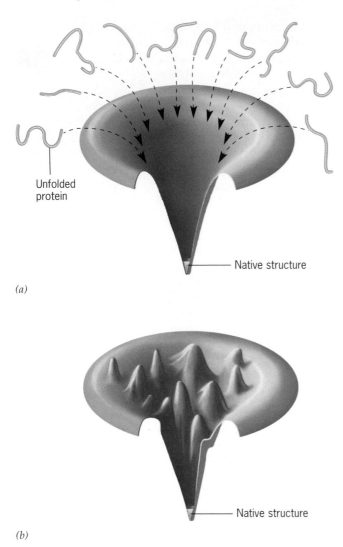

(a)

Unfolded protein

Native structure

(b)

Native structure

Figure 2.44 **Representation of the pathways of protein folding using energy landscape funnels.** Part *a* depicts the change in energy of a polypeptide that folds rapidly and directly into the native conformation. Part *b* depicts the change in energy state of a polypeptide that experiences a more complex folding pathway in which intermediates become trapped temporarily in nonproductive conformations.

Unfolded | Secondary structure | Molten globule | Native

Figure 2.45 **One possible series of steps in the protein folding process.** Curled segments represent α helices and arrows represent β strands.

and folding pattern. Molten globules lack many of the bonds between side chains that hold proteins in their final tertiary structure. The final steps to reach the native state include (1) the formation of noncovalent bonds between side chains leading to tighter packing and (2) the formation of covalent, disulfide bonds. According to an alternate scheme, the first major event in protein folding is the collapse of the polypeptide to form a compact structure, and only then does significant secondary structure develop. Recent studies indicate that for a given protein, one or the other of these pathways is usually taken, though an intermediate between these two schemes may also occur. Our understanding of protein folding has not yet advanced to the point where one can predict which route will be taken based simply on primary structure, but as newer computer programs are developed, this may become possible.

If amino acid sequence determines tertiary structure, then alterations in this sequence have the potential to change the way a protein folds, leading to a molecule with an abnormal tertiary structure. In fact, many mutations responsible for inherited disorders have been found to alter a protein's three-dimensional structure. In some cases, the consequences of protein misfolding can be fatal. Two examples of fatal neurodegenerative diseases that result from abnormal protein folding are discussed in the accompanying Human Perspective.

Not all proteins are able to assume their final tertiary structure by a simple process of self-assembly. This is not because the primary structure of these proteins lacks the required information for proper folding, but rather because proteins undergoing folding have to be prevented from interacting nonselectively with other molecules in the crowded compartments of the cell. Several families of proteins have evolved whose function is to "help" unfolded or misfolded proteins achieve their proper three-dimensional conformation. These "helper proteins" are called **molecular chaperones** and are discussed at length in the Experimental Pathways on page 76.

Protein Engineering Advances in molecular biology have created the opportunity to design and mass produce novel proteins that are different from those made by living organisms. It is possible with current DNA-synthesizing techniques to create an artificial gene that can be used in the production of a protein having any desired sequence of amino acids. The problem arises in knowing which of the virtually infinite variety of possible proteins one could manufacture might have some useful function. Consider, for example, a biotechnology company that wanted to manufacture a protein that would bind to the surface of the AIDS virus and remove it from an aqueous solution, such as the bloodstream. Assume that computer simulation programs could predict the shape such a protein should have to bind to the viral surface. What sequence of amino acids strung together would produce such a protein? The answer requires detailed insight into the rules governing the relationship between a protein's primary structure and its tertiary structure.

THE HUMAN PERSPECTIVE

UPDATE
For the very latest,
go to page U-4

Protein Misfolding Can Have Deadly Consequences

In April 1996 a paper was published in the medical journal *Lancet* that generated widespread alarm in the populations of Europe. The paper described a study of 10 persons afflicted with Creutzfeld-Jakob disease (CJD), a rare, fatal disorder that attacks the brain, causing a loss of motor coordination and dementia. Like numerous other diseases, CJD can occur as an inherited disease that runs in certain families or as a sporadic form that appears in individuals who have no family history of the disease. Unlike virtually every other inheritable disease, however, CJD can also be *acquired*. Until recently, persons who had acquired CJD had been recipients of organs or organ products that were donated by a person with undiagnosed CJD. The cases described in the 1996 *Lancet* paper had also been acquired, but the apparent source of the disease was contaminated beef that the infected individuals had eaten years earlier. The contaminated beef was derived from cattle raised in England that had contracted a neurodegenerative disease that caused them to lose motor coordination and develop demented behavior. The disease became commonly known as "mad cow disease." Patients who have acquired CJD from eating contaminated beef can be distinguished by several criteria from those who suffer from the classical forms of the disease. By February 2001, approximately 85 cases of CJD had been traced to contaminated European beef, and efforts are still underway to determine the scope of the epidemic.

A disease that runs in families can invariably be traced to a faulty gene, whereas diseases that are acquired from a contaminated source can invariably be traced to an infectious agent. How can the same disease be both inherited and infectious? The answer to this question has emerged gradually over the past several decades, beginning with observations by D. Carleton Gajdusek in the 1960s on a strange malady that once afflicted the native population of Papua, New Guinea. Gajdusek showed that these islanders were contracting a fatal neurodegenerative disease—which they called "kuru"—during a funeral ritual in which they ate the brain tissue of a recently deceased relative. Autopsies of the brains of patients who had died of kuru showed a distinct pathology, referred to as *spongiform encephalopathy*, in which certain brain regions were riddled with microscopic holes (vacuolations), causing the tissue to resemble a sponge.

It was soon shown that the brains of islanders suffering from kuru were strikingly similar in microscopic appearance to the brains of persons afflicted with CJD. This observation raised an important question: Did the brain of a person suffering from CJD, which was known to be an inherited disease, contain an infectious agent? In 1968, Gajdusek showed that when extracts prepared from a biopsy of the brain of a person who had died from CJD were injected into a suitable laboratory animal, that animal did indeed develop a spongiform encephalopathy similar to that of kuru or CJD. Clearly, the extracts contained an infectious agent, which at the time was presumed to be a virus.

In 1982, Stanley Prusiner of the University of California, San Francisco, published a paper suggesting that, unlike viruses, the agent responsible for CJD lacked nucleic acid and instead was composed solely of protein. He called the protein a *prion*. This "protein only" hypothesis, as it is called, was originally met with considerable skepticism, but subsequent studies by Prusiner and others have provided a large body of evidence to support the original conclusion. It was presumed initially that the prion protein was an external agent—some type of virus-like particle lacking nucleic acid. Contrary to this expectation, the prion protein was soon shown to be encoded by a gene (called *PRNP*) within the cell's own chromosomes. The gene is expressed within *normal* brain tissue and encodes a protein designated PrP^C (standing for prion protein cellular) that resides at the surface of nerve cells. The precise function of PrP^C remains unknown. A modified version of the protein (designated PrP^{Sc}, standing for prion protein scrapie) is present in the brains of humans with CJD. Unlike the normal PrP^C, the modified version of the protein accumulates within nerve cells, forming aggregates that kill the cells.

In their purified states, PrP^C and PrP^{Sc} have very different physical properties. PrP^C remains as a monomeric molecule that is soluble in salt solutions and is readily destroyed by protein-digesting enzymes. In contrast, PrP^{Sc} molecules interact with one another to form insoluble fibrils that are resistant to enzymatic digestion. Based on these differences, one might expect these two forms of the PrP protein to be composed of distinctly different sequences of amino acids, but this is not the case. The two forms can have identical amino acid sequences, but they differ in the way the polypeptide chain folds to form the three-dimensional protein molecule. Whereas a PrP^C molecule consists almost entirely of α-helical segments and interconnecting coils, approximately 45 percent of a PrP^{Sc} molecule consists of β sheet. The dramatic differences between these two forms of the protein are illustrated in Figure 1. PrP can be converted from a soluble, protease-sensitive conformation into insoluble, protease-insensitive aggregates in vitro by simply changing the conditions in the test tube.[1]

Under normal conditions within the cell, it would appear that the newly synthesized PrP polypeptide almost invariably folds into the PrP^C conformation. Persons who develop CJD as an inherited disease have a gene that encodes a mutant protein whose amino acid sequence is different from that of the general population. The mutant protein is presumed to be less stable in the PrP^C conformation than the normal version and more likely to fold into the abnormal β-rich conformation. Once formed, the β-rich proteins produce ag-

[1]At the time of this writing, none of these abnormal protein preparations produced in vitro have proven infectious. This is one of the strongest arguments used by researchers that do not accept the "protein only" hypothesis.

Figure 2 The microscopic appearance of brain tissue from a person who died of Alzheimer's disease. (Martin Rothker/Phototake.)

Figure 1 A comparison of the normal (PrP^C) and abnormal (PrP^Sc) prion proteins. These two proteins (shown in *a* and *b* respectively) are formed by polypeptide chains that can be identical in amino acid sequence, but they fold differently. As a result of the differences in folding, PrP^C remains soluble, whereas PrP^Sc produces insoluble aggregates that kill the cell. (The two molecules shown in this figure are called *conformers* because they differ only in conformation.) (After S. B. Prusiner, Trends Biochem. Sci. 21:483, 1996. Copyright 1996, with permission of Elsevier Science.)

gregates, which lead to disease. Approximately 20 different disease-causing mutations in the prion gene have been identified.

It is not hard to understand how a mutant polypeptide might fold into the abnormal PrP^Sc conformation, but how is such a protein able to act as an infectious agent? According to one major hypothesis, an abnormal prion molecule (PrP^Sc) can bind to a normal protein molecule (PrP^C) and act as a template that causes the normal protein to refold into the abnormal form. This conversion can be shown to occur in the test tube: addition of PrP^Sc to a preparation of PrP^C can apparently convert the PrP^C molecules into the PrP^Sc conformation. According to this hypothesis, the appearance of the abnormal protein in the body—whether as a result of a rare misfolding event in the case of sporadic disease or by exposure to contaminated surgical instruments—starts a chain reaction in which normal protein molecules in the cells are gradually converted to the abnormal prion form. Viewed in this way, prion diseases are seen to be a quirk of molecular fate: this particular protein happens to be susceptible to folding incorrectly, incorrect folding happens to be stimulated by the presence of incorrectly folded molecules, and the accumulation of this incorrectly folded protein happens to cause the death of the cells in which it occurs.

CJD is a rare disease caused by a protein with unique infective properties. Alzheimer's disease (AD), on the other hand, is a common disorder that strikes as many as 10 percent of individuals who are at least 65 years of age and perhaps 40

percent of individuals who are 80 years or older. Persons with AD exhibit memory loss, confusion, and a loss of reasoning ability. CJD and AD share a number of important features. Both are fatal neurodegenerative diseases that can occur in either an inherited or sporadic form. Like CJD, the brain of a person with Alzheimer's disease contains fibrillar deposits of an insoluble material referred to as *amyloid* (Figure 2). In both diseases, the toxic fibrillar deposits result from the self-association of a polypeptide composed predominantly of β sheet. There are also many basic differences between the two diseases: the proteins that form the disease-causing aggregates are totally unrelated, the parts of the brain that are affected are distinct, and the protein responsible for AD does not act like an infectious agent (i.e., it is *nontransmissable*).

Most of the evidence suggests that Alzheimer's disease is caused by the deposition of a molecule, called the *amyloid β-peptide* (*A*β), which is originally part of a larger protein called the *amyloid precursor protein* (*APP*), which spans the nerve cell membrane. The Aβ peptide is released from the APP molecule following cleavage by two specific enzymes, β-secretase and γ-secretase (Figure 3). The length of the Aβ peptide is somewhat variable. The predominant species has a length of 40 amino acids (designated as Aβ40), but a minor species with two additional hydrophobic residues (designated as Aβ42) is also produced. Both of these peptides can exist in a soluble form that consists predominantly of α helices, but Aβ42 has a tendency to refold itself into a conformation that contains considerable β-pleated sheet. It is the Aβ42 version of the molecule that has the potential to cause Alzheimer's disease, because it tends to self-associate and form the insoluble amyloid deposits that lead to nerve cell death. Persons who suffer from inherited forms of AD carry a mutation that leads to the increased production of the neurotoxic Aβ42 peptide. Overproduction of Aβ42 can be caused by mutations in the *APP* gene that encodes the amyloid precursor protein or the *PS1* gene that encodes γ-secretase. Individuals with one of these inherited mutations exhibit symptoms of the disease at an early age, typically in their 50s.

Figure 3 **Formation of the Aβ peptide.** The Aβ peptide is carved from the amyloid precursor protein (APP) as the result of cleavage by two enzymes, β-secretase and γ-secretase. It is interesting that APP and the two secretases are all proteins that span the membrane. Cleavage of APP occurs inside the cell (in the endoplasmic reticulum), and the Aβ product is ultimately secreted into the space outside of the cell. The γ-secretase can cut at either of two sites in the APP molecule, producing either Aβ40 or Aβ42 peptides, the latter of which is primarily responsible for production of the amyloid plaques seen in Figure 2. γ-secretase is unusual among proteolytic enzymes that it cleaves its substrate at a site within a membrane.

One of the best approaches to the development of treatments for human diseases is to find laboratory animals, particularly mice, that develop similar diseases, and use these animals to test the effectiveness of potential therapies. Animals that exhibit a disease that mimics a human disease are termed *animal models*. For whatever reason, the brains of aging mice show no evidence of the amyloid deposits found in humans and, up until 1995, there was no animal model for AD. Then, in that year, a group of researchers found that they could create a strain of mice that developed amyloid brain plaques by introducing a particular gene into the animal's genome. The gene in question was one of the mutant *APP* genes that causes AD in humans. In 1999, members of this same research group reported an astounding finding. They had discovered that the formation of amyloid plaques in these genetically engineered (*transgenic*) mice could be blocked by repeatedly injecting the animals with the very same substance that causes the problem, the Aβ42 peptide. Injection of Aβ42 peptide caused the animals to produce antibodies that bound specifically to the Aβ42 peptide produced within their brains as the result of cleavage of the mutant APP protein. In effect, the researchers were immunizing (i.e., vaccinating) the mice against the disease. When young (6-week old) mice were immunized with the Aβ42, they failed to develop the amyloid brain deposits as they grew older. When older (13-month-old) mice whose brains already contained extensive amyloid deposits were immunized with the peptide, a significant fraction of the deposits was cleared out of the nervous system. Further analysis suggested that the antibodies were passing from the bloodstream into the brain where they were activating a type of scavenger cell that ingested the amyloid deposits from the brain tissue. Not only did immunized mice lack amyloid brain deposits, they performed much better in studies designed to test their memory than did nonimmunized transgenic controls. Clinical tests of the safety of this vaccination procedure have begun, and at the time of this writing no reports of toxic side effects in AD patients have been reported.

The idea that humans may one day be vaccinated against AD has captured the attention of physicians and the media, but it is not the only therapy being tested. Several drugs are being developed that inhibit the enzymes that cut the Aβ peptide out of the APP precursor, thereby reducing the production of Aβ42. In an alternate approach, a number of small peptides have been synthesized that can bind specifically to the α-rich version of Aβ42 and prevent it from refolding into the β-rich version. These small peptides, which are called β-sheet breakers, have a sequence of amino acids that is similar to a stretch of hydrophobic residues in the Aβ peptide that are involved in abnormal folding. In addition to the hydrophobic residues, the β-sheet breaker contains a proline residue that inhibits the formation of a β sheet. If one of these β-sheet breakers is injected into the brain of a rat that has been treated to develop amyloid deposits, the continued formation of amyloid fibrils is blocked and existing amyloid deposits are reduced in size. Peptide drugs are not likely to be very effective because they are quickly destroyed in the body and are generally unable to reach the brain. But nonpeptide drugs with a similar structure have been designed and synthesized and may one day play a key role in the prevention of this dreaded disease.

Great strides have been made in the past few years in synthesizing artificial genes whose peptide products fold into relatively simple secondary structures, such as bundles of α helices or 2–3 stranded β sheets. Attempts to create more complex polypeptide structures from scratch, however, have proven much more difficult. Researchers are finding, for example, that it is difficult to put a particular sequence into the context of an entire protein. A similar stretch of amino acids can form different types of secondary structures in different proteins, depending on external influences from other regions of the molecule. These difficulties serve as a reminder of the high level of complexity of proteins and our limited understanding of the way that nature transforms a linear genetic message into a three-dimensional protein.

An alternate approach to the production of novel proteins has been to modify those that are already produced by cells. Recent advances in DNA technology have allowed in-

vestigators to isolate an individual gene from human chromosomes, to alter its information content in a precisely determined way, and to synthesize the modified protein with its altered amino acid sequence. This technique, which is called **site-directed mutagenesis**, has many different uses, both in basic research and in applied biology. If, for example, an investigator wants to know about the role of a particular residue in the folding or function of a polypeptide, the gene can be mutated in a way that substitutes an amino acid with different charge, hydrophobic character, or hydrogen-bonding properties. The ability of the modified polypeptide to attain its normal tertiary structure and function can then be determined. As we will see throughout this text, site-directed mutagenesis is proving invaluable in the analysis of the specific functions of minute parts of virtually every protein of interest to biologists.

Site-directed mutagenesis is also being used to modify the structure of clinically useful proteins to increase their effectiveness or reduce unwanted side effects. For example, the protein thrombin is an enzyme in the blood that, depending on circumstances, can act either to promote or to inhibit the formation of a blood clot. Researchers have been able to engineer a version of the thrombin molecule that retains its anticoagulant activity while losing nearly all of its procoagulant activity. This engineered protein shows promise as an anticlotting agent in the treatment of strokes and heart attacks.

Structure-Based Drug Design The production of new proteins is one clinical application of recent advances in molecular biology; another is the development of new drugs that act by binding to proteins, thereby inhibiting their activity. Drug companies have access to chemical "libraries" that contain millions of different organic compounds that have been synthesized over the past decades. One way to search for potential drugs is to expose the protein in question to combinations of these compounds and determine which compounds, if any, are capable of binding with high affinity. An alternate approach, which can be employed if the tertiary structure of a protein has been determined, is to use computers to design "virtual" drug molecules whose size and shape might allow them to fit into the apparent cracks and crevices of the protein, rendering it inactive. Once a potential inhibitor has been synthesized and shown to bind to the target protein, the three-dimensional structure of the protein–inhibitor complex can be examined by X-ray crystallography. This approach provides clues as to how the molecule might be modified or redesigned to bind more tightly to its protein target.

Using these procedures, new types of aspirin-like molecules have been designed that lack some of the undesirable side effects caused by aspirin (acetylsalicylate). Aspirin's anti-inflammatory powers are derived from its ability to inhibit an enzyme (cyclooxygenase-2, or COX-2) that is required in the synthesis of certain hormone-like compounds (prostaglandins) that promote inflammation, pain, and fever. Prolonged use of aspirin or other nonsteroidal anti-inflammatory drugs (NSAIDs, such as

ibuprofen and naproxen) can cause serious damage to the lining of the stomach because they inhibit a related enzyme (cyclooxygenase-1, or COX-1), which plays a role in maintaining the health of a variety of organs. Although these two enzymes are very similar, they have at least one important difference in their aspirin-binding site—an isoleucine at position 523 in COX-1 and a valine at the corresponding site in COX-2. Isoleucine has a bulkier side chain than valine, giving COX-1 a smaller drug-binding site than that of COX-2 (Figure 2.46a). Because

(a)

(b)

Figure 2.46 Structure-based design of COX-2-specific inhibitors. (*a*) This image shows the NSAID-binding pocket of both COX-1 and COX-2. Amino acid residues shown in orange are the same in both enzymes. The only significant difference between the two enzymes in this region occurs at position 523, where COX-1 has a bulkier isoleucine residue (purple). This residuc decreases the volume of the COX-1 pocket where aspirin and other NSAIDs bind. (*b*) Space-filling models of aspirin (left) and a COX-2-specific inhibitor (right). (A: FROM PHILIP NEEDLEMAN AND PETER C. ISAKSON, SCIENCE & MEDICINE 5:33, 1998, JAN/FEB.; B: REPRINTED WITH PERMISSION FROM S. W. ROWLINSON AND A. S. KALGUTKAR, SCIENCE 280:1191, 1998. COPYRIGHT © 1998 AMERICAN ASSOCIATION FOR THE ADVANCEMENT OF SCIENCE.)

of this difference, it has been possible to design aspirin-like compounds that bind COX-2 with much greater affinity than COX-1 (Figure 2.46*b*). The first two COX-2-specific inhibitors to become available (sold under the brand names of Celebrex and Vioxx) have proven safe and effective, particularly in patients with arthritis who must take large amounts of the drug. However, unlike aspirin, COX-2-specific inhibitors do not lower the risk of heart attack and stroke, a benefit obtained through inhibition of COX-1 in blood platelets.

Protein Adaptation and Evolution Adaptations are traits that improve the likelihood of survival of an organism in a particular environment. Proteins are biochemical adaptations that are subject to natural selection and evolutionary change in the same way as other types of characteristics, such as eyes or skeletons. This is best revealed by comparing the homologous proteins in organisms living under very different environmental conditions. For example, the proteins of halophilic (salt-loving) archaebacteria possess amino acid substitutions that allow them to maintain their solubility and function at very cytosolic high salt concentrations (up to 4 M KCl). Unlike its counterpart in other organisms, the surface of the halophilic version of the protein malate dehydrogenase (MDH), for example, is coated with aspartic and glutamic acid residues whose carboxyl groups can compete with the salt for water molecules (Figure 2.47).

Figure 2.47 Distribution of polar, charged amino acid residues in an enzyme (MDH) from a halophilic archaebacterium. Red balls represent acidic residues and blue balls represent basic residues. The surface of the enzyme is seen to be covered with acidic residues, which gives the protein a net charge of −156, and promotes its solubility in extremely salty environments. For comparison, a homologous protein from the dogfish, an ocean-dwelling shark, has a net charge of +16. (REPRINTED WITH PERMISSION FROM O. DYM, M. MEVARECH, AND J. L. SUSSMAN, SCIENCE 267:1345, 1995. COPYRIGHT © 1995 AMERICAN ASSOCIATION FOR THE ADVANCEMENT OF SCIENCE.)

Many, if not most, of the amino acid substitutions that occur in proteins over time are thought to be neutral changes, that is, random changes that do not affect the fitness of the molecule. A change from one nonpolar amino acid to another in the core of a protein, for example, is not likely to have an effect on either the structure or function of that molecule. In fact, homologous proteins isolated from different organisms may exhibit virtually identical shapes and folding patterns, but show strikingly divergent amino acid sequences. The greater the evolutionary distance between two organisms, the greater the difference in the amino acid sequences of their proteins. In some cases, only a few key amino acids located in a particularly critical portion of the protein will be present in all of the organisms from which that protein has been studied. For example, a comparison of 226 globin sequences reveals that only two residues are absolutely conserved in all of these polypeptides; one is a histidine residue that plays a key role in the binding and release of O_2. These observations indicate that the secondary and tertiary structures of proteins change much more slowly during evolution than their primary structures.

We have seen how evolution has produced different versions of proteins in different organisms, but it has also produced different versions of proteins in individual organisms. As more and more amino acid sequences and tertiary structures of proteins were reported, it became apparent that most proteins are members of **families** (or large *superfamilies*) of related molecules. The genes that encode the various members of a protein family are thought to have evolved from a single ancestral gene that underwent duplication to produce two or more copies of the gene (see Figure 10.26). Over long periods of evolutionary time, the nucleotide sequences of the various copies diverged from one another to generate proteins with related (*homologous*) amino acid sequences. As a result of this evolutionary process, an organism may possess many different versions of a particular protein, such as globin, myosin, or collagen. In most cases, different versions of the family, which are known as **isoforms**, are adapted to function in different tissues or at different stages of development. For example, humans possess six different genes encoding isoforms of the contractile protein actin. Two of these isoforms are found in smooth muscle, one in skeletal muscle, one in heart muscle, and two in virtually all other types of cells.

Nucleic Acids

Nucleic acids are macromolecules constructed as a long chain (**strand**) of monomers called **nucleotides**. Nucleic acids function primarily in the storage and transmission of genetic information, but they may also have structural or catalytic roles. There are two types of nucleic acids found in living organisms, **deoxyribonucleic acid (DNA)** and **ribonucleic acid (RNA)**. As noted in Chapter 1, DNA serves as the genetic material of all cellular organisms, though RNA carries out that role for many viruses. In cells, information stored in DNA is used to govern cellular

activities through the formation of RNA messages. In the present discussion, we will examine the basic structure of nucleic acids using RNA as the representative molecule.

Phosphate

(a)

Sugar phosphate backbone

(b)

Figure 2.48 **Nucleotides and nucleotide strands of RNA.**
(a) Nucleotides are the monomers from which strands of nucleic acid are constructed. A nucleotide consists of three parts: a sugar, a nitrogenous base, and a phosphate. The nucleotides of RNA contain the sugar ribose, which has a hydroxyl group bonded to the second carbon atom. In contrast, the nucleotides of DNA contain the sugar deoxyribose, which has a hydrogen atom rather than a hydroxyl group attached to the second carbon atom. Each nucleotide is polarized, having a 5′ end (corresponding to the 5′ side of the sugar) and a 3′ end. (b) Nucleotides are joined together to form strands by covalent bonds that link the 3′ hydroxyl group of one sugar with the 5′ phosphate group of the adjoining sugar.

We will look at the more complex double-stranded structure of DNA in Chapter 10, where it can be tied to its central role in the chemical basis of life.

Each nucleotide in a strand of RNA consists of three parts (Figure 2.48a): (1) a five-carbon sugar, ribose; (2) a nitrogenous base (so-called because nitrogen atoms form part of the rings of the molecule); and (3) a phosphate group. The sugar and nitrogenous base together form a *nucleoside*, so that the nucleotides of an RNA strand are also known as ribonucleoside monophosphates. The phosphate is linked to the 5′ carbon of the sugar, and the nitrogenous base is attached to the sugar's 1′ carbon. During the assembly of a nucleic acid strand, the hydroxyl group attached to the 3′ carbon of the sugar of one nucleotide becomes linked by an ester bond to the phosphate group attached to the 5′ carbon of the next nucleotide in the chain. Thus the nucleotides of an RNA (or DNA) strand are connected by sugar-phosphate linkages (Figure 2.48b), which are described as *3′-5′-phosphodiester bonds* because the phosphate atom is esterified to two oxygen atoms, one from each of the two adjoining sugars.

A strand of RNA (or DNA) contains four different types of nucleotides distinguished by their nitrogenous base. Two types of bases occur in nucleic acids: pyrimidines and purines (Figure 2.49). **Pyrimidines** are smaller molecules, consisting of a single ring; **purines** are larger, consisting of two rings. RNAs contain two different purines, **adenine** and **guanine**, and two different pyrimidines, **cytosine** and **uracil**. In DNA, uracil is replaced by **thymine**, a pyrimidine with an extra methyl group attached to the ring (Figure 2.49).

Although RNAs consist of a continuous single strand, they often fold back on themselves to produce molecules having extensive double-stranded segments

Figure 2.49 **Nitrogenous bases in nucleic acids.** Of the four standard bases found in RNA, adenine and guanine are purines, and uracil and cytosine are pyrimidines. In DNA, the pyrimidines are cytosine and thymine, which differs from uracil by a methyl group attached to the ring.

(a)

(b)

Figure 2.50 RNAs can assume complex shapes. (a) This ribosomal RNA is an integral component of the small ribosomal subunit of a bacterium. The RNA strand is folded back on itself in a highly ordered pattern so that most of the molecule is double stranded. (b) This hammerhead ribozyme, as it is called, is a small RNA molecule from a viroid (page 25). The RNA has folded back on itself to form a largely double-stranded molecule. (B: FROM WILLIAM G. SCOTT, ET AL., CELL 81:993, 1995; BY PERMISSION OF CELL PRESS.)

and complex tertiary structures. This is illustrated by the two RNAs shown in Figure 2.50. The RNA whose secondary structure is shown in Figure 2.50a is a component of the small subunit of the bacterial ribosome (see Figure 2.51). Ribosomal RNAs are not molecules that carry genetic information; rather, they serve as structural scaffolds on which the proteins of the ribosome can be attached and as elements that recognize and bind various soluble components required for protein synthesis. One of the ribosomal RNAs of the large subunit acts as the catalyst for the reaction by which amino acids are covalently joined during protein synthesis. RNAs having a catalytic role are called RNA enzymes, or **ribozymes**. Figure 2.50b depicts the tertiary structure of the so-called hammerhead ribozyme, which is able to cleave its own RNA strand. In both examples shown in Figure 2.50, the double-stranded regions are held together by hydrogen bonds between the bases. This same principle is responsible for holding together the two strands of a DNA molecule.

Our appreciation of RNA structural complexity has lagged behind that of proteins because it has not been possible to use X-ray crystallography to explore the structures of large RNA molecules. In recent years, techniques have been developed to determine the tertiary structures of small RNAs, such as the hammerhead ribozyme of Figure 2.50b, and it has become evident that RNA molecules can achieve a level of structural complexity approaching that of proteins.

Nucleotides are not only important as building blocks of nucleic acids, they also have important functions in their own right. Most of the energy being put to use at any given moment in any living organism is derived from the nucleotide **adenosine triphosphate (ATP)**. The structure of ATP and its key role in cellular metabolism are discussed in the following chapter. **Guanosine triphosphate (GTP)** is another nucleotide of enormous importance in cellular activities. GTP binds to a variety of proteins (called G proteins) and acts as a switch to turn on their activities (see Figure 15.10 for an example).

An RNA World Research over the past 50 years has revealed the remarkable feats accomplished by various types of proteins and the importance of DNA as the storehouse of genetic information. RNA, in contrast, has been relegated primarily to a role as messenger—an intermediate in the flow of genetic information from DNA (through RNA) to protein. This concept has changed over recent years as we have learned that certain RNAs, such as the one depicted in Figure 2.50b, catalyze essential chemical reactions. The discovery of these ribozymes has led to speculation that at one time, during an early stage of biological evolution, neither DNA nor proteins existed on Earth. Instead, RNA molecules performed double duty: they served as genetic material, and they catalyzed necessary reactions. Only at a later stage in evolution were these activities taken over by DNA and protein (see Section 11.3).

1. Which macromolecules are polymers? What is the basic structure of each type of monomer? How do the various monomers of each type of macromolecule vary among themselves?

2. Describe the structure of nucleotides and the manner in which these monomers are strung together to form a polynucleotide strand. Why would it be overly simplistic to describe RNA as a single-stranded nucleic acid?

3. Name three polysaccharides composed of polymers of glucose. How do these macromolecules differ from one another?

4. Describe the properties of three different types of lipid molecules. What are their respective biological roles?

5. What are the major properties that distinguish different amino acids from one another? What roles do these differences play in the structure and function of proteins?

6. What are the properties of glycine, proline, and cysteine that distinguish these amino acids from all the others?

7. How are the properties of an α helix different from a β strand? How are they similar? How do each of these secondary structures affect the properties of a protein, such as α-keratin or silk?

8. Given that proteins act as molecular machines, explain why conformational changes are so important in protein function.

2.6 THE FORMATION OF COMPLEX MACROMOLECULAR STRUCTURES

To what degree can the lessons learned from the study of protein architecture be applied to more complex structures in the cell? Can structures, such as membranes, ribosomes, and cytoskeletal elements, which consist of different types of subunits, also assemble by themselves? How far can subcellular organization be explained simply by having the pieces fit together to form the most stable arrangement? The assembly of cellular organelles is poorly understood, but it is apparent from the following examples that different types of subunits can self-assemble to form higher-order arrangements.

The Assembly of Tobacco Mosaic Virus Particles and Ribosomal Subunits

The most convincing evidence that a particular assembly process is self-directed is the demonstration that the assembly can occur outside the cell (in vitro) under physiologic conditions when the only macromolecules present

are those that make up the final structure. In 1955, Heinz Fraenkel-Conrat and Robley Williams of the University of California, Berkeley, demonstrated that TMV particles, which consist of one long RNA molecule (approximately 6600 nucleotides) wound within a helical capsule made of 2130 identical protein subunits (see Figure 1.22), were capable of self-assembly. In their experiments, they purified TMV RNA and protein separately, mixed them together under suitable conditions, and recovered mature, infective particles after a short period of incubation. Although the sequence of events in TMV assembly has become controversial in recent years, there is little disagreement that the two components contain all the information necessary for particle formation.

Ribosomes, like TMV particles, are made of RNA and protein. Unlike the simpler TMV, ribosomes contain several different types of RNA and a considerable collection of different proteins. All ribosomes, regardless of their source, are composed of two subunits of different size. Although ribosomal subunits are often depicted in drawings as symmetric structures, in fact they have a highly irregular shape, as indicated in Figure 2.51. The large (or 50S) ribosomal subunit of bacteria contains 2 molecules of RNA and approximately 32 different proteins. The small (or 30S) ribosomal subunit contains 1 molecule of RNA and 21 different proteins. The structure and function of the ribosome are discussed in detail in Section 11.6.

One of the milestones in the study of ribosomes came in the mid-1960s, when Masayasu Nomura and his

Figure 2.51 **Reconstruction of a ribosome from the cytoplasm of a wheat germ cell.** This reconstruction is based on high-resolution electron micrographs and shows the two subunits of the ribosome, the small (40S) subunit on the left and the large (60S) subunit on the right. The internal structure of a ribosome is discussed in Section 11.6. (FROM ADRIANA VERSCHOOR, ET AL., J. CELL BIOL. VOL. 133, COVER #3, 1996; BY COPYRIGHT PERMISSION OF THE ROCKEFELLER UNIVERSITY PRESS.)

coworkers at the University of Wisconsin succeeded in reconstituting complete, fully functional 30S subunits by mixing the 21 purified proteins of the small subunit with purified small-subunit ribosomal RNA. Apparently, the components of the small subunit contain all the information necessary for the assembly of the entire particle. Analysis of the intermediates that form at different stages during reconstitution in vitro indicates that subunit assembly occurs in a sequential step-by-step manner that closely parallels the process in vivo. As suggested in TMV assembly, the incorporation of individual proteins appears to change the conformation of the growing particle, making it more reactive to the attachment of additional proteins. At least one of the proteins of the small subunit (S16) appears to function solely in ribosome assembly; deletion of this protein from the reconstitution mixture greatly slowed the assembly process but did not block the formation of fully functional ribosomes. Many other proteins of the small subunit function primarily to stabilize the assembled structure. Reconstitution of the large subunit of the bacterial ribosome was accomplished in the following decade. It should be kept in mind that although it takes approximately 2 hours at 50°C to reconstitute the ribosome in vitro, the bacterium can assemble the same structure in a few minutes at temperatures as low as 10°C.

It may be that the bacterium uses something that is not available to the investigator who begins with purified components. Assembly of the ribosome within the cell, for example, may include the participation of accessory factors that function in protein folding, such as the chaperones described in the Experimental Pathways. In fact, the formation of ribosomes within a eukaryotic cell requires the transient association of several proteins that do not end up in the final particle, as well as the removal of approximately half the nucleotides of the large ribosomal RNA precursor (Section 11.3). As a result, the components of the mature eukaryotic ribosome no longer possess the information to reconstitute themselves in vitro.

REVIEW ?

1. What type of evidence suggests that bacterial ribosomal subunits are capable of self-assembly, but eukaryotic subunits are not?

2. What evidence would indicate that a particular ribosomal protein had a role in ribosome function but not assembly?

EXPERIMENTAL PATHWAYS

Chaperones: Helping Proteins Reach Their Proper Folded State

In 1962, F. M. Ritossa, an Italian biologist studying the development of the fruit fly *Drosophila*, reported a curious finding.[1] When the temperature at which fruit fly larvae were developing was raised from the normal 25°C to 32°C, a number of new sites on the giant chromosomes of the larval cells became activated. As we will see in Chapter 10, the giant chromosomes of these insect larvae provide a visual exhibit of gene expression. The results suggested that increased temperature induced the expression of new genes, a finding that was confirmed a decade later with the characterization of several proteins that appeared in larvae following temperature elevation.[2] It was soon found that this response, called the **heat shock response**, was not confined to fruit flies, but can be initiated in many different cells from virtually every type of organism—from bacteria to plants and mammals. Closer examination revealed that the proteins produced during the response were found not only in heat-shocked cells, but also at lower concentration in cells under normal conditions. What is the function of these so-called *heat-shock proteins (hsps)*? The answer to this question was gradually revealed by a series of seemingly unrelated studies.

We saw on page 74 that some complex, multisubunit structures, such as a bacterial ribosome or a tobacco mosaic virus particle, can self-assemble from purified subunits. It was demonstrated in the 1960s that the proteins that make up bacteriophage particles (see Figure 1.23c) also possess a remarkable ability to self-assemble, but they are generally unable to form a complete, functional virus particle by themselves in vitro. Experiments on phage assembly in bacterial cells confirmed that phages require bacterial help. It was shown in 1973, for example, that a certain mutant strain of bacteria, called *GroE*, could not support the assembly of normal phages. Depending upon the type of phage, the head or the tail of the phage particle was assembled incorrectly.[3,4] These studies suggested that a protein encoded by the bacterial chromosome participated in the assembly of viruses, even though this host protein was not a component of the final virus particles. As it obviously did not evolve as an aid for virus assembly, the bacterial protein required for phage assembly had to play some role in the cell's normal activities, but the precise role remained obscure. Subsequent studies revealed that the *GroE* site on the bacterial chromosome actually contains two separate genes, *GroEL* and *GroES*, that en-

code two separate proteins GroEL and GroES. Under the electron microscope, the purified GroEL protein appeared as a cylindrical assembly consisting of two discs. Each disc was composed of seven subunits arranged symmetrically around the central axis (Figure 1).[5,6]

Several years later, a study on pea plants hinted at the existence of a similar assembly-promoting protein in the chloroplasts of plants.[7] Rubisco is a large protein in chloroplasts that catalyzes the reaction in which CO_2 molecules taken up from the atmosphere are covalently linked to organic molecules during photosynthesis (Section 6.6). Rubisco comprises 16 subunits: 8 small subunits (molecular mass of 14,000 daltons) and 8 large subunits (55,000 daltons). It was found that large Rubisco subunits, synthesized inside the chloroplast, are not present in an independent state, but are associated with a huge protein assembly consisting of identical subunits of 60,000 daltons (60 kDa) molecular mass. In their paper, the researchers considered the possibility that the complex formed by the large subunit and the 60 kDa polypeptide was an intermediate in the assembly of a complete Rubisco molecule.

A separate line of investigation on mammalian cells also revealed the existence of proteins that appeared to assist the assembly of multisubunit proteins. Like Rubisco, antibody molecules consist of a complex of two different types of subunits, smaller light chains and larger heavy chains. Just as the large subunits of Rubisco become associated with another protein not found in the final complex, so too do the heavy chains of an antibody complex.[8] This protein, which associates with newly synthesized heavy chains, but not with heavy chains that are already bound to light chains, was named *binding protein*, or BiP. BiP was subsequently found to have a molecular mass of 70,000 daltons (70 kDa).

To this point, we have been discussing two lines of investigation: one concerned with the heat shock response and the other with proteins that promote protein assembly. These two fields came together in 1986, when it was shown that one of the proteins that figures most prominently in the heat shock response, a protein that had been named *heat-shock protein 70* (hsp70) because of its molecular mass, was identical to BiP, the protein implicated in the assembly of antibody molecules.[9]

Even before the discovery of the heat shock response, the structure of proteins was known to be sensitive to temperature, with a small rise in temperature causing these delicate molecules to begin to unfold. Unfolding exposes hydrophobic residues that were previously buried in the protein's core. Just as fat molecules in a bowl of soup are pushed together into droplets, so too are proteins with hydrophobic patches on their surface. Consequently, when a cell is heat shocked, soluble proteins become denatured and form aggregates. A report in 1985 demonstrated that, following temperature elevation, newly synthesized hsp70 molecules enter the nuclei and bind to aggregates of nuclear proteins, where they act like a molecular crowbar to promote disaggregation.[10] Because of their role in assisting the assembly of proteins by preventing undesirable interactions, hsp70 and related molecules were named **molecular chaperones**.[11]

It was soon demonstrated that the bacterial heat-shock protein GroEL and the Rubisco assembly protein in plants were homologous proteins. In fact, the two proteins share the same amino acids at nearly half of the more than 500 residues in their respective molecules.[12] The fact that the two pro-

Figure 1 A model of the GroEL complex built according to data from electron microscopy and molecular weight determination. The complex is seen to consist of two discs, each composed of seven identical subunits arranged symmetrically around a central axis. Subsequent studies showed the complex contains a large internal chamber. (FROM T. HOHN, ET AL., J. MOL. BIOL. 129:371, 1979. COPYRIGHT 1979, BY PERMISSION OF THE PUBLISHER ACADEMIC PRESS.)

teins—both members of the *Hsp60 chaperone family*—have retained so many of the same amino acids reflects their similar and essential function in the two types of cells. But what was that essential function? At this point it was thought that their primary function was to mediate the assembly of multisubunit complexes, such as bacteriophage particles or Rubisco. This view was changed by experiments studying molecular chaperones in mitochondria. It was known that newly made mitochondrial proteins produced in the cytosol had to cross the outer mitochondrial membranes in an unfolded, extended, monomeric form. A mutant was found that altered the activity of another member of the Hsp60 chaperone family that resided inside mitochondria. In cells containing this mutant chaperone, proteins that were transported into mitochondria failed to fold into their active forms.[13] Even proteins that consisted of a single polypeptide chain failed to fold into their native conformation. This finding changed the perception of chaperone function from a notion that they assist assembly of already-folded subunits into larger complexes, to our current understanding that they assist polypeptide chain folding.

The results of these and other studies indicated the presence in cells of two major families of molecular chaperones: the Hsp70 chaperones, such as BiP, and the Hsp60 chaperones (which are also called *chaperonins*), such as Hsp60, GroEL, and the Rubisco assembly protein. We will focus on the Hsp60 chaperonins, such as GroEL, which are best understood.

As first revealed in 1979, GroEL is a huge molecular complex of 14 polypeptide subunits arranged in two stacked rings resembling a double doughnut.[5,6] Fifteen years after these first electron micrographs were taken, the three-dimensional structure of the GroEL complex was determined by X-ray crystallography.[14] The study revealed the presence of a central cavity within the GroEL cylinder. Subsequent studies demonstrated that this cavity was divided into two separate chambers, one at each end of the complex. Each chamber was

Figure 2 Reconstructions of the GroEL based on high-resolution electron micrographs taken of specimens that had been frozen in liquid ethane and examined at −170°C. The image on the left shows the GroEL complex, and that on the right shows the GroEL complex with GroES, which appears as a dome on one end of the cylinder. It is evident that the binding of the GroES is accompanied by a marked change in conformation of the apical end of the proteins that make up the top GroEL ring (arrow), which results in a marked enlargement of the upper chamber. (REPRINTED WITH PERMISSION FROM S. CHEN, ET AL., COURTESY OF HELEN R. SAIBIL, NATURE 371:263, 1994. COPYRIGHT © 1994 MACMILLAN MAGAZINES LIMITED.)

situated within the center of one of the rings of the GroEL complex and was large enough to enclose a polypeptide undergoing folding.

Electron microscopic studies also provided information about the function of a second protein, GroES, which acts in conjunction with GroEL. Like GroEL, GroES is a ringlike protein with seven subunits arrayed symmetrically around a central axis. GroES, however, consists of only one ring, and its subunits are much smaller (10,000 daltons) than those of GroEL (60,000 daltons). GroES is seen as a cap or dome that fits on top of either end of a GroEL cylinder (Figure 2). The attachment of GroES to one end of GroEL causes a dramatic conformational change in the GroEL protein that markedly increases the volume of the enclosed chamber at that end of the complex.[15]

The importance of this conformational change has been revealed in remarkable detail by X-ray crystallographic studies in the laboratories of Arthur Horwich and Paul Sigler at Yale University.[16,17] As shown in Figure 3, the binding of the GroES cap is accompanied by a 60° rotation of the apical (red) domain of the subunits that make up the GroEL ring at that end of the GroEL cylinder. The attachment of GroES does more than trigger a conformational change that enlarges the GroEL chamber. Before attachment of GroES, the wall of the GroEL chamber has exposed hydrophobic residues that give the lining a hydrophobic character. Nonnative polypeptides also have exposed hydrophobic residues that eventually become buried in the interior of the native polypeptide. Because hydrophobic surfaces tend to interact, the hydrophobic lining of the GroEL cavity promotes the binding of nonnative polypeptides. Binding of GroES to GroEL buries the hydrophobic residues of the GroEL wall and exposes a number

(a)

(b)

Figure 3 Conformational change in GroEL. (*a*) The drawing on the left shows a surface view of the two rings that make up the GroEL chaperonin. The drawing on the right shows the tertiary structure of one of the subunits of the top GroEL ring. The polypeptide chain can be seen to fold into three domains. (*b*) When a GroES ring (arrow) binds to the GroEL cylinder, the apical domain of each GroEL subunit of the adjacent ring undergoes a dramatic rotation of approximately 60° with the intermediate domain (shown in green) acting like a hinge. The effect of this shift in parts of the polypeptide is a marked elevation of the GroEL wall and enlargement of the enclosed chamber. (REPRINTED WITH PERMISSION FROM Z. XU, A. L. HORWICH, AND P. B. SIGLER, NATURE 388:744, 1997. COPYRIGHT © 1997 MACMILLAN MAGAZINES LIMITED.)

of polar residues, thereby changing the character of the chamber wall. As a result of this change, the nonnative polypeptide that had been bound to the wall by hydrophobic interactions is displaced into the space within the chamber. Once freed from its attachment to the chamber wall, the polypeptide is given the opportunity to continue its folding in a pro-

Figure 4 A schematic illustration of the proposed steps that occur during the GroEL-GroES-assisted folding of a polypeptide. The GroEL is seen to consist of two chambers that have equivalent structures and functions and that alternate in activity. Each chamber is located within one of the two rings that make up the GroEL complex. The nonnative polypeptide enters one of the chambers (step 1) and binds to hydrophobic sites on the chamber wall. Binding of the GroES cap produces a conformational change in the wall of the top chamber, causing the enlargement of the chamber and release of the nonnative polypeptide from the wall into the encapsulated space (step 2). After about 15 seconds have elapsed, the GroES dissociates from the complex and the polypeptide is ejected from the chamber (step 3). If the polypeptide has achieved its native conformation, as has the molecule on the left, the folding process is complete. If, however, the polypeptide is only partially folded, or is misfolded, it will rebind the GroEL chamber for another round of folding. (*Note*: As indicated, the mechanism of GroEL action is driven by the binding and hydrolysis of ATP, an energy-rich molecule whose function is discussed at length in the following chapter.) (AFTER A. L. HORWICH, ET AL, PROC. NAT'L. ACAD. SCI. U.S.A. 96:11037, 1999.)

tected environment. After about 15 seconds, the GroES cap dissociates from the GroEL ring, and the polypeptide is ejected from the chamber. If the polypeptide has not reached its native conformation by the time it is ejected, it can rebind to the same or another GroEL, and the process is repeated. A model depicting some of the steps thought to occur during GroEL-GroES-assisted folding is shown in Figure 4.

It is estimated that as many as 50 percent of the nonnative soluble proteins of a bacterial cell can interact with GroEL. It was emphasized in this chapter that proteins interact with one another in a highly specific manner. Given this principle, how is it possible that a single protein—GroEL—can bind so many different polypeptides? The GroEL binding site consists of a hydrophobic surface formed largely by two α helices of the apical domain that is capable of binding virtually any sequence of hydrophobic residues that might be accessible in a partially folded or misfolded polypeptide.[18] A comparison of the crystal structure of the unbound GroEL molecule with that of GroEL bound to several different peptides revealed that the binding site on the apical domain of a GroEL subunit can locally adjust its positioning when bound to different partners. This finding indicates that the binding site has structural flexibility that allows it to adjust its own shape to fit the shape of the particular polypeptide with which it has to interact.

Given the flexibility of the GroEL binding site, how broad a spectrum of polypeptides can this chaperonin interact with? Ulrich Hartl and his colleagues at the Max Planck Institute in Germany have attempted to answer this question by incubating bacterial cells for brief periods with labeled amino acids, lysing the cells, and precipitating the GroEL-polypeptide complexes by addition of an anti-GroEL antibody.[19] Several hundred different labeled polypeptides were found to be present in these immunoprecipitates, a finding that directly confirms the proposal that a great variety of newly synthesized polypeptides interact with GroEL. Polypeptides containing a particular motif, called an αβ-fold, were particularly abundant among the labeled, GroEL-bound molecules. Such polypeptides contain α helices and buried β sheets with extensive hydrophobic surfaces. Polypeptides of this type are prone to aggregation if not properly folded and make ideal candidates for interaction with GroEL

Keep in mind that molecular chaperones do not convey information for the folding process but instead prevent proteins from veering off their correct folding pathway and finding themselves in misfolded or aggregated states. Just as Anfinsen discovered decades ago, the three-dimensional structure of a protein is determined by its amino acid sequence.

References

1. RITOSSA, F. 1962. A new puffing pattern induced by temperature shock and DNP in *Drosophila. Experientia* 18:571–573.

2. TISSIERES, A., MITCHELL, H. K., & TRACY, U. M. 1974. Protein synthesis in salivary glands of *Drosophila melanogaster:* Relation to chromosomal puffs. *J. Mol. Biol.* 84:389–398.

3. STERNBERG, N. 1973. Properties of a mutant of *Escherichia coli* defective in bacteriophage lambda head formation (groE). *J. Mol. Biol.* 76:1–23.

4. GEORGOPOULOS, C. P., ET AL. 1973. Host participation in bacteriophage lambda head assembly. *J. Mol. Biol.* 76:45–60.

5. HOHN, T., ET AL. 1979. Isolation and characterization of the host protein groE involved in bacteriophage lambda assembly. *J. Mol. Biol.* 129:359–373.

6. HENDRIX, R. W. 1979. Purification and properties of groE, a host protein involved in bacteriophage assembly. *J. Mol. Biol.* 129:375–392.

7. BARRACLOUGH, R. & ELLIS, R. J. 1980. Protein synthesis in chloroplasts. IX. *Biochim. Biophys. Acta* 608:19–31.

8. HAAS, I. G. & WABL, M. 1983. Immunoglobulin heavy chain binding protein. *Nature* 306:387–389.

9. MUNRO, S. & PELHAM, H. R. B. 1986. An Hsp70-like protein in the ER: Identity with the 78 kD glucose-regulated protein and immunoglobin heavy chain binding protein. *Cell* 46:291–300.

10. LEWIS, M. J. & PELHAM, H. R. B. 1985. Involvement of ATP in the nuclear and nucleolar functions of the 70kD heat-shock protein. *EMBO J.* 4:3137–3143.

11. ELLIS, J. 1987. Proteins as molecular chaperones. *Nature* 328:378–379.

12. HEMMINGSEN, S. M., ET AL. 1988. Homologous plant and bacterial proteins chaperone oligomeric protein assembly. *Nature* 333:330–334.

13. CHENG, M. Y., ET AL. 1989. Mitochondrial heat-shock protein Hsp60 is essential for assembly of proteins imported into yeast mitochondria. *Nature* 337:620–625.

14. BRAIG, K., ET AL. 1994. The crystal structure of the bacterial chaperonin GroEL at 2.8A. *Nature* 371:578–586.

15. CHEN, S., ET AL. 1994. Location of a folding protein and shape changes in GroEL-GroES complexes. *Nature* 371:261–264.

16. XU, Z., HORWICH, A. L., & SIGLER, P. B. 1997. The crystal structure of the asymmetric GroEL-GroES-(ADP)$_7$ chaperonin complex. *Nature* 388:741–750.

17. HORWICH, A. 2000. Working with Paul Sigler. *Nature Struct. Biol.* 7:269–270.

18. CHEN, L. & SIGLER, P. 1999. The crystal structure of a GroEL/peptide complex: plasticity as a basis for substrate diversity. *Cell* 99:757–768.

19. HOURY, W. A., ET AL. 1999. Identification of *in vivo* substrates of the chaperonin GroEL. *Nature* 402:147–154.

SYNOPSIS

Covalent bonds hold atoms together to form molecules. Covalent bonds are stable partnerships formed when atoms share their outer-shell electrons, each participant gaining a filled shell. Covalent bonds can be single, double, or triple depending on the number of pairs of shared electrons. If electrons in a bond are shared unequally by the component atoms, the atom with the greater attraction for electrons (the more electronegative atom) bears a partial negative charge, whereas the other atom bears a partial positive charge. Molecules that lack polarized bonds have a nonpolar, or hydrophobic, character, which makes them insoluble in water. Molecules that have polarized bonds have a polar, or hydrophilic, character, which makes them water soluble. Polar molecules of biological importance contain one or more electronegative atoms, usually O, N, S, or P. (p. 33)

Noncovalent bonds are formed by weak attractive forces between positively and negatively charged regions within the same molecule or between two nearby molecules. Noncovalent bonds play a key role in maintaining the structure of biological molecules and mediating their dynamic activities. Noncovalent bonds include ionic bonds, hydrogen bonds, and van der Waals forces. Ionic bonds form between fully charged positive and negative groups; hydrogen bonds form between a covalently bonded hydrogen atom (which bears a partial positive charge) and a covalently bonded nitrogen or oxygen atom (which bears a partial negative charge); van der Waals forces form between two atoms exhibiting a transient charge due to a momentary asymmetry in the distribution of electrons around the atoms. Nonpolar molecules or nonpolar portions of larger molecules tend to associate with one another in aqueous environments to form hydrophobic interactions. Examples of these various types of noncovalent interactions include the association of DNA and proteins by ionic bonds, the association of pairs of DNA strands by hydrogen bonds, and the formation of the hydrophobic core of soluble proteins as the result of hydrophobic interactions and van der Waals forces. (p. 36)

Water has unique properties on which life depends. The covalent bonds that make up a water molecule are highly polarized. As a result, water is an excellent solvent capable of forming hydrogen bonds with virtually all polar molecules. Water is also a major determinant of the structure of biological molecules and the types of interactions in which they can engage. The pH of a solution is a measure of the concentration of hydrogen (or hydronium) ions. Most biological processes are acutely sensitive to pH because changes in hydrogen ion concentration alter the ionic state of biological molecules. Cells are protected from pH fluctuations by buffers—compounds that react with hydrogen or hydroxyl ions. (p. 38)

Carbon atoms play a pivotal role in the formation of biological molecules. Each carbon atom is able to bond with up to four other atoms, including other carbon atoms. This property allows the formation of large molecules whose backbone consists of a chain of carbon atoms. Molecules consisting solely of hydrogen and carbon are called hydrocarbons. Most of the molecules of biological importance contain functional groups that include one or more electronegative atoms, making the molecule more polar, more water soluble, and more reactive. (p. 41)

Biological molecules are members of four distinct types: carbohydrates, lipids, proteins, and nucleic acids. Carbohydrates include simple sugars and larger molecules (polysaccharides) constructed of sugar monomers. Carbohydrates function primarily as a storehouse of chemical energy and as durable building materials for biological construction. Simple biological sugars consist of a backbone of three to seven carbon atoms, with each carbon linked to a hydroxyl group except one, which bears a carbonyl. Sugars with five or more carbon atoms self-react to form a ring-shaped molecule. Those carbon atoms along the sugar backbone that are linked to four different groups are sites of stereoisomerism, generating pairs of isomers that cannot be superimposed. The asymmetric carbon farthest from the carbonyl determines whether the sugar is D or L. Sugars are linked to one another by glycosidic bonds to form disaccharides, oligosaccharides, and polysaccharides. In animals, sugar is stored primarily as the branched polysaccharide glycogen, which provides a readily available energy source. In plants, glucose reserves are stored as starch, which is a mixture of unbranched amylose and branched amylopectin. Most of the sugars in both glycogen and starch are joined by $\alpha(1\rightarrow4)$ linkages. Cellulose is a structural polysaccharide manufactured by plant cells that serves as a major component of the cell wall. Glucose monomers in cellulose are joined by $\beta(1\rightarrow4)$ linkages, which are cleaved by cellulase, an enzyme that is absent in virtually all animals. Chitin is a structural polysaccharide composed of N-acetyl-glucosamine monomers. (p. 42)

Lipids are a diverse array of hydrophobic molecules having widely divergent structures and functions. Fats consist of a glycerol molecule esterified to three fatty acids. Fatty acids differ in chain length and the number and position of double bonds (sites of unsaturation). Fats are very rich in chemical energy; a gram of fat contains over twice the energy content of a gram of carbohydrate. Steroids are a group of lipids containing a characteristic four-ringed hydrocarbon skeleton. Steroids include cholesterol as well as numerous hormones (e.g., testosterone, estrogen, and progesterone) that are synthesized from cholesterol. Phospholipids are phosphate-containing lipid molecules that contain both a hydrophobic end and a hydrophilic end and play a pivotal role in the structure and function of cell membranes. (p. 49)

Proteins are macromolecules of diverse function consisting of amino acids linked by peptide bonds into polypeptide chains. Included among the diverse array of proteins are en-

zymes, structural materials, membrane receptors, gene regulatory factors, hormones, transport agents, and antibodies. The order in which the 20 different amino acids are incorporated into a protein is encoded in the sequence of nucleotides in DNA. All 20 amino acids share a common structural organization consisting of an α carbon bonded to an amino group, a carboxyl group, and a side chain of varying structure. In the present scheme, the side chains are classified into four categories: those that are fully charged at physiologic pH; those that are polar, but uncharged and capable of forming hydrogen bonds; those that are nonpolar and interact by means of van der Waals forces; and three amino acids (proline, cysteine, and glycine) that possess unique properties. (p. 51)

The structure of a protein can be described at four levels of increasing complexity. Primary structure is described by the amino acid sequence of a polypeptide; secondary structure by the three-dimensional structure (conformation) of sections of the polypeptide backbone; tertiary structure by the conformation of the entire polypeptide; and quaternary structure by the arrangement of the subunits if the protein consists of more than one polypeptide chain. The α helix and β-pleated sheet are both stable, maximally hydrogen-bonded secondary structures that are common in many proteins. The tertiary structure of a protein is highly complex and unique to each individual type of protein. Most proteins have an overall globular shape in which the polypeptide is folded to form a compact molecule in which specific residues are strategically situated to allow the protein to carry out its specific function. Analysis of large numbers of proteins has revealed that many consist of domains that maintain a structural and functional independence from one another. Using the technique of site-directed mutagenesis, researchers can learn about the role of specific amino acid residues by making specific substitutions. (p. 55)

The information required for a polypeptide chain to achieve its native conformation is encoded in its primary structure. Some proteins fold into their final conformation by themselves; others require the assistance of nonspecific chaperones, which prevent aggregation of partially folded intermediates. (p. 64)

Nucleic acids are primarily informational molecules that consist of strands of nucleotide monomers. Each nucleotide in a strand consists of a sugar, phosphate, and nitrogenous base. The nucleotides are linked by bonds between the 3' hydroxyl group of the sugar of one nucleotide and the 5' phosphate group of the adjoining nucleotide. Both RNA and DNA are assembled from four different nucleotides; nucleotides are distinguished by their bases, which can be a pyrimidine (cytosine or uracil/thymine) or a purine (adenine or guanine). DNA is a double-stranded nucleic acid, and RNA is generally single stranded, though the single strand is often folded back on itself to form double-stranded sections. Information in nucleic acids is encoded in the specific sequence of nucleotides that constitute a strand. (p. 71)

ANALYTIC QUESTIONS

1. Sickle cell anemia results from a substitution of a valine for a glutamic acid. What do you expect the effect might be if the mutation were to have placed a leucine at that site? An aspartic acid?

2. Of the following amino acids, glycine, isoleucine, and lysine, which would you expect to be the most soluble in an acidic aqueous solution? Which the least?

3. How many structural isomers could be formed from a molecule with the formula C_5H_{12}? C_4H_8?

4. Glyceraldehyde is the only three-carbon aldotetrose, and it can exist as two stereoisomers. What is the structure of dihydroxyacetone, the only ketotriose? How many stereoisomers does it form?

5. Bacteria are known to change the kinds of fatty acids they produce as the temperature of their external environment changes. What types of changes in the fatty acids would you expect as the temperature drops? Why would this be adaptive?

6. In the polypeptide backbone —C—C—N—C—C—N —C—C—NH$_2$, identify the α carbons.

7. Which of the following are true? Increasing the pH of a solution would (1) suppress the dissociation of a carboxylic acid, (2) increase the charge on an amino group, (3) increase the dissociation of a carboxylic acid, (4) suppress the charge on an amino group.

8. Which of the four classes of amino acids has side chains with the greatest hydrogen-bond-forming potential? Which has the greatest potential to form ionic bonds? Hydrophobic interactions?

9. If the three enzymes of the pyruvate dehydrogenase complex existed as physically separate proteins rather than as a complex, what effect might this have on the rate of reactions catalyzed by these enzymes?

10. Would you agree that neither ribonuclease nor myoglobin had quaternary structure? Why or why not?

11. How many different tripeptides are possible? How many carboxyl terminals of polypeptide chains are present in a molecule of hemoglobin?

12. You have isolated a pentapeptide composed of four glycine residues and one lysine residue that resides at the C-terminus of the peptide. Using the information provided in the legend of Figure 2.27, if the pK of the side chain of lysine is 10 and the pK of the terminal carboxyl group is 4, what is the structure of the peptide at pH 7? At pH 12?

13. The side chains of glutamic acid (pK 4.3) and arginine (pK 12.5) can form an ionic bond under certain conditions. Draw the relevant portions of the side chains and indicate whether or not an ionic bond could form at the following: (a) pH 4; (b) pH 7; (c) pH 12; (d) pH 13.

14. Would you expect a solution of high salt to be able to denature ribonuclease? Why or why not?

15. You have read in the Human Perspective that (1) mutations in the *PRNP* gene can make a polypeptide more likely to fold into the PrPSc conformation, thus causing CJD and (2) exposure to the PrPSc prion can lead to an infection that also causes CJD. How can you explain the occurrence of rare sporadic cases of the disease in persons who have no genetic propensity for it?

16. One of the arguments used by researchers who do not believe the "protein only" hypothesis in prion diseases is that the peak age group afflicted by CJD is 50 to 60 years of age and not older. In view of your answer to the previous question, why would the age of the victims of the disease relate to its underlying cause.

17. We saw on page 71 how evolution has led to the existence of protein families composed of related molecules with similar functions. A few examples are also known where proteins with very similar functions have primary and tertiary structures that show no evidence of evolutionary relationship. Subtilisin and trypsin, for example, are two protein-digesting enzymes (proteases) that show no evidence they are homologous despite the fact they utilize the same mechanism for attacking their substrates. How can this coincidence be explained?

18. Would you agree with the statement that many different amino acid sequences can fold into the same basic tertiary structure? What data can you cite as evidence for your position.

ADDITIONAL READING

General

GARRETT, R. H. & GRISHAM, C. M. 1998. *Biochemistry*, 2d ed. Saunders.

LEHNINGER, A. L., NELSON, D. L., & COX, M. M. 1993. *Principles of Biochemistry*, 2d ed. Worth.

MATHEWS, C. K., ET AL. 1999. *Biochemistry*, 3d ed. Benjamin-Cummings.

MYERS, R. A., ED. 1996. *Encyclopedia of Molecular Biology and Molecular Medicine*. 6 vols. VCH.

PASTERNAK, C. A. 1998. *The Molecules Within Us*. Plenum.

STRYER, L. 1995. *Biochemistry*, 4th ed. W. H. Freeman.

VOET, D., ET AL. 1999, *Fundamentals of Biochemistry*. Wiley.

Amino Acids and Proteins

BARRETT, G. C. & ELMORE, D. T. 1998. *Amino Acids and Peptides*. Cambridge.

BRANDEN, C. & TOOZE, J. 1999. *Introduction to Protein Structure*, 2d ed. Garland.

CREIGHTON, T. E. 1993. *Proteins: Structures and Molecular Properties*, 2d ed. W. H. Freeman.

DICKERSON, R. E. & GEIS, I. 1969. *The Structure and Action of Proteins*. Benjamin-Cummings.

KARPLUS, M. & MCCAMMON, J. A. 1986. The dynamics of proteins. *Sci. Am.* 254:42–51. (April)

KENDREW, J. C. 1963. Myoglobin and the structure of proteins. *Science* 139:1259–1266.

KYTE, J. 1995. *Structure in Protein Chemistry*. Garland.

PERUTZ, M. F. 1964. The hemoglobin molecule. *Sci. Amer.* 211:64–76. (Nov.)

SANGER, F. 1988. Sequences, sequences, and sequences. *Ann. Rev. Biochem.* 57:1–28.

SCHULZ, G. E. & SCHIRMER, R. H. 1990. *Principles of Protein Structure*, 2d ed. Springer-Verlag.

THOMPSON, E. O. P. 1955. The Insulin Molecule. *Sci. Amer.* 192:36–41. (May)

Human Perspective: Free Radicals as a Cause of Aging

BECKMAN, K. B. & AMES, B. N. 1998. The free radical theory of aging matures. *Physiol. Revs.* 78:547–581.

FREI, B., ET AL. 1999. Molecular and biological mechanisms of antioxidant action. *FASEB J.* 13:963–1024.

GUARENTE, L. 1999. Mutant mice live longer. *Nature* 402:243–244.

LITHGOW, G. J. & ANDERSEN, J. K. 2000. The real Dorian Gray mouse. *Bioess.* 22:410–413.

WALLACE, D. C. & MELOV, S. 1998. Radicals r'aging. *Nat. Gen.* 19:105–106.

WEINDRUCH, R. & SOHAL, R. S. 1997. Caloric intake and aging. *New Engl. J. Med.* 337:986–994.

WOOD, W. B. 1998. Aging of *C. elegans. Cell* 95:147–150.

Human Perspective: Protein Misfolding Can Have Deadly Consequences

BALTER, M. 1999. Prions: A lone killer or a vital accomplice. *Science* 286:660–662.

———— 2000. Tracking the human fallout from "Mad Cow disease." *Science* 289:1452–1454.

COHEN, F. E. & PRUSINER, S. B. 1998. Pathologic conformations of prion proteins. *Annu. Rev. Biochem.* 67:793–819.

DUFF, K. 1999. Curing amyloidosis: Will it work in humans? *Trends Neurosci.* 22:485–486.

HELMUTH, L. 2000. Further progress on a β-amyloid vaccine. *Science* 289:374.

Prusiner. S. B., ET AL. 1998. Prion protein biology. *Cell* 93:337–348.

SELKOE, D. J. & WOLFE, M. S. 2000. In search of gamma secretase:presenilin at the cutting edge. *Proc. Nat'l Acad. Sci. U. S. A.* 97:5690–5692.

SKOVRONSKY, D. M. & LEE, V. M. 2000. β-secretase revealed: starting gate for race to novel therapies for Alzheimer's disease. *Trends Pharmacol. Sci.* 21:161–163.

SOTO, C. 1999. Plaque busters: strategies to inhibit amyloid formation in Alzheimer's disease. *Mol. Med. Today* 5:343–350.

WADSWORTH, J. D. F., ET AL. 1999. Molecular biology of prion propagation. *Curr. Opin. Gen. Develop.* 9:338–345.

General Topics

ALBERTS, B. 1998. The cell as a collection of protein machines: Preparing the next generation of molecular biologists. *Cell* 92:291–294.

ANFINSEN, C. B. 1972. Principles that govern the folding of protein chains. *Science* 181:223–230.

BALDWIN, R. L. & ROSE, G. D. 1999. Is protein folding hierarchic? *Trends Biochem. Sci.* 24:26–33.

DANSON, M. J. & HOUGH, D. W. 1998. Structure, function, and stability of enzymes from the Archaea. *Trends Microbiol.* 6:307–314.

DE CHADAREVIAN, S. 1999. Protein sequencing and the making of molecular genetics. *Trends Biochem. Sci.* 24:203–206.

DILL, K. A. 1999. Polymer principles and protein folding. *Prot. Sci.* 8:1166–1180.

DILL, K. A. & CHAN, H. S. 1997. From Levinthal to pathways to funnels. *Nature Struct. Biol.* 4:10–19.

DOBSON, C. M. & PTITSYN, O. B., eds. 1999. Folding and Binding. *Curr. Opin. Struct. Biol.* 9:89–134.

ELLIS, R. J. 1999. Molecular chaperones: Pathways and networks. *Curr. Biol.* R137–139.

FERRY, G. 2000. *Dorothy Hodgkin: A Life.* Cold Spring Harbor.

FINK, A. L. 1999. Chaperone-mediated protein folding. *Physiol. Revs.* 79:425–449.

GERSTEIN, M. & CHOTHIA, C. 1999, Proteins in motion. *Science* 285:1682–1683.

KOSHLAND, D. E., Jr. 1998. Conformational changes: How small is big enough? *Nat. Med.* 4:1112–1114.

KREIL, G. 1997. D-amino acids in animal peptides. *Annu. Rev. Biochem.* 66:337–345.

MARTIN, T. W. & Derewenda, Z. S. 1999. The name is bond—H bond. *Nat. Struct. Biol.* 6:403–406.

NEEDLEMAN, P. & ISAKSON, P. C. 1999. Selective inhibition of cyclooxygenase 2. *Sci. & Med.* 5:26–35. (Jan./Feb.)

NOMURA, M. 1987. The role of RNA and protein in ribosome function: A review of early reconstitution studies and prospects for future studies. *Cold Spring Harbor Symp.* 52:653–663.

PENNISI, E. 1999. The race to the ribosome structure. *Science* 285:2048–2051.

PETSKO, G. A. 2000. Design by necessity. *Nature* 403:606–607.

POND, C. 1998. *The Fats of Life.* Cambridge.

RADFORD, S. E. & DOBSON, C. M. 1999. From computer simulations to human disease: Emerging themes in protein folding. *Cell* 97:291–298.

SHORTLE, D. 2000. Prediction of protein structure. *Curr. Biol.* 10:R49–R51.

VANE, J. R., ET AL. 1998. Cyclooxygenases 1 and 2. *Annu. Rev. Pharmacol. Toxicol.* 38:97–120.

WLODAWER, A. & VONDRASEK, J. 1998. Inhibitors of HIV-1 protease: a major success of structure-assisted drug design. *Annu. Rev. Biophys. Biomol. Str.* 27:249–284.

YEH, S.-R. & ROUSSEAU, D. L. 2000. Hierarchical folding of cytochrome c. *Nature Struct. Biol.* 7:443–445.

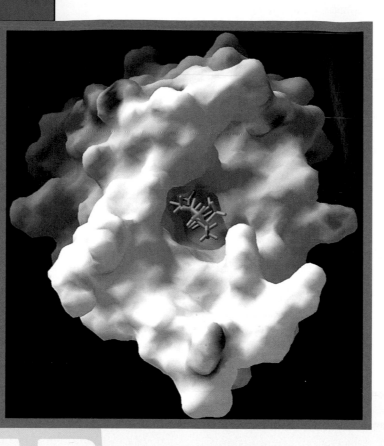

Bioenergetics, Enzymes, and Metabolism

he interrelationship between structure and function is evident at all levels of biological organization from the molecular to the organismal. We saw in the last chapter that proteins have an intricate three-dimensional structure that depends on particular amino acid residues being present in precisely the correct place. In this chapter, we will look more closely at one large group of proteins, the enzymes, and see how their complex architecture endows them with the capability to vastly increase the rate of biological reactions. To understand how enzymes are able to accomplish such feats, it is necessary to consider the flow of energy during a chemical reaction, which brings us to the subject of thermodynamics. A brief survey of the principles of thermodynamics also helps explain many of the cellular processes that will be discussed in this and following chapters, including the movement of ions across membranes, the synthesis of macromolecules, and the assembly of cytoskeletal networks. As we will see, the thermodynamic analysis of a particular system can reveal whether or not the events can occur spontaneously and, if not, provide a measure of the energy a cell must expend for the process to be accomplished. In the final section of this chapter, we will see how individual chemical reactions are linked together to

A model showing the surface of the enzyme Δ^5-3-ketosteroid isomerase with a substrate molecule (green) in the active site. The electrostatic character of the surface is indicated by color (red, acidic; blue, basic). (REPRINTED WITH PERMISSION FROM ZHENG RONG WU, ET AL., SCIENCE 276:417, 1997, COURTESY OF MICHAEL F. SUMMERS, UNIVERSITY OF MARYLAND, BALTIMORE COUNTY; COPYRIGHT 1997 AMERICAN ASSOCIATION FOR THE ADVANCEMENT OF SCIENCE.)

...rm metabolic pathways and how the flow of energy and ...aw materials through certain pathways can be controlled.

BIOENERGETICS

A living cell bustles with activity. Macromolecules of all types are assembled from raw materials, waste products are produced and excreted, genetic instructions flow from the nucleus to the cytoplasm, vesicles are moved from the Golgi complex to the plasma membrane, ions are pumped across cell membranes, and so forth. To maintain such a high level of activity, a cell must acquire and expend energy. The study of the various types of energy transformations that occur in living organisms is referred to as **bioenergetics**.

Energy is defined as the capacity to do work, that is, the capacity to change or move something. Energy exists in two alternate states: potential and kinetic. A rock perched on the edge of a cliff possesses **potential energy** because it has the potential to perform work. It has this potential because it exists in a field of force, in this case, a gravitational field. If the rock is pushed over the edge, then gravity can act on it and cause it to fall. While falling, the rock has **kinetic energy** and can accomplish work, for example, by lifting another object, as shown in Figure 3.1. Similarly, a resting nerve cell has potential energy when it maintains a high concentration of sodium ions on the outer side of its plasma membrane and a low concentration of sodium ions on the inner side. Like the flow of water through a dam, the opening of specific channels in the plasma membrane allows sodium ions to flow across the membrane into the cell. The directed movement of sodium ions into the cell is a form of kinetic energy that can be used to accomplish work, as occurs when a nerve impulse passes along the membrane of a nerve cell.

In both of the cases just described—the falling rock and the directed movement of sodium ions—as in all cases of measurement of energy at work, two factors must be considered: a potential factor and a capacity factor (Table 3.1). The potential factor is proportional to the intensity of the field of force, whereas the capacity factor provides some measure of the "size" of the subject being considered. In the case of the falling rock, the *potential factor* is the distance it will fall, and the *capacity factor* is the rock's mass. For the movement of charged ions, the potential factor is the voltage, and the capacity factor is the combined charge of the particles. The work, or energy released during these events, is a multiple of these two factors; as either one increases, so too does the amount of energy.

The Laws of Thermodynamics and the Concept of Entropy

Thermodynamics is the study of the changes in energy that accompany events in the universe. In the following pages, we will focus on a set of concepts that allow us to predict the direction that events will take and whether or not an input of energy is required to cause the event to happen. However, thermodynamic measurements provide no help in determining how rapidly a specific process will occur or the specific mechanism used by the cell to carry out the process.

The First Law of Thermodynamics **The first law of thermodynamics** is the law of conservation of energy. It states that energy can neither be created nor destroyed. Energy can, however, be converted (**transduced**) from one form to another. Electric energy is converted to mechanical energy when we plug in a clock, and chemical energy is converted to thermal energy when fuel is burned in an oil heater. Cells are also capable of energy transduction. As discussed in later chapters, the chemical energy present in

Rock has potential energy Falling rock has kinetic energy Rock has accomplished work

Figure 3.1 Accomplishing work. The rock perched on top of the cliff has potential energy. The energy available for work is proportional to the mass of the rock and the distance it can fall.

Once it leaves the edge of the cliff, the rock has kinetic energy that can be used to accomplish work, in this case by lifting both water and fish.

Table 3.1	Potential and Capacity Factors in Measuring Energy	
Type of Energy	**Potential Factor**	**Capacity Factor**
Mechanical		
falling rock	height	mass
compression	pressure	volume
stretching	tension	length
Electrical		
movement of ions	electric potential (voltage)	charge
Osmotic		
movement of water into a cell	solute concentration gradient	mass
Energy unavailable to do work	temperature	entropy

certain biological molecules, such as ATP, is converted to mechanical energy when organelles are moved from place to place in a cell, to thermal energy when heat is released during muscle cell contraction, or to electrical energy when ions flow across a membrane. The most important energy transduction in the biological world is the conversion of sunlight into chemical energy—the process of photosynthesis—which provides the fuel that directly or indirectly powers the activities of nearly all forms of life.[1]

[1]Several communities of organisms are known that are independent of photosynthesis. These include communitites that reside in the hydrothermal vents at the bottom of the ocean floor that depend on energy obtained by bacterial chemosynthesis.

A number of animals, including fireflies and luminous fish, are able to convert chemical energy back into light. Regardless of the transduction process, however, the total amount of energy in the universe remains constant.

To discuss energy transformations involving matter, we need to divide the universe into two parts: the *system* under study and the remainder of the universe, which we will refer to as the *surroundings*. A system can be defined in various ways: it may be a certain space in the universe or a certain amount of matter. For example, the system may be a living cell. The changes in a system's energy that occur during an event are manifested in two ways—as a change in the heat content of a system and in the performance of work. Even though the system may gain or lose energy, the first law of thermodynamics indicates that the loss or gain must be balanced by a corresponding gain or loss in the surroundings, so that the amount in the universe as a whole remains constant. The energy of the system is termed the *internal energy (E)*, and its change during a transformation is ΔE. One way to describe the first law of thermodynamics is that $\Delta E = Q - W$, where Q is the heat energy and W is the work energy.

Depending on the process, the internal energy of the system at the end can be greater than, equal to, or less than its internal energy at the start, depending on its relationship to its surroundings (Figure 3.2). In other words, ΔE can be positive, zero, or negative. Consider a system to be the contents of a reaction vessel. As long as there is no change in pressure or volume of the contents, there is no work being done by the system on its surroundings, or vice versa. In that case, the internal energy at the end of the

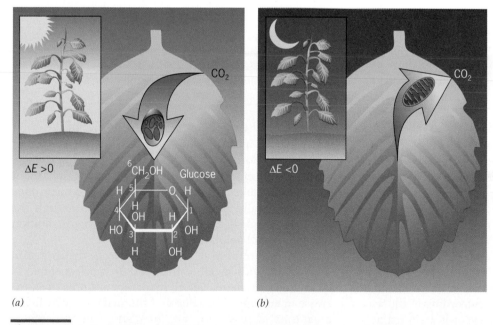

(a) *(b)*

Figure 3.2 A change in a system's internal energy. In this example, the system will be defined as a particular leaf of a plant. (*a*) During the day, sunlight is absorbed by photosynthetic pigments in the leaf's chloroplasts and used to convert CO_2 into carbohydrates, such as the glucose molecule shown in the drawing (which is subsequently incorporated into sucrose or starch).

As the cell continues to absorb light, its internal energy increases; the energy present in the remainder of the universe has to decrease. (*b*) At night, the energy relationship between the cell and its surroundings is reversed as the carbohydrates produced during the day are oxidized to CO_2 in the mitochondria and the energy is used to run the cell's nocturnal activities.

...sformation will be greater than that at the beginning if ...at is absorbed and less if heat is released. Reactions that ...ose heat are termed **exothermic**, and ones that gain heat are **endothermic**, and there are many reactions of both types. Because ΔE for a particular reaction can be positive or negative, it gives us no information as to the likelihood that a given event will occur. To determine the likelihood of a particular transformation, we need to consider some additional concepts.

The Second Law of Thermodynamics

The **second law of thermodynamics** expresses the concept that events in the universe have direction; they tend to proceed "downhill" from a state of higher energy to a state of lower energy. Thus, in any energy transformation, there is a decreasing availability of energy for doing additional work. Rocks fall off cliffs to the ground below, and once at the bottom, their ability to do additional work is reduced; it is very unlikely that they will lift themselves back to the top of the cliff. Similarly, opposite charges normally move together, not apart, and heat flows from a warmer to a cooler body, not the reverse. Such events are said to be spontaneous, a term that indicates they are thermodynamically favorable and can occur *without the input of external energy*.

The concept of the second law of thermodynamics was formulated originally for heat engines, and the law carried with it the idea that it is thermodynamically impossible to construct a perpetual-motion machine. In other words, it is impossible for a machine to be 100 percent efficient, which would be required if it were to continue functioning without the input of external energy. Some of the energy is inevitably lost as the machine carries out its activity. A similar relationship holds true for living organisms. For example, when a giraffe browses on the leaves of a tree or a lion preys on the giraffe, most of the chemical energy in the food never becomes available to the animal having the meal. The energy that is unavailable for doing additional work after an event has occurred has a potential factor and a capacity factor, as do other energy terms (see Table 3.1). The potential factor is temperature (in degrees) and the capacity factor is **entropy (S)**, which has the dimensions of energy per degree (e.g., calories per degree). The unavailable energy is equal to $T\Delta S$, where ΔS is the change in entropy between the initial and final states.

The loss of available energy during a process is a result of a tendency for the randomness, or disorder, of the universe to increase every time there is a transfer of energy. Entropy provides a measure of this disorder; it is also associated with the *random* movements of the particles of matter, which, because they are random, cannot be made to accomplish a *directed* work process. According to the second law of thermodynamics, every event is accompanied by an increase in the entropy of the universe. When a sugar cube is dropped into a cup of hot water, for example, there is a spontaneous shift of the molecules from an ordered state in the crystal to a much more disordered condition when the sugar molecules are spread throughout

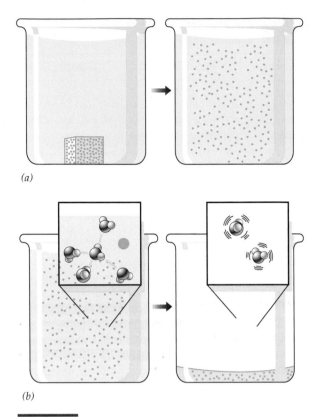

(a)

(b)

Figure 3.3 Events are accompanied by an increase in the entropy of the universe. (*a*) A sugar cube contains sucrose molecules in a highly ordered arrangement in which the freedom of movement of individual molecules is restricted. As the cube dissolves, the freedom of movement of the sucrose molecules is greatly increased, and their random movement causes them to become equally distributed throughout the available space. Once this occurs, there will be no further tendency for redistribution, and the entropy of the system is at a maximum. (*b*) Sugar molecules spread randomly through a solution can be returned to an ordered state, but only if the entropy of the surroundings is increased, as occurs when the more ordered water molecules of the liquid phase become disordered following evaporation.

the solution (Figure 3.3*a*). As the molecules of the sugar cube dissolve into solution, their freedom of movement increases, as does the entropy of the system. The change from the concentrated to the dispersed state results from the random movements of the molecules. The sugar molecules eventually spread themselves equally through the available volume because the state of uniform distribution is the most probable state.

The release of heat, for example, from the oxidation of glucose within a cell or from the friction generated as blood flows through a vessel, is another example of an increase in entropy. The release of thermal energy by living organisms increases the rate of random movements of atoms and molecules; it cannot be redirected to accomplish additional work. Since the energy of molecular and atomic movements increases with temperature, so too does the entropy. It is only at absolute zero (0° K), when all movements cease, that the entropy is zero.

(Removing the stray lines above.)

As with other spontaneous events, we must distinguish between the system and its surroundings. The second law of thermodynamics indicates only that the total entropy in the universe must increase; the disorder within one part of the universe (the system) can decrease at the greater expense of its surroundings. The dissolved sugar can decrease in entropy; it can be recrystallized by evaporating off the water (Figure 3.3b). The consequence of this process, however, is an increase in the entropy of the surroundings. The increased freedom of movement of the water molecules in the gaseous phase more than balances the decrease in freedom of the molecules of the sugar crystals.

Life operates on a similar principle. Living organisms are able to decrease their own entropy by increasing the entropy of their environment. Entropy is decreased in an organism when relatively simple molecules, such as amino acids, are ordered into more complex molecules, such as the protein myoglobin in a muscle cell. For this to occur, however, the entropy of the environment must increase, which is accomplished as complex, ordered molecules such as glycogen stored in liver or muscle tissue are converted into heat and smaller, less ordered compounds (such as CO_2 and H_2O) are released to the environment. It is this feature of metabolism that allows living organisms to maintain such a highly ordered and improbable state.

Another measure of the energy state of a living organism is provided by the information content of its macromolecules. Information is a subject that is difficult to define but easy to recognize. Information can be measured in terms of the ordered arrangement of a structure's subunits. For example, proteins and nucleic acids, in which the specific linear sequence of the subunits is highly ordered, are low in entropy and high in information content. Maintaining a state of high information content (low entropy) requires the input of energy. Consider just one molecule of DNA located in one cell in your liver. That cell has dozens of different proteins whose sole job is to patrol the DNA, looking for damage and repairing it (discussed further in Section 13.2). Nucleotide damage in an active cell can be so great that, without this expenditure of energy, the information content of DNA would rapidly deteriorate.

Free Energy

Together, the first and second laws of thermodynamics indicate that the energy of the universe is constant, but the entropy continues to increase toward a maximum. The concepts inherent in the first two laws were put together by the American chemist, J. Willard Gibbs, in 1878 into the expression $\Delta H = \Delta G + T\Delta S$, where ΔG is the change in **free energy**, that is, the change during a process in the energy available to do work; ΔH is the change in *enthalpy* or total energy content of the system (equivalent to ΔE for our purposes), T is the absolute temperature ($°K = °C + 273$), and ΔS is the change in the entropy of the system.

The equation states that the total energy change is equal to the sum of the changes in useful energy (ΔG) and energy that is unavailable to do further work ($T\Delta S$).

When rearranged and written as $\Delta G = \Delta H - T\Delta S$, the above equation provides a measure of the spontaneity of a particular process. It allows one to predict the direction in which a process will proceed and the extent to which that process will occur. All *spontaneous* energy transformations must have a negative ΔG; that is, the process must proceed toward a state of lower free energy. The magnitude of ΔG indicates the maximum amount of energy that can be passed on for use in another process. Processes that can occur spontaneously, that is, processes that are thermodynamically favored (have a $-\Delta G$), are described as **exergonic**. In contrast, if the ΔG for a given process is positive, then it cannot occur spontaneously. Such processes are thermodynamically unfavorable and are described as **endergonic**. As we will see, reactions that are normally endergonic can be made to occur by coupling them to energy-releasing processes.

The signs of ΔH and ΔS can be positive or negative, depending on the relation between the system and its surroundings. (Hence, ΔH will be positive if heat is gained by the system and negative if heat is lost; ΔS will be positive if the system becomes more disordered and negative if it becomes more ordered.) The counterplay between ΔH and ΔS is illustrated by the ice–water transformation. The conversion of water from the liquid to the solid state is accompanied by a decrease in entropy (ΔS is negative, as illustrated in Figure 3.4) and a decrease in enthalpy (ΔH is negative). In order for this transformation to occur (i.e., for ΔG to be negative), ΔH must be more negative than

Figure 3.4 When water freezes, its entropy decreases because the water molecules of ice exist in a more ordered state with less freedom of movement than in a liquid state. The decrease in entropy is particularly apparent in the formation of a snowflake.
(© NURIDSANY AND PERENNOU/PHOTO RESEARCHERS.)

| Table 3.2 | Thermodynamics of the Ice–Water Transformation | | | | |

Temperature (°C)	ΔE (cal/mol)	ΔH (cal/mol)	ΔS (cal/mol · °C)	$T\Delta S$ (cal/mol)	ΔG (cal/mol)
−10	−1,343	−1,343	−4.9	−1,292	−51
0	−1,436	−1,436	−5.2	−1,436	0
+10	−1,529	−1,529	−5.6	−1,583	+54

Source: I. M. Klotz, *Energy in Biochemical Reactions*, Academic Press, 1967.

$T\Delta S$, a condition that occurs only below 0°C. This relationship can be seen from Table 3.2, which indicates the values for the different terms if one mole of water were to be converted to ice at 10°C, 0°C, or −10°C. In all cases, regardless of the temperature, the energy level of the ice is less than that of the liquid (the ΔH is negative). However, at the higher temperature, the entropy term of the equation ($T\Delta S$) is more negative than the enthalpy term, and therefore, the free-energy change is positive and the process cannot occur spontaneously. At 0°C, the system is in equilibrium; at −10°C, the solidification process is favored, that is, the ΔG is negative.

Free-Energy Changes in Chemical Reactions Now that we have discussed the concept of free energy in general terms, we can apply the information to chemical reactions within the cell. All chemical reactions within the cell are reversible, and therefore, we must consider two reactions occurring simultaneously, one forward and the other in reverse. According to the law of mass action, the rate of a reaction is proportional to the concentration of the reactants. Consider, for example, the following hypothetical reaction:

$$A + B \rightleftharpoons C + D$$

The rate of the forward reaction is directly proportional to the product of the molar concentrations of A and B. The rate of the forward reaction can be expressed as $k_1[A][B]$, where k_1 is a rate constant for the forward reaction. The rate of the backward reaction equals $k_2[C][D]$. All chemical reactions proceed, however slowly, toward a state of equilibrium, that is, toward a point at which the rates of the forward and backward reactions are equal. At equilibrium, the same number of A and B molecules are converted into C and D molecules per unit time as are formed from them. At equilibrium, therefore,

$$k_1[A][B] = k_2[C][D]$$

which can be rearranged to

$$\frac{k_1}{k_2} = \frac{[C][D]}{[A][B]}$$

In other words, at equilibrium there is a predictable ratio of the concentration of products to the concentration of reactants. This ratio, which is equal to k_1/k_2, is termed the **equilibrium constant, K_{eq}.**

The equilibrium constant allows one to predict the direction (forward or reverse) in which the reaction is favored under a given set of conditions. Suppose, for example, that we are studying the above reaction, and we have just mixed the four components (A, B, C, D) so that each is present at an initial concentration of 0.5 M.

$$\frac{[C][D]}{[A][B]} = \frac{[0.5][0.5]}{[0.5][0.5]} = 1$$

The direction in which this reaction will proceed depends on the equilibrium constant. If the K_{eq} is greater than one, the reaction will proceed at a greater rate toward the formation of the products C and D than in the reverse direction. For example, if the K_{eq} happens to be 9.0, then the concentration of the reactants and products at equilibrium in this particular reaction mixture will be 0.25 M and 0.75 M, respectively.

$$\frac{[C][D]}{[A][B]} = \frac{[0.75][0.75]}{[0.25][0.25]} = 9$$

If, on the other hand, the K_{eq} is less than 1, the reverse reaction will proceed at a greater rate than the forward reaction, so that the concentration of A and B will rise at the expense of C and D. It follows from these points that the net direction in which the reaction is proceeding at any moment depends on the relative concentrations of all participating molecules and can be predicted if the K_{eq} is known.

Let us return to the subject of energetics. The ratio of reactants to products present at equilibrium is determined by the relative free-energy levels of the reactants and products. As long as the total free energy of the reactants is greater than the total free energy of the products, then ΔG has a negative value and the reaction proceeds in the direction of formation of products. The greater the ΔG, the farther the reaction is from equilibrium and the more work that can be performed by the system. As the reaction proceeds, the difference in free-energy content between the reactants and products decreases (ΔG becomes less negative), until at equilibrium the difference is zero ($\Delta G = 0$) and no further work can be obtained.

Because ΔG for a given reaction depends on the reaction mixture present at a given time, it is not a useful term to compare the energetics of various reactions. To place reactions on a comparable basis and to allow various types of calculations to be made, a convention has been adopted

to consider the free-energy change that occurs during a reaction under a set of **standard conditions**. For biochemical reactions, the conditions are arbitrarily set at 25°C (298°K) and 1 atm of pressure, with all the reactants and products present at 1.0 M concentration, except for water, which is present at 55.6 M, and H^+ at 10^{-7} M (pH 7.0).[2] The **standard free-energy change ($\Delta G^{\circ\prime}$)** describes the free energy released when reactants are converted to products under these standard conditions. It must be kept in mind that standard conditions do not prevail in the cell, and therefore, one must be cautious in the use of values for standard free-energy differences in calculations of cellular energetics.

The relationship between the equilibrium constant and standard free-energy change is given by the equation

$$\Delta G^{\circ\prime} = -RT\ln K'_{eq}$$

When the natural log (ln) is converted to \log_{10}, the equation becomes

$$\Delta G^{\circ\prime} = -2.303\,RT\log K'_{eq}$$

where R is the gas constant (1.987 cal/mol · °K) and T is the absolute temperature (298°K).[3] Recall that the log of 1.0 is zero. Consequently, it follows from the above equation that reactions having equilibrium constants greater than 1.0 have negative $\Delta G^{\circ\prime}$ values, which indicates that they can occur spontaneously under *standard conditions*. Reactions having equilibrium constants of less than 1 have positive $\Delta G^{\circ\prime}$ values and cannot occur spontaneously under standard conditions. In other words, given the reac-

Table 3.3	Relation Between $\Delta G^{\circ\prime}$ and K'_{eq} at 25°C
K'_{eq}	$\Delta G^{\circ\prime}$ (kcal/mol)
10^6	−8.2
10^4	−5.5
10^2	−2.7
10^1	−1.4
10^0	0.0
10^{-1}	1.4
10^{-2}	2.7
10^{-4}	5.5
10^{-6}	8.2

tion written as follows: $A + B \rightleftharpoons C + D$, if the $\Delta G^{\circ\prime}$ is negative, the reaction will go to the right when reactants and products are all present at 1.0 M concentration at pH 7. The greater the negative value, the farther to the right the reaction will proceed before equilibrium is reached. Under the same conditions, if the $\Delta G^{\circ\prime}$ is positive, the reaction will proceed to the left; that is, the reverse reaction is favored. The relation between $\Delta G^{\circ\prime}$ and K'_{eq} is shown in Table 3.3.

Free-Energy Changes in Metabolic Reactions One of the most important chemical reactions in the cell is the hydrolysis of ATP (Figure 3.5). In the reaction,

$$ATP + H_2O \rightarrow ADP + P_i$$

the standard free-energy difference between the products and reactants is −7.3 kcal/mol. Based on this information, it is evident that under standard conditions the hydrolysis of ATP is a highly favorable (exergonic) reaction,

[2] $\Delta G^{\circ\prime}$ indicates standard conditions include pH 7, whereas ΔG° indicates standard conditions in 1 M H^+ (pH 0.0). The designation K'_{eq} also indicates a reaction mixture at pH 7.

[3] The right-hand side of this equation is equivalent to the amount of free energy lost as the reaction proceeds from standard conditions to equilibrium.

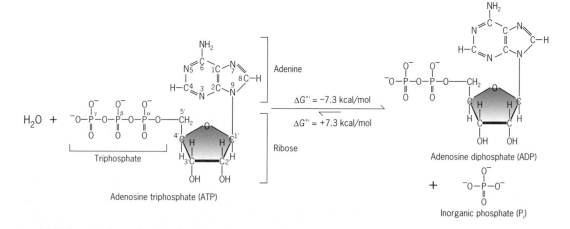

Figure 3.5 ATP hydrolysis. Adenosine triphosphate (ATP) is hydrolyzed as part of many biochemical processes. In most reactions, as shown here, ATP is hydrolyzed to ADP and inorganic phosphate (P_i), but in some cases (not shown) it is hydrolyzed to AMP, a compound with only one phosphate group, and pyrophosphate (PP_i). Both of these reactions have essentially the same $\Delta G^{\circ\prime}$ of −7.3 kcal/mol (30.5 kJ/mol).

that is, one that tends toward a large [ADP]/[ATP] ratio at equilibrium. There are several reasons why this reaction is so favorable, one of which is evident in Figure 3.5. The electrostatic repulsion created by four closely spaced negative charges on ATP^{4-} is partially relieved by formation of ADP^{3-}.

It is important that the difference between ΔG and $\Delta G^{\circ\prime}$ be kept clearly in mind. The $\Delta G^{\circ\prime}$ is a fixed value for a given reaction and indicates the direction in which that reaction would proceed were the system at standard conditions. Since standard conditions do not prevail within a cell, $\Delta G^{\circ\prime}$ values cannot be used to predict the direction in which a particular reaction is proceeding at a given moment within a particular cellular compartment. To do this, one must know the ΔG, which is determined by the concentrations of the reactants and products that are present at the time.

$$\Delta G = \Delta G^{\circ\prime} + 2.303RT\log\frac{[C][D]}{[A][B]}$$

$$\Delta G = \Delta G^{\circ\prime} + 2.303\,(1.987\text{ cal/mol}\cdot{}^{\circ}\text{K})\,(298{}^{\circ}\text{K})$$
$$\log\frac{[C][D]}{[A][B]}$$

$$\Delta G = \Delta G^{\circ\prime} + (1.4\text{ kcal/mol})\log\frac{[C][D]}{[A][B]}$$

where [A], [B], [C], and [D] are the actual concentrations at the time. Calculation of the ΔG reveals the direction in which the reaction in the cell is proceeding and how close the particular reaction in question is to equilibrium. For example, typical cellular concentrations of the reactants and products in the reaction for ATP hydrolysis might be [ATP] = 10 mM; [ADP] = 1 mM; [P_i] = 10 mM. Substituting these values into the equation,

$$\Delta G = \Delta G^{\circ\prime} + 2.303RT\log\frac{[ADP][P_i]}{[ATP]}$$

$$\Delta G = -7.3\text{ kcal/mol} + (1.4\text{ kcal/mol})\log\frac{[10^{-3}][10^{-2}]}{[10^{-2}]}$$

$$\Delta G = -7.3\text{ kcal/mol} + (1.4\text{ kcal/mol})(-3)$$

$$\Delta G = -11.5\text{ kcal/mol}$$

Thus, even though the $\Delta G^{\circ\prime}$ for the hydrolysis of ATP is -7.3 kcal/mol, the typical ΔG in the cell for this reaction is about -12 kcal/mol because a cell maintains a high [ATP]/[ADP] ratio.

Cells carry out many reactions with positive $\Delta G^{\circ\prime}$ values because the relative concentrations of reactants and products favor the progress of the reactions. This can happen in two ways. The first illustrates the important difference between ΔG and $\Delta G^{\circ\prime}$, and the second reveals the manner in which reactions with positive $\Delta G^{\circ\prime}$ values can be driven in the cell by the input of stored chemical energy.

Consider the reaction of glycolysis (see Figure 3.23) in which dihydroxyacetone phosphate is converted to glyceraldehyde 3-phosphate. The $\Delta G^{\circ\prime}$ for this reaction is

$+1.8$ kcal/mol; yet the formation of the product of this reaction takes place in the cell. The reaction proceeds because other cellular reactions maintain the ratio of the reactant to product above that defined by the equilibrium constant. As long as this condition holds, the ΔG will be negative and the reaction will continue spontaneously in the direction of formation of glyceraldehyde 3-phosphate. This brings up an important characteristic of cellular metabolism, namely, that specific reactions cannot be considered independently as if they were occurring in isolation in a test tube. Hundreds of reactions occur simultaneously in a cell. All these reactions are interrelated because a product of one reaction becomes a substrate for the next reaction in the sequence and so on throughout one metabolic pathway and into the next. To maintain the production of glyceraldehyde 3-phosphate at the expense of dihydroxyacetone phosphate, the reaction is situated within a metabolic pathway in such a way that the product is removed by the next reaction in the sequence at a sufficiently rapid rate that a favorable ratio of the concentrations of these two molecules is retained.

Coupling Endergonic and Exergonic Reactions Reactions with large positive $\Delta G^{\circ\prime}$ values are typically driven by the input of energy. Consider the formation of the amino acid glutamine from glutamic acid by the enzyme glutamine synthetase:

Glutamic acid + $NH_3 \rightarrow$
$$\text{glutamine} \qquad \Delta G^{\circ\prime} = +3.4\text{ kcal/mol}$$

This endergonic reaction takes place in the cell because glutamic acid is actually converted to glutamine in two sequential reactions, both of which are exergonic:

1st reaction: Glutamic acid + ATP \rightarrow
$$\text{glutamyl phosphate} + \text{ADP}$$
2nd reaction: Glutamyl phosphate + $NH_3 \rightarrow$
$$\text{glutamine} + P_i$$

Overall reaction: Glutamic acid + ATP + $NH_3 \rightarrow$
$$\text{glutamine} + \text{ADP} + P_i$$
$$\Delta G^{\circ\prime} = -3.9\text{ kcal/mol}$$

The formation of glutamine is said to be **coupled** to the hydrolysis of ATP. As long as the ΔG for ATP hydrolysis is more negative than the ΔG for glutamine synthesis from glutamic acid is positive, the "downhill" ATP hydrolysis reaction can be used to drive the "uphill" synthesis of glutamine. To couple the two chemical reactions, the product of the first reaction becomes the substrate for the second. The bridge between the two reactions—glutamyl phosphate in this case—is called the **common intermediate**. What happens, in essence, is that the exergonic hydrolysis of ATP is taking place in two steps. In the first step, glutamic acid acts as an acceptor of the phosphate group, which is displaced by NH_3 in the second step.

The hydrolysis of ATP can be used in the cell to drive reactions leading to the formation of molecules such as

glutamine because ATP levels are maintained at much higher levels (relative to those of ADP) than they would be at equilibrium. This can be demonstrated by the following calculation. As noted above, a typical cellular concentration of P_i might be 10 mM. To calculate the equilibrium ratio of [ATP]/[ADP] under these conditions, we can set ΔG at 0 and solve the following equation (taken from page 90) for [ADP]/[ATP]:

$$\Delta G = \Delta G°' + (1.4\,\text{kcal/mol}) \log \frac{[\text{ADP}][\text{P}_i]}{[\text{ATP}]}$$

$$0 = -7.3\,\text{kcal/mol} + (1.4\,\text{kcal/mol}) \log \frac{[\text{ADP}][10^{-2}]}{[\text{ATP}]}$$

$$0 = -7.3\,\text{kcal/mol} + (1.4\,\text{kcal/mol}) \left(\log\,10^{-2} + \log \frac{[\text{ADP}]}{[\text{ATP}]} \right)$$

$$+7.3\,\text{kcal/mol} = (1.4\,\text{kcal/mol})(-2) + (1.4\,\text{kcal/mol}) \log \frac{[\text{ADP}]}{[\text{ATP}]}$$

$$\log \frac{[\text{ADP}]}{[\text{ATP}]} = \frac{10.1\,\text{kcal/mol}}{1.4\,\text{kcal/mol}} = 7.2$$

$$\frac{[\text{ADP}]}{[\text{ATP}]} = 1.6 \times 10^7$$

Thus, at equilibrium, the ADP concentration would be expected to be more than 10^7 times that of ATP, but in fact, the ATP concentrations in most cells are 10 to 100 times that of ADP. This is a crucial point, since it's the relative concentrations of ATP and ADP that matter. If a cell were to contain an equilibrium mixture of ATP, ADP, and P_i, it wouldn't matter how much ATP was present, the cell would have no capacity to perform work.

ATP hydrolysis is used to drive most endergonic processes within the cell, including chemical reactions such as the one just described, the separation of charge across a membrane, the concentration of a solute, the movement of filaments in a muscle cell, and the generation of heat (Figure 3.6). ATP can be used for such diverse processes because its terminal phosphate group can be transferred to a variety of different types of molecules, including amino acids, sugars, lipids, and proteins. In most coupled reactions, the phosphate group is transferred in an initial step from ATP to one of these acceptors and is subsequently removed in a second step (see Figure 4.44 as an example).

Equilibrium Versus Steady-State Metabolism

As reactions tend toward equilibrium, the free energy available to do work decreases toward a minimum and the entropy increases toward a maximum. Thus, the farther a reaction is kept from its equilibrium state, the less its capacity to do work is lost to the increase in entropy. Cellular metabolism is essentially nonequilibrium metabolism; that is, it is characterized by nonequilibrium ratios of products to reactants. This does not mean that some reactions do not occur at or near equilibrium within cells. In fact, many of the reactions of a metabolic pathway may be near equilibrium (see Figure 3.24). However, at least one and often several reactions in a pathway are poised far from equilib-

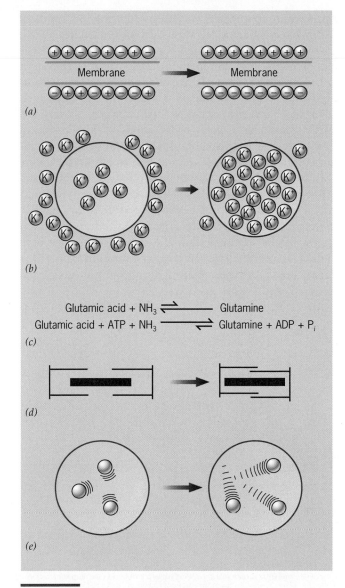

Figure 3.6 A few roles for ATP hydrolysis. In the cell, ATP may be used to (a) separate charge across a membrane; (b) concentrate a particular solute within the cell; (c) drive an otherwise unfavorable chemical reaction; (d) slide filaments across one another, as occurs during the shortening of a muscle cell; (e) increase the temperature of the cell, which results from an increase in the rate of molecular movements.

rium, making them essentially irreversible. These are the reactions that keep the pathway going in a single direction. These are also the reactions that are subject to cellular regulation, because the flow of materials through the entire pathway can be greatly increased or decreased by stimulating or inhibiting the activity of the enzymes that catalyze these reactions.

The basic principles of thermodynamics were formulated using nonliving, *closed* systems (no exchange of matter between the system and its surroundings) under reversible equilibrium conditions. The unique features of cellular metabolism require a different perspective. Cellular metabolism can maintain itself at irreversible, non-

equilibrium conditions, because unlike the environment within a test tube, the cell is an open system. Materials and energy are continually flowing into the cell from the bloodstream or culture medium. The extent of this input into cells from the outside becomes apparent if you simply hold your breath. We depend from minute to minute on an external source of oxygen because oxygen is a very important reactant in cellular metabolism. The continual flow of oxygen and other materials into and out of cells allows cellular metabolism to exist in a **steady state** (Figure 3.7). In a steady state, the concentrations of reactants and products remain relatively constant, even though the individual reactions are not necessarily at equilibrium. Because products from one reaction are acted upon as substrates of the next reaction, the concentrations of each metabolic intermediate can remain essentially constant as long as new substrates are brought in from the outside and terminal products are removed.

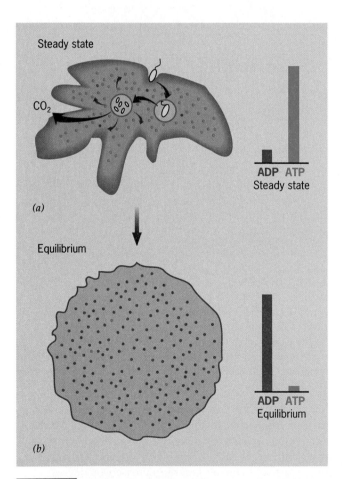

(a)

(b)

Figure 3.7 Steady state versus equilibrium. (a) As long as this amoeba can continue to take in nutrients from the outside world, it can harvest the energy necessary to maintain concentrations of compounds at a steady state, which may be far from equilibrium. The steady state concentrations of ATP and ADP are indicated by the colored dots and the histogram. (b) When the amoeba dies, the concentrations of ATP and ADP (as well as other biochemicals) shift toward their equilibrium ratios.

REVIEW **?**

1. Describe the differences between the first and second laws of thermodynamics and how, when taken together, they are able to describe the direction of events that take place in the universe.

2. How is the maintenance of the ordered living state consistent with the second law of thermodynamics?

3. Describe two examples in which the entropy of a system decreases and two examples in which the entropy of a system increases.

4. Discuss the differences between ΔG and $\Delta G^{\circ\prime}$; between ΔG° and $\Delta G^{\circ\prime}$; between the relative rates of the forward and reverse reactions when ΔG is negative, zero, or positive. What is the relationship between $\Delta G^{\circ\prime}$ and K^{\prime}_{eq}? How can a cell carry out a reaction that has a $+\Delta G^{\circ\prime}$?

5. How is it possible for a cell to maintain an [ATP]/[ADP] ratio greater than one? How does this ratio differ from that expected at equilibrium?

6. How is it that ice formation does not take place at temperatures above 0°C?

3.2 **ENZYMES AS BIOLOGICAL CATALYSTS**

At the end of nineteenth century, a debate was raging as to whether or not the process of ethanol formation required the presence of intact yeast cells. On one side of the debate was the organic chemist Justus von Liebig, who argued that the reactions of fermentation that produced the alcohol were no different from the types of organic reactions they had been studying in the test tube. On the other side was the biologist Louis Pasteur, who argued that the process of fermentation could only occur within the confines of an intact, highly organized, living cell.

In 1897, two years after Pasteur's death, Hans Büchner, a bacteriologist, and his brother Eduard, a chemist, were preparing "yeast juice"—an extract made by grinding yeast cells with sand grains and then filtering the mixture through filter paper. They wanted to preserve the yeast juice for later use. After trying to preserve the extract with antiseptics and failing, they attempted to protect the preparation from spoilage by adding sugar, the same procedure used to preserve jams and jellies. Instead of preserving the solution, the yeast juice produced gas from the sugar, and it bubbled continuously for days. After further analysis, Eduard discovered that fermentation was producing ethanol and bubbles of carbon dioxide. Büchner had shown that fermentation did not require the presence of intact cells.

It was soon found, however, that fermentation was very different from the types of reactions carried out by or-

ganic chemists. Fermentation required the presence of a unique set of catalysts that had no counterpart in the non-living world. These catalysts were called **enzymes** (after the Greek for "in yeast"). Enzymes are the mediators of metabolism, responsible for virtually every reaction that occurs in a cell. Without enzymes, metabolic reactions would proceed so slowly as to be imperceptible.

The first evidence that enzymes are proteins was obtained by James Sumner in 1926 when he crystallized the enzyme urease from jack beans and determined its composition. Though this finding was not greeted with much positive acclaim at the time, several other enzymes were soon shown to be proteins, and it became accepted for the next few decades that all biological catalysts were proteins. It has recently become evident, however, that certain biological reactions are catalyzed by RNA molecules. For the sake of clarity, the term "enzyme" is still generally reserved for protein catalysts, while the term "ribozyme" is used for RNA catalysts.

Even though enzymes are proteins, many of them are conjugated proteins; that is, they contain nonprotein components, called **cofactors**, which may be inorganic (metals) or organic (**coenzymes**). When present, cofactors are important participants in the functioning of the enzyme, often carrying out activities for which amino acids are not suited. In myoglobin, for example, as discussed in the previous chapter, the iron atom of a heme group is the site where oxygen is bound and stored until required by cellular metabolism.

The Properties of Enzymes

As is true of all catalysts, enzymes exhibit the following properties: (1) they are required only in small amounts; (2) they are not altered irreversibly during the course of the reaction, and therefore, each enzyme molecule can partic-ipate repeatedly in individual reactions; and (3) they have no effect on the thermodynamics of the reaction. This last point is particularly important. Enzymes do not supply energy for a chemical reaction and, therefore, do not determine whether a reaction is thermodynamically favorable (exergonic) or unfavorable (endergonic). Similarly, enzymes do not determine what the ratio of products to reactants is at equilibrium. These are inherent properties of the reacting chemicals. As catalysts, enzymes can only accelerate the rate at which a favorable chemical reaction proceeds.

There is no necessary relationship between the magnitude of the ΔG for a particular reaction and the rate at which that reaction takes place. The magnitude of ΔG informs us only of the difference in free energy between the beginning state and equilibrium. It is totally independent of the pathway or the time it takes to reach equilibrium. The oxidation of glucose, for example, is a very favorable process, as can be determined by the amount of energy released during its combustion. However, glucose crystals can be left out in room air virtually indefinitely without a noticeable conversion into less energetic materials. Glucose, in other words, is *kinetically* stable, even if it is *thermodynamically* unstable. Even if the sugar were to be dissolved, as long as the solution were kept sterile, it would not rapidly deteriorate. However, if one were to add a few bacteria, within a short period of time the sugar would be taken up by the cells and enzymatically degraded.

Enzymes are remarkably adept catalysts. Catalysts employed by organic chemists in the lab, such as acid, metallic platinum, and magnesium, generally accelerate reactions a hundred to a thousand times over the noncatalyzed rate. In contrast, enzymes typically increase the velocity of a reaction 10^8 to 10^{12} fold (Table 3.4). On the basis of these numbers, enzymes can accomplish in one second what would require from 3 years to 300,000 years if

Table 3.4	Catalytic Activity of a Variety of Enzymes		
Enzyme	**Nonenzymatic $t_{1/2}$[1]**	**Turnover number[2]**	**Rate enhancement[3]**
OMP decarboxylase	78,000,000 yr	39	1.4×10^{17}
Staphylococcal nuclease	130,000 yr	95	5.6×10^{14}
Adenosine deaminase	120 yr	370	2.1×10^{12}
AMP nucleosidase	69,000 yr	60	6.0×10^{12}
Cytidine deaminase	69 yr	299	1.2×10^{12}
Phosphotriesterase	2.9 yr	2,100	2.8×10^{11}
Carboxypeptidase A	7.3 yr	578	1.9×10^{11}
Ketosteroid isomerase	7 wk	66,000	3.9×10^{11}
Triosephosphate isomerase	1.9 d	4,300	1.0×10^{9}
Chorismate mutase	7.4 hr	50	1.9×10^{6}
Carbonic anhydrase	5 sec	1×10^{6}	7.7×10^{6}
Cyclophilin, human	23 sec	13,000	4.6×10^{5}

Source: A. Radzicka and R. Wolfenden, *Science* 267, 91, 1995. Copyright 1995 American Association for the Advancement of Science.
[1]The time that would elapse for half the reactants to be converted to product in the absence of enzyme.
[2]The number of reactions catalyzed by a single enzyme molecule per second when operating at a saturating substrate concentration.
[3]The increase in reaction rate achieved by the enzyme-catalyzed reaction over the noncatalyzed reaction.

the enzyme were absent. Even more remarkably, they accomplish this feat at the mild temperature and pH found within a cell. In addition, unlike the inorganic catalysts used by chemists, enzymes are highly specific in the reactants they can bind and the reaction they can catalyze. The reactants bound by an enzyme are called **substrates**. If, for example, the enzyme hexokinase is present in solution with a hundred low-molecular-weight compounds in addition to its substrate, glucose, only the glucose molecules will be recognized by the enzyme and undergo reaction. For all practical purposes, the other compounds may as well be absent. This type of specificity, whether between enzymes and substrates or other types of proteins and the substances that they bind, is crucial for maintaining the order required to sustain life.

In addition to their high level of activity and specificity, enzymes act as metabolic traffic directors in the sense that enzyme-catalyzed reactions are very orderly—the only products formed are the appropriate ones. This is very important because the formation of chemical by-products would rapidly take its toll on the life of a fragile cell. Finally, unlike other catalysts, the activity of enzymes can be regulated to meet the particular needs of the cell at a particular time. As will be evident in this chapter and the remainder of the text, the enzymes of a cell are truly a wondrous collection of miniature molecular machines.

Overcoming the Activation Energy Barrier

How are enzymes able to accomplish such effective catalysis? The first question to consider is why thermodynamically favorable reactions do not proceed on their own at relatively rapid rates in the absence of enzymes. Even ATP, whose hydrolysis is so favorable, is basically stable in the cell until it's broken down in a controlled enzymatic reaction. If this weren't the case, ATP would be of little use to a cell.

Chemical transformations require that certain covalent bonds be broken within the reactants. For this to occur, the reactants must contain sufficient kinetic energy (energy of motion) that they overcome a barrier, called the **activation energy** (E_A). This is expressed diagrammatically in Figure 3.8, where the activation energy is represented by the height of the curves. The reactants in a chemical reaction are often compared to an object resting on top of a cliff ready to plunge to the bottom. If left to itself, the object in all likelihood would remain there indefinitely. However, if someone were to come along and provide the object with sufficient energy to overcome the friction or other small barrier in its way and cause it to reach the edge of the cliff, it would spontaneously drop to the bottom. The object has the potential to fall to a lower energy state once the kinetic barriers have been removed.

In a solution at room temperature, molecules exist in a state of random movement, each possessing a certain amount of kinetic energy at a given instant. Among a population of molecules, their energy is distributed in a bell-shaped curve (Figure 3.9), some having very little energy and others much more. Molecules of high energy (activated molecules) remain as such only for a brief time, losing their excess energy to other molecules by collision. Let us consider a reaction in which one reactant molecule is split into two product molecules. If a given reactant molecule acquires sufficient energy to overcome the activation

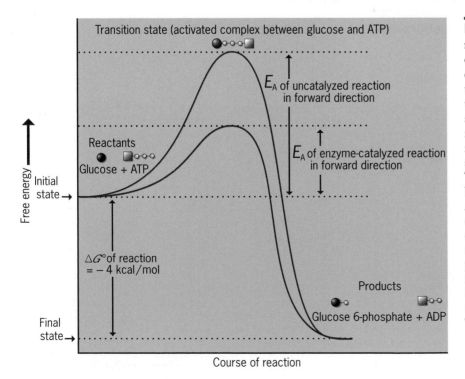

Figure 3.8 **Activation energy and enzymatic reactions.** Even though the formation of glucose 6-phosphate is a thermodynamically favored reaction ($\Delta G^{\circ\prime} = -4$ kcal/mol), the reactants must possess sufficient energy to achieve an activated state in which the atomic rearrangements necessary for the reaction can occur. The amount of energy required is called the activation energy (E_A) and is represented by the height of the curve. The activation energy is not a fixed value, but varies with the particular reaction pathway. E_A is greatly reduced when the reactants combine with an enzyme catalyst. (This diagram depicts a simple, one-step reaction mechanism. Many enzymatic reactions take place in two or more steps leading to the formation of intermediates (as in Figure 3.13). Each step in the reaction has a distinct E_A and a separate transition state.)

Figure labels:
Transition state (activated complex between glucose and ATP)
E_A of uncatalyzed reaction in forward direction
E_A of enzyme-catalyzed reaction in forward direction
Reactants
Glucose + ATP
Free energy
Initial state →
ΔG° of reaction = −4 kcal/mol
Products
Glucose 6-phosphate + ADP
Final state →
Course of reaction

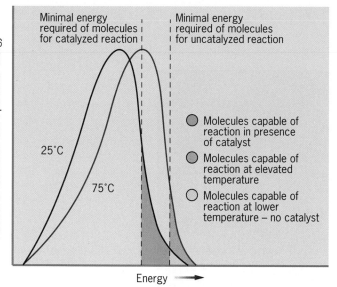

Figure 3.9 **The effect of lowering activation energy on the rate of a reaction.** The bell-shaped curves indicate the energy content of a population of molecules present in a reaction mixture at two different temperatures. The number of reactant molecules containing sufficient energy to undergo reaction is increased by either heating the mixture or adding an enzyme catalyst. Heat increases the rate of reaction by increasing the energy content of the molecules, while the enzyme does so by lowering the activation energy required for the reaction to occur.

barrier, then the possibility exists that it will split into the two product molecules. The rate of the reaction depends on the number of reactant molecules that contain the necessary kinetic energy at any given time. One way to increase the reaction rate is to increase the energy of the reactants. This is most readily done in the laboratory by heating the reaction mixture (Figure 3.9). In contrast, applying heat to an enzyme-mediated reaction leads to the rapid inactivation of the enzyme due to its denaturation.

Reactants, when they are at the crest of the energy hump and ready to be converted to products, are said to be at the **transition state** (Figure 3.8). At this point, the reactants have formed a fleeting, activated complex in which bonds are being formed and broken. We can illustrate the nature of a transition state structure by examining the interconversion of D and L stereoisomers of proline, a reaction catalyzed by the bacterial enzyme proline racemase.

D-proline Transition L-proline
 state

This reaction proceeds in either direction through the loss of a proton from the α carbon of the proline molecule. As

a result, the transition state structure contains a negatively charged carbanion in which the three bonds formed by the carbon atom are all in the same plane.

Unlike the difference in standard free energy for a reaction, the activation energy required to reach the transition state is not a fixed value but varies with the particular reaction mechanism utilized. Enzymes catalyze reactions by decreasing the magnitude of the activation energy barrier. Consequently, unlike catalysis by heat, enzymes cause their substrates to be very reactive without having to be raised to particularly high energy levels. A comparison between the percentage of molecules able to react in an enzyme-catalyzed reaction versus a noncatalyzed reaction is shown in Figure 3.9. Enzymes are able to lower activation energies by binding more tightly to the transition state than to the reactants, which stabilizes this activated complex, thereby decreasing its energy. The importance of the transition state can be demonstrated in numerous ways. For example:

- Compounds that resemble the transition state of a reaction tend to be very effective inhibitors of that reaction because they are able to bind so tightly to the catalytic region of the enzyme (illustrated on page 118).

- Antibodies normally do not behave like enzymes but, rather, simply bind to molecules with high affinity. However, antibodies that bind to a compound that resembles a transition state for a reaction are often able to act like enzymes and catalyze the breakdown of that compound.

As the transition state is converted to products, the affinity of the enzyme for the bound molecule(s) decreases and the products are expelled.

The Active Site and Molecular Specificity

As catalysts, enzymes accelerate bond-breaking and bond-forming processes. To accomplish this task, enzymes become actively involved in the activities that are taking place among the reactants. Enzymes do this by forming a complex with reactants, called an **enzyme-substrate (ES) complex** (Figure 3.10). In most cases, the association between enzyme and substrate is noncovalent, though many examples are known in which a transient covalent bond is formed.

That part of the enzyme molecule that is directly involved in binding the substrate is termed the **active site**. The active site and the substrate(s) have complementary shapes, enabling them to bind together with a high degree of precision, like pieces of a jigsaw puzzle. The binding of substrate to enzyme is accomplished by the same types of noncovalent interactions (ionic bonds, hydrogen bonds, hydrophobic interactions) that determine the structure of the protein itself. For example, the enzyme depicted in Figure 3.11a contains a number of positively charged residues strategically located to bind negatively charged atoms of the substrate. In addition to binding the sub-

(a) *(b)*

Figure 3.10 **Formation of an enzyme-substrate complex.**
(*a*) Schematic drawing of the reaction catalyzed by pyruvate kinase (see Figure 3.23) in which the two substrates, phosphoenolpyruvate and ADP, bind to the enzyme to form an enzyme-substrate (ES) complex, which leads to the formation of the products, pyruvate and ATP. (*b*) Computer-generated model of an enzyme-substrate complex between an RNA molecule (green) and the enzyme ribonuclease (purple). The RNA is bound within a cleft of the enzyme. (B: COURTESY OF ALEXANDER MCPHERSON AND STEPHEN KOSZELEK.)

strate, the active site contains a particular array of amino acid side chains that influence the substrate and lower the activation energy of the reaction. Often these influences are carried out in a hydrophobic environment within the enzyme, which enhances the reactivity of the side chains.

The active site is typically buried in a cleft or crevice that leads from the aqueous surroundings into the depths of the protein (Figure 3.11*b*). The amino acids that make up the active site are usually situated at distant points along the length of the extended polypeptide chain, but they are brought into close proximity to one another as the polypeptide folds into its final tertiary structure (Figure 3.11*c*). The structure of the active site not only accounts for the catalytic activity of the enzyme, but also for its *specificity*. As noted above, most enzymes are capable of binding only one or a small number of closely related biological molecules.

Mechanisms of Enzyme Catalysis

How is it that an enzyme can cause a reaction to occur hundreds of times a second when that same reaction might occur at undetectable rates in the enzyme's absence? The answer lies in the formation of the enzyme-substrate complex, which allows the substrate(s) to be taken out of solution and held onto the surface of the large catalyst

molecule. Once there, the substrate can be affected in a number of ways, as described next.

Substrate Orientation Suppose you were to place a handful of nuts and bolts into a bag and shake the bag for 15 minutes. It is very unlikely that any of the bolts would have a nut firmly attached to its end when you were finished. In contrast, if you were to pick up a bolt in one hand and a nut in the other, you could rapidly guide the bolt into the nut. By holding the nut and bolt in the proper orientation, you have greatly decreased the entropy of the system. Enzymes lower the entropy of their substrates in a similar manner.

Substrates bound to the surface of an enzyme are brought very close together in precisely the correct orientation to undergo reaction (Figure 3.12*a*). In contrast, when reactants are present in solution, they are free to undergo translational and rotational movements, and even those possessing sufficient energy do not necessarily undergo a collision that results in the formation of a transition-state complex.

Changing Substrate Reactivity Enzymes are composed of amino acids having a variety of different types of side chains (R groups), from fully charged to highly nonpolar.

(a)

(b)

Figure 3.11 The active site of an enzyme. (*a*) Diagrammatic representation of the active site of the enzyme ribulose bisphosphate carboxylase showing the various sites of interaction between the bound substrates (RuBP and CO_2) and certain amino acid side chains of the enzyme. In addition to determining the substrate-binding properties of the active site, these noncovalent interactions alter the properties of the substrate in ways that accelerate its conversion to products. (*b*) A computer-generated model of a cross section through a bacterial esterase depicting the electrostatic properties of the surface of the protein. The negatively charged surface (red) of the active site is seen buried deeply in the interior of the molecule (indicated by the yellow cone). Several side chains of the active site are shown in green. (*c*) An electron density map of the active site of a viral thymidine kinase with the substrate, deoxythymidine, shown in the center of the map (arrow). The edges of the grids provide an indication of the outer reaches of the electron orbitals of the atoms that make up the substrate and the side chains of the enzyme, thus portraying a visual representation of the space occupied by the atoms of the active site. (A: AFTER D. A. HARRIS, BIOENERGETICS AT A GLANCE, P. 88, BLACKWELL, 1995; B: REPRINTED WITH PERMISSION FROM Y. WEI, Z. S. DEREWENDA, ET AL., NATURE STR. BIOL. 2:220, 1995; C: REPRINTED WITH PERMISSION FROM D. G. BROWN, M. R. SANDERSON, ET AL., NATURE STR. BIOL. 2:878, 1995; B,C: COPYRIGHT 1995, MACMILLAN MAGAZINES LIMITED.)

(c)

When a substrate is bound to the surface of an enzyme, the distribution of electrons within that substrate molecule is influenced by the neighboring side chains of the enzyme (Figure 3.12*b*). This influence increases the reactivity of the substrate and stabilizes the transition-state complex formed during the reaction. These effects are accomplished without the input of external energy, such as heat.

There are several general mechanisms whereby the reactivity of substrates is increased by association with enzymes. Basically, these mechanisms are similar to those characterized by organic chemists studying the mechanisms of organic reactions in a test tube. For example, the rate of reactions can be greatly affected by changes in pH. Although enzymes cannot change the pH of their medium, they do contain numerous amino acids with acidic or basic side chains. These groups are capable of donating or accepting protons from the substrate, thereby altering the charge on the substrate, making it more reactive.

The active sites of many enzymes contain side chains with a partial positive or negative charge. Such groups are capable of interacting with a substrate to produce a temporary, covalent enzyme–substrate linkage. Chymotrypsin, an enzyme that digests food proteins within the small intestine, acts in this way. The series of reactions

Figure 3.12 Three mechanisms by which enzymes accelerate reactions: (*a*) maintaining precise substrate orientation, (*b*) changing substrate reactivity by altering its ionic structure, (*c*) exerting physical stress on bonds in the substrate to be broken.

that take place as chymotrypsin hydrolyzes a peptide bond in a substrate protein is shown in Figure 3.13. Three amino acids within the active site of the enzyme—a serine, a histidine, and an aspartic acid—play a prominent role. In Figure 3.13, the reaction is divided into two steps. In the first step, the electronegative oxygen atom of the side chain of a serine of the enzyme "attacks" a carbon atom of the substrate. As a result, the peptide bond of the substrate is hydrolyzed, and a covalent bond is formed between the serine and the substrate, displacing the remainder of the substrate as one of the products. As discussed in the figure legend, the ability of serine to carry out this reaction depends on a nearby histidine residue, which attracts the proton of serine's hydroxyl group, increasing the electronegativity of the group's oxygen atom. In the second step, the covalent bond between the enzyme and substrate is cleaved by a water molecule, returning the enzyme to its original unbonded state and releasing the remainder of the substrate as the second product.

Although amino acid side chains can engage in a variety of reactions, they are not well suited to donate or accept electrons. As we will see in the following section (and more fully in Chapters 5 and 6), the transfer of electrons is the central event in the oxidation-reduction reactions that play such a vital role in cellular metabolism. To catalyze these reactions, enzymes contain cofactors (metal ions or coenzymes) that increase the reactivity of substrates by removing or donating electrons.

Inducing Strain in the Substrate Although the active site of an enzyme may be complementary to its substrate(s), various studies reveal a shift in the relative positions of certain of the atoms of the enzyme once the substrate has bound. In many cases, the conformation shifts so that the complementary fit between the enzyme and reactants is improved (an induced fit), and the proper reactive groups of the enzyme move into place. An example of this type of **induced fit** is shown in Figure 3.14. Once in the "grip" of an enzyme, certain bonds within the substrate molecule may be placed under physical or electronic strain. This has the effect of destabilizing the substrate, causing it to adopt the transition state in which the strain is relieved (Figure 3.12*c*). An example of this mechanism is discussed at length in the Experimental Pathways on lysozyme catalysis at the end of the chapter.

Studying Conformational Changes and Catalytic Intermediates To fully understand the mechanism by which an enzyme catalyzes a particular reaction, it is necessary to describe the various changes in atomic and electronic structure in both the enzyme and the substrate(s) as the reaction proceeds. We saw in the last chapter how X-ray crystallographic techniques can reveal details of the structure of a large enzyme molecule. Since 40 to 60 percent of the volume of a typical protein crystal consists of trapped solvent, most crystallized enzymes retain a high level of enzymatic activity. It should be possible, therefore, to use X-ray diffraction techniques to study reaction mechanisms. There is one major limitation, namely, time. In a standard crystallographic study, enzyme crystals must be subjected to an X-ray beam for a period of hours or days while the necessary data is collected. The portrait that emerges from such studies captures the structure of the molecule averaged out over time. Recent innovations, however, have made it possible to use X-ray crystallographic techniques to observe the fleeting changes in the active site while an enzyme is catalyzing a single reaction cycle. This approach, which is called *time-resolved crystallography*, can include the following:

■ The use of ultra-high-intensity X-ray beams generated by a synchrotron, an instrument used by nuclear physicists to study subatomic particles. This can cut the X-ray exposure period to a matter of picoseconds, which is the same time scale required for an enzyme to catalyze a single chemical transformation.

(a)　　　　　　*(b)*

Figure 3.13 **The catalytic mechanism of chymotrypsin.** The reaction is divided into two steps. (*a*) The electronegative oxygen atom of a serine residue (Ser 195) in the enzyme, which carries a partial negative charge, carries out a nucleophilic attack on the carbonyl carbon atom, which carries a partial positive charge, splitting the peptide bond. The serine is made more reactive by a closely applied histidine residue (His 57) that draws the proton from the serine and subsequently donates the proton to the nitrogen atom of the cleaved peptide bond. Histidine is able to do this because its side chain is a weak base that is capable of gaining and losing a proton at physiological pH. (A stronger base, such as lysine, would remain fully protonated at this pH.) Part of the substrate forms a transient covalent bond with the enzyme by means of the serine side chain, while the remainder of the substrate is released. (It can be noted that the serine and histidine residues are situated 138 amino acids away from each other in the primary sequence but are brought together within the enzyme by the folding of the polypeptide. An aspartic acid, residue 102, which is not shown, also plays a role in catalysis by influencing the ionic state of the histidine.) (*b*) In the second step, the electronegative oxygen atom of a water molecule displaces the covalently linked substrate from the enzyme, regenerating the unbound enzyme molecule. As in the first step, the histidine plays a role in proton transfer; in this step, the proton is removed from water, making it a much stronger nucleophile. The proton is subsequently donated to the serine residue of the enzyme.

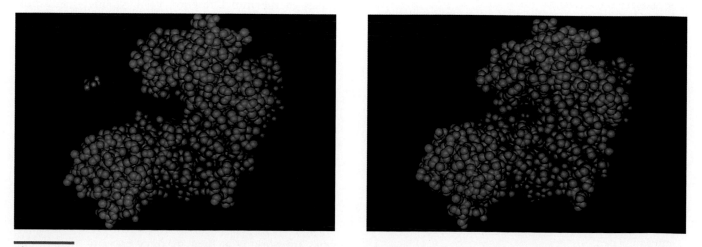

Figure 3.14 **An example of induced fit.** When a glucose molecule binds to the enzyme hexokinase, the protein undergoes a conformational change that encloses the substrate within the active site pocket. (COURTESY OF THOMAS A. STEITZ.)

Arg CD3

distal His

proximal His

—— Mb CO
—— Mb*CO
—— unlitgated Mb

F helix

Figure 3.15 **An example of time-resolved X-ray crystallography.** The active site of a myoglobin molecule shown in three different states: (1) with a carbon monoxide (CO) molecule bound to the protein's heme group (blue), (2) while the carbon monoxide (CO) molecule is released (red), and (3) with the site unoccupied (green). The differences in conformation among these three states can be seen by comparing the positions of the three colors. During these events, the iron atom is displaced approximately 0.35 Å out of the plane of the heme. This type of image is made possible by performing X-ray crystallography at very low temperature (20°K), which vastly slows the rate of CO rebinding. Release of CO from its binding site in the protein is induced simultaneously throughout the crystal by exposure to a brief flash of light. (COURTESY OF JOEL BERENDZEN.)

■ Cooling the enzyme crystals to temperatures within 20 to 40 degrees of absolute zero, which slows the reaction by a factor as high as 10 billion, greatly increases the lifetime of transient intermediates. An example of the data that can be collected using this approach is shown in Figure 3.15.

■ Use of techniques to synchronize a reaction throughout an entire crystal so that all of the enzyme molecules in the crystal are in the same stage of the reaction at the same time. For example, in the case of a reaction in which ATP is a substrate, the enzyme crystals can be infiltrated with ATP molecules that have been made nonreactive by linking them to an inert group (e.g., a nitrophenyl group) by a photosensitive bond. When the crystals are exposed to a brief pulse of light, all of the "caged" ATP molecules are released, initiating the reaction simultaneously in active sites throughout the crystal.

■ Use of enzymes bearing site-directed mutations (page 70) that impose kinetic barriers at specific stages of the reaction, leading to an increased lifetime for particular intermediates.

■ Generating a mutant version of the enzyme that stops catalyzing the reaction at a specific stage, thereby trapping the enzyme in an intermediate state.

By combining these techniques, researchers have been able to determine the three-dimensional structure of an enzyme at successive stages during a single catalyzed reaction. When these individual "snapshots" are put together in sequence, they produce a "movie" showing the various catalytic intermediates that appear and disappear as the reaction proceeds.

Enzyme Kinetics

Enzymes vary greatly in their ability to catalyze reactions. The catalytic activity of an enzyme is revealed by studying its **kinetics**, that is, the rate at which it catalyzes a reaction under various experimental conditions. In 1913, Leonor Michaelis and Maud Menten reported on the mathematical relationship between substrate concentration and the velocity of enzyme reactions as measured by the amount of product formed (or substrate consumed) in a given amount of time. This relationship can be expressed by an equation (given on page 101) that generates a hyperbola, as shown in Figure 3.16. Rather than considering the theoretical aspects of enzyme kinetics, we can obtain the same curve in a practical manner, as it is done for each enzyme studied. To determine the velocity of a reaction, an incubation mixture is set up at the desired temperature, containing all the ingredients required except one, which when added initiates the reaction. If, at the time the reaction begins, no product is present in the mixture, then the amount of product that appears over time provides a measure of the velocity of the reaction. There are complicating factors in this procedure. If the incubation time is too great, the substrate concentration becomes measurably reduced. In addition, as product appears, it can be converted back to substrate by the reverse reaction, which is also catalyzed by the enzyme. Ideally, what we want to determine is the *initial* velocity, that is, the velocity at the instant when no product has yet been formed. To obtain accurate values of the initial reaction velocity, brief incubation times and sensitive measuring techniques are employed.

To generate a curve such as that shown in Figure 3.16, the initial velocity is determined for a series of incubation mixtures that contain the same amount of enzyme but an increasing concentration of substrate. It is evident from this curve that the initial reaction velocity varies markedly with substrate concentration. At low substrate concentrations, enzyme molecules are subjected to relatively few collisions with substrate in a given amount of time. Consequently, the enzyme has "idle time"; that is, substrate molecules are rate limiting. At high substrate concentrations, enzymes are colliding with substrate molecules more rapidly than they can be converted to product. Thus, at high substrate concentrations, individual enzyme molecules are working at their maximal capacity; that is, enzyme molecules are rate limiting. Thus, as a greater and greater concentration of substrate is present in the reaction mixture, the enzyme approaches a state of *saturation*.

Figure 3.16 The relationship between the rate (velocity) of an enzyme-catalyzed reaction and the substrate concentration. Since each enzyme molecule is only able to catalyze a certain number of reactions in a given amount of time, the velocity of the reaction (typically expressed as moles of product formed per second) approaches a maximal rate as the substrate concentration increases. The substrate concentration at which the reaction is at half-maximal velocity ($V_{max}/2$) is called the Michaelis constant, or K_M.

The velocity at this theoretical saturation point is termed the **maximal velocity (V_{max})**.

The simplest measure of the catalytic activity of an enzyme is provided by its *turnover number*, which can be calculated from the V_{max}. The turnover number (or *catalytic constant, k_{cat}*, as it is also called) is the maximum number of molecules of substrate that can be converted to product by one enzyme molecule per unit time. A turnover number (per second) of 1 to 10^3 is typical for enzymes, although values as great as 10^6 (for carbonic anhydrase) are known (see Table 3.4). It is apparent from these values that relatively few molecules of enzyme can rapidly convert a relatively large number of substrate molecules into product.

The value of V_{max} is only one useful term obtained from a plot such as that of Figure 3.16; another is the **Michaelis constant (K_M)**, which is equal to the substrate concentration when the reaction velocity is one-half of V_{max}. As its name implies, the K_M is constant for a given enzyme and, thus, independent of substrate or enzyme concentration. The relationship between V_{max} and K_M is best appreciated by considering the Michaelis-Menten equation, which can be used to generate the plot shown in Figure 3.16.

$$V = V_{max}\frac{[S]}{[S] + K_M}$$

According to the equation, when the substrate concentration [S] is set at a value equivalent to K_M, then the velocity of the reaction (V) becomes equal to $V_{max}/2$, or one-half

the maximal velocity. Thus, $K_M = [S]$, when $V = V_{max}/2$. In contrast, when [S] is low relative to K_M, we can ignore the [S] term in the denominator of the equation, so that $V = V_{max} \cdot [S]/K_M$. Thus, at low substrate concentrations, the velocity of the reaction is directly proportional to the concentration of substrate. On the other hand, when [S] is high relative to K_M, then we can ignore the K_M term in the denominator of the equation, so that $V = V_{max}$; in other words, the velocity of the reaction is independent of the concentration of substrate.

In most cases, the value for K_M provides a measure of the affinity of the enzyme for the substrate. The higher the K_M, the greater the substrate concentration that is required to reach one-half V_{max} and, thus, the lower the affinity of the enzyme for that substrate. The K_M for most enzymes ranges between 10^{-1} and 10^{-7}, with a typical value about 10^{-4} M. Other factors that strongly influence enzyme kinetics are the pH and temperature of the incubation medium. Each enzyme has an optimal pH and temperature at which it operates at maximal activity (Figure 3.17).

To generate a hyperbolic curve such as that in Figure 3.16 and make an accurate determination of the values for V_{max} and K_M, a considerable number of points must be plotted. An easier and more accurate description is gained by plotting the reciprocals of the velocity and substrate concentration against one another, as formulated by Hans Lineweaver and Dean Burk. When this is done, the hyperbola becomes a straight line (Figure 3.18) whose x intercept is equal to $-1/K_M$, y intercept is equal to $1/V_{max}$, and slope is equal to K_M/V_{max}. The values of K_M and V_{max} are, therefore, readily determined by extrapolation of the line drawn from a relatively few points.

Enzyme Inhibitors **Enzyme inhibitors** are molecules that are able to bind to an enzyme and decrease its activity. The cell depends on inhibitors to regulate the activity of many of its enzymes; biochemists use inhibitors to study the properties of enzymes; and many biochemical companies produce enzyme inhibitors that act as drugs, antibiotics, or pesticides. Enzyme inhibitors can be divided into two types: reversible or irreversible. Reversible inhibitors, in turn, can be considered as competitive or noncompetitive.

Irreversible inhibitors are those that bind very tightly to an enzyme, often by forming a covalent bond to one of its amino acid residues. A number of nerve gases, such as diisopropylphosphofluoridate and the organophosphate pesticides, act as irreversible inhibitors of acetylcholinesterase, an enzyme that plays a crucial role in destroying acetylcholine, the neurotransmitter responsible for causing muscle contraction. With the enzyme inhibited, the muscle is stimulated continuously and remains in a state of permanent contraction. As discussed in the accompanying Human Perspective, the antibiotic penicillin acts as an irreversible inhibitor of a key enzyme in the formation of the bacterial cell wall.

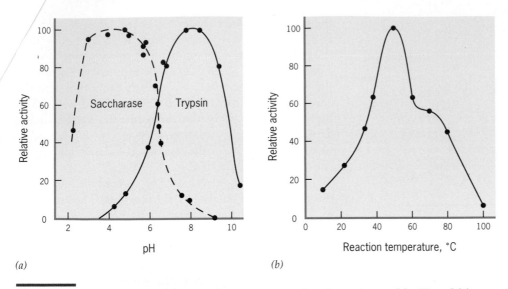

(a)

(b)

Figure 3.17 **Dependence of the rate of an enzyme-catalyzed reaction on (*a*) pH, and (*b*) temperature.** The shape of the curves and the optimal pH and temperature vary with the particular reaction. (*a*) Changes in pH affect the ionic properties of the substrate and the enzyme as well as the enzyme's conformation. (*b*) At lower temperatures, reaction rates rise with increasing temperature due to the increased energy of the reactants. At higher temperatures, this positive effect is offset by enzyme denaturation. (A: FROM E. A. MOELWYN-HUGHES, IN THE ENZYMES, J. B. SUMNER AND K. MYRBACK, EDS., VOL. 1, ACADEMIC PRESS, 1950; B: FROM K. HAYASHI ET AL., J. BIOCHEM. 64:93, 1968.)

Reversible inhibitors, on the other hand, bind only loosely to an enzyme, and thus are readily displaced. **Competitive inhibitors** are reversible inhibitors that compete with a substrate for access to the active site of an enzyme. Since substrates have a complementary structure to the active site to which they bind, competitive inhibitors must resemble the substrate to compete for the same binding site, but differ in a way that prevents them from being transformed into product (Figure 3.19). Analysis of the types of molecules that can compete with

Figure 3.19 **Competitive inhibition.** Because of their molecular similarity, competitive inhibitors are able to compete with the substrate for a binding site on the enzyme. The effect of a competitive inhibitor depends on the relative concentrations of the inhibitor and substrate.

Figure 3.18 **A Lineweaver-Burk plot** of the reciprocals of velocity and substrate concentration from which the values for the V_{max} and K_M are readily calculated.

the substrate for a binding site on the enzyme provides insight into the structure of the active site and the nature of the interaction between the natural substrate and its enzyme.

Competitive enzyme inhibition forms the basis of action of a wide variety of common drugs, as illustrated in the following example. Angiotensin converting enzyme (ACE) is a proteolytic enzyme that acts on a ten-residue peptide (angiotensin I) to produce an eight-residue peptide (angiotensin II). Elevated levels of angiotensin II are a major risk factor in the development of high blood pressure (*hypertension*). In the 1960s John Vane and his colleagues at the Eli Lilly Company began a search for compounds that could inhibit ACE. Previous studies had found that the venom of a Brazilian pit viper contained inhibitors of proteolytic enzymes, and it was found that one of the components of this venom, a peptide called teprotide

Teprotide (peptide bonds marked in red)

was a potent competitive inhibitor of ACE. Although teprotide was shown to lower blood pressure in hypertensive patients, it was not a very useful drug because it had a peptide structure and thus was rapidly degraded if taken orally. Subsequent efforts to develop nonpeptide inhibitors of the enzyme led researchers to synthesize a compound called captopril,

Captopril

which became the first useful antihypertensive drug that acted by binding to ACE.

The effectiveness of a competitive inhibitor depends on its relative affinity for the enzyme. Regardless, competitive inhibition can be overcome if the substrate/inhibitor ratio is great enough. In other words, if the number of collisions between the enzyme and inhibitor become insignificant relative to those between the enzyme and its substrate, then the effect of the inhibitor becomes minimal. Given a sufficient substrate concentration, it remains theoretically possible to achieve the enzyme's maximal velocity even in the presence of the competitive inhibitor.

In **noncompetitive inhibition**, the substrate and inhibitor do not compete for the same binding site; generally, the inhibitor acts at a site other than the enzyme's active site. The level of inhibition depends only on the concentration of the inhibitor, and increasing the concentration of the substrate cannot overcome it. Since, in the presence of a noncompetitive inhibitor, a certain fraction of the enzyme molecules are necessarily inactive at a given instant, the maximal velocity of the population of enzyme molecules cannot be reached. The effects on the kinetics of enzymes in the presence of noncompetitive and competitive inhibitors are shown in Figure 3.20. In one case, the V_{max} is lowered, and in the other, the K_M is increased. In both types, the slope (K_M/V_{max}) is increased relative to the uninhibited reaction.

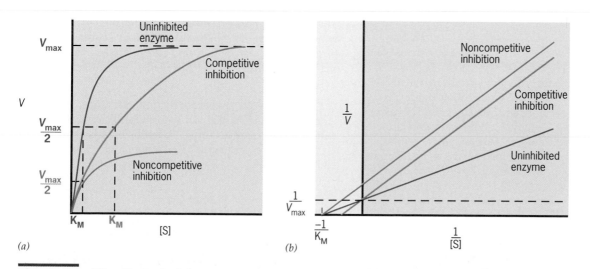

Figure 3.20 The effects of inhibitors on enzyme kinetics. The effect of both competitive and noncompetitive inhibitors is shown when the kinetics of the reaction are plotted as velocity of reaction versus substrate concentration (*a*) or its reciprocal (*b*). The noncompetitive inhibitor reduces V_{max} without affecting K_M, whereas the competitive inhibitor increases K_M without affecting V_{max}.

THE HUMAN PERSPECTIVE

UPDATE For the very latest, go to page U-7

The Growing Problem of Antibiotic Resistance

Not too long ago, it was widely believed that human health would no longer be threatened by serious bacterial infections. Bacterial diseases such as tuberculosis, pneumonia, gonorrhea, and dozens of others were being cured by administration of any one of a number of antibiotics—compounds that selectively kill bacteria without harming the human host in which they grow. In the wake of their success, pharmaceutical companies scaled back further research into the development of new types of antibiotics, turning their attention to other matters. It has become painfully evident, however, that the announcement of the death of infectious bacteria was premature.

The development of bacterial resistance provides an excellent example of natural selection; the widespread use (and misuse) of these drugs has killed off susceptible cells, leaving the rare resistant individuals to survive and repopulate the ranks. The result has been a marked upswing in the incidence of antibiotic-resistant strains of bacteria, including those responsible for pneumonia, tuberculosis, and septicemia. Many infectious disease specialists are predicting that the problem will become more acute in the coming years and that fatalities from once-curable diseases will increase sharply. Pharmaceutical companies are finally responding to this threat and are beginning to search for new types of antibiotics. Here, we will look briefly at the mechanism of action of antibiotics—particularly those that target enzymes, which are the subject of this chapter—and the development of bacterial resistance.

Most antibiotics are natural products, produced by microorganisms to kill other microorganisms. Antibiotics work because they interfere with certain bacterial activities without affecting those of eukaryotic cells. Several types of targets in bacterial cells have proven most vulnerable. These include the following:

1. *Enzymes involved in the synthesis of the bacterial cell wall.* Penicillin and its derivatives are structural analogues of the substrates of a family of transpeptidases that catalyze the final cross-linking reactions that give the cell wall its rigidity. If these reactions do not occur, a rigid cell wall fails to develop. Penicillin is an irreversible inhibitor of the transpeptidases; the antibiotic fits into the active site of these enzymes, forming a covalently bound complex that cannot be dislodged. The antibiotic vancomycin (which was originally derived from a microorganism living in soil samples taken from Borneo) inhibits transpeptidation by binding to the peptide substrate of the transpeptidase, rather than to the enzyme itself. Normally the transpeptidase substrate terminates in a D-alanine—D-alanine dipeptide. To become resistant to vancomycin, a bacterial cell must synthesize an alternate terminus that doesn't bind the drug, which is a roundabout process that requires the acquisition of several new enzymatic activities. As a result, vancomycin is the antibiotic to which bacteria have proven least able to develop resistance and thus is usually given as a last resort when other antibiotics have failed. Unfortunately, van-

comycin-resistant strains of several pathogenic bacteria, including *Enterococcus faecium* and *Staphylococcus aureus*, have emerged in recent years. These resistant strains often reside in hospital wards, which has made hospital-acquired infections increasingly difficult to treat. The selection for vancomycin-resistant strains is thought to have been driven by the widespread use of a related antibiotic (called avoparcin) in animal feed. Infectious disease specialists are urging hospitals to institute better hygiene programs and are pressuring legislators to ban the use of antibiotics in agriculture. In the meantime, researchers have modified the chemical structure of vancomycin to produce derivatives that may prove useful in attacking bacteria that have become resistant to the native form of the drug.

2. *Components of the system by which bacteria duplicate, transcribe, and translate their genetic information.* Although prokaryotic and eukaryotic cells have a similar system for storing and utilizing genetic information, there are many basic differences between the two types of cells that pharmacologists can take advantage of. Streptomycin and the tetracyclines, for example, bind to prokaryotic ribosomes, but not eukaryotic ribosomes. Quinolones, which are a rare example of wholly synthetic antibiotics (i.e., they are not based on natural products), inhibit the enzyme DNA gyrase, which is required for bacterial DNA replication.

3. *Enzymes that catalyze metabolic reactions specifically in bacteria.* Sulfa drugs, for example, are effective antibiotics because they closely resemble the compound *p*-aminobenzoic acid (PABA),

$$H_2N-\langle\rangle-COOH \qquad H_2N-\langle\rangle-SO_2-NH-R$$

PABA Sulfa drugs

which bacteria convert enzymatically to the essential coenzyme folic acid. Since humans lack a folic acid-synthesizing enzyme, they must obtain this essential coenzyme in their diet and, consequently, sulfa drugs have no effect on human metabolism.

Bacteria become resistant to antibiotics through a number of distinct mechanisms, many of which can be illustrated using penicillin as an example. Penicillin is a β-lactam; that is, it contains a characteristic four-membered β-lactam ring (shown in color).

Penicillin

As early as 1940, researchers discovered that certain bacteria possess an enzyme called β-lactamase (or penicillinase) that can open the lactam ring, rendering the compound harmless to the bacterium. At the time when penicillin was first introduced as an antibiotic during World War II, none of the major disease-causing bacteria possessed a gene for β-lactamase. This can be verified by examining the DNA of bacteria descended from laboratory cultures that were started in the pre-antibiotic era. Today, the β-lactamase gene is present in a wide variety of infectious bacteria, and the production of β-lactamase by these cells is the primary cause of penicillin resistance. How did these species acquire the gene?

The widespread occurrence of the β-lactamase gene illustrates how readily genes can spread from one bacterium to another, not only among the cells of a given species, but between species. There are several ways this can happen, including conjugation (shown in Figure 1.13), in which DNA is passed from one bacterial cell to another; transduction, in which a bacterial gene is carried from cell to cell by a virus; and transformation, in which a bacterial cell is able to pick up naked DNA from its surrounding medium. Pharmacologists have attempted to counter the spread of β-lactamase by synthesizing penicillin derivatives (e.g., cefuroxime) that are more resistant to the hydrolytic enzyme. As might be expected, natural selection leads rapidly to the evolution of bacteria whose β-lactamase can split the new forms of the antibiotic. As noted by Julian Davies of the University of British Columbia, "a single base change in a gene encoding a bacterial β-lactamase may render useless 100 million dollars worth of pharmaceutical research effort." One approach that has met with limited success is to treat patients with two separate drugs: a penicillin-like antibiotic to inhibit the transpeptidase and a separate enzyme inhibitor (e.g., clavulanic acid) to inhibit the β-lactamase.

Not all penicillin-resistant bacteria have acquired a β-lactamase gene. Some are resistant because they possess modifications in their cell walls that block entry of the antibiotic; others are resistant because they are able to selectively export the antibiotic once it has entered the cell; still others are resistant because they possess modified transpeptidases that fail to bind the antibiotic. Bacterial meningitis, for example, is caused by the bacterium *Neisseria meningitidis*, which has yet to display evidence it has acquired β-lactamase. Yet these bacteria are becoming resistant to penicillin because their transpeptidases have lost an affinity for the antibiotics.

The problem of drug resistance is not restricted to bacterial diseases, but has also become a major issue in the treatment of AIDS. Unlike bacteria, whose DNA-replicating enzymes operate at a very high level of accuracy, the replicating enzyme of the AIDS virus (HIV), called *reverse transcriptase*, makes a large number of mistakes, which leads to a high rate of mutation. This high error rate (about 1 mistake for every 10,000 bases duplicated), combined with the high rate of virus production ($>10^8$ virus particles produced in a person per day), makes it very likely that drug-resistant variants will emerge within an individual as the infection progresses. This problem is being combatted in two ways:

- By having patients take several different drugs targeted at different viral enzymes. This greatly reduces the likelihood that a variant will emerge that is resistant to all of the drugs.

- By designing drugs that interact with the most highly conserved portions of each targeted enzyme, that is, those portions where mutations are most likely to produce a defective enzyme. This point underscores the importance of knowing the structure and function of the target enzyme and the manner in which potential drugs interact with that target (page 70).

REVIEW ?

1. How is it possible to have a reaction characterized by a large ΔG and a small E_A? A large E_A and a small ΔG?

2. Explain how enzymes can be so specific toward the substrates they bind.

3. Examine one of the figures illustrating the steps in an enzymatically catalyzed reaction and describe what is occurring without reading the legend or the corresponding text.

4. Distinguish between turnover number and V_{max}.

3.3 METABOLISM

Metabolism is the collection of biochemical reactions that occur within a cell, which include a tremendous diversity of molecular conversions. Most of these reactions can be grouped into **metabolic pathways** containing a sequence of chemical reactions in which each reaction is catalyzed by a specific enzyme, and the product of one reaction is the substrate for the next. The enzymes constituting a metabolic pathway are usually confined to a specific region of the cell, such as the mitochondria or the cytosol. Increasing evidence suggests that the enzymes of a metabolic pathway are often physically linked to one another, a feature that allows the product of one enzyme to be delivered directly as a substrate to the active site of the next enzyme in the reaction sequence.

The compounds formed in each step along the pathway are **metabolic intermediates** (or **metabolites**) that lead ultimately to the formation of an **end product**. End products are molecules with a particular role in the cell, such as an amino acid that can be incorporated into a polypeptide, or a sugar that can be consumed for its energy content. The metabolic pathways of a cell are interconnected at various points so that a compound generated by one pathway may be shuttled in a number of directions depending on the requirements of the cell at the time. We will concentrate in this section on aspects of metabolism that lead to the transfer and utilization of chemical energy

within the cell, for this topic is one we will draw on throughout the book.

An Overview of Metabolism

Metabolic pathways can be divided into two broad types. **Catabolic pathways** lead to the disassembly of complex molecules to form simpler products. Catabolic pathways serve two functions: they make available the raw materials from which other molecules can be synthesized, and they provide chemical energy required for the many activities of a cell. As will be discussed at length, energy released by catabolic pathways is stored temporarily in two forms: as high-energy phosphates (primarily ATP) and as high-energy electrons (primarily in NADPH). **Anabolic pathways** lead to the synthesis of more complex compounds from simpler starting materials. Anabolic pathways are energy requiring and utilize chemical energy released by the exergonic catabolic pathways.

Figure 3.21 shows a greatly simplified profile of the ways in which the major anabolic and catabolic pathways are interconnected. Macromolecules are first disassembled (hydrolyzed) into the building blocks of which they are made (stage I, Figure 3.21). Once macromolecules have been hydrolyzed into their components—amino acids, sugars, and fatty acids—the cell can reutilize the building blocks directly to (1) form other macromolecules of the same class (stage I), (2) convert them into different compounds to make other products, or (3) degrade them further (stages II and III, Figure 3.21) and extract a measure of their free-energy content.

The pathways for degradation of the diverse building blocks of macromolecules vary according to the particular compound being catabolized. Ultimately, however, all of these molecules are converted into a small variety of compounds (stage II, Figure 3.21) that can be metabolized similarly. Thus, even though substances begin as macromolecules having a very different structure, they are converted by catabolic pathways to the same low-molecular-weight metabolites. For this reason, catabolic pathways are said to be convergent.

Remarkably, the chemical reactions and metabolic pathways described in this chapter are found in virtually every living cell, from the simplest bacterium to the most complex plant or animal. It is evident that these pathways appeared very early in, and have been retained throughout, the course of biological evolution.

Oxidation and Reduction: A Matter of Electrons

Both catabolic and anabolic pathways include key reactions in which electrons are transferred from one reactant to another. Reactions that involve a change in the electronic state of the reactants are called **oxidation-reduction (or redox) reactions**. Changes of this type involve the gain or loss of electrons. Consider the conversion of metallic iron (Fe^0) to the ferrous state (Fe^{2+}), by which the iron atom loses a pair of electrons, thereby attaining a more positive

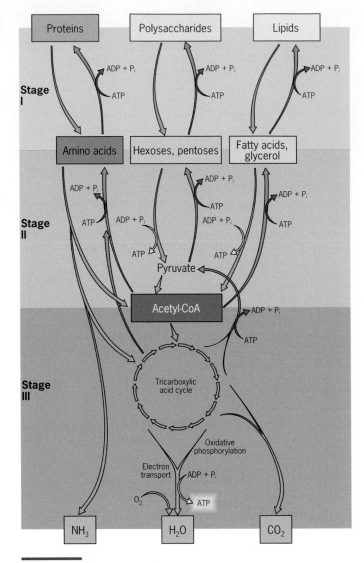

Figure 3.21 **Three stages of metabolism.** The catabolic pathways (green arrows downward) converge to form common metabolites and lead to ATP synthesis in stage III. The anabolic pathways (blue arrows upward) start from a few precursors in stage III and utilize ATP to synthesize a large variety of cellular materials. Metabolic pathways for nucleic acids are more complex and are not shown here. (FROM A. L. LEHNINGER, BIO-CHEMISTRY, 2D ED., 1975. WORTH PUBLISHERS, NEW YORK.)

state. When an atom loses one or more electrons, it is said to be **oxidized**. The reaction is reversible, meaning ferrous ions can be converted to metallic iron, a more negative state, by the acquisition of a pair of electrons. When an atom gains one or more electrons, it is said to be **reduced**.

For metallic iron to be oxidized, there must be some substance to accept the electrons that are released. Conversely, for ferrous ions to be reduced, there must be some substance to donate the necessary electrons. In other words, the oxidation of one reactant must be accompanied by the simultaneous reduction of some other reactant, and vice versa. One possible reaction involving iron might be

$$Fe^0 + Cu^{2+} \rightleftharpoons Fe^{2+} + Cu^0$$

The substance that is oxidized during an oxidation-reduction reaction, that is, the one that loses electrons, is called a **reducing agent**, and the one that is reduced, that is, the one that gains electrons, is called an **oxidizing agent**.

The oxidation or reduction of metals, such as iron or copper, involves the complete loss or gain of electrons. The same cannot occur with most organic compounds for the following reason: The oxidation and reduction of organic substrates during cellular metabolism involve carbon atoms that are covalently bonded to other atoms. As discussed in Chapter 2, when a pair of electrons is shared by two different atoms, the electrons are attracted more strongly to one of the two atoms of the polarized bond. In a C—H bond, the carbon atom has the strongest pull on the electrons; thus it can be said that the carbon atom is in a reduced state. In contrast, if a carbon atom is bonded to a more electronegative atom, as in a C—O or C—N bond, the electrons are pulled away from the carbon atom, which is thus in an oxidized state. Since carbon has four outer-shell electrons it can share with other atoms, it can exist in a variety of oxidation states. This is illustrated by the carbon atom in a series of one-carbon molecules (Figure 3.22) ranging from the fully reduced state in methane (CH_4) to the fully oxidized state in carbon dioxide (CO_2). The relative oxidation state of an organic molecule can be roughly determined by counting the number of hydrogen versus oxygen and nitrogen atoms per carbon atom. As we will see shortly, the oxidation state of the carbon atoms in an organic molecule provides a measure of the molecule's free-energy content.

The Capture and Utilization of Energy

The compounds that we use as chemical fuels to run our furnaces and automobiles are highly reduced organic compounds, such as natural gas (CH_4) and petroleum derivatives. Energy is released when these molecules are burned in the presence of oxygen, converting the carbons to more oxidized states, as in carbon dioxide and carbon monoxide gases. The degree of reduction of a compound is also a measure of its ability to perform chemical work within the cell. The more hydrogen atoms that can be stripped from a "fuel" molecule, the more ATP that ultimately can be produced. Carbohydrates are rich in chemical energy because they contain strings of (H—C—OH) units. Fats contain even greater energy per unit weight because they contain strings of more reduced (H—C—H) units. We will concentrate in the following discussion on carbohydrates.

As the sole building block of both starch and glycogen, glucose is a key molecule in the energy metabolism of both plants and animals. The free energy released by the complete oxidation of glucose is very large:

$$C_6H_{12}O_6 + 6\,O_2 \rightarrow 6\,CO_2 + 6\,H_2O$$
$$\Delta G^{\circ\prime} = -686 \text{ kcal/mol}$$

By comparison, the free energy required to convert ADP to ATP is relatively small:

$$ADP + P_i \rightarrow ATP + H_2O \qquad \Delta G^{\circ\prime} = +7.3 \text{ kcal mol}$$

It is evident from these numbers that the complete oxidation of a molecule of glucose to CO_2 and H_2O can release enough energy to produce a large number of ATPs. As we will see in Chapter 5, up to about 36 molecules of ATP are formed per molecule of glucose oxidized under conditions that exist within most cells. For this many ATP molecules to be produced, the sugar molecule is disassembled in many small steps. Those steps in which the free-energy

— Covalent bond in which carbon atom has greater share of electron pair
— Covalent bond in which oxygen atom has greater share of electron pair

Figure 3.22 The oxidation state of a carbon atom depends on the other atoms to which it is bonded. Each carbon atom can form a maximum of four bonds with other atoms. This series of simple, one-carbon molecules illustrates the various oxidation states in which the carbon atom can exist. In its most reduced state, the carbon is bonded to four hydrogens (forming methane); in its most oxidized state, the carbon atom is bonded to two oxygens (forming carbon dioxide).

difference between the reactants and products is relatively large can be coupled to reactions that lead to the formation of ATP.

There are basically two stages in the catabolism of glucose, and they are virtually identical in all aerobic organisms. The first stage, **glycolysis**, occurs in the soluble phase of the cytoplasm (the cytosol) and leads to the formation of pyruvate. The second stage is the **tricarboxylic acid** (or **TCA) cycle**, which occurs within the mitochondria of eukaryotic cells and the cytosol of prokaryotes and leads to the final oxidation of the carbon atoms to carbon dioxide. Most of the chemical energy of glucose is stored in the form of high-energy electrons, which are removed as substrate molecules are oxidized during both glycolysis and the TCA cycle. It is the energy of these electrons that is ultimately used to synthesize ATP. We will concentrate in the following pages on the steps during glycolysis, the first stage in the oxidation of glucose, which occurs without the involvement of oxygen. This is presumably the pathway of energy capture utilized by our early anaerobic ancestors and remains the major anabolic pathway utilized by anaerobic organisms living today. We will complete the story of glucose oxidation in Chapter 5 when we discuss the structure of the mitochondrion and its role in aerobic respiration.

Glycolysis and ATP Formation The reactions of glycolysis and the enzymes that catalyze them are shown in Figure 3.23. Before discussing the specific reactions, an im-

portant point concerning the thermodynamics of metabolism can be made. In an earlier discussion, the difference between ΔG and $\Delta G^{\circ\prime}$ was stressed; it is the ΔG for a particular reaction that determines its direction in the cell. Actual measurements of the concentrations of metabolites in the cell can reveal the value for ΔG of a reaction at any given time. Figure 3.24 shows the typical ΔG values measured for the reactions of glycolysis. In contrast to the $\Delta G^{\circ\prime}$ values of Figure 3.23, all but three reactions have ΔG values of nearly zero; that is, they are near equilibrium. The three reactions that are far from equilibrium, which makes them essentially irreversible in the cell, provide the driving force that moves the metabolites through the glycolytic pathway in a directed manner.

In 1905, two British chemists, Arthur Harden and William Young, were studying glucose breakdown by yeast cells, a process that generates bubbles of CO_2 gas. Harden and Young noted that the bubbling eventually slowed and stopped, even though there was plenty of glucose left to metabolize. Apparently, some other essential component of the broth was being exhausted. After experimenting with a number of substances, the chemists found that adding inorganic phosphates started the reaction going again. They concluded that the reaction was exhausting phosphate, the first clue that phosphate groups played a role in metabolic pathways. The importance of the phosphate group is illustrated by the initial reactions of glycolysis.

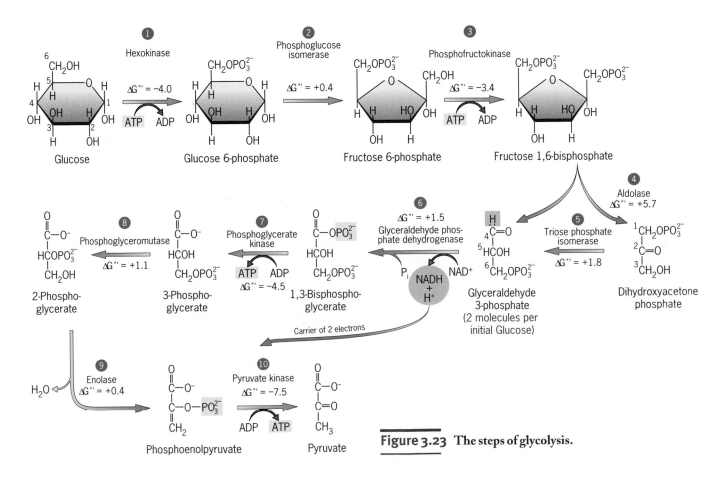

Figure 3.23 **The steps of glycolysis.**

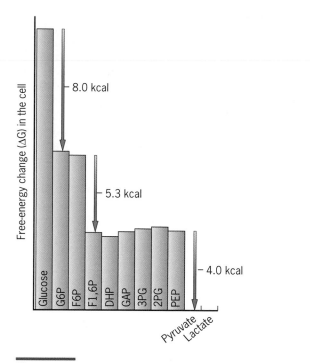

Figure 3.24 Free-energy profile of glycolysis in the human erythrocyte. All reactions are at or near equilibrium except those catalyzed by hexokinase, phosphofructokinase, and pyruvate kinase, which exhibit large differences in free energy. In the cell, all reactions must proceed with a decline in free energy; the slight increases in free energy depicted here for several steps must be regarded as deriving from errors in experimental measurements of metabolite concentrations. (FROM A. L. LEHNINGER, BIOCHEMISTRY, 2D ED., 1975. WORTH PUBLISHERS, NEW YORK.)

Glycolysis begins with the linkage of the sugar to a phosphate group (step 1, Figure 3.23) at the expense of one molecule of ATP. The use of ATP at this stage can be considered an energy investment—the cost of getting into the glucose oxidation business. Phosphorylation activates the sugar, making it capable of participating in subsequent reactions whereby phosphate groups are moved around and transferred to other acceptors. Glucose 6-phosphate is converted to fructose 6-phosphate and then to fructose 1,6-bisphosphate at the expense of a second molecule of ATP (steps 2, 3). The six-carbon bisphosphate is split into two three-carbon monophosphates (step 4), which sets the stage for the first exergonic reaction to which the formation of ATP can be coupled. Let us turn now to the formation of ATP.

ATP is formed in two basically different ways, both illustrated by one chemical reaction of glycolysis: the conversion of glyceraldehyde 3-phosphate to 3-phosphoglycerate (steps 6, 7). The overall reaction is an oxidation of an aldehyde to a carboxylic acid (as in Figure 3.22), and it occurs in two steps catalyzed by two different enzymes (Figure 3.25). The first of these enzymes requires a nonprotein cofactor (a coenzyme), called nicotinamide adenine dinucleotide (NAD), to catalyze the reaction. As will be evident in this and following chapters, NAD plays a key role

in energy metabolism by accepting and donating electrons. The first reaction (Figure 3.25a,b) is an oxidation-reduction in which two electrons and a proton (equivalent to a hydride ion, $:H^-$) are transferred from glyceraldehyde 3-phosphate (which becomes oxidized) to NAD^+ (which becomes reduced). The reduced form of the coenzyme is NADH (Figure 3.26). An enzyme that catalyzes this type of reaction is termed a **dehydrogenase**; the enzyme catalyzing the above reaction is *glyceraldehyde 3-phosphate dehydrogenase*. The NAD^+ molecule, which is derived from the vitamin niacin, acts as a loosely associated coenzyme to the dehydrogenase in position to accept the hydride ion (i.e., both the electrons and the proton). The NADH formed in the reaction is then released from the enzyme in exchange for a fresh molecule of NAD^+.

We will return to this reaction in a moment, but first we will continue with the consequences of the formation of NADH. NADH is considered a high-energy compound because of the ease with which it is able to transfer electrons to other electron-attracting molecules. (NADH is said to have a high electron-transfer potential relative to other electron acceptors in the cell, see Table 5.1). Electrons are usually transferred from NADH through a series of membrane-embedded electron carriers that constitute an *electron-transport chain*. As the electrons move along this chain, they move to a lower and lower free-energy state and are ultimately passed to molecular oxygen, reducing it to water. The energy released during electron transport is utilized to form ATP by a process called *oxidative phosphorylation*. Electron transport and oxidative phosphorylation will be explored in detail in Chapter 5.

In addition to the indirect route for ATP formation involving NADH and an electron-transport chain, the conversion of glyceraldehyde 3-phosphate to 3-phosphoglycerate includes a direct route for ATP formation. In the second step of this overall reaction (step 7, Figure 3.23 and Figure 3.25c), a phosphate group is transferred from 1,3-bisphosphoglycerate to ADP to form a molecule of ATP. The reaction is catalyzed by the enzyme *phosphoglycerate kinase*. This direct route of ATP formation is referred to as **substrate-level phosphorylation** because it occurs by transfer of a phosphate group from one of the substrates (in this case, 1,3-bisphosphoglycerate) to ADP. The remaining reactions of glycolysis (steps 8–10), which includes a second substrate-level phosphorylation of ADP (step 10), are indicated in Figure 3.23.

The substrate-level phosphorylation of ADP illustrates an important point about ATP. Its formation is not *that* endergonic. In other words, ATP is not so energetic a molecule that it cannot be readily formed by metabolic reactions. There are numerous phosphorylated molecules whose hydrolysis has a more negative $\Delta G°'$ than that of ATP. Figure 3.27 illustrates the relative $\Delta G°'$s for the hydrolysis of several phosphorylated compounds. Any donor higher on the scale can be used to phosphorylate any molecule lower on the scale, and the $\Delta G°'$ of this reaction will be equal to the difference between the two values

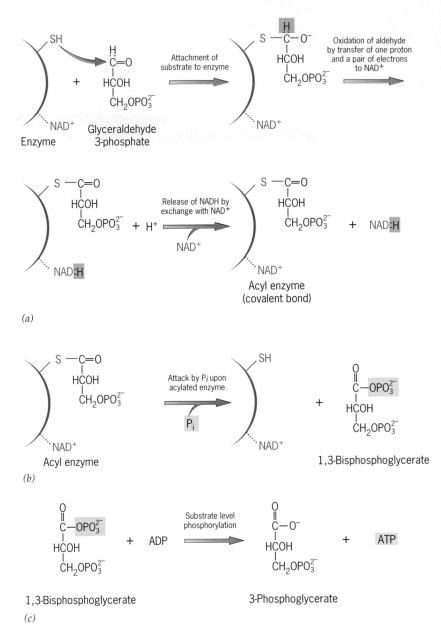

(a)

(b)

(c)

Figure 3.25 The transfer of energy during a chemical oxidation. The oxidation of glyceraldehyde 3-phosphate to 3-phosphoglycerate, which is an example of the oxidation of an aldehyde to a carboxylic acid, occurs in two steps catalyzed by two enzymes. The first reaction (parts *a* and *b*) is catalyzed by the enzyme glyceraldehyde phosphate dehydrogenase, which transfers a pair of electrons from the substrate to NAD^+. Once reduced, the NADH molecules are displaced by NAD^+ molecules from the cytosol. (*c*) The second reaction, which is catalyzed by the enzyme phosphoglycerate kinase, is an example of a substrate-level phosphorylation in which a phosphate group is transferred from a substrate molecule, in this case 1,3-bisphosphoglycerate, to ADP to form ATP.

given in the figure. For example, the $\Delta G^{\circ\prime}$ for the transfer of a phosphate group from 1,3-bisphosphoglycerate to ADP to form ATP is equal to −4.5 kcal/mol (−11.8 kcal/mol + 7.3 kcal/mol). This concept of **transfer potential** is useful for comparing any series of donors and acceptors regardless of the group being transferred, whether protons, electrons, oxygen, or phosphate groups. Those molecules higher on the scale, that is, ones with greater free energy (larger $-\Delta G^{\circ\prime}$s), are molecules with less affinity for the group being transferred than are ones lower on the scale. The less the affinity, the better the donor; the greater the affinity, the better the acceptor.

An important feature of glycolysis is that it can generate a limited number of ATP molecules even in the absence of oxygen. Neither the substrate-level phosphorylation of ADP by 1,3-bisphosphoglycerate nor a later one by phosphoenolpyruvate (step 10, Figure 3.23) requires mol-

ecular oxygen. Thus, glycolysis can be considered an **anaerobic pathway** to ATP production, indicating that it can proceed in the absence of molecular oxygen to continue to provide ATP. Two molecules of ATP are produced by substrate-level phosphorylation during glycolysis from each molecule of glyceraldehyde 3-phosphate oxidized to pyruvate. Since each molecule of glucose produces two molecules of glyceraldehyde 3-phosphate, four ATP molecules are generated per molecule of glucose oxidized to pyruvate. On the other hand, two molecules of ATP must be hydrolyzed to initiate glycolysis, leaving a net profit for the cell of two molecules of ATP per glucose oxidized. The net equation for the reactions of glycolysis can be written

Glucose + 2ADP + 2 P$_i$ + 2 NAD$^+$ →
 2Pyruvate + 2ATP + 2NADH + 2H$^+$ + 2H$_2$O

Figure 3.26 The structure of NAD⁺ and its reduction to NADH. When the 2′ OH of the ribose moiety (indicated by the purple colored screen) is covalently bonded to a phosphate group, the molecule is NADP⁺/NADPH, whose function is discussed later in the chapter.

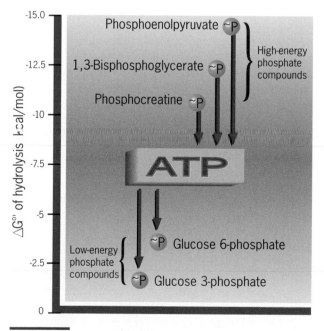

Figure 3.27 Ranking compounds by phosphate transfer potential. Those phosphate compounds higher on the scale (ones with a more negative $\Delta G°'$ of hydrolysis) have a lower affinity for their phosphate group than those compounds that are lower on the scale. As a result, compounds higher on the scale readily transfer their phosphate group to form compounds that are lower on the scale. Thus, phosphate groups can be transferred from 1,3-bisphosphate or phosphoenolpyruvate to ADP during glycolysis.

Pyruvate, the end product of glycolysis, is a key compound because it stands at the junction between anaerobic (oxygen-independent) and aerobic (oxygen-dependent) pathways. In the absence of molecular oxygen, pyruvate is subjected to fermentation, as discussed in the following section. When oxygen is available, pyruvate is further catabolized by aerobic respiration, as discussed in Chapter 5.

Anaerobic Oxidation of Pyruvate: The Process of Fermentation We have seen that glycolysis is able to provide a cell with a small net amount of ATP for each molecule of glucose converted to pyruvate. However, glycolytic reactions occur at rapid rates so that a cell is able to produce a significant amount of ATP using this pathway. In fact, a number of cells, including yeast cells, tumor cells, and muscle cells, rely heavily on glycolysis as a means of ATP formation. There is, however, a problem that must be confronted by such cells. One of the products of the oxidation of glyceraldehyde 3-phosphate is NADH. The formation of NADH occurs at the expense of one of the reactants, NAD⁺, which is in short supply in cells. Since NAD⁺ is a required reactant in this important step of glycolysis, it must be regenerated from NADH. If not, the oxidation of glyceraldehyde 3-phosphate can no longer take place, nor can any of the succeeding reactions of glycolysis. However, without oxygen, NADH cannot be oxidized to NAD⁺ by means of the electron-transport chain because oxygen is the final electron acceptor of the chain. Cells are able, however, to regenerate NAD⁺ by **fermentation**, the transfer of electrons from NADH to pyruvate, the end product of glycolysis, or to a compound derived from pyruvate (Figure 3.28). Like glycolysis, fermentation takes place in the cytosol of a eukaryotic cell. For most organisms, which depend on O₂, fermentation is a stopgap measure to regenerate NAD⁺ when O₂ levels are low, so that glycolysis can continue and ATP production can be maintained.

The product of fermentation varies from one type of cell or organism to another. When muscle cells are required to contract repeatedly, the oxygen level becomes too low to keep pace with the cell's metabolic demands. Under these conditions, skeletal muscle cells regenerate NAD⁺ by converting pyruvate to lactate. When oxygen once again becomes available in sufficient amounts, the lactate can be converted back to pyruvate for continued oxidation. Yeast cells have faced the challenges of anaerobic life with a different metabolic solution: they convert pyruvate to ethanol, as illustrated in Figure 3.28.

Although fermentation is a necessary adjunct to metabolism in many organisms, and the sole source of metabolic energy in some anaerobes, the energy gained by glycolysis alone is meager when compared with the complete oxidation of glucose to carbon dioxide and water. When a mole of glucose is completely oxidized, 686 kilocalories are released. In contrast, only 57 kilocalories are released when this same amount of glucose is converted to ethanol under standard conditions, and only 47 kilocalories are released when it is converted to lactate. In any case, only two

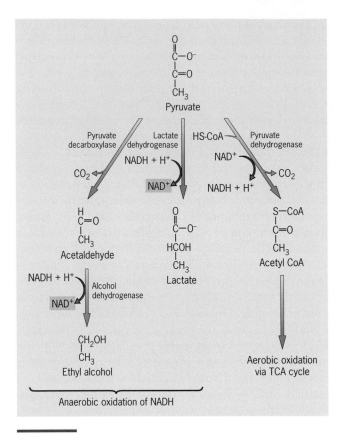

Figure 3.28 **Fermentation.** Most cells carry out aerobic respiration, which depends on molecular oxygen. If oxygen supplies should diminish, as occurs in a skeletal muscle cell undergoing strenuous contraction or a yeast cell living under anaerobic conditions, these cells are able to regenerate NAD^+ by fermentation. Muscle cells accomplish fermentation by formation of lactate, whereas yeast cells do so by formation of ethanol. Aerobic oxidation of pyruvate by means of the TCA cycle is discussed at length in Chapter 5.

molecules of ATP are formed per glucose oxidized by glycolysis and fermentation; over 90 percent of the energy is simply discarded in the fermentation product (as evidenced by the flammability of ethyl alcohol).

During the early stages of life on Earth, when oxygen had not yet appeared, glycolysis and fermentation were probably the primary metabolic pathways by which energy was extracted from sugars by primitive prokaryotic cells. Following the appearance of cyanobacteria, the oxygen of the atmosphere rose dramatically, and the evolution of an aerobic metabolic strategy became possible. As a result, the products of glycolysis could be completely oxidized, and much more ATP could be harvested. We will see how this is accomplished in Chapter 5 when we discuss the structure and function of the mitochondrion.

Reducing Power The energy required to synthesize complex biological molecules, such as proteins, fats, and nucleic acids, is derived largely from the ATP generated by glycolysis and electron transport. But many of these materials, par-

ticularly fats and other lipids, are more reduced than the metabolites from which they are built. Synthesis of fats requires the reduction of metabolites, which is accomplished by the transfer of high-energy electrons from NADPH, a compound similar in structure to NADH, but containing an additional phosphate group (described in the legend to Figure 3.26). A cell's reservoir of NADPH represents its **reducing power**, which is an important measure of the cell's usable energy content. The use of NADPH can be illustrated by one of the key reactions of photosynthesis:

1,3-Bisphosphoglycerate Glyceraldehyde 3-phosphate

In this reaction, a pair of electrons (together with a proton) are transferred from NADPH to the substrate 1,3-bisphosphoglycerate, reducing a carbon atom (indicated in red).

The oxidized form of NADPH, $NADP^+$, is formed from NAD^+ in the following reaction:

$$NAD^+ + ATP \rightleftharpoons NADP^+ + ADP$$

NADPH can then be formed by the reduction of $NADP^+$. Like NADH, NADPH is a high-energy compound because of its high electron-transfer potential. The transfer of free energy in the form of these electrons raises the acceptor to a more reduced, more energetic state.

The separation of reducing power into two distinct but related molecules, NADH and NADPH, reflects a separation of their primary metabolic roles. NADH and NADPH are recognized as coenzymes by different enzymes. Enzymes having a reductive role in anabolic pathways recognize NADPH as their coenzyme, whereas enzymes acting as dehydrogenases in catabolic pathways recognize NAD^+. Even though they are used differently, one coenzyme can reduce the other in the following reaction, which is catalyzed by the enzyme *transhydrogenase*:

$$NADH + NADP^+ \rightleftharpoons NAD^+ + NADPH$$

When energy is abundant, the production of NADPH is favored, providing a supply of electrons needed for biosynthesis of new macromolecules that are essential for growth. When energy resources are scarce, however, most of the high-energy electrons of NADH are "cashed in" for ATP, and only enough NADPH is generated to meet the cell's minimal biosynthetic requirements.

Metabolic Regulation

The amount of ATP present in a cell at a given moment is surprisingly small. A bacterial cell, for example, contains approximately one million molecules of ATP, whose half-

life is very brief (on the order of a second or two). With so limited a supply, it is evident that ATP is not a molecule in which a large total amount of free energy is stored. The energy reserves of a cell are stored instead as polysaccharides and fats. When the levels of ATP start to fall, reactions are set in motion to increase ATP formation at the expense of the energy-rich storage forms. Similarly, when ATP levels are high, reactions that would normally lead to ATP production are inhibited. Cells regulate these important energy-releasing reactions by controlling certain key enzymes in a number of metabolic pathways.

An enzyme's activity can be modified if its active site can be altered. Two common mechanisms for altering the shape of an enzyme's active site are covalent modification and allosteric modulation, both of which play key roles in regulating glucose oxidation.[4]

Altering Enzyme Activity by Covalent Modification

During the mid-1950s, Edmond Fischer and Edwin Krebs of the University of Washington were studying phosphorylase, an enzyme found in muscle cells that disassembles glycogen into its glucose subunits. The enzyme could exist in either an inactive or an active state. Fischer and Krebs prepared a crude extract of muscle cells and found that inactive enzyme molecules in the extract could be converted to active ones simply by adding ATP to the test tube. Further analysis revealed a second enzyme in the extract—a "converting enzyme," as they called it—that transferred a phosphate group from ATP to one of the 841 amino acids that make up the phosphorylase molecule. The presence of the phosphate group altered the shape of the active site of the enzyme molecule and increased its catalytic activity.

Subsequent research has shown that covalent modification of enzymes, as illustrated by the addition (or removal) of phosphates, is a general mechanism for changing the activity of enzymes. Enzymes that transfer phosphate groups to other proteins are called protein kinases and regulate such diverse activities as hormone action, cell division, and gene expression. The "converting enzyme" discovered by Krebs and Fischer was later named phosphorylase kinase, and its regulation is discussed in Section 15.2. There are two basically different types of protein kinases: one type adds phosphate groups to specific tyrosine residues in a substrate protein; the other type adds phosphates to specific serine or threonine residues in the substrate. The importance of protein kinases is reflected in the fact that approximately two percent of all the genes of a yeast cell (113 out of about 6200) encode members of this class of enzymes.

Altering Enzyme Activity by Allosteric Modulation Allosteric modulation is a mechanism whereby the activity of an enzyme is either inhibited or stimulated by a compound that binds to a site, called the **allosteric site**, that is spatially distinct from the enzyme's active site. Like the sequential collapse of a row of dominoes, the binding of a compound to the allosteric site sends a "ripple" through the protein, producing a defined change in the shape of the active site, which may be located on the opposite side of the enzyme or even on a different polypeptide within the protein molecule. Depending on the particular enzyme and the particular allosteric modulator, the change in shape of the active site may either stimulate or inhibit its ability to catalyze the reaction. Allosteric modulation illustrates the intimate relationship between molecular structure and function. Very small changes in the structure of the enzyme induced by the allosteric modulator can cause marked changes in enzyme activity.

Cells are highly efficient manufacturing plants that do not waste energy and materials producing compounds that are not utilized. One of the primary mechanisms cells use to shut down anabolic assembly lines is a type of allosteric modulation called **feedback inhibition**, in which the enzyme catalyzing the first committed step in a metabolic pathway is temporarily inactivated when the concentration of the end product of that pathway—an amino acid, for example—reaches a certain level. This is illustrated by the simple pathway depicted in Figure 3.29 in which two substrates, A and B, are converted to the end product E. As the concentration of the product E rises, it binds to the allosteric site of enzyme BC, causing a conformational change in the active site that decreases the enzyme's activity. Feedback inhibition exerts immediate, sensitive control over a cell's anabolic activity.

Separating Catabolic and Anabolic Pathways A brief consideration of the anabolic pathway leading to the synthesis of glucose (**gluconeogenesis**) will illustrate a few important aspects about synthetic pathways. Most cells are able to synthesize glucose from pyruvate at the same time they are oxidizing glucose as their major source of chemical energy. How are cells able to utilize these two opposing pathways? The first important point is that, even though enzymes can catalyze a reaction in both directions, the reactions of gluconeogenesis cannot proceed simply by the reversal of the reactions of glycolysis. The glycolytic pathway contains three thermodynamically irreversible reactions (Figure 3.24), and these steps must somehow be bypassed. Even if all the reactions of glycolysis could be reversed, it would be a very undesirable way for the cell to handle its metabolic activities, since the two pathways could not be controlled independently of one another. Thus, a cell could not shut down glucose synthesis and crank up glucose breakdown because the same enzymes would be active in both directions.

If one compares the pathway by which glucose is degraded to the pathway by which glucose is synthesized, it is apparent that some of the reactions are identical, although they are running in opposite directions, while oth-

[4]Metabolism is also regulated by controlling the concentrations of enzymes. The relative rates at which enzymes are synthesized and degraded are considered in subsequent chapters.

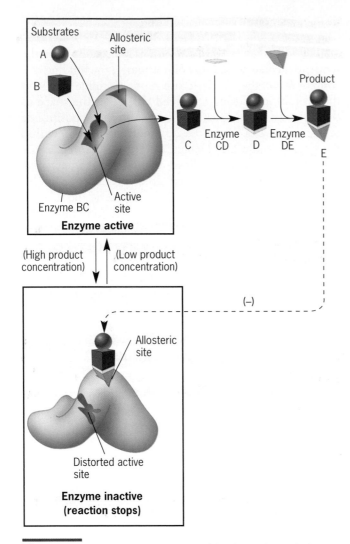

Figure 3.29 Feedback inhibition. The flow of metabolites through a metabolic pathway stops when the first enzyme of the pathway (enzyme BC) is inhibited by the end product of that pathway (compound E), which binds to an allosteric site on the enzyme. Feedback inhibition prevents a cell from wasting resources by continuing to produce compounds that are no longer required.

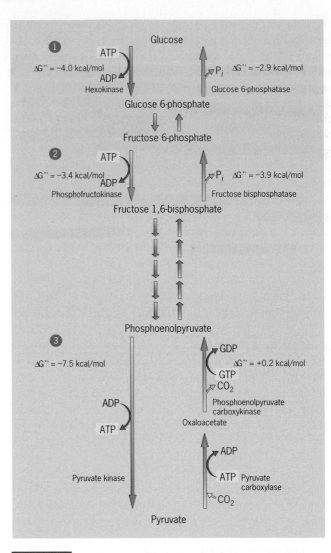

Figure 3.30 Glycolysis versus gluconeogenesis. Whereas most of the reactions are the same in the two pathways, even though they run in opposite directions, the three irreversible reactions of glycolysis (steps 1 to 3 here) are replaced in the gluconeogenic pathway by different, thermodynamically favored reactions.

ers are very different (steps 1 to 3, Figure 3.30). By using different enzymes to catalyze different key reactions in the two opposing pathways, a cell can solve both the thermodynamic and the regulatory problems inherent in being able to manufacture and degrade the same molecules.

We can see this better by looking more closely at key enzymes of both glycolysis and gluconeogenesis. As indicated in step 2 of Figure 3.30, *phosphofructokinase*, an enzyme of glycolysis, catalyzes the reaction

Fructose 6-phosphate + ATP \rightleftharpoons
\qquad Fructose 1,6-bisphosphate + ADP

which has a $\Delta G^{\circ\prime}$ of -3.4 kcal/mol, making it essentially irreversible. The reaction has such a large $-\Delta G^{\circ\prime}$ because

it is coupled to the hydrolysis of ATP. In gluconeogenesis, the formation of fructose 6-phosphate is catalyzed by the enzyme *fructose 1,6-bisphosphatase* by a simple hydrolysis:

Fructose 1,6-bisphosphate + H_2O \rightleftharpoons
\qquad Fructose 6-phosphate + P_i \qquad $\Delta G^{\circ\prime} = -3.9$ kcal/mol

The particular enzymes of glycolysis and gluconeogenesis described above are key regulatory enzymes in their respective pathways. Even though ATP happens to be a substrate of phosphofructokinase, ATP is also an allosteric inhibitor, while AMP is an allosteric activator. When ATP levels are high, the activity of the enzyme is decreased so that no additional ATP is formed by glycolysis. Conversely, when ADP and AMP levels are high rela-

tive to ATP, the activity of the enzyme is increased, which promotes the synthesis of ATP. In contrast, the activity of fructose 1,6-bisphosphatase, a key enzyme of gluconeogenesis, is inhibited by elevated levels of AMP.[5] As a result of these types of regulation, ATP levels generally do not fluctuate, but remain high despite great variations in demand.

We have concentrated in this chapter on the conservation of chemical energy in the form of ATP and its use in metabolism. The energy stored in ATP is used in a great variety of processes (e.g., see Figure 3.6), many of which will be discussed throughout this book. It can be noted here that ATP is not always used as described in this chapter to generate phosphorylated intermediates, such as glutamyl phosphate. In some cases, the phosphate is transferred to an amino acid residue of a protein to induce a conformational change, as occurs, for example, during the movement of sodium and potassium ions across the plasma membrane (see Figure 4.44).

[5]The activities of phosphofructokinase and fructose 1,6-bisphosphatase are also controlled by a compound, fructose 2,6-bisphosphate, which acts as an allosteric activator of the former enzyme and a competitive inhibitor of the latter enzyme. Thus, changes in the level of fructose 2,6-bisphosphate can act as a switch between the opposing pathways of glycolysis and gluconeogenesis. It is noteworthy that the synthesis of fructose 2,6-bisphosphate and its breakdown—two separate reactions—are catalyzed by a bifunctional enzyme, that is, an enzyme with two different active sites located in different domains of the same protein molecule.

REVIEW ?

1. Why are catabolic pathways described as convergent whereas anabolic pathways are described as divergent?

2. Compare the energy obtained by a cell that oxidizes glucose anaerobically and aerobically. What is the difference in the end products of these two types of metabolism?

3. Explain what is meant by phosphate transfer potential. How does the phosphate transfer potential of phosphoenolpyruvate compare with that of ATP? What does this mean thermodynamically (i.e., in terms of relative $\Delta G°'$s of hydrolysis)? What does it mean in terms of affinity for phosphate groups?

4. Why is reducing power considered a form of energy?

5. In the reaction in which glyceraldehyde 3-phosphate is oxidized to 3-phosphoglycerate, which compound is acting as the reducing agent? Which as the oxidizing agent? Which compound is being oxidized? Which is being reduced?

6. Which reactions of glycolysis are coupled to ATP hydrolysis? Which reactions involve substrate-level phosphorylation? Which reactions depend on either fermentation or aerobic respiration to continue?

EXPERIMENTAL PATHWAYS

Determining the Mechanism of Lysozyme Action

One day in 1922, while suffering from a cold, the Scottish bacteriologist Alexander Fleming discovered that a drop of nasal mucus, when added to a culture of bacteria, caused the lysis of the cells. The agent in the mucus responsible for killing the bacteria was discovered to be an enzyme that Fleming named lysozyme. Fleming's test of substances for their bacteria-killing activity was more than a casual interest. After watching hundreds of soldiers die from infected wounds during World War I, Fleming had set out on a lifelong search for an agent that would effectively kill microbes while, at the same time, remain relatively nontoxic to humans. Unlike penicillin, which Fleming discovered in 1928, lysozyme did not prove to have clinical value. Lysozyme has played an important role, however, in the study of enzyme mechanisms.

In 1965, David Phillips and his colleagues at Oxford University published a detailed model of the tertiary structure of lysozyme, the first enzyme whose three-dimensional structure had been determined using X-ray crystallography.[1,2] The lysozyme had been purified from the white of a hen's egg, where it protects the developing embryo from bacterial infection. Lysozyme lyses bacteria by hydrolyzing glycosidic bonds within the polysaccharide of the bacterial cell wall. The cell walls of sensitive bacteria (gram-positive bacteria) are composed of a copolymer of two alternating amino sugars, N-acetylglucosamine and N-acetylmuramic acid (Figure 1).

This normal substrate for lysozyme is a large polymeric polysaccharide. It was not feasible to study the enzyme as it was bound to its normal substrate, so Phillips and his colleagues set out to find a low-molecular-weight inhibitor that resembled the substrate but would not be hydrolyzed. The longest chain of sugars that they found would fit into the active site of lysozyme molecules when they were packed together in a crystal consisting of three linked units of N-acetylglucosamine, which we will refer to as (NAG)$_3$. When enzyme crystals were prepared in the presence of (NAG)$_3$ and then subjected to X-ray diffraction, the (NAG)$_3$ molecule

NAM NAG NAM NAG

Figure 1 Bacterial cell wall polysaccharide, the lysozyme substrate, consists of alternating *N*-acetylmuramic acid (NAM) and *N*-acetylglucosamine (NAG) residues.

was found to sit in a cleft within the egg-shaped enzyme, filling about one-half of the cleft's length. The trisaccharide was bound to the enzyme by hydrogen bonds and van der Waals forces. After constructing models of enzyme-substrate complexes, Phillips extrapolated the findings with (NAG)$_3$ to larger molecules and proposed that the active site of the enzyme resided in a cleft containing six subsites (A–F), each of which bound a single sugar along the polysaccharide chain. In other words, the cleft would normally be occupied by six adjacent sugar units of the cell wall polysaccharide (Figure 2).

Analysis of the structure of the enzyme–(NAG)$_3$ complex suggested a mechanism to account for the hydrolytic activity of the enzyme.[2,3] When six adjacent sugar units were bound in the cleft of the enzyme, one of the sugars (the fourth, or D, sugar of Figure 2) could not be readily accommodated in the available space. For this sugar to fit in the space adjacent to the tryptophan side chain, it would have to be forced out of its normal chair conformation (Figure 2.12e) and flattened into a shape approaching a half-chair or "sofa." Because this part of the substrate was subjected to physical strain, and for other reasons, Phillips proposed that the glycosidic bond that linked the sugars residing in subsites D and E (sugars 4 and 5) would be the bond that was hydrolyzed.

Close examination of the region of the enzyme in the vicinity of this glycosidic bond revealed that two amino acid residues approached within about 0.3 nm of the bond from either side. One residue was an aspartic acid and the other a glutamic acid, both of which have carboxyl-containing side chains. When the environment of these two residues was considered, it would be expected that the ionization states of the two carboxyl groups would be quite different. The environment of the glutamic acid is nonpolar, which would be expected to suppress the dissociation of its proton, while that of the aspartic acid is polar, which would promote the dissociation of the proton, leaving the aspartyl residue with a negative charge. Using this information based solely on the protein's structure, Phillips proposed the following catalytic mechanism (Figure 3).

In the first step of the proposed reaction mechanism, the splitting of the glycosidic bond is accomplished by the interaction of the bond with the nearby glutamic acid. The oxygen atom connecting sugars D and E is sufficiently electronegative to draw the proton from the undissociated carboxyl group of the glutamic acid, causing the acid-catalyzed hydrolysis of the bond in the substrate. The breaking of the bond by the proton leaves both the carbon and oxygen atoms of the substrate (the C1 and O5 atoms of sugar D, Figure 3)

Figure 2 Simplified model of a lysozyme molecule showing a hexasaccharide bound in the cleft of the enzyme. The locations of key amino acid residues of the enzyme are indicated.

with a partial positive charge. A molecule containing a shared positive charge between an O and C atom is called an *oxocarbonium ion*. The formation of the oxocarbonium ion is facilitated by the distortion of the sugar following its binding to the enzyme. This distorted, positively charged molecule corresponds to the substrate molecule as it exists at the transition state (page 95). It was noted that enzymes act by stabilizing the transition state, thereby favoring the formation of products. In the case of lysozyme, the positively charged oxo-

Figure 3 The mechanism for lysozyme action proposed by Phillips. The bond between the fourth (D) and fifth (E) sugar of the hexasaccharide residing in the cleft of the lysozyme molecule is cleaved by acid hydrolysis using a proton donated by the carboxyl group of the closely applied glutamic acid residue (Glu 35). The formation of the positively charged oxocarbonium ion at the C1 position of sugar D is facilitated by the distortion of the sugar shown in the figure and stabilized by the nearby aspartic acid residue (Asp52) of the enzyme. In the final step, the oxocarbonium ion reacts with an OH^- group from the solvent. (FROM D. VOET AND J. G. VOET, BIOCHEMISTRY, 2D ED. COPYRIGHT © 1995, JOHN WILEY & SONS, INC. REPRINTED BY PERMISSION OF JOHN WILEY & SONS, INC.)

Table 1	Relative Rates of Hydrolysis for N-Acetylglucosamine Oligosaccharides	
Saccharide	**Relative rate**	
$(NAG)_2$	0.003	
$(NAG)_4$	1.0	
$(NAG)_5$	4,000	
$(NAG)_6$	30,000	

Source: J. A. Rupley and V. Gates, *Proc. Nat'l. Acad. Sci. U.S.A.* 57:500, 1967.

was hydrolyzed by the enzyme as efficiently as a preparation of cell wall polysaccharide. The six-unit oligosaccharide is just large enough to fill the active site. As predicted, the hexasaccharide was split between the fourth and fifth sugar, generating products consisting of four and two sugar units.

$$(NAG)_6 \rightarrow (NAG)_4 + (NAG)_2$$

A key aspect of the model is the inability of the natural substrate to bind to the enzyme without being distorted. The distortion results from the presence of a bulky hydroxymethyl ($-CH_2OH$) group on carbon 6 of the fourth sugar of the substrate (see Figure 3). If the hydroxymethyl were to be removed, the substrate should be able to bind to the enzyme with greater ease. This is precisely what was found when the enzyme was incubated with a modified version of $(NAG)_4$ in which the fourth NAG was altered by removal of the CH_2OH group (the fourth sugar was an *N*-acetylxylosamine rather than *N*-acetylglucosamine). As predicted, this modified oligosaccharide bound 40 to 50 times more strongly to the enzyme than $(NAG)_4$.[5]

If the Phillips model for lysozyme is correct and only the fourth of six sugars in the active site is subjected to physical strain, then the binding of this sugar to its subsite in the enzyme should have different thermodynamic properties than the binding of the other sugars. It has been possible, by studying the kinetics in which dozens of different sugar compounds bind to lysozyme, to measure the favorability ($-\Delta G$) or unfavorability ($+\Delta G$) for the binding of each sugar in the substrate to its respective subsite.[6] When this is done, only the binding of a sugar residue to subsite D was found to be unfavorable (at least +2.9 kcal/mol), which, once again, supports the proposed model. The energy required to bind the fourth sugar is supplied by the energy released as the other sugars bind to the enzyme.

In Phillips' model, the distortion of the fourth sugar residue in the substrate places strain on the glycosidic bond, which aids in the bond's hydrolysis. This distorted sugar is an important element in the proposed structure of the substrate as it exists in the transition state. In 1948, Linus Pauling suggested that "enzymes are molecules that are complementary in structure to the activated complexes of the reactions they catalyze." In other words, it isn't the initial substrates to which the enzyme is most complementary, but rather the high-energy transition state that forms as the reactants are ready to be converted into products. If this is true, then inhibitors that resemble a transition state molecule should bind to an enzyme much more strongly than inhibitors that resemble the original substrate(s). Inhibitors of this type are called

carbonium ion is stabilized by the very close presence of a negatively charged aspartic acid residue in the enzyme. The reaction of the oxocarbonium ion with a hydroxyl ion from the solvent completes the hydrolysis. As a hallmark of a good model, the proposed mechanism of lysozyme made numerous predictions—which were soon put to the test.

Analysis of lysozyme's ability to hydrolyze substrates of various length provided early support for the model.[4] Oligosaccharides consisting of two, three, or four NAG units were not hydrolyzed by the enzyme (Table 1). When a five-unit substrate was offered, a low rate of hydrolysis was obtained. In contrast, a six-unit oligosaccharide (a hexamer)

transition state analogues (TSAs).[7] A transition state analogue for lysozyme, named TACL (tetra-*N*-acetylchitotetrose), was tested in 1972.[8] The analogue resembles (NAG)$_4$ except that its terminal sugar residue is oxidized to a δ-lactone

NAG
(at transition state)

TACL

that mimics the distorted conformation of the D sugar proposed by Phillips as a transition state structure. The binding of TACL to lysozyme was measured by its ability to inhibit the lysozyme-catalyzed lysis of a culture of sensitive bacterial cells. When tested near the pH where the enzyme is optimally active, it required more than 100 times the concentration of (NAG)$_4$ (an inhibitor that resembles the substrate) to cause the same level of inhibition of cell lysis as caused by TACL (an inhibitor that resembles the transition state). These results support the suggestion that the distorted substrate corresponds to a transition state. Calculations from the data indicated that TACL binds approximately 3600 times more strongly to the enzyme than does the unmodified tetrasaccharide (NAG)$_4$. These investigators concluded that the strain caused by substrate binding could increase the rate of catalysis by a factor of 10^3 to 10^4.

Another prediction made by Phillips' model is that two key amino acids, Asp 52 and Glu 35, should have very different ionization properties because of differences in the polarity of their environment. Lysozyme exhibits maximal activity at pH 5. At this pH, the model proposes that the carboxyl group of Glu 35 should be holding onto its proton (as —COOH), since this is the proton that is donated during acid catalysis. At this same pH, however, Asp 52 should be negatively charged (as —COO$^-$) and capable of stabilizing the positively charged oxocarbonium ion. The best measurements of the ionization state of these two carboxyl groups have come from a technique called circular dichroism (CD), which depends on the absorption of polarized light and is very sensitive to changes in the state of a protein. The CD spectrum of lysozyme shows a strong absorption band at a wavelength of 305 nm that can be traced to a specific tryptophan residue (Trp 108) in the protein. The Trp 108 is very close to Glu 35, which, in turn, is close to Asp 52. Because of these proximities, changes in the pH that alter the ionization state of the carboxyl groups of Glu 35 or Asp 52 have an effect on the CD spectrum derived from Trp 108. Comparisons of the CD data obtained at varying pHs indicate that the pK of Glu 35 (the pH at which half of the groups are protonated and half are ionized) is unusually high at 6.1 (a typical value would be 4.4).[9] The pK of Asp 52, in contrast, is about 3.4. As predicted by the model, the glutamyl carboxyl group would retain its proton at pH 5, while the aspartyl carboxyl would be negatively charged.

Another means to study key amino acids is to selectively modify those residues and measure the effect on the enzyme's catalytic activity. Once this has been done, the enzyme can be digested into fragments and the precise modification verified. The first report of the activity of a chemically modified lysozyme appeared in 1969.[10] These investigators found that when they treated lysozyme with an agent (triethyloxonium fluoroborate) that "removed" the carboxyl group of Asp 52 (by converting it into an ethyl ester), the modified enzyme lost all catalytic activity even though it continued to bind substrate with high affinity. Subsequent experiments in which either Asp 52 or Glu 35 was modified chemically confirmed the expectation that both of these residues must remain in their native state for the enzyme to remain active.[11]

With the development of new techniques in DNA technology, it has become possible to make deletions, additions, or substitutions to any polypeptide for which the gene has previously been isolated. Using this technique, called site-directed mutagenesis (SDM), the nucleotide sequence of the DNA is specifically altered so that an amino acid in the polypeptide is replaced by another amino acid of the investigator's choosing. Any amino acid in the protein can be replaced, and the investigator can be certain that all of the protein molecules produced possess the alteration. The first use of SDM in the study of lysozyme was published in 1989.[12] In this report, the roles of both Asp 52 and Glu 35 were studied by testing the activity of mutant proteins in which these amino acid residues had been replaced. As expected from previous studies, these mutant proteins had very little or no catalytic activity.

The early crystallographic studies of Blake, Phillips, and colleagues produced a relatively low-resolution portrait of the enzyme-substrate complex. More recent studies of the molecule at high resolution (1.5 Å) have confirmed the prediction that the sugar located in site D adopts a strained conformation, which allows it to fit into the available space. All of the atoms that make up the ring of the bound sugar lie in essentially the same plane, which is very different from their arrangement in the normal chair conformation.[13] This same study also found evidence that the Asp 52 residue was hydrogen bonded to an asparagine residue (Asn 46). This hydrogen bond would be expected to stabilize the negative charge on Asp 52, which would in turn promote the latter's ability to stabilize the oxocarbonium intermediate. Subsequent studies in which the Asp 52 and/or Asn 46 residues were mutated has provided support for this interpretation. When the Asn 46 residue is mutated to an alanine, the ability of the Asp 52 residue to bind to the substrate is disrupted and so too is the catalytic activity of the enzyme.[14]

We can conclude that the reaction mechanism proposed by Phillips more than 30 years ago has stood the test of time remarkably well.[15]

References

1. BLAKE, C. C. F., ET AL. 1965. Structure of hen egg-white lysozyme. *Nature* 206:757–761.
2. PHILLIPS, D. C. 1966. The three-dimensional structure of an enzyme molecule. *Sci. Am.* 215:78–90. (Nov.)
3. PHILLIPS, D. C. 1967. The hen egg-white lysozyme molecule. *Proc. Nat'l. Acad. Sci. U.S.A.* 57:484–495.

4. RUPLEY, J. A. & GATES, V. 1967. Studies on the enzymic activity of lysozyme. *Proc. Nat'l. Acad. Sci. U.S.A.* 57:496–510.

5. VAN EIKEREN, P. & CHIPMAN., D. M. 1972. Substrate distortion in catalysis by lysozyme: Interaction of lysozyme with oligosaccharides containing N-acetylxylosamine. *J. Am. Chem. Soc.* 94:4788–4790.

6. CHIPMAN, D. M. & SHARON, N. 1969. Mechanism of lysozyme action. *Science* 165:454–464.

7. WOLFENDEN, R. 1969. Transition state analogues for enzyme catalysis. *Nature* 223:704–705.

8. SECEMSKI, I., ET AL. 1972. A transition state analog for lysozyme. *J. Biol. Chem.* 247:4740–4748.

9. KURAMITSU, S., ET AL. 1974. Ionization constants of Glu 35 and Asp 52 in hen, turkey and human lysozymes. *J. Biochem.* 76:671–683.

10. PARSONS, S. M. & RAFTERY, M. A. 1969. The identification of aspartic acid residue 52 as being critical to lysozyme activity. *Biochemistry* 8:4199–4205.

11. KUROKI, R., ET AL. 1986. Chemical mutations of the catalytic carboxyl groups in lysozyme to the corresponding amides. *J. Biol. Chem.* 261:13571–13576.

12. MALCOLM, B. A., ET AL. 1989. Site-directed mutagenesis of the catalytic residues Asp-52 and Glu-35 of chicken egg white lysozyme. *Proc. Nat'l. Acad. Sci. U.S.A.* 86:133–137.

13. STRYNADKA, N. C. J. & JAMES, M. N. G. 1991. Lysozyme revisited: Crystallographic evidence for distortion of an N-acetylmuramic acid residue bound in site D. *J. Mol. Biol.* 220:401–424.

14. MATSUMURA, I. & KIRSCH, J. F. 1996. Synergistic contributions of asparagine 46 and aspartate 52 to the catalytic mechanism of chicken egg white lysozyme. *Biochemistry* 35:1890–1896.

15. JOHNSON, L. N. 1998. The early history of lysozyme. *Nature Struct. Biol.* 5:942–944.

SYNOPSIS

Energy is the capacity to do work. Energy can occur in various forms, including chemical, mechanical, light, electrical, and thermal energy, which are interconvertible. Whenever energy is exchanged, the total amount of energy in the universe remains constant, but there is a loss of free energy, that is, energy available to do additional work. The usable energy lost to entropy results from an increase in the randomness and disorder of the universe. Living organisms are systems of low entropy, maintained by the constant input of external energy that is ultimately derived from the sun. (p. 84)

All spontaneous (exergonic) energy transformations proceed from a state of higher free energy to a state of lower free energy; the ΔG must be negative. In a chemical reaction, ΔG is equivalent to the difference in the free-energy content between the reactants and products. The greater the ΔG, the farther the reaction is from equilibrium. As the reaction proceeds, ΔG decreases and reaches 0 at equilibrium. To compare the energy changes during different chemical reactions, free-energy differences between reactants and products are determined for a set of standard conditions and denoted as $\Delta G^{\circ\prime}$, thus $\Delta G^{\circ\prime} = -2.303\, RT \log K'_{eq}$. Reactions having equilibrium constants greater than one have negative $\Delta G^{\circ\prime}$ values. It must be kept in mind that $\Delta G^{\circ\prime}$ is a fixed value that describes the direction in which a reaction proceeds when the reaction mixture is at standard conditions. It is of no value in determining the direction of a reaction at a particular moment in the cell; this is governed by ΔG, which is dependent on the concentrations of reactants and products at the time. (p. 87)

The hydrolysis of ATP is a highly favorable reaction ($\Delta G^{\circ\prime} = -7.3$ kcal/mol) and can be used to drive reactions that would otherwise be unfavorable. The use of ATP hydrolysis to drive unfavorable reactions is illustrated by glutamine synthesis from glutamic acid and NH_3 ($\Delta G^{\circ\prime} = +3.4$ kcal/mol). The reaction is driven through the formation of a common intermediate, glutamyl phosphate. ATP hydrolysis can play a role in such processes because the cell maintains an elevated [ATP]/[ADP] ratio that is far above that at equilibrium, which illustrates that cellular metabolism operates under nonequilibrium conditions. This does not mean that every reaction is kept from equilibrium. Rather, certain key reactions in a metabolic pathway occur with large negative ΔG values, which makes them essentially irreversible within the cell and allows them to drive the entire pathway. The concentrations of reactants and products can be maintained at relatively constant nonequilibrium (steady state) values within a cell because materials are continually flowing into the cell from the external medium. (p. 89)

Enzymes are proteins that vastly accelerate the rate of specific chemical reactions by binding to the reactant(s) and increasing the likelihood that they will be converted to products. Like all true catalysts, enzymes are present in small amounts; they are not altered irreversibly during the course of the reaction; and they have no effect on the thermodynamics of the reaction. Enzymes, therefore, cannot cause an unfavorable reaction (one with a $+\Delta G$) to proceed in the forward direction, nor can they change the ratio of products to reactants at equilibrium. Enzymes are also characterized by specificity toward their substrates, highly efficient catalysis with virtually no undesirable by-products, and the opportunity for their catalytic activity to be regulated. (p. 92)

Enzymes act by lowering the activation energy (E_A)—the kinetic energy required by reactants to undergo reaction. As a result, a much greater percentage of reactant molecules possess the necessary energy to be converted to products in the presence of an enzyme. Enzymes lower E_A through the formation of an enzyme-substrate complex. The part of the enzyme that binds the substrate(s) is called the active site, which also contains the necessary amino acid side chains and/or cofactors to influence the substrates so as to facilitate the chemical transformation. Among the mechanisms that facilitate catalysis, enzymes are able to hold reactants in the proper orientation; they are able to make the substrates more reactive by influencing their electronic character; and they

are able to exert physical stress that weakens certain bonds within the substrate. (p. 94)

Metabolism is the collection of biochemical reactions that occur within the cell. These reactions can be grouped into metabolic pathways containing a sequence of chemical reactions in which each reaction is catalyzed by a specific enzyme. Metabolic pathways are divided into two broad types: catabolic pathways, in which compounds are disassembled and energy is released, and anabolic pathways that lead to the synthesis of more complex compounds using energy stored in the cell. Macromolecules of diverse structure are degraded by catabolic pathways into a relatively small number of low-molecular-weight metabolites that provide the raw materials from which anabolic pathways diverge. Both types of pathways include oxidation-reduction reactions in which electrons are transferred between substrates. (p. 105)

The state of reduction of an organic molecule, as measured by the number of hydrogens per carbon atom, provides a rough measure of the molecule's energy content. A mole of glucose releases 686 kcal when completely oxidized to CO_2 and H_2O, whereas the conversion of a mole of ADP to ATP requires only 7.3 kcal. Thus, the oxidation of a molecule of glucose can produce enough energy to generate a large number of ATP molecules. The first stage in the catabolism of glucose is glycolysis, during which glucose is converted to pyruvate with a net gain of two molecules of ATP and two molecules of NADH. The ATP molecules are produced by substrate-level phosphorylation in which a phosphate group is transferred from a substrate to ADP. The NADHs are produced by oxidation of an aldehyde to a carboxylic acid with the accompanying transfer of a hydride ion (a proton and two electrons) from the substrate to NAD^+. In the presence of O_2, most cells oxidize the NADH by means of an electron-transport chain, forming ATP by aerobic respiration. In the absence of O_2, NAD^+ is regenerated by fermentation in which the high-energy electrons of NADH are used to reduce pyruvate. NAD^+ regeneration is required for glycolysis to continue. (p. 107)

An enzyme's activity is commonly regulated by two mechanisms: covalent modification and allosteric modulation. Covalent modification is most often accomplished by transfer of a phosphate group from ATP to one or more serine, threonine, or tyrosine residues of the enzyme in a reaction catalyzed by a protein kinase. Allosteric modulators act by binding noncovalently to a site on the enzyme that is removed from the active site. Binding of the modulator alters the conformation of the active site, increasing or decreasing the enzyme's catalytic activity. The same compound degraded by a catabolic pathway may serve as the end product of an anabolic pathway. Glucose, for example, is degraded by glycolysis and synthesized by gluconeogenesis. While most of the enzymes are shared by the two pathways, three key enzymes are unique to each pathway, which allows the cell to regulate the two pathways independently and to overcome what would otherwise be irreversible reactions. (p. 113)

ANALYTIC QUESTIONS

1. How would you expect a drop in pH to affect a reaction catalyzed by chymotrypsin? by lysozyme? How might an increase in pH affect these two reactions?

2. Feedback inhibition typically alters the activity of the first enzyme of a metabolic pathway rather than one of the latter enzymes of the pathway. Why is this adaptive?

3. After reviewing the reactions of glutamine formation on page 90, explain why each of the following statements concerning the third (or overall) reaction is either true or false.

 a. If the reaction were written in reverse, its $\Delta G^{\circ\prime}$ would be +3.9 kcal/mol.

 b. If all reactants and products were at standard conditions at the beginning of an experiment, after a period of time, the [NH$_3$]/[ADP] ratio would decrease.

 c. As the reaction proceeds, the $\Delta G^{\circ\prime}$ moves closer to zero.

 d. At equilibrium, the forward and reverse reactions are equal and the [ATP]/[ADP] ratio becomes one.

 e. In the cell, it is possible for glutamine to be formed when the [glutamine]/[glutamic acid] ratio is greater than one.

4. You have just isolated a new enzyme and have determined the velocity of reaction at three different substrate concentrations. You find that the slope of the product versus time curve is the same for all three concentrations. What can you conclude about the conditions in the reaction mixture?

5. Lysozyme is a slow-acting enzyme, requiring approximately two seconds to catalyze a single reaction. What is the turnover number of lysozyme?

6. In the reaction $R \rightleftharpoons P$, if one mole of product (P) has the same free energy as one mole of reactant (R), what is the value for the K_{eq} of this reaction? What is the value of the $\Delta G^{\circ\prime}$?

7. What is meant in terms of concentration ratios when it is said that the ΔG of ATP hydrolysis in the cell is approximately -12 kcal/mol, whereas the $\Delta G^{\circ\prime}$ is -7.3 kcal/mol?

8. The enzymes under cellular regulation are those whose reactions typically proceed under nonequilibrium conditions. What would be the effect of allosteric inhibition of an enzyme that operated close to equilibrium?

9. In the reaction $A \rightleftharpoons B$, if the K'_{eq} is 10^3, what is the $\Delta G^{\circ\prime}$? What is the $\Delta G^{\circ\prime}$ if the K'_{eq} had been determined to be 10^{-3}? What is the K'_{eq} of the hexokinase reaction indicated in Figure 3.23 (step 1)?

10. The $\Delta G^{\circ\prime}$ of the reaction acetyl phosphate + ADP \rightleftharpoons acetate + ATP is -2.8 kcal/mol. Molecule for molecule, acetyl phosphate has (greater, lesser, equal) free energy than does ATP relative to its dephosphorylated compound; ADP has (greater, lesser, equal) affinity for phosphate than acetate. (Circle the correct answers.)

11. If the reaction XA + Y \rightleftharpoons XY + A has a $\Delta G^{\circ\prime}$ of +7.3 kcal/mol, could this reaction be driven in the cell by coupling it to ATP hydrolysis? Why or why not?

12. In a series of reactions, $A \rightarrow B \rightarrow C \rightarrow D$, it was determined that the equilibrium constant for the second reac-

tion (B → C) is 0.1. You would expect the concentration of C in a living cell to be: (1) equal to B, (2) one-tenth of B, (3) less than one-tenth of B, (4) ten times that of B, (5) more than ten times that of B. (Circle any correct answer.)

13. The reaction of compound X with compound Y to produce compound Z is an unfavorable reaction ($\Delta G°' = +5$ kcal/mol). Draw the chemical reactions that would occur if ATP was utilized to drive the reaction.

14. ATP has evolved as the central molecule in energy metabolism. Could 1,3-bisphosphoglycerate serve the same function? Why or why not?

15. Calculate the ΔG for ATP hydrolysis in a cell in which the [ATP]/[ADP] ratio had climbed to 100:1, while the P_i concentration remained at 10 mM. How does this compare to the ratio of [ATP]/[ADP] when the reaction is at equilibrium and the P_i concentration remains at 10 mM? What would be the value for ΔG when the reactants and products were all at standard conditions (1 M)?

16. Consider the reaction:

Glucose + P_i ⇌ Glucose 6-phosphate + H_2O
$$\Delta G°' = +3 \text{ kcal/mol}$$

What is the equilibrium constant, K'_{eq}, for this reaction? (*Note*: the concentration of water is ignored.) Does the positive $\Delta G°'$ in the above reaction mean that the reaction can never go spontaneously from left to right?

17. Under physiological conditions, [Glucose] = 5mM, [Glucose 6-phosphate] = 83mM, and [P_i] = 1mM. Under these conditions, will the reaction of problem 16 go spontaneously from left to right? If not, what would the concentration of glucose need to be for the reaction to go from left to right, if the concentrations of the other reactants and products are as stated above?

18. Consider the reaction:

glutamate + ammonia ⇌ glutamine + water
$$\Delta G°' = +3.4 \text{ kcal/mol}$$

If the concentration of ammonia is 10 mM, what is the ratio of glutamate/glutamine required for the reaction to proceed spontaneously from left to right at 25°C?

19. It should be clear that the synthesis of glutamine cannot occur by the reaction described in problem 18. The actual reaction couples glutamine synthesis to ATP hydrolysis:

glutamate + ammonia + ATP ⇌ glutamine + ADP + P_i

What is the $\Delta G°'$ for this reaction? Suppose that all reactants and products, except ammonia, are present at 10 mM. What concentration of ammonia would be required to drive the reaction to the right, that is, to drive the net synthesis of glutamine?

20. A noncompetitive inhibitor does not prevent the enzyme from binding its substrate. What will be the effect of increasing the substrate concentration in the presence of a noncompetitive inhibitor? Do you expect a noncompetitive inhibitor to affect the enzyme V_{max}? K_M? Explain briefly.

ADDITIONAL READING

Energy

HAROLD, F. M. 1986. *The Vital Force: A Study of Bioenergetics*. W. H. Freeman.

HARRIS, D. A. 1995. *Bioenergetics at a Glance*. Blackwell.

KLOTZ, I. M. 1986. *Introduction to Biomolecular Energetics*. Academic Press.

LEHNINGER, A. L. 1971. *Bioenergetics*, 2d ed. Benjamin-Cummings.

RACKER, E. 1976. *A New Look at Mechanisms in Bioenergetics*. Academic Press.

Enzymes and Metabolism

BLOW, D. M. 1997. The tortuous story of Asp . . . His . . . Ser: Structural analysis of α-chymotrypsin. *Trends Biochem. Sci.* 22:405–408.

BLOW, D. M. 2000. So do we understand how enzymes work? *Sructure* 8:R77-R81.

BOYER, P. D., ED. 1970–present. *The Enzymes*, 3d ed. Academic Press. (vols. 19, 20 on mechanisms of enzyme catalysis, 1990, 1992).

CUSHMAN, D. W. & ONDETTI, M. A. 1999. Design of angiotensin converting enzyme inhibitors. *Nat. Med.* 5:1110–1112.

DRESSLER, D. & POTTER, H. 1991. *Discovering Enzymes*. Scientific American Library.

FERSHT, A. 1998. *Structure and Mechanism in Protein Science: A Guide to Enzyme Catalysts and Protein Folding*. W.H. Freeman.

JENCKS, W. P. 1997. From chemistry to biochemistry to catalysis to movement. *Ann. Rev. Biochem.* 66:1–18.

KAUFMAN, B. T. 1993. Why NADP? *Trends Biochem. Sci.* 18:278.

KORNBERG, A. 1997. Centenary of the birth of modern biochemistry. *Trends Biochem. Sci.* 22:282–283.

KORNBERG, A. 1989. *For the Love of Enzymes*. Harvard.

KRAUT, J. 1988. How do enzymes work? *Science* 242:533–540.

KREBS, H. A. & A. MARTIN. 1981. *Reminiscences and Reflections*. Oxford.

LERNER, R. A., ET AL. 1998. Making enzymes. (on catalytic antibodies). *Harvey Lects.* 92:1–40.

LIPMANN, F. 1971. *Wanderings of a Biochemist*. Wiley.

PALMER, T. 1995. *Understanding Enzymes*. Prentice-Hall.

PHILLIPS, D. C. 1966. The three-dimensional structure of an enzyme molecule. *Sci. Am.* 215:78–90. (Nov.)

SCHRAMM, V. L. 1998. Enzymatic transition states and transition state analog design. *Annu. Rev. Biochem.* 67:693–720.

STROUD, R. M. 1974. A family of protein-cutting proteins. *Sci. Am.* 231:74–88. (July)

WALSH, C. 1979. *Enzymatic Reaction Mechanisms*. W. H. Freeman.

WALSH, C., ET AL. 2001. Reviews on biocatalysis. *Nature.* 409:226–268.

WESTHEIMER, F. H. 1987. Why nature chose phosphates. *Science* 235:1173–1178.

Antibiotic Resistance

ANDERSON, R. M. 1999. The pandemic of antibiotic resistance. *Nat. Med.* 5:147–149.

LEVY, S. B. 1998. The challenge of antibiotic resistance. *Sci. Am.* 278:46–53. (March)

TAN, Y.-T., ET AL. 2000 Molecular strategies for overcoming antibiotic resistance in bacteria. *Mol. Med. Today* 6:309–314.

WALSH, C. 2000. Molecular mechanisms that confer antibacterial drug resistance. *Nature* 406:775–781.

4

The Structure and Function of the Plasma Membrane

The outer walls of a house or car provide a strong, inflexible barrier that protects its human inhabitants from an unpredictable and harsh external world. You might expect the outer boundary of a living cell to be constructed of an equally tough and impenetrable barrier because it must also protect its delicate internal contents from a nonliving, and often inhospitable, environment. Yet cells are separated from the external world by a thin, fragile structure called the **plasma membrane** that is only 5 to 10 nm wide. It would require about ten thousand plasma membranes stacked one on top of the other to equal the thickness of a single page of this book.

Because it is so thin, no hint of the plasma membrane is detected when a section of a cell is examined under a light microscope. In fact, it wasn't until the late 1950s that techniques for preparing and staining tissue had progressed to the point where the plasma membrane could be resolved in the electron microscope. These early electron micrographs, such as those taken by J. D. Robertson of Duke University, portrayed the plasma membrane as a three-layered structure, consisting of two darkly staining outer layers and a lightly staining middle layer (Figure 4.1*a*). All membranes that were examined closely—whether they

A model of a fully hydrated lipid bilayer composed of phosphatidylcholine molecules (each containing two myristoyl fatty acids) penetrated by a single transmembrane helix composed of 32 alanine residues. (FROM LIYANG SHEN, DONNA BASSOLINO, AND TERRY STOUCH, BRISTOL-MYERS SQUIBB RESEARCH INSTITUTE, FROM BIOPHYSICAL JOURNAL, VOL. 73, P. 6, 1997.)

were plasma, nuclear, or cytoplasmic membranes (Figure 4.1*b*), or taken from plants, animals, or microorganisms—showed this same ultrastructure. In addition to providing a visual image of this critically important cellular structure, these electron micrographs touched off a vigorous debate as to the molecular composition of the various layers of a membrane, an argument that went to the very heart of the subject of membrane structure and function. We will return to the structure of membranes shortly, but first we will survey some of the major functions of membranes in the life of a cell (Figure 4.2). ■

4.1 AN OVERVIEW OF MEMBRANE FUNCTIONS

1. ***Compartmentalization.*** Membranes are continuous, unbroken sheets and, as such, inevitably enclose compartments. The plasma membrane encloses the contents of the entire cell, whereas the nuclear and cytoplasmic membranes enclose diverse intracellular spaces. The various membrane-bound compartments of a cell possess markedly different contents. Because of compartmentalization, specialized activities can proceed with a mini-

mum of outside interference and can be regulated independently of one another.

2. ***Scaffold for biochemical activities.*** Membranes not only enclose compartments but are also a distinct compartment themselves. As long as reactants are present in solution, their relative positions cannot be stabilized and their interactions are dependent on random collisions. Because of their construction, membranes provide the cell with an extensive framework or scaffolding within which components can be ordered for effective interaction.

3. ***Providing a selectively permeable barrier.*** Membranes prevent the unrestricted exchange of molecules from one side to the other. At the same time, membranes provide the means of communication between the compartments they separate. The plasma membrane, which encircles a cell, can be compared to a moat around a castle: both serve as a general barrier, yet both have gated "bridges" that promote the movement of desirable elements into and out of the enclosed living space.

4. ***Transporting solutes.*** The plasma membrane contains the machinery for physically transporting substances from one side of the membrane to another, often from a region where the solute is present at low concentration into a region where that solute is present at much higher concentration. The membrane's transport machinery allows a cell to accumulate substances, such as sugars and amino acids, that are necessary to fuel its metabolism and build its macromolecules. The plasma membrane is also able to transport specific ions, thereby establishing ionic gradients across itself. This capability is especially critical for nerve and muscle cells.

(a) 50 nm

Figure 4.1 The trilaminar appearance of membranes.
(*a*) Electron micrograph showing the three-layered (trilaminar) structure of the plasma membrane of an erythrocyte. The arrows denote the inner and outer edges of the membrane. (*b*) The outer edge of a differentiated muscle cell grown in culture showing the similar trilaminar structure of both the plasma membrane (PM) and the membrane of the sarcoplasmic reticulum (SR), a calcium-storing compartment of the cytoplasm. (A: Courtesy of J. D. Robertson; b: from Andrew R. Marks et al., J. Cell Biol. 114:305, 1991; by copyright permission of the Rockefeller University Press.)

(b) 0.1 μm

Figure 4.2 A summary of membrane functions in a plant cell.
(1) An example of membrane compartmentalization in which hydrolytic enzymes (acid hydrolases) are sequestered within the membrane-bound vacuole. (2) An example of the role of cytoplasmic membranes as a site of enzyme localization. The fixation of CO_2 by the plant cell is catalyzed by an enzyme that is associated with the outer surface of the thylakoid membranes of the chloroplasts. (3) An example of the role of membranes as a selectively permeable barrier. Water molecules are able to penetrate rapidly through the plasma membrane, causing the plant cell to fill out the available space and exert pressure against its cell wall. (4) An example of solute transport. Hydrogen ions, which are produced by various metabolic processes in the cytoplasm, are pumped out of plant cells into the extracellular space by a transport protein located in the plasma membrane. (5) An example of the involvement of a membrane in the transfer of information from one side to another (signal transduction). In this case, a hormone (e.g., abscisic acid) binds to the outer surface of the plasma membrane and triggers the release of a chemical message (such as IP_3) into the cytoplasm. In this case, IP_3 causes the release of Ca^{2+} ions from a cytoplasmic warehouse. (6) An example of the role of membranes in cell-cell communication. Openings between adjoining plant cells, called plasmodesmata, allow materials to move directly from the cytoplasm of one cell into its neighbors. (7) An example of the role of membranes in energy transduction. The conversion of ADP to ATP occurs in close association with the inner membrane of the mitochondrion.

5. *Responding to external signals.* The plasma membrane plays a critical role in the response of a cell to external stimuli, a process known as **signal transduction**. Membranes possess **receptors** that combine with specific molecules (or **ligands**) having a complementary structure. Different types of cells have membranes with different types of receptors and are, therefore, capable of recognizing and responding to different ligands in their environment. The interaction of a plasma membrane receptor with an external ligand may cause the membrane to generate a signal that stimulates or inhibits internal activities. For example, signals generated at the plasma membrane may tell a cell to manufacture more glycogen, to prepare for cell division, to move toward a higher concentration of a particular compound, to release calcium from internal stores, or possibly to commit suicide.

6. *Intercellular interaction.* Situated at the outer edge of every living cell, the plasma membrane of multicellular organisms mediates the interactions between a cell and its neighbors. The plasma membrane allows cells to recognize and signal one another, to adhere when appropriate, and to exchange materials and information.

7. *Energy transduction.* Membranes are intimately involved in the processes by which one type of energy is converted to another type (energy transduction). The most fundamental energy transduction occurs during photosynthesis when energy in sunlight is absorbed by membrane-bound pigments and converted into chemical energy contained in carbohydrates. Membranes are also involved in the transfer of chemical energy from carbohydrates and fats to ATP. In eukaryotes, the machinery for these energy conversions is contained within membranes of the chloroplasts and mitochondria.

We will concentrate in this chapter on the structure and functions of the plasma membrane, but remember that the principles discussed here are common to all cell membranes. Specialized aspects of the structure and functions of mitochondrial, chloroplast, cytoplasmic, and nuclear membranes will be discussed in Chapters 5, 6, 8, and 12, respectively.

4.2 A BRIEF HISTORY OF STUDIES ON PLASMA MEMBRANE STRUCTURE

The first insights into the chemical nature of the outer boundary layer of a cell were obtained by Ernest Overton of the University of Zürich during the 1890s. Overton knew that nonpolar solutes dissolved more readily in nonpolar solvents than in polar solvents, and that polar solutes had the opposite solubility property. Overton reasoned that a substance entering a cell from the medium would first have to dissolve in the outer boundary layer of that cell. To test the permeability of the outer boundary layer,

Stationary barrier · Lipids · Movable barrier

(a)

(b)

(c)

Figure 4.3 **The plasma membrane contains a lipid bilayer.** (*a*) Calculating the surface area of a lipid preparation. When a sample of phospholipids is dissolved in an organic solvent, such as hexane, and spread over an aqueous surface, the phospholipid molecules form a layer over the water that is a single molecule thick: a monomolecular layer. The molecules in the layer are oriented with their hydrophilic groups bonded to the surface of the water and their hydrophobic chains directed into the air. To estimate the surface area the lipids would cover if they were part of a membrane, the lipid molecules are compressed into the smallest possible area by means of movable barriers. Using this type of apparatus, Gorter and Grendel concluded that red blood cells contained enough lipid to form a layer over their surface that was two molecules thick: a bilayer. (*b*) As Gorter and Grendel first proposed, the core of a membrane contains a bimolecular layer of phospholipids oriented with their water-soluble head groups facing the outer surfaces and their hydrophobic fatty acid tails facing the interior. The structures of the head groups are given in Figure 4.6*a*. (*c*) Simulation of a fully hydrated lipid bilayer composed of the phospholipid phosphatidylcholine. Phospholipid head groups are orange, water molecules are blue & white, fatty acid chains are green. (c. FROM S.-W. CHIU, TRENDS IN BIOCHEM. SCI. 22:341, 1997, COPYRIGHT 1997, WITH PERMISSION FROM ELSEVIER SCIENCE.)

Overton placed plant root hairs into hundreds of different solutions containing solutes of varying solubility. He discovered that the more lipid soluble the compound, the more rapidly it would enter the root hair cells (see p. 151). He concluded that the dissolving power of the outer boundary layer of the cell matched that of a fatty oil.

The first proposal that cellular membranes might contain a lipid bilayer was made in 1925 by two Dutch scientists, E. Gorter and F. Grendel. These researchers extracted the lipid from human red blood cells and measured the amount of surface area the lipid would cover when spread over the surface of water (Figure 4.3*a*). Since mature mammalian red blood cells lack both nuclei and cytoplasmic organelles, the plasma membrane is the only lipid-containing structure and all of the lipids extracted from the cells can be assumed to have resided in the cells' plasma membranes. The ratio of the surface area of water covered by the extracted lipid to the surface area calculated for the red blood cells from which the lipid was extracted varied between 1.8 to 1 and 2.2 to 1. Gorter and Grendel guessed that the actual ratio was 2 : 1 and therefore concluded that the plasma membrane contained a bimolecular layer of lipids, that is, a **lipid bilayer** (Figure 4.3*b*). They also suggested that the polar groups of each molecular layer (or **leaflet**) were directed outward toward the aqueous environment, as shown in Figure 4.3*b*, *c*. This would be the thermodynamically favored arrangement because the polar head groups of the lipids could interact with surrounding water molecules, while the hydrophobic fatty acyl chains would be protected from the aqueous environment (Figure 4.3*c*). Thus, the polar head groups would face the cytoplasm on one edge and the blood plasma on the other. Even though Gorter and Grendel made several miscalculations (which fortuitously compensated for one another), they still arrived at the correct conclusion that membranes contain a lipid bilayer.

In the 1920s and 1930s, cell physiologists obtained evidence that there must be more to the structure of membranes than simply a lipid bilayer. It was found, for example, that lipid solubility was not the sole determining factor as to whether or not a substance could penetrate the plasma membrane. Similarly, the surface tensions of membranes were calculated to be much lower than those of pure lipid structures. This decrease in surface tension might be explained by the presence of protein in the membrane. Support for this idea came from the discovery that a protein film between a layer of water and lipid would greatly lower the tension at the interface between the two layers. In 1935,

Hugh Davson and James Danielli proposed that the plasma membrane was composed of a lipid bilayer that was lined on both its inner and outer surface by a layer of globular proteins. They revised their model in the early 1950s to account for the selective permeability of the membranes they had studied. In the revised version of their model (Figure 4.4a), Davson and Danielli suggested that, in addition to the outer and inner protein layers, the lipid bilayer was also penetrated by protein-lined pores, which could provide conduits for polar solutes and ions to enter and exit the cell.

Experiments conducted in the late 1960s led to a new concept of membrane structure, as detailed in the fluid-mosaic model proposed in 1972 by S. Jonathan Singer and Garth Nicolson of the University of California, San Diego (Figure 4.4b). In the **fluid-mosaic model**, which has served as the "central dogma" of membrane biology for three decades, the lipid bilayer remains the core of the membrane, but attention is now focused on the physical state of the lipid. Unlike previous models, the bilayer of a fluid-mosaic membrane is present in a fluid state, and in-

Figure 4.4 A brief history of the structure of the plasma membrane. (a) A revised 1954 version of the Davson-Danielli model showing the lipid bilayer, which is lined on both surfaces by a monomolecular layer of proteins that extend through the membrane to form protein-lined pores. (b) The fluid-mosaic model of membrane structure as initially proposed by Singer and Nicolson in 1972. Unlike previous models, the proteins penetrate the lipid bilayer. Although the original model shown here depicted a protein that was only partially embedded in the bilayer, lipid-penetrating proteins that have been studied span the entire bilayer. (c) A current representation of the plasma membrane showing the same basic organization as that proposed by Singer and Nicolson. We now know that the external surface of most membrane proteins, as well as a small percentage of the phospholipids, contain short chains of sugars, making them glycoproteins and glycolipids. Those portions of the polypeptide chains that extend through the lipid bilayer typically occur as α helices composed of hydrophobic amino acids. (A: FROM J. F. DANIELLI, COLLSTON PAPERS 7:8, 1954; B: REPRINTED WITH PERMISSION FROM S. J. SINGER AND G. L. NICOLSON, SCIENCE 175:720, 1972; COPYRIGHT 1972 AMERICAN ASSOCIATION FOR THE ADVANCEMENT OF SCIENCE.

dividual lipid molecules can move laterally within the plane of the membrane.

The structure and arrangement of membrane proteins in the fluid-mosaic model differ from that of previous models in that they occur as a "mosaic" of discontinuous particles that penetrate the lipid sheet (Figure 4.4b). Most importantly, the fluid-mosaic model presents cellular membranes as dynamic structures in which the components are mobile and capable of coming together to engage in various types of transient or semipermanent interactions. In the following sections, we will examine some of the evidence used to formulate and support this dynamic portrait of membrane structure and look at some of the recent data that bring the model up to date (Figure 4.4c).

REVIEW ?

1. Describe some of the important roles of membranes in the life of a eukaryotic cell.

2. Summarize some of the major steps leading to the current model of membrane structure.

4.3 THE CHEMICAL COMPOSITION OF MEMBRANES

Membranes are lipid-protein assemblies in which the components are held together in a thin sheet by noncovalent bonds. As noted above, the core of the membrane consists of a sheet of lipids arranged in a bimolecular layer (Figure 4.3b, c). The lipid bilayer serves primarily as a structural backbone of the membrane and provides the barrier preventing the random movements of water-soluble materials into and out of the cell. The proteins of the membrane, on the other hand, carry out most of the specific functions summarized in Figure 4.2. In addition to lipid and protein, membranes also contain carbohydrates (Figure 4.4c). The ratio of lipid to protein varies considerably (Table 4.1) depending on the type of cellular membrane (plasma vs. endoplasmic reticulum vs. Golgi), the type of organism (prokaryote vs. plant vs. animal), and the type of cell (cartilage vs. muscle vs. liver). For example, the inner mitochondrial membrane has a very high ratio of protein/lipid in comparison to the red blood cell plasma membrane, which is high in comparison to the membranes of the myelin sheath that form a multilayered wrapping around a nerve cell (Figure 4.5). To a large degree these differences can be correlated with the particular functions of these membranes. The inner mitochondrial membrane contains the protein carriers of the electron-transport chain, and relative to other membranes, lipid is diminished. In contrast, the myelin sheath acts primarily as electrical insulation for the nerve cell it encloses, a function that is best carried out by a thick lipid layer of high electrical resistance with a minimal content of protein.

Membrane Lipids

Membranes contain a wide diversity of lipids, all of which are **amphipathic**; that is, they contain both hydrophilic and hydrophobic regions. There are three main types of

Table 4.1	Protein and Lipid Content of Membranes		
Membrane	**Protein/lipid (wt/wt)**	**Cholesterol/polar lipid (mol/mol)**	**Major polar lipids***
Myelin	0.25	0.7–1.2	Cer, PE, PC
Plasma membranes			
Liver cell	1.0–1.4	0.3–0.5	PC, PE, PS, Sph
Ehrlich ascites	2.2		
Intestinal villi	4.6	0.5–1.2	
Erythrocyte ghost	1.5–4.0	0.9–1.0	Sph, PE, PC, PS
Endoplasmic reticulum	0.7–1.2	0.03–0.08	PC, PE, Sph
Mitochondrion			DPG, PC, PE, Plas
Outer membrane	1.2	0.03–0.09	
Inner membrane	3.6	0.02–0.04	
Retinal rods	1.5	0.13	PC, PE, PS
Chloroplast lamellae	0.8	0	GalDG, SL, PS
Bacteria			
Gram-positive	2.0–4.0	0	DPG, PG, PE, PGaa
Gram-negative		0	PE, PG, DPG, PA
Mycoplasma	2.3	0	
Halophilic	1.8	0	Ether analog PGP

*Abbreviations are: Cer, cerebrosides; DPG, diphosphatidylglycerol; GalDG, galactosyldiglyceride; PA, phosphatidic acid; PC, phosphatidylcholine; PE, phosphatidylethanolamine; Pgaa, amino acyl esters of phosphatidyglycerol; Plas, plasmalogen; SL, sulfolipid; Sph, sphingomyelin.

Source: E. D. Korn, Reproduced with permission from the *Annual Review of Biochemistry*, vol. 39, © 1969, by Annual Reviews Inc.

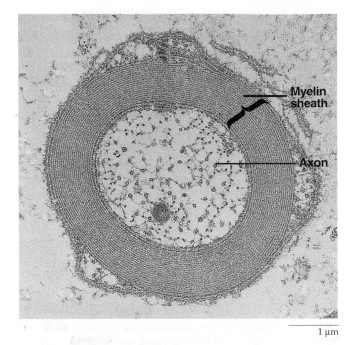

1 μm

Figure 4.5 **The myelin sheath**. Electron micrograph of a nerve cell axon surrounded by a myelin sheath consisting of concentric layers of plasma membrane that have an extremely low protein/lipid ratio. The myelin sheath insulates the nerve cell from the surrounding environment, which increases the velocity at which impulses can travel along the axon (discussed on page 169). The perfect spacing between the layers is maintained by interlocking protein molecules (called P_0) that project from each membrane. (FROM LEONARD NAPOLITANO, FRANCIS LEBARON, AND JOSEPH SCALETTI, J. CELL BIOL. 34:820, 1967; BY COPYRIGHT PERMISSION OF THE ROCKEFELLER UNIVERSITY PRESS.)

membrane lipids: phosphoglycerides, sphingolipids, and cholesterol.

Phosphoglycerides Most membrane lipids contain a phosphate group, which makes them **phospholipids**. Because most membrane phospholipids are built on a glycerol backbone, they are called **phosphoglycerides** (Figure 4.6a). Unlike triglycerides, which have three fatty acids (page 49) and are not amphipathic, membrane glycerides are *diglycerides*—only two of the hydroxyl groups of the glycerol are esterified to fatty acids; the third is esterified to a hydrophilic phosphate group. Without any additional substitutions beyond the phosphate and the two fatty acyl chains, the molecule is called *phosphatidic acid*, which is virtually absent in most membranes. Instead, membrane phosphoglycerides have an additional group linked to the phosphate, most commonly either choline (forming *phosphatidylcholine*, PC), ethanolamine (forming *phosphatidylethanolamine*, PE), serine (forming *phosphatidylserine*, PS), or inositol (forming *phosphatidylinositol*, PI). Each of these groups is small and hydrophilic and, together with the negatively charged phosphate to which it is attached, forms a highly water-soluble domain at one end of the molecule, called the **head group**. At physiological pH, the head groups of PS and PI have an overall negative charge, whereas those of PC and PE are neutral. In contrast, the fatty acyl chains are hydrophobic, unbranched hydrocarbons approximately 16 to 20 carbons in length (Figure 4.6). A membrane fatty acid may be fully saturated (i.e., lack double bonds), monounsaturated (i.e., possess one double bond), or polyunsaturated (i.e., possess more than one double bond). Phosphoglycerides often contain one unsaturated and one saturated fatty acyl chain. With fatty acid chains at one end of the molecule and a polar head group at the other end, phosphoglycerides exhibit a distinct amphipathic character.

Sphingolipids A less abundant class of membrane lipids, called **sphingolipids**, are derivatives of sphingosine, an amino alcohol that contains a long hydrocarbon chain (Figure 4.6b). Sphingolipids consist of sphingosine linked to a fatty acid (R of Figure 4.6b) by its amino group. This molecule is a *ceramide*. The various sphingosine-based lipids have additional groups esterified to the terminal alcohol of the sphingosine moiety. If the substitution is phosphorylcholine, the molecule is *sphingomyelin*, which is the only phospholipid of the membrane that is not built with a glycerol backbone. If the substitution is a carbohydrate, the molecule is a **glycolipid**. If the carbohydrate is a simple sugar, the glycolipid is called a **cerebroside**; if it is an oligosaccharide, the glycolipid is called a **ganglioside**. Since all sphingolipids have two long, hydrophobic hydrocarbon chains on one end and a hydrophilic region at the other, they are also amphipathic and basically similar in overall structure to the phosphoglycerides.

Glycolipids are interesting membrane components. Relatively little is known about them, yet tantalizing hints have emerged to suggest they play crucial roles in cell function. The nervous system is particularly rich in glycolipids. The myelin sheath pictured in Figure 4.5 contains a high content of a particular glycolipid, called galactocerebroside (shown in Figure 4.6b), which is formed when a galactose is added to ceramide. Mice lacking the enzyme that carries out this reaction exhibit severe muscular tremors and eventual paralysis. Interest in glycolipids has increased in recent years following the discovery that a group of fungal toxins, called fumonisins, act by inhibiting synthesis of these membrane components. Fumonisins disturb such diverse processes as cell growth, cell death, cell–cell interactions, and communication from the outside of the cell to the interior. Fumonisins are also carcinogenic; their presence in beer produced from moldy grains by African villagers has been linked to the high incidence of esophageal cancer in these individuals. Glycolipids also play a role in certain infectious diseases; the toxins that cause cholera and botulism both enter their target cell by first binding to cell-surface gangliosides, as does the influenza virus.

Cholesterol Another lipid component of certain membranes is the sterol **cholesterol** (see Figure 2.21), which in

Dioleoyl phosphatidic acid

Phosphatidic acid

Phosphatidyl-choline (lecithin)

Phosphatidyl-serine

Phosphatidyl-ethanolamine (cephalin)

Phosphatidyl-inositol

Diphosphatidyl-glycerol (cardiolipin)

(a)

Sphingosine

Ceramide

Sphingomyelin

A cerebroside

A Ganglioside (G$_{M2}$)

(b)

Figure 4.6 The chemical structure of membrane lipids. (*a*) The structures of phosphoglycerides (see also Figure 2.22). (*b*) The structures of sphingolipids. Sphingomyelin is a phospholipid; cerebrosides and gangliosides are glycolipids. A third membrane lipid is cholesterol, which is shown in the next figure. (R = fatty acyl chain). [The green portion of each lipid, which represents the hydrophobic tail(s) of the molecule, is actually much longer than the hydrophilic head group (see Figure 4.23).]

certain animal cells may constitute up to 50 percent of the lipid molecules in the plasma membrane. Cholesterol is absent from the plasma membranes of most plant and all bacterial cells. Cholesterol is smaller than the other lipids of the membrane and less amphipathic. As shown in Figure 4.7, cholesterol molecules are oriented with their small hydrophilic hydroxyl group toward the membrane surface and the remainder of the molecule embedded in the lipid bilayer. The hydrophobic rings of a cholesterol molecule are flat and rigid, and they interfere with the movements of the fatty acid tails of the phospholipids (page 141).

The Nature and Importance of the Lipid Bilayer Each type of cellular membrane has its own characteristic lipid composition, differing from one another in the types of lipids, the nature of the head groups, and the particular species of fatty acyl chain(s). Because of this structural variability, it is estimated that some biological membranes may contain over a hundred chemically distinct species of phospholipid. The percentages of some of the major types

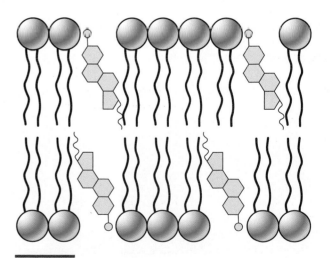

Figure 4.7 The cholesterol molecules of a membrane are oriented with their small hydrophilic end facing the external surface of the bilayer and the bulk of their structure packed in among the fatty acid tails of the phospholipids. The placement of the cholesterol molecules interferes with the tight packing of the phospholipids, which tends to increase the fluidity of the bilayer. Unlike most lipids of the membrane, cholesterol is often rather evenly distributed between the two layers (leaflets).

Table 4.2 Lipid Compositions of Some Biological Membranes*

Lipid	Human erythrocyte	Human myelin	Beef heart mitochondria	E. coli
Phosphatidic acid	1.5	0.5	0	0
Phosphatidylcholine	19	10	39	0
Phosphatidylethanolamine	18	20	27	65
Phosphatidylgycerol	0	0	0	18
Phosphatidylserine	8.5	8.5	0.5	0
Cardiolipin	0	0	22.5	12
Sphingomyelin	17.5	8.5	0	0
Glycolipids	10	26	0	0
Cholesterol	25	26	3	0

*The values given are weight percent of total lipid.
Source: C. Tanford, *The Hydrophobic Effect*, p. 109, copyright 1980, John Wiley & Sons, Inc. Reprinted by permission of John Wiley & Sons, Inc.

of lipids of a variety of membranes are given in Table 4.2. The lipids of a membrane are more than simple structural elements; they can have important effects on the biological properties of a membrane. Lipid composition can determine the physical state of the membrane (page 140) and influence the activity of particular membrane proteins. Membrane lipids also provide the precursors for highly active chemical messengers that regulate cellular function (Section 15.2).

Various types of measurements indicate that the combined fatty acyl chains of the lipid bilayer span a width of about 30 Å and that each row of head groups (with its adjacent shell of water molecules) adds another 15 Å (refer to the chapter-opening illustration on page 122). Thus, the entire lipid bilayer is only about 60 Å (6 nm) thick.

The presence in membranes of this thin double sheet of amphipathic lipid molecules has remarkable consequences for cell structure and function. Because lipids assemble into closed, bimolecular sheets, membranes are never seen to have a free edge; they are always continuous, unbroken structures. As a result, membranes form extensive interconnected networks within the cell. Because of the flexibility of the lipid bilayer, membranes are deformable and their overall shape can change, as occurs during locomotion (Figure 4.8*a*) or cell division (Figure 4.8*b*). The lipid bilayer is thought to facilitate the fusion or budding of membranes. For example, the events of secretion, in which cytoplasmic vesicles fuse to the plasma membrane, or of fertilization, where two cells fuse to form a single cell (Figure 4.8*c*), involve processes in which two separate

(a) *(b)* *(c)*

Figure 4.8 The dynamic properties of the plasma membrane. (*a*) The leading edge of a moving cell often contains sites where the plasma membrane displays undulating ruffles. (*b*) Division of a cell is accompanied by the deformation of the plasma membrane as it is pulled toward the center of the cell. Unlike most dividing cells, the cleavage furrow of this dividing

ctenophore egg begins at one pole and moves unidirectionally through the egg. (*c*) Membranes are capable of fusing with other membranes. This sperm and egg are in a stage leading to the fusion of their plasma membranes. (A: COURTESY OF JEAN PAUL REVEL; B: COURTESY OF GARY FREEMAN; C: COURTESY OF A. L. COLWIN AND L. H. COLWIN.)

membranes come together to become one continuous sheet (see Figure 8.31). The importance of the lipid bilayer in maintaining the proper internal composition of a cell, in separating electric charges across the plasma membrane, and in many other cellular activities will be apparent throughout this and subsequent chapters.

Another important feature of the lipid bilayer is its ability to self assemble, which can be demonstrated more easily within a test tube than a living cell. If, for example, a small amount of phosphatidylcholine is dispersed in an aqueous solution, the phospholipid molecules assemble spontaneously to form the walls of fluid-filled spherical vesicles, called **liposomes**. The walls of these liposomes consist of a continuous lipid bilayer that is organized in the same manner as that of the lipid bilayer of a natural membrane. Liposomes have proven invaluable in membrane research. Membrane proteins can be inserted into liposomes and their function studied in a much simpler environment than that of a natural membrane. Liposomes are also being tested as vehicles to deliver drugs or DNA molecules within the body. The drugs or DNA can be linked to the wall of the liposome or contained at high concentration within its lumen (Figure 4.9a). In these studies, the walls of the liposomes are constructed to contain specific proteins (such as antibodies or hormones) that allow the liposomes to bind selectively to the surfaces of particular target cells where the drug or DNA is intended to go. Most of the early clinical studies with liposomes met with failure because the injected vesicles were rapidly removed by phagocytic cells of the immune system. This obstacle has now been overcome with the devel-

opment of so-called "stealth liposomes" that contain an outer coating of a synthetic polymer that protects the liposomes from immune destruction (Figure 4.9b). Stealth liposomes containing doxorubicin have been approved for the treatment of Kaposi's sarcoma, and numerous clinical trials are underway to determine their effectiveness against other types of cancer.

Membrane Carbohydrates

The plasma membranes of eukaryotic cells contain carbohydrates that are covalently linked to both lipid and protein components (see Figure 4.4c). Depending on the species and cell type, the carbohydrate content of the plasma membrane ranges between 2 and 10 percent by weight. The plasma membrane of the red blood cell, for example, which is depicted in Figure 4.1a, contains approximately 52 percent protein, 40 percent lipid, and 8 percent carbohydrate. More than 90 percent of the membrane's carbohydrate is covalently linked to proteins to form glycoproteins; the remaining carbohydrate is covalently linked to lipids to form glycolipids, which were discussed on page 128. As indicated in Figure 4.4c, all of the carbohydrate of the plasma membrane faces outward into the extracellular space.[1] The carbohydrate of internal cellular membranes also faces away from the cytosol (the basis for this orientation is illustrated in Figure 8.14).

[1]It can be noted that even though phosphatidylinositol contains a sugar group (Figure 4.6), it is not considered to be part of the carbohydrate portion of the membrane in this discussion.

(a)

300 nm

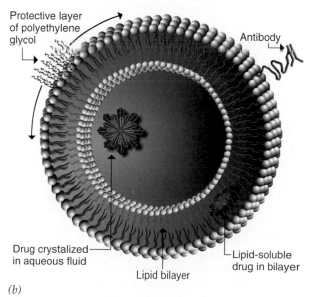

(b)

Figure 4.9 **Liposomes.** (a) These liposomal vesicles, which range in diameter from about 40 to 100 nm, contain the anticancer drug doxorubicin. At high concentration, doxorubicin sulfate forms insoluble gel fibers, seen as dark bands within the vesicles. The walls of the liposomes consist of a single lipid bilayer as shown in the inset. (b) A schematic diagram of a stealth liposome containing a hydrophilic polymer (such as polyethylene glycol) to protect it from destruction by immune cells, antibody molecules that target it to specific body tissues, a watersoluble drug enclosed in the fluid-filled interior chamber, and a lipid-soluble drug in the bilayer. (A: COURTESY OF P. M. FREDERIK.)

Figure 4.10 Two types of linkages that join sugars to a polypeptide chain. The *N*-glycosidic linkage between asparagine and *N*-acetylglucosamine is more common than the *O*-glycosidic linkage between serine or threonine and *N*-acetylgalactosamine.

Figure 4.11 Blood group antigens. Whether a person has type A, B, AB, or O blood is determined by a short chain of sugars covalently attached to membrane lipids and proteins of the red blood cell membrane. The oligosaccharides attached to membrane lipids (forming a ganglioside) that produce the A, B, and O blood types are shown here. A person with type AB blood has gangliosides with both the A and B structure. (Gal, galactose; GlcNAc, *N*-acetylglucosamine; Glu, glucose; Fuc, fucose; GalNAc, *N*-acetylgalactosamine.)

The carbohydrate of glycoproteins is present as short, branched oligosaccharides, typically having fewer than about 15 sugars per chain. In contrast to most high-molecular-weight carbohydrates (such as glycogen, starch or cellulose), which are polymers of a single sugar, the oligosaccharides attached to membrane proteins and lipids can display considerable variability in composition and structure. Oligosaccharides may be attached to several different amino acids by two major types of linkages (Figure 4.10). These carbohydrate projections play a role in mediating the interactions of a cell with its environment (Chapter 7) and sorting of membrane proteins to different cellular compartments (Chapter 8). The carbohydrates of the glycolipids of the red blood cell plasma membrane determine whether a person's blood type is A, B, AB, or O (Figure 4.11). A person having blood type A has an enzyme that adds an *N*-acetylgalactosamine to the end of the chain, whereas a person with type B blood has an enzyme that adds galactose to the chain terminus. People with AB blood type possess both enzymes, whereas people with O blood type lack enzymes capable of attaching either terminal sugar.

REVIEW **?**

1. Draw the basic structure of the major types of lipids found in cellular membranes. How do sphingolipids differ from glycerolipids? Which lipids are phospholipids? Which are glycolipids? How are these lipids organized into a bilayer?

2. What is a liposome? How are liposomes used in medical therapies?

3. What is an oligasaccharide? How are they linked to membrane proteins?

4.4 THE STRUCTURE AND FUNCTIONS OF MEMBRANE PROTEINS

Depending on the cell type and the particular organelle within that cell, a membrane may contain hundreds of different proteins. Each membrane protein has a defined orientation relative to the cytoplasm, so that the properties of one surface of a membrane are very different from those of the other surface. This asymmetry is referred to as membrane "sidedness." In the plasma membrane, for example, those parts of membrane proteins that interact with other cells or with extracellular ligands, such as hormones or growth factors, project outward into the extracellular space, whereas those parts of membrane proteins that interact with cytoplasmic molecules, such as G proteins or protein kinases (Chapter 15), project into the cytosol. Membrane proteins can be grouped into three distinct classes distinguished by the intimacy of their relationship to the lipid bilayer (Figure 4.12). These are

1. **Integral proteins** that penetrate the lipid bilayer. In

(a) Integral membrane proteins

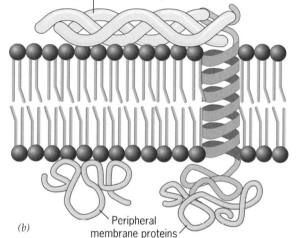

Peripheral membrane protein

(b) Peripheral membrane proteins

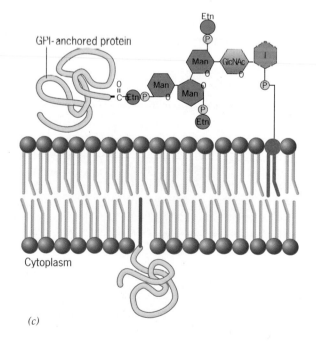

GPI-anchored protein

Etn

Man GlcNAc I

Man Man

Etn

Cytoplasm

(c)

Figure 4.12 **Three classes of membrane protein.** (*a*) Integral proteins typically contain one or more transmembrane helices (see Figure 5.3 for an exception). (*b*) Peripheral proteins are noncovalently bonded to the polar head groups of the lipid bilayer and/or to an integral membrane protein. (*c*) Lipid-anchored proteins are covalently bonded either to a phospholipid or a fatty acid that is embedded in one of the leaflets of the lipid bilayer. I, inositol; GlcN, *N*-acetylglucosamine; Man, mannose; Etn, ethanolamine.

fact, all integral proteins are probably **transmembrane proteins**; that is, they pass entirely through the lipid bilayer and thus have domains that protrude from both the extracellular and cytoplasmic sides of the membrane. Some integral proteins have only one membrane-spanning segment, whereas others are multispanning.

2. **Peripheral proteins** that are located entirely outside of the lipid bilayer, on the cytoplasmic or extracellular side, yet are associated with the surface of the membrane by noncovalent bonds.

3. **Lipid-anchored proteins** that are located outside the lipid bilayer, on either the extracellular or cytoplasmic surface, but are covalently linked to a lipid molecule that is situated within the bilayer.

Integral Membrane Proteins

Like the phospholipids of the bilayer, integral membrane proteins are also amphipathic, having both hydrophilic and hydrophobic portions. As discussed below, those portions of an integral membrane protein that reside within the lipid bilayer tend to have a hydrophobic character. Amino acid residues in these transmembrane domains form hydrophobic interactions with the fatty acyl chains of the bilayer, which seals the protein into the lipid "wall" of the membrane. As a result, the permeability barrier of the membrane is preserved and the protein is brought into direct contact with surrounding lipid molecules. Those portions of an integral membrane protein that project into either the cytoplasm or extracellular space tend to be more like the globular proteins discussed in Section 2.5. These nonembedded domains tend to have hydrophilic surfaces that interact with water-soluble substances (low-molecular-weight substrates, hormones, and other proteins) at the edge of the membrane. One group of integral membrane proteins—those that contain aqueous channels—has passageways through the lipid bilayer that are lined by hydrophilic residues. As will be discussed later, integral proteins need not be fixed structures but may be able to move laterally within the membrane itself.

Distribution of Integral Proteins: Freeze-Fracture Analysis The concept that proteins penetrate through membranes, rather than simply remaining external to the

..r, was derived primarily from the results of a technique called **freeze-fracture replication** (see Section 18.2). In this procedure, tissue is frozen solid and then struck with a knife blade, which fractures the block into two pieces. One of the favored paths taken by the fracture plane is between the two leaflets of the lipid bilayer (Figure 4.13*a*). Once the membranes are split in this manner, metals are deposited on their exposed surfaces to form a shadowed *replica*, which is viewed in the electron microscope (see Figure 18.16). As shown in Figure 4.13*b*, the replica resembles a road strewn with pebbles, which are called *membrane-associated particles*. Since the fracture plane passes through the center of the bilayer, most of these particles correspond to integral membrane proteins that extend at least halfway through the lipid core. When the fracture plane reaches a given particle, it goes around it rather than cracking it in half. Consequently, each protein (particle) separates with one half of the plasma membrane (Figure 4.13*c*), leaving a corresponding pit in the other half. One of the great values of the freeze-fracturing technique is that it allows an investigation of the microheterogeneity of the membrane. Localized differences in parts of the membrane stand out in these replicas and can be identified, as shown in the replica of a small portion of a sperm

cell in Figure 4.14. Unlike biochemical analyses, microscopic observations do not average out such individualities, but allow them to be appreciated.

Studying the Structure and Properties of Integral Membrane Proteins

Because of their hydrophobic transmembrane domains, integral membrane proteins are difficult to isolate in a soluble form. Removal of these proteins from the membrane normally requires the use of a detergent, such as the ionic (charged) detergent SDS (which denatures proteins) or the nonionic (uncharged) detergent Triton X-100 (which generally does not alter a protein's tertiary structure).

$$CH_3—(CH_2)_{11}—OSO_3^-Na^+$$

Sodium dodecyl sulfate (SDS)

Triton X-100

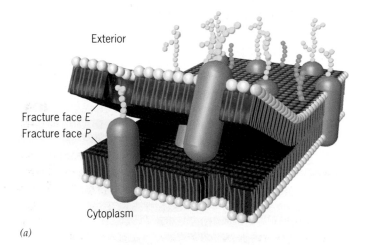

Exterior

Fracture face *E*
Fracture face *P*

Cytoplasm

(a)

Figure 4.13 Freeze fracture: a technique for investigating cell membrane structure. (*a*) When a block of frozen tissue is struck by a knife blade, a fracture plane runs through the tissue, often following a path that leads it through the middle of the lipid bilayer. The fracture plane goes around the proteins rather than cracking them in half, and they segregate with one of the two halves of the bilayer. The exposed faces within the center of the bilayer can then be covered with a metal deposit to form a metallic replica. These exposed faces are referred to as the E, or ectoplasmic face, and the P, or protoplasmic face. (*b*) Replica of a freeze-fractured human erythrocyte. The P fracture face is seen to be covered with particles approximately 8 nm in diameter. A small ridge (arrow) marks the junction of the particulate face with the surrounding ice. (*c*) This micrograph shows the surface of an erythrocyte that was frozen and then fractured, but rather than preparing a replica, the cell was thawed, fixed, and labeled

P

(b) 0.2 µm

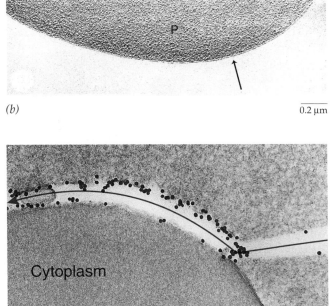

Cytoplasm

(c) 0.3 µm

with a marker for the carbohydrate groups that project from the external surface of the integral protein glycophorin (Figure 4.18). Thin sections of the labeled, fractured cell reveal that glycophorin molecules (black particles) have preferentially segregated with the outer half of the membrane. The red line shows the path of the fracture plane. (B: FROM THOMAS W. TILLACK AND VINCENT T. MARCHESI, J. CELL BIOL. 45:649, 1970; C: FROM PEDRO PINTO DA SILVA AND MARIA R. TORRISI, J. CELL BIOL. 93:467, 1982; B,C: BY COPYRIGHT PERMISSION OF THE ROCKEFELLER UNIVERSITY PRESS.)

Like membrane lipids, detergents are amphipathic, being composed of a polar end and a nonpolar hydrocarbon chain (see Figure 2.20). As a consequence of their struc-ture, detergents can substitute for phospholipids in stabilizing integral proteins while rendering them soluble in aqueous solution (Figure 4.15). Once the proteins have been solubilized by the detergent, various analytic procedures can be carried out to determine the protein's amino acid composition, molecular weight, amino acid sequence, and so forth.

Researchers have had great difficulty obtaining crystals of most integral membrane proteins for use in X-ray crystallography. Consequently, the structures of only a handful of membrane proteins have been determined at atomic resolution.[2] One of the first membrane proteins whose structure was determined at atomic resolution is shown in Figure 4.16. This protein—the bacterial photosynthetic reaction center—consists of three subunits containing 11 membrane-spanning α helices. While some of the technical difficulties in preparing membrane protein crystals have been overcome, researchers still rely largely on indirect approaches for solving the three-dimensional puzzle of membrane protein organization. We will examine some of these approaches in the following paragraphs.

Determining Membrane Sidedness Which parts of an integral plasma membrane protein project into the cytoplasm and which project outside the cell? This determination can be made experimentally using nonpenetrating agents that either label or modify the proteins. Consider what would happen if a preparation of intact cells were treated

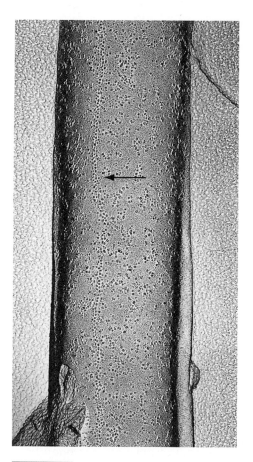

Figure 4.14 Heterogeneity in the distribution of integral membrane proteins. When the plasma membrane of a mammalian sperm tail is exposed during freeze fracture, it is seen to contain a double row of particles called the "zipper" (arrow). The freeze-fracture technique is ideal for revealing this type of microheterogeneity within a membrane. (COURTESY OF DANIEL S. FRIEND.)

[2]Many integral membrane proteins have a substantial portion that is present in the cytoplasm or extracellular space. In many cases, this soluble portion has been cleaved from its transmembrane domain, crystallized, and its tertiary structure determined. While this approach provides valuable data about the protein, it doesn't provide any information about its orientation within the membrane.

Figure 4.15 Solubilization of membrane proteins with detergents. The nonpolar end of the detergent molecules associates with the nonpolar residues of the protein, leaving the polar end of the detergent to dissolve the complex in aqueous solution. Nonionic detergents, as shown here, solubilize membrane proteins without disrupting their structure.

Figure 4.16 An integral protein as it resides within the plasma membrane. Tertiary structure of the photosynthetic reaction center of a bacterium as determined by X-ray crystallography. The protein contains three different membrane-spanning polypeptides, shown in yellow, light blue, and dark blue. (FROM G. FEHER, J. P. ALLEN, M. Y. OKAMURA, D. C. REES, REPRINTED WITH PERMISSION FROM NATURE 339:113, 1989; COPYRIGHT 1989, MACMILLAN MAGAZINES LIMITED.)

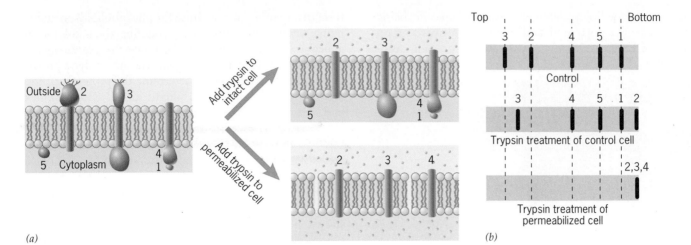

(a)

(b)

Figure 4.17 **Experimental procedure for determining the orientation of proteins within a plasma membrane.** The hypothetical membrane being studied contains five distinct proteins. (*a*) Treatment of intact cells with the nonpenetrating enzyme trypsin leads to digestion of those portions of the proteins that project into the external medium. In contrast, when a cell has been made permeable (by extraction with nonionic detergents or by exposure to a hypotonic medium), portions of membrane proteins that project into the cytoplasm are also accessible to proteolytic digestion. (*b*) Schematic diagram of the results of

SDS–polyacrylamide gel electrophoresis (SDS–PAGE) showing the pattern of bands obtained from the experiment described in *a*. The rate that a protein migrates during electrophoresis is inversely proportional to its molecular mass. Thus, removal of parts of a protein causes it to move more rapidly through the gel. Such proteins will be closer to the bottom of the gel at the end of the experiment than they would have been if they had not been attacked by the enzyme. An example of an SDS–PAGE gel is shown in Figure 4.32*c*.

with a proteolytic enzyme, such as trypsin, which is too large to penetrate the plasma membrane (Figure 4.17*a*, upper case). Those parts of the membrane's proteins that are situated on the external side of the lipid bilayer would be subjected to digestion by the added enzyme, but those parts within the bilayer or on the cytoplasmic face of the membrane could not be affected. The effect of the treatment can be monitored by extracting the proteins and subjecting them to SDS-polyacrylamide gel electrophoresis (discussed at length in Section 18.7). Those proteins that have had a significant portion of their structure digested migrate to a lower position in the gel than the same protein from an untreated membrane (Figure 4.17*b*). To determine whether a portion of a protein projects from the cytoplasmic face of the membrane, the cells can be made permeable by treatment with nonionic detergents or by osmotic shock (see Figure 4.32*b*). Under these conditions, the plasma membrane no longer acts as a barrier to the penetration of the proteolytic enzyme, so that the cytoplasmic portions of the protein are also subjected to digestion (lower portions of Figure 4.17*a,b*).

Identifying Transmembrane Domains Which segments of the polypeptide chain are actually embedded in the lipid bilayer? The identification of these **transmembrane domains** is generally predicted from the amino acid sequence of the protein, which is deduced from the nucleotide sequence of an isolated gene. Those segments of a protein predicted to span the membrane typically consist of a string of 20 to 30 predominantly nonpolar amino acids that adopt an α helical secondary structure (excep-

tions are discussed later).[3] The chemical structure of a single transmembrane helix is shown in Figure 4.18, which depicts the two-dimensional structure of glycophorin A, the major integral protein of the erythrocyte plasma membrane. Of the 20 amino acids that make up the lone α helix of glycophorin (amino acids 74 to 93 of Figure 4.18), all but three have hydrophobic side chains (or an H atom in the case of the glycine residues); the exceptions are serine and threonine, neither of which is charged.

Knowing the amino acid sequence of an integral membrane protein, we can usually identify the transmembrane segments using a **hydropathy plot**, in which each site along a polypeptide is assigned a value that provides a measure of the *hydrophobicity* of the amino acid at that site as well as that of its neighbors. This approach provides a "running average" of the hydrophobicity of short sections of the polypeptide, and guarantees that one or a few polar amino acids in a sequence do not alter the profile of the entire stretch. Hydrophobicity of amino acids can be determined using various criteria, such as their lipid solubility or the energy that would be required to transfer them from an aqueous into a lipid medium. A hydropathy plot

[3]It was noted on page 57 that the α helix is a favored conformation because it allows for a maximum number of hydrogen bonds to be formed between neighboring amino acid residues, thereby creating a highly stable (low-energy) configuration. This is particularly important for a membrane-spanning polypeptide that is surrounded by fatty acyl chains and, thus, cannot form hydrogen bonds with an aqueous solvent. Because each amino acid occupies 1.5 Å of the polypeptide's length, and the hydrophobic core of the lipid bilayer is 30 Å wide, it requires at least 20 amino acids to span this part of the membrane.

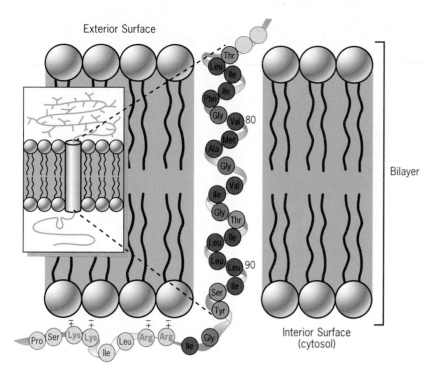

Exterior Surface

Bilayer

Interior Surface (cytosol)

Figure 4.18 Glycophorin A, an integral protein with a single transmembrane domain. The single α helix that passes through the membrane consists predominantly of hydrophobic residues. The four positively charged amino acid residues of the cytoplasmic domain of the membrane form ionic bonds with negatively charged lipid head groups. Carbohydrates are attached to a number of amino acid residues on the outer surface of the protein. All but one of the 16 oligosaccharides are small O-linked chains (the exception is a larger oligosaccharide linked to the asparagine residue at position 26). The role of glycophorin in the erythrocyte membrane is discussed on page 149.

for glycophorin A is shown in Figure 4.19. Transmembrane segments are usually identified as a jagged peak that extends well into the hydrophobic side of the spectrum. An educated guess concerning the orientation of the transmembrane segment within the bilayer can usually be made by examining the nature of the flanking amino acid residues. In most cases, as illustrated by glycophorin in Figure 4.18, those parts of the polypeptide at the cytoplas-

mic flank of a transmembrane segment tend to be more positively charged than those at the extracellular flank.

Not all transmembrane segments consist of hydrophobic α helices. The most obvious exceptions are found in proteins that contain an aqueous channel that allows the passage of ions or polar solutes through the lipid bilayer. The walls of these aqueous channels contain either (1) amphipathic helices in which one face of each helix consists of nonpolar amino acid residues that face the lipid bilayer, whereas the opposite face consists largely of polar residues facing the pore (Figure 4.20), or (2) a circle of

Figure 4.19 Hydropathy plot for glycophorin A, a single membrane-spanning protein. Hydrophobicity is measured by the free energy required to transfer each segment of the polypeptide from a nonpolar solvent to an aqueous medium. Values above the 0 line are energy requiring ($+\Delta G$s), indicating they consist of stretches of amino acids that have predominantly nonpolar side chains. Peaks that project above the red-colored line are interpreted as a transmembrane domain.

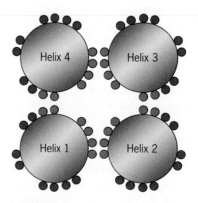

Figure 4.20 Schematic representation of four α helices that might occur as a bundle within a membrane. The small blue circles represent polar side chains that associate with one another in the center of the complex, whereas the small red circles represent nonpolar residues that associate with the surrounding lipid bilayer. Note that the side chains actually project from the backbone at various levels along the length of each helix; they have been compressed into a single plane for the purposes of the figure.

amphipathic β strands organized into a barrel as illustrated in Figure 5.3. To date, aqueous channels constructed of β barrels have only been found in the outer membranes of bacteria, mitochondria, and chloroplasts.

Determining Spatial Relationships Within an Integral Membrane Protein Suppose you have isolated a gene for an integral membrane protein and, based on its nucleotide sequence, determined that it contains four apparent membrane-spanning α helices. You might want to know how these helices are oriented relative to one another and which amino acid side chains of each helix face outward toward the lipid environment. Although these determinations are difficult to make without detailed structural models, considerable insight can be gained by site-directed mutagenesis, that is, by introducing specific changes into the gene that codes for the protein (Section 18.13). For example, site-directed mutagenesis can be employed to replace amino acid residues in neighboring helices with cysteine residues. As discussed on page 53, two cysteine residues are capable of forming a covalent disulfide bridge. If two transmembrane helices of a polypeptide each contain a cysteine residue, and the two cysteine residues are able to form a disulfide bridge with one another, then these helices must reside in very close proximity. The results of one site-directed cross-linking study on lactose permease, a sugar-transporting protein in bacterial cell membranes, is shown in Figure 4.21. It is found in this case that helix VII lies in close proximity to both helices I and II.

Determining spatial relationships between amino acids in a membrane protein can provide more than structural information; it can tell a researcher about some of the dynamic events that occur as a protein carries out its function. One way to learn about the distance between selected residues in a protein is to introduce chemical groups whose properties are sensitive to the distance that separates them. *Nitroxides* are chemical groups that contain an unpaired electron, which produces a characteristic spectrum when monitored by a technique called *electron paramagnetic resonance (EPR) spectroscopy*. A nitroxide group can be introduced at any site in a protein by first mutating that site to a cysteine and then attaching the nitroxide to the —SH group of the cysteine residue. Figure 4.22 shows how this technique was used to discover the conformational changes that occur in a membrane protein as its channel becomes activated in response to changes in the pH of the medium. The protein in question, a bacterial K$^+$ channel, is a tetramer composed of four identical subunits. The cytoplasmic opening to the channel is bounded by four transmembrane helices, one from each subunit of the protein. Figure 4.22a shows the EPR spectra that were obtained when a nitroxide was introduced near the cytoplasmic end of each transmembrane helix. The red line shows the spectrum obtained at pH 6.5 when the channel is in the closed state, and the blue line shows the spectrum at pH 3.5 when the channel is open. The shape of each line depends on the proximity of the nitroxides to one another. The spectrum is broader at pH 6.5 because the ni-

Figure 4.21 Determination of helix packing in a membrane protein by site-directed cross-linking. In these experiments, pairs of cysteine residues are introduced into the protein by site-directed mutagenesis, and the ability of the cysteines to form disulfide bridges is determined. Hydropathy plots and other data indicate that lactose permease has 12 transmembrane helices. It is found that a cysteine introduced at position 242 of helix VII can cross-link to a cysteine introduced at either position 28 or 29 of helix I. Similarly, a cysteine at position 245 of helix VII can cross-link to cysteines at either 52 or 53 of helix II. The proximity of these three helices is thus established. (REPRINTED FROM H. R. KABACK, J. VOSS, J. WU, CURR. OPIN. STRUCT. BIOL. 7:539, 1997, COPYRIGHT 1997, WITH PERMISSION FROM ELSEVIER SCIENCE.)

troxide groups on the four subunits are closer together at this pH, which decreases the intensity of their EPR signals. These results indicate that the activation of the channel is accompanied by an increased separation between the labeled residues of the four subunits (Figure 4.22b). An increase in diameter of the channel opening allows ions in the cytoplasm to reach the actual permeation pathway (shown in red) within the channel, which allows only the passage of K$^+$ ions (discussed on page 156).

Peripheral Membrane Proteins

Peripheral proteins are associated with the membrane by weak electrostatic bonds either to the hydrophilic head groups of lipids or to the hydrophilic portions of integral proteins protruding from the bilayer (refer to Figure 4.12b). Peripheral proteins can usually be solubilized by extraction with aqueous salt solutions. In actual fact, the distinction between integral and peripheral proteins is blurred because many membrane proteins consist of several polypeptides, some of which penetrate into the lipid bilayer and others that remain on the periphery.

The best-studied peripheral proteins are located on the internal (cytosolic) surface of the plasma membrane,

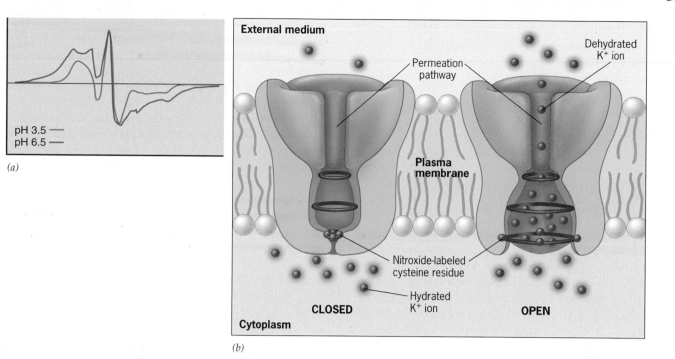

Figure 4.22 Use of EPR spectroscopy to monitor changes in conformation of a bacterial K⁺ ion channel as it opens and closes. (*a*) EPR spectra from nitroxides that have been attached to cysteine residues near the cytoplasmic end of the four transmembrane helices that line the channel. The cysteine residue in each helix replaces a glycine residue that is normally at that position. The shapes of the spectra depend on the distances between unpaired electrons in the nitroxides on different subunits. (Nitroxides are described as "spin-labels," and this technique is known as site-directed spin labeling). (*b*) A highly schematic model of the ion channel in the active and inactive states based on the data from part *a*. Opening of the channel is accompanied by the movement of the four nitroxide groups apart from one another. (A: FROM E. PEROZO ET AL., NAT. STRUCT. BIOL. 5:468, 1998.)

where they form a fibrillar network that acts as a membrane "skeleton" (see Figure 4.32*d*). These proteins provide mechanical support for the membrane and function as an anchor for integral membrane proteins. Other peripheral proteins on the internal plasma membrane surface function as enzymes, specialized coats (see Figure 8.24), or factors that transmit transmembrane signals (see Figure 15.23). Peripheral proteins associated with the external surface of the plasma membrane are typically part of the extracellular matrix, whose multiple functions are discussed in Chapter 7 (see Figure 7.5). Peripheral proteins typically have a dynamic relationship with the membrane, being recruited to the membrane or released from the membrane depending on prevailing conditions.

Lipid-Anchored Membrane Proteins

Two types of lipid-anchored membrane proteins are distinguished by the types of lipid anchors and the membrane surface on which they are exposed. A variety of proteins present on the external face of the plasma membrane are bound to the membrane by a short oligosaccharide linked to a molecule of glycophosphatidylinositol (GPI) that is embedded in the outer leaflet of the lipid bilayer (refer to Figure 4.12*c*). The presence of these GPI-anchored proteins was discovered when it was shown that certain membrane proteins could be released by a phospholipase that specifically recognized and cleaved inositol-containing phospholipids. The normal cellular scrapie protein PrP^C (page 67) is a GPI-linked molecule, as are various receptors, enzymes, and cell-adhesion proteins. A rare type of anemia, paroxysmal nocturnal hemoglobinuria, results from a deficiency in GPI synthesis that makes red blood cells susceptible to lysis.

Another group of proteins present on the *cytoplasmic* side of the plasma membrane is anchored to the membrane by long hydrocarbon chains embedded in the inner leaflet of the lipid bilayer (refer to Figure 4.12*c*). At least two proteins associated with the plasma membrane in this way (Src and Ras) have been implicated in the transformation of a normal cell to a malignant state.

REVIEW ?

1. Why are detergents necessary to solubilize membrane proteins? How might one determine the diversity of integral proteins that reside in a purified membrane fraction?

2. How can one determine: (1) membrane-sidedness, (2) the location of transmembrane segments in the amino acid sequence, or (3) the relative locations of transmembrane helices?

3. Describe the asymmetry of the plasma membrane in terms of each of its major components: proteins, lipids, and carbohydrates.

4. Describe the properties of the three classes of membrane proteins (integral, peripheral, and lipid anchored), how they differ from one another, and how they vary among themselves.

4.5 MEMBRANE LIPIDS AND MEMBRANE FLUIDITY

The physical state of the lipid of a membrane is described by its fluidity (or viscosity).[4] Consider a simple artificial bilayer made of phosphatidylcholine and phosphatidylethanolamine, whose fatty acids are largely unsaturated. If the temperature of the bilayer is kept relatively warm (e.g., 37°C), the lipid exists in a relatively fluid state (Figure 4.23a). At this temperature, the lipid bilayer is best described as a two-dimensional liquid crystal. As in a crystal, the molecules still retain a specified orientation; in this case, the long axes of the molecules remain essentially parallel, yet individual phospholipids can rotate around their axis or move laterally within the plane of the bilayer. If the temperature is slowly lowered, a point is reached where the bilayer distinctly changes (Figure 4.23b). The lipid is converted from its normal liquid-like state to a frozen crystalline gel in which the movement of the phospholipid fatty acid chains is greatly restricted. The temperature at which this change occurs is called the **transition temperature**.

⤷change from liquid to solid

[4]Fluidity and viscosity are inversely related; fluidity is a measure of the ease of flow, and viscosity is a measure of the resistance to flow.

The transition temperature of a membrane or lipid bilayer can be measured calorimetrically (Figure 4.24). The membrane suspension is placed in a *calorimeter*, an instrument that can measure the energy required to raise the temperature of a specimen. As in the case of any substance undergoing a transition from a solid to a liquid state, energy is absorbed by the system in order to break the intermolecular restraints without causing a corresponding increase in the temperature of the substance. The transition in phase is seen as a sharp peak in the top line of Figure 4.24.

The transition temperature of a particular bilayer depends on the ability of the lipid molecules to be packed together, which depends in turn on the particular lipids of which it is constructed. Saturated fatty acids have the shape of a straight, flexible rod. *Cis*-unsaturated fatty acids, on the other hand, have crooks in the chain (Figures 2.19, 4.23) because the carbon atoms sharing a double bond cannot rotate. Consequently, phospholipids with saturated chains pack together more tightly than those containing unsaturated chains. The greater the degree of unsaturation of the fatty acids of the bilayer, the *lower* the temperature before the bilayer gels. The introduction of one double bond in a molecule of stearic acid lowers the melting temperature almost 60°C (Table 4.3). Another factor that influences bilayer fluidity is fatty acid chain length. The shorter the fatty acyl chains of a phospholipid, the lower its melting temperature.

Different lipids can undergo their phase change over a very wide range of temperature. For example, various phosphatidylcholines can be synthesized and used to form bilayers whose transition temperatures range from below 0°C to greater than 60°C. The physical state of the membrane is also affected by cholesterol. Because of their orientation within the bilayer (Figure 4.7), cholesterol mole-

(a)

(b)

Figure 4.23 The structure of the lipid bilayer depends on the temperature. The bilayer shown here is composed of two phospholipids, phosphatidylcholine and phosphatidylethanolamine. (*a*) Above the transition temperature, the lipid molecules and their hydrophobic tails are free to move in certain directions, even though they retain a considerable degree of order.

(*b*) Below the transition temperature, the movement of the molecules is greatly restricted, and the entire bilayer can be described as a crystalline gel. (AFTER R. N. ROBERTSON, THE LIVELY MEMBRANES, CAMBRIDGE UNIVERSITY PRESS, 1983, REPRINTED WITH PERMISSION OF CAMBRIDGE UNIVERSITY PRESS.)

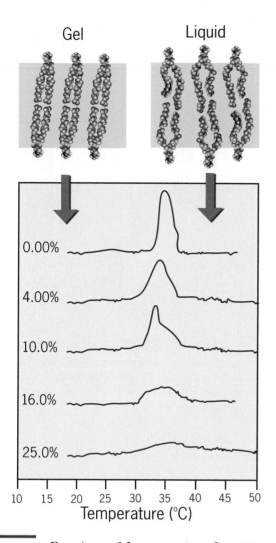

Figure 4.24 Experimental demonstration of transition temperature and the effect of cholesterol. Differential scanning calorimetry curves of model membranes containing different proportions of phosphatidylcholine and cholesterol. The mole percent cholesterol in the various preparations is given on the left side of the curves. Heat is absorbed by the lipid as it undergoes the transition from gel (left side of the peak) to the liquid crystal state (right side of the peak). This transition becomes blurred as the cholesterol content increases. (FROM F. M. LINSEISEN ET AL., CHEM. PHYS. LIPIDS 65:143, 1993, COPYRIGHT 1993, WITH PERMISSION FROM ELSEVIER SCIENCE.)

cules disrupt the close packing of fatty acyl chains and interfere with their mobility. The presence of cholesterol tends to abolish sharp transition temperatures (Figure

Table 4.3	Melting Points of the Common 18-Carbon Fatty Acids	

Fatty acid	cis Double bonds	M.p.(°C)
Stearic acid	0	70
Oleic acid	1	13
α-Linoleic acid	2	−9
Linolenic acid	3	−17

4.24) and create a condition of intermediate fluidity. In physiological terms, cholesterol tends to increase the durability and decrease the permeability of a membrane.

Recent studies suggest that cholesterol molecules may not be evenly distributed within the lipid bilayer but may be concentrated together with sphingolipids into specialized membrane domains. The transition temperature of cholesterol-sphingolipid patches is expected to be higher than that of surrounding phosphoglycerides. Consequently, cholesterol-sphingolipid domains might act like "rafts" floating on a fluid phosphoglyceride "sea." Lipid rafts may serve as platforms that recruit particular membrane proteins—most notably GPI-anchored proteins—giving these domains special properties that are not shared by the remainder of the membrane.

The Importance of Membrane Fluidity

What effect does the physical state of the lipid bilayer have on the biological properties of the membrane? Membrane fluidity provides a perfect compromise between a rigid, ordered structure in which mobility would be absent and a completely fluid, nonviscous liquid in which the components of the membrane could not be oriented and structural organization and mechanical support would be lacking. In addition, fluidity allows for interactions to take place within the membrane. For example, membrane fluidity makes it possible for clusters of membrane proteins to assemble at particular sites within the membrane and form specialized structures, such as intercellular junctions, light-capturing complexes, and synapses. Because of membrane fluidity, molecules that interact can come together, carry out the necessary reaction, and move apart.

Fluidity also plays a role in membrane assembly, a subject discussed in Chapter 8. Membranes arise only from preexisting membranes, and their growth is accomplished by the insertion of lipid and protein components into the fluid matrix of the membranous sheet. Many of the most basic cellular processes, including cell movement, cell growth, cell division, formation of intercellular junctions, secretion, and endocytosis, depend on the movement of membrane components and would probably not be possible if membranes were rigid, nonfluid structures.

Maintaining Membrane Fluidity

The internal temperature of most organisms (other than birds and mammals) fluctuates with the temperature of the external environment. Since it is essential for many activities that the membranes of a cell remain in a fluid state, cells respond to changing conditions by altering the types of phospholipids of which they are made. Maintenance of membrane fluidity is an example of homeostasis at the cellular level and can be demonstrated in various ways. For example, if the temperature of a culture of cells is lowered, the cells respond metabolically. The initial "emergency" response is mediated by enzymes that remodel membranes,

making the cell more cold resistant. Remodeling is accomplished by (1) desaturating single bonds in fatty acyl chains to form double bonds, and (2) reshuffling the chains between different phospholipid molecules to produce those that contain two unsaturated fatty acids, which greatly lowers the melting temperature of the bilayer. Desaturation of single bonds to form double bonds is catalyzed by enzymes called *desaturases*. Reshuffling is accomplished by *phospholipases*, which split the fatty acid from the glycerol backbone, and *acyltransferases*, which transfer them to a different phospholipid. In addition, the cell changes the types of phospholipids being synthesized in favor of ones containing more unsaturated fatty acids. As a result of the activities of these various enzymes, the physical properties of a cell's membranes are matched to the prevailing environmental conditions. Maintenance of fluid membranes by adjustments in fatty acyl composition has been demonstrated in a variety of organisms, including hibernating mammals, pond-dwelling fish whose body temperature changes markedly from day to night, cold-resistant plants, and bacteria living in hot water springs.

To verify that desaturation of fatty acids contributes to low-temperature tolerance, experiments have been conducted using strains of cyanobacteria that lack certain desaturases. When the growth rates of the wild-type and mutant strains incubated at 34°C are compared, little difference is seen. But when the temperature is dropped to 22°C, the doubling time of the mutant cells rises to 59 hours, compared with a doubling time of 22 hours for the wild type. When the gene that codes for this enzyme (*desA*) is experimentally introduced into the mutant cells, fatty acid desaturation is observed, and doubling time dramatically decreases.

The Asymmetry of Membrane Lipids

The lipid bilayer consists of two distinct leaflets, and there is no reason to assume that the lipid composition of the two halves should be identical. In fact, the lipids of the plasma membrane are distributed in a highly asymmetric pattern. One line of experiments that has led to this conclusion takes advantage of the fact that lipid-digesting enzymes cannot penetrate the plasma membrane and, consequently, are only able to digest lipids that reside in the outer leaflet of the bilayer. If intact human red blood cells are treated with a lipid-digesting phospholipase, approximately 80 percent of the phosphatidylcholine (PC) of the membrane is hydrolyzed, but only about 20 percent of the membrane's phosphatidylethanolamine (PE) and less than 10 percent of its phosphatidylserine (PS) are attacked. These data indicate that, compared to the inner leaflet, the outer leaflet has a relatively high concentration of PC (and sphingomyelin) and a low concentration of PE and PS (Figure 4.25). It follows that the lipid bilayer can be thought of as composed of two more-or-less stable, independent monolayers having different physical and chemical properties.

Figure 4.25 **The asymmetric distribution of phospholipids (and cholesterol) in the plasma membrane of human erythrocytes.** (SM, sphingomyelin; PC, phosphatidylcholine; PS, phosphatidylserine; PE, phosphatidylethanolamine; PI, phosphatidylinositol; Cl, cholesterol.)

The biological role of lipid asymmetry is not well understood, except in a few cases. All of the glycolipids are in the outer leaflet where they probably serve as receptors for extracellular ligands. Phosphatidylserine, which is concentrated in the inner leaflet, has a net negative charge at physiological pH, which makes it a candidate for binding positively charged lysine and arginine residues, such as those adjacent to the membrane-spanning α helix of glycophorin A in Figure 4.18. The appearance of PS on the outer surface of aging lymphocytes marks the cells for destruction by macrophages, whereas its appearance on the outer surface of platelets signals for blood coagulation. Phosphatidylinositol, which is concentrated in the inner leaflet, plays a key role in the transfer of stimuli from the plasma membrane to the cytoplasm (Section 15.2).

REVIEW ?

1. What is the importance of fatty acid unsaturation for membrane fluidity? Of enzymes that are able to desaturate fatty acids?
2. What is meant by a membrane's transition temperature and how can it be measured?
3. Why is membrane fluidity important to a cell?
4. How can the two sides of a lipid bilayer have different ionic charges?

4.6 THE DYNAMIC NATURE OF THE PLASMA MEMBRANE

It is apparent from the previous discussion that the lipid bilayer can exist in a relatively fluid state. The mobility of individual lipid molecules within the bilayer of the plasma membrane can be directly observed by linking the polar heads of the lipids with gold particles and following the movement of the particles under the microscope (Figure 4.26). A phospholipid can move laterally within the same leaflet with considerable ease. It is estimated that a phospholipid can diffuse from one end of a bacterium to the other end in a second or two. In contrast, the half-life of a phospholipid molecule moving across to the other leaflet is measured in hours to days. Thus, of all the possible motions that a phospholipid can make, its flip-flop to the other side of the membrane is the most restricted (Figure 4.27). This finding is not surprising. For flip-flop to occur, the hydrophilic head group of the lipid must pass through the internal hydrophobic sheet of the membrane, which is thermodynamically unfavorable. However, cells do contain enzymes, called *flippases*, that actively move certain phospholipids from one leaflet to the other. These enzymes may play a role in establishing lipid asymmetry and may also reverse the slow rate of passive transmembrane movement.

Because lipids provide the matrix in which integral proteins of a membrane are embedded, the physical state of the lipid is an important determinant of the mobility of integral proteins. The demonstration that integral proteins can move within the plane of the membrane was a cornerstone in the formulation of the fluid-mosaic model. The dynamic properties of membrane proteins have been revealed in several ways.

The Diffusion of Membrane Proteins After Cell Fusion

Cell fusion is a technique whereby two different types of cells, or cells from two different species, can be fused to produce one cell with a common cytoplasm and a single, continuous plasma membrane. Cells are induced to fuse with one another by making the outer surface of the cells "sticky" so that their plasma membranes adhere to one another. Cells are induced to fuse by addition of certain inactivated viruses that attach to the surface membrane, by adding the compound polyethylene glycol, or by a mild electric shock. Cell fusion has played an important role in cell biology and is currently used as part of the technique to prepare specific antibodies (Section 18.14).

The first experiments to demonstrate that membrane proteins could move within the plane of the membrane utilized cell fusion, and they were reported in 1970 by Larry Frye and Michael Edidin of Johns Hopkins University. In their experiments, mouse and human cells were fused, and the locations of specific proteins of the plasma membrane were followed once the two membranes had

Figure 4.26 Experimental determination of the mobility of phospholipids within the lipid bilayer of the plasma membrane. A small number of phospholipids of the outer (exoplasmic) leaflet of the plasma membrane of a living fibroblast were tagged with gold particles, and their subsequent movements under the microscope were followed on a video screen. The tagged lipids are shown to move randomly within the membrane as indicated by the tracks drawn over the cell surface. The bar represents 3 μm. (FROM GRETA M. LEE ET AL., J. CELL BIOL. 120:28, 1993; BY COPYRIGHT PERMISSION OF THE ROCKEFELLER UNIVERSITY PRESS.)

Figure 4.27 The possible movements of phospholipids in a membrane. The types of movements in which membrane phospholipids can engage and the approximate time scales over which they occur. Whereas phospholipids move from one leaflet to another (called flip-flop) at a very slow rate, phospholipids diffuse laterally within a leaflet rapidly.

become continuous. To follow the distribution of either the mouse membrane proteins or the human membrane proteins at various times after fusion, antibodies against

(a)

(b) 5 μm (c)

Figure 4.28 The use of cell fusion to reveal mobility of membrane proteins. (*a*) Outline of the experiment in which human and mouse cells were fused (steps 1–2) and the distribution of the proteins on the surface of each cell were followed in the hybrids over time (steps 3–4). Mouse membrane proteins are indicated by solid circles, human membrane proteins by open circles. Locations of human and mouse proteins in the plasma membranes of the hybrid cells were monitored by interaction with fluorescent red and fluorescent green antibodies, respectively. (*b*) Micrograph showing a fused cell in which mouse and human proteins are still in their respective hemispheres (equivalent to hybrid in step 3 of part *a*). (*c*) The effect of temperature on the diffusion of membrane proteins is revealed by examining the percentage of cells in which the proteins from the two species have become intermixed (forming a mosaic) at 40 minutes after fusion. The percentage of mosaic cells increases sharply as the temperature rises above 15°C, suggesting that the lipid undergoes a phase transition in that range from a gelated to a fluid state. (B,C: FROM L. D. FRYE AND MICHAEL EDIDIN, J. CELL SCI. 7:328, 334, 1970; BY PERMISSION OF THE COMPANY OF BIOLOGISTS LTD.)

one or the other type of protein were prepared and covalently linked to fluorescent dyes. The antibodies against the mouse proteins were complexed with a dye that fluoresces green and the antibodies against human proteins with one that fluoresces red. When the antibodies were added to fused cells, they bound to the human or mouse proteins and could be located under the fluorescence light microscope (Figure 4.28*a*). At the time of fusion, the plasma membrane appeared half human and half mouse; that is, the two protein types remained segregated in their own hemisphere (step 3, Figure 4.28*a,b*). As the time after fusion increased, the membrane proteins were seen to move laterally within the membrane into the opposite hemisphere. By about 40 minutes, each species' proteins were uniformly distributed around the entire hybrid cell membrane (step 4, Figure 4.28*a*). If the same experiment was performed at lower temperature, the viscosity of the lipid bilayer increased, and the mobility of the membrane

proteins decreased (Figure 4.28c). These early cell fusion experiments gave the impression that integral membrane proteins were capable of virtually unrestricted movements. As we will see shortly, subsequent studies made it apparent that membrane dynamics was a much more complex subject than first envisioned.

Patterns of Protein Mobility

Several techniques allow researchers to follow the movements of molecules in the membranes of living cells using the light microscope. In a technique called **fluorescence recovery after photobleaching (FRAP)**, which is illustrated in Figure 4.29*a*, integral membrane components in cultured cells are first labeled by linkage to a fluorescent dye. Membrane proteins can be labeled in an indiscriminate manner by treating cells with a nonspecific dye (such as fluorescein isothiocyanate) that reacts with all exposed

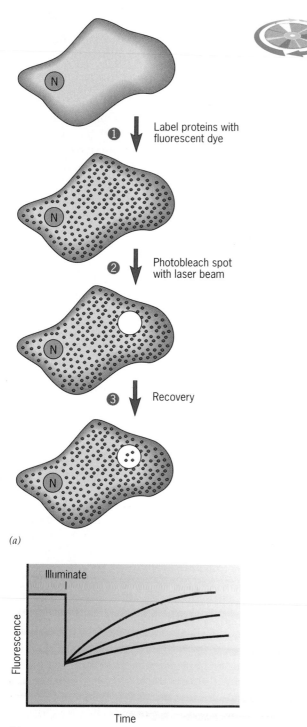

① Label proteins with fluorescent dye

② Photobleach spot with laser beam

③ Recovery

(a)

(b)

Figure 4.29 Measuring the diffusion rates of membrane proteins by fluorescence recovery after photobleaching (FRAP). (*a*) In this technique, a particular component of the membrane is first labeled with a fluorescent dye (step 1). A small region of the surface is then irradiated to bleach the dye molecules (step 2), and the recovery of fluorescence in the bleached region is followed over time (step 3). (N represents the cell nucleus.) (*b*) The rate of fluorescence recovery within the illuminated spot can vary depending on the protein(s) being followed. The rate of recovery is related to the diffusion coefficient of the fluorescent species.

protein molecules, or alternatively, a particular membrane protein can be labeled using specific probes, such as a fluorescent antibody. Once labeled, the cells are placed under the microscope and individually irradiated using a sharply focused laser beam, which bleaches irreversibly the fluorescent molecules in its path, leaving a circular spot (typically about 1 µm diameter) on the surface of the cell that is devoid of fluorescence. If the labeled proteins in the membrane are mobile, then the random movements of these molecules should produce a gradual reappearance of fluorescence in the irradiated circle. The rate of fluorescence recovery (Figure 4.29*b*) provides a direct measure of the rate of diffusion (expressed as a diffusion coefficient, *D*) of the mobile molecules. The extent of fluorescence recovery (expressed as a percentage of the original intensity) provides a measure of the percentage of the labeled molecules that are free to diffuse.

Early studies utilizing FRAP suggested that (1) membrane proteins moved much more slowly in plasma membranes than they would in a pure lipid bilayer and (2) that a significant fraction of membrane proteins (30 to 70 percent) were not free to diffuse back into the irradiated circle. But the FRAP technique has its drawbacks. FRAP can only follow the movement of a large population of labeled molecules as they diffuse over a relatively large distance (e.g., 1 µm). As a result, researchers using FRAP cannot distinguish between proteins that are truly immobile and ones that can only diffuse over a limited distance in the time allowed. To get around these limitations, alternate techniques have been developed that allow researchers to observe the movements of individual protein molecules over very short distances and to determine how they might be restrained.

In **single-particle tracking (SPT)**, individual membrane protein molecules are labeled, usually with antibody-coated gold particles (approximately 40 nm in diameter), and the movements of the labeled molecules are followed by computer-enhanced video microscopy (Section 18.1). The results of these studies often depend on the particular protein being investigated. For example,

- Some membrane proteins move randomly throughout the membrane (Figure 4.30, protein A), though generally at rates considerably less than would be measured in an artificial lipid bilayer. (If protein mobility were based strictly on physical parameters such as lipid viscosity and protein size, one would expect proteins to migrate with diffusion coefficients of approximately 10^{-8} to 10^{-9} cm²/sec rather than 10^{-10} to 10^{-12} cm²/sec, as is observed for molecules of this group.) The reasons for the reduced diffusion coefficient have been debated.

- Some membrane proteins fail to move and are considered to be immobilized (Figure 4.30, protein B).

- In some cases, a particular species of protein is found to move in a highly directed (i.e., nonrandom) manner toward one part of the cell or another. For example, a

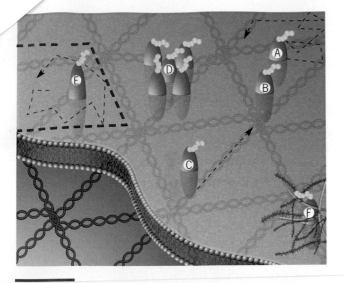

Figure 4.30 Patterns of movement of integral membrane proteins. Depending on the cell type and the conditions, integral membrane proteins can exhibit several different types of mobility. Protein A is capable of diffusing randomly throughout the membrane, though its rate of movement may be limited; protein B is immobilized as the result of its interaction with the underlying membrane skeleton; protein C is being moved in a particular direction as the result of its interaction with a motor protein at the cytoplasmic surface of the membrane; protein D is capable of diffusion, but its movements are restricted by other integral proteins of the membrane; protein E is capable of diffusion, but its movements are restricted by fences formed by proteins of the membrane skeleton; protein F is capable of diffusion, but its movement is restrained by extracellular materials.

particular membrane protein may tend to move toward the leading or the trailing edge of a moving cell (Figure 4.30, protein C).

■ In most studies, the largest fraction of protein species exhibit random (Brownian) movement within the membrane at rates consistent with free diffusion (diffusion coefficients about 5×10^{-9} cm^2/sec), but the molecules are unable to migrate more than a few tenths of a micron. The membrane appears to contain barriers that prevent extended movement. These barriers are discussed in the following sections.

Control of Membrane Motility It is apparent that plasma membrane proteins are not totally free to drift around randomly on the lipid "sea," but instead they are subjected to various influences that affect their mobility. Some membranes are crowded with proteins, so that the random movements of one molecule can be impeded by its neighbors (Figure 4.30, protein D). The strongest influences on an integral membrane protein are thought to be exerted from just beneath the membrane on its cytoplasmic face. The plasma membranes of many cells possess a fibrillar network, or "membrane skeleton," consisting of peripheral proteins situated on the cytoplasmic surface of the membrane. A certain proportion of a membrane's integral protein molecules are either tethered to the membrane skeleton (Figure 4.30, protein B) or otherwise restricted by it.

Information concerning the presence of membrane barriers has been obtained using an innovative technique that allows investigators to trap integral proteins and drag them through the plasma membrane with a known force. This technique, which uses an apparatus referred to as **optical tweezers**, takes advantage of the tiny optical forces that are generated by a focused laser beam. The integral proteins to be studied are tagged with antibody-coated beads, which serve as handles that can be gripped by the laser field. It is generally found that optical tweezers can drag integral proteins for a limited distance before the proteins encounter a barrier that causes them to be released from the laser's grip. As they are released, the protein typically springs rapidly backward suggesting that the barriers are elastic structures.

One approach to studying factors that affect membrane protein mobility is to genetically modify cells so they produce altered membrane proteins. Integral proteins whose cytoplasmic portions have been genetically deleted often move much greater *distances* than their intact counterparts, indicating that barriers reside on the cytoplasmic side of the membrane. These findings suggest that the membrane's underlying cytoskeleton forms a network of "fences" around portions of the membrane, creating microdomains that restrict the distance an integral protein can travel (Figure 4.30, protein E). Proteins move across the boundaries from one microdomain to another through breaks in the fences. Such openings are thought to appear and disappear along with the dynamic disassembly and reassembly of parts of the meshwork. Membrane domains may function primarily to keep specific combinations of proteins in close enough proximity to facilitate their interaction.

Integral proteins lacking that portion that would normally project into the extracellular space typically move at a much slower *rate* than the wild-type version of the protein. This finding suggests that the movement of a transmembrane protein through the bilayer is slowed by the presence of extracellular materials that can entangle the external portion of the protein molecule (Figure 4.30, protein F).

Membrane Domains and Cell Polarity For the most part, studies of membrane dynamics, such as those discussed above, are carried out on the relatively homogeneous plasma membrane situated at the free upper surface of a cell residing on a culture dish. Most membranes, however, exhibit distinct variations in protein composition and mobility, especially in cells whose various surfaces display distinct functions. For example, the epithelial cells that line the intestinal wall or make up the microscopic tubules of the kidney are highly polarized cells whose different surfaces carry out different functions

Apical plasma membrane
• regulation of nutrient and water intake
• regulated secretion
• protection

Glucose
Na+
Disaccharide
Monosaccharide

Lateral plasma membrane
• cell contact and adhesion
• cell communication

Basal membrane
• cell-substratum contact
• generation of ion gradients

(a)

(b)

0.5 μm

Figure 4.31 Differentiated functions of the plasma membrane of an epithelial cell. (*a*) The apical surface of this intestinal epithelial cell contains integral proteins that function in ion transport and hydrolysis of disaccharides, such as sucrose and lactose; the lateral surface contains integral proteins that function in intercellular interaction; and the basal surface contains integral proteins that function in the association of the cell with the underlying basement membrane. (*b*) Electron micrograph showing the location of sucrose-splitting enzymes (sucrase) in the microvilli of an intestinal epithelial cell. The location of the enzyme is revealed by labeled antibodies directed against the enzyme. (B: FROM J. A. M. FRANSEN ET AL., J. CELL BIOL. 115:50, 1991; BY COPYRIGHT PERMISSION OF THE ROCKEFELLER UNIVERSITY PRESS.)

(Figure 4.31). Studies indicate that the apical plasma membrane, which selectively absorbs substances from the lumen, possesses different enzymes than the lateral plasma membrane, which interacts with neighboring epithelial cells, or the basal membrane, which adheres to an underlying extracellular substrate (a *basement membrane*).

In other examples, the receptors for neurotransmitter substances are concentrated into regions of the plasma membrane located within synapses (see Figure 4.54), and receptors for low-density lipoproteins are concentrated into patches of the plasma membrane specialized to facilitate their internalization (see Section 8.8).

The Red Blood Cell: An Example of Plasma Membrane Structure

Of all the diverse types of membranes, the plasma membrane of the human erythrocyte (red blood cell) is the most studied and best understood (Figure 4.32). There are several reasons for the popularity of this membrane. The cells are inexpensive to obtain and readily available in huge numbers from whole blood. They are already present as single cells and need not be dissociated from a complex tissue. The cells are simple by comparison with other cell types, lacking nuclear and cytoplasmic membranes that inevitably contaminate plasma membrane preparations from other cells. In addition, purified, *intact* erythrocyte plasma membranes can be obtained simply by placing the cells in a dilute (hypotonic) salt solution. The cells respond to this osmotic shock by taking up water and swelling, a phenomenon termed *hemolysis*. As the surface area of each cell increases, the cell becomes leaky and the contents, composed almost totally of dissolved hemoglobin, flow out of the cell leaving behind a plasma membrane "ghost" (Figure 4.32*b*).

Once erythrocyte plasma membranes are isolated, the proteins can be solubilized and separated from one an-

Figure 4.32 The plasma membrane of the human erythrocyte. (*a*) Scanning electron micrograph of human erythrocytes. (*b*) Micrograph showing plasma membrane ghosts, which were isolated by allowing erythrocytes to swell and hemolyze as described in the text. (*c*) The results of SDS–polyacrylamide gel electrophoresis (SDS–PAGE) used to fractionate the proteins of the erythrocyte membrane, which are identified at the side of the gel. The gel has been stained for protein by Coomassie blue. (*d*) A model of the erythrocyte plasma membrane as viewed from the internal surface, showing the integral proteins embedded in the lipid bilayer and the arrangement of peripheral proteins that make up the membrane's internal skeleton. The band 3 dimer shown here is greatly simplified. (*e*) Electron micrograph showing the arrangement of the proteins of the inner membrane skeleton. (A: FROM FRANÇOIS M. M. MOREL, RICHARD F. BAKER, AND HAROLD WAYLAND, J. CELL BIOL. 48:91, 1971; B: COURTESY OF JOSEPH F. HOFFMAN; C: REPRODUCED, WITH PERMISSION, FROM V. T. MARCHESI, H. FURTHMAYR, AND M. TOMITA, ANNU. REV. BIOCHEM. VOL. 45; © 1976 BY ANNUAL REVIEWS INC.; D: AFTER D. VOET AND J. G. VOET, BIOCHEMISTRY, 2D ED., COPYRIGHT © 1995, JOHN WILEY & SONS, INC.; E: FROM SHIH-CHUN LIU, LAURA H. DERICK, AND JIRI PALEK, J. CELL BIOL. 104:527, 1987; A, E: BY COPYRIGHT PERMISSION OF THE ROCKEFELLER UNIVERSITY PRESS.)

other (fractionated), providing a better idea of the diversity of proteins within the membrane. Fractionation of membrane proteins can be accomplished using polyacrylamide gel electrophoresis (PAGE) in the presence of the ionic detergent sodium dodecyl sulfate (SDS). (The technique of SDS–PAGE is discussed in Section 18.7.) The SDS keeps the integral proteins soluble and, in addition, adds a large number of negative charges to the proteins with which it associates. Because the number of charged SDS molecules per unit weight of protein tends to be relatively constant, the molecules separate from one another according to their molecular weight. The largest proteins move the slowest through the molecular sieve of the gel. The major proteins of the erythrocyte membrane are separated into about a dozen major bands by SDS–PAGE (Figure 4.32c).

A closer look at the proteins of the erythrocyte plasma membrane will illustrate their various functions. Among the proteins are a variety of enzymes (including glyceraldehyde 3-phosphate dehydrogenase, one of the enzymes of glycolysis), transport proteins (for ions and sugars), and skeletal proteins (e.g., spectrin). Various proteins of the erythrocyte plasma membrane can be correlated with the transport of oxygen and carbon dioxide and with the physical stresses that the cells encounter as they circulate through the body. At the same time, the plasma membrane of the erythrocyte is presumed to be simpler than that of most other cells because it lacks proteins involved in cell signaling and cell–cell or cell–matrix interactions.

Integral Proteins of the Erythrocyte Membrane A model of the erythrocyte plasma membrane showing its major proteins is seen in Figure 4.32d. The most abundant integral proteins of this membrane are a pair of carbohydrate-containing, membrane-spanning proteins, called band 3 and glycophorin A. Band 3, which gets its name from its position in an electrophoretic gel (Figure 4.32c), is present as a dimer composed of two identical subunits. Each subunit spans the membrane at least a dozen times and contains a relatively small amount of carbohydrate (6–8 percent of the molecule's weight). Band 3 protein serves as a channel for the passive exchange of anions across the membrane. As blood circulates through the tissues, carbon dioxide becomes dissolved in the fluid of the bloodstream (the plasma) and undergoes the following reaction:

$$H_2O + CO_2 \rightarrow H_2CO_3 \rightarrow HCO_3^- + H^+$$

The bicarbonate ions (HCO_3^-) enter the erythrocyte in exchange for chloride ions, which leave the cell. In the lungs, where carbon dioxide is released, the reaction reverses and bicarbonate ions leave the erythrocyte in exchange for chloride ions. The reciprocal movement of HCO_3^- and Cl^- apparently occurs through a channel in the center of each band 3 dimer.

Glycophorin A was the first membrane protein to have its amino acid sequence determined. The arrangement of the polypeptide chain of glycophorin A within the plasma membrane is shown in Figure 4.18. (Other related glycophorins, B, C, D, and E, are also present in the membrane at much lower concentrations.) Unlike band 3, glycophorin A spans the membrane only once, and it contains an extensive carbohydrate portion consisting of 16 oligosaccharide chains that together make up about 60 percent of the molecule's weight. It is thought that the primary function of the glycophorins derives from the large number of negative charges borne on sialic acid, the sugar residue at the end of each carbohydrate chain. Because of these charges, red blood cells repel each other, which prevents the cells from clumping as they circulate through the body's tiny vessels. It is noteworthy that persons who lack both glycophorin A and B in their red blood cells show no ill effects from their absence. At the same time, the band 3 proteins in these individuals are more heavily glycosylated, which apparently compensates for the otherwise missing negative charges required to prevent cell-cell interaction. Glycophorin also happens to be the receptor utilized by the protozoan that causes malaria, providing a path for entry into the blood cell. Consequently, individuals whose erythrocytes lack glycophorin A and B are thought to be protected from acquiring malaria. Differences in glycophorin amino acid sequence determine whether a person has an MM, MN, or NN blood type.

The Erythrocyte Membrane Skeleton The peripheral proteins of the erythrocyte plasma membrane are located on its internal surface and constitute a fibrillar membrane skeleton (Figure 4.32d,e) that plays a major role in determining the biconcave shape of the erythrocyte and restraining the movement of the membrane's integral proteins. The major component of the skeleton is an elongated fibrous protein, called *spectrin*. Spectrin is a heterodimer approximately 100 Å long, consisting of an α and β subunit that curl around one another. Two such dimeric molecules are linked at their head ends to form a 200 Å-long tetrameric filament that is both flexible and elastic. Spectrin is attached to the internal surface of the membrane by means of noncovalent bonds to another peripheral protein, *ankyrin* (the green spheres of Figure 4.32d), which in turn is linked noncovalently to the cytoplasmic domain of a band 3 molecule. As evident in Figures 4.32d and e, spectrin filaments are organized into hexagonal or pentagonal arrays. This type of network is constructed by linking both ends of each spectrin filament to a cluster of proteins that includes *actin* and *tropomyosin*, which are typically involved in contractile activities. A number of genetic diseases (*hemolytic anemias*) characterized by fragile, abnormally shaped erythrocytes have been traced to mutations that alter the structure and function of ankyrin or spectrin.

If the peripheral proteins are removed from erythrocyte ghosts, the membrane becomes fragmented into small vesicles, indicating that the inner protein network is required to maintain the integrity of the membrane. Erythrocytes are circulating cells that are squeezed under

pressure through microscopic capillaries whose diameter is considerably less than that of the erythrocytes themselves. To traverse these narrow passageways, and to do so day after day, the red blood cell must be highly deformable, durable, and capable of withstanding shearing forces that tend to pull it apart. The spectrin-actin network gives the cell the strength, elasticity, and pliability necessary to carry out its demanding function. When first discovered, the membrane skeleton of the erythrocyte was thought to be a unique structure suited to the unique shape and mechanical needs of this cell type. However, as other cells are examined, similar types of membrane skeletons containing members of the spectrin and ankyrin families have been revealed, indicating that inner membrane skeletons are widespread.

REVIEW ?

1. Describe two techniques to measure the rates of diffusion of a specific membrane protein.
2. Compare and contrast the types of protein mobility depicted in Figure 4.30.
3. Discuss two major functions of the integral and peripheral proteins of the erythrocyte membrane.
4. Compare the rate of lateral diffusion of a lipid with that of flip-flop. What is the reason for the difference?

4.7 THE MOVEMENT OF SUBSTANCES ACROSS CELL MEMBRANES

Because the contents of a cell are completely surrounded by its plasma membrane, all communication between the cell and the extracellular medium must be mediated by this structure. In a sense, the plasma membrane has a dual function. On one hand, it must retain the dissolved materials of the cell so that they do not simply leak out into the environment, while on the other hand, it must allow the necessary exchange of materials into and out of the cell. The lipid bilayer of the membrane is ideally suited to prevent the loss of charged and polar solutes from a cell, including ions, sugars, and amino acids. Consequently, some provision must be made for the movement of nutrients, respiratory gases, hormones, waste products, and other compounds, in and out of the cell. There are basically two means for the movement of substances through a membrane: passively by diffusion or actively by an energy-coupled transport process. Both types of movements lead to the net flux of a particular ion or compound. The term **net flux** indicates that the movement of the substance into the cell (*influx*) and out of the cell (*efflux*) is not balanced, but that one exceeds the other.

Several different processes are known by which substances move across membranes: simple diffusion through

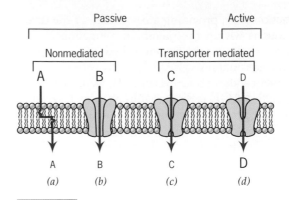

Figure 4.33 Four basic mechanisms by which solute molecules move across membranes. The relative sizes of the letters indicate the directions of the concentration gradients. (*a*) Simple diffusion through the bilayer, which always proceeds from high to low concentration. (*b*) Simple diffusion through an aqueous channel formed within an integral membrane protein or a cluster of such proteins. As in *a*, movement is always down a concentration gradient. (*c*) Facilitated diffusion in which solute molecules bind specifically to a membrane protein carrier (a facilitative transporter). As in *a* and *b*, movement is always from high to low concentration. (*d*) Active transport by means of a protein transporter with a specific binding site that undergoes a change in affinity driven with energy released by an exergonic process, such as ATP hydrolysis. Movement occurs against a concentration gradient.

the lipid bilayer; simple diffusion through an aqueous, protein-lined channel; facilitated diffusion; and active transport (Figure 4.33). We will consider each in turn, but first we will discuss the energetics of solute movement.

The Energetics of Solute Movement

Diffusion is a spontaneous process in which a substance moves from a region of high concentration to a region of low concentration, eventually eliminating the concentration difference between the two regions. As discussed on page 86, diffusion depends on the random thermal motion of solutes and is an exergonic process driven by an increase in entropy. We will restrict the following discussion to diffusion of substances across membranes.

The free-energy change when an uncharged solute (a nonelectrolyte) diffuses across a membrane depends on the magnitude of the concentration gradient, that is, the difference in concentration on each side of the membrane. The following relationship describes the movement of a nonelectrolyte *into* the cell:

$$\Delta G = RT \ln \frac{[C_i]}{[C_o]}$$

$$\Delta G = 2.303 \, RT \log_{10} \frac{[C_i]}{[C_o]}$$

where ΔG is the free-energy change (Section 3.1), R is the gas constant, T is the absolute temperature, and $[C_i]/[C_o]$

is the ratio of the concentration of the solute on the inside (i) and outside (o) surfaces of the membrane. At 25°C,

$$\Delta G = 1.4 \text{ kcal/mol} \cdot \log_{10} \frac{[C_i]}{[C_o]}$$

(see page 89 for RT values)

If the ratio of $[C_i]/[C_o]$ is less than 1.0, then the log of the ratio is negative, ΔG is negative, and the net influx of solute is thermodynamically favored (exergonic). If, for example, the external concentration of solute is 10 times the internal concentration, $\Delta G = -1.4$ kcal/mole. Thus, the maintenance of a tenfold concentration gradient represents a storage of 1.4 kcal/mol. As solute moves into the cell, the concentration gradient decreases, the stored energy is dissipated, and ΔG decreases until, at equilibrium, ΔG is zero. (To calculate ΔG for movement of a solute out of the cell, the term for concentration ratios becomes $[C_o]/[C_i]$).

If the solute is an electrolyte (a charged species), the overall charge difference between the two compartments must also be considered. As a result of the mutual repulsion of ions of like charges, it is thermodynamically unfavorable for an electrolyte to move across a membrane from one compartment into another compartment having a net charge of the same sign. Conversely, if the charge of the electrolyte is opposite in sign to the compartment into which it is moving, the process is thermodynamically favored. The greater the difference in charge (the potential difference or voltage) between the two compartments, the greater the difference in free energy. Thus, the tendency of an electrolyte to diffuse between two compartments depends on two gradients: a chemical gradient, determined by the concentration difference of the substance between the two compartments, and the electric potential gradient, determined by the difference in charge. Together these differences are combined to form an **electrochemical gradient**. The free-energy change for the diffusion of an electrolyte into the cell is

$$\Delta G = RT \ln \frac{[C_i]}{[C_o]} + zF\Delta E_m$$

where z is the charge of the solute, F is the Faraday constant (23.06 kcal/V·equivalent, where an equivalent is the amount of the electrolyte having one mole of charge), and ΔE_m is the potential difference (in volts) between the two compartments. We saw in the previous example that a tenfold difference in concentration of a nonelectrolyte across a membrane at 25°C generates a ΔG of -1.4 kcal/mol. Suppose the concentration gradient consisted of Na^+ ions, which were present at tenfold higher concentration outside the cell than in the cytoplasm. Because the voltage across the membrane of a cell is typically about -70 mV (page 166), the free-energy change for the movement of a mole of Na^+ ions into the cell under these conditions would be

$$\Delta G = -1.4 \text{ kcal/mol} + zF\Delta E_m$$
$$\Delta G = -1.4 \text{ kcal/mol} + (1)(23.06 \text{ kcal/V·mol})(-0.07\text{V})$$
$$= -3.1 \text{ kcal/mol}$$

Thus under the conditions just described, the concentration difference and the electric potential make similar contributions to the storage of free energy across the membrane.

The interplay between concentration and potential differences is seen in the diffusion of potassium ions (K^+) out of a cell. The efflux of the ion is favored by the K^+ concentration gradient, which has a higher K^+ concentration inside the cell, but hindered by the electrical gradient that its diffusion creates, which leaves a higher negative charge inside the cell. We will discuss this subject further when we consider the topic of membrane potentials and nerve impulses on page 166.

Diffusion of Substances Through Membranes

Two qualifications must be met before a nonelectrolyte can diffuse passively across a plasma membrane. The substance must be present at higher concentration on one side of the membrane than the other, and the membrane must be permeable to the substance. A membrane may be permeable to a given solute either (1) because that solute can pass directly through the lipid bilayer, or (2) because that solute can traverse an aqueous pore that spans the membrane and prevents the solute from coming into contact with the lipid molecules of the bilayer. Let us begin by considering the former route in which a substance must dissolve in the lipid bilayer on its way through the membrane.

Discussion of simple diffusion leads us to consider the polarity of a solute. One simple measure of the polarity (or nonpolarity) of a substance is its **partition coefficient**, which is the ratio of its solubility in a nonpolar solvent, such as octanol or a vegetable oil, to that in water under conditions where the nonpolar solvent and water are mixed together. Figure 4.34 shows the relationship between partition coefficient and membrane permeability of a variety of chemicals and drugs. It is evident that the greater the lipid solubility, the faster the penetration.

Another factor determining the rate of penetration of a compound through a membrane is its size. If two molecules have approximately equivalent partition coefficients, the smaller molecule tends to penetrate the lipid bilayer of a membrane more rapidly than the larger one. Very small, uncharged molecules penetrate very rapidly through cellular membranes. Consequently, membranes are highly permeable to small inorganic molecules, such as O_2, CO_2, NO, and H_2O, which are thought to slip between adjacent phospholipids.

In contrast, larger polar molecules, such as sugars, amino acids, and phosphorylated intermediates, exhibit poor membrane penetrability. As a result, the lipid bilayer of the plasma membrane provides an effective barrier that keeps these essential metabolites from diffusing out of the cell. Some of these molecules (e.g., sugars and amino acids) must enter cells from the bloodstream, but they cannot do so by simple diffusion. Instead, special mechanisms must be available to mediate their penetration

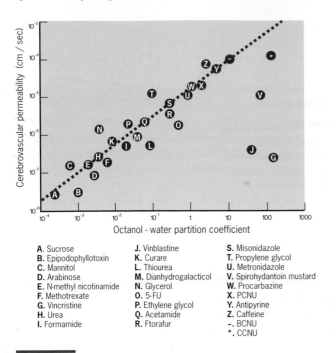

A. Sucrose
B. Epipodophyllotoxin
C. Mannitol
D. Arabinose
E. N-methyl nicotinamide
F. Methotrexate
G. Vincristine
H. Urea
I. Formamide
J. Vinblastine
K. Curare
L. Thiourea
M. Dianhydrogalacticol
N. Glycerol
O. 5-FU
P. Ethylene glycol
Q. Acetamide
R. Ftorafur
S. Misonidazole
T. Propylene glycol
U. Metronidazole
V. Spirohydantoin mustard
W. Procarbazine
X. PCNU
Y. Antipyrine
Z. Caffeine
-. BCNU
*. CCNU

Figure 4.34 **The relationship between partition coefficient and membrane permeability.** In this case, measurements were made of the penetration of a variety of chemicals and drugs across the plasma membranes of the cells that line the capillaries of the brain. Substances penetrate by passage through the lipid bilayer of these cells. The partition coefficient is expressed as the ratio of solubility of a solute in octanol to its solubility in water. Permeability is expressed as penetrance (P) in cm/sec. For all but a few compounds, such as vinblastine and vincristine, penetration is directly proportional to lipid solubility. (FROM N. J. ABBOTT AND I. A. ROMERO, MOLEC. MED. TODAY, 2:110, 1996. COPYRIGHT 1996, WITH PERMISSION FROM ELSEVIER SCIENCE.)

through the plasma membrane. The use of such mechanisms allows a cell to regulate the movement of substances across its surface barrier. We will return to this feature of membranes later.

The Diffusion of Water Through Membranes Water molecules move much more rapidly through a cell mem-

brane than do dissolved ions or small polar organic solutes, which are essentially nonpenetrating. Because of this difference in the penetrability of water versus solutes, membranes are said to be **semipermeable**. Water moves readily through a semipermeable membrane from a region of lower *solute* concentration to a region of higher *solute* concentration. This process is called **osmosis**, and it is readily demonstrated by placing a cell into a solution containing a nonpenetrating solute at a concentration different than that present within the cell itself.

When two compartments of different solute concentration are separated by a semipermeable membrane, the compartment of higher solute concentration is said to be **hypertonic** (or **hyperosmotic**) relative to the compartment of lower solute concentration, which is described as being **hypotonic** (or **hypoosmotic**). When a cell is placed into a hypotonic solution, the cell rapidly gains water by osmosis and swells (Figure 4.35*a*). Conversely, a cell placed into a hypertonic solution rapidly loses water by osmosis and shrinks (Figure 4.35*b*). These simple observations show that a cell's volume is controlled by the difference between the solute concentration inside the cell and that in the extracellular medium. The swelling and shrinking of cells in slightly hypotonic and hypertonic media are usually only temporary events. Within a few minutes, the cells recover and return to their original volume. In a hypotonic medium, recovery occurs as the cells lose ions, thereby reducing their internal osmotic pressure. In a hypertonic medium, recovery occurs as the cells gain ions from the medium. Once the internal solute concentration (which includes a high concentration of dissolved proteins) equals the external solute concentration, the internal and external fluids are **isotonic** (or **isosmotic**), and no net movement of water into or out of the cells occurs (Figure 4.35*c*).

Osmosis is an important factor in a multitude of bodily functions. Your digestive tract, for example, secretes several liters of fluid daily, which is reabsorbed osmotically by the cells that line your intestine. If this fluid weren't reabsorbed, as happens in cases of extreme diarrhea, you would face the prospect of rapid dehydration. Plants uti-

(a) Hypotonic solution

(b) Hypertonic solution

(c) Isotonic solution

Net water gain
Cell swells

Net water loss
Cell shrinks

No net loss or gain

Figure 4.35 **The effects of differences in the concentration of solutes on opposite sides of the plasma membrane.** (*a*) A cell placed in a hypotonic solution (one having a lower solute concentration than the cell) swells because of a net gain of water by osmosis. (*b*) A cell in a hypertonic solution shrinks because of a net loss of water by osmosis. (*c*) A cell placed in an isotonic solution maintains a constant volume because the inward flux of water is equal to the outward flux.

HYPOTONIC:
Normal turgor
pressure

Plasmolysis

H_2O

HYPERTONIC:
No turgor
pressure

(a)

(b)

Figure 4.36 **The effects of osmosis on a plant cell.** (a)
Aquatic plants living in freshwater are surrounded by a hypo-
tonic environment. Water therefore tends to flow into the cells,
creating turgor pressure. (b) If the plant is placed in a hypertonic
solution, such as seawater, the cell loses water, and the plasma
membrane pulls away from the cell wall. (© ED RESCHKE.)

lize osmosis in different ways. Unlike animal cells, which
are generally isotonic with the medium in which they are
bathed, plant cells are generally hypertonic compared to
their fluid environment. As a result, there is a tendency for
water to enter the cell, causing it to develop an internal
(*turgor*) pressure that pushes against its surrounding wall
(Figure 4.36a). If a plant cell is placed into a hypertonic
medium, its volume shrinks as the plasma membrane pulls
away from the surrounding cell wall, a process called **plas-
molysis** (Figure 4.36b). Turgor pressure provides support
for nonwoody plants and for the nonwoody parts of
plants, such as the leaves. The loss of water due to osmosis
causes plants to lose their support and wilt.

Many cells are much more permeable to water than
can be explained by simple diffusion through the lipid bi-

layer. The plasma membranes of such cells contain pro-
teins, called **aquaporins**, that allow the passive movement
of water from one side to the other. Aquaporins contain a
hydrophilic channel that is highly specific for water mole-
cules. Aquaporins are particularly prominent in cells, such
as those of the kidney tubule and plant roots, where the
passage of water plays a crucial role in the tissue's physio-
logical activities. The hormone vasopressin, which stimu-
lates water retention by the collecting ducts of the kidney,
acts by way of these channels. Some cases of the inherited
disorder *congenital nephrogenic diabetes insipidus* have been
traced to mutations in this aquaporin channel. Persons
suffering from this disease excrete huge quantities of urine
because their kidneys do not respond to vasopressin.

The Diffusion of Ions Through Membranes The lipid bi-
layer that constitutes the core of biological membranes is
highly impermeable to charged substances, including
small ions such as Na^+, K^+, Ca^{2+}, and Cl^-. Yet the rapid
movement (**conductance**) of these ions across membranes
plays a critical role in a multitude of cellular activities, in-
cluding formation and propagation of a nerve impulse, se-
cretion of substances into the extracellular space, muscle
contraction, regulation of cell volume, and the opening of
stomatal pores on plant leaves. In 1955, Alan Hodgkin
and Richard Keynes of Cambridge University first pro-
posed that cell membranes contain **ion channels**, that is,
openings in the membrane that are permeable to specific
ions. During the late 1960s and 1970s, Bertil Hille of the
University of Washington and Clay Armstrong of the
University of Pennsylvania began to obtain evidence for
the existence of such channels. The final "proof" emerged
through the work of Bert Sakmann and Erwin Neher at
the Max Planck Institute in the late 1970s and 1980s who
developed techniques to monitor the ionic current passing
through a single ion channel. This is accomplished using
very fine micropipette-electrodes made of polished glass
that are placed on the outer cell surface and sealed to the
membrane by suction. The voltage across the membrane
can be maintained (*clamped*) at any particular value, and
the current originating in the small patch of membrane
surrounded by the pipette can be measured (Figure 4.37).

Today, biologists have identified a bewildering variety
of ion channels, each formed by integral membrane pro-
teins that surround an aqueous pore. Most ion channels
are highly selective in allowing only one particular type of
ion to pass through the pore. As with the passive diffusion
of other types of solutes across membranes, the diffusion
of ions through a channel is always downhill, that is, from
a state of higher energy to a state of lower energy. Ion
channels are *bidirectional*: they allow the passage of ions in
either direction, with the net flux of the ion depending on
the electrochemical gradient. Comparisons of the amino
acid sequences of proteins that form different types of ion
channels in organisms as diverse as bacteria, plants, and
animals reveal that all ion channels are members of a small
number of giant *superfamilies*. Although the members of a
given superfamily may have very different ion selectivities,

(a)

(b) ———— 25 μm

Figure 4.37 Measuring ion channel conductance by patch-clamp recording. (*a*) In this technique, a highly polished glass micropipette is placed against a portion of the outer surface of a cell, and suction is applied to seal the rim of the pipette against the plasma membrane. Because the pipette is wired as an electrode (a *microelectrode*), a voltage can be applied across the patch of membrane enclosed by the pipette, and the responding flow of ions through the membrane channels can be measured. As indicated in the figure, the micropipette can enclose a patch of membrane containing a single ion channel, which allows inves-

tigators to monitor the opening and closing of a single gated channel, as well as its conductance at different applied voltages. (*b*) The micrograph shows patch-clamp recordings being made from a single photoreceptor cell of the retina of a salamander. One portion of the cell is drawn into a glass micropipette by suction, while a second micropipette-electrode (lower right) is sealed against a small patch of the plasma membrane on another portion of the cell. (B: FROM T. D. LAMB, H. R. MATTHEWS, AND V. TORRE, J. PHYSIOLOGY 372:319, 1986. REPRODUCED WITH PERMISSION.)

all are similar enough in amino acid sequence and overall structure to suggest that they are derived from a single protein that was present in a common ancestor living billions of years ago.

Most of the ion channels that have been identified can exist in either an open or a closed conformation; such channels are said to be **gated**. The opening and closing of the gates are subject to complex physiological regulation and can be induced by a variety of factors depending on the particular channel. Two major categories of gated channels are:

1. **Voltage-gated channels** whose conformational state depends on the difference in ionic charge on the two sides of the membrane.
2. **Ligand-gated channels** whose conformational state depends on the binding of a specific molecule (the ligand), which is usually not the solute that passes through the channel.

Some ligand-gated channels are opened (or closed) following the binding of a molecule to the outer surface of the channel; others are opened (or closed) following the binding of a ligand to the inner surface of the channel. For example, neurotransmitters, such as acetylcholine, act on

the outer surface of certain cation channels, while cyclic nucleotides, such as cAMP, act on the inner surface of certain calcium ion channels. We will focus in the following discussion on the structure and function of potassium ion channels because these are the best understood.

Genes that encode a variety of distinct voltage-gated K^+ (or Kv) channels have been isolated and the molecular anatomy of their proteins scrutinized. Members of this family of proteins have the basic architecture illustrated two-dimensionally in Figure 4.38*a*. The N- and C-terminal domains are situated on the cytoplasmic side of the membrane, whereas the middle portion of the polypeptide contains six membrane-spanning segments (S1–S6). A single Kv channel consists of four homologous polypeptides (subunits) arranged symmetrically around a central ion-conducting pore, as shown in Figure 4.38*b*.[5] The walls of the pore at its narrowest portion are lined by a segment of the polypeptide called H5 (or P) that connects the

———

[5]This four-fold symmetry is the apparent hallmark of all channels that are selective for specific cations. Whereas K^+ channels consist of four homologous polypeptides, Na^+ and Ca^{2+} channels are composed of a single giant polypeptide with four separate hydrophobic domains, each containing six membrane-spanning segments that are homologous to those of the K^+ channels.

S4 Segment

(a)

(b)

Figure 4.38 The structure of a voltage-gated K⁺ channel.
(*a*) A two-dimensional portrait of a K⁺ channel subunit showing its six transmembrane segments and a portion of the polypeptide (called H5 or P) that dips into the protein to form part of the channel's wall. The inset shows the sequence of amino acids of the S4 transmembrane segment of the *Drosophila* K⁺ *Shaker* ion channel, which contains seven positively charged side chains that are thought to serve as a voltage sensor. The positively charged side chains are situated at every third residue along the helix. This member of the Kv family is

called a *Shaker* channel because flies with certain mutations in the protein shake vigorously when anesthetized with ether. The *Shaker* channel was the first K⁺ channel to be identified and cloned in 1987. (*b*) Schematic model of the arrangement of the subunits of a voltage-gated K⁺ channel in the membrane. The pore is lined at its outer end by the H5 segments of each polypeptide and its cytoplasmic end primarily by the S6 helices. Each polypeptide ends in a dangling ball whose function is illustrated in the following figure. One of the four polypeptides that makes up the tetrameric channel is not shown here.

S5 and S6 transmembrane helices. The H5 segments from the four subunits dip in toward the center of the protein (see Figure 4.39*b*) forming a ring of residues just large enough to allow the passage of a K⁺ ion after it has been stripped of its surrounding shell of water molecules.

Voltage-gated potassium channels can exist in at least three distinct conformations: closed, open, and inactivated, which are related to one another as shown schematically in Figure 4.39*a*. As will be discussed later in the section on action potentials and nerve impulses (Section 4.8), Kv channels are opened by a change in voltage. The opening of the channel is regulated by the S4 transmembrane helix (Figure 4.38*a*, inset), which contains several posi-

Rest Open Inactivated

(a)

Figure 4.39 Conformational states of a voltage-gated K⁺ channel. (*a*) Schematic representation of a view down a K⁺ ion channel, perpendicular to the membrane, showing the channel in the closed (resting), open, and inactivated state. (*b*) Structural model of the ball-and-chain inactivation mechanism. The three-dimensional structure of the pore is shown in this model. (A: REPRINTED FROM NEURON, VOL. 20, C. M. ARMSTRONG AND B. HILLE, VOLTAGE-GATED ION CHANNELS AND ELECTRICAL EXCITABILITY, PAGE 377, COPYRIGHT 1998, WITH PERMISSION FROM ELSEVIER SCIENCE; B: REPRINTED WITH PERMISSION FROM C. ANTZ ET AL., NATURE 385:274, 1997; COPYRIGHT 1997, MACMILLAN MAGAZINES LIMITED; COURTESY HANS ROBERT KALBITZER.)

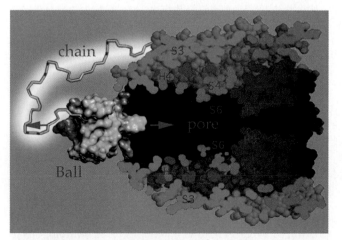

(b)

tively charged amino acid residues spaced along the polypeptide chain. This portion of the protein is postulated to act as a *voltage sensor*. Under resting conditions, the negative potential across the membrane (page 166) keeps the S4 helix in a conformation in which the pore is closed. A change in the potential to a more positive value (a depolarization, page 167) exerts an electric force on the S4 helix. This force is thought to cause the S4 helix to rotate in such a way that its positively charged residues move from a position where they were exposed to the cytoplasm to a new position where they are exposed to the outside of the cell. Movement of the positively charged residues of the S4 helix can be detected electrically as a small *gating current*. Movement of the S4 helix has also been followed in experiments in which a fluorescent group is attached to specific amino acids in this part of the protein. When the membrane containing these channels is depolarized, fluorescence appears at the cell surface, which indicates that the labeled amino acids have moved to a position facing the external medium.

The movement of the S4 helix in response to changes in voltage initiates a conformational change within the protein that leads to the opening of the mouth of the pore. This change in conformation may be similar to that illustrated in Figure 4.22*b*. Once the pore is opened, several thousand potassium ions can pass through the channel per millisecond, which is nearly the rate that would occur by free diffusion. Because of the large ion flux, the opening of a relatively small number of K^+ channels has significant impact on the electrical properties of the membrane. After the channel is open for a few milliseconds, the movement of K^+ ions is "automatically" stopped by a process of inactivation. Inactivation of the channel is accomplished by movement of the N-terminal portion of the polypeptide that is folded into a "ball on a chain." When one of the dangling balls (depicted in Figure 4.38*b*) swings into the cytoplasmic mouth of the channel (Figure 4.39*b*), the passage of ions is blocked, and the channel is inactivated. At a subsequent stage of the cycle, the ball is released, and the opening to the pore is closed as indicated schematically in Figure 4.39*a*.

The information concerning the K^+ channel described above was gleaned from many biochemical and biophysical studies conducted during the 1980s and 1990s. Finally, in 1998, the first three-dimensional image of a K^+ channel based on X-ray crystallography was published by Roderick MacKinnon and his colleagues at Rockefeller University. The channel chosen by these researchers was not the eukaryotic K^+ channel depicted in Figure 4.38, but a much simpler, more stable version that had been discovered a few years earlier in bacterial cells. This channel, which is called KcsA, contains only two transmembrane helices (M1 and M2) connected by a much shorter segment (P) that dips into the bilayer to form the lining of a narrow pore through which a dehydrated K^+ ion must pass (Figure 4.40). The M1-P-M2 portion of a KcsA channel is thought to be homologous to the S5-H5-S6 portion of the eukaryotic channel.

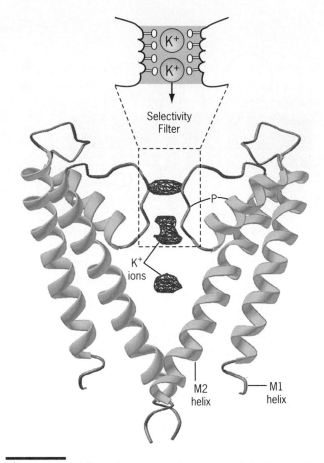

Figure 4.40 **Three dimensional structure of the bacterial KcsA channel and the selection of K^+ ions.** This K^+ ion channel consists of four subunits, two of which are shown here. Each subunit is comprised of M1 and M2 helices joined by a P (pore) segment consisting of a short helix and a nonhelical portion that lines the channel through which the ions pass. A portion of each P segment contains a conserved pentapeptide whose residues line the selectivity filter that screens selectively for K^+ ions. The oxygen atoms of the carbonyl groups of these residues project into the channel where they can interact selectively with K^+ ions (indicated by the red mesh objects) within the filter. As indicated in the top inset, the selectivity filter contains four rings of O atoms; each ring contains four O atoms, one donated by each subunit. The diameter of the rings is just large enough to interact with a K^+ ion, replacing its normal water of hydration. (FROM RODERICK MACKINNON, REPRINTED WITH PERMISSION FROM NATURE MED. 5:1108,1999; COPYRIGHT 1999, MACMILLAN MAGAZINES LIMITED.)

As in other cases in the field, the publication of a three-dimensional structure has served to (1) confirm basic conclusions about the molecule based on earlier, noncrystallographic data and (2) enable us to better understand the way the molecule works. In this case, the crystal structure of the KcsA channel has led to a much deeper understanding of the mechanism by which this channel (and its eukaryotic counterparts) selects overwhelmingly for K^+ ions over Na^+ ions, yet at the same time allows such a rapid conductance of K^+ ions through its pore. As shown in Figure 4.40, the P region of the

KcsA channel resides within the extracellular mouth of the channel. Within the P region is a pentapeptide—Thr-Val-Gly-Tyr-Gly—that is either identical or very similar to that found in the H5 regions of all known K^+ ion channels. Mutations within the conserved pentapeptide destroy the channel's ability to discriminate between K^+ and Na^+ ions. The crystal structure of the KcsA channel shows that the backbone carbonyl groups from the conserved pentapeptide (see the backbone structure on page 53) form the lining of the narrowest part of the pore (Figure 4.40, inset). This section of the pore is referred to as the *selectivity filter* because of its role in selecting K^+ ions. The conserved residues of the selectivity filter create four successive rings of carbonyl oxygen atoms (Figure 4.40, inset). Each ring, which contains an oxygen atom projecting from each subunit, has a diameter of approximately 3 Å, which is precisely that required to surround a K^+ ion (diameter 2.7 Å) that has lost its shell of hydration. It appears from these findings that the four electronegative O atoms of each ring substitute for the water molecules that are displaced as each K^+ ion enters the pore. Whereas the selectivity filter is a precise fit for a dehydrated K^+ ion, it is much larger than the diameter of a dehydrated Na^+ ion (1.9 Å). Consequently, the Na^+ ion cannot interact with the four carbonyl oxygen atoms necessary to stabilize its structure. As a result, the smaller Na^+ ions cannot overcome the higher energy barrier required to penetrate the pore. It is estimated that two K^+ ions can bind within the selectivity filter at the same time (Figure 4.40). The electrostatic repulsion between these closely bound dehydrated K^+ ions provides the mechanism for allowing the ions to move down their electrochemical gradient through the pore so rapidly.

Potassium channels come in many different varieties. It is remarkable that *C. elegans*, a nematode worm whose body consists of only about 1000 cells contains more than 90 different genes that encode K^+ channels. It is evident that a single cell—whether in a nematode, human, or plant—is likely to possess a variety of different K^+ channels that open and close in response to different voltages. In addition, the voltage required to open or close a particular K^+ channel can vary depending on whether or not the channel protein is phosphorylated, which in turn is regulated by hormones and other factors. It is apparent that ion channel function is under the control of a diverse and complex set of regulatory agents. The structure and function of a very different type of ion channel, the ligand-gated nicotinic acetylcholine receptor is the subject of the Experimental Pathways section at the end of this chapter.

Facilitated Diffusion

Substances always diffuse across a membrane from a region of higher concentration on one side to a region of lower concentration on the other side, but they do not always diffuse through the lipid bilayer or through a channel. Numerous examples are known in which the diffusing

Figure 4.41 Facilitated diffusion. A schematic model for the facilitated diffusion of glucose depicts the alternating conformation of a carrier that exposes the glucose binding site to either the inside or outside of the membrane. (AFTER S. A. BALDWIN AND G. E. LIENHARD, TRENDS BIOCHEM. SCI. 6:210, 1981.)

substance first binds selectively to a membrane-spanning protein, called a **facilitative transporter**,[6] that facilitates the diffusion process. The binding of the solute to the facilitative transporter on one side of the membrane is thought to trigger a conformational change in the protein, exposing the solute to the other surface of the membrane, from where it can diffuse down its concentration gradient. An example of this mechanism is illustrated in Figure 4.41. **Facilitated diffusion**, as this process is called, is similar in many ways to an enzyme-catalyzed reaction.

Because they operate passively, that is, without being coupled to an energy-releasing system, facilitated transporters can mediate the movement of solutes equally well in both directions. The direction of net flux depends solely on the relative concentration of the substance on the two sides of the membrane.

[6]The term "facilitative transporter" is used here to distinguish these proteins from the active transporters, discussed later in the chapter, whose activity is coupled to a process that releases energy. Technically, the term "transporter" applies to a membrane protein that can only bind a solute from one side of the membrane at a time and in which a conformational change acts as the mechanism for the movement of the substance through the membrane. This definition distinguishes transporters from channels, which, when open, can bind solutes from either side of the membrane at the same time. As more is learned about these proteins, the distinction between channels and transporters has become blurred.

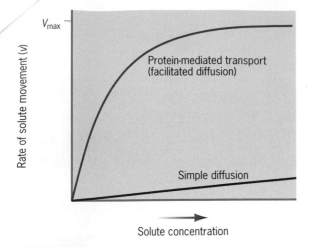

Figure 4.42 **The kinetics of facilitated diffusion as compared to that of simple physical diffusion.**

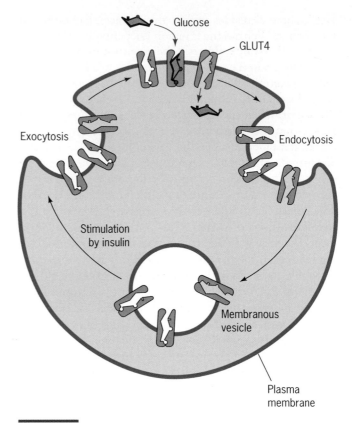

Figure 4.43 **The regulation of glucose uptake in muscle and fat cells by insulin.** Glucose transporters are stored in the walls of cytoplasmic vesicles that form by budding from the plasma membrane (endocytosis). When the insulin level increases, the cytoplasmic vesicles are translocated to the cell periphery. The vesicles fuse with the plasma membrane (exocytosis), delivering the transporters to the cell surface where they can take up glucose. (FROM D. VOET AND J. G. VOET, BIOCHEMISTRY, 2D ED., COPYRIGHT © 1995, JOHN WILEY & SONS, INC. REPRINTED BY PERMISSION OF JOHN WILEY & SONS, INC.)

Like enzymes, facilitative transporters are specific for the molecules they transport, discriminating, for example, between D and L stereoisomers (page 45). In addition, both enzymes and transporters exhibit saturation-type kinetics (Figure 4.42). Unlike ion channels, which can conduct millions of ions per second, most facilitative transporters can move only hundreds to thousands of solute molecules per second across the membrane. Another important feature of facilitative transporters is that, like enzymes and ion channels, their activity can be regulated. Facilitated diffusion is particularly important in mediating the entry and exit of polar solutes, such as sugars and amino acids, that do not penetrate the lipid bilayer. This is illustrated in the following section.

The Glucose Transporter: An Example of Facilitated Diffusion

Glucose is the body's primary source of direct energy, and most mammalian cells contain a membrane protein that facilitates the diffusion of glucose from the bloodstream into the cell (as depicted in Figure 4.41). A gradient favoring the continued diffusion of glucose into the cell is maintained by phosphorylating the sugar after it enters the cytoplasm, thus lowering the intracellular glucose concentration. Humans have at least five related proteins that act as facilitative glucose transporters (such genetic variants are referred to as *isoforms*). These isoforms, termed GLUT1 to GLUT5, are distinguished by the tissues in which they are located, as well as their kinetic and regulatory characteristics.

Insulin is a hormone produced by endocrine cells of the pancreas that plays a key role in maintaining proper blood sugar levels. An increase in blood glucose levels triggers the secretion of insulin, which stimulates the uptake of glucose into various target cells, most notably skeletal muscle and fat cells (adipocytes). Insulin-responsive cells share a common isoform of the facilitative glucose transporter, specifically GLUT4. When insulin levels are low,

these cells contain relatively few glucose transporters on their surface. Instead, the transporters are present within the membranes of cytoplasmic vesicles (Figure 4.43). When insulin levels rise in response to elevated blood glucose levels, the hormone acts on target cells to stimulate the translocation of the vesicles from the cytoplasm to the cell surface, where the transporters become incorporated into the plasma membrane and transport glucose into the cell.

Active Transport

Typically, the K^+ concentration inside a mammalian cell is about 100 mM, whereas that outside the cell is only about 5 mM. Consequently, there is a steep K^+ concentration gradient across the plasma membrane favoring the diffusion of K^+ out of the cell. Sodium ions are also distributed very unequally across the plasma membrane, but the gradient is oppositely disposed—the Na^+ concentra-

tion is about 150 mM outside the cell and 10–20 mM inside the cell. The concentration difference for Ca^{2+} is even greater; the typical cytosolic concentration of 10^{-7} M is 10,000 times less than that outside the cell. The ability of a cell to generate such steep concentration gradients across its plasma membrane cannot occur by either simple or facilitated diffusion. Rather, these gradients must be generated by **active transport**.

Like facilitated diffusion, active transport depends on integral membrane proteins that selectively bind a particular solute and move it across the membrane in a process driven by changes in the protein's conformation. Unlike facilitated diffusion, however, movement of a solute against a gradient requires the coupled input of energy. Consequently, the endergonic movement of ions or other solutes across the membrane against a concentration gradient is coupled to an exergonic process, such as the hydrolysis of ATP, the absorbance of light, the transport of electrons, or the flow of other substances down their gradients. Proteins that carry out active transport are often referred to as "pumps."

Coupling Active Transport to ATP Hydrolysis

In 1957, Jens Skou, a Danish physiologist, discovered an ATP-hydrolyzing enzyme in the nerve cells of a crab that was only active in the presence of both sodium and potassium ions (as well as Mg^{2+}, which acts as a cofactor by binding ATP). Skou proposed, and correctly so, that this enzyme, which was responsible for ATP hydrolysis, was the same protein that was active in transporting the two ions; the enzyme was called the Na^+/K^+-ATPase, or the *sodium–potassium pump*.

Unlike the protein-mediated movement of a facilitated diffusion system, which will carry the substance equally well in either direction, active transport drives the movement of ions in only one direction. A large number of studies have indicated that the ratio of Na^+/K^+ pumped by the Na^+/K^+-ATPase is not 1:1, but 3:2 (see Figure 4.44). In other words, for each ATP hydrolyzed, three sodium ions are pumped out as two potassium ions are pumped in. Because of this pumping ratio, the Na^+/K^+-ATPase is *electrogenic*, which means that it contributes directly to the separation of charge across the membrane.

The Na^+/K^+-ATPase is an example of a P-type ion pump. The "P" stands for phosphorylation, indicating that, during the pumping cycle, the hydrolysis of ATP leads to the transfer of the released phosphate group to an aspartic acid residue of the transport protein, which in turn causes an essential conformational change within the protein. Conformational changes are necessary to change the affinity of the protein for the two cations that are transported. Consider the activity of the protein. It must pick up sodium or potassium ions from a region of low concentration, which means that the protein must have a relatively high affinity for the ions. Then the protein must release the ions on the other side of the membrane into a much greater concentration of each ion. To do this, the

affinity of the protein for that ion must decrease. Thus, the affinity for each ion on the two sides of the membrane must be different. This is achieved by phosphorylation, which changes the shape of the protein molecule. The change in shape of the protein also serves to expose the ion binding sites to different sides of the membrane, as discussed in the following paragraph.

A scheme for the pumping cycle of the Na^+/K^+-ATPase is shown in Figure 4.44. When the protein binds three Na^+ ions on the inside of the cell (step 1) and becomes phosphorylated (step 2), it shifts from the E_1 conformation to the E_2 conformation (step 3). In doing so, the binding site becomes exposed to the extracellular compartment, and the protein loses its affinity for Na^+ ions, which are then released outside the cell. Once the three sodium ions have been released, the protein picks up two potassium ions (step 4), becomes dephosphorylated (step 5), and shifts back to the original E_1 conformation (step 6). In this state, the binding site opens to the internal surface of the membrane and loses its affinity for K^+ ions, leading to the release of these ions into the cell. The cycle is then repeated.

The importance of the sodium-potassium pump becomes evident when one considers that it consumes approximately one-third of the energy produced by most animal cells and two-thirds of the energy produced by nerve cells. Digitalis, a steroid obtained from the foxglove plant that has been used for 200 years as a treatment for congestive heart disease, binds to the Na^+/K^+-ATPase. Digitalis strengthens the heart's contraction by inhibiting the Na^+/K^+ pump, which leads to a chain of events that increases Ca^{2+} availability inside the muscle cells of the heart.

Other Ion Transport Systems

The sodium-potassium pump is found only in animal cells. This protein is thought to have evolved in primitive animals as the primary means to maintain cell volume and as the mechanism to generate the steep Na^+ and K^+ gradients that play such a key role in the formation of impulses in nerve and muscle cells. Plant cells have a H^+-transporting, P-type, plasma membrane pump. In plants, this proton pump plays a key role in the secondary transport of solutes (discussed later), in the control of cytosolic pH, and possibly in control of cell growth by means of acidification of the plant cell wall. Another well-studied P-type pump is the Ca^{2+}-ATPase present in both the plasma membrane and the membranes of the endoplasmic reticulum. This Ca^{2+} pump functions to actively transport calcium ions out of the cytosol into either the extracellular space or the lumen of the ER.

The epithelial lining of the stomach also contains a P-type pump, the H^+/K^+-ATPase, which secretes a solution of concentrated acid (up to 0.16 N HCl) into the stomach chamber. In the resting state, these pump molecules are situated in cytoplasmic membranes of the parietal cells of the stomach lining and are nonfunctional (Figure 4.45). When

Figure 4.44 Schematic concept of the Na⁺/K⁺-ATPase transport cycle. Sodium ions (1) bind to the protein on the inside of the membrane. ATP is hydrolyzed, and the phosphate is transferred to the protein (2), changing its conformation (3) and allowing sodium ions to be expelled to the external space. Potassium ions then bind to the protein (4), and then the phosphate group is removed (5), which causes the protein to snap back to its original conformation, moving the potassium ions to the inside of the cell (6). Unlike facilitated diffusion, the changes in the shape of the protein are driven by energy from ATP hydrolysis, which allows the transport system to move these ions against their electrochemical gradients. Note that the actual Na⁺/K⁺-ATPase is a tetramer consisting of two different membrane-spanning subunits: a larger α subunit, which carries out the transport activity, and a smaller β subunit, which functions primarily in the maturation and assembly of the pump within the membrane.

food enters the stomach, a hormonal message is transmitted to the parietal cells that causes the pump-containing membranes to move to the apical cell surface, where they fuse with the plasma membrane and begin secreting acid (Figure 4.45). In addition to functioning in digestion, stomach acid can also lead to heartburn. Prilosec is a widely prescribed drug that prevents heartburn by inhibiting the stomach's H⁺/K⁺-ATPase. Other acid-blocking heartburn medications (e.g., Zantac, Pepcid, and Tagamet) do not inhibit the H⁺/K⁺-ATPase directly, but block a receptor on the surface of the parietal cells, thereby stopping the cells from becoming activated by the hormone.

Unlike P-type pumps, V-type pumps utilize the energy of ATP without forming a phosphorylated protein intermediate. V-type pumps actively transport hydrogen ions across the walls of cytoplasmic organelles and vacuoles (hence the designation V-type). They occur in the membranes that line lysosomes, secretory granules, and plant cell vacuoles. V-type pumps have also been found in the plasma membranes of a variety of cells. For example, a V-type pump in the plasma membranes of kidney tubules helps maintain the body's acid-base balance by secreting protons into the forming urine.

Another diverse group of proteins that actively transport ions is the *ATP-binding cassette (ABC) transporters,* so-called because all of the members of this superfamily share a homologous ATP-binding domain. The best-studied ABC transporter is described in the accompanying Human Perspective.

Using Light Energy for Active Transport of Ions *Halobacterium salinarium* (previously *H. halobium*) is an archaebacterium that lives in extremely salty environments, such as that found in the Great Salt Lake. When grown under anaerobic conditions, the plasma membranes of these bacteria take on a purple color due to the presence of one particular protein, bacteriorhodopsin. As shown in Figure 4.46, *bacteriorhodopsin* contains retinal, the same prosthetic group present in rhodopsin, the light-absorbing protein of the rods of the vertebrate retina. The absorption of light energy by the retinal group induces a series of conformational changes in the protein that cause a proton to move from the

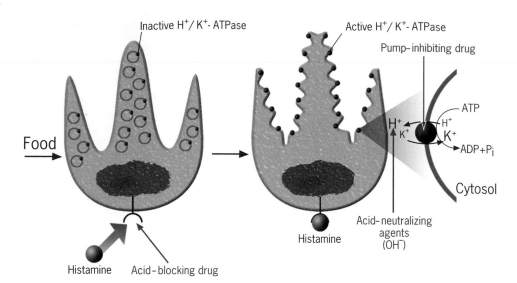

Figure 4.45 Control of acid secretion in the stomach. In the resting state, the H^+/K^+-ATPase molecules are present in the walls of cytoplasmic vesicles. Food entering the stomach triggers a cascade of hormone-stimulated responses in the stomach wall leading to the release of histamine, which binds to a receptor on the surface of the acid-secreting parietal cells. Binding of histamine to its receptor stimulates a response that causes the H^+/K^+-ATPase-containing vesicles to fuse to the plasma membrane, forming deep folds, or canaliculi. Once at the surface, the transport protein is activated and pumps protons into the stomach cavity against a concentration gradient (indicated by the size of the letters). The heartburn drug Prilosec blocks acid secretion by directly inhibiting the H^+/K^+-ATPase, whereas other acid-blocking medications interfere with activation of the parietal cells. Acid-neutralizing medications provide basic anions that combine with the secreted protons.

retinal group, through a channel in the protein, to the cell exterior (Figure 4.46). The proton donated by the photo-excited retinal is replaced by another proton transferred to the protein from the cytoplasm. In effect, this process results in the translocation of protons from the cytoplasm to the external environment, thereby generating a steep H^+ gradient across the plasma membrane. This gradient is subsequently used by an ATP-synthesizing enzyme to phosphorylate ADP, as described in the next chapter.

Figure 4.46 Bacteriorhodopsin: a light-driven proton pump. The protein contains seven membrane-spanning helices and a centrally located retinal group (shown in purple), which serves as the light-absorbing element (chromophore). Absorption of a photon of light causes a change in the electronic structure of retinal, leading to the transfer of a proton from the —NH^+ group to a closely associated, negatively charged aspartic acid residue (#85) (step 1). The proton is then released to the extracellular side of the membrane (step 2) by a relay system consisting of several amino acid residues (Asp82, Glu204, and Glu194). The spaces between these residues are filled with hydrogen-bonded water molecules that help shuttle protons along the pathway. The deprotonated retinal is returned to its original state (step 3) when it accepts a proton from an undissociated aspartic acid residue (Asp96) located near the cytoplasmic side of the membrane. Asp96 is then reprotonated by a H^+ from the cytoplasm (step 4). Asp85 is deprotonated (step 5) prior to receiving a proton from retinal in the next pumping cycle. As a result of these events, protons move from the cytoplasm to the cell exterior through a central channel in the protein. (FROM HARTMUT LUECKE ET AL., COURTESY OF JANOS K. LANYI, SCIENCE 286:255, 1999; COPYRIGHT 1999 AMERICAN ASSOCIATION FOR THE ADVANCEMENT OF SCIENCE).

THE HUMAN PERSPECTIVE

Defects in Ion Channels as a Cause of Inherited Disease

UPDATE For the very latest, go to page U-8

Several severe, inherited disorders have been traced to mutations in genes that encode ion channel proteins (Table 1). Most of the disorders listed in Table 1 affect the movement of ions across the plasma membranes of excitable cells (i.e., muscle, nerve, and sensory cells), reducing the ability of these cells to develop or transmit impulses (action potentials, page 167). In contrast, cystic fibrosis, the best-studied and most common inherited ion channel disorder, results from a defect in the ion channels of epithelial cells.

On average, 1 out of every 25 persons of Northern European descent carries one copy of the gene that can cause cystic fibrosis. Because they show no symptoms of the mutant gene, most heterozygotes are unaware that they are carriers. Consequently, approximately 1 out of every 2500 infants in this Caucasian population ($\frac{1}{25} \times \frac{1}{25} \times \frac{1}{4}$) is homozygous recessive at this locus and born with cystic fibrosis (CF). Although cystic fibrosis affects various organs, including the intestine, pancreas, sweat glands, and reproductive tract, the respiratory tract usually exhibits the most severe effects. Victims of CF produce a thickened, sticky mucus that is very hard to propel out of the airways. Afflicted individuals typically suffer from chronic lung infections and inflammation, which progressively destroy pulmonary function.

The gene responsible for cystic fibrosis was isolated in 1989. Once the sequence of the CF gene was determined and the amino acid sequence of the corresponding polypeptide was deduced, it was apparent that the polypeptide was a member of the ABC transporter superfamily (page 160). Like the other members in this superfamily, the polypeptide contains two domains situated within the lipid bilayer and two nucleotide-binding domains (NBDs) that project into the cytoplasm (Figure 1). Unlike the other members of the superfamily, the protein involved in CF includes a regulatory (R) domain containing several serine residues that can be phosphorylated by a protein kinase (called PKA) that is activated by the second messenger, cyclic AMP (Section 15.2). The protein was named *cystic fibrosis transmembrane conductance regulator (CFTR)*, an ambiguous term that reflected the fact that researchers weren't sure of its precise function. The question was answered after the protein was purified, incorporated into artificial lipid bilayers, and shown to act as a cyclic AMP-regulated chloride channel, not a transporter. The opening of the channel gate requires two independent events: the phosphorylation of the R domain and the binding and hydrolysis of ATP by one of the nucleotide-binding domains (Figure 1).

The cells of patients with cystic fibrosis exhibit an abnormally low efflux of chloride ions, as expected from a disease caused by a defect in a chloride ion channel. It remains uncertain, however, how this underlying molecular defect leads to the development of chronic lung infections. Numerous explanations have been offered, including the following:

1. Because the movement of water out of epithelial cells by osmosis follows the movement of salts, the decrease in Cl^- efflux could lead to a decrease in the volume of fluid that bathes the epithelial cells of the airways. A reduction in vol-

Table 1

Inherited disorder	Type of channel	Gene	Clinical consequences
Familial hemiplegic migraine (FHM)	Ca^{2+}	CACNL1A4	Migraine headaches
Episodic ataxia type-2 (EA-2)	Ca^{2+}	CACNL1A4	Ataxia (lack of balance and coordination)
Hypokalemic periodic paralysis	Ca^{2+}	CACNL1A3	Periodic myotonia (muscle stiffness) and paralysis
Episodic ataxia type-1	K^+	KCNA1	Ataxia
Benign familial neonatal convulsions	K^+	KCNQ2	Epileptic convulsions
Nonsyndromic dominant deafness	K^+	KCNQ4	Deafness
Long QT syndrome	K^+	HERG KVLQT1, or	Dizziness, sudden death from ventricular fibrillation
	Na^+	SCN5A	
Hyperkalemic periodic paralysis	Na^+	SCN4A	Periodic myotonia and paralysis
Liddle Syndrome	Na^+	β-ENaC	Hypertension (high blood pressure)
Myasthenia gravis	Na^+	nAChR	Muscle weakness
Dent's disease	Cl^-	CLCN5	Kidney stones
Myotonia congenita	Cl^-	CLC-1	Periodic myotonia
Cystic fibrosis	Cl^-	CFTR	Lung congestion and infections
Cardiac arrhythmias	Na^+ K^+ Ca^{2+}	many different genes	Irregular or rapid heartbeat

Figure 1 Proposal for the dual control of the CFTR dimer. Cyclic AMP activates PKA, a protein kinase, which phosphorylates serine residues of the R domain of the protein (step 1). The conformational change induced by phosphorylation facilitates the binding of ATP to one of the nucleotide binding domains (NBD1) of the protein (step 2). Hydrolysis of the bound ATP opens the chloride channel (step 3), allowing efflux of more than 10^6 Cl⁻ ions per second. The binding of a second ATP to NBD2 and its subsequent hydrolysis (step 4) leads to the closure of the channel.

ume of the surface liquid, and a resulting increase in viscosity of the secreted mucus, could impair the function of the cilia that are responsible for pushing bacteria out of the respiratory tract.

2. Among the bacteria that infect the airways of CF patients, *Pseudomonas aeruginosa* is the most prevalent and destructive species. This bacterium is rarely seen in the airways of individuals suffering from other types of lung disease, and it is not certain why CF patients are so susceptible to it. Studies indicate that *P. aeruginosa* binds to the extracellular end of the CFTR protein, and it is thought that this attachment may lead to ingestion and destruction of the bacterium by the epithelial cells. Individuals who lack the CFTR protein in their plasma membrane, as do many CF patients (see the following), may be unable to clear the bacterium from their respiratory tract.

In the past decade, researchers have identified more than 800 different mutations that give rise to cystic fibrosis, and the effect of many of these alterations on protein structure has been studied (Figure 2). Nearly 70 percent of the alleles responsible for cystic fibrosis in the United States contain the same genetic alteration (designated ΔF508)—they are all missing three base pairs of DNA that encode a phenylalanine at position 508, within one of the nucleotide-binding domains of the CFTR polypeptide. Subsequent research has revealed that CFTR polypeptides lacking this particular amino acid fail to be processed normally within the membranes of the endoplasmic reticulum and, in fact, never reach the surface of epithelial cells. As a result, CF patients who are homozygous for the ΔF508 allele completely lack the CFTR chloride channel in their plasma membranes and have a severe form of the disease. Other CF patients with less severe forms have mutant alleles that encode a CFTR that is able to reach the surface of cells but mediates a reduced chloride conductance. The mildest forms are characterized by infertility, with little or no damage to major organs.

Figure 2 Identifying the mutations responsible for cystic fibrosis. The location and type of more than 150 different CF mutations that have been identified by the Cystic Fibrosis Genetic Analysis Consortium are indicated below a schematic of the CFTR protein. (▲) In-frame deletion; (■) missense mutation; (●) nonsense mutation; (○) frame-shift mutation; and (▼) splicing mutation. Most of these classes of mutation are discussed in Chapter 11. This figure indicates the variety of alleles of a gene that can exist within a population. The numbers correspond to the coding segments (exons) that make up the gene. The Δ508 deletion common in persons of Northern European descent occurs in the 10th exon. (REPRINTED WITH PERMISSION FROM F. S. COLLINS, SCIENCE 256:775, 1992; COPYRIGHT 1992 AMERICAN ASSOCIATION FOR THE ADVANCEMENT OF SCIENCE.)

According to one estimate, the ΔF508 mutation had to have originated more than 50,000 years ago to have reached such a high frequency in the population. The fact that the CF gene has reached this frequency suggests that heterozygotes may receive some selective advantage over those lacking a copy of the defective gene. It has been proposed that CF heterozygotes may be protected from the effects of cholera, a disease that is characterized by excessive fluid secretion by the wall of the intestine. One difficulty with this proposal is that there is no record of cholera epidemics in Europe until the 1820s. An alternate proposal suggests that heterozygotes are protected from typhoid fever because the bacterium responsible for this disease adheres poorly to the wall of an intestine having a reduced number of CFTR molecules.

Ever since the isolation of the gene responsible for CF, the development of a cure by gene therapy—that is, by replacement of the defective gene with a normal version—has been a major goal of CF researchers. Cystic fibrosis is a good candidate for gene therapy because the worst symptoms of the disease result from the defective activities of epithelial cells that line the airways and, therefore, are accessible to substances that can be delivered by inhalation of an aerosol. Clinical trials have begun using several different types of delivery systems. In one group of trials, the normal *CFTR* gene is incorporated into the DNA of a defective adenovirus, a type of virus that normally causes upper respiratory tract infections. The recombinant virus particles are then allowed to infect the cells of the airway, delivering the normal gene to the genetically deficient cells. The primary disadvantage in using adenovirus is that the viral DNA (along with the normal *CFTR* gene) does not become integrated into the chromosomes of the infected host cell so that the virus must be readministered frequently. In addition, the virus induces an immune response that eliminates the virus and leads to lung inflammation. In other trials, the DNA encoding the normal *CFTR* gene has been linked to positively charged liposomes (page 131) that can fuse with the plasma membranes of the airway cells, delivering their DNA contents into the cytoplasm. Lipid-based delivery has an advantage over viruses in being less likely to stimulate a destructive immune response within the patient following repeated treatments, but has the disadvantage of being less effective in achieving genetic modification of target cells.

The results of early clinical trials are encouraging, especially those that rely on liposomes, which have not caused any significant side effects in patients after a single treatment. In most patients, the normal gene is able to enter a sufficient percentage of airway cells to at least partially correct the underlying Cl⁻ defect and to reduce the adherence of bacteria to lung tissue. The development of more effective DNA delivery systems, which are capable of genetically altering a greater percentage of airway cells, is probably required if a treatment for CF is to be achieved. It also remains to be shown whether repeated administration of the normal gene can control the disease over long periods of time.

Cotransport: Coupling Active Transport to Existing Ion Gradients The establishment of concentration gradients, such as those of Na^+, K^+, and H^+, provides a means by which free energy can be stored in the cell. The potential energy stored in ionic gradients is utilized by a cell in various ways to perform work, including the transport of other solutes. Consider the physiologic activity of the intestine. Within its lumen, enzymes hydrolyze high-molecular-weight polysaccharides into simple sugars, which are absorbed by the epithelial cells that line the intestine. The movement of glucose across the apical plasma membrane of the epithelial cells, against a concentration gradient, occurs by **cotransport** with sodium ions, as illustrated in Figure 4.47. The Na^+ concentration is kept very low within the cells by the action of a *primary* active transport system (the Na^+/K^+-ATPase), located in the basal and lateral plasma membrane, that pumps sodium ions out of the cell against a concentration gradient. The tendency for sodium ions to diffuse back across the apical plasma membrane down their concentration gradient is "tapped" by the epithelial cells to drive the cotransport of glucose molecules into the cell *against* a concentration gradient. The glucose molecules are said to be driven by **secondary active transport**.

In this case, the transport protein, called a *Na⁺/glucose cotransporter*, binds two sodium ions and one glucose molecule on the external surface of the apical plasma membrane. When the sodium ions are released on the inside of the cell into a solution of lesser sodium concentration, the conformation of the protein is changed so that it loses its affinity for the glucose molecule, which is then released into the cell. Once inside, the glucose molecules diffuse through the cell and are moved across the basal membrane by facilitated diffusion (page 158).

To appreciate the power of an ion gradient in accumulating other types of solutes in cells, we can briefly consider the energetics of the $Na^+/$glucose cotransporter. Recall from page 151 that the free-energy change for the movement of a mole of Na^+ ions into the cell is equal to -3.1 kcal/mol, and thus -6.2 kcal for two moles of Na^+ ions, which would be available to transport one mole of glucose uphill into the cell. Recall also from page 151 that the equation for the movement of a nonelectrolyte, such as glucose, across the membrane is

$$\Delta G = RT \ln \frac{[C_i]}{[C_o]}$$

$$\Delta G = 2.303\, RT \log_{10} \frac{[C_i]}{[C_o]}$$

Using this equation, we can calculate how steep a concentration gradient of glucose (X) that this cotransporter can generate. At 25°C,

$$-6.2 \text{ kcal/mol} = 1.4 \text{ kcal/mol} \cdot \log_{10} X$$

$$\log_{10} X = -4.43$$

$$X = \frac{1}{23,000}$$

Figure 4.47 Secondary transport: the use of energy stored in an ionic gradient. The Na^+/K^+-ATPase residing in the plasma membrane of the lateral surface maintains a very low cytosolic concentration of Na^+. The Na^+ gradient across the plasma membrane represents a storage of energy that can be tapped to accomplish work, such as the transport of glucose by a Na^+/glucose cotransporter (an SGLT) located in the apical plasma membrane. Once transported across the apical surface into the cell, the glucose molecules diffuse to the basal surface where they are carried by a glucose facilitative transporter (a GLUT) out of the cell and into the bloodstream. The relative size of the letters indicates the directions of the respective concentration gradients. Two Na^+ ions are transported for each glucose molecule; the $2:1$ Na^+/glucose provides a much greater driving force for moving glucose into the cell than a $1:1$ ratio.

This calculation indicates that the Na^+/glucose cotransporter is capable of transporting glucose into a cell against a concentration gradient greater than 20,000 fold.

Plant cells rely on secondary active transport systems to take up a variety of nutrients, including sucrose, amino acids, and nitrate. In plants, uptake of these compounds is coupled to the downhill, inward movement of H^+ ions rather than Na^+ ions. The secondary active transport of glucose into the epithelial cells of the intestine or of sucrose into a plant cell are examples of **symport**, in which the two transported species (Na^+ and glucose or H^+ and sucrose) move in the same direction. Numerous secondary transport proteins have been isolated that engage in **antiport**, in which the two transported species move in opposite directions. For example, numerous cells maintain a proper cytoplasmic pH by coupling the inward, downhill movement of Na^+ with the outward movement of H^+. Proteins that mediate antiport are usually called *exchangers*.

1. Compare and contrast the four basically different ways that a substance can move across the plasma membrane (as indicated in Figure 4.33).
2. Contrast the energetic difference between the diffusion of an electrolyte versus a nonelectrolyte across the membrane.
3. Describe the relationship between partition coefficient and particle size with regard to membrane permeability.
4. Explain the effects of putting a cell into a hypotonic, hypertonic, or isotonic medium.
5. Describe two ways in which energy is utilized to move ions and solutes against a concentration gradient.
6. How does the Na^+/K^+-ATPase illustrate the sidedness of the plasma membrane?
7. What is the role of phosphorylation in the mechanism of action of the Na^+/K^+-ATPase?
8. Because of its smaller size, one would expect Na^+ ions to be able to penetrate any pore large enough for a K^+ ion. How does the K^+ channel select for this specific ion.

4.8 MEMBRANE POTENTIALS AND NERVE IMPULSES

All organisms respond to external stimulation, a property referred to as *irritability*. Even a single-celled amoeba, if poked with a fine glass needle, responds by withdrawing its pseudopodia, rounding up, and moving off in another direction. Irritability in an amoeba depends on the same basic properties of membranes that lead to the formation and propagation of nerve impulses, which is the subject of the remainder of the chapter.

Nerve cells (or **neurons**) are specialized for the collection, conduction, and transmission of information, which is coded in the form of fast-moving electrical impulses. The basic parts of a neuron are illustrated in Figure 4.48). The nucleus of the neuron is located within an expanded region called the **cell body**, which is the metabolic center of the cell and the site where most of its material contents are manufactured. Extending from the cell bodies of most neurons are a number of minuscule extensions, called **dendrites**, which receive *incoming* information from external sources, typically other neurons. Also emerging from the cell body is a single, more prominent extension, the **axon**, which conducts *outgoing* impulses away from the cell body and toward the target cell(s). Impulses are generally initiated in the **axon hillock**, the region where the cell body merges into the axon. Although

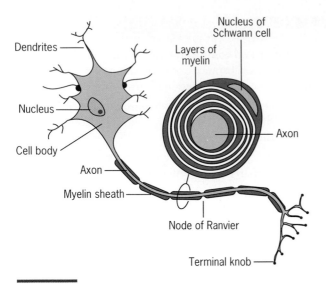

Figure 4.48 The structure of a nerve cell. Schematic drawing of a simple neuron with a myelinated axon. As the inset shows, the myelin sheath comprises individual Schwann cells that have wrapped themselves around the axon. The sites where the axon lacks myelin wrapping are called nodes of Ranvier. (Note: myelin-forming cells within the central nervous system are called oligodendrocytes rather than Schwann cells.)

(a) *(b)*

Figure 4.49 Measuring a membrane's resting potential. A potential is measured when a difference in charge is detected between the reference and recording electrodes. In (*a*), both electrodes are on the outside of the cell, and no potential difference (voltage) is measured. As one electrode penetrates the plasma membrane of the axon in (*b*), the potential immediately drops to −70 mV (inside negative), which approaches the equilibrium potential for potassium ions, that is, the potential that would result if the membrane were impermeable to all ions except potassium.

some axons may be only a few micrometers in length, others extend for many meters in the body of a large vertebrate, such as a giraffe or whale. Most axons split near their ends into smaller processes, each ending in a **terminal knob**—a specialized site where impulses are transmitted from neuron to target cell. Some neurons in the brain may end in thousands of terminal knobs, allowing these brain cells to communicate with thousands of potential targets. As discussed on page 127, most neurons in the vertebrate body are wrapped in a lipid-rich **myelin sheath**, whose function is described below.

The Resting Potential

A voltage or electric potential difference between two points, such as the inside and outside of the plasma membrane, results when there is an excess of positive ions at one point and an excess of negative ions at the other point. Voltages across plasma membranes can be measured by inserting one fine glass electrode into the cytoplasm of a cell, placing another electrode in the extracellular fluid outside the cell, and connecting the electrodes to a voltmeter, an instrument that measures a difference in charge between two points (Figure 4.49). When this experiment was first carried out on a giant axon of the squid, a potential difference of approximately 70 millivolts (mV) was recorded, the inside being negative with respect to the outside (indicated with a minus sign, −70 mV). The presence of a membrane potential is not unique to nerve cells; such potentials are present in all types of cells, the magnitude varying between about −15 and −100 mV. For

nonexcitable cells, that is, those other than neurons and muscle cells, this voltage is simply termed the **membrane potential**. In a nerve or muscle cell, this same potential is referred to as the **resting potential** because it is subject to dramatic change, as discussed in the following section.

The magnitude and direction of the voltage across the plasma membrane are determined by the differences in concentrations of ions on either side of the membrane and their relative permeabilities. As described earlier in the chapter, the Na^+/K^+-ATPase pumps Na^+ out of the cell and K^+ into the cell, thereby establishing steep gradients of these two ions across the plasma membrane. Because of these gradients, you might expect that potassium ions would leak out of the cell and sodium ions would leak inward through their respective ion channels (page 153). However, the vast majority of the ion channels that are open in the plasma membrane of a *resting* nerve cell are selective for K^+; they are often referred to as K^+ *leak channels*. It is interesting to note that, even after 50 years of studying these events, there is still controversy over the nature of the K^+ leak channels. In the past few years researchers have discovered a new family of K^+ channels that lack the S4 voltage sensor (page 156) and appear not to respond to changes in voltage. Studies on mammalian

brain tissue suggest that the K⁺ leak channels are members of this family. These proteins are called "twin-pore" channels because each subunit contains two ion pores rather than the single ion pore present in the subunits of other K⁺ channels (Figure 4.38).

Because K⁺ ions are the only charged species with significant permeability in a resting nerve cell, their outflow through the membrane leaves an excess of negative charges on the cytoplasmic side of the membrane. Although the concentration gradient across the membrane favors continued efflux of K⁺, the electrical gradient resulting from the excess negative charge on the inside of the membrane favors the retention of K⁺ ions inside the cell. When these two opposing forces are balanced, the system is at equilibrium, and there is no further *net* movement of K⁺ ions across the membrane. Using the following equation, which is called the Nernst equation, one can calculate the membrane potential (V_m) that would be measured at equilibrium if the plasma membrane of a nerve cell were permeable only to K⁺.[7] In this case, V_m would be equal to the potassium equilibrium potential (E_K):

$$E_K = 2.303 \frac{RT}{zF} \cdot \log_{10} \frac{[K^+_o]}{[K^+_i]}$$

For a squid giant axon, the internal $[K^+_i]$ is approximately 350 mM, while the external $[K^+_o]$ is approximately 10 mM; thus at 25°C (298°K) and $z = +1$ (for the univalent K⁺ ion),

$$E_K = 59 \log_{10} 0.028 = -91 \, \text{mV}$$

A similar calculation of the Na⁺ equilibrium potential (E_{Na}) would produce a value of approximately +55 mV. Because measurements of the voltage across the resting nerve membrane are similar in sign and magnitude (−70 mV) to the potassium equilibrium potential just calculated, the movement of potassium ions across the membrane is considered the most important factor in determining the resting potential. The difference between the calculated K⁺ equilibrium potential (−91 mV) and the measured resting potential (−70 mV, Figure 4.49) is due to a slight permeability of the membrane to Na⁺ and Cl⁻.

The Action Potential

Physiologists first learned about membrane potentials in nerve cells from studies on the giant axons of the squid. These axons, which are approximately 1 mm in diameter, carry impulses at high speeds, enabling the squid to escape rapidly from predators. If the membrane of a resting squid axon is stimulated by poking it with a fine needle or jolting it with a very small electric current, some of its sodium channels open, allowing a limited number of sodium ions to diffuse into the cell. This opportunity for positively

charged ions to move into the cell reduces the membrane potential, making it less negative. Because the reduction in membrane voltage causes a *decrease* in the polarity between the two sides of the membrane, it is called a **depolarization**.

If the stimulus causes the membrane to depolarize by only a few millivolts, say from −70 to −60 mV, the membrane rapidly returns to its resting potential as soon as the stimulus has ceased (Figure 4.50a, left box). If, however, the stimulus depolarizes the membrane beyond a certain point, called the **threshold**, which occurs at about −50 mV, then a new series of events is launched. The change in voltage causes the voltage-gated sodium channels to open. As a result, sodium ions diffuse freely into the cell (Figure 4.50a, middle box) down both their concentration and electric gradients. The increased permeability of the membrane to Na⁺ ions and the corresponding movement of positive charge into the cell causes the membrane to reverse potential briefly (Figure 4.50b), becoming positive at about +40 mV, which approaches the equilibrium potential for Na⁺ (Figure 4.49).

After approximately 1 msec, the sodium channels spontaneously inactivate, blocking further influx of Na⁺ ions. Inactivation results from the random diffusion of an attached ball into the opening of the channel pore in a manner similar to that described for K⁺ channels on page 156. Meanwhile, the change in membrane potential caused by Na⁺ influx triggers the opening of the voltage-gated potassium channels (Figure 4.50a, right box). As a result, potassium ions diffuse freely out of the cell down their steep concentration gradient. The decreased permeability of the membrane to Na⁺ and increased permeability to K⁺ causes the membrane potential to swing back to a negative value of about −80 mV, approaching that of the K⁺ equilibrium potential (Figure 4.49). The large negative membrane potential causes the voltage-gated potassium channels to close (see Figure 4.39a), which returns the membrane to its resting state. Collectively, these changes in membrane potential are called an **action potential** (Figure 4.50b). The entire series of changes during an action potential takes only about 5 msec in the squid axon and less than that in a myelinated mammalian nerve cell. Because the sodium channels cannot reopen for several milliseconds following their inactivation, the membrane enters a brief *refractory period* following an action potential in which it cannot be restimulated. Once the negative resting potential is reestablished, the inactivated Na⁺ ion channels close until they are once again stimulated to open by a threshold depolarization.

Although the action potential dramatically changes membrane voltage, only a minute percentage of the ions on the two sides of the membrane are involved in any given action potential. The dramatic changes in membrane potential seen in Figure 4.50b are not caused by changes in Na⁺ and K⁺ ion *concentrations* on the two sides of the membrane (which are insignificant). Rather, they are caused by the movements of charge in one direction or

[7]The Nernst equation is derived from the equation provided on page 151, paragraph 2, by setting ΔG at zero, which is the case when the movement of the ions is at equilibrium.

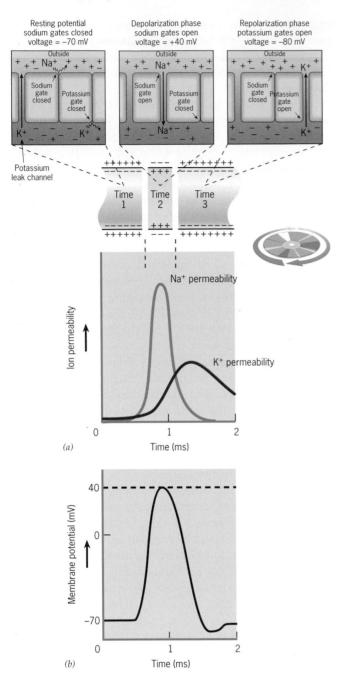

(a)

(b)

Figure 4.50 **Formation of an action potential.** (*a*) Time 1, upper left box: The membrane in this region of the nerve cell exhibits the resting potential, in which only the K⁺ leak channels are open and the membrane voltage is approximately −70 mV. Time 2, upper middle box, shows the depolarization phase: The membrane has depolarized beyond the threshold value, opening the voltage-regulated sodium gates, leading to an influx of Na⁺ (indicated in the permeability change in the lower graph). The increased Na⁺ permeability causes the membrane voltage to temporarily reverse itself, reaching a value of approximately +40 mV in the squid giant axon (time 2). It is this reversal of membrane potential that constitutes the action potential. Time 3, upper right box, shows the repolarization phase: Within a tiny fraction of a second, the sodium gates are inactivated and the potassium gates open, allowing potassium ions to diffuse across the membrane (lower part of the drawing) and establish an even more negative potential at that location (−80 mV) than that of the resting potential. Almost as soon as they open, the potassium gates close, leaving the potassium leak channels as the primary path of ion movement across the membrane and reestablishing the resting potential. (*b*) A summary of the voltage changes that occur during an action potential, as described in part *a*.

the other that result from the fleeting changes in permeability to these ions. Those Na⁺ and K⁺ ions that do change places across the membrane during an action potential are pumped back by the Na⁺/K⁺-ATPase. Even if the Na⁺/K⁺-ATPase is inactivated, a neuron can often continue to fire thousands of impulses before the ionic gradients established by the pump's activity are dissipated.

The movements of ions across the plasma membrane of nerve cells form the basis for neural communication. Certain *local* anesthetics, such as procaine and novocaine, act by closing ion channels in the membranes of sensory cells and neurons. As long as these ion channels remain closed, the affected cells are unable to generate action potentials and thus unable to inform the brain of events occurring at the skin or teeth.

Once the membrane of a neuron is depolarized to the threshold value, a full-blown action potential is triggered without further stimulation. This feature of nerve cell function is known as the **all-or-none law**. There is no in-between; subthreshold depolarization is incapable of triggering an action potential, whereas threshold depolarization automatically elicits a maximum response. It is also noteworthy that an action potential is not an energy-requiring process, but one that results from the flow of ions down their respective electrochemical gradients. Energy is required by the N⁺/K⁺-ATPase to generate the steep ionic gradients across the plasma membrane, but once that is accomplished, the various ions are poised to flow through the membrane as soon as their ion channel gates are opened.

Propagation of Action Potentials as an Impulse

Up to this point, we have restricted the discussion to events occurring at a particular site on the nerve cell membrane where experimental depolarization has triggered the action potential. Once an action potential has been initiated, it does not remain localized at a particular site but is *propagated* down the length of the cell as a **nerve impulse**. Action potentials are normally initiated at the axon hillock of the nerve cell and move from there to the opposite end of the cell. In the experimental case just described, the action potential has been triggered somewhere in the middle of the neuron, and it moves out from that site in both directions.

Figure 4.51 Propagation of an impulse results from the local flow of ions. An action potential at one site on the membrane depolarizes an adjacent region of the membrane, triggering an action potential at the second site. The action potential can only flow in the forward direction because the portion of the membrane that has just experienced an action potential remains in a refractory period.

Nerve impulses are propagated along a membrane because an action potential at one site has an effect on the adjacent site. The large depolarization that accompanies an action potential draws positive ions toward that site on the outer surface of the membrane and forces them away from that site on the inner surface (Figure 4.51). This flow of current causes the membrane in the region just ahead of the action potential to become depolarized itself. Because the depolarization accompanying the action potential is very large, the membrane in the adjacent region is readily depolarized to a level greater than the threshold value, and the sodium gates swing open in this adjacent region, generating another action potential. Thus, once triggered, a wave of action potential passes down the entire length of the neuron without any loss of intensity, arriving at its target cell with the same strength it had at its point of origin.

Because all impulses traveling along a neuron exhibit the same strength, stronger stimuli cannot produce "bigger" impulses than weaker stimuli. Yet, we are clearly capable of detecting differences in the strength of a stimulus. The ability to make sensory discriminations depends on several factors. For example, a stronger stimulus, such as scalding water, activates more nerve cells than does a weaker stimulus, such as warm water. It also activates "high-threshold" neurons that would remain at rest if the stimulus were weaker. Stimulus strength is also encoded in the pattern and frequency by which action potentials are launched down a particular neuron. In most cases, the stronger the stimulus, the greater the number of impulses generated.

Speed Is of the Essence The greater the diameter of an axon, the less the resistance to local current flow and the more rapidly an action potential at one site can activate adjacent regions of the membrane. Some invertebrates, such as squid and tube worms, have evolved giant axons that facilitate the animal's escape from danger. There is,

however, a limit to this evolutionary approach. Because the speed of conduction increases with the square root of the increase in diameter, an axon that is 480 μm in diameter can conduct an action potential only four times faster than one that is 30 μm in diameter.

During the evolution of vertebrates, an increase in conduction velocity was achieved when the axon became wrapped in a myelin sheath (see Figure 4.5). The myelin sheath is formed from accessory cells in the nervous system; these cells are known as oligodendrocytes in the central nervous system and Schwann cells in the peripheral nervous system (Figure 4.48). Composed of many layers of lipid-containing membranes, the myelin sheath is ideally suited to prevent the passage of ions across the plasma membrane. In addition, nearly all of the ion channels of a myelinated neuron reside in the unwrapped gaps, or *nodes of Ranvier*, between adjacent Schwann cells that make up the sheath (see Figure 4.48). Consequently, the nodes of Ranvier are the only sites where action potentials can be generated. An action potential at one node triggers an action potential at the next node (Figure 4.52), causing the impulse to jump from node to node without having to activate the intervening membrane. Propagation of an impulse by this mechanism is called **saltatory conduction**. Impulses are conducted along a myelinated axon at speeds up to 120 meters per second, which is nearly 20 times faster than the speed that impulses travel in an unmyelinated neuron of the same diameter.

The importance of myelination is dramatically illustrated by multiple sclerosis, a disease associated with deterioration of the myelin sheath that surrounds axons in various parts of the nervous system. Manifestations of the disease usually begin in young adulthood; victims experience weakness in their hands, difficulty in walking, and problems with their vision. The disease is characterized by progressive muscular dysfunction, often culminating in

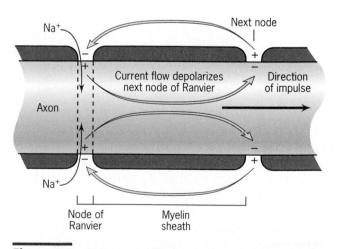

Figure 4.52 Saltatory conduction. During saltatory conduction, only the membrane in the nodal region of the axon becomes depolarized and capable of forming an action potential. This is accomplished as current flows directly from an activated node to the next resting node along the axon.

permanent paralysis (discussed in the Human Perspective of Chapter 17).

Neurotransmission: Jumping the Synaptic Cleft

Neurons are linked with their target cells at specialized junctions called **synapses**. Careful examination of a synapse reveals that the two cells do not make direct contact but are separated from each other by a narrow gap of about 20 to 50 nm. This gap is called the **synaptic cleft**. A **presynaptic cell** (a sensory cell or a neuron) conducts impulses toward a synapse, and a **postsynaptic cell** (a neuron, muscle, or gland cell) always lies on the receiving side

of a synapse. Figure 4.53 shows a number of synapses between the terminal branches of an axon and a skeletal muscle cell; synapses of this type are called **neuromuscular junctions** or *motor end plates*.

How does an impulse in a presynaptic neuron jump across the synaptic cleft and affect the postsynaptic cell? Studies carried out decades ago indicated that a chemical substance is involved in the transmission of an impulse from one cell to another. The very tips (terminal knobs) of the branches of an axon appear in the electron microscope to contain large numbers of **synaptic vesicles** (Figure 4.53, left insert) that serve as storage sites for the chemical transmitters that act on the postsynaptic cells. Two of the

Terminal knob of presynaptic neuron

Synaptic vesicles

Membrane of postsynaptic target cell

Synaptic cleft

Axon of nerve cell

Muscle fiber

Figure 4.53 The neuromuscular junction is where branches from a motor axon form synapses with the muscle fibers of a skeletal muscle. The left inset shows the synaptic vesicles residing within the terminal knob of the axon and the narrow synaptic cleft between the terminal knob and the postsynaptic target cell.

The right inset shows the terminal knob pressed closely against the muscle cell plasma membrane. Neurotransmitter molecules released from synaptic vesicles of the presynaptic neuron are binding to receptors on the surface of the muscle cell. (LEFT INSET FROM VU/T. REESE AND DON W. FAWCETT/VISUALS UNLIMITED.)

best-studied **neurotransmitters** are acetylcholine and norepinephrine,

Acetylcholine (ACh)

Norepinephrine

which transmit impulses to the body's skeletal and cardiac muscles.

The sequence of events during synaptic transmission can be summarized as follows (Figure 4.54). When an impulse reaches a terminal knob (step 1, Figure 4.54), the accompanying depolarization induces the opening of a number of voltage-gated Ca^{2+} channels in the plasma membrane of this part of the presynaptic nerve cell (step 2, Figure 4.54). Calcium ions are normally present at very low concentration within the neuron (about 100 nM), as in all cells. When the gates open, calcium ions diffuse from the extracellular fluid into the terminal knob of the neuron, causing the $[Ca^{2+}]$ to rise more than a thousand fold within localized microdomains near the channels. The elevated $[Ca^{2+}]$ triggers the rapid fusion of one or a few nearby synaptic vesicles with the plasma membrane, causing the release of neurotransmitter molecules into the synaptic cleft (step 3, Figure 4.54).

Once released from the synaptic vesicles, the neurotransmitter molecules diffuse across the narrow gap and bind selectively to receptor molecules that are concentrated directly across the cleft in the postsynaptic plasma membrane (step 4, Figure 4.54). A neurotransmitter molecule can have one of two opposite effects depending on the type of receptor on the target cell membrane to which it binds:

1. The bound transmitter can trigger the opening of cation-selective channels in the membrane, leading primarily to an influx of sodium ions and a less negative (more positive) membrane potential. This depolarization of the postsynaptic membrane *excites* the cell, making the cell more likely to respond to this or subsequent stimuli by generating an action potential of its own (Figure 4.54, steps 5a and 6).

2. The bound transmitter can trigger the opening of anion-selective channels, leading mainly to an influx of chloride ions, and a more negative (hyperpolarized) membrane potential. Hyperpolarization of the postsynaptic membrane makes it less likely the cell will generate an action potential because greater sodium influx is subsequently required to reach the membrane's threshold (Figure 4.54, step 5b).

Figure 4.54 The sequence of events during synaptic transmission with acetylcholine as the neurotransmitter. During steps 1–4, a nerve impulse reaches the terminal knob of the axon, calcium gates open leading to an influx of Ca^{2+}, and acetylcholine is released from synaptic vesicles and binds to receptors on the postsynaptic membrane. If the binding of the neurotransmitter molecules causes a depolarization of the post-synaptic membrane (as in 5a), a nerve impulse may be generated there (6). If, however, the binding of neurotransmitter causes a hyperpolarization of the postsynaptic membrane (5b), the target cell is inhibited, making it more difficult for an impulse to be generated in the target cell by other excitatory stimulation. The breakdown of the neurotransmitter by acetylcholinesterase is not shown.

Most nerve cells in the brain receive both excitatory and inhibitory signals from presynaptic neurons. It is the summation of these opposing influences that determine whether or not an impulse will be generated in the postsynaptic neuron.

All terminal knobs of a given neuron release the same neurotransmitter(s). However, a given neurotransmitter may have a stimulatory effect on one particular postsynaptic membrane and an inhibitory effect on another. Acetylcholine, for example, inhibits contractility of the heart but stimulates contractility of skeletal muscle. Within the brain, glutamate serves as the primary excitatory neurotransmitter and gamma-aminobutyric acid (GABA) as the primary inhibitory neurotransmitter. A number of general anesthetics, as well as valium and its derivatives, act by binding to the GABA receptor and enhancing the activity of the inhibitory neurotransmitter.

It is important that a neurotransmitter has only a short half-life following its release from a presynaptic neuron; otherwise the effect of the neurotransmitter would be extended and the postsynaptic neuron would not recover. A neurotransmitter is eliminated from the synapse in two ways: by enzymes that destroy neurotransmitter molecules in the synaptic cleft and by proteins that transport neurotransmitter molecules back to the presynaptic terminals—a process called *reuptake*. Because of the destruction or reuptake of neurotransmitter molecules, the effect of each impulse lasts no more than a few milliseconds.

Interfering with the destruction or reuptake of neurotransmitters can have dramatic physiological and behavioral effects. Acetylcholinesterase is an enzyme localized within the synaptic cleft that hydrolyzes acetylcholine. If this enzyme is inhibited by exposure to the nerve gas DFP, for example, the muscles of the body undergo violent contraction due to the continued presence of high concentrations of acetylcholine. Cocaine, on the other hand, interferes with the reuptake of the neurotransmitter dopamine that is released by certain nerve cells in a portion of the brain known as the limbic system. The sustained presence of dopamine in the synaptic clefts of the limbic system produces a short-lived feeling of euphoria. Amphetamines also act on dopamine-releasing neurons; they are thought to stimulate the excessive release of dopamine from presynaptic terminals and interfere with the reuptake of the neurotransmitter molecules from the synaptic cleft. Mice that have been genetically engineered to lack the dopamine transporter (DAT)—the protein responsible for dopamine reuptake—show behavior similar to that of normal mice that have been given cocaine or amphetamines. Administration of cocaine or amphetamines has no additional behavioral effects on animals lacking the

DAT gene. These studies have important implications in the treatment of Parkinson's disease, whose symptoms of motor dysfunction are caused by a marked drop in the levels of dopamine in certain parts of the brain. If drugs can be developed to block the dopamine transporter, the limited amount of dopamine released by the neurons of these patients should be more effective in producing a response in postsynaptic cells.

Synapses are more than simply connecting sites between adjacent neurons; they are key determinants in the routing of impulses through the nervous system. The billions of synapses in a complex mammalian nervous system act like gates stationed along the various pathways, allowing some pieces of information to pass from one neuron to another, while holding back other pieces or rerouting them in another direction. While synapses are often perceived as fixed, unchanging structures, they can display a remarkable dynamic quality that is often described as "synaptic plasticity." Synaptic plasticity is particularly important during infancy and childhood, when the neuronal circuitry of the brain achieves it mature configuration.

Synaptic plasticity is most readily observed in studies on neurons from the hippocampus, a portion of the brain that is vitally important in learning and short-term memory. The hippocampus is one of the major areas of the brain that is destroyed by Alzheimer's disease (page 68). When hippocampal neurons are repeatedly stimulated over a short period of time, the synapses that connect these neurons to their neighbors become "strengthened" by a process known as long-term potentiation (LTP), which may last for days or even weeks. Synapses that have undergone LTP are able to transmit weaker stimuli and evoke stronger responses in postsynaptic cells. These changes are thought to play a major role as memories are encoded in the neural circuits of the brain. There are numerous other reasons why the study of synapses is so important. For example, a number of diseases of the nervous system, including myasthenia gravis, Parkinson's disease, schizophrenia, and even depression, are thought to have their roots in synaptic dysfunction.

REVIEW ?

1. What is the role of the myelin sheath in conduction of an impulse?

2. Describe the steps between the time an impulse reaches the terminal knob of a presynaptic neuron and an action potential is initiated in a postsynaptic cell.

EXPERIMENTAL PATHWAYS

The Acetylcholine Receptor

In 1843, at the age of 30, Claude Bernard moved from a small French town, where he had been a pharmacist and an aspiring playwright, to Paris, where he planned to pursue his literary career. Instead, Bernard enrolled in medical school and went on to become the foremost physiologist of the nineteenth century. Among his many interests was elucidating the mechanism by which nerves stimulate the contraction of skeletal muscles. His studies included the use of curare, a highly toxic drug isolated from tropical plants and utilized for centuries by native South American hunters to make poisonous darts. Bernard found that curare paralyzed a skeletal muscle without interfering with either the ability of nerves to carry impulses to that muscle or the ability of the muscle to contract on direct stimulation. Bernard concluded that curare somehow acted on the region of contact between the nerve and muscle.

This conclusion was confirmed and extended by John Langley, a physiologist at Cambridge University. Langley was studying the ability of nicotine, another substance derived from plants, to stimulate the contraction of isolated frog skeletal muscles and the effect of curare in inhibiting nicotine action. In 1906, Langley concluded that "the nervous impulse should not pass from nerve to muscle by an electric discharge, but by the secretion of a special substance on the end of the nerve."[1] Langley proposed that this "chemical transmitter" was binding to a "receptive substance" on the surface of the muscle cells, the same site that bound nicotine and curare. These proved to be farsighted proposals.

Langley's suggestion that the stimulus from nerve to muscle was transmitted by a chemical substance was confirmed in 1921 in an ingenious experiment by the Austrian-born physiologist, Otto Loewi, the design of which came to Loewi during a dream. The heart rate of a vertebrate is regulated by input from two opposing (antagonistic) nerves. Loewi isolated a frog's heart with both nerves intact. When he stimulated the inhibitory (*vagus*) nerve, a chemical was released from the heart preparation into a salt solution, which was allowed to drain into the medium bathing a second isolated heart. The rate of the second heart slowed dramatically, as though its own inhibitory nerve had been activated.[2] Loewi called the substance responsible for inhibiting the frog's heart "Vagusstoff." Within a few years, Loewi had shown that the chemical and physiological properties of Vagusstoff were identical to acetylcholine, and he concluded that acetylcholine (ACh) was the substance released by the tips of the nerve cells that made up the vagus nerve.

In 1937, David Nachmansohn, a neurophysiologist at the Sorbonne, was visiting the World's Fair in Paris where he observed several living electric fish of the species *Torpedo marmorata* that were on display. These rays have electric organs that deliver strong shocks (40–60 volts) capable of killing potential prey. At the time, Nachmansohn was studying the enzyme acetylcholinesterase, which acts to destroy ACh after its release from the tips of motor nerves. Nachmanson was

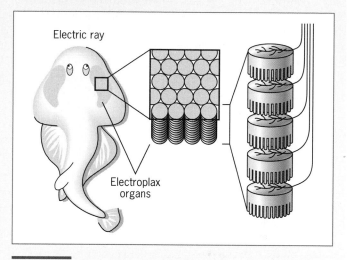

Figure 1 The electric organs of *Torpedo* consist of stacks of modified neuromuscular junctions located on each side of the body. (FROM Z. W. HALL, AN INTRODUCTION TO NEUROBIOLOGY, SINAUER ASSOCIATES, INC., SUNDERLAND, MA ©1992.)

aware that the electric organs of these fish were derived from modified skeletal muscle tissue (Figure 1), and he asked if he could have a couple of the fish for study once the fair had ended. The results of the first test showed the electric organ was an extraordinarily rich source of acetylcholinesterase.[3] It was also a very rich source of the nicotinic acetylcholine receptor (nAChR),* the receptor present on the postsynaptic membranes of skeletal muscle cells that binds ACh molecules released from the tips of a motor nerve. Finding an ideal system can prove invaluable in the study of a particular aspect of cell structure or function. As will be evident from the following discussion, the electric organs of fish have been virtually the sole source of material for the study of the nAChR.

The nAChR is an integral membrane protein, and it wasn't until the 1970s that techniques were developed for the isolation of such proteins. As discussed in Chapter 18, the purification of a particular protein requires a suitable assay to determine the amount of that protein present in any particular fraction. The ideal assay for nAChR was a compound that bound selectively and tightly to this particular protein. Such a compound was discovered in 1963 by Chen-Yuan Lee and

*The receptor is described as nicotinic because it can be activated by nicotine as well as by acetylcholine. This contrasts with muscarinic acetylcholine receptors of the parasympathetic nerve synapses, which can be activated by muscarine, but not nicotine, and are inhibited by atropine, but not curare. Smokers' bodies become accustomed to high levels of nicotine, and they experience symptoms of withdrawal when they stop smoking because the postsynaptic neurons that possess nAChRs are no longer stimulated at their usual level.

his colleagues of the National Taiwan University. The compound was α-bungarotoxin, a substance present in the venom of a Taiwanese snake. The α-bungarotoxin causes paralysis by binding tightly to the nAChRs on the postsynaptic membrane of skeletal muscle cells, blocking the response of the muscle to ACh.[4]

Equipped with labeled α-bungarotoxin to use in an assay, electric organs as a source, and a detergent capable of solubilizing membrane proteins, a number of investigators were able to isolate the acetylcholine receptors in the early 1970s. In one of these studies,[5] the membranes containing the nAChR were isolated by homogenizing the electric organs in a blender and centrifuging the suspension to pellet the membrane fragments. After attempting several procedures for solubilizing membrane proteins, it was found that Triton X-100 gave the best results. Triton X-100 is a nonionic detergent. Like other detergents, Triton X-100 has a long hydrophobic portion capable of replacing the phospholipids that normally surround the hydrophobic portions of a membrane protein, and a hydrophilic end that is soluble in water. Unlike most detergents, which bear a negatively charged carboxyl group, Triton X-100 lacks a charge (hence, is nonionic) and is thus less disruptive of protein structure. Membrane proteins were extracted from the membrane fragments using Triton X-100, and the mixture was passed through a column containing

tiny beads coated with a synthetic compound whose end bears a structural resemblance to ACh (Figure 2a). As the mixture of dissolved proteins passed through the column, two proteins that have binding sites for acetylcholine, nAChR and acetylcholinesterase (AChE), stuck to the beads. The remaining 90 percent of the protein in the extract failed to bind to the beads and simply passed through the column and was collected (Figure 2b). Once this group of proteins had passed through, a solution of 10^{-3} M flaxedil was passed through the column, which selectively removed the nAChR from the beads, leaving the AChE behind. Using this procedure, the acetylcholine receptor as measured by bungarotoxin binding was purified by more than 150 fold in a single step. This type of procedure is known as *affinity chromatography*, and its general use is discussed in Section 18.7.

The next step was to determine something about the structure of the acetylcholine receptor. Studies in the laboratory of Arthur Karlin at Columbia University determined that the nAChR was a pentamer, a protein consisting of five subunits. Each receptor contained two copies of a subunit called α, and one copy each of β, γ, and δ subunit. The subunits could be distinguished by extracting membrane proteins in Triton, purifying the nAChR by affinity chromatography, and then subjecting the purified protein to electrophoresis through a polyacrylamide gel (SDS-PAGE, as dis-

Figure 2 Steps used in the isolation of the nAChR. (*a*) Structure of a synthetic compound, CT5263, that was attached to sepharose beads to form an affinity column. The ends of the compound projecting from the beads resemble acetylcholine, causing both acetylcholinesterase (AChE) and the nicotinic acetylcholine receptor (nAChR) to bind to the beads. (*b*) When the Triton X-100 extract was passed through the column, both of the acetylcholine-binding proteins stuck to the beads, while the remaining dissolved protein (about 90 percent of the total protein in the extract) passed directly through the column. Subsequent passage of a solution of 10^{-3} M flaxedil through the column released the bound nAChR, without disturbing the bound AChE (which was subsequently eluted with 1 M NaCl). (FROM R. W. OLSEN, J.-C. MEUNIER, AND J.-P. CHANGEUX, FEBS LETT. 28:99, 1972.)

Figure 3 The top portion of the figure shows an SDS-polyacrylamide gel following electrophoresis of a preparation of the purified nAChR. The receptor consists of four different subunits whose molecular weights are indicated. Prior to electrophoresis, the purified receptor preparation was incubated with a radioactive compound (^3H-MBTA) that resembles acetylcholine and binds to the acetylcholine-binding site of the nAChR. Following electrophoresis, the gel was sliced into 1-mm sections and the radioactivity of each slice determined. All of the radioactivity was bound to the 39,000 dalton subunit, indicating this subunit contains the ACh binding site. The dotted line indicates the light absorbance of each fraction, which provides a measure of the total amount of protein present in that fraction. The heights of the peaks provide a measure of the relative amounts of each of the subunits in the protein. All of the subunits are present in equal numbers except the smallest subunit (the α subunit, which contains the ACh binding site), which is present in twice the number of copies. (FROM C. L. WEILL, M. G. MCNAMEE, AND A. KARLIN, BIOCHEM. BIOPHYS. RES. COMMUN. 61:1002, 1974.)

cussed in Section 18.7), which separates the individual polypeptides according to size (Figure 3).[6]

Another important milestone in the study of the nAChR was the demonstration that the purified receptor acted as both a site for binding of ACh and a channel for the passage of cations. It had been postulated years earlier by Jean-Pierre Changeux of the Pasteur Institute in Paris that the binding of ACh to the receptor caused a conformational change that opened an ion channel within the protein. The inward flux of Na^+ ions through the channel could then lead to a depolarization of the membrane and the activation of the muscle cell. During the last half of the 1970s, Changeux and his colleagues succeeded in incorporating purified nAChR molecules into artificial lipid vesicles.[7] Using vesicles containing various concentrations of labeled sodium and potassium ions, they demonstrated that binding of ACh to the receptors in the lipid bilayer initiated a flux of cations through the "membrane." It was evident that "the pure protein does indeed contain all the structural elements needed for the chemical transmission of an electrical signal—namely, an acetylcholine binding site, an ion channel and a mechanism for coupling their activity."

During the past decade, researchers have focused on determining the structure of the nAChR and the mechanism by which it becomes concentrated within one small part of the muscle membrane—that part adjacent to the tip of the motor neuron. Analysis of the structure has taken several different paths. In one approach, scientists used purified genes, determination of amino acid sequences, and site-directed mutagenesis to determine the specific parts of the polypeptides that span the membrane, or bind the neurotransmitter, or form the ion channel. These studies on the molecular anatomy of a protein are similar in principle to those described for the K^+ ion channel on page 154.

Another approach required the electron microscope. The first glimpses of the nAChR were seen in electron micrographs of the membranes of the electric organs (Figure 4).[8] The receptors appeared ring shaped, with a diameter of 8 nm and a central channel of 2-nm diameter, and protruded out from the lipid bilayer into the external space. An increasingly detailed model of the nAChR has been developed over the past decade as a result of the work of Nigel Unwin and his colleagues of the Medical Research Council of England.[9–11] Using a mathematical analysis of electron micrographs of frozen membranes from the electric organs, Unwin described the arrangement of the five subunits around a central channel (Figure 5). The ion channel consists of a narrow pore (7–8 Å diameter) lined by a wall composed of five α-helical segments, one from each of the surrounding subunits. The gate to the pore is postulated to be near the middle of its passage through the membrane, where each of the α helices bends inward (at the apex of the V-shaped bars in Figure 5a) to form a kink. The side chain of a leucine residue projects inward from each kink. In this model, the leucine residues from the five helices would form a tight ring that prevents the ions from crossing the membrane. The gate opens following the binding of an ACh molecule to a site located within a pocket on the external domain of each α subunit (Figure 5b).

To study the changes in the nAChR during channel opening, Unwin carried out the following experiment.[11] Preparations of nAChR-rich membranes were applied to a support grid that was allowed to fall into a bath of liquid nitrogen-cooled ethane, which freezes the membranes. Approximately 5 msec before they reached the surface of the

Figure 4 Electron micrograph of negatively stained, receptor-rich membranes from the electric organ of an electric fish showing the dense array of nAChR molecules. Each receptor molecule is seen as a small whitish circle with a tiny black dot in its center; the dot corresponds to the central channel, which has collected electron-dense stain. (FROM WERNER SCHIEBLER & FERDINAND HUCHO, EUR. J. BIOCHEM. 85:58, 1978.)

0.1 μm

freezing bath, the grids were sprayed with a solution of ACh, which bound to the receptors and triggered the conformational change required to open the channel. By comparing electron micrographs of nAChRs trapped in the open versus closed state, Unwin found that ACh binding triggers a conformational change in the extracellular domains of the receptor subunits, which is propagated down the protein, causing a change in orientation of the α helices that line the pore (Figure 6). Closure of the pore is attributed to formation of a kink in the helices, which is seen by comparison of the white and blue bars superimposed over the helix on the far right in Figure 6b.

(a)

Figure 5 (a) An electron density map of a slice through the nAChR obtained by analyzing electron micrographs of tubular crystals of *Torpedo* membranes embedded in ice. These analyses have allowed researchers to reconstruct the three-dimensional appearance of a single nAChR protein as it resides within the membrane. The continuous contours indicate lines of similar density greater than that of water. The two dark, bar-shaped lines represent α helices that line the channel at its narrowest point. (b) Schematic diagram of the nAChR showing the arrangement of the subunits and a cross-sectional representation of the protein. (A: FROM N. UNWIN, J. MOL. BIOL. 229:1118, 1993, COPYRIGHT 1993 BY PERMISSION OF THE PUBLISHER ACADEMIC PRESS.)

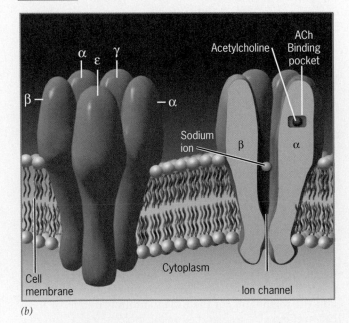

(b)

We saw on page 144 how the movement of integral membrane proteins can be restricted by peripheral components of the membrane. This finding is particularly well illustrated by the nAChR. During the early development of a skeletal muscle cell, acetylcholine receptors are distributed rather uniformly throughout the cell's plasma membrane. However, with the maturation of the muscle cell and the close approach of the tip of a motor neuron, these same receptor molecules become highly concentrated in the regions of the synapses, which account for only a very small fraction of the muscle cell surface. It is estimated that over 90 percent of the nAChR molecules are concentrated into less than 0.1 percent of a muscle cell's plasma membrane. How does the machinery required for neurotransmission become concentrated in the postsynaptic membrane of the muscle cell just across the cleft from the terminal knobs of the motor nerve?

During the mid-1980s, U. J. McMahan and his colleagues at Stanford University discovered a protein present in the fish electric organ that, when added to immature muscle cells in culture, would induce the clustering of nAChRs in the muscle cell membrane.[12] The protein was found concentrated outside the cells of the electric organ in the extracellular space of the synaptic clefts. In 1987, they purified this protein and named it agrin.[13] McMahan proposed that agrin was synthesized in the cell bodies of motor neurons and transported down their axons to the terminal knobs, where it was released into the extracellular space of the developing neuromuscular synapse. The agrin molecules would then bind to a specific agrin receptor on the outer surface of the plasma membrane of the adjacent muscle cell, triggering a cascade of events that would culminate in the clustering of nAChR molecules (and other components of the postsynaptic membrane) in the vicinity of the activated agrin receptor. Agrin would remain as part of the extracellular material of the synaptic cleft, where it would help maintain the organization of the postsynaptic membrane and ensure that newly synthesized AChRs became concentrated at the synaptic site.[14] This hypothesis was supported by various studies, including the use of antibodies that bind specifically to agrin. It was shown, for example, that when sections of a chick embryonic spinal cord were incubated with anti-agrin antibodies, the antibodies specifically labeled the cytoplasm of the motor neurons, indicating that these cells were the source of the agrin molecules. Anti-agrin antibodies were also localized in the synaptic cleft of neuromuscular junctions, as predicted by the hypothesis. When anti-agrin antibodies were added to cultures containing a mix of chick motor neurons and developing muscle cells, the antibodies blocked the normal aggregation of nAChR at sites of contact between the two types of cells. Additional insights into the formation of neuromuscular junctions have been gained from studies on agrin "knockout" mice, that is, mice that have been genetically engineered so that they are unable to synthesize agrin.[15] Agrin-deficient mice develop normally until the end of gestation, when they either die in the uterus or at birth. These mice exhibit no sign of muscular activity prior to their death, and examination of sections of muscle tissue reveals the absence of normal neuromuscular synapses.

The mechanism by which agrin induces nAChR aggregation remains poorly understood, but a number of steps in the process have been defined. One of the first events follow-

(a)

(b)

Figure 6 (*a*) The orientation of the five membrane-spanning helices in the open pore conformation. The helices, which are depicted as an electron-density map, are shown in white, with the lipid-facing rim of the protein in brown. The helices form a wall around the central pore of the channel. (*b*) The orientation of the helices in both the open- and closed-pore conformations. The position of the helices in the open pore is shown in white, as in *a*, with the white bar on the right indicating the path of one of the helices. The position of the helices in the closed pore is superimposed in blue. The difference in position between the white and blue images indicates the conformational change during pore closure. The blue line on the right shows the change in conformation of that helix, which develops a distinct bend or kink in the middle of the channel, closing the gate. (The front face of the rim has been removed to reveal the helices more clearly.)
(REPRINTED WITH PERMISSION FROM N. UNWIN, NATURE 373:41, 1995; COPYRIGHT 1995, MACMILLAN MAGAZINES LIMITED.)

ing addition of agrin to cultures of immature muscle cells is the phosphorylation of one or more tyrosine residues in the β subunit of the nAChR. This phosphorylation begins within 30 minutes of addition of agrin to the culture, long before any hint of nAChR aggregation is detected.[16] As will be discussed at length in Chapter 15, tyrosine phosphorylation is often a key step in the transmittal of signals across the plasma membrane. If tyrosine phosphorylation of the nAChR is blocked by addition of tyrosine kinase inhibitors, the receptors fail to cluster, suggesting that agrin-induced tyrosine phosphorylation of nAChR is a required step in its aggregation.[17,18]

Signaling pathways that are activated by tyrosine phosphorylation often lead to reorganization of elements of the cytoskeleton, particularly its actin filaments. We have seen in this chapter how the membrane skeleton, which is often linked to the actin cytoskeleton, plays a key role in the distribution of integral membrane proteins (Figure 4.32). Recent studies indicate that the signals generated at the cell surface following interaction betweeen agrin and its receptor lead to the polymerization of actin filaments that are directly responsible for the clustering of the nACh receptors. If actin filament formation is blocked by a drug that binds to actin monomers and prevents their incorporation into actin polymers, then agrin-induced clustering of nAChRs is prevented.[19]

References

1. LANGLEY, J. N. 1906. On nerve endings and on special excitable substances in cells. *Proc. R. Soc. London B Biol. Sci.* 78:170–194.

2. LOEWI, O. 1921. Uber humorale ubertragbarkeit der herznervenwirkung. *Pfluger's Arch.* 214:239–242. (A review of Loewi's work written by him in English can be found in *Harvey Lect.* 28:218–233, 1933.)

3. MARNAY, A. 1937. Cholinesterase dans l'organe electrique de la torpille. *Compte Rend.* 126:573–574. (A review of Nachmanson's work written in English can be found in his book, *Chemical and Molecular Basis of Nerve Action*, 2nd ed., Academic Press, 1975.)

4. CHANG, C. C. & LEE, C.-Y. 1963. Isolation of the neurotoxins from the venom of *Bungarus multicinctus* and their modes of neuromuscular blocking action. *Arch. Int. Pharmacodyn. Ther.* 144:241–257.

5. OLSEN, R. W., MEUNIER, J. C., & CHANGEUX, J. P. 1972. Progress in the purification of the cholinergic receptor protein from *Electrophorus electricus* by affinity chromatography. *FEBS Lett.* 28:96–100.

6. WEILL, C. L., McNAMEE, M. G., & KARLIN, A. 1974. Affinity-labeling of purified acetylcholine receptor from *Torpedo californica. Biochem. Biophys. Res. Commun.* 61:997–1003.

7. POPOT, J. L., CARTAUD, J., & CHANGEUX, J. P. 1981. Reconstitution of a functional acetylcholine receptor. *Eur. J. Biochem.* 118:203–214.

8. SCHIEBLER, W. & HUCHO, F. 1978. Membranes rich in acetylcholine receptor: Characterization and reconstitution to excitable membranes from exogenous lipids. *Eur. J. Biochem.* 85:55–63.

9. BRISSON, A. & UNWIN, N. 1984. Tubular crystals of acetylcholine receptor. *J. Cell Biol.* 99:1202–1211.

10. UNWIN, N. 1993. Acetylcholine receptor at 9 Å resolution. *J. Mol. Biol.* 229:1101–1124.

11. UNWIN, N. 1995. Acetylcholine receptor channel imaged in the open state. *Nature* 373:37–43.

12. GODFREY, E. W., ET AL. 1984. Components of *Torpedo* electric organ and muscle that cause aggregation of acetylcholine receptors on cultured muscle cells. *J. Cell Biol.* 99:615–627.

13. NITKIN, R. M., ET AL. 1987. Identification of agrin, a synaptic organizing protein from *Torpedo* electric organ. *J. Cell Biol.* 105:2471–2478.

14. McMAHAN, U. J. 1990. The agrin hypothesis. *Cold Spring Harbor Symp. Quant. Biol.* 55:407–418.

15. GAUTAM, M., ET AL. 1996. Defective neuromuscular synaptogenesis in agrin-deficient mutant mice. *Cell* 85:525–535.

16. WALLACE, B.G., ET AL. 1991. Agrin induces phosphorylation of the nicotinic acetylcholine receptor. *Neuron* 6:869–878.

17. WALLACE, B. G. 1994. Staurosporine inhibits agrin-induced acetylcholine receptor phosphorylation and aggregation. *J. Cell Biol.* 125:661–668.

18. HOCH, W. 1999. Formation of the neuromuscular junction: agrin and its unusual receptors. *Eur. J. Biochem.* 265:1–10.

19. DAI, Z., ET AL. 2000. The actin-driven movement and formation of acetylcholine receptor clusters. *J.Cell Biol.* 150:1321–1334.

SYNOPSIS

Plasma membranes are remarkably thin, delicate structures, yet they play a key role in many of a cell's most important functions. The plasma membrane separates the living cell from its environment; it provides a selectively permeable barrier that allows the exchange of certain substances, while preventing the passage of others; it contains the machinery that physically transports substances from one side of the membrane to another; it contains receptors that bind to specific ligands in the external space and relay the information to the cell's internal compartments; it mediates interactions with other cells; it provides a framework in which components can be organized; it is a site where energy is transduced from one type to another. (p. 122)

Membranes are lipid-protein assemblies in which the components are held together in a thin sheet by noncovalent bonds. The membrane is held together as a cohesive sheet by a lipid bilayer consisting of a bimolecular layer of amphipathic lipids, whose polar head groups face outward and hydrophobic fatty acyl tails face inward. Included among the lipids are phosphoglycerides, such as phosphatidylcholine; sphingosine-based lipids, such as the phospholipid sphingomyelin and the carbohydrate-containing cerebrosides and gangliosides (glycolipids); and cholesterol. The proteins of the membrane can be divided into three groups: integral proteins that penetrate into and through the lipid bilayer, with portions exposed on both the cytoplasmic and extracellular membrane surfaces; peripheral proteins that are present wholly outside the lipid bilayer, but are noncovalently associated with either the polar head groups of the lipid bilayer or with the surface of an integral protein; and lipid-anchored proteins that are outside the lipid bilayer but covalently linked to a lipid that is part of the bilayer. The transmembrane segments of integral proteins occur typically as an α helix, which may be predominantly hydrophobic if they simply span the bilayer or amphipathic if they line an internal aqueous channel. (p. 127)

Membranes are highly asymmetric structures whose two leaflets have very different properties. As examples, all of the membrane's carbohydrate chains face away from the cytosol; many of the integral proteins bear sites on their external surface that interact with extracellular ligands and sites on their internal surface that interact with peripheral proteins that form part of an inner membrane skeleton; and the phospholipid content of the two halves of the bilayer is highly asymmetric. The organization of the proteins within the membrane is best revealed in freeze-fracture replicas in which the cells are frozen, their membranes are split through the center of the bilayer by a fracture plane, and the exposed internal faces are visualized by formation of a metal replica. (p. 132)

The physical state of the lipid bilayer has important consequences for the lateral mobility of both phospholipids and integral proteins. The viscosity of the bilayer and the temperature at which it undergoes phase transition depend on the degree of unsaturation and the length of the fatty acyl chains of the phospholipids. Maintaining a fluid membrane is important for many cellular activities, including signal transduction, cell division, and formation of specialized membrane regions. The lateral diffusion of proteins within the membrane was originally demonstrated by cell fusion, and it can be quantitated by techniques that follow the movements of proteins tagged with fluorescent compounds or electron-dense markers. Measurement of the diffusion coefficients of integral proteins suggests that most are subject to restraining influences that inhibit their mobility. Proteins may be restrained by association with other integral proteins or with peripheral proteins located on the membrane surface. Because of these various types of restraint, membranes achieve a considerable measure of organizational stability in which particular membrane regions are differentiated from one another. (p. 140)

The erythrocyte plasma membrane contains two major integral proteins, band 3 and glycophorin A, and a well-defined inner skeleton composed of peripheral proteins. Each band 3 subunit spans the membrane at least a dozen times and contains an internal channel through which bicarbonate and chloride anions are exchanged. Glycophorin A is a heavily glycosylated protein of unknown function containing a single transmembrane domain consisting of a hydrophobic α helix. The major component of the membrane skeleton is the fibrous protein spectrin, which interacts with other peripheral proteins such as ankyrin to provide support for the membrane and restrain the diffusion of its integral proteins. (p. 147)

The plasma membrane is a selectively permeable barrier that allows solute passage by several mechanisms, including simple diffusion through either the lipid bilayer or membrane channels, facilitated diffusion, and active transport. Diffusion is an energy-independent process in which a solute moves down an electrochemical gradient, dissipating the free energy stored in the gradient. Small inorganic solutes, such as O_2, CO_2, and H_2O, penetrate the lipid bilayer readily, as do solutes with large partition coefficients (high lipid solubility). Ions and polar organic solutes, such as sugars and amino acids, require special transporters to enter or leave the cell. (p. 150)

Water moves by osmosis directly through the lipid bilayer of a semipermeable membrane from a region of lower solute concentration (the hypotonic compartment) to a region of higher solute concentration (the hypertonic compartment). Osmosis plays a key role in a multitude of physiological activities. In plants, for example, the influx of water generates turgor pressure against the cell wall that helps support nonwoody tissues. Ions diffuse through a plasma membrane by means of special protein-lined channels that are often specific for particular ions. Ion channels are usually gated and are controlled by either voltage or chemical ligands, such as neurotransmitters. (p. 152)

Facilitated diffusion and active transport involve integral membrane proteins that combine specifically with the solute to be transported. Facilitative transporters act without the input of energy and move solutes down a concentration gradient in either direction across the membrane. They are thought to act by changing conformation, which alternately exposes the solute binding site to the two sides of the membrane. The glucose transporter is a facilitative transporter whose presence in the plasma membrane is stimulated by increasing levels of insulin. Active transporters require the

input of energy and move ions and solutes against a concentration gradient. P-type active transporters, such as the Na^+/K^+-ATPase, are driven by the transfer of a phosphate group from ATP to the transporter, changing its affinity toward the transported ion. Secondary active-transport systems tap the energy stored in an ionic gradient to transport a second solute against a gradient. For example, the active transport of glucose across the apical surface of an intestinal epithelial cell is driven by the cotransport of Na^+ down its electrochemical gradient. (p. 157)

The resting potential across the plasma membrane is due largely to the limited permeability of the membrane to K^+ and is subject to dramatic change. The resting potential of a typical nerve or muscle cell is about −70 mV (inside negative). When the membrane of an excitable cell is depolarized past a threshold value, events are initiated that lead to the opening of the gated Na^+ channels and the influx of Na^+, which is measured as a reversal in the voltage across the membrane. Within milliseconds after they have opened, the Na^+ gates close and gated potassium channels open, which

leads to an efflux of K^+ and a restoration of the resting potential. The series of dramatic changes in membrane potential following depolarization constitutes an action potential. (p. 165)

Once an action potential has been initiated, it becomes self-propagating. Action potentials propagate because the depolarization that accompanies an action potential at one site on the membrane is sufficient to depolarize the adjacent membrane, which initiates an action potential at that site. In a myelinated axon, an action potential at one node in the sheath is able to depolarize the membrane at the next node, allowing the action potential to hop rapidly from node to node. When the action potential reaches the terminal knobs of an axon, the calcium gates in the plasma membrane open, allowing an influx of Ca^{2+}, which triggers the fusion of the membranes of neurotransmitter-containing secretory vesicles with the overlying plasma membrane. The neurotransmitter diffuses across the synaptic cleft and binds to receptors on the postsynaptic membrane, inducing either the depolarization or hyperpolarization of the target cell. (p. 168)

ANALYTIC QUESTIONS

1. What types of integral proteins would you expect to reside in the plasma membrane of an epithelial cell that might be absent from that of an erythrocyte? How do such differences relate to the activities of these cells?

2. Many different types of cells possess receptors that bind steroid hormones. Where in the cell do you think such receptors might reside? Where in the cell would you expect the insulin receptor to reside? Why?

3. When the trilaminar appearance of the plasma membrane was first reported, the images were taken as evidence to support the Davson-Danielli model of plasma membrane structure. Why do you think these micrographs might have been interpreted in this way?

4. Suppose you were planning to use liposomes in an attempt to deliver drugs to a particular type of cell in the body, for example, a fat or muscle cell. Is there any way you might be able to construct the liposome to increase its target specificity?

5. How is it that, unlike polysaccharides such as starch and glycogen, the oligosaccharides on the surface of the plasma membrane can be involved in specific interactions? How is this feature illustrated by determining a person's blood type prior to receiving a transfusion?

6. Trypsin is an enzyme that can digest the hydrophilic portions of membrane proteins, but it is unable to penetrate the lipid bilayer and enter a cell. Because of these properties, trypsin has been used in conjunction with SDS–PAGE to determine which proteins have an extracellular domain. Describe an experiment using trypsin to determine the sidedness of proteins of the erythrocyte membrane.

7. Would you describe the S4 transmembrane segment of the K^+ ion channel as a hydrophobic helix, an amphipathic helix, or some other type? Why?

8. Suppose you were culturing a population of bacteria at 15°C and then raised the temperature of the culture to 37°C. What effect do you think this might have on the fatty acid composition of the membrane? On the transition temperature of the lipid bilayer? On the activity of membrane desaturases?

9. Looking at Figure 4.6, which lipids would you expect to have the greatest rate of flip-flop? Which the least? Why? If you determined experimentally that phosphatidylcholine actually exhibited the greatest rate of flip-flop, how might you explain this finding? How would you expect the rate of phospholipid flip-flop to compare with that of an integral protein? Why?

10. What is the difference between a two-dimensional and a three-dimensional representation of a membrane protein? How are the different types of profiles obtained, and which is more useful? Why do you think there are so many more proteins whose two-dimensional structure is known?

11. If you were to inject a squid giant axon with a tiny volume of solution containing 0.1 M NaCl and 0.1 M KCl in which both the Na^+ and K^+ ions were radioactively labeled, which of the radioactively labeled ions would you expect to appear most rapidly within the seawater medium while the neuron remained at rest? After the neuron had been stimulated to conduct a number of action potentials?

12. It has been difficult to isolate proteins containing water channels (i.e., aquaporins) due to the high rate of diffusion of water through the lipid bilayer. Why would this make aquaporin isolation difficult? Is there any way you might be able to distinguish diffusion of water through the lipid bilayer versus that through aquaporins? The best approach to studying aquaporin behavior has been to express the aquaporin genes in frog oocytes. Is there any reason why the oocytes of a pond-dwelling amphibian might be particularly well suited for such studies?

13. How is it that diffusion coefficients measured for lipids within membranes tend to be closer to that expected for free diffusion than those measured for integral proteins in the same membranes?

14. Assume that the plasma membrane of a cell was suddenly permeable to the same degree to both Na^+ and K^+ and that each responded to a concentration gradient of the same magnitude. Would you expect these two ions to move across the membrane at the same rates? Why or why not?

15. Most marine invertebrates show no loss or gain of water by osmosis, whereas most marine vertebrates experience continual water loss in their high-salt environment. Speculate on the basis for this difference and how it might reflect different pathways of evolution of the two groups.

16. How would you expect the concentrations of solute inside a plant cell to compare to that of its extracellular fluids? Would you expect the same to be true of the cells of an animal?

17. What would be the consequence for impulse conduction if the Na^+ channels were able to reopen immediately after they had closed during an action potential?

18. What would be the value of the potassium equilibrium potential if the external K^+ concentration was 200 mM and the internal concentration was 10 mM at 25°C? at 37°C?

19. As discussed on page 164, the Na^+/glucose cotransporter transports two Na^+ ions for each glucose molecule. What if this ratio were 1:1 rather than 2:1; how would this affect the glucose concentration that the transporter could work against?

20. A transmembrane protein usually has the following features: (1) the portion that transits the membrane bilayer is at least 20 amino acids in length, all of which are nonpolar residues; (2) the portion that anchors the protein on the external face has two or more consecutive acidic residues; and (3) the portion that anchors the protein on the cytoplasmic face has two or more consecutive basic residues. Consider the transmembrane protein with the following sequence:

NH$_2$-MLSTGVKRKGAVLLILLFPWMVAGGPLFWLA ADESTYKGS-COOH

Draw this protein as it would reside in the plasma membrane. Make sure you label the N- and C-termini and the external and cytoplasmic faces of the membrane. (Single-letter code for amino acids is given in Figure 2.26.)

21. Many marine invertebrates, such as squid, have extracellular fluids that resemble seawater and therefore have much higher intracellular ion concentrations than mammals. For a squid neuron, the ionic concentrations are

ion	intracellular concentration		extracellular concentration	
K^+	410	mM	15	mM
Na^+	40	mM	440	mM
Cl^-	100	mM	560	mM
Ca^{2+}	2×10^{-4}	mM	10	mM
pH	7.6		8.0	

If the resting potential of the plasma membrane, V_m, is −70 mV, are any of the ions at equilibrium? How far out of equilibrium, in mV, is each ion? What is the direction of net movement of each ion through an open channel permeable to that ion?

22. The membrane potential of a cell is determined by the relative permeability of the membrane to various ions. When acetylcholine binds to its receptors on the postsynaptic muscle membrane, it causes a massive opening of channels that are equally permeable to sodium and potassium. Under these conditions,

$$V_m = (V_{K^+} + V_{Na^+})/2$$

If $[K^+_{in}] = 140$ mM and $[Na^+_{in}] = 10$ mM for the muscle cell, and $[Na^+_{out}] = 150$ mM and $[K^+_{out}] = 5$ mM, what is the membrane potential of the endplate region of an acetylcholine-stimulated muscle?

23. Transmembrane domains consist of individual α-helices or a β sheet formed into a barrel. Looking at Figures 2.30 and 2.31, why is a single α helix more suited to spanning the bilayer than a single β strand?

24. Knowing how the K^+ channel selects for K^+ ions, suggest a mechanism by which the Na^+ channel is able to select for its ion.

25. How would you compare the rate of movement of ions passing through a channel versus those transported actively by a P-type pump? Why?

ADDITIONAL READING

Membranes: Structure and Function

BROWN, R. E. 1998. Sphingolipid organization in membranes: What physical studies of model membranes reveal. *J. Cell Sci.* 111:1–9.

DOWHAN, W. 1997. Molecular basis for membrane phospholipid diversity: Why are there so many lipids? *Annu. Rev. Biochem.* 66: 199–232.

EDIDIN, M. 1992. Patches, posts, and fences: Proteins and plasma membrane domains. *Trends Cell Biol.* 2:376–380.

EDIDIN, M. 1997. Lipid microdomains in cell surface membranes. *Curr. Opin. Struct. Biol.* 7:528–532.

HUSTEDT, E. J. & BETH, A. H. 1999. Nitroxide spin-spin interactions:

Applications to protein structure and dynamics. *Annu. Rev. Biophys. Biomol. Str.* 28:129–153.

JACOBSON, K. & DIETRICH, C. 1999. Looking at lipid rafts. *Trends Cell Biol.* 9:87–91

LASIC, D. D. 1996. Liposomes. *Sci. & Med.* 3:34–43. (May)

LUX, S. E. & PALEK, J. 1995. Disorders of the red cell membrane, in *Blood: Principles and Practice of Hematology.* R. I. HANDIN, ET AL., EDS. Lippincott.

POPOT, J.-L. 2000. Helical membrane protein folding, stability, and evolution. *Annu. Rev. Biochem.* 69:881–922.

SCHULZ, G. E. 2000. β-barrel membrane proteins. *Curr. Opin. Struct. Biol.* 10:443–447.

SEXTON, M. J. & JACOBSON, K. 1997. Single-particle tracking: Applications to membrane dynamics. *Annu. Rev. Biophys. Biomol. Str.* 26: 373–399.

STORRIE, B. & KREIS, T. E. 1996. Probing the mobility of membrane proteins inside the cell. *Trends Cell Biol.* 6:321–324.

UNWIN, N. & HENDERSON, R. 1984. The structure of proteins in biological membranes. *Sci. Am.* 250:78–94. (Feb.)

VIEL, A. & BRANTON, D. 1996. Spectrin: On the path from structure to function. *Curr. Opin. Cell Biol.* 8:49–55.

YEAGLE, P. L. 1993. *The Membranes of Cells*, 2d ed. Academic Press.

Movement of Solutes Across Membranes

ANDERSEN, J. P., ED. 2000. Ion pumps. *Adv. Mol. Cell. Biol.*, vol. 23.

ARMSTRONG, C. M. 1998. The vision of the pore. *Science* 280:56–57.

ARMSTRONG, C. M. & HILLE, B. 1998. Voltage-gated ion channels and electrical excitability. *Neuron* 20:371–380.

BORGNIA, M., ET AL. 1999. Cellular and molecular biology of the aquaporin water channels. *Annu. Rev. Biochem.* 68:425–458.

CATTERALL, W. A. 2000. From ionic currents to molecular mechanisms; the structure and function of voltage-gated sodium channels. *Neuron* 26:13–25.

CHOE, S., ET AL. 1999. Towards the three-dimensional structure of voltage-gated potassium channels. *Trends Biochem. Sci.* 24:345–349.

CLAPHAM, D. E. 1999. Unlocking family secrets: K^+ channel transmembrane domains. *Cell* 97:547–550.

DEAMER, D. W., ET AL., eds. 1999. Membrane permeability. *Curr. Topics Membs.* vol. 48.

GENNIS, R. B. & EBREY, T. G. 1999. Proton pump caught in the act. *Science* 286:252–253.

HIGGINS, C. F. 1992. ABC transporters: From microorganisms to man. *Annu. Rev. Cell Biol.* 8:67–113.

HILLE, B. 1991. *Ionic Channels of Excitable Membranes*, 2d ed. Sinauer.

HILLE, B., ARMSTRONG, C. M., & MACKINNON, R. 1999. Ion channels: From idea to reality. *Nature Med.* 5:1105–1109.

HORN, R. 2000. A new twist to the saga of charge movement in voltage dependent ion channels. *Neuron* 25:511–514.

KÜHLBRANDT, W. 2000. Bacteriorhodopsin—the movie. *Nature* 406: 569–570.

KURACHI, Y., ET AL., eds. 1999. Potassium channels. *Curr. Topics Membs.* vol. 46.

LANG, F., ET AL. 1998. Functional significance of cell volume regulatory mechanisms. *Physiol. Revs.* 78:247–306.

LANYI, J. K., ED. 2000. Bacteriorhodopsin. *Biochim. Biophys. Acta* vol. 1460, #1.

NELSON, N. & HARVEY, W. R. 1999. Vacuolar and plasma membrane proton-ATPase. *Physiol. Revs.* 79:361–385.

SACHS, G., ET AL. 1995. The pharmacology of the gastric acid pump: The H^+, K^+-ATPase. *Annu. Rev. Pharm.* 35:277–305.

SANSOM, M. S. P. 2000. Potassium channels: Watching a voltage sensor tilt and twist. *Curr. Biol.* 10:R206–R209.

SANSOM, M. S. P. & LAW, R. J. 2001. Aquaporins—Channels without ions. *Curr. Biol.* R71–R73.

SHIN, Y.-K. 1998. K^+ channel gating mechanism proposed using EPR. *Nature Struct. Biol.* 5:418–420.

SKOU, J. C. 1989. Identification of the sodium pump as the membrane-bound Na^+/K^+-ATPase. *Biochim. Biophys. Acta* 1000: 435–446.

STEIN, W. D. 1990. *Channels, Carriers, and Pumps: An Introduction to Membrane Transport*. Academic Press.

YELLEN, G. 1998. Premonitions of ion channel gating. *Nature Struct. Biol.* 5:421.

ZAGROVIC, B. & ALDRICH, R. 1999. For the latest information, tune to channel KcsA. *Science* 285:59–61.

Ion Channels and Disease

ASHCROFT, F. M. 1999. *Ion Channels and Disease*. Academic.

ENNIST, D. L. 1999. Gene therapy for lung disease. *Trends Pharmacol.* 20:260–266.

FRIZELL, R. A. ED. 1999. Physiology of cystic fibrosis. *Physiol. Revs.* vol. 79, #1, supplement.

GUGGINO, W. B. 1999. Cystic fibrosis and the salt controversy. *Cell* 96:607–610.

JONGSMA, H. J. 1998. Sudden cardiac death: A matter of faulty ion channels? *Curr. Biol.* 8:R568–R571.

KING, L. S., ET AL. 2000. Aquaporins in health and disease. *Mol. Med. Today* 6:60–65.

LEHMANN-HORN, F. & JURKAT-ROTT, K. 1999. Voltage-gated ion channels and hereditary disease. *Physiol. Revs.* 79:1317–1372.

PRINCE, A. 1998. The CFTR advantage: Capitalizing on a quirk of fate. *Nature Med.* 4:663–664.

SAIER, M. H., Jr. 2000. Families of proteins forming transmembrane channels. *J. Memb. Biol.* 175:165–180.

SANSOM, C. 1999. A step towards gene therapy for CF. *Mol. Med. Today* 5:279.

WINE, J. J. 1999. Cystic fibrosis lung disease. *Sci. & Med.* 6:34–43. (May/June)

Nerve Impulses and Synaptic Transmission

AIDLEY, D. J. 1998. *The Physiology of Excitable Cells*. 4th ed. Cambridge.

CHANGEUX, J.-P. 1993. Chemical signaling in the brain. *Sci. Am.* 269:58–62. (Nov.)

JESSELL, T. M., KANDEL, E. R., ET AL. 1993. Synaptic transmission: Nine reviews. *Cell* 72: Suppl. 1–149.

HALL, Z. W. 1992. *An Introduction to Molecular Neurobiology*. Sinauer.

HARRISON, N. L. & FLOOD, P. 1998. Molecular mechanisms of general anesthetic action. *Sci. & Med.* 5:18–27. (May/June)

LEVITAN, I. B. & KACZMAREK, L. K. *The Neuron*. Oxford.

NEHER, E. 1998. Vesicle pools and Ca^{2+} microdomains: New tools for understanding their roles in neurotransmitter release. *Neuron* 20:389–399.

NICHOLLS, J. G., ET AL. 1992. *From Neuron to Brain*. Sinauer.

WAXMAN, S. G., ED. 1995. *The Axon*. Oxford.

WILLIAMS, J. T., ET AL. 2001. Cellular and synaptic adaptations mediating opioid dependence. *Physiol. Revs.* 81:299–343.

5

Aerobic Respiration and the Mitochondrion

uring the first two billion years or so that life existed on Earth, the atmosphere consisted largely of reduced molecules, such as molecular hydrogen (H_2), ammonia (NH_3), and H_2O. The Earth of this period was populated by anaerobes—organisms that captured and utilized energy by means of oxygen-independent (anaerobic) metabolism, such as glycolysis and fermentation (Figures 3.23, 3.28). Then, the cyanobacteria appeared—a new kind of organism that carried out a new type of photosynthetic process, one in which water molecules were split apart and molecular oxygen (O_2) was released. The cyanobacteria were immensely successful, and the oceans, lakes, and atmosphere were rapidly infused with the new gas.

As discussed on page 35, molecular oxygen can be a very toxic substance, taking on extra electrons and reacting with a variety of biological molecules. The presence of oxygen in the atmosphere must have been a powerful agent for natural selection. Over time, species evolved that were not only protected from the damaging effects of molecular oxygen, but possessed metabolic pathways that utilized the molecule to great advantage. Without the ability to use oxygen,

Micrograph of a mammalian fibroblast that has been fixed and stained with fluorescent antibodies that reveal the distribution of the mitochondria (green) and the microtubules of the cytoskeleton (red). The mitochondria do not appear as individual structures but rather as an extensive network or reticulum that extends through much of the cell. (FROM MICHAEL P. YAFFE, UNIVERSITY OF CALIFORNIA, SAN DIEGO, REPRINTED WITH PERMISSION FROM SCIENCE 283:1493, 1999. COPYRIGHT 1999, AMERICAN ASSOCIATION FOR THE ADVANCEMENT OF SCIENCE.)

organisms could only extract a limited amount of energy from their foodstuffs, excreting energy-rich products such as lactic acid and ethanol, which they were unable to metabolize further. In contrast, organisms that incorporated O_2 into their metabolism could completely oxidize such compounds to CO_2 and H_2O and, in the process, extract a much larger percentage of their energy content. These organisms that became dependent on oxygen were the Earth's first **aerobes**, and they eventually gave rise to all of the oxygen-dependent prokaryotes and eukaryotes living today. In eukaryotes, the utilization of oxygen as a means of energy extraction takes place in a specialized organelle, the **mitochondrion**. ∎

5.1 MITOCHONDRIAL STRUCTURE AND FUNCTION

Unlike most cytoplasmic organelles, mitochondria are large enough to be seen in the light microscope (Figure 5.1a), and their presence within cells has been known for over a hundred years. Even before 1900, mitochondria had been isolated from tissues by teasing the cells apart with a fine needle, and a number of the properties of these organelles had been described. It was shown, for example, that mitochondria are osmotically active; that is, they would swell in hypotonic media and shrink in hypertonic media. This property suggested that mitochondria were bounded by a semipermeable membrane not unlike that which surrounds the cell itself.

As with most other organelles, mitochondria possess recognizable morphologic characteristics while exhibiting considerable variability in appearance. When a tissue is fixed, sectioned, and examined under the electron microscope, its mitochondria typically appear as sausage-shaped organelles (Figure 5.1b), approximately 0.2 to 1.0 μm in cross-sectional diameter and 1 to 4 μm in length. More recent investigations have suggested that, at least in yeast and cultured mammalian cells, mitochondria are not present as discrete, individual organelles but rather as a highly branched, interconnected network or reticulum. This type of mitochondrial structure can be seen in the chapter opening micrograph of a fibroblast that had been fixed and stained with fluorescent antibodies (see also Figure 5.20). Whether or not this concept of a mitochondrial reticulum will apply to cells in general remains to be determined.

Mitochondria occupy approximately 15 to 20 percent of the volume of an average mammalian liver cell, and are even more prevalent in muscle cells, which require large amounts of ATP to fuel their contraction. Mitochondria are often associated with fatty acid-containing oil droplets from which they derive raw materials to be oxidized. A particularly striking arrangement of mitochondria occurs in sperm cells where they are often located in the midpiece, just behind the nucleus (Figure 5.1c). The movements of a sperm are powered by ATP produced in these mitochondria. Mitochondria are also prominent in many plant cells where they are the primary suppliers of ATP in nonphotosynthetic tissues, as well as being a source of ATP in photosynthetic leaf cells during periods of dark.

The internal structure of a mitochondrion is shown in the electron micrographs of Figure 5.1b and 5.2a, the three-dimensional reconstruction of Figure 5.2b, and the schematic diagram of Figure 5.2c. The role of mitochon-

Mitochondria

Cristae

Mitochondria

(a) *(b)* *(c)*

Figure 5.1 Mitochondria. (*a*) A living fibroblast viewed with a phase-contrast microscope. Mitochondria are seen as elongated, dark bodies. (*b*) Transmission electron micrograph of a thin section through a mitochondrion revealing the internal structure of the organelle, particularly the membranous folds (cristae) of the inner membrane. (*c*) Localization of mitochondria in the sperm midpiece surrounding the proximal portion of the flagellum. (A: COURTESY OF NORMAN K. WESSELS, B: COURTESY OF K. R. PORTER/PHOTO RESEARCHERS; C: DON W. FAWCETT/VISUALS UNLIMITED.)

(b) 1 μm

(a) 0.2 μm

Figure 5.2 The structure of a mitochondrion. (*a*) Scanning electron micrograph of a macerated mitochondrion, showing the internal matrix enclosed by folds of the inner membrane. (*b*) Three-dimensional reconstruction of a mitochondrion based on a series of micrographs taken with a high-voltage electron microscope of a single thick section of brown fat tissue that had been tilted at various angles. High-voltage instruments accelerate electrons at velocities that allow them to penetrate thicker tissue sections (up to 1.5 μm). This technique suggests that the cristae are present as flattened sheets (lamellae) that communicate with the intermembrane space by way of narrow tubular openings, rather than "wide open" gaps as is typically depicted. In this reconstruction, the inner mitochondrial membrane is shown in blue at the peripheral regions and in yellow when it penetrates into the matrix to form the cristae. (*c*) Schematic diagrams showing the generalized internal structure of a mitochondrion. (A: FROM K. TANAKA AND T. NAGURO, INT. REV. CYTOL. 68:111, 1980. B: FROM G. A. PERKINS, ET AL., J. BIOEN. BIOMEM. 30:436, 1998.)

(c)

dria as energy transducers is intimately tied to the membranes that are so prominent in electron micrographs of these organelles. Each mitochondrion contains an **outer membrane** and a complex **inner membrane** system. The outer membrane completely encloses the mitochondrion, serving as its outer boundary. Although part of the inner membrane may lie just inside the outer membrane, much of it occurs as deep folds or invaginations, called **cristae** (Figures 5.1*b*, 5.2). In some cells, as in those of the mammalian liver and brown adipose tissue, the cristae are wide sheets that cut across the entire width of the mitochondrion (Figure 5.2*b*). The infoldings of the inner membrane greatly increase the amount of surface available to house the machinery needed for aerobic respiration.

The membranes of the mitochondrion divide the organelle into two aqueous compartments, one within the interior of the mitochondrion, called the **matrix**, and a second between the outer and inner membrane, called the **intermembrane space**. The matrix has a gel-like consistency owing to the presence of a high concentration (up to 500 mg/ml) of water-soluble proteins. The intermembrane space appears rather limited in the micrographs and diagrams in Figures 5.1 and 5.2, but under conditions of active respiration, this compartment may become much more expansive.

Mitochondrial Membranes

The outer and inner membranes have very different properties. The outer membrane is composed of approximately 50 percent lipid by weight and contains a curious mixture of enzymes involved in such diverse activities as the oxidation of epinephrine, the degradation of tryptophan, and the elongation of fatty acids. In contrast, the inner membrane contains more than 100 different polypeptides and has a very high protein/lipid ratio (more than 3:1 by weight, which corresponds to about one protein molecule for every 15 phospholipids). The inner membrane is virtually devoid of cholesterol and rich in an unusual phospholipid, cardiolipin (diphosphatidylglycerol, see Figure 4.6

Figure 5.3 **Porins.** Gram-negative bacteria have a lipid-containing outer membrane outside of their plasma membrane as part of their cell wall. This outer membrane contains proteins, called porins, which consist of a barrel of β sheet and form an opening through which moderately sized molecules can penetrate. A variety of porins having different sized channels and selectivities are also found in the outer mitochondrial membrane in eukaryotic cells.

for the structure); both features are characteristic of bacterial plasma membranes, from which the inner mitochondrial membrane has presumably evolved. The outer mitochondrial membrane is thought to be derived from an outer membrane found as part of the cell wall of certain bacterial cells (Figure 5.3); both contain *porins*, integral proteins that have a relatively large internal channel (e.g., 2–3 nm) surrounded by a barrel of β strands. The porins of the outer mitochondrial membrane are not static structures but can undergo reversible closure in response to conditions within the cell. When the porin channels are wide open, the outer membrane is permeable to molecules up to about 5,000 daltons. Thus, molecules such as ATP, NAD, and coenzyme A, all of which have a molecular mass less than 1,000 daltons, can pass freely between the intermembrane space and the cytosol. In contrast, the inner mitochondrial membrane is highly impermeable; virtually all molecules and ions require special membrane transporters to gain entrance to the matrix.

Among the many proteins of the inner mitochondrial membrane are several involved in the uptake and release of calcium ions. Calcium ions are important triggers for cellular activities (Section 15.3) and recent studies have con-

firmed a role for mitochondria (along with the endoplasmic reticulum) in regulating the Ca^{2+} concentration of the cytosol.

As will be discussed in following sections, the composition and organization of the inner mitochondrial membrane are the keys to the bioenergetic activities of the organelle. In addition to containing a variety of transport systems, the inner membrane is the seat of most of the machinery required for the synthesis of ATP. The architecture of the inner membrane and the apparent fluidity of its bilayer facilitate the interactions of components that are required for ATP formation.

The Mitochondrial Matrix

In addition to various enzymes, the mitochondrial matrix also contains ribosomes (of considerably smaller size than those found in the cytosol) and several molecules of double-stranded, usually circular DNA (Figure 5.2c). Thus, mitochondria possess their own genetic material and the machinery to manufacture their own RNAs and proteins. This nonchromosomal DNA is important because it encodes a small number of mitochondrial polypeptides (13 in humans) that are tightly integrated into the inner mitochondrial membrane along with polypeptides encoded by genes residing within the nucleus. Human mitochondrial DNA also encodes 2 ribosomal RNAs and 22 tRNAs that are used in protein synthesis within the organelle.

Observations of living cells growing in culture have shown mitochondria to be dynamic organelles that change their shape, move from place to place within the cytoplasm, and undergo both fusion with one another and fission. Just as membranes arise from preexisting membranes

0.4 μm

Figure 5.4 **Mitochondrial fission.** Electron micrograph of mitochondria of an insect cell fixed while in the process of splitting. (FROM W. J. LARSEN, J. CELL BIOL. 47:379, 1970, BY COPYRIGHT PERMISSION OF THE ROCKEFELLER UNIVERSITY PRESS.)

Figure 5.5 **An overview of carbohydrate metabolism in eukaryotic cells.** The reactions of glycolysis generate pyruvate and NADH in the cytosol. In the absence of O_2, the pyruvate is reduced by NADH to lactate (or another product of fermentation, such as ethanol in yeast; see Figure 3.28 for details). The NAD^+ formed in the reaction is reutilized in the continuation of glycolysis. In the presence of O_2, the pyruvate moves into the matrix (facilitated by a membrane transporter), where it is decarboxylated and linked to coenzyme A (CoA), a reaction that generates NADH. The NADH produced during glycolysis donates its high-energy electrons to a compound that crosses the inner mitochondrial membrane. The acetyl CoA passes through the TCA cycle (as shown in Figure 5.7), which generates NADH and $FADH_2$. The electrons in these various NADH and $FADH_2$ molecules are passed along the electron-transport chain, which is made up of carriers that are embedded in the inner mitochondrial membrane, to molecular oxygen (O_2). The energy released during electron transport is used in the formation of ATP by a process discussed at length later in the chapter. If all of the energy from electron transport were to be utilized in ATP formation, approximately 36 ATPs could be generated from a single molecule of glucose.

and cells arise from preexisting cells, mitochondria arise by fission from preexisting mitochondria (Figure 5.4). We will return to discuss the structure and function of mitochondria, but first let us consider the role of these organelles in the basic oxidative pathways of eukaryotic cells, which is summarized in Figure 5.5. It might help to examine this overview figure and read the accompanying legend before moving on to the detailed descriptions of these pathways that follows.

REVIEW **?**

1. Describe the metabolic changes in oxidative metabolism that must have accompanied the evolution and success of the cyanobacteria.

2. Compare the properties of the inner and outer mitochondrial membranes; the intermembrane space and the matrix.

5.2 **OXIDATIVE METABOLISM IN THE MITOCHONDRION**

In Chapter 3 we described the initial stages in the oxidation of carbohydrates. Beginning with glucose, the first steps of the oxidation process are carried out by the enzymes of glycolysis, which are located in the cytosol (Figure 5.5). The 10 reactions that constitute the glycolytic pathway were illustrated in Figure 3.23; the major steps of the pathway are summarized in Figure 5.6. Only a small fraction of the free energy available in glucose is made available to the cell during glycolysis—enough for the net synthesis of only two molecules of ATP per molecule of glucose oxidized (Figure 5.6). Most of the energy remains stored in pyruvate. Each molecule of NADH produced during the oxidation of glyceraldehyde 3-phosphate (reaction 6, Figure 5.6) also carries a pair of high-energy electrons. The two products of glycolysis—pyruvate and NADH—can be metabolized in two very different ways,

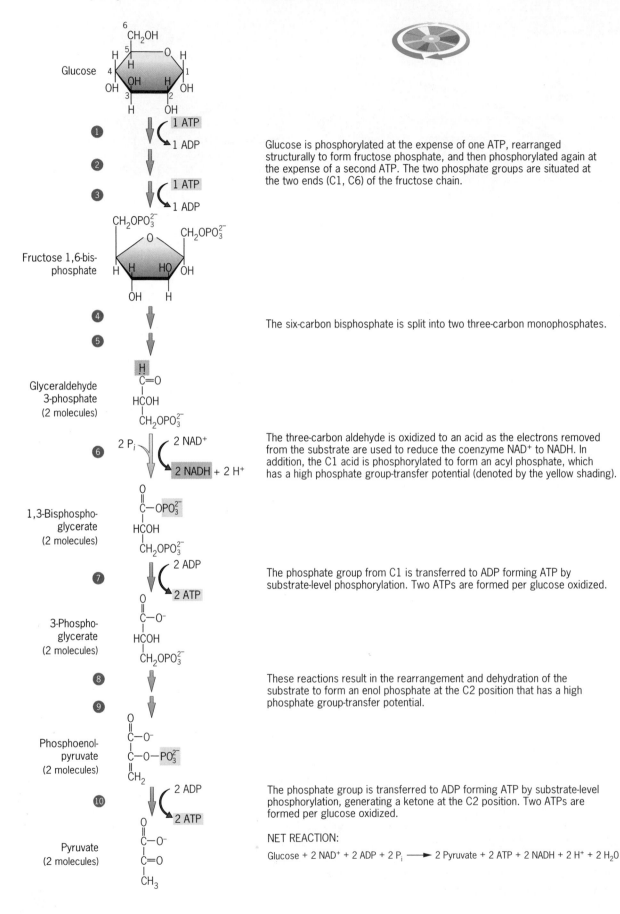

Glucose is phosphorylated at the expense of one ATP, rearranged structurally to form fructose phosphate, and then phosphorylated again at the expense of a second ATP. The two phosphate groups are situated at the two ends (C1, C6) of the fructose chain.

The six-carbon bisphosphate is split into two three-carbon monophosphates.

The three-carbon aldehyde is oxidized to an acid as the electrons removed from the substrate are used to reduce the coenzyme NAD^+ to NADH. In addition, the C1 acid is phosphorylated to form an acyl phosphate, which has a high phosphate group-transfer potential (denoted by the yellow shading).

The phosphate group from C1 is transferred to ADP forming ATP by substrate-level phosphorylation. Two ATPs are formed per glucose oxidized.

These reactions result in the rearrangement and dehydration of the substrate to form an enol phosphate at the C2 position that has a high phosphate group-transfer potential.

The phosphate group is transferred to ADP forming ATP by substrate-level phosphorylation, generating a ketone at the C2 position. Two ATPs are formed per glucose oxidized.

NET REACTION:

Glucose + 2 NAD^+ + 2 ADP + 2 P_i ⟶ 2 Pyruvate + 2 ATP + 2 NADH + 2 H^+ + 2 H_2O

Figure 5.6 An overview of glycolysis showing some of the key steps. These include the two reactions in which phosphate groups are transferred from ATP to the six-carbon sugar to produce fructose 1,6-bisphosphate (steps 1, 3); the oxidation and phosphorylation of glyceraldehyde 3-phosphate to produce 1,3-bisphosphoglycerate and NADH (step 6); and the transfer of phosphate groups from three-carbon phosphorylated substrates to ADP to produce ATP by substrate phosphorylation (steps 7, 10). Keep in mind that two molecules of glyceraldehyde 3-phosphate are formed from each molecule of glucose so reactions 6 through 10 shown here occur twice per glucose molecule oxidized.

depending on the type of cell in which they are formed and the presence or absence of oxygen.

In the presence of O_2, aerobic organisms are able to extract large amounts of additional energy from the pyruvate and NADH produced during glycolysis—enough to synthesize more than 30 additional ATP molecules. This energy is extracted in the mitochondrion (Figure 5.5). We will begin with pyruvate and return to consider the fate of the NADH later in the discussion. Each pyruvate molecule produced by glycolysis is transported across the inner mitochondrial membrane and into the matrix, where it is decarboxylated to form a two-carbon acetyl group ($-CH_3COO^-$). The acetyl group is transferred to *coenzyme A* (a complex organic compound derived from the vitamin pantothenic acid) to produce acetyl CoA.

$$\text{Pyruvate} + \text{HS---CoA} + \text{NAD}^+ \rightarrow$$
$$\text{Acetyl CoA} + CO_2 + \text{NADH} + H^+$$

The decarboxylation of pyruvate and transfer of the acetyl group to CoA (Figures 5.5, 5.7) is catalyzed by the giant multienzyme complex pyruvate dehydrogenase, whose structure was shown in Figure 2.41. The discovery of acetyl CoA by Fritz Lipmann in 1961 was the last piece in the puzzle of glucose oxidation.

The Tricarboxylic Acid (TCA) Cycle

Once formed, acetyl CoA is fed into a cyclic pathway, called the **tricarboxylic acid (TCA) cycle**, where the substrate is oxidized and its energy conserved. With the exception of succinate dehydrogenase, which is bound to the inner membrane, all the enzymes of the TCA cycle reside in the soluble phase of the matrix (Figure 5.5). The TCA cycle is also referred to as the Krebs cycle after the British biochemist Hans Krebs, who worked out the pathway in the 1930s. When Krebs first obtained sufficient evidence to support the idea of a metabolic cycle, he submitted the results to the British journal *Nature*. The paper was returned several days later accompanied by a letter of rejection. The editor had concluded it wasn't important enough to be published in the journal.

The first step in the TCA cycle is the condensation of the two-carbon acetyl group with a four-carbon oxaloacetate to form a six-carbon citrate molecule (Figure 5.7, step 12). During the cycle, the citrate molecule is decreased in chain length, one carbon at a time, regenerating the four-carbon oxaloactetate molecule, which can condense with another acetyl CoA. It is the two carbons that are removed during the TCA cycle (which are not the same ones that were brought in with the acetyl group) that are completely oxidized to carbon dioxide. During the TCA cycle, four reactions occur in which a pair of electrons are transferred from a substrate to an electron-accepting coenzyme. Three of the reactions reduce NAD^+ to NADH, one reaction reduces FAD (derived from the vitamin riboflavin) to $FADH_2$ (Figure 5.7). The net equation for the reactions of the TCA cycle can be written as

$$\text{Acetyl CoA} + 2\,H_2O + \text{FAD} + 3\,\text{NAD}^+ + \text{GDP} + P_i \rightarrow$$
$$2\,CO_2 + \text{FADH}_2 + 3\,\text{NADH} + 3\,H^+ + \text{GTP} + \text{HS---CoA}$$

The TCA cycle is a critically important metabolic pathway. If the position of the TCA cycle in the overall metabolism of the cell is considered (Figure 5.8, see also Figure 3.21), the metabolites of this cycle are found to be the same compounds generated by most of the cell's catabolic pathways. Acetyl CoA, for example, is an important end product of a number of catabolic paths, including the disassembly of fatty acids, which are degraded in two-carbon units (Figure 5.8a). These two-carbon compounds enter the TCA cycle as acetyl CoA. The catabolism of amino acids, the building blocks of proteins, also generates metabolites of the TCA cycle (Figure 5.8b), which enter the matrix by means of special transport systems in the inner mitochondrial membrane. It is apparent that all of a cell's energy-providing macromolecules (polysaccharides, fats, and proteins) are broken down into metabolites of the TCA cycle. Thus, the mitochondrion becomes the focus for the final energy-conserving steps in metabolism regardless of the nature of the starting material.

The Importance of Reduced Coenzymes in the Formation of ATP

It is evident from the net equation of the TCA cycle that the primary products of the pathway are the reduced coenzymes $FADH_2$ and NADH, which contain the electrons removed from various substrates as they are oxidized. NADH was also one of the products of glycolysis (along with pyruvate). Mitochondria are not able to import the NADH formed in the cytosol by glycolysis. Instead, the electrons of NADH are used to reduce a low-molecular-weight metabolite that can either (1) enter the mitochondrion (by a pathway called the malate-aspartate shuttle) and reduce NAD^+ to NADH, or (2) transfer its electrons to FAD (by a pathway called the glycerol phosphate shuttle, which is shown in Figure 5.9) to produce $FADH_2$.

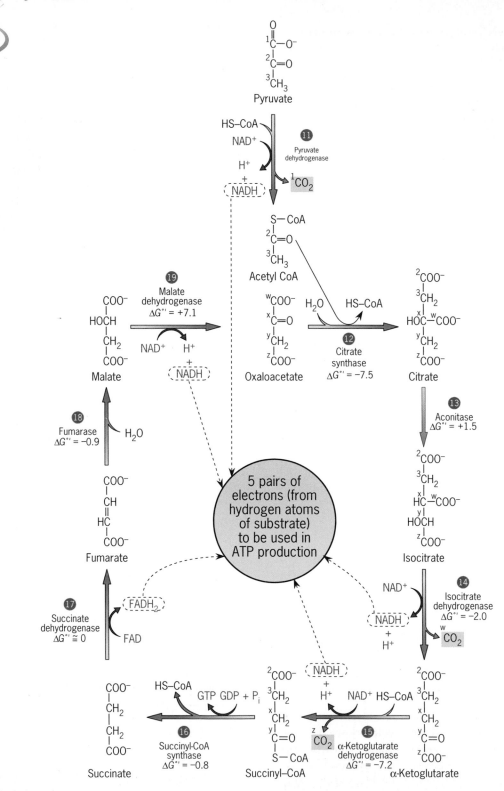

Figure 5.7 The tricarboxylic acid (TCA) cycle, also called the Krebs cycle after the person who formulated it, or the citric acid cycle after the first compound formed in it. The cycle begins with the condensation of oxaloacetate (OAA) and acetyl CoA (reaction 12). The carbons of these two compounds are marked with numbers or letters. The two carbons lost during passage through the cycle are derived from oxaloacetate. The standard free energies and names of enzymes are also provided.

Five pairs of electrons are removed from substrate molecules by pyruvate dehydrogenase and the enzymes of the TCA cycle. These high-energy electrons are transferred to NAD^+ or FAD and then passed down the electron-transport chain for use in ATP production. The reactions shown here begin with number 11 because the pathway continues where the last reaction of glycolysis (number 10 of Figure 5.6) leaves off.

(a)

(b)

Figure 5.8 Catabolic pathways generate compounds that are fed into the TCA cycle. (*a*) Oxidation of fatty acids. The first step in the oxidation of a fatty acid is its activation by linkage to the thiol (—SH) group of coenzyme A, which occurs after the fatty acyl group is transported across the inner mitochondrial membrane in association with a carrier protein. In the mitochondrion, the fatty acyl CoA molecule undergoes stepwise disassembly in which an acetyl CoA (shown in blue) is removed from the fatty acid chain with each round of the cycle. In addition to the molecule of acetyl CoA that is fed into the TCA cycle, each round of the fatty acid cycle produces one NADH and one FADH$_2$. It is apparent from examination of this series of reactions why fats are such a rich storehouse of chemical energy. (*b*) Input of amino acids into the TCA cycle.

Figure 5.9 The glycerol phosphate shuttle. In the glycerol phosphate shuttle, electrons are transferred from NADH to dihydroxyacetone phosphate (DHAP) to form glycerol 3-phosphate, which shuttles them into the mitochondrion. These electrons then reduce FAD at the inner mitochondrial membrane, forming FADH$_2$, which can transfer the electrons to a carrier of the electron-transport chain.

Now that we have accounted for the formation of NADH and FADH$_2$ by both glycolysis and the TCA cycle, we can turn to the steps that utilize these reduced coenzymes to produce ATP. The overall process can be divided into two steps, which can be summarized as follows:

Step 1. *High-energy electrons are passed from FADH$_2$ or NADH* to the first of a series of electron carriers that make up the electron-transport chain, which is located in the inner mitochondrial membrane. The terminal accep-

tor of electrons from the respiratory chain is molecular oxygen (O_2), which becomes reduced to water. Electrons pass along the respiratory chain in energy-releasing reactions. These reactions are coupled to energy-requiring conformational changes in electron carriers that move protons outward across the inner mitochondrial membrane. As a result, the energy released during electron transport is stored in the form of a proton gradient across the membrane.

Step 2. *The controlled movement of protons back across the membrane* through an ATP-synthesizing enzyme provides the energy required to phosphorylate ADP to ATP. The importance of proton movements in the formation of ATP was first proposed in 1961 by Peter Mitchell of the University of Edinburgh. The experiments that led to the acceptance of the **chemiosmotic mechanism**, as Mitchell called it, are discussed in the Experimental Pathways, which can be found at www.wiley.com/college/karp on the Web. Further analysis of the two steps summarized above will occupy us for much of the remainder of the chapter.

Each pair of electrons transferred from NADH to oxygen by means of the electron-transport chain releases sufficient energy to drive the formation of approximately three molecules of ATP. Each pair donated by $FADH_2$ releases sufficient energy for the formation of approximately two molecules of ATP. If one adds up all of the ATPs formed from one molecule of glucose completely catabolized by means of glycolysis and the TCA cycle, the net gain is 36 ATPs (which includes the GTP formed by each round of the TCA cycle, step 16, Figure 5.7). Keep in mind that the actual number of ATPs formed per molecule of glucose oxidized depends on the [ATP]/[ADP] ratio within the cell and the particular activities in which the cell is engaged. The relative importance of glycolysis versus the TCA cycle, that is, of anaerobic versus aerobic oxidative metabolism, in human skeletal muscle function is discussed in the accompanying Human Perspective.

REVIEW ?

1. How are the two products of glycolysis connected to the reactions of the TCA cycle?
2. Why is the TCA cycle considered to be the central pathway in the energy metabolism of a cell?
3. Describe the mechanism by which NADH produced by glycolysis is able to feed electrons into the TCA cycle.

5.3 THE ROLE OF MITOCHONDRIA IN THE FORMATION OF ATP

Mitochondria are often described as miniature power plants. Like power plants, mitochondria extract energy from organic materials and store it, temporarily, in the form of electrical energy. More specifically, the energy extracted from substrates is utilized to generate an ionic gradient across the inner mitochondrial membrane. An ionic gradient across a membrane represents a form of energy that can be tapped to perform work. We saw in Chapter 4 how intestinal cells utilize an ionic gradient across their plasma membrane to transport sugars and amino acids out of the intestinal lumen, while nerve cells use a similar gradient to conduct neural impulses. The use of ionic gradients as a form of energy currency requires several components, including a system to generate the gradient, a membrane capable of maintaining the gradient, and the machinery to tap the gradient so it can perform work.

Mitochondria utilize an ionic gradient across their inner membrane to drive numerous energy-requiring activities, most notably the synthesis of ATP. When ATP formation is driven by energy that is released from electrons removed during substrate oxidation, the process is called **oxidative phosphorylation**. Oxidative phosphorylation can be contrasted with substrate-level phosphorylation, as discussed on page 109, in which ATP is formed directly by transfer of a phosphate group from a substrate molecule to ADP. According to one estimate, oxidative phosphorylation accounts for the production of more than 2×10^{26} molecules (>160 kg) of ATP in our bodies per day. Unraveling the basic mechanism of oxidative phosphorylation has been one of the crowning achievements in the field of cell and molecular biology; filling in the remaining gaps continues to be an active area of research. To understand the mechanism of oxidative phosphorylation, it is first necessary to consider how substrate oxidation is able to release free energy.

Oxidation-Reduction Potentials

If one compares a variety of oxidizing agents, they can be ranked in a series according to their affinity for electrons: the greater the affinity, the stronger the oxidizing agent. Reducing agents can also be ranked according to their affinity for electrons: the lower the affinity (the more easily electrons are released), the stronger the reducing agent. To put this into quantifiable terms, reducing agents are ranked according to **electron-transfer potential**; those substances having a high electron-transfer potential, such as NADH, are strong reducing agents, whereas those with a low electron-transfer potential, such as H_2O, are weak reducing agents. Oxidizing and reducing agents occur as **couples**, such as NAD^+ and NADH, which differ in their number of electrons. Strong reducing agents are coupled to weak oxidizing agents, and vice versa. For example, NAD^+ (of the NAD^+-NADH couple) is a weak oxidizing agent, whereas O_2 (of the O_2-H_2O couple) is a strong oxidizing agent.

Because the movement of electrons generates a separation of charge, the affinity of substances for electrons can be measured by instruments that detect voltage (Figure 5.10). What is measured for a given couple is an

THE HUMAN PERSPECTIVE

The Role of Anaerobic and Aerobic Metabolism in Exercise

Muscle contraction requires large amounts of energy. Most of the energy is utilized in causing actin- and myosin-containing filaments to slide over one another, as discussed in Chapter 9. The energy that drives muscle contraction is derived from ATP. The rate of ATP hydrolysis increases more than 100-fold in a skeletal muscle undergoing maximal contraction compared to the same muscle at rest. It is estimated that the average human skeletal muscle has enough ATP available to fuel a 2 to 5 second burst of vigorous contraction. Even as ATP is hydrolyzed, it is important that additional ATP be produced; otherwise the ATP/ADP ratio would fall and with it the free energy available to fuel contraction. Muscle cells contain a store of creatine phosphate (CrP), one of the compounds with a phosphate transfer potential greater than that of ATP (see Figure 3.27) and thus can be utilized to generate ATP in the following reaction:

$$CrP + ADP \rightarrow Cr + ATP$$

Skeletal muscles typically have enough stored creatine phosphate to maintain elevated ATP levels for approximately 15 seconds. Because muscle cells have a very limited supply of both ATP and creatine phosphate, it follows that intense or sustained muscular activity requires the formation of additional quantities of ATP, which must be obtained by oxidative metabolism.

Human skeletal muscles consist of two general types of fibers (Figure 1): fast-twitch fibers, which can contract very rapidly (e.g., 15–40 msec), and slow-twitch fibers, which contract more slowly (e.g., 40–100 msec). Fast-twitch fibers are seen under the electron microscope to be nearly devoid of mitochondria, which indicates that these cells are unable to produce much ATP by aerobic respiration. Slow-twitch fibers, on the other hand, contain large numbers of mitochondria. These two types of skeletal muscle fibers are suited for different types of activities. For example, lifting weights or running sprints depends primarily on fast-twitch fibers, which are able to generate more force than their slow-twitch counterparts. Fast-twitch fibers produce nearly all of their ATP anaerobically as a result of glycolysis. Even though glycolysis only produces about five percent as much ATP per molecule of glucose oxidized compared with aerobic respiration, the reactions of glycolysis occur much more rapidly than those of the TCA cycle and electron transport; thus, the rate of anaerobic ATP production is actually higher than that achieved by aerobic respiration. The problems with producing ATP by glycolysis are the rapid use of the fiber's available glucose (stored in the form of glycogen) and the production of an undesirable end product, lactic acid. Let's consider this latter aspect further.

Recall that the continuation of glycolysis requires the regeneration of NAD$^+$, which occurs by fermentation (page 110). Muscle cells regenerate NAD$^+$ by reducing pyruvate—

the end product of glycolysis—to lactic acid. Most of the lactic acid diffuses out of the active muscle cells into the blood, where it is carried to the liver and converted back to glucose. Glucose produced in the liver is released into the blood, where it can be returned to the active muscles to continue to fuel the high levels of glycolysis. However, the formation of lactic acid is associated with a drop in pH within the muscle tissue (from about pH 7.00 to 6.35), which may produce the pain and cramps that accompany vigorous exercise. Increased acidity, together with the depletion of glycogen stores, probably accounts for the sensation of muscle fatigue that accompanies anaerobic exercises.

If, instead of trying to use your muscles to lift weights or run sprints, you were to engage in an aerobic exercise, such as bicycling or fast walking, you would be able to continue to perform the activity for much longer periods of time without feeling muscle pain or fatigue. Aerobic exercises, as their name implies, are designed to allow your muscles to continue to perform aerobically, that is, to continue to produce the necessary ATP by electron transport. Aerobic exercises depend largely on the contraction of the slow-twitch fibers of the skeletal muscles. Although these fibers generate less force, they can continue to function for long periods of time due to the continuing aerobic production of ATP without production of lactic acid.

Aerobic exercise is initially fueled by the glucose molecules stored as glycogen in the muscles themselves, but after a few minutes, the muscles depend more and more on free fatty acids released into the blood from adipose (fat) tissue. The longer the period of exercise, the greater the dependency on fatty acids. After 20 minutes of vigorous aerobic exercise, it is estimated that about 50 percent of the calories being con-

Figure 1 **Skeletal muscles** contain a mix of fast-twitch (or type II) fibers (darkly stained) and slow-twitch (or type I) fibers (lightly stained). (COURTESY OF DUNCAN MACDOUGALL.)

sumed by the muscles are derived from fat. Aerobic exercise, such as jogging, fast walking, swimming, or bicycling, is one of the best ways of reducing the body's fat content.

The ratio of fast-twitch to slow-twitch fibers varies from one particular muscle to another. For example, postural muscles in the back that are needed to allow a person to stand consist of a higher proportion of slow-twitch fibers than arm muscles used to throw or lift an object. The precise ratio of fast-twitch to slow-twitch fibers in a particular muscle is genetically determined and quite variable from person to person, allowing a particular individual to excel in certain types of physical activities. For example, world-class sprinters and weight lifters tend to have a higher proportion of fast-twitch fibers in their muscles than long-distance runners. In addition, training for sports such as weight lifting leads to a disproportionate enlargement of the fast-twitch fibers.

The muscle tissue of the heart must also increase its level of activity during vigorous exercise, but unlike skeletal muscle tissue, the heart can only produce ATP by aerobic metabolism. In fact, approximately 40 percent of the cytoplasmic space of a human heart muscle cell is occupied by ATP-producing mitochondria.

oxidation-reduction potential (or **redox potential**) relative to the potential for some standard couple. The standard couple has arbitrarily been chosen as hydrogen (H^+-H_2). As with free-energy changes, where the standard free-energy change, $\Delta G°$, was used, a similar assignment is used for redox couples. The standard redox potential, E_0, for a given couple is designated as the voltage produced by a half-cell (with only members of one couple present) in which each member of the couple is present at a standard concentration under standard conditions (as in Figure 5.10). Standard concentrations are 1.0 M for solutes and ions and 1 atm pressure for gases (e.g., H_2) at 25°C. The standard redox potential for the oxidation-reduction reaction involving hydrogen ($2H^+ + 2$ electrons

$\rightarrow H_2$) is 0.00 V. Table 5.1 gives the redox potentials of some biologically important couples. The value for the hydrogen couple in the table is not 0.00, but -0.42 V. This number represents the value when the concentration of H^+ is 10^{-7} M (pH of 7.0) rather than 1.0 M (pH of 0.0), which would be of little physiological use. When calculated at pH 7, the standard redox potential is indicated by the symbol E'_0 rather than E_0. The assignment of sign (positive or negative) to couples is arbitrary and varies among different disciplines. We will consider the assignment in the following way. Those couples whose reducing agents are better donors of electrons are assigned more negative redox potentials. For example, the standard redox potential for the NAD^+-NADH couple is -0.32 V (Table 5.1). Those couples whose oxidizing agents are better electron acceptors than NAD^+, that is, have greater affinity for electrons than NAD^+, have more positive redox potentials.

Just as any other spontaneous reaction is accompanied by a loss in free energy, so too are oxidation-reduction re-

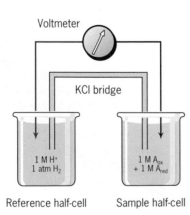

Voltmeter

KCl bridge

1 M H^+
1 atm H_2

1 M A_{ox}
+ 1 M A_{red}

Reference half-cell Sample half-cell

Figure 5.10 Measuring the standard oxidation-reduction potential. The sample half-cell contains the oxidized and reduced members of the couple, both present at 1 M. The reference half-cell contains a 1 M solution of H^+ that is in equilibrium with hydrogen gas at 1 atm pressure. An electric circuit is formed by connecting the half-cells through a voltmeter and a salt bridge. If electrons flow preferentially from the sample half-cell to the reference half-cell, then the standard redox potential (E_0) of the sample couple is negative; if the electron flow is in the opposite direction, the standard redox potential of the sample couple is positive. The salt bridge, which consists of a saturated KCl solution, provides a path for counter-ions to move between the half-cells and maintain electrical neutrality in the two compartments.

Table 5.1	Standard Redox Potentials of Selected Half-Reactions

Electrode equation	$E'_0(V)$
Succinate + CO_2 + $2H^+$ + $2e^- \rightleftharpoons$ α-ketoglutarate + H_2O	-0.670
Acetate + $2H^+$ + $2e^- \rightleftharpoons$ acetaldehyde	-0.580
$2H^+$ + $2e^- \rightleftharpoons H_2$	-0.421
α-Ketoglutarate + CO_2 + $2H^+$ + $2e^- \rightleftharpoons$ isocitrate	-0.380
Acetoacetate + $2H^+$ + $2e^- \rightleftharpoons$ β-hydroxybutyrate	-0.346
NAD^+ + $2H^+$ + $2e^- \rightleftharpoons$ NADH + H^+	-0.320
$NADP^+$ + $2H^+$ + $2e^-$ NADPH + H^+	-0.324
Acetaldehyde + $2H^+$ + $2e^- \rightleftharpoons$ ethanol	-0.197
Pyruvate + $2H^+$ + $2e^- \rightleftharpoons$ lactate	-0.185
Oxaloacetate + $2H^+$ + $2e^- \rightleftharpoons$ malate	-0.166
FAD + $2H^+$ + $2e^- \rightleftharpoons$ $FADH_2$ (in flavoproteins)	$+0.031$
Fumarate + $2H^+$ + $2e^- \rightleftharpoons$ succinate	$+0.031$
2 cytochrome $b_{K(ox)}$ + $2e^- \rightleftharpoons$ 2 cytochrome $b_{K(red)}$	$+0.030$
Ubiquinone + $2H^+$ + $2e^- \rightleftharpoons$ ubiquinol	$+0.100$
2 cythochrome C_{ox} + $2e^- \rightleftharpoons$ 2 cytochrome $c_{(red)}$	$+0.254$
2 cytochrome $a_{3(ox)}$ + $2e^- \rightleftharpoons$ 2 cytochrome $a_{3(red)}$	$+0.385$
$\frac{1}{2}O_2$ + $2H^+$ + $2e^- \rightleftharpoons H_2O$	$+0.816$

actions. The standard free-energy change during a reaction of the type

$$A_{(ox)} + B_{(red)} \rightleftharpoons A_{(red)} + B_{(ox)}$$

can be calculated from the standard redox potentials of the two couples involved in the reaction according to the equation

$$\Delta G^{\circ\prime} = -nF\Delta E'_0$$

where n is the number of electrons transferred in the reaction, F is the Faraday constant (23.063 kcal/V·mol), and $\Delta E'_0$ is the difference in volts between the standard redox potentials of the two couples. The greater the difference in standard redox potential between two couples, the farther the reaction proceeds under standard conditions to the formation of products before an equilibrium state is reached. Let's consider the reaction in which NADH, a strong reducing agent, is oxidized by molecular oxygen, a strong oxidizing agent.

$$NADH + \tfrac{1}{2}O_2 + H^+ \rightarrow H_2O + NAD^+$$

The standard redox potentials of the two couples can be written as

$$\tfrac{1}{2}O_2 + 2H^+ + 2e^- \rightarrow H_2O \qquad E'_0 = +0.82 \text{ V}$$
$$NAD^+ + 2H^+ + 2e^- \rightarrow NADH + H^+ \quad E'_0 = -0.32 \text{V}$$

The voltage change for the total reaction is equal to the difference between the two E'_0 values ($\Delta E'_0$):

$$\Delta E'_0 = +0.82 \text{ V} - (-0.32\text{V}) = 1.14\text{V}$$

which is a measure of the free energy released when NADH is oxidized by molecular oxygen under standard conditions. Substituting this value into the above equation,

$$\Delta G^{\circ\prime} = (-2)(23.063\text{kcal/V·mol})(1.14 \text{ V})$$
$$= -52.6 \text{ kcal/mol of NADH oxidized}$$

the standard free-energy difference ($\Delta G^{\circ\prime}$) is -52.6 kcal/mol. As with other reactions, the actual ΔG values depend on the relative concentrations of reactants and products (oxidized and reduced versions of the compounds) present in the cell at a given instant. Regardless, it would appear that the drop in free energy of a pair of electrons as they pass from NADH to molecular oxygen ($\Delta G^{\circ\prime} = -52.6$ kcal/mol) should be sufficient to drive the formation of several molecules of ATP ($\Delta G^{\circ\prime} = +7.3$ kcal/mol) even under conditions in the cell, where ATP/ADP ratios are much higher than those at standard conditions. The transfer of this energy from NADH to ATP within the mitochondrion occurs in a series of small, energy-releasing steps, which will be the primary topic of discussion in the remainder of the chapter.

Electrons are transferred to NAD^+ (or FAD) within the mitochondrion from several substrates of the TCA cycle, namely isocitrate, α-ketoglutarate, malate, and succinate. The first three of these intermediates have redox potentials of relatively high negative values (Table 5.1)—sufficiently high to transfer electrons to NAD^+ *under con-*

ditions that prevail in the cell.[1] In contrast, the oxidation of succinate to fumarate, which has a more positive redox potential, proceeds by the reduction of FAD, a coenzyme of greater electron affinity than NAD^+.

Electron Transport

Five of the nine reactions illustrated in Figure 5.7 are catalyzed by dehydrogenases, enzymes that transfer pairs of electrons from substrates to coenzymes. Four of these reactions generate NADH, one produces $FADH_2$. The NADH molecules, which are formed in the mitochondrial matrix, dissociate from their respective dehydrogenases and bind to NADH dehydrogenase, an integral protein of the inner mitochondrial membrane (see Figure 5.16). Unlike the other enzymes of the TCA cycle, succinate dehydrogenase, the enzyme that catalyzes the formation of $FADH_2$ (Figure 5.7, reaction 17), is a component of the inner mitochondrial membrane. In either case, the high-energy electrons associated with NADH or $FADH_2$ are transferred by NADH dehydrogenase or succinate dehydrogenase through a series of specific electron carriers that constitute the **electron-transport chain** (or **respiratory chain**) of the inner mitochondrial membrane.

Types of Electron Carriers

The electron-transport chain is composed of five types of membrane-bound electron carriers: flavoproteins, cytochromes, copper atoms, ubiquinone, and iron-sulfur proteins. With the exception of ubiquinone, all of the redox centers within the respiratory chain that accept and donate electrons are *prosthetic groups*, that is, non-amino acid components that are tightly associated with proteins.

- **Flavoproteins** consist of a polypeptide bound tightly to one of two related prosthetic groups, either flavin adenine dinucleotide (FAD) or flavin mononucleotide (FMN) (Figure 5.11a). The prosthetic groups of flavoproteins are derived from riboflavin (vitamin B_2), and each is capable of accepting and donating two protons and two electrons. The major flavoproteins of the mitochondria are NADH dehydrogenase of the electron-transport chain and succinate dehydrogenase of the TCA cycle.

- **Cytochromes** are proteins that contain *heme* prosthetic groups (such as that described for myoglobin on page 59). The iron atom of a heme undergoes reversible transition between the Fe^{3+} and Fe^{2+} oxidation states as a result of the acceptance and loss of a single electron

[1]As indicated in Table 5.1, the standard redox potential (E'_0) of the oxaloacetate-malate couple is more positive than that of the NAD^+-NADH couple. Thus, the oxidation of malate to oxaloacetate can proceed to the formation of oxaloacetate only when the ratio of products to reactants is kept below that of standard conditions. The ΔG of this reaction is kept negative by maintenance of elevated malate and/or NAD^+ levels relative to oxaloacetate or NADH levels in the region surrounding the active site of the enzyme. The situation is analogous to that for the formation of glyceraldehyde 3-phosphate from dihydroxyacetone phosphate (page 90).

Figure 5.11 **Structures of the oxidized and reduced forms of three types of electron carriers.** (*a*) FMN of NADH dehydrogenase, (*b*) the heme group of cytochrome *c*, and (*c*) ubiquinone (coenzyme Q). The heme groups of the various cytochromes of the electron-transport chain differ in the substitutions on the porphyrin rings (indicated by the blue shading) and type of linkage to the protein. Cytochromes can accept only one electron, whereas FMN and quinones can accept two electrons and two protons in successive reactions, as shown. FAD differs from FMN in having an adenosine group bonded to the phosphate.

(Figure 5.11*b*). There are at least five species of cytochromes present in the electron-transport chain, *a*, *a*$_3$, *b*, *c*, and *c*$_1$, that differ from one another by substitutions within the heme group (indicated by the blue shaded portions of Figure 5.11*b*) as well as the amino acid sequence of the polypeptide chain.

■ Three **copper atoms**, all located within a single protein of the inner mitochondrial membrane (see Figure 5.19), accept and donate a single electron as they alternate between the Cu^{+2} and Cu^{+1} oxidation states.

■ **Ubiquinone** (UQ, or coenzyme Q) is a lipid-soluble molecule containing a long hydrophobic chain composed of five-carbon isoprenoid units (Figure 5.11*c*). Like the flavoproteins, each ubiquinone is able to accept and donate two electrons and two protons. The partially reduced molecule is the free radical

ubisemiquinone, and the fully reduced molecule is ubiquinol (UQH$_2$). Ubiquinone remains within the lipid bilayer of the membrane, where it is capable of rapid lateral diffusion.

■ **Iron-sulfur proteins** are iron-containing proteins in which the iron atoms are not located within a heme group but instead are linked to inorganic sulfur atoms as part of an *iron-sulfur center*. The most common centers contain either two or four atoms of iron and sulfur—designated [2Fe-2S] and [4Fe-4S]—linked to the protein at cysteine residues (Figure 5.12). Even though a single center may have several iron atoms, the entire complex is capable of accepting and donating only a single electron. The redox potential of an iron-sulfur center depends on the hydrophobicity and charge of the amino acid residues that make up its local

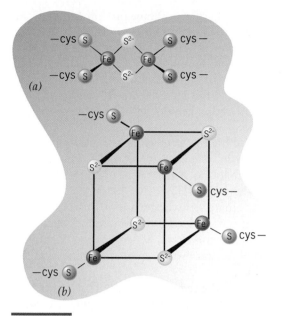

Figure 5.12 Iron-sulfur centers. Structure of a [2Fe-2S] (*a*) and a [4Fe-4S] (*b*) iron-sulfur center. Both types of iron-sulfur centers are joined to the protein by linkage to a sulfur atom (shown in orange) of a cysteine residue. Inorganic sulfide ions (S^{2-}) are shown in yellow. Both types of iron-sulfur centers accept only a single electron, whose charge is distributed among the various iron atoms.

Figure 5.13 The arrangement of several carriers in the electron-transport chain. The diagram illustrates the approximate redox potential of the carriers and the decline in free energy as electron pairs move along the respiratory chain to molecular oxygen. The numerous iron-sulfur centers are not indicated in this figure for the sake of simplicity. As discussed in the following section, each of the three electron transfers denoted by red arrows yields sufficient energy to move protons across the inner mitochondrial membrane, which in turn provides the energy required to generate ATP from ADP. (AFTER A. L. LEHNINGER, BIOCHEMISTRY, 2D ED., 1975, WORTH PUBLISHERS, NEW YORK.)

environment. As a group, iron-sulfur proteins have potentials ranging from about -700 mV to about $+300$ mV, corresponding to a major portion of the span over which electron transport occurs. Over a dozen distinct iron-sulfur centers have been identified within mitochondria.

The carriers of the electron-transport chain are arranged in order of increasing redox potential (Figure 5.13). Each carrier is reduced by the gain of electrons from the preceding carrier in the chain and is subsequently oxidized by the loss of electrons to the carrier following it. Thus, electrons are passed from one carrier to the next, losing energy as they move "downhill" along the chain. The final acceptor of this electron "bucket brigade" is O_2, which accepts the energy-depleted electrons and is reduced to water. The specific sequence of carriers that constitute the electron-transport chain was worked out by Britton Chance and coworkers at the University of Pennsylvania using a variety of inhibitors that blocked electron transport at specific sites along the route. The concept of these experiments is depicted by analogy in Figure 5.14. In each case, an inhibitor was added to cells, and the oxidation state of the various electron carriers in the inhibited cells was determined. In the case depicted in Figure 5.14, addition of antimycin A blocked electron transport at a site that left NADH, $FMNH_2$, QH_2, and cyt *b* in the reduced state, and left cytochromes *c* and *a* in the oxidized state. This result indicates that NAD, FMN, Q, and cyt *b* are situated on the "upstream" side of the block. In con-

Figure 5.14 Experimental use of inhibitors to determine the sequence of carriers in the electron-transport chain. In this hydraulic analogy, treatment of mitochondria with the inhibitor antimycin A leaves those carriers on the upstream (NADH) side of the point of inhibition in the fully reduced state and those carriers on the downstream (O_2) side of inhibition in the fully oxidized state. Comparison of the effects of several inhibitors revealed the order of the carriers within the chain. (AFTER A. L. LEHNINGER, BIOCHEMISTRY, 2D ED., 1975, WORTH PUBLISHERS, NEW YORK.)

Figure 5.15 **Electron-tunneling pathways for the yeast cytochrome c–cytochrome c peroxidase complex.** The heme group of cytochrome c is blue, and that of cytochrome c peroxidase (which is not a carrier of the mitochondrial electron-transport chain but provides an analogous electron acceptor for which high-resolution crystal structures are available) is red. Several defined pathways (yellow) exist for the movement of electrons from one heme to the other. Each of the pathways carries the electrons through several amino acid residues situated between the heme groups. (By Jeffrey J. Regan and J. N. Onuchic, from David N. Beratan et al., reprinted with permission from Science 258:1741, 1992. Copyright 1992, American Association for the Advancement of Science.)

trast, an inhibitor (e.g., rotenone) that acts between FMN and Q would leave only NADH and $FMNH_2$ in the reduced state. Thus, by identifying reduced and oxidized components in the presence of different inhibitors, the sequences of the carriers could be determined.

The tendency for electrons to be transferred from one carrier to the next depends on the potential difference between the two redox centers, but the rate of transfer depends upon the catalytic activities of the proteins involved. This distinction between thermodynamics and kinetics is similar to that discussed on page 93 regarding the activity of enzymes. Studies indicate that electrons may travel considerable distances (10–20 Å) between adjacent redox centers and that electrons probably flow through special "tunneling pathways" consisting of a series of covalent and hydrogen bonds that stretch across parts of several amino acid residues. An example of such a proposed pathway involving cytochrome c is shown in Figure 5.15.

Electron-Transport Complexes When the inner mitochondrial membrane is disrupted, the various electron carriers can be isolated as part of four distinct, asymmetric, membrane-spanning complexes, identified as complexes I, II, III, and IV (Figure 5.16). Two components of the electron-transport chain, cytochrome c and ubiquinone, are not part of any of the four complexes but exist independently within the membrane. Ubiquinone occurs as a pool of molecules dissolved in the lipid bilayer, and cy-

tochrome c is a peripheral membrane protein. Ubiquinone and cytochrome c are thought to move within or along the membrane, shuttling electrons between the large, relatively immobile protein complexes. Once within one of the large, multiprotein complexes, electrons travel along defined pathways (of the type illustrated in Figure 5.15) between adjacent redox centers whose relative positions are fixed in place.

When NADH is the electron donor, electrons enter the respiratory chain by means of complex I, which transfers electrons to ubiquinone generating ubiquinol (Figures 5.11, 5.16). When $FADH_2$ is the donor, electrons are passed from succinate dehydrogenase of the TCA cycle (which constitutes complex II) to ubiquinone, bypassing the upstream end of the chain, which has too negative a redox potential to accept the less energetic electrons of the flavin nucleotide.

If one examines the redox potentials of successive carriers in Figure 5.13, it is apparent that there are three places in which the transfer of electrons is accompanied by a major release of free energy (on the order of 200 mV). Each of these *coupling sites*, as they are called, occurs between carriers that are part of one of the three complexes, I, III, and IV. The free energy, released as electrons are passed across these three sites, is conserved by translocation of protons from the matrix across the inner membrane into the intermembrane space. These three protein complexes are often described as *proton pumps*. The translocation of protons by these electron-transporting complexes establishes the proton gradient that drives ATP synthesis. The ability of these complexes to act as independent proton-translocating units can be demonstrated by purifying each of them and incorporating them individually into artificial lipid vesicles. When given an appropriate electron donor, these protein-containing vesicles are capable of accepting electrons and using the released energy to pump protons across the vesicle membrane.

Great strides have been made in the past few years in elucidating the molecular architecture of all the protein complexes of the inner membrane. Researchers are no longer asking what these proteins look like but are using the wealth of new structural information to understand how they work. We will briefly examine the mammalian versions of each of the four electron-transport complexes, which together contain about 70 different polypeptides. The bacterial versions are considerably simpler than their mammalian counterparts and contain far fewer subunits. The additional subunits of the mammalian complexes do not contain redox centers and are thought to function in either the regulation or assembly of the complex rather than electron transport.

Complex I (or NADH Dehydrogenase) Complex I is the gateway to the electron-transport chain, catalyzing the transfer of a pair of electrons from NADH to ubiquinone (UQ) to form ubiquinol (UQH$_2$). The mammalian version of complex I is a huge L-shaped conglomerate, con-

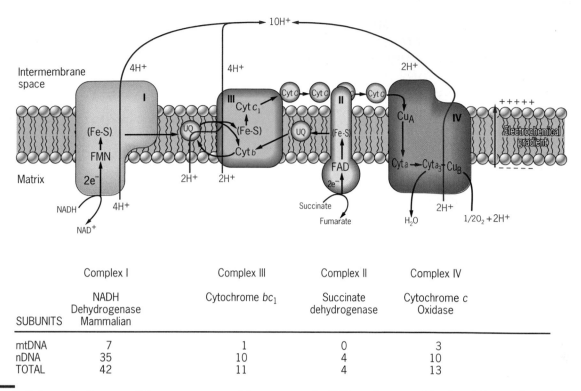

	Complex I	Complex III	Complex II	Complex IV
SUBUNITS	NADH Dehydrogenase Mammalian	Cytochrome bc_1	Succinate dehydrogenase	Cytochrome c Oxidase
mtDNA	7	1	0	3
nDNA	35	10	4	10
TOTAL	42	11	4	13

Figure 5.16 **Schematic diagram of the components of the electron-transport chain within the inner mitochondrial membrane.** The respiratory chain consists of four complexes of electron carriers and two other carriers (ubiquinone and cytochrome c) that are independently disposed within the membrane. Electrons enter the chain from either NADH (via complex I) or $FADH_2$ (a part of complex II). Electrons are passed from either complex I or II to ubiquinone (UQ), which exists as a pool within the lipid bilayer. Electrons are subsequently passed from reduced ubiquinone (ubiquinol) to complex III and then to the peripheral protein cytochrome c, which is thought to be mobile. Electrons are transferred from cytochrome c to com- plex IV (cytochrome oxidase) and then to O_2 to form H_2O. The sites of proton translocation from the matrix to the cytosol are indicated. The precise number of protons translocated at each site remains controversial; the number indicated is a general consensus. Keep in mind that the proton numbers shown are those generated by each pair of electrons transported, which is sufficient to reduce only one-half of an O_2 molecule. (The translocation of protons by complex III occurs by way of a Q cycle, whose details are shown in Figure 4 of the Experimental Pathway. The Q cycle can be divided into two steps, each of which leads to the release of two protons into the cytosol.)

taining 42 different subunits and representing a molecular mass of nearly one million daltons. Seven of the subunits—all hydrophobic, membrane-spanning polypeptides—are encoded by mitochondrial genes and are homologues of bacterial polypeptides. Complex I includes an FMN-containing flavoprotein that oxidizes NADH, at least six distinct iron-sulfur centers, and two bound molecules of ubiquinone. The passage of a pair of electrons through complex I is thought to be accompanied by the movement of four protons from the matrix into the intermembrane space.

Complex II (or Succinate Dehydrogenase) Complex II consists of several polypeptides, two of which make up succinate deyhdrogenase, a membrane-bound, FAD-containing enzyme that catalyzes a key reaction of the TCA cycle. Complex II provides a pathway for feeding lower-energy electrons (ones close to 0 mV) from succinate to FAD to ubiquinone forming ubiquinol. Electron transfer through complex II is not accompanied by proton translocation.

Complex III (or Cytochrome bc_1) Complex III catalyzes the transfer of electrons from ubiquinol to cytochrome c. Experimental determinations suggest that four protons are pumped across the membrane for every pair of electrons transferred through complex III. The protons are released into the intermembrane space in two separate steps powered by the energy released as a pair of electrons are separated from one another and passed along different pathways through the complex. Two protons are derived from the molecule of ubiquinol that entered the complex. Two additional protons are removed from the matrix and translocated across the membrane as part of a second molecule of ubiquinol. These steps are described in further detail in Figure 4 of the Experimental Pathways, which can be seen on the Web at the following address: www.wiley.com/college/karp. Three of the subunits of complex III contain redox groups: cytochrome b with two different heme groups, cytochrome c_1, and an iron-sulfur protein. Cytochrome b is the only polypeptide of the complex encoded by a mitochondrial gene.

Complex IV (Cytochrome c Oxidase) The final step in electron transport in a mitochondrion is the successive transfer of electrons from reduced cytochrome *c* to oxygen according to this reaction:

$$2 \text{ cyt } c^{2+} + 2 \text{ H}^+ + \tfrac{1}{2} \text{O}_2 \rightarrow 2 \text{ cyt } c^{3+} + \text{H}_2\text{O}$$

To reduce a whole molecule of O_2,

$$4 \text{ cyt } c^{2+} + 4 \text{ H}^+ + \text{O}_2 \rightarrow 4 \text{ cyt } c^{3+} + 2 \text{ H}_2\text{O}$$

The reduction of O_2 is catalyzed by complex IV, a huge assembly of polypeptides referred to as **cytochrome oxidase**. Cytochrome oxidase was the first component of the electron transport chain shown to act as a proton pump. This was demonstrated in the experiment depicted in Figure 5.17 in which the purified enzyme was incorporated into vesicles containing an artificial lipid bilayer (liposomes). Addition of reduced cytochrome *c* to the medium was accompanied by the expulsion of H^+ ions from the vesicles, which was measured as a drop in the surrounding pH. Whether inside a liposome or the inner mitochondrial membrane, translocation of protons is coupled to conformational changes generated by the release of energy that accompanies the transfer of electrons. For every molecule of O_2 reduced by cytochrome oxidase, eight protons are thought to be taken up from the matrix. Four of these protons are consumed in the formation of two molecules of water as indicated above, and the other four protons are translocated across the membrane and released into the intermembrane space (Figure 5.16). Consequently, the overall reaction can be written:

$$4 \text{ cyt } c^{2+} + 8 \text{ H}^+ \text{ matrix} + \text{O}_2 \rightarrow 4 \text{ cyt } c^{3+} + 2 \text{ H}_2\text{O} + 4\text{H}^+ \text{ outside}$$

It can be noted that a number of potent respiratory poisons, including carbon monoxide (CO), azide (N_3^-), and cyanide (CN^-), have their toxic effect by binding to the catalytic site of cytochrome oxidase. (Carbon monoxide also binds to the heme group of hemoglobin.)

A Closer Look at Cytochrome Oxidase In 1995, the three-dimensional structure of beef heart cytochrome oxidase was determined at 2.8 Å resolution. This enormous molecule consists of 13 subunits having a total molecular mass of 204,000 daltons, as depicted in Figure 5.18. The three largest polypeptides of the complex are encoded by the mitochondrial genome and contain all four of the protein's redox centers.

The mechanism of electron transfer through complex IV has been intensively studied, most notably in the laboratories of Mårten Wikström at the University of Helsinki and Gerald Babcock at Michigan State University. A major challenge for investigators is to explain how carriers that are only able to transfer single electrons can reduce a molecule of O_2 to two molecules of H_2O, a process that requires four electrons (together with four protons). Most importantly, the process must occur very efficiently because the cell is dealing with very dangerous substances; the "accidental" release of partially reduced oxygen species has the potential to damage virtually every macromolecule in the cell (page 35).

Figure 5.17 **Experimental demonstration that cytochrome oxidase is a proton pump.** When purified cytochrome oxidase is incorporated into the artificial bilayer of a liposome, the medium becomes acidified following the addition of reduced cytochrome *c*. This indicates that as electrons are transferred from cytochrome *c* to cytochrome oxidase, and O_2 is being reduced to water, protons are being translocated from the compartment within the vesicle to the external medium. This experiment was originally carried out in the late 1960s by Mårten Wikström and colleagues. (REPRINTED WITH PERMISSION FROM M. I. VERKHOVSKY, ET AL. NATURE 400:481, 1999. COPYRIGHT 1999 MACMILLAN MAGAZINES LIMITED.)

The movement of electrons between the redox centers of cytochrome oxidase is shown in Figure 5.19. Electrons are transferred one at a time from cytochrome *c* through a bimetallic copper center (Cu_A) of subunit II to a

Figure 5.18 **Three-dimensional structure of beef heart cytochrome oxidase.** As shown here, the protein consists of a dimer in which each monomer is composed of 13 different subunits. (BY SHINYA YOSHIKAWA, FROM T. TSUKIHARA, ET AL., REPRINTED WITH PERMISSION FROM SCIENCE 272:1136, 1996. COPYRIGHT 1996 AMERICAN ASSOCIATION FOR THE ADVANCEMENT OF SCIENCE.)

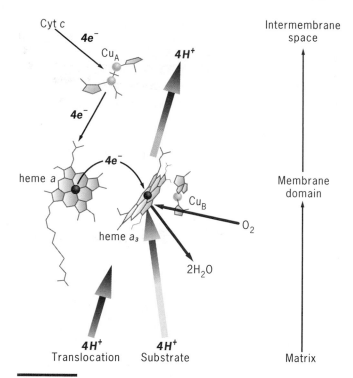

Cyt c

Cu$_A$

4e$^-$

4e$^-$

4e$^-$

heme a

heme a$_3$

Cu$_B$

O$_2$

2H$_2$O

Intermembrane space

4H$^+$

Membrane domain

Matrix

4H$^+$
Translocation

4H$^+$
Substrate

Figure 5.19 **The mechanism of action of cytochrome oxidase.** A model showing the flow of electrons through the four redox centers of cytochrome oxidase. Iron atoms are shown as red spheres, copper atoms as yellow spheres. Electrons are thought to be passed one at a time from cytochrome *c* to the dimeric copper center (Cu$_A$), then on to the heme group (Fe$_a$) of cytochrome *a*, then on to the binuclear redox center consisting of a second iron (of the heme group of cytochrome *a$_3$*) and copper ion (Cu$_B$). The structures and suggested orientations of the redox centers are indicated. (FROM M. WIKSTRÖM, ET AL., BIOCHIM. BIOPHYS. ACTA 1459:515, 2000.)

heme (heme *a*) of subunit I. From there, electrons are passed to a redox center located in subunit I that contains a second heme (heme *a$_3$*) and a second copper atom (Cu$_B$) situated less than 5 Å apart. Acceptance of the first two electrons reduces the *a$_3$*-Cu$_B$ *binuclear center* according to the reaction

$$Fe_{a_3}^{3+} + Cu_B^{2+} + 2e^- \rightarrow Fe_{a_3}^{2+} + Cu_B^{1+}$$

Once the binuclear center accepts its second electron, an O$_2$ molecule binds to the center and then accepts the pair of electrons, forming a peroxy anion O$_2^{2-}$.

$$Fe^{2+}-O \atop \phantom{Fe^{2+}-}O \, Cu^{1+} \quad \rightarrow \quad Fe^{3+}-O^- \atop \phantom{Fe^{3+}-}O^- \, Cu^{2+}$$

The reactive peroxy ion is the highest energy component in the reaction sequence and reacts quickly to extract a third electron, either from the heme itself or from a nearby amino acid residue. At the same time, the binuclear center accepts two protons from the matrix, cleaving the covalent O—O bond and reducing one of the O atoms.

$$Fe^{4+} = O^{2-} \quad Cu^{2+}-OH_2$$

The passage of a fourth electron and the entry of two additional protons from the matrix leads to the formation of two molecules of water:

$$Fe^{3+} \quad Cu^{2+} + 2 H_2O$$

For each proton removed from the matrix, an excess negative charge (in the form of an OH$^-$) is left behind, thus contributing directly to the electrochemical gradient across the inner mitochondrial membrane.

As noted above, protons taken up from the matrix are utilized in two very different ways. For every molecule of O$_2$ reduced to 2 H$_2$O by cytochrome oxidase: (1) four H$^+$ ions are consumed in this chemical reaction and (2) four additional H$^+$ ions are translocated across the inner mitochondrial membrane. The first four protons can be referred to as "substrate" protons and the latter four as "pumped" protons. Movement of both groups of protons contributes to the electrochemical gradient across the membrane.

With the publication of the three-dimensional crystal structure of cytochrome oxidase, it became possible to search for pathways through which substrate and pumped protons might move through the huge protein. Unlike other ions (e.g., Na$^+$ or Cl$^-$) that have to diffuse by themselves across the entire distance being traversed, H$^+$ ions can "move" through a channel by exchanging themselves with other protons present along the pathway. Such proton-conduction pathways can be identified because they consist of strings of acidic residues, hydrogen-bonded residues, and trapped water molecules. Researchers have identified at least two potential proton conduits—denoted as K and D channels—within the molecule. Unfortunately, static structural models, by themselves, can't reveal the dynamic movements that take place within the protein as it functions. As a result, considerable controversy remains as to how the energy released by O$_2$ reduction is coupled to proton translocation, whether or not translocated protons move through the vicinity of the binuclear center on their way from the matrix to the intermembrane space, and at what stages in the catalytic cycle that various proton movements occur.

REVIEW

1. Describe the steps by which the transport of electrons down the respiratory chain leads to the formation of a proton gradient.

2. Of the five types of electron carriers, which has the smallest molecular mass? Which has the greatest ratio of iron atoms to electrons carried? Which has a component that is located outside the lipid bilayer? Which are capable of accepting protons and electrons and which only electrons?

3. Describe the relationship between a compound's affinity for electrons and its ability to act as a reducing agent. What is the relationship between the

electron-transfer potential of a reducing agent and the ability of the other member of its couple to act as an oxidizing agent?

4. Look at Figure 5.11 and describe how the semi-quinone states of ubiquinone and FMN are similar.

5. What is meant by the binuclear center of cytochrome oxidase? How does it function in the reduction of O_2?

6. What are two different ways that cytochrome oxidase contributes to the proton gradient?

7. Why do some electron transfers result in a greater release of energy than other transfers?

5·4 TRANSLOCATION OF PROTONS AND THE ESTABLISHMENT OF A PROTON-MOTIVE FORCE

We have seen that the free energy released during election transport is utilized to move protons from the matrix to the intermembrane space. Translocation of protons across the inner membrane is electrogenic (i.e., voltage producing) because it results in a greater number of positive charges in the intermembrane space and cytosol and a greater number of negative charges within the matrix. Thus, there are two components of the proton gradient to consider. One of the components is the concentration difference between hydrogen ions on one side of the membrane versus the other side; this is a pH gradient (ΔpH). The other component is the voltage (Ψ) that results from the separation of charge across the membrane. A gradient that has both a concentration (chemical) and an electrical (voltage) component is an *electrochemical gradient* (page 151). The energy present in both components of the proton electrochemical gradient can be combined and expressed as the **proton-motive force (Δp)**, which is measured in millivolts. Thus,

$$\Delta p = \psi - 2.3\,(RT/F)\,\Delta pH$$

Because 2.3 RT/F is equal to 59 mV at 25°C, the equation[2] can be rewritten as

$$\Delta p = \psi - 59\,\Delta pH$$

The contribution to the proton-motive force made by the electric potential versus the pH gradient depends on the permeability properties of the inner membrane. For example, if the outward movement of protons during electron transport is accompanied by negatively charged chloride ions, then the electric potential (Ψ) is reduced without affecting the proton gradient (ΔpH). Measurements made in various laboratories suggest that actively respiring

mitochondria generate a proton-motive force of approximately 220 mV across their inner membrane. In mammalian mitochondria, approximately 80 percent of the free energy of Δp is represented by the voltage component and the other 20 percent by the proton concentration difference (approximately 0.5 to 1 pH unit difference). If the proton concentration difference were much greater than this it would likely affect the activity of cytoplasmic enzymes. The transmembrane voltage across the inner mitochondrial membrane can be visualized by the use of positively charged, lipid-soluble dyes that distribute themselves across membranes in proportion to the electric potential (Figure 5.20).

When cells are treated with certain lipid-soluble agents, most notably 2,4-dinitrophenol (DNP), they con-

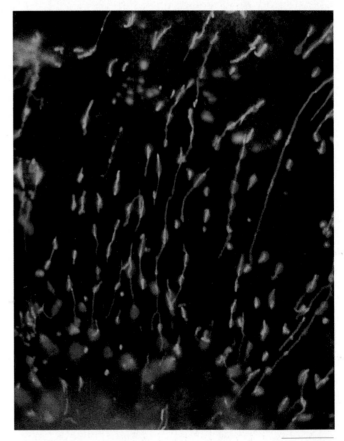

7 µm

Figure 5.20 Visualizing the proton-motive force. Fluorescence micrograph of a cultured cell stained with the fluorescent, cationic compound JC-1. When the cell is active, the voltage generated across the inner mitochondrial membrane (inside negative) leads to the accumulation of the lipid-soluble substance within the mitochondria, causing the organelles to fluoresce. At high membrane potentials, JC-1 forms aggregates, which changes its fluorescence properties. In the monomeric state, the dye fluoresces green, whereas in the aggregate state the dye fluoresces orange. The cell is seen to contain both orange- and green-fluorescing organelles, which suggests the presence of two different populations of mitochondria having different membrane potentials. (COURTESY OF LAN BO CHEN.)

[2]In other words, a difference of one pH unit, which represents a tenfold difference in H^+ concentration across the membrane, is equivalent to a potential difference of 59 millivolts, which is equivalent to a free-energy difference of 1.37 kcal/mol.

tinue to oxidize substrates without being able to generate ATP. In other words, DNP *uncouples* glucose oxidation and ADP phosphorylation. Because of this property, DNP has become widely used in the laboratory to inhibit ATP formation. During the 1920s, a few physicians even prescribed DNP as a diet pill. When exposed to the drug, the cells of obese patients continued to oxidize their stores of fat in a vain attempt to maintain normal ATP levels. The practice ceased with the deaths of a number of patients taking the drug. With the formulation of the chemiosmotic theory and the determination that mitochondria generate a proton gradient, the mechanism of DNP action became understood. This drug can uncouple oxidation and phosphorylation because it combines with protons, and because of its lipid solubility, it carries the protons across the inner mitochondrial membrane down their electrochemical gradient.

The maintenance of a proton-motive force requires that the inner mitochondrial membrane remain highly impermeable to protons. If not, the gradient established by electron transport is immediately dissipated by the diffusion of protons back into the matrix, leading to the release of energy as heat. It came as a surprise to discover that the inner mitochondrial membrane of certain cells contain proteins that act as a natural (endogenous) uncoupler. These proteins, called *uncoupling proteins* (or *UCPs*), are especially abundant in brown adipose tissue of mammals, which functions as a source of heat production. Adult humans are essentially devoid of brown adipose tissue, but UCPs are present in significant levels in the mitochondria of human skeletal muscle cells. It has been suggested that UCPs play a key role in determining a person's basal metabolic rate, which may help explain the physiological basis of obesity. Two persons having markedly different levels of UCP activity might ingest and oxidize the same amount of food, but the individual with the higher level of UCP would be expected to convert a greater percentage of the released energy into heat as opposed to fat. This notion has received dramatic support from a 2000 study in which mice were genetically engineered to produce high levels of a human UCP. Although these mice took in from 25 to 50 percent more food calories than control animals, they had approximately 50 percent less fat tissue as a result of an increased rate of energy consumption. These findings have led to a search for weight-loss drugs that act by stimulating UCP activity.

The experiments that led to the acceptance of the proton-motive force as an intermediate in the formation of ATP are discussed in the Experimental Pathways, which can be found at www.wiley.com/college/karp on the Web.

REVIEW ?

1. What are the two components of the proton-motive force, and how can their relative contribution vary from one cell to another?

2. What is the effect of dinitrophenol on ATP formation by mitochondria? Why is this the case?

5.5 THE MACHINERY FOR ATP FORMATION

Now that we have seen how the transport of electrons generates a proton electrochemical gradient across the inner mitochondrial membrane, we can turn to the molecular machinery that utilizes the energy stored in this gradient to drive the phosphorylation of ADP.

During the early 1960s, Humberto Fernandez-Moran of Massachussetts General Hospital was examining isolated mitochondria by the recently developed technique of negative staining. Fernandez-Moran discovered a layer of spheres attached to the inner (matrix) side of the inner membrane, projecting from the membrane and attached to it by stalks (Figure 5.21). A few years later, Efraim Racker of Cornell University isolated the inner membrane spheres, which he called **coupling factor 1**, or simply F_1. Racker discovered that the F_1 spheres behaved like an enzyme that hydrolyzed ATP, that is, an ATPase. At first glance this seems like a peculiar finding. Why should mitochondria possess a predominant enzyme that hydrolyzes the substance it is supposed to produce?

If one considers that the hydrolysis of ATP is the reverse reaction from its formation, the function of the F_1 sphere becomes more evident; it contains the catalytic site at which ATP formation normally occurs. Recall that

10 nm

Figure 5.21 The machinery for ATP synthesis. Electron micrograph of a small portion of a beef heart mitochondrion that has been air-dried and negatively stained. At magnifications of approximately one-half million, spherical particles are seen (arrow) attached by a thin stalk to the inner surface of the cristae membranes. (FROM HUMBERTO FERNANDEZ-MORAN ET AL., J. CELL BIOL. 22:71, 1964, BY COPYRIGHT PERMISSION OF THE ROCKEFELLER UNIVERSITY PRESS.)

1. Enzymes do not affect the equilibrium constant of the reaction they catalyze.

2. Enzymes are capable of catalyzing both the forward and reverse reactions.

Consequently, the direction of an enzyme-catalyzed reaction at any given time depends on the prevailing conditions. This is nicely shown in experiments with other ATPases, such as the Na^+/K^+-ATPase of the plasma membrane (page 159). When this enzyme was discussed in Chapter 4, it was described as an enzyme that used the energy obtained from ATP hydrolysis to export Na^+ and import K^+ against their respective gradients. In the cell this is the only function of the enzyme. However, under experimental conditions, this enzyme can catalyze the formation of ATP rather than its hydrolysis (Figure 5.22). To obtain such conditions, red blood cell ghosts (page 147) are prepared with a *very* high internal K^+ concentration and a *very* high external Na^+ concentration, higher than normally exists in the body. Under these conditions, K^+ moves out of the "cell" and Na^+ moves into the "cell." Both ions are moving down their respective gradients, rather than against them as would normally occur in a living cell. If ADP and P_i are present within the ghost, the movement of the ions causes ATP to be synthesized rather than hydrolyzed. Experiments such as this illustrate the reality of what would be expected on the basis of the theoretical reversibility of enzyme-catalyzed reactions. They also illustrate how an ionic gradient can be used to drive a reaction in which ADP is phosphorylated to ATP, which is precisely what occurs in the mitochondrion. The driving force is the proton-motive force established by electron transport.

The Structure of ATP Synthase

Although the F_1 sphere is the catalytic portion of the enzyme that manufactures ATP in the mitochondrion, it is not the whole story. The ATP-synthesizing enzyme, which is called **ATP synthase**, is a mushroom-shaped protein complex (Figure 5.23*a*) composed of two principal components: a spherical F_1 head (about 90 Å diameter) and a basal section, called F_0, embedded in the inner membrane. Recent high-resolution electron micrographs suggest that the two portions are connected by both a central and a peripheral stalk as shown in Figure 5.23*b*. A typical mammalian liver mitochondrion has roughly 15,000 copies of the ATP synthase. Homologous versions of the ATP synthase are found in the plasma membrane of aerobic bacteria, the thylakoid membrane of plant chloroplasts, and the inner membrane of mitochondria.

The F_1 portions of the bacterial and mitochondrial ATP synthases have similar structures; both contain five different polypeptides with a stoichiometry of $\alpha_3\beta_3\delta\gamma\epsilon$. The α and β subunits are arranged alternately within the F_1 head in a manner resembling the segments of an orange (Figure 5.23*b*; see also 5.25*b*). Two aspects can be noted

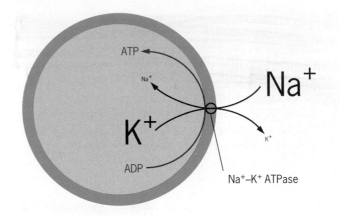

Figure 5.22 An experiment to drive ATP formation in membrane vesicles reconstituted with the Na^+/K^+-ATPase. By causing these vesicles to have very high internal K^+ concentration and very high external Na^+ concentration, the reaction is made to run in the opposite direction from that of its normal course in the plasma membrane. In the process, ATP is formed from ADP and P_i. The size of the letters indicates the direction of the concentration gradients.

for later discussion: (1) each F_1 contains three catalytic sites for ATP synthesis, one on each β subunit, and (2) the γ subunit runs from the outer tip of the F_1 head through the central stalk and makes contact with the F_0 basepiece. In the mitochondrial enzyme, all five polypeptides of F_1 are encoded by the nuclear DNA, synthesized in the cytosol, and imported posttranslationally (discussed in Section 8.9).

The F_0 portion of the bacterial ATP synthase resides within the outer cell membrane and consists of three different polypeptides with an apparent stoichiometry of ab_2c_{12} (Figure 5.23*b*).[3] In mitochondria, a more complex F_0 base is embedded in the inner mitochondrial membrane and contains a channel through which protons are conducted from the intermembrane space to the matrix. The presence of a channel in the F_0 base is best demonstrated by experiments in which the inner mitochondrial membrane is broken into fragments that form membrane vesicles, called *submitochondrial particles* (Figure 5.24*a*). Intact submitochondrial particles, which contain the ATP synthase embedded in the vesicle membrane, are capable of oxidizing substrates, generating a proton gradient, and synthesizing ATP (Figure 5.24*b*). If, however, the F_1 spheres are removed from the particles by treatment with urea, the vesicle membrane can no longer maintain a proton gradient despite continuing substrate oxidation and electron transport. Protons that are

[3] The number of c subunits in the F_0 base has been controversial, ranging from 9 to 14. An X-ray crystallographic study of the yeast enzyme suggests there are 10 copies in that species, but for the present purpose we will assume there are 12 copies as is generally accepted for the bacterial and mammalian enzymes. See S. J. Ferguson, *Curr. Biol. 10*:R804, 2000, for discussion of importance of c copy number.

Figure 5.23 **The structure of the ATP synthase.** (*a*) Electron micrograph of the isolated rat liver ATP synthase. (*b*) Schematic diagram of the bacterial ATP synthase. The enzyme consists of two major portions, called F_1 and F_0. The F_1 head consists of five different subunits in the ratio $3\alpha : 3\beta : 1\delta : 1\gamma : 1\epsilon$. The α and β subunits are organized in a circular array to form the spherical head of the particle; the γ subunit runs through the core of the ATP synthase from the tip of F_1 down to F_0 to form a central stalk; the ϵ subunit helps attach the γ subunit to the F_0 base. The F_0 base, which is embedded in the membrane, consists of three different subunits in the apparent ratio $1a : 2b : 12c$. As discussed later: the c subunits are thought to form a movable ring within the membrane; the paired b subunits of the F_0 base and the δ subunit of the F_1 head form a peripheral stalk that holds the α/β subunits in a fixed position; and the a subunit contains the proton channel that allows protons to traverse the membrane. (A: FROM J. W. SOPER, G. L. DECKER, AND P. L. PEDERSEN, J. BIOL. CHEM. 254:11173, 1979.)

translocated across the membrane during electron transport simply cross back through the "beheaded" ATP synthase, and the energy is dissipated.

The Basis of ATP Formation According to the Binding Change Mechanism

How does a proton electrochemical gradient provide the energy required to drive the synthesis of ATP? To answer this question, Paul Boyer of UCLA published an innovative hypothesis in 1979, called the **binding change mechanism,** which has since gained wide acceptance. Overall, the binding change hypothesis has several distinct components, which we will discuss in sequence.

1. The energy released by the movement of protons is not used to drive ADP phosphorylation directly but principally to change the binding affinity of the active site for the ATP product. We are used to thinking of cellular reactions occurring in an aqueous environment in which the concentration of water is 55 M and the reactants and products are simply dissolved in the medium.

Under these conditions, energy is required to drive the formation of the covalent bond linking ADP with inorganic phosphate to form ATP. It has been demonstrated, however, that once ADP and P_i are bound within the catalytic site of the ATP synthase, the two reactants readily condense to form a tightly bound ATP molecule without the input of additional energy. In other words, even though the reaction

$$\text{soluble ADP} + \text{soluble P}_i \rightarrow \text{soluble ATP} + H_2O$$

may require the input of considerable energy (7.3 kcal/mol under standard conditions), the reaction

$$\text{enzyme-bound ADP} + \text{enzyme-bound P}_i \rightarrow$$
$$\text{enzyme-bound ATP} + H_2O$$

has an equilibrium constant close to 1 ($\Delta G° = 0$) and thus can occur spontaneously without the input of energy (page 89). This does not mean that ATP can be formed from ADP without energy expenditure. Instead, the energy is required for the release of the tightly bound product from the catalytic site, rather than the phosphorylation event itself.

(a) 50 nm (b)

Figure 5.24 **ATP formation in experiments with submito-chondrial particles.** (*a*) Electron micrograph of submitochondrial particles, which are fragments of the inner mitochondrial membrane that have become closed vesicles with the F_1 spheres projecting outward into the medium. (*b*) Outline of an experiment showing that intact submitochondrial particles are capable of substrate oxidation, proton gradient formation (indicated by the separation of charge across the membrane), and ATP formation. In contrast, submitochondrial particles lacking the F_1 head are capable of substrate oxidation but are not able to maintain a proton gradient (indicated by the lack of charge separation), and thus they are unable to form ATP. (A: COURTESY OF EFRAIM RACKER.)

2. Each active site progresses successively through three distinct conformations that have different affinities for substrates and product. Recall that the F_1 complex has three catalytic sites, one on each of the three β subunits. When investigators examined the properties of the three catalytic sites in a *static* enzyme (one that was not engaged in enzymatic turnover), the different sites exhibited different chemical properties. Boyer proposed that, at any one time, the three catalytic sites are present in different conformations, which causes them to have different affinity for nucleotides. At any given instant, one site is in the "loose" or L conformation in which ADP and P_i are loosely bound; a second site is in the "tight" or T conformation in which nucleotides (ADP + P_i substrates, or ATP product) are tightly bound; and the third site is in the "open" or O conformation, which, because it has a very low affinity for nucleotides, allows release of ATP.

Although there were differences among the three catalytic sites in a static enzyme, Boyer obtained evidence that all of the ATP molecules produced by an active enzyme were synthesized by the same catalytic mechanism. It seemed, in other words, that all three of the catalytic sites of the enzyme were operating identically. To explain the apparent contradiction between the asymmetry of the enzyme's structure and the uniformity of its catalytic mechanism, Boyer proposed that each of the three catalytic sites passed sequentially through the same L, T, and O conformations (see Figure 5.26).

3. ATP is synthesized by rotational catalysis in which one part of the ATP synthase rotates relative to another part. To explain the sequential changes in the conformation of each of the catalytic sites, Boyer proposed that the α and β subunits, which form a hexagonal ring of subunits within the F_1 head (Figure 5.23), rotate *relative* to the central stalk. In this model, which is referred to as **rotational catalysis**, rotation is driven by the movement of protons through the membrane via the channel in the F_0 base.

Evidence to Support the Binding Change Mechanism and Rotary Catalysis The publication in 1994 of a detailed atomic model of the F_1 head by John Walker and his colleagues of the Medical Research Council in England provided a remarkable body of structural evidence to support Boyer's binding change mechanism. First, it revealed the structure of each of the catalytic sites in a static enzyme, confirming that they differed in their conformation and their affinity for nucleotides. Structures corresponding to the L, T, and O conformations were identified in the cat-

(a)

(b)

Figure 5.25 The structural basis of catalytic site conformation. (*a*) A section through the F$_1$ head shows the spatial organization of three of its subunits. The α-helical γ subunit is seen to project into the central cavity of the F$_1$, and between the α and β subunits on each side. The conformation of the catalytic site of the β subunit (shown on the left) is determined by its contact with the γ subunit. (*b*) A top view of the F$_1$ head showing the arrangement of the six α and β subunits (shown in

red and yellow) around the asymmetric γ subunit (shown in blue). The γ subunit is in position to rotate relative to the surrounding subunits. It is also evident that the γ subunit makes contact in a different way with each of the three β subunits, inducing each of them to adopt a different conformation. (REPRINTED WITH PERMISSION FROM J. P. ABRAHAMS, ET AL., COURTESY OF JOHN E. WALKER, NATURE 370:624, 627, 1994. COPYRIGHT 1994, MACMILLAN MAGAZINES LIMITED.)

alytic sites of the three β subunits. Secondly, it revealed that the γ subunit of the enzyme was perfectly positioned within the ATP synthase to transmit conformational changes from the F$_0$ membrane sector to the F$_1$ catalytic sites. The γ subunit could be seen to extend as a shaft from the F$_0$ sector through the stalk and into a central cavity within the F$_1$ sphere (Figure 5.25*a*), where it contacted each of the three β subunits differently (Figure 5.25*b*).

The apical end of the γ subunit is highly asymmetric, and at any given instant, different parts of the γ subunit interact with the different β subunits causing them to adopt different (L, T, and O) conformations. As it rotates, each binding site on the γ subunit interacts successively with the three β subunits of F$_1$. During a single catalytic cycle, the γ subunit rotates a full 360°, causing each catalytic site to pass sequentially through the L, T, and O conformations. This mechanism is illustrated schematically in Figure 5.26 and discussed in detail in the accompanying legend. As indicated in Figure 5.26*a*, condensation of ADP and P$_i$ to form ATP occurs while each subunit is in the T conformation. As indicated in Figure 5.23*b*, the ε subunit is associated with the "lower" portion of the γ subunit and the two subunits probably rotate together as a unit.

Direct evidence for the rotation of the γ subunit relative to the αβ subunits has been obtained in a variety of

experiments. If "seeing is believing," then the most convincing demonstration has come from the work of Masasuke Yoshida and colleagues at the Tokyo Institute of Technology and Keio University in Japan. These researchers devised an ingenious system to watch the enzyme catalyze the reverse reaction from that normally operating in the cell. First, they prepared a genetically engineered version of the working portion of the ATP synthase, consisting of 3 α subunits, 3 β subunits, and a γ subunit (α$_3$β$_3$γ) (Figure 5.27). This polypeptide complex was then fixed to a glass coverslip by its head, and a short, fluorescently labeled actin filament was attached to the end of the γ subunit that projected into the medium (Figure 5.27). When the preparation was incubated with ATP and observed under the microscope, the fluorescently labeled actin filaments were seen to rotate like microscopic propellors, powered by the energy released as ATP molecules were bound and hydrolyzed by the catalytic sites of the β subunits. Calculations of the energy expended and work performed suggest that nearly 100 percent of the chemical energy released by ATP hydrolysis is converted into mechanical energy used to turn the attached actin filament. These experiments demonstrate unequivocally that ATP synthase operates as a highly efficient rotary motor, the smallest such motor known to exist in a living organism.

(a)

(b)

Figure 5.26 The binding change mechanism for ATP synthesis. (*a*) Schematic drawing showing changes in a single catalytic site during a cycle of catalysis. At the beginning of the cycle, the site is in the open (O) conformation and substrates ADP and P$_i$ are entering the site. In step 1, the movement of protons through the membrane induces a shift to the loose (L) conformation in which the substrates are loosely bound. In step 2, the movement of additional protons induces a shift to the tight (T) conformation, in which the affinity for substrates increases, causing them to be tightly bound to the catalytic site. In step 3, the tightly bound ADP and P$_i$ spontaneously condense to form a tightly bound ATP; no change in conformation is required for this step. In step 4, the movement of additional protons induces a shift to the open (O) conformation, in which the affinity for ATP is greatly decreased, allowing the product to be released from the site. Once the ATP has dissociated, the catalytic site is available for substrate binding, and the cycle is repeated. (*b*) Schematic drawing showing changes at all three catalytic sites of the enzyme simultaneously. The movement of protons through the F$_0$ portion of the enzyme causes the rotation of the asymmetric γ subunit, which displays three different faces to the catalytic subunits. As the γ subunit rotates, it induces changes in the conformation of the catalytic site of the β subunits, causing each catalytic site to pass successively through the T, O, and L conformations.

Coverslip coated with Ni-NTA

Figure 5.27 Direct observation of rotational catalysis. To carry out the experiment, a modified version of a portion of the ATP synthase consisting of α$_3$β$_3$γ was prepared. Each β subunit was modified to contain 10 histidine residues at its N-terminus, a site located on the outer (matrix) face of the F$_1$ head. The side chains of histidine have a high affinity for a substance (Ni-NTA), which was used to coat the coverslip. The γ subunit was modified by replacing one of the serine residues near the end of the stalk with a cysteine residue, which provided a means to attach the fluorescently labeled actin filament. In the presence of ATP, the actin filament was observed to rotate counterclockwise (when viewed from the membrane side). At low ATP concentrations, the actin filaments could be seen to rotate in steps of 120°. (REPRINTED WITH PERMISSION FROM H. NOJI ET AL., COURTESY OF MASASUKE YOSHIDA, NATURE 386:300, 1997. COPYRIGHT 1997, MACMILLAN MAGAZINES LIMITED.)

Using the Proton Gradient to Drive the Catalytic Machinery: The Role of the F$_0$ Portion of ATP Synthase By 1997, a detailed understanding of the working machinery of the F$_1$ complex was in hand, but major questions about the structure and function of the membrane-bound F$_0$ portion of the enzyme remained to be answered. Most important among them were: What is the path taken by protons as they move through the F$_0$ complex and how does this movement lead to the synthesis of ATP? These ques-

tions have been the focus of the past few years of study of this remarkable enzyme. It had been presumed for several years that:

1. The c subunits of the F$_0$ base were assembled into a ring that resides within the lipid bilayer (as in Figure 5.23*b*),

2. The c ring is physically bound to the γ subunit of the stalk,

3. The "downhill" movement of protons through the membrane drives the rotation of the ring of c subunits, and

4. The rotation of the c ring of F_0 provides the twisting force (*torque*) that drives the rotation of the attached γ subunit, leading to the synthesis and release of ATP.

All of these presumptions have been confirmed by data collected over the past several years. We will look at each of these aspects in more detail.

A body of evidence, including X-ray crystallography and the type of disulfide cross-linking study depicted in Figure 4.21, has shown that the c subunits are indeed organized into a circle to form a ring-shaped complex. High-resolution electron micrographs indicate that the two b subunits and the single a subunit of the F_0 complex reside outside the ring of c subunits as shown in Figure 5.23*b*. The b subunits are thought to be primarily structural components of the ATP synthase. The two elongated b subunits form a *peripheral stalk* that connects the F_0 and F_1 portions of the enzyme (Figure 5.23*b*) and,

along with the δ subunit of F_1, are thought to hold the $\alpha_3\beta_3$ subunits in a fixed position while the γ subunit rotates within the center of the complex.

The rotation of the c ring of F_0 has been observed directly utilizing an experimental procedure similar to that employed to demonstrate rotation of the γ subunit of F_1 (Figure 5.27). In this case, the *E. coli* ATP synthase used in the study included the c subunits along with an attached fluorescently labeled actin filament (Figure 5.28*a*). When ATP was added to the medium and the bound enzymes were observed under the fluorescent microscope, the labeled filaments attached to the c subunits were seen to rotate (Figure 5.28*b*). While these experiments demonstrate the rotation of both the c ring and γ subunit, they provide no information on how these two rotary elements are connected to one another. Each c subunit is built like a hairpin: it contains two transmembrane helices connected by a hydrophilic loop that projects toward the F_1 head. The polar loops situated on the tops of the c subunits are thought to form a binding site for the bases of the γ and ϵ subunits, which together act like a "foot" that that is attached to the c ring (see Figure 5.23*b*). As a result of this attachment, rotation of the c ring drives rotation of the attached γ subunit.

The mechanism by which H^+ movements drive the rotation of the c ring is more complex and less well understood. Figure 5.29 provides a model as to how H^+ ions might flow through the F_0 complex. Keep in mind in the following discussion (1) that the subunits of the c ring move successively past a *stationary* a subunit and (2) that protons are picked up from the intermembrane space one at a time by each c subunit and carried *completely around a circle* before they are released into the matrix. In this model, each a subunit has two half-channels that are physically separated (offset) from one another. One half-channel leads from the intermembrane (cytosolic) space into the middle of the a subunit and the other leads from the middle of the a subunit into the matrix. It is proposed that each proton moves from the intermembrane space through the half-channel and binds to a negatively

(a)

(b)

3.5 µm

Figure 5.28 Direct observation of rotation of the c ring of F_0. (*a*) The experimental approach in this experiment is similar to that of the previous figure, but the entire ATP synthase molecule (F_1 and F_0) is attached to the coverslip prior to linkage of the c ring to an actin filament. Rotating subunits of the ATP synthase are shown in bright colors; nonrotating subunits are shown in bluish shades. (*b*) Series of photographs showing the rotation of the fluorescent actin filament in the presence of 5 mM ATP. (B: FROM YOH WADA, YOSHIHIRO SAMBONGI, MASAMITSU FUTAI, BIOCHEM. BIOPHYS. ACTA, 1459:503, 2000.)

Figure 5.29 A model in which proton diffusion is coupled to the rotation of the c ring of the F_0 complex. As discussed in the text, it is proposed in this model that each proton from the intermembrane space enters a half-channel within the a subunit and then binds to an aspartic acid residue that is accessible on one of the c subunits. Proton binding induces a conformational change that causes the ring to move by 30°. The bound proton is carried in a full circle by the rotation of the c ring and is then released into a second half-channel that opens into the matrix. A succession of protons that engage in this activity causes the c ring to rotate in a counter-clockwise direction as shown. Rotation in the opposite direction is probably prevented by a repulsive force exerted by a fully protonated arginine residue located on the a subunit.

charged aspartic acid residue situated at the surface of the adjoining c subunit. Binding of the proton to the carboxyl group generates a major conformational change in the c subunit that causes that subunit to rotate approximately 30° in a counterclockwise direction. The movement of the recently protonated c subunit brings the next subunit in the ring (which was protonated at an earlier step) into alignment with the second half channel of the a subunit. Once there, the aspartic acid releases its associated proton, which diffuses into the matrix. Following dissociation of the proton, the c subunit returns to its original conformation and is ready to accept another proton from the intermembrane space and repeat the cycle.

According to this model, the aspartic acid of each c subunit acts like a revolving proton carrier. A proton hops onto the carrier at a selected pick-up site and is then carried around a circle before being released at a selected drop-off site. Movement of the ring is driven by the con-

formational changes associated with the sequential protonation and deprotonation of the aspartic acid residue of each c subunit. If the c ring is composed of 12 subunits, the association/dissociation of 4 protons in the manner described would move the ring 120°. Movement of the c ring 120° would drive a corresponding rotation of the attached γ subunit 120°, which would lead to the release of one newly synthesized molecule of ATP by the F_1 complex. According to this stoichiometry, the translocation of 12 protons would lead to the full 360° rotation of the c ring and γ subunit, and the synthesis and release of 3 molecules of ATP.

Other Roles for the Proton-motive Force in Addition to ATP Synthesis

Although the production of ATP may be the most important activity of mitochondria, these organelles are engaged in numerous other activities that require the input of energy. Unlike most organelles that rely primarily on ATP hydrolysis to power their activities, mitochondria rely on an alternate source of energy—the proton-motive force. For example, the proton-motive force drives the uptake of ADP and P_i into the mitrochondria in exchange for ATP and H^+, respectively. These and other activities that take place during aerobic respiration are summarized in Figure 5.30. In other examples, the proton-motive force may be used as the source of energy to "pull" calcium ions into the mitochondrion; to drive the transhydrogenase reaction that leads to the production of NADPH, the cell's reducing power (page 112); and to cause specifically targeted polypeptides to enter the mitochondrion from the matrix (Section 8.9).

We saw in Chapter 3 how ATP levels play an important role in controlling the rate of glycolysis and the TCA cycle by regulating the activity of key enzymes. Within the mitochondrion, the ADP level is a key determinant of respiratory rate. When ADP levels are low, ATP levels are typically high, and there is no need for additional substrate to be oxidized to provide electrons for the respiratory chain. Under these conditions, ATP synthesis is low, so that protons are unable to reenter the mitochondrial matrix through the ATP synthase. This leads to a sustained build-up of the proton-motive force, which in turn inhibits the proton pumping reactions of the electron transport chain and the consumption of oxygen by cytochrome oxidase. When the ATP/ADP ratio decreases, oxygen consumption abruptly increases. Recent evidence suggests that respiratory control may also be mediated directly through binding of ADP and ATP to allosteric sites on cytochrome oxidase subunits.

REVIEW ?

1. Describe the basic structure of the ATP synthase and the mechanism by which it is able to synthesize ATP.

Figure 5.30 **Summary of the major activities during aerobic respiration in a mitochondrion.**

2. Describe the steps in ATP synthesis according to the binding change mechanism.

3. Describe some of the evidence that supports the binding change mechanism.

4. Describe a proposed mechanism whereby proton diffusion from the intermembrane space into the matrix drives the phosphorylation of ADP.

5.6 PEROXISOMES

Peroxisomes are simple, membrane-bound vesicles (Figure 5.31*a*) with a diameter of 0.1 to 1.0 μm that often contain a dense, crystalline core of oxidative enzymes. Peroxisomes (or *microbodies*, as they are also called) are multifunctional organelles, containing more than 50 enzymes involved in such diverse activities as the oxidation of very-long-chain fatty acids (VLCFAs, ones whose chain typically contains 24 to 26 carbons), and the synthesis of plasmalogens, which are an important class of phospholipids in brain tissue in which one of the fatty acids is linked to glycerol by an ether linkage rather than an ester linkage. The enzyme luciferase, which generates the light emitted by fireflies, is also a peroxisomal enzyme. Peroxi-

somes are considered in the present chapter because they share several properties with mitochondria: both mitochondria and peroxisomes form by splitting from preexisting organelles; both types of organelles import preformed proteins from the cytosol (section 8.9); and both engage in similar types of oxidative metabolism. In fact, at least one enzyme, alanine/glyoxylate aminotransferase, is found in the mitochondria of some mammals (e.g., cats and dogs) and the peroxisomes of other mammals (e.g., rabbits and humans).

These organelles were named "peroxisomes" because they are the site of synthesis and degradation of hydrogen peroxide (H_2O_2), a highly reactive and toxic oxidizing agent. Hydrogen peroxide is produced by a number of enzymes, including urate oxidase, glycolate oxidase, and amino acid oxidases, that utilize molecular oxygen to oxidize their respective substrates (Figure 5.31*b*). The H_2O_2 generated in these reactions is rapidly broken down by the enzyme catalase, which is present in high concentrations in these organelles. The importance of peroxisomes in human metabolism is evident from the discussion in the accompanying Human Perspective.

Peroxisomes are also present in plants. Plant seedlings contain a specialized type of peroxisome, called a **glyoxysome** (Figure 5.32). Plant seedlings rely on stored fatty acids to provide the energy and material to form a new

(a) 0.5 μm

(b)

Figure 5.31 The structure and function of peroxisomes. (*a*) Electron micrograph of purified peroxisomes isolated by centrifugation through a sucrose density gradient. An electron micrograph of a peroxisome within a plant cell is shown in Figure 6.22. (*b*) Peroxisomes contain enzymes that carry out the two-step reduction of molecular oxygen to water. In the first step, an oxidase removes electrons from a variety of substrates (RH$_2$), such as uric acid or amino acids. In the second step, the enzyme catalase converts the hydrogen peroxide formed in the first step to water. (A: FROM Y. FUJIKI, ET AL., J. CELL BIOL. 93:105, 1982; BY COPYRIGHT PERMISSION OF THE ROCKEFELLER UNIVERSITY PRESS.)

10 μm

Figure 5.32 Glyoxysome localization within plant seedlings. Light micrograph of a section through cotyledons from imbibed cotton seeds. The glyoxysomes, which are seen as small dark structures (arrow), have been made visible by a cytochemical stain for the enzyme catalase. (FROM KENT D. CHAPMAN AND RICHARD N. TRELEASE, J. CELL BIOL. 115:998, 1991; BY COPYRIGHT PERMISSION OF THE ROCKEFELLER UNIVERSITY PRESS.)

plant. One of the primary metabolic activities in these germinating seedlings is the conversion of stored fatty acids to carbohydrate. Disassembly of stored fatty acids generates acetyl CoA, which condenses with oxaloacetate (OAA) to form citrate, which is then converted into glucose by a series of enzymes of the glyoxylate cycle localized in the glyoxysome. The role of peroxisomes in the metabolism of leaf cells is discussed in Section 6.6.

REVIEW **?**

1. What are some of the major activities of peroxisomes? What is the role of catalase in these activities?

2. In what ways are peroxisomes similar to mitochondria? In what ways are they unique?

THE HUMAN PERSPECTIVE

UPDATE
For the very latest, go to page U-10

Diseases That Result from Abnormal Mitochondrial or Peroxisomal Function

Mitochondria

In 1962, Rolf Luft of the University of Stockholm in Sweden reported on a study of mitochondria isolated from a woman who had been suffering from long-standing fatigue and mus-

cle weakness and who exhibited an elevated metabolic rate and body temperature. The mitochondria from this patient were somehow released from their normal respiratory control. Normally, when ADP levels are low, isolated mitochon-

dria stop oxidizing substrate. In contrast, the mitochondria from this patient continued to oxidize substrate at a high rate, even in the absence of ADP to phosphorylate, producing heat rather than mechanical work. Since that initial report, a variety of disorders have been described that have been traced to abnormalities in mitochondrial structure and function. Muscle and nerve tissues tend to be the most seriously impacted in these disorders, because these are the tissues with the highest demand for ATP. Depending on the protein(s) affected, these conditions range in severity from diseases that lead to death during infancy; to disorders that produce seizures, blindness, deafness, and/or strokelike episodes; to mild conditions characterized by intolerance to exercise or nonmotile sperm. Patients with serious conditions generally have abnormal skeletal muscle fibers, which contain large peripheral aggregates of mitochondria (Figure 1*a*). Closer examination of the mitochondria reveals large numbers of abnormal inclusions that may cause the mitochondria to resemble a "parking lot" (Figure 1*b*).

More than 90 percent of the polypeptides of the respiratory chain are encoded by genes that reside in the nucleus, so one would expect that numerous mitochondrial disorders would be traced to nuclear mutations. Yet, it was not until 1995 that the first such disease-causing mutation was reported. The mutation occurred in the gene encoding the flavoprotein subunit of the TCA enzyme succinate dehydrogenase. A couple of other nuclear mutations have since been linked to mitochondrial diseases, but many more have been traced to mutations in mitochondrial DNA, or mtDNA. The most serious disorders usually stem from mutations (or deletions) that affect genes encoding mitochondrial transfer RNAs, which are required for the synthesis of all 13 polypeptides produced in human mitochondria. Because all of the genes carried in mtDNA encode proteins required for oxidative phosphorylation, one would expect all types of mtDNA mutations to produce a similar clinical phenotype. Clearly, this is not the case.

The inheritance of mitochondrial disorders contrasts in several ways with the Mendelian inheritance of nuclear genes. The mitochondria present in the cells of a human embryo are derived exclusively from mitochondria that were present in the egg at the time of conception without any contribution from the fertilizing sperm. Consequently, mitochondrial disorders are inherited maternally. In addition, the mitochondria of a cell can contain a mixture of normal (i.e., wild-type) and mutant mtDNA, a condition known as *heteroplasmy*. Studies indicate that mitochondrial disorders only arise when a preponderance of the mitochondria of a particular tissue contains defective genetic information. Because of this variability, different members of a family carrying the same mitochondrial DNA mutation can exhibit strikingly different symptoms. Those individuals carrying a higher percentage of mutant mitochondria suffer more severe disease.

Another difference between nuclear and mitochondrial DNA is that the former is protected from damage by a variety of DNA repair systems (Section 13.2), which are generally lacking in mitochondria. Mitochondrial DNA may also be subjected to high levels of mutagenic oxygen radicals (page 35). For these various reasons, mitochondrial DNA experiences more than 10 times the mutation rate of nuclear DNA.

Figure 1 Mitochondrial abnormalities in skeletal muscle.
(*a*) Ragged-red fibers. These degenerating muscle fibers are from a biopsy of a patient and show accumulations of red-staining "blotches" just beneath the cells' plasma membrane, which are due to abnormal proliferation of mitochondria. (*b*) Electron micrograph showing crystalline inclusions within the mitochondrial matrix from cells of a patient with abnormal mitochondria. (A: COURTESY OF DONALD R. JOHNS; B: FROM JOHN A. MORGAN-HUGHES AND D. N. LANDON, IN MYOLOGY, 2D ED., A. G. ENGEL AND C. FRANZINI-ARMSTRONG, EDS., MCGRAW-HILL, 1994.)

(*a*)

(*b*)

In fact, cells taken from older persons have an increased number of mtDNA mutations compared to the same cells from younger individuals. It has been suggested that such mutations could lead to abnormal mitochondrial function that in turn might cause the aging process itself (page 35).

Mitochondrial mutations are particularly likely to accumulate in cells that remain in the body for long periods of time, such as those of nerve and muscle tissue. A number of common neurological diseases with adult onset, most notably Parkinson's disease, might be a consequence of degenerative changes in mitochondrial function. This possibility originally came to light in the early 1980s when a number of young drug addicts reported to hospitals with the sudden onset of severe muscular tremors that are characteristic of advanced Parkinson's disease (PD) in elderly adults. It was discovered that these individuals had intravenously injected themselves with a synthetic heroin that was contaminated with a compound called MPTP. Subsequent studies revealed that MPTP caused damage to complex I of the mitochondrial respiratory chain, leading to the death of nerve cells in the same region of the brain (the substantia nigra) that is affected in patients with PD. When cells from the substantia nigra of PD patients were then studied, they too were found to have a marked and selective decrease in complex I activity. Recent data suggests that this problem stems from underlying defects in mtDNA of these brain cells.

Peroxisomes

Zellweger syndrome (ZS) is a rare inherited disease characterized by a variety of neurological, visual, and liver abnormalities leading to death during early infancy. In 1973, Sidney Goldfischer and his colleagues at Albert Einstein College of Medicine reported that liver and renal cells from these patients were lacking peroxisomes. Subsequent investigations revealed that peroxisomes were not entirely absent from the cells of these individuals, but, rather, they were present as empty membranous "ghosts," that is, organelles lacking the enzymes that were normally found in peroxisomes. It is not that these individuals are unable to synthesize peroxisomal enzymes, but rather that the enzymes fail to be imported into the peroxisomes and remain largely in the cytosol where they are unable to carry out their normal functions. Genetic studies of cells from ZS patients showed that the disorder can arise from mutations in at least 11 different genes, all encoding proteins involved in the uptake of peroxisomal enzymes from the cytosol. Studies of these human disorders received a boost when strains of yeast were discovered that exhibited similar defects in peroxisome formation. The genes responsible for these defects were soon isolated—initially from yeast and then from humans. The products of these genes function either as receptors for peroxisomal proteins or as part of the machinery for transporting these proteins across the peroxisomal membrane (discussed in Section 8.9). Disruption of either of these functions leads to the formation of empty peroxisomes.

Unlike Zellweger syndrome, which affects a wide variety of peroxisomal functions, a number of inherited diseases are characterized by the absence of a single peroxisomal enzyme. One of these single-enzyme deficiency diseases is adrenoleukodystrophy (ALD), which was the subject of the 1993 movie *Lorenzo's Oil*. Boys with the disease are typically unaffected until midchildhood, when symptoms of adrenal insufficiency and neurological dysfunction begin. The disease results from a defect in a membrane protein that transports very-long-chain fatty acids (VLCFAs) into the peroxisomes, where they are normally metabolized. In the absence of this protein, VLCFAs accumulate in the brain and destroy the myelin sheaths that insulate nerve cells. In *Lorenzo's Oil*, the parents of a boy stricken with ALD discover that a diet rich in certain fatty acids is able to retard the progress of the disease. Unfortunately, subsequent studies have not borne out these findings and the diet is no longer widely prescribed. A number of ALD patients have been successfully treated by bone marrow transplantation, which provides normal cells capable of metabolizing VLCFAs, and by administration of drugs (e.g., lovastatin) that may lower VLCFA levels. Clinical studies employing gene therapy are also being planned.

SYNOPSIS

Mitochondria are large organelles comprised of an outer porous membrane and a highly impermeable inner membrane that is thrown into folds (cristae) that contain much of the machinery required for aerobic respiration. The porosity of the outer membrane results from integral proteins called porins. The architecture of the inner membrane and the apparent fluidity of its bilayer facilitate the interactions of components that are required during electron transport and ATP formation. The inner membrane surrounds a gel-like matrix that contains, in addition to proteins, a genetic system that includes DNA, RNA, ribosomes, and all the machinery necessary to transcribe and translate genetic information. Many of the properties of mitochondria can be accounted for by their presumed evolution from ancient symbiotic bacteria. (p. 184)

The mitochondrion is the center of oxidative metabolism in the cell, converting the products from carbohydrate, fat, and protein catabolism into chemical energy stored in ATP. Pyruvate and NADH are the two products of glycolysis. Pyruvate is transported across the inner mitochondrial membrane where it is decarboxylated and combined with coenzyme A to form acetyl CoA, which is condensed with oxaloacetate to form citrate, which is fed into the TCA cycle. As it traverses the reactions of the TCA cycle, two of the carbons from the citrate are removed and released as CO_2, which represents the most highly oxidized state of the carbon atom. Electrons removed from substrates are transferred to FAD and NAD^+ to form $FADH_2$ and NADH. Fatty acids are broken down into acetyl CoA, which is fed into the TCA cycle, and all 20 amino acids are broken down into either pyruvate, acetyl CoA, or TCA cycle intermediates. Thus, the TCA cycle is the pathway on which the cell's major catabolic pathways converge. (p. 187)

Electrons transferred from substrates to $FADH_2$ and NADH are passed along a chain of electron carriers to O_2, releasing

energy that is used to generate an electrochemical gradient across the inner mitochondrial membrane. The controlled movement of protons back across the membrane through an ATP-synthesizing enzyme is used to drive the formation of ATP at the enzyme's catalytic site. Each pair of electrons from NADH releases sufficient energy to drive the formation of approximately three ATPs, whereas the energy released from a pair of electrons from $FADH_2$ accounts for the formation of approximately two ATPs. (p. 189)

The amount of energy released as an electron is transferred from a donor (reducing agent) to an acceptor (oxidizing agent) can be calculated from the difference in redox potential between the two couples. The standard redox potential of a couple is measured under standard conditions and compared to the H_2/H^+ couple. The standard redox potential of the $NADH/NAD^+$ couple is -0.32 V, reflecting the fact that NADH is a strong reducing agent, that is, one that readily transfers its electrons. The standard redox potential of the H_2O/O_2 couple is $+0.82$ V, reflecting the fact that O_2 is a strong oxidizing agent, that is, one that has a strong affinity for electrons. The difference between these two couples, which is equivalent to 1.14 V, provides a measure of the free energy released (52.6 kcal/mol) when a pair of electrons is passed from NADH along the entire electron-transport chain to O_2. (p. 192)

The electron-transport chain contains five different types of carriers: heme-containing cytochromes, flavin nucleotide-containing flavoproteins, iron-sulfur proteins, copper atoms and quinones. Flavoproteins and quinones are capable of accepting and donating hydrogen atoms, while cytochromes, copper atoms and iron-sulfur proteins are capable of accepting and donating only electrons. The carriers of the electron-transport chain are arranged in order of increasing redox potential. The various carriers are organized into four large, multiprotein complexes. Cytochrome c and ubiquinone are mobile carriers, shuttling electrons between the larger complexes. As pairs of electrons are passed through complexes I, III, and IV, a specific number of protons are translocated from the matrix across the membrane into the intermembrane space. The translocation of protons by these electron-transporting complexes establishes the proton gradient in which energy is stored. The last of the complexes is cytochrome oxidase, which transfers electrons from cytochrome c to O_2, reducing it to form water, a step which also removes protons from the matrix, contributing to the proton gradient. (p. 195)

The translocation of protons creates a separation of charge across the membrane, in addition to a difference in proton concentration. Consequently, the proton gradient has two components—a voltage and a pH gradient—the magnitudes of which depend on the movement of other ions across the membrane. Together the two components constitute a proton-motive force (Δp). In mammalian mitochondria, approximately 80 percent of the free energy of Δp is represented by the voltage and 20 percent by the pH gradient. (p. 202)

The enzyme that catalyzes the formation of ATP is a large multiprotein complex called ATP synthase. ATP synthase contains two distinct parts: an F_1 head that projects into the matrix and includes the catalytic sites, and an F_0 base that is embedded in the lipid bilayer and forms a channel through which protons are conducted from the intermembrane space into the matrix. In the binding change hypothesis of ATP formation, which is now generally accepted, the controlled movement of protons through the F_0 portion of the enzyme induces the rotation of the γ subunit of the enzyme, which runs through the stalk, connecting the F_0 and F_1 portions of the enzyme. Rotation of the γ subunit is brought about by rotation of the c ring of the F_0 base, which is induced by the movement of protons through half-channels in the a subunit. The rotation of the γ subunit induces changes in the conformation of the F_1 catalytic sites, driving the formation of ATP. Evidence indicates that the energy-requiring step is not the actual phosphorylation of ADP, but the release of the ATP product from the active site, which occurs in response to the induced changes in conformation. In addition to the formation of ATP, the proton-motive force also provides the energy required for a number of transport activities, including the uptake of ADP into the mitochondrion in exchange for the release of ATP to the cytosol, the uptake of phosphate and calcium ions, and the import of mitochondrial proteins. (p. 203)

Peroxisomes are membrane-bound cytoplasmic vesicles that carry out a number of diverse metabolic reactions, including the oxidation of urate, glycolate, and amino acids, generating H_2O_2. (p. 211)

ANALYTIC QUESTIONS

1. Consider the reaction $A:H + B \rightleftharpoons B:H + A$. If, at equilibrium, the ratio of [B:H]/[B] is equal to 2.0, one can conclude that (1) B:H is the strongest reducing agent of the four compounds; (2) the (A:H-A) couple has a more negative redox potential than the (B:H-B) couple; (3) none of the four compounds are cytochromes; (4) the electrons associated with B are higher energy than those associated with A. Which of the previous statements are true? Draw one of the half reactions for this redox reaction.

2. The membranous vesicle of a submitochondrial particle after removal of the F_1 spheres would be able to (1) oxidize NADH; (2) produce H_2O from O_2; (3) generate a proton gradient; (4) phosphorylate ADP. Which of the previous statements are true? How would this answer be different, if at all, if the objects being studied were intact submitochondrial particles that had been treated with dinitrophenol?

3. Protein A is a flavoprotein with a redox potential of -0.2 V. Protein B is a cytochrome with a redox potential of $+0.1$ V.

 a. Draw the half-reactions for each of these two electron carriers.

 b. Give the reaction that would occur if one were to mix reduced A molecules and oxidized B molecules.

 c. Which two compounds in the reaction of part b would be present at the highest concentration when the reaction reached equilibrium?

4. Of the following substances, ubiquinone, cytochrome c, NAD^+, NADH, O_2, H_2O, which is the strongest reducing agent? Which is the strongest oxidizing agent? Which has the greatest affinity for electrons?

5. If it were determined that the inner mitochondrial membrane was freely permeable to chloride ions, what effect might this have on the proton-motive force across the inner mitochondrial membrane?

6. Look at the drop in energy during electron transport shown in Figure 5.13. How would this profile be different if the original electron donor was $FADH_2$ rather than NADH?

7. Would you expect the import of P_i to lead to a decrease in the proton-motive force? Why?

8. In the chart of standard redox potentials in Table 5.1, the oxaloacetate-malate couple is less negative than the NAD^+-NADH couple. How are these values compatible with the transfer of electrons from malate to NAD^+ in the TCA cycle?

9. How many high-energy triphosphates are formed by substrate-level phosphorylation during each turn of the TCA cycle (consider the TCA cycle reactions only)? How many are formed as the result of oxidative phosphorylation? How many molecules of CO_2 are released? How many molecules of FAD are reduced? How many pairs of electrons are stripped from substrate?

10. Protons move in both directions across the inner mitochondrial membrane. They move in one direction as the result of electron transport. What causes them to move in the opposite direction?

11. The drop in $\Delta G^{\circ\prime}$ of a pair of electrons is -52.6 kcal/mol, and the $\Delta G^{\circ\prime}$ of ATP formation is $+7.3$ kcal/mol. If three ATPs are formed for each pair of electrons removed from substrate, can you conclude that oxidative phosphorylation is 21.9/52.6 or 42 percent efficient? Why or why not?

12. Would you expect metabolically active, isolated mitochondria to acidify or alkalinize the medium in which they were suspended? Would your answer be different if you were working with submitochondrial particles rather than mitochondria? Why?

13. Suppose the movement of three protons may be required for the synthesis of one ATP molecule. Calculate the energy released by the passage of three protons into the matrix (see p. 202 for information).

14. Would you expect isolated mitochondria to be able to carry out the oxidation of glucose with the accompanying production of ATP? Why or why not? What might be a good compound to add to a preparation of isolated mitochondria to achieve ATP synthesis?

15. Calculate the free energy released when $FADH_2$ is oxidized by molecular O_2 under standard conditions.

16. Using your understanding of the organization of the mitochondrion and how it generates ATP, answer the following questions. Give a brief explanation of each answer.

 a. What would be the effect on ATP production of adding a substance (protonophore) that makes the mitochondrial membrane leaky to protons?

 b. What would be the effect on pH of the mitochondrial matrix of reducing the oxygen supply?

 c. Suppose that the glucose supply to the cell is greatly reduced. What would be the effect on CO_2 production by the mitochondria?

17. Suppose that you are able to manipulate the potential of the inner membrane of an isolated mitochondrion. You measure the pH of the mitochondrial matrix and find it to be 8.0. You measure the bathing solution and find its pH to be 7.0. You clamp the inner membrane potential at $+59$ mV—that is, you force the matrix to be 59 mV positive with respect to the bathing medium. Under these circumstances, can the mitochondrion use the proton gradient to drive the synthesis of ATP? Explain your answer.

18. Suppose you were able to synthesize an ATP synthase that was devoid of the gamma subunit. How would the catalytic sites of the β subunits of such an enzyme compare to one another. Why?

19. In functional terms, the ATP synthase can be divided into two parts: a "stator" made up of subunits that do not move during catalysis, and a "rotor" made up of parts that do move. Which subunits of the enzyme make up each of these two parts. How are the immovable parts of the stator in F_0 related structurally to the immovable parts of the stator in F_1?

ADDITIONAL READING

Mitochondria Structure and Function

BABCOCK, G. T. 1999. How oxygen is activated and reduced in respiration. *Proc. Nat'l. Acad. Sci. U. S. A.* 96:12971–12973

BERRY, E. A., ET AL. 2000. Structure and function of cytochrome BC complexes. *Annu. Rev. Biochem.* 69:1005–1077.

BOYER, P. D. 1997. The ATP synthase—A splendid molecular machine. *Annu. Rev. Biochem.* 66:717–749.

BOYER, P. D. 1999. What makes ATP synthase spin? *Nature* 402:247–249.

BRANDT, U. 1997. Proton-translocation by membrane-bound NADH: ubiquinone-oxidoreductase. *Biochim. Biophys. Acta* 1318:79–91.

CAPALDI, R. A. 2000. The changing face of mitochondrial research. *Trends Biochem. Sci.* 25:212–214.

ELSTON, T., ET AL. 1998. Energy transduction in ATP synthase. *Nature* 391:510–513.

FILLINGAME, R. H. 1999. Molecular rotary motors. *Science* 286:1687–1688.

FREY, T. G. & MANNELLA, C. A. 2000. The internal structure of mitochondria. *Trends Biochem. Sci.* 25:319–324.

GARLID, K. D., ET AL. 2000. How do uncoupling proteins uncouple? *Biochim. Biophys. Acta* 1459:383–389.

Gennis, R. B. 1998. Cytochrome c oxidsase: One enzyme, two mechanisms. *Science* 280:1712–1713.

GRAY, M. W., ET AL. 1999. Reviews on mitochondria. *Science* 283:1476–1497.

HAROLD, F. M. 1986. *The Vital Force: A Study of Bioenergetics.* W. H. Freeman.

HARRIS, D. A. 1995. *Bioenergetics at a Glance.* Blackwell.

HERMANN, G. J. & SHAW, M. 1998. Mitochondrial dynamics in yeast. *Annu. Rev. Cell Dev. Biol.* 14:265–303.

JUNGE, W. 1999. ATP synthase and other motor proteins. *Proc. Nat'l. Acad. Sci. U. S. A.* 96:4735–4737.

KINOSITA, K., JR., ET AL. 1998. F_1-ATPase: A rotary motor made of a single molecule. *Cell* 93:21–24.

MICHEL, H., ET AL. 1998. Cytochrome C oxidase: structure and spectroscopy. *Annu. Rev. Biophys. Biomol. Struct.* 27:329–256.

NAKAMOTO, R. K., ET AL. 1999. Rotational coupling in the F_0F_1 ATP synthase. *Annu. Rev. Biophys. Biomol. Struct.* 28:205–234.

OSTER, G. & WANG, H. 1999. ATP synthase: Two motors, two fuels. *Structure* 7:R67–R72.

POMES, R., ET AL. 1998. Structure and dynamics of a proton shuttle in cytochrome c oxidase. *Biochim. Biophys. Acta* 1365:255–260.

ROUSSEAU, D. L. 1999. Two phases of proton translocation. *Nature* 400:412–413.

SCHEFFLER, I. E. 1999. *Mitochondria.* Wiley-Liss.

SLATER, E. C. 1981. The discovery of oxidative phosphorylation. *Trends Biochem. Sci.* 6:226–227.

SMITH, J. L. 1998. Secret life of cytochrome bc_1. *Science* 281:58–59.

WALKER. J., ED. 2000. The mechanism of F_1F_0-ATPase. *Biochim. Biophys. Acta* vol. 1458, #2–3.

WIKSTRÖM, M., ED. 1998. Cytochrome oxidase: Structure and mechanism. *J. Bioen. Biomemb.*. vol. 30, #1.

WIKSTRÖM, M., ET AL. 2000. The role of the D- and K-pathways of proton transfer in the function of the haem-copper oxidases. *Biochim. Biophys. Acta* 1459:514–520.

Peroxisomes

ERDMANN, R., ET AL. 1997. Peroxisomes: Organelles at the crossroads. Trends Cell Biol. 7:400–407.

MASTERS, C. and CRANE, D. 1995. *The Peroxisome.* Cambridge.

SUBRAMANI, S. 1998. Components involved in peroxisome import, biogenesis, proliferation, turnover, and movement. *Physiol. Revs.* 78:171–188.

TABAK, H. F., ET AL. 1999. Peroxisomes: simple in function but complex in maintenance. *Trends Cell Biol.* 9:447–453.

Human Perspectives

ANDERSEN, J. L., ET AL. 2000. Muscle, genes, and athletic performance. *Sci. Am.* 283:48–55. (Sept.)

DiMAURO, S., ET AL. 1998. Mitochondria in neuromuscular and neurodegenerative diseases. *Biochim. Biophys. Acta* 1366:197–233.

DUBOIS-DALCQ, M., ET AL. 1999. The neurobiology of X-linked adrenoleukodystrophy. *Trends Neurosci.* 22:4–12.

GOULD, S. J. & VALLE, D. 2000. Peroxisome biogenesis disorders. *Trends Gen.* 16:340–345.

POORTMANS, J. R., ed. 1989. *Principles of Exercise.* Karger.

POWERS, S. K. and HOWLEY, E. T. 1990. *Exercise Physiology: Theory and Application to Fitness and Performance.* Wm C. Brown.

SCHAPIRA, A. H. V., ED. 1999. Mitochondrial dysfunction in human disease. *Biochim. Biophys. Acta* vol. 1410:99–228.

SIMON, D. K. & JOHNS, D. R. 1999. Mitochondrial disorders: Clinical and genetic features. *Annu. Rev. Med.* 50:111–127.

WALLACE, D. C. 1997. Mitochondrial DNA in aging and disease. *Sci. Am.* 277:40–47. (August)

WANDERS, R. J. A. 1998. A happier sequel to *Lorenzo's oil? Nature Med.* 4:1245–1246.

6

Photosynthesis and the Chloroplast

The Earth's earliest life forms must have obtained their raw materials and energy from simple organic molecules dissolved in their aqueous environment. These organic molecules had to have formed *abiotically*, that is, as a result of nonbiological chemical reactions occurring in the primeval seas. Thus, just as we survive on nutrients taken in from our environment, so too must have the original life forms. Organisms that depend on an external source of organic compounds are called **heterotrophs**.

The number of heterotrophic organisms living on the primitive Earth must have been severely restricted, because the spontaneous production of organic molecules occurs very slowly. The evolution of life on Earth received a tremendous boost with the appearance of organisms that employed a new metabolic strategy. Unlike their predecessors, these organisms could manufacture their own organic nutrients from the simplest types of inorganic molecules, such as carbon dioxide (CO_2) and hydrogen sulfide (H_2S). Organisms capable of surviving on CO_2 as their principal carbon source are called **autotrophs**.

Light micrograph of a living leaf epidermal cell from Arabidopsis thaliana. *The cell contains a number of chloroplasts—the organelles that house the plant's photosynthetic machinery. Chloroplasts normally divide by binary fission in which a single constriction splits the organelle into two equal daughters. This cell is from a mutant plant that is characterized by asymmetric fission. The highly elongated chloroplast has initiated fission asymmetrically at several sites, as indicated by the multiple constrictions.* (COURTESY OF KEVIN D. STOKES AND STANISLAV VITHA, MICHIGAN STATE UNIVERSITY.)

The manufacture of complex molecules from CO_2 requires the input of large amounts of energy. Over the course of evolution, two main types of autotrophs have evolved that can be distinguished by their source of energy. **Chemoautotrophs** utilize the energy stored in inorganic molecules (such as ammonia, hydrogen sulfide, or nitrites) to convert CO_2 to organic compounds, while **photoautotrophs** utilize the radiant energy of the sun to accomplish this result. Because all chemoautotrophs are prokaryotes, and their relative contribution to the formation of biomass on Earth is small, we will not consider their metabolic activities further. Photoautotrophs, on the other hand, are responsible for capturing the energy that fuels the activities of nearly every organism on Earth. Photoautotrophs include plants and eukaryotic algae, various flagellated protists, and a variety of prokaryotes, including purple sulfur and nonsulfur bacteria, green bacteria, and cyanobacteria. All of these organisms carry out **photosynthesis**, a process in which energy from sunlight is transformed into chemical energy that is stored in carbohydrates and other organic molecules.

During photosynthesis, relatively low-energy electrons are removed from a donor compound and converted into high-energy electrons using the energy absorbed from light. These high-energy electrons are then used to synthesize reduced biological molecules, such as starch and oils. It is likely that the first groups of photoautotrophs, which dominated the Earth for two billion years, utilized hydrogen sulfide as their source of electrons for photosynthesis, carrying out the overall reaction

$$CO_2 + 2H_2S \xrightarrow{\text{light}} (CH_2O) + H_2O + 2S$$

where (CH_2O) represents a unit of carbohydrate. There are numerous bacteria living today that carry out this type of photosynthesis. But hydrogen sulfide today is neither abundant nor widespread, and consequently, organisms that depend on this compound as an electron source are limited to habitats such as sulfur springs and deep sea vents.

Approximately 3.5 billion years ago, a new type of photosynthetic prokaryote appeared on Earth that was able to utilize a much more abundant source of electrons, namely, water. Not only did the use of water allow these organisms—the cyanobacteria—to exploit a much more diverse array of habitats on Earth (see Figure 1.15), it produced a waste product of enormous consequence for all forms of life. The waste product was molecular oxygen (O_2), which is formed in the overall reaction

$$CO_2 + H_2O \xrightarrow{\text{light}} (CH_2O) + O_2$$

The switch from H_2S to H_2O as the substrate for photosynthesis is more difficult than switching one letter in the alphabet for another. The redox potential of the S-H_2S couple is -0.25 V as compared to $+0.816$ V for the $\frac{1}{2}$ O_2-H_2O couple (page 195). In other words, the sulfur atom in a molecule of H_2S has much less affinity for its electrons (and thus gives them up more readily) than does the oxygen atom in a molecule of H_2O. Thus, if an organism is going to carry out **oxygenic** (oxygen-releasing)

photosynthesis, it has to generate a very strong oxidizing agent as part of its photosynthetic metabolism in order to pull tightly held electrons from water. The switch from H_2S (or other reduced substrates) to H_2O as the electron source for photosynthesis required an overhaul of the photosynthetic machinery.

We will begin our analysis of photosynthesis by briefly examining the organelles in which the process takes place. ∎

6.1 CHLOROPLAST STRUCTURE AND FUNCTION

Photosynthesis in plants takes place in specialized cytoplasmic organelles, the **chloroplasts**, located predominantly in the mesophyll cells of leaves. The structure of a leaf and the arrangement of chloroplasts around the central vacuole of a mesophyll cell are shown in Figure 6.1. Chloroplasts of higher plants are generally lens-shaped (Figure 6.2), approximately 2 to 4 μm wide and 5 to 10 μm long, and typically numbering 20 to 40 per cell. Their dimensions make chloroplasts giants among organelles—as large as an entire mammalian red blood cell.

Chloroplasts were identified as the site of photosynthesis in 1881 in an ingenious experiment by the German biologist Theodor Engelmann. Engelmann showed that when cells of the green alga *Spyrogyra* were illuminated, actively moving bacteria would collect outside the cell near the site of the large ribbonlike chloroplast. The bacteria were utilizing the minute quantities of oxygen released by photosynthesis in the chloroplast for their aerobic respiration.

The outer covering of a chloroplast consists of an envelope composed of two membranes separated by a narrow space (Figure 6.2). Like the outer membrane of a mitochondrion, the outer membrane of a chloroplast envelope contains porins (page 186), which makes this membrane permeable to solutes of molecular mass as high as 10,000 daltons. In contrast, the inner membrane of the envelope is relatively impermeable; substances moving through this membrane do so only with the aid of a variety of transporters.

Much of the photosynthetic machinery of the chloroplast—including light-absorbing pigments, a complex chain of electron carriers, and an ATP-synthesizing apparatus—is part of an internal membrane system that is physically separate from the double-layered envelope. The internal membrane of the chloroplast, which contains the energy-transducing machinery, is organized into flattened membranous sacs, called **thylakoids**. Thylakoids are arranged in orderly stacks called **grana**, which resemble stacks of coins (Figures 6.2, 6.3). The space inside a thylakoid sac is the **lumen**, and the space outside the thylakoid and within the chloroplast envelope is the **stroma**, which contains the enzymes responsible for carbohydrate synthesis. The thylakoids of different grana are connected to one another by flattened membranous structures called **stroma thylakoids** (Figure 6.3). The differences in struc-

Figure 6.1 **The functional organization of a leaf.** The section of the leaf shows several layers of cells that contain chloroplasts distributed within their cytoplasm. These chloroplasts carry out photosynthesis, providing raw materials and chemical energy for the entire plant.

Figure 6.2 **The internal structure of a chloroplast.** (*a*) Transmission electron micrograph through a single chloroplast. The internal membrane is arranged in stacks of disklike thylakoids that are physically separate from the outer double membrane that forms the envelope. (*b*) Schematic diagram of a chloroplast showing the outer double membrane and the thylakoid membranes. (A: COURTESY OF LESTER K. SHUMWAY.)

ture and function between grana thylakoids and stroma thylakoids are discussed in the Experimental Pathways for this chapter, which can be found on the Web at www.wiley.com/college/karp.

Like the matrix of a mitochondrion, the stroma of a chloroplast contains small, double-stranded, circular DNA molecules and prokaryotic-like ribosomes. The chloroplast DNA encodes tRNAs, rRNAs, and up to about 100 different polypeptides, including many of the protein subunits that mediate the light reactions of the thylakoid membrane and the large subunit of the enzyme responsible for capturing CO_2 (page 234). The majority of proteins of a chloroplast are encoded by the DNA of the nucleus and synthesized in the cytosol. These proteins must be imported into the chloroplast by a specialized transport machinery (Section 8.9).

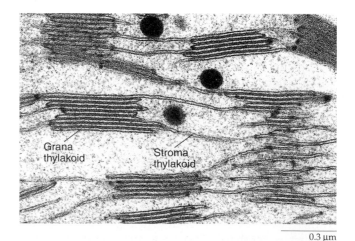

Figure 6.3 **Thylakoid membranes.** Electron micrograph of a section through a portion of a spinach chloroplast showing the stacked grana thylakoids and the unstacked stroma thylakoids. (FROM L. A. STAEHELIN, IN L. A. STAEHELIN, ED., ENCYCLOPEDIA OF PLANT PHYSIOLOGY, VOL. 19, P. 23, SPRINGER-VERLAG, 1986.)

Thylakoid membranes are unusual in having relatively little phospholipid and a high percentage of galactose-containing glycolipids such as the one shown below.

Monogalactosyl diacylglycerol

Both fatty acids of these lipids contain several double bonds, which makes the lipid bilayer of thylakoid membranes highly fluid. The fluidity of the lipid bilayer facilitates lateral diffusion of protein complexes through the membrane during photosynthesis.

As illustrated in the chapter opening micrograph, chloroplasts arise by fission from preexisting chloroplasts (or their nonpigmented precursors, which are called **pro-plastids**). Chloroplasts are believed to have evolved from an oxygen-producing, photosynthetic prokaryote (presumably a cyanobacterium) that had taken up a symbiotic existence inside a primitive, nonphotosynthetic pro-eukaryotic cell (page 26). As a result, chloroplasts and cyanobacteria share many basic characteristics, including porins, circular DNA, prokaryotic-type ribosomes, and a similar photosynthetic machinery, which will be discussed in detail in the following pages.

REVIEW ?

1. Describe the effect that the appearance of cyanobacteria is thought to have had on the metabolism of organisms?

2. Describe the organization of the membranes of a chloroplast. How does this organization differ from that of mitochondria?

3. Distinguish between stroma and lumen, envelope membrane and thylakoid membrane, grana thylakoid and stroma thylakoid.

6.2 AN OVERVIEW OF PHOTOSYNTHETIC METABOLISM

A major advance in our understanding of the chemical reactions of photosynthesis came in the early 1930s with a proposal by C. B. van Niel, who was then a graduate stu-dent at Stanford University. Consider the overall equation of photosynthesis as given earlier:

$$CO_2 + H_2O \xrightarrow{\text{light}} (CH_2O) + O_2$$

The prevailing belief in 1930 held that the energy of light was used to split CO_2, releasing molecular oxygen (O_2) and transferring the carbon atom to a molecule of water to form a unit of carbohydrate (CH_2O). In 1931, van Niel proposed an alternate mechanism based on his work with sulfur bacteria. It had been conclusively demonstrated that these organisms were able to reduce CO_2 to carbohydrate using the energy of light without the simultaneous production of molecular O_2. It was proposed that the reaction for sulfur bacteria was

$$CO_2 + 2 H_2S \xrightarrow{\text{light}} (CH_2O) + H_2O + 2S$$

Postulating a basic similarity in the photosynthetic processes of all organisms, van Niel proposed a general reaction to include all these activities:

$$CO_2 + 2 H_2A \xrightarrow{\text{light}} (CH_2O) + H_2O + 2A$$

For the production of a hexose, such as glucose, the reaction would be

$$6CO_2 + 12 H_2A \xrightarrow{\text{light}} C_6H_{12}O_6 + 6H_2O + 12A$$

Van Niel recognized that photosynthesis was essentially a process of oxidation-reduction. In the preceding reaction, H_2A is an electron donor (reducing agent) and can be represented by H_2O, H_2S, or other reduced substrates utilized by various types of bacteria. Carbon dioxide, however, is an oxidizing agent, which, in a plant cell, is reduced to form hexose in the following reaction:

$$6CO_2 + 12 H_2O \xrightarrow{\text{light}} C_6H_{12}O_6 + 6H_2O + 6O_2$$

In this scheme, each molecule of oxygen is derived not from CO_2 but from the breakdown of two molecules of H_2O, a process driven by the absorption of light. The role of water in the formation of molecular oxygen was clearly established in 1941 by Samuel Ruben and Martin Kamen of the University of California, Berkeley. These investigators carried out experiments using a specially labeled isotope of oxygen, ^{18}O, as a replacement for the common isotope, ^{16}O. One group of plants was given labeled $C[^{18}O_2]$ and unlabeled water, while another group was given unlabeled carbon dioxide and labeled $H_2[^{18}O]$. The investigators asked a simple question: Which of these two groups of plants released labeled $^{18}O_2$? The plants given labeled water produced labeled oxygen, demonstrating that O_2 produced during photosynthesis derived from H_2O. The plants given labeled carbon dioxide produced "regular" oxygen, confirming that O_2 was not being produced by a chemical splitting of CO_2. Contrary to popular belief, it wasn't carbon dioxide that was being split into its two atomic components, but water. Van Niel's hypothesis had been confirmed.

The van Niel proposal placed photosynthesis in a different perspective; it became, in essence, the reverse of respiration. Whereas respiration in mitochondria reduces

Figure 6.4 An overview of the energetics of photosynthesis and aerobic respiration.

oxygen to water, photosynthesis in chloroplasts oxidizes water to oxygen. The former process releases energy, so the latter process must require energy. An overview of the thermodynamics of photosynthesis and aerobic respiration is indicated in Figure 6.4. Many similarities between these two metabolic activities will be apparent in the following pages.

The events of photosynthesis can be divided into two series of reactions. During the first stage, the **light-dependent reactions**, energy from sunlight is absorbed and stored as chemical energy in two high-energy biological molecules: ATP and NADPH. During the second stage, the **light-independent reactions** (or "dark reactions"), carbohydrates are synthesized from carbon dioxide using the energy stored in the ATP and NADPH molecules produced in the light-dependent reactions. It is estimated that plant life on Earth converts approximately 500 trillion kg of CO_2 to carbohydrate while releasing about 450 trillion kg of O_2 each year.

We will begin with the light-dependent reactions, which are complex and remain incompletely understood.

REVIEW ?

1. In what ways is photosynthesis the reverse of respiration?

2. In what ways is nonoxygenic photosynthesis that uses H_2S as an electron source similar to oxygenic photosynthesis that uses H_2O as an electron source? In what ways are they different?

3. In general terms, how do the light-independent reactions differ from the light-dependent reactions? What are the primary products of the two types of reactions?

6.3 THE ABSORPTION OF LIGHT

Light travels in packets of energy called **photons**. The energy content of a photon depends on the wavelength of the light according to the equation

$$E = hc/\lambda$$

where h is Planck's constant (1.58×10^{-34} cal·sec), c is the speed of light in a vacuum, and λ is the wavelength of light. The shorter the wavelength, the higher the energy content. One mole (6.02×10^{23}) of photons of 680 nm wavelength, which is an important wavelength in photosynthesis, contains approximately 42 kcal of energy, which is equivalent to a change in redox potential of approximately 1.8 V (calculated by dividing 42 kcal by the Faraday constant of 23.06 kcal/V).

The absorption of light is the first step in any photochemical process. When a photon is absorbed by a molecule, an electron becomes sufficiently energetic to be pushed from an inner to an outer orbital. The molecule is said to have shifted from the **ground state** to an **excited state**. Because the number of orbitals in which an electron can exist is limited and each orbital has a specific energy

level, it follows that any given atom or molecule can absorb only light of certain specific wavelengths.

The excited state of a molecule is unstable and can be expected to last only about 10^{-9} second. There are several consequences that can befall an excited electron, depending on the circumstances. Consider a molecule of **chlorophyll**, which is the most important light-absorbing photosynthetic pigment. If the electron of an excited chlorophyll molecule drops back to a lower orbital, the energy it had absorbed must be released. If the energy is released in the form of heat or light (fluorescence), the chlorophyll has returned to the original ground state and the energy of the absorbed photon has not been utilized. This is precisely what is observed when a preparation of *isolated chlorophyll* in solution is illuminated: the solution becomes strongly fluorescent because the absorbed energy is reemitted at a longer (i.e., lower energy) wavelength. However, if the same experiment is performed on a preparation of *isolated chloroplasts*, only a faint fluorescence is observed, indicating that very little of the absorbed energy is dissipated. Instead, the energetic electrons of chlorophyll molecules are transferred to acceptors within the chloroplast membranes before they have a chance to drop back to lower energy orbitals. Thus, chloroplasts are able to harness the absorbed energy before it dissipates. It is estimated that transfer of an electron from chlorophyll can occur within a few picoseconds (1 ps = 10^{-12} second).

Photosynthetic Pigments

Pigments are colored molecules that absorb light of particular wavelength(s) within the visible spectrum. Leaves are green because their chloroplasts contain large quantities of the pigment chlorophyll, which absorbs most strongly in the blue and red, leaving the intermediate green wavelengths to be reflected to our eyes. The structure of chlorophyll is shown in Figure 6.5. Each molecule consists of two parts: (1) a porphyrin ring that functions in light absorption and (2) a hydrophobic phytol chain that keeps the chlorophyll embedded in the photosynthetic membrane. Unlike the red, iron-containing porphyrins (heme groups) of hemoglobin and myoglobin, the porphyrin in a chlorophyll molecule contains a magnesium atom. The alternating single and double bonds along the edge of the porphyrin ring delocalize electrons, which form a cloud around the ring (Figure 6.5). Systems of this type are said to be *conjugated* and are strong absorbers of visible light. The absorbed energy causes a redistribution of the electron density of the molecule, which in turn favors loss of an electron to a suitable acceptor. The conjugated bond system also broadens the absorption peaks, enabling individual molecules to absorb the energy of a range of wavelengths. These features are evident in an absorption spectrum of purified chlorophyll molecules (Figure 6.6). An **absorption spectrum** is a plot of the intensity of light absorbed relative to its wavelength. The range of wavelengths absorbed by photosynthetic pigments situated within the thylakoids is further increased because the

Figure 6.5 The structure of chlorophyll *a*. The molecule consists of a porphyrin ring (which in turn is constructed of four smaller pyrrole rings) with a magnesium ion at its center and a long hydrocarbon tail. The green shading around the edge of the porphyrin indicates the delocalization of electrons that forms a cloud. The structure of the magnesium-containing porphyrin of chlorophyll can be compared to the iron-containing porphyrin of a heme shown in Figure. 5.11. Chlorophyll *b* and bacteriochlorophyll *a* contain specific substitutions as indicated. For example, the —CH$_3$ group on ring II is replaced by a —CHO group in chlorophyll *b*. Chlorophyll *a* is present in all oxygen-producing photosynthetic organisms, but it is absent in the various sulfur bacteria. In addition to chlorophyll *a*, chlorophyll *b* is present in all higher plants and green algae. A third variety, chlorophyll *c*, is present in brown algae, diatoms, and certain protozoa. Bacteriochlorophyll is found only in green and purple bacteria, organisms that do not produce O$_2$ during photosynthesis.

pigments are noncovalently associated with a variety of different polypeptides.

Figure 6.6 **Absorption spectrum for several photosynthetic pigments of higher plants.** The background shows the colors that we perceive for the wavelengths of the visible spectrum. Chlorophylls absorb most strongly in the violet-blue and red regions of the spectrum, while carotenoids (e.g., β-carotene) also absorb into the green region. Red algae contain additional pigments (phycobilins) that absorb in the middle bands of the spectrum.

Figure 6.7 **Action spectrum for photosynthesis.** The action spectrum (represented by the red-colored line) indicates the efficiency with which light of various wavelengths is able to promote photosynthesis in the leaves of a plant. The black lines indicate the absorption spectra of each of the major photosynthetic pigments. The green line shows the combined absorption spectrum of all pigments.

Several classes of chlorophyll, differing from one another in the side groups attached to the porphyrin ring, occur among photosynthetic organisms. The structures of these pigments are indicated in Figure 6.5. Chlorophylls are the primary light-absorbing photosynthetic pigments, but terrestrial plants also contain orange and red *accessory pigments* called **carotenoids**, including β-carotene, which contain a linear system of conjugated double bonds:

[β - Carotene structure]

Carotenoids absorb light primarily in the blue and green region of the spectrum (Figure 6.6), while reflecting those of the yellow, orange, and red regions. Carotenoids produce the characteristic colors of carrots and oranges, and the leaves of some plants during the fall. Carotenoids have multiple functions: they act as secondary light collectors during photosynthesis, and they draw excess energy away from excited chlorophyll molecules and dissipate it as heat. If this excess energy were not absorbed by carotenoids, it could be transferred to oxygen, producing an excited and reactive form of the atom called singlet oxygen ($^1O^*$) that can destroy biological molecules and cause cell death. Mutant algal cells lacking carotenoids cannot survive in an aerobic environment because of oxygen-mediated destruction of their photosynthetic machinery.

Because the light falling on a leaf is composed of a variety of wavelengths, the presence of pigments with varying absorption properties ensures that a greater percentage of incoming photons will stimulate photosynthesis. This

can be seen by examining an **action spectrum** (Figure 6.7), which is a plot of the relative rate (or efficiency) of photosynthesis produced by light of various wavelengths. Unlike an absorption spectrum, which simply measures the wavelengths of light that are absorbed by particular pigments, an action spectrum identifies the wavelengths that are effective in bringing about a given physiological response. The action spectrum for photosynthesis follows the absorption spectrum of chlorophylls and carotenoids fairly closely, reflecting the participation of these pigments in the photosynthetic process.

REVIEW ?

1. What is the difference between an absorption spectrum and an action spectrum?
2. Compare the structure, absorption, and function of chlorophylls and carotenoids.

6.4 **PHOTOSYNTHETIC UNITS AND REACTION CENTERS**

In 1932, Robert Emerson and William Arnold of the California Institute of Technology carried out an experiment suggesting that not all of the chlorophyll molecules in a chloroplast were directly involved in conversion of light energy into chemical energy. Using suspensions of the green alga *Chlorella* and flashing lights of short duration (e.g., 10 msec) at saturating intensity, they determined the

minimum amount of light needed to produce maximal O_2 production during photosynthesis. Based on the number of chlorophyll molecules present in the preparation, they calculated that one molecule of oxygen was being released during a brief flash of light for every 2,500 molecules of chlorophyll present. It was later shown by Emerson that a minimum of eight photons (*quanta* of light) must be absorbed to produce one molecule of O_2, which meant that chloroplasts contain approximately 300 times as many chlorophyll molecules as would appear to be necessary to oxidize water and generate O_2.

One possible interpretation of this finding is that only a very small percentage of chlorophyll molecules is involved directly in photosynthesis. This is not the case, however. Rather, all 300 or so chlorophyll molecules act together as one **photosynthetic unit** in which only one member of the group—**the reaction-center chlorophyll**—actually transfers electrons to an electron acceptor. Even though the bulk of the pigment molecules do not participate *directly* in the conversion of light energy into chemical energy, they are responsible for light absorption; they form a **light-harvesting antenna** that absorbs photons of varying wavelength and transfers that energy (called *excitation energy*) very rapidly to the pigment molecule at the reaction center.

The transfer of excitation energy from one pigment molecule to another is very sensitive to the distance between the molecules. The chlorophyll molecules of an antenna are kept in close proximity (less than 1.5 nm separation) by noncovalent linkage to integral membrane polypeptides. These polypeptides modify the absorption properties of the bound chlorophyll molecules and provide a scaffolding that holds them in fixed orientations that favor energy transfer. A "rule" that operates among antenna pigments is that energy can only be transferred to an equal or less energy-requiring molecule. In other words, energy can only be passed to a pigment molecule that absorbs light of equal or longer wavelength (lower energy) than that absorbed by the donor molecule. As the energy "wanders" through a photosynthetic unit (Figure 6.8), it is repeatedly transferred to a pigment molecule that absorbs at a longer wavelength. The energy is ultimately transferred to the chlorophyll of the reaction center, which absorbs light of longer wavelength than any of its neighbors. It takes an estimated 200×10^{-12} seconds for energy to travel from the site of absorption to the reaction-center pigment. Once the energy is received by the reaction center, the electron excited by light absorption can be transferred to its acceptor.

Oxygen Formation: Coordinating the Action of Two Different Photosynthetic Systems

The evolution of organisms capable of using H_2O as an electron source was accompanied by major changes in the photosynthetic machinery. The reason for these changes is revealed by considering the energetics of oxygenic

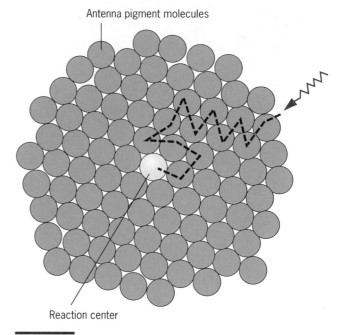

Antenna pigment molecules

Reaction center

Figure 6.8 The transfer of excitation energy. Energy is transferred randomly through a network of pigment molecules that absorb light of increasingly longer wavelength until the energy reaches the reaction-center chlorophyll, which transfers an excited electron to a primary acceptor, as described later in the chapter.

(O_2-releasing) photosynthesis. The O_2-H_2O couple has a standard redox potential of +0.82 V, whereas that of the $NADP^+$-NADPH couple is −0.32 V (Table 5.1, page 194). The difference between the redox potentials of these two couples (1.14 V) provides a measure of the minimal energy that must be absorbed by the system to remove an electron from H_2O and pass it to $NADP^+$ *under standard conditions*. However, cells don't operate under standard conditions, and the transfer of electrons from H_2O to $NADP^+$ requires more than the minimal input of energy. It is estimated that, during actual operations in the chloroplast, more than 2 V of energy are utilized to carry out this oxidation-reduction reaction. (This value of 2 V is estimated from the left-hand scale of Figure 6.9, which extends from below +1 V to above −1 V.) It was noted on page 222 that one mole of photons of 680 nm wavelength (red light) is equivalent to a change in redox potential of 1.8 V. Thus, while it is theoretically possible for one photon of red light to boost an electron to an energy level required to reduce $NADP^+$ under standard conditions (i.e., 1.14 V), the process is accomplished in the cell through the combined action of two different light-absorbing reactions. The light-absorbing reactions of photosynthesis occur in large pigment-protein complexes called **photosystems.** Two types of photosystems are required to catalyze the two light-absorbing reactions utilized in oxygenic photosynthesis. Each photosystem boosts electrons part of the way up the energy hill (Figure 6.9), not unlike

Figure 6.9 **An overview of the flow of electrons during the light-dependent reactions of photosynthesis.** The events depicted in this schematic drawing are described in detail in the following pages. The various carriers involved in photosynthetic electron transport are indicated in Figure 6.15*b*. The energy content of electrons is given in volts. To convert these values to calories, multiply by the Faraday constant, 23.06 kcal/V. For example, a difference of 2.0 V corresponds to an energy difference of 46 kcal/mol of electrons. This can be compared to the energy of red light (680 nm), which contains about 42 kcal/mol of photons.

the way a two-stage chairlift moves skiers up the side of a particularly long slope. One photosystem, **photosystem II (PSII)**, boosts electrons from an energy level below that of water to a midway point. The other photosystem, **photosystem I (PSI)**, raises electrons from a midway point to an energy level well above that of NADP$^+$. The two photosystems act in series, that is, one after the other. Even though they mediate distinctly different photochemical reactions, the two types of photosystems in plants, as well as those of bacterial cells, exhibit marked similarities in protein composition and overall architecture. These shared properties suggest that all photosynthetic reaction centers have evolved from a common ancestral structure that has been conserved for more than three billion years. Light-harvesting antenna complexes, in contrast, show great variability and adaptation to particular needs.

The reaction-center chlorophyll of photosystem II is referred to as **P680**, "P" standing for "pigment" and "680"

standing for the wavelength of light that this particular chlorophyll molecule absorbs most strongly. The reaction-center chlorophyll of photosystem I is referred to as **P700** for similar reasons. When sunlight strikes a thylakoid membrane, energy is absorbed by antenna pigments of both PSII and PSI and passed to the reaction centers of both photosystems. Electrons of both reaction-center pigments are boosted to an outer orbital, and each electron is transferred to a **primary electron acceptor**. The transfer of electrons out of the photosystems leaves the two reaction-center pigments missing an electron and thus positively charged. After losing their electrons, the reaction-center chlorophylls of PSII and PSI can be denoted as P680$^+$ and P700$^+$, respectively. Positively charged reaction centers act as attractants for electrons, which sets the stage for the flow of electrons along a chain of specific carriers. In oxygenic photosynthesis, where the two photosystems act in series, electron flow occurs along three

legs—from water to PSII, from PSII to PSI, and from PSI to NADP$^+$—an arrangement described as the **Z scheme**, which was first proposed by Robert Hill and Derek Bendall of the University of Cambridge. The broad outline of the Z scheme is illustrated in Figure 6.9; we will fill in the names of some of the specific components as we examine each of the major parts of the pathway. Like the members of the respiratory chain of mitochondria (Chapter 5), most of the electron carriers of the Z scheme are found as part of large membrane protein complexes (see Figure 6.15). As in mitochondria, electron transfer releases energy, which is used to establish a proton gradient, which in turn drives the synthesis of ATP. As discussed on page 236, ATP produced in the chloroplast is used primarily within the organelle in the synthesis of carbohydrates; ATP utilized outside the chloroplast is derived largely from that produced in plant cell mitochondria.

PSII Operations: Obtaining Electrons by Splitting Water

Photosystem II uses absorbed light energy for two interrelated activities: removing electrons from water and generating a proton gradient. The PSII of plant cells is a complex of more than 20 different polypeptides embedded in the thylakoid membrane. Two of these proteins, designated D1 and D2, are particularly important because together they bind the P680 reaction-center chlorophyll and the molecules active in electron transport (Figure 6.10).

The first step in PSII activation is the absorption of light by an antenna pigment. Most of the antenna pigments that collect light for PSII reside within a separate pigment-protein complex, called **light-harvesting complex II** or, simply, **LHCII**. LHCII proteins bind both chlorophylls and carotenoids and are situated outside the core of the photosystem (Figure 6.10). The pigments of the LHC are held in close contact with one another, facilitating rapid energy transfer toward the interior of the photosystem. As discussed in the Experimental Pathways , which can be accessed on the Web at www.wiley.com/college/karp, LHCII is not always associated with PSII but, under appropriate conditions, can migrate along the thylakoid membrane and become associated with PSI, serving as a light-harvesting complex for the PSI reaction center as well.

The Flow of Electrons from PSII to Plastoquinone Excitation energy is passed from the outer-antenna pigments of LHCII to a small number of inner-antenna chlorophyll molecules situated within the core of PSII. From there, the energy is ultimately passed to the PSII reaction-center pigment, which is apparently one of four closely spaced chlorophyll *a* molecules (Figure 6.10). The excited reaction-center pigment (P680*) responds (within about 3 picoseconds) by transferring a single photoexcited electron to a closely associated, chlorophyll-like pheophytin molecule (step 1, Figure 6.10), which is the primary electron acceptor. This electron transfer generates a positively charged donor (P680$^+$) and a negatively charged acceptor (Pheo$^-$). The importance of the formation of two oppositely charged species, P680$^+$ and Pheo$^-$, becomes more apparent when we consider the oxidation-reduction capabilities of these two species. P680$^+$ is electron-deficient and can accept electrons, making it an oxidizing agent. In contrast, Pheo$^-$ has an extra electron that it will readily lose, making it a reducing agent. This event—the light-driven formation of an oxidizing agent and a reducing agent—takes less than one billionth of a second and is the essential first step in photosynthesis.

Because they are oppositely charged, P680$^+$ and Pheo$^-$ exhibit an obvious attraction to one another. Interaction between the oppositely charged species is prevented by moving the charges farther apart, ultimately to opposite sides of the membrane, by passage through several different sites. Pheo$^-$ first transfers its electron (step 2, Figure 6.10) to a molecule of plastoquinone (designated PQ$_A$ in Figure 6.10) bound to the D2 protein near the outer (stromal) side of the membrane. Plastoquinone (PQ) is a lipid-soluble molecule (Figure 6.11) similar in structure to ubiquinone (see Figure 5.11c). The electron from PQ$_A$ is transferred (step 3, Figure 6.10) to a second plastoquinone (designated PQ$_B$ in Figure 6.10) to produce a semi-reduced form of the molecule (PQ$_B$·$^-$) that remains firmly bound to the D1 protein of the reaction center. With each of these transfers, the electron moves closer to the stromal side of the membrane.

The positively charged pigment (P680$^+$) is reduced back to P680 (as described below), which primes the reaction center for the absorption of another photon. The absorption of a second photon sends a second energized electron along the path from P680 to pheophytin to PQ$_A$ to PQ$_B$·$^-$, forming PQ$_B^{2-}$ (step 4, Figure 6.10), which combines with two protons to form plastoquinol, PQH$_2$ (step 5, Figure 6.10; Figure 6.11). The protons utilized in the formation of PQH$_2$ are derived from the stroma, causing a decrease in H$^+$ concentration of the stroma, which contributes to formation of the proton gradient. The reduced PQH$_2$ molecule dissociates from the D1 protein and diffuses into the lipid bilayer. The displaced PQH$_2$ is replaced by a fully oxidized PQ molecule derived from a small "pool" of plastoquinone molecules in the bilayer (step 6, Figure 6.10). We will follow the fate of the electrons (and protons) carried by PQH$_2$ in a following section.

The Flow of Electrons from Water to PSII The least understood part of the flow of electrons from water to NADP$^+$ is the first step between H$_2$O and PSII. The splitting of water is highly endergonic because of the stable association of its hydrogen and oxygen atoms. Splitting water in a laboratory requires the use of a strong electric current or temperatures approaching 2,000°C. Yet a plant cell can accomplish this feat on a snowy mountainside using only the energy of visible light.

We saw in the previous section how the absorption of light by PSII leads to the formation of two charged mole-

Figure 6.10 **The functional organization of photosystem II.**
(a) This drawing shows a simplified model of the huge protein-pigment complex, which catalyzes the light-driven oxidation of water and reduction of plastoquinone. The path taken by electrons is indicated by the yellow arrows. Events begin with the absorption of light by an antenna pigment in the outer light-harvesting complex (LHCII). Energy is transferred from LHCII through an inner antenna pigment-protein complex to the P680 reaction-center chlorophyll *a*, which recent studies suggest is present within a cluster of four chlorophyll *a* molecules. Absorption of this energy by P680 excites an electron, which is transferred to pheophytin (Pheo) (step 1), the primary electron acceptor of PSII. The electron subsequently passes to a plastoquinone PQ_A (step 2) and then to PQ_B (step 3) to form a negatively charged free radical ($PQ_B \cdot \,^-$). Absorption of a second photon sends a second electron along the same pathway, converting the acceptor to $PQ_B{}^{2-}$ (step 4). Two protons then enter from the stroma (step 5) generating PQH_2, which is released into the lipid bilayer and replaced by a new oxidized PQ_B molecule (step 6). As the above events are occurring, electrons are moving from

H_2O by way of Tyr_z to the positively charged reaction-center pigment (steps B and A). Thus, overall, PSII catalyzes the transfer of electrons from water to plastoquinone. The oxidation of 2 molecules of H_2O to release one molecule of O_2 generates two molecules of PQH_2. Because the oxidation of water releases protons into the thylakoid lumen and the reduction of $PQ_B{}^{2-}$ removes protons from the stroma, the operation of PSII makes a major contribution to formation of a H^+ gradient. (PSII is present in the membrane as a dimer, so that another copy of the reaction center and oxygen-evolving complex would be present behind the one shown here.) *(b)* A structural model of the PSII complex illustrated in part *a*. The oxygen-evolving complex is seen to be present as a peripheral membrane protein at the thylakoid surface. The yellow, orange, and purple helices in the center of the protein complex are part of the D1 and D2 subunits, the green regions to each side of D1/D2 comprise the inner antenna, and green helices at the far lateral ends correspond to the LHCII antenna complex. (B: REPRINTED WITH PERMISSION FROM JON NIELD, ET AL. NATURE 400:481, 1999, COURTESY OF JON NIELD AND JAMES BARBER. COPYRIGHT 1999 MACMILLAN MAGAZINES LIMITED.)

cules, $P680^+$ and $Pheo^-$. We have followed the route taken by the excited electron associated with $Pheo^-$; now we turn to the other species, $P680^+$, which is the most powerful oxidizing agent yet to be discovered in a biologi-

cal system. It is estimated that the redox potential of the oxidized form of P680 is $+1.17$ V (Figure 6.9), which is sufficiently strong to pull tightly held (low-energy) electrons from water (redox potential of $+0.82$ V), thus split-

Plastoquinone

1e⁻

1e⁻

2H⁺

Plastoquinol

Figure 6.11 Plastoquinone. The acceptance of two electrons and two protons reduces PQ (plastoquinone) to PQH_2 (plastoquinol). The intermediates are similar to those shown in Figure 5.11c for ubiquinone of the mitochondrion.

ting the molecule. The splitting of water during photosynthesis is called **photolysis**. The formation of one molecule of oxygen during photolysis is thought to require the *simultaneous* loss of four electrons from two molecules of water according to the reaction

$$2H_2O \rightarrow 4H^+{}_{lumen} + O_2 + 4e^-$$ (overall PSII reaction)

Yet one PSII reaction center can only generate one positive charge ($P680^+$), or oxidizing equivalent, at a time. This problem is apparently solved by the following mechanism, which was proposed around 1970 by Pierre Joliot and Bessel Kok. Closely associated with the D1 protein of PSII at its luminal surface is a complex of several peripheral subunits that bind a cluster of four manganese (Mn) ions (Figure 6.10). The Mn cluster accumulates four successive positive charges by transferring four electrons, one at a time, to the nearby $P680^+$.[1] The transfer of electrons from the Mn cluster to $P680^+$ (steps B and A of Figure 6.10) is accomplished by passage through a positively charged tyrosine residue of the D1 protein, denoted as $Tyr_z{}^+$ in the figure. After each electron is transferred to $P680^+$ regenerating P680, the pigment becomes reoxidized (back to $P680^+$) following the absorption of another photon by the photosystem. Thus, the accumulation of four positive charges (oxidizing equivalents) by the Mn cluster is driven by the successive absorption of four pho-

[1]There is controversy concerning the mechanism by which the Mn ions accumulate positive charge and whether additional elements of the proteins share this role.

Figure 6.12

Figure 6.12 **Measuring the kinetics of O_2 release.** The plot shows the response by isolated chloroplasts that have been kept in the dark to a succession of light flashes of very short duration. The amount of oxygen released peaks with every fourth flash of light. The first peak occurs after three flashes (rather than four) because most of the manganese cluster is present in the S_1 state (one oxidizing equivalent) when kept in the dark. The oscillations become damped as the number of flashes increases.

tons of light by the PSII photosystem. Once four charges have accumulated, the oxygen-evolving complex of PSII is capable of catalyzing the removal of 4 e^- from 2 H_2O molecules, as indicated in the following manner: This process can be written as follows:

where the subscript on the S indicates the number of positive charges (oxidizing equivalents) stored by the Mn cluster.

The protons produced in the photolysis reaction are retained in the thylakoid lumen (Figure 6.10), where they contribute to the proton gradient. The four electrons produced in the photolysis reaction serve to regenerate the fully reduced Mn cluster (S_0 state), while the O_2 is released as a waste product into the environment. Evidence for the accumulation of successive oxidizing equivalents was first obtained by exposure of algal cells to very brief (1 μsec) flashes of light. Virtually no oxygen is released after one or two flashes. Instead, the cells must be illuminated several times before measurable O_2 production occurs, indicating that the effect of individual photoreactions must accumulate before O_2 can be released (Figure 6.12).

From PSII to PSI It was described earlier how the successive absorption of two photons by the reaction center of PSII leads to the formation of a fully reduced PQH_2 molecule. Consequently, the production of a single O_2 molecule, which requires the absorption of four photons by PSII, leads to the formation of two molecules of PQH_2. PQH_2 is a mobile electron carrier capable of diffusing through the lipid bilayer of the thylakoid membrane and

Figure 6.13 **Electron transport between PSII and PSI.** The flow of a pair of electrons is indicated by the yellow arrow. Cytochrome b_6f operates in a manner very similar to that of cytochrome bc_1 in the mitochondria and engages in a Q cycle (not discussed in the text) that translocates four protons for every pair of electrons that moves through the complex. PQH$_2$ and PC are mobile carriers that can transport electrons between distantly separated photosystems.

Figure 6.14 **The functional organization of photosystem I.** Events begin with the absorption of light by an antenna pigment (not shown) and transfer of the energy to a P700 chlorophyll at the PSI reaction center. Absorption of energy by P700 causes the excitation of an electron and its transfer (step 1) to A_0, which is the primary electron acceptor of PSI. The electron subsequently passes to A_1 (step 2) and then to an iron-sulfur center named F_X (step 3). From F_X the electron is transferred through two more iron-sulfur centers (F_A and F_B) and finally to ferredoxin, a small iron-sulfur protein (step 5) that is peripheral to the PSI complex. When two different ferredoxin molecules have accepted an electron, they act together to reduce a molecule of NADP$^+$ to NADPH (step 6). The electron-deficient reaction-center pigment (P700$^+$) is reduced by an electron donated by plastocyanin (step A).

binding to a large multiprotein complex called *cytochrome b_6f* (Figure 6.13). Cytochrome b_6f is similar in structure and function to cytochrome bc_1 of the electron transport chain of mitochondria (page 199); both complexes share the same redox groups, both can be poisoned by some of the same inhibitors, and both engage in a Q cycle that translocates 4 H$^+$ for each pair of electrons. Because these protons had originally been derived from the stroma, their release into the lumen constitutes a movement of protons across the thylakoid membrane (see Figure 6.15*a*). The electrons from cytochrome b_6f are passed to another mobile electron carrier, a water-soluble, copper-containing, peripheral membrane protein called *plastocyanin*, situated on the luminal side of the thylakoid membrane (Figure 6.13). Plastocyanin carries electrons to the luminal side of PSI, where they are transferred to P700$^+$, the positively charged reaction-center pigment of PSI.

Keep in mind that all of the electron transfers described in this discussion are exergonic and occur as electrons are passed to carriers with increasing affinity for electrons (more positive redox potentials, page 192). The need for mobile electron carriers, such as PQH$_2$ and plastocyanin, became evident when it was discovered that the two types of photosystems (PSII and PSI) are not situated in close proximity to one another in the membrane but rather are separated by distances on the order of 0.1 μm. The experiments that led to this discovery are described in detail in the Chapter 6 Experimental Pathways, which can be found on the Web at www.wiley.com/college/karp.

PSI Operations: The Production of NADPH In 2000, the three-dimensional structure of PSI from a cyanobacterium was determined at 2.5Å resolution. This model has provided the structural basis for understanding the events that occur in this large pigment-protein complex in higher plants. The PSI core of the cyanobacterium contains 11

different polypeptides; in higher plants, the corresponding polypeptides are encoded by both the nuclear and chloroplast genomes. The photochemical events that take place in PSI begin with the absorption of photons by an antenna complex containing both chlorophyll and β-carotene molecules.[2]

Energy is passed from antenna pigments to the PSI reaction-center pigment P700, which is a chlorophyll *a* dimer (Figure 6.14). Following energy absorption, the excited reaction-center pigment P700* transfers an electron

[2]In the cyanobacterium, all the antenna pigments of PSI are bound to the same proteins that bind the redox centers required for electron transport. Higher plants have an additional antenna system, called LHCI, associated with the PSI core.

to a separate monomeric chlorophyll *a* molecule (designated A_0), which acts as the primary electron acceptor (step 1, Figure 6.14). As in PSII, absorption of light leads to the production of two charged species, in this case $P700^+$ and A_0^-. A_0^- is a very strong reducing agent with a redox potential of approximately -1.0 V, which is well above that needed to reduce $NADP^+$ (redox potential of -0.32 V). The electron hole in the $P700^+$ pigment is filled with an incoming electron donated by plastocyanin, as noted earlier.

The initial separation of charge in PSI is stabilized by passage of the electron from A_0^- through several "hands," beginning with a type of quinone called *phylloquinone* (designated A_1) and then three iron-sulfur clusters (designated F_X, F_B, and F_A) (steps 2–4, Figure 6.14). The oxidation of P700 to $P700^+$ occurs at the luminal side of the membrane. As indicated in Figure 6.14, the electron that is lost to the primary acceptor passes through PSI to the iron-sulfur centers bound at the stromal side of the membrane. The electron is subsequently transferred out of PSI to a small, water-soluble, iron-sulfur protein called *ferredoxin* (step 5, Figure 6.14) associated with the stromal surface of the membrane. The reduction of $NADP^+$ to form NADPH (step 6, Figure 6.14) is catalyzed by a large enzyme called ferredoxin-$NADP^+$ reductase, which contains an FAD prosthetic group capable of accepting and transferring two electrons (page 195). An individual ferredoxin molecule can donate only one electron, so that two ferredoxins act together in the reduction:

$$2\ Ferredoxin_{red} + H^+ + NADP^+ \xrightarrow{\text{Ferredoxin-NADP}^+ \text{ Reductase}}$$
$$2\ Ferredoxin_{ox} + NADPH$$

The removal of a proton from the stroma also adds to the proton gradient across the thylakoid membrane. We can write the overall reaction for PS I, based on the absorption of four photons as was done for PSII, as

$$4\ e^- + 2H^+_{stroma} + 2\ NADP^+ \rightarrow 2\ NADPH$$
$$\text{(overall PSI reaction)}$$

Not all electrons passed to ferredoxin inevitably end up in NADPH; alternate routes can be taken depending on the particular organism and the conditions at the time. For example, electrons from PSI can pass to various inorganic acceptors, which are thereby reduced. These paths for electrons can lead to the eventual reduction of nitrate (NO_3^-) to ammonia (NH_3), or of sulfate (SO_4^{2-}) to sulfhydryl (—SH), reactions that convert inorganic "wastes" into compounds necessary for life. Thus, the energy of sunlight is used not only to reduce the most oxidized carbon atoms (those in CO_2), but also to reduce highly oxidized forms of nitrogen and sulfur atoms as well.

If we look back over the entire process of electron transport that takes place during oxygenic photosynthesis (summarized in Figure 6.15), we can see that electrons travel from water to $NADP^+$ by the action of two light-absorbing photosystems. Events occurring in PSII generate a strong oxidizing agent capable of producing O_2 from water, whereas events in PSI generate a strong reducing agent capable of producing NADPH from $NADP^+$. These two events lie at the opposite ends of redox chemistry in living organisms. As noted on page 229, the production of one molecule of O_2 requires the removal of four electrons from two molecules of water. The removal of four electrons from water requires the absorption of four photons, one for each electron. At the same time, the reduction of one molecule of $NADP^+$ requires the transfer of two electrons. Thus, hypothetically, if only one photosystem were able to transfer electrons from H_2O to $NADP^+$, four photons would be sufficient to produce two molecules of NADPH. Because two photosystems are utilized in the cell, that number is doubled to eight. In other words, a total of eight moles of photons must be absorbed by the cell to generate one mole of molecular oxygen and two moles of NADPH. Thus, if we add the reactions of PSII (page 229) and PSI (left column), *ignoring the protons for a moment*, we arrive at an overall equation for the light reactions of

$$2\ H_2O + 2\ NADP^+ \rightarrow O_2 + 2\ NADPH$$
$$\text{(overall light reaction)}$$

In addition, the light reactions of photosynthesis establish a proton gradient across the thylakoid membrane that leads to the formation of ATP. The proton gradient forms as the result of the removal of H^+ from the stroma and the addition of H^+ to the thylakoid lumen. Contributions to the proton gradient (shown in Figure 6.15) arise from (1) the splitting of water in the lumen, (2) translocation from the stroma to the lumen by cytochrome b_6f, and (3) reduction of $NADP^+$ in the stroma.

Killing Weeds by Inhibiting Electron Transport

The light reactions of photosynthesis employ a considerable number of electron carriers, which serve as targets for a variety of different plant-killing chemicals (herbicides). A number of common herbicides, including diuron, atrazine, and terbutryn, act by binding to the D1 protein of PSII. We saw on page 227 how absorption of light by PSII leads to production of a PQH_2 molecule that is subsequently released from a site (the Q_B site) on the D1 protein and replaced by a PQ from the pool. The herbicides listed above act by binding to the open Q_B site after release of PQH_2, blocking electron transport through PSII. The herbicide paraquat has received attention in the news media because it is used to kill marijuana plants and because its residues are highly toxic to humans. Paraquat interferes with PSI function by competing with ferredoxin for electrons from the PSI reaction center. Electrons that are diverted to paraquat are subsequently used to reduce O_2, generating highly reactive oxygen radicals (page 35) that damage the chloroplasts and kill the plant.

1. What is the relationship between the energy content of a photon and the wavelength of light? How does the wavelength of light determine whether or not it will stimulate photosynthesis? How do the absorbance properties of photosynthetic pigments determine the direction in which energy is transferred within a photosynthetic unit?

2. What is the role in photosynthesis of the light-harvesting antenna pigments?

3. Describe the sequence of events following the absorption of a photon of light by the reaction-center pigment of photosystem II. Describe the comparable events in photosystem I. How are the two photosystems linked to one another?

4. Describe the difference in the redox potentials of the reaction-center pigments of the two photosystems.

5. Describe the process by which water is split during photolysis. How many photons must be absorbed by PSII for this to occur?

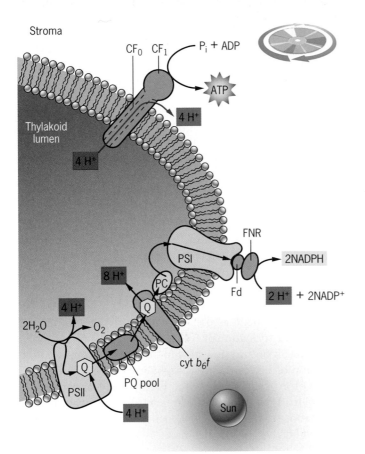

(a)

Figure 6.15 Summary of the light-dependent reactions. (a) Summary of the flow of electrons from H_2O to NADPH through the three transmembrane complexes. The estimated number of protons translocated through the membrane as the result of the oxidation of two molecules of water is shown. The ATP synthase of the thylakoid membranes is also shown and is discussed in the following section. Four protons are likely required for the synthesis of each molecule of ATP (page 210). The actual distribution of photosynthetic complexes within thylakoid membranes is discussed in the Experimental Pathways on the Web. (b) A detailed version of the Z scheme showing the energetics of the two photosystems and the various electron carriers involved in the light-dependent reactions. The dashed line shows the path of electron flow during cyclic photophosphorylation.

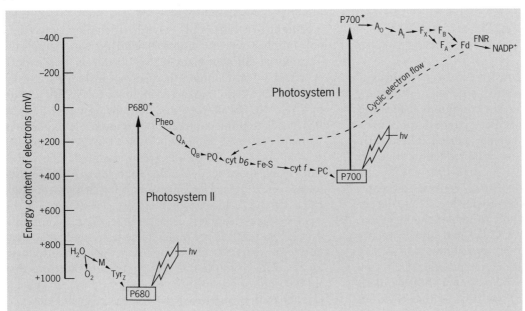

(b)

6.5 PHOTOPHOSPHORYLATION

The light-dependent reactions discussed in previous pages provide the energy required to reduce CO_2 to carbohydrate. In quantitative terms, conversion of one mole of CO_2 to one mole of carbohydrate (CH_2O) requires the input of three moles of ATP and two moles of NADPH (see Figure 6.18). We have seen how plant cells produce the NADPH required for the manufacture of carbohydrate; let us now examine how these same cells produce the required ATP.

The machinery for ATP synthesis in a chloroplast is virtually identical to that of a mitochondrion or a plasma membrane of an aerobic bacteria. As in those cases, the ATP synthase (Figure 6.15) consists of a head (called CF_1 in chloroplasts), which contains the catalytic site of the enzyme, and a base (CF_0), which spans the membrane and mediates proton movement. The two parts are connected by a stalk. The various parts of the ATP synthase are constructed of polypeptides that are homologous to those of the bacterial and mitochondrial enzymes, which were discussed at length in Chapter 5 (beginning on page 203).

ATP synthases are not uniformly dispersed throughout the thylakoid membrane, but are preferentially localized in the stroma thylakoids and in the exposed regions of the grana stacks. The CF_1 heads project outward into the stroma in keeping with the orientation of the proton gradient, which has its higher concentration within the thylakoid lumen (Figure 6.15). Thus, protons move from higher concentration in the lumen through the CF_0 base of the ATP synthase and into the stroma, thereby driving phosphorylation of ADP as described in Chapter 5 for the mitochondrion.

Measurements made during periods of maximal ATP synthesis suggest that differences in H^+ concentrations of 1,000 to 2,000 fold exist across the thylakoid membranes, corresponding to a pH gradient (ΔpH) of about 3 units. The movement of protons into the lumen during electron transport is compensated by movement of other ions, so that a significant membrane potential is not built up. Thus, unlike in mitochondria where the proton-motive force is expressed primarily as an electrochemical potential (page 202), the proton-motive force (Δp) acting in the chloroplasts is largely or exclusively due to a gradient of pH.

Noncyclic Versus Cyclic Photophosphorylation

The formation of ATP during the process of oxygenic photosynthesis is called **noncyclic photophosphorylation** because electrons move in a linear (i.e., noncyclic) path from H_2O to $NADP^+$ (Figure 6.15). During the 1950s, Daniel Arnon of the University of California, Berkeley, discovered that isolated chloroplasts were not only capable of synthesizing ATP from ADP but could do so even in the absence of added CO_2 or $NADP^+$. These experiments indicated that chloroplasts had a means for ATP formation that did not require most of the photosynthetic reac-

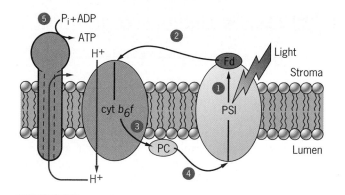

Figure 6.16 **Simplified scheme for cyclic photophosphorylation.** Absorption of light by PSI excites an electron, which is transferred to ferredoxin (step 1) and on to cytochrome b_6f (step 2), plastocyanin (step 3), and back to $P700^+$ (step 4). In the process, protons are translocated by cytochrome b_6f to form a gradient utilized for ATP synthesis (step 5).

tions that would have led to oxygen production, CO_2 fixation, or $NADP^+$ reduction. All that was necessary was illumination, chloroplasts, ADP, and P_i. The process Arnon had discovered was later called **cyclic photophosphorylation** and is illustrated in Figure 6.16. Cyclic photophosphorylation is carried out by PSI independent of PSII. The process begins with the absorption of a quantum of light by PSI and transfer of a high-energy electron to the primary acceptor. From there, the electron is passed along to ferredoxin, as is always the case, but rather than being transferred to $NADP^+$, the electron is passed back to the electron-deficient reaction center (as shown in Figure 6.16) to complete the cycle. During the flow of an electron around this course, sufficient free energy is released to translocate protons (estimated at two H^+/e^-) across the membrane by the cytochrome b_6f complex and to build a proton gradient capable of driving ATP synthesis. Cyclic photophosphorylation is thought to provide additional ATP required for carbohydrate synthesis (see Figure 6.18) and also for other ATP-requiring activities in the chloroplast (e.g., molecular chaperone involvement in protein import).

Now that we have seen how the light reactions of photosynthesis lead to the production of ATP and NADPH, which are the energy-rich compounds required for the synthesis of carbohydrates, we can turn to the so-called dark reactions that lead to carbohydrate formation.

REVIEW ?

1. Which steps in the light-dependent reactions are responsible for generating an electrochemical proton gradient across the thylakoid membrane?

2. To what degree is this gradient reflected in a pH gradient versus a voltage?

3. How does the proton gradient lead to the formation of ATP?

6.6 CARBON DIOXIDE FIXATION AND THE SYNTHESIS OF CARBOHYDRATE

After World War II, Melvin Calvin and his colleagues at the University of California, Berkeley, began what was to be a decade-long study of the enzymatic reactions by which carbon dioxide was assimilated into the organic molecules of the cell. Armed with the newly available, long-lived radioactive isotope of carbon (^{14}C) and a new technique, two-dimensional paper chromatography, they began the task of identifying all the labeled molecules produced when cells took up [^{14}C]O_2. The studies began with plant leaves but soon shifted to a simpler system, the green alga *Chlorella*. Algal cultures were grown in closed chambers in the presence of unlabeled CO_2, after which radioactive CO_2 was introduced by injection into the culture medium. After the desired period of incubation with the labeled CO_2, the algal suspension was drained into a container of hot alcohol, which had the combined effects of killing the cells immediately, stopping enzyme activity, and extracting soluble molecules. Extracts of the cells were then placed as a spot on chromatographic paper and subjected to two-dimensional chromatography. To locate the radioactive compounds at the end of the procedure, a piece of X-ray film was pressed against the chromatogram, and the plates were kept in the dark to expose the film. After photographic development, identification of radiolabeled compounds on the autoradiogram was made by comparison to known standards and by chemical analysis of the original spots. Let us consider some of their findings.

Carbohydrate Synthesis in C₃ Plants

Labeled CO_2 was converted to reduced organic compounds very rapidly. If the incubation period was very short (up to a few seconds), one radioactive spot on the chromatogram predominated (Figure 6.17). The compound that formed the spot was determined to be 3-phosphoglycerate (PGA), one of the intermediates of glycolysis. Calvin initially suspected that CO_2 was being covalently linked (or **fixed**) to a two-carbon compound to form the three-carbon PGA molecule. Because the first intermediate to be identified was a three-carbon molecule, plants that utilize this pathway to fix atmospheric CO_2 are referred to as **C₃ plants.**

After considerable investigation it became apparent that the initial acceptor was not a two-carbon compound, but a five-carbon compound, ribulose 1,5-bisphosphate (RuBP), which condensed with CO_2 to produce a six-carbon molecule. This six-carbon compound rapidly fragmented into two molecules of PGA, one of which contained the recently added carbon atom. Both the condensation of RuBP and the splitting of the six-carbon product (Figure 6.18*a*) are carried out in the stroma by a large multisubunit enzyme, *ribulose bisphosphate carboxylase*, which is known familiarly as **Rubisco**. Rubisco is capable of fixing only

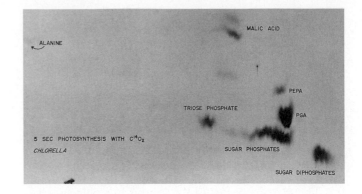

Figure 6.17 **Chromatogram showing the results of an experiment in which algal cells were incubated for 5 seconds in [^{14}C]O_2 prior to immersion in alcohol.** One spot, which corresponds to 3-phosphoglycerate (labeled PGA), contains most of the radioactivity. (COURTESY OF JAMES BASSHAM AND MELVIN CALVIN.)

about three molecules of CO_2 per second, which may be the worst turnover number of any known enzyme (see Table 3.4). To compensate for this inefficiency, as much as half of the protein of leaves is present as Rubisco. In fact, Rubisco constitutes the most abundant protein on Earth, accounting for nearly 10 kg for every human!

As the structure of the various intermediates was determined, together with the positions of the labeled carbon atoms, it became apparent that the pathway for conversion of CO_2 into carbohydrate was cyclic and complex. This pathway has become known as the **Calvin cycle**, and it occurs in cyanobacteria and all eukaryotic photosynthetic cells. A simplified version of the Calvin cycle is shown in Figure 6.18*b*. The cycle comprises three main parts: (1) carboxylation of RuBP to form PGA; (2) reduction of PGA to the level of a sugar (CH_2O) by formation of glyceraldehyde 3-phosphate (GAP) using the NADPH and ATP produced in the light-dependent reactions; and (3) the regeneration of RuBP, which also requires ATP. It can be seen in Figure 6.18*b* that for every 6 molecules of CO_2 fixed, 12 molecules of GAP are produced. (GAP is the spot marked triose phosphate on the chromatogram of Figure 6.17.) The atoms in 10 of these three-carbon GAP molecules are rearranged to regenerate 6 molecules of the five-carbon CO_2 acceptor, RuBP (Figure 6.18*b*). The other 2 GAP molecules can be considered as product. These GAP molecules can be exported to the cytosol in exchange for phosphate ions (see Figure 6.19) and used to synthesize the disaccharide sucrose. Alternatively, GAP can remain in the chloroplast where it is converted to starch. An overview of the entire process of photosynthesis, including the light reactions (light absorption, oxidation of water, reduction of NADP⁺, translocation of protons), phosphorylation of ADP, the Calvin cycle, and the synthesis of starch or sucrose, is shown in Figure 6.19.

Figure 6.18 Converting CO_2 into carbohydrate. (*a*) The reaction catalyzed by ribulose bisphosphate carboxylase (Rubisco) in which CO_2 is fixed by linkage to RuBP. The product rapidly splits into two molecules of 3-phosphoglycerate. (*b*) An abbreviated version of the Calvin cycle showing the fate of 6 molecules of CO_2 that are fixed by combination with 6 molecules of RuBP. (Numerous reactions have been deleted). The fixation of CO_2 is indicated in step 1. In step 2, the 12 PGA molecules are phosphorylated to form 12 1,3-bisphosphoglycerate (BPG) molecules, which are reduced in step 3 by electrons provided by

NADPH to form 12 molecules of glyceraldehyde 3-phosphate (GAP). In the case depicted here, 2 of the GAPs are drained away (step 4) to be used in the synthesis of sucrose in the cytosol, which can be considered the product of the light-independent reactions. The other 10 molecules are converted into 6 molecules of RuBP (step 5), which can act as the acceptor for 6 more molecules of CO_2. The regeneration of 6 RuBPs requires the hydrolysis of 6 molecules of ATP. The NADPH and ATP used in the formation of carbohydrate in the Calvin cycle are the two high-energy products of the light-dependent reactions.

Sucrose molecules that are formed in the cytosol from the GAPs of the Calvin cycle are transported out of the leaf cells and into the phloem, where they are carried to the various nonphotosynthetic organs of the plant. Just as glucose serves as the source of energy and organic building

blocks in most animals, sucrose serves an analogous role in most plants. Starch, on the other hand, is stored within chloroplasts as granules (see Figure 2.17*b*). Just as stored glycogen provides animals with readily available glucose in times of need, the starch stored in a plant's leaves provides

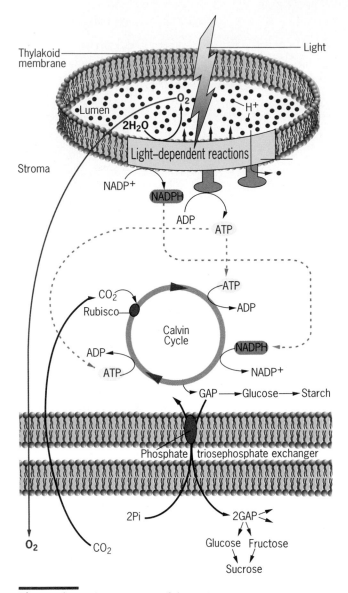

Figure 6.19 An overview of the various stages of photosynthesis.

Photorespiration One of the spots that appeared on the chromatograms from Calvin's early work on algal cells was identified as the compound glycolate, which was ignored (and correctly so) in formulating the Calvin cycle of Figure 6.18. While glycolate is not part of the Calvin cycle, it is a product of a reaction catalyzed by Rubisco. Approximately 20 years after it was found that Rubisco catalyzed the attachment of CO_2 to RuBP, it was discovered that Rubisco also catalyzed a second reaction in which O_2 is attached to RuBP to produce 2-phosphoglycolate (along with PGA) (Figure 6.20). Phosphoglycolate is subsequently converted to glycolate by an enzyme in the stroma. Glycolate produced in the chloroplast is transferred to the peroxisome and eventually leads to the release of CO_2 as described later (see Figure 6.22).

Because it involves the uptake of O_2 and the release of CO_2, this series of reactions is called **photorespiration**. Because photorespiration leads to the release of previously fixed CO_2 molecules, it is a waste of the plant's energy. In fact, photorespiration can account for the loss of up to 50 percent of newly fixed carbon dioxide by crop plants growing under conditions of high light intensity. Thus, as you might expect, a concerted effort has been underway for decades to breed plants that are less likely to engage in photorespiration. So far, these efforts have met with virtually no success.

Studies of the enzymatic activity of purified Rubisco show that the enzyme exhibits relatively little preference for CO_2 as a substrate over O_2. The reason for this lack of specificity is that neither CO_2 nor O_2 binds to the active site of the enzyme (see Figure 3.11a). Rather, the enzyme binds RuBP, which adopts the enediol form shown in Figure 6.20. This form of RuBP can then be attacked by either CO_2 or O_2. Viewed in this way, photorespiration appears to be an unavoidable consequence of the catalytic properties of Rubisco, an enzyme that is presumed to have evolved at a time when O_2 levels in the atmosphere were virtually nonexistent. Under today's atmospheric conditions, O_2 and CO_2 compete with one another, and the predominant direction of the Rubisco-catalyzed reaction is determined by the ratio of CO_2/O_2 that is available to the enzyme. When plants are grown in closed environments containing elevated levels of CO_2, they are capable of much more rapid growth by virtue of their elevated rates of CO_2 fixation. It is suggested that the rise in atmospheric CO_2 levels over the past century (from about 270 parts per million in 1870 to 380 ppm at present) is responsible for as much as 10 percent of the increased crop yield that has occurred during this period.[3] Two groups of plants, called C_4 and CAM plants, have overcome the negative effects of photorespiration by evolving metabolic mechanisms that increase the CO_2/O_2 ratio to which the Rubisco enzyme molecules are exposed.

it with sugars at night when the light-dependent reactions are not possible. It is evident from the reactions of Figure 6.18b that the synthesis of carbohydrate is an expensive activity. The conversion of 6 molecules of CO_2 to 1 six-carbon sugar molecule and the regeneration of RuBP requires 12 molecules of NADPH and 18 molecules of ATP. This large energy expenditure reflects the fact that CO_2 is the most highly oxidized, and thus least energetic, form in which carbon can occur.

Most of the enzymes of the Calvin cycle can be found in other pathways, including glycolysis. Thus, even though many of the intermediates formed in the Calvin cycle are also substrates in other pathways, their formation within the stroma of the chloroplasts isolates them from the enzymes of the cytosol. Once again, the importance of membranes in cell compartmentalization becomes evident.

[3]It can be noted that this rise in atmospheric CO_2 concentration is also thought to be responsible for global warming, that is, an overall rise in average temperatures on the planet. Studies suggest that CO_2 levels at the beginning of the Cenozoic era (about 65 million years ago) were much higher than they are today, as were global temperatures.

Figure 6.20 The reactions of photorespiration. Rubisco can catalyze two different reactions with RuBP as a substrate (shown in the enediol state within the plane, see Figure 6.18*a*). If the RuBP reacts with O_2 (step 1b), the reaction produces an oxygenase intermediate (step 2b) that breaks down into 3-PGA and 2-phosphoglycolate (step 3b). The subsequent reactions of phosphoglycolate are shown in Figure 6.22. The eventual out-come of these reactions is the release of CO_2, a molecule that the cell had previously expended energy to fix. In contrast, if the RuBP molecule reacts with CO_2 (step 1a), the reaction produces a carboxylase intermediate (step 2a) that breaks down into two molecules of PGA (step 3a), which continue through the Calvin cycle.

Carbohydrate Synthesis in C_4 Plants

In 1965, Hugo Kortschak reported that when sugarcane was supplied with $[^{14}C]O_2$, radioactivity first appeared in organic compounds containing four-carbon skeletons rather than the three-carbon PGA molecule found in other types of plants. Further analysis revealed that these four-carbon compounds (predominantly malate and oxaloac-etate) resulted from combination of CO_2 with phospho-enolpyruvate (PEP) and thus represented a second mechanism of fixation of atmospheric carbon dioxide (Figure 6.21). The enzyme responsible for linkage of CO_2 to PEP was named *phosphoenolpyruvate carboxylase*, and it cat-alyzes the first step of the **C_4 pathway.** Plants that utilize this pathway are referred to as **C_4 plants** and are represented primarily by tropical grasses. Before considering the fate of this newly fixed carbon atom, it is useful to examine the rea-sons why an alternate pathway of CO_2 fixation has evolved.

When a C_3 plant is placed in a closed chamber and its photosynthetic activity monitored, it is found that, once the plant reduces CO_2 levels in the chamber to approxi-mately 50 parts per million (ppm), the rate of CO_2 release by photorespiration equals the rate of CO_2 fixation by photosynthesis, so that net production of carbohydrate stops. In contrast, a plant that utilizes the C_4 pathway continues the net synthesis of carbohydrate until CO_2 lev-els have dropped to 1 to 2 ppm. The C_4 plants have this ability because PEP carboxylase continues to operate at much lower CO_2 levels than does Rubisco and is not in-hibited by O_2. But what is the value to a plant of being able to fix CO_2 at such low levels when the atmosphere invari-ably contains CO_2 at levels well above 300 ppm? The issue seems even more confusing because (1) the use of PEP re-quires an additional expenditure of ATP, and (2) plants have no direct biosynthetic pathway leading from the C_4 intermediates (malate and oxaloacetate) to carbohydrate.

The value of the C_4 pathway becomes apparent when C_4 plants are placed in a dry, hot environment similar to that in which many of them live. The most serious prob-lem faced by plants living in a hot, dry climate is loss of water through openings, called stomata, at the surfaces of their leaves. Although open stomata lead to loss of water, they also provide the channel through which CO_2 enters the leaves. C_4 plants are adapted to hot, arid environments because they are capable of closing their stomata to pre-vent water loss, yet they are still capable of maintaining sufficient CO_2 uptake to fuel their photosynthetic activity at a maximal rate. They are said to be high-efficiency pho-tosynthesizers because of their high levels of photosynthe-sis per unit of water lost. This is the reason why crabgrass, a C_4 plant, tends to take over a lawn, displacing the do-mestic C_3 grasses originally planted. Sugarcane, corn, and sorghum are the most important crop plants that utilize the C_4 pathway. Because most C_4 plants do not do nearly as well in cooler temperatures, their distribution is limited at higher southern and northern latitudes.

When the fate of the CO_2 fixed by means of the C_4 pathway is followed, it is found that the CO_2 group is soon

5 µm

Figure 6.21 **Structure and function in C_4 plants.** Electron micrograph of a transverse section through the leaf of a C_4 plant showing the spatial relationship between the mesophyll and bundle sheath cells. Superimposed on the micrograph are the reactions of CO_2 fixation that occur in each type of cell. In step 1, CO_2 is joined to PEP by the enzyme PEP carboxylase in a mesophyll cell that is located close to the leaf exterior. The four-carbon malate formed by the reaction is transported to the more centrally located bundle sheath cell (step 2), where the CO_2 is released. The CO_2 becomes highly concentrated in the bundle sheath cell, favoring the fixation of CO_2 by Rubisco to form 3-PGA (step 3), which can be circulated through the Calvin cycle. The pyruvate formed when CO_2 is released is shipped back to the mesophyll cell (step 4), where it is converted to PEP. Although the process requires the hydrolysis of ATP (step 5), the high CO_2/O_2 in the bundle sheath cell minimizes the rate of photorespiration. (ELECTRON MICROGRAPH COURTESY OF S. CRAIG.)

released, only to be captured by Rubisco and converted to metabolic intermediates of the C_3 pathway (Figure 6.21). The basis for this seemingly paradoxical metabolism becomes evident when the leaf anatomy of a C_4 plant is examined. Unlike those of C_3 plants, leaves of C_4 plants contain two concentric cylinders of cells. An outer cylinder is made of *mesophyll cells* and an inner cylinder of *bundle sheath cells* (Figure 6.21). The fixation of CO_2 to PEP occurs in the outer mesophyll cells. PEP carboxylase activity can continue, even when the leaf stomata are almost entirely closed and the CO_2 level in the cells is very low. Once formed, the C_4 products are transported to the thick-walled bundle sheath cells, which are sealed off from atmospheric gases. Once in the bundle sheath cells, the previously fixed CO_2 can be split from the C_4 carrier, producing a high CO_2 level in these interior cells—a level suitable for fixation by Rubisco. Carbon dioxide levels in the bundle sheath cells may be 100 times that of the mesophyll. Thus, the C_4 pathway provides a mechanism to drive CO_2 fixation by the less efficient C_3 pathway by "pumping" CO_2 into the bundle sheath. Once CO_2 is split from the four-carbon compound, the pyruvate formed returns to the mesophyll cells to be recharged as PEP (Figure 6.21). In addition to saving water, plants that utilize the C_4 pathway are able to generate high CO_2/O_2 ratios in the local Rubisco environment, thereby favoring the

process of CO_2 fixation over that of photorespiration. In fact, attempts to demonstrate photorespiration in the intact leaves of C_4 plants usually fail. C_4 plants are thought to have evolved during a relatively recent period (approximately 7–10 million years ago) when CO_2 levels were low.

Carbohydrate Synthesis in CAM Plants

Many desert plants, such as cacti, have another biochemical adaptation that allows them to survive in very hot, dry habitats. These plants are called **CAM plants** and they utilize PEP carboxylase in fixing atmospheric CO_2 just as C_4 plants.[4] Unlike C_4 plants, however, CAM species carry out light-dependent reactions and CO_2 fixation at different times of the day, rather than in different cells of the leaf. Whereas C_3 and C_4 plants open their stomata and fix CO_2 during the daytime, CAM plants keep their stomata tightly closed during the hot, dry daylight hours. Then, at night, when the rate of water vapor loss is greatly reduced, they open their stomata and fix CO_2 by means of PEP carboxylase. As more and more carbon dioxide is fixed during the night, the malate that is generated is transported across the tonoplast membrane into the cell's central vacuole. The

[4]CAM is an acronym for *crassulacean acid metabolism*, named for plants of the family Crassulaceae in which it was first discovered.

presence of malate (in the form of malic acid within the acidic vacuole) is evident by the sour "morning taste" of the plants. During the daylight hours, the stomata close, and the malic acid is moved into the cytoplasm. There it gives up its CO_2, which can be fixed by Rubisco under conditions of low O_2 concentration that exist when the stomata are closed. Carbohydrates are then synthesized using energy from ATP and NADPH generated by the light-dependent reactions.

Photoinhibition

Photosynthesis is a complex and finely tuned process. Just as too much O_2 leads to diminished photosynthetic output, so too does too much light. The negative effect of high light intensity on photosynthesis is called **photoinhibition** and is thought to result primarily from damage inflicted on photosystem II by the absorption of excess light. PSII operates at the highest oxidizing potential of any known biological system. The formation of a very strong oxidizing agent, and the ever-present danger of

forming highly toxic oxygen species, gives PSII the potential for its own self-destruction as the result of overexcitation of the system. Most of the damage appears to be targeted to the polypeptide (D1) that binds nearly all of the active redox centers of the photosystem. Chloroplasts contain a mechanism for the selective proteolytic degradation of D1, and its replacement by a newly-synthesized polypeptide molecule. This process of D1 turnover, which increases with increasing light intensity, appears to be the primary response to light-induced photoinhibition.

Peroxisomes and Photorespiration

Peroxisomes are cytoplasmic organelles whose role in oxidative metabolism in animal cells was discussed in Section 5.6 of the last chapter. Peroxisomes are also present in certain plant cells. Studies on the peroxisomes of leaf cells have revealed a striking example of interdependence among different organelles. The electron micrograph in Figure 6.22 shows a peroxisome of a leaf cell closely ap-

0.3 μm

Figure 6.22 The cellular basis of photorespiration. Electron micrograph of a portion of a leaf mesophyll cell of a tobacco plant showing a peroxisome (identified by its crystalline core) pressed against a pair of chloroplasts and close to a mitochondrion. The reactions of photorespiration that occur in each of these organelles are described in the text and shown superim-

posed on the organelles in which they occur. This series of reactions is referred to as the C_2 cycle. The last steps of the cycle—the conversion of serine to glycerate in the peroxisome and then to 3-PGA in the chloroplast—are not shown. (MICROGRAPH COURTESY OF SUE ELLEN FREDERICK AND ELDON H. NEWCOMB.)

posed to the surfaces of two adjacent chloroplasts and in close proximity to a nearby mitochondrion. This arrangement is not simply fortuitous but reflects an underlying biochemical relationship whereby the products of one organelle serve as substrates in another organelle. The reactions that take place in the different organelles are superimposed on the micrograph in Figure 6.22 and summarized below.

It was noted previously that chloroplasts engage in a process called photorespiration, which begins when RuBP reacts with O_2 rather than CO_2 to form 3-PGA and a two-carbon compound, phosphoglycolate (Figure 6.20). Once formed, phosphoglycolate is converted to glycolate, which is shuttled out of the chloroplast into a peroxisome. In the peroxisome, the enzyme glycolate oxidase converts glycolate to glyoxylate, which may then be converted to glycine and transferred to a mitochondrion. In the mitochondrion, two molecules of glycine (a two-carbon molecule) are converted to one molecule of serine (a three-carbon molecule), with the accompanying release of one molecule of CO_2. Thus, for every two molecules of phosphoglycolate produced by Rubisco, one carbon atom that

had previously been fixed is lost back to the atmosphere. The serine produced in the mitochondrion may be shuttled back to the peroxisome and converted to glycerate, which can be transported to the chloroplast and utilized in carbohydrate synthesis by means of the formation of 3-PGA. This series of reactions provides an excellent example of the interdependency of cellular organelles, a feature that is often lost in studies with isolated cellular structures.

REVIEW ?

1. Describe the basic plan of the Calvin cycle, indicating the reactions that require energy input. Why is it described as a cycle? Why does energy have to be expended in this type of pathway? What are the eventual products of the pathway?

2. Describe the major structural and biochemical differences between C_3 plants and C_4 plants. How do these differences affect the ability of these two types of plants to grow in hot, dry climates?

SYNOPSIS

The earliest life forms are presumed to have been heterotrophs that depended on organic molecules formed abiotically; these forms were eventually succeeded by autotrophs—organisms that were capable of surviving on CO_2 as their principal carbon source. The first autotrophs are thought to have carried out nonoxygenic photosynthesis in which compounds such as H_2S were oxidized as a source of electrons. The evolution of oxygenic photosynthesis, in which water is oxidized and O_2 is released, allowed the cyanobacteria to exploit a much wider array of habitats and set the stage for aerobic respiration. (p. 218)

Chloroplasts are large membrane-bound organelles that are thought to have evolved from photosynthetic prokaryotes. Chloroplasts are bounded by a double membrane made porous by the inclusion of porins in the lipid bilayer. Thylakoids are flattened membranous sacs that are arranged in orderly stacks, or grana. The thylakoids are surrounded by a fluid stroma that contains DNA, ribosomes, and the machinery required for gene expression. (p. 219)

The light-dependent reactions of photosynthesis begin with the absorption of photons of light by photosynthetic pigments, an event that pushes electrons to outer orbitals from which they can be transferred to an electron acceptor. The major light-absorbing pigments of plants are chlorophylls and carotenoids. Each chlorophyll molecule consists of a Mg^{2+}-containing porphyrin ring that acts in light absorption and a hydrocarbon tail (a phytol) that keeps the pigment embedded in the bilayer. Chlorophyll absorbs most strongly in the blue and red region of the visible spectrum and least strongly in the green. Carotenoids absorb most strongly in the blue and green region and least strongly in the red and orange. The action spectrum of photosynthesis, which shows the wavelengths of light that are able to stimulate photosyn-

thesis, follows the absorption spectrum of the pigments quite closely. Photosynthetic pigments are organized into functional units in which only one molecule—the reaction-center chlorophyll—transfers electrons to an electron acceptor. The bulk of the pigment molecules form a light-harvesting antenna that traps photons of differing wavelengths and transfers the energy of excitation to the pigment molecule at the reaction center. (p. 222)

The transfer of a pair of electrons from H_2O to $NADP^+$ under conditions operating in the chloroplast utilizes the input of approximately 2 V of energy, which is delivered by the combined action of two separate photosystems. Photosystem II (PSII) boosts electrons from an energy level below that of water to a midway point, while photosystem I (PSI) raises electrons to the highest energy level, above $NADP^+$. As photons are absorbed by each photosystem, energy is passed to their respective reaction-center pigments (P680 for PSII and P700 for PSI). The energy absorbed by the reaction-center chlorophylls serves to boost an electron to an outer orbital where it is transferred to a primary acceptor, producing a positively charged pigment ($P680^+$ and $P700^+$). (p. 225)

The path of noncyclic electron flow from water to PSII and then to PSI and $NADP^+$ is denoted in the form of a Z. The first leg of the Z is from H_2O to PSII. Most of the antenna pigments that collect light for PSII reside within a separate complex, called LHCII. Energy flows from LHCII to the PSII reaction center, where an electron is transferred from P680 to a primary acceptor, which is a chlorophyll-like pheophytin molecule. This transfer of electrons generates a strong oxidizing agent ($P680^+$) and a weak reducing agent ($Pheo^-$). The separation of charge in PSII is stabilized by moving the oppositely charged molecules farther apart; this is accomplished as the electron is transferred from pheophytin to the quinone

PQ_A and then PQ_B. The successive absorption of two photons by PSII leads to the transfer of two electrons to PQ_B to form PQ_B^{2-}, which then picks up two protons from the stroma, forming PQH_2 (reduced plastoquinone). The reduced plastoquinone leaves the reaction-center and is replaced by an oxidized plastoquinone that is capable of receiving additional electrons. As each electron is transferred from P680 to the primary acceptor and on to PQ_B, the hole in the positively charged reaction-center pigment (P680$^+$) is filled by an electron from a specific tyrosine residue (Tyr$_z$). Tyr$_z$ in turn receives electrons one at a time from a protein containing four manganese ions. As each electron is transferred to Tyr$_z$, the Mn-containing protein accumulates a positive charge. Once the protein has accumulated four positive charges, it is capable of removing four electrons from water, a reaction that generates O_2 and delivers four H^+ to the thylakoid lumen, which contributes to the proton gradient. (p. 227)

Electrons from reduced plastoquinone are transferred to the multiprotein complex cytochrome $b_6 f$, while the protons are released into the thylakoid lumen, thus contributing to the proton gradient. Electrons from cytochrome $b_6 f$ are passed to plastocyanin located on the luminal side of the thylakoid membrane and on to P700$^+$, the reaction-center pigment of PSI that had lost an electron following absorption of a photon. As each photon is absorbed by P700, the electron is transferred to a primary acceptor, A_0, and then through several iron-sulfur centers of the PSI reaction center to ferredoxin. Electrons are transferred from ferredoxin to NADP$^+$ forming NADPH, which requires a proton from the stroma, contributing to the proton gradient. In summary, noncyclic electron flow results in the oxidation of H_2O to O_2, with the transfer of electrons to NADP$^+$, forming NADPH, and the establishment of a H^+ gradient across the membrane. (p. 230)

The proton gradient established during the light-dependent reactions provides the energy required for the formation of ATP in the chloroplast, a process called photophosphorylation. The machinery for ATP synthesis in the chloroplast is virtually identical to that of the mitochondrion; the ATP synthase consists of a CF$_1$ headpiece that projects into the stroma and a CF$_0$ base that is embedded in the thylakoid membrane. Protons drive the formation of ATP as they move from higher concentration in the thylakoid lumen through the CF$_0$ base and into the stroma, thereby dissipating the H^+ gradient. ATP synthesis can occur in the absence of the oxidation of H_2O through a process of cyclic photophosphorylation that does not involve PSII. Light is absorbed by P700 of PSI, passed to ferredoxin, and returned to the electron-deficient PSI reaction center through cytochrome $b_6 f$. As electrons pass around the cyclic pathway, protons are

translocated into the thylakoid lumen and subsequently drive the formation of ATP. (p. 233)

During the light-independent reactions, the chemical energy stored in NADPH and ATP is used in the synthesis of carbohydrates from CO_2. CO_2 is converted to carbohydrate by the C_3 pathway (or Calvin cycle) in which CO_2 is fixed by RuBP carboxylase (Rubisco) to a five-carbon compound, RuBP, forming an unstable six-carbon intermediate, which splits into two molecules of 3-phosphoglyceric acid. NADPH and ATP are used to convert PGA molecules to glyceraldehyde phosphate (GAP). For every 6 molecules of CO_2 fixed, 2 molecules of GAP can be directed toward the formation of sucrose or starch, while the remaining 10 molecules of GAP can be used to regenerate RuBP for additional rounds of CO_2 fixation. (p. 234)

Rubisco can also catalyze a reaction in which O_2, rather than CO_2, is covalently joined to RuBP. This process, which is called photorespiration, leads to the formation of compounds that are metabolized in reactions that lead to the loss of CO_2. Because photorespiration involves the uptake of O_2 and the release of CO_2, it represents a waste of the plant's energy. The rate of photorespiration versus CO_2 fixation depends on the CO_2/O_2 ratio faced by Rubisco. Two groups of plants, called C_4 and CAM plants, possess mechanisms that increase this ratio. (p. 236)

C_4 and CAM plants possess an additional CO_2-fixing enzyme, called PEP carboxylase, that is able to operate at very low CO_2 concentrations. C_4 plants possess a unique leaf structure containing an outer cylinder of mesophyll cells and an inner cylinder of bundle sheath cells that are sealed off from atmospheric gases. PEP carboxylase operates in the mesophyll, where CO_2 is fixed to the three-carbon compound phosphoenolpyruvate (PEP) to form a four-carbon acid, which is transported to the bundle sheath, where it is decarboxylated. The CO_2 released in the bundle sheath accumulates to high concentration, favoring CO_2 fixation to RuBP and the formation of PGA and GAP by means of the Calvin cycle. CAM plants carry out the light-dependent and light-independent reactions at different times of the day. CAM plants keep their stomata closed during hot, dry daylight hours, thereby avoiding water loss. Then, during the night, they open their stomata and fix CO_2 by means of PEP carboxylase. The malic acid produced in these reactions is stored in the vacuole until daylight hours, when the compound is moved back into the chloroplast. There it gives up its CO_2, which can be fixed by Rubisco under conditions of low O_2 concentration and converted to carbohydrate using energy from the ATP and NADPH generated by the light-dependent reactions. (p. 237)

ANALYTIC QUESTIONS

1. Which of the two photosystems operates at the most negative redox potential? Which generates the strongest reducing agent? Which must absorb four photons during each round of noncyclic photophosphorylation?

2. Which type of plant (C_3, C_4, or CAM) might you expect to do best if exposed to continuous daylight under hot, dry conditions?

3. Of the following substances, PQH_2, reduced cytochrome b_6, reduced ferredoxin, NADP$^+$, NADPH, O_2, H_2O, Tyr$_z^+$, which is the strongest reducing agent? Which is the strongest oxidizing agent? Which has the greatest affinity for electrons? Which has the most energetic electrons?

4. Suppose you were to add the uncoupler dinitrophenol (DNP) to a preparation of chloroplasts carrying out photosynthesis. Which of the following activities would you expect to be affected? (1) absorption of light, (2) cyclic

photophosphorylation, (3) electron transport between PSII and PSI, (4) noncyclic photophosphorylation, (5) synthesis of PGA, (6) NADP$^+$ reduction.

5. Calculate the proton-motive force that would be formed across a thylakoid membrane that maintained a 10,000-fold difference in [H$^+$] and no electric potential difference. (The equation for proton-motive force is on page 202.)

6. Under what conditions would you expect a plant to be engaged most heavily in cyclic photophosphorylation?

7. Contrast the change in pH of the medium that occurs when isolated chloroplasts are carrying out photosynthesis compared to isolated mitochondria carrying out aerobic respiration.

8. It was noted in the previous chapter that the majority of the proton-motive force in mitochondria is expressed as a voltage. In contrast, the proton-motive force generated during photosynthesis is almost exclusively expressed as a pH gradient. How can you explain these differences?

9. Contrast the roles of PSI and PSII in generating the electrochemical gradient across the thylakoid membrane.

10. Would you agree that a C$_3$ plant has to expend more energy per molecule of CO$_2$ converted to carbohydrate than a C$_4$ plant? Why or why not?

11. In photosynthesis, the capture of light energy results in the release and subsequent transfer of electrons. From what molecules are the electrons originally derived? In what molecules do these electrons eventually reside?

12. How many ATP and NADPH molecules are required in the C$_3$ pathway to make one six-carbon sugar? If the synthesis of a molecule of ATP were to require four protons, would you expect that these relative requirements for ATP and NADPH could be met by noncyclic photophosphorylation in the absence of cyclic photophosphorylation?

13. If pheophytin and A$_0$ (a chlorophyll a molecule) are the primary electron acceptors of PSII and PSI, respectively, what are the primary electron donors of each photosystem?

14. Contrast the roles of three different metal atoms in the light-dependent reactions of photosynthesis.

15. Would you expect the greenhouse effect (an increase in the CO$_2$ content of the atmosphere) to have a greater effect on C$_4$ plants or C$_3$ plants? Why?

16. Would you expect the water content of the atmosphere to be an important factor in the success of a C$_3$ plant? Why?

17. It has been suggested that low CO$_2$ levels play an key role in keeping O$_2$ levels stable at 21%. How is it possible that CO$_2$ levels could affect O$_2$ levels in the atmosphere.

18. Suppose CO$_2$ levels were to rise to 600 ppm, which in fact they probably were about 300 million years ago. What effect would you suppose this might have regarding competition between C$_3$ and C$_4$ plants?

19. Suppose you were to place a C$_3$ plant under hot, dry conditions and provide it with radioactively labeled ^{18}O$_2$. In what compounds would you expect to find this label incorporated.

ADDITIONAL READING

General Topics

ARNON, D. I. 1984. The discovery of photosynthetic phosphorylation. *Trends Biochem. Sci.* 9:258–262.

BARBER, J. 1998. Photosystem two. *Biochim. Biophys. Acta* 1365:269–277.

BARBER, J. & KÜHLBRANDT, W. 1999. Photosystem II. *Curr. Opin. Struct. Biol.* 9:469–475.

BENDALL, D. S. & MANASSE, R. S. 1995. Cyclic photophosphorylation and electron transport. *Biochim. Biophys. Acta* 1229:23–38.

BOGORAD, L. 1981. Chloroplasts. *J. Cell Biol.* 91:256S–270S.

BREYTON, C. 2000. The cytochrome b_6f complex: Structural studies and comparison with the bc_1 complex. *Biochim. Biophys. Acta* 1459:467–474.

DEISENHOFER, J. & MICHEL, H. 1989. The photosynthetic reaction center from the purple bacterium, *Rhodopseudomonas viridis. Science* 245:1463–1473.

GOBECK, J. H. 1992. Structure and function of photosystem I. *Annu. Rev. Plant Physiol.* 43:293–324.

HANKAMER, B., ET AL. 1997. Structure and membrane organization of photosystem II in green plants. *Annu. Rev. Plant Physiol.* 48:641–671.

HARRIS, D. A. 1995. *Bioenergetics at a Glance*. Blackwell.

KRAUSS, N., ET AL. 1996. Photosystem I at 4 Å resolution represents the first structural model of a joint photosynthetic reaction center and core antenna system. *Nature Str. Biol.* 3:965–973.

KUHLBRANDT, W., ET AL. 1994. Atomic model of plant light-harvesting complex by electron crystallography. *Nature* 367:614–621.

LEE, A. G. 2000. Membrane lipids: it's only a phase. *Curr. Biol.* 10:R377–380.

NICHOLLS, D. G. & FERGUSON, S. J. 1992. *Bioenergetics 2*. Academic Press.

NIELD, J., ET AL., 2000. 3D map of the plant photosystem II supercomplex obtained by cryoelectron microscopy and single particle analysis. *Nature Struct. Biol.* 7:44–47.

NUGENT, J. H. A. 1996. Oxygenic photosynthesis: Electron transfer in photosystem I and photosystem II. *Eur. J. Biochem.* 237:519–531.

RAGHAVENDRA, A. S., ed. 1998. *Photosynthesis: A Comprehensive Treatise*. Cambridge.

RÖGNER, M., ET AL. 1996. How does photosystem II split water? *Trends Biochem. Sci.* 21:44–49.

TOMMOS, C. & BABCOCK, G. T. 2000. Proton and hydrogen currents in photosynthetic water oxidation. *Biochim. Biophys. Acta* 1458:199.

CO$_2$ Fixation, Photorespiration, and Plant Metabolism

MANN, C. C. 1999. Genetic engineers aim to soup up crop photosynthesis. *Science* 283:314–316.

MOORE, P. D. 1994. High hopes for C$_4$ plants. *Nature* 367:322–323.

NUGENT, J., ED. 2001. Reviews on photosythetic water oxidation. *Biochem. Biophys. Acta*, vol. 1503, #1.

PORTIS, A. R., Jr. 1992. Regulation of ribulose 1,5-bisphosphate carboxylase/oxygenase activity. *Annu. Rev. Plant Physiol.* 43:415–437.

TOLBERT, N. E. 1997. The C$_2$ oxidative photosynthetic carbon cycle. *Annu. Rev. Plant Physiol. Mol. Biol.* 48:1–25.

7

Interactions Between Cells and Their Environment

Although the plasma membrane constitutes the boundary between a living cell and its nonliving environment, materials present outside the plasma membrane play an important role in the life of a cell. Most cells in a multicellular plant or animal are organized into clearly defined tissues in which the component cells maintain a defined relationship with one another and with the extracellular materials that lie between the cells. Even those cells that have no fixed relationship within a solid tissue, such as white blood cells that patrol the body, must interact in highly specific ways with other cells and with extracellular materials with which they come in contact. These interactions regulate such diverse activities as cell migration, cell growth, and cell differentiation, and they determine the three-dimensional organization of tissues and organs that emerges during embryonic development.

Micrograph of a skin cell (a keratinocyte) on the bottom of a culture dish. The cell itself appears red-yellow after binding a substance (rhodamine-linked phalloidin) that stains the cell's cytoskeleton. The material that is stained green by a fluorescent antibody is a protein (laminin-5) that is normally present in the extracellular material on which these skin cells reside. This cell originally settled on the dish at the site indicated by the white arrow and then migrated over the laminin it secreted. Keratinocytes that cannot produce this laminin because of a genetic deficiency can migrate only one-fifth as rapidly as this wild-type cell. Laminin-5 is thought to play a role in the migration of keratinocytes following injury to the skin. (COURTESY OF DIANE FRANK, FRED HUTCHINSON CANCER RESEARCH CENTER, FROM BETH P. NGUYEN ET AL. CURR. OPIN. CELL BIOL. 12:557, 2000.)

Figure 7.1 An overview of how cells are organized into tissues and how they interact with one another and with their extracellular environment. In this schematic diagram of a section through human skin, the cells of the epidermis are seen to adhere to one another by specialized contacts. The basal layer of epidermal cells also adheres to an underlying, noncellular layer (the basement membrane). The dermis consists largely of extracellular elements that interact with each other and with the surfaces of scattered cells. The cells contain receptors that interact with extracellular materials.

We will focus in this chapter on the extracellular environment of cells and the types of interactions in which cells engage. Figure 7.1 shows a diagrammatic section of human skin and provides an overview of several of the subjects to be discussed in the chapter. The outer layer of skin (the epidermis) consists largely of closely packed cells attached to one another and to an underlying noncellular layer by specialized contacts (depicted in the top inset). These contacts provide a mechanism for cells to adhere to and communicate with one another. In contrast, the deeper layer of the skin (the dermis) consists largely of extracellular material, including a variety of distinct fibers that interact with each other in specific ways. A closer look at one of the scattered cells (fibroblasts) of the dermis shows that the outer surface of the plasma membrane contains receptors that mediate interactions between the cell and the components of its environment (depicted in the lower inset). These cell surface receptors interact not only with the external surroundings, but are connected at their internal end to various cytoplasmic proteins. Receptors with this type of dual attachment are well suited to transmit messages between the cell and its environment. ∎

7.1 THE EXTRACELLULAR SPACE

If we begin at the plasma membrane and move outward, we can examine the types of extracellular elements that surround various types of cells. It was noted in Chapter 4 that virtually all integral membrane proteins, as well as certain membrane lipids, bear short chains of sugars (oligosaccharides) that project outward from the plasma membrane (see Figure 4.4c). These carbohydrate projections form part of a layer closely applied to the outer surface of the plasma membrane called the **glycocalyx** (or *cell coat*) (Figure 7.2a). In addition to membrane carbohydrates, the glycocalyx may contain additional extracellular materials that have been secreted by the cell into the external space, where they remain closely associated with the cell surface. This extracellular material is very prominent in some types of cells, such as the epithelial cells that line the mammalian digestive tract (Figure 7.2b). The glycocalyx is thought to mediate cell–cell and cell–substratum interactions, provide mechanical protection to cells, and serve as a barrier to particles moving toward the plasma membrane.

The Extracellular Matrix

Many types of animal cells are surrounded by an **extracellular matrix (ECM)**—an organized network of extracellular materials that is present beyond the immediate vicinity of the plasma membrane (Figure 7.3). The ECM is more than an inert packing material; it often plays a key regulatory role in determining the shape and activities of the cell. For example, enzymatic digestion of the ECM

(a) 0.5 μm

(b)

Figure 7.2 The glycocalyx. (*a*) The basal surface of an ecto-dermal cell of an early chick embryo. Two distinct structures closely applied to the outer cell surface can be distinguished: an inner glycocalyx (GC) and an outer basal lamina (BL; or base-ment membrane). (*b*) Electron micrograph of the apical surface of an epithelial cell from the lining of the intestine, showing the extensive glycocalyx, which has been stained with the iron-con-taining protein ferritin. (A: FROM A. MARTINEZ PALOMO, INT. REV. CYTOL. 29:64, 1970; B: FROM S. ITO AND D. W. FAWCETT/PHOTO RE-SEARCHERS.)

Figure 7.3 Experimental determination of the thickness of the extracellular matrix. When cartilage cells (chondrocytes) are grown in culture, they produce an extracellular matrix whose outline can be made visible by adding a suspension of small par-ticles, such as fixed red blood cells (RBCs), as shown here. The extent of the extracellular matrix surrounding each cell can be measured by the thickness of the clear space that is not pene-trated by the particles (arrowhead). The bar represents 10 μm. (COURTESY OF GRETA M. LEE, BRIAN JOHNSTONE, KEN JACOBSON, AND BRUCE CATERSON.)

that surrounds cultured cartilage cells or mammary gland epithelial cells causes a marked decrease in the synthetic and secretory activities of the cells. Addition of extracellu-lar matrix materials back to the culture dish restores the differentiated state of the cells and their ability to produce their usual cell products (see Figure 7.29).

One of the best-defined extracellular matrices is the **basement membrane** (or **basal lamina**), a continuous sheet 50 to 200 nm thick that (1) surrounds muscle and fat cells, (2) underlies the basal surface of epithelial tissues, such as the epidermis of the skin (Figures 7.1 and 7.4*a*), or the lining of the digestive and respiratory tracts, and (3) underlies the inner endothelial lining of blood vessels. Basement membranes provide mechanical support for the attached cells, serve as a substratum for cell migration, separate adjacent tissues within an organ, and act as a bar-rier to the passage of macromolecules. In this latter capac-ity, basement membranes play an important role in pre-venting the passage of proteins out of the blood as it flows through the porous-walled capillaries of the body. This is particularly important in the kidney, where the blood is filtered under high pressure through a double-layered basement membrane that separates the capillaries of the glomerulus from the wall of the kidney tubules (Figure 7.4*b*). Kidney failure in long-term diabetics may result from an abnormal thickening of the basement membranes surrounding the glomeruli. Basement membranes also serve as a barrier to the invasion of tissues by errant cancer cells. The molecular organization of basement mem-branes is discussed later (see Figure 7.12).

Even though the extracellular matrix may take diverse forms in different tissues and organisms, it tends to be composed of similar proteins. Unlike most proteins pre-sent inside of cells, which are compact, globular mole-cules, most proteins of the extracellular space are ex-tended, *fibrous* species. Figure 7.5 shows a schematic overview of the interconnected three-dimensional net-work of the extracellular proteins that are discussed in the following sections. Among their diverse functions, the proteins of the ECM serve as scaffolds, girders, mortar, and wire. As noted throughout the following discussion, alterations in the amino acid sequence of extracellular pro-teins can lead to serious disorders. We will begin with one of the most important, and ubiquitous, molecules of the ECM, the glycoprotein collagen.

(a) 10 µm

(b)

Figure 7.4 The basement membrane (basal lamina).
(*a*) Scanning electron micrograph of human skin. The epidermis has pulled away from part of the basement membrane, which is visible beneath the epidermal cells. The dermis is a connective tissue layer consisting primarily of extracellular materials. (*b*) An unusually thick basement membrane is formed between the blood vessels of the glomerulus and the proximal end of the renal tubules of the kidney. This extracellular layer plays an important role in filtering the fluid that is forced out of the capillaries into the renal tubules during urine formation. The black dots within the glomerular basement membrane (GBM) are gold particles attached to antibodies that are bound to type IV collagen molecules in the basement membrane (CL, capillary lumen; P, podocyte of tubule). The bar represents 0.5 mm. (A: COURTESY OF KAREN HOLBROOK; B: FROM MICHAEL DESJARDINS AND M. BENDAYAN, J. CELL BIOL. 113:695, 1991; BY COPYRIGHT PERMISSION OF THE ROCKEFELLER UNIVERSITY PRESS.)

Collagen Collagens comprise a family of fibrous glycoproteins that are present only in extracellular matrices. Collagens are found throughout the animal kingdom and are noted for their high tensile strength, that is, their resistance to pulling forces. It is estimated that a collagen fiber 1 mm in diameter is capable of suspending a weight of 10

Figure 7.5 An overview of the macromolecular organization of the extracellular matrix. The proteins and polysaccharides shown in this illustration will be discussed in the following sections. The proteins depicted (fibronectin, collagen, and laminin) contain binding sites for one another, as well as binding sites for receptors (integrins) that are located at the cell surface. The proteoglycans are huge protein-polysaccharide complexes that occupy much of the volume of the extracellular space.

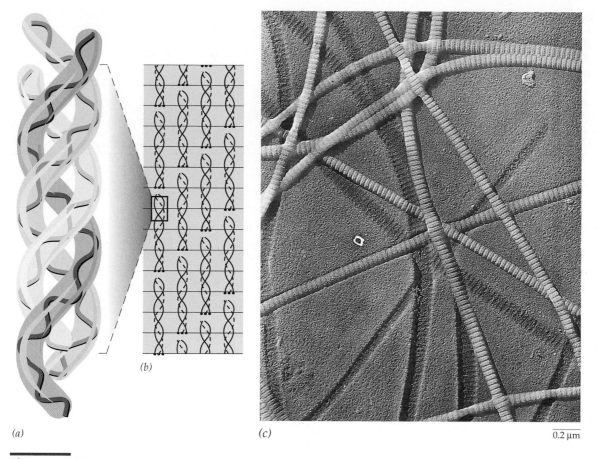

(a) (b) (c) 0.2 µm

Figure 7.6 **The structure of collagen.** This figure depicts several levels of organization of collagen. (*a*) The basic collagen monomer is a triple helix composed of three helical α chains. Some types of collagen contain three identical α chains and thus are homotrimers, while others are heterotrimers containing two or three different chains. (*b*) Collagen monomers become aligned in rows in which the molecules in one row are staggered relative to those in the neighboring row. (*c*) An electron micrograph of human collagen fibers showing their characteristic banding pattern, which results from the arrangement of the collagen monomers. The bands repeat along the fiber with a periodicity of 64 to 70 nm. (c: COURTESY OF JEROME GROSS AND FRANCIS O. SCHMITT.)

kg (22 lb) without breaking. Collagen is the single most abundant protein in the human body (constituting more than 25 percent of all protein), a fact that reflects the widespread occurrence of extracellular materials.

Collagen is produced primarily by fibroblasts, the cells found in various types of connective tissues, and also by smooth muscle cells and epithelial cells. Nineteen distinct types of collagen have been identified. Each collagen type is restricted to particular locations within the body, but two or more different types are often present together in the same ECM. Additional functional complexity is provided by mixing several collagen types within the same fiber. These "heterotypic" fibers are the biological equivalent of a metal alloy. It is likely that different structural and mechanical properties result from different mixtures of collagens in fibers. Although there are many differences among the members of the collagen family, all share at least two important structural features. (1) All collagen molecules are trimers consisting of three polypeptide chains, called α chains. (2) Along at least part of their length, the three polypeptide chains of a collagen mole-

cule are wound around each other to form a unique, rod-like triple helix (Figure 7.6*a*).

A number of the collagens, including types I, II, and III, are described as *fibrillar collagens* because they assemble into rigid, cable-like fibrils, which in turn assemble into thicker fibers that are typically large enough to be seen in the light microscope. The side-by-side packaging of the rows of collagen I molecules within a collagen fibril is depicted in Figure 7.6*b*. The individual collagen molecules of a fibril are not aligned in register but are staggered approximately one-quarter length relative to their neighbors. This staggered arrangement of the component molecules increases the mechanical strength of the complex and produces the banding patterns characteristic of collagen fibers (Figure 7.6*c*). The fibrils are further strengthened by covalent cross-links between lysine and hydroxylysine residues on adjacent collagen molecules. This cross-linking process continues through life and may contribute to the decreased elasticity of skin and increased brittleness of bones among the elderly.

Among the various components of the ECM, collagen molecules provide the insoluble framework that de-

1 μm

Figure 7.7 **The corneal stroma** consists largely of layers of collagen fibers in which the molecules of alternate layers are arranged at right angles to one another, resembling the structure of plywood. (COURTESY OF M. TAKUS.)

50 nm

Figure 7.8 **The type IV collagen network of the basement membrane.** Electron micrograph of a basement membrane from human amniotic tissue that had been extracted with a series of salt solutions to remove noncollagenous materials. The treatment leaves behind an extensive, branching, polygonal network of strands that form an irregular lattice. Evidence indicates that this lattice consists of type IV collagen molecules interlinked with one another in a complex three-dimensional array. A model of the basement membrane scaffold is shown in Figure 7.12. (B: FROM PETER D. YURCHENCO AND GEORGE C. RUBEN, J. CELL BIOL. 105:2561, 1987; BY COPYRIGHT PERMISSION OF THE ROCKEFELLER UNIVERSITY PRESS.)

termines many of the mechanical properties of the matrix. In fact, the properties of a particular tissue can often be correlated with the three-dimensional organization of its collagen molecules. For example, tendons, which connect muscles to bones, must resist tremendous pulling forces during times of muscular contraction. Tendons contain an ECM in which the collagen fibers are aligned parallel to the long axis of the tendon and thus parallel to the direction of the pulling forces. The cornea is also a remarkable tissue; it must serve as a durable, protective layer at the surface of the eyeball but must also be transparent so that light can pass through the lens to the retina. The thick, middle layer of the cornea is the stroma, which contains relatively short collagen fibrils that are organized into distinct layers. The layered structure of the stroma is similar to that of plywood: the fibrils of each layer are parallel to other fibrils in the layer but perpendicular to the fibrils in the layer on either side (Figure 7.7). This plywoodlike structure provides strength to this delicate tissue, while the uniformity of size and ordered packing of the fibers minimize the scattering of incoming light rays, thereby promoting the tissue's transparency.

Given their abundance and widespread distribution, it is not surprising that abnormalities in fibrillar collagen formation can lead to serious disorders. Burns or traumatic injuries to internal organs can lead to an accumulation of scar tissue, which consists largely of fibrillar collagen. Mutations in genes encoding type I collagen can produce *osteogenesis imperfecta*, a potentially lethal condition characterized by extremely fragile bones, thin skin, and weak tendons. Mutations in genes encoding type II collagen alter the properties of cartilage tissue, causing dwarfism and skeletal deformities. Mutations in a number of collagen genes can lead to a variety of distinct but related defects in collagen matrix structure (called *Ehler-Danlos syndromes*). Persons with one of these syndromes have hyperflexible joints and highly extensible skin.

Not all collagens form fibrils. One of the nonfibrillar collagens is type IV, whose distribution is restricted to basement membranes (page 245). Basement membranes are thin, supportive sheets, and the type IV collagen molecules are organized into a network that provides mechanical support and serves as a lattice for the deposition of other extracellular materials (see Figure 7.12). Unlike type I collagen, which consists of a long, uninterrupted triple helix, the type IV collagen trimer contains nonhelical segments interspersed along the molecule and globular domains at each end. The nonhelical segments give the molecule flexibility, while the globular ends serve as sites of interaction between molecules that give the complex its latticelike character (Figure 7.8). Mutations in type IV collagen genes have been identified in patients with *Alport syndrome*, an inherited kidney disease in which the glomerular basement membrane (Figure 7.4*b*) is disrupted.

Proteoglycans In addition to collagen, basement membranes and other extracellular matrices typically contain large amounts of a distinctive type of protein-polysaccharide complex called a **proteoglycan**. A proteoglycan (Figure 7.9*a*) consists of a core protein molecule (shown in

Figure 7.9 **The structure of a cartilage-type proteoglycan complex.** (*a*) Schematic representation of a single proteoglycan consisting of a core protein to which a large number of glycosaminoglycan chains (GAGs, shown in red) are attached. A proteoglycan from cartilage matrix (e.g., aggrecan) may contain about 30 keratan sulfate and 100 chondroitin sulfate chains. Proteoglycans found in basement membranes (e.g., perlecan and agrin) have only a few GAG chains attached to the core protein. (*b*) The structures of the repeating disaccharides that make up each of the GAGs shown in this figure. All GAGs bear large numbers of negative charges (indicated by the blue shading). (*c*) In the cartilage matrix, individual proteoglycans are linked to a nonsulfated GAG, called hyaluronic acid, to form a giant complex with a molecular mass of about 3,000,000 daltons. The box indicates one of the proteoglycans of the type shown in part *a*. (*d*) Electron micrograph of a proteoglycan complex, comparable to that illustrated in part *c*, that was isolated from cartilage matrix. (D: COURTESY OF LAWRENCE C. ROSENBERG.)

black in Figure 7.9*a*) to which chains of **glycosaminoglycans (GAGs)** are covalently attached (shown in red in the figure). Each glycosaminoglycan chain is composed of a repeating disaccharide; that is, it has the structure -A-B-A-B-A-, where A and B represent two different sugars. GAGs are highly acidic due to the presence of both sulfate and carboxyl groups attached to the sugar rings (Figure 7.9*b*). Proteoglycans of the extracellular matrix may be assembled into gigantic complexes by linkage of their core proteins to a molecule of *hyaluronic acid*, a nonsulfated GAG (Figure 7.9*c*). The microscopic appearance of one of these complexes, which can occupy a volume equivalent to that of a bacterial cell, is shown in Figure 7.9*d*.

Because of the negative charges borne on the sulfated GAGs, proteoglycans bind huge numbers of cations, which in turn bind large numbers of water molecules. As a result, proteoglycans form a porous, hydrated gel that fills the extracellular space like packing material (Figure 7.5) and resists crushing (compression) forces. This property complements that of the adjacent collagen molecules, which resist pulling forces and provide a scaffold for the proteoglycans. Together, collagens and proteoglycans give cartilage and other extracellular matrices strength and resistance to deformation.

Fibronectin The term "matrix" implies a structure made up of a network of interacting components. The term has proven very apt for the extracellular matrix, which contains a number of proteins, in addition to collagen and proteoglycans, that interact with one another in highly specific ways (Figure 7.5). **Fibronectin**, like several other proteins to be discussed in this chapter, consists of a linear array of distinct domains that gives each polypeptide a modular construction (Figure 7.10*a*). Each fibronectin polypeptide is constructed from a sequence of approximately 30 independently folding Fn domains, two of which are depicted in the top inset of Figure 7.10*a*. While Fn domains were first discovered in fibronectin, they are found as part of many other proteins, ranging from blood clotting factors to membrane receptors (see Figure 7.21) and other proteins of the ECM. As discussed on page 60, the presence of shared domains among diverse proteins strongly suggests that many present-day genes have arisen during evolution by the fusion of parts of separate ancestral genes. In fibronectin, the 30 or so structural domains combine to form 5 or 6 larger functional units, illustrated by the colored cylinders in Figure 7.10*a*. Each of the two polypeptide chains that make up a fibronectin molecule contains

1. Binding sites for other components of the ECM, such as collagens and proteoglycans. These binding sites facilitate interactions that link these diverse molecules into a stable, interconnected network (Figure 7.5).

2. Binding sites for receptors on the cell surface. These binding sites hold the cell in a stable attachment to the ECM (Figure 7.5). The importance of a fibronectin-

containing substratum for cell attachment is shown dramatically in the micrograph on the cover of this text. The cultured endothelial cell shown in this photo has adopted a shape unlike any it would have in the body because it has spread itself over the available surface provided by a square coat of fibronectin.

The importance of fibronectin and other extracellular proteins is particularly evident when tissues are undergoing dynamic activities, as occur during embryonic devel-

opment. Development is characterized by waves of cell migration during which different cells follow different routes from one part of the embryo to another (Figure 7.11*a*). Migrating cells are guided by proteins, such as fibronectin, that are contained within the molecular landscape through which they pass. For example, the cells of the neural crest, which migrate out of the developing nervous system into virtually all parts of the embryo, traverse pathways rich in fibrils composed of fibronectin (Figure 7.10*b*). When antibodies that bind to fibronectin are injected into embryos, neural crest cells are no longer able to interact with fibronectin molecules in the surrounding matrix and cellular movement is inhibited. The importance of fibronectin in neural crest cell migration is readily revealed on a culture dish as shown in Figure 7.10*c*.

(a)

(b)

(c)

Figure 7.10 **Structure and function of fibronectin.** (*a*) A human fibronectin molecule consists of two similar, but non-identical, polypeptides joined by a pair of disulfide bonds located near the C-termini. Each polypeptide is composed of a linear series of distinct modules, which are organized into several larger functional units, illustrated by the colored cylinders in this figure. Each of these functional units contains one or more binding sites for either a specific component of the ECM or for the surface of cells. Some of these binding activities are indicated by the labels. The cell-binding site of the polypeptide that contains the sequence arg-gly-asp, or RGD, is indicated. As discussed later in the chapter, this sequence binds specifically to a particular class of integral plasma membrane proteins (integrins) that are involved in cell attachment and signal transduction. The inset shows two of the 30 or so repeating Fn modules that comprise the polypeptide; the RGD sequence forms a loop of the polypeptide, protruding from one module. (*b*) A

section through a young chick embryo that has been treated with fluorescent antibodies against fibronectin. Fibronectin is present as fibrils in the basement membranes (dark red sites) that lie beneath the embryonic epithelia and provide a substratum over which cells migrate. (*c*) In this micrograph, neural crest cells are emigrating from a portion of the developing chick nervous system (at the left of the photo) onto a glass culture dish that contains strips of fibronectin-coated surface alternating with strips of bare glass. The boundary of the fibronectin-coated region is indicated by the white lines. It is evident that the cells remain exclusively on the fibronectin-coated regions. Cells that reach the glass substratum (arrowheads) tend to round up and lose their migratory capabilities. The arrow indicates the direction of migration. (B: COURTESY OF JAMES W. LASH; C: FROM GIOVANNI LEVI, JEAN-LOUP DUBAND, AND JEAN PAUL THIERY, INT. REV. CYTOL. 123:213, 1990.)

(a)

(b) 20 μm

Figure 7.11 The role of cell migration during embryonic development. (*a*) A summary of some of the cellular traffic occurring during mammalian development. The most extensive movements are conducted by the neural crest (shown in blue), which migrate out of the neural plate in the dorsal midline of the embryo and give rise to all the pigment cells of the skin (P), the sympathetic ganglia (SpG), adrenal medulla (AdM), and the cartilage of the embryonic skull (Mx, Md for maxillary and mandibular arches). The primordial germ cells (PGC) migrate from the yolk sac to the site of gonad (G) formation within the embryo. The progenitors of lymphoid cells are transported to the liver (L), bone marrow (Bm), thymus (Thy), lymph nodes (LN), and spleen (Sp). (Note: The "pathways" shown here connect the original sites of cells with their destinations; they do not accurately depict the actual routes of the cells.) (*b*) Micrograph of a section of a portion of the hind gut of a 10-day mouse embryo. The primordial germ cells (green) are seen to be migrating along the dorsal mesentery on their way to the developing gonad. The tissue has been stained with antibodies against the protein laminin (red), which is seen to be concentrated in the surface over which the cells are migrating. (A: AFTER AARON A. MOSCONA AND R. E. HAUSMAN, IN CELL AND TISSUE INTERACTIONS, J. W. LASH AND M. M. BURGER, EDS, RAVEN PRESS, 1977; B: FROM MARTIN I. GARCIA-CASTRO, ROBERT ANDERSON, JANET HEASMAN, AND CHRISTOPHER WYLIE, J. CELL BIOL. 138:475, 1997; BY COPYRIGHT PERMISSION OF THE ROCKEFELLER UNIVERSITY PRESS.)

Laminin Laminins are a family of extracellular glycoproteins that consist of three different polypeptide chains linked by disulfide bonds and organized into a molecule resembling a cross with three short arms and one long arm (Figure 7.5). Like fibronectin, extracellular laminins can greatly influence a cell's potential for migration, growth, and differentiation. For example, laminins play a critical role in the migration of primordial germ cells. These cells arise in the yolk sac, which is located outside the embryo itself, and then migrate by way of the bloodstream and embryonic tissues to the developing gonad, where they eventually give rise to sperm or eggs (Figure 7.11*a*). During their migration, the primordial germ cells traverse surfaces that are particularly rich in laminin (Figure 7.11*b*). Studies indicate that the primordial germ cells possess a cell-surface protein that adheres strongly to one of the subunits of the laminin molecule. It has also been demonstrated that certain cells migrate over a laminin-containing matrix that they secrete themselves. The chapter-opening micrograph on page 243 shows a cultured skin cell (or keratinocyte, shown in red) that is migrating over a layer of laminin (shown in green) secreted by the cell as it moves over the substratum. When these same types of cells are isolated from mice that have been geneti-

cally engineered to lack the genes for this particular type of laminin, their migratory ability is vastly diminished. The possible role of laminins in the dissemination of cancer cells is discussed in the Human Perspective on page 265.

In addition to binding tightly to cell-surface receptors, laminin can bind to other laminin molecules, to proteoglycans, and to other components of basement membranes. The laminin and type IV collagen molecules of basement membranes are thought to form separate, but interconnected, networks, as depicted in Figure 7.12. These interwoven networks give basement membranes both strength and flexibility. Basement membranes with this type of structure are not restricted to vertebrates but are found throughout the animal kingdom. Analysis of the genome sequences of fruit flies and nematodes, for example, reveal the presence of genes in these invertebrates that encode laminins, proteoglycans, as well as a type IV collagen.

The micrographs and diagrams presented in the first section of this chapter portray the ECM as a static structure, which fits with the fact that these materials exist outside of the living cell. In actual fact, however, the ECM can exhibit dynamic properties, both in space and over time. Spatially, for example, ECM fibrils can be seen to stretch several times their normal length as they are pulled

Figure 7.12 **A model of the basement membrane scaffold.**
Basement membranes contain two network-forming molecules,
collagen IV (pink), which was illustrated in Figure 7.8, and
laminin (green), which is indicated by the thickened cross-shaped
molecules. The collagen and laminin networks are connected by
entactin molecules (purple). (AFTER PETER D. YURCHENCO, YI-SHAN
CHENG, AND HOLLY COLOGNATO, J. CELL BIOL. 117:1132, 1992; BY
COPYRIGHT PERMISSION OF THE ROCKEFELLER UNIVERSITY PRESS.)

on by cells and contract when tension is relieved. Elonga-
tion of fibronectin molecules is thought to occur as the β
sheets that make up the fibronectin (Fn) domains (see
Figure 7.10a) become unfolded. Temporally, the compo-
nents of an ECM are subject to continual degradation and
reconstruction. These processes serve to renew the matrix
and to allow it to be remodeled during embryonic devel-
opment or following tissue injury. Even the calcified ma-
trix of our bones, which we think of as a stable, inert struc-
ture, is subject to continual restoration. The degradation
of ECM materials is accomplished largely by a family of
zinc-containing enzymes called **matrix metallopro-
teinases (MMPs)** that are either secreted into the extra-
cellular space or bound to the external surface of the
plasma membrane. Members of the family differ in the
types of extracellular proteins they are able to attack. As
might be expected from enzymes whose normal function
is to destroy extracellular materials, the excessive or inap-
propriate activity of MMPs is likely to cause disease. In
fact, MMPs have been implicated in a number of patho-
logical conditions, including arthritis, multiple sclerosis,
atherosclerosis, tooth decay, and tumor progression (page
265).

(page 265).

REVIEW ?

1. Distinguish between the glycocalyx, a basement
 membrane, and the extracellular matrix of cartilage
 tissue.

2. Contrast the roles of collagen, proteoglycans, and
 fibronectin in the extracellular space.

3. List a few of the functions of the extracellular ma-
 trix in animal tissues.

7.2 INTERACTIONS OF CELLS WITH NONCELLULAR SUBSTRATES

It was noted in the previous discussion that components of
the ECM, such as fibronectin, laminin, proteoglycans,
and collagen, are capable of binding to receptors situated
on the cell surface (as in Figure 7.5). The most important
family of receptors that attach cells to their extracellular
microenvironment is the integrins.

Integrins

Integrins are a family of integral membrane proteins
thought to be present on the surfaces of virtually all types
of vertebrate cells. Integrins are composed of two mem-
brane-spanning polypeptide chains, an α chain and a β
chain, that are noncovalently linked. At least 18 different
α subunits and 8 different β subunits have been identified.
Although more than a hundred possible pairings of α and
β subunits could theoretically occur, only about two dozen
different integrins have been identified on the surfaces of
cells, each with a specific distribution within the body.
Most cells possess a variety of different integrins, and con-
versely, most integrins are present on a variety of different
cell types.

Electron microscopic observations of integrin mole-
cules suggested that the two subunits are arranged so as to
form a globular extracellular head connected to the mem-
brane by a pair of rigid stalks (as was depicted in Figure
7.5). Both subunits have a single transmembrane helix and
a small cytoplasmic domain.[1] Studies over the past few
years have revealed the structure of the extracellular do-
mains of integrin subunits and the manner in which these
proteins are able to bind their ligands. Amino acid se-
quence analysis of several α subunits of integrins indicated
that the N-terminal extracellular portion of the subunit
consisted of seven repeating modules, each comprised of
approximately 60 amino acids. This portion of the
polypeptide is folded into a single domain of a type that
had been found in several other proteins and, because of
its flattened, circular shape, had been termed a seven-
bladed β propeller. A model of the β-propeller domain of
an integrin subunit is shown in Figure 7.13 with the seven
individual "blades" depicted in different colors. Three cal-
cium ions are seen to be present as part of blades 5, 6, and
7. These Ca^{2+} ions are probably required to maintain the
proper structure of the integrin rather than being involved
in any direct interaction with a bound ligand.

Alpha integrin subunits can be divided into two
groups based on the structure of their extracellular do-
main. In about half of the 18 or so α subunits the ligand
binding site is part of the seven-bladed β propeller. The

[1]One exception to this molecular architecture is the $β_4$ chain, which has an
extra thousand or so amino acids as part of its cytoplasmic domain. This
huge addition makes $β_4$ integrins capable of extending much more deeply
into the cytoplasm (see Figure 7.18).

Figure 7.13 **The proposed structure of the extracellular terminus of an α integrin subunit.** Ribbon diagram of a model for the seven-bladed β-propeller domain of the N-terminal region of an α_4 subunit. Each of the seven repeating modules (blades) (W1-W7) is shown in a different color and the three calcium ions are indicated as yellow spheres. (FROM TIMOTHY A. SPRINGER, PROC. NAT'L. ACAD. SCI. U.S.A. 94:69, 1997.)

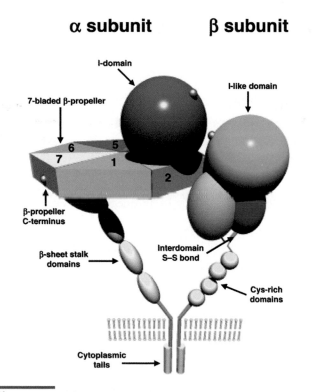

Figure 7.14 **The structure of an integrin.** Schematic representation of a heterodimeric integrin molecule. The β propeller domain is represented as a seven-sided disc oriented parallel to the plane of the lipid bilayer. The I domain of the α subunit, when present, resides at the upper surface of the β-propeller and is inserted between blades 2 and 3. The I (or I-like) domain of the β subunit is shown as a green sphere. The small orange spheres on the I domains represent the divalent Mg^{2+} or Mn^{2+} ions that play a role in ligand binding. (FROM JORDI BELLA AND HELEN M. BERMAN, STRUCT. 8:R122, 2000.)

remaining α integrin subunits have an "extra" module of about 200 amino acids, called an I domain, that loops out of the β propeller between the second and third blades. The I domain is postulated to form a separate globular element that sits "on top" of the β propeller of the α subunit facing the extracellular space (Figure 7.14). When present in an α subunit, the I domain contains the ligand binding site. The β subunits of integrins differ from the α subunits in (1) lacking a β-propeller and (2) having an I domain (or I-like domain) in all known cases (Figure 7.14). The ligand binding site of the β subunit is located within the I-like domain of the β subunit.

Studies indicate that certain integrins can exist on the surface of a cell in an inactive conformation. These integrins can be activated by events within the cell that alter the conformation of the cytoplasmic domains of the integrin's subunits. Changes in conformation of the cytoplasmic domains can be propagated through the molecule, increasing the integrin's affinity for a ligand. For example, the aggregation of platelets to form a blood clot (Figure 7.15) only occurs after the activation of $\alpha_{IIb}\beta_3$ integrins, which increases their affinity for fibrinogen. This type of alteration in affinity triggered by changes occurring inside the cell is often referred to as "inside-out" signaling.

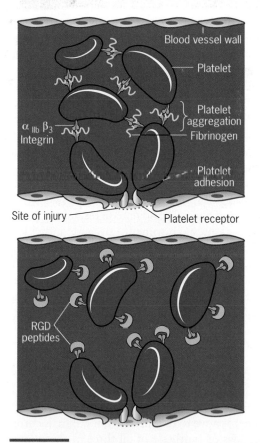

Figure 7.15 **Blood clots form when platelets adhere to one another through fibrinogen bridges that bind to the platelet integrins.** The presence of synthetic RGD peptides can inhibit blood clot formation by competing with the fibrinogen molecules for the RGD-binding sites on the integrins (page 255).

Integrins have been implicated in two major types of activities: adhesion of cells to their substratum (or to other cells) and transmission of signals from the external environment to the cell interior, a phenomenon known as "outside-in" signaling. Signals are transmitted across membranes by means of changes in conformation. The binding of the extracellular domain of an integrin to a ligand, such as fibronectin or laminin, can induce a conformational change at the cytoplasmic end of the integrin. Changes at the cytoplasmic end can, in turn, alter the way the integrin interacts with nearby cytoplasmic proteins, such as the enzyme focal adhesion kinase (FAK)(see Figure 7.17c). Thus as integrins bind to an extracellular ligand, they can activate cytoplasmic protein kinases, such as FAK, which can then phosphorylate other proteins, initiating a chain reaction. In some cases, the chain reaction leads all the way to the nucleus, where a specific group of genes may be activated.

Signals transmitted by integrins (and other cell-surface molecules) can influence many aspects of cell behavior, including motility, growth, and even the survival of the cell. The influence of integrins on cell survival is best illustrated by comparing normal and malignant cells. Most malignant cells are capable of growing while suspended in a liquid culture medium. Normal cells, in contrast, can only grow and divide if they are cultured on a solid substratum; if they are placed in suspension cultures, they die. Normal cells are thought to die in suspension culture because their integrins are not able to interact with extracellular substrates and, as a result, are not able to transmit life-saving signals to the interior of the cell. When cells become malignant, their survival no longer depends on integrin binding. The role of integrins in transmitting signals is one of the most active areas of research in cell biology and will be explored further on page 266 and in Chapter 15.

Table 7.1 lists a number of known integrins and the key extracellular ligands to which they bind. Linkage between integrins and these ligands mediates adhesion between cells and their environment. Because individual cells may express a variety of different integrins on their cell surface, such cells are capable of binding to a variety of different extracellular components (Figure 7.5). In many cases, it may be a combination of receptors recognizing different domains of an extracellular ligand that determines the effect the ECM has on cell behavior. For example, cells situated in the lower layer of the epidermis contain several β_1 integrins, including $\alpha_2\beta_1$, $\alpha_3\beta_1$, and $\alpha_5\beta_1$, that help attach these cells to components in the underlying basement membrane (Table 7.1). The detachment of these cells from the basement membrane, which is required for them to move to the surface of the skin, is correlated with the loss of synthesis of all three of these integrins. Like other integrins, the β_1 integrins are not only important in epidermal adhesion but play a role in regulating basic cellular activities. Normally, the basal cells of the epidermis are the only cells of this multilayered epithelium

| Table 7.1 | Classification of Integrin Receptors Based on Recognition of RGD Sequences |

RGD Recognition		**Non-RGD Recognition**	
Integrin Receptor	**Key Ligands**	**Integrin Receptor**	**Key Ligands**
$\alpha_3\beta_1$	Fibronectin	$\alpha_1\beta_1$	Collagen
$\alpha_5\beta_1$	Fibronectin	$\alpha_2\beta_1$	Collagen
$\alpha_v\beta_1$	Fibronectin		Laminin
		$\alpha_3\beta_1$	Collagen
$\alpha_{11b}\beta_3$	Fibronectin		Laminin
	von Willebrand		
	factor	$\alpha_4\beta_1$	Fibronectin
	Vitronectin		VCAM
	Fibrinogen	$\alpha_6\beta_1$	Laminin
$\alpha_v\beta_3$	Fibronectin	$\alpha_L\beta_2$	ICAM-1
	von Willebrand		
	factor		ICAM-2
	Vitronectin	$\alpha_M\beta_2$	Fibrinogen
$\alpha_v\beta_5$	Vitronectin		ICAM-1
$\alpha_v\beta_6$	Fibronectin		

Source: S. E. D'Souza, M. H. Ginsberg, and E. F. Plow, *Trends Biochem. Sci.* 16:249, 1991.

with the capacity to grow and divide. If the synthesis of the β_1 integrins does not stop when the cells leave the basal layer, the epidermal cells continue to divide, producing a condition resembling *psoriasis* in which the skin is abnormally thickened and inflamed.

Most of the extracellular proteins that bind to integrins do so because they contain the amino acid sequence arginine-glycine-aspartic acid (or, in abbreviated amino acid nomenclature, RGD). This tripeptide is present in the cell-binding sites of proteoglycans, fibronectin, laminin, collagen, and various other extracellular proteins. The cell-binding domain of fibronectin, with its extended RGD-containing loop, is shown in Figure 7.10. The binding site for the RGD loop is thought to reside on the I-like domain of the β subunit of specific integrins (such as those shown in the left column of Table 7.1)

As noted earlier, and shown on the cover of the text, many cells will adhere firmly to a culture dish coated with fibronectin. If the cells are added in the presence of a synthetic peptide containing the RGD sequence, the cells are no longer able to attach to the dish; the synthetic peptide is able to compete successfully with the RGD sequences in the fibronectin molecules for binding sites on the cell-surface integrins. It is estimated that about half of all integrins contain an RGD-binding site. Consequently, specific extracellular proteins bind to a variety of different integrins, and a single integrin may bind multiple ligands. Despite the apparent overlap, most integrins do appear to have unique functions, as knockout mice (Section 18.13) that lack different integrin subunits exhibit distinct phenotypes. For example, α_8 knockouts show kidney defects, α_4 knockouts exhibit heart defects, and α_5 knockouts display vascular defects.

The discovery of the importance of the RGD sequence has opened the door to new avenues for treatment of medical conditions that involve receptor-ligand interactions. The formation of a blood clot (*thrombus*) in a diseased artery can block the flow of blood to major organs and is one of the leading causes of heart attack and stroke. A blood clot starts with an aggregation of blood platelets, nonnucleated cells that circulate in the blood. The aggregation of platelets requires the interaction of a platelet-specific integrin ($\alpha_{IIb}\beta_3$) with soluble RGD-containing blood proteins, such as fibrinogen and von Willebrand factor, which act as linkers that hold the platelets together (Figure 7.15). Experiments with animals indicate that RGD-containing peptides can inhibit blood clot formation by preventing the platelet integrin from binding to blood proteins. These findings led to the design of a new class of nonpeptide, antithrombotic agents (e.g., Aggrastat) that resemble the RGD structure but bind only to the platelet integrin. A specific antibody (called ReoPro) directed against the $\alpha_{IIb}\beta_3$ integrin can prevent blood clots in patients undergoing high-risk vascular surgeries.

The cytoplasmic domains of integrins contain binding sites for a variety of cytoplasmic proteins, including several that act as adaptors to link the integrin to actin filaments of the cytoskeleton (see Figure 7.17). The role of integrins in making the connection between the ECM and the cytoskeleton is best seen in two specialized structures: focal adhesions and hemi-desmosomes.

Focal Adhesions and Hemidesmosomes: Anchoring Cells to Their Substratum

It is much easier to study the interaction of cells with the bottom of a culture dish than with an extracellular matrix inside an animal. Consequently, much of our knowledge in this area has been obtained from the study of cells adhering to various substrates in vitro. The stages in the attachment of a cell to the surface of a culture dish are shown in Figure 7.16. At first, the cell has a rounded morphology, as is generally true of animal cells suspended in aqueous medium. Once the cell makes contact with the substratum, it sends out projections that form increasingly stable attachments. Over time, the cell flattens and spreads itself out on the substratum.

When fibroblasts or epithelial cells spread onto the bottom of a culture dish, the lower surface of the cell is not pressed uniformly against the substratum. Instead, the cell is anchored to the surface of the dish only at scattered, discrete sites, called **focal adhesions** (or *focal contacts*) (Figure 7.17a). Focal adhesions are dynamic structures that can be rapidly disassembled if the adherent cell is stimulated to move or enter mitosis. The plasma membrane in the region of a focal adhesion contains large clusters of integrins (most often $\alpha_5\beta_1$) that connect the extracellular material that coats the culture dish to actin filaments of the cytoskeleton (Figure 7.17b,c). The role of focal adhesions as signaling complexes is discussed in Chapter 15 (see Figure

(a) 2.5 μm

(b) 2.5 μm

(c) 2.5 μm

(d) 2.5 μm

Figure 7.16 **Steps in the process of cell spreading.** Scanning electron micrographs showing the morphology of mouse fibroblasts at successive times during attachment and spreading on glass coverslips. Cells were fixed after (a) 30 minutes, (b) 60 minutes, (c) 2 hours, and (d) 24 hours of attachment. (FROM J. J. ROSEN AND L. A. CULP, EXP. CELL RES. 107:141, 1977.)

(a)

(b) 2.5 µm

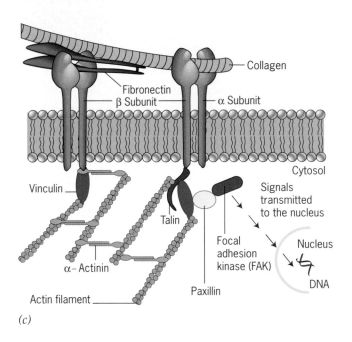

(c)

Figure 7.17 **Focal adhesions are sites where cells adhere to their substratum.** (*a*) This cultured cell has been stained with fluorescent antibodies to reveal the locations of the actin filaments (gray-green) and the integrins (red). The integrins are localized in small patches that correspond to the sites of focal adhesions. (*b*) The cytoplasmic surface of a focal adhesion of a cultured amphibian cell is shown here after the inner surface of the membrane was processed for quick-freeze, deep-etch analysis. Bundles of microfilaments are seen to associate with the inner surface of the membrane in the region of a focal adhesion. (*c*) Schematic drawing of a focal adhesion showing the interactions of integrin molecules with other proteins on both sides of the lipid bilayer. The binding of extracellular ligands, such as collagen and fibronectin, is thought to induce conformational changes in the cytoplasmic domains of the integrin that cause the integrins to become linked to actin filaments of the cytoskeleton. Linkages with the cytoskeleton lead, in turn, to the clustering of integrins at the cell surface. Linkages with the cytoskeleton are mediated by various actin-binding proteins, such as talin and α-actinin, that bind to the β subunit of the integrin. The cytoplasmic domains of integrins are also associated with protein kinases, such as FAK (focal adhesion kinase). The attachment of the integrin to an extracellular ligand can activate these protein kinases and start a chain reaction that transmits signals throughout the cell. (A: FROM MARGO LAKONISHOK AND CHRIS DOE, SCI. AM. P. 68, MAY 1997; B: FROM STEVEN J. SAMUELSSON, PAUL J. LUTHER, DAVID W. PUMPLIN, AND ROBERT J. BLOCH, J. CELL BIOL. 122:487, 1993; BY COPYRIGHT PERMISSION OF THE ROCKEFELLER UNIVERSITY PRESS.)

15. 28). Figure 7.17*c* also illustrates how the binding of an extracellular ligand, such as fibronectin or laminin, can activate protein kinases, such as FAK, that can transmit signals throughout the cell, including the cell nucleus.

Focal adhesions occur primarily in vitro. *Within the body,* the tightest attachment between a cell and its extracellular matrix is seen at the basal surface of epithelial cells where the cells are anchored to the underlying basement membrane by a specialized adhesive structure called a **hemidesmosome** (Figure 7.1 and 7.18). Hemidesmosomes contain a dense plaque on the inner surface of the plasma membrane with filaments coursing outward into the cytoplasm. Unlike the filaments of focal adhesions, which consist of actin, the filaments of the hemidesmosome are thicker and consist of the protein keratin. Keratin-containing filaments are classified as intermediate filaments, which serve primarily in a supportive function; they are distinguished from the thinner, actin-containing microfilaments (as discussed in detail in Chapter 9). The keratin filaments of the hemidesmosome are linked to the extracellular matrix by membrane-span-

ning integrins, including α₆β₄ (Figure 7.18*b*). These integrins also transmit signals from the ECM that affect the shape and activities of the attached epithelial cells.

The importance of hemidesmosomes is revealed by a rare disease, *bullous pemphigoid,* in which individuals produce antibodies that bind to proteins (the *bullous pemphigoid antigens*) present in these adhesive structures. Diseases caused by production of antibodies directed against one's own tissues (i.e., autoantibodies) are called autoimmune disorders and are responsible for a wide variety of conditions. The presence of autoantibodies causes the lower layer of the epidermis to lose attachment to the un-

Labels in figure (c): Collagen, Fibronectin, β Subunit, α Subunit, Cytosol, Vinculin, Talin, α-Actinin, Actin filament, Paxillin, Focal adhesion kinase (FAK), Signals transmitted to the nucleus, Nucleus, DNA

Labels in figure (b): Membrane attachment, Actin filaments

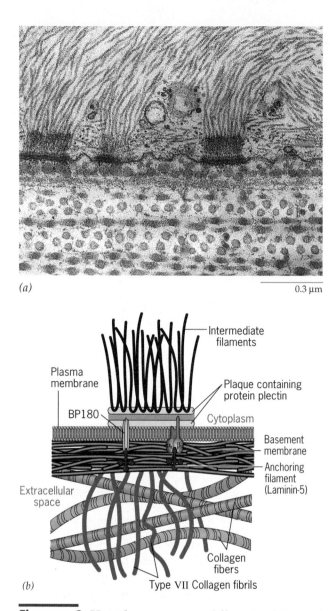

(a) 0.3 μm

Figure 7.18 Hemidesmosomes are differentiated sites at the basal surfaces of epithelial cells where the cells are attached to the underlying basement membrane. (*a*) Electron micrograph of several hemidesmosomes showing the dense plaque on the inner surface of the plasma membrane and the intermediate filaments projecting into the cytoplasm. (*b*) Schematic diagram showing the major components of a hemidesmosome connecting the epidermis to the underlying dermis. The $\alpha_6\beta_4$ integrin molecules of the epidermal cells are linked to cytoplasmic intermediate filaments by a protein called plectin that is present in the dark-staining plaque and to the basement membrane by anchoring filaments of a particular type of laminin. A second transmembrane protein (BP180) is also present in hemidesmosomes. The collagen fibers are part of the underlying dermis. (A: FROM DOUGLAS E. KELLY, J. CELL BIOL. 28:51, 1966; BY COPYRIGHT PERMISSION OF THE ROCKEFELLER UNIVERSITY PRESS.)

derlying basement membrane (and thus to the underlying connective tissue layer of the dermis). The leakage of fluid into the space beneath the epidermis results in severe blistering of the skin. A similar inherited blistering disease, *epidermolysis bullosa*, can occur in patients with genetic al-

terations in any one of a number of hemidesmosomal proteins, including the α_6 or β_4 integrin subunit or laminin. Other epithelia in the body, such as the lining of the gastrointestinal and urinary tracts, may also be affected in this disease.

REVIEW ?

1. Which types of integral membrane proteins play a key role in mediating interactions between the cell surface and the ECM? What is the structure of these proteins? Why do these interactions require the presence of Ca^{2+}?
2. How can a cell-surface protein be involved in both cell adhesion and transmembrane signal transduction?
3. List three characteristics that distinguish hemidesmosomes from focal adhesions.

7.3 INTERACTIONS OF CELLS WITH OTHER CELLS

Examination of a thin section through a major organ of an animal reveals a complex architecture involving a variety of different types of cells. Little is known about the mechanisms responsible for generating the complex three-dimensional cellular arrangements found within developing organs. It is presumed that this process depends heavily on *selective* interactions between cells of the same type, as well as between cells of a different type. Evidence indicates that cells can recognize the surfaces of other cells, interacting with some and ignoring others.

It is very difficult to study cellular interactions that occur within the minuscule organs of a developing embryo. Early attempts to learn about cell–cell recognition and adhesion were carried out by removing a developing organ from a chick or amphibian embryo, dissociating the organ's tissues to form a suspension of single cells, and determining the ability of the cells to reaggregate in culture. In experiments in which cells from two different developing organs were dissociated and mixed together, the cells would initially aggregate to form a mixed clump. Over time, however, the cells would move around within the aggregate and eventually "sort themselves out" so that each cell adhered only to cells of the same type (Figure 7.19). Once separated into a homogeneous cluster, these embryonic cells would often differentiate into many of the structures they would have formed within an intact embryo.

Little was known about the nature of the molecules that mediated cell–cell adhesion until the development of techniques for purifying integral membrane proteins and, more recently, for the isolation and cloning of the genes that encode these proteins. Dozens of different proteins

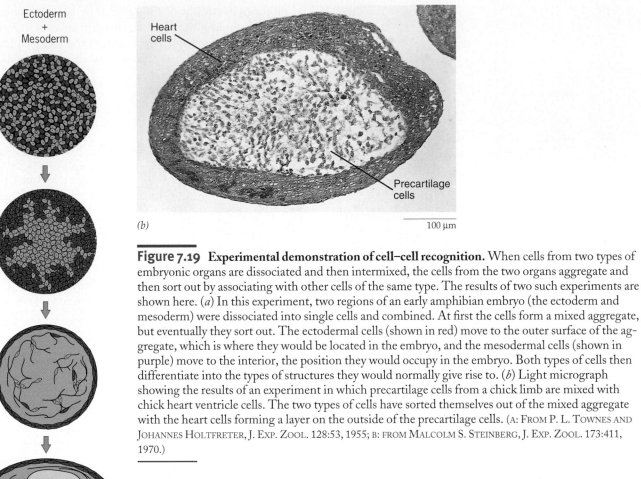

(a)

(b)

100 µm

Figure 7.19 Experimental demonstration of cell–cell recognition. When cells from two types of embryonic organs are dissociated and then intermixed, the cells from the two organs aggregate and then sort out by associating with other cells of the same type. The results of two such experiments are shown here. (*a*) In this experiment, two regions of an early amphibian embryo (the ectoderm and mesoderm) were dissociated into single cells and combined. At first the cells form a mixed aggregate, but eventually they sort out. The ectodermal cells (shown in red) move to the outer surface of the aggregate, which is where they would be located in the embryo, and the mesodermal cells (shown in purple) move to the interior, the position they would occupy in the embryo. Both types of cells then differentiate into the types of structures they would normally give rise to. (*b*) Light micrograph showing the results of an experiment in which precartilage cells from a chick limb are mixed with chick heart ventricle cells. The two types of cells have sorted themselves out of the mixed aggregate with the heart cells forming a layer on the outside of the precartilage cells. (A: FROM P. L. TOWNES AND JOHANNES HOLTFRETER, J. EXP. ZOOL. 128:53, 1955; B: FROM MALCOLM S. STEINBERG, J. EXP. ZOOL. 173:411, 1970.)

involved in cell adhesion have now been identified. Different arrays of these molecules on the surfaces of different types of cells are thought to be responsible for the specific interactions between cells within complex tissues. Four distinct families of integral membrane proteins play a major role in mediating cell-cell adhesion: (1) selectins, (2) certain members of the immunoglobulin superfamily (IgSF), (3) certain members of the integrin family, and (4) cadherins.

Selectins

During the 1960s, it was discovered that lymphocytes that had been removed from peripheral lymph nodes, radioactively labeled, and injected back into the body would return to the sites from which they were originally derived—a phenomenon called "lymphocyte homing." It was subsequently found that homing could be studied in vitro by allowing lymphocytes to adhere to frozen tissue sec-

tions of lymphoid organ tissues. Under these experimental conditions, the lymphocytes would selectively adhere to the endothelial lining of the venules (the smallest veins) of peripheral lymph nodes. Binding of lymphocytes to venules could be blocked by antibodies that bound to a specific glycoprotein on the lymphocyte surface. This lymphocyte glycoprotein was named LEU-CAM1 and later L-selectin.

Selectins are a family of integral membrane glycoproteins that recognize and bind to a particular arrangement of sugars in the oligosaccharides that project from the surfaces of other cells (page 131). The name of this class of cell-surface receptors is derived from the word "lectin," a term for a compound that binds to specific carbohydrate groups. Selectins possess a small cytoplasmic domain, a single membrane-spanning domain, and a large extracellular segment that consists of a number of separate domains, including an outermost domain that acts as the lectin (Figure 7.20). There are three known selectins: E-selectin, present on endothelial cells; P-selectin, present on platelets and endothelial cells; and L-selectin, present on all types of leukocytes (white blood cells). All three selectins recognize a particular grouping of sugars (Figure 7.20) that is found at the ends of the carbohydrate chains of certain complex glycoproteins. Binding of selectins to their carbohydrate ligands requires calcium. As a group, selectins mediate transient interactions between circulat-

(a)

(b)

(c)

(d)

Figure 7.20 Selectins. Schematic drawing showing the three types of known selectins (*a*). All of them recognize and bind to a similar carbohydrate ligand at the ends of oligosaccharide chains on glycoproteins, such as the one shown in (*b*). (*c*) The detailed structure of the carbohydrate ligand. The terminal fucose and sialic acid moieties are particularly important in selectin recognition, and the *N*-acetylglucosamine moiety is often sulfated. P-selectin, which is present on endothelial cells, also binds with high affinity to a different carbohydrate ligand called PSGL-1, which projects from the surface of leukocytes. (*d*) A model of the lectin domain of E-selectin. The yellow residues are those whose mutation significantly decreases the binding of the protein toward its ligand. The ligand (sLe^x) recognized by E-selectin is shown in blue, oriented as if it were ready to bind to the lectin domain of the selectin. The red groups of the ligand indicate the carbohydrate moieties that are crucial in binding to E-selectin; these include the carboxyl group of sialic acid, two hydroxyl groups of galactose, and three hydroxyl groups of fucose. The bar represents 10 Å. (D: FROM DAVID V. ERBE ET AL., J. CELL BIOL. 119:224, 1992; BY COPYRIGHT PERMISSION OF THE ROCKE- FELLER UNIVERSITY PRESS. IMAGE COURTESY OF LARRY LASKY, GENEN- TECH, INC.)

ing leukocytes and vessel walls at sites of inflammation and clotting. The role of selectins in inflammation is discussed further in the Human Perspective on page 264.

Immunoglobulins and Integrins

The elucidation of the structure of blood-borne antibody molecules in the 1960s was one of the milestones in our understanding of the immune response. Antibodies, which are a type of protein called an immunoglobulin (or Ig), were found to consist of polypeptide chains composed of a number of similar domains. Each of these Ig domains, as they are called, is composed of 70 to 110 amino acids organized into a tightly folded structure, as shown in the inset of Figure 7.21. With further study, it became apparent that Ig-type domains were present in a wide variety of proteins, which together constitute the **immunoglobulin superfamily**, or **IgSF**. Most members of the IgSF are involved in various aspects of immune function, but some of these proteins mediate calcium-*independent* cell–cell adhesion. In fact, the discovery of Ig-like domains in cell-adhesion receptors in invertebrates—animals that lack a classic immune system—suggests that Ig-like proteins originally evolved as cell-adhesion mediators and only secondarily took on their functions as effectors of the vertebrate immune system.

Most IgSF cell-adhesion molecules mediate the specific interactions of lymphocytes with cells required for an immune response (e.g., macrophages, other lymphocytes, and target cells). However, some IgSF members, such as VCAM (vascular cell-adhesion molecule), NCAM (neural cell-adhesion molecule), and L1, mediate adhesion between nonimmune cells. NCAM and L1, for example, play important roles in nerve outgrowth, synapse

Figure 7.21 Cell-adhesion molecules of the immunoglobulin (Ig) superfamily. One proposed model of cell-cell adhesion resulting from the specific interactions of the immunoglobulin (Ig) domains of two L1 molecules projecting from the surfaces of neighboring cells. Each L1 molecule contains a small cytoplasmic domain, a transmembrane segment, several segments that resemble the repeats found in fibronectin, and six Ig domains situated at the N-terminal portion of the molecule. The inset shows the structure of the two N-terminal Ig domains of VCAM, an IgSF molecule on the surface of endothelial cells. The Ig domains of VCAM and L1 have a similar three-dimensional structure consisting of two β sheets packed face-to-face. (INSET REPRINTED WITH PERMISSION FROM E. YVONNE JONES ET AL., NATURE 373:540, 1995. COPYRIGHT 1995 MACMILLAN MAGAZINES LTD.)

formation, and other events during the development of the nervous system. Like fibronectin and many other proteins involved in cell adhesion, the IgSF cell-adhesion molecules have a modular construction (Figure 7.21) and are composed of individual domains of similar structure to the domains in other proteins.

The importance of L1 in neural development has been revealed in several ways. In humans, mutations in the *L1* gene can have devastating consequences. In extreme cases, babies are born with a fatal condition of hydrocephalus ("water on the brain"). Children with less severe forms of the mutation typically exhibit mental retardation and difficulty in controlling limb movements (spasticity). Autopsies on patients that have died of an L1-deficiency disease reveal a remarkable condition: they are often missing two large nerve tracts, one that runs between the two halves of the brain and the other that runs between the brain and the spinal cord. The absence of such nerve tracts

suggests that L1 is involved in the growth of axons within the embryonic nervous system.

As discussed in Section 9.7, the tip of a growing axon is a highly motile portion called a **growth cone** that is responsible for guiding the axon to its correct target in the embryo. The growth cone is an exploratory structure. It creeps along the substratum, sensing and responding to substances in its environment that influence axonal growth in one direction or another. In vitro studies have shown (1) that growth cones of certain nerves contain the L1 protein (and other IgSF molecules), (2) that these growth cones will grow outward over an L1-containing substratum, and (3) that nerve outgrowth in these experiments is blocked by addition of antibodies directed against L1. These studies suggest that interaction between the extracellular domains of L1 molecules present on the cell surface and in the substratum generates signals within the growth cone that promote its extension.

Various types of proteins serve as ligands for IgSF cell-surface molecules. As described earlier, most integrins facilitate adhesion of cells to their substratum, but a few integrins mediate cell–cell adhesion by binding to proteins on other cells. For example, the integrin $\alpha_4\beta_1$ on the surface of leukocytes binds to VCAM, an IgSF protein on the endothelial lining of certain blood vessels.

Cadherins

The **cadherins** are a large family of glycoproteins that mediate Ca^{2+}-dependent cell adhesion and transmit signals from the ECM to the cytoplasm. Cadherins join cells of similar type to one another and do so predominantly by binding to the same cadherin present on the surface of the neighboring cell. This property of cadherins was first demonstrated by genetically engineering cells that were normally nonadhesive to express one of a variety of different cadherins. The cells were then mixed in various combinations and their interactions monitored. It was found that cells expressing one species of cadherin preferentially adhered to other cells expressing the same cadherin.

Cadherins are found on the surfaces of many different cell types in animals, with each particular member of the cadherin family having a specific distribution within the body. The best-studied cadherins are E-cadherin (epithelial), N-cadherin (neural), and P-cadherin (placental). These "classical" cadherins, as they are called, contain a relatively large extracellular segment consisting of five tandem domains, a single transmembrane segment, and a small cytoplasmic domain (Figure 7.22). The cytoplasmic domain is often associated with members of the **catenin** family of cytosolic proteins, which have a dual role: they tether the cadherins to the cytoskeleton (see Figure 7.26), and they transmit information to the cytoplasm.

The structural models of cadherin adhesion depicted in Figure 7.22 are derived from X-ray crystallographic studies on the extracellular domains of these molecules. These studies indicate that cadherins from the same cell surface associate laterally to form dimers. These studies also shed light on the role of calcium, which has been known for

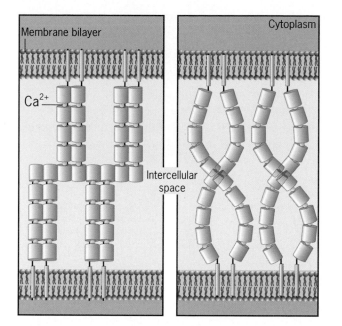

Figure 7.22 Cadherins and cell adhesion. Schematic representation of two cells adhering to one another as the result of interactions between similar types of cadherins projecting from the plasma membrane of each cell. Calcium ions (shown as small yellow spheres) are situated between the successive domains of the cadherin molecule where they play a critical role in maintaining the rigidity of the extracellular portion of the protein. The strength of the cell-cell adhesion is proportional to the number of cadherin molecules present in the cluster. This illustration shows two alternate structures by which cadherins from opposing cells interact; either by interdigitation (*a*) or head-to-head contact (*b*).

decades to be essential for cell-cell adhesion. As indicated in Figure 7.22, calcium ions form bridges between successive domains of a given molecule, not between molecules from different cells, as had long been presumed. These calcium ions maintain the extracellular portion of each cadherin in a rigid conformation required for cell adhesion. Adhesion between cells results from the interaction between the terminal domains of cadherins from opposing cells to form a "cell-adhesion zipper." Two alternate ways that this may happen are shown in Figure 7.22. The formation of an extensive network of interacting cadherins would be expected to hold apposing cells together with great strength.

Cadherin-mediated adhesion is thought to be responsible for the ability of like cells to "sort out" of mixed aggregates, as was illustrated in Figure 7.19. In fact, cadherins may be the single most important factor in molding cells into cohesive tissues in the embryo and holding them together in the adult. As discussed in the Human Perspective, the loss of cadherin function may play a key role in the spread of malignant tumors.

Embryonic development is characterized by change: change in gene expression, change in cell shape, change in cell motility, change in cell adhesion, and so forth. A number of morphogenetic events during embryonic development involve a group of cells changing from a mesenchyme

(loose, primarily nonadhesive cells) to an epithelium (tightly adherent, organized cell layer), or vice versa. The mesenchymal–epithelial transition is illustrated by the movement of mesodermal cells during gastrulation. Typically, these cells break away from a cohesive layer at the surface of the early gastrula and wander into the interior regions as mesenchymal cells (Figure 7.23*a*). After a period, many of these same cells regain adhesive properties and reassemble into an epithelium, such as the somites that form along the dorsal midline of the embryo (Figure 7.23*b*). At a later stage, cells in certain parts of the somite lose their adhesiveness and begin to wander again as a mesenchyme (Figure 7.23*c*), into the developing limb to become muscle or cartilage or beneath the developing epidermis to become dermal tissues. Cadherins (and other cell-adhesion molecules) play a key role in these events by changing the adhesive properties of cells. The aggregation of cells into an epithelium, such as the somite, is correlated with the appearance of N-cadherin on the surfaces of cells (Figure 7.23*d*), an event that would be expected to promote adhesion between cells. In contrast, the dispersion of cells from an epithelium is correlated with the disappearance of N-cadherin.

Whereas cadherins are typically distributed diffusely over the surface of cells, they also participate in the formation of specialized intercellular junctions, which are the subjects of the following section.

Linking Cells to Other Cells: Synapses, Adherens Junctions, and Desmosomes

Synaptic Junctions One of the great mysteries in cell biology lies in the mechanism by which the tip of a growing axon is able to hook up at the end of its journey with a particular target cell. During the development of the human eye, for example, each axon grows out of a specific site in the retina and eventually forms a **synaptic junction** with a specific cell (or group of cells) in the optic tectum of the brain. This type of synaptic relationship between neurons from the retina and optic tectum is necessary for us to comprehend spatial relationships in our visual field. But how does the tip of a retinal axon know which cell in the brain it is "supposed to select"? During the 1950s, Roger Sperry of the California Institute of Technology suggested that neural specificity was governed by a large family of cell surface molecules. A given axon and its target cell would have a specific pair of complementary surface molecules that would allow the growing axon to find a particular target cell among a large population of potential candidates. Over the years evidence has accumulated in favor of Sperry's proposal and has recently pointed to a unique family of cadherins, called *protocadherins*, as the molecules that carry the molecular code for mediating synaptic connections. Protocadherins have been localized to synapses within the brain (Figure 7.24) and it has been shown that different members of the family are found among different neurons. More than 50 different genes encoding protocadherins have been identified, and many more are likely to be discovered. As discussed in Chapter 17 in the case of antibodies, there are nu-

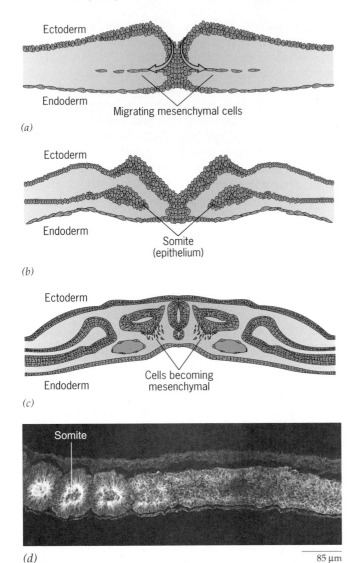

(a)

(b)

(c)

(d)

85 μm

Figure 7.23 **Cadherins and the epithelial-mesenchymal transformation.** (*a–c*) Stages in the early development of a chick embryo. (*a*) During gastrulation, cells in the upper layer of the embryo (the epiblast) migrate toward a groove in the center of the embryo, sink into the groove, and then migrate laterally as mesenchymal cells in the space beneath the epiblast. (*b*) Some of these mesenchymal cells collect into blocks of epithelial cells, forming the somites. (*c*) At a later stage of development, a part of the wall of each somite transforms into mesenchymal cells that migrate into various peripheral tissues. (*d*) A sagittal section through a chick embryo at the level of the somites. The anterior part of the embryo is shown on the left and the posterior part on the right. The anterior-posterior axis of a chick embryo is like a developmental time line; certain events occur first anteriorly, such as the formation of distinct somites, then at a later time in the posterior regions. In this photograph, the location of N-cadherin in the somite epithelium is revealed by immunofluorescence. The formation of somites, which is seen progressively from the anterior to the posterior portions of the embryo, is correlated with the synthesis of N-cadherin. (D: FROM KOHEI HATTA, SHIN TAGAKI, HAJIME FUJISAWA, AND MASATOSHI TAKEICHI, DEV. BIOL. 120:218, 1987.)

(a)

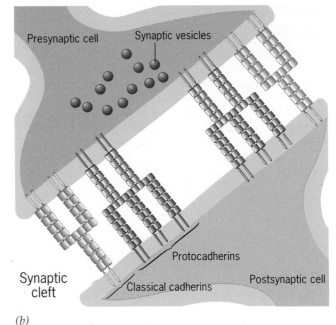

(b)

Figure 7.24 **The localization of cadherins in synaptic junctions.** (*a*) One technique to localize a protein within a subcellular region is to incubate tissue sections with labeled antibodies prepared against that protein. In this electron micrograph, a section through a synapse in a mouse brain has been incubated with gold-labeled antibodies (black grains) that bind to a protocadherin. The protocadherin is localized within the central region of the synaptic junction. (*b*) A model of a synaptic junction in which classical cadherins (which contain five extracellular domains) are situated at the edges of the synapse where they are thought to provide adhesion between the pre- and postsynaptic membranes. Protocadherins (which contain six extracellular domains) are present in the center of the synapse where they are thought to mediate specificity in the interaction between the two cells. (A: COURTESY OF TAKESHI YAGI, INSTITUTE FOR MOLECULAR CELLULAR BIOLOGY, OSAKA UNIVERSITY, JAPAN.)

(a)

Figure 7.25 A junctional complex. (a) Schematic diagram showing the junctional complex on the lateral surfaces of a simple columnar epithelial cell. The complex consists of a tight junction (zonula occludens), an adherens junction (zonula adherens), and a desmosome (macula adherens). Other desmosomes and gap junctions are located deeper along the lateral surfaces of the cells. Adherens junctions and tight junctions encircle the cell, whereas desmosomes and gap junctions are restricted to a particular site between adjacent cells. Hemidesmosomes are shown at the basal cell surface. (b) Electron micrograph of a junctional complex between two rat airway epithelial cells (TJ, tight junction; AJ, adherens junction; D, desmosome). (B: FROM EVELINE E. SCHNEEBERGER AND ROBERT D. LYNCH, AM. J. PHYSIOL. 262:L648, 1992.)

(b)

0.1 µm

merous mechanisms by which an organism can transform a few hundred genes into thousands of different protein species. This is the type of diversity that would be expected of a family of molecules that mediates synaptic specificity.[2] In addition to their role in mediating specificity, cadherins are thought to provide the adhesive glue that holds the presynaptic and postsynaptic membranes at the proper distance to facilitate synaptic transmission (Figure 4.54). The role of cadherins in the formation of synapses is likely to be a focus of research in the coming years.

Adherens Junctions and Desmosomes The cells of certain tissues, particularly epithelia and cardiac muscle, are notoriously difficult to separate from one another because they are held together tightly by specialized calcium-dependent adhesive junctions. There are two main types

of adhesive junctions: adherens junctions and desmosomes. In addition to adhesive junctions, epithelial cells often contain other types of cell junctions that are also located along their lateral surfaces near the apical lumen (Figure 7.25). When these junctions are arranged in a specific array, this assortment of surface specializations is

[2]The genes that encode these proteins have a striking organization that is suited for generating large numbers of similar polypeptides. The important discovery of these unique genetic loci by Qiang Wu and Tom Maniatis is described in *Cell* 97:779–790, 1999.

THE HUMAN PERSPECTIVE

The Role of Cell Adhesion in Inflammation and Metastasis

Inflammation is one of the primary responses to infection. If a part of the body were to become contaminated by bacteria, as might occur following a puncture wound to the skin, the site of injury would become a magnet for a variety of white blood cells. White blood cells (leukocytes) that would normally remain in the bloodstream instead are stimulated to traverse the endothelial layer that lines the smallest veins (venules) in the region and enter the tissue. Once in the tissue, the leukocytes move in response to chemical signals toward the invading microorganisms, which they ingest. Although inflammation is a protective response, it also produces negative side effects, such as fever, swelling due to fluid accumulation, redness, and pain.

Inflammation can also be triggered inappropriately. For example, damage to the tissues of the heart or brain can occur when blood flow to these organs is blocked during a heart attack or stroke. When blood flow to the organ is restored, circulating leukocytes may attack the damaged tissue, causing a condition known as *reperfusion damage*. An overzealous inflammatory response can also lead to asthma, toxic shock syndrome, and respiratory distress syndrome. A great deal of research has focused on questions related to these conditions: How are leukocytes recruited to sites of inflammation? Why do they stop flowing through the bloodstream and adhere to vessel walls? How do they penetrate the walls of the vessels? How can some of the negative side effects of inflammation be blocked without interfering with the beneficial aspects of the response? Answers to questions about inflammation have focused on three types of cell-adhesion molecules: selectins, integrins, and IgSF molecules.

A chain of events that has been proposed to occur during acute inflammation is shown in Figure 1. The first step begins as the walls of the venules respond to "signals" from nearby damaged tissue. The endothelial cells that line these venules become more adhesive to circulating neutrophils, a type of phagocytic leukocyte that carries out a rapid, nonspecific attack on invading pathogens. This change in adhesion is mediated by a temporary display of P- and E-selectins on the surfaces of the activated endothelial cells in the damaged area (step 2, Figure 1). When neutrophils encounter the selectins, they form transient adhesions that dramatically slow their movement through the vessel. At this stage, the neutrophils can be seen to roll slowly along the wall of the vessel. A number of biotechnology companies are attempting to develop anti-inflammatory drugs that act by interfering with binding of ligands to selectins, especially P-selectin. Anti-selectin antibodies block neutrophil rolling on selectin-coated surfaces in vitro and suppress inflammation and reperfusion damage in animals. A similar type of blocking effect has been attained using synthetic carbohydrates that compete with carbohydrate ligands on the surfaces of the neutrophils.

As the neutrophils interact with the inflamed venule endothelium, an activation process (triggered by various agents, including *platelet activating factor* released from the endothelium) causes an increase in the binding activity of integrins already situated on the neutrophil surface (step 3, Figure 1). The activated integrins bind with high affinity to IgSF molecules (ICAMs) on the surface of the endothelial cells, causing the neutrophils to stop their rolling and adhere firmly to the wall of the vessel (step 4). The bound neutrophils then change their shape and squeeze through the endothelial layer (a process called *extravasation*) into the damaged tissue (step 5). This cascade of events involving several different types of cell-adhesion molecules ensures that attachment of blood cells to the walls of blood vessels and subsequent penetration will occur only at sites where leukocyte invasion is required.

The importance of integrins in the inflammatory response is demonstrated by a rare disease called *leukocyte adhesion deficiency (LAD)*. Persons with this disease are unable to produce the β_2 subunit as part of a number of leukocyte integrins. As a result, the leukocytes of these individuals lack the

① Endothelial activation ② Neutrophil trapping ③ Neutrophil activation ④ Neutrophil adhesion ⑤ Neutrophil invasion

— ICAM — Selectin — Platelet activating factor — Integrin (unactivated) — Integrin (activated) Invasion

Figure 1 **Steps in the movement of neutrophils from the bloodstream during inflammation. Steps are described in the text.**

ability to adhere to the endothelial layer of venules, a step required for their exit from the bloodstream. These patients suffer from repeated life-threatening bacterial infections. The disease is best treated by bone marrow transplantation, which provides the patient with stem cells capable of forming normal leukocytes. Administration of antibodies against the β_2 subunit can mimic the effects of LAD, blocking the movement of neutrophils and other leukocytes out of blood vessels. Such antibodies might prove useful in preventing inflammatory responses associated with diseases such as asthma and rheumatoid arthritis or with reperfusion.

Cancer is a disease in which cells escape from the body's normal growth control mechanisms and proliferate in an unregulated manner. Were the malignant cells to remain in a single mass, as often occurs in a thyroid cancer or some types of skin cancer, most cancers would be readily cured by surgical removal of the diseased tissue. Most malignant tumors, however, spawn cells that are capable of leaving the primary tumor mass and entering the bloodstream or lymphatic channels, thereby initiating the growth of secondary tumors in other parts of the body. The spread of a tumor within the body is called **metastasis** and is the reason why cancer is such a devastating disease. Metastatic cells (cancer cells that are able to initiate the formation of secondary tumors) are thought to have special cell-surface properties that are not shared by most other cells in the tumor. For example:

1. Metastatic cells must be less adhesive than other cells to break free of the tumor mass.

2. They must be able to penetrate numerous barriers, such as the extracellular matrices of surrounding connective tissue and the basement membranes that line the blood vessels that carry them to distant sites.

3. They must be able to invade normal tissues if they are to form secondary colonies.

The penetration of extracellular matrices is accomplished largely by ECM-digesting enzymes, most notably the matrix metalloproteinases (MMPs) discussed on page 252. In some cases, the tumor cells secrete their own MMPs, but in most cases the tumor cells induce the synthesis and secretion of these enzymes by the surrounding "host" cells. Either way, these enzymes degrade the proteins and proteoglycans that stand in the way of tumor cell migration. Because of their apparent role in cancer and other diseases, MMPs have become a target of the pharmaceutical industry. A number of clinical trials of MMP inhibitors have been conducted in patients with advanced, inoperable pancreatic or lung cancers. To date, these inhibitors have not shown much promise in stopping tumor progression and, in some cases, have led to joint damage. Trials of MMP inhibitors in patients with earlier-stage cancers have shown more promise, but it is too early to tell if these drugs will ultimately prove useful.

Changes in the numbers and types of various cell-adhesion molecules—and thus the ability of cells to adhere to other cells or to extracellular matrices—have also been implicated in promoting metastasis. The major studies in this area have focused on E-cadherin, which is the predominant cell-adhesion molecule of epithelial cells (see Figure 7.26). In one survey of epithelial cell tumors (e.g., breast, prostate, and colon cancers), it was found that the greater the level of expression of E-cadherin, the less the cell's metastatic potential. Other studies suggest that the progressive loss of E-cadherin from the surfaces of cancer cells as they grow in a tumor increases the level of malignancy of the cells and worsens a patient's prognosis. It would appear that the presence of E-cadherin favors the adhesion of cells to one another and suppresses the dispersal of tumor cells to distant sites. In fact, when malignant cells are forced to express extra copies of the E-cadherin gene, the cells become much less capable of causing tumors when injected into host animals. The importance of E-cadherin is evident from a recent study of a family of native New Zealanders that had lost 25 members to stomach cancer over a 30-year period. Analysis of the DNA of family members has revealed that susceptible individuals carry mutations in the gene encoding E-cadherin.

called a **junctional complex**. The structures and functions of the two adhesive junctions of the complex are described in the following paragraphs, whereas discussion of the other types of epithelial junctions (tight junctions and gap junctions) is taken up later in the chapter.

Adherens junctions are found in a variety of sites within the body. They are particularly common in epithelia, such as the lining of the intestine, where they occur as a "belt" (or *zonulae adherens*) that encircles each of the cells near its apical surface, binding that cell to its surrounding neighbors (Figure 7.25a). The cells of an adherens junction are held together by calcium-dependent linkages formed between the extracellular domains of cadherin molecules that bridge the 30-nm gap between neighboring cells (Figure 7.26). As Figure 7.26 illustrates, the cytoplasmic domain of these cadherins is linked by catenins to a variety of cytoplasmic proteins, including actin filaments of the cytoskeleton. Thus, like the integrins of a focal adhesion, the cadherins of an adherens junctions (1) connect the external environment to the actin cytoskeleton and (2) provide a potential pathway for signals to be transmitted from the cell exterior to the cytoplasm. To give one example, the adherens junctions situated between endothelial cells that line the walls of blood vessels transmit signals that ensure the survival of the cells. Mice that are lacking an endothelial cell cadherin are unable to transmit these survival signals, and these animals die during embryonic development as a result of the death of the cells lining the vessel walls.

Desmosomes (or *maculae adherens*) are disk-shaped adhesive junctions approximately 1 μm in diameter (Figure 7.27a) that are found in a variety of tissues. Desmosomes are particularly numerous in tissues that are subjected to mechanical stress, such as cardiac muscle and the epithelial layers of the skin and uterine cervix. Like adherens junctions, desmosomes contain cadherins that link the two cells across a narrow (30 nm) extracellular gap. The cadherins of desmosomes have a different domain structure from that of adherens junctions and are referred to as *desmogleins* and

Figure 7.26 **The structure of an adherens junction.**
Schematic model for the molecular architecture of a zonula adherens. The cytoplasmic domain of each cadherin molecule is connected to the actin filaments of the cytoskeleton by linking proteins, including β-catenin and α-catenin. β-catenin has also been implicated as a key element in a signaling pathway leading from the cell surface to the cell nucleus, but that is probably not related to its presence in adherens junctions. Another member of the catenin family, p120 catenin, binds to a site on the cytoplasmic domain of the cadherin that is adjacent to the lipid bilayer. p120 catenin may regulate the strength of the junction and serve as a component of a signaling pathway.

desmocollins (Figure 7.27b). Dense cytoplasmic plaques on the inner surface of the plasma membranes serve as sites of anchorage for looping intermediate filaments similar to those of hemidesmosomes (Figure 7.18). The network of ropelike intermediate filaments provides structural continuity and tensile strength to the entire sheet of cells. Intermediate filaments are linked to the cytoplasmic domains of the desmosomal cadherins by additional proteins, as depicted in Figure 7.27b. The importance of cadherins in maintaining the structural integrity of an epithelium is illustrated by an autoimmune disease (*pemphigus vulgaris*) in which antibodies are produced against one of the desmogleins. The disease is characterized by a loss of epidermal cell-cell adhesion and severe blistering of the skin.

The Role of Cell-Adhesion Receptors in Transmembrane Signaling

Figure 7.28 provides a summary of some of the points that have been made in this chapter. The drawing depicts each of the four types of cell-adhesion molecules discussed and their interactions with both extracellular and cytoplasmic materials. One of the roles of integral membrane proteins

(a) 0.1 μm

(b)

Figure 7.27 **The structure of a desmosome.** (*a*) Electron micrograph of a desmosome from newt epidermis. (*b*) Schematic model of the molecular architecture of a desmosome. (A: FROM DOUGLAS E. KELLY, J. CELL BIOL. 28:51, 1966; BY COPYRIGHT PERMISSION OF THE ROCKEFELLER UNIVERSITY PRESS.)

is to transfer information across the plasma membrane, a process known as **transmembrane signaling**. While the subject will be explored in detail in Chapter 15, it can be noted that all four types of cell-adhesion molecules illustrated in Figure 7.28 have the potential to carry out this function. For example, integrins and cadherins can transmit signals from the extracellular environment and the cytoplasm by means of linkages with the cytoskeleton and with cytosolic regulatory molecules, such as protein kinases. The engagement of an integrin with its ligand can

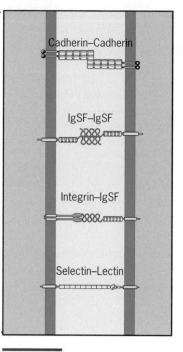

Figure 7.28 An overview of the types of interactions involving the cell surface. Four types of cell-cell adhesive interactions are shown, as well as two types of interactions between cells and extracellular substrata. Keep in mind that the various interactions depicted here do not occur on a single cell type but are shown in this manner for purposes of illustration. For example, interactions between selectins and lectins occur primarily between circulating leukocytes and the walls of blood vessels.

induce a variety of responses within a cell, including changes in cytoplasmic pH, Ca^{2+} concentration, protein phosphorylation, and gene expression. These changes, in turn, can alter a cell's growth potential, migratory activity, state of differentiation, or survival. This type of phenomenon is illustrated by the mammary gland epithelial cells shown in Figure 7.29. When these cells are removed from a mammary gland and grown on a bare culture dish, they lose their ability to synthesize milk proteins and appear as

(a) *(b)*

Figure 7.29 The role of extracellular proteins in maintaining the differentiated state of cells. (*a*) These mammary gland epithelial cells were removed from a mouse and cultured in the absence of an extracellular matrix. Unlike normal differentiated mammary cells, these cells are flattened and are not engaged in the synthesis of milk proteins. (*b*) When extracellular matrix molecules are added back to the culture, the cells regain their differentiated appearance and synthesize milk proteins. (COURTESY OF JOANNE EMERMAN)

flattened, undifferentiated cells (Figure 7.29*a*). When these same undifferentiated cells are cultured in the presence of certain extracellular molecules (e.g., laminin), they regain their differentiated appearance and become organized into milk-producing, gland-like structures (Figure 7.29*b*). Laminin is thought to stimulate mammary cells by binding to cell-surface integrins and activating kinases at the inner surface of the membrane (as in Figure 7.17*c*).

REVIEW ?

1. Distinguish between a hemidesmosome and a desmosome; a desmosome and an adherens junction.

2. Which type(s) of cell junctions contain actin filaments? Which contain(s) intermediate filaments? Which contain(s) integrins? Which contain(s) cadherins?

3. How do cadherins, IgSF members, and selectins differ at the molecular level in the way they mediate cell-cell adhesion?

7.4 TIGHT JUNCTIONS: SEALING THE EXTRACELLULAR SPACE

Biologists have known for decades that when certain types of epithelia, such as frog skin or the wall of the urinary bladder, are mounted between two compartments containing different solute concentrations, very little diffusion of ions or solutes is observed across the wall of the epithelium from one compartment to the other. Given the impermeability of plasma membranes, it is not surprising that solutes cannot diffuse freely through the cells of an epithelial layer. But why aren't they able to pass between cells by way of the *paracellular pathway* (as in Figure 7.30*a*)? The reason became apparent during the 1960s with the discovery of specialized contacts, called **tight junctions** (or *zonulae occludens*), between neighboring epithelial cells.

Tight junctions (TJs) are located at the very apical end of the junctional complex between adjacent epithelial cells (see Figure 7.25). An electron micrograph of a section through a TJ that has been cut to include the plasma membranes of the adjacent cells is shown in Figure 7.30*a*. The adjoining membranes make contact at intermittent points, rather than being fused over a large surface area. As indicated in Figure 7.30*b*, the points of cell–cell contact are sites where integral proteins of two adjacent membranes meet within the extracellular space.

Freeze fracture, which allows observation of the internal faces of a membrane, shows that the plasma membranes of a TJ contain interconnected strands (Figure 7.30*c*) that run mostly parallel to one another and to the

apical surface of the epithelium. The strands (or grooves in the opposite face of a fractured membrane) correspond to paired rows of aligned integral membrane proteins that are illustrated in the inset of Figure 7.30*b*. The integral proteins of TJs form continuous fibrils that completely encircle the cell like a gasket and make contact with neighboring cells on all sides (Figure 7.30*d*). As a result, TJs serve as a barrier to the free diffusion of water and solutes from the extracellular compartment on one side of an epithelial sheet to that on the other side. Tight junctions also serve as "fences" that help maintain the polarized character of epithelial cells (see Figure 4.31). They accomplish this by blocking the diffusion of integral proteins between the apical domain of the plasma membrane and its lateral and basal domains.

Not all TJs exhibit the same permeability properties. Part of the explanation can be seen under the electron microscope: TJs with several parallel strands (such as that in Figure 7.30*c*) tend to form better seals than junctions with only one or a couple of strands. But there is more to the story than numbers of strands. Some TJs are permeable to *specific* ions or solutes to which other TJs are impermeable. For example, all of the cells of a human kidney tubule are connected to their neighbors by TJs, but only one small region of the tubule has TJs that are permeable to magnesium (Mg^{2+}) ions. This region of the tubule known as the thick ascending limb (or TAL), is the site along the tubule where Mg^{2+} ions are reabsorbed from the tubular fluid back into the blood. Recent studies have shed considerable light on the molecular basis of TJ permeability.

Up until 1998 it was thought that all TJ strands were composed of a single protein, *occludin*. Then, in that year, it was found that cultured cells that lacked a gene for occludin, and thus could not produce the protein, were still able to form TJ strands of normal structure and function. Subsequent studies by S. Tsukita and his colleagues at the University of Kyoto led to the discovery of a family of proteins called *claudins* that form the major structural component of TJ strands. The electron micrograph of Figure 7.31 shows that occludin and claudin are present together within the linear fibrils of a TJ. At least 20 different claudins have been identified, and differences in the distribution of these proteins may explain differences in TJ permeability, such as the selective Mg^{2+} ion permeability of the TJs of the TAL. It is thought that the claudin-containing strands of the TAL have pores that are selectively permeable to Mg^{2+} ions. Support for this concept has come from the finding that one specific member of the claudin family, claudin-16, is expressed primarily in the TAL. The importance of claudin-16 in kidney function was revealed in 1999 on studies of patients suffering from a rare disease characterized by abnormally low Mg^{2+} levels in their blood. These patients were found to have mutations in both copies of their *claudin-16* gene. The Mg^{2+} level in their blood is low because tight junctions containing the abnormal protein are impermeable to Mg^{2+}. As a result, this important ion fails to exit the tubule and is simply excreted in the urine.

(a)

(c)

0.2 μm

(b)

(d)

Figure 7.30 Tight junctions. (*a*) Electron micrograph of a section through the apical region of two adjoining epithelial cells showing where the plasma membranes of the two cells come together at intermittent points within the tight junction. (*b*) A model of a tight junction showing the intermittent points of contact between integral proteins from two apposing membranes. Each of these contact sites extends as a paired row of proteins within the membranes, forming a barrier that blocks solutes from penetrating the space between the cells. (*c*) Freeze-fracture replica showing the E face of the plasma membrane of one of the cells in a region of a tight junction. The grooves in the E face are left behind after the integral membrane proteins are pulled from this half of the membrane. (*d*) Scanning electron micrograph of the apical surface of an epithelium showing the encircling nature of the tight junctions. (A: COURTESY OF DANIEL S. FRIEND; C: FROM PHILIPPA CLAUDE AND DANIEL A. GOODENOUGH, J. CELL BIOL. 58:390, 1973, BY COPYRIGHT PERMISSION OF THE ROCKEFELLER UNIVERSITY PRESS; D: COURTESY OF D. TARIN.)

Figure 7.31 The molecular composition of tight junction strands. Electron micrograph of a freeze-fracture replica of cells that had been joined to one another by tight junctions. The fracture faces were incubated with two types of gold-labeled antibodies. The smaller gold particles (arrowheads) reveal the presence of claudin molecules, whereas the larger gold particles (arrows) indicate the presence of occludin. These experiments demonstrate that both proteins are present in the same tight junction strands. Bar equals 0.15 μm. (FROM MIKIO FURUSE, HIROYUKI SASAKI, KAZUSHI FUJIMOTO, AND SHOICHIRO TSUKITA, J. CELL BIOLOGY 143:398, 1998, BY COPYRIGHT PERMISSION OF THE ROCKEFELLER UNIVERSITY PRESS.)

Tight junctions are also present between the endothelial cells that line the walls of capillaries. These junctions are particularly evident in the brain where they help form the *blood-brain barrier*, which prevents substances from passing from the bloodstream into the brain. Although small ions and even water molecules may not be able to penetrate the blood-brain barrier, cells of the immune system are able to pass across the endothelium through these junctions. These cells are thought to send a signal that opens the junction, allowing the cells to pass. While protecting the brain from unwanted solutes, the blood-brain barrier also prevents access of many drugs to the central nervous system. Consequently, a major goal of the pharmaceutical industry is to develop drugs that open the tight junctions of the brain to allow passage of therapeutic compounds.

REVIEW ?

1. What does freeze-fracture analysis tell you about the structure of a junction that can't be learned from examination of stained tissue sections?

2. How does the structure of a tight junction contribute to its function?

7·5 GAP JUNCTIONS AND PLASMODESMATA: MEDIATING INTERCELLULAR COMMUNICATION

Gap junctions are sites between animal cells that are specialized for intercellular communication. Electron micrographs reveal gap junctions to be sites where the plasma membranes of adjacent cells come very close to one another (within about 3 nm) but do not make direct contact. Instead, the cleft between the cells is spanned by very fine strands (Figure 7.32*a*) that are actually molecular "pipelines" that pass through the adjoining plasma membranes and open into the cytoplasm of the neighboring cells (Figure 7.32*b*).

Gap junctions have a simple molecular composition; they are composed entirely of an integral membrane protein called *connexin*. Connexins are clustered within the plasma membrane as a multisubunit complex, called a **connexon**, that completely spans the membrane (Figure 7.32*b*). Each connexon is composed of six connexin subunits arranged around a central opening, or *annulus*, that is approximately 16 Å in diameter at its extracellular surface (Figure 7.32*c*).

During the formation of gap junctions, the connexons in the plasma membranes of apposing cells become tightly linked to one another through extensive interactions of the extracellular domains of the connexin subunits. Once aligned, connexons in apposing plasma membranes form

complete intercellular channels connecting the cytoplasm of one cell with the cytoplasm of its neighbor (Figure 7.32*b*). The channels become clustered in specific regions of the membrane, forming gap-junction plaques that can be visualized when the membrane is split down the middle by freeze fracture (Figure 7.32*d*).

As discussed in the Experimental Pathways (see www.wiley.com/college/karp), gap junctions are sites of communication between the cytoplasms of adjacent cells. The existence of gap-junction intercellular communication (GJIC) is revealed through the passage of either ionic currents or low-molecular-weight dyes, such as fluorescein, from one cell to its neighbors (Figure 7.33). Mammalian gap junctions allow the diffusion of molecules having a molecular mass below approximately 1000 daltons. In contrast to the highly selective ion channels that connect a cell to the external medium (page 153), gap-junction channels are relatively nonselective; if the molecule is small enough, it passes through the open pipeline. Just as ion channels can be open or closed, gap junction channels are also thought to be gated. Closure of gap junction channels can be accomplished experimentally by several treatments, including elevation of the intracellular calcium concentration. Under normal circumstances, however, channel closure is probably triggered by phosphorylation of connexin subunits.

We saw in Chapter 4 how skeletal muscle cells are stimulated by chemicals released from the tips of nearby nerve cells. Stimulation of a cardiac or smooth muscle cell occurs by a different process involving gap junctions. The contraction of the mammalian heart is stimulated by an electrical impulse generated in a small region of specialized heart muscle called the *sinoatrial node*, which acts as the heart's pacemaker. The impulse spreads rapidly as a current of ions flows through gap junctions from one cardiac muscle cell to its neighbors, causing the cells to contract in synchrony. Similarly, the flow of ions through gap junctions that interconnect smooth muscle cells in the wall of the esophagus or intestine leads to the generation of coordinated peristaltic waves that move down the length of the wall.

Gap junctions can put a large number of cells of a tissue into intimate cytoplasmic contact. This has important physiological consequences, because a number of highly active regulatory substances, such as cyclic AMP and inositol phosphates (Chapter 15), are small enough to fit through gap-junction channels. As a result, gap junctions have the potential to integrate the activities of individual cells of a tissue into a functional unit. For example, if only a few cells near a particular blood vessel happen to be stimulated by a hormone, the stimulus can be rapidly transmitted to all cells of the tissue. Gap junctions also allow cells to cooperate metabolically by sharing key metabolites, such as ATP, sugar phosphates, amino acids, and many coenzymes, which are small enough to pass through these intercellular channels.

Connexins (Cx), the proteins of which gap junctions are constructed, are members of a multigene family. Ap-

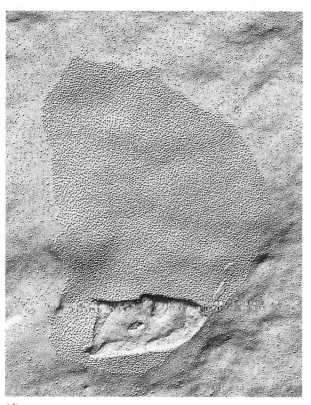

Figure 7.32 **The structure of a gap junction.** (*a*) Electron micrograph of a section through a gap junction perpendicular to the plane of the two adjacent membranes. The "pipelines" between the two cells are seen as electron-dense beads on the apposed plasma membranes. (*b*) Schematic model of a gap junction showing the arrangement of six connexin subunits to form a connexon, which contains half of the channel that connects the cytoplasm of the two adjoining cells. (*c*) A high-resolution image based on optical diffraction of a negatively stained, isolated membrane containing a plaque of gap junctions (as shown in *d*). The relative dimensions of the annulus and connexon are apparent, as is the hexagonal packing of the units within the plaque. (*d*) Freeze-fracture replica of a gap junction plaque showing the large numbers of connexons and their high degree of concentration. (A: FROM CAMILLO PERACCHIA AND ANGELA F. DULHUNTY, J. CELL BIOL. 70:419, 1976; BY COPYRIGHT PERMISSION OF THE ROCKEFELLER UNIVERSITY PRESS; C: COURTESY OF DANIEL GOODENOUGH; D: COURTESY OF DAVID ALBERTINI.)

proximately 20 different connexins with distinct tissue-specific distributions have been identified. Connexons composed of different connexins exhibit differences in conductance, permeability, and regulation. In some cases,

connexons in neighboring cells that are composed of different connexins are able to dock and form functional channels, whereas in other cases, they are not. These compatibility differences may play important roles in either promoting or preventing communication between different types of cells in an organ. For example, connexons joining cardiac muscle cells are composed of the connexin Cx43, whereas connexons joining the cells that make up the heart's electrical conduction system are composed of Cx40. Because these two connexins form incompatible connexons, the two types of cells are electrically insulated from one another even though they are still in physical

Figure 7.33 **Results of an experiment demonstrating the passage of low-molecular-weight solutes through gap junctions.** Dark-field micrograph showing the passage of fluorescein from one cell into which it was injected (X) to the surrounding cells. (FROM R. AZARNIA AND W. R. LOEWENSTEIN, J. MEMB. BIOL. 6:378, 1971.)

(a)
 50 nm

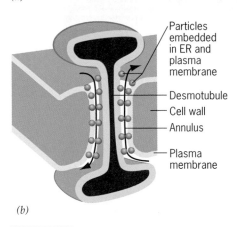

(b)

Figure 7.34 **Plasmodesmata.** (*a*) Electron micrograph of a section through a plasmodesma of a fern gametophyte. The desmotubule is seen to consist of membrane that is continuous with the endoplasmic reticulum (ER) of the cytoplasm on both sides of the plasma membrane. (*b*) Schematic drawing of a plasmodesma. (A: FROM LEWIS G. TILNEY, TODD J. COOKE, PATRICIA S. CONNELLY, AND MARY S. TILNEY, J. CELL BIOL. 112:740, 1991; BY COPYRIGHT PERMISSION OF THE ROCKEFELLER UNIVERSITY PRESS.)

contact. A number of inherited disorders have been associated with mutations in genes encoding connexins. Consequences of these disorders include deafness, blindness, or nerve degeneration.

Plasmodesmata

Although plants lack the specialized junctions found in animal tissues, most plant cells are connected to one another by plasmodesmata. **Plasmodesmata** (singular, **plasmodesma**) are cytoplasmic channels that pass through the cell walls of adjacent cells. A simple (that is, unbranched) plasmodesma is shown in Figure 7.34. Plasmodesmata are lined by plasma membrane and usually contain a dense central structure, the *desmotubule*, derived from the smooth endoplasmic reticulum of the two cells. Like gap junctions between animal cells, plasmodesmata serve as sites of intercellular communication, allowing the cells of a plant tissue to function as a metabolic unit.

Until recently, it was thought that plasmodesmata were impermeable to molecules greater than about 1 kDa. This conclusion was based on studies in which different-sized fluorescent dyes were injected into cells. More recent studies suggest that some plasmodesmata allow much larger molecules (up to 50 kDa) to pass between cells. It appears that younger tissues typically contain simpler plasmodesmata with larger passageways. As the tissue

matures, the plasmodesmata tend to become branched and their channels more restrictive. Unlike gap junctions, whose pipelines have a fixed opening, the plasmodesmatal pore is capable of dilation. The first insight into this dynamic property was gained from studies in the 1980s on plant viruses that spread from one cell to another through the plasmodesmata. It was found that the viruses encoded a *movement protein* that interacted with the wall of the plasmodesmata and increased the diameter of the pore. Subsequent studies have revealed that plant cells produce their own movement proteins that mediate the transport of proteins and RNAs from cell to cell.

1. Compare the arrangement of integral membrane proteins in a tight junction versus a gap junction.
2. How are plasmodesmata and gap junctions similar? How are they dissimilar? Would you expect a moderately sized protein to pass through a plasmodesma?

7.6 CELL WALLS

Because a lipid-protein plasma membrane of approximately 10-nm thickness can be expected to offer only minimal protection for a cell's contents, it is not surprising that "naked" cells are extremely fragile structures. The cells of nearly all organisms other than animals are enclosed in a protective outer envelope. Protozoa have a thickened outer coat, whereas bacteria, fungi, and plants have distinct **cell walls**. We will restrict our discussion to the cell walls of plants.

Plant cell walls perform numerous vital functions. As discussed on page 153, plant cells develop osmotic turgor pressure that pushes them against their surrounding wall. As a result, the wall gives the enclosed cell its characteristic polyhedral shape (Figure 7.35a) and provides mechanical support for the cell and, collectively, for the entire plant. Cell walls protect the cell against damage from mechanical abrasion and pathogens; they mediate cell–cell interactions; and they act as a primary barrier to the penetration of large molecular substances while allowing ions and small molecules to pass rapidly through. In recent years, it has become evident that, like the ECM at the surface of an animal cell, a plant cell wall can be a source of signals that alter the activities of the cells that it contacts. The fate of an epidermal root cell, for example, may be governed by the composition of the underlying cell wall.

(a)

(b)

100 nm

(c)

Hemicellulose

Cellulose microfibril

Glycoprotein

Pectin

Figure 7.35 **The plant cell wall.** (*a*) Electron micrograph of a plant cell surrounded by its cell wall. The middle lamella is a pectin-containing layer situated between adjacent cell walls. (*b*) Electron micrograph showing the cellulose microfibrils and hemicellulose cross-links of an onion cell wall after extraction of the nonfibrous pectin polymers. (*c*) Schematic diagram of a generalized plant cell wall. (A: COURTESY OF W. GORDON WHALEY; B: FROM M. C. MCCANN, B. WELLS, AND K. ROBERTS, J. CELL SCI. 96:329, 1990; BY PERMISSION OF THE COMPANY OF BIOLOGISTS LTD.)

Middle lamella Primary cell wall

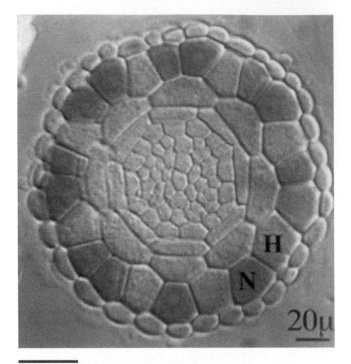

Figure 7.36 Cell walls are thought to provide information that directs the differentiation of adjacent cells. A cross-section through a root of *Arabidopsis thaliana*, showing the outer epidermal layer is composed of two different types of cells, hair cells (H) and nonhair cells (N). Just inside the epidermal layer is a layer of cortical cells. It can be seen that each of the H cells in the epidermal layer contacts the cell wall where it forms a junction between adjacent cortical cells. It is proposed that these cell wall junctions provide information that directs the differentiation of the adjoining epidermal cell into an H cell. (FROM FRED BERGER, JIM HASELOFF, JOHN SCHIEFELBEIN, AND LIAM DOLAN, CURR. BIOL. 8:423, 1998.)

As shown in Figure 7.36, two types of epidermal root cells can be distinguished: hair (H) cells that develop over cell-wall junctions and nonhair (N) cells that develop between cell-wall junctions. If a cell that is positioned to become an N cell is moved to a site overlying a cell wall junction, the fate of the cell is altered and it develops into an H cell.

Plant cell walls are often likened to fabricated materials such as reinforced concrete or fiberglass because they contain a fibrous element embedded in a nonfibrous, gel-like matrix. Cellulose, whose structure was described on page 48, provides the fibrous component of the cell wall, and proteins and pectin (described below) provide the matrix. Cellulose molecules are organized into **microfibrils** (Figure 7.35*b,c*) that confer rigidity to the cell wall and provide resistance to tensile (pulling) forces. Each microfibril is 5–10 nm in diameter and is composed of bundles of 30 to 60 cellulose molecules oriented parallel to one another and held together by hydrogen bonds. The walls of many plant cells are composed of layers in which the microfibrils of one layer are oriented at approximately 90° to those of adjacent layers (Figure 7.35*b*). This type of organization is similar to that of collagen fibers in the layers of the corneal stroma (see Figure 7.7).

Cellulose molecules are polymerized at the cell surface. Glucose subunits are added to the end of a growing cellulose molecule by a multisubunit enzyme called *cellulose synthase*. The subunits of the enzyme are organized into a six-membered ring, or rosette, which is embedded within the plasma membrane (Figure 7.37*a,b*). In contrast, materials of the matrix are synthesized within the cytoplasm (Figure 7.37*c*) and carried to the cell surface in secretory vesicles.

The matrix of the cell wall is composed of three types of macromolecules (Figure 7.35*c*).

1. **Hemicelluloses** are branched polysaccharides whose backbone consists of one sugar, such as glucose, and side chains of other sugars, such as xylose. Hemicellulose molecules bind to the surfaces of cellulose microfibrils, cross-linking them into a complex structural network.

2. **Pectins** are a heterogeneous class of negatively charged polysaccharides containing galacturonic acid. Like the glycosaminoglycans of animal cell matrices, pectins hold water and thus form an extensive hydrated gel that fills in the spaces between the fibrous elements. When a plant is attacked by pathogens, fragments of pectins released from the wall trigger a defensive response by the plant cell. Purified pectin is used commercially to provide the gel-like consistency of jams and jellies.

3. **Proteins,** whose functions are not well understood. One class, the *extensins*, provide structural support and may form a barrier to invading microorganisms.

The percentages of these various materials in cell walls are highly variable, depending on the type of plant, the type of cell, and the stage of the wall. Like the extracellular matrices of animal connective tissues, plant cell walls are dynamic structures that can be modified in response to changing environmental conditions.

Cell walls arise as a thin **cell plate** (described in Figure 14.37) that forms between the plasma membranes of newly formed daughter cells following cell division. The cell wall matures by the incorporation of additional materials that are assembled inside the cell and secreted into the extracellular space. In addition to providing mechanical support and protection from foreign agents, the cell wall of a young, undifferentiated plant cell must be able to grow in conjunction with the enormous growth of the cell it surrounds. The walls of growing cells are called **primary walls**, and they possess an extensibility that is lacking in the thicker **secondary walls** present around many mature plant cells. The transformation from primary to secondary cell wall occurs as the wall increases in cellulose content and, in most cases, incorporates a phenol-containing polymer called *lignin*. Lignin provides structural support. The lignin in the walls of water-conducting cells of the xylem provides the support required to move water through the plant. Lignin is also the major component of wood and, consequently, the most abundant organic molecule on Earth.

(a) $\overline{0.1\ \mu m}$

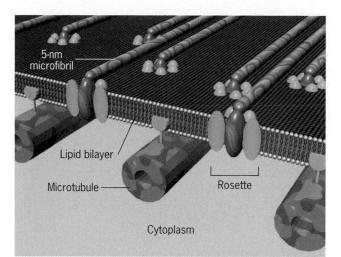

(b)

Figure 7.37 Synthesis of plant cell wall cellulose. (*a*) Freeze-fracture replica of the membrane of an algal cell. The rosettes are thought to represent the cellulose-synthesizing enzyme (cellulose synthase) situated within the plasma membrane. (*b*) A model of cellulose fibril deposition. Each rosette is thought to form a single microfibril that associates laterally with the microfibrils from other rosettes to form a larger fiber. The entire array of rosettes might move laterally within the membrane as it is pushed by the elongating cellulose molecules. Studies suggest that the direction of movement of the membrane rosettes is determined by oriented microtubules present in the cortical cytoplasm beneath the plasma membrane (discussed in Chapter 9). (*c*) Electron micrograph of a Golgi complex of a peripheral root cap cell stained with antibodies against a polymer of galacturonic acid, one of the major components of pectin. This material, like the other pectins and hemicellulose, is assembled in the Golgi complex. The antibodies have been linked to gold particles to make them visible as dark granules. The bar represents 0.25 μm. (A: FROM T. H. GIDDINGS, JR., D. L. BROWER, AND L. A. STAEHELIN, J. CELL BIOL. 84.332, 1980; C: FROM MARGARET LYNCH AND L. A. STAEHELIN, J. CELL BIOL. 118:477, 1991; ALL BY COPYRIGHT PERMISSION OF THE ROCKEFELLER UNIVERSITY PRESS.)

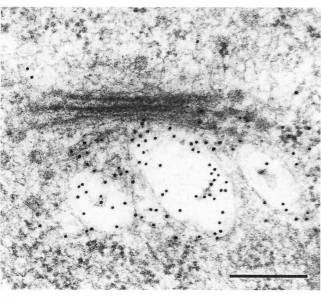

(c)

REVIEW ?

1. Describe the components that make up a plant cell wall and the role of each in the wall's structure and function.

2. Distinguish between cellulose and hemicellulose, a cellulose molecule and a microfibril, a primary cell wall and a secondary cell wall.

SYNOPSIS

The extracellular space extends outward from the outer surface of the plasma membrane and contains a variety of secreted materials that influence cell behavior. Epithelial tissues rest on a basement membrane, which consists of a thin, interwoven network of extracellular materials. Various types of connective tissue, including tendons, cartilage, and the

corneal stroma, contain an expansive extracellular matrix that gives the tissue its characteristic properties. (p. 244)

The major components of extracellular matrices include collagens, proteoglycans, and a variety of proteins, such as fibronectin, and laminin. Each of the proteins of the ECM has a modular construction composed of domains that contain binding sites for one another and for receptors on the cell surface. As a result, these various extracellular materials interact to form an interconnected network that is bound to the cell surface. Collagens are highly abundant, fibrous proteins that give the extracellular matrix the capability to resist pulling forces. Proteoglycans consist of protein and glycosaminoglycans and serve as an amorphous packing material that fills the extracellular space. (p. 246)

Integrins are cell-surface receptors involved in calcium-dependent interactions between cells and their substratum. Integrins are heterodimeric integral membrane proteins whose cytoplasmic domains may interact with cytoskeletal components, and whose extracellular domains may contain binding sites for various extracellular materials. Binding of an extracellular ligand to an integrin may send signals into the cell, triggering changes in cellular activities. (p. 252)

Cells attach to their substratum by means of cell-surface specializations such as focal adhesions and hemidesmosomes. Focal adhesions are sites of attachment where the plasma membrane contains clusters of integrins linked to actin-containing microfilaments of the cytoskeleton. Hemidesmosomes are sites of attachment where the plasma membrane contains clusters of integrins that are connected to the basement membrane on their outer surface and indirectly to keratin-containing intermediate filaments on their inner surface. (p. 255)

The adhesion of cells to other cells is mediated by several distinct families of integral membrane proteins, including selectins, integrins, cadherins, and members of the immunoglobulin superfamily (IgSF). Selectins bind to specific arrangements of carbohydrate groups projecting from the surfaces of other cells and mediate calcium-dependent, transient interactions between circulating leukocytes and the walls of blood vessels at sites of inflammation and clotting. Cell-adhesion molecules of the Ig superfamily mediate calcium-independent cell-cell adhesion. An IgSF protein on one cell may interact with an integrin or with the same or a different IgSF protein projecting from another cell. Cadherins mediate calcium-dependent cell-cell adhesion by binding to the same cadherin species on the opposing cell, thereby facilitating the formation of tissues composed of similar types of cells. (p. 257)

Strong adhesions between cells are facilitated by the formation of specialized adherens junctions and desmosomes. Adherens junctions encircle a cell near its apical surface, allowing contact between the cell and all of its surrounding neighbors. The plasma membranes of adherens junctions contain clusters of cadherins that are linked by means of intermediary proteins to the actin filaments of the cytoskeleton. Desmosomes occur as patches between cells and are characterized by dense cytoplasmic plaques on the inner surfaces of the membranes. Desmosomes are sites of concentration of cadherins, which are connected by means of intermediary proteins with intermediate filaments that loop through the cytoplasmic plaques. (p. 261)

Tight junctions are specialized sites of contact that block solutes from diffusing between the cells in an epithelium. A cross section through a tight junction shows the outer surfaces of adjacent cells come into direct contact at intermittent sites. Examination of the membranes by freeze fracture shows the sites to contain rows of aligned particles that form strands within the plasma membranes of the adjacent cells. (p. 268)

Gap junctions and plasmodesmata are specialized sites of communication between adjoining cells in animals and plants, respectively. The plasma membranes of a gap junction contain channels formed by a hexagonal array of connexin subunits forming a connexon. These channels connect the cytoplasm of one cell with the cytoplasm of the adjoining cell. The central channel of a connexon allows the direct diffusion of substances up to about 1000 daltons between cells. The passage of ionic currents through gap junctions plays a pivotal role in numerous physiological processes, including the spread of excitation through the cardiac muscle tissue of the heart. Plasmodesmata are cylindrical cytoplasmic channels that extend between adjacent plant cells directly through the intervening cell wall. These channels normally allow the free passage of solute molecules of approximately 1000 daltons, but they can be dilated by specific proteins to allow passage of macromolecules. (p. 270)

Plant cells are surrounded by a complex cell wall composed of a variety of secreted materials that provide mechanical support and protect the cell from potentially damaging influences. Cellulose microfibrils, which are composed of bundles of cellulose molecules and are secreted by enzymes residing within the plasma membrane, provide rigidity and tensile strength. Hemicellulose molecules cross-link the cellulose fibers, whereas pectins form an extensively cross-linked gel that fills the spaces between the fibrous elements of the cell wall. Like the extracellular matrices of animal cells, plant cell walls are dynamic structures that can be modified in response to changing environmental conditions. (p. 273)

ANALYTIC QUESTIONS

1. Cell adhesion can often be blocked in vitro by treating cells with specific agents. Which of the following substances would be expected to interfere with cell adhesion mediated by selectins? With cell adhesion mediated by L1 molecules? The substances are trypsin, which digests proteins; a peptide containing RGD; neuraminidase, which removes sialic acid from an oligosaccharide; collagenase, which digests collagen; hyaluronidase, which digests hyaluronic acid; EGTA, which binds Ca^{2+} ions in the medium.

2. What substance might you add to a culture dish to block the migration of neural crest cells? To block the outgrowth of an axon? To block the adhesion of fibroblasts to the substratum?

3. Mice lacking a gene for fibronectin do not survive early

embryonic development. Name two processes that might be disrupted in such embryos.

4. Suppose you found that molecule A, which had a molecular weight of 1500 daltons, was able to penetrate the channels of a gap junction, but molecule B, whose molecular weight was only 1200 daltons, was unable to diffuse between the same cells. How might these molecules be different so as to explain these results?

5. In what ways are the extracellular matrices of animals and the cell walls of plants similar in construction?

6. It was noted that two different autoimmune diseases, one producing antibodies against a component of hemidesmosomes and the other producing antibodies against a component of desmosomes, both cause severe blistering of the skin. Why do you think these two conditions have such similar symptoms?

7. Which of the various types of molecules that mediate cell adhesion is most likely to be responsible for the type of sorting out demonstrated by the cells of Figure 7.19? Why? How could you test your conclusion?

8. Follicle-stimulating hormone (FSH) is a pituitary hormone that acts on the follicle cells of the ovary to trigger the synthesis of cyclic AMP, which stimulates various metabolic changes. FSH normally has no effect on cardiac muscle cells. However, when ovarian follicle cells and cardiac muscle cells are grown together in a mixed culture, a number of the cardiac muscle cells are seen to contract following addition of FSH to the medium. How might this observation be explained?

9. Why do you think animal cells are able to survive without the types of cell walls that are found in nearly every other group of organisms?

10. Why would lowering the temperature of the medium in which cells were growing be expected to affect the ability of the cells to form gap junctions with one another?

11. Certain of the cell junctions occur as encircling belts, while others occur as discrete patches. How do these two types of structural arrangements correlate with the functions of the respective junctions?

12. Propose some mechanism that might explain how tobacco mosaic virus was able to alter the permeability of a plasmodesma. How might you be able to test your proposal?

13. The one type of vertebrate cell that is thought to lack integrins is the erythrocyte (red blood cell). Does this surprise you? Why or why not?

ADDITIONAL READING

The Extracellular Matrix

DE ARCANGELIS, A. & GEORGES-LABOUESSE, E. 2000. Integrin and ECM functions: Roles in vertebrate development. *Trends Gen.* 16:389–395.

BAUM, J. & BRODSKY, B. 1999. Folding of peptide models of collagen and misfolding in disease. *Curr. Opin. Struct. Biol.* 9:122–128.

BELLA, J. & BERMAN, H. M. 2000. Integrin-collagen complex: a metal-glutamate handshake. *Struct.* 8:R121–R126.

BISSELL, M. J. & NELSON, W. J., EDS. 1999. Cell-to-cell contact and extracellular matrix. *Curr. Opin. Cell Biol.* vol. 11, #5.

BORRADORI, L. & SONNENBERG, A. 1999. Structure and function of hemisdesmosomes: More than simple adhesion complexes. *J. Invest. Dermatol.* 112:411–418.

CAWSTON, T. 1998. Matrix metalloproteinases and TIMPs: Properties and implications for the rheumatic diseases. *Mol. Med. Today* 4:130–137.

HAY, E. D., ED. 1991. *Cell Biology of the Extracellular Matrix.* 2d ed. Plenum.

HEMLER, M. E. & RUTISHAUSER, U., EDS. 2000. Cell-to-cell contact and extracellular matrix. *Curr. Opin. Cell Biol.* 12:539–644.

HILEMAN, R. E., ET AL. 1998. Glycosaminoglycan-protein interactions. *BioEss.* 20:156–167.

IOZZO, R. V. 1998. Matrix proteoglycans: From molecular design to cellular function. *Annu. Rev. Biochem.* 67:609–652.

JONES, J. C. R., ET AL. 1998. Structure and assembly of hemidesmosomes. *BioEss.* 20:488–494.

KUCHARZ, E. J. 1992. *The Collagens: Biochemistry and Pathology.* Springer-Verlag.

LEITINGER, B. & HOGG, N. 2000. From crystal clear ligand binding to designer I domains. *Nature Struct. Biol.* 7:614–616.

PROCKOP, D. J. & KIVIRIKKO, K. I. 1995. Collagens: Molecular biology, diseases, and potential for therapy. *Annu. Rev. Biochem.* 64:403–434.

SHAUB, A. 1999. Unravelling the extracellular matrix. *Nature Cell Biol.* 1:E173–175.

SCHWARZBAUER, J. 1999. Basement membrane: Putting up the barriers. *Curr Biol.* 9:R242–244.

Cell Adhesion Molecules

BAZZONI, G. & HEMLER, M. E. 1998. Are changes in integrin affinity and conformation overemphasized? *Trends Biochem. Sci.* 23:30–34.

BITTAR, E., ET AL., EDS. 1999. The adhesive interactions of cells. *Adv. Mol. Cell Biol.* vol. 28.

CHOTHIA, C. & JONES, E. Y. 1997. The molecular structure of cell adhesion molecules. *Annu. Rev. Biochem.* 66:823–862.

GIANCOTTI, F. G. & RUOSLAHTI, E. 1999. Integrin signaling. *Science* 285:1028–1032.

GUMBINER, B. 2000. Regulation of cadherin adhesive activity. *J. Cell Biol.* 148:399–403.

HUMPHRIES, M. J. & NEWHAM, P. 1998. The structure of cell adhesion molecules. *Trends Cell Biol.* 8:78–83.

HYNES, R. O. 1999. Cell adhesion: Old and new questions. *Trends Cell Biol.* 9:M33–M37. (Dec.)

HYNES, R. O. & ZHAO, Q. 2000. The evolution of cell adhesion. *J. Cell Biol.* 150:F89–F95.

KAMIGUCHI, H., ET AL. 1998. Adhesion molecules and inherited diseases of the human nervous system. *Annu. Rev. Neurosci.* 21:97–125.

KENWRICK, S. & DOHERTY, P. 1998. Neural cell adhesion molecule L1: Relating disease to function. *BioEss.* 20:668–676.

KOCH, A., ET AL., 1999. Homophilic adhesion by cadherins. *Curr. Opin. Struct. Biol.* 9:275–281.

PRIMAKOFF, P. & MYLES, D. G. 2000. The Adam gene family: Surface proteins with adhesion and protease activity. *Trends Gen.* 16:83–87.

SHAPIRO, L. & COLMAN, D. R. 1999. The diversity of cadherins and implications for a synaptic adhesive code in the CNS. *Neuron* 23:427–430.

STEINBERG, M. S. 1996. Adhesion in development: an historical review. *Develop. Biol.* 180:377–388.

TAKEICHI, M. 1990. Cadherins: A molecular family important in selective cell–cell adhesion. *Annu. Rev. Biochem.* 59:237–252.

VESTWEBER, D. & BLANKS, J. E. 1999. Mechanisms that regulate function of the selectins and their ligands. *Physiol. Revs.* 79:181–213.

WALSH, F. S. & DOHERTY, P. 1997. Neural cell adhesion molecules of the immunoglobulin superfamily: Role in axon growth and guidance. *Annu. Rev. Cell Dev. Biol.* 13:425–456.

Inflammation and Metastasis

CHRISTOFORI, G. & SEMB, H. 1999. The role of the cell-adhesion molecule E-cadherin as a tumour-suppressor gene. *Trends Biochem. Sci.* 24:73–76.

GUILFORD, P. 1999. E-cadherin downregulation in cancer: Fuel on the fire. *Mol. Med. Today* 5:172–177.

HORWITZ, A. F. 1997. Integrins and health. *Sci. Am.* 276:68–75. (May)

LEFER, D. J. 2000. Pharmacology of selectin inhibitors in ischemia/ reperfusion states. *Annu. Rev. Pharmacol. Toxicol.* 40:283–294.

MATRISIAN, L. M. 1999. Extracellular proteinases in malignancy. *Curr. Biol.* 9:R776–778.

MCCAWLEY, L. J. & MATRISIAN, L. M. 2000. Matrix metalloproteinases: Multifunctional contributors to tumor progression. *Mol Med. Today* 6:149–156.

PERL, A.-K. 1998. A causal role for E-cadherin in the transition from adenoma to carcinoma. *Nature* 392:190–193.

RUOSLAHTI, E. 1996. How cancer spreads. *Sci. Am.* 275:72–77. (Sept.)

Cell Junctions

ABDULLA, S. 1998. Getting to the heart of gap junction pathology. *Mol. Med. Today* 4:192–193.

ANDERSON, J. M. & VAN ITALLIE, C. M. 1999. Tight junctions: Closing in on the seal. *Curr. Biol.* 9:R922–924.

CITI, S. & CORDENONSI, M. 1999. The molecular basis for the structure, function, and regulation of tight junctions. *Adv. Mol. Cell Biol.* 28:203–233.

DING, B. 1997. Cell-to-cell transport of macromolecules through plasmodesmata. *Trends Cell Biol.* 7:5–9.

HAGLER, D. J., Jr. & GODA, Y. 1998. Synaptic adhesion: The building blocks of memory? *Neuron* 20:1059–1062.

KOWALCZYK, A. P., ET AL. 1999. Desmosomes: Intercellular adhesive junctions specialized for attachment of intermediate filaments. *Int. Rev. Cytol.* 185:237–302.

KRAGLER, R., ET AL. 1998. Plasmodesmata: Dynamics, domains and patterning. *Ann. Bot.* 81:1–10.

PERACCHIA, C. , ED. 2000. Gap junctions: Molecular basis of cell communication in health and disease. *Curr. Topics Membs.* vol. 49.

PICKARD, B. G. & BEACHY, R. N. 1999. Intercellular connections are developmentally controlled to help more molecules through the plant. *Cell* 98:5–8.

SIMON, A. M. 1999. Gap junctions: More roles and new structural data. *Trends Cell Biol.* 9:169–170.

SIMON, A. M. & GOODENOUGH, D. A. 1998. Diverse functions of vertebrate gap junctions. *Trends Cell Biol.* 8:477–483.

STEVENSON, B. R. & KEON, B. H. 1998. The tight junction: Morphology and molecules. *Annu. Rev. Cell Dev. Biol.* 14:89–108.

TSUKITA, S. & FURUSE, M. 1999. Occludin and claudins in tight-junction strands: Leading or supporting players? *Trends Cell Biol.* 9:268–273.

TSUKITA, S. & FURUSE, M. 2000. Pores in the wall: Claudins constitute tight junction strands containing aqueous pores. *J. Cell Biol.* 149:13–16.

VAN ITALLIE, C. M. & ANDERSON, J. M. 2000. Molecular structure and regulation of tight junctions. *Curr. Topics. Membs.* 50:163–186.

WHITE, T. W. & PAUL, D. L. 1999. Genetic diseases and gene knockouts reveal diverse connexin functions. *Annu. Rev. Physiol.* 61:283–310.

WU, Q. & MANIATIS, T. 1999. A striking organization of a large family of human neural cadherin-like cell adhesion genes. *Cell* 97:779–790.

YAP, A. S., ET AL. 1997. Molecular and functional analysis of cadherin-based adherens junctions. *Annu. Rev. Cell Dev. Biol.* 13:119–146.

Cell Walls

CASSAB, G. T. 1998. Plant cell wall proteins. *Annu. Rev. Plant Physiol. Mol. Biol.* 49:281–309.

COSGROVE, D. J. 1997. Assembly and enlargement of the primary cell wall in plants. *Annu. Rev. Cell Dev. Biol.* 13:171–201.

DELMER, D. P. 1999. Cellulose biosynthesis. *Annu. Rev. Plant Physiol. Mol. Biol.* 50:245–276.

ROBERTS, K. 1994. The plant extracellular matrix: In a new expansive mood. *Curr. Opin. Cell Biol.* 6:688–594.

STRAUSS, E. 1998. When walls can talk, plant biologists listen. *Science* 282:28–29.

VARNER, J. E. & LIN. L.-S. 1989. Plant cell wall architecture. *Cell* 56:231–239.

Cytoplasmic Membrane Systems: Structure, Function, and Membrane Trafficking

Under the light microscope, the cytoplasm of many living cells appears relatively devoid of structure. Yet, even before the beginning of the twentieth century, examination of stained sections of tissue hinted at the existence of an extensive membranous network within the cytoplasm. It wasn't until the development of the electron microscope in the 1940s, however, that biologists began to appreciate the diverse array of membranous structures present in the cytoplasm of most eukaryotic cells. These early electron microscopists saw membrane-bound vesicles of varying diameter containing material of different electron density; long channels bounded by membranes that radiate through the cytoplasm to form an interconnected network of canals; and stacks of flattened membrane-bound sacs, called *cisternae*.

It became evident from these early electron microscopic studies and the biochemical investigations that followed that the cytoplasm of eukaryotic cells was subdivided into a variety of distinct compartments bounded by membrane barriers. As more types of cells were examined, it became apparent that the membranous structures of the cytoplasm formed distinct organelles that could be identified in diverse cells from yeast to multicellular plants and animals. The extent to which the cytoplasm of a eukaryotic cell is occupied by membranous structures is illustrated in the electron micrograph of a maize root cell shown in Figure 8.1.

Colorized scanning electron micrograph of a cell ingesting a synthetic particle.
(MICROGRAPH BY LENNART NILSSON/ALBERT BONNIERS FORLAG AB, THE INCREDIBLE MACHINE, NATIONAL GEOGRAPHIC SOCIETY.)

3 μm

Figure 8.1 Membrane-bound compartments of the cytoplasm. The cytoplasm of this root cap cell of a maize plant contains an array of membrane-bound organelles whose structure and function will be examined in this chapter. (COURTESY OF HILTON H. MOLLENHAUER.)

In the present chapter, we will examine the structure and functions of the endoplasmic reticulum, Golgi complex, endosomes, lysosomes, and vacuoles. Although each of these membranous organelles has its own distinct structure and function, taken together they form an **endomembrane system** in which the individual components function as part of a coordinated unit. Several other membranous organelles of the cytoplasm—mitochondria, peroxisomes, and chloroplasts—are not part of this interconnected system and were the subjects of previous chapters. ∎

8.1 AN OVERVIEW OF THE ENDOMEMBRANE SYSTEM

A multitude of studies, some of which are discussed in this chapter, have shown that most of the membranous organelles of the cytoplasm are part of a dynamic, integrated network in which materials are shuttled back and forth from one part of the cell to another. For the most part, materials are carried between organelles—from the Golgi complex to the plasma membrane, for example—in small, membrane-bounded **transport vesicles** that bud from a donor membrane compartment (Figure 8.2a). Transport vesicles move through the cytoplasm in a directed manner, often pulled by motor proteins that operate on tracks formed by microtubules of the cytoskeleton (see Figure 9.1a). When they reach their destination, the vesicles fuse with the membrane of the acceptor compartment, which receives the vesicle's soluble cargo as well as its membranous wrapper (Figure 8.2a). Repeated cycles of budding and fusion shuttle materials along pathways through the cell.

Several distinct pathways through the cytoplasm have been identified and are illustrated in the overview shown in Figure 8.2b. A **biosynthetic pathway** can be discerned in which proteins are synthesized in the endoplasmic reticulum, modified during passage through the Golgi complex, and transported from the Golgi complex to various destinations, such as the plasma membrane, a lysosome, or the large vacuole of a plant cell. This route is also referred to as the **secretory pathway**, as many of the proteins synthesized in the endoplasmic reticulum (as well as complex polysaccharides synthesized in the Golgi complex, page 301) are destined to be discharged (**secreted**) from the cell. Secretory activities of cells can be divided into two types: constitutive and regulated (Figure 8.2b). During **constitutive secretion**, materials are transported in secretory vesicles from their sites of synthesis and discharged into the extracellular space in a continual manner. Most cells engage in constitutive secretion, a process that contributes not only to the formation of the extracellular matrix (Section 7.1), but to the formation of the plasma membrane itself. During **regulated secretion**, materials to be secreted are stored in large, densely packed, membrane-bound **secretory granules** in the peripheral regions of the cytoplasm and are discharged only in response to an appropriate stimulus. Regulated secretion occurs, for example, in endocrine cells that release hormones, in pancreatic acinar cells that release digestive enzymes, and in nerve cells that release neurotransmitters.

Proteins, lipids, and complex polysaccharides are transported through the cell along the biosynthetic or secretory pathway. We will focus in the first part of the chapter on the synthesis and transport of proteins, as summarized in Figure 8.2b. During the discussion, we will consider several distinct classes of proteins. These include secretory proteins that are discharged from the cell, integral proteins of the various membranes depicted in Figure 8.2b, and soluble proteins that reside within the various compartments enclosed by the endomembranes (e.g., lysosomal enzymes).

Whereas materials move out of the cell by the secretory pathway, the endocytic pathway operates in the opposite direction. By following the **endocytic pathway**, materials move from the outer surface of the cell to compartments, such as endosomes and lysosomes, located within the cytoplasm (Figure 8.2b).

The movement of vesicles and enclosed materials along the various pathways of a cell is analogous to the movement of trucks carrying different types of cargo along the various arteries of a city. Both types of transport require defined *traffic patterns* to ensure that materials are accurately delivered to the appropriate sites. For example, protein trafficking within a salivary gland cell requires that the proteins of salivary mucus, which are synthesized in the endoplasmic reticulum, are specifically *targeted* for secretory granules, while lysosomal enzymes, which are also manufactured in the endoplasmic reticulum, are specifically targeted for a trip to a lysosome. Different organelles also contain different integral membrane proteins. Conse-

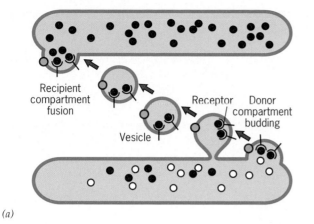

(a)

Figure 8.2 An overview of the biosynthetic/secretory and endocytic pathways that unite endomembranes into a dynamic, interconnected network. (*a*) Schematic diagram illustrating the process of vesicle transport by which materials are transported from a donor compartment to a recipient compartment. Vesicles form by membrane budding, whereby membrane proteins of the donor membrane are incorporated into the vesicle membrane and soluble proteins in the donor compartment are bound to specific receptors. When the transport vesicle subsequently fuses to another membrane, the proteins of the vesicle membrane become part of the recipient membrane, and the soluble proteins become sequestered within the lumen of the recipient compartment. (*b*) Materials follow the biosynthetic (or secretory) pathway from the endoplasmic reticulum, through the Golgi complex, and out to various locations including lysosomes, endosomes, secretory vesicles, secretory granules, vacuoles, and the plasma membrane. Materials follow the endocytic pathway from the cell surface to the interior by way of endosomes and lysosomes, where they are generally degraded by lysosomal enzymes.

(b)

quently, membrane proteins must also be targeted to particular organelles, such as a lysosome or Golgi cisterna. Proteins are targeted to predetermined cellular destinations through the presence of specific "addresses" or *sorting signals* that are encoded in the amino acid sequence of the proteins or in the attached oligosaccharides. The sorting signals, in turn, are recognized by specific membrane receptors (Figure 8.2*a*) or by the coats that form on the outer surface of transport vesicles (see Figure 8.25*b*).

Great advances have been made over the past decades in mapping the traffic patterns that exist in eukaryotic cells, identifying the specific addresses and receptors that govern the flow of traffic, and dissecting the machinery that ensures that materials are delivered to the appropriate sites in the cell. These subjects will be discussed in detail in the following pages. Motor proteins and cytoskeletal elements, which play key roles in the movements of transport vesicles and other cytomembranes, will be described in the following chapter. We will begin the study of cytomembranes by discussing a few of the most important experimental approaches that have led to our current understanding of the field.

REVIEW ❓

1. Compare and contrast the biosynthetic pathway with the endocytic pathway.
2. How are particular proteins targeted to particular subcellular compartments?

8.2 A FEW APPROACHES TO THE STUDY OF CYTOMEMBRANES

Early studies with the electron microscope provided biologists with a detailed portrait of the cytoplasm of cells but gave them little insight into the functions of the structures they were observing. Cells perform dynamic processes, yet the electron microscope portrays only static scenes. Determining the functions of cytoplasmic organelles required

the development of new techniques and the execution of innovative experiments. The early efforts in these areas were rewarded with the Nobel Prize in 1974 for three cell biologists: Christian De Duve, of the University of Louvain in Belgium, and Albert Claude and George Palade, of Rockefeller University. The experimental approaches described in the following sections have proven particularly useful in providing the foundation of knowledge on which current research on cytoplasmic membranes is based.

Insights Gained from Autoradiography

Among the hundreds of different cells in the body, the acinar cells of the pancreas have a particularly extensive endomembrane system. These cells function primarily in the synthesis and secretion of digestive enzymes. These enzymes are shipped by means of ducts from the pancreas, where they are synthesized, to the small intestine, where they degrade ingested food matter. Where are the secretory proteins of the pancreatic acinar cells synthesized, and how do they reach the surface of the cells where they are discharged? These questions are inherently difficult to answer because all of the steps in the process of secretion normally occur simultaneously within the cell. To follow the steps of a single cycle from start to finish, that is, from the synthesis of a secretory protein to its discharge from the cell, James Jamieson and George Palade utilized the technique of **autoradiography**.

As discussed in detail in Chapter 18, autoradiography provides a means to visualize biochemical processes by allowing an investigator to determine the location of radioactively labeled materials within a cell. In this technique, tissue sections containing radioactive isotopes are covered with a thin layer of photographic emulsion, which is exposed by radiation emanating from radioisotopes within the tissue. Sites in the cells containing radioactivity are revealed under the microscope by silver grains in the overlying emulsion (Figure 8.3).

To determine the sites where secretory proteins are synthesized, Palade and Jamieson incubated pieces of pancreatic tissue in a solution containing radioactive amino acids for a brief period of time. During this period, labeled amino acids were taken up by the living cells and incorporated into the digestive enzymes as they were being synthesized on ribosomes. The tissues were fixed, and the location of proteins that had been synthesized during the brief incubation with labeled amino acids was determined autoradiographically. Using this approach, the endoplasmic reticulum was discovered to be the site of synthesis of secretory proteins (Figure 8.3a).

To determine the intracellular path followed by secretory proteins from their site of synthesis to their site of discharge, Palade and Jamieson carried out an additional experiment. After incubating the tissue for a brief period in radioactive amino acids, they washed the tissue free of excess isotope and transferred the tissue to a medium containing only unlabeled amino acids. An experiment of this

(a)

Figure 8.3 **Autoradiography reveals the sites of synthesis and subsequent transport of secretory proteins.** (a) A section of a pancreatic acinar cell that had been incubated for three minutes in radioactive amino acids and then immediately fixed and prepared for autoradiography (see Section 18.4 for discussion of the technique). The black silver grains that appear in the emulsion following development are localized over the endoplasmic reticulum. (b–d) Diagrams of a sequence of autoradiographs showing the movement of labeled secretory proteins (represented by the silver grains in red) through a pancreatic acinar cell. When the cell is pulse labeled for 3 minutes and immediately fixed (as shown in a), radioactivity is localized in the endoplasmic reticulum (b). After a 3-minute pulse and 17-minute chase, radioactive label is concentrated in the Golgi complex and adjacent vesicles (c). After a 3-minute pulse and 117-minute chase, radioactivity is concentrated in the secretory granules and is beginning to be released into the pancreatic ducts (d). (e) Summary of the kinetics of labeling of the various compartments of the pancreatic acinar cell in the experiment described above. (A: COURTESY OF JAMES D. JAMIESON AND GEORGE PALADE.)

type is called a *"pulse-chase."* The *pulse* refers to the brief incubation with radioactivity during which labeled amino acids are incorporated into protein. The *chase* refers to the period when the tissue is exposed to the unlabeled medium, a period during which proteins are synthesized using nonradioactive amino acids. The longer the chase, the farther the radioactive proteins manufactured during the pulse will have traveled from their site of synthesis within the cell. Using this approach, one can ideally follow

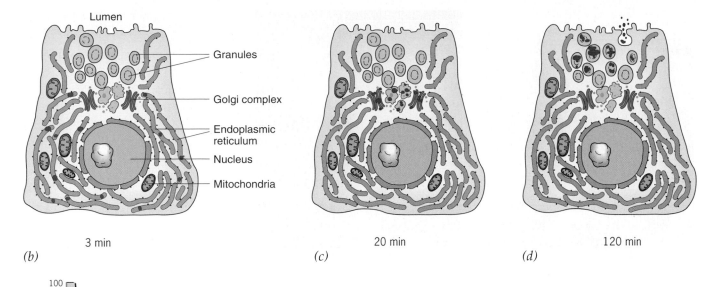

Lumen

Granules

Golgi complex

Endoplasmic reticulum

Nucleus

Mitochondria

(b) 3 min *(c)* 20 min *(d)* 120 min

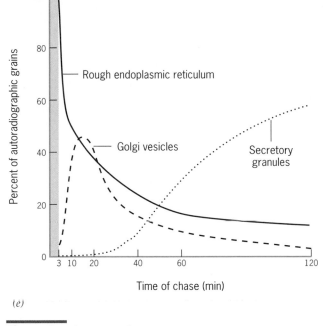

(e)

Figure 8.3 (continued)

the movements of newly synthesized molecules by observing a wave of radioactive material moving through the cytoplasm of cells from one location to the next until the process is complete. The results of these experiments—which first defined the biosynthetic (or secretory) pathway and tied a number of seemingly separate membranous compartments into an integrated functional unit—are summarized in Figure 8.3*b–e*.

Insights Gained from the Use of the Green Fluorescent Protein

The autoradiographic experiments described in the previous section require investigators to examine thin sections of different cells that have been been killed at various times after introduction of a radioactive label. In recent years, a new technology has been developed that allows researchers to follow with their own eyes the dynamic movements of specific proteins as they occur within a single living cell. This technology utilizes a gene isolated from certain jellyfish that encodes a small protein, called the **green fluorescent protein (GFP)**, which emits a green fluorescent light. To use this technique, DNA encoding GFP is fused to DNA encoding the protein to be studied and the resulting chimeric (i.e., composite) DNA is introduced into cells that can be observed under the microscope. Once inside the cells, the chimeric DNA expresses a chimeric protein consisting of GFP fused to the end of the protein to be studied. In most cases, the presence of GFP stuck to the end of a protein has little or no effect on the movement or function of that protein and, conversely, the protein under study has no effect on the fluorescence of the attached GFP. The structure of GFP and its fusion to a protein being studied is shown in molecular detail in Figure 9.16*c*.

Figure 8.4 shows a pair of micrographs depicting cells that contain a GFP fusion protein. In this case, the cells were infected with a strain of the vesicular stomatitis virus (VSV) in which one of the viral genes (*VSVG*) is fused to the *GFP* gene. Viruses are useful in these types of studies because they turn infected cells into factories for the production of one or a few viral proteins. When a cell is infected with the VSV virus, massive amounts of the VSVG protein are produced in the endoplasmic reticulum, traffic through the Golgi complex, and are transported to the plasma membrane of the infected cell where they are incorporated into viral envelopes. As in a radioactive pulse-chase experiment, the use of a virus allows investigators to follow a relatively synchronous wave of protein movement, in this case represented by a wave of green fluorescence that begins soon after infection. Synchrony can be enhanced, as was done in the experiment depicted in Figure 8.4, by use of a virus with a mutant VSVG protein that is unable to leave the ER of infected cells that are grown at an elevated temperature (e.g., 40°C). When the temperature is lowered to 32°C, the fluorescent GFP-VSVG protein that had accumulated in the ER (Figure 8.4*a*) moves

(a)

(b)

Golgi

Nucleus

ER

VSV

+

Viral genome
contains temperature-
sensitive *VSVG* gene
fused to *GFP* gene

(c)

Figure 8.4 **The use of green fluorescent protein (GFP) reveals the movement of proteins within a living cell.** (*a*) Fluorescence micrograph of a live cultured mammalian cell that had been infected with the VSV virus at 40°C. This particular strain of the VSV virus contained a *VSVG* gene that (1) was fused to a gene for the fluorescent protein GFP and (2) contained a temperature-sensitive mutation that prevented the newly synthesized VSVG protein from leaving the ER when kept at 40°C. The green fluorescence in this micrograph is restricted to the ER. (*b*) Fluorescence micrograph of a live infected cell that was held at 40°C to allow the VSVG protein to accumulate in the ER and then incubated at 32°C for 10 minutes. The fluorescent VSVG protein has moved on to the region containing the Golgi complex. (*c*) Schematic drawing showing the retention of the mutant VSVG protein in the ER at 40°C and its synchronous movement to the Golgi complex within ten minutes of incubation at the lower temperature. (A, B: FROM DANIEL S. CHAO, ET AL., COURTESY OF RICHARD H. SCHELLER, J. CELL BIOL. 144:873, 1999, BY COPYRIGHT PERMISSION OF THE ROCKEFELLER UNIVERSITY PRESS.)

synchronously to the Golgi complex (Figure 8.4*b*), where various processing events occur, and then on to the plasma membrane. Mutants of this type that function normally at a reduced (permissive) temperature, but not at an elevated (restrictive) temperature are described as *temperature-sensitive mutants*.

Insights Gained from the Biochemical Analysis of Subcellular Fractions

Electron microscopy, autoradiography, and the use of GFP provide information on the structure and function of cellular organelles but fail to provide much insight into the molecular composition of these structures. Techniques to break up (**homogenize**) cells and isolate particular types of

organelles were pioneered in the 1950s and 1960s by Albert Claude and Christian De Duve. When a cell is ruptured by homogenization, the cytoplasmic membranes become fragmented and the ends of the membrane fragments fuse to form spherical vesicles less than 100 nm in diameter. Vesicles derived from different organelles (nucleus, mitochondrion, plasma membrane, endoplasmic reticulum, and so forth) have different properties, which allow them to be separated from one another, an approach that is called **cell fractionation**.

Membranous vesicles derived from the endomembrane system (primarily the endoplasmic reticulum and Golgi complex) form a heterogeneous collection of similar sized vesicles referred to as **microsomes**. A rapid (and crude) purification of the microsomal fraction of a cell is

depicted in Figure 8.5*a*. The microsomal fraction can be further fractionated into smooth and rough membrane fractions (Figure 8.5*b,c*) by the gradient techniques discussed in Section 18.6. Once isolated, the biochemical composition of various fractions could be determined. It was found, for example, that vesicles derived from different parts of the Golgi complex contained enzymes that added different sugars to the end of a growing carbohy-

0.3 μm

Figure 8.5 **Isolation of a microsomal fraction by differential centrifugation.** (*a*) When a cell is broken by mechanical homogenization (step 1), the various membranous organelles become fragmented and form spherical membranous vesicles. Vesicles derived from different organelles can be separated by various techniques of centrifugation. In the procedure depicted here, the cell homogenate is first subjected to low-speed centrifugation to pellet the larger, denser particles, leaving the smaller vesicles (microsomes) in the supernatant (step 2). The microsomes can be removed from the supernatant by centrifugation at higher speeds for longer periods of time (step 3). A crude microsomal fraction of this type can be fractionated into different vesicle types in subsequent steps. (*b*) Electron micrograph of a smooth microsomal fraction in which the membranous vesicles lack ribosomes. (*c*) Electron micrograph of a rough microsomal fraction containing ribosome-studded membranes. (B,C: COURTESY OF J. A. HIGGINS AND R. J. BARNETT.)

drate chain of a glycoprotein or glycolipid. A specific enzyme could be isolated from the microsomal fraction and then used as an antigen to prepare antibodies against that enzyme. The antibodies could then be linked to materials, such as gold particles, that could be visualized in the electron microscope, and the localization of the enzyme in the membrane compartment could be determined. These studies detailed the role of the Golgi complex in the stepwise assembly of complex carbohydrates.

Insights Gained from the Use of Cell-Free Systems

Once techniques to fractionate membranous organelles were developed, researchers began to probe the capabilities of these crude subcellular preparations. They found that the isolated parts of a cell were capable of remarkable activities. These early **cell-free systems**—so called because they did not contain whole cells—provided a wealth of information about complex processes that were impossible to study using intact cells. During the 1960s, for example, George Palade, Philip Siekevitz, and their colleagues at Rockefeller University set out to learn more about the properties of the rough microsomal fraction (shown in Figure 8.5c). They found that they could strip a rough microsomal preparation of its attached particles, and the isolated particles (i.e., ribosomes) were capable of synthesizing proteins when provided with required ingredients from the cytosol. Under these conditions, the newly synthesized proteins were simply released by the ribosomes into the aqueous fluid of the test tube. When these same types of experiments were carried out using the complete rough microsomal fraction, the newly synthesized proteins were no longer released into the incubation medium but were trapped within the lumen of the membranous vesicles. It was apparent from these studies that the microsomal membrane was not required for the incorporation of amino acids into proteins but for sequestering newly synthesized secretory proteins within the ER cisternal space.

Over the past few decades, researchers have used cell-free systems to identify many of the proteins that are involved in specific steps of protein trafficking. Figure 8.6 shows a liposome with vesicles budding from its surface. As discussed on page 131, liposomes are vesicles whose walls consist of an artificial bilayer that is created in the laboratory from purified phospholipids. The buds and vesicles seen in Figure 8.6 were produced after the preparation of liposomes was incubated with purified proteins that normally comprise the coats of transport vesicles within the cell. Without the added protein, vesicle budding could not occur. Once a particular process such as vesicle budding is reconstituted in vitro, it becomes possible to study the specific roles of each of the proteins that is required to make that process happen.

Insights Gained from the Study of Genetic Mutants

A mutant is an organism (or cultured cell) whose chromosomes contain one or more genes that encode proteins that are altered from those of most of the organisms (or cells) in the population. When a protein encoded by a mutant gene is unable to carry out its normal function, the cell carrying the mutation exhibits a characteristic deficiency. Determining the precise nature of the deficiency provides information on the function of the normal protein. Study of the genetic basis of secretion has been carried out largely on yeast cells, primarily by Randy Schekman and colleagues at the University of California, Berkeley. Yeasts are particularly amenable to genetic studies because they have a small number of genes compared to other types of eukaryotes; they are small, single-celled organisms easily grown in culture; and they can be grown as haploids for the majority of their life cycle. A mutation in a single gene in a haploid cell produces an observable effect, because the cells lack a second copy of the gene that would mask the presence of the abnormal one.

In yeast, as in all eukaryotic cells, vesicles bud from the ER and travel to the Golgi complex, where they fuse with the Golgi cisternae (Figure 8.7a). To identify genes whose encoded protein is involved in this portion of the secretory pathway (i.e., *SEC* genes), researchers screen for mutant cells that exhibit an abnormal distribution of cytoplasmic membranes. An electron micrograph of a wild-type yeast cell is shown in Figure 8.7b. The cell depicted in Figure 8.7c has a mutation in a gene that encodes a protein involved in the formation of vesicles at the ER membrane (step 1, Figure 8.7a). In the absence of vesicle formation,

Figure 8.6 **The formation of coated vesicles in a cell-free system.** Electron micrograph of a liposome preparation that had been incubated with the components required to promote vesicle budding within the cell. The proteins in the medium have become attached to the surface of the liposomes and have induced the formation of protein-coated buds (arrows). (COURTESY OF LELIO ORCI AND RANDY SCHEKMAN.)

mutant cells accumulate an expanded endoplasmic reticulum. In contrast, the cell depicted in Figure 8.7*d* carries a mutation in a gene that encodes a protein involved in vesicle fusion (step 2, Figure 8.7*a*). When this gene product is defective, the mutant cells accumulate an excess number of unfused vesicles. Researchers have isolated dozens of different mutants, which taken as a group, exhibit disruptions in virtually every step of the secretory pathway. Many of the genes responsible for these defects have been cloned and sequenced, and the proteins they encode have been isolated. Isolation of proteins from yeast has launched successful searches for homologous proteins (i.e., proteins with related sequence) in mammalian systems.

One of the most important lessons learned from the use of all of these techniques is that the dynamic activities of the components of the endomembrane system are highly conserved. Not only do yeast, plant, insect, and human cells carry out similar processes, but they do so with remarkably similar proteins. It is evident that the structural diversity of cells belies their underlying molecular similarities. In many cases, proteins from widely divergent species are interchangeable. For example, cell-free systems derived from mammalian cells can utilize yeast proteins to facilitate vesicle transport. Conversely, yeast cells with genetic deficiencies that disrupt some phase of the biosynthetic pathway can often be "cured" by genetically engineering them to carry mammalian genes.

REVIEW ?

1. Describe the differences between an autoradiograph of a pancreas cell that had been incubated in labeled amino acids for 3 minutes and immediately fixed and one of a cell that had been labeled and then chased for 40 minutes.

2. How does the isolation of a mutant yeast that accumulates vesicles provide information on the process of protein trafficking?

3. How can GFP be used to study membrane dynamics"?

Figure 8.7 The use of genetic mutants in the study of secretion. (*a*) Schematic drawing of the first leg of the biosynthetic secretory pathway in the budding yeast (*Saccharomyces cerevisiae*). The steps are described below. (*b*) Electron micrograph of a section through a wild-type yeast cell containing a small amount of endoplasmic reticulum (ER). (*c*) A yeast cell bearing a mutation in the *sec12* gene whose product is involved in the formation of vesicles at the ER membrane (step 1, part *a*). Because vesicles cannot form, expanded ER cisternae accumulate in the cell. (*d*) A yeast cell bearing a mutation in the *sec17* gene, whose product is involved in vesicle fusion (step 2, part *a*). Because they cannot fuse with Golgi membranes, the vesicles (indicated by the arrowheads) accumulate in the cell. (The mutants depicted in *c* and *d* are temperature-sensitive mutants (page 283). As long as they are kept at the lower (permissive) temperature they are capable of normal growth and division.) (FROM CHRIS A. KAISER AND RANDY SCHEKMAN, CELL 61:724, 1990; BY COPYRIGHT PERMISSION OF CELL PRESS.)

(*a*)

(*b*)

(*c*)

(*d*)

8.3 THE ENDOPLASMIC RETICULUM

The **endoplasmic reticulum (ER)** is divided into two broad categories, the **rough endoplasmic reticulum (RER)** and the **smooth endoplasmic reticulum (SER)** (Figure 8.8). Both types of ER comprise a system of membranes that enclose a space, or lumen, that is separated from the surrounding cytosol. As will be evident in the following discussion, the composition of the **luminal** or **cisternal space** inside the ER membranes is quite different from that of the surrounding **cytosolic space**.

The rough ER has ribosomes bound to its cytosolic surface, whereas the smooth ER lacks associated ribosomes. The difference between these two types of membranous organelles, however, goes much deeper than this morphologic distinction. The RER is typically an extensive organelle composed primarily of interconnected flattened sacs (**cisternae**), as shown in Figure 8.9. As illustrated in Figure 8.2b, the rough ER is continuous with the outer membrane of the nuclear envelope, which also bears ribosomes on its cytosolic surface. The membranous elements of the SER are typically tubular (Figures 8.8 and 8.10) and form an interconnecting system of pipelines curving through the cytoplasm. When cells are homogenized, the SER fragments into smooth-surfaced vesicles, whereas the RER fragments into rough-surfaced vesicles (Figure 8.5b,c). Because the two types of vesicles have different densities, they can be readily separated by density gradient centrifugation (see Figure 18.23b).

Different types of cells contain markedly different amounts of either one type of ER or the other, depending on the activities of the cell. For example, cells that secrete large amounts of proteins, such as the cells of the pancreas or salivary glands, have extensive regions of RER (Figure 8.9b,c). We will return to the function of the RER shortly, but first we will describe the activities of the SER.

The Smooth Endoplasmic Reticulum

The SER is extensively developed in a number of cell types, including those of skeletal muscle, kidney tubules, and steroid-producing endocrine glands (Figure 8.10a). The specific proteins of the SER vary from one cell type to another depending on the organelle's particular functions, as illustrated by the following examples:

- Synthesis of steroid hormones in the endocrine cells of the gonad and adrenal cortex.

- Detoxification in the liver of a wide variety of organic compounds, including barbiturates and ethanol, whose chronic use can lead to proliferation of the SER in liver cells. Detoxification is carried out by a system of oxygen-transferring enzymes (oxygenases), including the *cytochrome P450*s. These enzymes are noteworthy for their lack of substrate specificity, being able to oxidize

Figure 8.8 **The endoplasmic reticulum.** Electron micrograph of part of a bat pancreatic cell showing both smooth and rough endoplasmic reticulum. (COURTESY OF KEITH R. PORTER.)

Smooth endoplasmic reticulum

Rough endoplasmic reticulum

thousands of different hydrophobic compounds and convert them into more hydrophilic, more readily excreted derivatives. The effects are not always positive. For example, the relatively harmless compound benzo[α]pyrene formed when meat is charred on a grill is converted into a potent carcinogen by the "detoxifying" enzymes of the SER.

- Release of glucose from glucose 6-phosphate in liver cells by the enzyme glucose 6-phosphatase. Large reserves of glycogen are stored in the liver as granules on the outside of SER membranes (Figure 8.10b). When the need for chemical energy arises, the enzyme phosphorylase converts glycogen to glucose 1-phosphate, which is then converted to glucose 6-phosphate in the cytoplasm. As long as the sugar remains in the phosphorylated state, it cannot leave the liver cell because the plasma membrane is impermeable to sugar phosphates. Glucose 6-phosphatase in the SER membranes removes the phosphate group, generating glucose molecules that ultimately move into the bloodstream where they are carried to the body's tissues.

- Sequestering calcium ions within the cisternal space. The SER contains a high concentration of calcium-binding proteins. The regulated release of Ca^{2+} from the SER triggers specific cellular responses, including the fusion of secretory vesicles with the plasma mem-

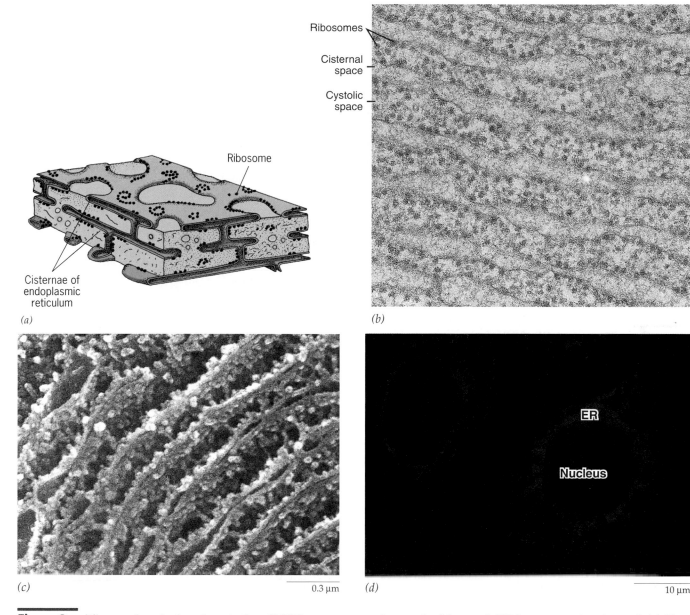

Ribosomes

Cisternal space

Cystolic space

Ribosome

Cisternae of endoplasmic reticulum

(a)

(b)

(c) 0.3 μm

(d) 10 μm

ER

Nucleus

Figure 8.9 **The rough endoplasmic reticulum (RER).**
(*a*) Schematic diagram showing the stacks of flattened cisternae that make up the rough ER. The cytosolic surface of the membrane contains bound ribosomes, which gives the cisternae their rough appearance. (*b*) Transmission electron micrograph of a portion of the rough ER of a pancreatic acinar cell. The division of the rough ER into a cisternal space (which is devoid of ribosomes) and a cytosolic space is evident. (*c*) Scanning electron micrograph of the rough ER in a pancreatic acinar cell. (*d*) Visualization of the rough ER in a whole cultured cell as revealed by imunofluorescent staining for the enzyme phosphate disulfide isomerase (PDI), an ER resident protein. (B: COURTESY OF S. ITO; C: FROM K. TANAKA, INT. REV. CYTOL. 68:101, 1980; D: FROM BRIAN STORRIE, RAINER PEPPERKOK, AND TOMMY NILSSON, TRENDS CELL BIOL. 10:388, 2000, COPYRIGHT 2000, WITH PERMISSION FROM ELSEVIER SCIENCE.)

brane and the contraction of skeletal muscle cells. This latter function of the SER is discussed in Chapter 9.

Functions of the Rough Endoplasmic Reticulum

Early investigations on the functions of the RER were carried out on cells that secrete large quantities of proteins, such as the acinar cells of the pancreas (Figure 8.3) or the mucus-secreting cells of the lining of the digestive tract (Figure 8.11). It is evident from the drawing and micrograph of Figure 8.11 that the organelles of these secretory epithelial cells are positioned in the cell in such a way as to produce a distinct polarity from one end to the other. The nucleus and an extensive array of RER cisternae are located near the basal surface of the cell, which faces the blood supply. The Golgi complex is located in the central region of the cell. The apical surface of the cell faces a duct that carries the secretory product out of the organ. The

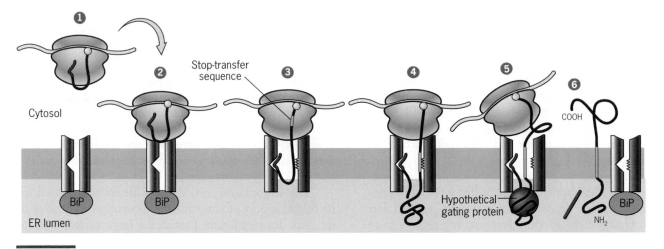

Figure 8.13 A model for the synthesis of an integral membrane protein that contains a single transmembrane segment and a signal sequence near the N-terminus of the nascent polypeptide. The SRP and the various components of the membrane that were shown in Figure 8.12 are also involved in the synthesis of integral proteins, but they have been omitted for simplicity. The nascent polypeptide enters the translocon just as if it were a secretory protein (steps 1–3). However, the entry of the hydrophobic stop-transfer sequence into the pore blocks further translocation of the nascent polypeptide through the channel (step 4). In the model depicted here, the ribosome remains attached to the translocon, but has tilted so as to allow the remaining portion of the polypeptide (the C-terminal portion) to be synthesized in the cytosol (step 5). It is also proposed that, because the ribosome is no longer tightly attached to the cytosolic end of the translocon, some other protein (labeled "hypothetical gating protein") must bind on the luminal side of the channel to maintain the membrane's permeability barrier. At some point, the channel apparently opens laterally, and the hydrophobic helix is inserted into the bilayer (step 6).

whose N-terminus is located on the cisternal side of the ER membrane and whose C-terminus is on the cytosolic side. Single-spanning membrane proteins can also have the opposite orientation with their N-terminus facing the cytosol. As noted on page 137, the most common determinant of membrane protein alignment is the presence of positively charged amino acid residues at the cytosolic end of a transmembrane segment (see Figure 4.18). During the synthesis of membrane proteins, the inner lining of the translocon is thought to orient the nascent polypeptide so that the more positive end faces the cytosol. In multi-spanning proteins (as shown in Figure 4.32d), sequential transmembrane segments have opposite orientations. For these proteins, every other transmembrane segment has to be rotated 180° before it can be released into the lipid bilayer. Evidence from studies with cell-free systems suggests that a translocon, by itself, is capable of properly orienting transmembrane segments. This is based on the finding that multi-spanning proteins are oriented properly when they are synthesized in cell-free systems using purified translocons that have been incorporated into artificial lipid bilayers (i.e., liposomes). Based on these types of studies it becomes evident that the translocon is not a simple passageway through the ER membrane, but a complex "machine" capable of recognizing various signal sequences and performing complex mechanical activities.

Membrane Biosynthesis in the ER Membranes do not arise *de novo*, that is, as new entities from pools of protein and lipid; rather, they are thought to arise only from pre-existing membranes. Membranes grow as newly synthesized proteins and lipids are inserted into existing membranes in the ER. As will be apparent in the following discussion, membrane components move from the ER to virtually every other compartment in the cell. As the membrane moves from one compartment to the next, its proteins and lipids are modified by enzymes that reside in the cell's various organelles. As a result of these modifications, the membranes that make up each of these compartments have their own unique composition and maintain their own distinct identity (see Table 4.1).

Keep in mind that cellular membranes are asymmetric: the two phospholipid layers (leaflets) of a membrane have different compositions (page 142). This asymmetry is initially established in the endoplasmic reticulum as lipids and proteins are inserted differently into the two layers. Asymmetry is maintained while the membrane passes through the cell by budding and fusion from one compartment to the next. As a result, components situated at the luminal surface of the ER membrane can be identified on the luminal surface of transport vesicles, the luminal surface of Golgi cisternae, and the external (exoplasmic) surface of the plasma membrane (Figure 8.14). Similarly, components situated at the cytosolic surface of the ER membrane maintain their orientation and can ultimately be found at the internal (cytoplasmic) surface of the plasma membrane.

Synthesis of Membrane Lipids Most membrane lipids are synthesized entirely within the endoplasmic reticulum. The major exceptions are (1) sphingomyelin and glyco-

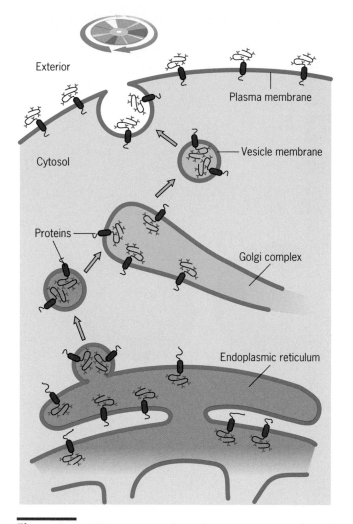

Figure 8.14 Maintenance of membrane asymmetry. As each protein is synthesized in the rough ER, it becomes inserted into the lipid bilayer in a predictable orientation determined by its amino acid sequence. This orientation is maintained throughout its travels in the endomembrane system, as illustrated in this figure. The carbohydrate chains, which are first added in the ER, provide a convenient way to assess membrane sidedness because they are always present on the cisternal side of the cytoplasmic membranes, which becomes the exoplasmic side of the plasma membrane following the fusion of vesicles with the plasma membrane.

The membranes of different organelles have markedly different lipid composition (Figure 8.15a), which indicates that changes take place as membrane flows through the cell. There are several factors that may contribute to such changes (Figure 8.15b):

1. Most membranous organelles contain enzymes that modify lipids already present within a membrane, converting one type of phospholipid (e.g., phosphatidylserine) to another (e.g., phosphatidylcholine) (step 1, Figure 8.15b).

2. When vesicles bud from a compartment (as in Figure 8.2a), some types of phospholipids may be preferentially included within the membrane of the forming vesicle, while other types may be left behind (step 2, Figure 8.15b).

3. Cells contain **phospholipid-transfer proteins** that can transport specific phospholipids *through the aqueous cytosol* from one membrane compartment to another (step 3, Figure 8.15b). These enzymes may facilitate the movement of specific phospholipids from the ER to other organelles, including mitochondria and chloroplasts.

Glycosylation in the Rough Endoplasmic Reticulum
Nearly all of the proteins produced on membrane-bound ribosomes—whether integral components of a membrane, soluble lysosomal or vacuolar enzymes, or parts of the extracellular matrix—become glycoproteins. Carbohydrate groups have key roles in the function of many glycoproteins, particularly as binding sites in their interactions with other macromolecules. The sequences of sugars that compose the oligosaccharides of glycoproteins are highly specific; if the oligosaccharides are isolated from a purified protein, the sequences are consistent and predictable. How is the sequence of sugars determined?

The addition of sugars to an oligosaccharide chain is catalyzed by a group of membrane-bound enzymes called **glycosyltransferases** that transfer a specific monosaccharide from an appropriate sugar donor to an appropriate sugar acceptor (Figure 8.16). The donor molecule is always a nucleotide sugar, such as CMP-sialic acid, GDP-mannose, or UDP-*N*-acetylglucosamine (Figure 8.16). The acceptor molecule that receives the transferred sugar is the growing end of the carbohydrate chain. The sequence in which sugars are transferred during assembly of an oligosaccharide depends on the sequence of glycosyltransferases that participate in the process. This in turn depends on the location of specific enzymes within the various membranes of the secretory pathway. Thus the arrangement of sugars in the oligosaccharide chains of a glycoprotein depends on the spatial localization of particular enzymes.

The initial steps during the addition of sugars to *N*-linked oligosaccharides (as opposed to *O*-linked oligosaccharides, see Figure 4.10) of both soluble proteins and integral membrane proteins are shown in Figure 8.17. The

lipids, whose synthesis begins in the ER and is completed in the Golgi complex, and (2) some of the unique lipids of the mitochondrial and chloroplast membranes, which are synthesized by enzymes that reside in those membranes. The enzymes involved in the synthesis of phospholipids are themselves integral proteins of the ER membrane with their active sites facing the cytosol. Newly synthesized phospholipids are inserted into the half of the bilayer facing the cytosol. Some of these lipid molecules later move into the opposite leaflet through the action of proteins (*flippases*) that actively translocate lipid molecules across the bilayer. Lipids are carried from the ER to the Golgi complex and plasma membrane as part of the bilayer that makes up the walls of transport vesicles.

ER = Endoplasmic reticulum
GC = Golgi complex
PM = Erythrocyte plasma membrane

PC = Phosphatidylcholine
PS = Phosphatidylserine
SM = Sphingomyelin

(a)

Figure 8.15 **Modifying the lipid composition of membranes.** (a) Histogram indicating the percentage of each of three phospholipids (phosphatidylcholine, phosphatidylserine, and sphingomyelin) in three different types of cellular membranes (the endoplasmic reticulum, Golgi complex, and plasma membrane). The percentage of each lipid species changes gradually as membrane flows from the ER to the Golgi to the plasma membrane. (b) Schematic diagram showing three distinct mechanisms that might explain how the phospholipid composition of one membrane in the endomembrane system can be different from another membrane in the system, even though the membranous compartments are spatially and temporally continuous. (1) The head groups of phospholipids of the bilayer are modified enzymatically; (2) the membrane of a forming vesicle contains a different phospholipid composition from the membrane it buds from; (3) phospholipids can be physically removed from one membrane and inserted into another membrane by phospholipid transfer proteins.

(b)

N-Acetylglucosamine | Uridine diphosphate

UDP–*N*–Acetylglucosamine

Mannose acceptor at growing end of oligosaccharide

N–Acetylglucosamine transferase

O (β1, 2 linkage)

Figure 8.16 **Example of a reaction catalyzed by a glycosyltransferase.** This large family of enzymes transfers sugars from a nucleotide sugar carrier (in this case UDP-*N*-acetylglucosamine) to the growing end of an oligosaccharide chain (which in this case has a mannose residue at its end). The reaction depicted here occurs in the Golgi complex.

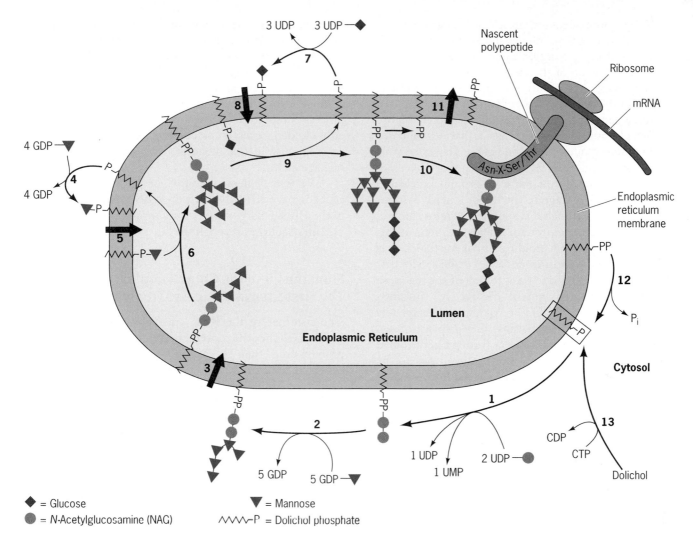

Figure 8.17 **The steps in the synthesis of the core portion of an *N*-linked oligosaccharide in the rough ER.** The first seven sugars (five mannose and two NAG residues) are transferred one at a time to the dolichol-PP on the cytosolic side of the ER membrane (steps 1 and 2). At this stage, the dolichol with its attached oligosaccharide apparently flips across the membrane (step 3), and the remaining sugars (four mannose and three glucose residues) are attached on the luminal side of the membrane. These latter sugars are attached one at a time on the cytosolic side of the membrane to the end of a dolichol phosphate molecule (as in steps 4 and 7), which then flips across the membrane (steps 5 and 8) and donates its sugar to the growing end of the oligosaccharide chain (steps 6 and 9). Once the oligosaccharide is completely assembled, it is transferred to an asparagine residue of the nascent polypeptide (step 10). The dolichol-PP flips back across the membrane (step 11) and is ready to begin accepting sugars again (steps 12 and 13). (FROM D. VOET AND J. G. VOET, BIOCHEMISTRY, 2D ED., COPYRIGHT 1995, JOHN WILEY & SONS, INC. REPRINTED BY PERMISSION OF JOHN WILEY & SONS, INC.)

basal, or core, segment of each carbohydrate chain is not assembled on the protein itself but put together independently on a lipid carrier and then transferred, as a block, to specific asparagine residues of the polypeptide. This lipid carrier, termed **dolichol phosphate**, is a hydrophobic molecule embedded within the lipid bilayer that is built from more than 20 isoprene units:

$$CH_3$$
$$H_2C{=}C{-}C{=}CH_2$$
$$H$$

Isoprene

Sugars are added to the dolichol phosphate molecule one at a time by membrane-bound glycosyltransferases, beginning with step 1 of Figure 8.17. This part of the glycosylation process is essentially invariant; in mammalian cells, it begins with the transfer of *N*-acetylglucosamine 1-phosphate, followed by the transfer of another *N*-acetylglucosamine, then nine mannose and three glucose units in the precise pattern indicated in Figure 8.17. This block of 14 sugars is then transferred by the enzyme oligosaccharyltransferase from dolichol phosphate to the nascent polypeptide (step 10, Figure 8.17) as the polypeptide is being translocated into the ER lumen.

Although some oligosaccharides, particularly in lower eukaryotes, remain essentially as shown in Figure 8.17, the evolution of more complex organisms was accompanied by diversification of the carbohydrate sequences attached to proteins. Modification of the core oligosaccharide begins in the ER with the enzymatic removal of two of the three terminal glucose residues (step 1, Figure 8.18). Each glycoprotein whose oligosaccharides contain a single remaining glucose then binds to a chaperone (calnexin or calreticulin)(step 2). The remaining glucose is enzymatically removed and the glycoprotein is released (step 3). This is a key stage in the life of a newly synthesized glycoprotein, whether it is a soluble secretory protein or an integral membrane protein. If a glycoprotein at this stage has not completed its folding or has misfolded, it is recognized by a "monitoring enzyme" (called GT) that adds a single glucose residue back to one of the mannose residues at the exposed end of the recently trimmed oligosaccharide (step 4). GT recognizes incompletely folded or misfolded proteins because they display exposed hydrophobic residues that are absent from properly folded proteins. Once the glucose residue has been added, the "tagged" glycoprotein is recognized by the same chaperones, which give the protein another chance to fold properly (step 5). After a period of time with the chaperone, the added glucose residue is removed and the "monitoring enzyme" checks it again to see if it has achieved its proper three-dimensional structure. If it is still partially unfolded or misfolded, another glucose residue is added and the process is repeated until, eventually, the glycoprotein has either folded correctly and continues on its way (step 6), or remains misfolded and is destroyed (step 7) as described on page 293.

We will pick up the story of protein glycosylation again on page 301, where the oligosaccharide that is added in the ER is enlarged as it passes through the Golgi complex on its journey through the biosynthetic pathway.

From the ER to the Golgi Complex: The First Step in Vesicular Transport

The cisternae of the RER are typically interconnected, which facilitates the movement of membrane and luminal proteins from their site of synthesis to sites that face the central regions of the cell. These faces of the RER cisternae are typically smooth-surfaced, that is, devoid of ribosomes (see Figure 8.10b arrows) and are referred to as *transitional elements*; they serve as exit sites for budding of the first transport vesicles in the biosynthetic pathway. The trip from the ER toward the Golgi complex can be followed visually in living cells by tagging secretory proteins with the green fluorescent protein (GFP) as described on page 283. Using this and other techniques it is found that soon after they bud from the ER membrane, the transport vesicles fuse with one another to form larger vesicles and interconnected tubules in the region between the ER and Golgi complex (Figure 8.19). This region has been named the **ERGIC** (*e*ndoplasmic *r*eticulum *G*olgi *i*ntermediate *c*ompartment), and the vesicular-tubular clusters that form there are called **VTCs** (see Figure 8.25a). Once formed, the VTCs move farther away from the ER toward the Golgi complex. The movement of two of these vesicular-tubular membranous carriers from the ERGIC to the Golgi complex is shown in Figure 8.19. Other studies indicate that movement occurs along tracks composed of microtubules.

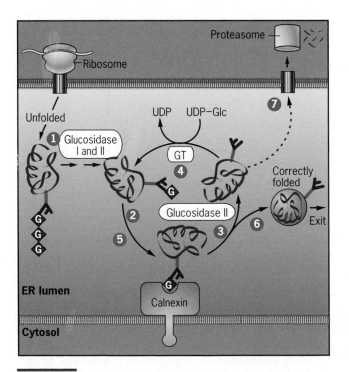

Figure 8.18 Quality control: ensuring that misfolded proteins do not leave the ER. Based on this proposed mechanism, misfolded proteins are recognized by a glucosyltransferase (GT) which adds a glucose to the end of the oligosaccharide chains. Glycoproteins containing monoglucosylated oligosaccharides are recognized by the chaperone calnexin and given an opportunity to achieve their correctly folded (native) state. If that does not occur after repeated attempts, the protein is translocated to the cytosol and destroyed. The steps are described in the text. (AFTER L. ELLGAARD ET AL. SCIENCE 286,1884, 1999; COPYRIGHT 1999 AMERICAN ASSOCIATION FOR THE ADVANCEMENT OF SCIENCE.)

REVIEW ?

1. What are the major morphological differences between the RER and SER? Major differences in their functions?

2. Describe the steps that occur between the time a ribosome attaches to a messenger RNA encoding a secretory protein and the time the protein leaves the RER.

3. How are newly synthesized integral proteins inserted into a membrane?

Figure 8.19 **Visualizing membrane traffic with the use of a fluorescent tag.** This series of photographs shows a small portion of a living mammalian cell that has been infected with the vesicular stomatitis virus (VSV) containing a *VSVG–GFP* chimeric gene (page 283). Once it is synthesized in the RER, the fusion protein emits a green fluorescence, which can be followed as the protein moves through the cell. In the series of photographs shown here, two vesicular-tubular carriers (VTCs) (arrows) containing the fluorescent protein have budded from the ER and are moving toward the Golgi complex (GC). The series of events depicted here took place over a period of 13 seconds. Bar represents 6 μm. (REPRINTED WITH PERMISSION FROM JOHN F. PRESLEY ET AL., NATURE 389:82, 1997. © COPYRIGHT 1997 MACMILLAN MAGAZINES LTD.)

4. Describe some of the ways that membranous organelles can maintain their unique compositions despite the continuous traffic of membranes and materials moving through them.

5. Describe how membrane asymmetry is maintained as membrane moves from the ER to the plasma membrane.

8.4 THE GOLGI COMPLEX

In the latter years of the nineteenth century, an Italian biologist, Camillo Golgi, was inventing new types of staining procedures that might reveal the organization of nerve cells within the central nervous system. One of these procedures employed a solution of silver nitrate applied to tissue that had been soaked in osmium and bichromate. In 1898, Golgi applied the stain to nerve cells from the cerebellum and discovered a darkly stained reticular network located near the cell nucleus. This network, which was later identified in other cell types and named the **Golgi complex**, helped earn its discoverer the Nobel Prize in 1906. The Golgi complex remained a center of controversy for decades between those who believed that the organelle existed in living cells and those who believed it was an *artifact*, that is, an artificial structure formed during preparation for microscopy. It wasn't until the Golgi complex was clearly identified in unfixed, freeze-fractured cells (see Figure 18.17) that its existence was verified beyond reasonable doubt.

The Golgi complex has a characteristic morphology consisting primarily of flattened, disklike, membranous cisternae with dilated rims and associated vesicles and tubules (Figure 8.20*a*). The cisternae, whose diameters are typically 0.5 to 1.0 μm, are arranged in an orderly stack, much like a stack of pancakes, and are curved so as to resemble a shallow bowl (Figure 8.20*b*).[4] Typically, a Golgi stack contains fewer than eight cisternae. An individual cell may contain from a few to several thousand stacks per cell, depending on the cell type. Individual Golgi stacks are often interconnected to form a ribbonlike complex. A closer look at an individual cisterna suggests that vesicles bud from a peripheral tubular domain of each cisterna (Figure 8.20*c*). As discussed later, many of these vesicles contain a distinct protein coat that is visible in Figure 8.20*c*.

The Golgi complex is divided into several functionally distinct compartments arranged along an axis from the *cis* or entry face closest to the ER to the *trans* or exit face at the opposite end of the the stack (Figures 8.20*a,b*). The *cis*-most face of the organelle is composed of an interconnected network of tubules referred to as the *cis* **Golgi network (CGN)**. The CGN is thought to function primarily as a sorting station that distinguishes between proteins to be shipped back to the ER (page 306) and those that are allowed to proceed to the next Golgi station. The bulk of the Golgi complex consists of a series of large, flattened cisternae, which are divided into *cis*, *medial*, and *trans* **cisternae** (Figure 8.20*a*). The *trans*-most face of the organelle contains a distinct network of tubules and vesicles called the *trans* **Golgi network (TGN)**. Like the CGN, the TGN is also a sorting station. Proteins are segregated in the TGN into different types of vesicles heading either to the plasma membrane or to various intracellular destinations.

Figure 8.21 provides visual evidence that the Golgi complex is not uniform in composition from one end to the other. Differences in composition of the membrane compartments from the *cis* to the *trans* face reflect the fact that the Golgi complex is primarily a "processing plant." Newly synthesized membrane proteins, as well as secretory and lysosomal proteins, leave the ER and enter the Golgi complex at its *cis* face and then pass across the stack to the *trans* face. As they progress along the stack, proteins that had been recently synthesized in the RER are sequentially modified in specific ways. For example, part of the protein's length may be trimmed by proteolytic enzymes; amino acids may be modified (e.g., hydroxylation of lysine and proline residues of a collagen molecule); and the protein's carbohydrates are modified by a series of stepwise enzymatic reactions, as discussed in the following section.

[4]A single Golgi stack in plant cells is sometimes called a **dictyosome**.

trans-Golgi network (TGN)

trans cisternae

medial cisternae

cis cisternae

cis-Golgi network (CGN)

(a)

trans

TGN

CGN

cis

(b)

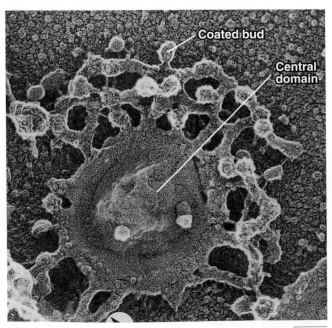

Coated bud

Central domain

(c)

0.1 μm

Figure 8.20 The Golgi complex. (*a*) Schematic model of a portion of a Golgi complex from an epithelial cell of the male rat reproductive tract. The *cis* elements are interrupted to permit visualization of the *medial* compartment. The elements of the *cis* and *trans* compartments are often discontinuous and appear as tubular networks. (*b*) Electron micrograph of a portion of a tobacco root cap cell showing a gradual increase in staining of the cisternae from the *cis* to the *trans* end of the Golgi stack. (*c*) Electron micrograph of a single isolated Golgi cisterna showing two distinct domains, a concave central domain and an irregular peripheral domain. The peripheral domain consists of a tubular network from which protein-coated buds are being pinched off. (A: FROM A. RAMBOURG AND Y. CLERMONT, EUR. J. CELL BIOL. 51:195, 1990; B: COURTESY OF THOMAS H. GIDDINGS; C: FROM PEGGY J. WEIDMAN AND JOHN HEUSER, TRENDS CELL BIOL. 5:303, 1995, COPYRIGHT 1995, WITH PERMISSION FROM ELSEVIER SCIENCE.)

(a) *(b)* *(c)* 0.15 μm

Figure 8.21 Regional differences in membrane composition across the Golgi stack. (*a*) Reduced osmium tetroxide preferentially impregnates the *cis* cisternae of the Golgi complex. (*b*) The enzyme mannosidase II, which is involved in trimming the mannose residues from the core oligosaccharide as described in the text, is preferentially localized in the *medial* cisternae. (*c*) The enzyme nucleoside diphosphatase, which splits dinucleotides (e.g., UDP) after they have donated their sugar, is preferentially localized in the *trans* cisternae. (A,C: FROM ROBERT S. DECKER, J. CELL BIOL. 61:603, 1974; B: FROM ANGEL VELASCO ET AL., J. CELL BIOL. 122:41, 1993; ALL BY COPYRIGHT PERMISSION OF THE ROCKEFELLER UNIVERSITY PRESS.)

Glycosylation in the Golgi Complex

The Golgi complex plays a key role in the assembly of the carbohydrate component of glycoproteins and glycolipids. When we left the topic of synthesis of *N*-linked carbohydrate chains on page 298, the glucose residues had just been removed from the ends of the core oligosaccharide.

As newly synthesized soluble and membrane glycoproteins pass through the *cis* and *medial* cisternae of the Golgi stack, most of the mannose residues are also removed from the core oligosaccharides, and other sugars are added sequentially by various glycosyltransferases (Figure 8.22) to produce a variety of different oligosaccharides.

cis Golgi | cis Golgi / medial Golgi | medial Golgi | medial Golgi | medial Golgi / trans Golgi | trans Golgi | trans Golgi

● *N*–Acetylglucosamine ▼ Mannose ▲ Galactose ▮ Sialic acid ◗ Fucose

Figure 8.22 Steps in the glycosylation of a mammalian *N*-linked oligosaccharide in the Golgi complex. Following the removal of the three glucose residues, various mannose residues are subsequently removed, while a variety of sugars (*N*-acetylglucosamine, galactose, fucose, and sialic acid) are added to the oligosaccharide by specific glycosyltransferases.

In the Golgi complex, as in the RER, the sequence in which sugars are incorporated into oligosaccharides is determined by the spatial arrangement of the specific glycosyltransferases that come into contact with the newly synthesized protein as it moves through the Golgi stack. The enzyme sialyltransferase, for example, which places a sialic acid at the terminal position of the chain in animal cells, is localized in the *trans* end of the Golgi stack, as would be expected if newly synthesized glycoproteins were continually moving toward this part of the organelle. Unlike the *N*-linked oligosaccharides, whose synthesis begins in the ER, those attached to proteins by *O*-linkages (Figure 4.10) are assembled wholly within the Golgi complex.

The Golgi complex is also the site of synthesis of most of a cell's complex polysaccharides, including the glycosaminoglycans found in the extracellular matrix of animal cells and the pectins and hemicellulose found in the cell walls of plants (see Figure 7.37c).

The Movement of Materials Through the Golgi Complex

Although the movement of materials through the Golgi complex has long been established, two contrasting views of the way this occurs have dominated the field for years. Up until the mid-1980s, it was generally accepted that Golgi cisternae were transient structures. It was supposed that Golgi cisternae formed at the *cis* face of the stack by

fusion of membranous carriers from the ER and ERGIC and that each cisterna physically moved from the *cis* to the *trans* end of the stack, changing in composition as it progressed. This is known as the **cisternal maturation model** because, according to the model, each cisterna "matures" into the next cisterna along the stack.

From the mid-1980s to the late-1990s, the maturation model of Golgi movement was largely abandoned and replaced by an alternate model, which proposed that the cisternae of a Golgi stack remain in place as stable compartments held together by a protein scaffold. In this latter model, which is known as the **vesicular transport model**, cargo (i.e., secretory, lysosomal, and membrane proteins) is shuttled through the Golgi stack, from the CGN to the TGN, in vesicles that bud from one membrane compartment and fuse with a neighboring compartment farther along the stack. The vesicular transport model is illustrated in Figure 8.23a, and its acceptance was based largely on two types of observations:

1. Each of the various Golgi cisternae of a stack has a distinct population of resident enzymes (Figure 8.21). How could the various cisternae have such different properties if each cisterna was giving rise to the next one in line, as suggested by the cisternal maturation model?

2. Large numbers of vesicles can be seen in electron micrographs to bud from the rims of Golgi cisternae. In

(a) Vesicular transport model

(b) Cisternal maturation model

(c) "Combined" model

Figure 8.23 Three models depicting the dynamics of transport through the Golgi complex. (*a*) In the vesicular transport model, cargo (dark dots) is carried in an anterograde direction by transport vesicles, while the cisternae themselves remain as stable elements. (*b*) In the cisternal maturation model, the cisternae progress gradually from a *cis* to a *trans* position and then disperse at the TGN. Transport vesicles carry resident Golgi enzymes (indicated by the colored vesicles) in a retrograde direction. The lens-shaped objects represent large cargo materials, such as procollagen complexes of fibroblasts. (*c*) In the "combined" model, the cisternae progress from *cis* to *trans*, at the same time that cargo (dark dots) is carried forward at a more rapid pace in anterograde vesicles and Golgi resident enzymes are carried backward in retrograde vesicles.

1983, James Rothman and his colleagues at Stanford University demonstrated, using cell-free preparations of Golgi membranes (page 286), that transport vesicles were capable of budding from one Golgi cisterna and fusing with another Golgi cisterna in vitro. This key experiment formed the basis for a hypothesis suggesting that inside the cell, cargo-bearing vesicles budded from *cis* cisternae and fused with cisternae situated at a more *trans* position in the stack.

Although both models of Golgi function continue to have their proponents, the consensus of opinion has shifted in the past few years back to the cisternal maturation model. Two of the major reasons for this shift can be noted:

■ Certain materials that are produced in the endoplasmic reticulum and travel through the Golgi complex can be shown to remain within the Golgi cisternae and never appear within Golgi-associated transport vesicles. For example, studies on fibroblasts indicate that large complexes of procollagen molecules (the precursors of extracellular collagen) move from the *cis* cisternae to the *trans* cisternae without ever leaving the cisternal lumen.

■ It was assumed until the mid-1990s that transport vesicles always moved in a "forward" (**anterograde**) direction, that is, from a *cis* origin to a more *trans* destination. But a large body of evidence has indicated that some vesicles move in a "backward" (**retrograde**) direction, that is, from a *trans* donor membrane to a *cis* acceptor membrane.

A current version of the cisternal maturation model is depicted in Figure 8.23*b*. Unlike the original versions of the cisternal maturation model, the version shown in Figure 8.23*b* acknowledges a role for transport vesicles, which have been clearly shown to bud from Golgi membranes. In this model, however, these transport vesicles do not shuttle cargo in an anterograde direction, but instead carry resident Golgi enzymes in a retrograde direction. This model explains how different Golgi cisternae in a stack can have a unique identity. An enzyme such as mannosidase I, for example, which removes mannose residues from oligosaccharides and is largely restricted to the *cis* cisternae, can be recycled backward in transport vesicles as each cisterna moves toward the *trans* end of the stack.

The two models of intra-Golgi transport are not mutually exclusive. In fact, recent evidence supports a scenario in which some cargo is transported by vesicles in an anterograde direction, even as *cis* cisternae are maturing into *trans* cisternae and Golgi enzymes are being recycled to *cis* cisternae. This is depicted in the "combined" model shown in Figure 8.23*c*. Vesicles may be utilized for rapid anterograde transport, at the same time cisternal maturation is moving other materials at a more leisurely pace. Depending upon the type of cell and organism, one process may be more predominant than the other. At this point, let us look more closely at the mechanism of vesicle transport, not only as it may apply to movement of materials within the Golgi complex, but also throughout the cell.

1. Describe the steps that occur as a soluble secretory protein, such as a digestive enzyme in a pancreas cell, moves from the RER to the *cis* Golgi cisterna; from the *cis* cisterna to the TGN.

2. What is the role of dolichol phosphate in the synthesis of membrane glycoproteins? How is the sequence of sugars attached to the protein determined?

3. How does the process of glycosylation in the Golgi complex differ from that in the RER?

4. How can the vesicular transport and cisternal maturation models of Golgi activity be reconciled?

8.5 TYPES OF VESICLE TRANSPORT AND THEIR FUNCTIONS

The biosynthetic pathway of a eukaryotic cell consists of a series of distinct membrane-bound organelles that function in the synthesis, modification, and delivery of soluble and membrane proteins to their appropriate destination in the cell. As was illustrated in Figure 8.2*a*, materials are carried between compartments by membrane-bound vesicles, typically 50–75 nm in diameter, that bud from donor membranes and fuse with acceptor membranes. If one scrutinizes electron micrographs for vesicles caught in the act of budding, one finds that most of these membranous buds are covered on their cytosolic surface by a "fuzzy," electron-dense layer. Further analysis reveals that the dark-staining layer consists of a **protein coat** formed from soluble proteins that assemble on the cytosolic surface of the donor membrane at sites where budding takes place. Each coated bud pinches off to form a **coated vesicle** such as those shown in Figure 8.24. (The discovery of coated vesicles is discussed in the Experimental Pathways at the end of the chapter.)

Protein coats have at least two distinct functions: (1) they act as a mechanical device that causes the membrane to curve and form a budding vesicle, and (2) they provide a mechanism for selecting the components to be carried by the vesicle. Selected components include (*a*) cargo to be transported and (*b*) the machinery required to target and dock the vesicle to an acceptor membrane (page 309). As discussed below, protein coats are able to make these selections by virtue of their specific affinity for the cytosolic "tails" of integral proteins that reside in the donor membrane (see Figure 8.25*b*).

Several distinct classes of coated vesicles have been identified; they are distinguished by the proteins that make up their coat, their appearance in the electron microscope, and their role in cell trafficking. The three best-studied coated vesicles are the following:

1. **COPII-coated vesicles** (Figure 8.24*a*) move materials from the ER "forward" to the ERGIC and Golgi

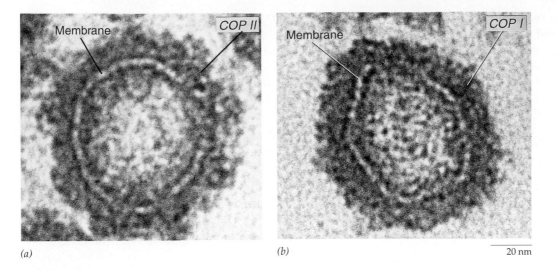

(a) (b) 20 nm

Figure 8.24 Coated vesicles. These high-power electron micrographs show the membranes of these vesicles to be covered on their outer (cytosolic) surface by a distinct protein coat. The first micrograph (*a*) shows a COPII-coated vesicle, whereas the second micrograph (*b*) shows a COPI-coated vesicle. (COURTESY OF RANDY SCHEKMAN AND LELIO ORCI.)

complex. (Recall from page 298 that the ERGIC is the intermediate compartment situated between the ER and Golgi complex.) (COP is an acronym for *co*at *p*roteins.)

2. **COPI-coated vesicles** (Figure 8.24*b*) move materials in a retrograde direction from the ERGIC and Golgi stack "backwards" to the ER. Other possible roles for COPII-coated vesicles are discussed below (and indicated in Figure 8.25*a*).

3. **Clathrin-coated vesicles** move materials from the TGN to endosomes, lysosomes, and plant vacuoles. They also move materials from the plasma membrane to cytoplasmic compartments along the endocytic pathway. They have also been implicated in trafficking from endosomes and lysosomes.

We will consider each type of coated vesicle in the following sections.[5] A summary of the various transport steps along the biosynthetic or secretory pathway that is mediated by each of these coated vesicles is indicated in Figure 8.25*a*.

COPII-Coated Vesicles: Transporting Cargo from the ER to the Golgi Complex

COPII-coated vesicles are the most recently discovered of the three major types. We are discussing them first because they mediate the first leg of the journey through the

biosynthetic pathway—from the ER to the ERGIC and CGN (Figure 8.25*a,b*). The COPII coat contains five protein subunits that were first identified in mutant yeast cells that were unable to carry out transport from the ER to the Golgi complex. Homologues of the yeast proteins were subsequently found in the coats of vesicles budding from the ER in mammalian cells. Antibodies to COPII-coat proteins block budding of vesicles from ER membranes but have no effect on movement of cargo at other stages of the secretory pathway.

COPII-coated vesicles are thought to select and concentrate certain components that they transport. Integral membrane proteins of the ER are selected because they interact specifically with COPII proteins of the vesicle coat (Figure 8.25*b*). Several types of membrane proteins are included in this group, including (1) enzymes that act at later stages in the biosynthetic pathway, such as the glycosyltransferases of the Golgi complex (indicated as an orange membrane protein in Figure 8.25*b*), (2) membrane proteins involved in the docking and fusion of the vesicle with the target compartment, and (3) membrane proteins that are able to bind soluble cargo (such as the secretory proteins, indicated by the red dots in Figure 8.25*b*). One of these cargo receptors (called ERGIC-53) binds to the mannose residues present on the oligosaccharides of lysosomal and secretory proteins in the ER (see Figure 8.17). Mutations in ERGIC-53 have been linked to an inherited bleeding disorder. Persons with this disorder fail to secrete certain coagulation factors that promote blood clotting.

The interaction between membrane proteins (e.g., ERGIC-53) and proteins in the COPII coat is mediated by signal sequences in the cytosolic tails of the membrane protein. Other types of soluble cargo (indicated by the green dots in Figure 8.25*b* are not selected at this stage

[5]Although emphasis in the discussion is placed on the protein molecules of the coat and vesicle, the phospholipids of the vesicle membrane should not be ignored. Considerable evidence indicates that phosphatidylinositol and its derivatives (phosphoinositides) play an important role in recruiting particular proteins to specific sites on membrane surfaces during membrane trafficking.

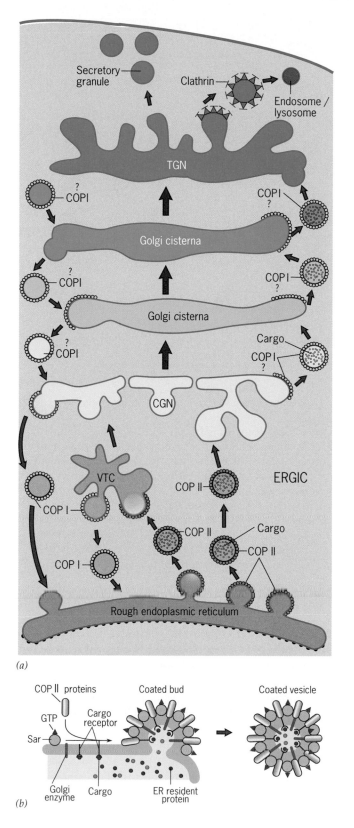

(a)

(b)

Figure 8.25 Movement of materials by vesicular transport between membranous compartments of the biosynthetic/secretory pathway. (*a*) The three different types of coated vesicles indicated in this schematic drawing are thought to have distinct transport roles. COPII-coated vesicles mediate transport from the ER to the ERGIC and Golgi complex. COPI-coated vesicles return proteins from the VTCs and Golgi complex to the ER. COPI-coated vesicles may also be involved in the transport of materials between the Golgi cisternae as anterograde and/or retrograde vesicles (indicated by the ? marks). Clathrin-coated vesicles mediate transport from the TGN to endosomes and lysosomes. Transport of materials along the endocytic pathway is not shown in this drawing. (*b*) Schematic drawing of the assembly of a COPII-coated vesicle. Binding of Sar-GTP to the ER membrane initiates the assembly of the COPII subunits to form the vesicle coat. Some cargo proteins of the ER lumen (red dots) bind to the luminal ends of transmembrane cargo receptors. These receptors are then concentrated within the coated vesicle through interaction of their cytosolic tails with components of the COPII coat. Other cargo proteins may simply be trapped in the vesicle (green dots) and carried by bulk flow. Some ER proteins (e.g., BiP) are excluded from the coated vesicles (blue dots).

vesicles and transported through the secretory pathway to the plasma membrane by bulk flow.

Among the COPII coat proteins is a small GTP-binding protein called Sar. Like other GTP-binding proteins, Sar plays a regulatory role, in this case regulating the assembly and disassembly of the vesicle coat. In its active state, Sar-GTP binds tightly to the ER membrane. As shown in Figure 8.25*b*, the presence of a membrane-bound Sar-GTP primes that site on the membrane to recruit the other proteins of the COPII coat to form a coated bud and subsequently a coated vesicle. Before the coated vesicle can fuse with a target membrane, the protein coat must be disassembled and its components released into the cytosol. Disassembly is thought to be triggered by hydrolysis of the bound GTP to produce a Sar-GDP subunit, which has decreased affinity for the vesicle membrane. Dissociation of Sar-GDP from the membrane is followed by the release of the other COPII subunits.

COPI-Coated Vesicles: Transporting Escaped Proteins Back to the ER

COPI-coated vesicles were first identified in experiments in which cells were treated with molecules similar in structure to GTP (GTP analogues) but, unlike GTP, cannot be hydrolyzed. Under these conditions, COPI-coated vesicles accumulated within the cell (Figure 8.26) and could be isolated from homogenized cells by density gradient centrifugation (Section 18.6). COPI-coated vesicles accumulate in the presence of a nonhydrolyzable GTP analogue because, like their COPII counterparts, the coat contains a GTP-binding protein, called *ARF1*, whose

and are present at the same concentration in the budding vesicle as in the ER lumen. Proteins that become enclosed in vesicles but are not specifically selected for inclusion are said to move by *bulk flow*. Some of the integral proteins in the ER membrane may also become trapped in budding

Figure 8.26 Accumulation of COPI-coated vesicles. The micrograph shows the Golgi complex of a permeabilized cell that had been treated with a nonhydrolyzable GTP analogue (GTPγS). Numerous COPI-coated vesicles and coated buds are visible. The COPI coat contains seven distinct proteins in addition to the GTP-binding protein ARF1, (FROM JAMES E. ROTHMAN AND LELIO ORCI, FASEB J. 4:1467, 1990.)

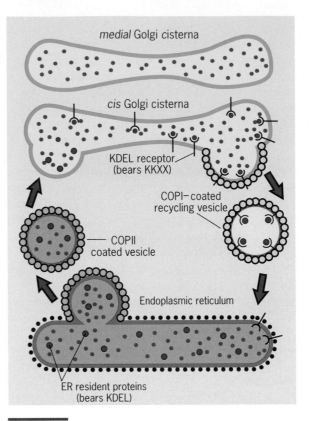

Figure 8.27 Retrieving ER proteins. Resident proteins of the ER contain amino acid sequences that lead to their retrieval from the Golgi complex if they are accidentally incorporated into a Golgi-bound transport vesicle. Soluble ER proteins bear the retrieval signal KDEL, whereas ER membrane proteins bear the signal KKXX. Retrieval is accomplished as soluble ER proteins bind to KDEL receptors residing in the membranous wall of *cis* Golgi compartments. The KDEL receptors, which bear the KKXX signal, bind in turn to proteins that compose the COPI coat, which allows the entire complex to be recycled back to the ER.

bound GTP must be hydrolyzed before the coat can disassemble. COPI-coated vesicles have been the subject of considerable controversy centered on whether or not they are involved in anterograde and/or retrograde transport between Golgi cisternae (indicated by the question marks in Figure 8.25a). Regardless, COPI-coated vesicles have been clearly shown to mediate the retrograde transport of proteins, particularly from the ERGIC and the Golgi complex back to the ER (Figure 8.25a). To understand the role of these retrograde COPI-coated vesicles, we have to consider a more general subject.

Retaining and Retrieving Resident ER Proteins If vesicles continually bud from membrane compartments, how does each compartment retain its unique composition? What determines, for example, whether a particular protein in the membrane of the ER remains in the ER or proceeds on to the Golgi complex? Studies suggest that proteins are maintained in an organelle by a combination of two mechanisms: (1) *retention* of resident molecules that are excluded from transport vesicles, and (2) *retrieval* of "escaped" molecules back to the compartment in which they normally reside. The mechanism by which proteins are retained in a particular membrane is not understood. According to one view, retained proteins become part of complexes that are too large to be incorporated into a budding transport vesicle. An alternate view is based on the microheterogeneity of membranes, that is, that membranes may possess different domains having dissimilar chemical composition and physical properties. It is possible that transported membrane proteins must reside in a particular ER membrane domain that can be "captured" by the COPII coat. The retrieval of escaped proteins is better understood.

Proteins that normally reside in the ER, those both in the lumen and in the membrane, contain short amino acid

sequences at their C-terminus that serve as **retrieval signals**, ensuring their return to the ER if they should be carried forward to the ERGIC or Golgi complex. The retrieval of "escaped" ER proteins from these compartments is accomplished by specific receptors that capture the molecules and return them to the ER in COPI-coated vesicles (Figure 8.25a, 8.27). Soluble proteins of the ER lumen (such as protein disulfide isomerase and the molecular chaperones that facilitate folding) typically possess the retrieval signal "lys-asp-glu-leu" (or KDEL in single-letter nomenclature). Soluble ER proteins bearing the KDEL signal are recognized and bound by an integral membrane protein, the *KDEL receptor*, whose cytosolic tail binds to the COPI coat (Figure 8.27), ensuring that these ER proteins will be returned to the ER. If the KDEL sequence is deleted from an ER protein, the protein is not returned to the ER compartment but instead is carried forward through the Golgi complex. Conversely, when a cell is genetically engineered to express a lysosomal or secretory protein that contains an added KDEL C-terminus, that

protein is returned to the ER rather than being sent on to its proper destination. Integral membrane proteins of the ER, such as the SRP receptor, also have a retrieval signal at their C-terminus (usually KKXX, where K is lysine and X is any amino acid), which binds to the COPI coat, facilitating their return to the ER. Each membrane compartment in the biosynthetic pathway may have its own unique retrieval signals, which helps explain how each compartment can maintain its unique complement of proteins despite the constant movement of vesicles in and out of that region.

Clathrin-Coated Vesicles: Sorting Lysosomal Proteins at the TGN

Despite all the discussion of transport vesicles, we have yet to examine how a particular protein that has been synthesized in the ER is targeted toward a particular cellular destination. It is important that a cell be able to distinguish among the various proteins that it manufactures. A pancreatic cell, for example, has to segregate newly synthesized digestive enzymes, which are packaged into secretory granules and ultimately secreted, from newly synthesized lysosomal enzymes, which are destined for the lysosomes. This is accomplished as the cell sorts proteins destined for different sites into different vesicles. Protein sorting occurs in the last of the Golgi compartments, the *trans* Golgi network (TGN), which functions as a major branchpoint in the movement of materials along the secretory pathway. The TGN is the site of assembly of clathrin-coated vesicles (the third and last type of coated vesicles to be discussed). Clathrin coats mediate cargo sorting at the TGN, and clathrin-coated vesicles carry hydrolytic enzymes and membrane proteins from there to endosomes, lysosomes and plant vacuoles.

The structure of clathrin-coated vesicles is described in detail on page 320 in connection with endocytosis, a process that is better understood than budding at the TGN. It will suffice at this point to note that the coats of these vesicles contain (1) an outer honeycomblike lattice composed of the protein clathrin, which forms a structural scaffold, and (2) an inner shell composed of protein complexes called **adaptors**, which cover the surface of the vesicle membrane that faces the cytosol (Figure 8.28). The first step in vesicle formation is the recruitment of the adaptors to the cytosolic surface of the TGN, a step that also requires the presence of the small GTP-binding protein ARF1, which surprisingly is the same protein that controls COPI-coat formation (page 305). The presence of the adaptors triggers the subsequent assembly of the clathrin lattice, which drives membrane budding.

As their name implies, adaptors provide the connection between two different types of components (Figure 8.28, inset). On their outer (cytosolic) surface, the adaptors bind to clathrin molecules, holding the clathrin scaffolding onto the surface of the vesicle. On their inner surface, the adaptors bind to sorting signals in the cytosolic tails of integral membrane proteins. Some of the se-

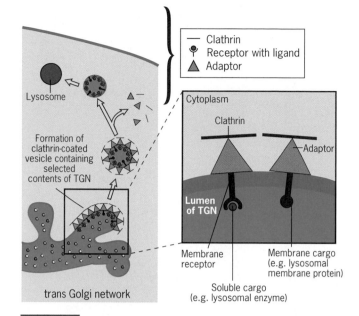

Figure 8.28 The formation of clathrin-coated vesicles at the TGN. Clathrin-coated vesicles of the TGN transport membrane proteins whose cytosolic tails are capable of binding to a particular type of adaptor (AP1) complex. The adaptors in turn possess binding sites for the surrounding cytosolic network of clathrin molecules. Clathin coated vesicles that form at the TGN carry membrane proteins that will become components of the lysosome membrane and membrane receptors that bind soluble lysosomal enzymes. A more realistic structure of the adaptor complex is shown in Figure 8.43.

lected integral membrane proteins serve as receptors that bind to specific cargo molecules within the vesicle lumen (Figure 8.28, inset). As a result of these interactions with adaptors, specific membrane proteins of the TGN and soluble proteins within the TGN lumen become concentrated into clathrin-coated vesicles. Once the vesicle has budded from the TGN, the clathrin coat is lost and the uncoated vesicle proceeds to its destination, which may be an endosome, lysosome, or plant vacuole. The operation of the various components of a clathrin-coated vesicle is illustrated by the sorting of lysosomal proteins.

Sorting of Lysosomal Proteins Lysosomal proteins are synthesized on membrane-bound ribosomes of the ER and carried to the *cis* Golgi cisternae along with other types of proteins. Once in the Golgi cisternae, soluble lysosomal enzymes are recognized by enzymes that catalyze the two-step addition of a phosphate group to certain mannose sugars of the *N*-linked carbohydrate chains (Figure 8.29a). Thus, unlike other glycoproteins sorted at the TGN, lysosomal enzymes possess phosphorylated mannose residues, which act as recognition signals. This mechanism of protein sorting was discovered through studies on cells from humans that lacked one of the enzymes involved in phosphate addition (discussed in the Human Perspective on page 315). Lysosomal enzymes

(a)

(b)

Figure 8.29 **The mechanism by which lysosomal enzymes are targeted to lysosomes.** (*a*) Lysosomal enzymes are recognized by an enzyme in the *cis* cisternae that transfers a phosphorylated *N*-acetylglucosamine from a nucleotide sugar donor to one or more mannose residues of *N*-linked oligosaccharides. The glucosamine moiety is then removed in a second step by a second enzyme, leaving mannose 6-phosphate residues as part of the oligosaccharide chain. (*b*) Schematic diagram showing the pathways followed by a lysosomal enzyme (black) from its site of synthesis in the ER to its delivery to a lysosome. The mannose residues of the lysosomal enzyme are phosphorylated in the Golgi cisternae (step 1) and then selectively incorporated into a clathrin-coated vesicle at the TGN (step 2). The mannose 6-phosphate receptors are thought to have a dual role: they interact specifically with the lysosomal enzymes on the luminal side of the vesicle, and they interact specifically with adaptors on the cytoplasmic surface of the vesicle (step 3). The mannose 6-phosphate receptors separate from the enzymes (step 4) and are returned to the Golgi complex (step 5). Mannose 6-phosphate receptors are also present in the plasma membrane, where they are able to capture lysosomal enzymes that are secreted into the extracellular space and return the enzymes to a pathway that directs them to a lysosome (step 6).

site sides of the membrane. While a part of the MPR projecting into the lumen of the TGN recognizes and binds a lysosomal enzyme, a part of the MPR projecting into the cytosol binds specifically with an adaptor-clathrin complex that assembles on the cytosolic surface of the TGN membrane (Figure 8.29*b*). These interactions ensure that lysosomal enzymes will become enclosed in clathrin-coated vesicles. Once it has budded, a clathrin-coated vesicle must be targeted to a particular destination. Coated vesicles containing lysosomal enzymes, for example, are ultimately targeted to a lysosome. The MPRs dissociate from the lysosomal enzymes and are returned to the TGN as shown in Figure 8.29*b*.

Lysosomal proteins are not the only materials that are exported from the TGN. As indicated in Figure 8.2, membrane proteins destined for the plasma membrane and secretory materials destined for export from the cell are transported from the TGN in secretory vesicles and secretory granules. For many years it was assumed that these membranous carriers must also be surrounded by a protein coat, but no such coat has yet been identified. According to a recent model, membranous carriers are produced as the TGN fragments into vesicles and tubules of various size. This concept fits with the cisternal maturation model, which proposes that the cisternae of the Golgi complex move continually toward the TGN, where they would have to disperse to allow continued maturation of the Golgi stack. According to one view, secretory vesicles don't have to be coated because the secretory materials they contain are present by "default." In other words, all of the other materials of the TGN are selectively removed, leaving only those materials that are targeted for the plasma membrane or discharge into the extracellular space.

carrying this mannose 6-phosphate signal are recognized and captured by **mannose 6-phosphate receptors (MPRs)**, which are integral membrane proteins concentrated within clathrin-coated pits of the TGN (Figure 8.29*b*).

Mannose 6-phosphate receptors are thought to span the TGN membrane with different binding sites on oppo-

In the case of cells that contain large, densely packed secretory granules, such as the pancreas cell of Figure 8.3*a*, the secretory product is thought to form aggregates that are trapped as secretory granules that bud off from the rims of the *trans* Golgi cisternae and TGN. We can now turn to the mechanism by which vesicle targeting is accomplished.

Targeting Vesicles to a Particular Compartment

Vesicle fusion requires specific interactions between different membranes. Vesicles from the ER, for example, fuse with the ERGIC or *cis* Golgi network and not with a *trans* cisterna. Selective fusion is one of the factors that ensures a highly directed flow through the membranous compartments of the cell. Despite a major research effort, we still do not understand the mechanisms by which cells target vesicles to particular compartments. It is thought that a vesicle contains specific proteins associated with its membrane that govern the targeting and fusion potential of that vesicle. To understand the nature of these proteins, we will consider the steps that occur between vesicle budding and vesicle fusion.

1. *Movement of the vesicle toward the specific target compartment.* In many cases, membranous vesicles must move considerable distances through the cytoplasm before reaching their eventual target. These types of movement are thought to be directed by microtubules, which act like railroad tracks carrying cargo containers along a defined pathway to a predetermined destination. The membranous carriers seen in Figure 8.19, for example, were observed to move from the ERGIC to the Golgi complex over microtubules.

2. *Tethering vesicles to the target compartment.* Microscopic studies indicate that vesicles are often "tethered" to a presumed target compartment, such as a Golgi cisterna, by extended, fibrous proteins (Figure 8.30*a*). It is hypothesized that tethering is an early stage in the process of vesicle fusion that requires specificity between the vesicle and target compartment. Part of this specificity may be conferred by a large family of small GTP-binding proteins called **Rabs**. Different members of the Rab family are found associated with different membrane compartments. Rabs are thought to recruit cytosolic tethering proteins to membrane surfaces (Figure 8.30*a*) and are likely to have other regulatory functions that remain unclear.

3. *Docking vesicles to the target compartment.* At some point during the process leading to vesicle fusion, the membranes of the vesicle and target compartment become tightly apposed to one another as the result of an interaction between the cytosolic regions of integral proteins of the two membranes. The key proteins that engage in these interactions are called **SNAREs**, and they constitute a large family of integral membrane proteins whose members are localized to specific subcellular compartments. SNAREs are divided into two categories,

v-SNAREs, which become incorporated into the membranes of transport vesicle during budding, and **t-SNAREs**, which are located in the membranes of target compartments (Figure 8.30*b*). The best-studied SNAREs are those that mediate docking of synaptic vesicles with the presynaptic membrane during the release of neurotransmitters (page 171). In this case, the plasma membrane of the nerve cell contains two t-SNAREs, syntaxin and SNAP-25, whereas the synaptic vesicle membrane contains a single v-SNARE, synaptobrevin. As the synaptic vesicle and presynaptic membrane approach one another, the t- and v-SNARE molecules from apposing membranes interact with one another to form four-stranded helical bundles as shown in Figure 8.31*a*.

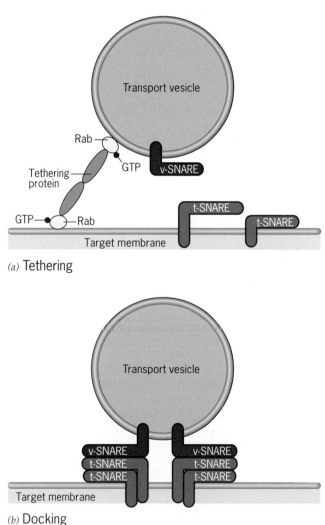

(a) **Tethering**

(b) **Docking**

Figure 8.30 **Steps in the targeting of transport vesicles to target membranes.** (*a*) According to this model, Rab proteins on the vesicle and target membrane are involved in the recruiting of one or more tethering proteins that mediate initial contact between the two membranes. (*b*) During the docking stage leading up to membrane fusion, the v-SNAREs in the vesicle membrane interact with the t-SNAREs in the target membrane to form a four-stranded α-helical bundle that brings the two membranes into intimate contact (see next figure).

(a)

(b)

(c)

Figure 8.31 A model of the interactions between v- and t-SNAREs leading to membrane fusion and exocytosis. (*a*) The synaptic vesicle has become docked to the plasma membrane through the formation of four-stranded bundles comprising α helices donated by syntaxin (red), synaptobrevin (blue), and SNAP-25 (green) with hypothetical transmembrane helices (yellow). (*b*) A speculative transition state in the fusion of the two membranes. A small water-filled cavity is shown at the center of the transmembrane helix bundle. (*c*) The transmembrane helices that previously resided in the two separate membranes are now present in the same bilayer and a fusion pore has opened between the vesicle and target membrane. The neurotransmitter contents of the vesicle can now be discharged by exocytosis. (FROM PEHR A. B. HARBURY, STRUCTURE 6:1490, 1998.)

Each bundle consists of two α helices donated by SNAP-25 and one α helix donated by both syntaxin and synaptobrevin. Together, these parallel α helices form a tightly

interwoven coiled coil (page 61) that pulls the two apposing lipid bilayers into very close association (Figure 8.30*b*, 8.31*a*). The formation of similar four-stranded helical bundles is thought to occur among other SNAREs at other sites in the cell where membranes are destined to fuse. It is interesting to note that the SNAREs of the synaptic vesicle and presynaptic membrane are the targets of two of the most potent bacterial toxins; those responsible for botulism and tetanus. These toxins act as proteases that cleave the SNAREs and, as a result, block the release of neurotransmitters.

4. *Fusion between the vesicle and target membranes.* When artificial lipid vesicles (liposomes) containing purified t-SNAREs are mixed with liposomes containing a purified v-SNARE, the two types of vesicles fuse with one another, but not with themselves. This finding indicates that t- and v-SNAREs constitute the minimal machinery required for membrane fusion of different compartments (Figure 8.31*b,c*). However, much evidence suggests that, while interaction between v- and t-SNAREs is required for membrane fusion, it is not sufficient by itself to bring about such fusion *within a cell*. Numerous studies indicate that additional proteins besides the SNAREs must be present in or associated with the membrane of a vesicle before that vesicle can fuse with a target membrane. Once fusion occurs, the SNAREs that previously projected from separate membranes become situated in the same membrane (Figure 8.31*c*). Dissociation of the four-stranded SNARE complex is achieved by a doughnut-shaped, cytosolic protein called NSF that fits around the SNARE bundle and twists it apart.

Now that we have described the events that occur during the fusion of a vesicle with a target membrane, we can return to the question: How is the specificity of this interaction determined? During the mid-1990s, SNAREs were thought to be the primary, or even sole, determinants of membrane-fusion specificity. It was presumed that v-SNAREs of a vesicle membrane could only bind to a target membrane that contained a specific complementary (cognate) t-SNARE. A v-SNARE in an ER-derived vesicle, for example, could engage a t-SNARE in a *cis* Golgi cisterna, but not a t-SNARE in a lysosomal membrane. Although the role of SNAREs in specific vesicle targeting remains controversial, a considerable body of evidence suggests that interactions between SNAREs are not as specific as previously thought. According to current opinion, the ability of a particular vesicle and target membrane to fuse is determined by a combination of factors, including tethering proteins, Rabs, and SNAREs. The contribution of each of these elements to target specificity remains unclear.

Exocytosis The best studied examples of vesicle fusion are those that occur during the regulated secretion of a secretory vesicle or synaptic vesicle with the plasma membrane. In these cases, membrane fusion produces an open-

ing through which the contents of the granule are released into the extracellular space (Figure 8.31). This process of membrane fusion and content discharge is called exocytosis, and it is usually triggered by a local increase in the concentration of calcium ions. It was noted on page 171 that the arrival of a nerve impulse at the terminal knob of a neuron leads to an increase in the influx of Ca^{2+} and the subsequent discharge of neurotransmitter molecules by exocytosis. In this case, fusion is mediated by a calcium binding protein (synaptotagmin) present in the membrane of the synaptic vesicle. In other types of cells, exocytosis is generally triggered by release of Ca^{2+} from cytoplasmic stores. Injection of calcium solutions into secretory cells leads to the wholesale exocytosis of secretory product.

The steps in exocytosis are not well understood. Contact between the membranes is thought to lead to the formation of a small, protein-lined "fusion pore" (Figure 8.31c) that rapidly dilates to form an opening for discharge. Regardless of the mechanism, when a cytoplasmic vesicle fuses with the plasma membrane, the luminal surface of the vesicle membrane becomes part of the outer surface of the plasma membrane, whereas the cytosolic surface of the vesicle membrane becomes part of the inner surface of the plasma membrane (Figure 8.14).

REVIEW ?

1. What determines the specificity of interaction between a transport vesicle and the membrane compartment with which it will fuse?
2. Describe the steps that ensure that a lysosomal enzyme will be targeted to a lysosome rather than a secretory vesicle.
3. Contrast the roles of COPI- and COPII-coated vesicles in protein trafficking.

8.6 LYSOSOMES

Lysosomes are an animal cell's digestive organelles. A typical lysosome contains approximately 50 different hydrolytic enzymes (Table 8.1) produced in the rough ER and targeted for these organelles. Taken together, lysosomal enzymes can hydrolyze virtually every type of biological macromolecule. The low-molecular-weight products of enzymatic digestion can then be transported across the lysosomal membrane into the cytosol. The enzymes of a lysosome share an important property: all have their optimal activity at an acid pH and thus are **acid hydrolases**. The pH optimum of these enzymes is suited to the low pH of the lysosomal compartment, which is approximately 4.6. The high internal proton concentration is maintained by a proton pump (an H^+-ATPase) present in the organelle's boundary membrane. Lysosomal mem-

Table 8.1 A Sampling of Lysosomal Enzymes

Enzyme	Substrate
Phosphatases	
acid phosphatase	phosphomonoesters
acid phosphodiesterase	phosphodiesters
Nucleases	
acid ribonuclease	RNA
acid deoxyribonuclease	DNA
Proteases	
cathepsin	proteins
collagenase	collagen
GAG-hydrolyzing enzymes	
iduronate sulfatase	dermatan sulfate
β-glactosidase	keratan sulfate
heparan N-sulfatase	heparan sulfate
α-N-acetylglucosaminidase	heparan sulfate
Polysaccharidases and oligosaccharidases	
α-glucosidase	glycogen
fucosidase	fucosyl oligosaccharides
α-mannosidase	mannosyl oligosaccharides
sialidase	sialyl oligosaccharides
Sphingolipid hydrolyzing enzymes	
ceramidase	ceramide
glucocerebrosidase	glucosylceramide
β-hexosaminidase	G_{M2} ganglioside
arylsulfatase A	galactosylsulfatide
Lipid hydrolyzing enzymes	
acid lipase	triacylglycerols
phospholipase	phospholipids

branes contain a variety of acidic, highly glycosylated, integral proteins that probably protect the membrane from attack by the enclosed enzymes.

Although lysosomes house a predictable collection of enzymes, their appearance in electron micrographs is neither distinctive nor uniform. For example, lysosomes range in size from relatively large structures (over 1 μm in diameter) to very small vesicles (25–50 nm in diameter). Figure 8.32 shows a small portion of a Kupffer cell, a phagocytic cell in the liver that engulfs aging red blood cells. The lysosomes of a Kupffer cell exhibit an irregular shape and variable electron density, illustrating how difficult it is to identify these organelles on the basis of morphology alone.

The presence within a cell of what is, in essence, a bag of destructive enzymes suggests a number of possible

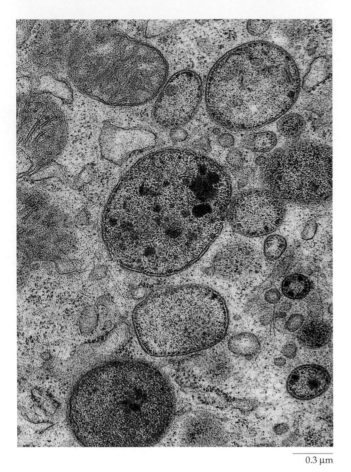

0.3 µm

Figure 8.32 Lysosomes. Portion of a phagocytic Kupffer cell of the liver showing at least 10 lysosomes of highly variable size. (FROM HANS GLAUMANN ET AL., J. CELL BIOL. 67:887, 1975, BY COPYRIGHT PERMISSION OF THE ROCKEFELLER UNIVERSITY PRESS.)

functions. The best-studied role of lysosomes is the breakdown of materials brought into the cell from the extracellular environment. Many single-celled organisms ingest food particles, which are then enzymatically disassembled in the lysosome. The resulting nutrients pass through the lysosomal membrane into the cytosol. In mammals, phagocytic cells, such as macrophages and neutrophils, function as scavengers that ingest debris and potentially dangerous microorganisms (page 316). Ingested bacteria are generally inactivated by the low pH of the lysosome and then digested enzymatically. It is estimated that a macrophage actively engaged in phagocytosis has as many as a thousand lysosomes.

Lysosomes do more than destroy materials that enter the cell from the external environment. One of the more unusual lysosomal activities occurs during fertilization, where lysosomal enzymes are active outside the cell. The head of a sperm contains a packet of lysosomal enzymes, called the *acrosome*. As the sperm reaches the surroundings of an egg, the membrane bounding the acrosome fuses with the cell's plasma membrane, releasing the stored lysosomal enzymes (Figure 8.33). The released enzymes digest their way through the outer covering of the egg, leaving a hole through which the advancing sperm can reach the egg surface.

Lysosomes also play a key role in organelle **turnover**, that is, the destruction of the cell's own organelles and their replacement. During this process, which is called **autophagy**, an organelle, such as the mitochondrion shown in Figure 8.34, is surrounded by a double membrane derived from an ER cisterna. The ER membrane then fuses with a lysosome to produce an *autophagolysosome*. It is not

Acrosome

Lysosomal enzymes

(a) (b)

Figure 8.33 The acrosomal reaction. (*a*) The acrosome is situated at the anterior end of this sea urchin sperm, just in front of the nucleus. (*b*) When the plasma membrane of the sperm makes contact with materials surrounding the egg, the sperm undergoes a reaction that releases lysosomal enzymes stored in the acrosome, which digest a path leading to the egg surface. (COURTESY OF GLEN L. DECKER, D. B. JOSEPH, AND WILLIAM J. LENNARZ.)

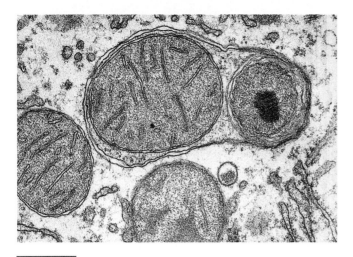

Figure 8.34 Autophagy. This electron micrograph shows a mitochondrion and a peroxisome enclosed in a double membrane wrapper derived from the ER. This autophagic vacuole would have fused with a lysosome and its contents digested. (FROM DON FAWCETT AND DANIEL FRIEND/PHOTO RESEARCHERS)

uncommon when surveying an electron micrograph to see a mitochondrion or other organelle trapped in an autophagolysosome. It is calculated that one mitochondrion undergoes autophagy every 10 minutes or so in a mam-

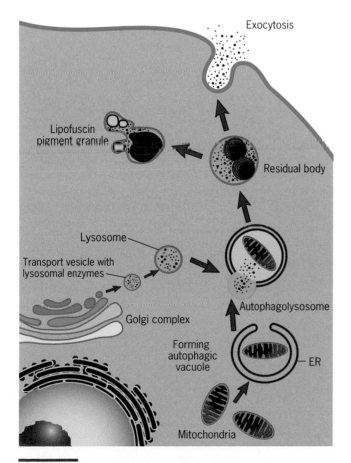

Figure 8.35 A summary of the autophagic pathway. The steps are described in the text.

malian liver cell. If a cell is deprived of nutrients, a marked increase in autophagy is observed. Under these conditions, the cell acquires energy to maintain its life by cannibalizing its own organelles.

The role of lysosomes in autophagy is summarized in Figure 8.35. Once the digestive process in the autophagolysosome has been completed, the organelle is termed a *residual body*. Depending on the type of cell, the contents of the residual body may be eliminated from the cell by exocytosis, or they may be retained within the cytoplasm indefinitely as a *lipofuscin granule*. Lipofuscin granules increase in number as an individual becomes older; accumulation is particularly evident in long-lived cells such as neurons, where these granules are considered a major characteristic of the aging process. The role of lysosomes in various diseases is discussed in the accompanying Human Perspective.

REVIEW ?

1. Describe three distinct roles of lysosomes.
2. Describe the events that occur during the autophagic destruction of a mitochondrion.

8.7 PLANT CELL VACUOLES

As much as 90 percent of the volume of many plant cells is occupied by a single membrane-bound, fluid-filled central **vacuole** (Figure 8.36). While simple in structure, plant vacuoles carry out a wide spectrum of essential functions. Many of the cell's solutes and macromolecules, including ions, sugars, amino acids, proteins, and polysaccharides, are stored temporarily in the vacuole. Vacuoles may also store a host of toxic compounds. Some of these compounds (such as cyanide-containing glycosides and glucosinolates) are part of an arsenal of chemical weapons that are released when the cell is injured by an herbivore or fungus. Other toxic compounds are simply the byproducts of metabolic reactions; because plants lack the type of excretory systems found in animals, they utilize their vacuoles to isolate these byproducts from the rest of the cell. Some of these compounds, such as digitalis, have proven to have important clinical value.

The membrane that bounds the vacuole, the **tonoplast**, contains a number of active transport systems that pump ions into the vacuolar compartment to a concentration much higher than that in the cytoplasm or the extracellular fluid. Because of the vacuole's high ionic concentration, water enters the organelle by osmosis. Hydrostatic (turgor) pressure exerted by the vacuole not only provides mechanical support for the soft tissues of a plant (page 153), it also stretches the cell wall during cell growth.

Plant vacuoles are also sites of intracellular digestion, not unlike lysosomes, which are absent in plants. In fact, plant vacuoles have some of the same acid hydrolases

(a)

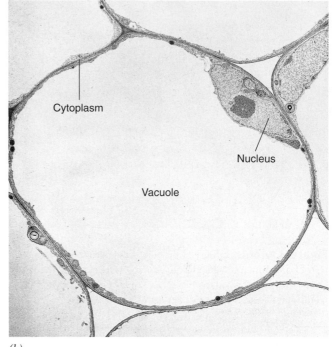

(b)

Figure 8.36 **Plant cell vacuoles.** (*a*) Each of the cylindrical leaf cells of the aquatic plant *Elodea* contains a large central vacuole surrounded by a layer of cytoplasm containing the chloroplasts that are visible in the micrograph. (*b*) Transmission electron micrograph of a soybean cortical cell showing the large central vacuole and thin layer of surrounding cytoplasm. (A: FROM M. I. WALKER/PHOTO RESEARCHERS; B: FROM M. F. BROWN/VISUALS UNLIMITED.)

found in lysosomes. The pH of the vacuole is maintained at a low value by a V-type H^+-ATPase (page 159) within the tonoplast that pumps protons into the vacuolar fluid. Another feature that plant vacuoles share with lysosomes is their participation as an endpoint in the cell's biosynthetic pathway. Like the proteins of the lysosome, many of the proteins of a plant vacuole are synthesized on membrane-bound ribosomes of the RER, transported through the Golgi complex, and sorted at the *trans* face of the Golgi before being targeted to the vacuole.

REVIEW ?

1. Describe three distinct roles of plant vacuoles.
2. In what ways is a plant vacuole similar to a lysosome? In what ways is it different?

8.8 CELLULAR UPTAKE OF PARTICLES AND MACROMOLECULES

In a previous chapter, we considered the movement of low-molecular-weight solutes directly through the plasma membrane. But how are cells able to take in materials that

are too large to penetrate this membrane regardless of its permeability properties? To meet this requirement, mechanisms have evolved that bring about the enclosure of material from the extracellular environment. Extracellular materials are enclosed within vesicles derived from folds or invaginations of the plasma membrane. The uptake of extracellular materials into cytoplasmic vesicles is divided into two separate categories, phagocytosis and endocytosis, which occur by different mechanisms. **Phagocytosis** is the uptake of particulate matter, and **endocytosis** is the uptake of fluid, dissolved solutes, and suspended macromolecules.[6]

Phagocytosis

Phagocytosis ("cell eating") is carried out extensively by a few types of cells specialized for the uptake of relatively large particles (>0.5 μm diameter) from the environment.

[6]The terminology in this area has undergone changes in recent years. In 1963, Christian de Duve introduced the term *endocytosis* to include two types of activities, the ingestion of particles (phagocytosis) and the uptake of fluid and solutes (pinocytosis). Because it has become clear in recent years that phagocytosis and pinocytosis are fundamentally different activities, the term *pinocytosis* is being used less often. In the present discussion, endocytosis is employed to describe the uptake of both fluid and dissolved or suspended molecules and is distinguished from phagocytosis.

THE HUMAN PERSPECTIVE

Disorders Resulting from Defects in Lysosomal Function

Considering that the hydrolytic enzymes of a lysosome can digest the entire contents of a cell, it is not unexpected that disturbances in lysosome function can have profound effects on human health. For example, a miners' disease known as *silicosis* results from the uptake of silica fibers by phagocytic cells in the lungs. The fibers become enclosed within lysosomes but cannot be digested; instead, they cause the lysosomal membrane to leak, spilling the contents of digestive enzymes into the cell and damaging the tissue of the lungs. A similar result occurs when asbestos fibers are taken up by scavenging cells, resulting in the disease *asbestosis*. Both conditions can be debilitating and may even be fatal. The debilitating symptoms of inflammatory joint diseases, such as rheumatoid arthritis, result in part from the release of lysosomal enzymes from immune cells into the synovial fluids of the joints.

Our understanding of the mechanisms by which proteins are targeted to particular organelles began with the discovery that the mannose 6-phosphate residues in lysosomal enzymes act as an "address" for delivery of the proteins to lysosomes. The discovery of the lysosome "address" was made in studies of patients with a rare and fatal inherited condition known as *I-cell disease*. Many cells in these patients contain lysosomes that are bloated with undegraded materials. Materials accumulate in the lysosomes because of the absence of hydrolytic enzymes. When fibroblasts from these patients were studied in culture, it was found that lysosomal enzymes are synthesized at normal levels but are secreted into the medium and not targeted to lysosomes. Further analysis revealed that the secreted enzymes lacked the mannose phosphate residues that are present on the corresponding enzymes of cells from normal individuals. The I-cell defect was soon traced to a deficiency in an enzyme (*N*-acetylglucosamine phosphotransferase) required for mannose phosphorylation (see Figure 8.29*a*).

In 1965, W. G. Hers of the University of Louvain in Belgium offered an explanation as to how the absence of a seemingly unimportant lysosomal enzyme, α-glucosidase, could lead to the development of a fatal inherited condition known as Pompe disease. Hers suggested that, in the absence of α-glucosidase, undigested glycogen accumulated in lysosomes, causing swelling of the organelles and irreversible damage to the cells and tissues. Diseases of this type, characterized by the deficiency of a lysosomal enzyme and the corresponding accumulation of undegraded substrate (Figure 1), are called **lysosomal storage disorders**, and over 30 have been described. Those diseases resulting from an accumulation of undegraded sphingolipids are listed in Table 1. The symptoms of lysosomal storage disorders can range from very severe to barely detectable, depending on the degree of enzyme dysfunction.

Among the best-studied lysosomal storage disorders is Tay-Sachs disease, which results from a deficiency of the en-

Figure 1 Lysosomal storage disorders. Electron micrograph of a section through a portion of a neuron of a person with a lysosomal storage disorder characterized by an inability to degrade G_{M2} gangliosides. These cytoplasmic vacuoles stain for both lysosomal enzymes and the ganglioside, indicating they are lysosomes in which undigested glycolipids have accumulated. (COURTESY OF KINUKO SUZUKI.)

zyme β-*N*-hexosaminidase A, an enzyme that degrades the ganglioside G_{M2} (Figure 4.6). G_{M2} is a major component of the membranes of brain cells, and in the absence of the hydrolytic enzyme, the ganglioside accumulates in the cytoplasm of brain cells (Figure 1), causing dysfunction. In its severe form, which strikes during infancy, the disease is characterized by progressive mental and motor retardation, as well as skeletal, cardiac, and respiratory abnormalities. The disease is very rare in the general population but reaches an incidence up to 1 in 3600 newborns among Jews of eastern European ancestry. The incidence of the disease has dropped dramatically in this ethnic population in recent years as the result of identification of carriers, genetic counseling of parents at risk, and prenatal diagnosis by amniocentesis. In fact, all of the known lysosomal storage disorders can be diagnosed prenatally.

In the past few years, the prospects for treatment of lysosomal storage disorders have improved with the demonstration that the symptoms of Gaucher's disease, a deficiency of

| Table 1 | **Sphingolipid Storage Diseases** | | |

Disease	Enzyme Deficiency	Principal Storage Substance	Consequences
G_{M1} Gangliosidosis	G_{M1} β-Galactosidase	Ganglioside G_{M1}	Mental retardation, liver enlargement, skeletal involvement, death by age 2
Tay-Sachs disease	Hexosaminidase A	Ganglioside G_{M2}	Mental retardation, blindness, death by age 3
Fabry's disease	α-Galactosidase A	Trihexosylceramide	Skin rash, kidney failure, pain in lower extremities
Sandhoff's disease	Hexosaminidases A and B	Ganglioside G_{M2} and globoside	Similar to Tay-Sachs disease but more rapidly progressing
Gaucher's disease	Glucocerebrosidase	Glucocerebroside	Liver and spleen enlargement, erosion of long bones, mental retardation in infantile form only
Niemann-Pick disease	Sphingomyelinase	Sphingomyelin	Liver and spleen enlargement, mental retardation
Farber's lipogranulomatosis	Ceramidase	Ceramide	Painful and progressively deformed joints, skin nodules, death within a few years
Krabbe's disease	Galactocerebrosidase	Galactocerebroside	Loss of myelin, mental retardation, death by age 2
Sulfatide lipidosis	Arylsulfatase A	Sulfatide	Mental retardation, death in first decade

the lysosomal enzyme glucocerebrosidase, can be alleviated by *enzyme replacement therapy*. Infants with Gaucher's disease accumulate large quantities of glucocerebroside lipids in the lysosomes of their macrophages, causing liver and spleen enlargement and anemia. Initial attempts to correct the disease by infusing a solution of the normal human enzyme into the bloodstream were unsuccessful because the enzyme was taken up by liver cells, which are not seriously affected by the deficiency. To target macrophages, the enzyme was purified from human placental tissue and treated with three different glycosidases to remove terminal sugars on the enzyme's oligosaccharide chains, which exposed underlying mannose residues (Figure 8.22). Following infusion into the bloodstream, this modified enzyme (marketed under the name Ceredase) is recognized by mannose receptors on the surface of the macrophages and rapidly taken up by receptor-mediated endocytosis (page 319). Because lysosomes are the natural target site of materials brought into the macrophage by endocytosis, the enzymes are efficiently delivered to the precise sites in the cell where the deficiency is manifested. Thousands of victims of this disease have been successfully treated in this way. Enzyme replacement therapy, gene therapy, and bone marrow transplantation are currently being studied as treatments for several other lysosomal storage disorders.

Single-celled heterotrophic organisms, such as amoebas and ciliates, make their livelihood by trapping food particles and smaller organisms and enclosing them within folds of the plasma membrane. The folds fuse to produce a vacuole (or *phagosome*) that pinches off inwardly from the plasma membrane (Figure 8.37*a*). The phagosome fuses with a lysosome, and the material is digested within the resulting *phagolysosome*.

In most animals, phagocytosis is a protective mechanism rather than a mode of feeding. Mammals possess a variety of "professional" phagocytes, including macrophages and neutrophils, that wander through the blood and tissues phagocytizing invading organisms, damaged and dying cells, and debris. These materials are recognized and bound by receptors on the surface of the phagocyte prior to uptake. Mammalian phagocytosis is markedly enhanced by a number of blood-borne factors, called **opsonins,** that coat the particle to be ingested. Once inside the phagocyte, microorganisms may be killed by lysosomal enzymes or by oxygen free radicals generated within the lumen of the phagosome. The process of particle engulfment is depicted in Figure 8.37*b*. The steps in the digestion of engulfed materials, which are similar to those in which cytoplasmic organelles are digested by autophagy, are illustrated in Figure 8.38. The engulfment of particulate material by phagocytosis is driven by contractile activities of the actin-containing microfilaments that underlie the plasma membrane.

Not all bacteria ingested by phagocytic cells are destroyed. In fact, some species "hijack" the phagocytic machinery to promote their own survival in the body. *Mycobacterium tuberculosis*, the agent responsible for tuberculosis, for example, is taken into the cytoplasm of a macrophage by phagocytosis, but the phagosomes in which the bacteria are enclosed fail to fuse with a lysosome. The engulfed organism inhibits the membrane fusion that would lead to its destruction and instead multiplies within the cell. In contrast, the bacterium responsible for Q fever, *Coxiella burnetii*, becomes enclosed in a phagosome that does fuse with a lysosome, but neither the acidic environment nor the lysosomal enzymes can destroy the pathogen. *Listeria monocytogenes*, a bacterium that causes meningitis, produces a phospholipase that destroys the integrity of the lysosomal membrane, allowing the bacterium to escape into the cell's cytosol (see Figure 9.69).

Figure 8.37 **Phagocytosis.** (*a*) Schematic diagram of the steps in the engulfment, digestion, and absorption of materials taken into an amoeba by phagocytosis. (*b*) The process of engulfment as illustrated by a polymorphonuclear leukocyte ingesting a yeast particle. (B: FROM JANET BOYLES AND DOROTHY F. BAINTON, CELL 24:906, 1981, BY PERMISSION OF CELL PRESS.)

Endocytosis

Endocytosis can be divided broadly into two categories: bulk-phase endocytosis and receptor-mediated endocytosis. **Bulk-phase endocytosis** is the nonspecific uptake of extracellular fluids. Any molecules, large or small, that happen to be present in the enclosed fluid also gain entry into the cell. Bulk-phase endocytosis can be visualized by adding a substance to the culture medium, such as the dye lucifer yellow or the enzyme horseradish peroxidase, that is taken up by cells nonspecifically. Bulk-phase endocytosis occurs continually in certain types of cells where it may function primarily to convert plasma membrane into cytoplasmic membrane. This conversion is required in cells that have been engaged in secretion and have had large numbers of secretory vesicles fuse with the plasma membrane. **Receptor-mediated endocytosis (RME)** brings about the uptake of specific extracellular macromolecules (ligands) following their binding to receptors on the external surface of the plasma membrane. Receptor-mediated endocytosis is discussed later and in the Experimental Pathways at the end of the chapter.

Both bulk-phase and receptor-mediated endocytosis can occur at a remarkably rapid rate—as much as half the surface membrane can be internalized in as little as 30 minutes. Despite the rapid inward movement of plasma membrane, there is no shrinkage of the cell surface, nor is there any immediate need for the synthesis of new membrane components; the membrane is simply cycled between the surface and the cell interior so that membrane is added to the surface as fast as it is removed.

The Endocytic Pathway Molecules taken into a cell by endocytosis are routed through a well-defined endocytic pathway (Figure 8.39*a*). The endocytic pathway begins with a dynamic network of tubules and vesicles known collectively as **endosomes**. The fluid in the lumen of an

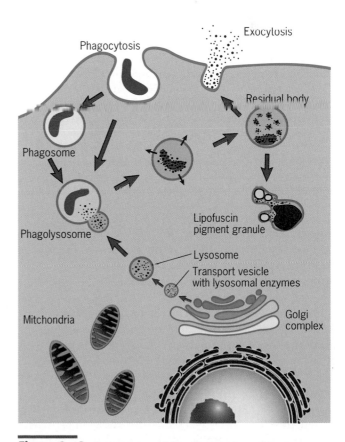

Figure 8.38 **A summary of the phagocytic pathway.**

(a)

(b)

(c)

1 μm

Figure 8.39 **The endocytic pathway.** (*a*) Schematic diagram showing the movement of materials from the extracellular space to the early endosomes, where sorting is thought to occur. Membrane receptors are usually sent back to the plasma membrane, whereas materials within the vesicle are transferred to the late endosomes. Materials (and hydrolytic enzymes) are transferred from late endosomes to lysosomes by a number of routes. Mannose 6-phosphate receptors (MPRs) carry lysosomal enzymes to late endosomes and are then shuttled back to the TGN for further transport duties. (*b,c*) Experimental demonstration of the movement of materials from early endosomes to late endosomes. The cell depicted in *b* had been incubated for

five minutes with the enzyme horseradish peroxidase (HRP) and then immediately fixed. HRP is taken into the cell by endocytosis and can be localized by electron microscopic cytochemistry. The reaction product of the ingested HRP is seen as an electron-dense deposit that is restricted to the early endosomes (arrows), which can take the shape of tubules, vesicles, or cisternae. The cell depicted in *c* had been incubated for an additional 30 minutes prior to fixation, which is sufficient time for the ingested enzyme to have moved to the larger, vesicular-shaped late endosomes (arrowheads). (B,C: FROM FENG GU ET AL., COURTESY OF JEAN GRUENBERG, J. CELL BIOL. 139:1189, 1997; BY COPYRIGHT PERMISSION OF THE ROCKEFELLER UNIVERSITY PRESS.)

endosome is acidified due to the activity of an H⁺-ATPase (a proton pump) in the boundary membrane. Endosomes are divided into two distinct classes, **early endosomes**, which are typically located near the peripheral region of the cell (Figure 8.39*b*), and **late endosomes**, which are typically located closer to the nucleus (Figure 8.39*c*). Early and late endosomes can be distinguished from one another on the basis of such properties as their buoyant density (which allows them to be isolated in different fractions on a density gradient), pH, and protein composition.

Materials that are taken up by endocytosis are transported in endocytic vesicles to early endosomes, where they are sorted (Figure 8.39*a,b*). Integral proteins of the vesicle membrane dissociate from their bound ligands in

the acidified endosomal environment and are then concentrated into specialized tubular compartments of the early endosome, which represent recycling centers. Vesicles that bud from these tubules carry membrane proteins back to the plasma membrane. In contrast, dissolved materials and bound ligands that have been released from their receptors become concentrated into a more spherical sorting compartment before being dispatched to late endosomes (Figure 8.39*a,c*). The transfer of materials from early to late endosomes probably occurs by means of specialized endosomal carrier vesicles (ECVs). Alternatively, early endosomes may simply mature into late endosomes.

Molecules that reach the late endosomes are presumably directed to a lysosome, which is the terminal compartment of the endocytic pathway. Movement of materi-

Figure 8.40 Receptor-mediated endocytosis. This sequence of micrographs shows the steps in the uptake of yolk lipoproteins by the hen oocyte. (*a*) The proteins to be taken into the cell are concentrated on the extracellular surface of an indented region of the plasma membrane, forming a coated pit. The cytosolic surface of the plasma membrane of the coated pit is covered with a layer of bristly, electron-dense material consisting of the protein clathrin. (*b*) The coated pit has sunk inward to form a coated bud. (*c*) The plasma membrane is about to pinch off as a vesicle containing the yolk protein on its luminal (previously extracellular) surface and clathrin on its cytosolic surface. (*d*) A coated vesicle that is no longer attached to the plasma membrane. The next step in the process is the release of the clathrin coat. (FROM M. M. PERRY AND A. B. GILBERT, J. CELL SCIENCE 39:257, 1979, BY PERMISSION OF THE COMPANY OF BIOLOGISTS, LTD.)

als from late endosomes to lysosomes may occur by several routes, including (1) maturation of late endosomes into lysosomes, (2) fusion of late endosomes with preexisting lysosomes, or (3) vesicle transport between late endosomes and lysosomes that remain in fixed positions. In addition to receiving material from the early endosomes, late endosomes also receive newly synthesized lysosomal enzymes from the TGN. Once they have delivered their cargo, the receptors that carried the lysosomal enzymes (page 308) are recycled back to the TGN for additional rounds of transport (Figure 8.39*a*).

Receptor-Mediated Endocytosis and the Role of Coated Pits Receptor-mediated endocytosis (RME) provides a means for the selective and efficient uptake of macromolecules that may be present at relatively low concentrations in the extracellular fluid. Cells have receptors for the uptake of many different types of ligands, including hormones, growth factors, enzymes, and plasma proteins. Substances that enter a cell by means of RME become bound to receptors that collect in specialized areas of the plasma membrane, known as **coated pits**. Receptors are concentrated in coated pits at 10–20 times their level in the remainder of the plasma membrane. Coated pits (Figure 8.40*a*) are recognized in electron micrographs as sites where the surface is indented and the plasma membrane is covered on its cytoplasmic face by a bristly, electron-dense protein composed of clathrin and adaptors, the same proteins found in clathrin-coated vesicles at the TGN (page 307). Coated pits invaginate into the cytoplasm (Figure 8.40*b*) and then pinch free of the plasma membrane to form coated vesicles (Figure 8.40*c,d*). To understand the mechanism of coated vesicle formation, we need to examine the molecular structure of the clathrin coat.

Figure 8.41 shows a coated pit as seen from the extracellular and cytoplasmic surfaces of the plasma membranes of cells that had been engaged in receptor-mediated endocytosis. When viewed from its cytoplasmic surface (Figure 8.41*b,c*), the bristly coat appears to consist of a network of hexagons resembling a honeycomb. The geometric construction of the coat is derived from the structure of its clathrin building blocks. Each clathrin molecule consists

(a) (b) (c) 0.1 µm

Figure 8.41 **Coated pits.** (*a*) Electron micrograph of a replica formed on the extracellular surface of a whole, freeze-dried fibroblast that had been incubated with LDL-cholesterol. Particles of LDL-cholesterol are visible as droplets located on the *extracellular surface* of the coated pit. (*b*) Electron micrograph of a replica formed on the *cytosolic surface* of a coated pit of a fibroblast. The coat is composed of a flattened network of clathrin-containing polygons associated with the inner surface of the plasma membrane. (*c*) Electron micrograph of the cytosolic surface showing the invaginated plasma membrane surrounded by a clathrin lattice that has assumed a hemispheric shape. (A–C: FROM JOHN HEUSER AND LOUISE EVANS, J. CELL BIOL. 84:568, 1980; BY COPYRIGHT PERMISSION OF THE ROCKEFELLER UNIVERSITY PRESS.)

of three heavy chains and three light chains, joined together to form a three-legged assembly called a *triskelion* (Figure 8.42). A triskelion provides a remarkably adaptable module with which a cell can construct different types of polygonal lattices. During invagination, the relatively flat lattice of the coated pit becomes curved around the forming vesicle (Figure 8.41*c*). Curvature of the clathrin coat is facilitated by conversion of some of the hexagons of the flattened lattice into pentagons. A scaffold composed of triskelions is well suited to undergo the type of structural rearrangement required to transform a flat lattice into a cagelike structure. In fact, purified clathrin triskelions self-assemble to form empty cages similar to those that enclose a coated vesicle. The overlapping arrangement of the three-legged clathrin triskelions within the wall of the vesicle coat, and the interaction between the clathrin lattice and the adaptors, are shown in Figure 8.43.

Figure 8.41*c* shows the structure of a coated pit that has undergone invagination and is nearing the stage when it would have pinched off to become a coated vesicle. The formation of a clathrin-coated endocytic vesicle requires an additional protein—a GTP-binding protein called **dynamin**—that self-assembles into a helical collar around the neck of the invaginated coated pit (Figure 8.44). Several models have been put forward to explain the mechanism of dynamin activity. According to the model depicted in Figure 8.44, steps 3–4, hydrolysis of the bound GTP by the polymerized dynamin molecules induces a conformational change in the dynamin helix that severs the coated vesicle from the plasma membrane. According to this model, dynamin acts as a structural protein capable of generating mechanical forces. According to another model, dynamin acts more like other G proteins described in this chapter by switching on the activity of a separate effector protein that severs the vesicle.

Like the clathrin-coated vesicles that bud from the TGN (page 307), the coated vesicles that form during endocytosis also contain a layer of adaptor complexes situated

Heavy chain

Light chain

50 nm

Figure 8.42 **Clathrin triskelions.** Electron micrograph of a metal-shadowed preparation of clathrin triskelions. Inset shows the triskelion is composed of three heavy chains. The inner "leg" of each heavy chain is linked to a smaller light chain. (REPRINTED WITH PERMISSION FROM ERNST UNGEWICKELL AND DANIEL BRANTON, NATURE 289:421, 1981. © COPYRIGHT 1981, MACMILLAN MAGAZINES LTD.)

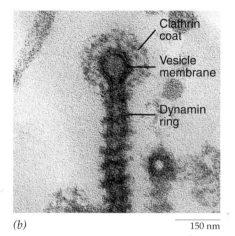

(a)

(b)

150 nm

Figure 8.43 Molecular assembly of a coated vesicle.
Schematic drawing of the incorporation of triskelions and adaptors into the outer clathrin coat of a coated vesicle. The sides of the polygons are formed by parts of the legs of overlapping triskelia. The N-terminus of each clathrin heavy chain forms a "hook" that projects toward the surface of the membrane where it engages an adaptor. Each adaptor consists of four different polypeptide subunits. Both the hooks and adaptors are situated at the vertices of the polyhedrons. (Note: not all of the triskelia of the lattice are shown in this figure; if they were, every vertex would have a clathrin hook and associated adaptor.) (REPRODUCED, WITH PERMISSION, AFTER S. SCHMID, ANNUAL REVIEW OF BIOCHEMISTRY, VOL. 66, © 1997 BY ANNUAL REVIEWS.)

between the clathrin lattice and the surface of the vesicle facing the cytosol. The adaptors specifically engage the cytoplasmic tails of the plasma membrane receptors (see Figure 8.39). As a result of their interaction with adaptors, the receptors (and their bound cargo molecules) are enclosed within the forming coated vesicle rather than being left behind in the plasma membrane. The interaction between adaptors and membrane receptors is mediated by *internalization signals* present in the cytoplasmic tail of the membrane protein as discussed further in the Experimental Pathways. The AP2 adaptors employed in coated vesicle formation at the plasma membrane contain polypeptides different from but related to the AP1 adaptors found at the TGN. A third type of adaptor (AP3) functions in the endosomal-lysosomal pathway and new AP-related adaptors that mediate unknown events continue to be identified.

Within a minute of its formation, the coated endocytic vesicle loses its clathrin coat and becomes a smooth-surfaced vesicle that typically fuses with an early endosome (page 318). In most cases, the low pH of the early

Figure 8.44 The role of dynamin in the formation of clathrin-coated vesicles. (*a*) The clathrin lattice of the coated pit (step 1) undergoes rearrangement to form an invaginated vesicle connected to the overlying plasma membrane by a stalk (step 2). At this point, the dynamin subunits, which are concentrated in the region, undergo polymerization to form a ring around the stalk (step 3). Changes in the conformation of the ring, which are thought to be induced by GTP hydrolysis (step 4), lead to fission of the coated vesicle from the plasma membrane and disassembly of the dynamin ring (step 5a). If vesicle budding occurs in the presence of GTPγS, a nonhydrolyzable analogue of GTP, dynamin polymerization continues beyond formation of a simple collar, producing a narrow tubule constructed from several turns of the dynamin helix (step 5b). (*b*) Electron micrograph showing a coated vesicle forming in the presence of GTPγS, which corresponds to the stage depicted in step 5b of part *a*. (A: FROM P. DE CAMILLI ET AL. CURRENT OPIN. NEUROBIOL. VOL. 5, P. 562, 1995; B: REPRINTED WITH PERMISSION FROM KOHJI TAKEI, ET AL., NATURE VOL. 374, COVER 3/9/95. © COPYRIGHT 1995 MACMILLAN MAGAZINES LTD.)

endosomal compartment alters the ionic state of the receptor molecules, causing the release of their ligands. Once dissociated from their ligands, the receptors are shuttled back to the plasma membrane for additional rounds of endocytosis (see Figure 8.39).

LDLs and Cholesterol Metabolism Among many examples of receptor-mediated endocytosis, the first studied and

Figure 8.45 **LDL cholesterol.** Each particle consists of approximately 1500 cholesterol ester molecules, surrounded by a mixed monomolecular layer of phospholipids and cholesterol, and a single molecule of the protein apolipoprotein B, which interacts specifically with the LDL receptor projecting from the plasma membrane.

best understood is one that provides animal cells with exogenous cholesterol. Cholesterol is used by animal cells as an essential part of their plasma membranes and as a precursor to steroid hormones. Cholesterol is a hydrophobic molecule that is transported in the blood as part of huge lipoprotein complexes, such as the **low-density lipoprotein (LDL)** shown in Figure 8.45. Each LDL particle contains a central core of about 1500 cholesterol molecules esterified to long-chain fatty acids. The core is surrounded by a single layer of phospholipids, unesterified cholesterol molecules, and a single copy of a large protein,

called *apolipoprotein B-100*, which binds specifically to LDL receptors on the surfaces of cells.

LDL receptors are transported to the plasma membranes of cells, where they become concentrated in coated pits, even in the absence of ligand. As a result, the receptors are on the cell surface ready to take up the blood-borne lipoproteins if they should become available. Once the LDL is bound to a coated pit (Figure 8.41*a*), the pit invaginates to form a coated vesicle, the clathrin coat is disassembled, and the LDL receptors are recycled back to the plasma membrane, as depicted in Figure 8.39*a*. Meanwhile, the LDL particles are delivered to lysosomes, where the protein component is degraded and the cholesterol is released for use by the cell in membrane assembly or other metabolic processes (e.g., steroid hormone formation). The number of LDL receptors present on a given cell is modulated by the cell's metabolic needs. If a cell is actively growing and synthesizing large amounts of membrane or is a specialized endocrine cell that is producing steroid hormones, the number of receptors increases and greater numbers of LDL particles are taken in from the surrounding medium. The steps that led to the discovery of receptor-mediated endocytosis and LDL internalization are described in detail in the accompanying Experimental Pathways.

The level of LDL in the blood correlates strongly with the development of atherosclerosis, a condition characterized by the narrowing of major arteries. Arterial occlusion is a complex and poorly understood process that includes the deposition of LDL-containing plaques within the inner walls of the vessels (Figure 8.46). In addition to reducing blood flow, atherosclerotic plaques act as sites for the formation of blood clots, which can block the flow of blood through a vessel. Blood clots that block the coronary arteries are the leading cause of myocardial infarction (heart attack). Lowering blood LDL levels is most readily accomplished by administration of drugs

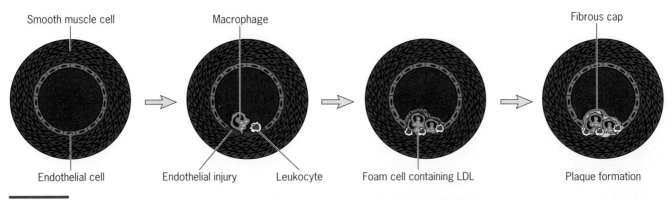

Figure 8.46 **A model of atherosclerotic plaque formation.** Plaque formation is initiated by various types of injury to the endothelial cells that line the vessel, including damage inflicted by oxygen free radicals that chemically alter the LDL-cholesterol particles. The injured endothelium acts as an attractant for white blood cells (leukocytes) and macrophages, which migrate beneath the endothelium and remain there. The macrophages

ingest the oxidized LDL, which becomes deposited in the cytoplasm as cholesterol-rich fatty droplets. These cells are referred to as macrophage foam cells. Substances released by the macrophages stimulate the proliferation of smooth muscle cells, which produce a dense, fibrous connective tissue matrix (fibrous cap) that bulges into the arterial lumen.

called *statins* (e.g., lovastatin and pravastatin) that block HMG CoA reductase, a key enzyme in the synthesis of cholesterol (page 327). When blood cholesterol levels are lowered, the frequency of heart attack is reduced.

LDLs are not the only cholesterol-transporting agents in the blood. **HDLs (high-density lipoproteins)** have a similar construction but contain a different protein (apolipoprotein A-I) and play a different physiological role in the body. LDL serves primarily to carry cholesterol molecules from the liver, where they are synthesized and packaged, through the blood to the body's cells. HDL carries cholesterol in the opposite direction. Excess cholesterol is transferred from the plasma membrane of the body's cells directly to circulating HDL particles, which carries cholesterol to the liver. Just as high blood levels of LDL are associated with increased risk of heart disease, high blood levels of HDL are associated with decreased risk. But the matter is not this simple. For example, HDL particles are present in different sizes, and persons with higher levels of smaller-sized HDL are at greater risk of coronary disease. In addition, there is an enzyme (cholesteryl ester transfer protein, CETP) that moves cholesterol molecules from HDL to LDL. CETP has become a focus of research following the discovery of a population of Japanese families whose members routinely live more than 100 years and carry mutations in the *CETP* gene. A vaccine has been developed against the CETP protein, which has shown promise in elevating HDL levels in test animals with coronary artery disease. Clinical tests of the vaccine in humans are underway.

REVIEW ?

1. Describe the steps that occur between the binding of an LDL particle to the plasma membrane of a cell and the entry of cholesterol molecules into the cytosol.

2. Describe the molecular structure of clathrin and the relationship between its structure and function.

8.9 POSTTRANSLATIONAL UPTAKE OF PROTEINS BY PEROXISOMES, MITOCHONDRIA, AND CHLOROPLASTS

We have seen in this chapter that protein trafficking within the cell is governed by (1) sorting signals, such as the signal peptide of secreted proteins or mannose phosphate groups of lysosomal enzymes, and (2) receptors that recognize these signals and deliver the protein containing them to the proper compartment. Four of the cell's major organelles—nuclei, mitochondria, chloroplasts, and peroxisomes—import proteins through one or more outer boundary membranes. As in the case of the rough ER, proteins that are imported by these organelles contain amino acid sequences that serve as addresses that are recognized by receptors at the organelle's outer membrane. Unlike the rough ER, which generally imports its proteins cotranslationally, the proteins of these other organelles are imported posttranslationally, that is, following their complete synthesis in the cytosol.

The import of proteins into the nucleus is a specialized topic and will be discussed separately in Section 12.1. Import of proteins into peroxisomes, mitochondria, and chloroplasts will be discussed in the following sections.

Uptake of Proteins into Peroxisomes

Peroxisomes are very simple organelles having only two subcompartments in which an imported protein can be placed: either the boundary membrane or the internal matrix (page 211). Proteins destined for a peroxisome possess a *peroxisomal targeting signal*, either a *PTS* for a peroxisomal matrix protein or an *mPTS* for a peroxisomal membrane protein. Several different PTSs, mPTSs, and PTS receptors have been identified. PTS receptors apparently bind to peroxisome-destined proteins in the cytosol and shuttle them to the peroxisome membrane prior to import. Unlike mitochondria and chloroplasts, whose imported proteins must assume an unfolded state, peroxisomes are somehow able to import peroxisomal matrix proteins in their native, folded conformation. The mechanism by which peroxisomes are able to accomplish this daunting assignment remains a mystery.

Uptake of Proteins into Mitochondria

Mitochondria have four subcompartments into which proteins can be delivered: an outer mitochondrial membrane (OMM), inner mitochondrial membrane (IMM), intermembrane space, and matrix (see Figure 5.24). Although mitochondria synthesize a few of their own integral membrane polypeptides (13 in mammals), the vast majority of the organelle's proteins are encoded by the nuclear genome, synthesized in the cytosol, and imported posttranslationally. The targeting signal of most mitochondrial preproteins (called the *presequence*) consists of a segment located at the N-terminus of the molecule that contains a number of positively charged residues. The N-terminal targeting sequence is ultimately removed following import by a mitochondrial processing protease.

Before a preprotein can enter a mitochondrion, several events are thought to take place. First, the preprotein must be presented to the mitochondrion in a relatively extended, or unfolded, state. Several different molecular chaperones have been implicated in preparing polypeptides for uptake into mitochondria, including ones that specifically direct mitochondrial preproteins to the cytosolic surface of the OMM (Figure 8.47a).

The OMM contains a protein import complex, the **TOM complex**, which includes a receptor that binds mitochondrial preproteins and a protein-lined channel through which unfolded polypeptides are translocated

(a)

(b)

0.2 μm

Figure 8.47 **Importing proteins into a mitochondrion.**
(*a*) Proposed steps taken by a protein imported into the mitochondrial matrix. The polypeptide is targeted to a mitochondrion by a targeting sequence at the N-terminus, which is ultimately removed within the matrix. Hsp70 and mHsp70 are molecular chaperones. The cytosolic Hsp70 unfolds the polypeptide prior to its entry into the mitochondrion, whereas the mHsp70 plays a role in the movement of the polypeptide into the matrix as discussed in the text. Once the protein has fully entered the matrix, it is transferred to Hsp60 where refolding occurs. Hsp60 is a cylindrical chamber whose structure and function is similar to the bacterial GroEL described on page 76. (*b*) Electron micrograph of a mitochondrion that had been importing mitochondrial precursor proteins in vitro. Sites where proteins were being imported were revealed by labeling the proteins with gold particles that are visible in the electron microscope. The particles are localized at sites where the inner and outer membrane are in close proximity. (A: AFTER D. VOET AND J. G. VOET, BIOCHEMISTRY, 2D ED. WILEY, 1995; B: FROM MARTIN SCHWAIGER, VOLKER HERZOG, AND WALTER NEUPERT, J. CELL BIOL. 105:243, 1987; BY COPYRIGHT PERMISSION OF THE ROCKEFELLER UNIVERSITY PRESS.)

into or through the outer membrane (Figure 8.47*a*). Proteins that are destined for the intermembrane space are simply translocated through the TOM complex into this compartment. However, most of the proteins that pass through the OMM are bound for the mitochondrial ma-

trix, which requires that they pass through a second barrier—the inner mitochondrial membrane (IMM). The preprotein is translocated from the cytosol into the mitochondrial matrix at sites where the outer and inner mitochondrial membranes come into close proximity (Figure

8.47*b*). Like the OMM, the IMM also contains a protein import complex, called the **TIM complex**, which includes a receptor for the preprotein and a channel for its translocation through the membrane.[7] Movement into the matrix is powered by the electric potential across the IMM acting on the positively charged targeting signal; if the potential is dissipated by addition of a drug such as DNP (page 203), translocation ceases and the polypeptide remains trapped within the membrane, spanning both the TOM and TIM complexes.

As it enters the matrix, a polypeptide interacts with mitochondrial chaperones (e.g., mHsp70, Figure 8.47*a*) that mediate entry into the aqueous compartment. Two alternate mechanisms have been proposed to explain the general action of chaperones involved in the movement of proteins across membranes, which is a widespread phenomenon. According to one view, these proteins act as force-generating motors that use energy derived from ATP hydrolysis to "pull" the unfolded polypeptide through the translocation pore. According to the alternate view, these chaperones aid in the diffusion of the polypeptide across the membrane. Diffusion is a random process in which a molecule can move in any available direction. Consider what would happen if an unfolded polypeptide had entered a translocation pore in the mitochondrial membrane and had "poked its head" into the matrix. Then consider what would happen if a chaperone residing on the inner surface of the membrane were able to bind the protruding polypeptide in such a way that it blocked the diffusion of the polypeptide back through the pore and into the cytosol, but did not block its diffusion further into the matrix. As the polypeptide moved further into the matrix, it would be bound repeatedly by the chaperone and at each stage prevented from diffusing backward. This mechanism of chaperone action is referred to as *biased diffusion*, and the chaperone is said to be acting as a "Brownian ratchet"; the term "Brownian" implies random diffusion, and a "ratchet" is a tool that allows movement in only one direction. This aspect of chaperone action is a major subject of current speculation.

Uptake of Proteins into Chloroplasts

Chloroplasts have six subcompartments into which proteins can be delivered: an inner and outer envelope membrane and intervening intermembrane space, as well as the stroma, thylakoid membrane, and lumen (Figure 8.48). Chloroplast and mitochondrial import mechanisms exhibit many similarities, although their translocation machineries have evolved independently. As in the mitochondria,

1. the vast majority of chloroplast proteins are imported from the cytosol,

2. the outer and inner envelope membranes contain distinct translocation complexes (*Toc* and *Tic complexes*, respectively) that work together during import,

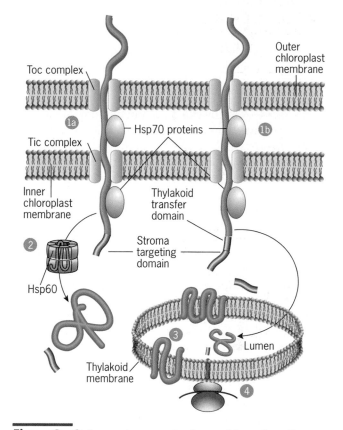

Figure 8.48 Importing proteins into a chloroplast. Proteins encoded by nuclear genes are synthesized in the cytosol and imported through protein-lined pores in both membranes of the outer chloroplast envelope (step 1). Proteins destined for the stroma (step 1a) contain a stroma-targeting domain at their N-terminus, whereas proteins destined for the thylakoid (step 1b) contain both a stroma-targeting domain and a thylakoid-transfer domain at their N-terminus. Stromal proteins remain in the stroma (step 2) following translocation through the outer envelope and removal of their single targeting sequence. The presence of the thylakoid transfer domain causes thylakoid proteins to be translocated either into or completely through the thylakoid membrane (step 3). Many of the proteins of the thylakoid membrane are encoded by chloroplast genes and synthesized by chloroplast ribosomes that are bound to the outer surface of the thylakoid membrane (step 4).

3. chaperones aid in the unfolding of the polypeptides in the cytosol and folding of the proteins in the chloroplast, and

4. preproteins destined for the chloroplast are synthesized with a removable N-terminal sequence (termed the *transit peptide*).

The transit peptide does more than simply target a polypeptide to a chloroplast: it provides an "address" that localizes the polypeptide to one of several possible subcompartments within the organelle (Figure 8.48). All proteins translocated through the chloroplast envelope contain a *stroma targeting domain* as part of their transit peptide, which guarantees that the polypeptide will enter the stroma. Once in the stroma, the stroma targeting domain

[7]Integral membrane proteins targeted to the IMM are translocated into the membrane by a separate TIM complex, which is not discussed.

is removed by a processing protease located in that compartment. Those polypeptides that belong in a thylakoid membrane or thylakoid lumen bear an additional segment in their transit peptide, called the *thylakoid transfer domain*, that dictates entry into the thylakoids. Several distinct pathways have been identified that translocate different proteins across the thylakoid membrane into the lumen. These pathways exhibit striking similarities to transport systems in bacterial cells, the presumed ancestors of chloroplasts. Many of the proteins that reside within the thylakoid membrane are encoded by chloroplast genes and synthesized on membrane-bound ribosomes of the chloroplast as illustrated in Figure 8.48, step 4.

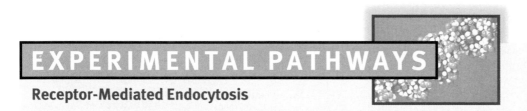

EXPERIMENTAL PATHWAYS

Receptor-Mediated Endocytosis

Embryonic development begins with the fusion of a minuscule sperm and a much larger egg. Eggs develop from oocytes, which accumulate yolk that has been synthesized elsewhere in the body of the female. How are the high-molecular-weight yolk proteins able to enter the oocyte? In 1964, Thomas Roth and Keith Porter of Harvard University reported on the mechanism by which yolk proteins might be taken into the oocytes of mosquitoes.[1] They noticed that during stages of rapid oocyte growth, there was a dramatic increase in the number of pitlike depressions seen on the surface of the oocyte. The pits, which were formed by invaginations of the plasma membrane, were covered on their inner surface by a bristly coat. In a farsighted proposal, Roth and Porter postulated that yolk proteins were specifically adsorbed onto the outer surface of the membrane of the coated pits, which would then invaginate as coated vesicles. The coated vesicles would lose their bristly coats and fuse with

one another to produce the large, membrane-bound yolk bodies characteristic of the mature oocyte.

The first insight into the structure of coated vesicles was obtained in 1969 by Toku Kanaseki and Ken Kadota of the University of Osaka.[2] Electron microscopic examination of a crude vesicle fraction isolated from guinea pig brains showed that coated vesicles were covered by a polygonal basketwork (Figure 1). These investigators suggested that the coatings were an apparatus to control the infolding of the plasma membrane during vesicle formation.

The first studies of the biochemical nature of the vesicle coat were published in 1975 by Barbara Pearse of the Medical Research Council in Cambridge, England.[3] Pearse developed a procedure in which membrane vesicles from pig brains were centrifuged through a succession of sucrose density gradients until a purified fraction of coated vesicles was obtained. Protein from the coated vesicles was solubilized and

(a)

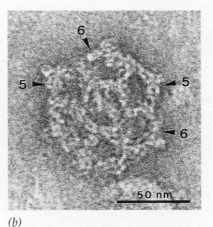

(b)

Figure 1 *(a)* A handmade model of an "empty basket" that would form the surface lattice of a coated vesicle. (b) A high-power electron micrograph of an empty basket. (A–B: FROM TOKU KANASEKI AND KEN KADOTA, J. CELL BIOL. 42:216, 1969, BY COPYRIGHT PERMISSION OF THE ROCKEFELLER UNIVERSITY PRESS.)

fractionated by SDS–polyacrylamide gel electrophoresis (SDS–PAGE, Section 18.7). The results indicated that the coat contained one predominant protein species having a molecular mass of approximately 180,000 daltons. Pearse named the protein clathrin. She found the same protein (based on molecular mass and peptide mapping) in preparations of coated vesicles that were isolated from several different types of cells obtained from more than one animal species.[4]

As the research described above was being conducted, an apparently independent line of research was begun in 1973 in the laboratories of Michael Brown and Joseph Goldstein of the University of Texas Medical School in Dallas. Brown and Goldstein had become interested in the inherited condition *familial hypercholesterolemia (FH)*. People who were homozygous for the defective gene (the *FH* allele) had profoundly elevated levels of serum cholesterol (800 mg/dl versus 200 mg/dl for a normal person), invariably developed severely blocked (atherosclerotic) arteries and usually died from heart attack before the age of 20. At the time, very little was known about the fundamental physiological or biochemical defects in this disorder.

Brown and Goldstein began their studies of FH by examining cholesterol metabolism in cultured fibroblasts derived from the skin of normal and FH-afflicted individuals. They found that the rate-controlling enzyme in cholesterol biosynthesis, HMG CoA reductase, could be inhibited in normal fibroblasts by cholesterol-containing lipoproteins (such as LDL) placed in the medium (Figure 2).[5] Thus, the addition of LDL to the culture medium in which normal fibroblasts were growing led to a decreased level of activity of HMG CoA reductase and a corresponding decrease in the synthesis of cholesterol by the fibroblasts. When HMG CoA reductase levels were measured in FH-derived fibroblasts, they were found to be 40 to 60 times that of normal fibroblasts.[6] In addition, enzyme activity in the FH fibroblasts was totally unaffected by the presence of LDL in the medium (Figure 3).

How could lipoproteins in the medium affect the activity of an enzyme in the cytoplasm of cultured cells? To answer this question, Brown and Goldstein initiated studies on the interaction between the cells and the lipoproteins. They added radioactively labeled LDL to culture dishes containing a single layer of fibroblasts derived from either FH-afflicted or normal human subjects.[7] The normal fibroblasts bound the labeled LDL molecules with high affinity and specificity, but the mutant cells showed virtually no ability to bind these lipoprotein molecules (Figure 4). These results indicated that normal cells have a highly specific receptor for LDL and that this receptor was defective or missing in cells from patients with FH.

To visualize the process of receptor binding and internalization, Brown and Goldstein teamed up with Richard Anderson, who had been studying cellular structure with the electron microscope. The group incubated fibroblasts from normal and FH subjects with LDL that had been covalently linked to the iron-containing protein ferritin. Because of the iron atoms, ferritin molecules are able to scatter a beam of electrons and thus can be visualized in the electron microscope. When normal fibroblasts were incubated with LDL-ferritin at

Figure 3 Fibroblast cells from either a control subject (closed circles) or a patient with homozygous FH (open circles) were grown in dishes containing 10 percent fetal calf serum. On day 6 (which corresponds to 0 hours on the graph), the medium was replaced with fresh medium containing 5 percent human lipoprotein-deficient plasma. At the indicated time, extracts were prepared and HMG CoA reductase activity was measured. If we look at the control cells, it is apparent that at the beginning of the monitoring period the cells have very little enzyme activity, because the medium had contained enough cholesterol-containing lipoproteins that the cells did not need to synthesize their own. Once the medium was changed to the lipoprotein-deficient plasma, the cells were no longer able to use cholesterol from the medium and thus increased the amount of enzyme within the cell. In contrast, the cells from the FH patients showed no response to either the presence or absence of lipoproteins in the medium. (FROM J. L. GOLDSTEIN AND M. S. BROWN, PROC. NAT'L. ACAD. SCI. U. S. A. 70:2805, 1973.)

Figure 2 HMG CoA reductase activity in normal fibroblasts was measured following addition of the lipoprotein fraction of calf serum (open squares), unfractionated calf serum (closed circles), or nonlipoprotein fraction of calf serum (open triangles). It is evident that the lipoproteins greatly depress the activity of the enzyme, while the nonlipoproteins have little effect. (FROM M. S. BROWN ET AL., PROC. NAT'L. ACAD. SCI. U. S. A. 70:2166, 1973.)

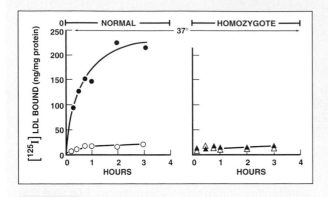

Figure 4 Time course of radioactive [^{125}I]-labeled LDL binding to cells from a normal subject (circles) and a homozygote with FH (triangles) at 37°C. The cells were incubated in a buffer containing 5 mg/ml [^{125}I] LDL in the presence (open circles and triangles) and absence (closed circles and triangles) of 250 mg/ml of nonradioactive LDL. It is evident that in the absence of added nonradioactive LDL, the normal cells bind significant amounts of the labeled LDL, while the cells from the FH patients do not. The binding of labeled LDL is greatly reduced in the presence of nonradioactive LDL because the nonlabeled lipoproteins compete with the labeled ones for binding sites. Thus, the binding of the lipoprotein to the cells is specific (i.e., it is not the result of some nonspecific binding event). (FROM M. S. BROWN AND J. L. GOLDSTEIN, PROC. NAT'L. ACAD. SCI. U. S. A. 71:790, 1974.)

4°C, a temperature at which ligands can bind to the cell surface but cannot be internalized, LDL-ferritin particles were seen to be bound to the cell surface. Close examination revealed that the LDL particles were not randomly scattered over the cell surface but were localized to short (0.5 μm) segments of the plasma membrane where the membrane was indented and coated by a "fuzzy" material (Figure 5).[8] These segments of the membrane were the coated pits originally described by Roth and Porter and had since been seen in a variety of cell types. Although cells from patients with FH were found to have a similar number of coated pits on their surface, no LDL-ferritin was bound to these mutant cells. The results supported the proposal that the mutant FH allele encoded a receptor that was unable to bind LDL. Subsequent electron microscopic studies on the internalization of LDL-ferritin revealed the endocytic pathway by which these lipoprotein particles were internalized, as described in the text.[9]

Based on these results, the group postulated that the rapid internalization of receptor-bound LDL is strictly dependent on the localization of LDL receptors in coated pits. It follows from this postulate that, if an LDL receptor failed to become localized within a coated pit, it would be unable to deliver its bound ligand to cellular lysosomes and thus would be unable to affect cholesterol biosynthesis within the cell. At about this time, an LDL receptor with a different type of mutation was discovered. LDL receptors bearing this new defect (known as the J. D. mutation after the patient in which it occurred) bound normal amounts of radioactively labeled LDL, yet the receptor-bound lipoprotein failed to be internal-

ized and consequently was not delivered to cytoplasmic lysosomes for processing.[10] Anderson and coworkers postulated that the LDL receptor was a transmembrane protein that normally became localized in coated pits, because its cytoplasmic domain was specifically bound by a component of the coated pits, possibly clathrin (but later identified as a likely subunit of the adaptor, as discussed below). Because of a defect in its cytoplasmic domain, the J. D. mutant receptor was unable to become localized in a cell's coated pits. People with this mutation exhibit the same phenotype as patients whose receptors are unable to bind LDL.

Subsequent studies determined that the normal LDL receptor is a transmembrane glycoprotein of 839 amino acids, with the 50 amino acids at the C-terminal end of the protein extending inward from the membrane as a cytoplasmic domain. Analysis of the J. D. mutant receptor revealed that the protein contained a single amino acid substitution: a tyrosine residue normally located at position 807 was replaced by a cysteine.[11] This single alteration in amino acid sequence obliterated the ability of the protein to become concentrated in coated pits.

Over the next few years, attention turned to the amino acid sequences of the cytoplasmic tails of other receptors that became localized in coated pits. Was there a common "internalization signal?" It was found that a number of different receptors subject to receptor-mediated endocytosis contained an essential tyrosine residue, most often situated within the sequence YXXφ, in which Y is the tyrosine, X can be any amino acid, and φ is an amino acid with a bulky hydrophobic side chain. Studies have shown that the YXXφ sequence of the receptor binds to one of the polypeptides (the μ subunit) of the AP2 adaptors (Figure 8.43).[12] X-ray crystallographic studies have revealed the nature of the interaction between the adaptor and internalization signal.[13] Included within the μ subunit, which consists almost totally of β sheets, are two hydrophobic pockets, one that binds the tyrosine residue and the other that

Figure 5 Electron micrograph showing the binding of LDL to coated pits of human fibroblasts. The LDL is made visible by conjugating the particles to electron-dense, iron-containing ferritin. (FROM R.G.W. ANDERSON, M. S. BROWN, AND J. L. GOLDSTEIN, CELL 10:356, 1977, BY COPYRIGHT PERMISSION OF CELL PRESS.)

Figure 6 X-ray crystallographic structure of the internalization signal of a membrane receptor (shown in purple) bound to the μ chain of the AP adaptor (shown in gray). The tyrosine (Y) and leucine (L) residues of the receptor fit into hydrophobic binding pockets of the μ subunit. (COURTESY OF DAVID J. OWEN AND PHILLIP R. EVANS.)

binds the bulky hydrophobic side chain of the internalization signal (Figure 6). This type of interaction is reminiscent of the connection between a two-pronged plug at the end of an electric cord and the socket into which it fits. Once the residues of the internalization signal have become bound to the adaptor, additional attractive forces are generated through the formation of hydrogen bonds between the backbone of the receptor and one of the β strands of the μ subunit. As a result of these various intermolecular contacts, the adaptor is trapped in coated pits prior to endocytosis.[14]

References

1. ROTH, T. F. & PORTER, K. R. 1964. Yolk protein uptake in the oocyte of the mosquito *Aedes aegypti. J. Cell Biol.* 20:313–332.

2. KANASEKI, T. & KADOTA, K. 1969. The "vesicle in a basket." *J. Cell Biol.* 42:202–220.

3. PEARSE, B.M.F. 1975. Coated vesicles from pig brain: Purification and biochemical characterization. *J. Mol. Biol.* 97:93–96.

4. PEARSE, B.M.F. 1976. Clathrin: A unique protein associated with the intracellular transfer of membrane by coated vesicles. *Proc. Nat'l. Acad. Sci. U. S. A.* 73:1255–1259.

5. BROWN, M. S., DANA, S. E., & GOLDSTEIN, J. L. 1973. Regulation of HMG CoA reductase activity in human fibroblasts by lipoproteins. *Proc. Nat'l. Acad. Sci. U. S. A.* 70:2162–2166.

6. GOLDSTEIN, J. L. & BROWN, M. S. 1973. Familial hypercholesterolemia: Identification of a defect in the regulation of HMG CoA reductase activity associated with overproduction of cholesterol. *Proc. Nat'l. Acad. Sci. U. S. A.* 70:2804–2808.

7. BROWN, M. S. & GOLDSTEIN, J. L. 1974. Familial hypercholesterolemia: Defective binding of lipoproteins to cultured fibroblasts associated with impaired regulation of HMG CoA reductase activity. *Proc. Nat'l. Acad. Sci. U. S. A.* 71:788–792.

8 ANDERSON, R.G.W., GOLDSTEIN, J. L., & BROWN, M. S. 1976. Localization of low density lipoprotein receptors on plasma membrane of normal human fibroblasts and their absence in cells from a familial hypercholesterolemia homozygote. *Proc. Nat'l. Acad. Sci. U. S. A.* 73:2434–2438.

9. ANDERSON, R.G.W., BROWN, M. S., & GOLDSTEIN, J. L. 1977. Role of the coated endocytic vesicle in the uptake of receptor-bound low density lipoprotein in human fibroblasts. *Cell* 10:351–364.

10. ANDERSON, R.G.W., GOLDSTEIN, J. L., & BROWN, M. S. 1977. A mutation that impairs the ability of lipoprotein receptors to localise in coated pits on the cell surface of human fibroblasts. *Nature* 270:695–699.

11. DAVIS, C. G., ET AL. 1986. The J. D. mutation in familial hypercholesterolemia: Amino acid substitution in cytoplasmic domain impedes internalization of LDL receptors. *Cell* 45:15–24.

12. OHNO, H., ET AL. 1995. Interaction of tyrosine-based sorting signals with clathrin-associated proteins. *Science* 269:1872–1874.

13. OWEN, D. J. & EVANS, P. R. 1998. A structural explanation for the recognition of tyrosine-based endocytic signals. *Science* 282:1327–1332.

14. PEARSE, B. M. F., ET AL. 2000. Clathrin coat construction in endocytosis. *Curr. Opin. Struct. Biol.* 10:220–228.

SYNOPSIS

The cytoplasm of eukaryotic cells contains a system of membranous organelles, including the endoplasmic reticulum, Golgi complex, and lysosomes, that are functionally and structurally related to one another and to the plasma membrane. These various membranous organelles are part of a dynamic, integrated network in which materials are shuttled back and forth as part of transport vesicles that form by budding from one compartment and fuse with another one. A biosynthetic (secretory) pathway moves proteins from their site of synthesis in the ER through the Golgi complex to their final destination (an organelle, the plasma membrane, or the extracellular space), whereas an endocytic pathway moves material in the opposite direction from the plasma membrane or extracellular space to the cell interior. (p. 280)

The endoplasmic reticulum (ER) is a system of tubules, cisternae, and vesicles that divides the cytoplasm into a luminal space within the ER membranes and a cytosolic space outside the membranes. The ER is divided into two broad types: the rough ER (RER), which occurs typically as flattened cisternae and whose membranes have attached ribosomes, and the smooth ER (SER), which occurs typically as tubular compartments and whose membranes lack ribosomes. The functions of the SER include the synthesis of steroid hormones, detoxification of a wide variety of organic compounds, and sequestration of calcium ions. The functions of the RER include the synthesis of proteins to be secreted, lysosomal proteins, integral membrane proteins, and membrane lipids. (p. 288)

Proteins to be synthesized on membrane-bound ribosomes of the RER are recognized by a hydrophobic signal sequence, which is usually situated near the N-terminus of the nascent polypeptide. As the signal sequence emerges from the ribosome, it is bound by a signal recognition particle (SRP) that arrests further synthesis and mediates the binding of the complex to the RER membrane. Following binding, the SRP is released from the membrane, and the nascent polypeptide passes through a protein-lined pore in the ER membrane and into the ER lumen. Lysosomal and secretory proteins are translocated into the lumen in their entirety, whereas membrane proteins become embedded in the lipid bilayer by virtue of one or more hydrophobic transmembrane sequences. Proteins that are not folded correctly are translocated back into the cytosol and destroyed. Once a newly synthesized protein is situated in either the lumen or the membrane of the RER, the protein can be moved from that location to specific destinations along the biosynthetic pathway. (p. 290)

Most of the lipids of a cell's membranes are also synthesized in the ER and moved from that site to various destinations. Phospholipids are synthesized at the cytosolic face of the ER and inserted into the outer leaflet of the ER membrane. Some of these molecules are subsequently flipped into the opposite leaflet. The lipid composition of membranes is subject to modification in a number of ways. (p. 294)

The addition of sugars (glycosylation) to the asparagine residues of proteins begins in the rough ER and continues in the Golgi complex. The sequences of sugars that make up the oligosaccharide chains of glycoproteins are determined by the types and locations of members of a large family of glycosyltransferases. The carbohydrate chains are assembled in the ER one sugar at a time and then transferred as a unit from the dolichol carrier to an asparagine residue of the polypeptide. Almost as soon as the carbohydrate block is transferred, it begins to be modified—first by removal of the terminal glucose residues. (p. 295)

The Golgi complex functions as a processing plant, modifying the membrane components and cargo synthesized in the ER before it moves on to its target destination. The Golgi complex is also the site of synthesis of the complex polysaccharides that make up the matrix of the plant cell wall. Each Golgi complex consists of a stack of flattened, platelike cisternae with dilated rims and associated vesicles and tubules. Materials enter the stack at the *cis* face and are modified as they move by vesicle transport toward the opposite, or *trans*, face. Once the proteins reach the *trans* Golgi network (TGN) at the end of the stack, they are ready to be sorted

and targeted for delivery to their ultimate cellular or extracellular destination. (p. 299)

Most if not all of the vesicles that transport material through the endomembrane system are encased initially in a protein coat. Several distinct types of coated vesicles have been identified. COPII-coated vesicles carry materials from the ER to the Golgi complex. COPI-coated vesicles carry materials in the opposite (retrograde) direction, from the Golgi back to the ER, and may be involved in numerous other transport steps. Clathrin-coated vesicles carry materials from the TGN to the endosomes, lysosomes, and central vacuole (in plants). Clathrin-coated vesicles are also part of the endocytic pathway, carrying materials from the plasma membrane to endosomes and lysosomes. (p. 303)

Each compartment along the biosynthetic or endocytic pathway has a characteristic protein composition. Resident proteins tend to be retained in that particular compartment and retrieved if they escape to other compartments. Vesicles that bud from a donor compartment are thought to possess specific proteins in their membrane (members of a family of v-SNAREs) that recognize a complementary protein (members of a family of t-SNAREs) located in the target (acceptor) compartment. The docking of the two membranes is mediated by the SNAREs. (p. 306)

The sorting of proteins in the biosynthetic pathway occurs to a large extent in the last of the Golgi compartments, the TGN. The TGN is the source of clathrin-coated vesicles, which contain particular membrane proteins that target the vesicle to a particular destination. Clathrin-coated vesicles carry specific cargoes by virtue of adaptors that form a layer between the outer clathrin coat and the vesicle membrane. Lysosomal enzymes produced in the rough ER are sorted in the TGN and targeted for lysosomes. As the various lysosomal enzymes pass through the Golgi cisternae, they are recognized by enzymes that produce phosphorylated mannose residues on the core oligosaccharide. The mannose phosphate residues are recognized by membrane receptors in the TGN, which causes the enzymes to be packaged into vesicles that are targeted to a lysosome. (p. 307)

Lysosomes are membrane-bound organelles that contain a diverse array of acid hydrolases capable of digesting virtually every type of biological macromolecule. Among their numerous roles, lysosomes degrade materials, such as bacteria and debris, that are brought into the cell by phagocytosis; they degrade aging cytoplasmic organelles by a process called autophagy; and they digest various macromolecules that are delivered by means of endosomes by receptor-mediated endocytosis. (p. 311)

The vacuoles of plants carry out a diverse array of activities. Vacuoles serve as a temporary storehouse for solutes and macromolecules; they contain toxic compounds used in defense; they provide a container for cellular wastes; they maintain a hypertonic compartment that exerts turgor pressure against the cell wall; and they are sites of intracellular digestion mediated by acid hydrolases. (p. 313)

Phagocytosis is the uptake of particulate matter, and endocytosis is the uptake of fluid, dissolved solutes, and suspended macromolecules. Phagocytosis can function as a feeding mechanism or as a cellular system of defense. In receptor-mediated endocytosis, specific ligands are bound to

plasma membrane receptors. The receptors collect in pits of the membrane that are coated on their cytoplasmic surface by a polygonal scaffold of clathrin molecules. Rearrangement of the clathrin lattice facilitates the formation of coated vesicles. The vesicles lose their coat, and deliver their contents to a lysosome. (p. 314)

ANALYTIC QUESTIONS

1. Which, if any, of the following polypeptides would be expected to lack a signal peptide: collagen, acid phosphatase, hemoglobin, ribosomal proteins, glycophorin, proteins of the tonoplast?

2. Suppose you suspected that a patient might be suffering from I-cell disease (page 315). How could you decide if this diagnosis was accurate using cells that were cultured from the patient?

3. In what cellular compartment would a glycoprotein be expected to have its greatest mannose content? Its greatest N-acetylglucosamine content? Its greatest sialic acid content?

4. Chondroitin sulfate is a glycosaminoglycan. Would you expect to find this material within the lumen of the RER? Why or why not?

5. What are two proteins you would expect to find as integral components of the RER membrane that would be absent from the SER? What are two proteins of the SER that would not be present in the RER?

6. Suppose you wanted to study the process of regulated secretion in a cell that lacked mature secretory granules, that is, vesicles containing secretory material that was ready to be discharged. How might you obtain a cell that lacked such granules?

7. It was described on page 316 how glucocerebrosidase could be prepared that bore mannose residues at the end of its surface oligosaccharides rather than the usual sugar, sialic acid. A similar version of glucocerebrosidase with exposed mannose residues is being produced by recombinant DNA technology using a special line of cells. What characteristics do you think these cells might exhibit? Clue: Information to answer this question is provided in Figure 8.22.

8. Autoradiography depends upon particles emitted from radioactive atoms striking a photographic emulsion that lies on top of the tissue section. When the emulsion is developed, the site where the particle struck the emulsion appears as a silver grain, as in Figure 8.3α. How do you think the thickness of the section might affect the resolution of the technique, that is, the ability to locate the precise site in the cell where radioactivity is incorporated?

9. In which part of a cell would you expect the following compounds to be first incorporated: [^3H]leucine, [^3H]sialic acid, [^3H]mannose, [^3H]choline, [^3H]glucuronic acid (a precursor of GAGs), [^3H]pregnenolone (a precursor of steroid hormones), [^3H]rhamnose in a plant cell (rhamnose is a precursor of pectins)?

10. Which of the following cells would you expect to be engaged most heavily in bulk-phase endocytosis: (a) an erythrocyte, (b) a pancreatic acinar cell, (c) a macrophage? Why?

11. Would you expect the properties of the cisternal side of Golgi membranes to be more similar to the extracellular or cytosolic side of the plasma membrane? Why?

12. Which compartment(s) of a cell is associated with each of the following: clathrin, calcium ions in a skeletal muscle cell, dolichol phosphate, ribonuclease and lipase, digitalis, LDL receptors, COPI proteins, COPII proteins, unbound SRPs?

13. If a slice of pancreatic tissue had been incubated in [^3H]leucine continuously for 2 hours, where in the acinar cells would you expect to find incorporated radioactivity?

14. If you were to add a drug that interfered with the ability of ribosomes to bind to mRNA, what effect would this be expected to have on the structure of the RER?

15. Not all receptors that carry out RME are located in coated pits prior to binding ligand, yet they too become concentrated in coated pits prior to internalization. How do you suppose the binding of a ligand to a receptor would cause it to become concentrated into a coated pit?

16. It was noted in the text that lysosomes lack a distinctive appearance in the electron microscope. How might you determine whether a particular vacuole is in fact a lysosome?

17. Can you design any experiments to test the opposing models of Golgi function depicted in Figure 8.23? Suppose you were able to inject a relatively large, inert, fluorescent bead into the lumen of a *cis* cisterna of the Golgi complex of a cultured cell. How might you be able to use this procedure to evaluate the cisternal maturation versus vesicular transport models?

ADDITIONAL READING

Cytomembranes and Membrane Trafficking

ALLAN, B. B. & BALCH, W. E. 1999. Protein sorting by directed maturation of Golgi compartments. *Science* 285:63–66.

ARIDOR, M. & BALCH, W. E. 1999. Integration of endoplasmic reticulum signaling in health and disease. *Nature Med.* 5:745–751.

BERGER, E. G., ET AL. 1998. Reviews on Golgi complex. *Trends Cell Biol.* vol. 8, #1.

BRITTLE, E. E. & WATERS. M. G. 2000. ER-to-Golgi traffic: This bud's for you. *Science* 289:403–404.

BRODSKY, J. L. & MCCRACKEN, A. A. 1999. ER protein quality control and proteasome-mediated degradation. *Sem. Cell & Develop. Biol.* 10:507–513.

DeCamilli, P. & Warren, eds. 1999. Membranes and sorting. *Curr. Opin. Cell Biol.* vol. 11, #4.

Donaldson, J. G. & Lippincott-Schwartz, J. 2000. Sorting and signaling at the Golgi complex. *Cell* 101:693–696.

Ellgaard, L., et al. 1999. Setting the standards: Quality control in the secretory pathway. *Science* 286:1882–1888.

Featherstone, C. 1998. Coming to grips with the Golgi. *Science* 282:2172–2174.

Hegde, R. S. & Lingappa, V. R. 1999. Regulation of protein biogenesis at the endoplasmic reticulum membrane. *Trends Cell Biol.* 9:132–137.

Herrmann, J. M., et al. 1999. Out of the ER—outfitters, escorts, and guides. *Trends Cell Biol.* 9:5–7, 1999.

Johnson, A. E. & van Waes., M. A. 1999. The translocon: A dynamic gateway at the ER membrane. *Annu. Rev. Cell Dev. Biol.* 15:799–842.

Kirchausen, T. 1999. Adaptors for clathrin-mediated traffic. *Annu. Rev. Cell Dev. Biol.* 15:705–732.

Ladinsky, M. S., et al. 1999. Golgi structure in three dimensions: Functional insights from the normal rat kidney cell. *J. Cell Biol.* 144:1135–1149.

Lewin, D. & Mellman, I. 1998. Sorting out adaptors. *Biochim. Biophys. Acta* 1401:129–145.

Lippincott-Schwartz, J., et al. 2000. Secretory protein trafficking and organelle dynamics. *Annu. Rev. Cell Dev. Biol.* 16:557–590.

Matouschek, A. 2000. Recognizing misfolded proteins in the endoplasmic reticulum. *Nature Struct. Biol.* 7:265–266.

Mellman, I. & Warren, G. 2000. The road taken: Past and future foundations of membrane traffic. *Cell* 100:99–112.

Munro, S. & Kaiser, C., eds. 2000. Membranes and sorting: Vesicles and beyond. *Curr. Opin. Cell Biol.* 12:443–523.

Pelham, H.R.B. & Rothman, J. E. 2000. The debate about transport in the Golgi—two sides of the same coin? *Cell* 102:713–719.

Pilon, M. & Schekman, R. 1999. Protein translocation: How Hsp70 pulls it off. *Cell* 97:679–682.

Scales, S. J., et al. 2000. Coat proteins regulating membrane traffic. *Int. Revs. Cytol.* 195:67–144.

Schekman, R. & Orci, L. 1996. Coat proteins and vesicle budding. *Science* 271:1526–1533.

Schmid, S. L. 1997. Clathrin-coated vesicle formation and protein sorting. *Annu. Rev. Biochem.* 66:511–548.

Smith, C J. & Pearse, B.M.F. 1999. Clathrin: Anatomy of a coat protein. *Trends Cell Biol.* 9:335–338.

Springer, S., et al. 1999. A primer on vesicle budding. *Cell* 97:145–148.

Storrie, B., et al., eds. 1998. Reviews on biosynthetic pathway. *Biochim. Biophys. Acta* 1404:1–270.

Storrie, B., et al. 2000. Breaking the COPI monopoly on Golgi recycling. *Trends Cell Biol.* 10:385–391.

Ungewickell, E. 1999. Clathrin: A good view of a shapely leg. *Curr. Biol.* 9:R32–R35.

Walter, P., et al. 2000. SRP—Where the RNA and membrane worlds meet. *Science* 287:1212–1213.

Warren, G. & Mellman, I. 1999. Bulk flow redux? *Cell* 98:125–127.

Membrane Fusion and Exocytosis

Bock, J. B. & Scheller, R. H. 1999. SNARE proteins mediate lipid bilayer fusion. *Proc. Nat'l Acad. Sci. U. S. A.* 96:12227–12229.

Fernandez-Chacón, R. & Südhof, T. C. 1999. Genetics of synaptic vesicle formation. *Annu. Rev. Physiol.* 61:753–776.

Jahn, R. & Südhof, T. C. 1999. Membrane fusion and exocytosis. *Annu. Rev. Biochem.* 68:863–911.

Lang, J. 1999. Molecular mechanisms and regulation of insulin exocytosis as a paradigm of endocrine secretion. *Eur. J. Biochem.* 259:3–17.

Lin, R. C., & Sheller, R. 2000. Mechanisms of synoptic vesicle exocytosis. *Annu. Rev. Cell Dev. Biol.* 16:19–50.

Pfeffer, S. R. 1999. Transport-vesicle targeting: Tethers before SNAREs. *Nature Cell Biol.* 1:E17–E22.

Scales, S. J., et al. 2000. The specifics of membrane fusion. *Nature* 407:144–146.

Lysosomes and Lysosomal Diseases

Chavany, C. & Jendoubi, M. 1998. Biology and potential strategies for the treatment of G_{M2} gangliosidoses. *Mol. Med. Today* 4:158–165.

Chen, Y.-T. & Amalfitano, A. 2000. Towards a molecular therapy for glycogen storage disease type II (Pompe disease). *Mol. Med. Today* 6:245–251.

Holtzmann, E. 1989. *Lysosomes*. Academic Press.

Neufeld, E. F. 1991. Lysosomal storage diseases. *Annu. Rev. Biochem.* 60:257–280.

Phagocytosis and Endocytosis

Acton, S. L., et al. 1999. The HDL receptor SR-BI: A new therapeutic target for atherosclerosis? *Mol. Med. Today* 5:518–524.

Aderem, A. & Underhill, D. M. 1999. Mechanisms of phagocytosis in macrophages. *Annu. Rev. Immunol.* 17:593–623.

Kirchauser, T. 2000. Cathrin. *Annu. Rev. Biochem.* 69:699–728.

Krieger, M. 1999. Charting the fate of the "good cholesterol:" Identification and characterization of the high-density lipoprotein receptor SR-BI. *Annu. Rev. Biochem.* 68:523–558.

Lowe, M. 2000. Membrane transport: Tethers and TRAPPs. *Curr. Biol.* 10:R407–R409.

Lusis, A. J. 2000. Atherosclerosis. *Nature* 407:233–241.

Marsh, M. & McMahon, H. T. 1999. The structural era of endocytosis. *Science* 285:215–220.

Mukherjee, S., et al. 1997. Endocytosis. *Physiol. Revs.* 77:759–803.

Pearse, B. M. F., et al. 2000. Clathrin coat construction in endocytosis. *Curr. Opin. Struct. Biol.* 10:220–228.

Schmid, S. L. & Cullis, P. R. 1998. Endosome marker is fat not fiction. *Nature* 392:135–136.

Tjelle, T. E., et al. 2000. Phagosome dynamics and function. *BioEss.* 22:255–263.

Importing Proteins into Organelles

Bauer, M. F., et al. 2000. Protein translocation into mitochondria: The role of TIM complexes. *Trends Cell Biol.* 10:25–31.

Chen, X. & Schnell, D. J. 1999. Protein import into chloroplasts. *Trends Cell Biol.* 9:222–227.

Cline, K. & Henry, R. 1996. Import and routing of nucleus-encoded chloroplast proteins. *Annu. Rev. Cell Biol.* 12:1–26.

Koehler, C. M., et al. 1999. How membrane proteins travel across the mitochondrial intermembrane space. *Trends Biochem. Sci.* 24:428–432.

May, T. & Soll, J. 1999. Chloroplast precursor protein translocon. *FEBS Lett.* 452:52–56.

Pfanner, N. 1998. Mitochondrial import: Crossing the aqueous intermembrane space. *Curr. Biol.* 8:R262–R265.

Subramani, S., et al. 2000. Import of peroxisomal matrix and membrane proteins. *Annu. Rev. Biochem.* 69:399–418.

The Cytoskeleton and Cell Motility

he skeleton of a vertebrate is a familiar organ system consisting of hardened elements that support the soft tissues of the body and play a key role in mediating bodily movements. Eukaryotic cells also possess a "skeletal system"—a **cytoskeleton**—which has analogous functions. The cytoskeleton is composed of three well-defined filamentous structures—microtubules, microfilaments, and intermediate filaments—which together form an elaborate interactive network. Each of the three filaments is a polymer of protein subunits with distinct mechanical properties. **Microtubules** are rigid tubes composed of subunits of the protein tubulin. **Microfilaments** are solid, thinner structures composed of the protein actin. **Intermediate filaments** are tough, ropelike fibers composed of a variety of related proteins. Although the components of the cytoskeleton appear stationary in micrographs, in reality they are highly dynamic structures capable of rapid and dramatic reorganization. We will consider each of the three major cytoskeletal elements separately in this chapter, after we briefly survey their major activities.

The terminus of a growing axon from the sea hare Aplysia. *Actin filaments, which are required for the motile activities of the elongating axon, are shown in blue; microtubules, which are located behind the growing tip, are shown in red. The photograph is a composite of separate video images that have been digitally superimposed.* (FROM P. FORSCHER AND S. J. SMITH, J. CELL BIOL. 107:1514, 1988; BY COPYRIGHT PERMISSION OF THE ROCKEFELLER UNIVERSITY PRESS. COURTESY OF PAUL FORSCHER, YALE UNIVERSITY.)

9.1 OVERVIEW OF THE MAJOR FUNCTIONS OF THE CYTOSKELETON

An overview of the major activities of the cytoskeleton in three different nonmuscle cells is shown in Figure 9.1. The cells depicted in this schematic illustration include a polarized epithelial cell, the tip of an elongating nerve cell, and a dividing cultured cell. As will be discussed in this chapter, the cytoskeleton of these cells functions as

1. A dynamic scaffold providing structural support that can determine the shape of the cell and resist forces that tend to deform it. The flat, rounded shape of many cultured cells, for example, depends on a radial array of microtubules in the cell's cytoplasm (see Figure 9.10).

2. An internal framework responsible for positioning the various organelles within the interior of the cell. This function is particularly evident in polarized epithelial cells, such as those depicted in Figure 8.11, in which organelles are arranged in a defined pattern along an axis from the apical to the basal end of the cell.

3. A network of highways that direct the movement of materials and organelles within cells. Examples of this function include the delivery of mRNA molecules to specific parts of a cell, the movement of membranous carriers from the endoplasmic reticulum to the Golgi complex, and the transport of vesicles containing neurotransmitters down the length of a nerve cell. Figure 9.2 shows a small portion of a cultured cell, and it is evident that most of the fluorescent green organelles, which are peroxisomes labeled with the fluorescent protein GFP (page 283), are closely associated with the microtubules (red) of the cell's cytoskeleton. Microtubules are the rails over which the peroxisomes are transported.

4. The force-generating apparatus that moves cells from one place to another. Single-celled organisms move either by "crawling" over the surface of a solid substratum or by propelling themselves through their aqueous environment with the aid of specialized locomotor

Key to Cytoskeletal Functions

(1) Structure and Support **(2)** Intracellular Transport **(3)** Contractility and Motility **(4)** Spatial Organization

(a) **Epithelial cell**

(b)

(c)

Nerve cell

Dividing cell

Figure 9.1 An overview of the structure and functions of the cytoskeleton. Schematic drawings of (*a*) an epithelial cell, (*b*) a nerve cell, and (*c*) a dividing cell. The microtubules of the epithelial and nerve cells function primarily in support and organelle transport, whereas the microtubules of the dividing cell form the mitotic spindle required for chromosome segregation. Intermediate filaments provide structural support for both the epithelial cell and nerve cell. Microfilaments support the microvilli of the epithelial cell and are an integral part of the motile machinery involved in nerve cell elongation and cell division.

5 μm

Figure 9.2 An example of the role of microtubules in transporting organelles. The peroxisomes of this cell (shown in green and indicated by arrows) are closely associated with microtubules of the cytoskeleton (shown in red). Peroxisomes appear green because they contain a peroxisomal protein fused to the green fluorescent protein (page 283). Microtubules appear red because they are stained with a fluorescently labeled antibody. (FROM E. A. C. WIEMER ET AL., J. CELL BIOL. 136:78, 1997, COURTESY OF S. SUBRAMANI; BY COPYRIGHT PERMISSION OF THE ROCKEFELLER UNIVERSITY PRESS.)

organelles (cilia and flagella) that protrude from the cell's surface. Multicellular animals have a variety of cells that are capable of independent locomotion, including sperm, white blood cells, and fibroblasts (Figure 9.3). The tip of a growing axon is also highly motile (Figure 9.1), and its movement resembles a crawling blood cell.

Figure 9.3 The flexibility of the cytoskeleton is apparent in this mouse fibroblast that is migrating over the 90° edge of a coverslip. Bar, 30 μm. (FROM GUENTER ALBRECHT-BUEHLER, INT. REV. CYTOL. 120:211, 1990.)

5. A site for anchoring mRNAs and facilitating their translation into polypeptides (see Figure 12.56).

6. An essential component of the cell's division machinery. Cytoskeletal elements make up the apparatus responsible for separating the chromosomes during mitosis and meiosis and for splitting the parent cell into two daughter cells during cytokinesis. These events are discussed at length in Chapter 14.

9.2 THE STUDY OF THE CYTOSKELETON

The cytoskeleton is currently one of the most actively studied subjects in cell biology, due in large part to the development of techniques that allow investigators to pursue a coordinated morphological, biochemical, and molecular approach. As a result, we know a great deal about the families of proteins that make up the cytoskeleton; how the subunits are organized into their respective fibrous structures; the molecular "motors" that generate the forces that allow the cytoskeleton to engage in motile activities; and the dynamic qualities that control the spatial organization, assembly, and disassembly of the various elements of the cytoskeleton. A few of the most important approaches to the study of the cytoskeleton will be briefly surveyed.

The Use of Fluorescence Microscopy

Traditionally, the light microscope has been a tool used to learn about the organization of cells within fixed, stained tissues and to observe living cells as they engage in various types of motile activities. In the past couple of decades, a revolution in light microscopy has taken place, enabling the instrument to provide information about cellular structures that are much too small to be visualized by standard optical techniques. The fluorescence microscope (Section 18.1) has played a leading part in this revolution in light microscopy.

The fluorescence microscope has been used to study the dynamics of the cytoskeleton. In one approach, the protein subunits of cytoskeletal structures (e.g., tubulin or keratin) are made fluorescent by covalent linkage to a small fluorescent dye. The labeled subunits are then microinjected into a living cell where they are incorporated into the polymeric form of the protein, such as a microtubule or an intermediate filament. The location of the fluorescent structure can be followed over time as the cell participates in its normal activities. Thus, unlike most other high-resolution techniques, the cell remains alive during observation. Figure 9.4 shows the distribution of microtubules in the elongating axon of a neuron that had been injected with fluorescently labeled tubulin. Alternatively, fluorescently labeled proteins can be produced within a cell, as in Figure 9.2, by fusing them with the green fluorescent protein as described in the previous chapter (page 283).

5 μm

Figure 9.4 The distribution of microtubules in an elongating axon of a living nerve cell. The cell was injected with a small volume of brain tubulin that had been covalently linked to the fluorescent dye rhodamine. Once in the cytoplasm, the fluorescent tubulin subunits become incorporated into microtubules whose positions can be followed under the fluorescence microscope as the cell carries out its normal activities. (FROM JAMES H. SABRY ET AL., J. CELL BIOL. 115:383, 1991, COURTESY OF DAVID BENTLEY; BY COPYRIGHT PERMISSION OF THE ROCKEFELLER UNIVERSITY PRESS.)

Fluorescence microscopy can also be used to reveal the location within a cell of a protein present in very low concentration. This type of experiment is best carried out with fluorescently labeled antibodies that bind with high affinity to the protein being sought. Antibodies are particularly useful because they can distinguish between closely related variants (isoforms) of a protein. The protein can be localized by injecting the labeled antibody into a living cell, which may also reveal the function of the target protein, because attachment of the antibody often destroys the ability of the protein to perform its normal activity. Alternatively, the location of the target protein may be determined by adding the fluorescent antibody to fixed cells or tissue sections, as illustrated in Figure 9.5. The resolution obtained in these types of images has become much better in recent years with the development of the confocal fluorescence microscope, an instrument that allows one to focus at a chosen level in a cell without background interference from fluorescent structures at other levels in the specimen (Section 18.1).

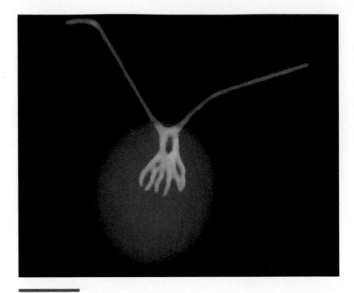

Figure 9.5 The localization of a protein within a cell using fluorescent antibodies. This algal cell was stained with fluorescent antibodies (yellow-green color) directed against a protein called centrin. Centrin is seen to be localized within the cell's flagella and the rootlike structure at the base of the flagella. The red color of the cell results from autofluorescence of chlorophyll molecules of this photosynthetic alga. (FROM MARK A. SANDERS AND JEFFREY L. SALISBURY OF THE MAYO CLINIC, FROM CELL 70:533, 1992; BY PERMISSION OF CELL PRESS.)

The Use of Video Microscopy and Focused Laser Beams for In Vitro Motility Assays

The power of the light microscope can be greatly increased using a video camera, tape recorder, and television monitor. The image captured by a video camera possesses exceptional contrast and can be enhanced by computer processing. These features allow objects ordinarily below the resolution of the light microscope, such as 25-nm microtubules or 40-nm membrane vesicles, to be observed and photographed. For example, video microscopy can be used to observe the growth or shrinkage of individual microtubules as they gain or lose subunits in vivo (see Figure 9.28).

The advent of video microscopy has also led to the development of **in vitro motility assays**, making it possible to detect the activity of individual protein molecules acting as molecular motors. In one type of assay, microtubules are attached to a glass coverslip. Then, microscopic beads containing attached motor proteins are placed directly onto the microtubules using focused laser beams. The laser beams are shone through the objective lens of a microscope, producing a weak attractive force near the point of focus. Because it can grasp microscopic objects, this apparatus is referred to as **optical tweezers**. Under appropriate conditions, with ATP present as an energy source, the movements of a bead along a microtubule can be followed on a television screen (Figure 9.6). Focused laser beams can also be used to "trap" a single

(a) (b) (c) (d)

(e)

Figure 9.6 **Using video microscopy to follow the activities of molecular motors.** In this video sequence, (*a*) a bead, which is coated with the motor protein kinesin, is captured out of suspension using a laser-based optical tweezers; (*b*) the bead is deposited on a microtubule (the reflection of laser light at the coverglass-water interface partially obscures the image); (*c,d*) the bead binds to the microtubule and begins moving along the microtubular track. (*e*) Schematic drawing of the movement of one of these beads along a microtubule. (FROM STEVEN M. BLOCK, L. S. B. GOLDSTEIN, AND BRUCE J. SCHNAPP. REPRINTED WITH PERMISSION FROM NATURE 348:349, 1990. © COPYRIGHT 1990, MACMILLAN MAGAZINES LTD.)

bead and determine the minute forces (measured as a few piconewtons, pN) generated by a single motor protein as it "tries" to move the bead against the force exerted by the optical trap.

The development of techniques for working with single molecules has coincided with the creation of a new field of mechanical engineering called **nanotechnology**. The goal of nanotechnologists is the development of tiny "nanomachines" (in the 10–100 nm size range) capable of performing specific activities in a submicroscopic world. There are many perceived uses of such machines, including a role in medicine. It is envisioned that, one day, such machines might be introduced into the human body where they could carry out a specific task, such as finding and destroying individual cancer cells. As it happens, Nature has already evolved molecular machines, including the motor proteins that will be discussed in this chapter. Not surprisingly, a number of nanotechnology labs have begun to use these motor proteins to move molecular cargoes unlike any found within living organisms.

The Use of Genetically Engineered Cells

One of the best ways to learn about the function of a particular polypeptide is to study the phenotype of cells in which that polypeptide is either absent or nonfunctional. Since the development of recombinant DNA technologies, researchers no longer have to wait for random mutations to accomplish this goal. Instead, they can create DNA molecules with the desired alteration and then incorporate the altered DNA into the genomes of laboratory animals or cultured cells. The techniques to carry out these procedures are discussed in Chapter 18.

Now that the genomes of several eukaryotes have been sequenced, we are in a position to study the role of any of the genes that might be required for a particular function, such as the operation of the cytoskeleton. Analysis of the genome of the fruit fly, for example, indicates that these insects possess genes that encode approximately 100 proteins that either function as motor proteins or interact with them. Any one of these genes can be isolated, modified, and genetically inactivated. The effects of gene inactivation on an organism or cell are usually studied in one of two experimental approaches: through the formation of

1. knockout animals—most often mice—that lack a particular gene. In some cases, knockout mice may show little or no effects of a missing protein, which tells the researcher very little about the possible function of that molecule, besides the fact that it is nonessential. In other cases, mice lacking a particular gene may show highly specific defects, which strongly suggests that the missing protein plays an important role in the process that is defective. Knockout mice often die during early stages of development, but cells from these abnormal embryos can be isolated and cultured, allowing the molecular deficiency to be identified (see Figure 9.17). For example, mice that lack a motor protein called cytoplasmic dynein fail to develop beyond about 8 days. When cells from these embryos were studied, the Golgi complex was found to be fragmented and dispersed throughout the cytoplasm. These findings suggest that cytoplasmic dynein plays an essential role in positioning the Golgi complex within the cell.

2. cells that overexpress a *dominant negative mutant* protein, that is, cells that produce large amounts of a nonfunctional protein. Cells of this type are generally produced by transfection, that is, by causing the cells to take up the altered DNA and incorporating it into their chro-

mosomes. Once the cells have been genetically modified, the overexpressed mutant protein competes with the normal protein produced at lower levels in the cell, causing the cell to exhibit the mutant phenotype. An example of this experimental strategy is illustrated in Figure 9.7. Figure 9.7a shows a control pigment cell from the amphibian *Xenopus* that has been treated with a hormone that induces dispersion of the pigment granules into the cell's peripheral processes. In the whole animal, this response has the effect of lightening the skin. The pigment cell in Figure 9.7b, which was treated with the same hormone, contains an overexpressed mutant version of a motor protein called kinesin II. The failure of the pigment granules to disperse in this cell implicates kinesin II as the motor protein responsible for the outward movement of these granules.

REVIEW ?

1. Describe two examples in which each of the following methodologies have contributed to our understanding of the nature of the cytoskeleton: fluorescence microscopy, in vitro motility assays, and genetic engineering.

2. List some of the major functions of the cytoskeleton.

9.3 MICROTUBULES

Structure and Composition

As the name implies, microtubules are hollow, tubular structures, and they occur in nearly every eukaryotic cell. Microtubules are components of a diverse array of structures, including the mitotic spindle of dividing cells and the core of cilia and flagella. Microtubules have an outer diameter of 24 nm, a wall thickness of approximately 5 nm, and may extend across the length or breadth of a cell. The wall of a microtubule is composed of globular proteins arranged in longitudinal rows, termed **protofilaments**, that are aligned parallel to the long axis of the tubule (Figure 9.8a). When viewed in cross section, microtubules are seen to consist of 13 protofilaments arrayed in a circular pattern within the wall (Figure 9.8b).

Each protofilament is assembled from dimeric building blocks consisting of one α-tubulin and one β-tubulin globular subunit. The two types of tubulin subunits have a similar three-dimensional structure and fit tightly together as shown in Figure 9.8c. The tubulin dimers are organized in a linear array along the length of each protofilament, as shown in Figure 9.8d. Because each assembly unit contains two nonidentical components (a heterodimer), the protofilament is asymmetric, with an α-tubulin at one end and a β-tubulin at the other end. All of

(a)

(b)

Figure 9.7 **Expression of a mutant motor protein inhibits the dispersion of pigment granules in a pigment cell.** (a) A control pigment cell from the amphibian *Xenopus*, showing the black pigment granules dispersed into the elongated cellular processes. (b) A pigment cell that is overexpressing a gene for a mutant motor protein. The mutant protein acts as a competitive inhibitor of the normal protein produced in the cell and inhibits

the dispersion of the pigment granules. This result points to a role of this motor protein in the outward transport of these membrane-bounded granules. Bar = 20 μm. (FROM M. CAROLINA TUMA ET AL., COURTESY OF VLADIMIR GELFAND, J. CELL BIOL. 143, 1551, 1998; BY COPYRIGHT PERMISSION OF THE ROCKEFELLER UNIVERSITY PRESS.)

(a) 100 nm *(b)* *(c)*

4 nm

8 nm

5cam

(d)

Figure 9.8 The structure of microtubules. (*a*) Electron micrograph of negatively stained micro-tubules from brain showing the globular subunits that make up the protofilaments. The bumps at the surface of the microtubules are microtubule-associated proteins (MAPs), discussed later. (*b*) Electron micrograph of a cross section through a microtubule of a *Juniperus* root tip cell revealing the 13 subunits arranged within the wall of the tubule. The microtubules of these plant cells are most abundant in a cor-tical zone about 100 nm thick just beneath the plasma membrane (seen in the lower right of the micro-graph). (*c*) A ribbon model showing the three-dimensional structure of the αβ-tubulin heterodimer. Note the complementary shapes of the subunits at their interacting surfaces. The α-tubulin subunit has a bound GTP, which is not hydrolyzed and is nonexchangeable. The β-tubulin subunit has a bound GDP, which is exchanged for a GTP prior to assembly into a polymer (page 352). The plus end of the dimer is at the top. (*d*) Diagram of a longitudinal section of a microtubule shown in the B-lattice, which is the structure thought to occur in the cell. The wall consists of 13 protofilaments composed of αβ-tubulin heterodimers. Adjacent protofilaments are not aligned in register, but are staggered about 1 nm so that the tubulin molecules form a helical array around the circumference of the microtubule. The helix is interrupted at one site where α and β subunits make lateral contacts. This produces a "seam" that runs the length of the microtubule. (A: FROM LINDA A. AMOS, J. CELL BIOL. 72:645, 1977; BY COPYRIGHT PERMISSION OF THE ROCKEFELLER UNIVERSITY PRESS; B: REPRINTED WITH PERMISSION OF MYRON C. LEDBETTER, J. AGR. FOOD CHEM. 13:406, 1965; © COPYRIGHT 1965, AMERICAN CHEMICAL SOCIETY; C: COURTESY OF EVA NOGALES AND KENNETH DOWNING.)

the protofilaments of a microtubule have the same polar-ity. Consequently, the entire polymer has polarity. One end of a microtubule is known as the **plus (or fast-grow-ing) end**, the other as the **minus (or slow-growing) end**. The endmost row of subunits at the plus end of a micro-tubule are β tubulins, whereas subunits at the tip of the minus end are α tubulins. As discussed later in the chap-ter, the structural polarity of microtubules is an important factor in the growth of these structures and their ability to participate in directed mechanical activities.

Microtubule-Associated Proteins

Although microtubules can assemble in vitro from puri-fied tubulin, those obtained from cells typically contain additional proteins, called **microtubule-associated proteins** (or **MAPs**). Most of the MAPs that have been identified are found only in brain tissue, but one of these

proteins, called MAP4, has a wide distribution in non-neuronal mammalian cells. MAPs typically have one do-main that attaches to the side of a microtubule and an-other domain that extends outward as a filament from the microtubule's surface. The binding of MAP2 to the sur-face of a microtubule is depicted in Figure 9.9. Some MAPs can be seen in electron micrographs as cross-bridges connecting microtubules to each other. Other MAPs increase the stability of microtubules, alter their rigidity, or influence the rate of their assembly. The activ-ity of the various MAPs is controlled primarily by the ad-dition and removal of phosphate groups from particular amino acid residues by protein kinases and phosphatases, respectively. An abnormally high level of phosphorylation of one particular MAP, called *tau*, has been implicated in the development of several fatal neurodegenerative dis-eases. The brain cells of people with these diseases contain strange, tangled filaments (called *neurofibrillary tangles*)

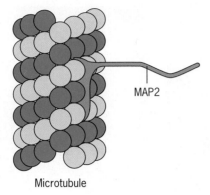

Figure 9.9 **Microtubule-associated proteins (MAPs).** Schematic diagram of a MAP2 molecule bound to the surface of a microtubule. Each MAP2 molecule contains three tubulin-binding sites connected by short segments of the polypeptide chain. The binding sites are spaced at sufficient distance to allow the MAP2 molecule to attach to three separate tubulin subunits in the wall of the microtubule. The tails of the MAP molecules project outward, allowing them to interact with other cellular components.

15 μm

Figure 9.10 **Localization of microtubules** of a large, flat, cultured mouse cell is revealed by fluorescent anti-tubulin antibodies. Microtubules are seen to extend from the perinuclear region of the cell in a radial array. Individual microtubules can be followed and are seen to curve gradually as they conform to the shape of the cell. (FROM MARY OSBORN AND KLAUS WEBER, CELL 12:563, 1977; BY PERMISSION OF CELL PRESS.)

consisting of tau molecules that are excessively phosphorylated and unable to bind to microtubules. The neurofibrillary filaments are thought to contribute to the death of nerve cells. Persons with one of these diseases, a type of dementia called FTDP-17, carry mutations in the *tau* gene, indicating that alteration of the tau protein is the primary cause of this disorder.

Microtubules as Structural Supports and Organizers

Microtubules are stiff enough to resist forces that might compress or bend the fiber. This property enables them to provide mechanical support much as steel girders support a tall office building. The distribution of cytoplasmic microtubules in a cell helps determine the shape of that cell. In cultured animal cells, the microtubules extend in a radial array outward from the area around the nucleus, giving these cells their round, flattened shape (Figure 9.10). In contrast, the microtubules of columnar epithelial cells are typically oriented with their long axis parallel to the long axis of the cell (Figure 9.1*a*). This configuration suggests that microtubules help support the elongated shape of the cell.

The role of microtubules as skeletal elements is evident in certain highly elongated cellular processes, such as the axons of nerve cells and the axopods of heliozoan protists. The axon of a nerve cell is filled with microtubules that are oriented parallel to the long axis of the axon (Figure 9.11). In mature axons, these microtubules serve as tracks for the movement of vesicles and other cytoplasmic particles (discussed on page 342). In the developing embryo, microtubules play a key role in maintaining the extended shape of the axon as it slowly grows out of the central nervous system into the peripheral embryonic tissues.

A similar process of axonal outgrowth can be followed in cell culture. If a nerve cell with a growing axon is treated with microtubule-disrupting drugs, such as nocodazole or colchicine, the axonal process ceases its outgrowth and typically collapses back toward the rounded cell body. A striking example of the role of microtubules in the maintenance of cell shape is provided by the long, slender axopodial processes of the heliozoan protists (Figure 9.12). Each axopod contains a core structure composed of large numbers of microtubules arranged in a spiral with individual microtubules traversing the entire length of the process.

In plant cells, microtubules play an indirect role in maintaining cell shape through their influence on the formation of the cell wall. During interphase, most of a plant cell's microtubules are located just beneath the plasma membrane (as in Figure 9.8*b*), forming a distinct cortical zone. The microtubules of the cortex are thought to influence the movement of the cellulose-synthesizing enzymes located in the plasma membrane (see Figure 7.37). As a result, the cellulose microfibrils of the innermost layer of the cell wall are assembled in an orientation that is parallel to the underlying microtubules of the cortex (Figure 9.13). The orientation of these cellulose microfibrils plays an important role in determining the growth characteristics of the cell and thus its shape. In most cells, newly synthesized cellulose microfibrils and coaligned microtubules are arranged perpendicular to the long axis of the cell (transversely), as shown in Figure 9.13. Because the cellulose microfibrils resist lateral expansion, the turgor pressure exerted by the cell is directed to the ends of the cell, causing cell elongation.

Microtubules are also thought to play a role in maintaining the internal organization of cells. Treatment of

(a)

Figure 9.11 Microtubules of an axon and their role in organelle movement. (*a*) Light micrograph of a living, isolated chick neuron. (*b*) Electron micrograph of a portion of the same cell shown in *a* (the box indicates EM coverage). The particle located in this area of the light micrograph was moving until the moment of fixation. The electron micrograph reveals that the particle corresponds to a pair of vesicles, including a lysosome. The numerous parallel microtubules function as tracks in these types of movements. (FROM A. C. BREUER, C. N. CHRISTIAN, M. HENKART, AND P. G. NELSON, J. CELL BIOL. 65:568, 1965; BY COPYRIGHT PERMISSION OF THE ROCKEFELLER UNIVERSITY PRESS.)

(b) 0.25 µm

Microtubules

(b)

(a)

Figure 9.12 Microtubules as supporting rods. (*a*) This protist has tentacles projecting from its cell body. (*b*) The tentacles are supported by an elaborate spiral array of microtubules. (A: FROM PETER PARKS/ANIMALS ANIMALS; B: COURTESY OF MANFRED HAUSER.)

(a) *(b)* 20 µm

Figure 9.13 Spatial relationship between microtubule orientation and cellulose deposition in plant cells. Wheat mesophyll cell that was double stained for microtubules (*a*) and cell wall cellulose (*b*). It is evident that the cortical microtubules and cellulose microfibrils are coaligned. (FROM GEORG JUNG AND WOLFGANG WERNICKE, PROTOPLASMA 53:145, 1990.)

cells with microtubule-disrupting drugs can seriously affect the location of membranous organelles, particularly the Golgi complex. The Golgi complex is typically located near the center of an animal cell, just outside the nucleus. Treatment of cells with nocodazole or colchicine can disperse the Golgi elements into the peripheral regions of the cell. When the drug is removed and microtubules reassemble, the Golgi membranes return to their normal position in the cell interior.

Microtubules as Agents of Intracellular Motility

Living cells bristle with activity as macromolecules and organelles move in a directed manner from place to place. While this hustle and bustle can be appreciated by watching the movement of particulate materials in a living cell, the underlying processes are usually difficult to study because most cells lack a highly ordered cytoskeleton. We know, for example, that the transport of materials from one membrane compartment to another depends on the presence of microtubules because disruption of these cytoskeletal elements brings the movements to a halt. We will begin our study of intracellular motility by focusing on nerve cells whose intracellular movements rely on a highly organized arrangement of microtubules and other cytoskeletal filaments.

Axonal Transport The axon of an individual motor neuron may stretch from the spinal cord to the tip of a finger or toe. The manufacturing center of this neuron is its cell body, a rounded portion of the cell that resides within the ventral portion of the spinal cord. The cell body contains the nucleus, endoplasmic reticulum, and Golgi complex that are necessary for synthesizing proteins and other molecules, whereas the axon is devoid of this capacity. When labeled amino acids are injected into the cell body, they are incorporated into labeled proteins that move into the axon and gradually travel down its length. Most materials, including neurotransmitter molecules, are compartmentalized within membranous vesicles in the ER and Golgi complex of the cell body and then transported down the length of the axon (Figures 9.11, 9.14a). Different materials move at different rates, with the fastest axonal transport occurring at rates of 5 μm per second (400 mm per day). Structures and materials traveling from the cell body toward the terminals of a neuron are said to move in an *anterograde* direction. Other structures, including endocytic vesicles that form at the axon terminals and carry regulatory factors from target cells, move in the opposite, or *retrograde*, direction—from the synapse toward the cell body.

Axons are filled with cytoskeletal structures, including bundles of microfilaments, intermediate filaments, and microtubules interconnected in various ways (Figure 9.14b). Evidence suggests that both anterograde and retrograde movements are mediated primarily by microtubules. Using video microscopy, investigators can follow

(a)

(b)

Figure 9.14 Axonal transport. (*a*) Schematic drawing of a nerve cell showing the movement of vesicles down the length of an axon along tracks of microtubules. Vesicles move in both directions within the axon. (*b*) Schematic drawing of the organization of the microtubules and intermediate filaments (neurofilaments) within an axon. Vesicles containing transported materials are attached to the microtubules by cross-linking proteins, including motor proteins, such as kinesin and dynein.

individual vesicles as they move along the microtubules of an axon, either toward or away from the cell body (Figure 9.15). The involvement of microtubules can be verified by fixing the cell and observing the same region in the electron microscope. When this is done, the moving vesicle is seen in close association with a microtubule (as in Figure 9.11). But how are these vesicles able to make their way along these cytoskeletal elements? It has been known for several decades that microtubules are primarily passive structures that serve as tracks for a large number of **motor proteins** that generate the forces required to move objects within a cell. The study of motor proteins has become a major focus in cell and molecular biology, and a great deal of information has been collected about the nature of these molecules and their mechanism of action, which are the subjects of the following section.

Motor Proteins That Traverse the Microtubular Cytoskeleton

The motor proteins of a cell convert chemical energy (stored in ATP) into mechanical energy, which is used to move cellular cargo attached to the motor. Types of cellu-

(a) *(b)* *(c)*

Figure 9.15 Visualizing axonal transport. *(a–c)* These video micrographs show the progression of a membranous organelle along a branched axon. The cell body is far out of the field to the upper left, while the termini (growth cones, page 390) are out of the field to the bottom right. The position of the organelle is indicated by the arrowheads. The organelle being followed (an autophagic vacuole) moves in the retrograde direction across the branch point and continues to move toward the cell body. Bar, 10 μm. (FROM PETER J. HOLLENBECK, J. CELL BIOL. 121:307, 1993; BY COPYRIGHT PERMISSION OF THE ROCKEFELLER UNIVERSITY PRESS.)

lar cargo transported by these proteins include vesicles, mitochondria, lysosomes, chromosomes, and other cytoskeletal filaments. A single cell may contain dozens of different motor proteins, each presumably specialized for the movement of a particular type of cargo in a particular region of the cell.

Collectively, motor proteins can be grouped into three broad families: myosins, kinesins, and dyneins. Kinesins and dyneins move along microtubules, whereas myosins move along microfilaments. No protein motor is known that uses intermediate filament tracks. Motor proteins move unidirectionally along their cytoskeletal track in a stepwise manner, from one binding site to the next As the protein moves along, it undergoes a series of conformational changes that constitute a *mechanical cycle*. The steps of the mechanical cycle are coupled to the steps of a *chemical cycle*, which provide the energy necessary to fuel the movement. The steps in the chemical cycle include the binding of an ATP molecule to the motor, the hydrolysis of the ATP, the release of the products (ADP and P$_i$) from the motor, and the binding of a new molecule of ATP. The binding and hydrolysis of a single ATP molecule typically moves the motor a few nanometers along its track. As the motor protein moves to successive sites along the cytoskeletal polymer, the mechanical and chemical cycles are repeated over and over again, pulling the cargo over considerable distances. Recent reports on the atomic structure of several motor proteins have generated hypotheses as to how the working parts of these molecules perform complex mechanochemical activities. We will begin by examining the molecular structure and functions of kinesins, which are the smallest and best understood microtubular motors.

Kinesins In 1985, a motor protein was isolated from squid giant axons, which is responsible for moving vesicles and organelles from the cell body toward the synaptic terminals. This motor protein, which was named **kinesin**, is a tetramer constructed from two identical heavy chains and two identical light chains (Figure 9.16*a*). A kinesin molecule has several parts, including a pair of globular heads that bind a microtubule and act as ATP-hydrolyzing, force-generating "engines." Each head is connected to a neck, a rodlike stalk, and a fan-shaped tail that binds cargo to be hauled (Figure 9.16*a,b*).

When purified kinesin molecules are attached to beads and observed in an in vitro motility assay (page 336), the beads move along individual microtubules toward the plus end of the polymer (see Figure 9.6). Kinesin is therefore said to be a *plus end–directed microtubular motor*. Because all of the microtubules of an axon are oriented with their minus ends facing the cell body and their plus ends facing the synaptic terminals, kinesin is presumed to mediate anterograde axonal transport. The discovery of kinesin and its role in axonal transport are discussed in the Experimental Pathways, which can be found on the Web at www.wiley.com/college/karp.

A single kinesin molecule moves along a single protofilament of a microtubule at a velocity proportional to the ATP concentration (up to a maximal velocity of about 1 μm per second). At low ATP concentrations, kinesin molecules travel slowly enough for observers to conclude that the protein moves in distinct steps (Figure 9.16*b*). Each step is approximately 8 nm in length, which is also the spacing between tubulin dimers along a protofilament. It would appear that kinesin molecules move over a microtubular track two globular subunits (or one tubulin heterodimer) at a time. The movement of kinesin molecules, both in vitro and in vivo, is highly **processive**, meaning that the motor protein tends to move along an individual microtubule for considerable distances (over one μm) without falling off. A two-headed kinesin

(b)

(c)

Figure 9.16 **Kinesin.** (*a*) Structure of a kinesin molecule, which consists primarily of two heavy chains that wrap around each other as a coiled coil in the region of the stalk. The force-generating heads bind to the microtubule, and the tail binds to the cargo being transported. Having a molecular mass of approximately 380 kDa, kinesin is considerably smaller than the other motor proteins, myosin (muscle myosin, 520 kDa) and dynein (over 1 thousand kDa). (*b*) Schematic diagram of a kinesin molecule moving a vesicle along a microtubular track. This type of movement is not unlike that of a person walking through a garden on a linear path of stepping stones. (*c*) The

conformational changes that occur in the head (blue) and neck (red) portions of a monomeric kinesin molecule that drive the movement of the protein along a microtubule (yellow contour map). Rather than being connected to a second head, the neck of this truncated kinesin molecule is attached to a GFP molecule (green). The swinging movement of the neck would normally drive the forward motion of the partner head, enabling the dimer to walk towards the plus end of a protofilament. (C: FROM RYAN B. CASE ET AL., COURTESY OF RONALD D. VALE, CURR. BIOL., VOL. 10, #3 COVER, 2000.)

molecule can accomplish this feat because at least one of the heads is attached to the microtubule at all times (Figure 9.16*b*).

The two heads of a kinesin molecule behave in a coordinated manner, so that they are always present at different stages in their chemical and mechanical cycles at any given time. When one head binds to the microtubule, the interaction induces a conformational change in the adjacent neck region of the motor protein that swings the other head forward toward its binding site on the next tubulin dimer of the protofilament. These changes in conformation are illustrated in Figure 9.16*c*, which depicts the head and neck portions of a monomeric kinesin heavy chain associated with a microtubule. The force generated by the catalytic activity of the head leads to the swinging movement of the neck which, in this illustration, is attached to a molecule of GFP (the green barrel-shaped protein) rather than to a second kinesin head. As a result of these conformational changes, a kinesin molecule "walks" along a microtubule (Figure 9.16*b*), hydrolyzing one molecule of ATP with each step. A motor protein

with this capability is well adapted for independent, long-distance transport of small parcels of cargo.

The kinesin molecule discovered in 1985, which is often referred to as "conventional kinesin," is only one member of a superfamily of related proteins, called **KRPs** (kinesin-related proteins) or **KLPs** (kinesin-like proteins). Based on the analysis of genome sequences, it is estimated that mammals produce more than 50 different KLPs. The heads of KLPs have related amino acid sequences, reflecting their common evolutionary ancestry and their similar role in moving along microtubules. In contrast, the tails of KLPs have diverse sequences, reflecting the variety of cargo the different proteins haul.

Like kinesin, most KLPs move toward the plus end of the microtubule to which they are bound. One small subfamily of KLPs, however, including the widely studied Ncd protein of *Drosophila*, moves in the opposite direction, that is, toward the minus end of the microtubular track. One might expect that the heads of plus end-directed and minus end-directed KLPs would have a different structure because they contain the catalytic core of

the motor domains. But the heads of the two proteins are virtually indistinguishable. Instead, differences in directional movement are determined by differences in the structure of the adjacent neck regions of the two proteins. When the head of a minus end-directed Ncd molecule is joined to the neck and stalk portions of a kinesin molecule, the hybrid protein moves toward the plus end of the track. Thus, even though the hybrid has a catalytic domain that normally would move toward the minus end of a microtubule, as long as it is joined to the neck of a plus end motor, it moves in the plus direction.

A third subfamily of kinesin-like proteins are incapable of movement. The KLPs of this latter group, such as XKCM1 discussed on page 353, are thought to destabilize microtubules rather than acting as microtubular motors.

Kinesin-Mediated Organelle Transport We saw in Chapter 8 how vesicles move from one membranous compartment, such as the Golgi complex, to another, such as a lysosome. The routes followed by cytoplasmic vesicles and organelles are largely defined by microtubules (see Figure 9.1), and members of the kinesin superfamily are strongly implicated as force-generating agents that drive the movement

of this membrane-bounded cargo. In most cells, as in axons, microtubules are aligned with their plus ends pointed away from the center of the cell. Therefore, kinesin and kinesin-like proteins tend to move vesicles and organelles (e.g., peroxisomes and mitochondria) in an outward direction toward the cell's plasma membrane. This is illustrated in the micrographs in Figure 9.17. The pair of micrographs on the left shows a cell that was isolated from a normal 9.5 day mouse embryo and stained to reveal the location of its microtubules (green) and mitochondria (orange). The pair of micrographs on the right shows a cell that was isolated from a 9.5 day mouse embryo lacking both copies of the gene that encodes the KIF5B kinesin heavy chain. The mitochondria of the KIF5B-deficient cell are absent from the peripheral regions of the cell, as would be expected if this plus end-directed kinesin is responsible for the outward movement of membranous organelles (see also Figure 9.18c). The role of kinesin-like proteins in cell division is discussed in Chapter 14.

Cytoplasmic Dynein The first microtubule-associated motor was discovered in 1963 as the protein responsible for the movement of cilia and flagella. The protein was

Figure 9.17 Alteration in the phenotype of a cell lacking a member of the kinesin superfamily.
(*a, c*) Control cell from the extraembryonic tissues of a normal 9.5 day mouse embryo stained in *a* for microtubules (green) and in *c* for mitochondria (yellow-orange). A significant fraction of the cell's mitochondria are located along microtubules in the peripheral regions of the cell. (*b, d*) A corresponding cell obtained from an embryo that lacked both copies of the gene encoding the kinesin-like protein KIF5B. All of the mitochondria are clustered in the central region of the cell, suggesting that KIF5B is responsible for transporting mitochondria in an outward direction. (FROM YOSUKE TANAKA ET AL., COURTESY OF NOBUTAKA HIROKAWA, CELL 93:1150, 1998, BY PERMISSION OF CELL PRESS.)

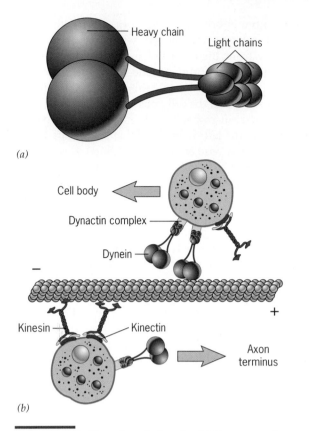

(a)

(b)

Figure 9.18 Cytoplasmic dynein. (*a*) Structure of a cytoplasmic dynein molecule, which consists of two large, force-generating globular heads (each composed of a dynein heavy chain), a stalk, and a number of smaller subunits at the base of the molecule. These latter subunits are presumed to mediate the binding of the motor protein to the cargo being transported. (*b*) Schematic diagram of two vesicles moving in opposite directions along the same microtubule, one powered by kinesin moving toward the plus end of the track, and the other by cytoplasmic dynein moving toward the minus end of the track. In the model shown here, each vesicle contains both types of motor

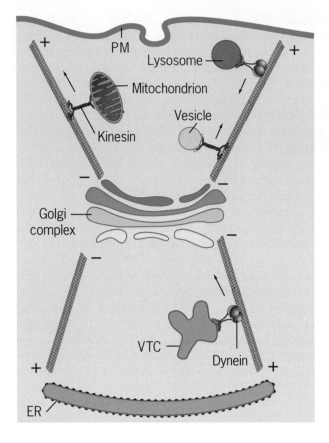

(c)

proteins, but the kinesin molecules are inactivated in the upper vesicle and the dynein molecules are inactivated in the lower vesicle. Both motor proteins are attached to the vesicle membrane by an intermediary: kinesin is attached by the transmembrane protein kinectin, and dynein by a soluble protein complex called dynactin. (*c*) Schematic illustration of kinesin-mediated and dynein-mediated transport of vesicles and organelles in a nonpolarized, cultured cell.

named **dynein**. The existence of cytoplasmic forms was almost immediately suspected, but it took over 20 years before a dyneinlike protein was purified and characterized from mammalian brain tissue. A related protein was soon identified in a variety of nonneural cells, and like kinesin, **cytoplasmic dynein**, as it is called, is now known to be a ubiquitous motor protein in eukaryotic cells.

Cytoplasmic dynein is a huge protein (molecular mass of approximately 1.5 million daltons) composed of two identical heavy chains and a variety of intermediate and light chains. Each dynein heavy chain forms a large globular head—an order of magnitude larger than a kinesin head—which acts as a force-generating engine (Figure 9.18*a*). In vitro motility assays indicate that, unlike conventional kinesin, cytoplasmic dynein moves along a microtubule toward the polymer's minus end (Figure 9.18*b*). A body of evidence suggests at least two roles for cytoplasmic dynein:

1. A force-generating agent in the positioning of the

spindle and movement of chromosomes during mitosis (discussed in Chapter 14).

2. A minus end–directed microtubular motor for the positioning of the Golgi complex and the movement of vesicles and organelles through the cytoplasm.

In nerve cells, cytoplasmic dynein has been implicated in the retrograde movement of cytoplasmic organelles and the anterograde movement of microtubules (see Analytic Question 19). In fibroblasts and other nonneural cells, cytoplasmic dynein is thought to transport membranous organelles from peripheral locations toward the center of the cell (Figure 9.18*c*). Dynein-driven cargo includes endosomes, lysosomes, and ER-derived vesicles heading toward the Golgi complex. Cytoplasmic dynein does not interact directly with membrane-bounded cargo, but requires an intervening multisubunit complex, called ***dynactin***. Dynactin may regulate dynein activity and help bind the motor protein to the microtubule.

According to the current model shown in Figure 9.18*c*, which is probably overly simplistic, kinesin and cytoplasmic dynein move similar materials in opposite directions over the same railway network. As indicated in Figure 9.18*b*, individual organelles may bind kinesin and dynein simultaneously, although presumably only one of them is active at any given time. As discussed on page 371, myosin may also be present on some of these organelles.

Microtubule-Organizing Centers (MTOCs)

The function of a microtubule within a living cell depends on its location and orientation, which makes it important to understand why a microtubule assembles in one place as opposed to another. The assembly of microtubules from αβ-tubulin dimers occurs in two distinct phases: a slower phase of *nucleation*, in which a small portion of the micro-

tubule is initially formed, and a more rapid phase of *elongation*. The nucleation of microtubules in vivo is associated with a variety of specialized structures that, because of their role in microtubule nucleation and organization, are called **microtubule-organizing centers** (or **MTOCs**). The best-studied MTOC is the centrosome.

Centrosomes In animal cells, the microtubules of the cytoskeleton typically form in association with the **centrosome**, a complex structure that contains two barrel-shaped **centrioles** surrounded by amorphous, electron-dense **pericentriolar material** (or **PCM**) (Figure 9.19*a,b*). Centrioles are cylindrical structures about 0.2 mm in diameter and typically about twice as long. Centrioles contain nine evenly spaced fibrils, each of which appears in cross section as a band of three microtubules, designated the A, B, and C tubules. Only the A tubule is a complete microtubule (Fig-

(a) (b) (c) 0.3 μm (d) 500 nm

Figure 9.19 Structure of the centrosome. (*a*) Schematic diagram of a centrosome showing the paired centrioles; the surrounding pericentriolar material (PCM); and microtubules emanating from the PCM, where nucleation occurs. (*b*) Electron micrograph of a cross section of a centriole showing the pinwheel arrangement of the nine peripheral fibrils, each of which consists of one complete microtubule and two incomplete microtubules. (*c*) Electron micrograph showing two pairs of centrioles. Each pair consists of a longer parental centriole and a smaller daughter centriole (arrow), which is undergoing elonga-

tion in this phase of the cell cycle (discussed in Section 14.2). (*d*) Electron micrographic reconstruction of a 1.0 M potassium iodide-extracted centrosome, showing the PCM to consist of a loosely organized fibrous lattice. (A: AFTER S. J. DOXSEY ET AL., CELL 76:643, 1994; BY PERMISSION OF CELL PRESS; B: COURTESY OF B. R. BRINKLEY; C: FROM J. B. RATTNER AND STEPHANIE G. PHILLIPS, J. CELL BIOL. 57:363, 1973; BY COPYRIGHT PERMISSION OF THE ROCKEFELLER UNIVERSITY PRESS; D: FROM BRADLEY J. SCHNACKENBERG ET AL., COURTESY OF ROBERT E. PALAZZO, PROC. NAT'L. ACAD. SCI. U.S.A. 95:9298, 1998.)

ure 9.19*a,b*) and it is connected to the center of the centriole by a radial spoke. The three microtubules of each triplet are arranged in a pattern that gives a centriole a characteristic pinwheel appearance. Centrioles are nearly always found in pairs, with the members situated at right angles to one another (Figure 9.19*a,c*). Extraction of isolated centrosomes with 1.0 M potassium iodide removes about 90 percent of the protein of the PCM, leaving behind a spaghetti-like scaffold of insoluble fibers (Fig. 9.19*d*). Examination of a series of sections through a cell shows that centrosomes are sites where large numbers of microtubules converge.

The origin of the microtubules of the cytoskeleton of a cultured animal cell can be best studied by depolymerizing the microtubules with cold temperature or chemicals, such as nocodazole or colchicine, and then following the reassembly of the microtubules after the cells are warmed or the chemicals are removed. The disassembly and reassembly of microtubules can be followed by fixing cells at various times and staining them with fluorescent antitubulin antibodies. Within a few minutes after removal of the inhibiting conditions, one or two bright fluorescent spots are seen in the cytoplasm of each cell. Within 15 to 30 minutes (Figure 9.20*a*), the number of labeled filaments radiating out of these foci increases dramatically. When these same cells are sectioned and examined in the electron microscope, the newly formed microtubules are found to radiate outward from a centrosome. Close examination shows that the microtubules do not actually penetrate into the centrosome and make contact with the centrioles, but terminate in the dense pericentriolar material that resides at the centrosome periphery. It is this material

that initiates the formation of microtubules (page 349). Although centrioles are not involved in microtubule nucleation, they probably play a role in recruiting the surrounding PCM during assembly of the centrosome.

The centrosome is typically situated near the center of a cell, just outside of the nucleus (as in Figure 9.20*a*). In columnar epithelial cells, the centrosome migrates from its position in the central region of the cell to a site in the apical region, just beneath the cortex. The microtubules of the cytoskeleton emanate from this site, extending toward the nucleus and basal surface of the cell (see Figure 9.1). Regardless of their location, centrosomes are sites of microtubule nucleation. The polarity of these microtubules is always the same: the minus end is associated with the centrosome, and the plus (or growing) end is situated at the opposite tip (Figure 9.20*b*). Thus, even though microtubules are nucleated at the MTOC, they are elongated at the opposite end of the polymer.

Not all of the microtubules of an animal cell are associated with a centrosome. The microtubules of an axon, for example, are not associated with the centrosome, which is located in the neuron's cell body. It is thought, however, that axonal microtubules are initially formed at the centrosome and then released from that MTOC and transported into the axon by motor proteins. Certain animal cells, including mouse oocytes, lack centrosomes entirely, yet they are still capable of forming complex microtubular structures, such as the meiotic spindle.

Basal Bodies and Other MTOCs Centrosomes are not the only MTOCs in cells. For example, the microtubules

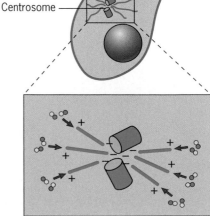

Centrosome

Nucleus

(a) 10 μm *(b)*

Figure 9.20 Microtubule nucleation at the centrosome.
(*a*) Fluorescence micrograph of a cultured fibroblast that had been exposed to colcemid to bring about the disassembly of the cell's microtubules and was then allowed to recover for 30 minutes before treatment with fluorescent anti-tubulin antibodies. The bright starlike structure marks the centrosome, a micro-

tubule-organizing center (MTOC) from which newly assembled microtubules have begun to grow in all directions. (*b*) Schematic diagram of the regrowth of microtubules showing the addition of subunits at the plus end of the polymer away from the centrosome. (A: FROM MARY OSBORN AND KLAUS WEBER, PROC. NAT'L ACAD. SCI. U. S. A., 73:869, 1976.)

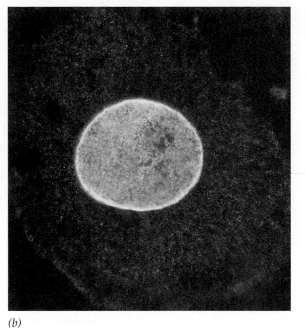

(a) 4 μm (b)

Figure 9.21 MTOC of a plant cell. (*a*) Localization of microtubules of a *Haemanthus* endosperm cell during interphase (between mitotic divisions). The microtubules are seen to emanate from the surface of the nucleus. (*b*) The same cell seen in *a* that has been stained with an antibody directed against purified calf centrosomes. The outer surface of the plant cell nucleus appears brightly fluorescent, indicating the presence of proteins homologous to those of animal cell centrosomes. (COURTESY OF ANNE-MARIE LAMBERT.)

in a cilium or flagellum originate from a structure called a **basal body**, which resides at the base of the projecting organelle (see Figure 9.33). Basal bodies are identical in structure to centrioles, and in fact, basal bodies and centrioles can give rise to one another. The flagellum of a sperm cell, for example, originates as a basal body derived from a centriole that had been part of the meiotic spindle of the spermatocyte from which the sperm arose. Conversely, the sperm basal body typically becomes a centriole during the first mitotic division of the fertilized egg.

Plant cells lack both centrosomes and centrioles, and their MTOCs are more dispersed than those of animal cells. In plant endosperm cells, for example, the primary MTOC consists of patches of material situated at the outer surface of the nuclear envelope from which microtubules of the cytoskeleton emerge (Figure 9.21). Microtubule nucleation is also thought to occur throughout the plant cell cortex.

Microtubule Nucleation Regardless of their diverse appearance, all MTOCs play similar roles in all cells; they control the number of microtubules, their polarity, the number of protofilaments that make up their walls, and the time and location of their assembly. In addition, all MTOCs share a common protein component—a type of tubulin discovered in the mid-1980s, called **γ-tubulin**.[1]

Unlike the α- and β-tubulins, which make up about 2.5 percent of the protein of a nonneural cell, γ-tubulin constitutes only about 0.005 percent of the cell's total protein. Fluorescent antibodies to γ-tubulin stain all types of MTOCs, including the pericentriolar material of centrosomes (Figure 9.22*a*), suggesting that γ-tubulin is a critical component in microtubule nucleation. This conclusion is supported by other studies. For example, microinjection of anti-γ-tubulin antibodies into a living cell blocks the reassembly of microtubules following their depolymerization by drugs or cold temperature. Similarly, genetically engineered fungal cells that are unable to synthesize γ-tubulin are also unable to assemble normal microtubules.

To understand the mechanism of microtubule nucleation, researchers have focused recently on the structure and composition of the pericentriolar material (PCM) at the periphery of centrosomes. The insoluble fibers of the PCM (Figure 9.19*d*) are thought to serve as attachment sites for ring-shaped structures that have the same diameter as microtubules (25 nm) and contain γ-tubulin. These ring-shaped structures were discovered when centrosomes were purified and incubated with gold-labeled antibodies that bound to γ-tubulin. The gold particles were seen to be clustered in semicircles or rings situated at the minus ends of microtubules (Figure 9.22*b*). These are the ends of the microtubules that are embedded in the PCM of the centrosome where nucleation occurs. Similar γ-tubulin ring complexes (called γ-*TuRCs*) have been isolated from

[1]Two other tubulin isoforms, δ-tubulin and ε-tubulin, have also been identified in centrosomes, but their function has not been determined.

(a)

(b)

30 nm

(c)

Figure 9.22 The role of γ-tubulin in centrosome function. (*a*) A dividing fibroblast that has been double stained with antibodies against γ-tubulin (red) and β-tubulin (green). Orange staining is due to the coincidence of the two types of tubulin, which occurs in the two centrosomes located at opposite poles of a dividing cell. (*b*) A reconstruction based on electron micrographs of a portion of a centrosome that had been incubated with purified tubulin in vitro and then labeled with antibodies against γ-tubulin. The antibodies were linked to gold particles to make them visible (as white dots) in the reconstruction. During the incubation with tubulin, the centrosome served as an MTOC to nucleate microtubules whose minus ends are seen to be labeled with clusters of gold, which are often arranged in semicircles or rings. The accompanying drawing shows the outline of the microtubule seen in the micrograph. (*c*) A model for γ-tubulin function during the assembly of microtubules. Nucleation begins as αβ-tubulin dimers bind to an open ring of γ-tubulin molecules, which are held in place by material contributed by the MTOC. This establishes the orientation of tubulin dimers in protofilaments and thus defines microtubule polarity. In the model shown here, the α-tubulin subunits are associated with the γ-tubulin rings at the minus end of the microtubule. (A: COURTESY OF M. KATHERINE JUNG AND BERL R. OAKLEY; B: REPRODUCED WITH PERMISSION FROM MICHELLE MORITZ ET AL., NATURE 378:639, 1995. PHOTO COURTESY OF DAVID A. AGARD. COPYRIGHT 1995 BY MACMILLAN MAGAZINES LIMITED.)

cell extracts and shown to nucleate microtubule assembly in vitro. These and other studies have suggested the model shown in Figure 9.22*c* in which a helical array of 13 γ-tubulin subunits forms an open, ring-shaped template on which the first row of αβ-tubulin dimers assemble. In this model, only the α-tubulin of the heterodimer can bind to a ring of γ-subunits. In this way, the polarity of the entire microtubule is established.

The Dynamic Properties of Microtubules

Although all microtubules appear quite similar morphologically, there are marked differences in their stability. Microtubules of the mitotic spindle or the cytoskeleton are extremely labile, that is, sensitive to disassembly. Microtubules of mature neurons are much less labile, and those of cilia and flagella are highly stable. Microtubules of this latter group are stabilized by the attachment of MAPs and by enzymatic modification (e.g., acetylation) of specific amino acid residues within the tubulin subunits. Living cells can be subjected to a variety of treatments that lead to the disassembly of labile cytoskeletal

microtubules without disrupting other cellular structures. Disassembly can be induced by cold temperature; hydrostatic pressure; elevated Ca^{2+} concentration; and a variety of chemicals, including colchicine, vinblastine, vincristine, nocodazole, and podophyllotoxin. The drug taxol disrupts the dynamic activities of microtubules by a very different mechanism. Taxol binds to the microtubule polymer, inhibiting its disassembly, thereby preventing the cell from assembling new microtubular structures as required. Many of these compounds, including taxol, are used in chemotherapy against cancer because they preferentially kill tumor cells. It had been assumed for years that tumor cells were particularly sensitive to these drugs because of their high rate of cell division. But recent research has revealed there is more to the story. As discussed in Chapter 14, normal cells have a mechanism (or checkpoint) that stops them from dividing in the presence of

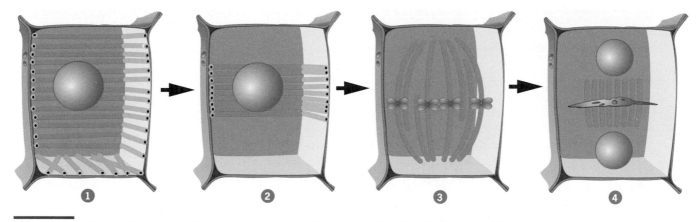

Figure 9.23 **Four major arrays of microtubules present during the cell cycle of a plant cell.** The organization of the microtubules at each stage are described in the text. (AFTER R. H. GODDARD ET AL., PLANT PHYSIOL. 104:2, 1994.)

drugs, such as vinblastine or taxol, that alter the mitotic spindle. As a result, normal cells usually arrest their division activities until the drug has been eliminated from the body. In contrast, many cancer cells lack this mitotic checkpoint and attempt to complete their division even in the absence of a functioning mitotic spindle. This usually results in the death of the tumor cell.

The lability of cytoskeletal microtubules reflects the fact that they are polymers formed by the noncovalent association of dimeric building blocks. The microtubules of the cytoskeleton are normally subject to depolymerization and repolymerization as the requirements of the cell change from one time to another. The dynamic character of the microtubular cytoskeleton is well illustrated by plant cells. If a typical plant cell is followed from one mitotic division to the next, four distinct arrays of microtubules appear, one after another (Figure 9.23).

1. During most of interphase, the microtubules of a plant cell are distributed widely throughout the cortex, as depicted in Figure 9.23, stage 1.

2. As the cell approaches mitosis, the microtubules disappear from most of the cortex, leaving only a single transverse band, called the *preprophase band*, that encircles the cell like a belt (Figure 9.23, stage 2). The preprophase band marks the site of the future division plane.

3. As the cell progresses into mitosis, the preprophase band is lost and microtubules reappear in the form of the mitotic spindle (Figure 9.23, stage 3).

4. After the chromosomes have been separated, the mitotic spindle disappears and is replaced by a bundle of microtubules called the *phragmoplast* (Figure 9.23, stage 4), which plays a role in the formation of the cell wall that separates the two daughter cells (Section 14.2).

These dramatic changes in the spatial organization of microtubules are thought to be accomplished by a combination of two separate mechanisms: (1) the rearrangement of existing microtubules and (2) the disassembly of existing microtubules and reassembly of new ones in different regions of the cell. In the latter case, the microtubules that make up the cortical array are formed from the same subunits that a few minutes earlier were part of the phragmoplast or, before that, the mitotic spindle.

Insight into factors that influence the rates of microtubule growth and shrinkage has come primarily from studies of microtubule assembly and disassembly in vitro.

The Study of Microtubule Dynamics in Vitro The first successful approach to the in vitro assembly of microtubules was taken in 1972 by Richard Weisenberg of Temple University. Reasoning that cell homogenates should possess all the macromolecules required for the assembly process, Weisenberg obtained tubulin polymerization in brain homogenates at 37°C by adding Mg^{2+}, GTP, and EGTA (which binds Ca^{2+}, an inhibitor of polymerization). Weisenberg found that microtubules could be disassembled and reassembled over and over simply by lowering and raising the temperature of the incubation mixture. Figure 9.24 shows three microtubules that were assembled in the test tube from purified tubulin. It can be seen that one of the three microtubules contains only 11 protofilaments (as indicated by its thinner diameter). It is not unexpected that microtubules assembled in vitro have abnormal protofilament numbers because they lack a 13-subunit template (Figure 9.22c) normally provided by the γ-tubulin ring complexes in vivo. In vitro assembly of microtubules is aided greatly by the addition of pieces of microtubules or structures that contain microtubules (Figure 9.25), which serve as templates for the addition of free subunits. As occurs in vivo, tubulin subunits are added primarily to the plus end of the existing polymer.

Early in vitro studies established that GTP is required for microtubule assembly. Assembly of tubulin dimers requires that a GTP molecule be bound to the β-tubulin

Figure 9.24 Microtubules assembled in the test tube. Electron micrograph of frozen, unfixed microtubules that had polymerized in vitro. The individual globular tubulin subunits are visible, as are the protofilaments into which the subunits are organized. Note that the middle of the three microtubules contains only 11 protofilaments. (COURTESY OF R. H. WADE, INSTITUT DE BIOLOGIE STRUCTURALE, GRENOBLE, FRANCE.)

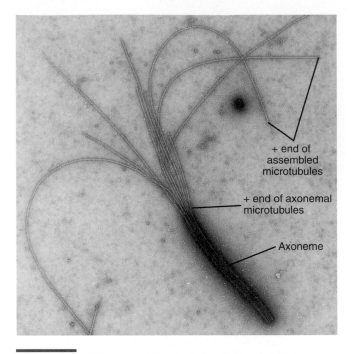

Figure 9.25 The assembly of tubulin onto existing microtubular structures. Electron micrograph showing the in vitro assembly of brain tubulin onto the plus ends of the microtubules of an axoneme from the flagellum of *Chlamydomonas*. (COURTESY OF L. I. BINDER AND JOEL L. ROSENBAUM.)

subunit.[2] GTP hydrolysis is not required for the actual incorporation of the dimer onto the end of a microtubule. Rather, the GTP is hydrolyzed to GDP shortly *after* the dimer is incorporated into a microtubule, and the resulting GDP remains bound to the assembled polymer. After a dimer is released from a microtubule during disassembly and enters the soluble pool, the GDP is replaced by a new GTP. This nucleotide exchange "recharges" the dimer, allowing it to serve once again as a building block for polymerization.

Because it includes GTP hydrolysis, assembly of a microtubule is not an inexpensive cellular activity. Why has such a costly polymerization pathway evolved? The most likely answer is that it allows a cell to control the rate of two opposing reactions—assembly and disassembly—independently. A dimer being added to a microtubule has a bound GTP, whereas a dimer being released from a microtubule has a bound GDP. These two types of subunits have different conformations, and consequently, they participate in different reactions. Because of these differences, the ends of growing and shrinking microtubules are seen under the electron microscope to have different structures. When a microtubule is growing, the plus end is present as an open sheet to which GTP-dimers are added (step 1, Figure 9.26). During times of rapid microtubule growth, tubulin dimers are added more rapidly than their GTP can be hydrolyzed. The presence of a cap of tubulin-

GTP dimers at the ends of the protofilaments is thought to favor the addition of more subunits and the growth of the microtubule. However, microtubules with open ends, as in step 1, Figure 9.26, are thought to undergo a spontaneous reaction that leads to closure of the tube (steps 2, 3). In this model, tube closure is accompanied by the hydrolysis of the bound GTP, which changes the conformation of the tubulin dimers. The resultant mechanical strain destabilizes the microtubule. The strain is released as the protofilaments curl outward from the tubule and undergo catastrophic depolymerization (step 4). Microtubules can shrink remarkably fast, especially in vivo, which allows a cell to disassemble its microtubular cytoskeleton very rapidly.

The Study of Microtubule Dynamics in Vivo The dynamic character of the microtubular cytoskeleton inside a cell is best revealed by microinjecting labeled tubulin into a nondividing cultured cell. The labeled subunits are rapidly incorporated into the preexisting microtubules of the cytoskeleton, even in the absence of any obvious morphologic change (Figure 9.27). When a cell containing fluorescently labeled microtubules is observed over time in a fluorescence microscope, the dynamic character of these polymers can be directly observed (Figure 9.28). At any given moment, some microtubules are growing in length, while others are shrinking. Observations of living cells suggest that both growth and shrinkage in vivo occur pre-

[2]A molecule of GTP is also bound to the α-tubulin subunit, but it is not exchangeable and it is not hydrolyzed following subunit incorporation. The positions of the guanine nucleotides in the αβ-tubulin heterodimer are shown in Figure 9.8c.

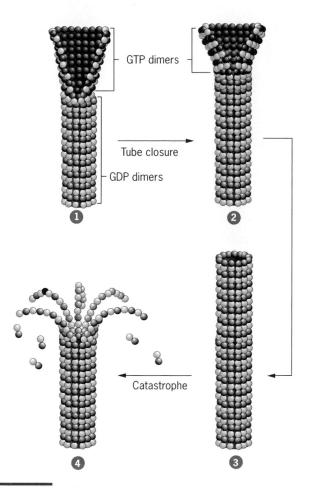

Figure 9.26 **The structural cap model of dynamic instability.** According to the model, the growth or shrinkage of a microtubule depends on the state of the tubulin dimers at the microtubule tip. Tubulin-GTP dimers are depicted in red. Tubulin-GDP dimers are depicted in blue. (step 1) In a growing microtubule, the tip consists of an open sheet containing tubulin-GTP subunits. (step 2) The tube has begun to close, forcing the hydrolysis of the bound GTP. (step 3) The tube has closed to its end, leaving only tubulin-GDP subunits. (step 4) The strain resulting from GTP hydrolysis is released as the protofilaments curl outward from the tubule and undergo catastrophic shrinkage. (FROM A. A. HYMAN AND E. KARSENTI, CELL 84:402, 1996; BY PERMISSION OF CELL PRESS.)

Figure 9.27 **Microtubule dynamics in living cells.** This cultured fibroblast was injected with a small volume of tubulin that had been covalently linked to biotin, a small molecule whose location in the cell is readily determined using fluorescent antibiotin antibodies. Approximately one minute following injection, the cells were fixed, and the location of biotinylated tubulin that had been incorporated into insoluble microtubules was determined. It is evident from this fluorescence micrograph that, even during periods as short as one minute, tubulin subunits are widely incorporated at the growing ends of cytoskeletal microtubules. (FROM MARC KIRSCHNER, J. CELL BIOL. 102, COVER OF NO. 3, 1986; BY COPYRIGHT PERMISSION OF THE ROCKEFELLER UNIVERSITY PRESS.)

dominantly at the plus end of the polymer—the end located opposite the centrosome (or other MTOC).

If individual microtubules are followed over time, they appear to switch back and forth unpredictably between growing and shrinking phases. Because shrinking occurs more rapidly than elongation, most microtubules disappear from the cell in a matter of minutes and are replaced by new microtubules that grow out from the centrosome. In 1984, Timothy Mitchison and Marc Kirschner of the University of California, San Francisco, proposed that microtubule behavior could be explained by a phenomenon they termed **dynamic instability**. Dynamic instability refers to the fact that growing and shrinking microtubules can coexist in the same region of a cell, and that a given microtubule can switch back and forth between growing and shortening phases (Figure 9.28). Cells can regulate the rates of microtubule growth or shrinkage, and the frequency of interconversion between the two phases, by altering prevailing conditions in the cytoplasm.

Microtubule dynamics has been best studied in extracts from amphibian eggs, which contain factors that promote both the assembly and disassembly of mitotic spindles. These eggs typically contain a MAP (XMAP215) that enhances microtubule assembly and a kinesin-like protein (XKCM1) that enhances microtubule disassembly. The average length of microtubules in an egg extract depends on the relative activities of these (and other) regulatory proteins. The rate of disassembly in a cell may also be enhanced by proteins such as katanin (the word for Samurai sword in Japanese), which sever microtubules into fragments, thereby increasing the number of free ends available for depolymerization. By regulating the equilibrium between assembly and disassembly of microtubules, cells can respond rapidly to changing conditions that require a remodeling of the structure of the microtubular cytoskeleton. Unlike the microtubules of the

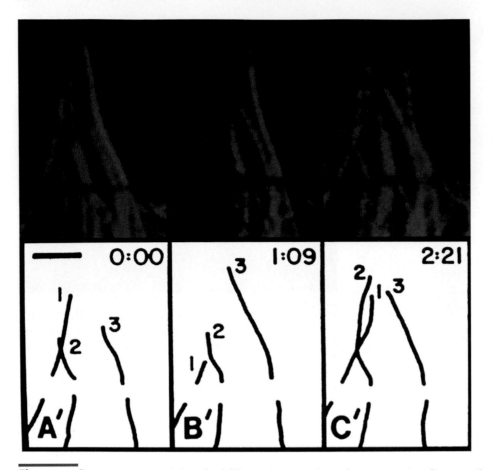

Figure 9.28 Dynamic instability. (*a–c*) These photographs and accompanying diagrams illustrate the changes in length of individual microtubules within living cells. The cells were injected with rhodamine-labeled tubulin, and the fluorescence was followed with video microscopy. A reference mark was produced across the microtubules by photobleaching. During the period of observation (indicated in the upper right of drawings), the end of microtubule 1 shrank and then grew with respect to its bleached domain, microtubule 2 grew, and microtubule 3 grew and then shrank. Length changes were due to the gain and loss of subunits distal to the bleached domain on each microtubule. These photographs illustrate the dynamic instability of microtubules. Bar, 3 μm. (FROM PAUL J. SAMMAK AND GARY G. BORISY, REPRINTED WITH PERMISSION FROM NATURE 332: 725, 1988; © COPYRIGHT 1988 BY MACMILLAN MAGAZINES LTD.)

cytoskeleton, the microtubules of the organelles to be discussed in the following section lack dynamic activity and, instead, are highly stable.

Cilia and Flagella: Structure and Function

Anyone who has placed a drop of pond water under the lens of a microscope and tried to keep a protozoan from swimming out of the field of view can appreciate the activity of cilia and flagella. **Cilia** and **flagella** are hairlike, motile organelles that project from the surface of a variety of eukaryotic cells. Bacteria also possess structures referred to as *flagella*, but prokaryotic flagella are simple filaments that bear no evolutionary relationship to their eukaryotic counterparts (see Figure 1.14). The discussion that follows pertains only to the eukaryotic organelles.

Cilia and flagella are two versions of the same structure and are best distinguished by their patterns of move-

ment. A cilium can be likened to an oar as it moves the cell in a direction perpendicular to the cilium itself. In its power stroke, the cilium is maintained in a rigid state (Figure 9.29*a*) as it pushes against the surrounding medium. In its recovery stroke, the cilium becomes flexible, offering little resistance to the medium. Cilia tend to occur in large numbers on a cell's surface, and their beating activity is usually coordinated (Figure 9.29*b*). In multicellular organisms, cilia move fluid and particulate material through various tracts (Figure 9.29*c*). In humans, for example, the ciliated epithelium lining the respiratory tract propels mucus and trapped debris away from the lungs.

Recent studies have revealed a fascinating role for cilia in determining the basic plan of the vertebrate body, including that of humans. When you look into a mirror, you see a relatively symmetrical organism, one whose left half is essentially a mirror image of the right half. A surgeon,

on the other hand, sees a strikingly asymmetric organism when they open a person's thoracic or abdominal cavity. The stomach, heart, and spleen, for example, are shifted to the left side of the body, whereas the liver is located on the right side. On occasion, a physician will see a patient in which the left-right asymmetry of the visceral organs is reversed (a condition called *situs inversus*). The basic body plan of a mammal is laid down during gastrulation in association with a structure called the *embryonic node*. Each cell that makes up the node has a single cilium. Studies on a strain of mouse embryos that developed *situs inversus* with high frequency revealed that the cells of the embryonic node possessed abnormal cilia. Examination of normal embryos revealed that the cilia of their nodal cells did not beat in typical back-and-forth fashion (as in Figure

9.29*a*), but instead *rotated* in a counterclockwise direction. The counterclockwise rotation of these cilia moves the surrounding fluid to the left side of the embryonic midline, as was demonstrated by tracking the movement of microscopic fluorescent beads. It is proposed that the fluid moved by the nodal cilia contains morphogenetic substances (i.e., substances that direct embryonic development) that become concentrated on the left side of the embryo, leading to the eventual formation of different organs on different sides of the midline.

Flagella are usually longer than cilia, and cells have fewer of them (Figure 9.30). Flagella exhibit a variety of different beating patterns (*waveforms*), depending on the cell type. For example, the single-celled alga pictured in Figure 9.30*a* pulls itself forward by waving its two flagella in an asymmetric manner that resembles the breast stroke of a human swimmer (Figure 9.30*b*). This same algal cell can also push itself through the medium using a symmetric beat resembling that of a sperm (shown in Figure 9.34). The degree of asymmetry in the pattern of the beat in the algal cell is regulated by the internal calcium ion concentration.

The Structure of Cilia and Flagella

An electron micrograph of a cross section of a cilium or flagellum is one of the most familiar images in cell biology (Figure 9.31*a*). The entire ciliary or flagellar projection is

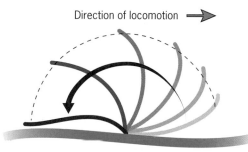

Direction of locomotion

Power stroke

Recovery stroke

(*a*)

(*b*) 10 µm

(*c*) 3.5 µm

Figure 9.29 Ciliary beat. (*a*) The various stages in the beat of a cilium. (*b*) The cilia on the surface of a ciliated protozoan beat in metachronal waves in which the cilia in a given row are in the same stage of the beat cycle, but those in adjacent rows are in a different stage. RS, cilia in recovery stroke; ES, cilia in effective

power stroke. (*c*) Cilia on the surface of the lip (fimbrium) of a mouse oviduct. (B: REPRINTED WITH PERMISSION FROM G. A. HORRIDGE AND S. L. TAMM, SCIENCE 163:818, 1969, © COPYRIGHT 1969, AMERICAN ASSOCIATION FOR THE ADVANCEMENT OF SCIENCE; C: COURTESY OF ELLEN R. DIRKSEN.)

(a)

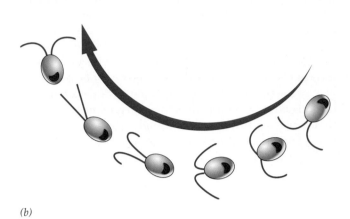

(b)

Figure 9.30 A eukaryotic flagellum. (*a*) The unicellular bi-flagellate alga *Chlamydomonas reinhardtii*. The two flagella appear green due to a fluorescent antibody directed against a major flagellar membrane protein. The red color results from autofluorescence of the cell's chlorophyll. Unlike many flagellated organisms, *Chlamydomonas* does not need its flagella to survive and reproduce, so that mutant strains exhibiting various types of flagellar defects can be cultured. (*b*) The forward movement of *Chlamydomonas* is accomplished by an asymmetric waveform that resembles the breast stroke. A different type of flagellar waveform is shown in Figure 9.34. (A: COURTESY OF ROBERT BLOODGOOD, UNIVERSITY OF VIRGINIA.)

covered by a membrane that is continuous with the plasma membrane of the cell. The core of the cilium, called the **axoneme**, contains an array of microtubules that run longitudinally through the entire organelle. In nearly every case, the axoneme consists of nine peripheral doublet microtubules surrounding a central pair of single microtubules. This axonemal structure is known as the "9 + 2 array." All microtubules of the axoneme have the same po-

(a) 50 nm

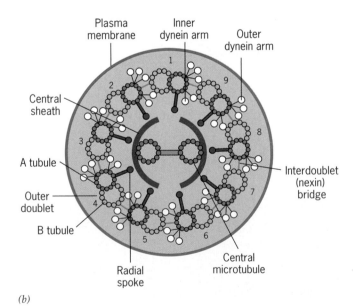

(b)

Figure 9.31 The structure of a ciliary or flagellar axoneme.
(*a*) Cross section through a sperm axoneme. The peripheral doublets are seen to consist of a complete and an incomplete microtubule, whereas the two central microtubules are complete. The dynein arms are seen as "fuzzy" projections from the wall of the complete microtubule. (*b*) Schematic diagram of an axoneme of a protist showing the structure of the microtubular fibers, the two types of dynein arms (three-headed outer arms and two-headed inner arms), the nexin links between the doublets, the central sheath surrounding the central microtubules, and the radial spokes projecting from the outer doublets toward the central sheath. (Note: The cilia and flagella of animals typically contain two-headed outer arms.) (A: COURTESY OF LEWIS G. TILNEY AND K. FUJIWARA.)

larity: their plus ends are at the tip of the projection, and their minus ends at the base. Each peripheral doublet consists of one complete microtubule, the *A tubule* , and one incomplete microtubule, the *B tubule*, the latter containing 10 or 11 subunits rather than the usual 13.

Not all eukaryotic organisms have cilia and flagella; they are generally absent, for example, among fungi, nematodes, and insects. However, where they do occur, cilia and flagella nearly always exhibit the same 9 + 2 array depicted in Figure 9.31a, another of the many reminders that all living eukaryotes have evolved from a common ancestor. Although the 9 + 2 pattern has been highly conserved, there are a number of notable evolutionary departures from this structural mode. For example, a 9 + 1

pattern occurs in flatworms, and a 9 + 0 pattern in the Asian horseshoe crab, eel, and mayfly. Some flagella that lack central elements are motile; others are not.

The basic structure of the axoneme was first described in 1954 by Don Fawcett and Keith Porter of Harvard University. As the resolving power of electron microscopes improved, certain of the less obvious components became visible (Figure 9.31b). The central tubules were seen to be enclosed by projections that form the *central sheath*, which is connected to the A tubules of the peripheral doublets by a set of *radial spokes*. The doublets are connected to one another by an *interdoublet bridge* composed of an elastic protein, nexin. Of particular importance was the observation that a pair of "arms"—an *inner arm* and an *outer arm*—project from the A tubule in a clockwise direction (when viewing the cilium from the base toward the tip, as in Figure 9.31a). A longitudinal section, that is, one cut through the axoneme parallel to its long axis, reveals the continuous nature of the microtubules and the discontinuous nature of the other elements (Figure 9.32a). The radial spokes, for example, occur typically in groups of three with a major repeat of 96 nm (Figure 9.32b).

As noted earlier, a cilium or flagellum emerges from a *basal body* (Figure 9.33a) similar in structure to the centriole of Figure 9.19a. Like a cilium or flagellum, a basal

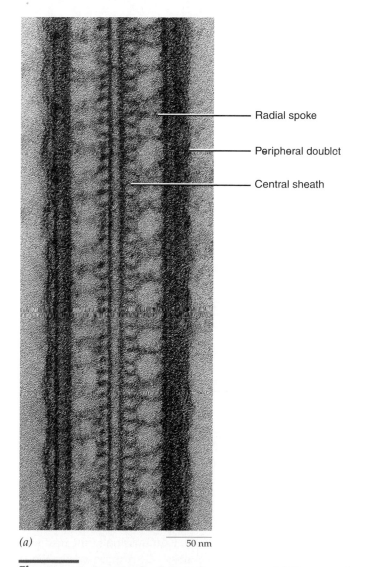

(a) 50 nm

- Radial spoke
- Peripheral doublet
- Central sheath

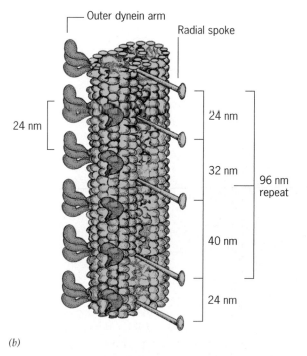

(b)

- Outer dynein arm
- Radial spoke
- 24 nm
- 24 nm
- 32 nm
- 40 nm
- 24 nm
- 96 nm repeat

Figure 9.32 Longitudinal view of an axoneme. (*a*) Electron micrograph of a median longitudinal section of a straight region of a cilium. The radial spokes are seen joining the central sheath with the A microtubule of the doublet. (*b*) Schematic diagram of a longitudinal section of a flagellar doublet. Radial spokes emerge in groups of three, which repeat (at 96 nm in this case) along the length of the microtubule. Outer dynein arms are spaced at intervals of 24 nm. (A: FROM FRED D. WARNER AND PETER SATIR, J. CELL BIOL. 63:41, 1974; BY COPYRIGHT PERMISSION OF THE ROCKEFELLER UNIVERSITY PRESS)

(a)　　　　　　　　　　　　　　*(b)*

Figure 9.33 Basal bodies and axonemes. (*a*) Electron micrograph of a longitudinal section through the basal bodies of a number of cilia at the apical surface of the epithelial cells of a rabbit oviduct. (*b*) Schematic diagram of the structural relationship between the microtubules of the basal body and the axoneme of a cilium or flagellum. (A: COURTESY OF R. G. W. ANDERSON.)

body contains nine peripheral fibers, but each consists of three microtubules rather than two. The A tubule is complete, while the B and C tubules are both incomplete. There are no central microtubules in either the basal body or centriole. The A and B tubules of the basal body elongate to form the doublets of the cilium or flagellum (Figure 9.33*b*). If a cilium or flagellum is sheared from the surface of a living cell, a new organelle is regenerated as an outgrowth of the basal body.

The Dynein Arms The machinery for ciliary and flagellar locomotion resides within the axoneme. This is illustrated in the experiment shown in Figure 9.34 in which a sperm tail axoneme devoid of its overlying membrane is still capable of normal, sustained beating in the presence of added Mg^{2+} and ATP. The greater the ATP concentration, the faster the beat frequency of these "reactivated" organelles.

The protein responsible for conversion of the chemical energy of ATP into the mechanical energy of ciliary locomotion was isolated in the 1960s by Ian Gibbons of Harvard University. Gibbons' experiments provide an elegant example of the relationship between structure and function in biological systems and the means by which the relationship can be revealed by experimental analysis. Using various solutions capable of solubilizing different components, Gibbons carried out a chemical dissection of cilia from the protozoan *Tetrahymena* (Figure 9.35). An

intact cilium is shown in cross section in step 1. The dissection began by dissolving the enclosing plasma membrane in the detergent digitonin (step 2). The isolated axonemes of the insoluble fraction were then immersed in a solution containing EDTA, a compound that binds and removes (or chelates) divalent ions. When EDTA-treated axonemes were examined in the electron microscope, the central tubules were missing, as were the arms projecting from the A tubules (step 3). Simultaneous with the loss of these structures, the insoluble axonemes lost their ability to hydrolyze ATP, while the supernatant gained it. The ATPase found in the supernatant was a huge protein (up to 2 million daltons) that Gibbons named *dynein* (from the words *dyne*, meaning "force," and *protein*). This protein is now referred to as **ciliary** (or **axonemal**) **dynein** to distinguish it from cytoplasmic dynein, a related protein involved in organelle transport (page 345). When the insoluble parts of the axoneme were mixed with the soluble protein in the presence of Mg^{2+}, much of the ATPase activity once again became associated with the insoluble material in the mixture (step 4). Examination of the insoluble fraction showed that the arms had reappeared on the A tubules of the axonemes. Gibbons concluded that the arms contained an ATPase that released the energy required for locomotion.

In subsequent investigations it was found that treatment of isolated sperm axonemes with high salt (0.6 M

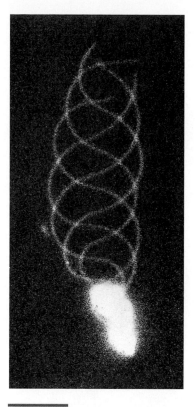

Figure 9.34 **A sea urchin sperm reactivated with 0.2 mM ATP following demembranation with the detergent Triton X-100.** This multiple-exposure photomicrograph was obtained with five flashes of light and shows the reactivated flagellum at different stages of its beat. (FROM CHARLES J. BROKAW AND T. F. SIMONICK, J. CELL BIOL. 75:650, 1977; BY COPYRIGHT PERMISSION OF THE ROCKEFELLER UNIVERSITY PRESS.)

NaCl) would selectively remove the outer arms, leaving the inner arms in place (step 5). When ATP was added to axonemes lacking outer arms, they beat at about half the rate of the intact axoneme, though they beat with a normal waveform.

Electron microscopic examination of the outer arms released by high salt from *Tetrahymena* axonemes showed them to contain three globular heads attached by slender stems to a common base (Figure 9.36*a,b*). These heads function as ATP-hydrolyzing cross-bridges of the type illustrated in Figure 9.36*c*. In the intact axoneme, the base of the protein is tightly anchored to the outer surface of the A tubule, with the globular heads projecting toward the B tubule of the neighboring doublet (Figure 9.36*b*, 9.37). The inner arms consist of a number of distinct types containing either one or two globular heads.

The Mechanism of Ciliary and Flagellar Locomotion As will be discussed later in this chapter, the contraction of muscle results from the sliding of actin filaments over adjacent myosin filaments. The sliding force is generated by ratchetlike cross-bridges that reside in the head of the myosin molecule. With the muscle system as a model, it was proposed that ciliary movement could be explained by

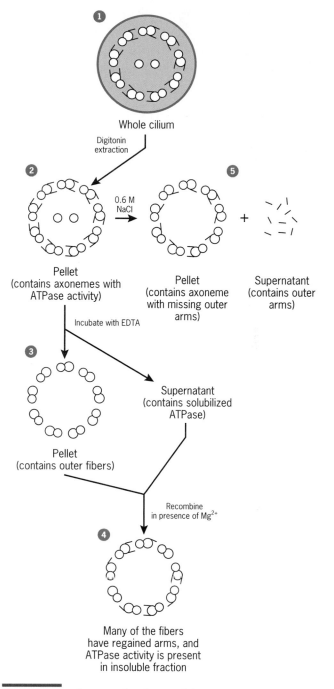

Figure 9.35 **Steps in the chemical dissection of cilia from the protozoan** *Tetrahymena.* The numbered steps are described in the text.

the sliding of adjacent microtubular doublets relative to one another. According to this proposal, the dynein arms act as swinging cross-bridges that generate the forces required for ciliary or flagellar movement. In essence, the dynein arms that project from one doublet "walk" along the wall of an adjacent doublet, causing the two doublets to slide relative to one another. The sequence of events is shown in Figure 9.37. In step 1, the dynein arms anchored

(b)

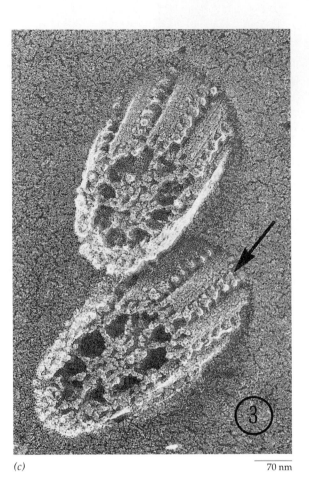

(c) ———— 70 nm

Figure 9.36 Axonemal dynein. (*a*) Platinum replica of a rotary-shadowed, outer-arm flagellar dynein molecule prepared by rapid-freeze, deep-etch electron microscopy. Each of the three heavy chains forms a prominent globular head with an extension that functions in linking the dynein arm to the neighboring doublet. An interpretive drawing is shown on the right. (*b*) A schematic drawing of an outer arm from the alga *Chlamydomonas*. The protein consists of three heavy chains that are named α, β, and γ, two intermediate chains (green), and a number of light chains (orange). The motor protein is attached to the A tubule of the doublet by one of the intermediate chains. (*c*) Electron micrograph of a replica of *Chlamydomonas* axonemes that had been rapidly frozen, fractured, and deep etched. A row of dynein arms is indicated by the arrow. (A: COURTESY OF JOHN E. HEUSER AND URSULA W. GOODENOUGH; C: FROM U. W. GOODENOUGH AND J. E. HEUSER, J. CELL BIOL. 95:800, 1982; BY COPYRIGHT PERMISSION OF THE ROCKEFELLER UNIVERSITY PRESS.)

along the A tubule of the lower doublet attach to binding sites on the B tubule of the upper doublet. In step 2, the dynein molecules undergo a conformational change that causes the lower doublet to slide toward the basal end of the upper doublet. In step 3, the dynein arms have detached from the B tubule of the upper doublet. In step 4 the arms have reattached to the upper doublet so that another cycle can begin (step 4).

Sliding on one side of the axoneme alternates with sliding on the other side so that a part of the cilium or flagellum bends first in one direction and then in the opposite direction (Figure 9.38*a*). This requires that, at any given time, the dynein arms on one side of the axoneme are active, while those on the other side are inactive. As a result of this difference in dynein activity, the doublets on the inner side of the bend (at the top and bottom of Figure 9.38*a*) extend beyond those on the opposite side of the axoneme. The differences in the relative positions of the doublets depicted in Figure 9.38*a* can be verified by electron microscopic examination of the tips of axonemes in different stages of the bending cycle (Figure 9.38*b*).

Evidence has steadily accumulated in favor of the sliding-microtubule theory and the role of the dynein arms. Microtubule sliding in beating flagellar axonemes

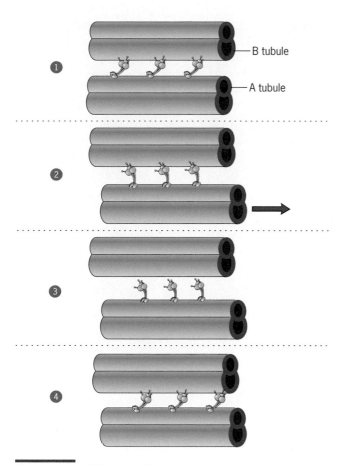

- B tubule
- A tubule

Figure 9.37 **The role of dynein arms in generating the force that drives ciliary or flagellar motility.** The steps are described in the text.

(a)

Figure 9.38 **The sliding-microtubule mechanism of ciliary or flagellar motility.** (*a*) Schematic diagram of the sliding of neighboring microtubules relative to one another. When the cilium is straight, all the outer doublets end at the same level (*center*). Cilium bending occurs when the doublets on the inner side of the bend slide beyond those on the outer side (*top and bottom*). The force responsible for the sliding of neighboring microtubules was shown in the previous figure. (*b*) Electron micrograph of the tip of a splayed axoneme of a motile cilium. The axoneme attaches to the grid and opens at random in the process of splaying out. As predicted by the sliding microtubule model, the outer tip of doublet 9 extends beyond that of doublet 5. This amount of tip displacement (0.36 μm) corresponds to a bend of just over 100°. (A: FROM D. VOET AND J. G. VOET, BIOCHEMISTRY, 2D ED., COPYRIGHT © 1995 JOHN WILEY AND SONS, INC. REPRINTED BY PERMISSION OF JOHN WILEY & SONS, INC. B: FROM W. S. SALE AND P. SATIR, J. CELL BIOL. 71:598, 1976; BY COPYRIGHT PERMISSION OF THE ROCKEFELLER UNIVERSITY PRESS.)

(b) 0.3 μm

6 μm

Figure 9.39 **Experimental demonstration of microtubule sliding.** Sea urchin sperm were demembranated, reactivated with ATP, and photographed by multiple exposure as in Figure 9.34. In this experiment, gold beads were attached to the exposed outer doublet microtubules, where they could serve as markers for specific sites along the microtubules. As the flagellum underwent beating, the relative positions of the beads were monitored. As shown here, the beads moved farther apart and then closer together as the flagellum undulated, indicating that the doublets are sliding back and forth relative to one another. (FROM CHARLES J. BROKAW, J. CELL BIOL. 114, COVER OF NO. 6, 1991; BY COPYRIGHT PERMISSION OF THE ROCKEFELLER UNIVERSITY PRESS.)

has been directly visualized, as illustrated in the photographs of Figure 9.39. In this experiment, isolated axonemes were incubated with tiny gold beads that became attached to the exposed outer surface of the peripheral doublets. The beads served as fixed markers of specific sites on the doublets. The relative position of the beads was then monitored as the axonemes were stimulated to beat by addition of ATP. As the axonemes bent back and forth, the distances between the beads on different doublets increased and decreased in an alternating manner (Figure 9.39) as would be expected if neighboring doublets slid up and down over one another.

Regulating Ciliary and Flagellar Locomotion If one considers that cilia beat 10 to 40 times per second, that each stroke has a precise form, and that ciliary motion is usually coordinated so that thousands of cilia move together, it becomes clear that this locomotor activity must be tightly regulated. Regulation of ciliary and flagellar locomotion begins with the regulation of dynein arm activity. As noted above, not all dynein arms are active at the same time; if they were, the organelle would remain in a "frozen," or paralyzed, state. The central pair of microtubules and the radial spokes are thought to determine which dynein arms are active at any given instant. In a number of species studied, the central pair rotates as the cilium or flagellum beats. According to one hypothesis, as it rotates, the central pair periodically sweeps across each radial spoke (see Figure 9.31b), causing the spoke to send a signal to the dynein arm on the adjacent A tubule, which activates the arm to undergo its swinging movement. Studies suggest that the activation or inactivation of the dynein arm is accomplished by the removal or addition of phosphate groups from one or more of the polypeptides that make up the huge motor protein.

REVIEW **?**

1. Describe three functions of microtubules.
2. Contrast the apparent roles of kinesin and cytoplasmic dynein in axoplasmic transport.
3. What is an MTOC? Describe the structure of three different MTOCs and how each of them functions.
4. What is the role of GTP in the assembly of microtubules?

5. Contrast the movements of a flagellum and a cilium. Describe the mechanism by which these organelles are able to undergo their bending movements.

9.4 INTERMEDIATE FILAMENTS

For many years, electron microscopy had revealed the presence in numerous cell types of solid, smooth-surfaced, unbranched filaments that had a diameter of approximately 10 nm, which is intermediate between that of microtubules and microfilaments. Because of their relative dimensions, these structures were called **intermediate filaments** (or **IFs**). To date, intermediate filaments have only been identified with certainty in animal cells. IFs radiate through the cytoplasm of a wide variety of animal cells and are often interconnected to the other types of cytoskeletal filaments by thin, wispy cross-bridges (Figure 9.40). In most cells, these cross-bridges consist of a huge, elongated protein called *plectin* that can exist in several different isoforms. Each plectin molecule has a binding site for an intermediate filament at one end and, depending upon the isoform, a binding site for another intermediate filament, microtubule, or microfilament at the other end.

Unlike microfilaments and microtubules, IFs are a chemically heterogeneous group of structures that, in humans, are encoded by at least 50 different genes. The polypeptide subunits of IFs can be divided into six major classes based on their tissue distribution (Table 9.1) as well as biochemical, genetic, and immunologic criteria.

Figure 9.40 Cytoskeletal elements are connected to one another by protein crossbridges. Electron micrograph of a replica of a small portion of the cytoskeleton of a fibroblast after selective removal of actin filaments. Individual components have been digitally colorized to assist visualization. Intermediate filaments (blue) are seen to be connected to microtubules (red) by long wispy crossbridges consisting of the fibrous protein plectin (green). Plectin is localized by antibodies linked to colloidal gold particles (yellow). (COURTESY OF TATYANA SVITKINA AND GARY BORISY.)

Most, if not all, of these polypeptides have a similar arrangement of domains that allows them to form similar looking filaments. Most notably, the polypeptides of IFs all contain a central, rod-shaped, α-helical domain of sim-

Table 9.1 Properties and Distribution of the Major Mammalian Intermediate Filament Proteins

IF Protein	Sequence Type	Average Molecular Mass ($\times 10^{-3}$)	Estimated Number of Polypeptides	Primary Tissue Distribution
Keratin	I	40–56.5	15	Epithelia
Keratin	II	53–67	15	Epithelia
Vimentin	III	57	1	Mesenchymal cells
Desmin	III	53–54	1	Muscle
Glial fibrillary acidic protein (GFAP)	III	50	1	Glial cells astrocytes
Peripherin	III	57	1	Peripheral neurons
Neurofilament proteins				Neurons of central and peripheral nerves
NF-L	IV	62	1	
NF-M	IV	102	1	
NF-H	IV	110	1	
Lamin proteins				All cell types (Nuclear envelopes)
Lamin A	V	70	1	
Lamin B	V	67	1	
Lamin C	V	60	1	
Nestin	VI	240	1	Neuronal stem cells

Adapted from Kathryn Albers and Elaine Fuchs, *Int. Rev. Cytol.* 134:244, 1992.

Figure 9.41 A model of intermediate filament assembly and architecture. Each monomer has a pair of globular terminal domains separated by a long α-helical region (step 1). Pairs of monomers associate in parallel orientation with their ends aligned to form dimers (step 2). Depending on the type of intermediate filament, the dimers may be composed of identical monomers (homodimers) or nonidentical monomers (heterodimers). Dimers in turn associate in an antiparallel, staggered fashion to form tetramers (step 3), which are thought to be the basic subunit in the assembly of intermediate filaments. The organization of the tetrameric subunits within the filament is shown in step 4.

ilar length and homologous amino acid sequence. This domain is flanked on each side by globular domains of variable size and sequence (step 1, Figure 9.41). Two such polypeptides spontaneously interact as their α-helical rods wrap around each other in a coiled coil to form a ropelike dimer approximately 45 nm in length (step 2). Because the two polypeptides are aligned parallel to one another in the same orientation, the dimer has polarity, with one end defined by the C-termini of the polypeptides and the opposite end by their N-termini.

Intermediate Filament Assembly and Disassembly

The basic unit of IF assembly is thought to be a tetramer formed by two dimers that become aligned side-by-side in a staggered fashion with their N- and C-termini pointing in opposite (antiparallel) directions, as shown in Figure 9.41, step 3. Because the dimers point in opposite directions, the tetramer itself lacks polarity. Tetramers associate with one another both side to side and end to end to form poorly described intermediates that assemble into

the final filament (step 4). Because the tetrameric building blocks lack polarity, so too does the assembled filament, which is another feature that distinguishes IFs from other cytoskeletal elements.

Like the collagen fibers of the extracellular matrix, which are also composed of staggered subunits, IFs are highly resistant to tensile (pulling) forces. The outer layer of the skin is made up of the remnants of dead epidermal cells and consists almost entirely of a dense mat of keratin-containing intermediate filaments. These filaments make the skin airtight, watertight, and resistant to bacteria and to many chemicals.

Intermediate filaments tend to be more resistant to chemical disruption than other types of cytoskeletal elements and more difficult to solubilize. However, IFs can be solubilized in harsh ionic detergents such as SDS. Once dissolved, IFs can be depolymerized and repolymerized repeatedly in vitro, indicating that the subunits possess all of the information necessary for their self-assembly. Because of their insolubility, IFs were initially thought to be permanent, unchanging structures, so it came as a surprise to find that they behave dynamically in vivo. When labeled keratin subunits are injected into cultured skin cells, they are rapidly incorporated into existing IFs. Surprisingly, the subunits are not incorporated at the ends of the filament, as might have been expected by analogy with microtubule and microfilament assembly, but rather into the filament's interior (Figure 9.42). Initially, the filaments become labeled at scattered sites along their length, but within an hour or so, the entire intermediate filament network is labeled. These results suggest that epidermal cells contain a pool of keratin subunits that, like microtubule and microfilament subunits, are in a dynamic equilibrium with the polymerized form. Similar findings have been obtained on the IFs of neurons. Assembly and disassembly of most types of IFs are controlled by phosphorylation and dephosphorylation of the subunits. For example, phosphorylation of vimentin filaments by protein kinase A leads to their complete disassembly.

Types and Functions of Intermediate Filaments

The IFs found in epithelial cells (including epidermal cells, hepatocytes, and pancreatic acinar cells) are composed of two types of keratins, the type I (or acidic) keratins and the type II (or neutral and basic) keratins. Keratin-containing IFs always consist of heterodimers containing one member of each type of keratin polypeptide. The keratin filaments of epithelial cells form an elaborate cagelike network around the nucleus and radiate through the cytoplasm (Figure 9.42b). Many of them terminate in the cytoplasmic plaques of desmosomes and hemidesmosomes that attach these cells to other cells and to the underlying basement membrane (pages 265 and 256).

The cytoplasm of neurons contains loosely packed bundles of intermediate filaments whose long axes are oriented parallel to that of the nerve cell axon (see Figure 9.14b). These IFs, or **neurofilaments**, as they are called,

Figure 9.42 Experimental demonstration of the dynamic character of intermediate filaments. These photographs show the results of an experiment in which biotin-labeled type I keratin was microinjected into cultured epithelial cells and localized 20 minutes later using immunofluorescence. The photograph in *a* shows the localization of the injected biotinylated keratin (as revealed by anti-biotin antibodies) that had become incorporated into filaments during the 20-minute period following injection. The photograph in *b* shows the distribution of intermediate filaments in the cell as revealed by anti-keratin antibodies. The dotlike pattern of fluorescence in *a* indicates that the injected subunits are incorporated into the existing filaments at sites throughout their length, rather than at their ends. (Compare with a similar experiment with labeled tubulin in Figure 9.27.) Bar, 10 mm. (FROM RITA K. MILLER, KAREN VIKSTROM, AND ROBERT D. GOLDMAN, J. CELL BIOL. 113:848, 1991; BY COPYRIGHT PERMISSION OF THE ROCKEFELLER UNIVERSITY PRESS.)

are composed of at least three distinct proteins: NF-L, NF-H, and NF-M, all of the type IV group of Table 9.1. Unlike the polypeptides of other IFs, NF-H and NF-M have sidearms that project outward from the neurofilament. These sidearms are thought to maintain the proper spacing between the parallel neurofilaments of the axon (see Figure 9.14*b*). In the early stages of differentiation when the axon is growing toward a target cell, it contains very few neurofilaments but large numbers of supporting microtubules. Once the nerve cell has become fully extended, it becomes filled with neurofilaments that provide support as the axon increases dramatically in diameter. This view of neurofilament function is based largely on studies of a naturally occurring mutant of Japanese quail that fails to produce neurofilaments. Quails bearing this mutation possess axons that are markedly thinner than normal. This gene alteration is called the *quiver* mutation because the bird displays uncontrolled muscle tremors.

Recent efforts to probe IF function have relied on genetically engineered mice that fail to produce a particular IF polypeptide (a gene knockout, page 337) or produce an altered IF polypeptide. These studies have revealed the importance of intermediate filaments in particular cell types. For example, mice carrying deletions in the gene encoding

K14, a type I keratin polypeptide normally synthesized by cells of the basal epidermal layer, have serious health problems. These mice are so sensitive to mechanical pressure that even mild trauma, such as that occurring during passage through the birth canal or during nursing by the newborn, can cause severe blistering of the skin or tongue. This phenotype bears strong resemblance to a rare skin-blistering disease in humans, called *epidermolysis bullosa simplex (EBS)*.[3] Subsequent analysis of EBS patients has shown that they carry mutations in the gene that encodes the homologous K14 polypeptide (or the K5 polypeptide, which forms dimers with K14). These studies confirm the role of IFs in imparting mechanical strength to cells situated in epithelial layers. Similarly, knockout mice that fail to produce the desmin polypeptide exhibit serious cardiac and skeletal muscle abnormalities. Desmin plays a key structural role in maintaining the alignment of the myofibrils of a muscle cell, and the absence of these IFs makes the cells extremely fragile. In 1998 it was discovered that an inherited human disease, named *desmin-related myopathy*, was

[3]As mentioned in Chapter 7, similar types of blistering diseases can be caused by defects in proteins of the hemidesmosome, which holds the basal layer of the epidermis to the basement membrane.

caused by mutations in the gene that encodes desmin. Persons with this disorder suffer from skeletal muscle weakness, cardiac arrhythmias, and eventual congestive heart failure. Not all IF polypeptides have such essential functions. For example, mice that lack the vimentin gene, which is expressed in fibroblasts, macrophages, and white blood cells, show no obvious abnormalities, even though the affected cells lack cytoplasmic IFs. It is evident from these studies that IFs have tissue-specific functions, which are more important in some cells than in others.

Other studies have utilized transgenic mice whose nerve cells produce three to four times the normal quantity of NF-L, one of the polypeptides of neurofilaments. Overexpression of this protein leads to the gradual accumulation of greater-than-normal numbers of neurofilaments, which impedes the normal movement of materials and organelles down the axon. The consequence is nerve cell degeneration and muscle atrophy (Figure 9.43). Accumulation and ab-

normal assembly of neurofilaments are also seen in the motor nerves of patients with certain degenerative neuromuscular diseases, including ALS (or Lou Gehrig disease), but it is not clear whether these cytoskeletal abnormalities are a cause or a byproduct of the disease. A small fraction of ALS patients has been found to harbor a mutation in the NF-H gene, suggesting that disruption of neurofilament structure is one of several primary causes of this disorder.

REVIEW ?

1. Give some examples that reinforce the suggestion that intermediate filaments are important primarily in tissue-specific functions rather than in basic activities that are common to all cells.
2. Compare and contrast microtubule assembly and intermediate filament assembly.

Figure 9.43 The effect of overexpression of neurofilament polypeptides. Light micrograph of a silver-stained section of a portion of the spinal cord of a transgenic mouse that was overexpressing the gene encoding the NF-L polypeptide. The motor neurons of these mice are abnormal and contain swirls of neurofilaments with ballooned cell bodies. (FROM Z. S. XU, L. C. CORK, J. W. GRIFFIN, AND D. W. CLEVELAND, CELL 73, COVER OF NO. 2, 1993; BY COPYRIGHT PERMISSION OF CELL PRESS.)

9.5 MICROFILAMENTS

Cells are capable of remarkable motility. The neural crest cells in a vertebrate embryo leave the developing nervous system and migrate across the entire width of the embryo, forming such diverse products as the pigment cells of the skin, teeth, and the cartilage of the jaws (see Figure 7.11). Legions of white blood cells patrol the tissues of the body searching for debris and microorganisms. Certain parts of cells can also be motile; broad projections of epithelial cells at the edge of a wound act as motile devices that pull the sheet of cells over the damaged area, sealing the wound. Similarly, the leading edge of a growing axon sends out microscopic processes that survey the substratum and guide the cell toward a synaptic target. All of these various examples of motility share at least one component: they all depend on the presence of microfilaments, the third major type of cytoskeletal element. Microfilaments are also involved in intracellular motile processes, such as the movement of vesicles and organelles, phagocytosis, and cytokinesis.

Microfilaments are approximately 8 nm in diameter and composed of globular subunits of the protein **actin**. In the presence of ATP, actin monomers polymerize to form a stiff filament composed of two strands of actin molecules wound around each other in a double helix (Figure 9.44). The terms "actin filament," "F-actin," and "microfilament" are basically synonyms for this type of two-stranded filament, although the term "F-actin" is often reserved for actin polymers formed in vitro. Because each actin subunit has polarity and all the subunits of an actin filament are pointed in the same direction, the entire microfilament has polarity. Depending on the type of cell and the activity in which it is engaged, actin filaments can be organized into highly ordered arrays, loose ill-defined networks, or tightly anchored bundles.

Actin was identified more than 50 years ago as one of the major contractile proteins of muscle cells. Since then,

actin has been identified as a major protein in virtually every type of eukaryotic cell that has been examined. Higher plant and animal species possess a number of actin-coding genes whose encoded products are specialized for different types of motile processes. The structure of actin has been remarkably conserved throughout the course of evolution. For example, the amino acid sequences of actin molecules from a yeast cell and from rabbit skeletal muscle are 88 percent identical. In fact, actin molecules from diverse sources can copolymerize to form hybrid filaments.

The identification of actin filaments in a given cell can be made with certainty using a cytochemical test that takes advantage of the fact that actin filaments, regardless of their source, will interact in a highly specific manner with the protein myosin. To facilitate the interaction, purified myosin (obtained from muscle tissue) is cleaved into fragments by a proteolytic enzyme. One of these fragments, called S1 (see Figure 9.48), binds to the actin molecules all along the microfilament. In addition to identifying the filaments as actin, the bound S1 fragments reveal the filament's polarity. When S1 fragments are bound, one end of the microfilament appears *pointed* like an arrowhead, while the other end looks *barbed*. An example of this arrowhead "decoration" is shown in the microvilli of intestinal epithelial cells of Figure 9.45. The orientation of the arrowheads formed by the S1-actin complex provides information as to the direction in which the microfilaments are likely to be moved by a myosin motor protein. Actin

(a) *(b)* 100 nm

Figure 9.44 Actin filament structure. (*a*) A model of an actin filament. The actin subunits are represented in three colors to distinguish the consecutive subunits more easily. The subdomains in one of the actin subunits are labeled 1, 2, 3, and 4. (*b*) Electron micrograph of a replica of an actin filament showing its double-helical architecture. (A: FROM MICHAEL F. SCHMID ET AL., COURTESY OF WAH CHIU, J. CELL BIOL. 124:346, 1994; BY COPYRIGHT PERMISSION OF THE ROCKEFELLER UNIVERSITY PRESS; B: FROM ROBERT H. DEPUE, JR. AND ROBERT V. RICE, J. MOL. BIOL. 12:302, 1965. COPYRIGHT 1965, BY PERMISSION OF THE PUBLISHER ACADEMIC PRESS.)

200 nm

Figure 9.45 Determining the location and polarity of actin filaments using the S1 subunit of myosin. Electron micrograph of a replica showing the microfilament bundles in the core of the microvilli of an intestinal epithelial cell. The cell had been fixed, treated with S1 myosin fragments, freeze fractured, and deep etched to expose the filamentous components of the cytoplasm. The intermediate filaments (IFs) at the bottom of the micrograph do not contain actin and therefore do not bind the S1-myosin fragments. These intermediate filaments originate at the desmosomes of the lateral surfaces of the cell. (FROM N. HIROKAWA, L. G. TILNEY, K. FUJIWARA, AND J. E. HEUSER, J. CELL BIOL. 94:430, 1982; BY COPYRIGHT PERMISSION OF THE ROCKEFELLER UNIVERSITY PRESS.)

can also be localized by light microscopy using fluorescently labeled S1 fragments or fluorescently labeled anti-actin antibodies (see Figure 9.68a).

Microfilament Assembly and Disassembly

Before it is incorporated into a filament, an actin monomer binds a molecule of ATP. Actin is an ATPase, just as tubulin is a GTPase, and the role of ATP in actin assembly is similar to that of GTP in microtubule assembly (page 352). The ATP associated with the actin monomer is hydrolyzed to ADP at some time after it is incorporated into the growing actin filament. When cells are assembling actin filaments at a rapid rate, the end of the filament contains a cap of actin-ATP subunits, which hinders the disassembly of the filament and favors continued assembly.

When microfilaments are incubated with a high concentration of labeled actin-ATP subunits in vitro, both ends of the microfilament become labeled, but one end incorporates monomers at a rate 5 to 10 times that of the other end. Decoration with the S1 myosin fragment reveals that the barbed (or plus) end of the microfilament is the fast-growing end, while the pointed (or minus) end is the slow-growing tip (Figure 9.46a). Conversely, the pointed end is the preferential site of depolymerization. Because actin-ATP subunits tend to add to a filament's plus end and actin-ADP subunits tend to leave from its minus end, individual monomers move down the length of an actin filament in vitro—a process known as "treadmilling" (Figure 9.46b). Whether treadmilling occurs in vivo has been difficult to establish because it can occur without any change in the length of the filaments.

Cells maintain a dynamic equilibrium between the monomeric and polymeric forms of actin, just as they do with microtubules. As discussed on page 382, the rate of assembly and disassembly of actin filaments in the cell can be influenced by a number of different proteins. Changes in the local conditions in a particular part of the cell can push the equilibrium toward either the assembly or the disassembly of microfilaments. By controlling this equilibrium, the cell can reorganize its microfilament cytoskeleton. Such reorganization is required for dynamic processes such as cell locomotion, changes in cell shape, and cytokinesis. These activities will be explored in the following pages.

As noted above, actin filaments play a role in nearly all of a cell's motile processes. The involvement of these filaments is most readily demonstrated by treating the cells with one of the cytochalasins, a group of compounds extracted from mold cells, or with phalloidin, a compound extracted from poisonous mushrooms. Cytochalasins promote depolymerization of microfilaments, whereas phalloidin increases microfilament stability so that the cell can no longer use the structures in its dynamic activities. Microfilament-mediated processes rapidly grind to a halt when cells are exposed to either of these compounds. The effect of cytochalasin D on the very fine motile processes (*filopodia*) of embryonic sea urchin cells is shown in Figure 9.47.

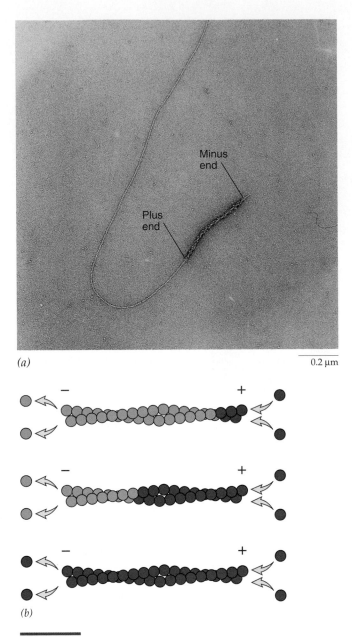

(a) 0.2 μm

(b)

Figure 9.46 Actin assembly in vitro. (*a*) Electron micrograph of a short actin filament that was labeled with S1 myosin and then used to nucleate actin polymerization. The addition of actin subunits occurs much more rapidly at the barbed (plus) end than at the pointed (minus) end of the existing filament. (*b*) Schematic diagram of the preferential addition of actin subunits at the plus end of a microfilament and their preferential loss from the minus end in an in vitro assay. As a result, subunits treadmill through the filament in vitro. (A: COURTESY OF M. S. RUNGE AND T. D. POLLARD.)

Myosin: The Molecular Motor for Actin Filaments

We have previously examined the structure and actions of two molecular motors—kinesin and dynein—that operate over tracks of microtubules. To date, all of the motors known to operate in conjunction with actin filaments are members of the myosin superfamily. With one known ex-

Figure 9.47 The effect of cytochalasin on structures containing actin filaments. A pair of sea urchin mesenchyme cells with fine filamentous processes (filopodia) after 30 seconds (*a*) and 5 minutes (*b*) of exposure to cytochalasin D. The drug causes dissolution of the cell processes, which are composed primarily of actin filaments. (FROM GERALD KARP AND MICHAEL SOLURSH, DEV. BIOL. 112:281, 1985.)

ception (page 372), myosins move toward the plus ends of microfilaments. If the myosin molecules are restrained and unable to move, as in muscle cells, myosin activity will cause the attached actin filaments to move with their minus end leading the way.

Myosin was first isolated from mammalian skeletal muscle tissue and subsequently from a wide variety of eukaryotic cells, including protists, higher plants, nonmuscle cells of animals, and vertebrate cardiac and smooth muscle tissues. All myosins share a characteristic motor (head) domain. The head domain contains a site that binds an actin filament and a site that binds and hydrolyzes ATP to drive the myosin motor. Whereas the head domains of myosins are similar, the tail domains are highly divergent. Myosins also contain a variety of low-molecular-weight (light) chains. Based on these differences in construction, myosins are generally divided into two large groups, the **conventional** (or **type II**) **myosins** and the **unconventional myosins**. The unconventional myosins are subdivided into at least 14 different classes (type I and types III-

XV), two of which are apparently unique to plants. Each type of myosin is presumed to have its own specialized functions, and several types may be present together in the same cell. Of the various myosins, the type II molecules are the best understood.

Conventional (Type II) Myosins Proteins of the myosin II class are found in various types of muscle tissues and also in a variety of nonmuscle cells. Among their nonmuscle activities, type II myosins are required for splitting a cell in two during cell division and for generating tension at focal adhesions. An electron micrograph of a pair of myosin II molecules is shown in Figure 9.48*a*. Each myosin II molecule is composed of six polypeptide chains—one pair of heavy chains and two pairs of light chains—organized in such a way as to produce a highly asymmetric protein (Figure 9.48*a*). Examination of the molecule in Figure 9.48*b* shows it to consist of (1) a pair of globular heads that contain the catalytic site of the molecule; (2) a pair of necks, each consisting of a single, unin-

(*a*) 20 nm (*b*)

Figure 9.48 Structure of a myosin II molecule. (*a*) Electron micrograph of negatively stained myosin molecules. The two heads and tail of each molecule are clearly visible. (*b*) A myosin II molecule (molecular mass of 520,000 daltons) consists of one pair of heavy chains and two pairs of light chains, which are named as indicated. The paired heavy chains consist of a rod-shaped tail in which portions of the two polypeptide chains wrap around one another to form a coiled coil and a pair of globular heads. When treated with a protease under mild conditions, the molecule is cleaved at the junction between the neck and tail, which generates the S1 fragment. (A: FROM S. A. BURGESS, M. L. WALKER, H. D. WHITE, AND J. TRINICK, J. CELL BIOL. 139:676, 1997; BY COPYRIGHT PERMISSION OF THE ROCKEFELLER UNIVERSITY PRESS.)

terrupted α helix and two associated light chains; and (3) a single, long, rod-shaped tail formed by the intertwining of long α-helical sections of the two heavy chains to form an α-helical coiled coil (page 61).

Isolated myosin heads (S1 fragments of Figure 9.48*b*) that have been immobilized on the surface of a glass coverslip are capable of sliding attached actin filaments in in vitro assays such as that shown in Figure 9.49. Thus, all of the machinery required for motor activity is contained in a single head domain. The mechanism of action of the

myosin head is discussed on page 377. The fibrous tail portion of a myosin II molecule plays a structural role, allowing the protein to form filaments. Myosin molecules assemble so that the ends of the tails point toward the center of the filament and the globular heads point away from the center (Figures 9.50 and 9.57). As a result, the filament is described as *bipolar*, indicating a reversal of polarity at the filament's center. As described in the following section, the myosin II filaments that assemble in skeletal muscle cells are highly stable components of the contractile apparatus. However, the smaller myosin II filaments that form in most nonmuscle cells often display transient construction, assembling when and where they are needed and then disassembling after they have acted.

Unconventional Myosins　In 1973, Thomas Pollard and Edward Korn of the National Institutes of Health described a unique myosin-like protein that was extracted from the protist *Acanthamoeba*. Unlike muscle myosin, this smaller, unconventional myosin had only a single head and was unable to assemble into filaments in vitro. The protein became known as myosin I. Other unconventional myosins have since been isolated from a variety of cell types and classified into numerous classes based on their amino acid sequences. A schematic structure of a myosin I molecule is shown in Figure 9.51. Like myosin

(a)

(b)

Figure 9.49 In vitro motility assay for myosin. (*a*) Schematic drawing in which myosin heads are bound to a silicone-coated coverslip, which is then incubated with a preparation of actin filaments. (*b*) Results of the experiment depicted in *a*. Two video images were taken 1.5 sec apart and photographed as a double exposure on the same frame of film. The dashed lines with arrowheads show the movement of the actin filaments over the myosin heads during the brief period between exposures. (FROM T. YANAGIDA, ADV. BIOPHYSICS 26:82, 1990.)

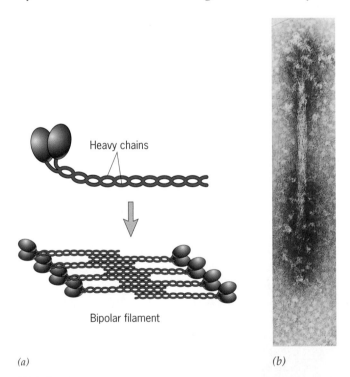

(a)　　　　　　　　　　　　　　　　　　(b)

Figure 9.50 Structure of a bipolar myosin II filament. (*a*) Schematic diagram of the staggered arrangement of the individual myosin molecules in a myosin II filament. (*b*) Electron micrograph of a bipolar myosin filament formed in vitro. The heads of the filament are seen at both ends, leaving a smooth section in the center of the filament. (B: COURTESY OF HUGH HUXLEY.)

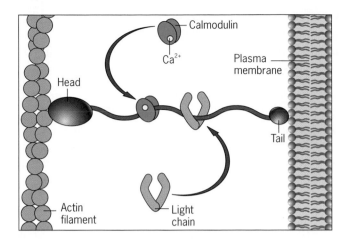

Figure 9.51 Myosin I. A highly schematic diagram of a myosin I molecule. This particular molecule has four light chains, two of which are calmodulin, a ubiquitous calcium-binding protein. The head of the molecule binds to an actin filament, while the tail is thought to bind to lipid head-groups of the plasma membrane. (AFTER J. S. WOLENSKI, TRENDS CELL BIOL. 5:311, 1995. COPYRIGHT 1995, WITH PERMISSION FROM ELSEVIER SCIENCE.)

Figure 9.52 The mechanical cycle of myosin V, a two-headed unconventional myosin involved in organelle transport. This series of negatively stained electron micrographs shows individual myosin V molecules that were caught at different stages in their mechanical cycle as each of them was "walking" along an actin filament. The circular arrow shows the sequence as a molecule might move from left to right along the actin filament towards its barbed (plus) end. The micrograph at the bottom center shows a stage in the cycle where both heads are attached to the actin filament with a spacing of approximately 36 nm, which is equivalent to the helical repeat of the filament. (FROM MATTHEW L. WALKER ET AL., COURTESY OF PETER J. KNIGHT. REPRINTED WITH PERMISSION FROM NATURE 405:804, 2000. © COPYRIGHT 2000, MACMILLAN MAGAZINES LTD.)

IIs, unconventional myosins are able to bind to actin filaments in vitro and cause them to move in the presence of ATP.

Class I myosins remain the best studied unconventional myosins. The best insight into the functions of class I myosins has come from studies on genetically engineered cells of the slime mold *Dictyostelium*, whose single myosin II gene can be deleted or rendered nonfunctional. Cells mutated this way express only myosin I, yet they remain capable of a remarkable variety of actin-based processes, including normal locomotion and phagocytosis. These cells are not, however, capable of normal cell division because the splitting of the cytoplasm (cytokinesis) requires myosin II.

The steps taken by one type of unconventional myosin—myosin V—have been revealed in a series of electron micrographs that capture the molecule at various stages in its mechanical cycle (Figure 9.52). Myosin V is a dimeric myosin that, unlike other myosin molecules that have been studied, moves processively along actin filaments. Processivity is thought to be due to the high affinity of the myosin head for the actin filament, which ensures that each head remains attached to the filament until the second head reattaches. Myosin V is also noteworthy for the length of its neck, which at 23 nm is about three times that of myosin II. Because of its long neck, myosin V can take very large steps (over 30 nm). This is very important for a motor protein that moves processively along an actin filament made up of *helical* strands of subunits. The actin double helix repeats itself about every 13 subunits (36 nm), which is just about the step size of a myosin V molecule (Figure 9.52). As a result of its giant steps, myosin V can apparently walk along this repeat in a straight path even though the underlying "roadway" spirals 360° between its "feet."

Several unconventional myosins (including myosin I, V, and VI) are associated with various types of cytoplasmic vesicles. Some vesicles have been shown to contain microtubule-based motors (kinesins and/or cytoplasmic dynein) *and* microfilament-based motors (unconventional myosins) and, in fact, the two types of motors may be physically linked to one another. The movement of vesicles and other membranous carriers over long distances within cells occurs on microtubules, as previously described. However, once they approach the end of the microtubule, these membranous vesicles are often thought to switch over to microfilament tracks for the journey through the actin-rich periphery of the cell (Figure 9.53).

Cooperation between microtubules and microfilaments has been best studied in pigment cells (Figure

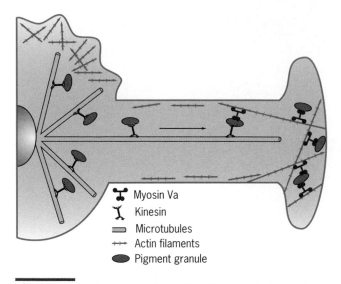

Myosin Va
Kinesin
Microtubules
Actin filaments
Pigment granule

Figure 9.53 **The contrasting roles of microtubule- and microfilament-based motors in organelle transport.** Most organelle transport is thought to be mediated by members of the kinesin and dynein families, which carry their cargo over relatively long distances. It is thought that some vesicles also carry myosin motors, such as myosin Va, which transport their cargo over microfilaments, including those present in the peripheral (cortical) regions of the cell. The two types of motors may act in a cooperative manner as illustrated here in the case of a pigment cell in which pigment granules move back and forth within extended cellular processes. (AFTER X. WU ET AL., J. CELL BIOL. 143:1915, 1998, BY COPYRIGHT PERMISSION OF THE ROCKEFELLER UNIVERSITY PRESS.)

9.53). In mammals, pigment granules (melanosomes) are transported into fine peripheral processes of the pigment cell by myosin Va and then transferred to hair follicles where they become incorporated into a developing hair. Mice lacking myosin Va activity are unable to transfer melanosomes into hair follicles, causing their coat to have a much lighter color. The importance of myosin Va in organelle transport became more apparent when it was discovered that mice lacking this motor protein also suffer fatal neurologic dysfunction, a presumed result of defective axonal transport. Humans lacking a normal gene encoding myosin Va also exhibit partial albinism (lack of skin coloration) and suffer neurologic dysfunction.

Another unconventional myosin, type VIIa, is localized in the hair cells of the cochlea of the inner ear (Figure 9.54). Hair cells are so-named because of the bundle of stiff, hairlike stereocilia that project from the apical surface of the cell. Displacement of the stereocilia by mechanical stimuli leads to the generation of nerve impulses that we perceive as sound. Each stereocilium consists of a bundle of actin filaments that contains myosin VIIa molecules as depicted in the drawing of Figure 9.54a and the micrographs of Figure 9.54b,c. Mutations in myosin VIIa are responsible for several types of deafness in humans as well as Usher 1B syndrome, which is characterized by

both deafness and blindness. The morphologic effects of mutations in the myosin VIIa gene on the hair cells of the inner ear of mice are shown in Figures 9.54d,e. As in humans, mice that are homozygous for the mutant allele encoding this motor protein are deaf. Myosin VI, another organelle transporter, has the distinction of being the only known myosin that moves in a "reverse direction," that is, toward the pointed (minus) end of an actin filament.

Now that we have described the basics of actin and myosin structure and function, we can see how these two proteins interact to mediate a complex cellular activity.

REVIEW ?

1. Compare and contrast the characteristics of microtubule assembly versus actin filament assembly.
2. Compare the structure of a fully assembled microtubule, actin filament, and intermediate filament.
3. Describe three functions of actin filaments.
4. Contrast the structure and function of conventional and unconventional myosins in cellular activities.

9.6 MUSCLE CONTRACTILITY

Skeletal muscles derive their name from the fact that most of them are anchored to bones that they move. They are under voluntary control and can be consciously commanded to contract. Skeletal muscle cells have a highly unorthodox structure. A single, cylindrically shaped muscle cell is 10 to 100 μm thick, up to 40 mm long, and contains hundreds of nuclei. Because of these properties, a skeletal muscle cell is more appropriately called a **muscle fiber**. Muscle fibers have multiple nuclei because each fiber is a product of the fusion of large numbers of mononucleated myoblasts (premuscle cells) in the embryo. Fusion of mononucleated myoblasts is readily demonstrated in the culture dish, even when the myoblasts are derived from distantly related animals (Figure 9.55).

Skeletal muscle cells may have the most highly ordered internal structure of any cell in the body. A longitudinal section of a muscle fiber (Figure 9.56) reveals a cable made up of hundreds of thinner, cylindrical strands, called myofibrils. Each myofibril consists of a repeating linear array of contractile units, called sarcomeres. Each sarcomere in turn exhibits a characteristic banding pattern, which gives the muscle fiber a **striated** appearance. Examination of stained muscle fibers in the electron microscope shows the banding pattern to be the result of the partial overlap of two distinct types of filaments, **thin filaments** and **thick filaments** (Figure 9.57a). Each sarcomere extends from one Z line to the next Z line and contains several

1 μm

(a)

(b)

(c)

(d)

(e)

Figure 9.54 The role of myosin VIIa in hearing. (*a*) Drawing of a hair cell of the cochlea. The inset shows a portion of several of the apical stereocilia, which are composed of a tightly grouped bundle of actin filaments. Stereocilia are thought to be linked to one another by myosin VIIa molecules. (*b*) Transmission electron micrograph of a cross-section of a stereocilium showing it is composed of a dense bundle of actin filaments. (*c*) Fluorescence micrograph of a hair cell from the bullfrog inner ear. The stereocilia appear red due to labeling by rhodamine-labeled phalloidin, which binds to actin filaments. The localization of myosin VIIa is indicated in green. The orange bands near the bases of the stereocilia indicate a concentration of myosin VIIa that may link adjacent stereocilia as depicted in *a*. (*d*) Scanning electron micrograph of the hair cells of the cochlea of a control mouse. The stereocilia are arranged in V-shaped rows. (*e*) A corresponding micrograph of the hair cells of a mouse with mutations in the gene encoding myosin VIIa, which causes deafness. The stereocilia of the hair cells exhibit a disorganized arrangement. (A: AFTER T. HASSON CURR. BIOL. 9:R839, 1999, WITH PERMISSION FROM ELSEVIER SCIENCE; B: COURTESY OF A. J. HUDSPETH, R. A. JACOBS, AND P. G. GILLESPIE; C: FROM PETER GILLESPIE AND TAMA HASSON, J. CELL BIOL. VOL. 137, COVER #6, 1997, BY COPYRIGHT PERMISSION OF THE ROCKEFELLER UNIVERSITY PRESS; D,E: FROM TIM SELF ET AL., COURTESY OF KAREN P. STEEL, DEVELOP. 125:560, 1998; BY PERMISSION OF THE COMPANY OF BIOLOGISTS LTD.)

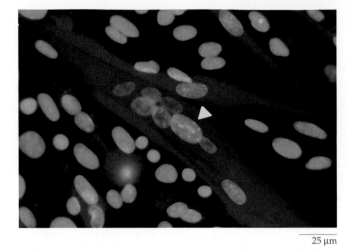

25 μm

Figure 9.55 Demonstration in cell culture that skeletal muscle cells arise by cell fusion. The muscle cells depicted here (called myotubes when formed in culture) are seen to contain two different types of nuclei. These myotubes arise from the fusion of two types of mononucleated muscle cell precursors (myoblasts), one type derived from a mouse cell line and the other from quail. The triangle points to a mouse nucleus in a hybrid myotube. Muscle fibers in the body arise by a similar process of myoblast fusion. (FROM M. CRESCENZI, D. H. CROUCH, AND F. TATO, J. CELL BIOL. 125:1142, 1994; BY COPYRIGHT PERMISSION OF THE ROCKEFELLER UNIVERSITY PRESS.)

dark bands and light zones. A sarcomere has a pair of lightly staining *I bands* located at its outer edges, a more densely staining *A band* located between the outer I bands, and a lightly staining *H zone* located in the center of the A band. A densely staining *M line* lies in the center of the H zone. The I band contains only thin filaments, the H zone only thick filaments, and that part of the A band on either side of the H zone represents the region of overlap and contains both types of filaments.

Cross sections through the region of overlap show that the thin filaments are organized in a hexagonal array around each thick filament, and that each thin filament is situated between two thick filaments (Figure 9.57*b*). Longitudinal sections show the presence of projections from the thick filaments at regularly spaced intervals. The pro-

Figure 9.56 The structure of skeletal muscle. Schematic diagram of the levels of organization of skeletal muscle from an entire body muscle to a single sarcomere. The first inset shows a light micrograph of a multinucleated muscle fiber. The second inset shows an electron micrograph of a section through a single sarcomere with the bands lettered. (MICROGRAPHS COURTESY OF GERALDINE F. GAUTHIER.)

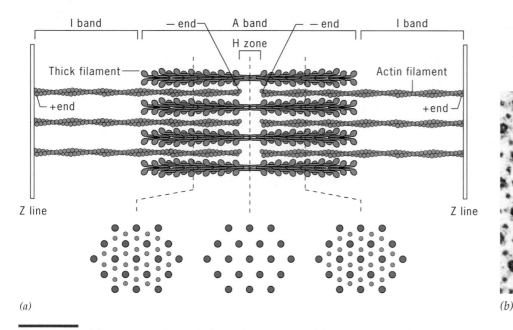

(a)

(b)

Figure 9.57 The contractile machinery of a sarcomere. (*a*) Schematic diagram of a part of a myofibril showing the overlapping array of thin actin-containing (orange) and thick myosin-containing (purple) filaments. The small transverse projections on the myosin fiber represent the heads (cross-bridges). A more detailed drawing of the thin filaments is shown in Figure 9.59. (*b*) Electron micrograph of a cross section through a portion of an insect flight muscle showing the hexagonal arrangement of the thin filaments around each thick filament. (MICROGRAPH BY J. AUBER/PHOTO RESEARCHERS.)

jections represent cross-bridges capable of forming attachments with neighboring thin filaments.

The Sliding Filament Model of Muscle Contraction

All skeletal muscles operate by shortening; there is no other way they can perform work. The units of shortening are the sarcomeres, whose combined decrease in length accounts for the decrease in length of the entire muscle. The most important clue to the mechanism underlying muscle contraction came from observations of the banding pattern of the sarcomeres at different stages in the contractile process. As a muscle fiber shortened, the A band in each sarcomere remained essentially constant in length, while the H band and I bands decreased in width and then disappeared altogether. As shortening progressed, the Z lines on both ends of the sarcomere moved inward until they contacted the outer edges of the A band (Figure 9.58).

Based on these observations, two groups of British investigators, Andrew Huxley and Ralph Niedergerke, and Hugh Huxley and Jean Hanson, proposed a far-reaching model to account for muscle contraction. They proposed that the shortening of individual sarcomeres did not result from the shortening of the filaments, but rather from their sliding over one another. The sliding of the thin filaments toward the center of the sarcomere would result in the observed increase in overlap between the filaments and the decreased width of the I and H bands (Figure 9.58). The

sliding-filament model was rapidly accepted, and evidence in its favor continues to accumulate.

The Composition and Organization of Thick and Thin Filaments

In addition to actin, the thin filaments of a skeletal muscle contain two other proteins, **tropomyosin** and **troponin** (Figure 9.59). Tropomyosin is an elongated molecule (approximately 40 nm long) that fits securely into the grooves between the two actin chains of the thin filament. Each rod-shaped tropomyosin molecule is associated with seven actin subunits along the thin filament (Figure 9.59). Troponin is a globular protein complex composed of three subunits, each having an important and distinct role in the overall function of the molecule. Troponin molecules are spaced approximately 40 nm apart along the thin filament and contact both the actin and tropomyosin components of the filament. The actin filaments of each half sarcomere are aligned with their barbed ends linked to the Z line.

Each thick filament is composed of several hundred myosin molecules together with small amounts of other proteins. Like the filaments that form in vitro (see Figure 9.50), the polarity of the thick filaments of muscle cells is reversed at the center of the sarcomere. The center of the filament is composed of the opposing tail regions of the myosin molecules and is devoid of heads. The myosin heads project from each thick filament along the remainder of its length due to the staggered position of the myosin molecules that make up the body of the filament (see Figure 9.50).

(a)

(b)

Figure 9.58 The shortening of the sarcomere during muscle contraction. (*a*) Schematic diagram showing the difference in the structure of the sarcomere in a relaxed and contracted muscle. During contraction, the myosin cross-bridges make contact with the surrounding thin filaments, and the thin filaments are forced to slide toward the center of the sarcomere. Cross-bridges

work asynchronously, so that only a fraction are active at any given instant. (*b*) Electron micrographs of longitudinal sections through a relaxed (*top*) and contracted (*bottom*) sarcomere. The micrographs show the disappearance of the H-zone as the result of the sliding of the thin filaments toward the center of the sarcomere. (B: FROM JAMES E. DENNIS/PHOTOTAKE.)

Figure 9.59 The molecular organization of the thin filaments. Each thin filament consists of a double-helical array of actin subunits with rod-shaped tropomyosin molecules situated in the grooves and troponin molecules spaced at defined intervals, as described in the text. The positional changes in these proteins that trigger contraction are shown in Figure 9.64.

The third most abundant protein of the skeletal muscle fiber is **titin**, whose molecular mass of almost 3 million daltons (27,000 amino acids) and length of more than 1 μm make it the largest protein yet to be discovered. Titin molecules originate at the M line and extend along the myosin filament, continuing past the A band and terminating at the Z line (Figure 9.60). Titin is a highly elastic protein that stretches like a bungee cord as certain domains within the molecule become unfolded. Titin is

thought to prevent the sarcomere from becoming pulled apart during muscle stretching. Titin also maintains myosin filaments in their proper position within the center of the sarcomere during muscle contraction.

The Molecular Basis of Contraction Following the formulation of the sliding-filament hypothesis, attention turned to the heads of the myosin molecules as the force-generating components of the muscle fiber. During a contraction, each myosin head extends outward and binds tightly to the thin filament, forming the cross-bridges seen between the two types of filaments (Figure 9.58). The heads from a single myosin filament interact with six surrounding actin filaments. While it is bound tightly to the actin filament, the myosin head undergoes a conformational change (described below) that moves the thin actin filament approximately 5 to 15 nm toward the center of the sarcomere.[4] Muscle myosin is not a processive motor protein like kinesin (page 343) because it remains in contact with its track (in this case the thin filament) for

[4]There has been considerable controversy over the working distance of a single skeletal muscle cross-bridge. The values of 5 to 15 nm used here cover most of the range of reported values.

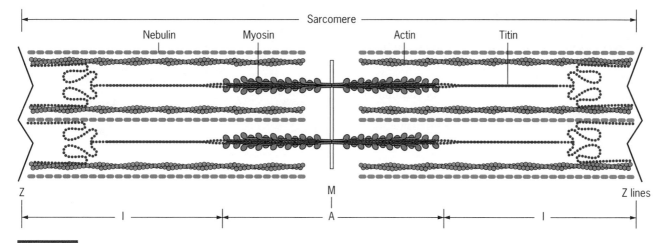

Figure 9.60 The arrangement of titin molecules within the sarcomere. These huge elastic molecules stretch from the end of the sarcomere at the Z line to the M band in the sarcomere center. Titin molecules are thought to maintain the thick filaments in the center of the sarcomere during contraction. The nebulin molecules (which are not discussed in the text) are thought to act like a "molecular ruler" by regulating the number of actin monomers that are allowed to assemble into a thin filament. (AFTER T. C. S. KELLER, CURR. OPIN. CELL BIOL. 7:33, 1995.)

only a small fraction (less than 5 percent) of the overall cycle. However, each thin filament is contacted by a team of a hundred or so myosin heads that beat out of synchrony with one another (Figure 9.58a). Consequently, the thin filament undergoes continuous motion during each contractile cycle. It is estimated that a single thin filament in a muscle cell can be moved several hundred nanometers during a period as short as 50 milliseconds.

Muscle physiologists have long sought to understand how a single myosin molecule can move an actin filament 5 to 15 nm in a single power stroke. The publication of the atomic structure of the S1 myosin fragment in 1993 by Ivan Rayment, Hazel Holden, and their colleagues at the University of Wisconsin was accompanied by a proposal to explain its mechanism of action. In their proposal, the energy released by ATP hydrolysis induces a small (0.5 nm) conformational change within the head while the head is tightly bound to the actin filament. The small movement of the head is then amplified approximately 20-fold by the swinging movement of an adjoining α-helical neck (Figure 9.61). According to this hypothesis, the elongated neck acts as a rigid "lever arm" causing the attached actin filament to slide a much greater distance (approximately 10 nm) than would otherwise be possible. The two light chains, which

Figure 9.61 Model showing the neck of the myosin molecule acting as a swinging lever arm. (a) During the power stroke, the neck of the myosin molecule moves through a rotation of approximately 90°, which would produce a movement of the actin filament of approximately 10 nm. (b) A model of the power stroke of the myosin motor domain consisting of the catalytic core (or head) and adjoining neck (or lever arm). An attached actin filament is shown in gray/brown. The long helical neck is shown in two positions, both before and after the power stroke (depicted as the upper dark blue and lower light blue necks, respectively). It is this displacement of the neck region that is thought to power muscle movement. The essential and regulatory light chains, which wrap around the neck, are shown in yellow and magenta, respectively. (B: FROM MALCOLM IRVING ET AL. REPRINTED WITH PERMISSION FROM NATURE STRUCT. BIOL. 7:482, 2000. © COPYRIGHT 2000, MACMILLAN MAGAZINES LTD.)

are wrapped around the neck, are thought to provide rigidity for the lever. Support for the "lever-arm" proposal was obtained through a series of experiments by James Spudich and colleagues at Stanford University. These researchers constructed genes that encoded altered versions of the myosin molecule, which contained necks of different length. The genetically engineered myosin molecules were then tested in an in vitro motility assay with actin filaments, similar to that depicted in Figure 9.49. As predicted by the "lever-arm" hypothesis, the apparent length of the power stroke of the myosin molecules was directly proportional to the length of their necks. Myosin molecules with shorter necks generated smaller displacements, whereas those with longer necks generated greater displacements.

The Energetics of Filament Sliding Like the other motor proteins kinesin and dynein, myosin converts the chemical energy of ATP into the mechanical energy of filament sliding. Each cycle of mechanical activity of the myosin cross-bridge takes about 50 msec and is accompanied by a cycle of ATPase activity, as illustrated in the model shown in Figure 9.62. According to this model, the cycle begins as a molecule of ATP binds to the myosin head, an event that induces the dissociation of the cross-bridge from the actin filament (Figure 9.62, step 1). Binding of ATP is followed by its hydrolysis, which occurs before the myosin head makes contact with the actin filament. The products of ATP hydrolysis, namely ADP and P_i, remain bound to the active site of the enzyme, while the energy released by hydrolysis is absorbed by the protein as a whole (Figure 9.62, step 2). At this point, the cross-bridge is in an energized state, analogous to a stretched spring capable of spontaneous movement. The energized myosin then attaches to the actin molecule (step 3) and releases its bound phosphate, which triggers a large conformational change driven by the stored free energy (step 4). This conformational change shifts the actin filament toward the center of the sarcomere. This movement represents the power stroke of the myosin head as shown in Figure 9.61. The release of the bound ADP (step 5) is followed by the binding of a new ATP molecule so that a new cycle can begin. In the absence of ATP, the myosin head remains tightly bound to the actin filament. The inability of myosin cross-bridges to detach in the absence of ATP is the basis for the condition of *rigor mortis*, which ensues following death.

Excitation-Contraction Coupling Muscle fibers are organized into groups termed *motor units*. All the fibers of a motor unit are jointly innervated by branches from a single motor neuron and contract simultaneously when stimulated by an impulse transmitted along that neuron. The point of contact of a terminus of an axon with a muscle fiber is called a **neuromuscular junction** (Figure 9.63; see also Figure 4.53 for a closer view of the structure of the synapse). The neuromuscular junction is a site of transmission of the nerve impulse from the axon across a synaptic cleft to the muscle fiber, whose plasma membrane is also excitable and capable of conducting an action potential.

Figure 9.62 A schematic model of the actinomyosin contractile cycle. The movement of the thin filament by the force-generating myosin head occurs as the result of a linkage between a mechanical cycle involving the attachment, movement, and detachment of the head and a chemical cycle involving the binding, hydrolysis, and release of ATP, ADP, and P_i. In this model, the two cycles begin in step 1 with the binding of ATP to a cleft in the myosin head, causing the detachment of the head from the actin filament. The hydrolysis of the bound ATP (step 2) energizes the head, causing it to bind weakly to the actin filament (step 3). The release of P_i causes a tighter attachment of the myosin head to the thin filament and the power stroke (step 4) that moves the thin filament toward the center of the sarcomere. The release of the ADP (step 5) sets the stage for another cycle. (AFTER M. Y. JIANG AND M. P. SHEETZ, BIOESS. 16:532, 1994.)

The steps that link the arrival of a nerve impulse at the muscle plasma membrane to the shortening of the sarcomeres deep within the muscle fiber constitute a process referred to as **excitation–contraction coupling**. Unlike a neuron, where an action potential remains at the cell surface, the impulse generated in a skeletal muscle cell is propagated into the interior of the cell along membranous folds called **transverse (T) tubules** (Figure 9.63). The T tubules terminate in very close proximity to a system of cytoplasmic membranes that make up the **sarcoplasmic reticulum (SR)**, which forms a membranous sleeve around the myofibril. Approximately 80 percent of the integral protein of the SR membrane consists of Ca^{2+}-ATPase molecules whose function is to transport Ca^{2+} out of the cytosol and into the lumen of the SR, where it is stored until needed.

The importance of calcium in muscle contraction was first shown by L. V. Heilbrunn of the University of Pennsylvania in 1947 when he injected calcium into a muscle fiber and found it to stimulate the fiber to contract. In the relaxed state, the Ca^{2+} levels within the cytoplasm are very low (ap-

Figure 9.63 **The functional anatomy of a muscle fiber.** Calcium is housed in the elaborate network of internal membranes that make up the sarcoplasmic reticulum (SR). When an impulse arrives by means of a motor neuron, it is carried into the interior of the fiber along the membrane of the transverse tubule to the SR. The calcium gates of the SR open, releasing calcium into the cytosol. The binding of calcium ions to troponin molecules of the thin filaments leads to the events described in the following figure and the contraction of the fiber.

proximately 2×10^{-7} M)—below the threshold concentration required for contraction. With the arrival of an action potential by way of the transverse tubules, calcium channels in the SR membrane are opened, and calcium diffuses out of the SR compartment and over the short distance to the myofibrils. As a result, the intracellular Ca^{2+} levels rise to about 5×10^{-5} M. To understand how elevated calcium levels trigger contraction in a skeletal muscle fiber, it is necessary to reconsider the protein makeup of the thin filaments.

When the sarcomere is relaxed, the tropomyosin molecules of the thin filaments (see Figure 9.59) block the myosin-binding sites on the actin molecules. The position of the tropomyosin within the groove is under the control of the attached troponin molecule. When Ca^{2+} levels rise, these ions bind to one of the subunits of troponin (troponin C), causing a conformational change in another subunit of the troponin molecule. Like the collapse of a row of dominoes, the movement of troponin is transmitted to the adjacent tropomyosin, which moves approximately 1.5 nm closer to the center of the filament's groove (from position b to a in Figure 9.64). This shift in position of the tropomyosin exposes the myosin-binding sites on the adjacent actin molecules, allowing the myosin heads to attach to the thin filaments. Each troponin molecule controls the position of one tropomyosin molecule, which in turn controls the binding capacity of seven actin monomers in the thin filament.

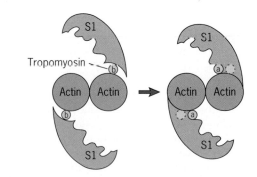

Figure 9.64 **The role of tropomyosin in muscle contraction.** Schematic diagram of the steric hindrance model in which the myosin-binding site on the thin actin filaments is controlled by the position of the tropomyosin molecule. When calcium levels rise, the interaction between calcium and troponin (not shown) leads to a movement of the tropomyosin from position b to position a, which exposes the binding site on the thin filament to the myosin head.

THE HUMAN PERSPECTIVE

The Molecular Basis of Muscular Dystrophy

The muscular dystrophies comprise a collection of inherited neuromuscular diseases characterized by muscle-fiber degeneration and resulting muscle weakness. One of these diseases, called *Duchenne muscular dystrophy* after the person who described it in 1861, is a common, debilitating disorder that strikes about 1 in 3,300 males. This is the disease that is most often associated with the name "muscular dystrophy," and it is the primary subject of this section. But there are several other rare types of muscular dystrophy that have revealed interesting insights into the relationship between the cytoskeleton, plasma membrane, and the extracellular matrix. We will begin with Duchenne muscular dystrophy (DMD).

The gene responsible for DMD is named *dystrophin*, and it has the distinction of being the largest gene so far identified in the mammalian genome. The gene stretches for more than 2.3 million bases along the X chromosome, a length that requires 16 hours to be transcribed into a single mRNA! Only about 0.6 percent of the gene actually codes for amino acids; the remaining 99.4 percent consists of noncoding introns (Section 11.3).

The protein encoded by the dystrophin gene resides in the membrane skeleton of striated muscle cells. Like the related spectrin molecules of the erythrocyte cytoskeleton (page 149), dystrophin molecules are rod-shaped dimers that lie just beneath the plasma membrane (Figure 1). On their cytoplasmic side, dystrophin molecules are attached to actin filaments, and on their membrane side, they are attached to a cluster of *dystrophin-associated proteins (DAPs)* that are part of the plasma membrane. The DAPs, in turn, are linked on their extracellular surface to components of the basement membrane that surround these contractile cells. Taken together,

these various proteins form a functional pathway linking the extracellular matrix with the internal cytoskeleton. In addition, dystrophin provides structural support for the plasma membrane as the muscle fiber repeatedly contracts and relaxes.

DMD is a severe form of muscular dystrophy, which is characterized at the cellular level by the virtual absence of dystrophin molecules from both skeletal and cardiac muscle cells. A less severe form, called *Becker muscular dystrophy*, also occurs, in which the protein is present but is abnormal and/or greatly reduced in amount. The symptoms of DMD

(a)

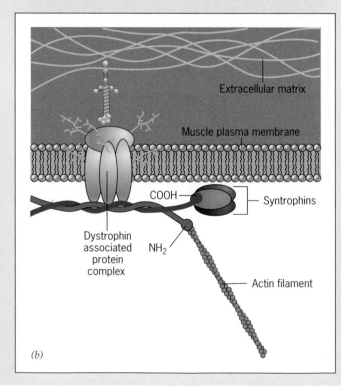

(b)

Figure 1. (*a*) Fluorescence micrograph of a portion of a mouse skeletal muscle cell that was labeled with antibodies against dystrophin. The location of dystrophin at the inner surface of the plasma membrane is shown by the larger reddish-colored areas (large arrows). Dystrophin is concentrated where the Z lines at the ends of the sarcomeres contact the membrane. The Z lines are faintly labeled because anti-dystrophin antibodies show a weak ability to bind to a related protein, α-actinin, which is concentrated in the Z lines (small arrows). Bar, 10 mm. (*b*) Schematic diagram of the manner in which dystrophin links the extracellular matrix to the actin cytoskeleton. Dystrophin molecules are attached to both actin filaments of the cytoplasm and to integral membrane proteins (DAPs). The membrane proteins are in turn attached to laminin molecules of the extracellular matrix. Mutations in each of these various components can result in a form of muscular dystrophy. (A: FROM VOLKER STRAUB, REGINALD E. BITTNER, JEAN J. LEGER, AND THOMAS VOIT, J. CELL BIOL. 119:1186, 1992; BY COPYRIGHT PERMISSION OF THE ROCKEFELLER UNIVERSITY PRESS.)

In figure (b) the following labels appear: Extracellular matrix; Muscle plasma membrane; COOH; Syntrophins; NH₂; Actin filament; Dystrophin associated protein complex

usually appear in childhood when the patient has difficulty performing certain motor tasks, such as climbing stairs or rising from a prone position. The disease progresses with increasing muscular weakness and debilitation, finally taking the patient's life as the result of respiratory or cardiac failure.

Patients with DMD show dramatic abnormalities in both cardiac and skeletal muscle tissue. At the light microscopic level, some of the muscle fibers are seen to be degenerating (necrotic) and often infiltrated by macrophages of the immune system. At the electron microscopic level, segments of the plasma membrane are seen to be missing from the cell surface so that the extracellular basement membrane remains as the primary encasement of the muscle fiber. As expected, destruction of the plasma membrane is accompanied by marked internal changes in the cell, including the dilation of the sarcoplasmic reticulum, swelling of mitochondria, increased proteolytic digestion of cell materials, and rupture of myofibrils. Damage to the plasma membrane is apparently caused by mechanical stress exerted on it as the muscle contracts. This view is supported by the fact that the greatest damage occurs to those muscles in the body that are subjected to the greatest mechanical stress.

Dystrophin is not the only gene that, when mutated, can cause muscular dystrophy. *Congenital muscular dystrophy (CMD)* is a rare form of the disease that develops in early infancy and strikes boys and girls with equal frequency. The disease has been traced to a deficiency in one of the subunits of the laminin molecule that forms a key component of the muscle cell basement membrane (Figure 1). Another form of the disease, called *severe childhood autosomal recessive muscular dystrophy (SCARMD)*, results from the absence of one of the subunits of the DAP complex (Figure 1). It is noteworthy that patients with DMD, CMD, and SCARMD all exhibit severe forms of the disease, indicating that all three components of the molecular chain—dystrophin, DAP, and laminin—are required for muscle function.

At the present time, there is no treatment to stem the degenerative changes that occur with any of the muscular dystrophies, and the best hopes of patients lie in the development of gene therapy in which the defective gene is replaced by a normal version. Fortunately, several animal models of muscular dystrophy have been developed, including the *mdx* mouse, which lacks a functional dystrophin gene. Unlike its human counterparts, these dystrophin-deficient mice show only mild symptoms. Regardless, they are still useful models for testing new molecular-based therapies and screening

drugs that might have therapeutic value. Studies on *mdx* mice have turned up one drug—an antibiotic called gentamicin—that causes the animals to synthesize dystrophin and protects them from muscle damage. *mdx* mice fail to produce dystrophin because their dystrophin genes contain a mutation that leads to premature termination during synthesis of the protein. Gentamicin reverses the effects of this mutation by allowing the ribosome to read through the defect on the mRNA and produce a full-sized dystrophin molecule. It is estimated that approximately 15 percent of DMD patients fail to synthesize dystrophin because of a similar type of genetic defect. Clinical trials of gentamicin on members of this population are planned.

Another approach that has been tried in clinical trials on human patients—without measurable success—is cell transplantation therapy. In these trials, normal, undifferentiated muscle cells (myoblasts) were isolated from muscle tissue of close relatives and injected into particular muscles of the patient. It was hoped that the donor myoblasts would fuse with the genetically deficient cells of the patient and provide genetic messages for the production of normal dystrophin throughout the entire muscle fiber. The survival rate of the injected cells was very low, and little or no evidence of the production of dystrophin was observed.

In recent years, a new approach to cell transplantation therapy has begun using injected bone marrow cells, rather than myoblasts. Blood forming tissues, such as bone marrow, contain stem cells that are capable of differentiating into muscle tissue (discussed on page 17). When bone marrow cells from a healthy donor mouse are injected intravenously into an *mdx* mouse, a small percentage of the introduced cells make their way into muscles throughout the recipient's body and differentiate into dystrophin-producing muscle cells. To date, the effect on the muscle tissues of these dystrophic animals is not dramatic enough to provide much encouragement to human DMD patients, but it does represent a start toward a new type of therapy. It is hoped, one day, that bone marrow cells can be removed from a DMD patient, provided with a normal copy of the dystrophin gene in vitro, and then reintroduced into the patient to give rise to normal muscle tissue. Alternatively, DMD patients might receive bone marrow cells from a healthy, compatible donor. Such cells would contain the normal dystrophin gene without having to be genetically altered, but the use of bone marrow from another individual raises problems of immune rejection that have to be considered.

Once stimulation from the innervating motor nerve fiber ceases, the Ca^{2+} channels in the SR membrane close, and the Ca^{2+}-ATPase removes excess calcium from the cytosol. As the Ca^{2+} concentration decreases, these ions dissociate from their binding sites on troponin, which causes the tropomyosin molecules to move back to a position where they block the actin-myosin interaction. The process of relaxation can be thought of as a competition for calcium between the transport protein of the SR membrane and troponin. The transport protein has a greater affinity for the ion, so it preferentially removes it from the cytosol, leaving the troponin molecules without bound calcium.

REVIEW ?

1. Describe the structure of the sarcomere of a skeletal muscle myofibril and the changes that occur during its contraction

2. Describe the steps that occur between the time that a nerve impulse is transmitted across a neuromuscular junction to the time that the muscle fiber actually begins to shorten. What is the role of calcium ions in this process?

9.7 NONMUSCLE MOTILITY

Skeletal muscle cells are an ideal system for the study of contractility and movement; the interacting contractile proteins are present in high concentration and are part of defined cellular structures. The study of nonmuscle motility is more challenging because the critical components tend to be present in less ordered, more labile, transient arrangements, and they are typically restricted to a thin *cortical zone* just beneath the plasma membrane. The cortical zone is an active region of the cell, responsible for such processes as the ingestion of extracellular materials, the extension of processes during cell movement, and the constriction of a single animal cell into two cells during cell division. All of these processes are dependent on the assembly of microfilaments in the cortex.

In the following pages, we will consider a number of examples of nonmuscle contractility and motility that depend on actin filaments and, in some cases, myosinlike motor proteins. First, however, it is important to survey the factors that govern the rates of assembly, numbers, lengths, and spatial patterns of actin filaments.

Actin-Binding Proteins

Purified actin can polymerize in vitro to form actin filaments, but such filaments cannot interact with one an-

other or perform useful activities. Under the microscope, they resemble the floor of a barn covered by straw. In contrast, actin filaments in living cells are organized into a variety of patterns, including various types of bundles, thin (two-dimensional) networks, and complex three-dimensional gels (Figure 9.65). The organization and

Figure 9.65 The organization of actin filaments within a cell. As described later in the chapter, cells move across a substratum by extending various types of processes. This electron micrograph of the leading edge of a motile fibroblast shows the high density of actin filaments. These filaments are seen to be organized into two distinct arrays: as bundles in which the filaments are arranged parallel to one another (arrows) and as a cross-linked meshwork in which the filaments are arranged in various directions. (COURTESY OF J. VICTOR SMALL.)

Table 9.2 Actin-Binding Proteins

Proteins	Relative Molecular Mass (kDa)	Source
Nucleating proteins		
Arp2/3 complex	7 subunits	Widespread
Monomer-sequestering proteins		
Thymosins	5	Widespread
Monomer-polymerizing proteins		
Profilin	12–15	Widespread
End-blocking proteins		
β-Actinin	35–37	Kidney, muscle
Cap Z	32, 34	Muscle
Capping protein	28–31	*Acanthamoeba*
Network-forming proteins		
Filamin	250	Smooth muscle
Actin-binding protein (ABP)	250	Platelets, macrophages
Gelactin	23–28	Amoebae
Bundling proteins		
Fimbrin	68	Intestine
Villin	95	Intestine
Fascin	57	Sea urchin eggs
α-Actinin	95	Muscle
Filament-severing proteins		
Gelsolin	90	Widespread
Brevin	93	Blood plasma
Filament-depolymerizing proteins		
Cofilin	21	Widespread
ADF	19	Widespread
Depactin	18	Sea urchin eggs
Membrane-binding proteins		
Dystrophin	427	Skeletal muscle
Vinculin	130	Widespread
Ponticulin	17	*Dictyostelium*

Note: Many of the actin-binding proteins can have more than one function depending on conditions.

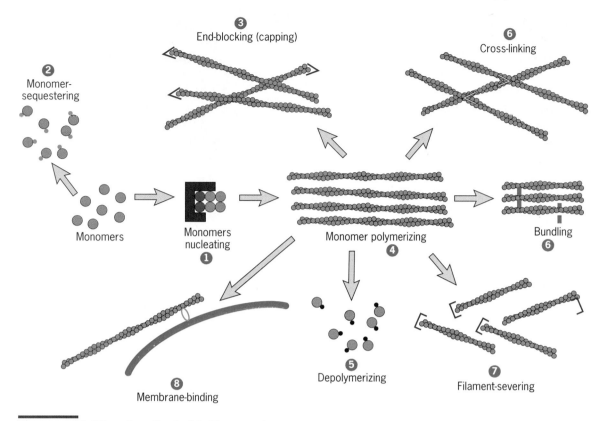

Figure 9.66 The roles of actin-binding proteins.

behavior of actin filaments inside cells are determined by a remarkable variety of **actin-binding proteins** that affect the assembly or disassembly of the actin filaments, their physical properties, and their interactions with one another and with other cellular organelles. More than 100 different actin-binding proteins belonging to numerous families have been isolated from different cell types. Actin binding proteins can be divided into several categories based on their presumed function in the cell (Table 9.2 and Figure 9.66).[5]

1. *Nucleating proteins.* The slowest step in the formation of an actin filament is the first step, nucleation, which requires that at least two or three actin monomers come together to begin formation of the polymer. As with microtubules, the formation of an actin filament is accelerated by the presence of a preexisting template to which subunits can be added (as in Figure 9.46a). Cells contain a protein complex, called the *Arp2/3 complex*, that can nucleate actin filaments both in vitro and in vivo. "Arp" stands for "actin-related proteins," which are proteins that share considerable sequence homology with actins, but are not considered "true" actins. An Arp2/3 complex consists of two Arps (Arp2 and Arp3) and five other pro-

teins. It is likely that the two Arps are bound together within the complex in such a way as to form a template to which actin monomers can be added, analogous to the way that γ-tubulin forms a template for microtubule nucleation (Figure 9.22).

2. *Monomer-sequestering proteins.* Thymosins (e.g., thymosin β₄) are proteins that bind to actin monomers (often called *G-actin*) and prevent them from polymerizing. Proteins with this activity are described as actin monomer-sequestering proteins. These proteins are believed responsible for the relatively high concentration of actin monomers in nonmuscle cells (50–200 μM). Without monomer-sequestering proteins, conditions within the cytoplasm would favor the near complete polymerization of soluble actin monomers into filaments. Because of their ability to bind G-actin and stabilize the monomer pool, changes in the concentration or activity of monomer-sequestering proteins can shift the monomer-polymer equilibrium in a certain region of a cell and determine whether polymerization or depolymerization is favored at the time.

3. *End-blocking (capping) proteins.* Proteins of this group regulate the length of actin filaments by binding to one or the other end of the filaments, forming a cap. If the fast-growing, barbed end of a filament is capped, depolymerization may proceed at the opposite end, resulting in the disassembly of the filament. The thin filaments of striated muscle are capped at their barbed end at the Z line by a protein called "capZ" and at their pointed end by

[5]It should be noted that some of these proteins can carry out more than one of the types of activities listed, depending on the concentration of the actin-binding protein and the prevailing conditions (for example, the concentration of Ca²⁺ and H⁺). Most studies of these proteins are conducted in vitro, and it is often difficult to extend the results to activities within the cell.

the protein tropomodulin. If the tropomodulin cap is disturbed by microinjection of antibodies into a muscle cell, the thin filaments add additional actin subunits at their newly exposed pointed end and undergo dramatic elongation into the middle of the sarcomere.

4. *Monomer-polymerizing proteins*. Profilin is a protein that binds to actin-ATP monomers. For years, profilin was thought to be a sequestering protein that, like thymosin, inhibited actin polymerization. But profilin isn't present at high enough concentration in cells to bind enough actin to act as an effective sequestering agent. Recent evidence suggests that rather than inhibiting polymerization, profilin probably promotes the addition of the bound monomer to the barbed end of a growing actin filament following removal of a capping protein from that site. Profilin has been shown to stimulate actin polymerization during cell locomotion.

5. *Actin filament-depolymerizing proteins*. Members of the cofilin family of proteins (including cofilin, ADF, and depactin) bind to the pointed end of actin filaments and greatly enhance their depolymerization into monomers. These proteins play a role in the rapid turnover of actin filaments at sites of rapid changes in cytoskeletal structure. They are essential for cell locomotion, phagocytosis, and cytokinesis.

6. *Cross-linking proteins*. Proteins of this group are able to alter the three-dimensional organization of a population of actin filaments. Each of these proteins has two or more actin-binding sites and therefore can cross-link two or more separate actin filaments. Certain of these proteins (e.g., ABP280 and filamin) have the shape of a long, flexible rod and promote the formation of loose networks of filaments interconnected at near right angles to one another (as in Figure 9.65). Regions of the cytoplasm containing such networks have the properties of an elastic gel that resists local mechanical pressures. Other cross-linking proteins (e.g., villin and fimbrin) have a globular shape and promote the bundling of actin filaments into tightly knit, parallel arrays. Such arrays are found in the microvilli that project from certain epithelial cells (Figure 9.67) and in the hairlike *stereocilia* (Figure 9.54) that project from receptor cells of the inner ear. Bundling filaments together adds to their rigidity, allowing them to act as a supportive internal skeleton for these cytoplasmic projections.

7. *Filament-severing proteins*. Proteins of this class have the ability to bind to the side of an existing filament and break it in two. Because they reduce the length of the filaments, proteins of this class decrease the viscosity of the cytoplasm. Gelsolin, the first of these proteins to be identified, was discovered through its ability to liquify (solate) gelated cytoplasmic extracts. Severing proteins may also promote the incorporation of actin monomers by creating additional free barbed ends, or they may cap the fragments they generate.

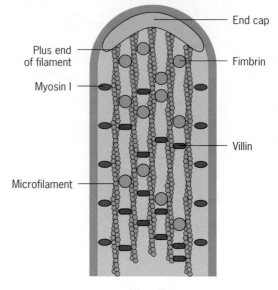

Microvillus

Figure 9.67 Actin filaments and actin-binding proteins in a microvillus. Microvilli are present on the apical surface of epithelia that function in absorption of solutes, such as the lining of the intestine and wall of the kidney tubule. Each microvillus contains about 25 actin filaments that are maintained in a highly ordered arrangement by the bundling proteins villin and fimbrin. The role of myosin I, which is present between the plasma membrane of the microvillus and the peripheral actin filaments, remains unclear.

8. *Membrane-binding proteins*. Much of the contractile machinery of nonmuscle cells lies just beneath the plasma membrane. During numerous activities, the forces generated by the contractile proteins act on the plasma membrane, causing it to protrude outward (as occurs, for example, during cell locomotion) or to invaginate inward (as occurs, for example, during phagocytosis or cytokinesis). These activities are generally facilitated by linking the actin filaments to the plasma membrane indirectly, by means of attachment to a peripheral membrane protein. Two examples were described in previous chapters: the inclusion of short actin polymers into the membrane skeleton of erythrocytes (see Figure 4.32*d*), and the attachment of actin filaments to the membrane at focal adhesions and adherens junctions (see Figures 7.17 and 7.26). Proteins that link membranes to actin include vinculin, members of the ERM family (ezrin, radixin, and moesin), and members of the spectrin family (including dystrophin, which was discussed in the Human Perspective).

Examples of Nonmuscle Motility and Contractility

Actin filaments, often working in conjunction with myosin motors, are responsible for a variety of dynamic activities in nonmuscle cells, including cytokinesis,

phagocytosis, cytoplasmic streaming (the directed bulk flow of cytoplasm that occurs in certain large plant cells), vesicle trafficking, blood platelet activation, lateral movements of integral proteins within membranes, cell-substratum interactions, cell locomotion, axonal outgrowth, and changes in cell shape. Nonmuscle motility and contractility are illustrated by the following examples.

Cell-Substratum Interactions and the Role of Stress Fibers When cells are allowed to settle to the bottom of a culture dish, they flatten and spread themselves onto the substratum. Attachment to the substratum often involves the formation of tightly adherent focal adhesions (described on page 255). Just beneath the plasma membrane

(a)

(b)

Figure 9.68 Stress fibers. (*a*) The distribution of stress fibers in a cultured fibroblast as revealed by fluorescent anti-actin antibodies (green). (*b*) Schematic drawing of a stress fiber showing the periodic nature of the non-actin components and the antiparallel arrangement of the actin filaments. (MICROGRAPH COURTESY OF ELIAS LAZARIDES; B FROM G. LANGANGER, ET AL., J. CELL BIOL. 102:208, 1986; BY COPYRIGHT PERMISSION OF THE ROCKEFELLER UNIVERSITY PRESS.)

of the focal adhesion, actin filaments become aligned in closely packed bundles, termed **stress fibers** (Figure 9.68*a*). More than any other structure in a nonmuscle cell, stress fibers most closely resemble the myofibrils of skeletal muscle tissue. In addition to actin, stress fibers contain a number of other proteins characteristic of muscle cells, including myosin II, tropomyosin, filamin, and α-actinin. These latter proteins are present discontinuously along the length of a stress fiber, giving it a banded appearance (Figure 9.68*b*), which resembles the sarcomere of a skeletal muscle fiber. In contrast, the actin filaments run as continuous elements through the stress fiber, but staining with heavy S1 myosin indicates that actin filaments are oriented in both directions within a given bundle.

Given this type of antiparallel organization, one might expect a stress fiber to be a contractile structure. This expectation is borne out in studies of stress fibers that have been isolated from cells by laser beam microsurgery: isolated stress fibers contract in the presence of ATP, which raises an interesting question. Why should these potentially contractile stress fibers be present in adherent cells that are not engaged in any type of locomotor activity? In fact, when a sedentary cultured cell is stimulated to move, one of the first visible changes is often the loss of its stress fibers. It is thought that the stress fibers of a tightly adherent cell undergo *isometric* contraction, that is, contraction that creates tension without a change in length. This type of contraction helps maintain the tension between cell and substratum. Within the body, stress fibers are found in cells that are associated with extracellular surfaces and are exposed to shearing forces acting to pull them from their substratum.

Actin Polymerization as a Force-Generating Mechanism Some types of cell motility occur solely as the result of actin polymerization and do not involve the activity of myosin. Consider the example of *Listeria monocytogenes*, a bacterium that infects macrophages and can cause encephalitis or food poisoning. *Listeria* is propelled like a rocket through the cytoplasm of an infected cell by the polymerization of actin monomers just behind the bacterium (Figure 9.69). How is the bacterial cell able to induce the formation of actin filaments at a particular site on its surface? Questions about location are important in the study of any type of motile process because motility depends on a cell being able to assemble the necessary machinery at a particular site and a particular time. *Listeria* can accomplish this feat because it contains a protein called ActA that is present only at one end of the bacterium. When ActA is exposed within the host cytoplasm, it recruits and activates a number of host proteins that work together to direct the process of actin polymerization. The process of *Listeria* propulsion has been reconstituted in vitro, which has allowed researchers to demonstrate conclusively that actin polymerization by itself, that is, without the participation of myosin motors, is capable of providing the force required for motility.

(a) 5 μm

(b)

Figure 9.69 **Cell motility may be based solely on actin polymerization.** (*a*) Fluorescence micrograph of a portion of a cell infected with the bacterium *L. monocytogenes*. The bacteria appear as red stained objects just in front of the green-stained filamentous actin tails. (*b*) Electron micrograph of a cell infected with the same bacterium as in *a*, showing the actin filaments that form behind the bacterial cell and push it through the cytoplasm. The actin filaments have a bristly appearance because they have been decorated with myosin heads. Bar at the upper left, 0.1 μm. (A: COURTESY OF PASCALE COSSART; B: FROM LEWIS G. TILNEY ET AL., J. CELL BIOL. 118:77, 1992; BY COPYRIGHT PERMISSION OF THE ROCKEFELLER UNIVERSITY PRESS.)

Recent studies have indicated that the same events that occur during *Listeria* propulsion are utilized for normal cellular activities, ranging from the propulsion of cytoplasmic vesicles to the movement of cells themselves, which is the subject of the following section.

Cell Locomotion Cell locomotion is required for many activities in higher vertebrates, including wound healing, protection against infection, and blood clotting. Cell locomotion is difficult to observe within the body because migratory cells cannot usually be distinguished from their opaque surroundings. Consequently, most researchers who study cell locomotion have turned to systems in which single cells can be followed as they crawl over the bottom of a culture dish, as seen in Figure 9.70.

Cell locomotion shares properties with other types of locomotion, for example, walking. As you walk, your body performs a repetitive series of activities: first, a leg is extended in the direction of locomotion; second, the bottom of your foot makes contact with the ground, which acts as a point of temporary adhesion; third, force is generated by the muscles of your legs that moves your entire body forward past the stationary foot, all the while generating traction against the point of adhesion; fourth, your foot—

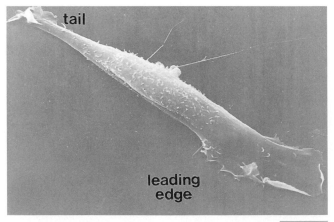

15 μm

Figure 9.70 **Scanning electron micrograph of a mouse fibroblast crawling over the surface of a culture dish.** The leading edge of the cell is spread into a flattened lamellipodium whose structure and function are discussed later in the chapter. (FROM GUENTER ALBRECHT-BUEHLER, INT. REV. CYTOL. 120:194, 1990.)

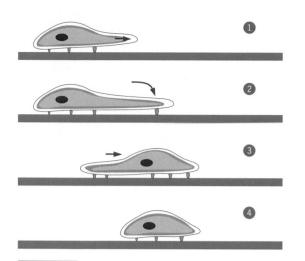

Figure 9.71 The repetitive sequence of activities that occurs as a cell crawls over the substratum. Step 1 shows the protrusion of the leading edge of the cell in the form of a lamellipodium. Step 2 shows the adhesion of the lower surface of the lamellipodium to the substratum, an attachment that is mediated by integrins residing in the plasma membrane. The cell uses this attachment to grip the substratum. Step 3 shows the movement of the bulk of the cell forward over the site of attachment, which remains relatively stationary. This movement is accomplished by a contractile (traction) force exerted against the substratum. Step 4 shows the cell after the attachments with the substratum have been severed and the rear of the cell has been pulled forward.

(a)

(b)

Figure 9.72 The leading edge of a motile cell. (*a*) The leading edge of this motile fibroblast is flattened against the substratum and spread out into a veil-like lamellipodium. (*b*) Scanning electron micrograph of the leading edge of a cultured cell, showing the ruffled membranes of the lamellipodium. (A: COURTESY OF J. VICTOR SMALL. B. FROM JEAN PAUL REVEL, SYMP. SOC. EXP. BIOL. 28:447, 1974.)

which is now behind your body rather than in front of it—is lifted from the ground in anticipation of your next step. Even though motile cells may assume very different shapes as they crawl over a substratum, they display a similar sequence of activities (Figure 9.71). (1) Movement is initiated by the protrusion of a part of the cell surface in the direction in which the cell is to move. (2) A portion of the lower surface of the protrusion attaches to the substratum, forming temporary sites of anchorage. (3) The bulk of the cell moves forward over the adhesive contacts, which eventually become part of the rear of the cell. (4) The cell breaks its rear contacts with the surface, causing retraction of the trailing edge, or "tail."

Cells That Crawl over the Substratum Anyone who has watched an amoeba crawl along the surface of a microscope slide has had the opportunity to observe one of the most dramatic types of cell locomotion. As the organism moves, portions of the cell surface are pushed outward by a column of cytoplasm that flows through the interior of the cell toward the periphery. The broad, rounded protrusions formed during ameboid movement are called **pseudopodia**. As the cytoplasm streams into the advancing pseudopodia, the amoeba slowly moves in that direction.

Locomotion of single cells of higher organisms is not usually accomplished by obvious cytoplasmic flow, but it is associated with protrusions of the cell surface at the leading edge of the cell. A quick look at a mammalian fibroblast moving along the surface of a culture dish reveals a type of locomotion very different from that of an amoeba. As it moves, a fibroblast typically flattens itself close to the substratum and becomes fan-shaped, with a broadened frontal end and a narrow "tail" (see Figure 9.70). Its movement is erratic and jerky, sometimes advancing and other times withdrawing. On a good day, a fibroblast may move a distance of about 1 mm. The key to the fibroblast's locomotion is seen by examining its leading edge, which is extended out from the cell as a broad, flattened, veil-like protrusion, called a **lamellipodium** (Figure 9.72*a*). Lamellipodia are typically devoid of particulate structures, and the outer edge often exhibits an undulating motion, giving it a ruffled appearance (Figure 9.72*b*). As a lamellipodium is extended from the cell, it adheres to the underlying substratum at specific points, providing temporary sites of anchorage for the cell to pull itself forward.

We saw on page 385 how the polymerization of actin monomers can provide the force that propels a *Listeria* bacterium through the cytoplasm. This type of intracellular movement is accomplished without the involvement of molecular motors. A similar type of actin-polymerization mechanism is thought to provide the motile force required for the protrusion of the leading edge of a lamellipodium. Suppose we begin with a rounded white blood cell that receives a chemical signal coming from one particular direction where the body has been wounded. This is the type of stimulus that will cause the cell to move in that direction. Once the stimulus is received at the plasma membrane, it triggers the localized polymerization of actin. Just as *Listeria* has a protein (ActA) that activates polymerization at the bacterial cell surface, mammalian cells have a family of proteins (the WASP family) that are thought to activate the Arp2/3 complex at the site of stimulation near the plasma membrane. As discussed on page 385 and illustrated in Figure 9.66, the Arp2/3 complex plays a key role in nucleating the formation of actin filaments. WASP was discovered as the product of a gene responsible for Wiskott-Aldrich syndrome. Patients with this disorder have a crippled immune system because their white blood cells lack a functional WASP protein and consequently fail to respond to chemotactic signals.

Figure 9.73 depicts a model for the major steps in formation of a lamellipodium that would move a cell in a particular direction. A stimulus is received at one end of the cell (step 1, Figure 9.73) that leads to the activation of Arp2/3 protein complexes by a member of the WASP family of proteins (step 2). Activated Arp2/3 complexes serve as nucleating sites for the formation of actin filaments

(step 3). Polymerization of actin monomers onto the free barbed ends of the growing filaments is promoted by profilin molecules, which are bound to the actin monomers (page 384). Once actin filaments have formed, Arp2/3 complexes bind to the sides of these filaments (step 4) and nucleate the formation of additional actin filaments that form as branches (step 5). The Arp2/3 complexes remain at the pointed ends, which are situated at the branchpoints. Meanwhile, growth of the barbed ends of older filaments is blocked by the addition of capping protein (step 5). In contrast, addition of actin monomers to the barbed ends of more recently formed filaments of the network pushes the membrane of the lamellipodium outward in the direction of the attractive stimulus (steps 5,6). As newer filaments are growing by addition of monomers to their barbed ends, the older capped filaments are undergoing disassembly from their pointed ends (step 6). Actin-ADP monomers released from the disassembling filaments are recharged by conversion into profilin-actin-ATP monomers, which can be reutilized in the assembly of new actin filaments.

Figure 9.74 illustrates some of the major structural features of cell locomotion. The electron micrograph of Figure 9.74 shows the branched, crosslinked nature of the filamentous actin network that resides just beneath the plasma membrane of an advancing lamellipodium. The circular insets in Figure 9.74 show a succession of actin-filament branches, with Arp2/3 complexes highlighted by immunogold labeling. The Arp2/3 complexes are seen to

Figure 9.73 A Proposed mechanism for the movement of a nonmuscle cell in a directed manner. A stimulus is received at the cell surface (step 1) that leads to the activation of Arp2/3 complex by a member of the WASP family (step 2). The activated Arp2/3 complexes initiate the nucleation of actin polymerization, which leads to the formation of actin filaments (step 3). Once filaments have formed, Arp2/3 complexes attach to the side of the filaments (step 4) and initiate side branches that extend outward (step 5) at an angle of about 70° relative to the existing filaments to which they are anchored. As these fila-

ments polymerize, they are thought to push the plasma membrane outward, resulting in extension of the leading edge of the lamellipodium. Meanwhile, the barbed end of previously formed filaments are bound by a capping protein, which prevents further growth of these filaments in an inappropriate direction. Eventually, the pointed end of the older, preexisting actin filaments undergo depolymerization, releasing actin-ADP subunits. These subunits become recharged by ATP/ADP exchange and bind profilin molecules, which makes them ready to engage in actin polymerization.

Figure 9.74 The structural basis of lamellipodial extension.
Electron micrograph of a replica of the cytoskeleton at the leading
edge of a motile mouse fibroblast. The actin filaments are seen to
be arranged in a branched network, which has been colorized to
indicate individual "trees." The circular insets show a succession of
Y-shaped junctions between branched actin filaments. Arp2/3
complexes are localized at the base of each branch by antibodies
linked to colloidal gold particles (yellow). (FROM TATYANA M. SVITK-
INA AND GARY G. BORISY, FROM J. CELL BIOL., VOL. 145, #5, 1999, BY
COPYRIGHT PERMISSION OF THE ROCKEFELLER UNIVERSITY PRESS.)

reside at the Y-shaped junctions where the newly poly-
merized filaments have branched off of preexisting fila-
ments.

Lamellipodial movement is a dynamic process. As
actin filament polymerization and branching continue at
the very front edge of the lamellipodium, actin filaments
are depolymerizing toward the rear of the lamellipodium
(step 6, Figure 9.73). Thus taken as a whole, the entire
actin-filament array undergoes a type of treadmilling
(page 368) in which actin monomers are added to barbed
ends of the array at its front end and lost from pointed
ends of the array toward the rear.

According to current opinion, actin polymerization is
responsible for extending the leading edge of a cell (step 1,
Figure 9.71), whereas myosin (in conjunction with actin
filaments) generates the contractile (traction) forces that
pull the remainder of the cell forward (step 3, Figure 9.71).
These contrasting roles of actin and myosin are best illus-
trated in studies on fish keratocytes (skin cells). These cells
have been a favored system for studying locomotion be-
cause their gliding movement depends on the formation of
a very broad, thin lamellipodium. Figure 9.75 shows a
moving keratocyte that has been fixed and stained for actin
(Figure 9.75a) and myosin II (Figure 9.75b). As expected
from the previous discussion, the advancing edge of the
lamellipodium is filled with actin. Myosin, on the other

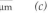

(a) (b) 10 μm (c)

**Figure 9.75 The roles of actin and myosin in the lamellipo-
dial-based movement of fish keratocytes.** (a–b) Fluorescence
micrographs of a fish keratocyte (skin cell) moving over a cul-
ture dish by means of a broad, flattened lamellipodium. The
arrow shows the direction of movement. The distribution of fil-
amentous actin is revealed in part a, which shows the localiza-
tion of fluorescently labeled phalloidin, which binds only to
actin filaments. The distribution of myosin in the same cell is
revealed in part b, which shows the localization of fluorescent
anti-myosin antibodies. It is evident that the body of the lamel-

lipodium contains actin filaments, but is virtually devoid of
myosin. Myosin is concentrated, instead, in a band that lies just
behind the lamellipodium, where it merges with the body of the
cell. (c) A schematic drawing depicting the filamentous actin
network of the lamellipodium and the actin-myosin interac-
tions toward the rear of the lamellipodium. The actin network
is shown in red, myosin molecules in blue. (BY ALEXANDER B.
VERKHOVSKY, FROM TATYANA M. SVITKINA ET AL., J. CELL BIOL.
139:397, 1997, BY COPYRIGHT PERMISSION OF THE ROCKEFELLER UNI-
VERSITY PRESS.)

hand, is concentrated in a band where the rear of the lamellipodium joins the remainder of the cell. Electron micrographs of this region show the presence of clusters of small, bipolar myosin II filaments bound to the actin network (Figure 9.75c). Forces generated by these myosin molecules are presumed to pull the bulk of the cell along behind the advancing lamellipodium. Myosin I is also thought to generate forces for cell locomotion in some organisms.

Axonal Outgrowth In 1907, Ross Harrison of Yale University performed one of the classic experiments in biology. Harrison removed a small piece of tissue from the developing nervous system of a frog embryo and placed the fragment into a tiny drop of lymphatic fluid. Harrison watched the tissue under a microscope over the next few days and found that the nerve cells not only remained healthy, but many of them sprouted processes that grew out into the surrounding medium. Not only was this the first time that cells had been kept alive in tissue culture, the experiment provided strong evidence that axons develop by a process of active outgrowth and elongation.

The tip of an elongating axon is very different from the remainder of the cell. Although the bulk of the axon shows little outward evidence of motile activity, the tip, or **growth cone**, resembles a highly motile, crawling fibroblast. Close examination of a living growth cone reveals several types of locomotor protrusions: a broad, flattened lamellipodium that creeps outward over the substratum; short, stiff *microspikes* (Figure 9.76a) that point outward to the edge of the lamellipodium; and highly elongated *filopodia* that extend and retract in a continuous display of motile activity. Fluorescence microscopy shows all of these structures to be filled with actin filaments (Figure 9.76b). Microtubules, on the other hand, do not penetrate into the tip of the growth cone but fill the core of the axon leading toward the tip (Figure 9.76c). In neurons, as in other types of cells, microfilaments function primarily in

motile activities, and microtubules function primarily in support and transport.

The growth cone is a highly motile region of the cell that explores its environment and elongates the axon. Within the embryo, the axons of developing neurons grow along defined paths, following certain topographical features of the substratum or responding to the presence of

(a)

(b)

(c) 10 μm

Figure 9.76 The structure of a growth cone: the motile tip of a growing axon. (a) A video image of a live growth cone. The terminus is spread into a flattened lamellipodium that creeps forward over the substratum. Rodlike microspikes (arrows) can be seen within the transparent veil of the lamellipodium, and fine processes called filopodia (arrowheads) can be seen projecting ahead of the leading edge of the lamellipodium. Bar, 5 μm. (b) The same growth cone shown in a after the cell has been fixed and stained for microfilaments using fluorescent phalloidin. Actin filaments are concentrated in the basal region of the lamellipodium and aligned with the microspikes. Filamentous actin is also seen along the outer edge of the lamellipodium, which had been creeping outward over the substratum. Actin filaments are also present in filopodia but are not resolved in this micrograph. (c) The same growth cone as in a and b stained for microtubules by immunofluorescence. Microtubules are concentrated in the central domain of the axon, but are largely absent from the lamellipodium. The outer edge of the lamellipodium is indicated by the asterisks in c. (FROM PAUL FORSCHER AND STEPHEN J. SMITH, J. CELL BIOL. 107:1508, 1988.)

certain chemicals that diffuse into their path. The lamellipodia and filopodia from the growth cone respond to the presence of these physical and chemical stimuli, causing the pathfinding axons to grow in directions that lead them to the target organ they will innervate.

Changes in Cell Shape During Embryonic Development

Each part of the body has a characteristic shape and internal architecture that arises during embryonic development: the spinal cord is basically a hollow tube, the kidney consists of microscopic tubules, each lung is composed of microscopic

air spaces, and so forth. Numerous cellular activities are necessary for the development of the characteristic morphology of an organ, including programmed changes in cell shape. Changes in cell shape are brought about largely by changes in the orientation of cytoskeletal elements within the cells. One of the best examples of this phenomenon is seen in the early stages of the development of the nervous system.

Toward the end of gastrulation in vertebrates, the outer (ectodermal) cells situated along the embryo's dorsal surface elongate and form a tall epithelial layer called the neural plate (Figure 9.77a,b). The cells of the neural plate elongate

Figure 9.77 Early stages in the development of the vertebrate nervous system. (*a–d*) Schematic drawings of the changes in cell shape that cause a layer of flattened ectodermal cells at the middorsal region of the embryo to roll into a neural tube. The initial change in height of the cells is thought to be driven by the orientation and elongation of microtubules, whereas the rolling of the plate into a tube is thought to be driven by contractile forces generated by actin filaments at the apical ends of the cells. (*e*) Scanning electron micrograph of the dorsal surface of a chick embryo as its neural plate is being folded to form a tube. (E: COURTESY OF KATHRYN W. TOSNEY.)

as microtubules become oriented with their long axes parallel to that of the cell (inset, Figure 9.77*b*). Following elongation, the cells of the neural epithelium become constricted at one end, causing the cells to become wedge shaped and the entire layer of cells to curve inward (Figure 9.77*c*). This latter change in cell shape is brought about by the contraction of a band of microfilaments that assemble in the cortical region of the cells just beneath the apical cell membrane (inset, Figure 9.77*c*). Eventually, the curvature of the neural tube causes the outer edges to contact one another forming a cylindrical, hollow tube (Figure 9.77*d,e*) that will give rise to the animal's entire nervous system.

REVIEW ?

1. List the various types of actin-binding proteins and one function of each type.
2. Describe the steps taken by a mammalian cell crawling over a substratum.
3. Describe the role of actin filaments in the activities of the growth cone of a neuron.

SYNOPSIS

The cytoskeleton is composed of three distinct types of fibrous structures: microtubules, intermediate filaments, and microfilaments (actin filaments), which participate in a number of cellular activities. Collectively, the elements of the cytoskeleton function as a structural support that helps maintain the shape of the cell; as an internal framework responsible for positioning the various organelles within the cell interior; as part of the machinery required for the movement of materials and organelles within cells; as force-generating elements responsible for the movement of cells from one place to another; and as sites for anchoring mRNAs and facilitating their translation into polypeptides. (p. 333)

Microtubules are hollow, cylindrical structures 25 nm in diameter that are assembled from the protein tubulin and, in addition to the cytoskeleton, form part of the mitotic spindle, centrioles, and the core of cilia and flagella. Microtubules are polymers assembled from αβ-tubulin heterodimers that are arranged in rows, or protofilaments. Many of the properties of microtubules, including their flexibility, stability, and capabilities for interaction, are influenced by members of a group of microtubule-associated proteins (MAPs). Because of their rigidity, microtubules often act in a supportive capacity not unlike the way that steel girders provide support for a tall building. Microtubules also function in such diverse activities as influencing the deposition of cellulose in a plant cell wall; maintaining the position of the membranous organelles of the biosynthetic pathway; and the movement of vesicles and other materials between the cell body and axon terminals of a nerve cell. (p. 338)

Three families of motor proteins have been identified and characterized: kinesins and dyneins, which move along microtubules, and myosins, which move along microfilaments. These motor proteins are able to convert chemical energy stored in ATP into mechanical energy that is used to move cellular cargo attached to the motor. Forces are generated as conformational changes in the motor protein are coupled to a chemical cycle involving the binding and hydrolysis of nucleotides and the release of the bound products. (p. 342)

In most cases, kinesin and cytoplasmic dynein move materials along microtubules in opposite directions. Both kinesin and cytoplasmic dyneins are large motor proteins with globular heads that interact with microtubules and act as force-generating engines and an opposite end that binds to specific types of cargo to be transported. Kinesin moves materials toward the plus end of a microtubule, cytoplasmic dynein toward the minus end. Kinesin has been implicated in the movement of ER-derived vesicles, endosomes, lysosomes, and secretory granules and has been shown to be the primary motor protein mediating anterograde transport in an axon (from cell body to axon terminus). (p. 343)

The nucleation of microtubules in vivo occurs in association with a variety of MTOCs (microtubule organizing centers). In animal cells, the microtubules of the cytoskeleton typically form in association with the centrosome, a complex structure that contains two barrel-shaped centrioles surrounded by amorphous, electron-dense pericentriolar material. Microtubules typically radiate outward from the pericentriolar material, which contains the components required for microtubule nucleation. The microtubules that form the fibers of a cilium or flagellum originate from a basal body, which has essentially the same structure as a centriole. Centrosomes, basal bodies, and other MTOCs, such as the outer surface of the nuclear envelope of plant cells, share a common protein component called γ-tubulin, which plays a key role in nucleation of microtubules. (p. 347)

The microtubules of the cytoskeleton are dynamic polymers that are subject to shortening, lengthening, disassembly, and reassembly. Disassembly of the microtubular cytoskeleton can be induced by a number of agents including colchicine, cold temperature, and elevated Ca^{2+} concentration. At any given moment, some microtubules of the cytoskeleton are growing in length while others are shrinking. When individual microtubules are followed over time, they are seen to switch back and forth between growing and shrinking phases, a phenomenon known as dynamic instability. Both growth and shrinkage occur predominantly, if not exclusively, at the plus end of the

polymer—the end opposite the MTOC. Tubulin dimers that are assembled into a microtubule contain a GTP molecule that is hydrolyzed soon after its incorporation into the polymer. During periods of rapid assembly, the hydrolysis of GTP bound to incorporated tubulin dimers lags behind the incorporation of new dimers, so that the end of the microtubule contains a cap of tubulin-GTP dimers that favors the addition of more subunits and the growth of the microtubule. Assembly and disassembly may also be governed by Ca^{2+} concentration and the presence of specific MAPs. (p. 350)

Cilia and flagella contain a core structure, the axoneme, that is composed of an array of microtubules that support the organelle as it projects from the cell surface and provide the machinery for generating the forces required for locomotion. In cross section, an axoneme is seen to consist of nine outer doublet microtubules (a complete A microtubule and an incomplete B microtubule) surrounding a pair of single microtubules. A pair of arms projects from the A microtubule of each doublet. The arms are composed of ciliary dynein, a motor protein that is responsible for using the energy released by ATP hydrolysis to generate the forces required for ciliary or flagellar bending. This is accomplished as the dynein arms of one doublet attach to the B microtubule of the neighboring doublet and then undergo a conformational change that causes the A microtubule to slide a perceptible distance. Sliding on one side of the axoneme alternates with sliding on the other side so that a part of the cilium or flagellum first bends in one direction and then in the opposite direction. Microtubule sliding has been demonstrated directly by attaching beads to the doublets of demembranated axonemes and following their relative movements during reactivation. (p. 354)

Intermediate filaments (IFs) are ropelike cytoskeletal structures approximately 10 nm in diameter that, depending on the cell type, may be composed of a variety of different protein subunits capable of assembling into similar types of filaments. Unlike microtubules, intermediate filaments are composed of symmetric building blocks (tetrameric subunits), which assemble into filaments that lack polarity. IFs are resistant to tensile forces and relatively insoluble; yet, like the other two types of cytoskeletal elements, they are dynamic structures that rapidly incorporate fluorescently labeled subunits that are injected into a cell. Assembly and disassembly are thought to be controlled primarily by phosphorylation and dephosphorylation. IFs are thought to provide mechanical stability to cells and to be required for specialized, tissue-specific functions. (p. 363)

Actin filaments (or microfilaments) are 8 nm in diameter, are composed of a double-helical polymer of the protein actin, and play a key role in virtually all types of contractility and motility within cells. Depending on the type of cell and the activity, actin filaments can be organized into highly ordered arrays, loose ill-defined networks, or tightly held bundles. Actin filaments are often identified by their ability to bind to the S1 myosin fragment, which also reveals the polarity of the filament. Although both ends of an actin filament can gain or lose subunits, the barbed (or plus end) of the filament is the preferred site of subunit addition, and the pointed (or minus end) is the preferred site of subunit loss. To be incorporated onto the growing end of a filament, an actin subunit must be bound to an ATP, which is hydrolyzed soon after its incorporation. Cells maintain a dynamic equilibrium

between the monomeric and polymeric forms of actin, which can be altered by changes in a variety of local conditions. The role of actin filaments in a particular process is most readily tested by treating the cells with cytochalasin, which promotes the depolymerization of the filament, or with phalloidin, which prevents it from disassembling and participating in dynamic activities. (p. 366)

The forces responsible for microfilament-dependent processes may be generated by the assembly of the actin filament or, more often, as the result of interaction with the motor protein myosin. The propulsion of certain bacteria through the cytoplasm of an infected phagocyte is a process driven by actin polymerization. Myosins are generally divided into two classes, conventional (type II) myosins and nonconventional (types I and III-XIV) myosins. Myosin II is the molecular motor that generates force in the various types of muscle tissues and also a variety of nonmuscle activities, including cytokinesis. Myosin II molecules all contain a long, rod-shaped tail attached at one end to two globular heads. The heads are the sites that bind the actin filament, hydrolyze ATP, and undergo the conformational changes required for force generation. The neck is thought to act as a lever arm that amplifies conformational changes of the head. The fibrous tail mediates the assembly of myosin into bipolar filaments. Most unconventional myosins have a single head and variable tail domains; they have been implicated in cell motility and organelle transport. (p. 368)

The contraction of a skeletal muscle fiber results from the sliding of actin-containing thin filaments toward the center of the individual sarcomeres of a myofibril, and it is driven by forces generated in the myosin cross-bridges that extend outward from the thick filaments. Changes that occur during the shortening of a muscle fiber are reflected in changes in the banding pattern of the sarcomeres as the Z lines at the ends of the sarcomere are moved toward the outer edges of the A bands. Contraction is triggered as an impulse penetrates into the interior of the muscle fiber along the membranous transverse tubules, stimulating the release of Ca^{2+} from storage sites in the sarcoplasmic reticulum (SR). The binding of calcium ions to the troponin molecules of the thin filaments causes a change in conformation that moves the tropomyosin molecules to a position that exposes the myosin-binding sites on the actin subunits of the thin filament. The subsequent interaction between myosin and actin triggers filament sliding. (p. 372)

Nonmuscle motility and contractility depend on some of the same proteins found in muscle cells, but they are arranged in less highly ordered, more labile and transient configurations. Nonmuscle motility depends on actin, usually in conjunction with myosin. The organization and behavior of actin filaments are determined by a variety of actin-binding proteins that affect the assembly of the actin filaments, their physical properties, and their interactions with one another and with other cellular organelles. Included among the list are proteins that sequester actin monomers and prevent their polymerization; proteins that form a cap at one end of an actin filament, which may block filament growth or lead to filament disassembly; proteins that cross-link actin filaments into bundles, loose meshworks, or three-dimensional gels; proteins that sever actin filaments; and proteins that link actin filaments to the inner surface of the plasma membrane. (p. 382)

Examples of nonmuscle motility and contractility include the maintenance of tension by the action of stress fibers, the crawling of cells over a substratum, and axonal outgrowth. Stress fibers consist of bundles of actin filaments with sarcomerelike distribution of actin-associated proteins. Stress fibers are thought to generate tension between a cell and its substratum by generating an isometric contraction. Cell crawling is typically accomplished by a flattened, veil-like projection, or lamellipodium, that forms at the cell's leading edge. As a lamellipodium extends from the cell, it adheres to the underlying substratum at specific points, providing temporary sites of anchorage for the cell to crawl over. The protrusion of a lamellipodium is associated with the assembly of actin filaments and their association with various types of actin-binding proteins. The forces required for lamellipodial protrusion are thought to result from actin polymerization. The tip of an elongating axon consists of a growth cone, which resembles a highly motile, crawling fibroblast and contains several types of locomotor protrusions, including a lamellipodium, microspikes, and filopodia. The growth cone acts both to explore the environment and to elongate the axon along an appropriate pathway. (p. 385)

ANALYTIC QUESTIONS

1. If a myofibril were pulled so that the sarcomeres increased in length by approximately 50 percent, what effect would you expect this to have on the contractile ability of the myofibril? Why? What effects would this have on the H, A, and I bands?

2. What are three different radioactively labeled substances that you might inject into a cell that would label the microtubules of a cell without labeling the other cytoskeletal elements?

3. What are two types of nonmuscle motility that might be unaffected by antibodies against both myosin I and myosin II? Why?

4. A centriole contains _____ complete microtubules, and a cilium contains _____ complete microtubules.

5. Some members of the kinesin family contain only a single heavy chain. How would you expect their activity in an in vitro motility assay to differ from that of kinesin itself?

6. List three things you could do that would shift the dynamic equilibrium of an in vitro preparation of tubulin and microtubules toward the formation of microtubules. List three treatments that would shift the equilibrium in the opposite direction.

7. It was noted that a demembranated ciliary or flagellar axoneme is able to beat with a normal frequency and pattern. Can you conclude that the plasma membrane is not important in ciliary or flagellar function?

8. Because cytoplasmic vesicles are seen to move in both directions within an axon, can you conclude that some microtubules are oriented with their plus end facing the axon terminus and others oriented with the opposite polarity?

9. Would you agree with the statement that the centrosome plays a key role in determining the rates of lengthening and shortening of the microtubules of an animal cell? Why or why not?

10. If you were comparing the molecular structure of kinesin and dynein, which part (heads or tails) would you expect to be most similar between them? Why?

11. Figure 9.31*a* shows the appearance of a cross section of a ciliary axoneme cut at an interior portion of the cilium. How might the image of a cross section differ if it had been cut very near the tip of a cilium at the beginning of the recovery stroke?

12. If an individual kinesin molecule can move at a rate of 800 nm/sec in an in vitro motility assay, what is the maximal turnover rate (ATP molecules hydrolyzed per second) by one of the molecule's motor domains?

13. Why do you suppose more can be learned about microtubule dynamics by injecting fluorescent tubulin into a cell than radioactively labeled tubulin? Can you think of a question that would be better answered by radioactively labeled tubulin?

14. Suppose you discovered that a mouse lacking copies of the kinesin gene appeared to show no ill effects and lived to a ripe old age. What could you conclude about the role of kinesin in intracellular locomotion?

15. Which type of vertebrate tissue would you expect to be an excellent source of tubulin? of actin? of keratin? Which protein would you expect to be the least soluble and most difficult to extract? What types of protein would you expect as contaminants in a preparation of tubulin? Which in a preparation of actin?

16. Actin is one of the most evolutionarily conserved proteins. What does this tell you about the structure and function of this protein in eukaryotic cells?

17. It was noted on page 35 that mutations in a gene whose product destroys free radicals have been implicated in ALS. How can this be compatible with the finding that accumulations of neurofilaments appear to play a causative role in this disease?

18. Myosin action (Figure 9.61) differs from that of kinesin (Figure 9.16) in that one of the kinesin heads is always in contact with a microtubule, whereas both myosin heads become completely detached from the actin filament. How are these differences correlated with the two types of motor activities in which these proteins engage?

19. The microtubules of an axon are thought to be nucleated at the centrosome, then severed from their nucleation site and moved into the axon. It was noted in the text that cytoplasmic dynein is responsible for the retrograde movement of organelles in axons, yet this same motor is thought to mediate the anterograde movement of microtubules in these same cellular processes. How is it possible that the same minus end-directed motor can be involved in both anterograde and retrograde movements.

ADDITIONAL READING

General Cytoskeleton

BERNS, M. W. 1998. Laser scissors and tweezers. *Sci. Am.* 278:62–67. (April)

BORISY, G. G. & CARLIER, M.-F., EDS. 2000. Cytoskeleton. *Curr. Opin. Cell Biol.* vol. 12, #1.

GOLDSTEIN, L. S. B. & THERIOT, J. A. 2001. Cytoskeleton. *Curr. Opin. Cell Biol.* vol. 13, #1.

INOUÉ, S., ET AL. 1999. A half-century of advances in microscopy. *FASEB J.* 13:S179–S283.

MEHTA, A. D., ET AL. 1999. Single-molecule biomechanics with optical methods. *Science* 283:1689–1695.

SALMON, E. D. & WAY, M., EDS. 1999. Cytoskeleton. *Curr. Opin. Cell Biol.* vol. 11, #1.

SCHEIBEL, E., ET AL. 2000. Reviews on cytoskeleton. *Nature Cell Biol.* vol. 2, #1.

Microtubules

AMOS, L. A. 2000. Focusing-in on microtubules. *Curr. Opin. Struct. Biol.* 10:236–241.

BAAS, P. W. 1999. Microtubules and neuronal polarity: Lessons from mitosis. *Neuron* 22:23–31.

CAPDEVILA, J. & BELMONTE, J. C. I. 2000. Knowing left from right: The molecular basis of laterality defects. *Mol. Med. Today* 6:112–118.

DESAI, A. & MITCHISON, T. J. 1997. Microtubule polymerization dynamics. *Annu. Rev. Cell Dev. Biol.* 13:83–117.

DREWES, G., ET AL. 1998. MAPs, MARKs, and MT dynamics. *Trends Biochem. Sci.* 23:307–311.

ERICKSON, H. P. 2000. Gamma-tubulin nucleation: Template or protofilament? *Nature Cell Biol.* 2:E93–E96.

GIBBONS, I. R. 1981. Cilia and flagella of eukaryotes. *J. Cell Biol.* 91:107S–124S.

JENG, R. & STEARNS, T. 1999. Gamma-tubulin complexes: Size does matter. *Trends Cell Biol.* 9:339–342.

MARGOLIS, R. L. & WILSON, L. 1998. Microtubule treadmilling: What goes around comes around. *BioEss.* 20:830–836.

NOGALES, E. 2000. Structural insights into microtubule function. *Annu. Rev. Biochem.* 69:277–302.

PALAZZO, R. E. 1999. The centrosome. *Science & Med.* 6:32–41 (March/April)

SATIR, P. 1999. The cilium as a biological nanomachine. *FASEB J.* 13:S235–S237.

URBANI, L. & STEARNS, T. 1999. The centrosome. *Curr. Biol.* 9:R315–R317.

VAUGHN, K. C. & HARPER, J. D. 1998. Microtubule-organizing centers and nucleating sites in land plants. *Int. Rev. Cytol.* 181:75–98.

WIESE, C. & ZHENG, Y. 1999. Gamma-tubulin complexes and their interactions with microtubule-organizing centers. *Curr. Opin. Struct. Biol.* 9:250–256.

Intermediate Filaments

BRADY, S. T. 2000. Neurofilaments run sprints not marathons. *Nature Cell Biol.* 2:E43–E45.

COULOMBE, P. A., ET AL. 2000. The "ins" and "outs" of intermediate filament organization. *Trends Cell Biol.* 10:420–428.

FUCHS, E. & CLEVELAND, D. W. 1998. A structural scaffolding of intermediate filaments in health and disease. *Science* 279:514–519.

FUCHS, E. & YANG, Y. 1999. Crossroads on cytoskeletal highways. *Cell* 98:547–550.

HERRMANN, H. & AEBI, U. 2000. Intermediate filaments and their associates. *Curr. Opin. Cell Biol.* 12:79–90.

HOUSEWEART, M. K. & CLEVELAND, D. W. 1999. Cytoskeletal linkers: New MAPs for old destinations. *Curr. Biol.* 9:R864–R866.

KLYMKOWSKY, M. W. 1999. Weaving a tangled web: The intermediate cytoskeleton. *Nature Cell Biol.* 1:E121–123.

WICHE, G. 1998. Role of plectin in cytoskeletal organization and dynamics. *J. Cell Sci.* 111:2477–2486.

Molecular Motors

AMOS, L. A. 2000. Kinesin sticks its neck out. *Nature Cell Biol.* 2:E15–E16.

BLOCK, S. M. 1998. Leading the procession: New insights into kinesin motors. *J. Cell Biol.* 140:1281–1284.

COOKE, R. 1999. Myosin structure: Does the tail wag the dog. *Curr. Biol.* 9:R773–R775.

CRAMER, L. P. 2000. Myosin VI: Roles for a minus end-directed actin motor in cells. *J. Cell Biol.* 150:F121–F126.

CROSS, R. & SCHOLEY, J. 1999. Kinesin: The tail unfolds. *Nature Cell Biol.* 1:E119–E121.

CROSS, R. A. & CARTER, N. J. 2000. Molecular motors. *Curr. Biol.* 10:R177–R179.

CROSS, R. A. 2001. Molecular motors: Kinesin's string variable. *Curr. Biol.* 11:R147–R150.

CYR, R. J. 1994, Microtubules in plant morphogenesis: Role of the cortical array. *Annu. Rev. Cell Biol.* 10:153–180.

ENDOW, S. A. 1999. Determinants of molecular motor directionality. *Nature Cell Biol.* 1:E163–E167.

ENDOW, S. A. & FLETTERICK, R. J. 1998. Reversing a "backwards" motor. *BioEss.* 20:108–112.

GOLDSTEIN, L. S. B. & PHILP, A. V. 1999. The road less traveled: Emerging principles of kinesin motor utilization. *Annu. Rev. Cell Dev. Biol.* 15:141–185.

HASSON, T. 1999. Molecular motors: Sensing a function for myosin VIIa. *Curr. Biol.* 9:R838–R841.

HAYS, T. & LI, M. 2001. Kinesin transport: Driving kinesin in the neuron. *Curr. Biol.* 11:R136–R139.

HUXLEY, A. F. 1998. Biological motors: Energy storage in myosin motors. *Curr. Biol.* 8:R485–R488.

HUXLEY, A. F. 1998. Support for the lever arm. *Nature* 396:317–318.

IRVING, M. & GOLDMAN, Y. E. 1999. Another step ahead for myosin. *Nature* 398:463–465.

KNIGHT, A. E. & MOLLOY, J. E. 1999. Coupling ATP hydrolysis to mechanochemical work. *Nature Cell Biol.* 1:E87–E89.

MOOSEKER, M. S. & CHENEY, R. E. 1995. Unconventional myosins. *Annu. Rev. Cell Dev. Biol.* 11:633–675.

SCHLIWA, M. 1999. Myosin steps backwards. *Nature* 401:431–432.

TITUS, M.A. 2000. Cytoskeleton: Getting to the point with myosin VI. *Curr. Biol.* 10:R294–297.

VALE, R. D. 1999. Millennial musings on molecular motors. *Trends Biochem. Sci.* 24:M38–M42. (Dec.)

VALE, R. D. & MILLIGAN, R. A. 2000. The way things move: Looking under the hood of molecular motor proteins. *Science* 288:88–95.

VALLEE, R. B. & GEE, M. A. 1998. Make room for dynein. *Trends Cell Biol.* 8:490–494.

WADE, R. H. & KOZIELSKI, F. 2000. Structural links to kinesin directionality and movement. *Nature Struct. Biol.* 7:456–460.

YANAGIDA, T. & IWANE, A. H. 2000. A large step for myosin. *Proc. Nat'l Acad. Sci. U. S. A.* 97:9357–9359.

Actin Filaments and Actin-Binding Proteins

AYSCOUGH, K. R. 1998 In vivo functions of actin-binding proteins. *Curr. Opin. Cell Biol.* 10:102–111.

BAMBURG, J. R., ET AL. 1999. Putting a new twist on actins: ADF/cofilin modulates actin dynamics. *Trends Cell Biol.* 9:364–370.

CHEN, H., ET AL. 2000. Regulating actin-filament dynamics in vivo. *Trends Biochem. Sci.* 25:19–23.

DEROSIER, D. J. & TILNEY, L. G. 2000. F-actin bundles are derivatives of microvilli: What does this tell us about how bundles might form? *J. Cell Biol.* 148:1–6.

HOLT, M. R. & KOFFER, A. 2001. Cell motility: Proline-rich proteins promote protrusions. *Trends Cell Biol.* 11:38–46.

MACHESKY, L. M. & INSALL, R. H. 1999. Signaling to actin dynamics. *J. Cell Biol.* 146:267–272.

POLLARD, T. D., ET AL. 2000. Molecular mechanisms controlling actin filament dynamics in nonmuscle cells. *Annu. Rev. Biophys. Biomol. Struct.* 29:545–576.

SCHAFER, D. A. & SCHROER, T. A. 1999. Actin-related proteins. *Annu. Rev. Cell Dev. Biol.* 15:341–363.

Muscle Contractility

COOKE, R. 1997. Actomyosin contraction in striated muscle. *Physiol. Revs.* 77:671–697.

EBASHI, S. 1991. Excitation-contraction coupling and the mechanism of muscle contraction. *Annu. Rev. Physiol.* 53:1–16.

FRANZINI-ARMSTRONG, C. 1999. The sarcoplasmic reticulum and the control of muscle contraction. *FASEB J.* 13:S266–S270.

GEEVES, M. A. & HOLMES, K. C. 1999. Structural mechanism of muscle contraction. *Annu. Rev. Biochem.* 68:687–728.

GORDON, A. M., ET AL. 2000. Regulation of muscle contraction in striated muscle. *Physiol. Revs.* 80:853–924.

HOLMES, K. C. 1998. A powerful stroke. *Nature Struct. Biol.* 5:940–942.

HUXLEY, H. 1996. A personal view of muscle and motility mechanisms. *Annu. Rev. Physiol.* 58:1–19.

HUXLEY, H. E. 1998. Getting to grips with contraction: The interplay of structure and biochemistry. *Trends Biochem. Sci.* 23:84–87.

SQUIRE, J. M. & MORRIS, E. P. 1998. A new look at thin filament regulation in vertebrate skeletal muscle. *FASEB J.* 12:761–771.

TRINICK, J. & TSKHOVREBOVA, L. 1999. Titin: A molecular control freak. *Trends Cell Biol.* 9:377–380.

Nonmuscle Motility

BECKERLE, M. C. 1998. Spatial control of actin filament assembly: Lessons from *Listeria*. *Cell* 95:741–748.

BRAY, D. 1992. *Cell Movement*. Garland.

BROWN, S. S. 1999. Cooperation between microtubule- and actin-based motor proteins. *Annu. Rev. Cell Dev. Biol.* 15:63–80.

DRAMSI, S. & COSSART, P. 1998. Intracellular pathogens and the actin cytoskeleton. *Annu. Rev. Cell Dev. Biol.* 14:137–166.

KELLEHER, J. F. & TITUS, M. A. 1998. Intracellular motility: How can we all work together? *Curr. Biol.* 8:R394–R397.

MACHESKY, L. M. 1999. Rocket-based motility: A universal mechanism? *Nature Cell Biol.* 1:E29–E31.

MACHESKY, L. M. & COOPER, J. A. 1999. Bare bones of the cytoskeleton. *Nature* 401:542–543.

MAHADEVAN, L. & MATSUDAIRA, P. 2000. Motility powered by supramolecular springs and ratchets. *Science* 288:95–99.

PENNISI, E. 2000. Kinesin movements revealed. *Science* 287:23–24.

SVITKINA, T. M. & BORISY, G. G. 1999. Progress in protrusion: The telltale scar. *Trends Biochem. Sci.* 24:432–436.

ZIGMOND, S. H. 1998. Actin cytoskeleton: The Arp2/3 complex gets to the point. *Curr. Biol.* 8:R654–R657.

The Human Perspective on Muscular Dystrophy

CHAMBERLAIN, J. 1999. The dynamics of dystroglycan. *Nature Gen.* 23:256–258.

GUSSONI, E., ET AL. 1999. Dystrophin expression in the *mdx* mouse restored by stem cell transplantation. *Nature* 401:390–394.

PENNISI, E. 1998. Bone marrow cells may provide muscle power. *Science* 279:1456.

QU, Z., ET AL. 1998. Development of approaches to improve cell survival in myoblast transfer therapy. *J. Cell Biol.* 142:1257–1267.

The Nature of the Gene and the Genome

ur concept of the gene has undergone a remarkable evolution as biologists have learned more and more about the nature of inheritance. The earliest studies revealed genes to be discrete factors that were retained throughout the life of an organism and then passed on to each of its progeny. Over the following half-century, these hereditary factors were shown to reside on chromosomes and to consist of DNA, a macromolecule with extraordinary properties. An overview of some of the early milestones along this journey of discovery are illustrated in Figure 10.1. In the half-century that has passed since the identification of DNA as the genetic material, researchers have learned a great deal about the complex mechanisms that cells employ to utilize their store of genetic information. These discoveries are the subject of the present and the following three chapters.

Model of DNA built by James Watson and Francis Crick at Cambridge University, 1953. (COURTESY SCIENCE & SOCIETY PICTURE LIBRARY, SCIENCE MUSEUM, LONDON.)

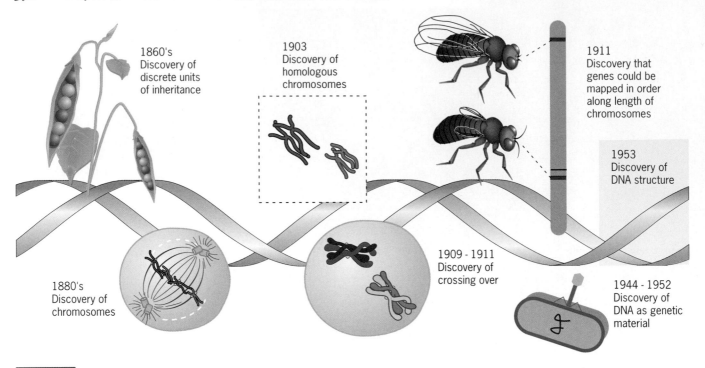

Figure 10.1 **An overview depicting several of the most important early discoveries on the nature of the gene.** Each of these discoveries is discussed in the present chapter.

10.1 THE CONCEPT OF A GENE AS A UNIT OF INHERITANCE

The science of genetics began with the work of Gregor Mendel in the 1860s. Mendel's laboratory was a small garden plot on the grounds of the Austrian monastery to which he belonged. We don't know exactly what motivated Mendel to begin his studies, but he evidently had a clear experimental plan in mind: his goal was to mate, or *cross*, pea plants having different inheritable characteristics and to determine the pattern by which these characteristics were transmitted to the offspring. Mendel chose the garden pea for a number of practical reasons, not the least of which was that he could purchase a variety of seeds that would give rise to plants with distinct characteristics. Mendel chose to focus on seven clearly definable traits, including plant height and flower color, each of which occurred in two alternate and clearly identifiable forms (Table 10.1). Mendel crossbred plants through several generations and counted the number of individuals having various characteristics. After several years of research, Mendel drew the following conclusions, which are expressed below in modern genetic terminology.

1. The characteristics of the plants were governed by distinct factors (or units) of inheritance, which were later termed **genes**. An individual plant possessed two copies of a gene that controlled the development of each trait, one derived from each parent. The two genes could be either identical to one another or nonidentical. Alternate forms of the two genes are called

alleles. For each of the seven traits studied, one of the two alleles was dominant over the other. When both were present together in the same plant, the existence of the recessive allele was masked by the dominant one.

2. Each reproductive cell (or *gamete*) produced by a plant contained only one copy of a gene for each trait. A particular gamete could have either the recessive or the dominant allele for a given trait, but not both. Each plant arose by the union of a male and female gamete. Consequently, one of the alleles that governed each trait in a plant was inherited from the female parent, and the other allele was inherited from the male parent.

3. Even though the pair of alleles that governed a trait remained together throughout the life of an individual plant, they became separated (or *segregated*) from one

Table 10.1	Seven Traits of Mendel's Pea Plants	
Trait	**Dominant Allele**	**Recessive Allele**
Height	Tall	Dwarf
Seed color	Yellow	Green
Seed shape	Round	Angular (wrinkled)
Flower color	Purple	White
Flower position	Along stem	At stem tips
Pod color	Green	Yellow
Pod shape	Inflated	Constricted

another during the formation of the gametes. This finding formed the basis of Mendel's "law of segregation."

4. The segregation of the pair of alleles for one trait had no effect on the segregation of alleles for another trait. A particular gamete, for example, could receive a paternal gene governing seed color and a maternal gene governing seed shape. This finding formed the basis of Mendel's "law of independent assortment."

Mendel presented the results of his work to the members of the Natural History Society of Brünn, Austria; the minutes of the meeting record no discussion of his presentation. Mendel's experiments were published in the Brünn society's journal in 1865, where they generated no interest whatsoever until 1900, sixteen years after his death. In that year, three European botanists *independently* reached the same conclusions, and all three rediscovered Mendel's paper, which had been sitting on the shelves of numerous libraries throughout Europe for 35 years.

10.2 CHROMOSOMES: THE PHYSICAL CARRIERS OF THE GENES

Although Mendel provided convincing evidence that inherited traits were governed by discrete factors, or genes, his studies were totally unconcerned with the physical nature of these elements or their location within the organism. Mendel was able to carry out his entire research project without ever observing anything under a microscope. During the time between Mendel's work and its rediscovery, a number of biologists became concerned with this other aspect of heredity—its physical basis within the cell.

The Discovery of Chromosomes

By the 1880s, a number of European biologists were closely following the activities of cells and using rapidly improving light microscopes to observe newly discernible cell structures. None of these scientists was aware of Mendel's work, yet they realized whatever it was that governed inherited characteristics would have to be passed from cell to cell and from generation to generation. This, by itself, was a crucial realization; all the genetic information needed to build and maintain a complex plant or animal had to fit within the boundaries of a single cell. Observations of dividing cells by the German biologist Walther Flemming in the early 1880s revealed that the cytoplasmic contents were shuttled into one daughter cell or the other as a matter of chance, depending on the plane through which the furrow happened to divide the cell. In contrast, the cell appeared to go to great lengths to divide the nuclear contents equally between the daughters. During cell division, the material of the nucleus became organized into visible "threads," which were named **chromosomes**, meaning "colored bodies."

At about this time, the process of fertilization was observed, and the roles of the two gametes—the sperm and the egg—were described (Figure 10.2). Even though the sperm is a tiny cell, it was known to be as important genetically as the much larger egg. What was it that these two very different cells had in common? The most apparent feature was the nucleus and its chromosomes. The importance of the chromosomes contributed by the male became evident in a study by the German biologist Theodore Boveri on sea urchin eggs that had been fertilized by two sperm rather than just one, as is normally the case. This condition, referred to as *polyspermy*, is characterized by disruptive cell divisions and the early death of the embryo. Why should the presence of one extra tiny sperm nucleus within the very large egg have such drastic consequences? The second sperm donates an extra set of chromosomes and an extra centriole (page 349). These additional components lead to abnormal cell divisions in the embryo during which daughter cells receive variable numbers of chromosomes. Boveri concluded that the orderly process of normal development is "dependent upon a particular combination of chromosomes; and this can only mean that the individual chromosomes must possess different qualities." This was the first evidence of a *qualitative* difference among chromosomes.

The processes occurring after fertilization were followed most closely in the roundworm *Ascaris*, whose few chromosomes are large and were as readily observed in the nineteenth century as in introductory biology laboratories today. In 1883, it was noted by the Belgian biologist

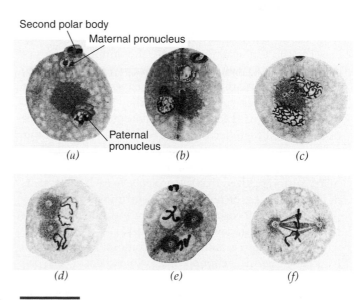

Figure 10.2 Events occurring in the roundworm *Ascaris* following fertilization, as reported in a classic nineteenth-century investigation. Both the male and female gamete are seen to contain two chromosomes. Fusion of the sperm and egg nuclei (called pronuclei) in the egg cytoplasm (between *e* and *f*) produces a zygote containing four chromosomes. The second polar body shown in *a* is a product of the previous meiosis as described in Section 14.3. (FROM T. BOVERI, JENAISCHE ZEIT 22:685, 1888.)

Edouard van Beneden that the cells of the worm's body had four large chromosomes, but that the male and female nuclei present in the egg just after fertilization (before the two nuclei fuse with one another) had but two chromosomes apiece (Figure 10.2). About this same time, the process of meiosis was described, and in 1887 the German biologist August Weismann proposed that meiosis included a "reduction division" during which the chromosome number was reduced by half prior to formation of the gametes. If a reduction division did not occur, and each gamete contained the same number of chromosomes as an adult cell, then the union of two gametes would double the number of chromosomes in the cells of the progeny. The number of chromosomes would double from one generation to the next, which obviously does not and cannot occur.[1]

Chromosomes as the Carriers of Genetic Information

The rediscovery of Mendel's work and its confirmation had an immediate influence on research in cell biology. Whatever their physical nature, the carriers of the hereditary units would have to behave in a manner consistent with Mendelian principles. In 1903, Walter Sutton, a graduate student at Columbia University, published a paper that pointed directly to the chromosomes as the physical carriers of Mendel's genetic factors. Sutton followed the development of sperm cells in the grasshopper, which, like *Ascaris*, has large, easily observable chromosomes. The cells that give rise to sperm are called spermatogonia, and they can undergo two very different types of cell division. A spermatogonium can divide by mitosis to produce more spermatogonia, or it can divide by meiosis to produce cells that differentiate into sperm (see Figure 14.40). On examination of the mitotic stages of grasshopper spermatogonia, Sutton counted 23 chromosomes. Careful examination of the shapes and sizes of the 23 chromosomes suggested that they were present in "look-alike" pairs. Eleven pairs of chromosomes could be distinguished together with one additional chromosome, termed an *accessory chromosome* (and later shown to be a sex-determining X chromosome), that did not have a mate. Sutton realized that the presence in each cell of pairs of chromosomes, or **homologous chromosomes**, as they were soon called, correlated perfectly with the pairs of inheritable factors uncovered by Mendel.

When Sutton examined the chromosomes in cells just beginning meiosis, he found that the members of each pair of chromosomes were associated with one another, forming a complex called a *bivalent*. Eleven bivalents were visible, each showing a cleft along its length where the two homologous chromosomes were associated (Figure 10.3). The ensuing first meiotic division separated the two homologous chromosomes into separate cells. This was the reduction di-

Figure 10.3 Homologous chromosomes. Sutton's drawing of the homologous chromosomes of the male grasshopper that have associated during meiotic prophase to form bivalents. Eleven pairs of homologous chromosomes (*a–k*) and an unpaired X chromosome were observed. (FROM W. S. SUTTON, BIOL. BULL. 4:24, 1902.)

vision proposed 15 years earlier by Weismann on theoretical grounds. Here also was the physical basis for Mendel's proposal that hereditary factors exist in pairs that remain together throughout the life of an individual and then separate from one another upon formation of the gametes. The reduction division observed by Sutton explained several of Mendel's other findings: gametes could contain only one version (allele) of each gene; the number of gametes containing one allele was equal to the number containing the other allele; and two gametes that united at fertilization would produce an individual with two alleles for each trait. But many questions remained unanswered. For instance, how were the genes organized within the chromosomes, and could the location of specific genes be determined?

The Chromosome as a Linkage Group As clearly as Sutton saw the relationship between chromosome behavior and Mendel's alleles, he saw one glaring problem. Mendel had examined the inheritance of seven traits and found that each was inherited independently of the others. This observation formed the basis of Mendel's law of independent assortment. But if genes are packaged together on chromosomes, like beads on a string, then genes should be passed from parent to offspring in packages, just as intact chromosomes should be passed to the next generation. Genes on the same chromosome should act as if they are *linked* to one another; that is, they should be part of the same **linkage group**.

How is it that all seven of Mendel's traits assorted independently? Were they all on different linkage groups, that is, different chromosomes? As it happens, the garden pea has seven different pairs of homologous chromosomes.

[1]Readers who have forgotten some of the events that occur during meiosis should read through Section 14.3, which discusses this pivotal event in the life cycle of eukaryotic organisms.

The genes that govern each of the traits on which Mendel reported either occur on different chromosomes or are so far apart from one another on the same chromosome that they act independently (page 403). Whether Mendel owed his findings to good fortune or simply to a lack of interest in any traits that did not fit his predictions remains unclear.

Sutton's prophecy of linkage groups was soon transformed from speculation to fact. Within a couple of years, two traits in sweet peas (flower color and pollen shape) were shown to be linked, and other evidence for chromosomal linkage rapidly accumulated.

Genetic Analysis in *Drosophila*

Research in genetics soon came to focus on one particular organism, the fruit fly, *Drosophila* (Figure 10.4). Fruit flies have a generation time (from egg to maturity) of 14 days and can produce up to 1,000 eggs in a lifetime. In addition, fruit flies are very small, so a large number could be kept on hand; they were easy to house and breed; and they were very inexpensive to maintain. In 1909, fruit flies seemed like the ideal organism to Thomas Hunt Morgan of Columbia University, and he began what was to be the start of a new era in genetic research. There was one major disadvantage in beginning work with this insect—there was only one "strain" of fly available, the **wild type** (Figure 10.4). Whereas Mendel had simply purchased varieties of pea seeds, Morgan had to generate his own varieties of fruit flies. Morgan expected that variants from the wild type would appear if he bred sufficient numbers of flies. Within a year, and after viewing thousands of flies, he found his first **mutant**, that is, an individual having an inheritable characteristic that distinguished it from the wild type. The mutant had white eyes rather than the normal red-colored ones.

By 1915, Morgan and his students had found 85 different mutants with a wide variety of affected structures. It was apparent that on rare occasions a spontaneous change, or **mutation**, occurred within a gene, altering it permanently, so that the altered trait could be passed from generation to generation. The demonstration of a spontaneous, inheritable alteration in a gene had consequences far beyond the study of *Drosophila* genetics. It suggested a mechanism for the origin of variation that exists within populations—evidence for a vital link in the theory of evolution. If variants of genes could arise spontaneously, then natural selection could act upon the mutants and new species could slowly emerge.

Whereas mutations are a necessary occurrence for evolution, they are a tool for geneticists, a marker providing contrast to the wild-type condition. As *Drosophila* mutants were isolated, they were bred and crossbred and kept as stocks within the lab. As expected, the 85 mutations did not all assort independently; instead, Morgan found that they belonged to four different linkage groups, one of which contained very few mutant genes (only two in 1915). This finding correlated perfectly with the observation that *Drosophila* cells have four pairs of homologous chromosomes, one of which is very small (Figure 10.5). Little doubt remained that genes reside on chromosomes.

Crossing Over and Recombination

Even though the association of genes into linkage groups was confirmed, linkage between alleles on the same chromosome was *incomplete*. In other words, alleles of two different genes, such as short wings and black body (as in Figure 10.7), that were originally present together on a given chromosome did not always remain together during the production of gametes. Thus maternal and paternal characteristics that were inherited by an individual on separate, homolo-

Figure 10.4 The fruit fly *Drosophila melanogaster*. Photograph of a wild-type male and female fruit fly. (© Vu/R. Calentine/Visuals Unlimited.)

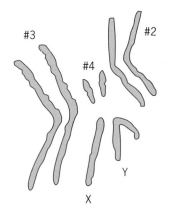

Figure 10.5 Fruit flies have four pairs of homologous chromosomes, one of which is very small. The two dissimilar homologues are the chromosomes that determine sex. As in humans, male fruit flies are XY and females are XX.

gous chromosomes could be reshuffled in such a way that they ended up on the same chromosome of a gamete. Conversely, two characteristics that were inherited together on the same chromosome from one parent could become separated from one another and end up in separate gametes.

In 1911, Morgan offered an explanation for the "breakdown" in linkage. Two years earlier, F. A. Janssens had observed that the homologous chromosomes of bivalents became wrapped around each other during an early stage of meiosis (Figure 10.6). Janssens had proposed that this interaction between maternal and paternal chromosomes resulted in the breakage and exchange of pieces. Capitalizing on this observation and proposal, Morgan suggested that this phenomenon, which he termed **crossing over** (or **genetic recombination**), could account for the appearance of offspring (*recombinants*) having unexpected combinations of genetic traits. An example of crossing over is shown in Figure 10.7.

Analyses of the offspring from a large number of crosses between adults carrying a variety of alleles on the same chromosome indicated that (1) the percentage of recombination between a given pair of genes on a chromosome, such as eye color and wing length, was constant from one experiment to another, and (2) the percentages of recombination between different pairs of genes, such as between eye color and wing length versus eye color and body color, could be very different.

Figure 10.6 Visualizing sites of crossing over. Homologous chromosomes wrap around each other during meiosis, as seen in this micrograph of a meiotic cell of a lily. The points at which the homologues are crossed are termed *chiasmata* (arrows) and (as discussed in Chapter 14) are sites at which crossing over had occurred at an earlier stage. (COURTESY OF A. H. SPARROW.)

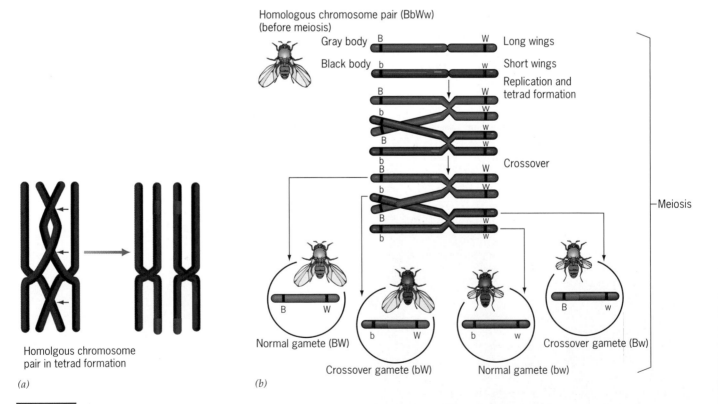

(a) *(b)*

Figure 10.7 Crossing over provides the mechanism for reshuffling alleles between maternal and paternal chromosomes. (*a*) Bivalent (tetrad) formation during meiosis, showing three possible crossover intersections (chiasmata, indicated by red arrows). (*b*) Simplified representation of a single crossover in a *Drosophila* heterozygote (*BbWw*) at chromosome number 2 and the resulting gametes. If either of the crossover gametes participates in fertilization, the offspring will have a chromosome with a combination of alleles that was not present on a single chromosome from either parent.

The fact that a given pair of genes gave the same frequency of recombination in every cross strongly suggested that the position of the genes along the chromosome (their **locus**) was fixed and did not vary from one fly to the next. If the locus of each gene is fixed, then the frequency of recombination between two genes provides a measure of the distance that separates those two genes. The more room between two sites on a chromosome for breakage to occur, the more likely it is that a break would occur between those two sites and, therefore, the greater the recombination frequency. In 1911, Alfred Sturtevant, an undergraduate at Columbia University who was working in Morgan's laboratory, realized that recombination frequencies could be used to map the relative positions of individual genes along a given chromosome. An example of the underlying principles of this mapping procedure is illustrated in Figure 10.7. In this example, the genes for wing length and body color in *Drosophila* are situated at a considerable distance from one another on the chromosome and, therefore, are likely to become unlinked by an intervening breakage and crossover. In contrast, the genes for eye color and body color are situated very close to one another on the chromosome and consequently are less likely to become unlinked. Using recombination frequencies, Sturtevant—who became one of the century's most prominent geneticists—began to construct detailed maps of the serial order of genes along each of the four chromosomes of the fruit fly. Recombination frequencies have since been used to prepare chromosomal maps from viruses and bacteria as well as a large variety of eukaryotic species.

Mutagenesis and Giant Chromosomes

During the early period of genetics, the search for mutants was a slow, tedious procedure dependent on the *spontaneous* appearance of altered genes. Using a special strain of fruit flies designed to reveal the presence of recessive alleles, Herman Muller of Indiana University found in 1927 that flies subjected to a sublethal dose of X rays displayed more than 100 times the spontaneous mutation rate exhibited by nonirradiated controls. This finding had important consequences. On the practical side, the use of mutagenic agents, such as X rays and ultraviolet radiation, generated a great increase in the number of mutants available for genetic research. The finding also pointed out the hazard of the increasing use of radiation in the industrial and medical fields.

The rediscovery in 1933 of giant chromosomes in certain insect cells by Theophilus Painter of the University of Texas illustrates a basic characteristic of biological systems. There is such remarkable variability among organisms, not only at the obvious macroscopic levels, but also at the cellular and subcellular levels, that often one particular type of cell may be much better suited for a particular type of research than all the others.

Cells from the larval salivary gland of *Drosophila* contain chromosomes that are about 100 times thicker than the chromosomes found in most other cells of the organism (Figure 10.8*a*). During larval development, these cells

Single DNA molecule

Band

(*a*)

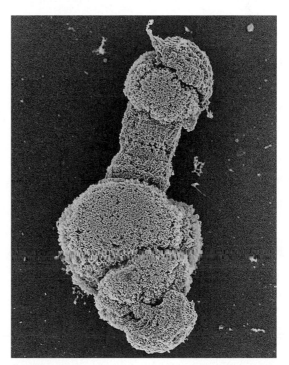

(*b*)

Figure 10.8 **Giant polytene chromosomes of larval insects.** (*a*) These giant polytene chromosomes from the salivary gland of a larval fruit fly show thousands of distinct, darkly staining bands. Many of the bands have been identified as the loci of particular genes. The inset shows how polytene chromosomes consist of a number of individual DNA strands. The bands on the chromosomes correspond to sites where the DNA is more tightly compacted. (*b*) Scanning electron micrograph of a giant polytene chromosome from a *Chironomus* larva showing how specific sites are expanded to form a "puff." Chromosome puffs are sites where DNA is being transcribed. (A: FROM BIOLOGICAL PHOTO SERVICE; B: COURTESY OF TERRY D. ALLEN AND CLAUS PELLING, J. CELL SCIENCE, COVER OF VOL.93, PART 4, 1989.)

cease dividing, but they keep growing. DNA replication continues, providing the additional genetic material needed to support the high level of secretory activity of these enormous cells. The duplicated DNA strands remain attached to each other in perfect side-by-side alignment (inset of Figure 10.8a), producing giant chromosomes with as many as 1,024 times the number of DNA strands of normal chromosomes.

These unusual **polytene chromosomes**, as they are called, are rich in visual detail, displaying about 5,000 bands when stained and examined microscopically. The banding pattern is essentially constant from one individual to the next, but considerable differences are observed between the chromosomes from flies of different species of the *Drosophila* genus. Painter soon found that individual bands could be correlated with specific genes. The relative positions of these genes on the giant chromosomes agreed with those predicted on the basis of genetic maps prepared from recombination frequencies, thus providing visual confirmation of the validity of the entire mapping procedure.

Giant insect chromosomes have been useful in other ways. Comparisons of banding patterns among polytene chromosomes of different species have provided an unparalleled opportunity for the investigation of evolutionary changes at the chromosome level. In addition, these chromosomes are not inert cellular objects but dynamic structures in which certain regions become "puffed out" at particular stages of development (Figure 10.8b). These chromosome puffs are sites where DNA is being transcribed into RNA, providing one of the best systems available for the direct visualization of gene expression (see Figure 18.21a).

REVIEW ?

1. What is a linkage group? What is its relationship to a chromosome? How can one determine the number of linkage groups in a species?

2. What is a genetic marker? How do mutations act as genetic markers?

3. What is meant by incomplete linkage? What does this have to do with pairing of homologous chromosomes during meiosis?

4. How do polytene chromosomes of an insect differ from normal chromosomes?

10.3 THE CHEMICAL NATURE OF THE GENE

Classical geneticists uncovered the rules governing the transmission of genetic characteristics and the relationship between genes and chromosomes. In his acceptance speech for the Nobel Prize in 1934, T. H. Morgan stated,

"At the level at which the genetic experiments lie, it does not make the slightest difference whether the gene is a hypothetical unit or whether the gene is a material particle." By the 1940s, however, a new set of questions was being considered, foremost of which was, "What is the chemical nature of the gene?" The experiments that provided the answer to this question are outlined in the accompanying Experimental Pathways. Let us go on to consider some of the information that was gathered once it was evident that genes were made of DNA.

The Structure of DNA

To understand the workings of a complex macromolecule—whether it is a protein, polysaccharide, lipid, or nucleic acid—it is essential to know how that molecule is constructed. The mystery of DNA structure was investigated by a number of laboratories in both the United States and England in the early 1950s and was solved by James Watson and Francis Crick at Cambridge University in 1953. Before describing their proposed DNA structure, let us consider the facts available at the time.

Base Composition The basic building block of DNA was known to be a **nucleotide** (Figure 10.9a,b) consisting of the five-carbon sugar **deoxyribose** to which one phosphate is esterified at the 5′ position of the sugar ring and one nitrogenous base is attached at the 1′ site.[2] There are two types of nitrogenous bases present in a nucleic acid, **pyrimidines**, which contain a single ring, and **purines**, which contain two rings (Figure 10.9c). DNA contains two different pyrimidines, **thymine (T)** and **cytosine (C)**, and two different purines, **guanine (G)** and **adenine (A)**. The nucleotides were known to be covalently linked to one another to form a linear polymer, or *strand*, with a backbone composed of alternating sugar and phosphate groups joined by *3′,5′-phosphodiester bonds* (Figure 10.9c). The bases attached to each sugar were thought to project from the backbone like a column of stacked shelves.

A nucleotide has a polarized structure: one end, where the phosphate is located, is called the *5′ end* (pronounced "five-prime end"), while the other end is the *3′ end* (Figure 10.9b). Because each of the stacked nucleotides in a strand

[2]It is useful to introduce a bit of terminology at this point. A molecule consisting simply of one of the four nitrogenous bases of Figure 10.9 linked to a pentose sugar moiety is known as a nucleoside. If the sugar is deoxyribose, the nucleoside is a deoxyribonucleoside. There are four major deoxyribonucleosides distinguished by the attached base: deoxyadenosine, deoxyguanosine, deoxythymidine, and deoxycytidine. If the nucleoside has one or more attached phosphate groups (generally at the 5′ position, but alternatively at the 3′ position), the molecule is a nucleotide. There are nucleoside 5′-monophosphates, nucleoside 5′-diphosphates, and nucleoside 5′-triphosphates, depending on the number of phosphates in the molecule. Examples of each are deoxyadenosine 5′-monophosphate (dAMP), deoxyguanosine 5′-diphosphate (dGDP), and deoxycytidine 5′-triphosphate (dCTP). A similar set of nucleosides and nucleotides involved in RNA metabolism contain the sugar ribose rather than deoxyribose. The nucleotides employed in energy metabolism, such as adenosine triphosphate (ATP), are ribose-containing molecules.

(a)

(b)

(c)

faces the same direction, the entire strand has a direction. One end is the 3′ end, the other is the 5′ end (Figure 10.9c). X-ray diffraction analysis indicated that the distance between the nucleotides of the stack was 3.4 Å (0.34 nm) and suggested the presence of a large structural repeat every 3.4 nm.

As discussed on page 432, DNA was thought for many years to consist of a simple repeating tetranucleotide (e.g.,—ATGCATGCATGC—), which made it unlikely to serve as an informational macromolecule. In 1950, Erwin Chargaff of Columbia University made an important finding that delivered the final blow to the tetranucleotide theory and provided vital information to Watson and Crick. Chargaff, believing that the sequence of nucleotides of the DNA molecule held the key to its importance, determined the relative amounts of each base in various samples of DNA, that is, the **base composition** of the samples. Base composition analysis of a DNA sample was accomplished by hydrolyzing the bases from their attached sugars, separating the bases in the hydrolysate by paper chromatography, and determining the amount of material in each of the four spots to which the four bases migrated.

If the tetranucleotide theory were correct, each of the four bases in a DNA sample would be present as 25 percent of the total number. Chargaff found that the ratios of the four component bases were quite variable from one type of organism to another, often being very different from the 1:1:1:1 ratio predicted by the tetranucleotide theory. For example, the A:G ratio of the DNA of a tubercle bacillus was 0.4, while the A:G ratio of human DNA was 1.56. It made no difference which tissue was used as the source of the DNA; the base composition remained essentially constant for that species. Amid this great variability in base composition of different species' DNA, an important numerical relationship was discovered. The number of purines always equaled the number of pyrimidines in a given sample of DNA. More specifically, the number of adenines always equaled the number of thymines, and the number of guanines always equaled the number of cytosines. In other words, Chargaff discovered the following rules of DNA base composition:

$$[A] = [T], [G] = [C], [A] + [T] \neq [G] + [C]$$

Figure 10.9 The chemical structure of DNA. (*a*) Model of a DNA nucleotide containing the base thymine; the molecule is deoxythymidine 5′-monophosphate (dTMP). The netlike cage represents the electron density of the atoms that make up the molecule. (*b*) Chemical structure of a DNA nucleotide containing the base adenosine; the molecule is deoxyadenosine 5′-monophosphate (dAMP). A nucleotide is composed of a nucleoside linked to a phosphate; the nucleoside portion of the molecule (i.e., deoxyadenosine) is enclosed by the dashed line. (*c*) The chemical structure of a small segment of a single DNA strand showing all four nucleotides. (A: REPRINTED WITH PERMISSION FROM ARNON LAVIE ET AL., NATURE STR. BIOL. 4:604, 1997. © COPYRIGHT 1997 BY MACMILLAN MAGAZINES LIMITED.)

Chargaff's findings put the DNA molecule in a new light, giving it specificity and individuality from one organism to another. The significance of the base equivalencies, however, remained obscure.

The Watson-Crick Proposal

When protein structure was discussed in Chapter 2, the importance of secondary and tertiary structure as determinants of the protein's activity was stressed. Similarly, information about the three-dimensional organization of DNA was needed if its biological activity was to be understood. Using X-ray diffraction data (obtained by Rosalind Franklin and Maurice Wilkins at King's College London) and models constructed from cutouts of the four types of nucleotides, Watson and Crick proposed a structure of DNA that included the following components (Figure 10.10):

1. The molecule is composed of two chains of nucleotides. This conclusion followed on the heels of an erroneous proposal by Linus Pauling, who had suggested that DNA was composed of three nucleotide strands.

2. The two chains spiral around each other to form a pair of right-handed helices. In a right-handed helix, an observer looking down the central axis of the molecule would see that each strand follows a clockwise path, moving away from the observer. The helical nature of DNA was revealed in the pattern of spots produced by X-ray diffraction.

3. The two chains composing one double helix run in opposite directions; that is, they are *antiparallel*. Thus, if one chain is aligned in the $5' \rightarrow 3'$ direction, its partner must be aligned in the $3' \rightarrow 5'$ direction.

4. The –sugar–phosphate–sugar–phosphate– backbone of each strand is located on the outside of the molecule with the two sets of bases projecting toward the center. The phosphate groups give the molecule a negative charge.

5. The bases occupy planes that are approximately perpendicular to the long axis of the molecule and are, therefore, stacked one on top of another like a pile of plates. Hydrophobic interactions and van der Waal forces (page 37) between the stacked, planar bases provide stability for the entire DNA molecule. Together, the helical turns and planar base pairs cause the molecule to resemble a spiral staircase. This manner of construction is evident in the chapter opening photograph, which shows the original Watson-Crick model.

6. The two strands are held together by hydrogen bonds between each base of one strand and an associated base on the other strand. Because individual hydrogen bonds are weak and easily broken, the DNA

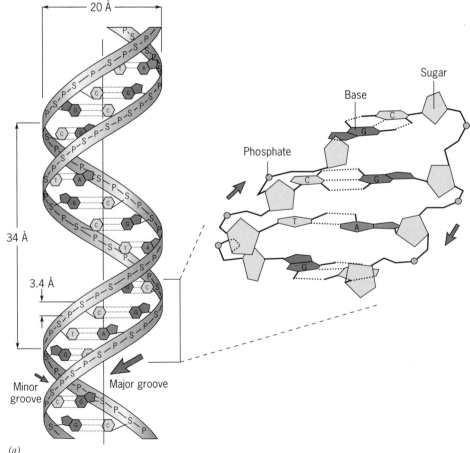

(a)

Figure 10.10 The double helix. (*a*) Schematic representation of the DNA double helix. (*b*) The Watson-Crick base pairs. (*c*) Space-filling model of the B form of DNA, showing how the bases fill the space between the two backbones. (Atoms are shown at a size that includes their van der Waals radii.) (*d*) Electron micrograph of DNA being released from the ruptured head of a T2 bacteriophage. This linear DNA molecule (note the two free ends) has a molecular mass of approximately 1.2×10^8 daltons (180,000 base pairs) and measures 68 μm in length, approximately 600 times longer than the phage head in which it is contained. (B: FROM D. VOET AND J. G. VOET, BIOCHEMISTRY, 2D ED., COPYRIGHT © 1995, JOHN WILEY & SONS, INC. REPRINTED BY PERMISSION. AFTER S. ARNOTT ET AL., ACTA CRYST. B25:2196, 1969; C: COURTESY OF NELSON MAX; D: COURTESY OF A. K. KLEINSCHMIDT ET AL., BIOCHIM. BIOPHYS. ACTA 61:861, 1962.)

Major groove

Minor groove

Major groove

Minor groove

(b)

(c)

Figure 10.10 (continued)

strands can become separated during various activities. But the strengths of hydrogen bonds are additive, and the large numbers of hydrogen bonds holding the strands together make the double helix a stable structure.

7. The distance from the phosphorus atom of the backbone to the center of the axis is 1 nm (thus the width of the double helix is 2 nm).

8. A pyrimidine in one chain is always paired with a purine in the other chain. This arrangement produces a molecule that is 2 nm wide along its entire length.

9. The nitrogen atoms linked to carbon 4 of cytosine and carbon 6 of adenine are predominantly in the amino (NH_2) configuration (Figure 10.9c) rather than the imino (NH) form. Similarly, the oxygen atoms linked to carbon 6 of guanine and carbon 4 of thymine are predominantly in a keto (C=O) configuration rather than the enol (C—OH) form. These structural restrictions on the configurations of the bases suggested that adenine was the only purine structurally capable of bonding to thymine and that guanine was the only purine capable of bonding to cytosine. Therefore, the only possible pairs were A-T and G-C (Figure 10.10b), which fit perfectly with

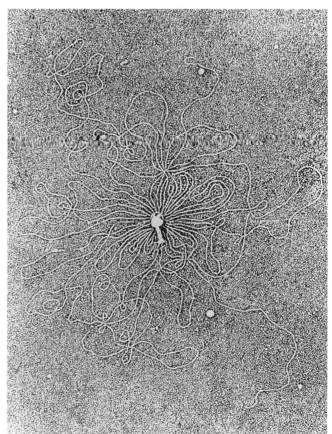

(d)

the previous base composition analysis by Chargaff. A-T pairs were joined by two hydrogen bonds, G-C pairs by three hydrogen bonds.

10. The spaces between adjacent turns of the helix form two grooves of different width—a wider *major groove* and a more narrow *minor groove*—that spiral around the outer surface of the double helix. Proteins that bind to DNA often contain domains that fit into these grooves. In some cases, a protein bound in a groove is able to read the sequence of nucleotides along the DNA without having to separate the strands.

11. The double helix makes one complete turn every 10 residues (3.4 nm), or 150 turns per million daltons of molecular mass.

12. Because an A on one strand is always bonded to a T on the other strand, and a G is always bonded to a C, the nucleotide sequences of the two strands are always fixed relative to one another. This relationship between the two chains of the double helix is referred to as **complementarity**. For example, A is complementary to T, AGC is complementary to TCG, and one entire chain is complementary to the other. As we shall see, complementarity is of overriding importance in nearly all the activities and mechanisms in which nucleic acids are involved.

The Importance of the Watson-Crick Proposal From the time biologists first considered DNA as the genetic material, there were three primary functions it was expected to fulfill (Figure 10.11):

1. *Storage of genetic information.* As genetic material, DNA must contain a stored record of instructions that determine all the heritable characteristics that an organism exhibits. In molecular terms, DNA must contain the information for the specific order of amino acids in all the proteins that are synthesized by an organism.

2. *Self-duplication and inheritance.* DNA must contain the information for its own replication (duplication). DNA replication allows genetic instructions to be transmitted from one cell to its daughter cells and from one individual to its offspring.

3. *Expression of the genetic message.* DNA is more than a storage center; it is also a director of cellular activity. Consequently, the information encoded in DNA has to be expressed in some form that can take part in events that are taking place within the cell. More specifically, the information in DNA must be used to direct the order by which specific amino acids are incorporated into a polypeptide chain.

The Watson-Crick model of DNA structure suggested a way in which the first two of these three genetic functions might be accomplished. The model strongly confirmed the suspicion that the information content of DNA resided in the linear sequence of its bases. A given segment of DNA would correspond to each gene. The specific sequence of nucleotides in that segment would

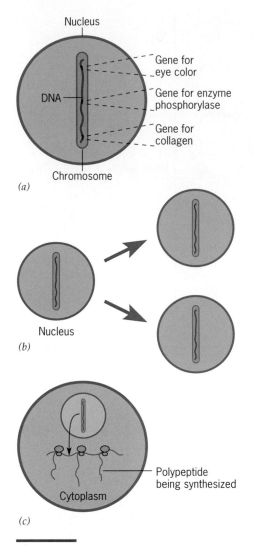

Figure 10.11 Three functions required of the genetic material. (*a*) DNA must contain the information that encodes inheritable traits. (*b*) DNA must contain the information that directs its own duplication. (*c*) DNA must contain the information that directs the assembly of specific proteins.

dictate the sequence of amino acids in a corresponding polypeptide. A change in the linear sequence of nucleotides within that segment would correspond to an inheritable mutation in that gene.

As for the second function, Watson and Crick's initial publication of DNA structure included a proposal as to how such a molecule might replicate. They suggested that during replication, the hydrogen bonds holding the two strands of the DNA helix were sequentially broken, causing the gradual separation of the strands much like the separation of two halves of a zipper. Each of the separated strands, with its nitrogenous bases exposed, would then serve as a **template** directing the order in which complementary nucleotides become assembled to form the complementary strand. When complete, the process would generate two double-stranded DNA molecules that were (1) identical to one another and (2) identical to the original DNA molecule. According to the Watson-Crick pro-

posal, each DNA double helix would contain one strand from the original DNA molecule and one newly synthesized strand (see Figure 13.1). As we will see in Chapter 13, the Watson-Crick proposal fared quite well in predicting the mechanism of DNA replication. Of the three primary functions listed above, only the mechanism by which DNA governs the assembly of a specific protein remained a total mystery.

Not only was the elucidation of DNA structure important in its own right, but it provided the stimulus for investigating all the activities in which the genetic material must take part. Once the model for its structure was accepted, any theory of a genetic code, DNA synthesis, or information transfer had to be consistent with that structure.

DNA Supercoiling In 1963, Jerome Vinograd and his colleagues at the California Institute of Technology discovered that two closed, circular DNA molecules of identical molecular mass could exhibit very different rates of sedimentation during centrifugation (Section 18.11). Further analysis indicated that the DNA molecule sedimenting more rapidly had a more compact shape because the molecule was twisted upon itself (Figure 10.12*a,b*), much like a rubber band in which the two ends are twisted in opposite directions or a tangled telephone cord after extended use. DNA in this state is said to be **supercoiled**. Because supercoiled DNA is more compact than its relaxed counterpart, it occupies a smaller volume and moves more rapidly in response to centrifugal force or an electric field (Figure 10.12*c*).

Supercoiling is best understood if one pictures a length of double-helical DNA lying free on a flat surface. A molecule in this condition has the standard number of 10 base pairs per turn of the helix and is said to be *relaxed*. The DNA would still be relaxed if both ends of the two strands were simply fused to form a circle. Consider, however, what would happen if the molecule were to be twisted *before* the ends were sealed. If the length of DNA were twisted in a direction opposite to that in which the duplex was wound, the molecule would tend to unwind. An *underwound* DNA molecule has a greater number of base pairs per turn of the helix (Figure 10.13). Because the molecule is most stable with 10 residues per turn, it tends to resist the strain of becoming underwound by becoming twisted upon itself into a supercoiled conformation (Figure 10.13).

DNA is referred to as *negatively supercoiled* when it is underwound and *positively supercoiled* when it is overwound. Circular DNAs found in nature (e.g., mitochondrial, viral, bacterial) are invariably negatively supercoiled. Supercoiling is not restricted to small, circular DNAs but also occurs in linear, eukaryotic DNA. For example, negative supercoiling plays a key role in allowing the DNA of the chromosomes to be compacted so as to fit inside the confines of a microscopic cell nucleus (Section 12.1). Because negatively supercoiled DNA is underwound, it exerts a force that helps separate the two strands of the helix, which is required during both replication (DNA synthesis) and transcription (RNA synthesis).

(a) (b) (c)

Figure 10.12 Supercoiled DNA. (*a–b*) Electron micrographs showing the differences in conformation between a relaxed, circular molecule of phage DNA (*a*) and the same type of molecule in a supercoiled state (*b*). (*c*) When a mixture of relaxed and supercoiled SV40 DNA molecules are subjected to gel electrophoresis, the highly compact, supercoiled form (seen at the bottom of the gel) moves much more rapidly than the relaxed form. The DNA molecules are visualized by staining the gel with ethidium bromide, a fluorescent molecule that inserts itself into the double helix. (A, B: COURTESY OF JAMES C. WANG; C: FROM WALTER KELLER, PROC. NAT'L. ACAD. SCI. U. S. A. 72:2553, 1975.)

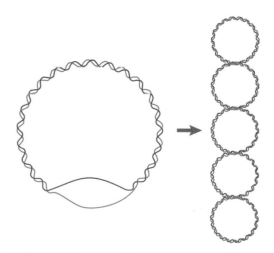

Figure 10.13 Underwound DNA. The DNA molecule at the left is underwound; that is, it has more than an average of 10 base pairs per turn of a helix. An underwound molecule spontaneously assumes a negatively supercoiled conformation, as shown in the drawing on the right.

Certain enzymes in both prokaryotic and eukaryotic cells are able to change the supercoiled state of a DNA duplex. These enzymes are called **topoisomerases** because they change the *topology* of the DNA. Cells contain a variety of topoisomerases, which can be divided into two classes. *Type I topoisomerases* change the supercoiled state of a DNA molecule by creating a transient break in one strand of the duplex. A model for the mechanism of action of human topoisomerase I is shown in Figure 10.14*a*. The enzyme cleaves one strand of the DNA and then allows the intact, complementary strand to undergo a controlled

(a)

(c)

(b)

(d)

Figure 10.14 **DNA Topoisomerases.** (*a*) A model depicting the action of human topoisomerase I. The enzyme has cut one of the strands of the DNA, which is rotating about a phosphodiester bond in the intact strand. The cut strand will then be resealed. (Note: The drawing depicts a type IB topoisomerase; type IA enzymes found in bacteria act by a different mechanism.) (*b*) A molecular model based on X-ray crystallography depicting the action of topoisomerase II. In step 1, the dimeric enzyme is in an "open" conformation ready to bind the G-DNA segment, so named because it will form the gate through which the T-DNA (or transported DNA) segment will pass. In step 2, the enzyme has undergone a conformational change as it binds the G-segment. In steps 3 and 4, the enzyme binds a molecule of ATP, the G-segment is cleaved, and the T-segment is passed through the open "gate." The bracketed stage represents a hypothetical intermediate carrying out the step in which the T-segment is transported through the G-segment. At this stage,

both cut ends of the G segment are covalently bound to the enzyme. In step 5, the two ends of the G-segment are rejoined, and the T-segment is released. ATP hydrolysis and release of ADP and P$_i$ are proposed to occur as the starting state is regenerated. (*c*) The types of reactions that can be catalyzed by topoisomerases. Part 1 shows supercoiling-relaxation reactions; part 2 shows knotting-unknotting reactions; part 3 shows catenation-decatenation reactions. (*d*) Electron micrograph of a pair of interconnected (catenated) circular DNA molecules. Molecules of this type collect in bacteria lacking a specific topoisomerase. (A: FROM DNA TOPOLOGY AND ITS BIOLOGIC EFFECTS, N. COZZARELLI AND J. C. WANG, EDS., COLD SPRING HARBOR LABORATORY PRESS, 1990; B: REPRINTED WITH PERMISSION FROM J. M. BERGER ET AL., NATURE 379:231, 1997. © COPYRIGHT 1997, MACMILLAN MAGAZINES LTD; D: FROM NICHOLAS COZZARELLI, CELL VOL. 71, COVER #2, 1992, BY PERMISSION OF CELL PRESS.)

rotation, which relaxes the supercoiled molecule. Topoisomerase I is essential for processes such as DNA replication and transcription to prevent excessive supercoiling from developing as the complementary strands of a DNA duplex separate and unwind (as in Figure 13.6). *Type II topoisomerases* make a transient break in both strands of a DNA duplex. Another segment of the DNA molecule (or a separate molecule entirely) is then transported through the break, and the severed strands are resealed. As expected, this complex reaction mechanism is accompanied by a series of dramatic conformational changes that are depicted in the model in Figure 10.14*b*. These enzymes are capable of some remarkable "tricks." In addition to being able to supercoil and relax DNA (Figure 10.14*c*, panel 1), type II topoisomerases can tie a DNA molecule into knots or untie a DNA knot (Figure 10.14*c*, panel 2). They can also cause a population of independent DNA circles to become interlinked (*catenated*), or separate interlinked circles into individual components (Figure 10.14*c,d*). Topoisomerase II is required to unlink DNA molecules before duplicated chromosomes can be separated during mitosis. Human topoisomerase II is a target for a number of drugs (e.g., etoposide and doxorubicin) used to kill rapidly dividing cancer cells. These drugs bind to the enzyme and prevent the cleaved DNA strands from being resealed.

REVIEW ?

1. What is meant by saying that a DNA strand has polarity? That the two strands are antiparallel? That the molecule has a major groove and a minor groove? That the strands are complementary to one another?
2. What was the importance of Chargaff's analysis of DNA base composition in determining the structure of the double helix?
3. How do underwound and overwound DNAs differ from one another? How do the two types of topoisomerases differ?

10.4 THE STRUCTURE OF THE GENOME

DNA is a macromolecule—an assemblage of a large number of atoms bonded together in a defined arrangement whose three-dimensional structure can be described by techniques such as X-ray crystallography. But DNA is also an information repository, which is a property more difficult to describe in simple molecular terms. If, as noted above, a gene corresponds to a particular segment of DNA, then the sum of the genetic information that an individual organism inherits from its parents is equivalent to the sum of all of the DNA segments present in the fertil-

ized egg at the start of life. All the individuals that make up a species population share the same set of genes, even though different individuals invariably possess slightly different versions (alleles) of many of these genes. Thus each species of organism has a unique content of genetic information, which is known as its **genome**. For humans, the genome is essentially equivalent to all of the genetic information that is present in a complete set of human chromosomes (includes 22 different autosomes and both the X and Y sex chromosomes).

The Complexity of the Genome

To understand how the complexity of the genome is determined, it is first necessary to consider one of the most important properties of the DNA double helix: its ability to separate into its two component strands, a property termed **denaturation**.

DNA Denaturation As first suggested by Watson and Crick, the two strands of a DNA molecule are held together by weak, noncovalent bonds. When DNA is dissolved in saline solution and the solution is slowly warmed, a temperature is reached when strand separation begins. Within a few degrees, the process is generally complete and the solution contains single-stranded molecules that are completely separated from their original partners. The progress of thermal denaturation (or *DNA melting*) is usually monitored by following the increase in absorbance of the dissolved DNA. The nitrogenous bases of a nucleic acid molecule absorb ultraviolet radiation with an absorbance maximum near 260 nm (Figure 10.15). Once the two DNA strands have separated, the hydrophobic interactions that result from base stacking are greatly decreased, which changes the electronic nature of

Figure 10.15 The absorption spectrum of thymine. All four nucleotides of DNA, and consequently DNA itself, absorb maximally near 260 nm. The absorbance at 260 nm is generally used to determine nucleic acid concentrations. The ratio of absorbance at 260/280 nm is one measure of the purity of a nucleic acid preparation because contaminating proteins generally absorb most strongly at 280 nm. The 260/280 ratio of purified DNA is about 2.

Figure 10.16 Thermal denaturation of DNA. A thermal denaturation curve for native bacteriophage T6 DNA in 0.3 M sodium citrate. The "melting" of the DNA (strand separation) occurs over a narrow range of temperature, particularly for the simpler DNAs of small viruses. The temperature corresponding to half the increase in absorbance is termed the T_m (FROM J. MARMUR AND P. DOTY, J. MOL. BIOL. 3:593, 1961, COPYRIGHT 1961, BY PERMISSION OF THE PUBLISHER ACADEMIC PRESS.)

the bases and increases their absorbance of ultraviolet radiation. The rise in absorbance that accompanies DNA denaturation is shown in Figure 10.16. The temperature at which the shift in absorbance is half completed is termed the *melting temperature (T_m)*. The higher the GC content (%G + %C) of the DNA, the higher the T_m. This increased stability of GC-containing DNA reflects the presence of the extra hydrogen bond between the bases as compared with AT pairs. The electron micrograph in Figure 10.17 shows that even within a single DNA molecule, AT-rich sections melt before GC-rich regions.

DNA Renaturation The separation of the two strands of the DNA duplex by heat is not an unexpected finding, but the *reassociation* of single strands into stable double-stranded molecules seems much less likely. However, in 1960, Julius Marmur and coworkers at Harvard University found that if they slowly cooled a solution of bacterial DNA that had been thermally denatured, the DNA regained the properties of the double helix; it absorbed less ultraviolet light and once again behaved like genetic material in being able to transform bacterial cells (discussed on page 433). Similar results were obtained by heating the DNA to 100°C to denature it, rapidly dropping the temperature of the solution to approximately 25°C below the T_m, and allowing the DNA to incubate at this temperature for a period of time. It became apparent from these studies

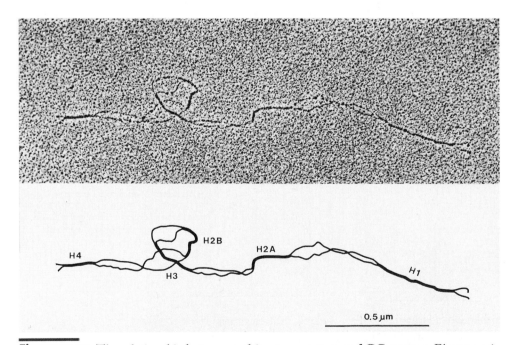

Figure 10.17 The relationship between melting temperature and GC content. Electron micrograph showing a segment of sea urchin DNA containing the genes for the five types of histone molecules (discussed in Section 12.1 and indicated in the accompanying map). Those regions of DNA that undergo strand separation under the conditions employed are seen as thinner, single-stranded segments. These regions correspond to segments that are rich in AT base pairs that lie between the DNA segments that code for the histone proteins. (COURTESY OF R. PORTMANN AND M. L. BIRNSTIEL, COPYRIGHT RESERVED.)

that complementary single-stranded DNA molecules were capable of reassociating, an event termed **renaturation**, or **reannealing**. This finding has proved to be one of the most valuable observations ever made in molecular biology. On one hand, reannealing has served as the basis for an investigation into the complexity of the genome, a subject discussed in the following section. On the other hand, reannealing has led to the development of a methodology called *nucleic acid hybridization*, in which complementary strands of nucleic acids from different sources can be mixed to form double-stranded (hybrid) molecules. Examples of the types of question that can be answered by allowing single-stranded nucleic acids to hybridize are discussed later in the chapter (page 416) and illustrated in Figure 10.22.

The Complexity of Viral and Bacterial Genomes Several factors determine the rate of renaturation of a given preparation of DNA. They are (1) the ionic strength of the solution, (2) the temperature, (3) the concentration of DNA, (4) the period of incubation, and (5) the size of the interacting molecules. With these factors in mind, consider what might happen if one were to compare the rate of renaturation of three different DNAs, each constituting the entire genome of (1) a small virus, such as SV40 (5.4×10^3 nucleotide pairs); (2) a larger virus, such as T4 (1.8×10^5 nucleotide pairs); and (3) a bacterial cell, such as *E. coli* (4.5×10^6 nucleotide pairs). The primary difference between these DNAs is their length. To compare their renaturation, it is important that the reacting molecules are of equivalent length, typically about 1,000 base pairs. DNA molecules can be fragmented into pieces of this length in various ways, one of which is to force them through a tiny orifice at high pressure.

When these three types of DNA—all present at the same lengths and DNA concentration (e.g., mg/ml)—are allowed to reanneal, they do so at distinctly different rates. The smaller the genome, the faster the renaturation. The reason becomes apparent when one considers the concentration of complementary sequences in the three preparations. Because all three preparations have the same amount of DNA in a given volume of solution, it follows that the smaller the genome size, the greater the number of genomes present in a given weight of DNA and the greater the chance of collision between complementary fragments. This is illustrated by the hypothetical case shown in Figure 10.18. The same reannealing profiles are observed, regardless of whether the three DNAs reanneal in separate tubes or together in the same solution (Figure 10.18*b,c*). This illustrates an important principle of hybridization reactions: the rate of reaction between molecules of complementary sequence is not affected by the presence of molecules having unrelated sequences.

The Complexity of Eukaryotic Genomes The renaturation of viral and bacterial DNAs occurs along single, symmetrical curves (Figure 10.19). Renaturation occurs along

such curves because all of the sequences are present (with the exception of a very few sequences in bacterial DNA) at the same concentration. Consequently, every nucleotide sequence in the population is as likely to find a partner in a given time as any other sequence. This is the result that would have been predicted from the various gene mapping studies that suggested the DNA of a chromosome contained one gene after another in a linear array. These results on viral and bacterial genomes contrast sharply with those obtained on mammalian DNA by Roy Britten and David Kohne of the California Institute of Technology. Unlike those of the simpler genomes, the DNA fragments from a mammalian genome reanneal at markedly different rates (Figure 10.20). These differences reflect the fact that the various nucleotide sequences in a preparation of eukaryotic DNA fragments are present at markedly different concentrations. This was the first insight that eukaryotic DNA is not just a simple progression of one gene after another, as in a bacterium or virus, but has a much more complex organization.

When DNA fragments from plants and animals are allowed to reanneal, the curve typically shows three more-or-less distinct steps (Figure 10.20), which correspond to the reannealing of three broad classes of DNA sequences. The three classes reanneal at different rates because they differ as to the number of times their nucleotide sequence is repeated within the population of fragments. The greater the number of copies of a particular sequence in the genome, the greater is its concentration and the faster it reanneals. The three classes are termed the **highly repeated fraction**, the **moderately repeated fraction**, and the **nonrepeated fraction**.

Highly Repeated DNA Sequences The highly repeated fraction, which consists of sequences present in at least 10^5 copies per genome, typically constitutes about 10 percent of the total DNA of vertebrates. This fraction of the genome reanneals so rapidly that its progress can be followed only in dilute solutions where the concentration of the reactants is very low. Highly repeated sequences are typically short (a few hundred nucleotides at their longest) and present in clusters in which the given sequence repeats itself over and over again without interruption. A sequence arranged in this end-to-end manner is said to be present *in tandem*. Highly repeated sequences fall into several overlapping categories, including satellite DNAs, minisatellite DNAs, and microsatellite DNAs.

■ **Satellite DNAs.** Satellite DNAs consist of short sequences (about 5 to a few hundred base pairs in length) that are repeated a large number of times in tandem to form very large clusters each containing up to several million base pairs of DNA. In many species, the base composition of these DNA segments is sufficiently different from the bulk of the DNA that fragments containing the sequence can be separated into a distinct "satellite" band during density gradient centrifugation

(a)

Reannealing experiment

(1) Extract DNA from each of 3 hypothetical organisms shown above.
(2) Dissolve each DNA in a volume of buffer (e.g., 0.12 M sodium phosphate) so that final concentration is 5 mg/ml.
(3) Shear DNAs B and C to 10^3 bps (∴ A, B, and C at same length).
(4) Denature as shown above.
(5) Incubate each solution at 60°C and follow the progress of the reannealing as shown by the curves below.

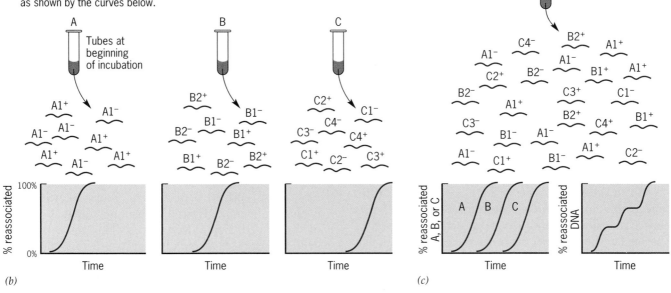

(b)

(c)

Figure 10.18 A hypothetical experiment on the renaturation of DNA from three different organisms. (*a*) The genome of each of the organisms. Each DNA preparation is sheared to fragments 1,000 base pairs in length and denatured into single strands by heat. Each copy of a genome from A yields one pair of complementary fragments; each from B yields two pairs of complementary fragments; and each from C yields four pairs of complementary fragments. (*b*) A comparison of the rates of re-

annealing of the three different DNAs when all are present at the same DNA concentration. (*c*) The reannealing of DNA when all three preparations are present in the same mixture. The plot on the left indicates the reannealing curve of the DNA from each of the three organisms. The plot on the right follows the reannealing of total DNA. In all the hypothetical curves of this figure, the time variable of the abscissa is plotted on a log scale.

(hence the name *satellite* DNA). A species may have more than one satellite DNA sequence; *Drosophila virilis,* for example, has three different satellite sequences, each seven nucleotides long and all very similar in sequence, indicating a common genetic origin. The lo-

calization of satellite DNA within the centromeres of chromosomes is described on page 416 and in further detail in Section 12.1. Despite decades of research, the precise function of satellite DNA remains a mystery.

Base pairs in genome

C_0t (mole X sec/liter)

Figure 10.19 The kinetics of renaturation of viral and bacterial DNAs. The curves show the renaturation of sheared strands of DNA from two viruses (MS-2 and T4) and a bacterium (*E. coli*). (The formation of double-stranded DNA is plotted against C_0t, which is a term that combines two variables, time of incubation and DNA concentration. A solution containing a high concentration of DNA incubated for a short time will have the same C_0t as one of low concentration incubated for a correspondingly longer time; both will have the same percentage of reannealed DNA.) The genome size, that is, the number of nucleotide base pairs in one copy of total genetic information of the organism, is indicated by the arrows near the upper numerical scale. The shape of each of the renaturation curves is very simple and occurs with a single slope. However, the time over which renaturation occurs is very different and depends on the concentration of complementary fragments, which in turn depends on the size of the genome. The larger the genome, the lower the concentration of complementary fragments in solution, and the greater the time required for renaturation to be complete.

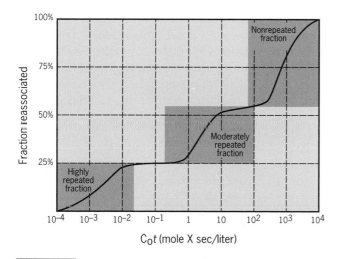

C_0t (mole X sec/liter)

Figure 10.20 An idealized plot showing the kinetics of renaturation of eukaryotic DNA. When single-stranded DNA is allowed to reanneal, three classes of fragments can usually be distinguished by their frequency of repetition within the genome: a highly repeated DNA fraction, a moderately repeated DNA fraction, and a nonrepeated (single-copy) DNA fraction. (Note: This is an idealized plot: the three classes of sequences are not as clearly separated in an actual renaturation curve).

- **Minisatellite DNAs.** Minisatellite sequences range from about 12 to 100 base pairs in length and are found in clusters containing as many as 3,000 repeats. Thus,

minisatellite sequences occupy considerably shorter stretches of the genome than do satellite sequences. For an unknown reason, minisatellites tend to be unstable, and the number of copies of a particular sequence often increases or decreases from one generation to the next. As a result, the length of a particular minisatellite locus is highly variable in the population, even among members of the same family. Because they are so variable (or **polymorphic**), minisatellite sequences are used to identify individuals in criminal or paternity cases through the technique of DNA fingerprinting (Figure 10.21). Changes in minisatellite sequences can alter the expression (transcription) of nearby genes, which in turn can have serious consequences. For example, changes at certain minisatellite loci have been implicated in causing cancer and diabetes.

- **Microsatellite DNAs.** Microsatellites are the shortest sequences (1 to 5 base pairs long) and are present in small clusters of about 50 to 100 base pairs in length. Microsatellites are scattered quite evenly through the DNA—at least 30,000 different microsatellite loci are present in the human genome. DNA replicating enzymes have trouble copying regions of the genome that contain these small, repetitive sequences, which causes them to exhibit an extremely high mutation rate. Because of their variability within the population, microsatellite DNAs have been used to analyze the relationships between different ethnic human populations,

Figure 10.21 DNA fingerprinting. In this technique, which is used widely to determine the identity of an individual from a sample of DNA, the DNA is digested by treatment with specific restriction nucleases (as in Figure 10.35), and the DNA fragments are subjected to gel electrophoresis. The location in the gel of DNA fragments containing specific DNA sequences is determined using radioactively labeled probes with sequences complementary to those being sought. The DNA fragments that bind these probes have variable lengths from one person to the next because of the presence of variable numbers of short repeated sequences in the genome. The autoradiograph shown in this figure was used in a criminal case in which the defendant was charged with the fatal stabbing of a young woman. The bloodstains on the pants and shirt of the defendant were compared to the known blood standards from the victim and defendant. DNA from the bloodstains on the defendant's clothing did not match his own blood standard, but they did match that of the victim. The lanes contain DNA from the following sources: 1, 2, 3, 9, and 10 are control DNA samples that serve as quality control checks; 4, the defendant's blood; 5, bloodstains from the defendant's pants; 6 and 7, bloodstains from the defendant's shirt; and 8, the victim's blood. (COURTESY OF CELL-MARK DIAGNOSTICS, INC. GERMANTOWN, MD.)

as illustrated in the following example. Many anthropologists argue that the modern human species arose in Africa. If this is true, then members of different African populations should exhibit greater DNA sequence variation than human populations living on other continents because genomes of African populations have had longer to diverge. The argument for African genesis has received support from studies on human DNA sequences. In one study that analyzed 60 different microsatellite loci, it was found that members of African populations had a significantly greater ge-

netic divergence than Asian or European populations. Changes in the numbers of copies of certain microsatellite sequences are responsible for several debilitating inherited diseases, which are discussed in the accompanying Human Perspective.

Once it became apparent that eukaryotic genomes contain large numbers of copies of short DNA sequences, researchers sought to learn where these DNA sequences are located in the chromosomes. For example, are satellite DNA sequences clustered in particular parts of a chromosome, or scattered uniformly from one end to the other? It was noted above that the discovery of DNA reannealing led to the development of a vast nucleic acid hybridization methodology (discussed at length in Chapter 18). The ability to locate satellite DNA sequences illustrates the analytic power of nucleic acid hybridization.

In the reannealing experiments described on page 412, complementary DNA strands were allowed to bind to one another while colliding randomly in solution. An alternate experimental protocol called **in situ hybridization** was developed by Mary Lou Pardue and Joseph Gall of Yale University and was used initially to determine the location of satellite DNA. The term *in situ* means "in place," which refers to the fact that the DNA of the chromosomes is kept in place while it is allowed to react with a particular preparation of labeled DNA. In early in situ hybridization studies, the DNA to be detected (the probe DNA) was radioactively labeled and was localized by autoradiography (as in Figure 11.9). The resolution of the technique has been increased using probe DNAs (or RNAs) that are labeled with fluorescent dyes and then localized with a fluorescence microscope (as in Figure 10.22). This latter technique is called **fluorescence in situ hybridization (FISH)** and has become so refined that it can be used to map the locations of specific sequences along single DNA fibers.

To carry out nucleic acid hybridization, both interactants must be single stranded. In the experiment depicted in Figure 10.22, the chromosomes from a mitotic cell are spread on a slide, and the DNA is made single stranded by treating the chromosomes with a hot salt solution that causes the DNA strands to separate and remain apart. During the ensuing hybridization step, the denatured chromosomes are incubated with a solution of labeled, single-stranded satellite DNA, which binds selectively to complementary strands of immobilized satellite DNA located in the chromosomes. Following the incubation period, the soluble, unhybridized satellite DNA is washed away or digested, and the sites of the bound, labeled DNA fragments are revealed. As shown in Figure 10.22, and described in the accompanying legend, satellite DNA is localized in the centromeric regions of the chromosome (see Figure 12.17). Other examples using FISH technology are shown in Figures 10.23 and 10.25.

Moderately Repeated DNA Sequences The moderately repeated fraction of the DNA can vary from about 20 to 80

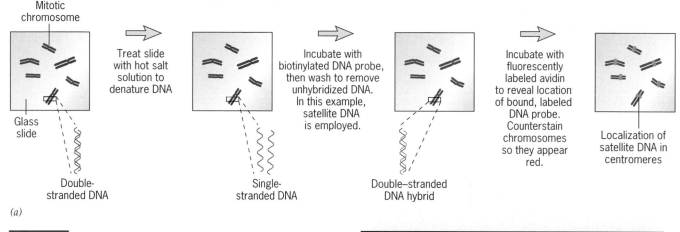

Mitotic
chromosome

Treat slide
with hot salt
solution to
denature DNA

Incubate with
biotinylated DNA probe,
then wash to remove
unhybridized DNA.
In this example,
satellite DNA
is employed.

Incubate with
fluorescently
labeled avidin
to reveal location
of bound, labeled
DNA probe.
Counterstain
chromosomes
so they appear
red.

Localization of
satellite DNA in
centromeres

Glass
slide

Double-
stranded DNA

Single-
stranded DNA

Double–stranded
DNA hybrid

(a)

Figure 10.22 **Fluorescence in situ hybridization and the localization of satellite DNA.** (*a*) The steps that are taken in carrying out fluorescence in situ hybridization. In this technique, the probe DNA is labeled without using radioactive isotopes; instead, certain of the nucleotides in the DNA are linked covalently to a small organic molecule, usually biotin. Following hybridization, the location of the bound, biotin-labeled DNA can be visualized by treatment of the preparation with fluorescently labeled avidin, a protein that binds to biotin with very high affinity. (Alternatively, the bound biotin-labelled DNA can be incubated with nonlabeled avidin, and the locations of the avidin-biotin complex can then be visualized with a fluorescently labeled anti-avidin antibody.) Chromosomes in these preparations usually appear red because they have been counterstained with propidium iodide. (*b*) Localization of α-satellite DNA at the centromere of human chromosomes. The location of the bound, biotin-labeled, satellite DNA is revealed by the yellow fluorescence, which stands out against a background of red, counterstained chromosomes. Fluorescence appears only at the site where each chromosome is constricted, which marks the location of the centromere. (B: FROM HUNTINGTON F. WILLARD, TRENDS GENET. 6:414, 1990.)

(b)

percent of the total DNA, depending on the organism. This fraction includes sequences that are repeated within the genome anywhere from a few times to tens of thousands of times. Included in the moderately repeated DNA fraction are sequences that code for known gene products, either RNAs or proteins, and those that lack a coding function.

1. *Repeated DNA Sequences with Coding Functions.* This fraction of the DNA includes the genes that code for transfer RNAs and ribosomal RNAs as well as those that code for an important group of chromosomal proteins, the histones. The repeated sequences that code for each of these products are typically identical to one another and located in tandem array. It is essential that the genes encoding the ribosomal and transfer RNAs be present in multiple numbers because these RNAs are required in large amounts and their synthesis does not benefit from the extra amplification step that occurs for protein-coding genes in which each mRNA acts as a template for the re-

peated synthesis of a polypeptide. Even though the production of histones does involve an intermediary messenger RNA, so many copies of these proteins are required during early development that several hundred DNA templates must be present.

2. *Repeated DNAs That Lack Coding Functions.* The bulk of the moderately repeated DNA fraction does not encode any type of product. Rather than occurring as clusters of tandem sequences, the members of these families are scattered (i.e., *interspersed*) throughout the genome as individual elements. Most of these repeated sequences can be grouped into two classes that are referred to as SINEs (*short interspersed elements*) or LINEs (*long interspersed elements*). SINEs are typically less than 500 base pairs long, whereas LINEs are more than 1,000 base pairs in length. The nucleotide sequences of these elements vary greatly from one species to the next; two families of repeated sequences found in humans (*Alu*, which is a SINE, and L1, which is a LINE) are discussed on page 425.

THE HUMAN PERSPECTIVE

Diseases That Result from Expansion of Trinucleotide Repeats

For decades, biologists believed that genes were always transmitted from generation to generation as stable entities. On rare occasion, a change occurs in the nucleotide sequence of a gene in the germ line, creating a mutation that is subsequently inherited. This is one of the basic tenets of Mendelian genetics. Then in 1991, several laboratories reported a new type of "dynamic mutation" in which the nucleotide sequence of particular genes changed dramatically between parents and offspring. In each case, these mutations affected genes that contained a repeating trinucleotide unit (e.g., CCG or CAG) as part of their sequence. In most members of the population, these particular genes contain a relatively small (but variable) number of repeated trinucleotides, and they are transmitted from one generation to the next without a significant change in the number of repeats. In contrast, a small fraction of the population possesses a mutant version of the gene that contains a larger number of these repeating units. Unlike the normal version, the mutant alleles are highly unstable, and the number of repeating units tends to increase as the gene passes from parents to offspring. When the number of repeats increases beyond a critical number, the individual inheriting the mutant allele develops a serious disease.

At the present time, a dozen or so distinct diseases have been attributed to expansion of trinucleotide repeats. These diseases fall into two basic categories, which we will consider in turn. Type I diseases are all neurodegenerative disorders that result from the expansion of the number of repeats of a CAG trinucleotide within the coding portion of the mutant gene (Figure 1). We can illustrate the nature of these disorders by considering the most prevalent and widely studied of the group, namely Huntington's disease (HD). HD is characterized by involuntary, uncoordinated movements; changes in

personality, including depression and irritability; and gradual intellectual decline. Symptoms usually begin in the third to fifth decade of life and increase in severity until death.

The normal *HD* gene, which is stably transmitted, contains between 6 and 35 copies of the CAG trinucleotide. The protein encoded by the *HD* gene is called **huntingtin**, and its precise function remains unclear. CAG is a triplet that codes for the amino acid glutamine (see Figure 11.41 for triplet coding assignments). Thus the normal huntingtin polypeptide contains a stretch of 6 to 35 glutamine residues—a polyglutamine tract—as part of its primary structure. We think of polypeptides as having a highly defined primary structure, and most of them do, but huntingtin is normally polymorphic with respect to the length of its polyglutamine tract. The protein appears to function normally as long as the tract length remains below approximately 35 glutamine residues. But if this number is exceeded, the protein takes on new properties, and the person is predisposed to developing HD.

HD (and the other trinucleotide degenerative diseases) exhibit a number of unusual characteristics. Unlike most inherited diseases, HD is a dominant genetic disorder, which means that a person with the mutant allele will develop the disease regardless of whether or not he or she has a normal *HD* allele. In fact, persons who are homozygous for the *HD* allele are no more seriously affected than heterozygotes. This observation indicates that the mutant huntingtin polypeptide causes the disease, not because it fails to carry out a particular function, but because it acquires some type of toxic property, which is referred to as a *gain-of-function* mutation. This interpretation is supported by studies with mice. Mice that are engineered to carry the mutant human *HD* allele (in addition to their own normal alleles) develop a neurodegenerative dis-

Figure 1 Trinucleotide repeat sequences and human disease. The top shows a generalized gene that is transcribed into an mRNA with several distinct portions (purple boxes), including a 5′ noncoding portion called the 5′ UTR (5′ untranslated region), a coding exon that carries the information for the amino acid sequence of the polypeptide, and a 3′ noncoding portion (the 3′ UTR). The introns in the DNA are not represented in the mature mRNA. The general location of the trinucleotide responsible for each of four different diseases (fragile X syndrome, Friedreich's ataxia,

Huntington's disease, and myotonic dystrophy) is indicated by the location of each pyramid. The number of repeats responsible for the normal (red), carrier (orange), and disease (yellow) conditions for each disease-causing gene is indicated. Genes responsible for type I diseases, such as Huntington's, do not exhibit the intermediate "carrier" state in which an individual possesses an unstable allele but is not affected. (AFTER J.-L. MANDEL, NATURE 386:768, 1997. © COPYRIGHT 1997, MACMILLAN MAGAZINES LTD.)

ease resembling that found in humans. The presence of the one abnormal allele is sufficient to cause disease.

Another unusual characteristic of HD and the other CAG disorders is a phenomenon called genetic anticipation, which means that, as the disease is passed from generation to generation, its severity increases and it strikes at an increasingly earlier age. This was once a very puzzling feature of HD but is now readily explained by the fact that the number of CAG repeats (and glutamine residues in the protein) in a mutant allele often increases dramatically from one generation to the next. Anticipation only occurs when the gene is transmitted paternally, that is, from a father to his offspring. Expansion probably occurs selectively in men because male germ cells (spermatogonia) undergo many more mitotic divisions before they differentiate into sperm than do female germ cells (oogonia) before they differentiate into eggs. Because expansion is thought to occur during DNA replication, the greater the number of cell divisions, the greater the likelihood of trinucleotide expansion.

The molecular basis of HD remains unclear. The function of a polypeptide is often revealed when the gene is isolated and sequenced, as was the case with the identification of the gene for cystic fibrosis (page 162). Unfortunately, the HD sequence provides little information concerning the possible function of the encoded protein. It is evident that when the polyglutamine tract of huntingtin exceeds 35 residues, the protein molecules associate with one another to form highly insoluble, protease-resistant aggregates, not unlike those seen in the brains of Alzheimer's victims (page 68). In HD, the aggregates are seen in both the nuclei and cytoplasm of certain neurons within the striatum of the brain; these are the neurons that subsequently degenerate and die. Whether or not mutant huntingtin aggregates are actually responsible for the death of the affected neurons has been hotly debated. Some researchers argue that the mutant protein itself is the toxic, cell-killing agent and that aggregation of the aberrant molecule may actually protect the cell by sequestering the harmful substance. This view is bolstered by the finding that mice expressing a shortened version of a protein with an expanded polyglutamine tract go on to develop the disease, even though protein aggregation is greatly reduced.

The most encouraging animal studies on HD have been performed on a genetically engineered strain of mice that carry a mutant huntingtin gene whose expression can be turned on or off as desired. Animals that express the mutant gene from the time of birth develop unmistakable symptoms of neurodegeneration at an early age. Remarkably, both the cellular and behavioral symptoms of the disease can be reversed by treating the affected animals in a manner that switches off the expression of the mutant gene. These findings provide hope that, when a treatment is developed for the disease, those individuals who already exhibit its symptoms can still be restored to full health.

Type II trinucleotide repeat diseases differ from the Type I diseases in a number of ways. Type II diseases (1) arise from the expansion of a variety of trinucleotides, not only CAG, (2) the trinucleotides involved are present in a part of the gene that does not encode amino acids (Figure 1), (3) the trinucleotides are subject to massive expansion into thousands of repeats, and (4) the diseases affect numerous parts of the body, not only the brain. The best studied of the Type II diseases is fragile X syndrome, so-named because the mutant X chromosome is especially susceptible to damage. Fragile X syndrome is characterized by mental retardation as well as a number of physical abnormalities. The disease is caused by a dynamic mutation in a gene called *FMR1*, whose product has not been precisely defined. A normal allele of this gene contains anywhere from about 5 to 55 copies of a specific trinucleotide (CGG) that is repeated in a part of the gene that corresponds to the 5′ noncoding portion of the messenger RNA (Figure 1; see also Section 12.3). A person can carry up to about 200 copies of this triplet without showing adverse effects. However, once the number of copies rises above about 60, the locus becomes very unstable and the copy number tends to increase rapidly into the thousands. Thus, persons with an *FMR1* gene containing 60 to 200 copies of the triplet exhibit a normal phenotype but are carriers for the transmission of a highly unstable chromosome to their offspring. If the repeat number in the offspring rises above about 200, the individual is almost always mentally retarded. Unlike an abnormal *HD* allele, which causes disease as the result of a gain of function, an abnormal *FMR1* allele causes disease as the result of a loss of function; *FMR1* alleles containing an expanded CGG number are selectively inactivated so that the gene is not transcribed or translated. Although there is no effective treatment for any of the diseases caused by trinucleotide expansion, the risk of transmitting or possessing a mutant allele can be assessed through genetic screening.

Nonrepeated DNA Sequences As initially predicted by Mendel, classical studies on the inheritance patterns of visible traits led geneticists to conclude that each gene was present in one copy per haploid set of chromosomes. When denatured eukaryotic DNA is allowed to reanneal, a significant fraction of the fragments (70 percent of human DNA fragments of 1,000 base-pair length) are very slow to find partners, so slow in fact that they are presumed to be present in a single copy per genome. This fraction comprises the *nonrepeated* (or *single-copy*) DNA sequences, which includes the genes that exhibit Mendelian patterns of inheritance. Because they are present in a single copy in the genome, nonrepeated genes always localize to a particular site on a particular chromosome (Figure 10.23).

Considering the variety of different sequences present, the nonrepeated fraction contains by far the greatest amount of genetic information. Included within the nonrepeated fraction are the DNA sequences that code for virtually all proteins other than histones. Even though these sequences are not present in multiple copies, it has become increasingly evident that genes that code for polypeptides are usually members of a family of related genes. This is true for the globins, actins, myosins, collagens, tubulins, integrins, and probably most other proteins in a eukaryotic cell. Each member of a multigene

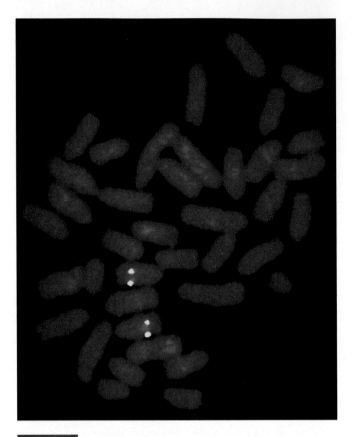

Figure 10.23 Chromosomal localization of a nonrepeated DNA sequence. These mitotic chromosomes were prepared from a dividing mouse cell and incubated with a purified preparation of biotin-labeled DNA encoding one of the nuclear lamin proteins (lamin B$_2$), which is encoded by a nonrepeated gene. The locations of the bound, labeled DNA appear as bright dots. The lamin gene is present on the homologues of chromosome 10. Each chromosome contains two copies of the gene because the DNA had been replicated prior to the cells entering mitosis. (FROM MONIKA ZEWE ET AL., COURTESY OF WERNER FRANKE, EUR. J. CELL BIOL. 56:349, 1991.)

family is encoded by a different but related nonrepeated sequence.

REVIEW ?

1. What is a genome? How does the complexity of bacterial genomes differ from that of eukaryotic genomes?

2. What is meant by the term DNA denaturation? How does denaturation depend on the GC content of the DNA? How does this variable affect the T_m?

3. What is a microsatellite DNA sequence? What role do these sequences play in human disease?

4. Which fraction of the genome contains the most information? Why is this true?

10.5 THE STABILITY OF THE GENOME

Because DNA is the genetic material, we tend to think of it as a conservative molecule whose information content changes only very slowly over long periods of evolutionary time. Studies indicate, however, that the sequence organization of the genome is capable of rapid change, not only from generation to generation, but within the lifetime of a given individual.

Duplication and Modification of DNA Sequences

We can begin an examination of genome stability with the well-established finding that eukaryotic organisms contain families of genes that code for polypeptides having related amino acid sequences. It is presumed that all the members of a multigene family arose from a single ancestral copy that became repeated over the course of evolution. The first step in the formation of a protein family is the duplication of some region of the chromosome, an event that could occur as DNA is synthesized but is more likely to be produced by a process of *unequal crossing over*, as depicted in Figure 10.24. Unequal crossing over occurs when a pair of homologous chromosomes come together during meiosis in such a way that they are not perfectly aligned. As a result of the misalignment, genetic exchange between the homologues causes one chromosome to acquire an extra segment of DNA (a **duplication**) and the other chromosome to lose a DNA segment (a **deletion**). If the duplication of a particular sequence is repeated in subsequent generations, a cluster of tandemly repeated segments is generated at a localized site within that chromosome (Figure 10.25). Once a gene has been duplicated, nucleotide substitutions can modify the repeated sequences in different ways, producing a multigene family whose encoded proteins vary in amino acid sequence. The production of a multigene family is illustrated by the evolution of the globin genes.

Evolution of Globin Genes Hemoglobin is a tetramer composed of four globin polypeptides (see Figure 2.40*b*). Examination of globin genes, whether from a mammal or a fish, reveals a characteristic organization. Each of these genes is constructed of three exons and two introns. Exons are parts of genes that code for amino acids in the encoded polypeptide, while introns do not; they are noncoding intervening sequences. The subject of exons and introns is discussed in detail in Section 11.3 (see Figure 11.24). For the present purposes, we will simply use these gene parts as markers of evolution. Examination of genes encoding certain globinlike polypeptides, such as the plant protein leghemoglobin and the muscle protein myoglobin, reveals the presence of four exons and three introns. This is proposed to represent the ancestral form of the globin gene. It is thought that the modern globin polypeptide arose from the ancestral

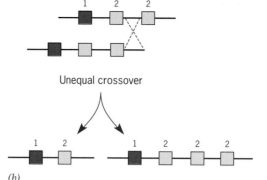

(a) (b)

Figure 10.24 Unequal crossing over between duplicated genes provides a mechanism for generating changes in gene number. (*a*) The initial state shown has two related genes (1 and 2). In a diploid individual, gene 1 on one homologue can align with gene 2 on the other homologue during meiosis. If a

crossover occurs during this misalignment, half the gametes will be missing gene 2 and half will have an extra gene 2. (*b*) As unequal crossing over continues to occur during meiotic divisions in subsequent generations, a tandemly repeated array of DNA sequences will gradually evolve.

form as the result of the fusion of two of the globin exons (step 1, Figure 10.26) some 800 million years ago.

A number of primitive fish are known that have only one globin gene (step 2), suggesting that these fish di-

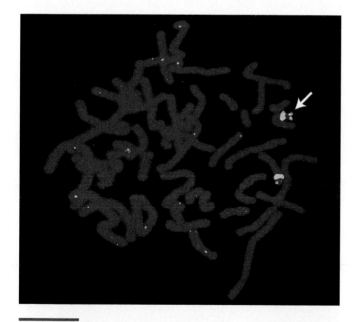

Figure 10.25 Location of a region in human chromosomes containing a recently duplicated gene. Fluorescence in situ hybridization using human mitotic chromosomes that had been incubated with biotin-labeled DNA from a gene implicated in polycystic kidney disease. The arrow indicates a region of chromosome 16 that contains several copies of the bound, labeled DNA. Further analysis indicated the presence in this region of three genes (*HG-A*, *HG-B*, and *HG-C*) whose nucleotide sequence is similar enough to bind the same labeled DNA probe. The fact that the sequences have not diverged significantly indicates that their duplication occurred in recent ancestors. (FROM CHRISTOPHER J. WARD ET AL., CELL 77:883, 1994; COURTESY OF PETER C. HARRIS, BY PERMISSION OF CELL PRESS.)

verged from other vertebrates prior to the first duplication of the globin gene (step 3). Following this duplication approximately 500 million years ago, the two copies diverged by mutation (step 4) to form two distinct globin types, an α type and a β type. In subsequent steps, the α and β forms are thought to have become separated from one another by a process that moved them to separate chromosomes (step 5). Each gene then underwent subsequent duplications and divergence (step 6), generating the arrangement of globin genes that exists today in humans (step 7). The α-globin genes of the human genome are clustered on chromosome 16, while the β-globin genes are clustered on chromosome 11.

When the DNA sequences of globin gene clusters were analyzed, researchers found "genes" whose sequences are homologous to those of functional globin genes, but which have accumulated severe mutations that render them nonfunctional. Genes of this type, which appear to be evolutionary relics, are known as **pseudogenes**, and their occurrence in genomes is widespread. Examples of pseudogenes are found in both the human α- and β-globin gene clusters of Figure 10.26. Another point that is evident from examination of the two globin clusters of human chromosomes is how much of the DNA consists of noncoding sequences, either as introns within genes or as spacers between genes. In fact, less than 15 percent of the DNA located in these regions actually codes for the amino acid sequences of the globin polypeptides; the remainder consists of noncoding regions, most of which have no known function. Studies of DNA sequences from different species indicate that different parts of the globin gene region evolve at very different rates. The DNA of the coding regions is highly conserved, while that of the noncoding regions is much more variable. Moreover, changes that occur in the coding regions are almost exclusively single base substitutions, while those in the noncoding portions often include additions and deletions of DNA segments.

Figure 10.26 A pathway for the evolution of globin genes.
Exons are shown in red, introns in yellow. The evolutionary
steps depicted in the diagram are discussed in the text. The
arrangement of α- and β-globin genes on human chromosomes
16 and 11 (shown without their introns on the right) are the
products of several hundred million years of evolution. As dis-
cussed in Chapter 2, hemoglobin molecules consist of two pairs
of polypeptide chains—one pair is always a member of the α-
globin subfamily, and the other pair is always a member of the
β-globin subfamily. Specific combinations of α- and β-globins
are found at different stages of development (see Figure 12.44).
The α- and β-globin chains that are observed in embryonic,
fetal, and adult hemoglobins are indicated.

In some instances, the polypeptides encoded by dif-
ferent members of a family have evolved divergent func-
tions, even though their amino acid sequences still show
their ancestral relationship. For example, growth hor-
mone and prolactin are pituitary hormones with clearly re-
lated amino acid sequences, but they evoke completely
different responses in target cells. Even where there is very

little change in amino acid sequence of a particular protein
over long periods of evolutionary time, as in the case of the
actins, there are always substantial differences in nu-
cleotide sequence between corresponding genes of dis-
tantly related organisms. The nucleotide substitutions
that have been tolerated are ones that do not alter the se-
quence of amino acids in the polypeptide chain.[3]

Mobile Genetic Elements

If one looks at repeated sequences that have arisen during
the normal course of evolution, one finds that repeats are
sometimes present in tandem arrays, sometimes present
on two or a few chromosomes (as in the case of the globin
genes of Figure 10.26), and sometimes dispersed through-
out the genome. If we assume that all members of a family
of repeated sequences arose from a single copy, then how
can individual members become dispersed among differ-
ent chromosomes?

The first person to suggest that genetic elements were
capable of moving from one site in the genome to another
site was Barbara McClintock, a geneticist working with
maize (corn) at the Cold Spring Harbor Laboratories in
New York. Genetic traits in maize are often revealed as
changes in the patterns and markings in leaf and kernel
coloration (Figure 10.27). In the late 1940s, McClintock
found that certain mutations were very unstable, appear-

[3]As discussed in Chapter 11, the same amino acid may be encoded by a num-
ber of different nucleotide triplets (codons). Consequently, the nucleotide
sequence can change without changing the encoded amino acid sequence.

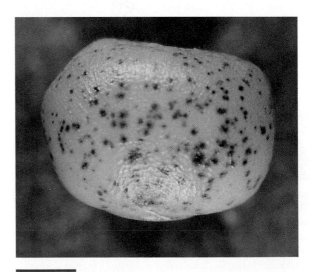

**Figure 10.27 Visible manifestations of transposition in
maize.** Kernels of corn are typically uniform in color. The spots
on this kernel result from a mutation in a gene that codes for an
enzyme involved in pigment production. Mutations of this type
can be very unstable, arising or disappearing during the period
in which a single kernel develops. These unstable mutations ap-
pear and disappear as the result of the movement of transpos-
able elements during the period of development. (COURTESY OF
VENKATESAN SUNDARESAN, COLD SPRING HARBOR LABORATORY.)

ing and disappearing from one generation to the next or even during the lifetime of an individual plant. After several years of careful study, she concluded that certain genetic elements were moving from one place in a chromosome to an entirely different site. She called this genetic rearrangement **transposition**, and the mobile genetic elements **transposable elements**. Meanwhile, molecular biologists working with bacteria were finding no evidence of transposition. In their studies, genes appeared as stable elements situated in a linear array on the chromosome that remained constant from one individual to another and from one generation to the next. As a result, McClintock's findings were largely ignored.

Then, in the late 1960s, several laboratories discovered that certain DNA sequences in bacteria were capable of moving from one place in the genome to another. These transposable elements were called **transposons**. Transposons were subsequently found to encode a protein, or *transposase*, that single-handedly catalyzes the excision of a transposon from a donor DNA site and its subsequent insertion at a target DNA site. Recent studies on

bacterial transposition indicate that this "cut-and-paste" mechanism is mediated by two separate transposase subunits that bind to specific sequences at the two ends of the transposon (Figure 10.28, step 1). The two subunits then come together to form an active dimer (step 2) that catalyzes a series of reactions leading to the excision of the transposon (step 3). The transposase—transposon complex then binds to a target DNA (step 4) where the transposase catalyzes the reactions required to integrate the transposon into its new residence (step 5).

Sequence analysis of transposons indicated that the sequence present at one end of the element is repeated at the opposite end but in an opposite orientation (i.e., as an inverted repeat) (red segments in Figure 10.29). These terminal repeats are recognized by transposases and are required for transposition. Repeating sequences are found at the ends of all types of transposable elements, both in bacteria and eukaryotes. In addition, integration of the element creates a small duplication in the target DNA that flanks the transposed element at the site of insertion (green segments in Figure 10.29). Target site duplications serve as "footprints" to identify sites in the genome that are or have been occupied by transposable elements.

As originally demonstrated by McClintock, eukaryotic genomes also contain large numbers of transposable elements. Approximately 40 percent of the DNA in a human cell nucleus is derived from transposable elements. The vast majority of these elements are incapable of moving from one place to another; they have either been crippled by mu-

Figure 10.28 Transposition of a bacterial transposon by a "cut-and-paste" mechanism. As discussed in the text, the two ends of this bacterial Tn5 transposon are brought together by the dimerization of a pair of subunits of the transposase. Both strands of the double helix are cleaved at each end, which excises the transposon as part of a complex with the transposase. The transposon-transposase complex is "captured" by a target DNA, and the transposon is inserted in such a way as to produce a small duplication that flanks the transposed element. (AFTER D. R. DAVIES ET AL., SCIENCE 289:77, 2000,. COPYRIGHT © 2000 AMERICAN ASSOCIATION FOR THE ADVANCEMENT OF SCIENCE.)

Figure 10.29 Sequence organization of a transposon. The ends of the mobile transposable element contain a sequence (red) that is inversely repeated. Integration of the transposon into the host DNA creates a duplication that is seen as a short direct repeat (green) on both sides of the element. The nucleotide sequences depicted here are those of Tn3, a bacterial transposon that codes for three proteins, including the transposase that allows it to be integrated at new sites in the DNA.

tation or their movement is suppressed by the cell. However, when transposable elements do change position, they insert almost randomly within the target DNA. In fact, many transposable elements can insert themselves within the center of a protein-coding gene. Several examples of such an occurrence have been documented in humans, including a number of cases of hemophilia caused by a mobile genetic element that had "jumped" into the middle of one of the key blood-clotting genes. It is estimated that approximately 1 out of 500 human mutations is the result of the insertion of a transposable element.

Several different types of eukaryotic transposable elements are known, as are several different mechanisms of transposition (Figure 10.30). In some cases, as described above, the transposable element is excised from the DNA at the donor site and inserted at a distant target site (Figure 10.30a). This "cut-and-paste" mechanism is utilized by members of the *mariner* family of transposons, which are found throughout the plant and animal kingdoms. In other cases, the transposable element is replicated, and the DNA copy is inserted into a target site leaving the donor site unchanged (Figure 10.30b). Most often, however, transposition in eukaryotic cells involves an RNA intermediate (Figure 10.30c). The DNA of the gene is transcribed, producing an RNA, which is then "reverse transcribed" by reverse transcriptase, producing a comple-

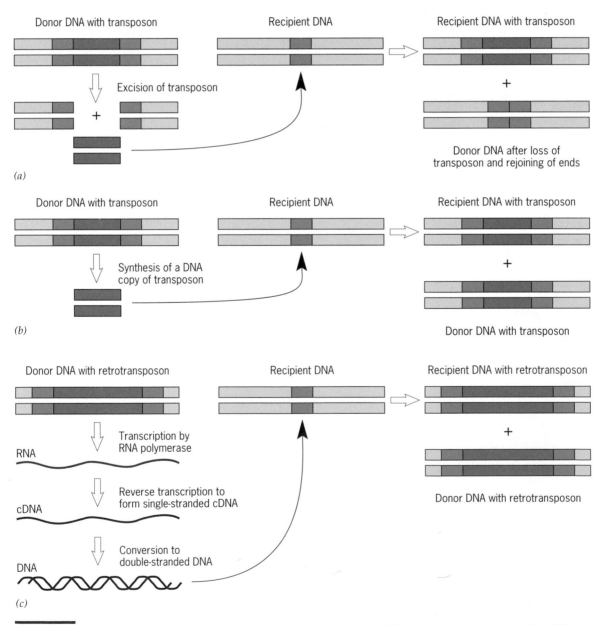

Figure 10.30 Three alternate pathways by which transposable elements move from place to place within the genome. (*a*) Nonreplicative transposition, (*b*) replicative transposition, and (*c*) transposition by means of an RNA intermediate. This latter pathway is virtually identical to that by which retroviruses integrate DNA copies of their genome into the host DNA.

mentary DNA. The DNA copy is made double stranded and then integrated into a target DNA site. In many cases, the transposable element itself contains the sequence that codes for a reverse transcriptase. Transposable elements that require reverse transcriptase for their movement are called **retrotransposons**. Retroviruses, such as the virus responsible for AIDS, are thought to have evolved from retrotransposons by acquiring genes that allowed them to leave the cell and become infectious (e.g., genes that encode for envelope proteins). The reverse event, in which retroviruses have given rise to retrotransposons, is also likely to have happened.

The Role of Mobile Genetic Elements in Evolution It was noted on page 416 that moderately repeated DNA sequences constitute a significant portion of the human genome. Unlike the highly repeated fraction of the genome (satellite, minisatellite, and microsatellite DNA), whose sequences reside in tandem and arise by gene duplication, most of the moderately repeated sequences of the genome are interspersed and arise by transposition. In fact, the two most common families of moderately repeated sequences in human DNA are transposable elements; they are the *Alu* and L1 families, both of which transpose by means of RNA intermediates. The human genome is estimated to contain about 500,000 copies of L1 (about 15 percent of nuclear DNA), but the vast majority of these are incomplete elements and are not capable of transposition. A full-length, transposable L1 sequence (at least 6,000 base pairs in length) encodes a unique protein with two catalytic activities: a reverse transcriptase that makes a DNA copy of the RNA and an endonuclease that cleaves the target DNA prior to insertion. L1 elements are found in all types of eukaryotes and may have been present in the earliest eukaryotic cells.

Even more abundant than L1 are the *Alu* sequences, which are interspersed at roughly one million different sites throughout the human genome. *Alu* is a family of short, related sequences about 300 base pairs in length. The *Alu* sequence closely resembles that of the 7S RNA present in the signal recognition particles found in conjunction with membrane-bound ribosomes (page 292). It is presumed that during the course of evolution this cytoplasmic RNA was copied into a DNA sequence by reverse transcriptase and integrated into the genome. The tremendous amplification of the *Alu* sequence is thought to have occurred by retrotransposition using the reverse transcriptase and endonuclease encoded by L1 sequences. Given its prevalence in the human genome, one might expect the *Alu* sequence to be repeated in genomes throughout the rest of the animal kingdom, but this is not the case. Studies of the genomes of various mammals indicate that the *Alu* sequence first appeared as a transposable element in the genome of higher primates about 60 million years ago and has been increasing in copy number at an average rate of about one copy every 100 years ever since.

When something new like a transposable element is discovered in an organism, a biologist's first question is usually: What is its function? Many researchers who study transposition believe that transposable elements are primarily "junk." According to this view, a transposable element is a type of genetic parasite that can invade a host genome from the outside world, spread within that genome, and be transmitted to offspring-so long as it doesn't have serious adverse effects on the ability of the host to survive and reproduce. Even if this is the case, it doesn't mean that transposable elements cannot make positive contributions to eukaryotic genomes. Regardless of its origin, once a DNA sequence is present in a genome, it has the *potential* to be "put to use" in some manner during the course of evolution. This potential has been realized in a number of ways:

1. DNA sequences that were originally derived from transposable elements are found as essential parts of the regulatory regions that flank a number of eukaryotic genes. The expression of the gene encoding human parathyroid hormone is controlled, in part, by relics of a transposable element.

2. In some cases, transposable elements can carry adjacent parts of the host genome with them as they move from one site to another. As a result, two unlinked segments of the host genome can be brought together to form a new, composite segment. This may be a primary mechanism in the evolution of proteins that are composed of domains derived from different ancestral genes (as in Figure 2.36).

3. In some cases, transposable elements appear to have given rise to genes, themselves. The enzyme telomerase, which plays a key role in replicating the DNA at the ends of chromosomes (see Figure 12.19), is thought to be derived from a reverse transcriptase encoded by an ancient retrotransposon. It is also speculated that the enzymes involved in the rearrangement of antibody genes (see Figure 17.15) are derived from a transposase encoded by an ancient DNA transposon. If this is in fact the case, our ability to ward off infectious diseases is a direct consequence of transposition.

One point is clear: transposition has had a major impact on altering the genetic composition of organisms. This point is dramatically illustrated by a transposable element of *Drosophila melanogaster* called the P element. Examination of laboratory fruit flies descended from individuals trapped by T. H. Morgan and his colleagues at the start of the century are devoid of the P element. In contrast, every member of the species caught in the "wild" today carries multiple copies of this transposable element (Figure 10.31). This element is thought to have been introduced into the genome of a single *D. melanogaster* within the past 80 years (probably by transmission from an individual of another *Drosophila* species) and has rapidly spread through the entire species population. Transmission of genetic material from one species to another—whether between dif-

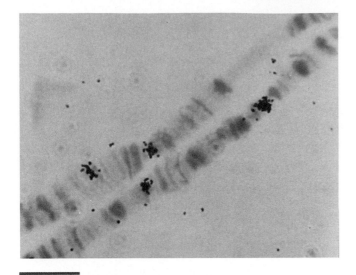

Figure 10.31 P elements in *Drosophila.* A portion of a pair of *Drosophila* polytene homologues in which the locations of the P elements are indicated in the autoradiograph by the silver grains overlying sites of radioactively labeled RNA that is complementary in sequence to the P transposable element. The P element is seen to be inserted at different sites on the two homologous chromosomes. Insertion of a P element at one of the sites has caused a mutation in the gene residing there. (FROM KATE LOUGHNEY, R. KREBER, AND BARRY GANETZKY, CELL 58:1143, 1989; BY PERMISSION OF CELL PRESS.)

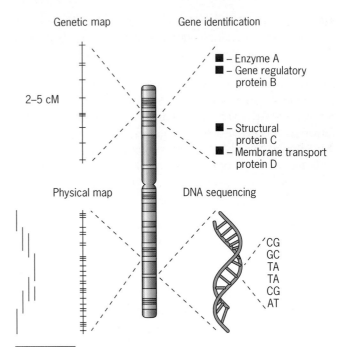

Figure 10.32 Types of information obtained from genetic analysis. Genetic maps provide information on the relative positions of genetic markers as determined by recombination frequencies. One centimorgan (cM) is equal to a 1 percent chance that two markers will be separated by recombination during meiosis. Physical maps are constructed from overlapping DNA fragments obtained by restriction enzyme digestion. Gene identification has become very important in clinical research as increasing numbers of genes responsible for causing human diseases have been identified. DNA sequencing provides information on the sequence of base pairs that constitute specific regions of chromosomes. (AFTER F. S. COLLINS, SCIENCE 262:45, 1993. COPYRIGHT © 1993 AMERICAN ASSOCIATION FOR THE ADVANCEMENT OF SCIENCE.)

ferent fruit flies or different types of vertebrates—is likely mediated by parasites that pick up DNA fragments from one host and transfer it to a subsequent host.

It is interesting to note that, only a couple of decades ago, molecular biologists considered the genome to be a stable repository of genetic information. Now it seems remarkable that organisms can maintain themselves from one day to the next in the face of this large-scale disruption by genetic rearrangement. For her discovery of transposition, Barbara McClintock was the sole recipient of a Nobel Prize in 1983, approximately 35 years after her initial discovery.

REVIEW ?

1. Describe the course of evolutionary events that are thought to have given rise to multiple gene families, such as those that encode the globins. How could these events give rise to pseudogenes? How could they give rise to proteins having entirely different functions?

2. Describe three mechanisms by which genetic elements are able to move from one site in the genome to another.

3. Describe the impact that transposable elements have had on the structure of the human genome over the past 50 million years.

10.6 MOLECULAR MAPS OF THE GENOME

For decades, geneticists have mapped genes along a chromosome by determining the frequency of crossing over between genes (page 403). The closer two genes reside on a chromosome, the less likely their alleles will become separated from one another during crossing over. When used for genetic mapping, different alleles of a gene act as signposts (*genetic markers*) that help reveal the location of nearby sites along the chromosome. Genes that are known to exist in several common allelic forms, such as those that determine blood or tissue type, are said to be *polymorphic.* Polymorphic genes make good genetic markers because particular alleles of these genes can be followed from one generation to the next in a family being studied. The assignment of genetic markers to relative positions on a chromosome produces a **genetic map** (Figure 10.32) of that chromosome.

During the past few years, variations in DNA sequences within the human population have replaced polymorphic genes in the construction of genetic maps. The most detailed maps depend on the identification of **single nucleotide polymorphisms (SNPs)**, which are sites in the genome where alternate base pairs are found at high frequency in the population. Even though most of these sites reside in regions of the genome that have no known biological function, they make excellent genetic markers because they are polymorphic and extremely abundant. It is estimated that there are approximately three million common SNPs in the human genome, or approximately one every thousand base pairs. By February 2001, 1.4 million SNPs had been identified and placed in a public database. Now that highly detailed genetic maps have been created, the localization and isolation of a gene can be made quite rapidly by determining which genetic markers it tends to be inherited with.

SNPs have important uses beyond their value as genetic markers. Many of these single nucleotide differences are found within protein-coding genes and presumably affect the functions of the encoded proteins in subtle ways. It is likely, for example, that SNPs play an important role in our susceptibility to complex diseases that involve the activities of a number of different genes. Researchers are hoping to scan the genomes of large numbers of peoples to identify SNPs that are associated with particular diseases. It may be possible, one day, to determine if a person is likely to develop heart disease, or Alzheimer's disease, or a particular type of cancer by identifying which nucleotides are present at key positions within their genome. In addition, SNPs may determine how a person reacts to a particular drug, whether they are greatly helped by it or develop serious side effects (Figure 10.33). The pharmaceutical industry is hoping that SNP data will eventually lead to an era of "customized drug therapy," allowing physicians to prescribe specific drugs that are tailored to each individual patient based on their "genetic profile."

Another type of chromosomal map has as its goal the determination of the nucleotide sequence of the DNA that makes up each chromosome. This type of map is the subject of the Human Perspective: Sequencing the Human Genome. To reach this goal, the huge DNA molecules that make up each chromosome have been dissected into overlapping fragments of manageable size. Such fragments can then be cloned and sequenced (Section 18.13), and their order in the chromosomes determined. A map that utilizes identifiable, overlapping DNA fragments rather than genetic markers is called a **physical map** (Figure 10.32). Construction of a physical map begins with dissection of the DNA, a feat that is accomplished using restriction endonucleases.

(a)

(b)

Figure 10.33 The potential use of single nucleotide polymorphisms (SNPs) in prescribing medication. (*a*) A hypothetical SNP map of a small portion of a human chromosome. Each site along the DNA that contains an SNP (designated by the short green line) is indicated. (*b*) A hypothetical example of the way that SNP maps could be used as a guide to prescribing drugs. The top line indicates the bases that appear at SNP sites in several individuals that respond well to a particular drug, whereas the middle line indicates the bases that appear at these same sites in several individuals that have adverse reactions. The bottom line indicates those nucleotides that predict a successful outcome using the drug. In a real example, thousands of different sites in the genome would have to be screened and the data subjected to statistical analysis to determine which SNPs and which nucleotides at those sites are predictive of success or failure.

Restriction Endonucleases

During the 1970s, it was found that bacteria contained nucleases that would recognize short nucleotide sequences within duplex DNA and cleave the DNA backbone at specific sites on both strands of the duplex. These enzymes are called **restriction endonucleases**, or simply **restriction enzymes**. They were given this name because they function in bacteria to destroy viral DNAs that might enter the cell, thereby *restricting* the potential growth of the viruses. The bacterium protects its own DNA from nucleolytic attack by methylating the bases at susceptible sites, a chemical modification that blocks the action of the enzyme.

Enzymes from several hundred different prokaryotic organisms have been isolated that, together, recognize over 100 different nucleotide sequences. The sequences recognized by most of these enzymes are four to six nucleotides long and are characterized by a particular type of

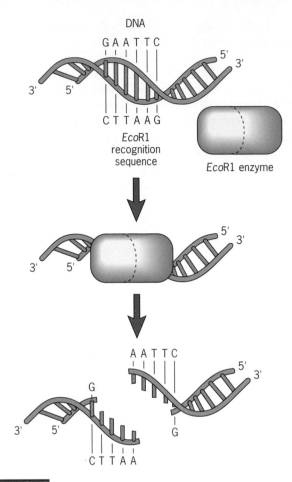

Figure 10.34 The cleavage of DNA by the restriction enzyme *Eco*R1.

Enzyme Recognition sequence Enzyme Recognition sequence

R. *Hind*II
3'–N–C–A–R ↓ Y–T–G–N–5'
5'–N–G–T–Y · R–A–C–N–3'

R. *Eco*RI
3'–N–C–T–T–·–A–A↓G–N–5'
5'–N–G↑A–A · T–T–C–N–3'

R. *Bsu*R
3'–N–N–C–C ↓ G–G–N–N–5'
5'–N–N–G–G · C–C–N–N–3'

R. *Eco*RII
3'–N–N–G–G–·–C–C↓N–N–5'
5'–N–N–C↑C–·–G–G–N–N–3'

R. *Hpa*I
3'–N–C–A–A ↓ T–T–G–N–5'
5'–N–G–T–T · A–A–C–N–3'

R. *Hind*III
3'–N–T–T–C–·–G–A↓A–N–5'
5'–N–A↑A–G–·–C–T–T–N–3'

Figure 10.35 Cleavage of double-stranded DNA molecules by most restriction enzymes occurs at sequences of four to six specific nucleotides. The sequences recognized by several of these enzymes are illustrated along with the site on each strand where the cut is made (indicated by the arrows). Those enzymes on the left break both strands at the same point, whereas those on the right cleave the strands at staggered sites. N is any nucleotide (opposite nucleotide must be complementary), R is a purine, and Y is a pyrimidine. The central dot indicates the point of twofold rotational symmetry. (REPRINTED FROM K. MURRAY, ENDEAVOUR 35:130, 1976, WITH KIND PERMISSION FROM ELSEVIER SCIENCE LTD.)

internal symmetry. Consider the particular sequence recognized by the enzyme *Eco*R1 (Figure 10.34):

$$3'—\text{CTT}\overset{\downarrow}{\text{A}}\text{AG}—5'$$
$$5'—\text{GA}\underset{\uparrow}{\text{A}}\text{TTC}—3'$$

This segment is said to have *twofold rotational symmetry* because it can be rotated 180° without change in base sequence. Thus, if one reads the sequence in the same direction (3' to 5' or 5' to 3') on either strand, the same order of bases is observed. A sequence with this type of symmetry is called a *palindrome*. When the enzyme *Eco*R1 attacks this palindrome, it breaks each strand at the same site *in the sequence*, which is indicated by the arrows between the A and G residues. The red dots indicate the methylated bases in this sequence that protect the host DNA from enzymatic attack. Figure 10.35 gives the sequences recognized by a number of restriction enzymes. Some of these enzymes cleave bonds directly opposite one another on the two strands producing blunt ends, whereas others, such as *Eco*R1, make staggered cuts.

The discovery and purification of restriction enzymes have been invaluable in the advances made by molecular biologists in recent years. Since a particular sequence of four to six nucleotides occurs quite frequently simply by chance, any type of DNA is susceptible to fragmentation by these enzymes. The use of restriction enzymes allows the DNA of the human genome, or that of any other organism, to be dissected into a precisely defined set of specific fragments. Once the DNA from a particular individual is digested with one of these enzymes, the fragments generated can be fractionated on the basis of length by gel electrophoresis (Figure 10.36*a*). Different enzymes cleave the same preparation of DNA into different sets of fragments, and the sites within the genome that are cleaved by various enzymes can be identified and ordered into a restriction map (which is one type of physical map), such as that shown in Figure 10.36*b*. Because of their ability to carve DNA into a predictable collection of fragments, restriction enzymes are finding application in numerous human activities, including diagnostic medicine, criminal prosecution, and sequencing the human genome (see the accompanying Human Perspective).

REVIEW ?

1. Describe the steps one would take in preparing a restriction map of a portion of the DNA using a particular restriction enzyme.

2. How does a genetic map of the genome differ from a physical map?

(a)

(b)

Figure 10.36 The construction of a restriction map of the small circular genome of the DNA tumor virus polyoma. (a) Autoradiographs of ^{32}P-labeled DNA fragments that have been subjected to gel electrophoresis. The gel on the left shows the pattern of DNA fragments obtained after a complete digestion of the polyoma genome with the enzyme *Hpa*II. To determine how these eight fragments are pieced together to make up the intact genome, it is necessary to treat the DNA in such a way that overlapping fragments are generated. Overlapping fragments can be produced by treating the intact genome with a second enzyme that cleaves the molecule at different sites, or by treating the genome with the same enzyme under conditions where the DNA is not fully digested as it was in the left gel. The two gels on the right represent examples of partial digests of the polyoma genome with *Hpa*II. The middle gel shows the fragments generated by partial digestion of the superhelical circular DNA, and the gel on the right shows the *Hpa*II fragments formed after the circular genome is converted into a linear molecule by *Eco*R1 (an enzyme that makes only one cut in the circle). (b) The restriction map of the linearized polyoma genome based on cleavage by *Hpa*II. The eight fragments from the complete digest are shown along the DNA at the top. The overlapping fragments from the partial digest are shown in their ordered arrangement below the map. (Fragments L and M migrate to the bottom of the gel in part *a*, right side.) (From Beverly E. Griffin, Mike Fried, and Alison Cowie, Proc. Nat'l. Acad. Sci. U. S. A. 71:2078, 1974.)

THE HUMAN PERSPECTIVE

Sequencing the Human Genome

UPDATE
For the very latest, go to page U-12

In 1986, a group of scientists from around the world gathered in the United States to discuss the possibility that all of the genetic information stored in human chromosomes might be decoded as the result of a giant, international, scientific collaboration. Within a few years, governmental agencies in the United States, Europe, and Asia had decided to proceed with the effort, which has become known as the Human Genome Project (HGP). The term *genome* refers to the information stored in all the DNA of a single representative set of chromosomes. The human genome contains approximately 3 billion base pairs. If each base pair in the DNA were equivalent to a single letter on this page, the information contained in the human genome would produce a book approximately 1 million pages long!

During the first few years of the Human Genome Project, researchers prepared detailed genetic and physical maps of each human chromosome (page 426). Once these maps were completed, DNA from different human chromosomes was isolated and cleaved into large overlapping fragments of approximately 150,000 base pairs that could be cloned as "artificial chromosomes" in bacteria (Section 18.14). The large fragments were then dissected into smaller overlapping fragments that were cloned and sequenced. As the sequences were generated, they were entered into a public database known as GenBank, which could be accessed by anyone using the Internet. Once the sequences of the overlapping fragments had been determined, the fragments were arranged in linear order to reassemble each of the larger fragments, which in turn could be arranged in linear order to reassemble the entire DNA molecule of that chromosome.

In 1999, reports of the nucleotide sequences of the two smallest human autosomes—chromosomes number 21 and 22—were published. The sequences of these two chromosomes were reported in their "finished" form, which means that (1) each site had been sequenced on average about 10 times to ensure a very high degree of accuracy, and (2) the two sequences contained a minimal number of small gaps, none larger than 150,000 base pairs. The gaps, in this case, represented regions of the two chromosomes that for one reason or another did not allow themselves to be cloned and thus were impossible to sequence using the available technology.

The next major announcement from the HGP came in June, 2000, with the report of a "rough draft" of the entire human genome. This "rough draft" differed from the "finished" versions reported for chromosomes 21 and 22 in that (1) it was less accurate because individual sites had only been sequenced an average of about four times over, and (2) there were a large number of gaps. Altogether it is estimated that the rough draft reported in June, 2000, covered approximately 90 percent of the human genome, excluding certain regions containing repeated DNA sequences whose positions are difficult to assign. The publication of a "rough draft" was not part of the original HGP plan, but was decided upon when it became clear that a private company, Celera Ge-

nomics, was rushing to complete the task of sequencing the human genome before the HGP. Rather than sequencing the genome chromosome-by-chromosome, and clone-by-clone, as the HGP was doing, Celera was using a "shotgun" approach. They shredded the entire genome into millions of tiny, overlapping pieces, which were then sequenced and assembled into larger molecules with the aid of powerful supercomputers and information available in the public database. The announcement of the rough draft in June, 2000, was made jointly by both Celera and the HGP. A "finished" version of the genome is expected by 2002–2003, but even that version is expected to contain a significant number of gaps.

By itself, a million pages of As, Ts, Gs, and Cs have little inherent value, but they contain the raw data from which invaluable information can be obtained. The task of "mining" the nucleotide sequence falls on the shoulders of those who will use computers to identify the genes, predict their function, and eventually determine how their expression is controlled through regulatory sequences that lie outside the gene itself (Chapter 12). Preliminary findings concerning the information content of the human genome were reported in February, 2001, by groups representing both the HGP and Celera Genomics. Up until these publications, it had been widely assumed that the human genome contained at least 50,000 different genes and maybe as many as 150,000. Instead, our genome contains only about 30,000 different genes, or roughly 50–75 percent more than that of a microscopic nematode worm. It would appear that the difference in complexity between humans and worms is less a matter of the amount of genetic information that we inherit than what we do with that information as we develop. As will be discussed in Chapters 12 and 17, a single gene present in the genome can generate a large number of gene products as the result of processes such as alternate splicing (Figure 12.53) and DNA rearrangement (Figure 17.15). It is possible that greater differences between organisms will emerge when these types of "gene-enhancing" mechanisms are further explored. This expectation is consistent with the fact that human genes tend to be more complex (i. e., contain more exons) than those of flies and worms. It was also revealed in these studies that more than a hundred genes in the human genome closely resemble bacterial genes and have presumably been acquired from bacteria during the period of vertebrate evolution. A similar process of lateral gene transfer was discussed on page 28 in connection with the evolution of eukaryotic cells.

Eventually, analysis of the genome sequence is expected to identify those genes that cause or contribute to human diseases—including such complex diseases as cancer, hypertension, and mental illness—and determine the changes in sequence that trigger the disease. In addition to its importance in clinical research, this type of information can be used to screen individuals who are at high risk for a particular disease because of their family history. At the present time, screening procedures are limited to a small number of dis-

eases (e.g., cystic fibrosis, Huntington's disease, familial Alzheimer's disease, sickle cell anemia, and familial breast cancer) for which mutations in a single gene can either cause the disease or place a person at very high risk for developing the disease. New technologies are being developed that will allow DNA samples to be screened simultaneously for polymorphisms (page 427) in a number of different genes. This should allow estimation of a person's risk for developing a particular disease for which the genetic contributions are more complex. For example, a person might learn that he or she has a number of alleles at several different loci that, taken together, makes them susceptible to developing heart disease later in life. Using this information, persons at risk might want to follow diet and exercise regimes that reduce their risk of developing this condition.

The Human Genome Project has caused considerable debate within the nonscientific community. Some critics argue that knowledge of human DNA sequences opens the door to further invasion of our privacy. It is feasible, for example, that insurers or employers could demand information about the genetic makeup of potential employees or insurees to determine their predisposition toward particular illnesses. A num-

ber of states have already passed legislation to bar insurance companies from using this type of information. Another battle has raged over whether or not human DNA sequences should be patented. Thousands of patent applications have been filed and many have been granted. A few companies hold patents on hundreds of different genes, though it is not yet evident the extent to which these patents will restrict basic research on the structure and function of the encoded proteins. The most heated current controversy centers on how much an applicant must know about the encoded protein of a gene before a patent on that gene should be granted. Individuals, companies, and universities are filing applications for patents on genes whose function is not known, although most applicants can argue a likely function based on sequence similarity (homology) to other proteins that are better understood. Many scientists believe that it isn't the discovery of a gene that should be subject to patent, but a specific invention in which that gene is put to use. In other words, a patent should be awarded to an applicant who can specify a clear and valuable use for the encoded protein. At the same time, other applicants might receive patents for the same gene if they could specify a different and novel use.

EXPERIMENTAL PATHWAYS

The Chemical Nature of the Gene

Three years after Gregor Mendel presented the results of his work on inheritance in pea plants, Friedrich Miescher graduated from a Swiss medical school and traveled to Tubingen, Germany, to spend a year studying under Ernst Hoppe-Seyler, one of the foremost chemists (and possibly the first biochemist) of the period. Miescher was interested in the chemical contents of the cell nucleus. To isolate material from cell nuclei with a minimum of contamination from cytoplasmic components, Miescher needed cells that had large nuclei and were easy to obtain in quantity. He chose white blood cells, which he obtained from the pus in surgical bandages that were discarded by a local clinic. Miescher treated the cells with dilute hydrochloric acid to which he added an extract from pig's stomach that removed protein (the stomach extract contained the proteolytic enzyme pepsin). The residue from this treatment was composed primarily of isolated cell nuclei that settled to the bottom of the vessel. Miescher then extracted the nuclei with dilute alkali. The alkali-soluble material was further purified by precipitation with dilute acid and reextraction with dilute alkali. Miescher found that the alkaline extract contained a substance that had properties unlike any previously discovered: the molecule was very large, acidic, and rich in phosphorus. He called the material "nuclein." His year up, Miescher returned home to Switzerland while Hoppe-Seyler, who was cautious about the findings, repeated the work. With the results confirmed, Miescher published the findings in 1871.[1]

Back in Switzerland, Miescher continued his studies of the chemistry of the cell nucleus. Living near the Rhine River, Miescher had ready access to salmon that had swum upstream and were ripe with eggs or sperm. Sperm were ideal cells to study nuclei. Like white blood cells, they could be obtained in large quantity, and 90 percent of their volume was occupied by nuclei. Miescher's nuclein preparations from sperm cells contained a higher percentage of phosphorus (almost 10 percent by weight) than those from white blood cells, indicating they had less contaminating protein. In fact, they were the first preparations of relatively pure DNA. The term "nucleic acid" was coined in 1889 by Richard Altmann, one of Miescher's students, who worked out the methods of purifying protein-free DNA from various animal tissues and yeast.[2]

During the last two decades of the nineteenth century, numerous biologists focused on the chromosomes, describing their behavior during and between cell division (page 399). One way to observe chromosomes was to stain these cellular structures with dyes. A botanist named E. Zacharias discovered that the same stains that made the chromosomes visible also stained a preparation of nuclein that had been extracted using Miescher's procedure of digestion with pepsin in an HCl medium. Furthermore, when the pepsin-HCl-extracted cells were subsequently extracted with dilute alkali, a procedure that was known to remove nuclein, the cell residue (which included the chromosome remnants) no longer contained stainable material. These and other results

pointed strongly to nuclein as a component of the chromosomes. In one remarkably farsighted proposal, Otto Hertwig, who had been studying the behavior of the chromosomes during fertilization, stated in 1884: "I believe that I have at least made it highly probable that nuclein is the substance that is responsible not only for fertilization but also for the transmission of hereditary characteristics."[3] Ironically, as more was learned about the properties of nuclein, the less it was considered as a candidate for the genetic material.

During the fifty years that followed Miescher's discovery of DNA, the chemistry of the molecule and the nature of its components were described. Some of the most important contributions in this pursuit were made by Phoebus Aaron Levene, who immigrated from Russia to the United States in 1891 and eventually settled in a position at the Rockefeller Institute in New York. It was Levene who finally solved one of the most resistant problems of DNA chemistry when he determined in 1929 that the sugar of the nucleotides was 2-deoxyribose.[4] To isolate the sugar, Levene and E. S. London placed the DNA into the stomach of a dog through a surgical opening and then collected the sample from the animal's intestine. As it passed through the stomach and intestine, various enzymes of the animal's digestive tract acted on the DNA, carving the nucleotides into their component parts, which could then be isolated and analyzed. Levene summarized the state of knowledge on nucleic acids in a monograph published in 1931.[5]

While Levene is credited for his work in determining the structure of the building blocks of DNA, he is also credited as having been the major stumbling block in the search for the genetic material. Through this period, it became increasingly evident that proteins were very complex and exhibited great specificity in catalyzing a remarkable variety of chemical reactions. DNA, on the other hand, was thought to be composed of a monotonous repeat of its four nucleotide building blocks. The major proponent of this view of DNA, which was called the tetranucleotide theory, was Phoebus Levene. Because chromosomes consisted of only two components—DNA and protein—there seemed little doubt at that time that protein was the genetic material.

Meanwhile, as the structure of DNA was being worked out, a seemingly independent line of research was being carried out in the field of bacteriology. During the early 1920s, it was found that a number of species of pathogenic bacteria could be grown in the laboratory in two different forms. Virulent bacteria, that is, bacterial cells capable of causing disease, formed colonies that were smooth, dome shaped, and regular. In contrast, nonvirulent bacterial cells grew into colonies that were rough, flat, and irregular (Figure 1).[6] The British microbiologist J. A. Arkwright introduced the terms smooth (S) and rough (R) to describe these two types. Under the microscope, cells that formed S colonies were seen to be surrounded by a gelatinous capsule, whereas the capsule was absent from cells in the R colonies. The bacterial capsule helps protect a bacterium from its host's defenses, which explains why the R cells, which lacked these structures, were unable to cause infections in laboratory animals.

Because of its widespread impact on human health, the bacterium responsible for causing pneumonia (*Streptococcus pneumoniae*, or simply pneumococcus) has long been a focus of attention among microbiologists. In 1923, Frederick Grif-

Figure 1 The large glistening colonies on the right are virulent S-type pneumococci that resulted from the transformation of nonvirulent R-type pneumococci (smaller colonies on the left) by DNA from heat-killed S pneumococci. (FROM O. T. AVERY, C. M. MACLEOD, AND M. MCCARTY, J. EXP. MED. 79:153, 1944; BY COPYRIGHT PERMISSION OF ROCKEFELLER UNIVERSITY PRESS.)

fith, a medical officer at the British Ministry of Health, demonstrated that pneumococcus also grew as either S or R colonies and, furthermore, that the two forms were interconvertible; that is, on occasion an R bacterium could revert to an S form, or vice versa.[7] Griffith found, for example, if he injected exceptionally large numbers of R bacteria into a mouse, the animal frequently developed pneumonia and produced bacteria that formed colonies of the S form.

It had been shown earlier that pneumococcus occurred as several distinct types (types I, II, and III) that could be distinguished from one another immunologically. In other words, antibodies could be obtained from infected animals that would react with only one of the three types. Moreover, a bacterium of one type never gave rise to cells of another type. Each of the three types of pneumococcus could occur as either the R or S form.

In 1928, Griffith made a surprising discovery while injecting various bacterial preparations into mice.[8] Injections of large numbers of heat-killed S bacteria or small numbers of living R bacteria, by themselves, were harmless to the mouse. However, if he injected both of these preparations together into the same mouse, it contracted pneumonia and died. Virulent bacteria could be isolated from the mouse and cultured. To extend the findings, he injected combinations of bacteria of different types (Figure 2). Initially, eight mice were injected with heat-killed, type I, S bacteria together with a small inoculum of a live, type II, R strain. Two of the eight animals contracted pneumonia, and Griffith was able to isolate and culture virulent, type I, S bacterial cells from the infected mice. Because there was no possibility that the heat-killed bacteria had been brought back to life, Griffith concluded that the dead type I cells had provided something to the live, nonencapsulated, type II cells that *transformed* them into an encapsulated type I form. The transformed bacteria continued to produce type I cells when grown in culture; thus the change was stable and permanent. Griffith went on to show that type I R strains could be permanently transformed into either type II or type

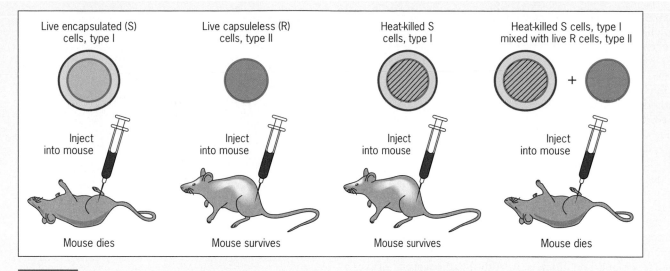

Figure 2 Outline of the experiment by Griffith of the discovery of bacterial transformation.

III S strains, and vice versa. In all cases, transformation appeared to be type specific, predictable, and inheritable.

Griffith's finding of transformation was rapidly confirmed by several laboratories around the world, including that of Oswald Avery, an immunologist at the Rockefeller Institute, the same institution where Levene was working. Avery had initially been skeptical of the idea that a substance released by a dead cell could alter the appearance of a living cell, but he was convinced when Martin Dawson, a young associate in his lab, confirmed the results.[9] Dawson went on to show that transformation need not occur within a living animal host. A crude extract of the dead S bacteria, when mixed in bacterial culture with a small number of the nonvirulent (R) cells in the presence of anti-R serum, was capable of converting the R cells into the virulent S form. The transformed cells were always of the type (I, II, or III) characteristic of the dead S cells.[10]

The next major step was taken by J. Lionel Alloway, another member of Avery's lab, who was able to solubilize the transforming agent. This was accomplished by rapidly freezing and thawing the killed donor cells, then heating the disrupted cells, centrifuging the suspension, and forcing the supernatant through a porcelain filter whose pores blocked the passage of bacteria. The soluble, filtered extract possessed the same transforming capacity as the original heat-killed cells.[11]

For the next decade, Avery and his colleagues focused their attention on purifying the substance responsible for transformation and determining its identity. As remarkable as it may seem today, no other laboratory in the world was pursuing the identity of the "transforming principle," as Avery called it. Progress on the problem was slow.[12] Eventually, Avery and his coworkers, Colin MacLeod and Maclyn McCarty, succeeded in isolating a substance from the soluble extract that was active in causing transformation when present at only 1 part per 600 million. All the evidence suggested that the active substance was DNA: (1) it exhibited a host of chemical properties that were also characteristic of DNA, (2) no other type of material could be detected in the preparation, and (3) tests of various enzymes indicated that only those enzymes that digested DNA inactivated the transforming principle.

The paper published in 1944 was written with scrupulous caution and made no dramatic statements that genes were made of DNA rather than protein.[13] The paper drew remarkably little attention. Maclyn McCarty, one of the three authors, describes an incident in 1949 when he was asked to speak at Johns Hopkins University along with Leslie Gay, who had been testing the effects of the new drug Dramamine for the treatment of sea sickness. The large hall was packed with people, and "after a short period of questions and discussion following [Gay's] paper, the president of the Society got up to introduce me as the second speaker. Very little that he said could be heard because of the noise created by people streaming out of the hall. When the exodus was complete, after I had given the first few minutes of my talk, I counted approximately thirty-five hardy souls who remained in the audience because they wanted to hear about pneumococcal transformation or because they felt they had to remain out of courtesy." But Avery's awareness of the potential of his discovery was revealed in a letter he wrote in 1943 to his brother Roy, also a bacteriologist:

If we are right, & of course that's not yet proven, then it means that nucleic acids are not merely structurally important but functionally active substances in determining the biochemical activities and specific characteristics of cells—& that by means of a known chemical substance it is possible to induce predictable and hereditary changes in cells. This is something that has long been the dream of geneticists. . . . Sounds like a virus—may be a gene. But with mechanisms I am not now concerned—one step at a time. . . . Of course the problem bristles with implications. . . . It touches genetics, enzyme chemistry, cell metabolism & carbohydrate synthesis—etc. But today it takes a lot of well documented evidence to convince anyone that the sodium salt of deoxyribose nucleic acid, protein free, could possibly be endowed with such biologically active & specific properties & that evidence we are now trying to get. It's lots of fun to blow bubbles,—but it's wiser to prick them yourself before someone else tries to.

Many articles and passages in books have dealt with the reasons why Avery's findings were not met with greater ac-

claim. Part of the reason may be due to the subdued manner in which the paper was written and the fact that Avery was a bacteriologist, not a geneticist. Some biologists were persuaded that Avery's preparation must have been contaminated with minuscule amounts of protein and that the contaminant, not the DNA, was the active transforming agent. Others questioned whether studies on bacteria published in a medical journal had any relevance to the field of genetics, and they viewed the phenomenon of transformation as a bacterial peculiarity.

During the years following the publication of Avery's paper, the climate in genetics changed in an important way. The existence of the bacterial chromosome was recognized, and a number of prominent geneticists turned their attention to these prokaryotes. These scientists believed that knowledge gained from the study of the simplest cellular organisms would shed light on the mechanisms that operate in the most complex plants and animals. In addition, the work of Erwin Chargaff and his colleagues on the base composition of DNA shattered the notion that DNA was a molecule consisting of a simple repetitive series of nucleotides (discussed on page 405).[14] This finding awakened researchers to the possibility that DNA might have the properties necessary to fulfill a role in information storage.

Seven years after the publication of Avery's paper on bacterial transformation, Alfred Hershey and Martha Chase of the Cold Spring Harbor Laboratories in New York turned their attention to an even simpler system—bacteriophages, or viruses that infect bacterial cells. By 1950, researchers recognized that viruses also had a genetic program. Viruses injected their genetic material into the host bacterium, where it directed the formation of new virus particles inside the infected cell. Within a matter of minutes, the infected cell broke open, releasing new bacteriophage particles, which then infected neighboring host cells.

It was clear that the genetic material directing the formation of viral progeny had to be either DNA or protein because these were the only two molecules the virus contained. Electron microscopic observations showed that, during the infection, the bulk of the bacteriophage remains outside the cell, attached to the cell surface by tail fibers (Figure 3). Hershey and Chase reasoned that the virus's genetic material must possess two properties. First, if the material were to direct the development of new bacteriophages during infection, it must pass into the infected cell. Second, if the material carries inherited information, it must be passed on to the next generation of bacteriophages. Hershey and Chase prepared two batches of bacteriophages to use for infection (Figure 4). One batch contained radioactively labeled DNA ([^{32}P]DNA); the other batch contained radioactively labeled protein ([^{35}S]protein). Because DNA lacks sulfur (S) atoms, and protein usually lacks phosphorus (P) atoms, these two radioisotopes provided distinguishing labels for the two types of macromolecules. Their experimental plan was to allow one or the other type of bacteriophage to infect a population of bacterial cells, wait a few minutes, and then strip the empty viruses from the surfaces of the cells. After trying several methods to separate the bacteria from the attached phage coats, they found this was best accomplished by subjecting the infected suspension to the spinning blades of a Waring blender. Once the virus particles had been detached from the cells, the bac-

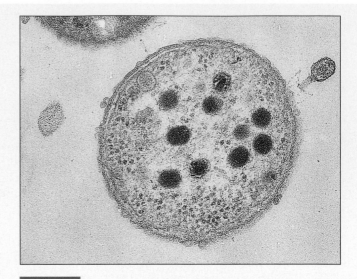

Figure 3 Electron micrograph of a bacterial cell infected by T4 bacteriophage. A phage is seen attached by its tail fibers to the outer surface of the bacterial cell, while new phage heads are being assembled in the host cell's cytoplasm. (COURTESY OF JONATHAN KING AND ERIKA HARTWIG.)

teria could be centrifuged to the bottom of the tube, leaving the empty viral coats in suspension.

By following this procedure, Hershey and Chase determined the amount of radioactivity that entered the cells versus that which remained behind in the empty coats. They found that when cells were infected with protein-labeled bacteriophages, the bulk of the radioactivity remained in the empty coats. In contrast, when cells were infected with DNA-labeled bacteriophages, the bulk of the radioactivity passed inside the host cell. When they monitored the radioactivity passed onto the next generation, they found that less than 1 percent of the labeled protein could be detected in the progeny, whereas approximately 30 percent of the labeled DNA could be accounted for in the next generation.

The publication of the Hershey-Chase experiments in 1952,[15] together with the abandonment of the tetranucleotide theory, removed any remaining obstacles to the acceptance of DNA as the genetic material. Suddenly, tremendous new interest was generated in a molecule that had largely been ignored. The stage was set for the discovery of the double helix.

References

1. MIESCHER, J. F. 1871. *Hoppe-Seyler's Med. Chem. Untersuchungen.* 4:441.
2. ALTMANN, R. 1889. *Anat. u. Physiol., Physiol. Abt.* 524.
3. Taken from MIRSKY, A. E. 1968. The discovery of DNA. *Sci. Amer.* 218:78–88. (June)
4. LEVENE, P. A. & LONDON, E. S. 1929. The structure of thymonucleic acid. *J. Biol. Chem.* 83:793–802.
5. LEVENE, P. A. & BASS, L. W. 1931. *Nucleic Acids.* The Chemical Catalog Co., NY.
6. ARKWRIGHT, J. A. 1921. Variation in bacteria in relation to agglutination both by salts and by specific serum. *J. Path. Bact.* 24:36–60.
7. GRIFFITH, F. 1923. The influence of immune serum on the

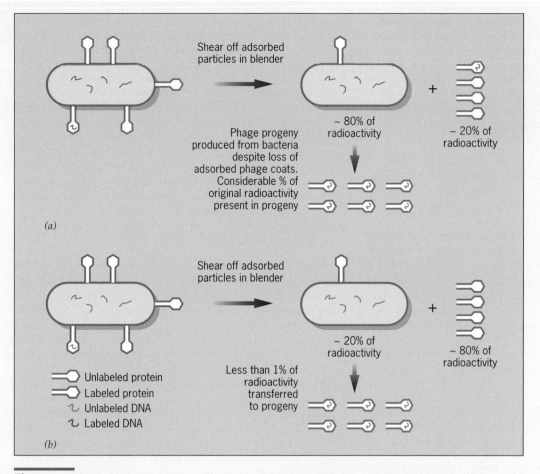

Figure 4 Outline of the Hershey-Chase experiment showing that bacterial cells infected with phage containing ^{32}P-labeled DNA (red DNA molecules) became radioactively labeled and produced labeled phage progeny. In contrast, bacterial cells infected with phage containing ^{35}S-labeled protein (red phage coats) did not become radioactively labeled and produced only nonlabeled progeny.

biological properties of pneumococci. *Rep. Public Health Med. Subj.* 18:1–13.

8. GRIFFITH, F. 1928. The significance of pneumococcal types. *J. Hygiene* 27:113–159.

9. DAWSON, M. H. 1930. The transformation of pneumoccal types. *J. Exp. Med.* 51:123–147.

10. DAWSON, M. H. & SIA, R. H. P. 1931. In vitro transformation of pneumococcal types. *J. Exp. Med.* 54:701–710.

11. ALLOWAY, J. L. 1932. The transformation in vitro of R pneumococci into S forms of different specific types by use of filtered pneumococcus extracts. *J. Exp. Med.* 55:91–99.

12. MCCARTY, M. 1985. *The Transforming Principle: Discovering That Genes Are Made of DNA.* Norton.

13. AVERY, O. T., MACLEOD, C. M., & MCCARTY, M. 1944. Studies on the chemical nature of the substance inducing transformation of pneumococcal types: Induction of transformation by a desoxyribonucleic acid fraction isolated from pneumococcus type III. *J. Exp. Med.* 79:137–158.

14. CHARGAFF, E. 1950. Chemical specificity of nucleic acids and mechanism of their enzymic degradation. *Experentia* 6:201–209.

15. HERSHEY, A. D. & CHASE, M. 1952. Independent functions of viral protein and nucleic acid in growth of bacteriophage. *J. Gen. Physiol.* 36:39–56.

SYNOPSIS

Chromosomes are the carriers of genetic information. A number of early observations led biologists to consider the genetic role of chromosomes. These included observations of the precision with which chromosomes are divided between the daughter cells during cell division; the realization that the chromosomes of a species remained constant in shape and number from one division to the next, the finding that embryonic development required a particular complement of chro-

mosomes, and the observation that the chromosome number is divided in half prior to the formation of the gametes and is doubled following the union of a sperm and egg at fertilization. Mendel's findings provided biologists with a new set of criteria for identifying the carriers of the genes. Sutton's studies of gamete formation in the grasshopper revealed the existence of homologous chromosomes, the association of homologues during the cell divisions that preceded gamete formation, and the separation of the homologues during the first of these meiotic divisions. (p. 399)

If genes are packaged together on chromosomes that are passed from parents to offspring, then genes on the same chromosome should be linked to one another to form a linkage group. The existence of linkage groups was confirmed in various systems, particularly in fruit flies where dozens of mutations were found to assort into four linkage groups that correspond in size and number to the chromosomes present in the cells of these insects. At the same time, it was discovered that linkage was incomplete; that is, the alleles originally present on a given chromosome did not necessarily remain together during the formation of the gametes, but could be reshuffled between maternal and paternal homologues. This phenomenon, which Morgan called crossing over, was found to be the result of breakage and reunion of segments of homologous chromosomes that occurred while the homologues were physically associated during the first meiotic division. Analyses of offspring from matings between adults carrying a variety of mutations on the same chromosome indicated that the frequency of recombination between two genes provided a measure of the relative distance that separates those two genes. Thus, recombination frequencies could be used to prepare detailed maps of the serial order of genes along each of the chromosomes of a species. Genetic maps of the fruit fly based on recombination frequencies were verified independently by examination of the locations of various bands in the giant polytene chromosomes found in certain larval tissues of these insects. (p. 400)

Experiments discussed in the Experimental Pathways provided conclusive evidence that DNA was the genetic material. DNA is a helical molecule consisting of two chains of nucleotides running in opposite directions with their backbones on the outside and the nitrogenous bases facing inward. Adenine-containing nucleotides on one strand always pair with thymine-containing nucleotides on the other strand, likewise for guanine- and cytosine-containing nucleotides. As a result, the two strands of a DNA molecule are complementary to one another. Genetic information is encoded in the specific linear sequence of nucleotides that makes up the strands. The Watson-Crick model of DNA structure suggested a mechanism of replication that included strand separation and the use of each strand as a template that directed the order of nucleotide assembly during construction of the complementary strand. The mechanism by which DNA governed the assembly of a specific protein remained a total mystery. The DNA molecule depicted in Figure 10.10 is in a relaxed state having 10 base pairs per turn of the helix. DNA within a cell tends to be underwound (contains a greater number of base pairs per turn) and is said to be negatively supercoiled, a condition that facilitates the separation of strands during replication and transcription. The supercoiled state of DNA is altered by topoisomerases, enzymes that are able to cut, rearrange, and reseal DNA strands. (p. 404)

All of the genetic information present in a single haploid set of chromosomes of an organism constitutes that organism's genome. The variety of DNA sequences that make up the genome and the numbers of copies of these various sequences describe the complexity of the genome. Understanding the complexity of a genome has grown out of early studies showing that the two strands that make up a DNA molecule can be separated by heat; when the temperature of the solution is lowered, complementary single strands are capable of reassociating to form stable, double-stranded DNA molecules. Analysis of the kinetics of this reassociation process provides a measure of the concentration of complementary sequences, which in turn provides a measure of the variety of sequences that are present within a given quantity of DNA. The greater the number of copies of a particular sequence in the genome, the greater is its concentration and the faster it reanneals. (p. 411)

When DNA fragments from eukaryotic cells are allowed to reanneal, the curve typically shows three steps, which correspond to the reannealing of three different classes of DNA sequences. The highly repeated fraction consists of short DNA sequences that are repeated in great number; these include satellite DNAs situated at the centromeres of the chromosomes, minisatellite DNAs, and microsatellite DNAs. The latter two groups tend to be highly variable, causing certain inherited diseases and forming the basis of DNA fingerprint techniques. The moderately repeated fraction includes DNA sequences that encode ribosomal and transfer RNAs and histone proteins, as well as various sequences with noncoding functions. The nonrepeated fraction contains protein-coding genes, which are present in one copy per haploid set of chromosomes. (p. 413)

The sequence organization of the genome is capable of change, either slowly over the course of evolution or rapidly as the result of transposition. The genes encoding eukaryotic proteins are often members of multigene families whose members show evidence that they have evolved from a common ancestral gene. The first step in this process is thought to be the duplication of a gene, which probably occurs primarily by unequal crossing over. Once duplication has occurred, nucleotide substitutions would be expected to modify various members in different ways, producing a family of repeated sequences of similar but nonidentical structure. The globin genes, for example, consist of clusters of genes located on two different chromosomes. Each cluster contains a number of related genes that code for globin polypeptides produced at different stages in the life of an animal. The clusters also contain pseudogenes, which are homologous to the globin genes, but are nonfunctional. The duplication of genes over the course of generations can be followed in the laboratory by exposing cultured cells to drugs that select for cells that have increased copies of a particular protein. (p. 420)

Certain DNA sequences are capable of moving rapidly from place to place in the genome by transposition. These transposable elements are called transposons, and they are capable of integrating themselves randomly throughout the genome. The best-studied transposons occur in bacteria. They are characterized by inverted repeats at their termini, an internal segment that codes for a transposase required for their integration, and the formation of short repeated sequences in the target DNA that flank the element at the site of

integration. Eukaryotic transposable elements are capable of moving by several mechanisms. Some are replicated, and the copy is inserted into a target site, leaving the donor site unchanged. In other cases, the element is excised from the donor site and inserted at a target site. In most cases, the element is transcribed into RNA, which is copied by a reverse transcriptase encoded by the element, and the DNA copy is integrated into the target site. The moderately repeated fraction of human DNA contains two large families of transposable elements, the *Alu* and L1 families. (p. 422)

ANALYTIC QUESTIONS

1. How would Mendel's results have differed if two of the seven traits he studied were encoded by genes that resided close to one another on one of the chromosomes of a pea plant?

2. Sutton was able to provide visual evidence for Mendel's law of segregation. Why would it have been impossible for him visually to confirm or refute Mendel's law of independent assortment?

3. Genes X, Y, and Z are all located on one chromosome. Draw a simple map showing the gene order and relative distances among the genes X, Y, and Z using the data below:

Crossover Frequency	Between These Genes
36%	X-Z
10%	Y-Z
26%	X-Y

4. Alleles on opposite ends of a chromosome are so likely to be separated by crossing over between them that they segregate independently. How would one be able to determine, using genetic crosses, that these two genes belong to the same linkage group? How might this be established using nucleic acid hybridization?

5. Where in the curve of Figure 10.20 would you expect to see the reannealing of the DNA that codes for ribosomal RNA? Suppose you had a pure preparation of DNA fragments whose sequences code for ribosomal RNA. Draw the reannealing of this DNA superimposed over the curve for total DNA shown in Figure 10.20.

6. Would you agree with the following statement? In an average population of eukaryotic DNA fragments of 200 nucleotides average length, a restriction enzyme would be expected to attack most of the fragments. If not, why not?

7. You have been given a sample of DNA extracted from an unknown source. After fragmenting and denaturing it, you find that it has an unusual reannealing curve in which only 10 percent of the DNA is made up of nonrepeated (single-copy) DNA sequences. (1) Draw this reannealing curve. (2) Suppose you wanted a preparation of pure, nonrepeated DNA sequences that you could use in a subsequent experiment. How might you go about obtaining such a preparation? (3) Draw the reannealing curve of this pure preparation of nonrepeated DNA superimposed over that of the curve from part 1.

8. Suppose you had two solutions of DNA, one single stranded and the other double stranded, with equivalent absorbance at 260 nm. How would the concentrations of DNA compare in these two solutions?

9. What type of labeling pattern would you expect after in situ hybridization between mitotic chromosomes and labeled myoglobin mRNA? Between the same chromosomes and labeled histone DNA?

10. According to Chargaff's determination of base composition, which of the following would characterize any sample of DNA? (1) [A] + [T] = [G] + [C]; (2) [A]/[T] = 1; (3) [G] = [C]; (4) [A] + [G] = [T] + [C]. If the C content of a preparation of double-stranded DNA is 15 percent, what is the A content?

11. If 30 percent of the bases on a single strand of DNA are T, then 30 percent of the bases on that strand are A. True or false? Why?

12. Would you agree with the statement that the T_m represents the temperature at which 50 percent of the DNA molecules in a solution are single stranded?

13. Suppose you were to find a primate that had a β-globin gene with only one intervening sequence. Do you think this animal may have evolved from a primitive ancestor that split away from other animals prior to the appearance of the second intron?

14. In 1996, a report was published in the journal *Lancet* on the level of trinucleotide expansion in the DNA of blood donors. These researchers found that the level of trinucleotide expansion decreased significantly with age. Can you provide any explanation for these findings?

ADDITIONAL READING

General References

ADAMS, R. L. P., ET AL. 1986. *The Biochemistry of the Nucleic Acids.* Chapman & Hall.

JUDSON, H. F. 1979. *The Eighth Day of Creation.* Simon & Schuster.

MOORE, J. A. 1972. *Heredity and Development*, 2d ed. Oxford.

WATSON, J. D., ET AL. 1987. *Molecular Biology of the Gene.* 4th ed. Benjamin/Cummings.

DNA Structure and Topology

FELSENFELD, G. 1985. DNA. *Sci. Am.* 253:58–67. (Oct.)

FORTUNE, J. M. & OSHEROFF, N. 2000. Topoisomerase II as a target for anticancer drugs. *Prog. Nucleic Acid Res. & Mol. Biol.* 64:221–253.

KECK, J. L. & BERGER, J. M. 1999. Enzymes that push DNA around. *Nature Struct. Biol.* 6:900–902.

MCCARTY, M. 1985. *The Transforming Principle.* Norton.

MIRSKY, A. E. 1968. The discovery of DNA. *Sci. Am.* 218:78–88. (June).

MONDRAGON, A. & DiGATE, R. 1999. The structure of *Escherichia coli* DNA topoisomerase II. *Struct.* 7:1373–1383.

OLBY, R. 1974. *The Path to the Double Helix.* University of Washington Press.

OSHEROFF, N., ET AL. 1998. Reviews on topoisomerases. *Biochim. Biophys. Acta,* vol. 1400.

STASIAK, A. 2000. DNA topology: Feeling the pulse of a topoisomerase. *Curr. Biol.* 10:R526–R528.

WANG, J. C. 1996. DNA topoisomerases. *Annu. Rev. Biochem.* 65:635–692.

WATSON, J. D. 2000. *A Passion for DNA: Genes, Genomes and Society.* Oxford.

WATSON, J. D. & CRICK, F. C. 1953. Molecular structure of nucleic acids. A structure for deoxyribose nucleic acid. *Nature* 171:737–738.

The Complexity of the Genome

BEVAN, M. & MURPHY, G. 1999. The small, the large and the wild: The value of comparison in plant genomics. *Trends Gen.* 15:211–214.

BRENT, R. 2000. Genomic biology. *Cell* 100:169–183.

BRITTEN, R. J. & KOHNE, D. E. 1970. Repeated segments of DNA. *Sci. Am.* 222:24–31. (April)

DJIAN, P. 1998. Evolution of simple repeats in DNA and their relation to human disease. *Cell* 94:155–160.

KIDWELL, M. G. & LISCH, D. R. 1998. Transposons abound. *Nature* 393:22–23.

LANDER, E. S. 1999. Genes and genomes. *Harvey Lects.* 93:35–48.

LANDER, E. S. & WEINBERG, R. A. 2000. Genomics—Journey to the center of biology. *Science* 287:1777–1782.

MOXON, E. R. & WILLS, C. 1999. DNA microsatellites: Agents of evolution? *Sci. Am.* 280:72–77. (Jan.)

VOYTAS, D. E. & NAYLOR, G. J. P. 1998. Rapid flux in plant genomes. *Nat. Gen.* 20:6–7.

Transposition

EICKBUSH, T. E. 1999. Exon shuffling in retrospect. *Science* 283:1465–1466.

FEDOROFF, N. 1992. Barbara McClintock: The geneticist, the genius, the woman. An obituary. *Cell* 71:181–182.

FINNEGAN, D. J. 1989. Eukaryotic transposable elements and genome evolution. *Trends Genet.* 5:103–107.

FLAVELL, A. J. 1999. Long terminal repeat retrotransposons jump between species. *Proc. Nat'l. Acad. Sci. U.S.A.* 96:12211–12212.

FURANO, A. V. 2000. The biological properties and evolutionary dynamics of mammalian LINE-1 retrotransposons. *Prog. Nucleic Acid Res. & Mol. Biol.* 64:255–94.

KAZAZIAN, H. H. 2000. L1 retrotransposons shape the mammalian genome. *Science* 289:1152–1153.

KAZAZIAN, H. H. & MORAN, J. V. 1998. The impact of L1 retrotransposons on the human genome. *Nature Gen.* 19:19–24.

KUMAR, A. & BENNETZEN, J. L. 1999. Plant retrotransposons. *Annu. Rev. Gen.* 33:479–532.

PENNISI, E. 1998. How the genome readies itself for evolution. *Science* 281:1131–1134.

SHEDLOCK, A. M. & OKADA, N. 2000. SINE insertions: Powerful tools for molecular systematics. *BioEss.* 22:148–160.

Human Perspective: Trinucleotide Expansion

FERRIGNO, P. & SILVER, P. A. 2000. Polyglutamine expansions: Proteolysis, chaperones and the dangers of promiscuity. *Neuron* 26:9–12.

ORR, H. T. & ZOGHBI, H. Y. 2000. Reversing neurodegeneration: A promise unfolds. *Cell* 101:1–4.

PERUTZ, M. F. 1999. Glutamine repeats and neurodegenerative diseases: Molecular aspects. *Trends Biochem. Sci.* 24:58–63.

REDDY, P. H., ET AL. 1999. Recent advances in understanding the pathogenesis of Huntington disease. *Trends Neurosci.* 22:248–255.

RICHARDS, R. I. & SUTHERLAND, G. R. 1997. Dynamic mutation: Possible mechanisms and significance in human disease. *Trends Biochem. Sci.* 22:432–436.

SISODIA, S. S. 1998. Nuclear inclusions in glutamine repeat disorders: Are they pernicious, coincidental, or beneficial? *Cell* 95:1–4.

Human Perspective: Sequencing the Human Genome

BOTSTEIN, D., ET AL. 1997. Yeast as a model organism. Science 277:1259–1260.

BROWN, K., ET AL. 2000. Articles on the Human Genome Project. *Sci. Am.* 283:50–69. (July)

DENNIS, C., ET AL. 2001. The human genome. *Nature* 409:813–958.

DHAND, R., ET AL. 2000. Functional genomics. *Nature* 405:819–865.

DOOLITTLE, R. F. 1998. Microbial genomes opened up. *Nature* 392:339–342.

DUNHAM, I. 2000. Genomics—The new rock and roll. *Trends Gen.* 16:456–461.

HARRIS, R. F. 2000. Patenting genes: Is it necessary and is it evil? *Curr. Biol.* 10:R174–R175.

HODGKIN, J. 2000. A view of mount *Drosophila. Nature* 404:442–443.

PENNISI, E. 2000. Finally the book of life and instructions for navigating it. *Science* 288:2304–2307.

PENNISI, E., ET AL., 2001. The human genome. *Science* 291:1177–1351.

REEVES, R. H. 2000. Recounting a genetic story. (sequencing human chromosome #21) *Nature* 405:283–284.

11

Expression of Genetic Information: From Transcription to Translation

I n many ways, progress in biology over the past century is reflected in our changing concept of the gene. As the result of Mendel's work, biologists learned that genes are discrete elements that govern the appearance of specific traits. Given the opportunity, Mendel might have argued for the concept that one gene determines one trait. Boveri, Weismann, Sutton, and their contemporaries discovered that genes have a physical embodiment as parts of the chromosome. Morgan, Sturtevant, and their coworkers demonstrated that genes have specific addresses—they reside at particular locations on particular chromosomes, and these addresses remain constant from one individual of a species to the next. Griffith, Avery, Hershey, and Chase demonstrated that genes were composed of DNA, and Watson and Crick solved the puzzle of DNA structure, which explained how this remarkable macromolecule could encode heritable information.

Although the formulations of these concepts were important steps along the path to genetic understanding, none of them addressed the mechanism by which the information stored in a gene is put to work in governing cellular activities. This is the major topic to be discussed in the present chapter. Let us begin with additional insights into the nature of a gene, which bring us closer to its role in the expression of inherited traits.

A model of the large subunit of a prokaryotic ribosome at 2.4Å resolution as determined by X-ray crystallography. This view looks into the subunit's active site cleft, which consists entirely of RNA. RNA is shown in gray, protein in gold, and the active site is revealed by a bound inhibitor (green). (COURTESY OF THOMAS A. STEITZ, YALE UNIVERSITY.)

439

11.1 THE RELATIONSHIP BETWEEN GENES AND PROTEINS

The first meaningful insight into gene function was gained by Archibald Garrod, a Scottish physician who reported in 1908 that the symptoms exhibited by persons with certain rare inherited diseases were caused by the absence of specific enzymes. One of the diseases investigated by Garrod was *alcaptonuria*, a condition readily diagnosed because the urine becomes dark on exposure to air. Garrod found that persons with alcaptonuria lacked an enzyme in their blood that oxidized homogentisic acid, a compound formed during the breakdown of the amino acids phenylalanine and tyrosine. As homogentisic acid accumulates, it is excreted in the urine and darkens in color when oxidized by air. Garrod had discovered the relationship between a genetic defect, a specific enzyme, and a specific metabolic condition. He called such diseases "inborn er-

rors of metabolism." As seems to have happened with other early observations of basic importance in genetics, Garrod's findings went unappreciated for decades.

The idea that genes direct the production of enzymes was resurrected in the 1940s by George Beadle and Edward Tatum of the California Institute of Technology. They studied *Neurospora*, a tropical bread mold that normally grows in a very simple medium containing a single organic carbon source (e.g., a sugar), inorganic salts, and biotin (a B vitamin). Because it needs so little to live on, *Neurospora* was presumed to synthesize all of its required metabolites. Beadle and Tatum reasoned that an organism with such broad synthetic capacity should be very sensitive to enzymatic deficiencies, which should be easily detected using the proper experimental protocol. A simplified outline of their protocol is given in Figure 11.1.

Beadle and Tatum's plan was to irradiate mold spores and screen them for mutations that caused cells to lack a

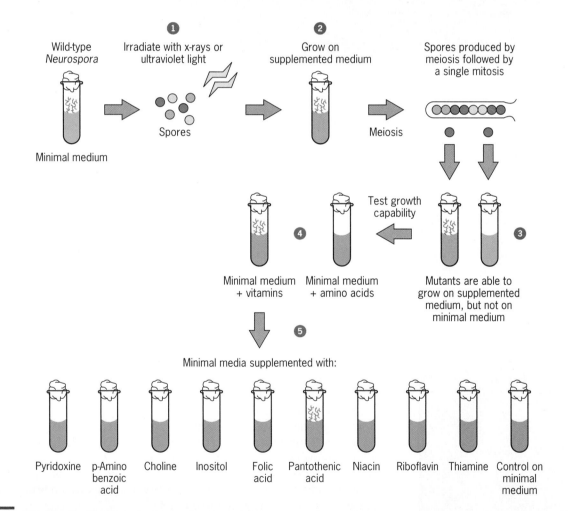

Figure 11.1 The Beadle-Tatum experiment for the isolation of genetic mutants in *Neurospora*. Spores were irradiated to induce mutations (step 1) and allowed to grow into colonies in tubes containing supplemented medium (step 2). Individual spores produced by the colonies were then tested for their ability to grow on the minimal medium (step 3). Those that failed to grow were mutants, and the task was to determine the mutant gene. In the example shown in step 4, a sample of mutant cells was found to grow in the minimal medium supplemented with vitamins, but they would not grow in medium supplemented with amino acids. This observation indicates a deficiency in an enzyme leading to the formation of a vitamin. In step 5, growth of these same cells in minimal medium supplemented with one or another of the vitamins indicates the deficiency resides in a gene involved in the formation of pantothenic acid (part of coenzyme A).

particular enzyme. To screen for such mutations, irradiated spores were tested for their ability to grow in a *minimal medium* that lacked the essential compounds known to be synthesized by this organism (Figure 11.1). If a spore is unable to grow in minimal medium but can grow in a medium supplemented with a particular coenzyme (e.g., pantothenic acid of coenzyme A), then the researchers can conclude that the cells have an enzymatic deficiency that prevents them from synthesizing this essential compound.

Beadle and Tatum began by irradiating over a thousand cells. Two of the cells proved unable to grow on the minimal medium: one needed pyridoxine (vitamin B$_6$), and the other needed thiamine (vitamin B$_1$). Eventually, the progeny of about 100,000 irradiated spores were tested, and dozens of mutants were isolated. Each mutant had a gene defect that resulted in an enzyme deficiency that prevented the cells from catalyzing a particular metabolic reaction. The results were clear-cut: a gene carries the information for the construction of a particular enzyme. This conclusion became known as the "one gene—one enzyme" hypothesis. Once it was learned that enzymes are often composed of more than one polypeptide chain, each of which is encoded by its own gene, the concept became modified to "one gene—one polypeptide." Although this relationship remains a close approximation of the basic function of a gene, it too has had to be modified owing to the discovery that a single gene often generates a variety of related polypeptides as the result of alternate splicing (discussed in Section 12.3).

What is the molecular nature of the defect in a protein caused by a genetic mutation? The answer to this question came in 1956, when Vernon Ingram of Cambridge University reported on the molecular consequence of the mutation that causes sickle cell anemia. Hemoglobin consists of four large polypeptides. Although they were too large to be sequenced by the technology of that period, Ingram devised a shortcut. He cleaved preparations of both the normal and sickle cell forms of hemoglobin into a number of specific pieces using the proteolytic enzyme trypsin. He then analyzed these peptide fragments by paper chromatography to determine if any of the products of trypsin digestion of the normal and sickled hemoglobin types were different. Of the 30 or so peptides in the mixture, one migrated differently in the two preparations (similar to that shown in Figure 2.29); this one difference was apparently responsible for all of the symptoms associated with the disease. Once the peptides had been separated, Ingram had only one small fragment to sequence rather than an entire protein. The difference in the proteins proved to be the substitution of a single amino acid, valine, in the sickle cell hemoglobin for a glutamic acid in the normal molecule. Ingram had demonstrated that a mutation in a single gene had caused a single substitution in the amino acid sequence of a single polypeptide.

An Overview of the Flow of Information Through the Cell

We have seen how the relationship between genetic information and amino acid sequence was established, but this knowledge by itself provided no clue as to the mechanism by which the specific polypeptide chain is generated. As we now know, there is an intermediate between a gene and its polypeptide; the intermediate is **messenger RNA (mRNA)** whose existence was first demonstrated in 1961 by François Jacob and Jacques Monod of the Pasteur Institute in Paris. A messenger RNA is assembled as a complementary copy of one of the two DNA strands that make up a gene. The synthesis of an RNA from a DNA template is called **transcription**. Because its nucleotide sequence is complementary to that of the gene from which it is transcribed, the mRNA retains the same information as the gene itself. An overview of the role of mRNA in the flow of information through a eukaryotic cell is illustrated in Figure 11.2.

Figure 11.2 An overview of the flow of information in a eukaryotic cell. The DNA of the chromosomes located within the nucleus contains the entire store of genetic information. Selected sites on the DNA are transcribed into RNAs (step 1), which are processed into messenger RNAs (step 2). The messenger RNAs are transported out of the nucleus (step 3) into the cytoplasm, where they are translated into polypeptides by ribosomes that move along the mRNA (step 4). Following translation, the polypeptide folds to assume its native conformation (step 5).

The use of messenger RNA allows the cell to separate information storage from information utilization. While the gene remains stored away in the nucleus as part of a huge, immobile DNA molecule, its information can be imparted to a much smaller, mobile RNA that passes into the cytoplasm. Once in the cytoplasm, the mRNA can serve as a template to direct the incorporation of amino acids in a particular order encoded by the nucleotide sequence of the DNA and mRNA. The use of a messenger RNA also allows a cell to greatly amplify its synthetic output. One DNA molecule can serve as the template in the formation of many mRNA molecules, each of which can be used in the formation of a large number of polypeptide chains.

Proteins are synthesized in the cytoplasm by a complex process called **translation**. Translation requires the participation of dozens of different components, including ribosomes. Ribosomes are nonspecific components of the translation machinery. These complex, cytoplasmic "machines" can be programmed, like a computer, to translate the information encoded by any mRNA. This is why bacterial cells can be used as "pharmaceutical factories" to churn out proteins encoded by human mRNAs. As discussed on page 74, a functioning ribosome consists of a large and a small subunit. The ribosome is assembled from these subunits at the time it begins to synthesize a particular polypeptide and then is disassembled when synthesis has been completed. Ribosomes consist of RNA and protein (as illustrated on page 447). The RNAs of a ribosome are called **ribosomal RNAs** (or **rRNAs**), and like mRNAs, each is transcribed from one of the DNA strands of a gene. Rather than functioning in an informational capacity, rRNAs are adapted to recognize and bind other molecules, to provide structural support, and to catalyze the chemical reaction in which amino acids are covalently linked to one another.

Transfer RNAs (or **tRNAs**) constitute a third major class of RNA that is required during protein synthesis. Transfer RNAs are required to translate the information in the mRNA nucleotide code into the amino acid "alphabet" of a polypeptide. Although tRNAs and rRNAs exhibit long half-lives within the cell (typically measured in days), mRNAs are much less stable, having half-lives typically measured in hours.

Both rRNAs and tRNAs owe their activity to their complex secondary and tertiary structures. Thus, unlike DNA, which has a similar overall structure regardless of the source, many RNAs fold into complex three-dimensional shapes, which are markedly different from one type of RNA to another. Thus, like proteins, RNAs carry out a diverse array of functions because of their different shapes. As with proteins, the folding of RNA molecules follows certain rules. Whereas protein folding is driven by the withdrawal of hydrophobic residues into the interior, RNA folding is driven by the formation of regions having complementary base pairs (Figure 11.3). As seen in Figure 11.3, base-paired regions typically form double-stranded (and double-helical) "stems," which are connected to single-stranded "loops." Unlike DNA, which consists exclusively of standard Watson-Crick (G-C, A-T) base pairs, RNAs often contain nonstandard

Figure 11.3 Two-dimensional structure of a bacterial ribosomal RNA showing the extensive base-pairing between different regions of the single strand. The expanded section shows the base sequence of a stem and loop, including a nonstandard base pair (G-U) and a modified nucleotide, methyl adenosine. One of the helices is shaded differently because it plays an important role in ribosome function as discussed on page 481. Examples of the three-dimensional structures of RNAs are shown in Figures 11.31 and 11.43.

base pairs (inset, Figure 11.3) and modified nitrogenous bases. These unorthodox regions of the molecule often serve as recognition sites by proteins and other RNAs.

The roles of mRNAs, rRNAs, and tRNAs are explored in detail in the following sections of this chapter. For background material on the structure of RNA, review page 72.

REVIEW ?

1. How were Beadle and Tatum able to conclude that a gene coded for a specific enzyme?
2. How was Ingram able to determine the amino acid substitution in the hemoglobin of persons with sickle cell anemia without sequencing the protein?

11.2 TRANSCRIPTION: THE BASIC PROCESS

Transcription is a process in which a DNA strand provides the information for the synthesis of an RNA strand. The enzymes responsible for transcription in both prokaryotic and eukaryotic cells are called **DNA-dependent RNA polymerases**, or simply **RNA polymerases**. These enzymes

are able to incorporate nucleotides, one at a time, into a strand of RNA whose sequence is complementary to one of the DNA strands, which serves as the **template**.

The first step in the synthesis of an RNA is the association of the polymerase with the DNA template. This brings up a matter of more general interest, namely, the specific interactions of two very different macromolecules, proteins and nucleic acids. Just as different proteins have evolved to bind different types of substrates and catalyze different types of reactions, so too have some of them evolved to recognize and bind to specific sequences of nucleotides in a strand of nucleic acid. The site on the DNA to which an RNA polymerase molecule binds prior to initiating transcription is called the **promoter**. RNA polymerases are not capable of recognizing promoters on their own but require the help of additional proteins called **transcription factors**, which are particularly important in the initiation of transcription of eukaryotic genes. In addition to providing a binding site for the polymerase, the promoter contains the information that determines which of the two DNA strands is transcribed and the site at which transcription begins.

The polymerase moves along the template DNA strand toward its 5′ end (i.e., in a 3′ → 5′ direction). As the polymerase progresses, the DNA is temporarily unwound, and the polymerase assembles a complementary strand of RNA that grows from its 5′ terminus in a 3′ direction (Figure 11.4a,b). As indicated in Figure 11.4c, RNA polymerase catalyzes the reaction

$$RNA_n + NPPP \rightarrow RNA_{n+1} + PP_i$$

in which ribonucleoside triphosphate precursors (NPPPs) are hydrolyzed into nucleoside monophosphates as they are polymerized into a covalent chain. Reactions leading to the synthesis of nucleic acids (and proteins) are inherently different from those of intermediary metabolism discussed in Chapter 3. Whereas some of the reactions leading to the formation of small molecules, such as amino acids, may be close enough to equilibrium that a considerable reverse reaction can be measured, those reactions leading to the synthesis of nucleic acids and proteins must occur under conditions in which there is virtually no reverse reaction. This condition is met during transcription by a second reaction

$$PP_i \rightarrow 2\, P_i$$

catalyzed by a different enzyme, a pyrophosphatase, in which the pyrophosphate (PP_i) produced in the first reaction is hydrolyzed to inorganic phosphate (P_i). The hydrolysis of pyrophosphate releases a large amount of free energy and makes the incorporation of nucleotides essentially irreversible.

As the polymerase moves along the DNA template, it incorporates complementary nucleotides into the growing RNA chain. A nucleotide is incorporated into the RNA strand if it is able to form a proper (Watson-Crick) base pair with the nucleotide in the DNA strand being transcribed. This can be seen in Figure 11.4c where the incoming adenosine 5′-triphosphate pairs with the thymine-containing nu-

cleotide of the template. Once the polymerase has moved past a particular stretch of DNA, the double helix reforms (as in Figure 11.4a,b). Consequently, the RNA chain does not remain associated with its template as a DNA-RNA hybrid (except for about eight nucleotides just behind the site where the polymerase is operating). A bacterial RNA polymerase is capable of incorporating 50 to 100 nucleotides into a growing RNA molecule per second, and most genes in a cell are transcribed simultaneously by numerous polymerases. The electron micrograph of Figure 11.4d shows a molecule of phage DNA with a number of bound RNA polymerase molecules.

RNA polymerases are capable of forming prodigiously long RNAs. Consequently, the enzyme must remain attached to the DNA over long stretches of template (the enzyme is said to be *processive*). At the same time, the enzyme must be associated loosely enough that it can move from nucleotide to nucleotide of the template. It is difficult to study certain properties of RNA polymerases, such as processivity, using biochemical methodologies that tend to average out differences between individual protein molecules. Consequently, researchers have developed techniques to follow the activities of single RNA polymerase molecules similar to those used to study individual cytoskeletal motors. Two examples of such studies are depicted in Figure 11.5. In both of these examples, a single RNA polymerase is attached to the surface of a glass coverslip and allowed to transcribe a DNA molecule containing a fluorescent bead covalently linked to one of its ends. The movement of the fluorescent bead can be monitored under a fluorescence microscope. In Figure 11.5a, the bead is free in solution and its range of movement is proportional to the length of the DNA between the polymerase and the bead. As the polymerase transcribes the template, the connecting DNA strand is shortened, and the movement of the bead is restricted. This type of system allows investigators to study the rate of transcription of an individual polymerase and to determine if the polymerase transcribes the DNA in a steady or discontinuous movement. In Figure 11.5b, the bead at the end of the DNA molecule being transcribed is trapped by a focused laser beam (page 336). The minute force exerted by the laser trap can be varied, until it is just sufficient to stop the polymerase from continuing to transcribe the DNA. Measurements carried out on RNA polymerase molecules in the act of transcription indicate that these enzymes can move with a force more than twice that of a myosin molecule. The energy required to drive the movements of the polymerase is derived from the hydrolysis of the NPPP precursors as they are incorporated into the growing RNA chain.

Even though polymerases contain relatively powerful motors, these enzymes do not necessarily move in a steady, continuous fashion but may pause at certain locations along the template for varying periods. In some cases, a stalled polymerase must digest away the 3′ end of the newly synthesized transcript and resynthesize the missing portion before it is able to resume its movement. A number of elongation factors have been identified that enhance the enzyme's ability to traverse these roadblocks.

(a)

(b)

(c)

(d)

Figure 11.4 Chain elongation during transcription. (*a*) A schematic model of the elongation of a newly synthesized RNA molecule during transcription. The polymerase covers approximately 35 base pairs of DNA, the transcription bubble composed of single-stranded (melted) DNA contains about 17 base pairs, and the segment present in a DNA-RNA hybrid includes about 8 base pairs. (*b*) Structural model of an RNA polymerase in the act of transcription elongation based on electron microscopic analyses. The DNA is thought to lie in a groove within the polymerase and clamped at its downstream end by a pair of jaws that project from the enzyme. The nascent RNA is thought to exit from the enzyme's active site through a tunnel (labeled RNA exit

channel). (*c*) Chain elongation occurs as a result of an attack by the 3′ OH of the nucleotide at the end of the growing strand on the 5′ α-phosphate of the incoming nucleoside triphosphate. The pyrophosphate released is subsequently cleaved, which drives the reaction in the direction of polymerization. The geometry of base-pairing between the nucleotide of the template strand and the incoming nucleotide determines which of the four possible nucleoside triphosphates is incorporated into the growing RNA chain at each site. (*d*) Electron micrograph of several RNA polymerase molecules bound to a phage DNA template. (B: AFTER J. GELLES AND R. LANDICK, CELL 93:13,1998, COPYRIGHT 1998, WITH PERMISSION OF ELSEVIER SCIENCE; D: COURTESY OF R. C. WILLIAMS.)

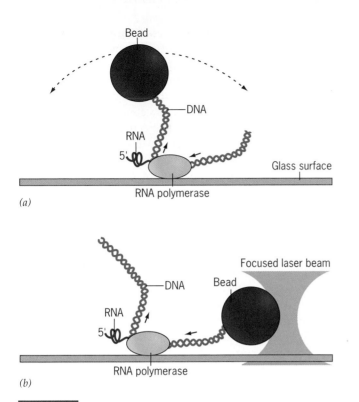

(a)

(b)

Figure 11.5 Examples of experimental techniques to follow the activities of single RNA polymerase molecules. (*a*) In this protocol, the RNA polymerase molecule is firmly attached to the underlying coverslip and allowed to transcribe a DNA molecule containing a fluorescent bead at the upstream end. The arrows indicate the movement of the DNA through the polymerase. The rate of movement and progression of the polymerase can be followed by observing the position of the bead over time using a fluorescence microscope. (*b*) In this protocol, the attached polymerase is transcribing a DNA molecule with a bead at its downstream end. As in *a*, the arrows indicate direction of DNA movement. The bead is caught in an optical (laser) trap, which delivers a known force that can be varied by adjusting the laser beam. This type of apparatus can measure the forces generated by a transcribing polymerase. (REPRINTED FROM J. GELLES AND R. LANDICK, CELL 93:15,1998, COPYRIGHT 1998, WITH PERMISSION FROM ELSEVIER SCIENCE.)

At this point in the discussion, it is useful to examine the differences in the process of transcription between prokaryotic and eukaryotic cells.

Transcription in Prokaryotes

Bacteria, such as *E. coli*, contain a single type of RNA polymerase composed of four subunits that are tightly associated to form a *core enzyme*.[1] If the core enzyme is purified from bacterial cells and added to a solution of bacterial DNA molecules and ribonucleoside triphosphates, the enzyme binds to the DNA and synthesizes RNA. The RNA molecules produced by a purified polymerase, how-

ever, are not the same as those found within cells because the core enzyme has attached to random sites in the DNA, sites that it would normally have ignored in vivo. If, however, a purified accessory polypeptide called *sigma factor* (σ) is added to the RNA polymerase before it attaches to DNA, transcription begins at selected locations (Figure 11.6). Attachment of σ factor to the core enzyme in-

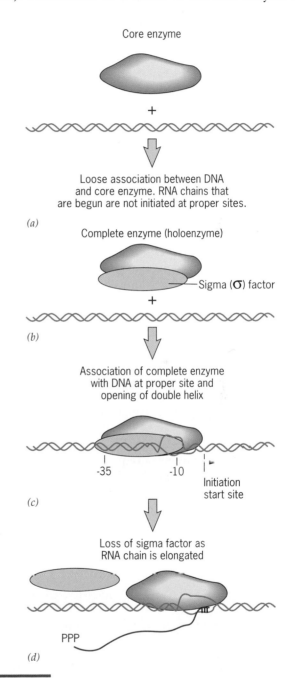

Core enzyme

+

(a) Loose association between DNA and core enzyme. RNA chains that are begun are not initiated at proper sites.

Complete enzyme (holoenzyme)

Sigma (σ) factor

(b) +

(c) Association of complete enzyme with DNA at proper site and opening of double helix

-35 -10

Initiation start site

(d) Loss of sigma factor as RNA chain is elongated

PPP

Figure 11.6 Initiation of transcription in prokaryotes. (*a*) In the absence of the σ factor, the core enzyme cannot interact with the DNA at specific initiation sites. (*b–d*) When the core enzyme is associated with the σ factor, the complete enzyme is able to recognize and bind to the promoter region of the DNA, separate the strands of the DNA double helix, and initiate transcription at the proper start site (see Figure 11.7). The σ factor then dissociates from the core enzyme, which is capable of transcription elongation.

[1]Archaea, the other major group of prokaryotes along with bacteria (page 12), also possess a single RNA polymerase, but it has a very different subunit composition from that described here and more closely resembles a eukaryotic RNA polymerase.

creases the enzyme's affinity for promoter sites in DNA and decreases its affinity for DNA in general. X-ray crystallographic analysis of the core bacterial RNA polymerase (see Figure 11.4*b*) reveals a molecule shaped like a "crab claw" with a broad internal channel that holds the DNA double helix and a pair of jaws that grips the DNA as it enters the enzyme. Once bound to the promoter, the enzyme separates (or *melts*) the two DNA strands in the region surrounding the start site, which makes the template strand accessible to the enzyme's active site. Initiation of transcription appears to be a difficult undertaking because an RNA polymerase typically makes several unsuccessful attempts to assemble an RNA transcript. Once approximately ten nucleotides have been successfully incorporated into a growing transcript, the enzyme undergoes a major change in conformation, leading to the release of the sigma factor and the promoter DNA. This conformational change stabilizes the polymerase, converting the enzyme into an *elongation complex* that can move processively along the DNA of a gene.

Bacterial promoters are located in the region of a DNA strand just preceding the initiation site of RNA synthesis (Figure 11.7). The nucleotide at which transcription is initiated is denoted as +1 and the preceding nucleotide as −1. Those portions of the DNA preceding the initiation site (toward the 3′ end of the template) are said to be *upstream* from that site. Those portions of the DNA succeeding it (toward the 5′ end of the template) are said to be *downstream* from that site. Analysis of the DNA sequences just upstream from a large number of bacterial genes reveals that two short stretches of DNA are similar from one gene to another. One of these stretches is centered at approximately 35 bases upstream from the initiation site and typically occurs as the sequence TTGACA (Figure 11.7). This TTGACA sequence is called a **consensus sequence**, which indicates that it is the most common version of a conserved sequence, but that some variation occurs from one gene to another. This site in the promoter (known as the −35 region) is recognized by the σ factor associated with the

polymerase. Bacterial cells possess a variety of different σ factors that recognize different versions of the −35 sequence. The σ^{70} is known as the "housekeeping" σ factor, because it initiates transcription of most genes. Alternative σ factors initiate transcription of a small number of specific genes that participate in a common response. For example, when *E. coli* cells are subjected to a sudden rise in temperature, a new σ factor is synthesized that recognizes a different promoter sequence and leads to the coordinated transcription of a battery of *heat-shock genes*. The products of these genes protect the proteins of the cell from thermal damage (page 75).

The second conserved sequence is found approximately 10 bases upstream from the initiation site and occurs as the consensus sequence TATAAT (Figure 11.7). This site in the promoter, named the *Pribnow box* after its discoverer, is responsible for identifying the precise nucleotide at which transcription begins. Although most mutations in promoter regions reduce the rate of transcription, certain base substitutions in both the −35 and −10 regions can greatly increase this rate. Thus promoters are more than simply recognition regions; they can act as important control sites in regulating the rate of gene expression.

Just as transcription is initiated at specific points in the chromosome, it also terminates when a specific nucleotide sequence is reached. In some cases, a protein called *rho* is required for termination. Rho has enzymatic activity that may peel the 3′ end of the RNA transcript away from the DNA to which it is bound. In most cases, the polymerase stops transcription when it reaches a *terminator sequence* and releases the completed RNA chain without requiring additional factors.

REVIEW ?

1. What is the role of a promoter in gene expression? Where are the promoters for prokaryotic polymerases located?

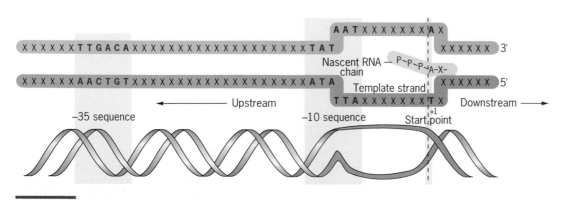

Figure 11.7 The basic elements of a promoter region in the DNA of the bacterium *E. coli*. The key regulatory sequences required for initiation of transcription are found in regions located at −35 and −10 base pairs from the site at which transcription is initiated. The initiation site marks the boundary between the + and − sides of the gene.

2. Describe the steps during initiation of transcription in prokaryotes. What is the role of the σ factor? What is the nature of the reaction in which nucleotides are incorporated into a growing RNA strand? How is the specificity of nucleotide incorporation determined? What is the role of pyrophosphate hydrolysis?

11.3 TRANSCRIPTION AND RNA PROCESSING IN EUKARYOTIC CELLS

Eukaryotic cells have three distinct transcribing enzymes, each responsible for synthesizing a different group of RNAs. **RNA polymerase I** synthesizes the larger (28S, 18S, and 5.8S) ribosomal RNAs; **RNA polymerase II** synthesizes messenger RNAs and most small nuclear RNAs (discussed later); and **RNA polymerase III** synthesizes low-molecular-weight RNAs, including the various transfer RNAs and the 5S ribosomal RNA. No prokaryote has been found with multiple RNA polymerases, whereas the simplest eukaryotes (yeast) have the same three types that are present in mammalian cells. This difference in number of RNA polymerases is another sharp distinction between prokaryotic and eukaryotic cells (see footnote on page 445).

The three types of eukaryotic RNA polymerases can be distinguished on the basis of their sensitivity to α-amanitin, a highly toxic octapeptide (8 linked amino acids). α-Amanitin is isolated from a common poisonous mushroom, *Amanita phalloides*, which is also the source of the microfilament toxin, phalloidin (page 368). RNA polymerase II activity is very sensitive to α-amanitin, whereas RNA polymerase I is not affected by the compound. RNA polymerase III is inhibited to a lesser degree than RNA polymerase II. A person poisoned by ingestion of these mushrooms shows no immediate symptoms, but liver function deteriorates over the following days due to the lack of production of new mRNAs necessary to direct continuing protein synthesis. In severe cases, the only way to save a person's life may be a liver transplant operation.

Another distinction between transcription in prokaryotes and eukaryotes is the requirement in eukaryotes for a large variety of accessory proteins, or transcription factors. These proteins play a role in virtually every aspect of the transcription process, from the binding of the polymerase to the DNA template, to the initiation of transcription, to its elongation and termination. Although transcription factors are crucial for the operation of all three types of eukaryotic RNA polymerases, they will only be discussed in regard to the synthesis of mRNAs by RNA polymerase II (page 455).

All three major types of RNAs—mRNAs, rRNAs, and tRNAs—are derived from precursor RNA molecules that are considerably longer than the final RNA product. The initial precursor RNA is equivalent in length to the full length of the DNA transcribed and is called the **primary transcript**, or **pre-RNA**. The corresponding segment of DNA from which a primary transcript is transcribed is called a **transcription unit**. Primary transcripts do not exist within the cell as naked RNA but become associated with proteins even as they are synthesized. Primary transcripts typically have a fleeting existence, being **processed** into smaller, functional RNAs by a series of "cut-and-paste" reactions.

RNA processing requires a variety of small RNAs (90 to 300 nucleotides long) and their associated proteins. The RNAs are called **small nuclear RNAs (snRNAs)** because of their small size and the fact that they function in the nucleus. In the following sections, we will examine the activities associated with the transcription and processing of each of the major eukaryotic RNAs.

Ribosomal RNAs

Eukaryotic cells contain millions of ribosomes, each consisting of several molecules of rRNA together with dozens of ribosomal proteins. The composition of a mammalian ribosome is shown in Figure 11.8. Ribosomes are so numerous that more than 80 percent of the RNA in most

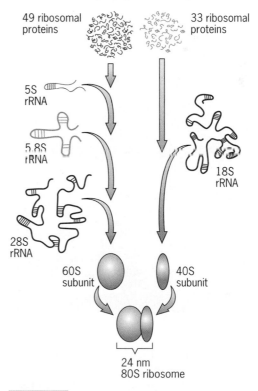

Figure 11.8 The macromolecular composition of a mammalian ribosome. The drawing shows the components present in each of the subunits of a mammalian ribosome. The synthesis and processing of the rRNAs and the assembly of the ribosomal subunits are discussed in the following pages. (FROM D. P. SNUSTAD ET AL., PRINCIPLES OF GENETICS, COPYRIGHT © 1997, JOHN WILEY & SONS, INC. REPRINTED BY PERMISSION OF JOHN WILEY & SONS, INC.)

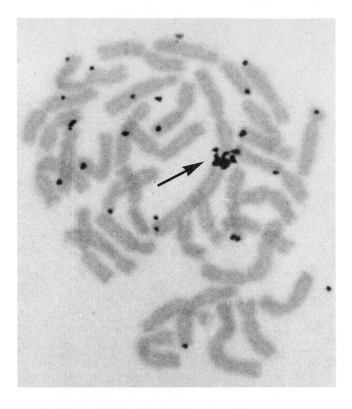

Figure 11.9 **Experimental localization by in situ hybridization of the DNA that codes for ribosomal RNA in the amphibian *Xenopus*.** Ribosomal DNA was isolated and used as a template to synthesize complementary, radioactively labeled RNA in vitro. The labeled RNA was then hybridized to the denatured DNA of a preparation of *Xenopus* chromosomes, and the location of the bound radioactivity was determined by autoradiography. The DNA that codes for rRNA is located in one or more sites where nucleoli form (hence the sites are called nucleolar organizers). Because the nucleolar organizers of both members of a homologous pair of *Xenopus* chromosomes are associated with one another, the radioactivity appears localized at one site on the set of chromosomes (arrow). (FROM MARY LOU PARDUE, COLD SPRING HARBOR SYMP. QUANT. BIOL. 38:476, 1973.)

cells consists of ribosomal RNA. To furnish the cell with such a large number of transcripts, the DNA sequences encoding rRNA are normally repeated hundreds of times. This DNA, called **rDNA**, is typically clustered in one or a few regions of the genome (Figure 11.9). The human genome has five rDNA clusters, each on a different chromosome. In a nondividing (interphase) cell, the clusters of rDNA are gathered together as part of one or more irregularly shaped nuclear structures, called **nucleoli** (singular, **nucleolus**), that function in producing ribosomes (Figure 11.10*a*). Nucleoli disappear during mitosis and then reap-

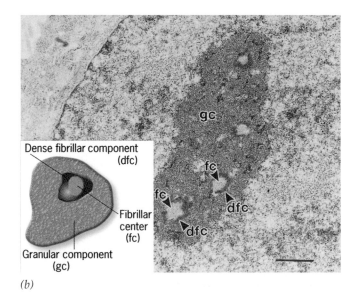

(a)

(b)

Figure 11.10 **The nucleolus.** (*a*) Light micrograph of two human HeLa cells transfected with a gene for a ribosomal protein fused to the green fluorescent protein (GFP). The fluorescent ribosomal protein can be seen in the cytoplasm where it is synthesized and ultimately functions, and in the nucleoli (white arrows), where it is assembled into ribosomes. (*b*) Electron micrograph of a section of part of a nucleus with a nucleolus. Three distinct regions can be distinguished morphologically. The bulk of the nucleolus consists of a granular component (gc). Embedded within the granules are fibrillar centers (fc) that are surrounded by a more dense fibrillar component (dfc). The inset shows a schematic drawing of these parts of the nucleolus. According to one current model, the fc contains the DNA that codes for ribosomal RNA, and the dfc contains the nascent pre-rRNA transcripts and associated proteins. According to this model, transcription of the pre-rRNA precursor takes place at the border between the fc and dfc. The gc is thought to contain ribosomal subunits in various stages of assembly. Bar, 1 μm.
(A: FROM C. E. LYON AND A. I. LAMOND, CURR. BIOL. 10:R323,2000; B: FROM PAVEL HOZAK ET AL., J. CELL SCIENCE 107:641, 1994; BY PERMISSION OF THE COMPANY OF BIOLOGISTS, LTD.)

pear in the nuclei of the daughter cells around those parts of the genome that contain ribosomal RNA genes. Consequently, regions of a chromosome that contain rDNA have been called *nucleolar organizers*.

The bulk of a nucleolus is composed of ribosomal subunits that give the nucleolus a granular appearance (Figure 11.10*b*). Embedded within this granular mass are one or more rounded cores consisting primarily of fibrillar material. As discussed in the legend of Figure 11.10, the fibrillar material is thought to consist of rDNA templates and nascent rRNA transcripts. In the following sections, we will examine the process by which these rRNA transcripts are synthesized.

Synthesizing the rRNA Precursor Oocytes are typically very large cells (e.g., 100 μm in diameter in mammals); those of amphibians are generally enormous (up to 2.5

mm in diameter). During the growth of amphibian oocytes, the amount of rDNA in the cell is greatly increased as are the number of nucleoli (Figure 11.11*a*). The selective amplification of rDNA is required to provide the large numbers of ribosomes that are required by the fertilized egg to begin embryonic development. Because these oocytes contain hundreds of nucleoli, each actively manufacturing rRNA, they are ideal subjects for the study of rRNA synthesis and processing.

Our understanding of rRNA synthesis was greatly advanced in the late 1960s with the development of techniques by Oscar Miller, Jr. of the University of Virginia to visualize "genes in action" with the electron microscope. To carry out these studies, the fibrillar cores of oocyte nucleoli were gently dispersed to reveal the presence of a large circular fiber. When one of these fibers was examined in the electron microscope, it was seen to resemble a

Nucleoli

(a)

(b) 2 μm

(c) 0.5 μm

Figure 11.11 The synthesis of ribosomal RNA. (*a*) Light micrograph of an isolated nucleus from a *Xenopus* oocyte stained to reveal the hundreds of nucleoli. (*b*) Electron micrograph of a segment of DNA isolated from one of the nucleoli from a *Xenopus* oocyte. The DNA (called rDNA) contains the genes that code for the two large ribosomal RNAs, which are carved from a single primary transcript. Numerous genes are shown, each in the process of being transcribed. Transcription is evident by the fibrils that are attached to the DNA. These fibrils consist of nascent RNA and associated protein. The stretches of DNA between the transcribed genes are nontranscribed spacers. The arrows indicate sites where transcription is initiated. (*c*) A closer view of two nucleolar genes being transcribed. The length of the nascent rRNA primary transcript increases with increasing distance from the point of initiation. The RNA polymerase molecules at the base of each fibril can be seen as dots. (A: FROM DAVID D. BROWN AND IGOR B. DAWID, SCIENCE 160:272, 1968; © COPYRIGHT 1968 AMERICAN ASSOCIATION FOR THE ADVANCEMENT OF SCIENCE. B–C: COURTESY OF OSCAR L. MILLER, JR. AND BARBARA R. BEATTY.)

chain of Christmas trees (Figure 11.11*b,c*). The electron micrographs of Figure 11.11 reveal a number of aspects of nucleolar activity and ribosomal RNA synthesis.

1. The micrograph in Figure 11.11*b* shows several distinct genes for ribosomal RNA situated one after the other along a single DNA molecule, thus revealing the tandem arrangement of the repeated rRNA genes.

2. The micrograph in Figure 11.11*b* shows a static image of dynamic events that occur in the nucleolus. We can interpret this photograph to provide a great deal of information about the process of rRNA transcription. Each of the 100 or so fibrils emerging from the DNA as a branch of a Christmas tree is a nascent rRNA transcript caught in the act of elongation. The dark granule at the base of each fibril, visible in the higher magnification photograph of Figure 11.11*c*, is the RNA polymerase I molecule synthesizing that transcript. The length of the fibrils increases gradually from one end of the Christmas tree trunk to the other. The shorter fibrils are RNA molecules of fewer nucleotides that are attached to polymerase molecules bound to the DNA closer to the transcription initiation site. The longer the fibril, the closer the transcript is to completion. The length of DNA between the shortest and longest RNA fibrils corresponds to a single transcription unit. The promoter lies just upstream from the transcription initiation site. The high density of RNA polymerase molecules along each transcription unit (about 1 every 100 base pairs of DNA) reflects the high rate of rRNA synthesis in the nucleoli of these oocytes.

3. It can be seen from the electron micrographs that the RNA fibrils contain clumps and associated particles. These particles consist of RNA and protein that work together to convert the rRNA precursors to their final rRNA products and assemble them into ribosomal subunits.

4. It can be noted from Figure 11.11*b* that the region of the DNA fiber between adjacent transcription units is devoid of nascent RNA chains. Because this region of the ribosomal gene cluster is not transcribed, it is referred to as the **nontranscribed spacer** (Figure 11.12). Nontranscribed spacers are present between various types of tandemly repeated genes, including those of tRNAs and histones.

Processing the rRNA Precursor Eukaryotic ribosomes have four distinct ribosomal RNAs, three in the large subunit and one in the small subunit. In humans, the large subunit contains a 28S, 5.8S, and 5S RNA molecule, and the small subunit contains an 18S RNA molecule.[2] Three of these rRNAs (the 28S, 18S, and 5.8S) are carved by various nucleases from a single primary transcript (called the **pre-rRNA**). The 5S rRNA is synthesized from a separate RNA precursor outside the nucleolus. We will begin this discussion with the pre-rRNA.

Two of the peculiarities of the pre-rRNA, as compared with other RNA transcripts, are the large number of methylated nucleotides and pseudouridine residues. By the time a human pre-rRNA precursor is first cleaved, over 100 methyl groups have been added to ribose groups in the molecule, and approximately 95 of its uridine residues have been chemically converted to pseudouridine

[2]The S value (or Svedberg unit) refers to the sedimentation coefficient of the RNA; the larger the number, the more rapidly the molecule moves through a field of force during centrifugation, and (for a group of chemically similar molecules) the larger the size of the molecule. The 28S, 18S, 5.8S, and 5S RNAs consist of nucleotide lengths of about 5,000, 2,000, 160, and 120 nucleotides, respectively.

Figure 11.12 The rRNA transcription unit. The top drawing depicts the appearance of a portion of the DNA from a nucleolus as it is transcribed into rRNA. The lower drawings illustrate one of the transcription units that codes for rRNA in *Xenopus* and the mouse. Those parts of the DNA that code for the mature rRNA products are shown in blue. Regions of transcribed spacer, that is, portions of the DNA that are transcribed, but whose corresponding RNAs are degraded during processing, are shown in yellow. The nontranscribed spacer, which lies between the transcription units, contains the promoter region at the 5′ side of the gene and an enhancer (a region of DNA that increases the rate of transcription). (AFTER B. SOLLNER-WEBB AND E. B. MOUGEY, TRENDS BIOCHEM. SCI. 16:59, 1991.)

(see Figure 11.15*a*). All of these modifications occur after the nucleotides are incorporated into the nascent RNA, that is, *posttranscriptionally*. The altered nucleotides are located at specific positions and are clustered within portions of the molecule that have been conserved during vertebrate evolution. All of the nucleotides in the pre-rRNA that are altered remain as part of the final products, while unaltered sections are discarded during processing. The function of the methyl groups and pseudouridines is unclear. It is speculated that these modified nucleotides may protect parts of the pre-rRNA from enzymatic cleavage, promote folding of rRNAs into their final three-dimensional structures, and/or promote interactions of rRNAs with other molecules.

Because rRNAs are so heavily methylated, their synthesis can be followed by incubating cells with radioactively labeled methionine, a compound that serves in most cells as a methyl group donor. The methyl group is trans-

ferred enzymatically from methionine to nucleotides in the pre-rRNA. When [^{14}C]-methionine is provided to cultured mammalian cells for a short period of time, a considerable fraction of the incorporated radioactivity is present in a 45S RNA molecule, which corresponds to a length of about 13,000 nucleotides. The 45S RNA is cleaved into smaller molecules that are subsequently trimmed down to the 28S, 18S, and 5.8S rRNA molecules. The combined length of the three mature rRNAs is approximately 7,000 nucleotides, or a little more than half of the primary transcript.

The various steps in the processing pathway from the 45S pre-rRNA to the mature rRNAs can be followed by incubating mammalian cells very briefly with labeled methionine and then chasing the cells in nonlabeled medium for varying lengths of time (Figure 11.13). As noted above, the first species to become labeled in this type of experiment is the 45S primary transcript, which is seen as a

Figure 11.13 Kinetic analysis of rRNA synthesis and processing. A culture of mammalian cells was incubated for 10 minutes in [^{14}C]methionine and was then chased in unlabeled medium for various times as indicated in each panel. After the chase, cells were washed free of isotope and homogenized, and nucleolar and cytoplasmic fractions were prepared. The RNA was extracted from each fraction and analyzed by ultracentrifugation, which separates RNAs according to size (the larger the size, the closer the RNA is to the bottom of the tube, which corresponds to fraction 1). The continuous line represents the UV absorbance of each cellular fraction, which provides a measure of the amount of RNA of each size class. This absorbance profile does not change with time. The dotted line shows the radioac-tivity at various times during the chase. The graphs of nucleolar RNA (upper profiles) show the synthesis of the 45S rRNA precursor and its subsequent conversion to a 32S molecule, which is a precursor to the 28S rRNA. The other major product of the 45S precursor leaves the nucleus very rapidly and, therefore, does not appear prominently in the nucleolar RNA. The lower profiles show the time course of the appearance of the mature rRNA molecules in the cytoplasm. The 18S rRNA appears in the cytoplasm well in advance of the larger 28S species, which correlates with the rapid exodus of the former from the nucleolus. (FROM H. GREENBERG AND S. PENMAN, J. MOL. BIOL. 21:531, 1966, COPYRIGHT 1966, BY PERMISSION OF THE PUBLISHER ACADEMIC PRESS.)

peak of radioactivity (red dotted line) in the nucleolar RNA fraction after 10 minutes. After about an hour, the 45S RNA has disappeared from the nucleolus and is largely replaced by a 32S RNA, which is one of the two major products produced from the 45S primary transcript. The 32S RNA is seen as a distinct peak in the nucleolar fraction from 40 minutes to 150 minutes. The 32S RNA is a precursor to the mature 28S rRNA. The other major product of the 45S pre-rRNA leaves the nucleolus quite rapidly and appears in the cytoplasm as the mature 18S rRNA (seen in the 40 minute cytoplasmic fraction). After a period of two or more hours, nearly all of the radioactivity has left the nucleolus, and most has accumulated in the 28S and 18S rRNAs of the cytoplasm. The radioactivity in the 4S RNA peak of the cytoplasm represents methyl groups that have been transferred to small tRNA molecules. One of the pathways in the processing of a rRNA primary transcript is indicated in Figure 11.14.

The Role of the snoRNAs The processing of the pre-rRNA is accomplished with the help of a large number of **small, nucleolar RNAs** (or **snoRNAs**) that are packaged with particular proteins to form particles called **snoRNPs (small, nucleolar ribonucleoproteins)**. Electron micro-

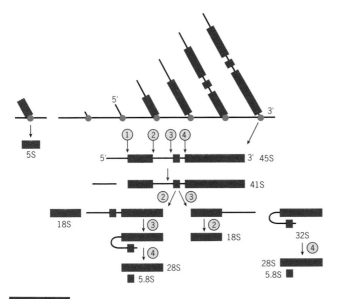

Figure 11.14 Scheme for the processing of mammalian ribosomal RNA. The primary transcript for the rRNAs is a 45S molecule of about 13,000 bases. The four principal cleavage events during processing of this pre-rRNA are indicated by the circled numbers. Cleavage of the primary transcript at site 1 removes the 5′ terminal leader sequence and produces a 41S intermediate. The second cleavage can occur at either site 2 or 3 depending on the type of cell. Cleavage at site 3 generates the 32S intermediate seen in curves of the previous figure. During the final processing steps, the 28S and 5.8S sections are separated from one another, and the ends of the various intermediates are trimmed down to their final mature size. (FROM R. P. PERRY, J. CELL BIOL. 91:29s, 1981, BY COPYRIGHT PERMISSION OF THE ROCKEFELLER UNIVERSITY PRESS.)

graphs indicate that snoRNPs begin to associate with the rRNA precursor before it is fully transcribed. The first RNP particle to attach to a pre-rRNA transcript contains the U3 snoRNA, which binds to the 5′ terminus of the precursor molecule. This RNP particle can be seen as a ball situated at the outer end of the nascent RNA fibrils in Figure 11.11c, where it catalyzes the removal of the 5′ end of the transcript (Figure 11.14). The other enzymatic cleavages shown in Figure 11.14 are thought to be carried out by different snoRNPs.

U3 and several other snoRNAs were discovered many years ago because they are present in relatively large quantities (about 10^6 copies per cell). More recently, a different class of snoRNAs present at lower concentration (about 10^4 copies per cell) was discovered. These low-abundance snoRNAs contain relatively long stretches (10 to 21 nucleotides) that are complementary to sections of the rRNA transcript. For this reason they have been called *antisense snoRNAs*. Each antisense snoRNA binds to a specific portion of the pre-rRNA to form an RNA-RNA duplex, which serves as a recognition site for enzymes that modify a particular nucleotide in the pre-rRNA sequence. Antisense snoRNAs can be divided into two groups based on their function and similarities in nucleotide sequence. Members of one group (called the *box C/D snoRNAs*) determine which nucleotides will have their ribose moieties methylated, whereas members of the other group (called the *box H/ACA snoRNAs*) determine which uridines are converted to pseudouridines. The structures of nucleotides modified by these two reactions are shown in Figure 11.15a. Taken together, there are as many as 200 different antisense snoRNAs, one for each site in the pre-RNA that is ribose-methylated or pseudouridylated. If the gene encoding one of these snoRNAs is deleted, one of the nucleotides of the pre-rRNA fails to be enzymatically modified. The mechanism of action of the two types of antisense snoRNAs is illustrated in Figure 11.15b,c. The remarkable discovery that these RNAs are encoded within the intervening sequences of other genes is discussed on page 470.

The nucleolus is the site not only of rRNA processing, but also of assembly of the two ribosomal subunits. Consequently, two types of proteins become associated with the RNA as it is processed: those proteins that remain in the ribosomal subunits, and those proteins of the nucleolus that have a transient interaction with the rRNA intermediates and are required only for processing. The proteins required for processing include enzymes that cleave the rRNA precursors and proteins that protect sites from cleavage.

Synthesis and Processing of the 5S rRNA A 5S rRNA, about 120 nucleotides long, is present as part of the large ribosomal subunit of both prokaryotes and eukaryotes. In eukaryotes, the 5S rRNA molecules are encoded by a large number of identical genes that are separate from the other rRNA genes and are located outside the nucleolus. The 5S

(a)

(b)

(c)

Figure 11.15 **Modifying the pre-rRNA.** (*a*) The most frequent modifications to nucleotides in a pre-rRNA are the conversion of a uridine to a pseudouridine and the methylation of a ribose at the 2′ site of the sugar. To convert uridine to pseudouridine, the N1—C1′ bond is cleaved, and the uracil ring is rotated through 180°, which brings the C5 of the ring into place to form the new bond with C1′ of ribose. These chemical modifications are thought to be catalyzed by protein components of the snoRNPs, but the enzymes have not yet been identified. (*b*) The formation of an RNA–RNA duplex between the U20 snoRNA and a portion of the pre-rRNA that leads to 2′ ribose methylation. In each case, the methylated nucleotide in the rRNA is hydrogen bonded to a nucleotide of the snoRNA that is located five base pairs from box D. Box D, which contains the invariable sequence CUGA, is present in all snoRNAs that guide ribose methylation. (*c*) The formation of an RNA–RNA duplex between U68 snoRNA and a portion of the pre-rRNA that leads to conversion of a uridine to pseudouridine (ψ). Pseudouridylation occurs at a fixed site relative to a hairpin fold in the snoRNA. The snoRNAs that guide pseudouridylation have a common ACA 3′ terminus. (B: AFTER J.-P. BACHELLERIE AND J. CAVAILLÉ, TRENDS IN BIOCH. SCI. 22:258, 1997, COPYRIGHT 1997, WITH PERMISSION FROM ELSEVIER SCIENCE; C: AFTER P. GANOT ET AL., CELL 89:802, 1997.)

tion of the promoter was first clearly demonstrated by introducing modified 5S rRNA genes into host cells and determining the ability of that DNA to serve as templates for the host polymerase III. It was found that the entire 5′ flanking region could be removed and the polymerase would still transcribe the DNA starting at the normal initiation site. If, however, the deletion included the central part of the gene (from about nucleotide 50 to 80 of a 120 base-pair gene), the polymerase would not transcribe the DNA or even bind to it. If the internal promoter from a 5S rRNA gene is inserted into another region of the genome, the new site becomes a template for transcription by RNA polymerase III.

Transfer RNAs

Plant and animal cells are estimated to have approximately 50 different species of transfer RNA, each encoded by a DNA sequence that is repeated a number of times within the genome. The degree of repetition varies with the organism: yeast cells have a *total* of about 275 tRNA genes, fruit flies about 850, and humans about 1,300. Transfer RNAs are synthesized from genes that are found in small clusters scattered around the genome. A single cluster typically contains multiple copies of *different* tRNA genes, and conversely, the DNA sequence encoding a given tRNA is typically found in more than one cluster. The DNA within a cluster (**tDNA**) consists largely of nontranscribed spacer sequences with the tRNA coding sequences situated at irregular intervals in a tandemly repeated arrangement (such as that of Figure 11.16).

rRNA genes are organized in tandem array, with the transcribed portions alternating with nontranscribed spacers to form repeating units analogous to the repeating units encoding the 45S rRNA precursor. The 5S rRNA genes are transcribed by RNA polymerase III. The 5′ end of the primary transcript is identical with that of the mature 5S rRNA, but the 3′ end usually contains extra nucleotides that are removed during processing. Following its synthesis, the 5S rRNA is transported to the nucleolus to join the other components involved in the assembly of ribosomal subunits.

RNA polymerase III is unusual among the three polymerases in that it binds to a promoter site located within the transcribed portion of the gene rather than a site upstream from the gene's 5′ flank. The internal posi-

Figure 11.16 The arrangement of genes that code for transfer RNAs in *Xenopus*. A 3.18 kilobase fragment of DNA, showing the arrangement of various tRNA genes and spacers. (FROM S. G. CLARKSON ET AL., IN D. D. BROWN, ED., DEVELOPMENTAL BIOLOGY USING PURIFIED GENES, ACADEMIC PRESS, 1981.)

Like the 5S rRNA, tRNAs are transcribed by RNA polymerase III, and the promoter sequence lies within the coding section of the gene rather than being located at its 5′ flank. The primary transcript of a transfer RNA molecule is larger than the final product, and pieces on both the 5′ and 3′ sides of the precursor tRNA (and a small interior piece in some cases) must be trimmed away. One of the enzymes involved in pre-tRNA processing is an endonuclease called ribonuclease P, which is present in both bacterial and eukaryotic cells and consists of both RNA and protein subunits. It is the RNA subunit of ribonuclease P that catalyzes cleavage of the pre-tRNA substrate, a subject discussed in the Experimental Pathways, which can be found at the end of the chapter. All mature tRNAs have the triplet sequence CCA at their 3′ end. These three nucleotides are added enzymatically after the tRNA is processed, and they play a key role in protein synthesis.

Messenger RNAs

When eukaryotic cells are incubated for a short period (30 minutes) in [³H]uridine or [³²P]phosphate and immediately killed, most of the radioactivity is incorporated into a large group of RNA molecules that have the following properties: (1) they have large molecular weights (up to about 80S, or 50,000 nucleotides); (2) as a group, they are represented by RNAs of diverse (heterogeneous) nucleotide sequence; and (3) they are found only in the nucleus. Because of these properties, these RNAs are referred to as **heterogeneous nuclear RNAs (hnRNAs)**, and they are indicated by the red radioactivity line in Figure 11.17*a*. When cells that have been incubated in [³H]uridine or [³²P]phosphate for a brief pulse are placed into unlabeled medium, chased for an hour or more before they are killed and the RNA extracted, the amount of radioactivity in the large nuclear RNAs drops sharply and appears instead in much smaller mRNAs found in the cytoplasm (red line of Figure 11.17*b*). These early experiments suggested that the large, rapidly labeled hnRNAs were precursors to the smaller cytoplasmic mRNAs. This interpretation has been fully substantiated by a large body of research over the past 35 years.

It is important to note that the blue and red lines in Figure 11.17 follow a very different course. The blue lines, which indicate the optical density (i.e., the absorbance of

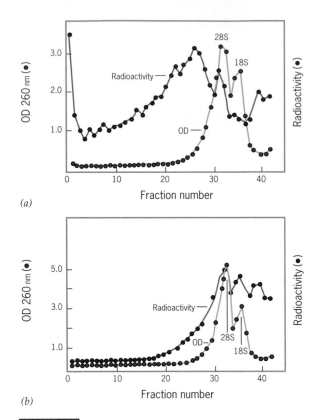

Figure 11.17 The formation of heterogeneous nuclear RNA (hnRNA) and its conversion into smaller mRNAs. (*a*) Curves showing the sedimentation pattern of total RNA extracted from duck blood cells after exposure to [³²P]phosphate for 30 minutes. The larger the RNA, the farther it travels during centrifugation and the closer to the bottom of the tube it ultimately lies. The absorbance (blue line) indicates the total amount of RNA in different regions of the centrifuge tube, whereas the red line indicates the corresponding radioactivity. It is evident that most of the newly synthesized RNA is very large, much larger than the stable 18S and 28S rRNAs. These large RNAs are the hnRNAs. (*b*) The absorbance and radioactivity profiles of the RNA were extracted from cells that were pulse labeled for 30 minutes as in part *a*, but then chased for 3 hours in the presence of actinomycin D, which prevents the synthesis of additional RNA. It is evident that the large hnRNAs have been processed into smaller RNA products. (FROM G. ATTARDI ET AL., J. MOL. BIOL. 20:160, 1966, COPYRIGHT 1966, BY PERMISSION OF THE PUBLISHER ACADEMIC PRESS.)

UV light by each fraction), provide information about the amount of RNA in each fraction following centrifugation. It is evident from the blue lines that most of the RNA in the cell is present as 18S and 28S rRNA (along with tRNAs and snRNAs that stay at the top of the tube). The red lines, which indicate the radioactivity in each fraction, provide information about the number of radioactive nucleotides incorporated into the RNA during the brief pulse. It is evident from these graphs that neither the hnRNA (Figure 11.17*a*) nor the mRNA (Figure 11.17*b*) constitute a significant fraction of the RNA in the cell. If

they did, there would be a greater correspondence between the blue and red lines. Thus, even though the mRNAs (and their heterogeneous nuclear RNA precursors) constitute only a small percentage of the total RNA of most eukaryotic cells, they constitute a large percentage of the RNA that is being synthesized by that cell at any given moment (Figure 11.17*a*). The reason that there is little or no evidence of the hnRNAs and mRNAs in the optical density plots of Figure 11.17 is that these RNAs are degraded after relatively brief periods of time. This is particularly true of the hnRNAs, which have half-lives of only a few minutes. In contrast, the rRNAs and tRNAs have half-lives that are measured in days or weeks and thus gradually accumulate to become the predominant species in the cell. The half-lives of mRNAs vary depending on the particular species, ranging from about 15 minutes to a period of days.

The Machinery Used in mRNA Transcription All eukaryotic mRNA precursors are synthesized by RNA polymerase II, an enzyme composed of 12 different subunits that is remarkably conserved from yeast to mammals. Initiation of transcription by RNA polymerase II occurs in cooperation with a number of **general transcription factors (GTFs)**, whose precise roles remain to be determined. These proteins are referred to as "general" transcription factors because the same ones are required for transcription of a diverse array of genes in a wide variety of different organisms. Promoters for RNA polymerase II lie to the 5′ side of each transcription unit. In the vast majority of the genes studied, a critical portion of the promoter (called a *core promoter element*) lies between 24 and 32 bases upstream from the site at which transcription is initiated (Figure 11.18*a*). This region contains a consensus sequence that is either identical or very similar to the oligonucleotide 5′-TATAAA-3′ and is known as the **TATA box**. The TATA box of the DNA is the site of assembly of a **preinitiation complex** that contains the GTFs and the polymerase. The preinitiation complex must assemble before transcription of the gene can be initiated.

The first step in assembly of the preinitiation complex is binding of a protein, called *TATA-binding protein (TBP)*, that specifically recognizes the TATA box of eukaryotic promoters. Thus, as in prokaryotic cells, a purified eukaryotic polymerase is not able to recognize a promoter directly and cannot initiate accurate transcription on its own. TBP is present as a subunit of a much larger protein complex called TFIID (*t*ranscription *f*actor for polymerase *II, f*raction *D*).[3] X-ray crystallography has revealed that binding of TBP to a polymerase II promoter causes a dramatic distortion in conformation of the DNA.

(a)

(b)

Figure 11.18 The TATA box and its interaction with the TATA-binding protein (TBP). (*a*) Nucleotide sequence of the region just upstream from the site where transcription is initiated in three different eukaryotic genes. The TATA box is indicated by the blue shading. Many eukaryotic promoters contain a second, less highly conserved core promoter element called the initiator (Inr), which includes the site where transcription is initiated (shown in orange). Other promoter elements are depicted in Figure 12.32. (*b*) A model of the structure formed by DNA and three of the GTFs, TBP of TFIID, TFIIA, and TFIIB. Interaction between the TATA box and TBP bends the DNA approximately 80° and allows TFIIB to bind to the DNA both upstream and downstream of the TATA box. (B: FROM GOURISANKAR GHOSH AND GREGORY D. VAN DUYNE, STRUCTURE 4:893, 1996.)

As shown in Figure 11.8*b*, TBP inserts itself into the minor groove of the double helix, bending the DNA molecule more than 80° at the site of DNA-protein interaction.

Binding of TFIID sets the stage for the assembly of the complete preinitiation complex, which may occur in a stepwise manner as depicted in Figure 11.19, or in a single step with the binding of a large, preassembled complex (or *holoenzyme*) that contains the polymerase and the remaining GTFs. Interaction of three of the GTFs (TBP of TFIID, TFIIA, and TFIIB) with DNA is shown in Figure 11.18*d*. The presence of these three GTFs bound to the promoter provides a platform for the binding of the huge, multisubunit polymerase with its attached TFIIF (Figure 11.19). Once the RNA polymerase-TFIIF is in position, another pair of GTFs (TFIIE and TFIIH) join the complex and transform the polymerase into an active, transcribing machine. TFIIH is the only GTF known to possess enzymatic activities. Two of the subunits of TFIIH function as DNA unwinding enzymes (helicases),

[3]TBP is actually a universal transcription factor that mediates the binding of all three eukaryotic RNA polymerases. TBP is present as one of the subunits of three different proteins. As a subunit of TFIID, TBP promotes the binding of RNA polymerase II. As subunits of the proteins SL1 or TFIIIB, TBP promotes the binding of RNA polymerases I and III, respectively.

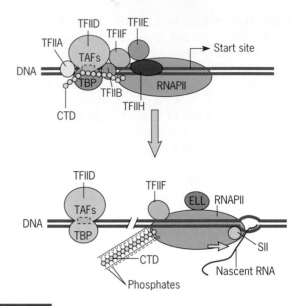

Figure 11.20 **Initiation of transcription by RNA polymerase II is associated with phosphorylation of the C-terminal domain (CTD).** Phosphorylation of the polymerase is thought to provide the trigger for the separation of the transcriptional machinery from the general transcription factors and/or the promoter DNA. Escape from the promoter is associated with major changes in the conformation of the polymerase, which converts it from an enzyme that "has trouble" synthesizing RNA to one that is highly processive. During the elongation phase of transcription, the polymerase moves along the DNA, unwinding the DNA duplex ahead of the active site, and rewinding the duplex in its wake. The 8–10 most recently incorporated nucleotides of the nascent RNA remain bound to the template strand in the form of an RNA-DNA hybrid. TFIIF may reassociate with the polymerase if it becomes stalled. SII and ELL are two of a number of elongation factors that are thought to be associated with the polymerase as it moves along the DNA.

Figure 11.19 **A model of the steps in the assembly of the preinitiation complex for RNA polymerase II at the TATA box of a eukaryotic promoter.** The polymerase itself is denoted RNAP II; the other components are the various general transcription factors required in assembly of the complete complex. TFIID includes the TBP subunit, which specifically binds to the TATA box, and a number of other subunits, which collectively are called TBP-associated factors (TAFs). TFIIB is thought to provide a binding site for RNA polymerase. TFIIF is bound to the entering polymerase. TFIIH contains 9 subunits, 3 of which possess enzymatic activities.

which separate the DNA strands of the promoter, allowing the polymerase access to the template strand. Another subunit of TFIIH functions as a protein kinase to phosphorylate RNA polymerase, as described below. Once transcription begins, one of the GTFs (TFIID) may be left behind at the promoter, while others are released from the complex (Figure 11.20). As long as TFIID remains bound to the promoter, additional RNA polymerase molecules may be able to attach to the promoter site and initiate additional rounds of transcription without delay.

The carboxyl-terminal domain (CTD) of the largest subunit of RNA polymerase II has an unusual structure; it consists of a sequence of seven amino acids (—Tyr—Ser—Pro—Thr—Ser—Pro—Ser—) that is repeated over and over. In mammals, the CTD consists of 52 repeats of this heptapeptide. Of the seven amino acids, all but the two prolines are prime candidates for phosphorylation by protein kinases. Studies indicate that the RNA polymerase that assembles into the preinitiation complex is not phosphorylated, whereas the same enzyme engaged in transcription is heavily phosphorylated; all of the phosphate groups are localized in the CTD (Figure 11.20). It appears that RNA polymerase II is phosphorylated either just before or soon after the initiation of transcription. Phosphorylation of the CTD is catalyzed by the protein kinase activity of TFIIH. Phosphorylation of the polymerase may act as the trigger that uncouples the enzyme from the GTFs and/or the promoter DNA, allowing the

enzyme to escape from the preinitiation complex and move down the DNA template. An RNA polymerase engaged in elongation may be associated with a number of large accessory proteins, only two of which (ELL and SII) are indicated in Figure 11.20. According to some estimates, an elongating RNA polymerase is part of a huge complex consisting of over 50 components and a total molecular mass of more than 3 million daltons (see Figure 11.37). In all likelihood, it is not the transcriptional machinery that moves along the template, but the template that moves through an immobilized transcriptional machine.

Together, RNA polymerase II and its GTFs are sufficient to promote a low, basal level of transcription from most promoters under in vitro conditions. As will be discussed at length in Chapter 12, a variety of *specific* transcription factors are able to bind at numerous sites in the regulatory regions of the DNA. These specific factors can determine (1) whether or not a preinitiation complex assembles at a particular promoter and/or (2) the rate at which the polymerase initiates new rounds of transcription from that promoter. Before discussing the pathway by which mRNAs are produced, let us first describe the structure of mRNAs so that the reasons for some of the processing steps will be clear.

The Structure of mRNAs Messenger RNAs share certain properties:

1. They contain a continuous sequence of nucleotides encoding a specific polypeptide.

2. They are found in the cytoplasm.

3. They are attached to ribosomes when they are translated.

4. Most mRNAs contain a significant noncoding segment, that is, a portion that does not direct the assembly of amino acids. For example, approximately 25 percent of each globin mRNA consists of noncoding, nontranslated regions (Figure 11.21). Noncoding portions are found on both the 5′ and 3′ ends of a messenger RNA and contain sequences that have important regulatory roles (Section 12.3).

5. Eukaryotic mRNAs have special modifications at their 5′ and 3′ termini that are found neither on prokaryotic messages nor on tRNAs or rRNAs. The 3′ end of nearly all eukaryotic mRNAs has a string of 50 to 250 adenosine residues that form a poly(A) tail, while the 5′ end has a methylated guanosine cap (Figure 11.21).

We will return shortly to describe how mRNAs are provided with their specialized 5′ and 3′ termini. First, however, it is necessary to take a short detour to understand how mRNAs are formed in the cell.

Split Genes: An Unexpected Finding Almost as soon as hnRNAs were discovered, it was proposed that this group of rapidly labeled nuclear RNAs were precursors to cyto-

Figure 11.21 Structure of the human β-globin mRNA. The mRNA contains a 5′ methylguanosine cap, a 5′ and 3′ noncoding region that flank the coding segment, and a 3′ poly(A) tail. The lengths of each segment are given in numbers of nucleotides. The length of the poly(A) tail is variable. It typically begins at a length of about 250 nucleotides and is gradually reduced in length, as discussed in Chapter 12. The structure of the 5′ cap is shown.

plasmic mRNAs (page 454). The major sticking point was the difference in size between the two RNA populations: the hnRNAs were several times the size of the mRNAs (Figure 11.22). Why would cells synthesize huge molecules that were precursors to much smaller versions? Early studies on the processing of ribosomal RNA had shown that mature RNAs were carved from larger precursors. Recall that large segments were removed from both the 5′ and 3′ sides of various rRNA intermediates (Figure 11.14) to yield the final, mature rRNA products. It was thought that a similar pathway might account for the processing of hnRNAs to mRNAs. But mRNAs constitute such a diverse population that it was impossible to follow the steps in the processing of a single mRNA species as had been done for the rRNAs. The problem was solved by an unexpected discovery.

It was assumed by molecular biologists until 1977 that a continuous linear sequence of nucleotides in a messenger RNA is complementary to a continuous sequence of nucleotides in one strand of the DNA of a gene. Then in that

Figure 11.22 The difference in size between hnRNAs and mRNAs. (*a,b*) Electron micrographs of metal-shadowed preparations of poly(A)-mRNA (*a*) and poly(A)-hnRNA (*b*) molecules. Representative size classes of each type are shown. Reference molecule is ΦX-174 viral DNA. (*c*) Size distribution of hnRNA and mRNA from mouse L cells as determined by density-gradient sedimentation. The red line represents rapidly labeled hnRNA, whereas the purple line represents mRNA that was isolated from polyribosomes after a four-hour labeling period. The abscissa has been converted from fraction number (indicated by the points) to molecular size by calibration of the gradients. (FROM JOHN A. BANTLE AND WILLIAM E. HAHN, CELL 8:145, 1976; BY PERMISSION OF CELL PRESS.)

year, a remarkable new finding was made by Phillip Sharp and his colleagues at MIT and Richard Roberts, Louise Chow, and their colleagues at the Cold Spring Harbor Laboratories in New York. These groups found that the mRNAs they were studying were transcribed from segments of DNA that were separated from one another along the template strand.

The first observations of major importance were made during the analysis of transcription of the adenovirus genome. Adenovirus is a pathogen capable of infecting a variety of mammalian cells. It was found that a number of different adenovirus mRNAs had the same 150- to 200-nucleotide 5′ terminus. One might expect that this leader sequence represents a repeated stretch of nucleotides located near the promoter region of each of the genes for these mRNAs. However, further analysis revealed that the 5′ leader sequence is not complementary to a repeated sequence and, moreover, is not even complementary to a continuous stretch of nucleotides in the template DNA. Instead, the leader is transcribed from three distinct and separate segments of DNA (represented by blocks x, y, and z in Figure 11.23). The regions of DNA between these blocks, referred to as **intervening sequences** (I_1 to I_3 in Figure 11.23), are somehow missing in the corresponding mRNA. It could have been argued that the presence of intervening sequences is a peculiarity of viral genomes, but the basic observation was soon extended to cellular genes themselves.

The presence of intervening sequences in nonviral, cellular genes was first reported in 1977 by Alec Jeffreys and Richard Flavell of the University of Amsterdam. These researchers discovered an intervening sequence of approximately 600 bases located directly within a part of the globin gene that coded for the amino acid sequence of the globin polypeptide (Figure 11.24). The basis for this finding is discussed in the legend accompanying the figure. Intervening sequences were soon found in other genes, and it became apparent that the presence of genes with intervening sequences—called **split genes**—is the rule not the exception. Those parts of a split gene that contribute to the mature RNA product are called **exons**, whereas the intervening sequences are called **introns**. Split genes are widespread among eukaryotes, both in the simpler eukaryotes, such as yeast and protists, and throughout the animal and plant kingdoms. The introns of simpler eukaryotes, however, tend to be fewer in number and smaller in size than those of plants and animals. Introns are found in all types of genes, including those that code for tRNAs, rRNAs, as well as mRNAs.

The discovery of genes with introns immediately raised the question as to how such genes were able to produce messenger RNAs lacking these sequences. One likely possibility was that cells produce a primary transcript that corresponds to the entire transcription unit, and that those portions of the RNA corresponding to the introns in the DNA are somehow removed. If this were the case, then the segments corresponding to the introns should be present in the primary transcript. Such an explanation would also

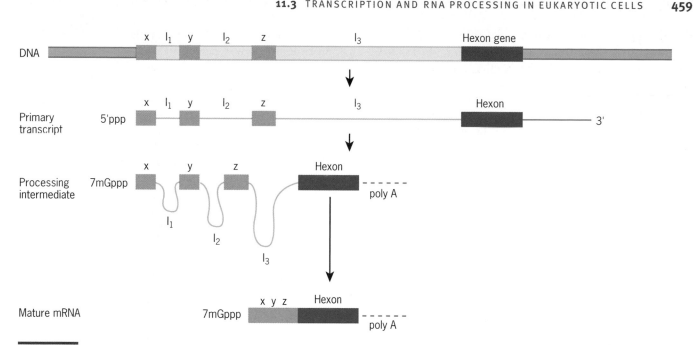

Figure 11.23 **The discovery of intervening sequences (introns).** A portion of the adenovirus genome is shown at the top. The noncontinuous sequence blocks labeled x, y, and z appear in a continuous arrangement in the mRNAs that code for a variety of polypeptides, such as the hexon protein. As discussed later in the text, the conversion of the primary transcript to the mRNA involves the removal (excision) of the intervening sequences, or introns (I_1 to I_3), and ligating the remaining portions to produce a continuous RNA molecule (bottom). The steps by which this occurs are shown in Figure 11.35.

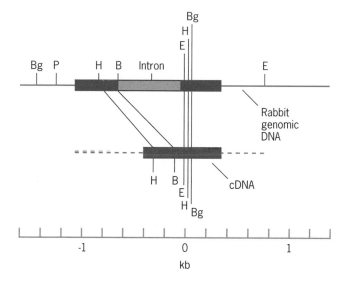

Figure 11.24 **The discovery of introns in a eukaryotic gene.** As discussed in Chapter 10, bacteria contain restriction enzymes that recognize and cleave DNA molecules at the site of certain nucleotide sequences. The drawing shows a map of restriction enzyme cleavage sites in the region of the rabbit β-globin gene (upper) and the corresponding map of a cDNA prepared from the β-globin mRNA (lower). (A cDNA is a DNA made in vitro by reverse transcriptase using the mRNA as a template. Thus, the cDNA has the complementary sequence to the mRNA. cDNAs had to be used for this experiment because restriction enzymes don't cleave RNAs.) The letters indicate the sites at which various restriction enzymes cleave the two DNAs. The upper map shows that the globin gene contains a restriction site for the enzyme *Bam*H1 (B) located approximately 700 base pairs from a restriction site for the enzyme *Eco*R1 (E). When the globin cDNA was treated with these same enzymes (lower map), the corresponding B and E sites were located only 67 nucleotides apart. It is evident that the DNA prepared from the genome has a sizable region that is absent from the corresponding cDNA (and thus absent from the mRNA from which the cDNA was produced). Complete sequencing of the globin gene later showed it to contain a second smaller intron. (FROM A. J. JEFFREYS AND R. A. FLAVELL, CELL 12:1103, 1977; BY PERMISSION FROM CELL PRESS.)

provide a reason why hnRNA molecules are so much larger than the mRNA molecules they ultimately produce.

Research on nuclear RNA had proceeded by this time to the point where the size of a few mRNA precursors (pre-mRNAs) had been determined. The globin sequence, for example, was found to be present within a nuclear RNA molecule that sediments at 15S, unlike the final globin mRNA, which has a sedimentation coefficient of 10S. An ingenious technique (known as R-loop formation) was employed by Shirley Tilghman, Philip Leder, and their coworkers at the National Institutes of Health to determine the physical relationship between the 15S and 10S globin RNAs and provide information on the transcription of split genes.

Recall from page 412, that single-stranded, complementary DNA strands can bind specifically to one another. Single-stranded DNA and RNA molecules can also bind

to one another as long as their nucleotide sequences are complementary; this is the basis of the technique of DNA-RNA hybridization discussed in Section 18.12 (the DNA-RNA complex is called a *hybrid*). Tilghman and her coworkers used the electron microscope to examine a fragment of DNA containing the globin gene that had been hybridized to the 15S globin RNA. The hybrid was seen to consist of a continuous, double-stranded, DNA-RNA complex (the red dotted and blue lines of Figure 11.25*a*). In contrast, when the same DNA fragment was incubated with the mature 10S globin mRNA, a large segment of the DNA in the center of the coding region was seen to bulge out to form a double-stranded loop (Figure 11.25*b*). The loop resulted from a large intron in the DNA that was not complementary to any part of the smaller globin message. It was apparent that the 15S RNA does indeed contain segments corresponding to the introns of the genes that are removed during formation of the 10S mRNA.

At about the same time, a similar type of hybridization experiment was performed between the DNA that codes for ovalbumin, a protein found in hen's eggs, and its corresponding mRNA. The ovalbumin DNA and mRNA hybrid contains seven distinct loops corresponding to seven introns (Figure 11.26). Taken together, the introns accounted for about three times as much DNA as that present in the eight combined coding portions (exons). Subsequent studies have revealed that individual exons are typically less than 300 nucleotides. Individual introns typically range from 1,000 to 100,000 nucleotides, which is why hnRNA molecules are typically much longer than mRNAs. One mouse gene extends for over 20 times the length needed to code for its corresponding message, and the type I collagen gene contains over 50 introns.

These and other findings provided strong evidence for the proposition that mRNA formation in eukaryotic cells occurs by the removal of internal sequences of ri-

(a)

(b)

Figure 11.25 Visualizing an intron in the globin gene. Electron micrographs of hybrids formed between (*a*) 15S globin precursor RNA and the DNA of a globin gene and (*b*) 10S globin mRNA and the same DNA as in *a*. The red colored lines indicate the positions of the RNA molecules. The precursor mRNA is equivalent in length and sequence to the DNA of the globin gene, but the 10S mRNA is missing a portion that is present in the DNA of the gene. These results suggest that the 15S RNA is processed by removing an internal RNA sequence and rejoining the flanking regions. (FROM SHIRLEY M. TILGHMAN ET AL., PROC. NAT'L ACAD. SCI. U. S. A. 75:1312, 1978.)

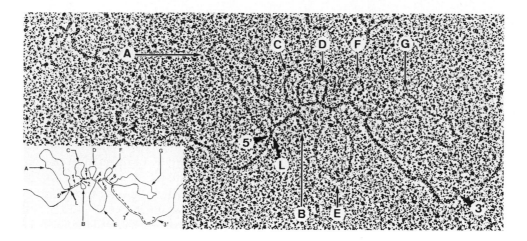

Figure 11.26 Introns in the ovalbumin gene. Electron micrograph of a hybrid formed between ovalbumin mRNA and a fragment of genomic chicken DNA containing the ovalbumin gene. The hybrid shown in this micrograph is similar in nature to that of Figure 11.25*b*. In both cases, the DNA contains the entire gene sequence because it was isolated directly from the genome.

In contrast, the RNA has been completely processed, and those portions that were transcribed from the introns have been removed. When the genomic DNA and the mRNA are hybridized, those portions of the DNA that are not represented in the mRNA are thrown into loops. The loops of the seven introns (A–G) can be distinguished. (COURTESY OF PIERRE CHAMBON.)

bonucleotides from a much larger pre-mRNA. Let us turn to the steps by which this occurs.

The Processing of Eukaryotic Messenger RNAs RNA polymerase II assembles a primary transcript that is complementary to the DNA of the entire transcription unit. This transcript is then processed in the nucleus to produce a mature mRNA that is transported to the cytoplasm.

Electron microscopic examination of transcriptionally active genes indicates that RNA transcripts become associated with proteins and larger particles while they are still in the process of being synthesized (Figure 11.27). These particles, which consist of proteins and ribonucleoproteins, include the agents responsible for converting the primary transcript into a mature messenger. This conversion process requires addition of a 5′ cap and 3′ poly(A)

(a) *(b)*

Figure 11.27 Pre-mRNA transcripts are processed as they are synthesized (cotranscriptionally). (*a*) Electron micrograph of a nonribosomal transcription unit showing the presence of ribonucleoprotein particles attached to the nascent RNA transcripts. (*b*) Interpretive tracing of the micrograph shown in part *a*. The dotted line represents the chromatin strand, the solid lines represent ribonucleoprotein (RNP) fibrils, and solid circles represent RNP particles associated with the fibrils. The RNP particles are not distributed randomly along the nascent transcript, but rather are bound at specific sites where RNA processing is taking place. (FROM ANN L. BEYER, OSCAR L. MILLER, JR., AND STEVEN L. MCKNIGHT, CELL 20:78, 1980; BY PERMISSION OF CELL PRESS.)

tail to the ends of the transcript, and removal of any intervening introns.

5′ Caps and Poly(A) Tails The 5′ ends of all RNAs initially possess a triphosphate derived from the first nucleoside triphosphate incorporated at the site of initiation of RNA synthesis. Once the 5′ end of an mRNA precursor has been synthesized, several enzyme activities act on this end of the molecule (Figure 11.28). In the first step, the last of the three phosphates is removed, converting the 5′ terminus to a diphosphate (step 1, Figure 11.28). Then, a GMP is added in an *inverted* orientation so that the 5′ end of the guanosine is facing the 5′ end of the RNA chain (step 2, Figure 11.28). As a result, the first two nucleosides are joined by a 5′–5′ triphosphate bridge. Finally, the ter-

minal, inverted guanosine is methylated at the 7 position on its guanine base, while the nucleotide on the internal side of the triphosphate bridge is methylated at the 2′ position of the ribose (step 3, Figure 11.28). The 5′ end of the RNA now contains a **methylguanosine cap** (shown in greater detail in Figure 11.21). These enzymatic modifications at the 5′ end of the primary transcript occur very quickly, while the RNA molecule is still in its very early stages of synthesis. In addition to methylation of the terminal nucleotides of the mRNA precursor, various internal nucleotide residues may or may not be methylated as well. The methylguanosine cap at the 5′ end of an mRNA serves several functions: it prevents the 5′ end of the mRNA from being digested by exonucleases, it aids in the transport of the mRNA out of the nucleus, and it plays an important role in the initiation of mRNA translation (page 480).

As noted above, the 3′ end of an mRNA contains a string of adenosine residues that forms a **poly(A) tail**. As a number of mRNAs were sequenced, it became evident that the poly(A) tail invariably begins approximately 15 nucleotides downstream from the sequence AAUAAA. This sequence in the primary transcript serves as a recognition site for the assembly of a complex of proteins that carry out the processing reactions at the 3′ end of the mRNA (Figure 11.28). Studies suggest that the assembled poly(A) processing complex is physically associated with the RNA polymerase as it synthesizes the primary transcript (see Figure 11.37).

Included among the proteins of the processing complex is an endonuclease that cleaves the pre-mRNA downstream from the recognition site (Figure 11.28, top). Following cleavage by the nuclease, an enzyme called *poly(A) polymerase* adds 250 or so adenosines without the need of a template (steps a–c, Figure 11.28). As discussed

Figure 11.28 Steps in the addition of a 5′ methylguanosine cap and a 3′ poly(A) tail to a pre-mRNA. The 5′ end of the nascent pre-mRNA binds to a capping enzyme with two active sites that catalyze different reactions: a phosphohydrolase that removes the terminal phosphate group (step 1) and a guanylyl transferase that adds a guanine residue in a reverse orientation, by means of a 5′-to-5′ linkage (step 2). In step 3, different methyltransferases add a methyl group to the terminal guanosine cap and to the ribose of the nucleotide that had been at the end of the nascent RNA. A very different series of events occur at the 3′ end of the pre-mRNA, where a large protein complex is assembled. First, an endonuclease cleaves the RNA strand, generating a new 3′ end. In steps a–c, poly(A) polymerase adds adenosine residues to the 3′ end without the involvement of a DNA template. The growing poly(A) tail associates with multiple copies of a protein (PABII) that stimulates continued addition of adenosine residues by the polymerase. A typical mammalian mRNA contains 200 to 250 adenosine residues in its completed poly(A) tail; the number is considerably less in lower eukaryotes. (AFTER D. A. MICKLOS AND G. A. FREYER, DNA SCIENCE, CAROLINA BIOLOGICAL SUPPLY CO.)

in Section 12.3, the poly(A) tail protects the mRNA from premature degradation by exonucleases. Poly(A) tails have also proven useful to molecular biologists interested in isolating mRNA. When a mixture of cellular RNAs is passed through a column containing bound synthetic poly(T), the mRNAs bind to the column and are removed from solution, while the more abundant tRNAs, rRNAs, and snRNAs that lack the poly(A) tail pass through the column along with the aqueous solvent.

RNA Splicing: Removal of Introns from a Pre-mRNA The key steps in the processing of a pre-mRNA are shown in Figure 11.29. In addition to formation of the 5′ cap and poly(A) tail, which have already been discussed, those parts of a primary transcript that correspond to the inter-

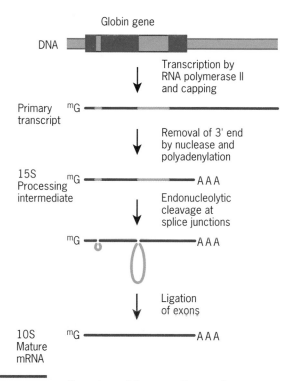

Globin gene

DNA

Transcription by RNA polymerase II and capping

Primary transcript ᵐG

Removal of 3′ end by nuclease and polyadenylation

15S Processing intermediate ᵐG ——— A A A

Endonucleolytic cleavage at splice junctions

ᵐG ——— A A A

Ligation of exons

10S Mature mRNA ᵐG ——————— A A A

Figure 11.29 Overview of the steps during the processing of the globin mRNA. Introns are shown in violet, whereas dark blue portions of the gene indicate the positions of the exons, that is, the DNA sequences that are represented in the mature messenger RNA.

vening DNA sequences (the introns) must be removed by a complex process known as **RNA splicing**. To splice an RNA, breaks in the strand must be introduced at the 5′ and 3′ edges (or **splice sites**) of each intron, and the exons situated on either side of the splice sites must be covalently joined (ligated). It is imperative that the splicing process occur with absolute precision, because the addition or loss of a single nucleotide at any of the splice junctions would cause the resulting mRNA to be mistranslated.

How does the same basic splicing machinery recognize the exon-intron boundaries in thousands of different pre-mRNAs? Examination of hundreds of junctions between exons and introns in eukaryotes ranging from yeast to insects to mammals revealed the presence in splice sites of a conserved nucleotide sequence of ancient evolutionary origin. In marked contrast, the sequences of the internal portions of introns tend to be highly divergent. The sequence most commonly found at the exon-intron borders within mammalian pre-mRNA molecules is shown in Figure 11.30. The G/GU at the 5′ end of the intron (*the 5′ splice site*) and the AG/G at the 3′ end of the intron (*the 3′ splice site*) are present in the vast majority of eukaryotic pre-mRNAs.[4] In addition, the adjacent regions of the intron contain preferred nucleotides, as indicated in Figure 11.30, which play an important role in splice site recognition. Studies indicate that specific sequences within the exons (called *exonic enhancers*) also play a role in intron recognition. Changes in DNA sequence within either a splice site or an adjacent exon can block the excision of an intron, leading to the accumulation of unspliced precursors. Certain types of inherited anemias (called thalassemias), for example, are commonly caused by mutations in the splice sites of the globin gene.

Understanding the mechanism of RNA splicing has come about through an appreciation of the remarkable capabilities of RNA molecules. The first evidence that RNA molecules were capable of catalyzing chemical reactions was obtained in 1982 by Thomas Cech and colleagues at the University of Colorado. As discussed at

[4]Approximately one percent of introns have AT and AC dinucleotides at their 5′ and 3′ ends (rather than GU and AG). Many of these AT/AC introns are processed by a different type of spliceosome, called a U12 spliceosome, because it contains a U12 snRNA in place of the U2 snRNA of the major spliceosome. U12 spliceosomes are absent from yeast and nematodes, but present in plants, insects, and vertebrates.

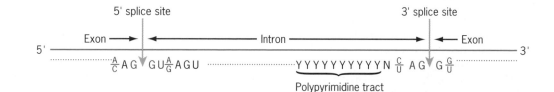

5′ splice site 3′ splice site

Exon ———→ ←———————————— Intron ————————————→ ←——— Exon

5′ ··········· A_C A G ▼ G U A_G A G U ············ Y Y Y Y Y Y Y Y Y Y Y N C_U A G ▼ G G_U ··········· 3′

Polypyrimidine tract

Figure 11.30 Nucleotide sequences at the splice sites of pre-mRNAs. The nucleotide sequences shown in the regions of the splice sites are based on analysis of a large number of pre-mRNAs and are therefore referred to as *consensus* sequences. The bases shown in orange are virtually invariant, those in black represent the preferred base at that position. N represents any of the four nucleotides, Y represents a pyrimidine. The polypyrimidine tract near the 3′ splice site typically contains between 10 and 20 pyrimidines.

Figure 11.31 The structure of group I introns. The complex three-dimensional structure of the *Tetrahymena* group I intron is indicated in this 1990 model, which has recently been confirmed by X-ray crystallography. The two exons that flank the intron are shown in white. As a result of the folding of the intron, the two exons are brought into close proximity, allowing them to be ligated after removal of the intron. Base pairs are indicated as phosphate-phosphate bridges between paired strands. (COURTESY OF E. WESTHOF.)

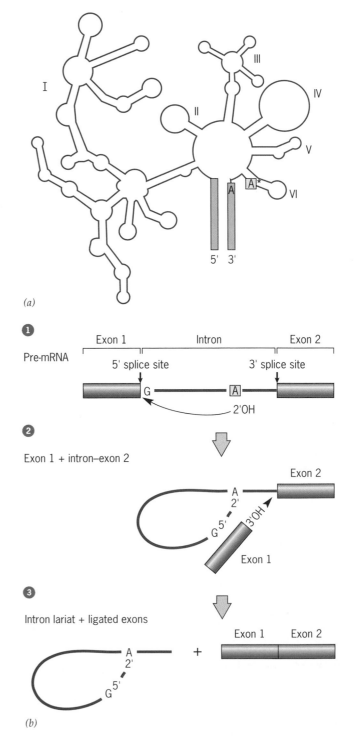

Figure 11.32 The structure and self-splicing pathway of group II introns. (*a*) Two-dimensional structure of a group II intron. The intron folds into six characteristic domains. The asterisk indicates the adenosine nucleotide that bulges out of domain VI and forms the lariat structure as described in the text. The two ends of the intron become closely applied to one another as indicated by the proximity of the two intron-exon boundaries. (*b*) Steps in the self-splicing of group II introns. In step 1, the 2′ OH of an adenosine within the intron (asterisk in domain VI of part *a*) carries out a nucleophilic attack on the 5′ splice site, cleaving the RNA and forming an unusual 2′–5′ phosphodiester bond with the first nucleotide of the intron. This branched structure is described as a lariat. In step 2, the free 3′ OH of the displaced exon attacks the 3′ splice site, which cleaves the RNA at the other end of the intron. As a result of this reaction, the intron is released as a free lariat, and the 3′ and 5′ ends of the two flanking exons are ligated (step 3). A similar pathway is followed in the splicing of introns from pre-mRNAs, but rather than occurring by self-splicing, these steps require the aid of a number of additional factors.

length in the Experimental Pathways for this chapter, these researchers found that the ciliated protozoan *Tetrahymena* synthesized an rRNA precursor (a pre-rRNA) that was capable of splicing itself. In addition to revealing the existence of RNA enzymes, or **ribozymes**, these experiments changed the thinking among biologists about the relative roles of RNA and protein in the mechanism of RNA splicing.

The intron in the *Tetrahymena* pre-rRNA is an example of a *group I intron*. Group I introns are found most commonly in fungal and plant mitochondria, plant chloroplasts, and in the nuclear RNA of lower eukaryotes, such as *Tetrahymena*. Though the nucleotide sequence of group I introns may be quite variable, all of them form very similar three-dimensional structures (Figure 11.31). It is this complex folded structure, together with several well-placed patches of conserved nucleotides, that allows these introns to catalyze the reactions necessary to excise themselves from a larger RNA molecule.

Another type of self-splicing intron, called a *group II intron*, was subsequently discovered in fungal mitochondria and plant chloroplasts. Group II introns fold into a complex structure (Figure 11.32*a*) that is very different from that of group I introns. Unlike group I introns, group II introns undergo self-splicing by passing through an intermediate stage, called a *lariat* (Figure 11.32*b*) because it resembles the type of rope used by cowboys to catch runaway calves. The first step in group II intron splicing is the cleavage of the 5′ splice site (step 1, Figure 11.32*b*), followed by formation of a lariat by means of a covalent bond between the 5′ end of the intron and an adenosine residue near the 3′ end of the intron (step 2). The subsequent cleavage of the 3′

splice site releases the lariat and allows the cut ends of the exon to be covalently joined (ligated) (step 3).

The steps that occur during the removal of introns from pre-mRNA molecules in animal cells are similar to those followed by group II introns. The primary difference is that the pre-mRNA is not able to splice itself, requiring instead a host of nuclear snRNAs and their associated proteins (Figure 11.33). As each large hnRNA molecule is transcribed, it becomes associated with a variety of proteins to form an **hnRNP (heterogeneous nuclear ribonucleoprotein)**, which represents the substrate for the processing reactions that follow. Processing occurs as each intron of the pre-mRNA becomes associated with a macromolecular complex called a **spliceosome** (Figure 11.34). Each spliceosome consists of a variety of proteins and a number of distinct ribonucleoprotein particles, called **snRNPs** (pronounced "snurps") because they are composed of snRNAs (small nuclear RNAs) bound to specific proteins. Spliceosomes are not present within the nucleus in a prefabricated state, but rather are assembled as their component snRNPs bind to the pre-mRNA. Once the spliceosome machinery is assembled, the snRNPs carry out the reactions that cut the introns out of the transcript and paste the ends of the exons together.

Our understanding of the steps in RNA splicing has been achieved largely through studies of cell-free extracts that can accurately splice pre-mRNAs in vitro. Some of the major steps in the assembly of a spliceosome and removal of an intron are indicated in Figure 11.35 and described in some detail in the accompanying legend. Taken together, removal of an intron requires several snRNP particles: the U1snRNP, U2 snRNP, U5 snRNP, and the U4/U6 snRNP, which contains the U4 and U6 snRNAs bound together.

The events described in Figure 11.35 provide excellent examples of the complex and dynamic interactions

Figure 11.33 Visual demonstration of the presence of small nuclear ribonucleoproteins (snRNPs) associated with nascent transcripts. Giant polytene chromosomes from the salivary glands of *Drosophila melanogaster* were stained with a fluorescent antibody against the protein of snRNP particles. The red colored fluorescent antibodies bind selectively to material present in the puffed regions of the chromosomes, which are sites where transcription is occurring. (FROM ERIKA L. MATUNIS, MICHAEL J. MATUNIS, AND GIDEON DREYFUSS, J. CELL BIOL. 121:226, 1993; BY COPYRIGHT PERMISSION OF THE ROCKEFELLER UNIVERSITY PRESS.)

Figure 11.34 The spliceosome. Electron micrographs of several isolated spliceosomes. Note that each of these large complexes can be seen to consist of several distinct subunits. Inset (top left) shows a particle with a thin RNA filament and a smaller particle at the end of the filament. (COURTESY OF JACK GRIFFITH, UNIVERSITY OF NORTH CAROLINA AT CHAPEL HILL.)

Figure 11.35 Schematic model of the assembly of the splicing machinery and some of the steps that occur during splicing. Step 1 shows the portion of the pre-mRNA to be spliced. In step 2, the first of the splicing components, U1 snRNP, has become attached at the 5′ splice site of the intron. The nucleotide sequence of U1 snRNA is complementary to the 5′ splice site of the pre-mRNA, and evidence indicates that U1 snRNP initially binds to the 5′ side of the intron by the formation of specific base pairs between the splice site and U1 snRNA (see inset A). The U2 snRNP is next to enter the splicing complex, binding to the pre-mRNA (as shown in inset A) in a way that causes a specific adenosine residue (dot) to bulge out of the surrounding helix (step 3). This is the site that later becomes the branch point of the lariat. The next step is the binding of the U4/U6 and U5 snRNPs to the pre-mRNA with the accompanying displacement of U1 (step 4). The assembly of a spliceosome involves a series of dynamic interactions between the pre-mRNA and specific snRNAs and among the snRNAs themselves. As they enter the complex with the pre-mRNA, the U4 and U6 snRNAs are extensively base-paired to one another (inset B). The U4 snRNA is subsequently stripped away from the duplex,

and the regions of U6 that were paired with U4 become base-paired to a portion of the U2 snRNA (inset C). Another portion of the U6 snRNA is situated at the 5′ splice site (inset C), having displaced the U1 snRNA that was previously bound there (inset A). It is thought that U6 is a ribozyme and that U4 is an inhibitor of its catalytic activity. Once the U1 and U4 snRNA have been displaced, the U6 snRNA is in position to catalyze the two chemical reactions required for intron removal. The first reaction (indicated by the arrow in inset C) results in the cleavage of the 5′ splice site, forming a free 5′ exon and a lariat intron–3′ exon intermediate (step 5). The free 5′ exon is thought to be held in place by its association with the U5 snRNA of the spliceosome, which also interacts with the 3′ splice site (step 5). The first cleavage reaction at the 5′ splice site is followed by a second cleavage reaction at the 3′ splice site, which excises the lariat intron and simultaneously joins the ends of the two neighboring exons (step 6). Following splicing, the snRNPs must be released from the pre-mRNA, the original associations between snRNAs must be restored, and the snRNPs must be reassembled at the sites of other introns.

that can occur between RNA molecules. In several of the steps depicted in Figure 11.35, an snRNA that is originally base-paired to one RNA molecule is stripped of that partner and base-paired to a different RNA molecule. Of the various snRNAs that participate in RNA splicing, U6 is most likely to act as a ribozyme, making both cuts in the pre-mRNA required for intron removal.

In addition to its snRNA, each snRNP contains a dozen or more proteins. One family, called *Sm proteins*, are present in all of the snRNPs. Sm proteins bind to one another and to a conserved site on each snRNA (except U6 snRNA) to form the core of the snRNP. Sm proteins were first identified because they are the targets of antibodies produced by patients with the autoimmune disease systemic lupus erythematosus. The other proteins of the snRNPs are unique to each particle. For example, the rearrangements among RNA molecules that occur during the assembly of a spliceosome are probably mediated by ATP-consuming, RNA helicases (enzymes that unwind double-stranded RNAs) present within the snRNPs. At least eight different helicases have been implicated in splicing of pre-mRNAs in yeast.

The fact that (1) pre-mRNAs are spliced by the same pair of chemical reactions that occur as group II introns splice themselves, and (2) the snRNAs that are required for splicing pre-mRNAs closely resemble parts of the group II introns (Figure 11.36), strongly suggest that the snRNAs are the catalytically active components of the snRNPs, not the proteins. The proteins likely serve supplementary roles: to maintain the proper three-dimensional structure of the snRNA, to drive changes in snRNA conformation, to transport spliced mRNAs to the nuclear envelope, and to select the splice sites to be used during the processing of a particular pre-mRNA.

The proteins of the snRNPs are not the only proteins involved in mRNA processing. Another family of RNA-

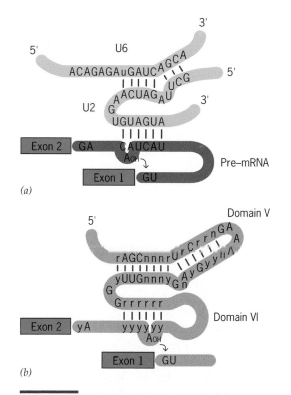

Figure 11.36 Proposed structural similarity between the splicing reactions carried out by the spliceosome on pre-mRNAs and the self-splicing reactions of group II introns. (*a*) The structural interactions between an intron of a premRNA molecule (flanked by exons 1 and 2) and two snRNAs (U2 and U6) that are required for splicing. (*b*) The structure of a group II intron showing the arrangement of critical parts that become aligned during self-splicing (as indicated in Figure 11.32). The resemblance of the parts of the group II intron to the combined RNAs in part *a* is evident. Invariant residues are shown in upper case letters; conserved purines and pyrimidines are denoted by r and y, respectively. Variable residues are denoted by n. (AFTER A. M. WEINER, CELL 72:162, 1993; BY PERMISSION OF CELL PRESS.)

Figure 11.37 **Schematic representation of a mechanism for the coordination of transcription, capping, polyadenylation, and splicing.** In this simplified model, the C-terminal domain of the large subunit of the RNA polymerase (page 456) serves as a scaffold for the organization of factors involved in processing pre-mRNAs, including those for capping, polyadenylation, and intron removal. The double-headed arrows indicate potential physical and functional interactions between factors. In addition to the proteins depicted here, the polymerase is probably associated with a host of transcription factors. (AFTER E. J. STEINMETZ, CELL 89:493, 1997, BY PERMISSION OF CELL PRESS.)

binding proteins, called *SR proteins* because of their large number of serine (S) and arginine (R) dipeptides, are thought to form interacting networks that span the intron/exon borders and help recruit snRNPs to the splice sites. Positively charged SR proteins may also bind electrostatically to the negatively charged phosphate groups that are added to the CTD of the polymerase as transcription is initiated (page 456). As a result, the assembly of the splicing machinery at an intron is thought to occur in conjunction with the synthesis of the intron by the polymerase. Based on current evidence, most of the machinery required for RNA processing travels with the polymerase as part of a giant "mRNA factory" (Figure 11.37).

Because most genes contain a number of introns, the splicing reactions depicted in Figure 11.35 must occur repeatedly on a single primary transcript. Evidence suggests that introns are removed in a preferred order, generating specific processing intermediates whose size lies between that of the primary transcript and the mature mRNA. An example of the intermediates that form during the nuclear processing of the ovomucoid mRNA in cells of the hen's oviduct is shown in Figure 11.38.

Evolutionary Implications of Split Genes and RNA Splicing

The discovery that RNAs are capable of catalyzing chemical reactions has had enormous impact on our view of biological evolution. Ever since the discovery of DNA as the genetic material, biologists have wondered which came first, protein or DNA. The dilemma arose from the seem-

Figure 11.38 **Processing the ovomucoid pre-mRNA.** The photograph shows a technique called a Northern blot in which extracted RNA (in this case from the nuclei of hen oviduct cells) is fractionated by gel electrophoresis and blotted onto a membrane filter. The immobilized RNA on the filter is then incubated with a radioactively labeled cDNA (in this case a cDNA made from the ovomucoid mRNA) to produce bands that reveal the position of RNAs containing the complementary sequence. The mature mRNA encoding the ovomucoid protein is 1,100 nucleotides long and is shown at the bottom of the blot. It is evident that the nucleus contains a number of larger-sized RNAs that also contain the nucleotide sequence of the ovomucoid mRNA. The largest RNA on the blot has a length of 5,450 nucleotides, which corresponds to the size of the ovomucoid transcription unit; this RNA is presumably the primary transcript from which the mRNA is ultimately carved. Other prominent bands contain RNAs with lengths of 3,100 nucleotides (which corresponds to a transcript lacking introns 5 and 6), 2,300 nucleotides (a transcript lacking introns 4, 5, 6, and 7), and 1,700 nucleotides (a transcript lacking all introns except number 3). (COURTESY OF BERT O'MALLEY.)

ingly nonoverlapping functions of these two types of macromolecules. Nucleic acids store information, whereas proteins catalyze reactions. With the discovery of ribozymes in the early 1980s, it became apparent that one type of molecule—RNA—could do both.

These findings have fueled the belief that both DNA and protein were probably absent at an early stage in the evolution of life. During this period, RNA molecules performed double duty: they served as genetic material, and

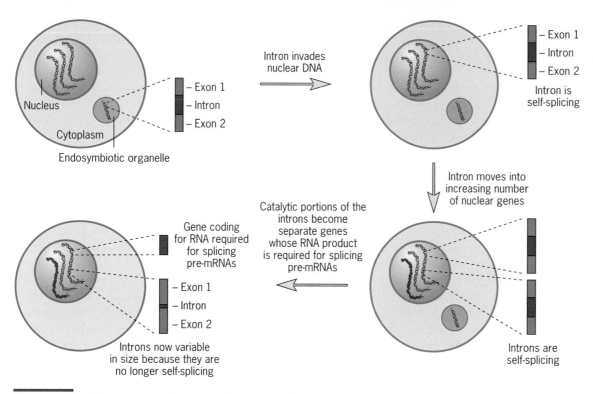

Figure 11.39 A hypothesis that may explain the evolutionary origin of introns in eukaryotic DNA. The proposed steps in the evolution of a self-splicing intron in an ancestral prokaryotic symbiont into introns in the eukaryotic genome. The steps are discussed in the text.

they catalyzed chemical reactions. They also had to be able to replicate themselves. Life at this stage is described as "an RNA world." Only at a later stage in evolution were the functions of catalysis and information storage taken over by protein and DNA, respectively, leaving RNA to function primarily as a go-between in the flow of genetic information. Many researchers believe that splicing provides an example of a legacy from an ancient RNA world.

Insights into the ancestry of RNA splicing have arisen from the discovery of group II introns in purple bacteria and cyanobacteria, which are thought to be close relatives of the ancestors of mitochondria and chloroplasts, respectively. This finding supports the hypothesis that group II introns are the source from which pre-mRNA introns have arisen. The hypothesis is outlined in Figure 11.39. According to this hypothesis, introns were originally present in the endosymbiotic organelles that resided inside primitive eukaryotic cells. Over time, the introns were transferred from the organelle genome to the nuclear genome where they took up permanent residence. The transfer of genes from mitochondria and chloroplasts to the cell nucleus is well documented, so the movement of DNA in this direction is not unexpected. Once in the nucleus, the introns are thought to have moved from place to place by transposition, a phenomenon discussed on page 422 of the last chapter. In fact, some introns of modern nuclear genes are still capable of acting like mobile genetic elements.

Because introns were originally capable of self-splicing, their presence in the middle of genes would not have caused a problem. Sequences corresponding to introns would have simply excised themselves from the primary RNA transcripts. Over time, the catalytic portions of introns moved from the interior of protein-coding genes to separate locations in the genome. The RNAs encoded by these "new" genes continued to participate in the splicing process. Over additional time, the RNAs encoded by these genes evolved into the snRNAs and their catalytic activity became dependent on proteins. Together, the snRNAs and proteins evolved into the snRNP components of the spliceosome. As this occurred, the internal nucleotides of introns no longer had any function, which explains their variable length and divergent nucleotide sequences.

Although the presence of introns may have created an added burden for cells, because they had to remove these intervening sequences from their transcripts, introns are not without their virtues. As we will see in the following chapter, RNA splicing is one of the steps along the path to mRNA formation that is subject to cellular regulation. Many primary transcripts can be processed by two or more pathways so that a sequence that acts as an intron in one pathway becomes an exon in an alternate pathway. As a result of this process, called *alternate splicing*, the same gene can code for more than one polypeptide.

It was also discovered that certain RNAs—namely, the snoRNAs required for rRNA processing (page 452)—

are encoded by introns rather than exons. Most of these snoRNAs are contained within the introns of genes that code for certain polypeptides involved in ribosome assembly and function. When these genes are transcribed and the introns are removed from the primary transcripts, rather than being discarded, some of these introns are processed into snoRNAs. Several genes have been discovered in which the roles of the introns and exons are essentially reversed. These genes are transcribed into primary transcripts whose introns are processed into snoRNAs, while the exons are degraded without ever giving rise to an mRNA.

The presence of introns is also thought to have had a major impact on biological evolution. When the amino acid sequence of a protein is examined, it is often found to contain sections that are homologous to parts of several other proteins (see Figure 2.36, page 61 for an excellent example). Proteins of this type are encoded by genes that are almost certainly composites made up of parts of other genes. The movement of genetic "modules" among different genes—a process called **exon shuffling**—is greatly facilitated by the presence of introns, which act like inert spacer elements between exons. Genetic rearrangements require breaks in DNA molecules, which can occur within introns without introducing mutations that might impair the organism. Over long periods of time, exons can be shuffled independently in various ways, allowing a nearly infinite number of combinations in search for new and useful coding sequences. As a result of exon shuffling, evolution need not occur only by the slow accumulation of point mutations, but it might also move ahead by "quantum leaps" with new proteins appearing within a single generation.

Creating New Ribozymes in the Laboratory

The main sticking point in the minds of many biologists concerning the feasibility of an RNA world in which RNA acted as the sole catalyst is that, to date, only a few reactions have been found to be catalyzed by *naturally occurring* RNAs. These include the cleavage and ligation of phosphodiester bonds required for RNA splicing and the formation of peptide bonds during protein synthesis. Are these the only types of reactions that RNA molecules are capable of catalyzing, or has their catalytic repertoire been sharply restricted by the evolution of more efficient protein enzymes? Several groups of researchers are currently exploring the catalytic *potential* of RNA by creating new RNA molecules in the laboratory. Although these experiments can never prove that such RNA molecules existed in ancient organisms, they provide proof of the *principle* that such RNA molecules could have existed.

In one approach, researchers are creating catalytic RNAs from scratch without any preconceived design as to how the RNA should be constructed. The RNAs are produced by allowing automated DNA-synthesizing machines to assemble DNAs with *random* nucleotide sequences. Transcription of these DNAs produces a population of RNAs whose nucleotide sequences are also randomly determined. Once a population of RNAs is obtained, individual members can be selected from the population by virtue of particular properties they possess. This approach has been described as "molecular evolution in the laboratory."

In one group of studies, synthetic RNAs have been selected for their ability to bind to a particular ligand, such as a nucleoside triphosphate or a peptide, by affinity chromatography (Section 18.7). In these experiments, the ligands are linked to beads, which are then used to pack a chromatography column. When a preparation of RNA is allowed to seep through the column, those RNAs that are able to bind to the immobilized ligand are pulled out of solution, while the vast majority of RNA molecules simply pass through unaffected. Once a subpopulation of RNAs has been selected for their binding properties, their numbers can be increased under conditions that introduce random substitutions (mutations) into their base sequence. This new population of RNAs can then be subjected to another round of selection under more stringent conditions. With each round of selection, those RNAs that "pass the test" bind with greater and greater affinity to the immobilized ligand.

It was expected that some of the RNAs that bind with high affinity to a ligand might possess catalytic activities capable of modifying that ligand. This expectation has been met in numerous experiments carried out in several laboratories over the past decade. In one study, researchers initially selected for RNAs that bound to ATP and, subsequently, for a subpopulation that hydrolyzed the ATP and transferred the phosphate group to the end of another RNA. The transfer of a phosphate group from ATP to a suitable acceptor is one of the most important classes of reactions carried out by protein enzymes in living cells.

Several other reactions of basic biological importance can be catalyzed by ribozymes that have been assembled in the laboratory. For example, one study has reported on ribozymes that function like RNA polymerases; they are capable of incorporating several nucleotides onto the end of an existing RNA using a complementary strand as a template. This finding provides support for the concept of an RNA world because ancient life forms that depended on RNA genomes had to have some mechanism for self-replication.

Other studies have described the synthesis of ribozymes that are capable of adding an amino acid onto the 3′ end of an RNA. In one study, a ribozyme was obtained that would transfer a specific amino acid onto the 3′ end of a targeted tRNA. This is the same basic reaction carried out by aminoacyl-tRNA synthetases, which are enzymes that link amino acids to tRNAs as required for protein synthesis (page 478). It is speculated that amino acids may have been used initially as adjuncts (cofactors) to enhance catalytic reactions carried out by ribozymes. Over time, ribozymes presumably evolved that were able to string amino acids together to form small proteins, which were more versatile catalysts than their RNA predecessors. As

we will see later in this chapter, ribosomes—the ribonucleoprotein machines responsible for protein synthesis—are essentially ribozymes at heart, which provides strong support for this evolutionary scenario.

As proteins took over a greater share of the workload in the primitive cell, the RNA world was gradually transformed into an "RNA-protein world." At a later point in time, RNA was presumably replaced by DNA as the genetic material, which propelled life forms into the present "DNA-RNA-protein world." The evolution of DNA may have required only two types of enzymes: a ribonucleotide reductase to convert ribonucleotides into deoxyribonucleotides and a reverse transcriptase to transcribe RNA into DNA. The fact that RNA catalysts do not appear to be involved in either DNA synthesis or transcription supports the idea that DNA was the last member of the DNA-RNA-protein triad to appear on the scene.

Somewhere along the line of evolutionary progress, a code had to evolve that would allow the genetic material to specify the sequence of amino acids to be incorporated into a given protein. The nature of this code is the subject of the next section of the chapter.

REVIEW ?

1. Describe the difference between a primary transcript, a transcription unit, a mature mRNA, and a processing intermediate.

2. Draw a representation of an electron micrograph of rDNA being transcribed. Label the nontranscribed spacer, the transcribed spacer, the RNA polymerase molecules, the U3 snRNP, and the promoter.

3. Compare the organization of the genes that code for the large rRNAs, the 5S RNA, and the tRNAs within the vertebrate genome.

4. What is a split gene? How was the existence of split genes discovered?

5. What is the relationship between hnRNAs and mRNAs? How was this relationship uncovered?

6. What are the general steps in the processing of a pre-mRNA into an mRNA? What is the role of the snRNAs and the spliceosome?

7. What is meant by the term "RNA world"? What type of evidence argues for its existence?

11.4 ENCODING GENETIC INFORMATION

Once the structure of DNA had been revealed in 1953, it became evident that the sequence of amino acids in a polypeptide was specified by the sequence of nucleotides in the DNA of a gene. It seemed very unlikely that the DNA could serve as a direct, physical template for the assembly of a protein. Instead, it was widely assumed that the information stored in the nucleotide sequence was present in some type of **genetic code**. With the discovery of messenger RNA as an intermediate in the flow of information from DNA to protein, attention turned to the manner in which a sequence written in a ribonucleotide "alphabet" would code for a sequence in an "alphabet" consisting of amino acids.

The Properties of the Genetic Code

One of the first models of the genetic code was presented by the physicist George Gamow, who proposed that each amino acid in a polypeptide was encoded by three sequential nucleotides. In other words, the code words, or **codons**, for amino acids were nucleotide triplets. Gamow arrived at this conclusion by a bit of armchair logic. He reasoned that it would *require* at least three nucleotides for each amino acid to have its own unique codon. Consider the number of words that can be spelled using an alphabet containing four different letters corresponding to the four possible bases that can be present at a particular site in DNA (or mRNA). There are 4 possible one-letter words, 16 (4^2) possible two-letter words, and 64 (4^3) possible three-letter words. Because there are 20 different amino acids (words) that have to be specified, codons must contain at least 3 successive nucleotides (letters). The triplet nature of the code was soon verified in a number of different genetic experiments.

In addition to proposing that the code was triplet, Gamow also suggested that it was *overlapping*. Even though this suggestion proved to be incorrect, it raises an interesting question concerning the genetic code. Consider the following sequence of nucleotides:

$$—AGCAUCGCAUCGA—$$

If the code is overlapping, then the ribosome would move along the mRNA one nucleotide at a time, recognizing a new codon with each move. In the preceding sequence, AGC would specify an amino acid, GCA would specify the next amino acid, CAU the next, and so on. In contrast, if the code is *nonoverlapping*, each nucleotide along the mRNA would be part of one, and only one, codon. In the preceding sequence, AGC, AUC, and GCA would specify successive amino acids.

A conclusion as to whether the code was overlapping or nonoverlapping could be inferred from studies of mutant proteins, such as the mutant hemoglobin responsible for sickle cell anemia. In sickle cell anemia, as in most other cases that were studied, the mutant protein was found to contain a single amino acid substitution. If the code were overlapping, a change in one of the base pairs in the DNA would be expected to affect three consecutive codons (Figure 11.40) and, therefore, three consecutive amino acids in the corresponding polypeptide. If, however, the code is nonoverlapping and each nucleotide is part of only one codon, then only one amino acid replacement would be expected. These and other data indicated that the code was nonoverlapping.

THE HUMAN PERSPECTIVE

UPDATE

For the very latest, go to page U-15

Clinical Applications of Ribozymes and Antisense Oligonucleotides

Medical scientists are continually searching for "magic bullets," therapeutic compounds that fight a particular disease in a highly specific manner without toxic side effects. Let us consider two major types of diseases—viral infections and cancer—that have become targets for a new type of molecular "magic bullet." Viruses are able to ravage an infected cell because they synthesize messenger RNAs that code for viral proteins that disrupt cellular activities. Most cells that become cancerous contain mutations in certain genes (called oncogenes), causing them to produce mutant mRNAs that are translated into abnormal versions of cellular proteins. Consider what might happen if a patient with one of these diseases could be treated with a drug that destroyed or inhibited the specific mRNAs transcribed from the viral genome or mutant cancer gene while, at the same time, ignoring all the other mRNAs of the cell. Two strategies have been developed in recent years that have this goal in mind.

Ribozymes

We have seen in this chapter that a number of ribozymes are able to cleave a phosphodiester bond linking nucleotides in an RNA strand. One of the simplest RNA-cleaving ribozymes is called the *hammerhead ribozyme*, which is found in certain plant pathogens known as viroids (page 25). Figure 1 shows the structure of a hammerhead ribozyme that is complexed to an RNA substrate and the site on the substrate that is cleaved by the ribozyme. It can be seen from this figure that the catalytic site of the ribozyme is flanked by a pair of arms (stems I and III of Figure 1) that form complementary base pairs with the substrate RNA. Only those RNAs that bind to the ribozyme by this type of complementary interaction are cleaved by the ribozyme. Hammerhead ribozymes can be custom designed so that the base sequence of the flanking arms is complementary to any desired mRNA target. The flanking arms can be extended in length to form 12 to 16 base pairs with the substrate, which makes it very unlikely that the ribozyme will be "accidentally" complementary to any other cellular mRNA in the cell. In most studies, the gene that encodes the ribozyme is inserted into the genome of a defective retrovirus, which carries the ribozyme gene into the cell where it becomes integrated into the host cell's chromosomes.

It was shown a number of years ago that white blood cells isolated from patients with AIDS could be genetically engineered to carry a DNA sequence that encodes a ribozyme capable of recognizing and cleaving HIV mRNA. The gene for the ribozyme is inserted into the DNA vector at a site adjacent to a strong promoter, which ensures that the gene is actively transcribed in the host cell. Tests have indicated that ribozymes produced in cultured HIV-infected cells reduce the amount of virus to undetectable levels. Just as important, the ribozyme is effective in destroying the viral RNAs without damaging the cell's normal RNAs. Clinical trials of ribozymes have begun on patients that test positively for the HIV virus but do not yet

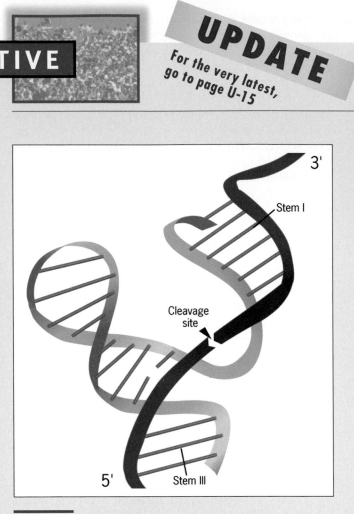

Figure 1 Custom-designed ribozymes. The ribozyme shown here in green, called a hammerhead ribozyme, binds to RNAs having a complementary nucleotide sequence and catalyzes a reaction that cleaves the backbone of the substrate RNA (shown in red).

show symptoms of AIDS. Cells are harvested from the blood of the patient, genetically engineered to contain a gene for the ribozyme, and then infused back into the bloodstream. It is hoped that the engineered cells will prove resistant to the virus and preserve the patient's crippled immune system. Two sets of clinical trials have begun. In one case, investigators are modifying fully differentiated T lymphocytes, the cells that carry out immune reactions that are inhibited in AIDS patients. In the other case, investigators are modifying hematopoietic stem cells (page 17), which are undifferentiated cells that give rise to T lymphocytes. Unlike T lymphocytes, stem cells are capable of proliferation so, conceivably, a single infusion of these modified stem cells could render a person resistant to further immune destruction. Preliminary reports indicate that T lymphocytes carrying the ribozyme gene have a survival advantage within the body of test patients.

Antisense Oligonucleotides

RNA-cleaving ribozymes provide one strategy to disrupt the function of a specific mRNA. In another strategy (Figure 2),

single-stranded nucleic acids are introduced that selectively bind to the target mRNA, interfering with the ability of the mRNA to be translated into protein and also promoting its degradation. This latter approach utilizes **antisense oligodeoxyribonucleotides (ODNs)**, which are short synthetic DNA molecules 7 to 30 nucleotides in length that can be made complementary to any transcript of known sequence. They are described as "antisense" because they are complementary to the target mRNA, which contains the "sense." The nucleotides used to synthesize the ODNs are chemically modified to make the synthetic polymer resistant to nuclease digestion either in the bloodstream or within a cell. Most studies employ an ODN in which one of the oxygen atoms of each nucleotide is replaced by a sulfur atom to form what is called a phosphorothioate oligonucleotide (or simply an "S-oligo").

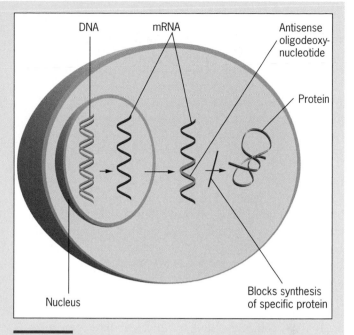

Figure 2 **Antisense oligonucleotides.** Antisense oligonucleotides bind to a portion of a specific, targeted mRNA and stop the RNA from being translated into the encoded protein.

Various approaches have been used to get cells to take up ODNs from the extracellular medium. The best results have been obtained by encapsulating the substances within liposomes (page 131) or linking them to positively charged lipids, which are able to diffuse across cell membranes.

As with the RNA-cleaving ribozymes discussed above, researchers initially tested antisense oligonucleotides on diseased cells grown in culture to determine their effectiveness and cellular toxicity. These in vitro experiments were followed by animal studies, which confirmed the potential use of antisense ODNs as therapeutic agents, and finally clinical trials. In 1998, the first antisense ODN was approved for use by the FDA. This drug (trade name Vitravene) is capable of combatting a type of eye infection caused by cytomegalovirus (CMV) that is prevalent in AIDS patients. To treat the disease, a concentrated preparation of an ODN complementary to the viral mRNA is injected into the vitreous humor of the eye, from which it is taken up by the infected cells of the retina.

Antisense ODNs are currently being tested in a number of clinical trials on patients with advanced cancers in which the malignant cells possess a mutated oncogene. Patients with certain types of leukemia or lymphomas can be treated by cell infusion. In these studies, bone marrow cells are withdrawn from the patients, treated with antisense ODNs in vitro, and then reintroduced into the body. Patients with various types of solid tumors have been treated by intravenous infusion of the ODN. Preliminary results indicate that a fraction of patients respond well to these treatments, whereas the majority of patients show little or no response. It is too early to tell whether antisense ODNs will find widespread application in cancer therapy.

Base Sequence	Codons		
Original sequence . . . AGCATCG . . .	Overlapping code . . ., AGC, GCA, CAT, ATC, TCG, . . .		Nonoverlapping code . . ., AGC, ATC, . . .
Sequence after single-base substitution . . . AGAATCG . . .	Overlapping code . . ., AGA, GAA, AAT, ATC, TCG, . . .		Nonoverlapping code . . ., AGA, ATC, . . .

Figure 11.40 **The distinction between an overlapping and nonoverlapping genetic code.** The effect on the information content of an mRNA by a single base substitution depending on whether the code is overlapping or nonoverlapping. The affected codons are underlined in red.

Given a triplet code that can specify 64 different amino acids and the reality that there are only 20 amino acids to be specified, the question arises as to the function of the extra 44 triplets. If any or all of these 44 triplets code for amino acids, then at least some of the amino acids would be specified by more than one codon. A code of this type is said to be *degenerate*. As it turns out, the code is highly degenerate, as nearly all of the 64 possible codons specify amino acids. Those that do not (3 of the 64) have a special "punctuation" function—they are recognized by the ribosome as termination codons and cause reading of the message to stop.

The degeneracy of the code was originally predicted by Francis Crick on theoretical grounds when he considered the great range in the base composition among the DNAs of various bacteria. It had been found, for example, that the G + C content could range from 20 percent to 74 percent of the genome, whereas the amino acid composition of the proteins from these organisms showed little overall variation. This suggested that the same amino acids were being encoded by different base sequences, which would make the code degenerate.

Identifying the Codons By 1961, the general properties of the code were known, but not one of the coding assignments of the specific triplets had been discovered. At the time, most geneticists thought it would take 5 to 10 years to decipher the entire code. But a breakthrough was made by Marshall Nirenberg and Heinrich Matthaei when they developed a technique to synthesize their own artificial genetic messages and then determine what kind of a protein they encoded. The first message they tested was a polyribonucleotide consisting exclusively of uridine; the message was called *poly(U)*. When poly(U) was added to a test tube containing a bacterial extract with all 20 amino acids and the materials necessary for protein synthesis (ribosomes and various soluble factors), the system followed the artificial messenger's instructions and manufactured a polypeptide. The assembled polypeptide was analyzed and found to be polyphenylalanine—a polymer of the amino acid phenylalanine. Nirenberg and Matthaei had thus shown that the codon UUU specifies phenylalanine.

Over the next four years, the pursuit was joined by a number of researchers, and synthetic mRNAs were constructed to test the amino acid specifications for all 64 possible codons. The result was the universal decoder chart for the genetic code shown in Figure 11.41. The decoder chart lists the nucleotide sequence for each of the 64 possible codons in an mRNA. Instructions for reading the chart are provided in the accompanying figure legend. The codon assignments given in Figure 11.41 are essentially universal: regardless of the source of a cell—whether a bacterium, yeast, mushroom, redwood tree, or human—the same codons specify the same amino acids. The primary exceptions to the universality of the genetic code occur in the codons of mitochondrial mRNAs. For example, in human mitochondria, UGA is read as tryptophan rather than stop, AUA is read as methionine rather than

isoleucine, and AGA and AGG are read as stop rather than arginine. Even with these discrepancies, the similarities of the mitochondrial and nuclear genetic codes far outweigh their differences, and it is generally believed that one is derived from the other and that all organisms present on Earth today share a common evolutionary origin.

Examination of the codon chart in Figure 11.41 indicates that the amino acid assignments are distinctly nonrandom. If one looks for the codon boxes for a specific amino acid, they tend to be clustered within a particular portion of the chart. Clustering reflects the similarity in codons that specify the same amino acid. As a result of this similarity in codon sequence, spontaneous mutations causing single base changes in a gene often will not produce a change in the amino acid sequence of the corresponding protein. The "safeguard" aspect of the code goes beyond its degeneracy. Codon assignments are such that *similar* amino acids tend to be specified by similar codons. For example, codons of the various hydrophobic amino acids (depicted as brown boxes in Figure 11.41) are clustered in the first two columns of the chart. Consequently, a mutation that results in a base substitution in one of these codons is likely to substitute one hydrophobic residue for another. In addition, the greatest similarities between amino acid-related codons occur in the first two nucleotides of the triplet, whereas the greatest variability occurs in the third nucleotide. For example, glycine is encoded by four codons, all of which begin with the nucleotides GG. The explanation for this phenomenon is revealed in the following section in which the role of the transfer RNAs is described.

R E V I E W **?**

1. Explain what is meant by stating that the genetic code is triplet and nonoverlapping? What did the finding that DNA base compositions varied greatly among different organisms suggest about the genetic code? How was the identity of the UUU codon established?

2. Distinguish between the effects of a base substitution in the DNA on a nonoverlapping and an overlapping code.

11.5 DECODING THE CODONS: THE ROLE OF TRANSFER RNAS

Nucleic acids and proteins are like two languages written with different types of letters. This is the reason why protein synthesis is referred to as *translation*. Translation requires that information encoded in the nucleotide sequence of an mRNA be decoded and used to direct the sequential assembly of amino acids into a polypeptide chain. Decoding the information in an mRNA is accomplished by transfer RNAs, which act like adaptors. On one

Figure 11.41 The genetic code. This universal decoder chart lists each of the 64 possible mRNA codons and the corresponding amino acid specified by that codon. To use the chart to translate the codon UGC, for example, find the first letter (U) in the indicated row on the left. Follow that row to the right until you reach the second letter (G) indicated at the top; then find the amino acid that matches the third letter (C) in the row on the right. UGC specifies the insertion of cysteine. Each amino acid (except two) has two or more codons that order its insertion, which makes the code degenerate. A given amino acid tends to be encoded by related codons. This feature reduces the likelihood that base substitutions will result in changes in the amino acid sequence of a protein. Amino acids with similar properties also tend to be clustered. Amino acids with acidic side chains are shown in red, those with basic side chains in blue, those with polar uncharged side chains in green, and those with hydrophobic side chains in brown. As discussed in the following section, decoding in the cell is carried out by tRNAs, a few of which are illustrated schematically in the right side of the figure.

hand, each tRNA is linked to a specific amino acid (as an aa-tRNA), while on the other hand, that same tRNA is able to recognize a particular codon in the mRNA. The interaction between successive codons in the mRNA and specific aa-tRNAs leads to the synthesis of a polypeptide with an ordered sequence of amino acids. To understand how this occurs, we must first consider the structure of tRNAs.

The Structure of tRNAs

In 1965, after seven years of work, Robert Holley of Cornell University reported the first base sequence of an RNA molecule, that of a yeast transfer RNA that carries the amino acid alanine (Figure 11.42a). This tRNA is composed of 77 nucleotides, 10 of which are modified from the standard four nucleotides of RNA (A, G, C, U), as indicated by the shading in the figure.

Over the following years, a number of other tRNA species were purified and sequenced, and a number of distinct similarities present in all the different tRNAs became evident (Figure 11.42b). All tRNAs consisted of roughly the same small number of nucleotides—between 73 and 93—and all had a significant percentage of unusual bases that were found to result from enzymatic modifications of one of the four standard bases *after* it had been incorporated into the RNA chain, that is, posttranscriptionally. In addition, all tRNAs had sequences of nucleotides

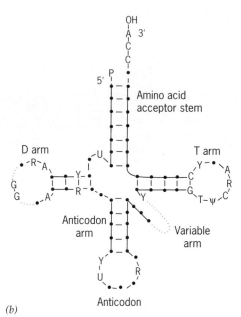

(a)

(b)

Figure 11.42 Two-dimensional structure of transfer RNAs.
(a) The nucleotide sequence of the cloverleaf form of a yeast
tRNA^Ala. The amino acid becomes linked to the 3′ end of the
tRNA, whereas the opposite end bears the anticodon, in this
case IGC. The function of the anticodon is discussed later. In
addition to the four bases, A, U, G, and C, this tRNA contains
ψ, pseudouridine; T, ribothymidine; mI, methyl inosine; I, ino-
sine; me₂G, dimethyl guanosine; D, dihydrouridine; and meG,
methyl guanosine. The sites of the 10 modified bases in this

tRNA are indicated by the color shading. (b) Generalized rep-
resentation of tRNA in the cloverleaf form. The bases common
to all tRNAs (both prokaryotic and eukaryotic) are indicated by
letters; R is an invariant purine, Y is an invariant pyrimidine, ψ
is an invariant pseudouridine. The greatest variability among
tRNAs occurs in the V (variable) arm, which can range from 4
to 21 nucleotides. There are two sites of minor variability in the
length of the D arm.

in one part of the molecule that were complementary to
sequences located in other parts of the molecule. Because
of these complementary sequences, the various tRNAs
become folded in a similar way to form a structure that can
be drawn in two dimensions as a cloverleaf. The base-
paired stems and unpaired loops of the tRNA cloverleaf
are shown in Figure 11.42. The amino acid carried by the
tRNA always becomes attached to the A at the 3′ end of
the molecule. The unusual bases, which are concentrated
in the loops, disrupt hydrogen bond formation in these re-
gions and serve as potential recognition sites for various
proteins.

Up to this point we have considered only the sec-
ondary, or two-dimensional, structure of these adaptor
molecules. Transfer RNAs fold into a unique and defined
tertiary structure. X-ray diffraction analysis has shown
that tRNAs are constructed of two double helices
arranged in the shape of an L (Figure 11.43). Those bases
that are found at comparable sites in all tRNA molecules
(the *invariant* bases of Figure 11.42b) are particularly im-
portant in generating the L-shaped tertiary structure. The
common shapes of tRNAs reflects the fact that all of them
take part in a similar series of reactions during protein syn-
thesis. However, each tRNA has unique features that dis-
tinguish it from other tRNAs. As discussed in the follow-
ing section, it is these unique features that make it possible

for an amino acid to become attached to the appropriate
(cognate) tRNA.

Transfer RNAs translate a sequence of mRNA
codons into a sequence of amino acid residues. Informa-
tion contained in the mRNA is decoded through the for-
mation of base pairs between complementary sequences in
the transfer and messenger RNAs (see Figure 11.50). The
part of the tRNA that participates in this specific interac-
tion with the codon of the mRNA is a stretch of three se-
quential nucleotides, called the **anticodon**, that is located
in the middle loop of the tRNA molecule (Figure 11.42a).
This loop is invariably composed of seven nucleotides, the
middle three of which constitute the anticodon. The anti-
codon is located at one end of the L-shaped tRNA mole-
cule opposite from the end at which the amino acid is at-
tached (Figure 11.43b).

Given the fact that 61 different codons can specify an
amino acid, a cell might be expected to have at least 61 dif-
ferent tRNAs, each with a different anticodon comple-
mentary to one of the codons of Figure 11.41. But recall
that the greatest similarities among codons that specify
the same amino acid occur in the first two nucleotides of
the triplet, whereas the greatest variability in these same
codons occurs in the third nucleotide of the triplet. Con-
sider the 16 codons ending in U. In every case, if the U is
changed to a C, the same amino acid is specified (first two

(a)

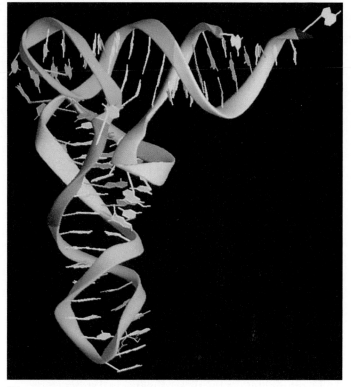

(b)

Figure 11.43 Three-dimensional structure of a tRNA. (*a*) Two-dimensional structure of a yeast phenylalanyl-tRNA with the various regions of the molecule color-coded to match part *b*. (*b*) Three-dimensional structure of tRNA$^{\text{Phe}}$ derived from X-ray crystallography. The amino acceptor (AA) arm and the

TψC (T) arm form a continuous double helix, and the anticodon (AC) arm and the D arm form the other partially continuous double helix. These two helical columns meet to form an L-shaped molecule. (B: COURTESY OF MIKE CARSON.)

lines of each box in Figure 11.41). Similarly, in most cases, a switch between an A and a G at the third site is also without effect on amino acid determination. The interchangeability of the base of the third position led Francis Crick to propose that the same transfer RNA may be able to recognize more than one codon. His proposal, termed the *wobble hypothesis*, suggested that the steric requirement between the anticodon of the tRNA and the codon of the mRNA is very strict for the first two positions but is more flexible at the third position. As a result, two codons that specify the same amino acid and differ only at the third po-

sition should use the same tRNA in protein synthesis. Once again, a Crick hypothesis proved to be correct.

The rules governing the wobble at the third position of the codon are as follows (Figure 11.44): U of the anticodon can pair with A or G of the mRNA; G of the anticodon can pair with U or C of the mRNA; and I (inosine, which is derived from guanine in the original tRNA molecule) of the anticodon can pair with U, C, or A of the mRNA. The wobble hypothesis has been verified by showing that certain purified tRNA molecules can bind to more than one codon for a particular amino acid. As a re-

mRNA codons 5'

Figure 11.44 The wobble in the interaction between codons and anticodons. In some cases, the nucleotide at the 5′ end of the tRNA anticodon is capable of pairing with more than one nucleotide at the 3′ end (third position) of the mRNA codon.

Consequently, more than one codon can use the same tRNA. The rules for pairing in the wobble scheme are indicated in the figure as well as in the text.

sult of the wobble, the six codons for leucine, for example, require only three tRNAs.

Amino Acid Activation It is critically important during polypeptide synthesis that each transfer RNA molecule be attached to the correct (cognate) amino acid. Amino acids are covalently linked to the 3′ ends of their cognate tRNA(s) by an enzyme called an **aminoacyl-tRNA synthetase** (Figure 11.45). Each amino acid is recognized by a specific aminoacyl-tRNA synthetase, which is capable of "charging" any of the tRNAs that are appropriate for that amino acid (i.e., any tRNA whose anticodon recognizes one of the various codons specifying that amino acid, as indicated in Figure 11.41). Aminoacyl-tRNA synthetases provide an excellent example of the specificity of protein-nucleic acid interactions. Certain common features must exist among all tRNA species specifying a given amino acid to allow a single aminoacyl-tRNA synthetase to recognize all of these tRNAs while, at the same time, discriminating against all of the tRNAs for other amino acids. Information concerning the structural features of tRNAs that cause them to be selected or rejected as substrates has come primarily from two sources:

1. Determination of the three-dimensional structure of these enzymes by X-ray crystallography, which allows investigators to identify which sites on the tRNA make direct contact with the protein. As illustrated in Figure 11.45, the two ends of the tRNA—the acceptor stem and anticodon—are particularly important for recognition by these enzymes.

2. Experiments to determine the changes in a tRNA that cause the molecule to be aminoacylated by a noncognate synthetase. It is found, for example, that a specific base pair in a tRNAAla (the G-U base pair involving the third G from the 5′ end of the molecule in Figure 11.42a) is the primary determinant of its interaction with the alanyl-tRNA synthetase. Insertion of this specific base pair into the acceptor stem of a tRNAPhe or a tRNACys is sufficient to cause these tRNAs to be recognized by alanyl-tRNA synthetase and to be aminoacylated with alanine.

Aminoacyl-tRNA synthetases carry out the following two-step reaction:

1st step: ATP + amino acid → aminoacyl-AMP + PP$_i$

2nd step: aminoacyl-AMP + tRNA → aminoacyl-tRNA + AMP

In the first step, the energy of ATP activates the amino acid by formation of an adenylated amino acid, which is bound to the enzyme. This is the primary energy-requiring step in the chemical reactions leading to polypeptide synthesis. Subsequent events, such as the transfer of the amino acid to the tRNA molecule (step 2 above), and eventually to the growing polypeptide chain, are thermodynamically favorable. The PP$_i$ produced in the first reaction is subsequently hydrolyzed to P$_i$, further driving the

Figure 11.45 Three-dimensional portrait of the interaction between a tRNA and its aminoacyl-tRNA synthetase. The crystal structure of *E. coli* glutaminyl-tRNA synthetase (in blue) complexed with tRNAGln (in red and yellow). The enzyme recognizes this specific tRNA and discriminates against others through interactions with the acceptor stem and anticodon of the tRNA. (FROM THOMAS A. STEITZ, SCIENCE VOL 246, COVER OF 12/1/89 ISSUE, © COPYRIGHT BY THE AMERICAN ASSOCIATION FOR THE ADVANCEMENT OF SCIENCE.)

overall reaction toward the formation of products. As we will see later, energy is expended during protein synthesis, but it is not used in the formation of peptide bonds. In the second step, the enzyme transfers its bound amino acid to the 3′ end of a cognate tRNA. Should the synthetase happen to place an inappropriate amino acid on a tRNA, a proofreading mechanism of the enzyme is activated and the bond between the amino acid and tRNA is severed.[5]

The aminoacyl-tRNA synthetases play a key role in translating the nucleotide sequence of the mRNA into the amino acid sequence of the polypeptide because they determine which amino acid is linked to a particular tRNA. Consequently, the amino acid itself plays no direct role in determining where it is incorporated in the polypeptide. This was demonstrated in an early experiment by Fritz Lipmann and coworkers when they chemically altered an amino acid *after* it had already been attached to its specific tRNA (Figure 11.46). In this experiment, they first prepared tRNACys charged with cysteine. They subsequently converted the attached cysteines to alanine and allowed the substituted amino acids to be incorporated. The recognition group of the tRNA, that is, its anticodon, remained unaltered and, therefore, continued to recognize the same nucleotides of the mRNA it would have if the

[5]Not all aa-tRNA synthetases possess this type of proofreading mechanism. Some of these enzymes remove an incorrect amino acid by hydrolyzing the aminoacyl-AMP linkage following the first reaction step. The two amino acids most difficult to distinguish are valine and leucine, which differ by a single methylene group (Figure 2.26). The leucyl-tRNA synthetase employs both types of proofreading mechanisms to ensure accurate aminoacylation.

Figure 11.46 Experimental demonstration that the specificity of amino acid incorporation is determined by the tRNA, not the attached amino acid. The alteration of a cysteine to an alanine after it has been attached to a tRNACys does not affect incorporation of the aa-tRNA into the polypeptide chain. As a result, alanine is incorporated into sites normally occupied by cysteine residues.

chemical alteration of the amino acid had not been made. Consequently, wherever a cysteine was encoded in the message, an alanine was incorporated into the polypeptide chain.

REVIEW **?**

1. Why are tRNAs referred to as adaptor molecules? What aspects of their structure do tRNAs have in common?

2. Describe the nature of the interaction between tRNAs and aminoacyl-tRNA synthetases. Describe the nature of the interaction between tRNAs and mRNAs. What is the wobble hypothesis?

11.6 TRANSLATING GENETIC INFORMATION

Protein synthesis, or **translation**, may be the most complex synthetic activity in a cell. The assembly of a protein requires all the various tRNAs with their attached amino acids, ribosomes, a messenger RNA, numerous proteins having different functions, cations, and GTP. The complexity is not surprising considering that protein synthesis requires the incorporation of 20 different amino acids in the precise sequence dictated by a coded message written in a language that uses different symbols. In the following discussion, we will draw most heavily on translation mechanisms as they operate in bacterial cells, which is best understood. The process is remarkably similar in eukaryotic cells. The primary difference is that translation in eukaryotic cells involves a larger number of soluble (nonribosomal) protein factors.

The synthesis of a polypeptide chain can be divided into three rather distinct activities: **initiation** of the chain, **elongation** of the chain, and **termination** of the chain. We will consider each of these activities in turn.

Initiation

Once it attaches to an mRNA, a ribosome always moves along the mRNA from one codon to the next, that is, in consecutive blocks of three nucleotides. To ensure that the proper triplets are read, the ribosome attaches to the mRNA at a precise site, termed the **initiation codon**, which is specified as AUG. Binding to this codon automatically puts the ribosome in the proper **reading frame** so that it correctly reads the entire message from that point on. For example, in the following case,

—CUAGUUACAUGCUCCAGUCCGU—

the ribosome moves from the initiation codon, AUG, to the next three nucleotides, CUC, and so on along the entire line.

The basic steps in the initiation of translation in prokaryotic cells are illustrated in Figure 11.47.

Step 1: Bringing the Small Ribosomal Subunit to the Initiation Codon As seen in Figure 11.47, an mRNA does not bind to an intact ribosome, but to the small and large subunits in separate stages. The first major step of initiation is the binding of the small ribosomal subunit to the first AUG sequence (or one of the first) in the message, which serves as the initiation codon.[6] How does the small subunit select the initial AUG codon as opposed to some internal one? Bacterial mRNAs possess a specific sequence of nucleotides (called the Shine-Dalgarno sequence after its discoverers) that resides 5 to 10 nucleotides before the initiation codon. The Shine-Dalgarno sequence is complementary to a sequence of nucleotides near the 3′ end of the 16S *ribosomal RNA* of the small ribosomal subunit.

rRNA — — A C C U C C U U U A 3′
 ⋮ ⋮ ⋮ ⋮ ⋮ ⋮
mRNA — — — G G A G G A — — — 5′

[6]GUG is also capable of serving as an initiation codon and is found as such in a few natural messages. When GUG is used, *N*-formylmethionine is still used to form the initiation complex despite the fact that internal GUG codons specify valine.

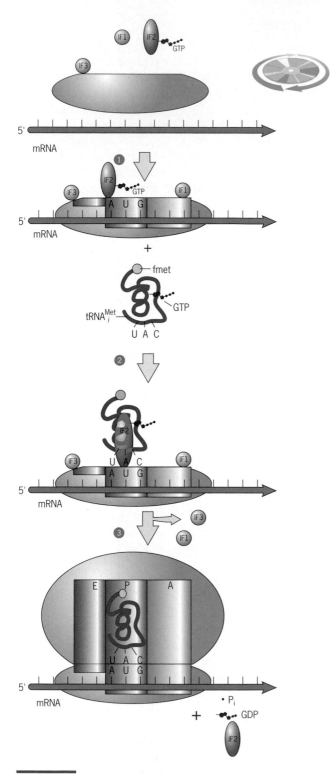

Figure 11.47 Initiation of protein synthesis in prokaryotes.
In step 1, initiation of translation begins with the association of the 30S ribosomal subunit with the mRNA at the AUG initiation codon, a step that requires IF1 and IF3. The 30S ribosomal subunit binds to the mRNA at the AUG initiation codon as the result of an interaction between a complementary nucleotide sequence on the rRNA and mRNA, as discussed in the text. In step 2, the formylmethionyl-tRNA becomes associated with the mRNA and the 30S ribosomal subunit complex by binding to IF2-GTP. In step 3, the 50S subunit joins the complex, GTP is hydrolyzed, and IF2-GDP is released.

Attachment of the 30S subunit at the AUG initiation codon results from an interaction between these complementary sequences on the mRNA and rRNA.

Initiation Requires Initiation Factors Several of the steps outlined in Figure 11.47 require the help of soluble proteins, called **initiation factors** (designated as IFs in prokaryotes and eIFs in eukaryotes). Prokaryotic cells require three initiation factors—IF1, IF2, and IF3—which attach to the 30s subunit (step 1, Figure 11.47). IF2 is a GTP-binding protein required for attachment of the first aminoacyl-tRNA. IF3 may prevent the large (50S) subunit from joining prematurely to the 30S subunit, and also facilitate entry of the initial aa-tRNA. IF1 facilitates attachment of the 30S subunit to the mRNA and may prevent the aa-tRNA from entering the wrong site on the ribosome.

Step 2: Bringing the First aa-tRNA into the Ribosome If one examines the codon assignments (Figure 11.41), it can be seen that AUG is more than just an initiation codon; it is also the only codon for methionine. In fact, methionine is always the first amino acid to be incorporated at the N-terminus of a nascent polypeptide chain. (In prokaryotes, the initial methionine bears a formyl group, making it *N*-formylmethionine.) In many cases, the methionine (or *N*-formylmethionine) is subsequently removed enzymatically. Cells possess two distinct methionyl-tRNAs: one used to initiate protein synthesis (designated $tRNA_i^{Met}$) and a different one (designated $tRNA^{Met}$) to incorporate methionyl residues into the interior of a polypeptide. The aminoacylated initiator tRNA enters the preinitiation complex by binding to both the AUG codon of the mRNA and the IF2 initiation factor (step 2, Figure 11.47). IF1 and IF3 are released.

Step 3: Assembling the Complete Initiation Complex Once the initiator tRNA is bound to the AUG codon, the large subunit joins the complex and the GTP bound to IF2 is hydrolyzed (step 3, Figure 11.47). GTP hydrolysis probably drives a conformational change in the ribosome that is required for the release of IF2-GDP.

Initiation of Translation in Eukaryotes Eukaryotic cells require at least 10 initiation factors comprising a total of at least 25 polypeptide chains. As indicated in Figure 11.48, several of these eIFs (e.g., eIF1, eIF1A, and eIF3) bind to the 40S subunit, which prepares the subunit for binding to the mRNA. The initiator $tRNA_i^{Met}$ also binds to the 40S subunit prior to its interaction with the mRNA. The initiator tRNA enters the subunit in association with eIF2-GTP, which is homologous to the bacterial initiation factor IF2-GTP. Once these events have occurred, the small ribosomal subunit with its associated initiation factors and charged tRNA (which together form a 43S preinitiation complex) is ready to find the 5′ end of the mRNA, which bears the methylguanosine cap (page 462).

Figure 11.48 Initiation of protein synthesis in eukaryotes. As discussed in the text, initiation begins with the union of two complexes, one (called the 43S complex) contains the 40S ribosomal subunit bound to several initiation factors (eIFs) and the initiator tRNA, whereas the other contains the 5′ end of the mRNA bound to a separate group of initiation factors. Once the 43S complex has bound to the end of the mRNA, it scans along the message until it reaches the AUG initiation codon.

The 43S complex is originally recruited to the mRNA with the help of a cluster of initiation factors that are already bound to the mRNA (Figure 11.48). Among these factors: (1) eIF4E binds to the 5′ cap of the eukaryotic mRNA; (2) eIF4A moves along the 5′ end of the message removing any double-stranded regions that would interfere with the movement of the 43S complex along the mRNA; and (3) eIF4G serves as a linker between the 5′ capped end and the 3′ polyadenylated end of the mRNA (Figure 11.48). Thus EIF4G, in effect, converts a linear mRNA into a circular message. The reasons for the circularization of eukaryotic mRNAs are not entirely clear.

Once the 43S complex binds to the 5′ end of the mRNA, the complex then scans along the message until it reaches a recognizable sequence of nucleotides (typically 5′–CCACCAUGC–3′) that contains the AUG initiation codon. Once the 43S complex reaches the AUG codon, eIF2-GTP is hydrolyzed, eIF2-GDP (and other associated eIFs) are released, and the large (60S) subunit joins the complex to complete initiation.

The Role of the Ribosome Now that we have reached a point where we have assembled a complete ribosome, we can look more closely at the structure and function of this multisubunit structure. Ribosomes are molecular machines, similar in some respects to the molecular motors described in Chapter 9. During translation, a ribosome undergoes a repetitive cycle of mechanical changes that is driven by energy released by GTP hydrolysis. Unlike myosin or kinesin, which simply move along a physical track, ribosomes move along an mRNA "tape" containing encoded information. Ribosomes, in other words, are *programmable* machines: the information stored in the mRNA determines the sequence of aminoacyl-tRNAs the ribosome uses in translation. Another feature that distinguishes ribosomes from many other cellular machines is the importance of its component RNAs. Ribosomal RNAs play major roles in selecting tRNAs, ensuring accurate translation, binding protein factors, and polymeriz-

ing amino acids (discussed in the Experimental Pathways at the end of the chapter).

The past few years have seen great progress in our understanding of the structure of the bacterial ribosome. Earlier studies using high-resolution electron microscopic imaging techniques revealed the ribosome to be a highly irregular structure with bulges, lobes, channels, and bridges (see Figure 2.51). During the 1990s, major advances were made in the crystallization of ribosomes, and by the end of the decade the first reports of the X-ray crystallographic structure of bacterial ribosomes had appeared. Figures 11.49*a* and *b* show the overall structure of the two ribosomal subunits of a prokaryotic ribosome as revealed by X-ray crystallography.

Each ribosome has three sites for association with transfer RNA molecules. These sites, termed the **A (aminoacyl) site**, the **P (peptidyl) site**, and the **E (exit) site**, receive each tRNA in successive steps of the elongation cycle, as described in the following section. The positions of tRNAs bound to the A, P, and E sites of both the small and large ribosomal subunit are shown in Figure 11.49 *a,b*. The tRNAs bind within these sites and span the gap between the two ribosomal subunits (Figure 11.49*c*). The anticodon ends of the bound tRNAs contact the small subunit, which plays a key role in decoding the information contained in the mRNA. In contrast, the amino acid-carrying ends of the bound tRNAs contact the large subunit, which plays a key role in catalyzing peptide bond formation. Other features revealed by recent high-resolution structural studies include the following:

1. The interface between the small and large subunits forms a relatively spacious cavity (Figure 11.49*c*) that is lined almost exclusively by RNA. The side of the small subunit that faces this cavity is lined along its length by a single continuous double-stranded RNA helix. This helix is shaded in the two-dimensional structure of the 16S rRNA in Figure 11.3. The surfaces of the two subunits that face one another contain the binding sites for the mRNA and incoming tRNAs and, thus, are of key importance for the function of the ribosome. The fact that these surfaces consist largely of RNA supports the proposal that primordial ribosomes were composed exclusively of RNA (page 471).

2. The active site, where amino acids are covalently linked to one another, also consists of RNA. This catalytic portion of the large subunit resides in a deep cleft, which protects the newly formed peptide bond from hydrolysis by the aqueous solvent.

3. A tunnel runs completely through the large subunit beginning at the active site. This tunnel is thought to provide a passageway for translocation of the elongating polypeptide through the ribosome (Figure 11.49*c*).

4. Most of the proteins of the ribosomal subunits have multiple RNA-binding sites and are ideally situated to stabilize the complex tertiary structure of the rRNAs.

Figure 11.49 Model of the bacterial ribosome based on X-ray crystallographic data, showing tRNAs bound to the A, P, and E sites of the two ribosomal subunits. (*a–b*) Views of the 30S and 50S subunits, respectively, with the three bound tRNAs shown at the interface between the subunits. (*a'–b'*) Drawings corresponding to the structures shown in parts *a* and *b*. The drawing in *a'* of the 30S subunit shows the approximate locations of the anticodons of the three tRNAs and their interaction with the complementary codons of the mRNA. The drawing in *b'* of the 50S subunit shows the tRNA sites from the opposite direction. The amino acid acceptor ends of the tRNAs of the A and P sites are in close proximity within the peptidyl transfer site of the subunit, where peptide bond formation occurs. The binding sites for the elongation factors EF-Tu and EF-G are located on the protuberance at the right side of the subunit. (*c*) Drawing of the 70S prokaryotic ribosome showing the space between the two subunits that is spanned by each tRNA molecule and the channel within the 50S subunit through which the nascent polypeptide exits the ribosome. (A–B: FROM JAMIE H. CATE ET AL., COURTESY OF HARRY F. NOLLER, SCIENCE 285:2100, 1999; A'–B',C: FROM A. LILJAS, SCIENCE 285:2078,1999; © COPYRIGHT 1999 AMERICAN ASSOCIATION FOR THE ADVANCEMENT OF SCIENCE.

Elongation

The basic steps in the process of elongation of translation in prokaryotic cells are illustrated in Figure 11.50.

Step 1: Aminoacyl-tRNA Selection With the charged initiator tRNA in place within the P site, the ribosome is available for entry of the second aminoacyl-tRNA into the vacant A site, which is the first step in elongation (step 1, Figure 11.50*a*). Before the second aminoacyl-tRNA can effectively bind to the exposed mRNA in the A site, it must combine with a protein elongation factor linked to GTP. This particular elongation factor is called **EF-Tu** (or Tu) in prokaryotes and eEF1α in eukaryotes. EF-Tu is required to deliver aminoacyl-tRNAs to the A site of the ribosome Although any aminoacyl-tRNA—Tu–GTP complex can enter the site, only one whose anticodon is complementary to the mRNA codon situated in the A site will be trapped there by the ribosome. Studies indicate that the rRNA molecule of the small subunit plays a key role in recognizing a proper codon-anticodon interaction. Once the proper aminoacyl-tRNA—Tu–GTP is bound to the mRNA codon, the GTP is hydrolyzed and the Tu–GDP complex released, leaving the aa-tRNA bound in the ribosome's A site. Regeneration of Tu–GTP from the released Tu–GDP requires another elongation factor, EF-Ts.

Step 2: Peptide Bond Formation At the end of the first step, the two amino acids, attached to their separate tRNAs, are juxtaposed in such a position that they can react with one another (Figure 11.49*a'*, *b'*). The second step in the elongation cycle is the formation of a peptide bond between these two amino acids (step 2, Figure 11.50*a*). Peptide bond formation is accomplished as the amino group of the aa-tRNA in the A site reacts with the carbonyl group of the amino acid bound to the tRNA of the P site, displacing the P-site tRNA (Figure 11.50*b*). As a result of this reaction, the tRNA bound to the second codon in the A site has an attached dipeptide, whereas the tRNA in the P site is deacylated. Peptide bond formation occurs spontaneously without the input of external energy. The reaction is catalyzed by **peptidyl transferase**, a component of the large subunit of the ribosome. For years it was assumed that the peptidyl transferase was one of the proteins of the ribosome. Then, as the catalytic powers of RNA became apparent, attention shifted to the ribosomal RNA as the catalyst for peptide bond formation. It has now been shown that peptidyl transferase activity does indeed reside in the large ribosomal RNA molecule of the large ribosomal subunit. In other words, peptidyl transferase is a ribozyme, a subject discussed in the Experimental Pathways at the end of the chapter.

Step 3: Translocation The formation of the first peptide bond leaves one end of the tRNA molecule of the A site

(b)

Figure 11.50 Steps in the elongation of the nascent polypeptide during translation in prokaryotes. (*a*) In step 1, an aminoacyl-tRNA whose anticodon is complementary to the second codon of the mRNA enters the empty A site of the ribosome. The binding of the tRNA is accompanied by the release of GDP-Tu. In step 2, peptide bond formation is accomplished by the transfer of the nascent polypeptide chain from the tRNA in the P site to the aminoacyl-tRNA of the A site, forming a peptidyl-tRNA in the A site and a deacylated tRNA in the P site. The reaction is catalyzed by a part of the rRNA acting as a ribozyme. In step 3, the binding of EF-G and the hydrolysis of its associated GTP results in the translocation of the ribosome relative to the mRNA. Translocation is accompanied by the movement of the deacylated tRNA and peptidyl-tRNA into the E and P sites, respectively. In step 4, the deacylated tRNA leaves the ribosome, and a new aminoacyl-tRNA enters the A site. (*b*) Peptide bond formation and the subsequent displacement of the deacylated tRNA.

still attached to its complementary codon on the mRNA and the other end of the molecule attached to a dipeptide (step 2, Figure 11.50*a*). The tRNA of the P site is now devoid of any linked amino acid. In the following step, called **translocation**, the ribosome and the mRNA move relative to one another. In essence, the ribosome moves three nucleotides (one codon) along the mRNA in the $5' \rightarrow 3'$ direction (step 3, Figure 11.50*a*). Translocation is accompanied by the movement of the tRNA-dipeptide from the A site to the P site of the ribosome, still hydrogen-bonded to the second codon of the mRNA, and the movement of the deacylated tRNA from the P site to the E site. Translocation is promoted by another GTP-bound elongation factor (**EF-G** in prokaryotes and eEF2 in eukaryotes). It is worth noting that, under certain conditions, translocation can occur in vitro in the absence of EF-G and GTP. This finding demonstrates that translocation, with all of its complex conformational changes, is an in-

trinsic property of the ribosome whose rate is greatly accelerated by EF-G and GTP hydrolysis.

Step 4: Releasing the Deacylated tRNA In the final step (step 4, Figure 11.50a), the deacylated tRNA leaves the ribosome, emptying the E site.

For each cycle of elongation, at least two molecules of GTP are hydrolyzed: one during aminoacyl-tRNA selection and one during translocation. Each cycle of elongation takes about one-twentieth of a second, most of which is probably spent sampling aa-tRNAs from the surrounding cytosol.

Once the peptidyl-tRNA has moved to the P site by translocation, the A site is once again open to the entry of another aminoacyl-tRNA, in this case one whose anticodon is complementary to the third codon (step 4, Figure 11.50a). Once the third charged tRNA is associated with the mRNA in the A site, the dipeptide from the tRNA of the P site is displaced by the aa-tRNA of the A site, forming the second peptide bond and, as a result, a tripeptide attached to the tRNA of the A site. The tRNA in the P site is once again devoid of an amino acid. Peptide bond formation is followed by translocation of the ribosome to the fourth codon and release of the deacylated tRNA, whereupon the cycle is ready to begin again.

We have seen in this section how the ribosome moves three nucleotides (one codon) at a time along the mRNA. The particular sequence of codons in the mRNA that is utilized by a ribosome (i.e., the reading frame) is fixed at the time the ribosome attaches to the initiation codon at the beginning of translation. Some of the most destructive mutations are ones in which a single base pair is either added to or deleted from the DNA. Consider the effect of an addition of a single nucleotide in the following sequence:

$$\text{—AUG CUC CAG UCC GU— } \rightarrow$$
$$\text{—AUG CUC GCA GUC CGU—}$$

The ribosome moves along the mRNA in the incorrect reading frame from the point of mutation through the remainder of the coding sequence. Mutations of this type are called **frameshift mutations**. Frameshift mutations lead to the assembly of an entirely abnormal sequence of amino acids from the point of the mutation to the end of the polypeptide. Ironically, after more than two decades in which it was assumed that the ribosome always moved from one triplet to the next, several examples have been discovered in which the mRNA contains a recoding signal that causes the ribosome to change its reading frame, either backing up one nucleotide (a shift to the −1 frame) or skipping a nucleotide (a shift to the +1 frame).

Termination

As noted on page 474, 3 of the 64 trinucleotide codons are stop codons whose function is to terminate polypeptide

assembly rather than encode an amino acid. No tRNAs exist whose anticodons are complementary to a stop codon.[7] When a ribosome reaches one of these codons, UAA, UAG, or UGA, the signal is read to stop further elongation and release the polypeptide associated with the last tRNA.

Termination requires the presence of **release factors**. Bacteria have three release factors: RF1, which recognizes UAA and UAG stop codons; RF2, which recognizes UAA and UGA stop codons; and RF3, which is not codon-specific but increases the activity of the other factors. Eukaryotic cells have two release factors, eRF1 and eRF3, which work together and recognize all of the stop codons. Studies have uncovered a novel example of molecular mimicry in which the release factor proteins superficially resemble a tRNA. Because of this resemblance, the release factors are able to enter the A site of the ribosome. A tripeptide within the release factor substitutes for the anticodon of a tRNA and interacts directly with the stop codon.

As with initiation and elongation factors, one of the release factors (RF3 or eRF3) carries a bound GTP that is subsequently hydrolyzed. Once translation stops, the completed polypeptide is severed from its attachment to the last tRNA in the P site, and both the release factor and deacylated tRNA are released from the ribosome. Once termination is complete, the ribosome separates from the message and dissociates into its large and small subunits in preparation for another round of translation.

Because the three termination codons can be readily formed by single base changes from many other codons, one might expect mutations to arise that produce stop codons *within* a gene. Mutations of this type, termed **nonsense mutations**, have been studied for decades and are known to be responsible for a significant fraction of inherited conditions in humans. In some cases, nonsense mutations result in the synthesis of partial polypeptides due to premature termination of the translation process. In most cases, mRNAs containing such mutations are only translated once or a few times before they are recognized and destroyed by a process called *nonsense-mediated decay (NMD)*. As a result, nonsense mutations usually cause a severe form of an inherited disease due to the absence of the encoded polypeptide.

[7]There is one minor exception to this statement. It is stated in this chapter that codons dictate the incorporation of 20 different amino acids. In actual fact, there is a 21st amino acid, called selenocysteine, that is incorporated into a very small number of polypeptides. Selenocysteine is a rare amino acid that contains the metal selenium, and it occurs in a dozen or so proteins in mammals. Selenocysteine has its own tRNA, called tRNA$^{\text{Sec}}$, but lacks its own aa-tRNA synthetase. This unique tRNA is recognized by the seryl-tRNA synthetase, which attaches a serine to the 3′ end of the tRNA$^{\text{Sec}}$. Following attachment, the serine is altered enzymatically to a selenocysteine. Selenocysteine is encoded by UGA, which is one of the three stop codons. Under most circumstances UGA is read as a termination signal. In a few cases, however, the UGA is followed by a folded region of the mRNA that binds a special elongation factor that causes the ribosome to recruit a tRNA$^{\text{Sec}}$ into the A site rather than a termination factor.

(a)

(b)

(c)

Figure 11.51 Polyribosomes. (*a*) Schematic drawing of a polyribosome (polysome). (*b*) Electron micrograph of a grazing section through the outer edge of a rough ER cisterna. The ribosomes are aligned in loops and spirals, indicating their attachment to mRNA molecules to form polysomes. (*c*) Electron micrograph of metal-shadowed polysomes isolated from reticulocytes engaged in hemoglobin synthesis. Most of these polysomes have between four and six ribosomes. (B: COURTESY OF E. YAMADA; C: COURTESY OF ALEXANDER RICH.)

Polyribosomes When a messenger RNA in the process of being translated is examined in the electron microscope, a number of ribosomes are invariably seen to be attached along the length of the mRNA thread. This complex of ribosomes and mRNA is called a **polyribosome**, or **polysome** (Figure 11.51*a*). Each of the ribosomes initially assembles from its subunits at the initiation codon and then moves from that point toward the 3′ end of the mRNA until it reaches a termination codon. As soon as each ribosome moves a sufficient distance along the message from the initiation codon, another ribosome attaches to the mRNA and begins its translation activity. The simultaneous translation of the same mRNA by numerous ribosomes greatly increases the rate of protein synthesis within the cell. Figures 11.51*b* and *c* show eukaryotic polysomes as seen with the electron microscope. The micrograph in Figure 11.51*b* shows polysomes bound to the cytosolic surface of the ER membrane that were engaged in synthesis of membrane and/or organelle proteins (page 290). The polysomes seen in Figure 11.51*c* were free in the cytosol of a reticulocyte where they were synthesizing the soluble protein hemoglobin.

Now that we have described the basic events of translation, it is fitting to close the chapter with pictures of the process, one taken from a prokaryotic cell (Figure 11.52*a*) and the other from a eukaryotic cell (Figure 11.52*b*). Unlike eukaryotic cells, in which transcription occurs in the nucleus and translation occurs in the cytoplasm with many intervening steps, the corresponding activities in prokaryotic cells are tightly coupled. Thus protein synthesis in bacterial cells begins on mRNA templates well before the mRNA has been completely synthesized. The synthesis of an mRNA proceeds in the same direction as the movement of ribosomes translating that message, that is, from the 5′ to 3′ end. Consequently, as soon as an RNA molecule has begun to be produced, the 5′ end is available for attachment of ribosomes. The micrograph in Figure 11.52*a* shows DNA being transcribed, nascent mRNAs being synthesized, and ribosomes that are translating each of the nascent mRNAs. Nascent protein chains are not visible in the micrograph of Figure 11.52*a*, but they are visible in the micrograph of Figure 11.52*b*, which shows a single polyribosome isolated from a silk gland cell of a silk moth. The silk protein being synthesized is visible because of its large size and fibrous nature. The development of techniques to visualize transcription and translation by Oscar Miller, Jr. have provided a fitting visual demonstration of processes that had been previously described in biochemical terms.

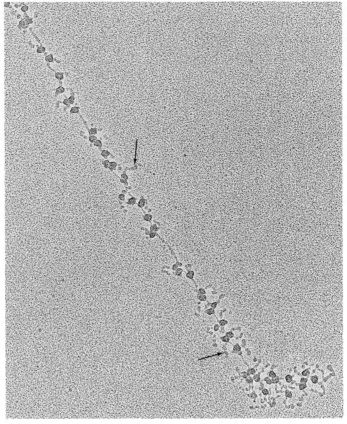

(a) (b)

Figure 11.52 Visualizing transcription and translation.
(*a*) Electron micrograph of portions of an *E. coli* chromosome engaged in transcription. The DNA is seen as faint lines running the length of the photo, while the nascent mRNA chains are seen to be attached at one of their ends, presumably by an RNA polymerase molecule. The particles associated with the nascent RNAs are ribosomes in the act of translation; in bacteria, transcription and translation occur simultaneously. The RNA molecules increase in length as the distance from the initiation site increases. (*b*) Electron micrograph of a polyribosome isolated from cells of the silk gland of the silk worm, which produces large quantities of the silk protein fibroin. This protein is large enough to be visible in the micrograph (arrows point to nascent polypeptide chains). (A: REPRINTED WITH PERMISSION FROM OSCAR L. MILLER, JR., BARBARA A. HAMKALO, AND C. A. THOMAS, SCIENCE 169:392, 1970; © COPYRIGHT 1970 AMERICAN ASSOCIATION FOR THE ADVANCEMENT OF SCIENCE, B: COURTESY OF STEVEN L. MCKNIGHT AND OSCAR L. MILLER, JR.)

REVIEW ?

1. Describe some of the ways that the initiation step of translation differs from the elongation steps of translation.

2. How does the effect of a nonsense mutation differ from that of a frameshift mutation? Why?

3. What is a polyribosome? How does its formation differ in prokaryotes and eukaryotes?

4. During translation elongation, it can be said that an aminocyl-tRNA enters the A site, a peptidyl-tRNA enters the P site, and a deacylated tRNA enters the E site. Explain how each of these events occurs.

EXPERIMENTAL PATHWAYS

The Role of RNA as a Catalyst

Research in biochemistry and molecular biology through the 1970s had solidified our understanding of the roles of proteins and nucleic acids. Proteins were the agents that made things happen in the cell, the enzymes that accelerated the rates of chemical reactions within organisms. Nucleic acids, on the other hand, were the informational molecules in the cell, storing genetic instructions in their nucleotide sequence. The division of labor between protein and nucleic acids seemed as well defined as any distinction that had been drawn in the biological sciences. Then, in 1981, a paper was published that began to blur this distinction.[1]

Thomas Cech and his coworkers at the University of Colorado had been studying the process by which the ribosomal RNA precursor synthesized by the ciliated protozoan *Tetrahymena thermophila* was converted to mature rRNA molecules. The *T. thermophila* pre-rRNA contained an intron of about 400 nucleotides that had to be excised from the primary transcript before the adjoining segments could be linked together (ligated).

Cech had previously found that nuclei isolated from the cells were able to synthesize the pre-rRNA precursor and carry out the entire splicing reaction. Splicing enzymes had yet to be isolated from any type of cell, and it seemed that *Tetrahymena* might be a good system in which to study these enzymes. The first step was to isolate the pre-rRNA precursor in an intact state and determine the minimum number of nuclear components that had to be added to the reaction mixture to obtain accurate splicing. It was found that when isolated nuclei were incubated in a medium containing a low concentration of monovalent cations (5 mM NH_4^+), the pre-rRNA molecule was synthesized but the intron was not excised. This allowed the researchers to purify the intact precursor, which they planned to use as a substrate to assay for splicing activity in nuclear extracts. They found, however, that when the purified precursor was incubated by itself at higher concentrations of NH_4^+ in the presence of Mg^{2+} and a guanosine phosphate (e.g., GMP or GTP), the intron was spliced from the precursor (Figure 1).[1] Nucleotide-sequence analysis confirmed that the small RNA that had been excised from the precursor was the intron with an added guanine-containing nucleotide at its 5′ end. The additional nucleotide was shown to be derived from the GTP that was added to the reaction mixture.

Splicing of an intron is a complicated reaction that requires recognition of the sequences that border the intron, cleavage of the phosphodiester bonds on both sides of the intron, and ligation of the adjoining fragments. All efforts had been made to remove any protein that might be clinging to the RNA before it was tested for its ability to undergo splicing. The RNA had been extracted with detergent-phenol, centrifuged through a gradient, and treated with a proteolytic enzyme. There were only two reasonable explanations: either splicing was accomplished by a protein that was bound very tightly to the RNA, or the pre-rRNA molecule was capable of splicing itself. The latter idea was not easy to accept.

Figure 1 Purified ^{32}P-labeled *Tetrahymena* ribosomal RNA, transcribed at different $(NH_4)_2SO_4$ concentrations, was analyzed by polyacrylamide gel electrophoresis. The numbers at the top indicate the concentration of ammonium sulfate. Two groups of samples are shown, "native" and heat-denatured. Samples of the latter group were boiled for five minutes in buffer and cooled on ice to dissociate any molecules that might be held together by hydrogen bonding between complementary bases. The right two columns contain bacterial 16S and 23S rRNAs, which provide markers of known size with which the other bands in the gel can be compared. It can be seen from the positions of the bands that, as the ammonium sulfate concentration rises, the presence of a small RNA appears whose size is equal to that of the isolated intron (IVS, intervening sequence). These data provided the first indication that the rRNA was able to excise the intron without the aid of additional factors. (FROM T. R. CECH ET AL., CELL 27:488, 1981; BY PERMISSION OF CELL PRESS.)

To resolve the question of the presence of a protein contaminant, Cech and coworkers utilized an artificial system that could not possibly contain nuclear splicing proteins.[2] The DNA that encodes the rRNA precursor was synthesized in *E. coli*, purified, and used as a template for in vitro transcription by a purified bacterial RNA polymerase. The pre-rRNA synthesized in vitro was then purified and incubated by itself in the presence of monovalent and divalent ions and a guanosine compound. Because the RNA had never been in a cell, it could not possibly be contaminated by cellular splicing enzymes. Yet the isolated pre-rRNA underwent the precise splicing reaction that would have occurred in the cell. The RNA had to have spliced itself.

As a result of these experiments, RNA was shown to be capable of catalyzing a complex, multistep reaction. Calculations indicated that the reaction had been speeded approximately 10 billion times over the rate of the noncatalyzed reaction. Thus, like a protein enzyme, the RNA was active in small concentrations, was not altered by the reaction, and

Figure 2 The results of polyacrylamide gel electrophoresis of reaction mixtures that had contained the precursor of tyrosinyl-tRNA (pTyr) and the precursor of another RNA called 4.5S RNA (p4.5). We will focus only on pTyr, which is normally processed by ribonuclease P into two RNA molecules, Tyr and 5'-Tyr (which is the 5' end of the precursor). The positions at which these 3 RNAs (pTyr, Tyr, and 5'-Tyr) migrate during electrophoresis are indicated on the left side of the gel. Lane 1 shows the RNAs that appear in the reaction mixture when pTyr (and p4.5) are incubated with the complete ribonuclease P. Very little of the pTyr remains in the mixture; it has been converted into the two products (Tyr and 5'-Tyr). Lane 5 shows the RNAs that appear in the reaction mixture when pTyr is incubated with the purified protein component of the molecule. The protein shows no ability to cleave the tRNA precursor as evidenced by the absence of bands where the two products would migrate. In contrast, when pTyr is incubated with the purified RNA component of the molecule (lane 7), the pTyr is processed as efficiently as if the intact ribonucleoprotein molecule had been used. (FROM CECILIA GUERRIER-TOKADA ET AL. CELL 35:850, 1983; BY PERMISSION OF CELL PRESS.)

was able to greatly accelerate a chemical reaction. The only distinction between this RNA and "standard proteinaceous enzymes" was that the RNA acted on itself rather than on an independent substrate. Cech called the RNA a "ribozyme."

In 1983, a second, unrelated example of RNA catalysis was discovered.[3] Sidney Altman of Yale University and Norman Pace of the National Jewish Hospital in Denver were collaborating on the study of ribonuclease P, an enzyme involved in the processing of a transfer RNA precursor in bacteria. The enzyme was unusual in being composed of both protein and RNA. When incubated in buffers containing high concentrations (60 mM) of Mg^{2+}, the purified RNA subunit was found capable of removing the 5' end of the tRNA precursor (lane 7, Figure 2) just as the entire ribonuclease P molecule would have done inside the cell. The reaction products of the in vitro reaction included the processed, mature tRNA molecule. In contrast, the isolated protein subunit of the enzyme was devoid of catalytic activity (lane 5, Figure 2).

To eliminate the possibility that a protein contaminant was actually catalyzing the reaction, the RNA portion of ribonuclease P was synthesized in vitro from a recombinant DNA template. As with the RNA that had been extracted

Figure 3 A molecular model of a portion of the catalytic RNA subunit of bacterial ribonuclease P (white) and its bound substrate, precursor tRNA (red). The site on the precursor tRNA where cleavage by the ribozyme occurs is indicated by a yellow sphere. (COURTESY OF MICHAEL E. HARRIS AND NORMAN R. PACE.)

from bacterial cells, this artificially synthesized RNA, without any added protein, was capable of accurately cleaving the tRNA precursor.[4] Unlike the rRNA processing enzyme studied by Cech, the RNA of ribonuclease P acted on another molecule as a substrate rather than on itself. Thus, it was demonstrated that ribozymes have the same catalytic properties as proteinaceous enzymes. A model of the interaction between the catalytic RNA subunit of ribonuclease P and a precursor tRNA substrate is shown in Figure 3.

The demonstration that RNA could catalyze chemical reactions created the proper atmosphere to reinvestigate an old question: Which component of the large ribosomal subunit is the peptidyl transferase, that is, the catalyst for peptide bond formation (page 482)? During the 1970s, a number of independent findings raised the possibility that ribosomal RNAs might be doing more than simply acting as a scaffold to hold ribosomal proteins in the proper position so they could catalyze translation. Included among the evidence were the following types of data:

1. Certain strains of *E. coli* carry genes that encode bacteria-killing proteins called colicins. One of these toxins, colicin E3, was known to inhibit protein synthesis in sensitive bacterial cells. Ribosomes that had been isolated from colicin E3-treated cells appeared perfectly normal by most criteria, but they were unable to support protein synthesis in vitro. Further analysis of such ribosomes revealed that the defect resides in the rRNA, not the ribosomal proteins. The 16S RNA of the small subunit is cleaved by the colicin about 50 nucleotides from its 3' end, which renders the entire subunit unable to support protein synthesis.[5]

2. Treatment of large ribosomal subunits with ribonuclease T_1, an enzyme that cleaves bonds between adjacent RNA nucleotides, was found to destroy the subunit's ability to carry out the peptidyl transferase reaction.[6]

3. A wide variety of studies on antibiotics that inhibited peptide bond formation, including chloramphenicol, carbomycin, and erythromycin, suggested that these drugs act on the ribosomal RNA, not the protein. For example, it was found that ribosomes became resistant to the effects of chloramphenicol as the result of substitutions in the bases in the ribosomal RNA.[7]

4. Ribosomal RNAs were shown to have a highly conserved base sequence, much more so than the amino acid sequences of ribosomal proteins. Some of the regions of the ribosomal RNAs are virtually unchanged in ribosomes isolated from bacteria, plants, and animals, as well as ribosomes isolated from mitochondria and chloroplasts. The fact that rRNA sequences are so highly conserved suggests that the molecules have a crucial role in the function of the ribosome.[8,9] In fact, the 1975 paper by C. R. Woese and coworkers states, "Since little or no correlation exists between these conserved regions and known ribosomal protein binding sites, the implication is strong that large areas of the RNA are directly involved in ribosomal function."[8]

Following the discovery of catalytic RNAs by Cech's and Altman's laboratories, investigation into the roles of ribosomal RNAs intensified. A number of studies by Harry Noller and his colleagues at the University of California, Santa Cruz, pinpointed the site in the ribosomal RNA that resides in or around the peptidyl transferase center. In one study, it was shown that transfer RNAs bound to the ribosome protect specific bases in the rRNA of the large subunit from attack by specific chemical agents. Protection from chemical attack is evidence that the tRNA must be situated very close to the rRNA bases that are shielded.[10] Protection is lost if the 3' end of the tRNA (the end with the CCA bonded to the amino acid) is removed. This is the end of the tRNA involved in peptide bond formation, which would be expected to reside in close proximity to the peptidyl transferase site.

Attempts to ascribe a particular function to *isolated* ribosomal RNA had always met with failure. It was presumed that, even if ribosomal RNAs did have specific functions, the presence of ribosomal proteins were most likely required to keep the rRNAs in their proper conformation. Considering that the ribosomal proteins and rRNAs have coevolved for billions of years, the two molecules would be expected to be interdependent. Despite this expectation, the catalytic power of isolated rRNA was finally demonstrated by Noller and coworkers in 1992.[11] Working with particularly stable ribosomes from *Thermus aquaticus*, a bacterium that lives at high temperatures, Noller treated preparations of the large ribosomal subunit with a protein-extracting detergent (SDS), a protein-degrading enzyme (proteinase K), and several rounds of phenol, a protein denaturant. Together, these agents removed at least 95 percent of the protein from the ribosomal subunit, leaving the rRNA behind. Most of the 5 percent of the protein that remained associated with the rRNA was demonstrated to consist of small fragments of the ribosomal proteins. Yet, despite the removal of nearly all of the protein, the rRNA retained 80 percent of the peptidyl transferase activity of the intact subunit. The catalytic activity was blocked by chloramphenicol and by treatment with ribonuclease. When the RNA was subjected to additional treatments to remove the small amount of remaining protein, the preparation lost its catalytic activity. Because it is very unlikely that the remaining protein had any catalytic importance, it is generally agreed that these experiments demonstrated that peptidyl transferase is a ribozyme.

This conclusion has gained further support from recent X-ray crystallographic studies by Thomas Steitz, Peter Moore, and their colleagues at Yale University on the structure of the large ribosomal subunit of the prokaryote *Haloarcula marismortui*, a model of which is shown in the chapter opening image on page 439. To identify the peptidyl transferase site within the large ribosomal subunit, these researchers soaked crystals of these subunits with CCdA-phosphate-puromycin, a substance that inhibits peptide bond formation by binding tightly to the peptidyl transferase active site. Determination of the structure of these subunits at atomic resolution reveals the location of the bound inhibitor, and thus the location of the peptidyl transferase site.[12] It was found in this study that the active-site inhibitor is bound within a cleft of the subunit that is surrounded entirely by conserved nucleotide residues of the 23S rRNA. In fact, there isn't a single amino acid side-chain of any of the ribosomal proteins within about 18 Å of the site where a peptide bond is synthesized. It would seem hard to argue with the conclusion that the ribosome is a ribozyme.

References

1. CECH, T. R., ZAUG, A. J., & GRABOWSKI, P. J. 1981. In vitro splicing of the ribosomal RNA precursor of Tetrahymena. *Cell* 27:487–496.

2. KRUGER, K., ET AL.1982. Self-splicing RNA: Autoexcision and autocyclization of the ribosomal RNA intervening sequence of Tetrahymena. *Cell* 31:147–157.

3. GUERRIER-TOKADA, C., ET AL.1983. The RNA moiety of ribonuclease P is the catalytic subunit of the enzyme. *Cell* 35:849–857.

4. GUERRIER-TOKADA, C., ET AL.1984. Catalytic activity of an RNA molecule prepared by transcription in vitro. *Science* 223.285–286.

5. BOWMAN, C. M., ET AL.1971. Specific inactivation of 16S ribosomal RNA produced by colicin E3 in vivo. *Proc. Nat'l. Acad. Sci. U.S.A.* 68:964–968.

6. CERNA, J., RYCHLIK, I., & JONAK, J. 1975. Peptidyl transferase activity of Escherichia coli ribosomes digested by ribonuclease T$_1$. *Eur. J. Biochem.* 34:551–556.

7. KEARSEY, S. & CRAIG, I. W. 1981. Altered ribosomal RNA genes in mitochondria from mammalian cells with chloramphenicol resistance. *Nature* 290:607–608.

8. WOESE, C. R., ET AL.1975. Conservation of primary structure in 16S ribosomal RNA. *Nature* 254:83–86.

9. NOLLER, H. F. & WOESE, C. R. 1981. Secondary structure of 16S ribosomal RNA. *Science* 212:403–411.

10. MOAZED, D. & NOLLER, H. F. 1989. Interaction of tRNA with 23S rRNA in the ribosomal A, P, and E sites. *Cell* 57:585–597.

11. NOLLER, H. F., HOFFARTH, V., & ZIMNIAK, L. 1992. Unusual resistance of peptidyl transferase to protein extraction procedures. *Science* 256:1416–1419.

12. NISSEN, P., ET AL. 2000. The structural basis of ribosome activity in peptide bond synthesis. *Science* 289:920–930.

SYNOPSIS

Our understanding of the relationship between genes and proteins came as the result of a number of key observations. The first important observation was made by Garrod, who concluded that persons suffering from inherited metabolic diseases were missing specific enzymes. Later, Beadle and Tatum induced mutations in the genes of Neurospora and identified the specific metabolic reactions that were affected. These studies led to the concept of "one gene—one enzyme" and subsequently to the more refined version of "one gene—one polypeptide chain." The molecular consequence of a mutation was first described by Ingram, who showed that the inherited disease of sickle cell anemia results from a single amino acid substitution in a globin chain. (p. 440)

The first step in gene expression is the transcription of a DNA template strand by an RNA polymerase. Polymerase molecules are directed to the proper site on the DNA by binding to a promoter region, which in almost every case lies just upstream from the site at which transcription is initiated. The polymerase moves in the 3' to 5' direction along the template DNA strand, assembling a complementary, antiparallel strand of RNA that grows from its 5' terminus in a 3' direction. At each step along the way, the enzyme catalyzes a reaction in which ribonucleoside triphosphates (NTPs) are hydrolyzed into nucleoside monophosphates as they are incorporated. The reaction is further driven by hydrolysis of the released pyrophosphate. Prokaryotes possess a single type of RNA polymerase that can associate with a variety of different σ factors that determine which genes are transcribed. The site at which transcription is initiated is determined by a nucleotide sequence located approximately 10 bases upstream from the initiation site. (p. 442)

Eukaryotic cells have three distinct RNA polymerases (I, II, and III), each responsible for the synthesis of a different group of RNAs. Approximately 80 percent of a cell's RNA consists of ribosomal RNA (rRNA). Ribosomal RNAs (with the exception of the 5S species) are synthesized by RNA polymerase I, transfer RNAs and 5S rRNA by RNA polymerase III, and mRNAs by RNA polymerase II. All three types of RNA are derived from primary transcripts that are longer than the final RNA product. RNA processing requires a variety of small nuclear RNAs (snRNAs). (p. 447)

Three of the four eukaryotic rRNAs (the 5.8S, 18S, and 28S rRNAs) are synthesized from a single transcription unit, consisting of DNA (rDNA) that is localized within the nucleolus, and are processed by a series of nucleolar reactions. Nucleoli from amphibian oocytes can be dispersed to reveal actively transcribed rDNA, which takes the form of a chain of "Christmas trees." Each of the trees is a transcription unit whose smaller branches represent shorter RNAs that are at an earlier stage of transcription, that is, closer to the site where RNA synthesis was initiated. Analysis of these complexes reveals the tandem arrangement of the rRNA genes, the presence of nontranscribed spacers separating the transcription units, and the presence of associated ribonucleoprotein (RNP) particles that are involved in processing the transcripts. The steps in the processing of rRNA have been studied by exposing cultured mammalian cells to labeled precursors, such as [^{14}C]methionine, whose methyl groups are transferred to a number of nucleotides of the pre-rRNA. The presence of the

methyl groups is thought to protect certain sites on the RNA from cleavage by nucleases and to aid in the folding of the RNA molecule. Approximately half of the 45S primary transcript is removed during the course of formation of the three mature rRNA products. The nucleolus is also the site of assembly of the two ribosomal subunits. (p. 449)

The precursors of the 5S rRNA and the tRNAs are synthesized by RNA polymerase III, whose promoter is located within the transcribed portion of the gene rather than in the upstream flanking region. Both types of RNA are synthesized on stretches of DNA in which the genes are organized in tandem array, with the transcribed portions alternating with nontranscribed spacers. Following processing, 5S rRNAs are transported to the nucleolus, where they are assembled with other components into the large ribosomal subunits. Transfer RNA genes are located in clusters that contain multiple copies of different genes. (p. 452)

Kinetic studies of rapidly labeled RNAs first suggested that mRNAs were derived from much larger precursors. When eukaryotic cells are incubated for one to a few minutes in [^3H]uridine or other labeled RNA precursors, most of the label is incorporated into a group of RNA molecules of very large molecular weight and diverse nucleotide sequence that are restricted to the nucleus. These RNAs are referred to as heterogeneous nuclear RNAs (or hnRNAs). When cells that have been incubated briefly with [^3H]uridine are chased in a medium containing unlabeled precursors for an hour or more, radioactivity appears in smaller cytoplasmic mRNAs. This and other evidence, such as the presence of 5' caps and 3' poly(A) tails on both hnRNAs and mRNAs, led to the conclusion that hnRNAs were the precursors of mRNAs. (p. 454)

Pre-mRNAs are synthesized by RNA polymerase II in conjunction with a number of general transcription factors that allow the polymerase to recognize the proper DNA sites and to initiate transcription at the proper nucleotide. In the majority of the genes studied, the promoter lies between 24 and 32 bases upstream from the site of initiation in a region containing the TATA box. The TATA box is recognized by the TATA-binding protein (TBP), whose binding to the DNA initiates the assembly of a preinitiation complex. Phosphorylation of a portion of the RNA polymerase leads to the disengagement of the polymerase and the initiation of transcription. (p. 455)

One of the most important revisions in our concept of the gene came in the late 1970s with the discovery that the coding regions of a gene did not form a continuous sequence of nucleotides. The first observations in this regard were made on studies of transcription of the adenovirus genome in which it was found that the terminal portion of a number of different messenger RNAs are composed of the same sequence of nucleotides that is encoded by several discontinuous segments in the DNA. The regions between the coding segments are called intervening sequences, or introns. A similar condition was soon found to exist for cellular genes, such as those that code for β-globin and ovalbumin. In each case, the regions of DNA that encode portions of the polypeptide (the exons) are separated from one another by noncoding regions (the introns). Subsequent analysis indicated that the entire gene is transcribed into a primary transcript. The regions corresponding to

the introns are subsequently excised from the pre-mRNA, and adjacent coding segments are ligated together. This process of excision and ligation is called RNA splicing. (p. 457)

The major steps in the processing of primary transcripts into mRNA include the addition of a 5′ cap, formation of a 3′ end, addition of a 3′ poly(A) tail, and removal of introns. The formation of the 5′ cap occurs by stepwise reactions in which the terminal phosphate is removed, a GMP is added in an inverted orientation, and methyl groups are transferred to both the added guanosine and the first nucleotide of the transcript itself. The final 3′ end of the mRNA is generated by cleavage of the primary transcript at a site just downstream from an AAUAAA recognition site and the addition of adenosine residues one at a time by poly(A) polymerase. Removal of the introns from the primary transcript depends on the presence of invariant residues at both the 5′ and 3′ splice sites on either side of each intron. Splicing is accomplished by a multicomponent spliceosome that contains a variety of proteins and ribonucleoprotein particles (snRNPs) that assemble in a stepwise fashion at the site of intron removal. Insight into the evolution of splicing mechanisms has been gained from studies of the self-splicing reactions of group I and group II introns of RNAs in lower eukaryotes, mitochondria, and chloroplasts. These studies strongly suggest that the snRNAs of the spliceosomes, not the proteins, are the catalytically active components of the snRNPs. One of the apparent evolutionary benefits of split genes is the ease with which exons can be shuffled within the genome, generating new genes from portions of preexisting ones. (p. 461)

Information for the incorporation of amino acids into a polypeptide is encoded in the sequence of triplet codons in the mRNA. In addition to being triplet, the genetic code is nonoverlapping and degenerate. In a nonoverlapping code, each nucleotide is part of one, and only one, codon, and the ribosome must move along the message three nucleotides at a time. The ribosome attaches to the mRNA at the initiation codon, AUG, which automatically puts the ribosome in the proper reading frame so that it correctly reads the entire message. A triplet code constructed of 4 different nucleotides can have 64 (4^3) different codons. The code is degenerate because many of the 20 different amino acids have more than one codon. Of the 64 possible codons, 61 specify an amino acid, while the other 3 are stop codons that cause the ribosome to terminate translation. The codon assignments are essentially universal, and their sequences are such that base substitutions in the mRNA tend to minimize the effect on the properties of the polypeptide. (p. 471)

Information in the nucleotide alphabet of DNA and RNA is decoded by transfer RNAs during the process of transla-

tion. Transfer RNAs are small RNAs (73 to 93 nucleotides long) that share a similar, L-shaped, three-dimensional structure and a number of invariant residues. One end of the tRNA bears the amino acid, and the other end contains a three-nucleotide anticodon sequence that is complementary to the triplet codon of the mRNA. The steric requirements of complementarity between codon and anticodon are relaxed at the third position of the codon to allow different codons that specify the same amino acid to use the same tRNA. It is essential that each tRNA is linked to the proper (cognate) amino acid, that is, the amino acid specified by the mRNA codon to which the tRNA anticodon binds. Linkage of tRNAs to their cognate amino acids is accomplished by a group of enzymes called aminoacyl-tRNA synthetases. Each enzyme is specific for one of the 20 amino acids and is able to recognize all of the tRNAs to which that amino acid must be linked. Formation of the aminoacyl-tRNAs is the primary energy-requiring step in the entire process of polypeptide assembly. (p. 474)

Protein synthesis is a complex synthetic activity involving all the various tRNAs with their attached amino acids, ribosomes, messenger RNA, a number of proteins, cations, and GTP. The process is divided into three distinct activities: initiation, elongation, and termination. The primary activities of initiation include the precise attachment of the small ribosomal subunit to the initiation codon of the mRNA, which sets the reading frame for the entire translational process; the entry into the ribosome of a special initiation $tRNA_i^{Met}$; and the assembly of the translational machinery. During elongation, a cycle of tRNA entry, peptide bond formation, and tRNA exit repeats itself with every amino acid incorporated. Aminoacyl-tRNAs enter the A site, where they bind to the complementary codon of the mRNA. As each tRNA enters, the nascent polypeptide attached to the tRNA of the P site is transferred to the amino acid on the tRNA of the A site, forming a peptide bond. Peptide bond formation is catalyzed by a portion of the large rRNA acting as a ribozyme. In the last step of elongation, the ribosome translocates to the next codon of the mRNA, as the deacylated tRNA of the P site is transferred to the E site, and the deacylated tRNA that was in the E site is released from the ribosome. Both initiation and elongation require GTP hydrolysis. Translation is terminated when the ribosome reaches one of the three stop codons. After a ribosome assembles at the initiation codon and moves a short distance toward the 3′ end of the mRNA, another ribosome generally attaches to the initiation codon so that each mRNA is translated simultaneously by a number of ribosomes, which greatly increases the rate of protein synthesis within the cell. The complex of an mRNA and its associated ribosomes is a polyribosome. (p. 479)

ANALYTIC QUESTIONS

1. Look at the codon chart of Figure 11.41; which codons would you expect to have a unique tRNA, that is, one that is used only for that codon? Why is it that many codons do not have their own unique tRNA?

2. Proflavin is a compound that inserts itself into DNA and causes frameshift mutations (page 484). How would the effect on the amino acid sequence of a proflavin-induced

mutation differ between an overlapping and a nonoverlapping code?

3. You have just isolated a new drug that has only one effect on cell metabolism; it totally inhibits the breakdown of pre-rRNA to ribosomal RNA. After treating a culture of cells with this drug, you give the cells [^3H]-uridine for 2 minutes, and then grow the cells in the presence of the drug in unla-

beled medium for 4 hours before extracting the RNA and centrifuging it through a sucrose gradient. Draw the curves you would obtain by plotting both absorbance at 260 nm and radioactivity against the fraction number of the gradient. Label the abscissa using S values of RNA.

4. Using the same axes as in the previous question, draw the profile of radioactive RNA that you would obtain after a culture of cells had been incubated for 48 hours in [³H]uridine without any inhibiting drugs present.

5. Suppose you were to construct a synthetic RNA from a repeating dinucleotide (e.g., AGAGAGAG) and then use this RNA as a messenger to synthesize a polypeptide in an in vitro protein-synthesizing system, such as that used by Nirenberg and Matthei to produce polyphenylalanine. What type of polypeptide would you make from this particular polynucleotide? Would you expect to have more than one type of polypeptide produced? Why or why not?

6. Suppose that you found an enzyme that incorporated nucleotides randomly into a polymer without the requirement for a template. How many different codons would you be able to find in synthetic RNAs made using two different nucleotide precursors (e.g., CTP and ATP)? (An enzyme called polynucleotide phosphorylase catalyzes this type of reaction and was used in studies that identified codons.)

7. Draw the parts of a 15S globin pre-mRNA, labeling the noncoding portions.

8. What is the minimum number of GTPs that you would need to synthesize a pentapeptide?

9. On the same set of axes as in Figure 10.20, draw two reannealing curves, one for DNA extracted from *Xenopus* brain tissue and the other for DNA extracted from *Xenopus* oocytes.

10. Would you agree with the following statement? The discovery that sickle cell anemia resulted from a single amino acid change was proof that the genetic code was nonoverlapping. Why or why not?

11. Thalassemia is a disease characterized by mutations that convert amino acid codons into stop codons. Suppose you were to compare the polypeptides synthesized in

vitro from mRNAs purified from a wide variety of thalassemia patients. How would you expect these polypeptides to compare? Refer to the codon chart of Figure 11.41; how many amino acid codons can be converted into stop codons by a single base substitution?

12. Do you think it would be theoretically possible to have a genetic code with only two letters, A and T? If so, what would be the minimum number of nucleotides required to make a codon?

13. During elongation of a polypeptide chain, (1) the tRNA moves from one codon of the mRNA to the next, (2) the ribosome moves closer to the 3' end of the mRNA, (3) the growing polypeptide chain is passed from the tRNA of the P site to the amino acid present in the A site, (4) tRNAs enter the empty P site. Circle the correct answers.

14. How is it possible that mRNA synthesis occurs at a greater rate in bacterial cells than any other class, yet very little mRNA is present within the cell?

15. If a codon for serine is 5'-AGC-3', the anticodon for this triplet would be 5'- — — — -3'. How would the wobble affect this codon-anticodon interaction?

16. One of the main arguments that proteins evolved before DNA (i.e., that the RNA world evolved into an RNA-protein world rather than an RNA-DNA world) is based on the fact that the translational machinery involves a large variety of RNAs (e.g., tRNAs, rRNAs), whereas the transcriptional machinery shows no evidence of RNA involvement. Can you explain how such an argument about the stages of early evolution might be based on these observations?

17. Frameshift mutations and nonsense mutations were described on page 484. It was noted that nonsense mutations often lead to the destruction by NMD of an mRNA containing a premature termination codon. Would you expect mRNAs containing frameshift mutations to be subject to NMD?

18. The arrowheads in Figure 11.16 indicate the direction of transcription of the various tRNA genes. What does this drawing tell you about the template activity of each strand of a DNA molecule within a chromosome?

ADDITIONAL READING

Genetic Code

CRICK, F. H. C. 1966. The genetic code III. *Sci. Am.* 215:55–62. (Oct.)

KHORANA, H. G. 1966. Polynucleotide synthesis and the genetic code. *Harvey Lects.* 62:79–106.

KNIGHT, R. D., ET AL. 1999. Selection, history and chemistry: The three faces of the genetic code. *Trends Biochem. Sci.* 24:241–247.

NIRENBERG, M. W. 1963. The genetic code II. *Sci. Am.* 208:80–94. (March)

VOGEL, G. 1998. Tracking the history of the genetic code. *Science* 281:329–331.

Transcription and Promoter Function

BURATOWSKI, S. 2000. Snapshots of RNA polymerase II transcription initiation. *Curr. Opin. Cell Biol.* 12:320–325.

BURLEY, S. K. & ROEDER, R. G. 1996. Biochemistry and structural biology of transcription factor IID (TFIID). *Annu. Rev. Biochem.* 65:769–799.

CONAWAY, J. W. & CONAWAY, R. C. 1999. Transcription elongation and human disease. *Annu. Rev. Biochem.* 68:301–319.

CONAWAY, J. W., ET AL. 2000. Control of elongation by RNA polymerase II. *Trends Biochem. Sci.* 25:375–380.

DREYFUSS, G. & STRUHL, K., EDS. 1999. Nucleus and gene expression. *Curr. Opin. Cell Biol.*, vol. 11, #3.

GEIDUSCHEK, E. P. & BARTLETT, M. S. 2000. Engines of gene expression. *Nature Struct. Biol.* 7:437–439.

GELLES, J. & LANDICK, R. 1998. RNA polymerase as a molecular motor. *Cell* 93:13–16.

MOONEY, R. A. & LANDICK, R. 1999. RNA polymerase unveiled. *Cell* 98:687–690.

NUDLER, E. 1999. Transcription elongation: Structural basis and mechanisms. *J. Mol. Biol.* 288:1–12.

REEDER, R. H. 1999. Regulation of RNA polymerase I transcription in yeast and vertebrates. *Prog. Nucleic Acid Res. Mol. Biol.* 62:293–327.

STILLMAN, B., ET AL., 1998. Mechanisms of transcription. *Cold Spring Harbor Symp. Quant. Biol.* vol. 63.

VON HIPPEL, P. H. 1998. An integrated model of the transcription complex in elongation, termination, and editing. *Science* 281:660–665.

Nucleolus and rRNA Synthesis

BACHELLERIE, J.-P. & CAVAILLÉ, J. 1997. Guiding ribose methylation of rRNA. *Trends Biochem. Sci.* 22:257–261.

CARMO-FONSECA, M., ET AL. 2000. To be or not to be in the nucleolus. *Nature Cell Biol.* 2:E107–E112.

MILLER, O. L., JR. 1973. The visualization of genes in action. *Sci. Am.* 228:34–42. (March)

OLSON, M. O. J., ET AL. 2000. The nucleolus: An old factory with unexpected capabilities. *Trends Cell Biol.* 10:189–196.

SCHEER, U. & HOCK, R. 1999. Structure and function of the nucleolus. *Curr. Opin. Cell Biol.* 11:385–390.

SMITH, C. M. & STEITZ, J. A. 1997. Sno storm in the nucleolus: New roles for myriad small RNPs. *Cell* 89:669–672.

Ribozymes

CECH, T. R. 1986. RNA as an enzyme. *Sci. Am.* 255:64–75. (Nov.)

FEDOR, M. J. 1998. Ribozymes. *Curr. Biol.* 8:R441–R443.

JOYCE, G. F. 1992. Directed molecular evolution. *Sci. Am.* 267:90–97. (Dec.)

SCOTT, W. G. & KLUG, A. 1996. Ribozymes: Structure and mechanism in RNA catalysis. *Trends Biochem. Sci.* 21:220–224.

WESTHOF, E. 1999. Chemical diversity in RNA cleavage. *Science* 286:61–62.

WESTHOF, E. & MICHEL, F. 1998. Ribozyme architectural diversity made visible. *Science* 282:251–252.

WILSON, D. S. & SZOSTAK, J. W. 1999. In vitro selection of functional nucleic acids. *Annu. Rev. Biochem.* 68:611–647.

RNA Processing and Nucleic Acid Evolution

BARTEL, D. P. & UNRAU, P. J. 1999. Constructing an RNA world. *Trends Biochem. Sci.* 24:M9–M12. (Dec.)

BLENCOWE, B. J. 2000. Exonic splicing enhancers: Mechanism of action, diversity and role in human genetic diseases. *Trends Biochem. Sci.* 25.106–110.

COHEN, J. 1995. The culture of credit. *Science* 268:1706–1711. (on the discovery of split genes)

FREELAND, S. J., ET AL. 1999. Do proteins predate DNA? *Science* 286:690–692.

GESTELAND, R. F., ET AL., EDS. 1999. *The RNA World*, 2d ed. Cold Spring Harbor Laboratory Press.

HURST, L. D. 1994. The uncertain origin of introns. *Nature* 371:381–382.

KELLER, W. 1999. In the beginning, there was RNA. *Science* 285:668–669.

LEWIS, J. D. & TOLLERVEY, D. 2000. Like attracts like: Getting RNA processing together in the nucleus. *Science* 288:1385–1389.

MURRAY, H. L. & JARRELL, K. A. 1999. Flipping the switch to an active spliceosome. *Cell* 96:599–602.

NEWMAN, A. 1998. RNA splicing. *Curr. Biol.* 8:R903–R905.

PROUDFOOT, N. 2000. Connecting transcription to messenger RNA processing. *Trends Biochem. Sci.* 25:290–293.

REED, R. 2000. Mechanisms of fidelity in pre-mRNA splicing. *Curr. Opin. Cell Biol.* 12:340–345.

SHARP, P. A. 1994. Split genes and RNA splicing. *Cell* 77:805–815.

STALEY, J. P. & GUTHRIE, C. 1998. Mechanical devices of the spliceosome: Motors, clocks, springs, and things. *Cell* 92: 315–326.

Human Perspective on Ribozymes and Antisense Oligonucleotides

COHEN, J. S. & HOGAN, M. E. 1994. The new genetic medicines. *Sci. Am.* 271:76–82. (Dec.)

GALDERISI, U., ET AL. 1999. Antisense oligonucleotides as therapeutic agents. *J. Cell. Physiol.* 181:251–257.

JAMES, H. A. & GIBSON, I. 1998. Ribozymes. *Science & Med.* 5:26–35. (March/April)

Transfer RNA and tRNA Synthetases

ARNEZ, J. G. & MORAS, D. 1997. Structure and function considerations of the aminoacylation reaction. *Trends Biochem. Sci.* 22:211–216.

IBBA, M. & SÖLL, D. 2000. Aminoacyl-tRNA synthesis. *Annu. Rev. Biochem.* 69:617–650.

RICH, A. & KIM, S. H. 1978. The three-dimensional structure of transfer RNA. *Sci. Am.* 238:52–62. (Jan.)

WOESE, C. R., ET AL. 2000. Aminoacyl-tRNA synthetases, the genetic code, and the evolutionary process. *Microbiol. Mol. Biol. Revs,* 64:202–236.

Translation

CATE, J. H., ET AL. 1999. X-ray crystal structure of 70S ribosome functional complexes. *Science* 285:2095–2104.

CECH, T. R. 2000. The ribosome is a ribozyme. *Science* 289:878–879.

DEVER, T. E. 1999. Translation initiation: Adept at adapting. *Trends Biochem. Sci.* 24:398–403.

DOUDNA, J. A. 1999. Ribosomes: Cashing in on crystals. *Curr. Biol.* 9:R731–R734.

GINGRAS, A.-C., ET AL. 1999. eIF4 initiation factors: Effectors of mRNA recruitment to ribosomes and regulators of translation. *Annu. Rev. Biochem.* 68:913–963.

GREEN, R. 2000. Ribosomal translocation: EF-G turns the crank. *Curr. Biol.* 10:R369–R373.

HENTZE, M. & KULOZIK, A. E. 1999. A perfect message: RNA surveillance and nonsense-mediated decay. *Cell* 96:307–310.

IBBA, M. & SÖLL, D. 1999. Quality control mechanisms during translation. *Science* 286:1893–1897.

KURLAND, C. G. 1992. Translational accuracy and the fitness of bacteria. *Ann. Rev. Genet.* 26:29–50.

LILJAS, A. 1999. Function is structure. *Science* 285:2077–2078.

MANKIN, A. S. & LIEBMAN, S. W. 1999. Baby, don't stop. (on premature termination). *Nature Gen.* 23:8–10.

NAKAMURA, Y., ET AL. 2000. Mimicry grasps reality in translation termination. *Cell* 101:349–352.

PENNISI, E. 1999. The race to the ribosome structure. *Science* 285:2048–2051.

SACHS, A. B. & VARANI, G. 2000. Eukaryotic translation initiation: There are (at least) two sides to every story. *Nature Struct. Biol.* 7:356–361.

The Cell Nucleus and the Control of Gene Expression

I n spite of their obvious differences in form and function, the cells
that make up a multicellular organism contain a complete set of
genes. A cartilage cell and a nerve cell show little outward similar-
ity, yet both contain the same complement of genes. The genetic information
present in a specialized eukaryotic cell can be compared to a book of blueprints
prepared for the construction of a large, multipurpose building. During the
construction process, all of the blueprints will probably be needed, but only a
small subset of this information has to be consulted during the work on a par-
ticular floor or room. The same is true for a fertilized egg, which contains a
complete set of genetic instructions that is faithfully replicated and distributed
to each cell of a developing organism. As a result, differentiated cells carry
much more genetic information than they will ever have to use. Consequently,
cells contain mechanisms that allow them to express their genetic information
selectively, following only those instructions that are relevant for that cell at

*Scanning force micrograph of a bacterial DNA molecule (shown in blue) with a
regulatory protein (called NtrC, shown in yellowish tones) bound at a site that is
upstream from a gene that encodes glutamine synthetase. Phosphorylation of NtrC
leads to activation of transcription of the regulated gene. A scanning force microscope
measures the height of various parts of the specimen at the atomic level, which is
converted into the colors shown in the key on the right.* (FROM I. ROMBEL ET AL., COURTESY
OF S. KUSTU, UNIVERSITY OF CALIFORNIA, BERKELEY, COLD SPRING HARBOR SYMP. QUANT. BIOL.
63:160, 1998.)

that particular time. In this chapter, we will explore some of the ways in which cells control gene expression, and thereby ensure that certain proteins are synthesized, while others are not. First, however, we will begin by describing the structure and properties of the eukaryotic cell nucleus in which most of the regulatory machinery is contained.

12.1 THE NUCLEUS OF A EUKARYOTIC CELL

Considering its importance in the storage and utilization of genetic information, the nucleus of a eukaryotic cell has a rather undistinguished morphology (Figure 12.1). The contents of the nucleus are present as a viscous, amorphous mass of material enclosed by a complex *nuclear envelope* that forms a boundary between the nucleus and cytoplasm. Included within the nucleus of a typical interphase (i.e., nonmitotic) cell are (1) the chromosomes, which are present as highly extended nucleoprotein fibers, termed *chromatin*; (2) the *nuclear matrix*, which is a protein-containing fibrillar network; (3) one or more *nucleoli*, which are irregular shaped electron-dense structures that function in the synthesis of ribosomal RNA and the assembly of ribosomes (discussed on page 447); and (4) the *nucleoplasm*, the fluid substance in which the solutes of the nucleus are dissolved.

The Nuclear Envelope

The separation of a cell's genetic material from the surrounding cytoplasm may be the single most important feature that distinguishes eukaryotes from prokaryotes, which makes the appearance of the nuclear envelope a landmark in biological evolution. The **nuclear envelope** consists of several distinct components (Figure 12.2*a*). The core of the nuclear envelope consists of two cellular membranes arranged parallel to one another and separated by 10 to 50 nm. The membranes of the nuclear envelope serve as a barrier that keeps ions, solutes, and macromolecules from passing between the nucleus and cytoplasm. The two membranes are fused at sites forming circular pores that contain complex assemblies of proteins (discussed in the next section). The average mammalian cell contains approximately 3,000 nuclear pores. The outer membrane is generally studded with ribosomes and is often seen to be continuous with the membrane of the rough endoplasmic reticulum (Figure 12.2*a*).

The inner surface of the nuclear envelope is bound to a dense filamentous meshwork, called the **nuclear lamina** (Figure 12.3. The nuclear lamina provides mechanical support to the nuclear envelope and serves as a site of attachment for chromatin fibers at the nuclear periphery (Figure 12.2*b*). The filaments of the nuclear lamina are approximately 10 nm in diameter and composed of polypeptides, called *lamins*. Lamins are members of the same superfamily of polypeptides that assemble into the 10-nm intermediate filaments of the cytoplasm (page

(a) 1 μm

(b)

Figure 12.1 **The cell nucleus.** (*a*) Electron micrograph of an interphase HeLa cell nucleus showing a pair of nucleoli and scattered chromatin. Heterochromatin (page 505) is evident around the entire inner surface of the nuclear envelope. Two prominent nucleoli are visible, and clumps of chromatin can be seen scattered throughout the nucleoplasm. (*b*) Schematic drawing showing some of the major components of the nucleus. (A: FROM WERNER W. FRANKE, INT. REV. CYTOL. (SUPPL.) 4:130, 1974.)

363). Mutation in one of the lamin genes (*LMNA*) is responsible for several inherited human diseases, including a rare form of muscular dystrophy (called EMD2). The basis for these diseases is unclear. As in the cytoplasm, the integrity of the intermediate filaments that make up nuclear lamina is regulated by phosphorylation and dephosphorylation. The disassembly of the nuclear lamina prior to mitosis is thought to be induced by phosphorylation of the lamins by a specific kinase (Section 14.2).

(a)

(a) _____ 2 μm

(b) _____ 2 μm

(b) _____ 0.2 μm

Figure 12.2 The nuclear envelope. (*a*) Schematic diagram showing the double membrane, nuclear pore complex, nuclear lamina, and the continuity of the outer membrane with the rough endoplasmic reticulum. (*b*) Electron micrograph of a section through a portion of the nuclear envelope of an onion root tip cell. Note the double membrane (NM) with intervening space, nuclear pore complexes (NPC), and associated heterochromatin (HC) that does not extend into the region of the nuclear pores. (B: FROM WERNER W. FRANKE ET AL. J. CELL BIOL. 91:47S, 1981; BY COPYRIGHT PERMISSION OF THE ROCKEFELLER UNIVERSITY PRESS.)

The Structure of the Nuclear Pore Complex and its Role in Nucleocytoplasmic Exchange

The nuclear envelope is the barrier between the nucleus and cytoplasm, and nuclear pores are the gateways across that barrier. Unlike the plasma membrane, which prevents passage of macromolecules between the cytoplasm and the extracellular space, the nuclear envelope is a hub of activity for the movement of RNAs and proteins in both directions between the nucleus and cytoplasm. The replication and transcription of genetic material within the nucleus require the participation of large numbers of proteins that are synthesized in the cytoplasm and transported across the nuclear envelope. Conversely, the mRNAs, tRNAs, and ribosomal subunits that are manufactured in the nucleus must be transported through the nuclear envelope in the opposite direction. Some components, such as the snRNAs of the spliceosome (page 465), move in both directions; they are synthesized in the nucleus, assembled into RNP particles in the cytoplasm, and then shipped back to the nucleus where they function in mRNA processing. To appreciate the magnitude of the traffic between the two major cellu-

Figure 12.3 The nuclear lamina. (*a*) Nucleus of a cultured human cell that has been stained with fluorescently labeled antibodies to reveal the nuclear lamina (red), which lies on the inner surface of the nuclear envelope. The nuclear matrix (page 516) is stained green. (*b*) Electron micrograph of a freeze-dried, metal-shadowed nuclear envelope of a *Xenopus* oocyte that has been extracted with the nonionic detergent Triton X-100. The lamina appears as a rather continuous meshwork comprising filaments oriented roughly perpendicular to one another. Inset shows a well-preserved area from which nuclear pores have been mechanically removed. (A: FROM H. MA, A. J. SIEGEL, AND R. BEREZNEY, J. CELL BIOL. 146:535, 1999; BY COPYRIGHT PERMISSION OF THE ROCKEFELLER UNIVERSITY PRESS; B: REPRINTED WITH PERMISSION FROM U. AEBI, J. COHN, L. BUHLE, AND L. GERACE, NATURE 323:561, 1986; © COPYRIGHT 1986 MACMILLAN MAGAZINES LIMITED.)

lar compartments, consider a HeLa cell, which is estimated to contain about 10,000,000 ribosomes. To support its growth, a single HeLa cell nucleus must import approximately 560,000 ribosomal proteins and export approximately 14,000 ribosomal subunits every minute.

(a)

(b)

0.1 μm

Figure 12.4 The movement of materials through the nuclear pore. (a) Electron micrograph of the nuclear-cytoplasmic border of an amoeba taken a few minutes after injection with colloidal gold particles. These particles are seen to pass through the center of the nuclear pores on their way from the cytoplasm to the nucleus. The inset shows a portion of the nuclear envelope at higher magnification. Arrows indicate nuclear pores. (b) Electron micrograph of a section through the nuclear envelope of an insect cell showing the movement of granular material (presumed to be a ribosomal subunit) through a nuclear pore. (A: COURTESY OF C. M. FELDHERR; B: FROM BARBARA J. STEVENS AND HEWSON SWIFT, J. CELL BIOL. 31:72, 1966; BY COPYRIGHT PERMISSION OF THE ROCKEFELLER UNIVERSITY PRESS.)

How do all of these materials pass through the nuclear envelope? To begin to answer this question, Carl Feldherr and his colleagues at the University of Florida injected gold particles of varying sizes into cells and observed their passage through the nuclear envelope with the electron microscope. As illustrated in Figure 12.4a, these particles move from the cytoplasm into the nucleus by passing single-file through the center of the nuclear pores (shown in the inset to Fig. 12.4a). Electron micrographs of cells fixed in the normal course of their activities have also shown that particulate material can pass through a nuclear pore. An example is shown in Figure 12.4b, in which granular material presumed to consist of a ribosomal subunit is seen squeezing through one of these pores.

Given the fact that materials as large as gold particles and ribosomal subunits can penetrate nuclear pores, one might assume that these pores are merely open channels, but just the opposite is true. Nuclear pores contain a complex, basketlike apparatus called the **nuclear pore complex (NPC)** that appears to fill the pore like a stopper, projecting into both the cytoplasm and nucleoplasm. The structure of the NPC is seen in the electron micrographs of Figure 12.5 and the models of Figure 12.6. The NPC is a huge, supramolecular complex—approximately 30 times the mass of a ribosome—that exhibits octagonal symmetry due to the eightfold repetition of a number of structures (Figure 12.6). Depending on the species, NPCs contain anywhere from 30 to 50 or more different proteins, which are called *nucleoporins*. Many of the nucleoporins have been purified and their position within the NPC located in the electron microscope using gold-labeled antibodies (Fig. 12.7a). A molecular map of a yeast NPC indicating the distribution of its component proteins is shown in Figure 12.7b. It can be seen from this figure that most nucleoporins are symmetrically positioned, being present on both the cytoplasmic and nuclear side of the structure.

When low-molecular-weight solutes are injected into the cytoplasm of a cell, they rapidly penetrate the nuclear pores by simple diffusion. Such solutes are thought to diffuse through the slots between the spokes that connect the cytoplasmic and nuclear rings of the NPC (Figure 12.6a). The ability of larger molecules (most notably proteins and ribonucleoproteins) to pass from the cytoplasm into the nucleus correlates with their usual location within the cell. When a nonnuclear protein, such as bovine serum albumin, is radioactively labeled and injected into the cytoplasm, it remains in the cytoplasm. In contrast, when the same experiment is carried out with a nuclear protein, such as nucleoplasmin, the labeled protein rapidly enters the nucleus.

In 1982, Robert Laskey and his coworkers at the Medical Research Council of England found that nucleoplasmin, one of the more abundant nuclear proteins of amphibian oocytes, contains a stretch of amino acids near its C-terminus that functions as a **nuclear localization signal (NLS)**. This sequence enables a protein to pass through the nuclear pores and enter the nucleus. The best-studied, or "classical" NLSs, consist of one or two

(a) 0.2 μm

(b)

NEL

(c) 0.6 μm

Figure 12.5 Scanning electron micrographs of the nuclear pore complex from isolated nuclear envelopes of an amphibian oocyte. (*a*) The cytoplasmic face of the nuclear envelope showing the peripheral cytoplasmic granules of the nuclear pore complex. (*b*) The nuclear face of the nuclear envelope showing the basketlike appearance of the inner portion of the complex. (*c*) The nuclear face of the envelope showing the distribution of the NPCs and places where intact patches of the nuclear lamina (NEL) are retained. The lamina is attached to the innermost aspect of the NPC baskets. In all of these micrographs, the isolated nuclear envelopes were fixed, dehydrated, dried, and metal coated. (FROM M. W. GOLDBERG AND T. D. ALLEN, J. CELL BIOL. 119:1431, 1992; BY COPYRIGHT PERMISSION OF THE ROCKEFELLER UNIVERSITY PRESS.)

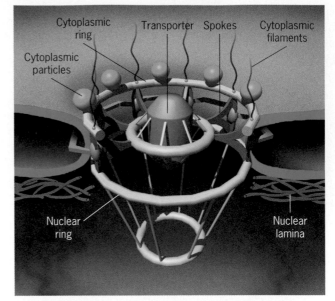

Cytoplasmic ring Transporter Spokes Cytoplasmic filaments

Cytoplasmic particles

Nuclear ring Nuclear lamina

(a)

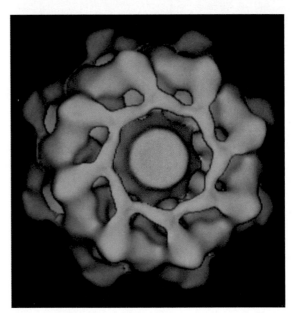

(b)

Figure 12.6 A model of a vertebrate nuclear pore complex (NPC). (*a*) A three-dimensional representation of a vertebrate NPC as it is situated within the nuclear envelope. This elaborate structure contains at least 50 different polypeptides (called *nucleoporins*, or *nups*) and consists of several parts, including a spoke-ring assembly (drawn as two yellow rings and rose-colored spokes), a nuclear basket that resembles a fish trap, a central "plug" or transporter, and eight cytoplasmic filaments. The cytoplasmic ring has been cut open to reveal the transporter. The translocation of materials through the complex is described in Figure 12.8. (*b*) A three-dimensional model based on high-resolution electron microscopy showing the nuclear pore complex as it would appear from the cytoplasm. (A: AFTER CHRISTOPHER W. AKEY AND MICHAEL RADERMACHER, J. CELL BIOL. 122:15, 1993; BY COPYRIGHT PERMISSION OF THE ROCKEFELLER UNIVERSITY PRESS; B: COURTESY OF CHRISTOPHER AKEY.)

(a)

(b)

Figure 12.7 The molecular architecture of the yeast nuclear pore complex. To identify the locations of each of the 30 or so nucleoporins that comprise the yeast nuclear pore complex, researchers prepared a subcellular fraction that was highly enriched in NPCs, extracted the nucleoporins, purified the individual proteins by column chromatography and gel electrophoresis, and prepared antibodies against each of the purified components. (*a*) Electron micrograph showing the location of two of the nucleoporins within the NPC using gold-labeled antibodies. One of these nucleoporins (NIC96) is situated symmetrically on both sides of the NPC, whereas the other nucleoporin (NUP159) is present only on the cytoplasmic side of the structure. (*b*) A map showing the positions of each of the nucleoporins relative to the central axis (R) of the yeast NPC. The positions of individual proteins, which are indicated by the circles, were determined by statistical analysis of the distribution of gold particles in electron micrographs. Most of the proteins are located symmetrically on both the cytoplasmic and nuclear sides of the complex (represented as grey and green

circles). Nucleoporins present strictly on the nuclear or cytoplasmic side of the NPC are indicated by the blue and red circles, respectively. The purple circles correspond to integral *membrane* proteins that associate with the NPC. The background shows a cross-section of the NPC structure. (FROM MICHAEL P. ROUT ET AL., J. CELL BIOL. VOL. 148, COVER OF #4, 2000; BY COPYRIGHT PERMISSION OF THE ROCKEFELLER UNIVERSITY PRESS.)

short stretches of positively charged amino acids. The T antigen encoded by the virus SV40, for example, contains an NLS identified as –pro–lys–lys–lys–arg–lys–val–. If one of the basic amino acids in this sequence is replaced by a nonpolar amino acid, the protein fails to become localized in the nucleus. Conversely, if this NLS is fused to a non-nuclear protein, such as serum albumin, and injected into the cytoplasm, the modified protein becomes concentrated in the nucleus. Thus, targeting of proteins to the nucleus is similar in principle to trafficking of other proteins that are destined for segregation within a particular organelle, such as a mitochondrion or a peroxisome (page 323). In all of these cases, the proteins possess a specific "address" that is recognized by a specific receptor that mediates its transport into the organelle.

The study of nuclear transport has been a very active area of research the past few years, driven by the development of in vitro systems capable of selectively importing proteins and RNPs into the nucleus. Using these systems, researchers can identify which proteins are required for the nuclear import of a particular macromolecule. These efforts have identified a family of proteins (called *karyopherins*) that function as **transport receptors**, moving

macromolecules across the nuclear envelope. Within this family, *importins* move macromolecules from the cytoplasm into the nucleus and *exportins* move macromolecules in the opposite direction.

Figure 12.8*a* depicts some of the major steps that occur during the nuclear import of a protein, such as nucleoplasmin, that contains a classical NLS. Import begins as the NLS-containing cargo protein binds to a heterodimeric, soluble NLS receptor, called *importin α/β*, that resides in the cytoplasm (step 1, Figure 12.8*a*). The NLS transport receptor is thought to escort the protein cargo to the outer surface of the nucleus where it docks with the cytoplasmic filaments that extend from the outer ring of the NPC (step 2). Figure 12.8*b* shows a number of gold particles bound to these filaments; these particles were coated with an NLS-containing nuclear protein that is being transported through the nuclear pore complex. In the model depicted in Figure 12.8*a*, the cytoplasmic filaments bend toward the nucleus, delivering the receptor-cargo complex to specific binding sites on the NPC (step 3, Figure 12.8*a*). Translocation of protein cargo through the pore complex is thought to be accompanied by changes in conformation of the transporter, the large,

(a)

(b)

0.4 μm

Figure 12.8 Importing proteins from the cytoplasm into the nucleus. (*a*) Proposed steps in nuclear protein import, as described in the text. The protein bearing a nuclear localization signal (NLS) binds to the heterodimeric receptor (importin α/β) (step 1) forming a complex that associates with a cytoplasmic filament (step 2). The protein complex is moved through the nuclear pore (step 3) and into the nucleoplasm where it interacts with Ran-GTP and dissociates (step 4). The importin β subunit is transported back to the cytoplasm in association with Ran-GTP (step 5), where the Ran-GTP is hydrolyzed. Ran-GDP is subsequently transported back to the nucleus and importin α back to the cytoplasm. (*b*) Nucleoplasmin is a protein present in high concentration in the nucleoplasm of *Xenopus* oocytes. When gold particles are coated with nucleoplasmin and injected into the cytoplasm of a *Xenopus* oocyte, they are seen to bind to the cytoplasmic filaments (CF) projecting from the outer ring of the nuclear pore complex. Several particles are also seen in transit through the pore (NP) into the nucleus. (A: BASED ON A MODEL BY M. OHNO ET AL., CELL 92:327, 1998, B: FROM W. D. RICHARDSON ET AL., CELL 52:662, 1988; BY PERMISSION OF CELL PRESS, PHOTOGRAPH COURTESY OF A. D. MILLS.)

pluglike structure situated in the center of the NPC (Figure 12.6*b*), which is not shown in Figure 12.8. These changes in conformation may open an aqueous channel through the center of the transporter, allowing protein cargo (and its bound receptor) to enter the nucleoplasm.

Now that we've gotten the bound cargo through the NPC and into the nuclear compartment, we need to introduce another key player, a GTP-binding protein called **Ran**. Like other GTP-binding proteins, such as ARF (page 305) and EF-Tu (page 482) discussed in earlier chapters, Ran can exist in an active GTP-bound form or an inactive GDP-bound form. Ran's role in regulating nucleocytoplasmic transport depends on the fact that the cell maintains a high concentration of Ran-GTP in the nucleus and a very low concentration of Ran-GTP in

the cytoplasm. The steep gradient of Ran-GTP across the nuclear envelope depends on the compartmentalization of certain accessory proteins (see Figure 15.3 for further discussion). One of these accessory proteins (named *Ran-GAP1*) resides in the cytoplasm where it promotes the conversion of Ran-GTP to Ran-GDP, thus maintaining the low cytoplasmic level of Ran-GTP. Another accessory protein (named *RCC1*) is sequestered in the nucleus where it promotes the conversion of Ran-GDP to Ran-GTP, thus maintaining the high nuclear level of Ran-GTP. As we will see, the nuclear/cytoplasmic gradient of Ran-GTP is crucial in determining the direction in which a particular cargo molecule is transported.

We can now return to our description of the classical NLS import pathway. When the importin-cargo complex

arrives in the nucleus, it is met by a molecule of Ran-GTP, which binds to the complex and causes its disassembly as indicated in step 4, Figure 12.8*a*. The imported cargo is released into the nucleoplasm, and one portion of the NLS receptor (the importin β subunit) is shuttled back to the cytoplasm together with the bound Ran-GTP (step 5). Once in the cytoplasm, the GTP molecule bound to Ran is hydrolyzed, releasing Ran-GDP from the importin β subunit. Ran-GDP is returned to the nucleus, where it is converted back to the GTP-bound state for additional rounds of activity. Importin α is transported back to the cytoplasm by one of the exportins.

Less is known about the export of macromolecules from the nucleus through the NPC into the cytoplasm. Most of the traffic moving in this direction consists of various types of RNA molecules—mRNAs, rRNAs, and tRNAs—that are synthesized in the nucleus and function in the cytoplasm. In most cases, these RNAs move through the NPC as ribonucleoproteins (RNPs).[1] The protein components of the RNP carry amino acid sequences (called *nuclear export signals*, or *NESs*) that are recognized by transport receptors (exportins) that carry them through the nuclear envelope to the cytoplasm. Messenger RNAs, for example, are bound in the nucleoplasm by certain hnRNPs (page 465) that move with them into the cytoplasm. Only mature (i.e., fully processed) mRNAs are capable of nuclear export; if an mRNA still contains an unspliced intron, that RNA is retained in the nucleus.

Ran-GTP plays a key role in the escort of macromolecules from the nucleus, just as it does in their import from the cytoplasm. Recall that Ran-GTP is essentially confined to the nucleus. Whereas Ran-GTP induces the *disassembly* of import complexes, as shown in step 4 of Figure 12.8*a*, Ran-GTP promotes the *assembly* of export complexes. Messenger RNAs, for example, leave the nucleus as an mRNA—hnRNP—exportin—Ran-GTP complex. Once in the cytoplasm, the GTP is hydrolyzed, the mRNA is released, and the proteins (hnRNP, Ran-GDP, and exportin) shuttle back into the nucleus to escort additional mRNAs.

Chromosomes

Chromosomes seem to appear out of nowhere at the beginning of mitosis and disappear once again when cell division has ended. The appearance and disappearance of chromosomes provided cytologists with one of their most important, yet challenging questions: what is the nature of the chromosome in the nonmitotic cell? When viewed in the electron microscope, most subcellular organelles reveal a structure that provides considerable insight into their organization and function. However, electron mi-

crographs of thin sections cut through the nucleus generally reveal little of the nature of the interphase chromosome (see Figure 12.1*a*).

Packaging the Genome An average human cell contains about 6 billion base pairs of DNA divided among 46 chromosomes (the value for a diploid, unreplicated number of chromosomes). Each unreplicated chromosome contains a single, continuous DNA molecule; the larger the chromosome, the longer the DNA it contains. Given that each base pair is about 0.34 nm in length, 6 billion base pairs would constitute a DNA molecule fully 2 m long. In addition, the DNA in a cell binds a large amount of water (about six water molecules per base pair), expanding its volume even further. How is it possible to fit 2 meters of hydrated DNA into a nucleus only 10 μm in diameter and, at the same time, maintain the DNA in a state that is accessible to enzymes and regulatory proteins? Just as important, how is the single DNA molecule of each chromosome organized so that it does not become hopelessly tangled with the molecules of other chromosomes? The answers lie in the remarkable manner in which a DNA molecule is packaged.

It has long been known that chromosomes are composed of DNA and associated protein, which together are called **chromatin**. There are two types of proteins in chromatin: histones and nonhistone chromosomal proteins. **Histones** are small, well-defined, basic proteins, whereas **nonhistone chromosomal proteins** include a large number of widely diverse structural, enzymatic, and regulatory proteins, most of which have yet to be characterized.

Nucleosomes: The Lowest Level of Chromosome Organization The orderly packaging of eukaryotic DNA depends on histones, a remarkable group of proteins that possess an unusually high content of the basic amino acids arginine and lysine. Histones are divided into five classes, which can be distinguished by their arginine/lysine ratio (Table 12.1). Histones are further diversified by a variety of modifications, such as phosphorylation and acetylation, that occur after synthesis of the polypeptide (i.e., posttranslationally). The amino acid sequences of histones, particularly

| Table 12.1 | Calf Thymus Histones |

Histone	Number of Residues	Mass (kDa)	% Arg	% Lys	UEP* ($\times 10^{-6}$ year)
H1	215	23.0	1	29	8
H2A	129	14.0	9	11	60
H2B	125	13.8	6	16	60
H3	135	15.3	13	10	330
H4	102	11.3	14	11	600

*Unit evolutionary period: the time for a protein's amino acid sequence to change by 1% after two species have diverged.

[1]Transfer RNAs are a major exception to this statement. Evidence indicates that tRNAs are recognized directly by a transport receptor called exportin-t that carries them to the cytoplasm.

H3 and H4, have undergone very little change over long periods of evolutionary time. The H4 histones of both peas and cows, for example, contain 102 amino acids, and their sequences differ at only two amino acid residues. Why are histones so highly conserved? One reason is histones interact with the backbone of the DNA molecule, which is identical in all organisms. Another reason is nearly all of the amino acids in a histone molecule are engaged in an interaction with another molecule, either DNA or another histone. As a result, very few amino acids in a histone can be replaced with other amino acids without severely affecting the function of the protein.

In the early 1970s, it was found that when chromatin was treated with nonspecific nucleases, most of the DNA was converted to fragments of approximately 200 base pairs in length. In contrast, a similar treatment of *naked* DNA (i.e., DNA devoid of proteins) produced a randomly sized population of fragments. This finding suggested that chromosomal DNA was protected from enzymatic attack, except at certain periodic sites along its length. It was presumed that the proteins associated with the DNA were providing the protection. In 1974, using the data from nuclease digestion and other types of information, Roger Kornberg of Harvard University proposed an entirely new structure for chromatin. Kornberg proposed that DNA and histones are organized into repeating subunits, called **nucleosomes**. We now know that each nucleosome contains a *nucleosome core particle* con-

sisting of 146 base pairs of supercoiled DNA (page 409) wrapped almost twice around a disc-shaped complex of eight histone molecules (Figure 12.9*a*). The histone core of each nucleosome consists of two copies each of histones H2A, H2B, H3, and H4 assembled into an octamer, as discussed below. The remaining histone—type H1—resides outside the nucleosome core particle. The H1 histone is referred to as a *linker histone* because it binds to part of the *linker DNA* that connects one nucleosome core particle to the next.

The exact location of H1 has been the focus of considerable debate the past few years. According to early models of the nucleosome, the H1 protein was situated symmetrically so that it contacted the DNA helix as it both entered and exited the nucleosome core particle (dotted line of Figure 12.9*a*). Recent studies, however, suggest that the H1 protein is situated asymmetrically, contacting one end of the DNA and leaving the other end unprotected (as indicated by the orange colored protein in Figure 12.9*a*). Together the H1 protein and the histone octamer interact with about 168 base pairs of DNA. H1 histone molecules can be selectively removed from the chromatin fibers by subjecting the preparation to solutions of low ionic strength. When H1-depleted chromatin is observed under the electron microscope, the nucleosome core particles and naked linker DNA can be seen as separate elements, which together appear like "beads on a string" (Figure 12.9*b*).

(a)

(b)

Figure 12.9 Nucleosomal organization of chromatin.
(*a*) Schematic diagram showing the structure of a nucleosome core particle and an associated histone H1 molecule. The core particle itself consists of approximately 1.8 turns (146 base pairs) of negatively supercoiled DNA wrapped around eight core histone molecules (two each of H2A, H2B, H3, and H4). The H1 linker histone binds an additional 20 or so base pairs of DNA. Recent evidence indicates the H1 molecule is situated

asymmetrically, as shown. The dotted line depicts the earlier view of a symmetrically positioned H1. (*b*) Electron micrograph of chromatin fibers released from the nucleus of a *Drosophila* cell in a buffer of low ionic strength. The nucleosome core particles are approximately 10 nm in diameter and are connected by short strands of linker DNA, which are approximately 2 nm in diameter. (B: COURTESY OF OSCAR L. MILLER, JR.)

(a) (b) (c)

Figure 12.10 The three-dimensional structure of a nucleosome as revealed by X-ray crystallography. (*a*) A nucleosome core particle viewed down the central axis of the DNA superhelix, showing the position of each of the eight histone molecules of the core octamer. The histones are seen to be organized into four dimeric complexes, two H2A-H2B heterodimers and two H3-H4 heterodimers. The H3-H4 dimers are organized into a tetramer. Each histone dimer binds 27 to 28 base pairs of DNA, with contacts occurring where the minor groove of the DNA faces the histone core. (*b*) The disc shape of the nucleosome core particle is evident in this view perpendicular to the central axis. (*c*) A simplified, schematic model of half of a nucleosome core particle, showing one turn (73 base pairs) of the DNA superhelix and four core histone molecules. The four different histones are shown in separate colors, as indicated by the key. Each core histone is seen to consist of (1) a globular region, called the "histone fold," consisting of three α helices (represented by the cylinders) and (2) a flexible, extended N-terminal tail (indicated by the letter N) that projects out of the histone disk and past the DNA double helix. The intermittent points of interaction between the histone molecules and the DNA are indicated by white hooks. The dotted lines indicate the outermost portion of the histone tails; these flexible tails lack a defined tertiary structure and therefore do not appear in the X-ray structures shown in *a* and *b*. (REPRINTED WITH PERMISSION FROM KAROLIN LUGER ET AL., NATURE 389:251, 1997, COURTESY OF TIMOTHY J. RICHMOND. C: AFTER DRAWING BY D. RHODES. © COPYRIGHT 1997 BY MACMILLAN MAGAZINES LIMITED.)

Our understanding of DNA packaging has been greatly advanced in recent years by dramatic portraits of the nucleosome core particle obtained by X-ray crystallography (Figure 12.10). The eight histone molecules that comprise a nucleosome core particle are organized into four heterodimers: two H2A-H2B dimers and two H3-H4 dimers (Figure 12.10*a,c*). The two H3-H4 dimers are associated with one another in the center of the core particle to form a tetramer (Figure 12.10*b*). The two H2A-H2B dimers are positioned on each side of the (H3-H4)$_2$ tetramer. Each of the four heterodimers binds approximately 30 base pairs of DNA. This binding is mediated in part by ionic bonds between the negatively charged phosphates of the DNA backbone and the positively charged lysine and arginine residues of the histones. The two molecules contact at sites where the minor groove of the DNA faces inward toward the histone core, which occurs at approximately 10 base-pair intervals (Figure 12.10*c*). In between these points of contact, the two molecules are seen to be separated by considerable space, which might provide access to the DNA for transcription factors and other DNA-binding proteins.

The interaction between histones and DNA is primarily structural and is relatively independent of nucleotide sequence. For example, DNA that is not normally associated with histones, such as that of a bacteriophage or a synthetic, double-stranded polynucleotide, is incorporated into nucleosome subunits in vitro when incubated with purified histones derived from plants or animals. These types of experiments also show that nucleosomes can self-assemble.

Even though the binding of histones to DNA is not sequence specific, it does not necessarily follow that the nucleosome core particles are situated at random sites within a given gene. It has been shown, in fact, that certain parts of genes exhibit a constant, predictable relationship to nearby nucleosome core particles. The electron micrograph in Figure 12.11 shows the "chromosome" of an SV40 virus in which the nucleosome core particles are distributed periodically except in one region, which is devoid of associated histones. This site is known to contain the origin of replication for the viral chromosome.

Several factors can affect the location of nucleosomes along the DNA. For example, the binding of a nonhistone protein to a specific site in the DNA can influence the positioning of nearby nucleosome particles. The curving of DNA around the histone core is also a factor in nucleosome positioning because AT-rich segments of DNA are

50 nm

Figure 12.11 Demonstration of the positioning of nucleosomes on DNA. Electron micrograph of an SV40 "chromosome" shows the characteristic gap (top part of photo) in the distribution of nucleosome core particles. (FROM S. SARAGOSTI, G. MOYNE, AND M. YANIV, CELL 20:67, 1980; BY PERMISSION OF CELL PRESS.)

preferred at sites where the minor DNA groove faces inward toward the histone octamer, and GC-rich segments are preferred where the major DNA groove faces inward (Figure 12.10c). These distinctions are correlated with differences in the ease of bending between AT-rich and GC-rich segments. As a result, nucleosomes tend to form where there is an optimal interspersion of AT-rich and GC-rich regions. The positioning of nucleosomes along the DNA is important because it can determine whether that stretch of DNA is accessible to various types of regulatory proteins.

We began this section by wondering how a nucleus 10 μm in diameter can pack 200,000 times this length of DNA within its boundaries. The assembly of nucleosomes is the first important step in the compaction process. With a nucleotide-nucleotide spacing of 0.34 nm, the 200 base pairs of a single 10-nm nucleosome would stretch nearly 70 nm if fully extended. Consequently, it is said that the packing ratio of the DNA of nucleosomes is approximately 7:1.

Higher Levels of Chromatin Structure A DNA molecule wrapped around nucleosome core particles of 10-nm diameter is the lowest level of chromatin organization. Chromatin does not, however, exist in this relatively extended, beads-on-a-string state within the cell. Electron micrographs of sections through nuclei reveal large numbers of tiny dots, approximately 30 nm in diameter—three times the size of a nucleosome. These dots are chromatin fibers that have been cut in cross section. When chromatin is released from nuclei and prepared at physiological ionic strength, a fiber of similar (30-nm) thickness is observed

(Figure 12.12a). The structure of the 30-nm fiber remains a subject of debate. Two models in which the nucleosomal filament is coiled into the higher-order, thicker fiber are shown in Figure 12.12b,c. Regardless of how it is accomplished, the assembly of the 30-nm fiber increases the DNA-packing ratio an additional 6-fold, or about 40-fold altogether.

Maintenance of the 30-nm fiber depends on the interaction between histone molecules of neighboring nucleosomes. Linker histones and core histones have both been implicated in higher-order chromatin structure. If, for example, H1 linker histones are selectively extracted from compacted chromatin, the 30-nm fibers uncoil to form the thinner, more extended beaded filament shown in Figure 12.9b. Adding back H1 histone leads to restoration of the higher-order structure. Core histones of adjacent nucleosomes may interact with one another by means of their long, flexible tails (represented by the dotted lines of Figure 12.10c). Structural studies indicate, for example, that the N-terminal tail of an H4 histone from one nucleosome core particle can reach out and make extensive contact with the H2A/H2B dimer of an adjacent particle. This type of interaction could also mediate the folding of the nucleosomal filament into a thicker fiber.

The next stage in the hierarchy of DNA packaging is thought to occur as the 30-nm chromatin fiber is gathered into a series of large, supercoiled loops, or domains. The DNA loops typically begin and end with AT-rich sequences tethered to a variety of proteins that are part of an organized nuclear scaffold or matrix (discussed on page 515). Included among these proteins is a type II topoisomerase that presumably regulates the degree of DNA supercoiling. The topoisomerase would also be expected to untangle the DNA molecules of different loops when they become intertwined. Other proteins located at the bases of the loops may function as insulators that prevent transcription of a gene in one domain from affecting the expression of a gene in an adjacent domain (page 528).

Normally, loops of chromatin fibers are spread out within the nucleus and cannot be visualized, but their presence can be revealed under certain circumstances. For instance, when isolated *mitotic* chromosomes are treated with solutions that extract histones, the histone-free DNA can be seen to extend outward as loops from a protein scaffold (Figure 12.13a). Similar types of loops can be visualized in the interphase polytene chromosomes of insect cells (see Figure 10.8) and the meiotic *lampbrush chromosomes* of amphibian oocytes (Figure 12.13b), indicating that they are not simply peculiarities of mitotic chromosomes.

The mitotic chromosome represents the ultimate in chromatin compactness (Figure 12.13a); 1 μm of mitotic chromosome length typically contains approximately 1 cm of DNA, which represents a packing ratio of 10,000:1. This compaction occurs by a poorly understood process that is discussed in Section 14.2. An overview of the various levels of chromatin organization, from the nucleoso-

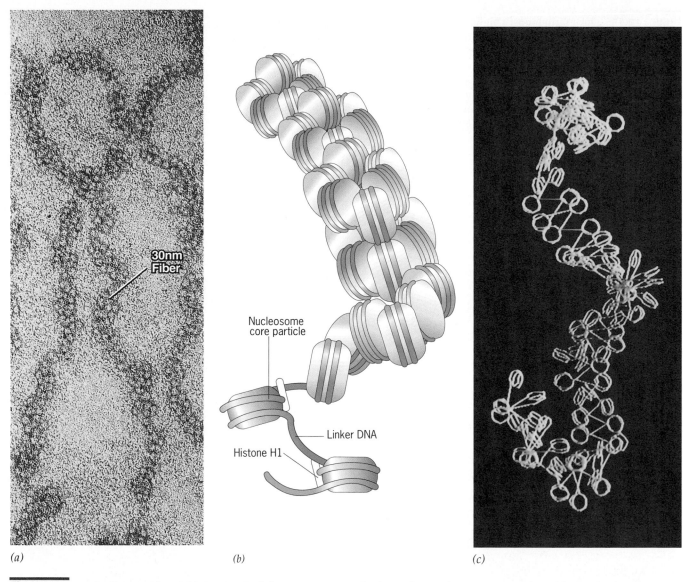

(a) *(b)* *(c)*

Figure 12.12 The 30-nm fiber: a higher level of chromatin structure. (*a*) Electron micrograph of a 30-nm chromatin fiber released from a nucleus following lysis of the cell in a hypotonic salt solution. (*b*) The "solenoid" model of the 30-nm chromatin fiber. In this model, the 30-nm fiber exhibits a regular, helical organization consisting of 6–8 nucleosomes per turn. The fiber is represented (from bottom to top) as it might form with increasing salt concentrations. (*c*) An alternative model based on high-resolution electron microscopy of frozen, hydrated, unfixed chromatin. In this model, the 30-nm fiber exhibits an irregular, zigzag organization, and the linker DNA is present in an extended state between consecutive core particles (represented by circles). (A: COURTESY OF BARBARA HAMKALO AND JEROME B. RATTNER; C: FROM R. A. HOROWITZ ET AL., COURTESY OF C. L. WOODCOCK, J. CELL BIOL. 125:8, 1994; BY COPYRIGHT PERMISSION OF THE ROCKEFELLER UNIVERSITY PRESS.)

mal filament to a mitotic chromosome, is depicted in Figure 12.14.

Heterochromatin and Euchromatin After mitosis has been completed, most of the chromatin in highly compacted mitotic chromosomes returns to its diffuse interphase condition. Approximately 10 percent of the chromatin, however, generally remains in a condensed, compacted form throughout interphase. This compacted chromatin is seen at the periphery of the nucleus in Figure 12.1*a*. Chromatin that remains compacted during inter-phase is called **heterochromatin** to distinguish it from **euchromatin**, which returns to a dispersed state. When a radioactively labeled RNA precursor such as [³H]uridine is given to cells that are subsequently fixed, sectioned, and autoradiographed, the clumps of heterochromatin remain unlabeled, indicating that they have little or no transcriptional activity.

Heterochromatin is divided into two classes, **constitutive heterochromatin** and **facultative heterochromatin**, depending on whether the chromatin is permanently or transiently compacted. Constitutive hete-

Figure 12.13 Chromatin loops: a higher level of chromatin structure. (*a*) Electron micrograph of a mitotic chromosome that had been treated with a solution of dextran sulfate to remove histones. The histone-depleted chromosome displays loops of DNA that are attached at their bases to a residual protein scaffold. (*b*) The looped chromatin domains are also apparent in this micrograph of a lampbrush chromosome isolated from an amphibian oocyte. (A: FROM JAMES R. PAULSON AND U. K. LAEMMLI, CELL 12:823, 1977; BY PERMISSION OF CELL PRESS; B: COURTESY OF JOSEPH G. GALL.)

(*a*) 2 μm

(*b*) 30 μm

rochromatin remains in the compacted state in all cells at all times and, thus, represents DNA that is permanently silenced. In mammalian cells, the bulk of the constitutive heterochromatin is found in and around the centromere of each chromosome (page 508) and in a few other sites, such as the distal arm of the Y chromosome in male mammals. In many plants, the ends of the chromosomes (telomeres) also consist of blocks of constitutive heterochromatin. The DNA of constitutive heterochromatin consists primarily of highly repeated sequences (page 413) and contains relatively few genes. In fact, when genes that are normally active move into a position adjacent to heterochromatin (having changed position as the result of transposition or translocation), they tend to become inactivated themselves, a phenomenon known as *position effect*. It is thought that heterochromatin contains compo-

nents whose influence can spread outward a certain distance, affecting nearby genes.

Unlike the constitutive variety, facultative heterochromatin is chromatin that has been specifically inactivated during certain phases of an organism's life. An example of facultative heterochromatin can be seen by comparing cells of a female mammal to those of a male. The cells of males have a tiny Y chromosome and a much larger X chromosome. Because the X and Y chromosomes have very few genes in common, males have only a single copy of those genes that are carried on the sex chromosomes. Although cells of females contain two X chromosomes, only one of them is transcriptionally active. The other X chromosome remains condensed as a heterochromatic clump (Figure 12.15*a*) called a *Barr body* after the researcher who discovered it in 1949. Formation of a Barr body ensures that the

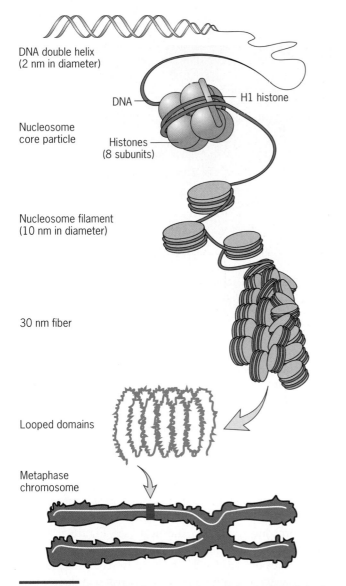

DNA double helix
(2 nm in diameter)

DNA ——— ——— H1 histone

Nucleosome
core particle

Histones
(8 subunits)

Nucleosome filament
(10 nm in diameter)

30 nm fiber

Looped domains

Metaphase
chromosome

Figure 12.14 Levels of organization of chromatin. Naked DNA molecules are wrapped around histones to form nucleosomes, which represent the lowest level of chromatin organization. Nucleosomes are organized into 30-nm fibers, which in turn are organized into looped domains. When cells prepare for mitosis, the loops become further compacted into mitotic chromosomes (see Figure 14.13).

cells of both males and females have the same number of active X chromosomes and thus synthesize equivalent amounts of the products encoded by X-linked genes.

X Chromosome Inactivation Based on her studies of the inheritance of coat color in mice, the British geneticist Mary Lyon proposed the following in 1961:

1. Heterochromatization of the X chromosome in female mammals occurs during early embryonic development and leads to the inactivation of the genes on that chromosome.

2. Heterochromatization in the embryo is a random

process in the sense that the paternally derived X chromosome and the maternally derived X chromosome stand an equal chance of becoming inactivated *in any given cell.* Consequently, at the time of inactivation, the paternal X can be inactivated in one cell of the embryo, and the maternal X can be inactivated in a neighboring cell. From that time on, however, the same X chromosome is inactive in all the descendants of that particular cell.

3. Reactivation of the heterochromatized X chromosome occurs in germ cells prior to the onset of meiosis. Consequently, both X chromosomes are active during oogenesis, and all of the gametes receive a euchromatic X chromosome.

The Lyon hypothesis was soon confirmed. Because maternally and paternally derived X chromosomes may contain different alleles for the same trait, adult females are in a sense *genetic mosaics*, where different alleles function in different cells. X-chromosome mosaicism is reflected in the patchwork coloration of the fur of some mammals, including calico cats (Figure 12.15*b*). Pigmentation genes in humans are not located on the X chromosome, hence the absence of "calico women." Mosaicism due to X inactivation can be demonstrated in women, nonetheless. For example, if a narrow beam of red or green light is shone into the eyes of a woman who is a heterozygous carrier for red-green color blindness, patches of retinal cells with defective color vision can be found interspersed among patches with normal vision.

The mechanism responsible for X inactivation has been a focus of attention since a 1992 report suggesting that inactivation is initiated by a noncoding RNA molecule—rather than a protein—that is transcribed from one of the genes (called *XIST* in humans) on the X chromosome that becomes inactivated. The *XIST* RNA does not diffuse into the nucleoplasm, but accumulates along the length of the chromosome, just before chromosomal inactivation.[2] The *XIST* gene is required to initiate inactivation, but not to maintain it from one cell generation to the next. This is based on the discovery of certain tumor cells in women that contain an inactivated X chromosome whose *XIST* gene is deleted. X inactivation is thought to be maintained by DNA methylation, as discussed on page 534.

The Structure of the Mitotic Chromosome The relatively dispersed state of the chromatin of an interphase cell favors interphase activities, such as replication and transcription. In contrast, the chromatin of a mitotic cell exists in its most highly condensed state, which favors the delivery of an intact "package" of DNA to each daughter cell. Mitotic chromosomes have also proven useful to biologists and physicians because they contain a complete set of genetic material of a cell and are made readily visible by simple techniques.

[2]A number of genes on the chromosome escape inactivation by an unknown mechanism. The "escapees" include genes that are also present on the Y chromosome, which ensures that they are expressed equally in both sexes.

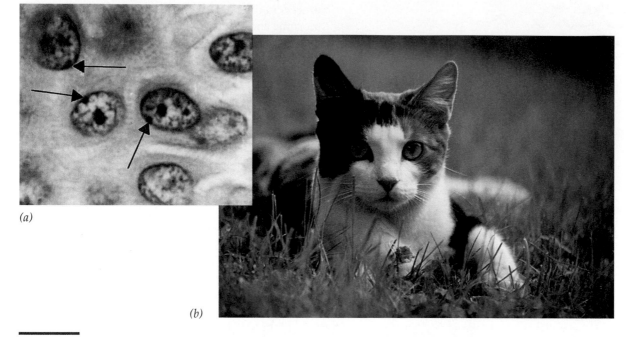

(a)

(b)

Figure 12.15 **The inactive X chromosome: an example of facultative heterochromatin.** (*a*) The inactivated X chromosome in the nucleus of a woman's cells appears as a darkly staining heterochromatic structure, called a Barr body (arrows). (*b*) A calico cat. Random inactivation of either X chromosome in different cells during early embryonic development creates a mosaic of tissue patches. Each patch comprises the descendants of one cell that was present in the embryo at the time of inactivation. These patches are visually evident in calico cats, which have an allele for black coat color residing on one X chromosome and an allele for yellow coat color on the other X. This explains why male calico cats are virtually nonexistent: because all cells in the male have the same coat color allele. (A: COURTESY OF MURRAY L. BARR; B: FROM JEAN PRAGEN/STONE.)

When a chromosome undergoes compaction during mitotic prophase, it adopts a distinct and predictable shape determined primarily by the length of the DNA molecule in that chromosome and the position of the centromere (discussed later). The mitotic chromosomes of a dividing cell can be displayed by the technique depicted in Figure 12.16*a*. In this technique, a dividing cell is broken open, and the mitotic chromosomes from the cell's nucleus settle and attach to the surface of the slide over a very small area (as in Figure 12.16*b*). The chromosomes displayed in Figure 12.16*b* have been prepared using a newly developed staining methodology in which chromosome preparations are incubated with fluorescently labeled DNA probes that bind specifically to particular chromosomes. By using various combinations of DNA probes and computer-aided visualization techniques, each chromosome can be "painted" with a different fluorescent color, making it readily identifiable to the trained eye. In addition to providing a colorful image, this technique delivers enhanced resolution, which allows clinical geneticists to discover chromosomal aberrations that might otherwise be missed (see Figure 2 of the accompanying Human Perspective).

If the individual chromosomes are cut out of a photograph such as that of Figure 12.16*b*, they can be matched up into homologous pairs (23 in humans) and ordered according to decreasing size, as shown in Figure 12.16*c*. A preparation of this type is called a **karyotype**. The chromosomes shown in the karyotype of Figure 12.16*c* have been prepared using a staining procedure that gives the chromosomes a cross-banded appearance. The pattern of these bands is highly characteristic for each chromosome of a species and provides a basis to identify chromosomes and compare them from one species to the next (see Figure 3 in the Human Perspective). Karyotypes are routinely prepared from cultures of blood cells and used to screen individuals for chromosomal abnormalities. As discussed in the accompanying Human Perspective, extra, missing, or grossly altered chromosomes can be detected in this manner.

Centromeres Each chromosome depicted in Figure 12.16 contains a site where the outer surfaces are markedly indented. The site of the constriction marks the **centromere** of the chromosome (Figure 12.17). It was mentioned earlier that the centromeres of chromosomes contain constitutive heterochromatin. In humans the centromeric DNA contains a 171-base pair, AT-rich sequence (called α-*satellite DNA*) that is tandemly repeated thousands of times per centromere. Centromeric DNA binds specific proteins, including those that serve as attachment sites for the microtubules that separate chromosomes during cell division (see Figure 14.16).

It is interesting that comparisons of centromeric DNA sequences and centromeric proteins among diverse organisms indicate that the proteins are much more highly conserved than the DNA sequences to which they bind. This finding suggests that the DNA sequence itself may not be that important a determinant in centromere structure and function, a conclusion that is strongly supported

Figure 12.16 Human mitotic chromosomes and karyotypes.
(*a*) Procedure used to obtain preparations of mitotic chromosomes for microscopic observation from leukocytes in the peripheral blood. (*b*) Photograph of a cluster of mitotic chromosomes that spilled out of the nucleus of a single dividing human cell. The DNA of each chromosome has hybridized to an assortment of fluorescently labeled probes that are specific for that chromosome. Different chromosomes bind different combinations of these dyes and, consequently, emit light of different wavelengths. Emission spectra from different chromosomes are converted into distinct and recognizable display colors by computer processing. Pairs of homologous chromosomes can be identified by searching for chromosomes of the same color and size. (*c*) The chromosomes of a human male arranged in a karyotype. Karyotypes are prepared from a photograph of chromosomes released from a single nucleus. Each chromosome is cut out of the photograph, and homologues are paired according to size, as shown. (B: FROM E. SCHRÖCK ET AL., COURTESY OF THOMAS RIED, SCIENCE 273:495, 1996, © COPYRIGHT 1993 AMERICAN ASSOCIATION FOR THE ADVANCEMENT OF SCIENCE; C: FROM CNRI/SCIENCE PHOTO LIBRARY/PHOTO RESEARCHERS.)

Figure 12.17 Each mitotic chromosome has a centromere whose site is marked by a distinct indentation. Scanning electron micrograph of a mitotic chromosome. The centromere (C) contains short, highly repeated DNA sequences (satellite DNA) and a protein-containing structure called the kinetochore that serves as a site for the attachment of spindle microtubules during mitosis and meiosis (discussed in Chapter 14). (FROM JEROME B. RATTNER, BIOESS. 13:51, 1991.)

by the following studies on humans. Approximately 1 in every 2,000 humans is born with cells that have an excess piece of chromosomal DNA that forms an additional diminutive chromosome, called a *marker chromosome*. In some cases, marker chromosomes are devoid of α-satellite DNA, yet they still contain a primary constriction and a functional centromere that allows the duplicated chromosomes to be separated normally into daughter cells at each cell division. Clearly, some other DNA sequence in these "chromosomes" becomes "selected" as the binding site for centromeric proteins. The centromere appears at the same site on a marker chromosome in all of the person's cells, indicating that the property is transmitted to the daughter chromosomes during cell division.

These observations indicate that α-satellite DNA is not absolutely required for the development of a centromere. They also raise a larger issue. Not all inherited characteristics are strictly dependent on DNA sequences. Inheritance of this type is referred to as **epigenetic** as opposed to *genetic*. The inactivation of the X chromosome discussed on page 507 is another example of an epigenetic phenomenon: the two X chromosomes can have identical DNA sequences, but one is inactivated and the other is not. Furthermore, the state of inactivation is transmitted from one cell to its daughters. Biologists have discussed epigenetic phenomena for decades, but have struggled to understand their underlying mechanisms. We do not know, for example, what properties of a particular region of DNA cause it to become selected as a centromere and then "mark" it as a centromere from that time on.

Telomeres Each chromosome contains a single, continuous, double-stranded DNA molecule. The tips of each DNA molecule are composed of an unusual stretch of repeated sequences called a **telomere**, which forms a cap at each end of the chromosome. Human telomeres contain the sequence $\frac{\text{TTAGGG}}{\text{AATCCC}}$ repeated from about 500 to 5,000 times (Figure 12.18*a*). Unlike most repeated sequences that vary considerably from species to species, even those

(*a*)

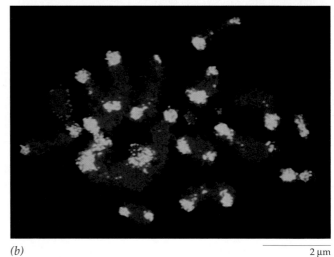

(*b*) ————————— 2 μm

Figure 12.18 Telomeres. (*a*) In situ hybridization of a DNA probe containing the sequence TTAGGG, which localizes to the telomeres of human chromosomes. (*b*) Demonstration that certain proteins bind specifically to telomeric DNA. These chromosomes were prepared from a meiotic nucleus of a yeast cell and incubated with the protein RAP1, which was subsequently localized at the telomeres by a fluorescent anti-RAP1 antibody. Blue areas indicate DNA staining, yellow areas represent anti-RAP1 antibody labeling, and red shows RNA stained with propidium iodide. Humans have a homologous telomere protein called hRAP1. (A: FROM J. MEYNE, IN R. P. WAGNER, CHROMOSOMES: A SYNTHESIS, COPYRIGHT © 1993. REPRINTED BY PERMISSION OF WILEY-LISS, INC., A SUBSIDIARY OF JOHN WILEY & SONS, INC.; B: FROM FRANZ KLEIN ET AL., J. CELL BIOL. 117:940, 1992, COURTESY OF SUSAN M. GASSER; BY COPYRIGHT PERMISSION OF THE ROCKEFELLER UNIVERSITY PRESS.)

that are closely related, the same telomere sequence is found throughout the vertebrates, and similar sequences are found in most other organisms. This similarity in sequence suggests that telomeres have a conserved function in diverse organisms. A number of DNA-binding proteins have been identified that bind specifically to the telomere sequence. The protein that is bound to the chromosomes in Figure 12.18*b* plays a role in regulating telomere length in yeast.

As discussed in Chapter 13, the DNA polymerases that replicate DNA do not initiate the synthesis of a strand of DNA but only add nucleotides to the 3′ end of an existing strand. Replication is initiated at the 5′ end of each newly synthesized strand by synthesis of a short RNA primer that is subsequently removed (Figure

12.19*a*). Because of this mechanism, the 5′ end of each newly synthesized strand is missing a short segment of DNA that is present at the 3′ end of the complementary template strand. As a result, the strand with the 3′ end overhangs the strand with the 5′ end. Rather than existing as an unprotected single-stranded terminus, the overhanging strand is "tucked back" into the double-stranded portion of the telomere to form a loop as illustrated in Figure 12.19*b*. This conformation may protect the telomeric end of the DNA from proteins that normally recognize single-stranded DNA and trigger a DNA repair response.

If cells were not able to replicate the ends of their DNA, the chromosomes would be expected to become shorter and shorter with each round of cell division (Figure 12.19*a*). This predicament has been called "the end-

(a)

(b)

(c)

Figure 12.19 The end-replication problem and the role of telomerase. (*a*) When the DNA of a chromosome is replicated, the 5′ ends of the newly synthesized strands (red) contain a short segment of RNA (green), which had functioned as a primer for synthesis of the adjoining DNA. Once this RNA is removed, the 5′ end of the DNA becomes shorter relative to that of the previous generation. (*b*) Electron micrographs reveal that the single-stranded overhang does not remain as a free extension, but invades the duplex as shown here, displacing one of the strands, which forms a loop. (*c*) The mechanism of action of telomerase. The enzyme contains an RNA molecule that is complementary to the end of the G-rich strand, which extends past the C-rich strand, forming an overhang. The RNA binds

to the protruding end of the G-rich strand (step 1) and then serves as a template for the addition of nucleotides onto the 3′ terminus of the strand (step 2). After a segment of DNA is synthesized, the telomeric RNA slides to the new end of the strand being elongated (step 3) and serves as the template for the incorporation of additional nucleotides (step 4). The gap in the complementary strand is filled by the cell's conventional replication machinery. (The TTGGGG sequence depicted in this drawing is that of the ciliated protist *Tetrahymena*, which is the organism in which telomerase was first discovered.) (C: AFTER C. W. GREIDER AND E. H. BLACKBURN, REPRINTED WITH PERMISSION FROM NATURE 337:336, 1989; © COPYRIGHT 1989 MACMILLAN MAGAZINES LIMITED.)

replication problem" The primary mechanism by which organisms solve the "end-replication problem," came to light in 1984 when Elizabeth Blackburn and Carol Greider of the University of California, Berkeley, discovered a novel enzyme, called **telomerase**, that can add new repeat units to the 3' end of the overhanging strand (Figure 12.19*c*). Telomerase presumably acts on free 3' ends at a stage when they are not part of a loop. Once the 3' end of the strand has been lengthened, a conventional DNA polymerase can use the newly synthesized 3' segment as a template to return the 5' ends of the complementary strand to its previous length. Telomerase is a reverse transcriptase that synthesizes DNA using an RNA template. Unlike most reverse transcriptases, the enzyme itself contains the RNA that serves as its template (Figure 12.19*c*).

Telomeres are very important parts of a chromosome: they are required for the complete replication of the chromosome; they form caps that protect the chromosomes from nucleases and other destabilizing influences; they facilitate interaction between the ends of the chromosome and the nuclear envelope in some types of cells; and they prevent the ends of chromosomes from fusing with one another. The importance of telomeres in preventing chromosome fusion is apparent in the micrograph of Figure 12.20, which shows the effect of a lack of telomerase activity on chromosome stability. Recent experiments suggest additional roles, which have made telomeres a focus of current research.

Suppose a researcher were to take a small biopsy of your skin, isolate a population of fibroblasts from the dermis, and allow these cells to grow in an enriched culture medium. The fibroblasts would divide every day or so in culture and eventually cover the dish. If a fraction of these cells were removed from the first dish and replated on a second dish, they would once again proliferate and cover the second dish. You might think you could subculture these cells indefinitely—as was thought in the first half of the last century—but you would be wrong. After about 50 to 80 population doublings, the cells stop dividing and eventually die. If the average length of the telomeres in the fibroblasts at the beginning and end of the experiment are compared, one finds a dramatic decrease in the length of the telomeres over time in culture. A similar decrease in telomere length is found in somatic cells from an elderly adult as compared with the telomeres in the corresponding cells from an infant or young child. Telomeres shrink because most cells in adult mammals lack telomerase and are unable to prevent the loss of their chromosome ends.[3] With each cell division, the telomeres of the chromosomes become shorter and shorter. Telomere shortening is thought to continue to a critical point, called "crisis," when the cells exhibit extensive chromosome abnormalities and stop dividing.

Figure 12.20 The importance of telomerase in maintaining chromosome integrity. The chromosomes in this micrograph are from the cell of a knockout mouse that lacks a functional gene for the enzyme telomerase. The telomeres appear as red spots following in situ hybridization with a fluorescently labeled telomere probe. It can be seen that some chromosomes lack telomeres entirely, and that a number of chromosomes have fused to one another at their ends. The occurrence of chromosome fusion, which does not occur in a normal cell, reveals the importance of telomeres in maintaining chromosome integrity. (FROM MARIA A. BLASCO, ET AL., COURTESY OF CAROL W. GREIDER, CELL, VOL. 91, COVER #1, 1997, BY PERMISSION OF CELL PRESS.)

The importance of telomere shortening in cellular aging was confirmed in a 1998 study in which genetically engineered cells were forced to express an active telomerase enzyme. In this study, control cells that lacked telomerase proliferated for a period and then underwent aging and death, as expected. In contrast, the cells that expressed telomerase continued to proliferate for hundreds of extra divisions. Not only did the telomerase-expressing cells continue to divide, they did so without showing signs of the aging process seen in control cultures. This report led some researchers to suggest that cellular aging—and human aging as well—might be prevented by causing cells to express this normally inactive enzyme. But giving cells this property could have serious consequences.

According to one current view, telomere shortening plays a key role in protecting the body from cancer. Malignant cells, by definition, are cells that have escaped the body's normal growth control and continue to divide indefinitely. How is it that malignant tumor cells can divide repeatedly without leading to cell death? Unlike normal cells that lack detectable telomerase activity, approximately 90 percent of human tumors consist of cells that contain an active telomerase enzyme.[4] It is speculated that the growth of tumors is accompanied by intense selection for cells in which the expression of telomerase has been reactivated. Cells that fail to express telomerase die out, whereas cells that express the enzyme are "immortalized."

[3]It is noteworthy that, unlike somatic cells, the germ cells of the gonads retain telomerase activity, and the telomeres of the chromosomes do not shrink as the result of cell division. Consequently, each offspring begins life as a zygote that contains telomeres of maximal length.

[4]The other 10 percent or so have an alternate mechanism based on genetic recombination that maintains telomere length in the absence of telomerase.

THE HUMAN PERSPECTIVE

Chromosomal Aberrations

In addition to mutations that alter the information content of a single gene, chromosomes may be subjected to more extensive alterations that occur most commonly during cell division. Pieces of a chromosome may be lost or segments may be exchanged between different chromosomes. Because these chromosomal aberrations follow chromosomal breakage, their incidence is increased by exposure to agents that damage DNA, such as viral infection, X rays, or reactive chemicals. In addition, the chromosomes of some individuals contain "fragile" sites that are particularly susceptible to breakage. Persons with certain rare inherited conditions, such as Bloom syndrome, Fanconi anemia, and ataxia-telangiectasia, have unstable chromosomes that exhibit a greatly increased tendency toward breakage.

The consequences of a chromosomal aberration depend on the genes that are affected and the type of cell in which it occurs. If the aberration occurs in a somatic (nonreproductive) cell, the consequences are generally minimal because only a few cells of the body are usually affected. On rare occasion, however, a cell carrying an aberration may be transformed into a malignant cell, which can grow into a cancerous tumor.

Chromosomal aberrations that occur during meiosis—particularly as a result of abnormal crossing over—can be transmitted to the next generation. When an aberrant chromosome is inherited through a gamete, all cells of the offspring will have the aberration, and the individual generally does not survive embryonic development. There are several types of chromosomal aberrations, including the following:

- Inversions. Sometimes a chromosome breaks in two places, and the segment between the breaks becomes resealed into the chromosome in reverse. This aberration is called an inversion. As many as 1 percent of humans carry an inversion that can be detected during chromosome karyotyping. A chromosome bearing an inversion usually contains all the genes of a normal chromosome, and therefore, the individual is not adversely affected. However, if a cell with a chromosome inversion enters meiosis, the aberrant chromosome cannot pair properly with its normal homologue because of differences in the order of their genes. In such cases, chromosome pairing is usually accompanied by a loop (Figure 1). If crossing over occurs within the loop, as shown in the figure, the gametes generated by meiosis either possess an additional copy of certain genes (a duplication) or are missing those genes (a deletion). When a gamete containing an altered chromosome fuses with a normal gamete at fertilization, the resulting zygote has a chromosome imbalance and is usually nonviable.

- Translocations. When all or a piece of one chromosome becomes attached to another chromosome, the aberration is called a translocation (Figure 2). Like inversions, a translocation that occurs in a somatic cell generally has little effect on the functions of that cell or its progeny. However, certain translocations increase the likelihood that the cell will

become malignant. The best-studied example is the Philadelphia chromosome, which is found in the malignant cells (but not the normal cells) of individuals with certain forms of leukemia. The Philadelphia chromosome, which is named for the city in which it was discovered in 1960, is a shortened version of human chromosome number 22. For years, it was thought that the missing segment represented a simple deletion, but with improved techniques for observing chromosomes, the missing genetic piece was found translocated to another chromosome (number 9). Chromosome number 9 contains a gene (c-*abl*) that encodes a protein kinase that plays a role in cell proliferation. As a result of translocation, one small end of this protein is replaced by about 600 extra amino acids encoded by a gene (*bcr*) carried on the translocated piece of chromosome number 22. This novel "fusion protein" apparently retains the catalytic activity of the original version but is no

Figure 1 The effect of inversion. Crossing over between a normal chromosome (purple) and one containing an inversion (green) is usually accompanied by formation of a loop. The chromosomes that result from the crossover contain duplications and deficiencies, which are shown in the chromosomes at first meiotic division in the lower part of the figure.

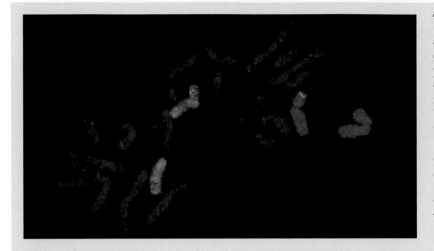

Figure 2 A translocation. Micrograph shows a set of human chromosomes in which chromosome 12 (bright blue) has exchanged pieces with chromosome 7 (red). The affected chromosomes have been made fluorescent by in situ hybridization with a large number of DNA fragments that are specific for each of the two chromosomes involved. The use of these "stains" makes it very evident when one chromosome has exchanged pieces with another chromosome. (COURTESY OF LAWRENCE LIVERMORE NATIONAL LABORATORY, FROM A TECHNIQUE DEVELOPED BY JOE GRAY AND DAN PINKEL.)

longer subject to the cell's normal regulatory mechanisms. The bcr-abl fusion protein is thought to play a role in the progression of chronic myelogenous leukemia (CML).

Like inversions, translocations cause problems during meiosis. A chromosome altered by translocation has a different genetic content from its homologue. As a result, the gametes formed by meiosis will either contain extra copies of genes or be missing genes. Translocations have been shown to play an important role in evolution, generating large-scale changes that may be pivotal in the branching of separate evolutionary lines from a common ancestor. Such a genetic incident probably happened during our own recent evolutionary history. A comparison of the 23 pairs of chromosomes in human cells with the 24 pairs of chromosomes in the cells of chimpanzees, gorillas, and orangutans reveals a striking similarity. Close examination of the two ape chromosomes that have no counterpart in humans reveals that together they are equivalent, band for band, to human chromosome number 2 (Figure 3). At some point during the evolution of humans, an entire chromosome was translocated to another, creating a single fused chromosome and reducing the haploid number from 24 to 23.

■ Deletions. A deletion occurs when a portion of a chromosome is missing. As noted above, zygotes containing a chromosomal deletion are produced when one of the ga-

metes is the product of an abnormal meiosis. Forfeiting a portion of a chromosome often results in a loss of critical genes, producing severe consequences, even if the individual's homologous chromosome is normal. Most human embryos that carry a significant deletion do not develop to term, and those that do exhibit a variety of malformations. The first correlation between a human disorder and a chromosomal deletion was made in 1963 by Jerome Lejeune, a French geneticist who had earlier discovered the chromosomal basis of Down syndrome. Lejeune discovered that a baby born with a variety of facial malformations was missing a portion of chromosome 5. A defect in the larynx (voice box) caused the infant's cry to resemble the sound of a suffering cat. Consequently, the scientists named the disorder cri du chat syndrome, meaning cry-of-the-cat syndrome.

■ Duplications. A duplication occurs when a portion of a chromosome is repeated. The role of duplications in the formation of multigene families was discussed on page 420. More substantial chromosome duplications create a condition in which a number of genes are present in three copies rather than the two copies normally present (a condition called partial trisomy). Cellular activities are very sensitive to the number of copies of genes, and thus extra copies of genes can have serious deleterious effects.

Chromosome #2
Human
Chimpanzee
Gorilla
Orangutan

Figure 3 Translocation and evolution. If the only two ape chromosomes that have no counterpart in humans are hypothetically fused, they match human chromosome number 2, band for band.

This does not mean that activation of telomerase, by itself, causes cells to become malignant. As discussed in Chapter 16, the development of cancer is a multistep process in which cells typically develop abnormal chromosomes, changes in cell adhesion, and the ability to invade normal tissues. Unlimited cell division is only one property of cancer cells. In fact, when normal cells are forced to express telomerase, as described above, they divide indefinitely, but they aren't transformed into cancer cells. If telomerase-expressing "normal" cells are injected into a mouse, they do not develop into tumors, as they would if they were cancer cells.

The discovery that cancer cells express telomerase has launched a search for specific inhibitors of this enzyme in the hope that they might arrest the growth of tumors. Further incentive for this approach has come from a study in which cells derived from several different human tumors were genetically engineered to carry an inactive, mutant form of telomerase. When expressed at high levels, the mutant version of the enzyme acted in a *dominant negative* manner, overriding the activity of the normal version of telomerase present in these cells. As a result, the modified cancer cells gradually lost their telomeres and stopped dividing when grown in culture. The primary stumbling block to the use of telomerase inhibitors in the treatment of cancer is the number of times the cells would have to divide in the presence of the inhibitor before they would lose their telomeres and die.

The Nucleus as an Organized Organelle

Examination of the cytoplasm of a eukaryotic cell under the electron microscope reveals the presence of a diverse array of membranous organelles and cytoskeletal elements. Examination of the nucleus, on the other hand, typically reveals little more than scattered clumps of chromatin and one or more irregular nucleoli. As a result, researchers were left with the impression that the nucleus is simply a "sack" of randomly positioned components. With the development of new types of microscopic techniques, most notably that of fluorescence in situ hybridization (FISH, discussed on page 417) and confocal microscopy (Section 18.1), it became possible to localize specific DNA and RNA sequences within the interphase nucleus. It is now evident that the nucleus is an ordered compartment. For example, the chromatin fibers of a given interphase chromosome are not strewn through the nucleus like a bowl of spaghetti, but are concentrated into a specific territory that does not overlap extensively with the territories of other chromosomes. In the micrograph shown in Figure 12.21, the chromatin that comprises human chromosome number 18 occupies a territory near the periphery of the nucleus, whereas the chromatin of chromosome number 19 is localized more centrally within the organelle. This difference in nuclear location may be related to the levels of activity of these two chromosomes: chromosome number 18 happens to be relatively devoid of

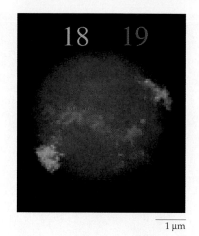

1 μm

Figure 12.21 Localizing specific chromosomes within an interphase nucleus. The micrograph shows the nucleus of a human lymphocyte (stained blue) that was subjected to dual-label FISH (fluorescence in situ hybridization) to visualize two different chromosomes, numbers 18 and 19, which appear as green and red, respectively. Each chromosome occupies a defined territory that does not overlap with that of any other labeled chromosome. (FROM JENNY A. CROFT, ET AL., COURTESY OF WENDY A. BICKMORE, J. CELL BIOL. 145:1119, 1999; BY COPYRIGHT PERMISSION OF THE ROCKEFELLER UNIVERSITY PRESS.)

genes, whereas chromosome number 19 happens to be rich in protein-coding sequences, many of which are presumably transcribed in these cells. A similar disposition is observed in the nuclei of women, where the inactive X chromosome is located at one edge of the nucleus, whereas the active X chromosome is situated internally (Figure 12.15a). Different parts of chromosomes may also have predictable locations. In some plant nuclei, for example, the centromeres and telomeres of the chromosomes are clustered near the nuclear envelope.

Another example of nuclear organization is illustrated in Figure 12.22a. This micrograph shows a cell that has been stained with a fluorescent antibody against one of the protein factors involved in pre-mRNA splicing. Rather than being spread uniformly throughout the nucleus, the processing machinery is concentrated within 20 to 50 irregular domains, referred to as "speckles." According to current opinion, these speckles function as dynamic storage depots that supply splicing factors for use at nearby sites of transcription. The green dot in the nucleus in Figure 12.22a is a viral gene that is being transcribed near one of the speckles. The micrographs of Figure 12.22b show a trail of splicing factors extending from a speckle domain toward a nearby site where pre-mRNA synthesis has recently been activated. The various components of the nucleus, including nucleoli and speckles, are thought to be ordered within the compartment by a complex, interactive network of filaments that make up the nuclear matrix.

The Nuclear Matrix When isolated nuclei are treated with nonionic detergents and high salt (e.g., 2 M NaCl), which remove lipids and nearly all of the histone and non-

(a) 2 µm

(b) 1.5 µm

Figure 12.22 Nuclear compartmentalization of the cell's mRNA processing machinery. (*a*) The nucleus of a cell stained with fluorescent antibodies against one of the factors involved in processing pre-mRNAs. The mRNA processing machinery is localized to approximately 30 to 50 discrete sites, or "speckles." The cell shown in this micrograph had been infected with cytomegalovirus, whose genes (shown as a green dot) are being transcribed near one of these domains. (*b*) Cultured cells were transfected with a virus, and transcription was activated by addition of cyclic AMP. Images are seen at various times after transcriptional activation. The site of transcription of the viral genome in this cell is indicated by the arrows. This site was revealed at the end of the experiment by hybridizing the viral RNA to a fluorescently labeled probe (indicated by the white arrow in the fourth frame). Pre-mRNA splicing factors (orange) form a trail from existing speckles in the direction of the genes being transcribed. (A: FROM TOM MISTELI AND DAVID L. SPECTOR, CURR. OPIN. CELL BIOL. 10:324, 1998; B: REPRINTED WITH PERMISSION FROM TOM MISTELI, JAVIER F. CÁCERES, AND DAVID L. SPECTOR, NATURE 387:525, 1997, COPYRIGHT 1997, MACMILLAN MAGAZINES LIMITED.)

(a) 5 µm

(b) 0.2 µm

Figure 12.23 The nuclear matrix. (*a*) Electron micrograph of a nucleus isolated in the presence of detergent and 2 M salt, which leaves the nuclear matrix surrounded by a halo of DNA loops. Arrow denotes the outer limit of the DNA loops. (*b*) Electron micrograph of a portion of a mouse fibroblast that has been extracted with detergent and depleted of its chromatin and DNA by treatment with nucleases and high salt. The nucleus (N) is seen to consist of a residual filamentous matrix whose elements terminate at the site of the nuclear envelope. The cytoplasm (C) contains a different cytoskeletal matrix whose structure is discussed in Chapter 9. (A: REPRINTED WITH PERMISSION FROM D. A. JACKSON, S. J. MCCREADY, AND P. R. COOK, NATURE 292:553, 1981; © COPYRIGHT 1981 MACMILLAN MAGAZINES LIMITED; B: FROM DAVID G. CAPCO, KATHERINE M. WAN, AND SHELDON PENMAN, CELL 29:851, 1982; BY PERMISSION OF CELL PRESS.)

histone proteins of the chromatin, the DNA is seen as a halo surrounding a residual nuclear core (Figure 12.23*a*). If the DNA fibers are subsequently digested with DNase, the structure that remains possesses the same shape as the original nucleus but is composed of a network of thin protein-containing fibrils crisscrossing through the nuclear space (Figure 12.23*b*). This insoluble fibrillar network is called the **nuclear matrix**.

The nuclear matrix serves as more than a skeleton to maintain the shape of the nucleus or a scaffold on which to organize loops of chromatin (page 504); it also serves to anchor much of the machinery that is involved in the various activities of the nucleus, including transcription, RNA processing, and replication. For instance, if cells are incubated with fluorescently or radioactively labeled RNA or DNA precursors for a brief period, nearly all of the newly synthesized nucleic acid is found to be associated with the fibrils of the nuclear matrix.

1. Describe the components that make up the nuclear envelope. What is the relationship between the nuclear membranes and the nuclear pore complex? How does the nuclear pore complex regulate bidirectional movement of materials between the nucleus and cytoplasm?

2. What is the relationship between the histones and DNA of a nucleosome core particle? How was the existence of nucleosomes first revealed? How are nucleosomes organized into higher levels of chromatin? What is the role of the H1 histone in this hierarchy?

3. What is the difference in structure and function between heterochromatin and euchromatin? Between constitutive and facultative heterochromatin? Between an active and inactivated X chromosome in a female mammalian cell?

4. What is the difference in structure and function between the centromeres and telomeres of a chromosome?

5. Describe some of the observations that suggest that the nucleus is an ordered compartment.

12.2 CONTROL OF GENE EXPRESSION IN PROKARYOTES

A bacterial cell lives in direct contact with its environment, which may change dramatically in chemical composition from one moment to the next. At certain times, a particular compound may be present, while at other times that compound is absent. Consider the consequences of transferring a culture of bacteria from a minimal medium to one containing either (1) lactose or (2) tryptophan.

1. Lactose is a disaccharide (see Figure 2.16) composed of glucose and galactose whose oxidation can provide the cell with metabolic intermediates and energy. The first step in the catabolism (i.e., degradation) of lactose is the hydrolysis of the bond (a β-galactoside linkage) that joins the two sugars, a reaction catalyzed by the enzyme β-galactosidase. When the bacterial cells were growing under the minimal conditions, the cells had no need for β-galactosidase. Under minimal conditions, an average cell contains fewer than five copies of β-galactosidase and a single copy of the corresponding mRNA. Within a few minutes after adding lactose to the culture medium, cells contain approximately 1,000 times the number of β-galactosidase molecules. The presence of lactose has induced the synthesis of this enzyme (Figure 12.24).

2. Tryptophan is an amino acid required for protein synthesis. In the absence of this compound in the medium, a bacterium must expend energy synthesiz-

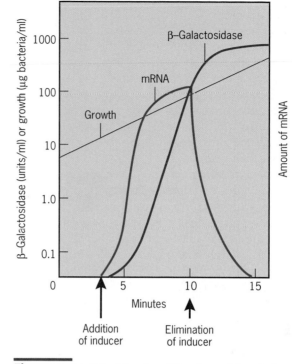

Figure 12.24 **The kinetics of β-galactosidase induction in *E. coli*.** When an appropriate inducer (a β-galactoside) is added, the production of mRNA for the enzyme β-galactosidase begins very rapidly, followed within a minute or so by the appearance of the enzyme, whose concentration increases rapidly. Removal of the inducer leads to a precipitous drop in the level of the mRNA, which reflects its rapid degradation. The amount of enzyme then levels off because new molecules are no longer synthesized.

ing this amino acid. Cells growing in the absence of tryptophan contain the enzymes, and their corresponding mRNAs, needed for tryptophan manufacture. If, however, this amino acid should become available in the medium, the cells no longer have to synthesize their own tryptophan, and within a few minutes, the production of the enzymes of the tryptophan pathway stops. In the presence of tryptophan, the genes that encode these enzymes are repressed.

The Bacterial Operon

In bacteria, the genes that encode the enzymes of a metabolic pathway are usually clustered together on the chromosome in a functional complex called an **operon**. All of the genes of an operon are coordinately controlled by a mechanism first described in 1961 by Francois Jacob and Jacques Monod of the Pasteur Institute in Paris. A typical bacterial operon consists of structural genes, a promoter region, an operator region, and a regulatory gene (Figure 12.25).

■ **Structural genes** code for the enzymes themselves. The structural genes of an operon usually lie adjacent to one another, and the RNA polymerase moves from one structural gene to the next, transcribing all of the genes

Figure 12.25 Organization of a bacterial operon. The enzymes that make up a metabolic pathway are encoded by a series of structural genes that reside in a contiguous array within the bacterial chromosome. All of the structural genes of an operon are transcribed into a continuous mRNA, which is translated into separate polypeptides. Transcription of the structural genes is controlled by a repressor protein that, when bound to the operator site of the DNA, blocks movement of the RNA polymerase from the promoter to the structural genes.

into a single mRNA. This extended mRNA is then translated into the various individual enzymes of the metabolic pathway. Consequently, turning on one gene turns on all the enzyme-producing genes of an operon.

- The **promoter** is the site where the RNA polymerase binds to the DNA prior to beginning transcription (discussed on page 443).

- The **operator**, which typically resides adjacent to or overlapping with the promoter (see Figure 12.27), serves as the binding site for a protein, called the **repressor**. The repressor is an example of a **gene regulatory protein**—a protein that recognizes a specific sequence of base pairs within the DNA and binds to that sequence with high affinity. As will be evident in the remaining sections of this chapter, DNA-binding proteins, such as bacterial repressors, play a predominant role in determining whether or not a particular segment of the genome is transcribed.

- The **regulatory gene** encodes the repressor protein.

The key to operon expression lies with the repressor. When the repressor binds the operator (Figure 12.26), the promoter is shielded from the polymerase, and transcription of the structural genes is prevented. The capability of the repressor to bind the operator and inhibit transcription depends on the conformation of the protein, which is regulated allosterically by a key compound in the metabolic pathway, such as lactose or tryptophan, as described shortly. It is the concentration of this key metabolic substance that determines if the operon is active or inactive at any given time.

The *lac* Operon The interplay among these various elements is illustrated by the *lac operon*—the cluster of genes that regulates production of the enzymes needed to degrade lactose in bacterial cells. The *lac* operon is an example of an **inducible operon**, that is, one in which the presence of a key metabolic substance (in this case, lactose) induces transcription of the structural genes (Figure

Figure 12.26 Gene regulation by operons. Inducible and repressible operons work on a similar principle: if the repressor is able to bind to the operator, genes are turned off; if the repressor is inactivated and unable to bind to the operator, genes are expressed. (*a*) In an inducible operon, (1) the inducer (in this case, the disaccharide lactose) binds to the repressor protein and (2) prevents its attachment to the operator (O). (3) Without the repressor in the way, RNA polymerase attaches to the promoter (P) and (4) transcribes the structural genes. Thus, when the lactose concentration is high, the operon is induced, and the needed sugar-digesting enzymes are manufactured. (5) As the sugar is metabolized, its concentration dwindles (6), causing bound inducer molecules to dissociate from the repressor, which then regains its ability to attach to the operator and prevent transcription. Thus, when the inducer concentration is low, the operon is repressed, preventing synthesis of unneeded enzymes. (*b*) In a repressible operon, such as the *trp* operon, the repressor, by itself, is *unable* to bind to the operator, and the structural genes coding for the enzymes are actively transcribed. The enzymes of the *trp* operon catalyze the reactions in which this essential enzyme is synthesized. (1) When available, tryptophan molecules act as corepressors by binding to the *inactive* repressor and (2) changing its shape, allowing it to attach to the operator, (3) preventing transcription of the structural genes. Thus, when tryptophan concentration is high, the operon is repressed, preventing overproduction of tryptophan. (4) When the tryptophan concentration is low, most repressor molecules lack a corepressor and therefore fail to attach to the operator. (5) Genes are transcribed, (6) enzymes are synthesized, and (7) the needed end product (tryptophan) is manufactured (8).

12.26*a*). The *lac* operon contains three tandem structural genes: the *z* gene, which encodes β-galactosidase; the *y* gene, which encodes galactoside permease, a protein that promotes the entry of lactose into the cell; and the *a* gene, which encodes thiogalactoside acetyltransferase, an enzyme whose physiological role is unclear. If lactose is present in the medium, the disaccharide enters the cell where it binds to the *lac* repressor, changing the conformation of

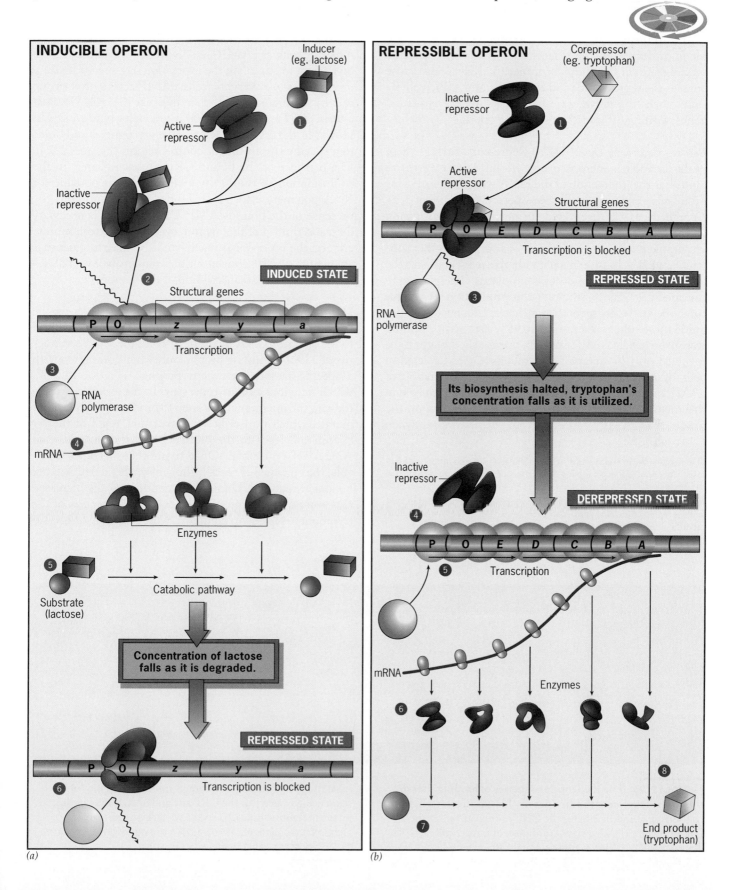

(a) INDUCIBLE OPERON

(b) REPRESSIBLE OPERON

the repressor and making it unable to attach to the DNA of the operator. In this state, the structural genes are transcribed, the enzymes are synthesized, and the lactose molecules are catabolized. Thus, in an inducible operon, such as the *lac* operon, the repressor protein is able to bind to the DNA only in the absence of lactose, which functions as the **inducer**.[5] As the concentration of lactose in the medium decreases, the disaccharide dissociates from its binding site on the repressor molecule. Release of lactose enables the repressor to bind to the operator, which physically blocks the polymerase from reaching the structural genes, turning off transcription of the operon.[6]

Positive Control by Cyclic AMP Repressors, such as those of the lac and trp operons, exert their influence by *negative control*, as the interaction of the DNA with this protein inhibits gene expression. The *lac* operon is also under *positive control*, as was discovered during an early investigation of a phenomenon called the *glucose effect*. If bacterial cells are supplied with glucose as well as a variety of other substrates, such as lactose or galactose, the cells catabolize the glucose and ignore the other compounds. The glucose in the medium acts to suppress the production of various catabolic enzymes, such as β-galactosidase, needed to degrade these other substrates. In 1965, a surprising finding was made: cyclic AMP (cAMP), previously thought to be involved only in eukaryotic metabolism, was detected in cells of *E. coli*. It was found that the concentration of cAMP in the cells was related to the presence of glucose in the medium; the higher the glucose concentration, the lower the cAMP concentration. Furthermore, when

cAMP was added to the medium in the presence of glucose, the catabolic enzymes that were normally absent were suddenly synthesized by the cells.

Although the exact means by which glucose lowers the concentration of cAMP has still not been elucidated, the mechanism by which cAMP overcomes the effect of glucose is well understood. As might be expected, a small molecule such as cAMP (see Figure 15.6) could not serve on its own to stimulate the expression of a specific battery of genes. As in eukaryotic cells, cAMP acts in prokaryotic cells by binding to a protein, in this case the *cAMP receptor protein (CRP)*. By itself, CRP is unable to bind to DNA. However, the cAMP-CRP complex recognizes and binds to a specific site in the *lac* control region (Figure 12.27). The presence of the bound CRP causes a change in the conformation of the DNA, which makes it possible for RNA polymerase to transcribe the *lac* operon. Thus the presence of the bound cAMP-CRP complex is necessary for transcription of the operon, even when lactose is present and the repressor is inactivated. As long as glucose is abundant, cAMP concentrations remain below that required to promote transcription of the operon.

The *trp* Operon In a **repressible operon**, such as the tryptophan (or *trp*) operon, the repressor is unable to bind to the operator DNA by itself. Instead, the repressor is active as a DNA-binding protein only when complexed with a specific factor, such as tryptophan (Figure 12.26*b*), which functions as a **co-repressor**. In the absence of tryptophan, the operator site is open to binding by RNA polymerase, which transcribes the structural genes of the *trp* operon and leads to the production of the enzymes that synthesize tryptophan. Once tryptophan becomes available, the enzymes of the tryptophan synthetic pathway are no longer required. Under these conditions, the increased concentration of tryptophan leads to the formation of the tryptophan-repressor complex, which blocks transcription.

[5]The actual inducer is allolactose, which is derived from and differs from lactose by the type of linkage joining the two sugars. This feature is ignored in the discussion.

[6]It is noteworthy that, even though the *lac* operon was first described in 1961 and the repressor was first purified in 1966, it was not until 1996 that technical difficulties were overcome and the three-dimensional, crystal structure of the *lac* repressor was finally solved.

Figure 12.27 The nucleotide sequence of binding sites in the control region of the *lac* operon. The promoter region contains the binding site for the CRP protein as well as the RNA polymerase. The site of initiation of transcription is denoted as +1, which is approximately 40 nucleotides upstream from the site at which translation is initiated. Regions of sequence symmetry in the CRP site and operator are indicated by the horizontal line. (REPRINTED WITH PERMISSION FROM R. C. DICKSON ET AL., SCIENCE 187:32, 1975; © COPYRIGHT 1975 AMERICAN ASSOCIATION FOR THE ADVANCEMENT OF SCIENCE.)

1. Describe the cascade of events responsible for the sudden changes in gene expression in a bacterial cell following the addition of lactose. How does this compare with the events that occur in response to the addition of tryptophan?

2. What is the role of cyclic AMP in the synthesis of β-galactosidase?

12.3 CONTROL OF GENE EXPRESSION IN EUKARYOTES

The proteins synthesized by a bacterial cell are largely determined by the conditions in which the cell is growing. The problems confronting eukaryotic cells are more complex. In addition to possessing a genome containing much more information, complex plants and animals are composed of many different types of cells. Vertebrates are composed of hundreds of different cell types, each far more complex than a bacterial cell and each requiring a distinct battery of proteins that enable it to carry out specialized activities.

There was a time when biologists thought that cells achieved a particular differentiated state by retaining those parts of the chromosomes required for the functions of that cell type while eliminating those parts of the chromosomes that would not be needed. This idea that differentiation was accompanied by the loss of genetic information was finally put to rest in the 1950s and 1960s by a series of key experiments in both plants and animals demonstrating that differentiated cells retain the genes required to become any other cell in that organism. For example, Frederick Steward and his colleagues at Cornell University demonstrated that individual root cells could be isolated from a mature plant and, under appropriate laboratory conditions, induced to grow into a fully developed plant that contained all the various cell types normally present.

Although single cells from an adult animal are not able to give rise to new individuals, the nuclei from such cells have been shown to contain all the information necessary for the development of a new individual. This was most dramatically demonstrated in 1997 when Ian Wilmut and his colleagues at a Scottish research institute reported the first cloning of a mammal—a lamb they named Dolly. To achieve this controversial feat, the researchers prepared two types of cells: (1) unfertilized sheep's eggs whose chromosomes had been removed, and (2) cultured cells derived from the mammary gland (udder) of an adult sheep. Each of the enucleated eggs was then fused with one of the cultured cells (Figure 12.28). Cell fusion was accomplished by bringing the two types of cells into contact and subjecting them to a brief electric pulse, which also served to stimulate the egg to begin em-

bryonic development. This procedure accomplishes, in essence, the transplantation of a nucleus from an adult cell into an egg lacking its own genetic material. Using the genetic instructions supplied by its new nucleus, one egg developed into a healthy lamb containing all of the fully differentiated cells normally found in this animal. It is

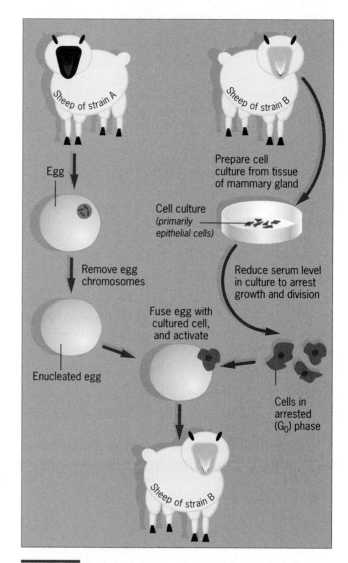

Figure 12.28 The cloning of animals demonstrates that nuclei retain a complete complement of genetic information. In this experiment, an enucleated egg from a sheep of one breed was fused with a mammary gland cell from a female of another breed. The activated egg developed into a healthy lamb. Because all of the genes in the newborn lamb had to have been derived from the transplanted nucleus (which was demonstrated by use of genetic markers), this experiment confirms the widely held belief that differentiated cells retain all of the genetic information originally present in the zygote. (The primary difficulty that is generally encountered in nuclear transplantation experiments occurs when the nucleus from an active somatic [nongerm] cell is suddenly plunged into the cytoplasm of a relatively inactive egg. To avoid damaging the donor nucleus, the cultured cells were forced into a quiescent state [called G_0] by drastically reducing the content of serum in the culture medium.)

evident from these experiments that nuclei of specialized cells, such as those from a plant root or an animal gland, contain the genetic information required for differentiation into many different types of cells. Results from a wide variety of experiments indicate that cells from most organisms do not lose genetic material as they undergo differentiation, but rather they repress those genes whose products confer the character of another cell type. Those cells that become liver cells, for example, do so because they express a specific set of "liver genes," while those cells that develop into neurons do so because they express a specific set of "nerve genes."

The average bacterial cell contains enough DNA to encode approximately 3,000 polypeptides, of which about one-third are typically expressed at any particular time. Compare this number to that of a human cell that contains enough DNA (6 billion base pairs) to encode several million different polypeptides. Even though most of this DNA does not actually contain protein-coding information, mammalian genomes are thought to contain approximately 30,000 protein-coding genes (page 430). Of this number, it is estimated that a typical mammalian cell manufactures approximately 5,000 different polypeptides at any given time. Many of these polypeptides, such as the enzymes of glycolysis and the electron carriers of the respiratory chain, are synthesized by virtually all the cells of the body. At the same time, each cell type synthesizes proteins that are unique to that differentiated state. It is these proteins, more than any other components, that give the cell its unique characteristics. Because of the tremendous amount of DNA found in a eukaryotic cell and the large number of different proteins being assembled, regulating eukaryotic gene expression is an extraordinarily complex process and is only beginning to be understood.

Consider the situation that faces a cell developing into a red blood cell inside the marrow of a human bone. Hemoglobin accounts for more than 95 percent of its protein, yet the genes that code for the hemoglobin polypeptides represent less than one-millionth of its total DNA. Not only does the cell have to find this genetic needle in the chromosomal haystack, it has to regulate its expression to such a high degree that production of these few polypeptides becomes the dominant synthetic activity of the cell. Because the chain of events leading to the synthesis of a particular protein consists of a number of discrete steps, there are several levels at which control might be exercised. Regulation of gene expression in eukaryotic cells occurs primarily at three distinct levels, as illustrated in the overview of Figure 12.29:

1. **Transcriptional-level** control mechanisms determine whether a particular gene can be transcribed and, if so, how often.

2. **Processing-level** control mechanisms determine the path by which the primary mRNA transcript (pre-mRNA) is processed into a messenger RNA that can be translated into a polypeptide.

3. **Translational-level** control mechanisms determine

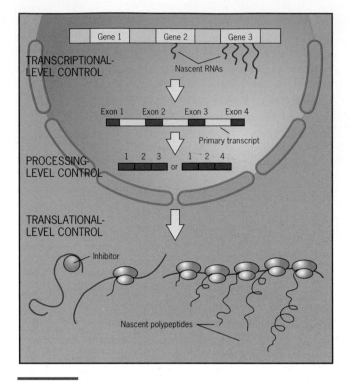

Figure 12.29 Overview of the levels of control of gene expression. Transcriptional-level controls operate by determining which genes are transcribed and how often; processing-level controls operate by determining which parts of the primary transcripts become part of the pool of cellular mRNAs; and translational-level controls regulate whether or not a particular mRNA is translated and, if so, how often and for how long.

whether a particular mRNA is actually translated and, if so, how often and for how long a period.

In the following sections of the chapter, we will consider each of these regulatory strategies in turn.

Transcriptional-Level Control

As in prokaryotic cells, differential gene transcription is the single most important mechanism by which eukaryotic cells selectively synthesize proteins at any given time. A large body of evidence indicates that different genes are expressed by cells at different stages of embryonic development, by cells in different tissues, and by cells that are exposed to different types of stimuli. The most direct evidence of widescale transcriptional level control comes from studies that use **DNA microarrays** (or "**DNA chips**") to provide visual portraits of the genes expressed in a particular cell population at a given time. Using this new technology, the expression of thousands of genes can be monitored in a single experiment.

The use of a DNA microarray to compare the population of mRNAs present in yeast cells under two different growth conditions is illustrated in Figure 12.30. Part *a* of this figure provides an outline of the basic steps taken in this type of experiment; these steps are described briefly as follows:

1. DNA fragments representing individual genes to be studied are generated using techniques discussed in Chapter 18 (PCR, Figure 18.46; DNA cloning, Figure 18.40). In the case depicted in step a of Figure 12.30a, each well on the plate holds a small volume containing a specific, cloned yeast gene. Different wells contain different genes. The cloned DNAs are then spotted, one at a time, in an ordered array on a glass slide by an instrument that delivers a few nanoliters of a concentrated DNA solution onto a specific spot on the slide (step b). A completed DNA microarray is shown in step c. Using this technique, DNA fragments from thousands of different genes can be spotted at known locations on a glass surface.

2. Meanwhile, the mRNAs present in the cells being studied are purified (step 1, Figure 12.30) and converted to a population of fluorescently labeled, complementary DNAs (cDNAs)(step 2). The method for preparation of cDNAs is also described in Chapter 18 (see Figure 18.41). In the example illustrated in Figure 12.30, and discussed below, preparations of cDNAs from two different cell populations are prepared, one labeled with a fluorescent green dye and the other with a fluorescent red dye. The two preparations of labeled cDNAs are mixed (step 3) and then incubated with the slide containing the immobilized DNA (step 4).

3. Those DNAs in the microarray that have hybridized to a labeled cDNA are identified by examination of the slide under the microscope (step 5). Any spot in the microarray that exhibits fluorescence represents a gene that has been transcribed in the cells being studied.

Figure 12.30b shows the actual results of this experiment. Each spot on the microarray of Figure 12.30b contains an immobilized DNA fragment from a different gene in the yeast genome. Taken together, the spots contain DNA from all of the roughly 6,200 protein-coding genes present in a yeast cell. As discussed above, this particular DNA microarray was hybridized with a mixture of two different cDNA populations. One population of cDNAs, which was labeled with a green fluorescent dye, was prepared from the mRNAs of yeast cells grown in the presence of a high concentration of glucose. Yeast cells growing under these conditions obtain their energy by glycolysis and fermentation, rapidly converting the glucose into ethanol (Section 3.3). The other population of cDNAs, which was labeled with a red fluorescent dye, was prepared from the mRNAs of yeast cells that were growing aerobically in a medium rich in ethanol but lacking glucose. Cells growing under these conditions obtain their energy by oxidative phosphorylation, which requires the enzymes of the TCA cycle (page 190). The two cDNA populations were mixed as described above, hybridized to the DNA in the microarray, and the slide examined in a fluorescence microscope. Genes that are active in one or the other growth media appear as either green or red spots in the microarray (step 5, Figure 12.30a, 12.30b). Those spots that remain devoid of color in Figure 12.30 repre-

sent genes that are not transcribed in these cells in either growth medium, whereas those spots that have a yellow fluorescence represent genes that are transcribed in cells in both types of media. A close-up of a small portion of the microarray is shown in the inset.

The experimental results shown in Figure 12.30b identify those genes transcribed in yeast cells under two different growth conditions. But just as importantly, they also provide information about the abundance of individual mRNAs in the cells, which is proportional to the intensity of a spot's fluorescence. A spot that displays a very strong green fluorescence, for example, represents a gene whose transcripts are abundant in yeast cells grown in glucose but is repressed in yeast cells grown in ethanol. Concentrations of individual mRNAs can vary by more than a hundredfold in yeast cells. The technique is so sensitive that an mRNA can be detected at a level of less than one copy per cell.[7]

Figure 12.30c shows the changes in concentration of glucose and ethanol over the course of an experiment. The glucose is rapidly metabolized by the yeast cells, leading to the disappearance of the sugar within a few hours. The ethanol produced by glucose fermentation is then gradually metabolized by the yeast cells over the next five days until it eventually disappears from the medium. Figure 12.30d shows the changes in the level of expression of the genes encoding the enzymes of the TCA cycle during the course of this experiment. The levels of expression of each gene (labeled on the right) are shown in intervals of time (labeled at the top), using shades of red to represent increases in expression, and shades of green to represent decreases in expression. It is evident that genes encoding these enzymes are induced when the cells adapt to growth on a carbon source (ethanol) that is metabolized by aerobic respiration, and then repressed when the ethanol is exhausted. Studies utilizing DNA microarrays are currently being conducted on a wide variety of phenomena, such as cell division and the transformation of a normal cell into a malignant cell, to determine which genes are normally turned on or off at various times during the event. Now that a number of eukaryotic genomes have been sequenced, researchers have an unlimited variety of genes whose expression can be monitored under different conditions.

DNA microarrays have many potential uses in addition to providing a visual portrait of gene expression. For example, DNA microarrays can be used to determine the degree of genetic variation in human populations or to identify the alleles for particular genes that a person carries in their chromosomes. It is hoped, one day, that this latter information may advise a person of the diseases to which they may be susceptible during their life, giving them an opportunity to take early preventative measures.

[7]Differences in mRNA abundance likely reflect differences in mRNA stability (page 543) as well as differences in rates of transcription. Consequently, results obtained from DNA microarrays cannot be interpreted *solely* on the basis of transcriptional control.

(a)

(b)

(c)

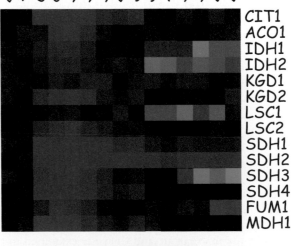

(d)

Figure 12.30 The production of DNA microarrays and their use in monitoring gene transcription. (*a*) Steps in the construction of a DNA microarray. Preparation of the cDNAs (i.e., DNAs that represent the mRNAs present in a cell) used in the experiment are shown in steps 1–3. Preparation of the DNA microarray is shown in steps a–c. The cDNA mixture is incubated with the microarray in step 4 and a hypothetical result is shown in step 5. The intensity of the color of the spot is proportional to the number of cDNAs bound there. (*b*) Results of an experiment performed using a mixture of cDNAs representing mRNAs transcribed from yeast cells in the presence of glucose (green-labeled cDNAs) and after the glucose has been depleted (red-labeled cDNAs). Those spots displaying yellow fluorescence correspond to genes that are expressed under both growth conditions. The inset at the lower right shows a close-up of a small portion of the microarray. The details of this experiment are discussed in the text. (*c*) Plot showing the changes in glucose and ethanol concentrations in the media and in cell density during the experiment. Initially, the yeast cells consume glucose, then the ethanol they have produced by fermentation, and finally they cease growth after exhausting both sources of chemical energy. (*d*) Changes in expression of genes encoding TCA-cycle enzymes during the course of the experiment. Each horizontal row depicts the level of expression of a particular gene at different time periods. The names of the genes are provided on the right (see Figure 5.7 for the enzymes). Bright red squares indicate the highest level of gene expression, bright green squares indicate the lowest level of expression. Expression of genes encoding TCA cycle enzymes is seen to be induced as the cells begin metabolizing ethanol and repressed when the ethanol has been exhausted. (B–D: COURTESY OF PATRICK O. BROWN. SEE JOSEPH DERISI ET AL., SCIENCE 278:680, 1997 AND TRACY L. FEREA AND PATRICK O. BROWN, CURR. OPIN. GEN. DEVELOP. 9:715, 1999.)

The Role of Transcription Factors in Regulating Gene Expression Great progress has been made in the past decade elucidating how certain genes are transcribed in a particular cell, while all other genes remain inactive. Transcriptional control is governed by the actions of a large number of proteins, called **transcription factors**. As discussed in Chapter 11, these proteins can be divided into two functional classes: general transcription factors that bind at core promoter sites in association with RNA polymerase, and specific transcription factors that bind to various regulatory sites of particular genes and either stimulate or inhibit the transcription of those genes. Unless otherwise specified, we will be discussing the action of specific transcription factors in this section.

Even though we know the detailed structure of a number of transcription factors and how they interact with their target DNA sequences, a unified picture of their mechanism of operation remains a distant goal. It is apparent, for example, that a single gene may be controlled by many different DNA regulatory sites that bind a variety of different regulatory proteins. Conversely, a single DNA-binding protein may become attached to numerous sites around the genome, thereby controlling the expression of a host of different genes. The control of gene

transcription is complex and is influenced by various factors, including the affinity of transcription factors for particular DNA sequences and the ability of transcription factors bound at nearby sites on the DNA to interact directly with one another. An example of this type of cooperative interaction between neighboring transcription factors is illustrated in Figure 12.31 and discussed further on page 533. In a sense, the regulatory region of a gene can be thought of as a type of integration center for that gene's expression. Cells exposed to different stimuli respond by synthesizing different transcription factors, which bind to different sites in the DNA. The extent to which a given gene is transcribed presumably depends upon the particular combination of transcription factors bound to its upstream regulatory elements.

The complexity inherent in the control of gene transcription can be illustrated by examining the DNA in and around a single gene. In the following discussion, we will focus on the gene that codes for phosphoenolpyruvate carboxykinase (PEPCK). PEPCK is one of the key enzymes of gluconeogenesis, the metabolic pathway that converts pyruvate to glucose (see Figure 3.30). The enzyme is synthesized in the liver when glucose levels are low, as might occur, for example, when a considerable period of time has passed since one's last meal. Conversely, synthesis of the

Figure 12.31 Interactions between transcription factors bound to different sites in the regulatory region of a gene. This image depicts the tertiary structure of two separate transcription factors, NFAT-1 (green) and AP-1 (whose two subunits are shown in red and blue) bound to the DNA upstream from a cytokine gene involved in the immune response. Cooperative interaction between these two proteins alters the expression of the cytokine gene. The gray bar traces the path of the DNA helix, which is seen to be bent as the result of this protein-protein interaction. (FROM TOM K. KERPPOLA, STRUCTURE 6:550, 1998.)

Figure 12.32 Identifying promoter sequences required for transcription. Line 1 shows the nucleotide sequence of one strand of the PEPCK gene promoter. The TATA box, CAAT box, and GC box are indicated. The other five lines show the results of experiments in which particular regions of the promoter have been deleted (indicated by black boxes) prior to transfecting cells with the DNA. The observed level of tran-scription of the PEPCK gene in each of these cases is indicated at the right. Deletions that remove all or part of the three boxes cause a marked decrease in the level of transcription, whereas deletions that affect other regions have either a lesser or no ef-fect. (DATA TAKEN FROM STUDIES BY RICHARD W. HANSON AND DARYL K. GRANNER AND THEIR COLLEAGUES.)

enzyme drops sharply after ingestion of a carbohydrate-rich meal, such as pasta. The level of synthesis of PEPCK mRNA is controlled by a variety of different transcription factors, including a number of receptors for hormones that are involved in regulating carbohydrate metabolism. The keys to understanding the regulation of PEPCK gene ex-pression lie in (1) unraveling the functions of the numer-ous DNA regulatory sequences that reside upstream from the gene itself, (2) identifying the transcription factors that bind these sequences, and (3) elucidating the signal-ing pathways that activate the machinery responsible for selective gene expression (discussed in Chapter 15).

The closest (most proximal) upstream regulatory se-quence is the TATA box, which is a major element of the gene's promoter (Figure 12.32). As discussed on page 455, a promoter is a regulatory region upstream from a gene that regulates the initiation of transcription. For our purposes, we will divide the eukaryotic promoter into separate re-gions, which are not that well delineated. The region that stretches from roughly the TATA box to the transcription start site is designated the *core promoter*. As discussed in Chapter 11, the core promoter is the site of assembly of a preinitiation complex consisting of RNA polymerase II and a number of general transcription factors that are re-quired before a eukaryotic gene can be transcribed. The TATA box is not the only short sequence that is shared by large numbers of genes. Two other promoter sequences, called the CAAT box and GC box, which are located far-ther upstream from the gene, are often required for a poly-merase to initiate transcription of that gene. The CAAT and GC boxes bind transcription factors (e.g., NF1 and SP1) that are found in many tissues and widely employed in gene expression. Whereas the TATA box determines

the site of initiation of transcription, the CAAT and GC boxes regulate the frequency with which the polymerase transcribes the gene. The TATA, CAAT, and GC boxes are typically located within 100–150 base pairs upstream from the transcription start site. Because of their proximity to the start of the gene, these shared sequences can be re-ferred to as **proximal promoter elements**, and are indi-cated in line 1 of Figure 12.32.

How do researchers learn which sites in a promoter region have these functions? Two general strategies are usually employed to make these determinations.

■ **Deletion mapping.** In this procedure, DNA molecules are prepared that contain deletions of various parts of the gene's promoter (Figure 12.32). Cells are then transfected with the altered DNA molecules; that is, they are caused to take up the DNA from the medium. Finally, the ability of the cells to transcribe the trans-fected DNA is determined. In many cases, the deletion of a few nucleotides has little or no effect on the tran-scription of an adjacent gene. However, if the deletion falls within any of the three boxes described above, the level of transcription tends to be sharply decreased (as in lines 2, 4, and 5 of Figure 12.32). Deletions in other parts of the promoter have less of an effect on tran-scription (lines 3 and 6 of Figure 12.32).

■ **DNA footprinting.** When a transcription factor binds to a DNA sequence, it protects that sequence from di-gestion by nucleases. Researchers take advantage of this property by isolating chromatin from cells and treating it with DNA-digesting enzymes, such as DNase I, that destroy those sections of the DNA that are not protected by bound transcription factors. Once the chromatin has

been digested, the bound protein is removed, and the DNA sequences that had been protected are identified. Once they have been identified, the DNA sequences can be used in subsequent experiments to isolate the transcription factor that recognizes each sequence.

In many cases, DNAs that contain intact promoters are capable of supporting a basal level of transcription in vitro in the presence of RNA polymerase II and a complete complement of general transcription factors. This basal level of transcription is then modulated (stimulated or inhibited) by specific transcription factors (activators or repressors) bound at other sites on the DNA. Whether or not genes are transcribed at basal levels *in the cell* in the absence of transcriptional activators is a matter of debate.

Activation of Transcription in Eukaryotes Among the various hormones that affect the expression of the PEPCK gene are insulin, thyroid hormone, glucagon, and glucocorticoids, all of which act by means of specific transcription factors that bind to the DNA. The sites at which these transcription factors bind to the region flanking the PEPCK gene are called *response elements* and are shown in Figure 12.33. In the following discussion, we will focus on the glucocorticoids, a group of steroid hormones (e.g., cortisol) synthesized by the adrenal gland in response to stress. Glucocorticoids stimulate PEPCK gene expression by binding to a specific DNA sequence, called a *glucocorticoid response element (GRE)*. When a glucocorticoid hormone enters a target cell, it binds to a glucocorticoid receptor protein in the cytosol, changing the conformation

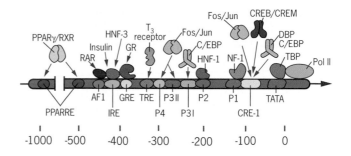

Figure 12.33 Regulating transcription from the rat PEPCK gene. Transcription of this gene, like others, is controlled by a variety of transcription factors that interact with specific DNA sequences located in a regulatory region upstream from the gene's coding region. Included within this region is a glucocorticoid response element (GRE) that, when bound by a glucocorticoid receptor, stimulates transcription from the promoter. Also included within the regulatory region are binding sites for a thyroid hormone receptor (labeled TRE); a protein that binds cyclic AMP, which is produced in response to the hormone glucagon (labeled CRE-1); and the hormone insulin (labeled IRE). A number of other transcription factors are also seen to bind to regulatory sites in this promoter. (FROM S. E. NIZIELSKI ET AL., J. NUTRITION 126: 2699 1996, © AMERICAN SOCIETY FOR NUTRITIONAL SCIENCES.)

of the protein. The change in protein structure is thought to expose a nuclear localization signal (page 497), which facilitates translocation of the receptor into the nucleus (Figure 12.34). The ligand-bound receptor binds to a GRE located upstream from the PEPCK gene, activating transcription of the gene. The same GRE sequence is lo-

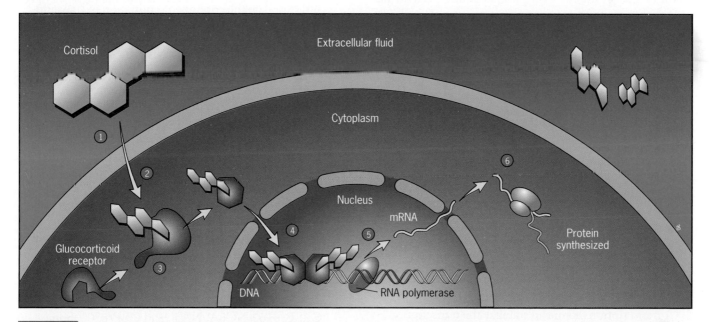

Figure 12.34 Steps in the activation of a gene by a steroid hormone, such as the glucocorticoid cortisol. The hormone enters the cell from the extracellular fluid (step 1), diffusing through the lipid bilayer (step 2) and into the cytoplasm, where it binds to a glucocorticoid receptor (step 3). Binding of the hormone changes the conformation of the receptor, causing it to translocate into the nucleus, where it acts as a transcription factor and binds to a glucocorticoid response element (GRE) of the DNA (step 4). As discussed on page 530, binding of two adjacent receptor molecules leads to the formation of a dimer, which activates transcription of the DNA (step 5), leading to the synthesis of specific proteins in the cytoplasm (step 6).

cated upstream from different genes on different chromosomes. Consequently, a single stimulus (elevated glucocorticoid concentration) can simultaneously activate a number of genes whose function is necessary for a comprehensive response to stress.

The GRE situated upstream from the PEPCK gene, and the other response elements illustrated in Figure 12.33, are generally considered to be part of the promoter; they are referred to as *distal promoter elements* to distinguish them from the proximal elements situated closer to the gene (Figure 12.32). The expression of most genes is also regulated by even more distant DNA elements called **enhancers**. Enhancers are often distinguished from promoter elements by a unique property: they can be moved experimentally from one place to another within a DNA molecule, or even inverted (rotated 180°), without affecting the ability of a bound transcription factor to stimulate transcription. Deletion of an enhancer can decrease the level of transcription by 100-fold or more. Some enhancers are located thousands or even tens of thousands of base pairs upstream or downstream from the gene whose transcription they stimulate. The human β-globin gene cluster, for example, contains an enhancer that governs transcription of the various β-globin genes from a site located tens of thousands of base pairs upstream from most of them.[8]

Even though enhancers and promoters may be separated by large numbers of nucleotides, enhancers are thought to stimulate transcription by influencing events that occur at the core promoter. Enhancers and core promoters can be brought into close proximity because the intervening DNA can form loops. The ability of DNA to be held in loops by interactions between bound proteins has been demonstrated numerous times; an example is shown in Figure 12.35. If enhancers can interact with promoters over such long distances, what is to prevent an enhancer from binding to an inappropriate promoter located even farther downstream on the DNA molecule? Recent studies indicate that a promoter and its enhancers are, in essence, "cordoned off" from other promoter/enhancer elements by specialized boundary sequences called **insulators**. Insulator sequences are thought to bind certain proteins that prevent interactions between enhancers and promoters across that boundary region (see Figure 12.36a).

A transcription factor bound to an enhancer has the potential to stimulate transcription in several ways:

1. A transcription factor can recruit general transcription factors (GTFs) and RNA polymerase II to the core promoter (Figure 12.36a). These events lead to the assembly of the preinitiation complex and the initiation of RNA synthesis (page 455).

2. A transcription factor can stabilize the transcription machinery previously assembled at the core promoter

Figure 12.35 A mechanism whereby transcription factors bound at distant sites can influence gene expression. Electron micrograph showing the binding of the progesterone receptor to DNA upstream from the uteroglobin gene, causing the formation of a loop. The bar equals 50 nm. (B: REPRINTED WITH PERMISSION FROM BERNARD THÉVENY ET AL., COURTESY OF EDWIN MILGROM, FROM NATURE 329:79, 1987; © COPYRIGHT 1987 MACMILLAN MAGAZINES LIMITED.)

and stimulate the continued reinitiation of transcription. It was noted on page 456 that TFIID, one of the GTFs required for initiation of transcription, consists of a number of subunits denoted as TAFs. Some transcription factors apparently influence events at the core promoter by interacting with one or more of these TFIID subunits. A known example of two different transcription factors acting cooperatively through different TFIID subunits is shown in Figure 12.36b.

3. It is becoming increasingly evident that most transcription factors do not interact directly with the proteins that make up the transcription machinery bound at the core promoter. Instead, transcription factors influence events at the core promoter through an intermediary known as a **coactivator** (Figure 12.36c) that does not bind to the DNA. Transcriptional coactivators are present in the nucleus as surprisingly large complexes containing as many as 15–20 different subunits. Examples of coactivators include DRIP/TRAP, SRB/MED, CRSP, p160/NCoA, NAT, and CBP/p300. Some subunits are thought to be shared among different coactivators, whereas other subunits are probably unique to particular coactivators. It has been shown in a few cases that a particular subunit of a coactivator is required for the action of a particular transcription factor. Given the large number of transcription factors en-

[8]There is wide variation in the terminology used to describe the various types of regulatory elements that control gene transcription. The terms employed here—core promoter, proximal promoter elements, distal promoter elements, and enhancers—are not universally employed, but they describe elements that are usually present and often capable of being distinguished.

Figure 12.36 Models for the mechanism of transcriptional activation. In all of these models, interactions among proteins are facilitated by the looping of the intervening DNA as shown in Figure 12.35. (*a*) In this model, two transcription factors bound at an upstream enhancer have recruited the proteins that comprise the preinitiation complex required for transcription to be initiated. (*b*) As discussed in Chapter 11, the general transcription factor TFIID contains the TATA-binding protein (TBP) and a number of other subunits, which are collectively referred to as TAFs (TBP-associated factors). In this example of transcriptional activation in *Drosophila*, two different transcription factors, bicoid (BCD) and hunchback (HB), interact specifically with different TFIID subunits (the numbers indicate the molecular mass of each of the TAFs that make up a TFIID protein). (*c*) In this model, transcription factors bound at an upstream enhancer promote the assembly of the preinitiation complex through an intermediate protein complex called a coactivator. The mechanism of action of coactivators is discussed on page 537 and illustrated in Figure 12.47. (B: AFTER C. P. VERRIJZER AND R. TJIAN, TRENDS BIOCHEM. SCI. 21:339, 1996, COPYRIGHT 1996, WITH PERMISSION OF ELSEVIER SCIENCE.)

coded in the genome, each coactivator is presumably capable of operating in conjunction with a wide variety of different transcription factors. The functions of coactivators have been intensively studied in the past few years, though they are still poorly understood. As discussed later in the chapter (and illustrated in Figure 12.47), some coactivators play a key role in converting chromatin from a state that is relatively inaccessible to the transcription machinery to a state that is much more transcription "friendly."

It is evident from the number of regulatory sites depicted for the PEPCK gene in Figure 12.33 that control of gene expression at the transcriptional level can become very complex. To understand how transcription factors interact with the DNA, with each other, and with the proteins bound to the core promoter, we will briefly examine the structure of transcriptional regulatory proteins.

The Structure of Transcription Factors The three-dimensional structure of numerous DNA-protein complexes has been determined by X-ray crystallography and NMR spectroscopy, providing a basic portrait of the way that these two giant macromolecules interact with one another. Like most proteins, transcription factors contain different domains that mediate different aspects of the protein's function. Transcription factors typically contain at least two domains: a **DNA-binding domain** that binds to a specific sequence of base pairs in the DNA, and an **activation domain** that activates transcription by interacting with other proteins (as in Figure 12.36). In addition, many transcription factors contain a surface that promotes the binding of the protein with another protein of identical or similar structure to form a dimer. The formation of dimers has proved to be a common feature of many different types of transcription factors and, as discussed below, is thought to play an important role in regulating gene expression. We can begin our analysis of transcription factors with the glucocorticoid receptor whose role in stimulating PEPCK gene expression has already been described.

The Glucocorticoid Receptor Glucocorticoids (such as cortisol) are steroid hormones, secreted by the outer portion of the adrenal gland, that promote the conversion of amino acids to glucose and its uptake by the brain. Secretion of these hormones is highest during periods of stress, as occurs during starvation or following a severe physical injury. For a cell to respond to glucocorticoids, it must possess a specific receptor capable of binding the hormone. The glucocorticoid receptor (GR) is a member of a family of nuclear receptors (including receptors for thyroid hormone, retinoic acid, and estrogen) that are thought to have evolved from a common ancestral protein. Like the other members of the family, the GR has three distinct domains: a *ligand-binding domain* that binds the steroid hormone, a *DNA-binding domain* that binds to a specific DNA sequence, and an *activation domain* that binds to other proteins and activates transcription. An example of a

DNA sequence that binds the glucocorticoid receptor (i.e., a glucocorticoid response element, or GRE) is

$$5'\text{-}A\,G\,A\,A\,C\,A\,n\,n\,n\,T\,G\,T\,T\,C\,T\text{-}3'$$
$$3'\text{-}T\,C\,T\,T\,G\,T\,n\,n\,n\,A\,C\,A\,A\,G\,A\text{-}5'$$

where n can be any nucleotide. (A symmetrical sequence of this type is called a *palindrome* because the two strands have the same 5′ to 3′ sequence.) The GRE is seen to consist of two defined stretches of nucleotides separated by three undefined nucleotides. The twofold nature of a GRE is important because pairs of GR polypeptides bind to the DNA to form dimers in which each subunit of the dimer binds to one half of the DNA sequence indicated above (Figure 12.37).

The importance of the GRE in mediating a hormone response is most clearly demonstrated by introducing one of these sequences into the upstream region of a gene that normally does not respond to glucocorticoids. When cells containing DNA that has been engineered in this way are exposed to glucocorticoids, transcription of the gene downstream from the transplanted GRE is initiated. Each GR subunit contains two distinctive α-helical loops oriented perpendicular to each other. One of the α helices, called the "recognition helix," projects into the major groove of the DNA where it recognizes and binds to the GRE sequence (Figure 12.37). The other α helix mediates binding to another GR molecule to form a dimer. As a result of dimerization, two recognition helices, one from each subunit of the dimer, are symmetrically extended into two adjacent major grooves of the target DNA.

Transcription Factor Motifs Comparison of a large number of transcription factors has revealed that DNA-binding domains of most transcription factors can be grouped into several broad classes whose members possess related structures (*motifs*) that interact with DNA sequences. The existence of several families of DNA-binding proteins indicates that evolution has found a number of different solutions to the problem of constructing polypeptides that can bind to the DNA double helix. As we will see shortly, most of these motifs contain a segment (often an α-helix, as in Figure 12.37) that is inserted into the major groove of the DNA, where it recognizes the sequence of base pairs that line the groove. Binding of the protein to the DNA is accomplished by a combination of van der Waals (hydrophobic) forces, ionic bonds, and hydrogen bonds between amino acid residues and various parts of the DNA, including the backbone.

Among the most common motifs that occur in eukaryotic DNA-binding proteins are the zinc finger, the helix-loop-helix, the leucine zipper, and the HMG box. Each provides a structurally stable framework on which the specific DNA-recognizing surfaces of the protein can be properly positioned to interact with the double helix. Such motifs are found in a wide variety of transcription factors that regulate diverse types of cellular activities in organisms ranging from fungi to plants and animals.

1. *The zinc finger motif.* A variety of transcription factors

contain a motif called the *zinc finger*. The zinc ion of each finger is coordinated to two cysteines and two histidines. The two cysteine residues are part of a two-stranded β-sheet on one side of the finger, and the two histidine residues are part of a short α-helix on the opposite side of the finger (inset, Figure 12.38a). These proteins typically have a number of such fingers that act independently of

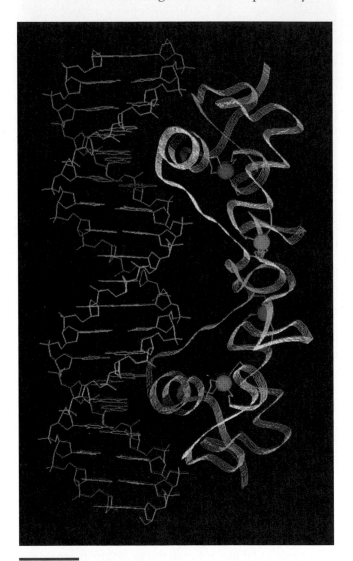

Figure 12.37 **Interaction between a transcription factor and its DNA target sequence.** A model of the interaction between the two DNA-binding domains of the dimeric glucocorticoid receptor (GR) and the target (GRE) DNA. It is noted in the text that both the glucocorticoid and estrogen receptors belong to the same family of proteins. Those residues of the GR that are essential for discriminating between a glucocorticoid response element and an estrogen response element (ERE) are indicated in red. The corresponding portion of the DNA (represented by two A-T base pairs in the center of each of the GRE half-sites) to which these residues bind are also indicated in red. Residues in the dimeric GR that are important for protein-protein interactions are colored green. The four zinc ions (two per monomer) are shown as purple spheres. (REPRINTED WITH PERMISSION FROM T. HÄRD ET AL., COURTESY OF ROBERT KAPTEIN, SCIENCE 249:159, 1990; © COPYRIGHT 1990 AMERICAN ASSOCIATION FOR THE ADVANCEMENT OF SCIENCE.)

one another and are spaced apart so as to project into successive major grooves in the target DNA, as illustrated in Figure 12.38*a*. The first zinc finger protein to be discovered, TFIIIA, has nine zinc fingers (Figure 12.38*b*). Other zinc finger transcription factors include Egr, which is involved in activating genes required for cell division, and GATA, which is involved in cardiac muscle development. Comparison of a number of zinc finger proteins indicates that the motif provides the structural framework for a wide variety of amino acid sequences that recognize a diverse set of DNA sequences.

2. *The helix-loop-helix (HLH) motif.* As the name implies, this motif is characterized by two α-helical segments separated by an intervening loop. The HLH domain is often preceded by a stretch of highly basic amino acids whose positively charged side chains contact the DNA and determine the sequence specificity of the transcription factor. Proteins with this basic-HLH (or bHLH) motif always occur as dimers, as illustrated in the example of the transcription factor MyoD in Figure 12.39. The two subunits of the dimer are usually encoded by different genes, making the protein a heterodimer.

(*a*)

(*b*)

(*b*)

Figure 12.38 Transcription factors that bind by means of a zinc finger motif. (*a*) A model of the complex between a protein with five zinc fingers (called GLI) and DNA. Each of the zinc fingers is colored differently; the DNA is a darker blue. Cylinders and ribbons highlight the α helices and β sheets, respectively. The inset shows the structure of a single zinc finger. (*b*) A model of TFIIIA bound to the DNA of the 5S RNA gene. TFIIIA is required for the transcription of the 5S rRNA gene by RNA polymerase III. (A: REPRINTED WITH PERMISSION FROM NIKOLA PAVLETICH AND CARL O. PABO, SCIENCE 261:1702, 1993; © COPYRIGHT 1993 AMERICAN ASSOCIATION FOR THE ADVANCEMENT OF SCIENCE; B: AFTER K. R. CLEMENS ET AL., PROC. NAT'L. ACAD. SCI. U.S.A. 89:10825, 1992.)

Figure 12.39 A transcription factor with a basic helix-loop-helix (bHLH) motif. (*a*) MyoD, a dimeric transcription factor involved in triggering muscle cell differentiation. MyoD is a bHLH protein that binds to the DNA by an associated basic region. The 14-base-pair binding site in the DNA is shown in light blue. The basic region of each MyoD monomer is red, whereas the helix-loop-helix region of each MyoD monomer is shown in brown. The DNA bases bound by the transcription factor are indicated in yellow. (*b*) A sketch of the dimeric MyoD complex in the same orientation as in part *a*. The α helices are represented as cylinders. (FROM P. C. M. MA ET AL., COURTESY OF CARL O. PABO, CELL 77:453, 1994; BY PERMISSION OF CELL PRESS.)

A-A homodimer B-B homodimer A-B heterodimer

Figure 12.40 Increasing the DNA-binding specificities of transcription factors through dimerization. In this model of an HLH protein, three different dimeric transcription factors that recognize different DNA-binding sites can be formed when the two subunits associate in different combinations.

Heterodimerization greatly expands the diversity of regulatory factors that can be generated from a limited number of polypeptides (Figure 12.40). Suppose, for example, that a cell were to synthesize five different bHLH-containing polypeptides that were capable of forming heterodimers with one another in any combination; then 32 (2^5) different transcription factors that recognize 32 different DNA sequences could conceivably be formed. In actuality, combinations between polypeptides are probably restricted, not unlike the formation of heterodimeric integrin molecules (page 252).

HLH-containing transcription factors play a key role in the differentiation of certain types of tissues, including skeletal muscle. This is illustrated in the photograph of Figure 12.41, which shows a transgenic mouse embryo that contains the regulatory region of the myogenin gene placed upstream from a bacterial β-galactosidase gene. The β-galactosidase gene is commonly employed to test the tissue-specific expression of a gene because the presence of the enzyme β-galactosidase is easily revealed by the blue color produced in a simple histochemical test. In the example depicted in Figure 12.41, the synthesis of the bacterial enzyme can only occur following the binding of a specific bHLH transcription factor to the regulatory region of the myogenin gene. It is evident in the photograph that myogenin activation occurs only in portions of the embryo (the myotome of the somites) that will give rise to muscle tissue.

HLH-containing transcription factors also participate in the control of cell proliferation and have been implicated in the formation of certain cancers. It was noted on page 513 that chromosome translocation can generate abnormal genes whose expression causes the cell to become cancerous. Genes encoding at least four different bHLH proteins (Myc, SCL, LYL-1, and E2A) have been found in chromosome translocations that lead to the development of specific cancers. The most prevalent of these cancers is Burkitt's lymphoma in which the c-*myc* gene on chromosome 8 is translocated to a locus on chromosome 14 that contains an enhancer for a gene encoding part of an antibody molecule. The overexpression of the c-*myc* gene in its new location is thought to be a significant factor in development of this lymphoma.

3. *The leucine zipper motif.* This motif gets its name from the fact that leucines occur every seventh amino acid along an α helix of 30 to 40 residues. Because an α helix repeats every 3.5 residues, all of the leucines along the he-

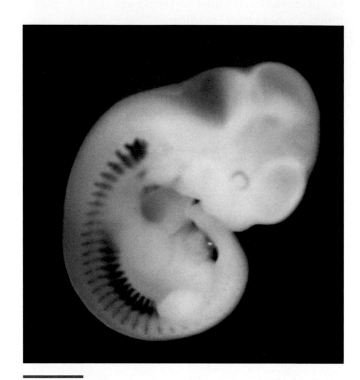

Figure 12.41 Experimental demonstration of the tissue-specific expression of a transcription factor involved in muscle cell differentiation. Transcription of the myogenin gene is activated specifically in those parts of this 11.5-day-old mouse embryo that give rise to muscle tissue. The activation of transcription is indicated by the blue-stained cells using a technique described in the text. (FROM T. C. CHENG ET AL., COURTESY OF ERIC N. OLSON, J. CELL BIOL. 119:1652, 1992; BY COPYRIGHT PERMISSION OF THE ROCKEFELLER UNIVERSITY PRESS.)

lical stretch of polypeptide face the same direction. Two α helices of this type are capable of zipping together along their length to form a coiled-coil (page 61), in which the leucines of one helix are pressed against the leucines of the other helix. Thus, like most other transcription factors, proteins with a leucine zipper motif exist as dimers. The leucine zipper motif can bind DNA because it contains a stretch of basic amino acids on one side of the leucine-containing α helix. Together the basic segment and leucine zipper are referred to as a *bZIP* motif. Thus, like bHLH proteins, the α-helical portions of bZIP proteins are important in dimerization, while the stretch of basic amino acids allows the protein to recognize a specific nucleotide sequence in the DNA. AP1, whose struc-

ture and interaction with DNA are shown in Figure 12.31, is an example of a bZIP transcription factor. AP1 is a heterodimer whose two subunits (Fos and Jun, shown in red and blue in Figure 12.31, respectively) are encoded by the genes c-*fos* and c-*jun*. Both of these genes play an important role in cell proliferation and, when mutated, can cause a cell to become malignant. Mutations in either of these genes that prevent the proteins from forming heterodimers also prevent the proteins from binding to DNA, indicating the importance of dimer formation in regulating their activity as transcription factors.

4. *The HMG-box motif.* The HMG box is named after a group of abundant proteins, called high mobility group (HMG) proteins, in which it was first discovered. The motif consists of three α helices organized into a boomerang-shaped motif capable of binding DNA, as shown in Figure 12.42*a*. Transcription factors that possess HMG boxes are referred to as "architectural" factors; they activate transcription by bending the DNA, which is thought to promote the interaction of other transcription factors bound at nearby sites (Figure 12.42*b*). The HMG protein called SRY, which is shown in Figure 12.42*a*, has a key role in sexual differentiation in human males. The SRY protein, which is encoded by a gene on the Y chromosome, activates transcription of genes in a pathway leading to differentiation of the testes. Mutations in the SRY gene that render the encoded protein unable to bind to DNA lead to a condition known as "sex reversal." Individuals with this phenotype have an XY pair of sex chromosomes but develop into females.

A different HMG protein called UBF, which is shown in Figure 12.42*b*, activates transcription of rRNA genes by RNA polymerase I. UBF binds to the DNA as a dimer whose two subunits contain a total of 10 HMG boxes. By interacting with successive sites along the DNA, the HMG boxes distort the DNA helix in a way that causes it to loop around the protein. The looping of the DNA brings two DNA regulatory sequences that are normally separated by 120 base pairs into close proximity, where they can be bound cooperatively by one or more transcription factors.

Repression of Transcription in Eukaryotes As is evident from Figures 12.25 and 12.26, control of transcription in prokaryotic cells relies heavily on repressors, which are proteins that bind to the DNA and block transcription of a nearby gene. Although research in eukaryotes has focused primarily on factors that activate or enhance transcription of specific genes, it is evident that eukaryotic cells also possess negative regulatory mechanisms. A number of regulatory proteins act as repressors by binding to specific promoter elements, blocking the assembly of the preinitiation complex required to initiate transcription (page 455). Some transcription factors act as an activator of certain genes and a repressor of others. The glucocorticoid receptor (GR), for example, which was described above as a transcriptional activator, can bind to DNA elements (called *negative GREs*, or *nGREs*) and inhibit the

(a)

(b)

Figure 12.42 **HMG proteins bend the DNA.** (*a*) A model showing a portion of the human HMG protein SRY (green color) bound to DNA (blue/red color). Binding of the protein induces a large bend in the DNA, which closely follows the concave binding surface of the HMG box. Bending occurs as the protein contacts the minor groove of the DNA and inserts the side chain of an isoleucine residue between a specific base pair. (*b*) UBF is a dimeric transcription factor that bends the DNA, making it a suitable template for RNA polymerase I. The binding of RNA polymerase I requires a number of general transcription factors, not unlike those required for binding of RNA polymerase II to the TATA box of its promoters. (A: FROM MILTON H. WERNER ET AL., COURTESY OF G. M. CLORE AND A. M. GRONENBORN, CELL 81:707, 1995; BY PERMISSION OF CELL PRESS.)

transcription of the associated gene. Other repressors act by inhibiting the binding or function of transcriptional activators. Thus, as in prokaryotes, the pattern of transcription of particular genes at a particular time in the life of a eukaryotic cell may depend on the balance between positive-acting and negative-acting regulatory factors.

Although transcription factors are the primary agents that regulate gene expression, they are not the only mechanism. As we will see in the following sections, regulation of transcription can also be exerted by modification of the

Figure 12.43 DNA methylation. (*a*) Examples of the most common methylated nucleotide sequences in DNA. Colored dots indicate the position of the methyl groups, all of which occur at 5′-CpG-3′ dinucleotides. (*b*) Examples of CG-rich islands in the promoter regions of three mammalian genes. The lengths of the islands are indicated by the width of the red boxes. The widths of the exons are indicated in black; it is evident that the intervening sequences constitute the great bulk of each gene, as is usually the case. (FROM A. P. BIRD, TRENDS GENET. 3:343, 1987.)

DNA itself and by changes made in the way that DNA is wrapped into nucleosomes.

The Role of DNA Methylation Examination of the DNA of mammals and other vertebrates indicates that as many as 1 out of 100 nucleotides bears an added methyl group, which is always attached to carbon 5 of a cytosine. Methyl groups are added to the DNA by a small family of enzymes called *methyltransferases* encoded in humans by *DNMT* genes. This simple chemical modification is thought to serve as a mark or "tag" that allows certain regions of the DNA to be identified and utilized differently from other regions. In mammals, virtually all of the methylcytosine residues are part of a 5′-CpG-3′ dinucleotide within a symmetrical sequence, such as those indicated in Figure 12.43*a*. Such sequences do not occur randomly in the DNA, but tend to be concentrated in CG-rich "islands" that are often located in promoter regions that regulate gene expression (Figure 12.43*b*).

Methylation of promoter DNA is strongly correlated with gene repression.[9] For instance, the level of methylation

in the region lying upstream of the γ-globin gene in the DNA in fetal liver is greatly decreased compared with the same gene in other fetal tissues. This change correlates with the high level of transcription of the γ-globin gene in the liver during fetal development (Figure 12.44). Similarly, when methylated DNA templates are introduced (transfected) into cultured cells, they stop being transcribed after a brief period. If the same DNAs are introduced in an unmethylated state, they continue to be expressed.

Most evidence suggests that DNA methylation serves more to maintain a gene in an inactive state than as a mechanism for initial inactivation. As an example, inactivation of the genes on one of the X chromosomes of female mammals (page 507) occurs prior to a wave of DNA methylation that is thought to convert the DNA into a more permanently repressed condition. Inhibition of transcription is mediated by proteins, such as MeCP2 and MBD1, that recognize and bind to DNA sequences containing methylated cytosine residues. Once bound to the DNA, these proteins recruit protein complexes that convert the chromatin into an inactive state (see Figure 12.51). MeCP2 is a likely inhibitor of transcription in the fetal brain, because children born with mutations in the *MECP2* gene suffer from Rett syndrome, which is characterized by mental retardation and autism.

[9]The majority of the added methyl groups are present on transposable elements, which were discussed on page 423. Methylation is thought to maintain these elements in an inactive state. We will restrict the present discussion to the role of DNA methylation in the control of gene expression.

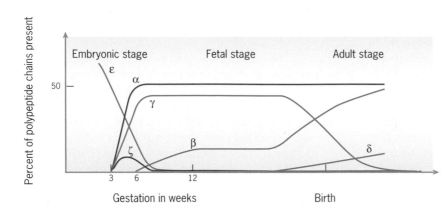

Figure 12.44 Transcription of different members of the globin gene family at different stages of human development. Hemoglobin molecules are formed by the association of two α-like globins with two β-like globins. Different members of each of these two subfamilies are synthesized at different stages of development, as indicated in the figure. Expression of α-like globin genes is indicated by the red lines, and expression of β-like globin genes is indicated by the blue lines.

Figure 12.45 Changes in DNA methylation levels during mammalian development. The DNA of the zygote is substantially methylated. During cleavage, the genome undergoes global demethylation. After implantation, the DNA is subjected to new (de novo) methylation in cells that give rise to the embryo, whereas the DNA of the primordial germ cells, which give rise to the gametes in the adult, remains unmethylated. The DNA of the germ cells becomes methylated at later stages of gamete formation. The general level of methylation is maintained in the somatic (nongerm) cells at a high level throughout the remainder of development and adulthood. (FROM R. JAENISCH, TRENDS GENET. 13:325, 1997, COPYRIGHT 1997, WITH PERMISSION OF ELSEVIER SCIENCE.)

that occur during the life of a mammal are indicated in Figure 12.45. The first major change in the level of methylation occurs during the first few divisions of the zygote, when enzymes sweep along the DNA, removing nearly all of the methylation "tags" that were inherited from the previous generation. Then, at about the time the embryo implants into the uterus, a wave of new (or *de novo*) methylation spreads through the cells, establishing a new pattern of methylation throughout the DNA. We don't know the signals that determine whether a gene is targeted for methylation or spared at this time. Regardless, once the pattern of DNA methylation is established within a cell, it is passed on to the DNA of each daughter cell by a process referred to as maintenance methylation. It is presumed that promoter regions of genes that encode tissue-specific products (such as the γ-globin gene discussed above) become selectively demethylated prior to the differentiation of that cell.

DNA methylation is not a universal mechanism for inactivating gene expression. DNA methylation does not, for example, occur in yeast or in many invertebrates, including the fruit fly and nematode. Plant DNA, in contrast, is often heavily methylated, and studies on cultured plant cells indicate that, as in animals, DNA methylation is associated with gene inactivation. In one experiment, plants treated with compounds that interfere with DNA methylation produced a greatly increased number of flower stalks. Moreover, flowers that developed on these stalks had a markedly altered morphology (Figure 12.46).

One of the most dramatic examples of the importance of methylation in gene expression occurs as part of a phe-

Although DNA methylation is thought to inhibit transcription on a rather permanent basis, it is not irreversible. The remarkable shifts in DNA-methylation levels

Figure 12.46 Experimental demonstration of the importance of DNA methylation in plant development. Plants of the species *Arabidopsis thaliana* were genetically engineered to carry a gene for an antisense oligonucleotide (page 472) that interfered with the synthesis of an enzyme involved in DNA methylation (a methyltransferase). The photographs show the effects of this treatment on a strain whose DNA methylation levels were reduced 71 percent. Control plants are shown in parts *a*, *c*, and *e*, whereas antisense-treated plants are shown in

b, *d*, and *f*. A comparison of *a* and *b* shows the treated plants produce a much greater number of leaves; comparison of *c* and *d* shows the treated plants produce about five times the number of flower stalks; and comparison of *e* and *f* shows that the flowers of the treated plant contain many more stamens than the controls. (REPRINTED WITH PERMISSION FROM MICHAEL J. RONEMUS ET AL., SCIENCE 273:655, 1996, COURTESY OF STEPHEN L. DELLAPORTA; © COPYRIGHT 1996 AMERICAN ASSOCIATION FOR THE ADVANCEMENT OF SCIENCE.)

nomenon known as genomic imprinting, which is unique to mammals.

Genomic Imprinting It had been assumed until the mid-1980s that the set of chromosomes inherited from a male parent was functionally equivalent to the corresponding set of chromosomes inherited from the female parent. But, as with many other long-standing assumptions, this proved not to be the case. Instead, certain genes are either active or inactive during early mammalian development depending solely on whether they were brought into the zygote by the sperm or the egg. For example, during early human development, the gene that encodes a fetal growth factor called IGF2 is only active on the chromosome transmitted from the male parent. In contrast, the gene that encodes a specific potassium channel (KVLQT1) is only active on the chromosome transmitted from the female parent. Genes of this type are said to be **imprinted** according to their parental origin. Imprinting is an excellent example of an epigenetic phenomenon (page 510) because the differences between the alleles are inherited from one's parents but are not based on differences in DNA sequence. It is estimated that the mammalian genome contains more than 100 imprinted genes located in several different chromosomal clusters.

Genes are thought to become imprinted as the result of selective DNA methylation of one of the two alleles. This conclusion is based on the finding that (1) the maternal and paternal versions of imprinted genes differ consistently in their degree of methylation, and (2) mice that lack a key methyltransferase (Dnmt1) are unable to maintain the imprinted state of the genes they inherit. The methylation state of imprinted genes is not affected by the waves of demethylation and remethylation that sweep through the early embryo (Figure 12.45). Consequently, the same alleles that are inactive due to imprinting in the fertilized egg will be inactive in the cells of the fetus and most adult tissues. The major exception occurs in the germ cells, where the imprints inherited from the parents are erased during early development and then reestablished when that individual produces his or her own gametes. Some mechanism must exist whereby specific genes (e.g., *KVLQT1*) are marked for inactivation during sperm formation, whereas other genes (e.g., *IGF2*) are marked for inactivation during formation of the egg.

Disturbances in imprinting patterns have been implicated in a number of rare human genetic disorders, particularly those involving a cluster of imprinted genes residing on chromosome 15. Prader-Willi syndrome (PWS) is an inherited neurological disorder characterized by mental retardation, obesity, and underdevelopment of the gonads. The disorder appears when chromosome 15 inherited from the father carries a deletion in a small region containing the imprinted genes. Because the paternal chromosome carries a deletion of one or more genes and the maternal chromosome carries the inactive, imprinted version of the homologous region, the individual lacks a functional copy of the gene(s). Cases are also known where an allele fails to become imprinted during gamete formation in the parent so that both alleles of the offspring are active. When both alleles of the *IGF2* gene are active, excessive amounts of the encoded growth factor are produced in the fetus, leading to the enlargement of certain organs and an increased risk of certain cancers.

What possible role could genomic imprinting play in the development of an embryo? Although there are several different thoughts on this question, there is no definitive answer. According to one researcher, genomic imprinting is a "phenomenon in search of a reason," which is where we will leave the matter.

Chromatin Structure and Transcription As discussed on page 502, the DNA in a eukaryotic nucleus is not present in a naked state but is wrapped around histone complexes to form nucleosomes. The discovery of nucleosomes in the 1970s raised an important question that has still not been satisfactorily answered: how are nonhistone proteins (such as transcription factors and RNA polymerases) able to interact with DNA that is tightly associated with core histones? In fact, a large body of evidence suggests that incorporation of DNA into nucleosomes *does* inhibit the initiation of transcription. Recall from page 455 that the initiation of transcription requires the assembly of a large protein complex consisting of a number of general transcription factors and a huge multisubunit RNA polymerase. Nucleosomes are thought to block the assembly of this transcriptional machinery at the core promoter. How do cells overcome this inhibitory effect that results from chromatin structure?

Transcription is accomplished through the collaboration of many different types of proteins. Among these proteins are molecules that act like "tools" to pry open regions of the DNA that are blocked by nucleosomes, thereby allowing access to the DNA. In the mid-1960s, Vincent Allfrey and his colleagues at Rockefeller University found that genes that were actively transcribed tended to be bound by histones that were more highly acetylated than histones bound to inactive genes. Over the years, this correlation between histone acetylation and gene expression was repeatedly confirmed, but it wasn't until the mid-1990s that the significance of histone acetylation became appreciated, as the enzymes responsible for this histone modification were purified and characterized.

Each of the histone molecules of the nucleosome core has a flexible N-terminal tail that extends outside the core particle and past the DNA helix. (The tails of these histones can be seen in the model of Figure 12.10c.) Acetyl groups are added to specific lysine residues on the N-terminal tails of the core histones by a family of enzymes called **histone acetyltransferases (HATs)**. Acetylation of these lysine residues is thought to accomplish two functions: (1) neutralize the positive charge of the lysine residue, thereby decreasing the strength of the DNA-histone interaction, and/or (2) destabilize the interac-

tions between the histone tails and various proteins involved in maintaining higher order chromatin structure. Either of these actions would be expected to increase access to the DNA, which should facilitate the assembly of the preinitiation complex and promote the initiation of transcription.

As discussed on page 528, coactivators are proteins or multiprotein complexes that function as adapters linking transcriptional activators bound at upstream sites on the DNA to the transcription machinery bound at the core promoter (see Figure 12.36c). In the late 1990s, it was discovered that a number of coactivators possessed HAT activity. If the HAT activity of these coactivators was eliminated by mutation, so too was their ability to stimulate transcription. The discovery that coactivators contain HAT activity provided a crucial link between histone acetylation, chromatin structure, and gene activation. As indicated in the model depicted in Figure 12.47, transcriptional activators, such as the glucocorticoid receptor, are thought to bind to their response elements on the DNA, even when that region of the DNA is wrapped by nucleosomes. Once bound to the GRE, the glucocorticoid receptor recruits a coactivator (e.g., CBP) to a region of the chromatin that is targeted for transcription. Once bound to the target region, the HAT activity of the coactivator acetylates the core histones bound to the promoter DNA, causing the release of the nucleosome core particles (as shown in Figure 12.47), or at least a loosening of the histone-DNA interaction. In either case, the HAT activity of the coactivator makes the DNA of the targeted promoter accessible to the general transcription factors and RNA polymerase, which assemble at the site where transcription

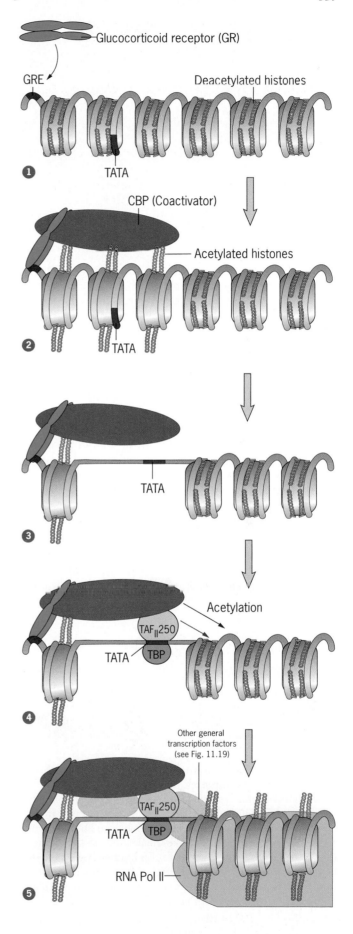

Figure 12.47 A model for the role of coactivators in regulating transcription by histone acetylation. Transcription factors, such as glucocorticoid receptor (GR), bind to the DNA and recruit coactivators, such as CPB, which possess histone acetyltransferase activity. (These enzymes transfer acetyl groups from an acetyl CoA donor to the amino groups of specific lysine residues.) Step 1 of this drawing depicts a region of a chromosome that is in a repressed state because of the association of its DNA with deacetylated histones. In step 2, the GR is bound to the GRE, the coactivator CBP is bound to the GR, and histones of the nucleosome core particles in the regions both upstream and downstream from the TATA box are being acetylated. In step 3, the acetylated histones have dissociated from the DNA. In step 4, TFIID binds to the open region of the DNA. Although it wasn't mentioned in the text, one of the subunits of TFIID (called TAF$_{II}$250) also possesses histone acetyltransferase activity as indicated by the red arrow. Together, CPB and TAF$_{II}$250 disrupt additional nucleosomes to allow transcription to be initiated. In step 5, the remaining nucleosomes of the promoter have been acetylated, the RNA polymerase is bound to the promoter, and transcription is about to begin. (BASED ON A MODEL BY P. A. WADE ET AL., TRENDS BIOCHEM. SCI. 22:129, 1997)

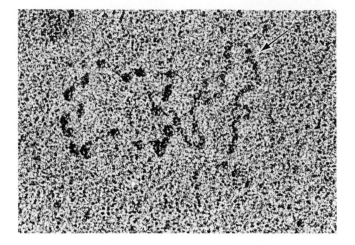

Figure 12.48 Demonstration of the transcription of DNA in a region containing nucleosomes. This SV40 "chromosome" contains nucleosomes, which are seen as dark spheres, as well as a nascent RNA chain (arrow) indicating that the DNA is being transcribed by an RNA polymerase. (COURTESY OF PIERRE CHAMBON.)

will be initiated. Once transcription is initiated, an RNA polymerase appears capable of transcribing a DNA template that is packaged into nucleosomes (Figure 12.48).

CBP serves as a coactivator in the transcription of a wide variety of different genes. Mutations that diminish the HAT activity of CBP have consequences on transcription in both yeast and mammalian cells. The importance of CBP in cellular growth control is illustrated by the fact that certain tumor viruses (e.g., adenovirus) produce proteins

Figure 12.49 Models for the mechanism of action of chromatin remodeling complexes, such as SWI/SNF. SWI/SNF (or the human counterpart BRG1) is recruited to chromatin by transcription factors, such as the glucocorticoid receptor that bind to specific response elements (green box on DNA). In one model, SWI/SNF has caused a key nucleosome to slide along the DNA, thereby exposing the TATA binding site, where the preinitiation complex can be assembled. In the other model, SWI/SNF has caused the reorganization of the histone octamer of several nucleosomes. Although the TATA box is not completely free of histone association, it is now able to bind the proteins of the preinitiation complex.

that specifically inactivate CBP and the discovery that chromosome translocations involving CBP are associated with the development of monocytic leukemia in humans.

Histone acetylation is not the only mechanism eukaryotic cells utilize to make targeted DNA templates more accessible to the transcriptional machinery. Organisms as diverse as yeast and mammals possess a variety of **chromatin remodeling complexes** that use the energy released by ATP hydrolysis to disrupt nucleosomes and allow the binding of transcription factors to promoter regions. The best-studied chromatin remodeling "machines" are members of the SWI/SNF family. The founding member of the family was originally discovered in mutant yeast cells that exhibited altered transcription of certain genes involved in mating type *switching* (SWI) and *sucrose nonfermentation* metabolism (SNF). The SWI/SNF complex consists of 9–12 subunits and has a molecular mass of about two million daltons. The SWI/SNF complex is recruited to specific promoters by transcription factors that have bound to upstream regulatory sites. Once recruited to a promoter, chromatin remodeling complexes are thought to act in two different ways (Figure 12.49):

1. by changing the conformation of nucleosome core particles thereby disrupting DNA-histone interactions. This may cause the DNA to become partially unwrapped from the histone octamer. Recognition sites on the DNA of a "remodeled" nucleosome may be more accessible to binding regulatory proteins.

2. by promoting the mobility of the histone octamer so that it slides along the DNA to new positions, thereby exposing DNA sequences that were previously covered by histones. The movement of histone octamers along the DNA is inhibited by the presence of linker histones (page 502), which may also be removed from chromatin by these remodeling complexes.

Regardless of their mechanism of action, chromatin remodeling complexes are thought to be recruited by transcriptional activators (e.g., glucocorticoid receptor) and coactivators (e.g., CBP) to help make targeted promoter sites accessible to the transcriptional machinery (Figure 12.49).

The actions of histone acetylation and chromatin remodeling complexes give transcriptionally active chromatin certain distinctive properties that distinguish it from inactive chromatin. Active chromatin is more accessible to DNA-digesting enzymes, such as DNase I, which cleaves the DNA at certain sites. These *DNase I hypersensitive sites*, as they are called, are typically found in the regulatory regions of genes that are being actively transcribed, but they are absent from the same regions of genes that are transcriptionally silent. Hypersensitive sites are presumed to be regions of the DNA from which nucleosomes have been displaced and, thus, are available for interaction with gene regulatory proteins (Figure 12.50).

The state of acetylation of chromatin is a dynamic property; just as there are enzymes to add acetyl groups,

PEPCK Promoter

Figure 12.50 **Nuclease hypersensitive sites are located in regions that are more accessible to enzymes.** A map of a portion of the phosphoenolpyruvate carboxykinase (PEPCK) gene, showing the locations of nucleosomes. The site of initiation of transcription is indicated by the black arrow. The sites on the DNA that are sensitive to cleavage by two different nucleases, DNase I (red arrows) and MNase (green arrows), are shown. These nuclease hypersensitive sites occur at sites where the chromatin is more accessible to enzymes, which is particularly true of the promoter region, which contains eight different sites of nuclease cleavage.

there are also enzymes to remove them. Removal of acetyl groups is accomplished by **histone deacetylases (HDACs)**. Just as HATs are associated with transcriptional activation, HDACs are associated with transcriptional repression. The HDACs are present as subunits of larger complexes described as *corepressors*. Corepressors (e.g., Sin3 and N-CoR) are similar to coactivators, except they have the opposite function; they lead to transcriptional repression of the targeted gene rather than activation.

It was noted above (page 534) that methylation of DNA regulatory regions is associated with repression of transcription. Studies indicate that HDACs are guided to particular regions of a chromosome by the methylation patterns of the DNA. It is thought that methylated DNA binding proteins (e.g., MeCP2) recruit corepressor complexes containing an HDAC, whose activity leads to chromatin compaction and gene repression (Figure 12.51. The correlation between transcriptional repression, DNA methylation, and core histone deacetylation can be seen by comparing the inactive X chromosome of female cells, which contains deacetylated histones, to the active X chromosome, whose histones exhibit a normal level of acetylation (Figure 12.52).

Figure 12.51 **The role of DNA methylation and histone deacetylase activity in the repression of transcription.** When actively transcribed, the histone tails are acetylated. Methylation of the CG islands of the gene's promoter leads to the binding of MeCP2, which recognizes the methylated cytosines, and the subsequent recruitment of a corepressor (e.g., Sin3) that has histone deacetylase (HDAC) enzymatic activity. Removal of the acetyl groups from the histone tails leads to chromatin inactivation.

Figure 12.52 **Experimental demonstration of a correlation between transcriptional activity and histone acetylation.** This metaphase chromosome spread has been doubly labeled with fluorescent antibodies to (1) acetylated histone H4, which fluoresce green, and (2) bromodeoxyuridine (BrdU), which fluoresce red. As discussed in the following chapter, BrdU is a thymidine analogue that is incorporated into DNA during replication. The inactivated X chromosome (arrow) replicates late in the S phase of the cell cycle, so it can be selectively labeled with BrdU, as was done here. It is evident that all of the chromosomes except the inactivated X stain brightly with the antibody against the acetylated histone. (FROM P. JEPPESEN AND B. M. TURNER, COVER OF CELL VOL. 74, NO. 2, 1993; BY PERMISSION OF CELL PRESS.)

Processing-Level Control

We have seen throughout this book that proteins are often encoded by members of multigene families. The various genes that constitute a multigene family are presumed to have arisen during evolution from a single ancestral gene that underwent repeated duplication. Over time, the duplicates underwent sequence divergence, generating a family of genes that encode homologous proteins with related functions. The formation of multigene families is an evolutionary mechanism that generates protein diversity. Protein diversity can also be generated within an individual organism by a process called **alternate splicing**, which regulates gene expression at the level of RNA processing.

Alternate splicing is a widespread mechanism by which a single gene can encode two or more related proteins. As discussed in the previous chapter, most eukaryotic genes (and thus their primary transcripts) contain numerous introns and exons. In many cases, there is more than one pathway by which a particular primary transcript can be processed. The particular processing pathway that is followed may depend upon the particular stage of development, or the particular cell type or tissue being considered. It is estimated that transcripts from as many as 35 percent of human genes may be subject to alternate splicing. Furthermore, it is not uncommon for a primary transcript to be alternately spliced to produce several different polypeptides. Given these numbers, it becomes apparent that alternate splicing plays a very important role in gene expression. Consequently, the "functional" number of genes in a genome may be several times the number identified by DNA sequencing alone.

In the simplest case, a specific segment can either be spliced out of the transcript or retained as part of the final mRNA. An example of this type of alternate splicing occurs during synthesis of fibronectin, a protein found in both blood plasma and the extracellular matrix (page 249). Fibronectin produced by fibroblasts and retained in the matrix contains two extra peptides compared to the version of the protein produced by liver cells and secreted into the blood (Figure 12.53). The extra peptides are encoded by portions of the pre-mRNA that are retained during processing in the fibroblast but are removed during processing in the liver cell.

In most cases, the proteins produced from a given gene by alternate splicing are identical along most of their length but differ in key regions that may affect such important properties as their cellular location, the types of ligands they can bind, or the kinetics of their catalytic activity. Antibody molecules, for example, can be either membrane bound or soluble depending on which of two alternate exons are situated at the 3′ end of their mRNA. A number of transcription factors are produced from genes that can be alternately spliced, producing variants that can determine the pathway of differentiation that is taken by a cell. In the fruit fly, for example, the developmental pathway that leads an embryo into becoming a male or female is determined by alternate splicing of the transcripts from certain genes.

Alternate splicing can become very complex, allowing a wide variety of different combinations of possible exons in the final mRNA product. The mechanism by which a particular exon is included or excluded depends primarily on whether or not specific 3′ and 5′ splice sites are selected by the splicing machinery as sites to be cleaved (page 463). Some splice sites are described as "weak," which indicates that they can be bypassed by the splicing machinery under certain conditions. The recognition and use of weak splice sites is governed by sequences in the RNA, called **splicing enhancers**, that are located within the exons whose inclusion is regulated. Splicing enhancers serve as binding sites for specific regulatory proteins. If a particular regulatory protein is produced in a cell, that protein is available to bind to the splicing enhancer and to recruit the necessary splicing factors to a nearby weak 3′ or 5′ splice site. Use of these splice sites results in the inclusion of the exon into the mRNA. A model of how this may work is depicted in Figure 12.54. If the regulatory protein is not produced in the cell, the neighboring splice sites are not recognized and the exon is excised along with the flanking introns.

Translational-Level Control

Translational-level control encompasses a wide variety of regulatory mechanisms affecting the translation of mRNAs previously transported from the nucleus to the cytoplasm. Subjects considered under this general regulatory umbrella include (1) localization of mRNAs to certain sites within a cell; (2) whether or not an mRNA is translated and, if so, how often; and (3) the half-life of the mRNA, a property that determines how long the message is translated.

Figure 12.53 Alternative splicing of the fibronectin gene. The fibronectin gene consists of a number of exons shown in the top drawing (the introns shown in black are not drawn to scale). Two of these exons encode portions of the polypeptide called EIIIA and EIIIB, which are included in the protein produced by fibroblasts, but which are excluded from the protein produced in the liver. The difference is due to alternative splicing; those portions of the pre-mRNA that encode these two exons are excised from the transcript in liver cells. The sites of the missing exons are indicated by the arrows in the liver mRNA.

Figure 12.54 A model for the role of splicing enhancers in regulating alternate splicing. In this case, exon 2 contains several splicing enhancers that bind specific regulatory proteins (typically SR proteins, page 468). These bound regulatory proteins recruit key splicing factors (U2AF) and U1 snRNP to nearby 3′ and 5′ splice sites, respectively. (U2AF plays a direct role in recruiting U2 snRNP to the branchpoint site [BPS], as is required for formation of the lariat, page 465). If U2AF and U1 snRNP were not recruited to the splice sites on each side of exon 2, this exon would not be recognized and, instead, would be excised as part of the intron. (AFTER K. J. HERTEL ET AL., CURR. OPIN. CELL BIOL. 9:351, 1997.)

Translational-level control mechanisms generally operate through interactions between specific mRNAs and various proteins present within the cytoplasm. It was noted on page 457 that mRNAs contain noncoding segments, called **untranslated regions (UTRs)**, at both their 5′ and 3′ ends. The 5′ UTR extends from the methylguanosine cap at the start of the message to the AUG initiation codon, whereas the 3′ UTR extends from the termination codon at the end of the coding region to the end of the poly(A) tail that is attached to nearly all eukaryotic mRNAs (see Figure 11.21). For many years the untranslated sections of the message were largely ignored, but it has become evident in recent years that the UTRs contain nucleotide sequences used by the cell to mediate translational-level control. In following sections, we will consider three distinct aspects of translational-level control: mRNA localization, mRNA translation, and mRNA stability.

Cytoplasmic Localization of mRNAs Information required to start a fertilized animal egg on the path toward formation of an embryo is laid down within the oocyte during oogenesis. We will briefly consider the fruit fly, whose egg, larval, and adult stages are depicted in Figure 12.55a. The development of the anterior-posterior (head-abdomen) axis of a larval fly and subsequent adult is foreshadowed by localization of specific mRNAs along this same axis in the oocyte. For example, mRNAs transcribed from the *bicoid* gene become preferentially localized at the anterior end of the oocyte, while mRNAs transcribed from the *oskar* gene become localized at the opposite end (Figure 12.55b,c). The protein encoded by *bicoid* mRNA plays a critical role in the development of the head and thorax, whereas the protein encoded by *oskar* mRNA is re-

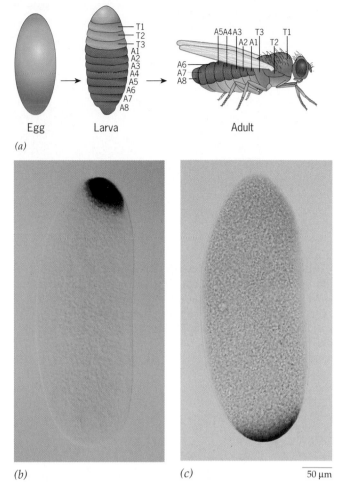

Figure 12.55 Localization of mRNAs in the cytoplasm of a *Drosophila* egg. (a) Schematic drawings showing three stages in the life of a fruit fly: the egg, larva, and adult. The segments of the thorax and abdomen are indicated. (b) Localization of *bicoid* mRNA at the anterior pole of an early cleavage stage by in situ hybridization. (c) Localization of *oskar* mRNA at the posterior pole of a comparable stage to that shown in b. Both of these localized RNAs play an important role in the development of the anterior-posterior axis of the fruit fly. (B: COURTESY OF DANIEL ST JOHNSTON; C: COURTESY OF ANTOINE GUICHET AND ANNE EPHRUSSI.)

quired for the formation of germ cells, which develop at the posterior end of the larva. Localizing mRNAs is more efficient than localizing their corresponding proteins because each mRNA can be translated into large numbers of protein molecules.

The information that governs the cytoplasmic localization of an mRNA resides in the 3′ UTR. This can be demonstrated using fruit flies that carry a foreign gene whose coding region is coupled to a DNA sequence that encodes the 3′ UTR of either the *bicoid* or *oskar* mRNA. When the foreign gene is transcribed during oogenesis, the mRNA becomes localized in the site determined by the 3′ UTR. Localization of mRNAs is mediated by specific proteins that recognize localization sequences (called *zipcodes*) in this region of the mRNA.

0.1 μm

Figure 12.56 Localization of mRNA to actin filaments in the cytoplasm. The black dots in this electron micrograph are gold particles at a site of intersection between actin filaments where a poly(A) mRNA molecule is localized (arrowhead). To obtain this micrograph, human fibroblasts were extracted with the nonionic detergent Triton X-100 and then incubated with a gold-labeled probe that binds specifically to the poly(A) tails of mRNAs. (FROM GARY J. BASSELL ET AL., J. CELL BIOL. 126:869, 1994, BY COPYRIGHT PERMISSION OF THE ROCKEFELLER UNIVERSITY PRESS.)

Microtubules, and the motor proteins that use them as tracks, play a key role in transporting mRNAs to particular locations. The localization of *bicoid* mRNAs in a fruit fly oocyte, for example, is disrupted by agents such as colchicine that depolymerize microtubules. Microfilaments, on the other hand, are thought to anchor mRNAs after they have arrived at their destination. As illustrated in Figure 12.56, individual mRNA molecules are often found at intersections between microfilaments of the cytoskeletal network. These filaments may provide a solid surface on which the genetic information of the cytoplasm can be organized. *Bicoid* and *oskar* mRNAs are thought to be anchored to the cytoskeleton by the same protein, Staufen, which binds to a region of double-stranded RNA in their 3′ UTRs. The localization of mRNAs is not restricted to eggs and oocytes but occurs in all types of polarized cells. For instance, actin mRNAs are localized near the leading edge of a migrating fibroblast, which is the site where actin molecules are needed for locomotion.

The Control of mRNA Translation The mRNAs stored in an unfertilized egg are templates for proteins synthesized during early stages of development; they are not utilized for protein synthesis in the egg itself. Messenger RNAs that are stored in the egg for later use are often referred to as "masked" mRNAs because they are rendered inactive by association with inhibitory proteins. Consider the case of an unfertilized sea urchin egg. If a suspension of these eggs is incubated with a radioactively labeled amino acid, very little radioactivity is incorporated into

(a)

(b)

Figure 12.57 Experimental demonstration of the activation of "masked" mRNAs following fertilization of the eggs of sea urchins. (*a*) The cumulative incorporation of [^{14}C]leucine by unfertilized and fertilized eggs of the sea urchin. The 0 time marks the point of fertilization, which is followed after a brief lag by a marked elevation in the rate of protein synthesis. (*b*) Rates of incorporation of [^{14}C]valine into fertilized eggs of the sea urchin in the presence or absence of actinomycin D, an inhibitor of RNA synthesis. The initial increase in protein synthesis that follows fertilization (*a*) is not inhibited by actinomycin D. These results indicate that protein synthesis in the period following fertilization does not depend on newly synthesized mRNA templates, but rather is taking place on mRNAs present in the egg at the time of fertilization. In contrast, the second rise in protein synthesis beginning at about 10 hours postfertilization does require new mRNA templates because it is inhibited by the drug. (A: FROM D. EPEL, PROC. NAT'L. ACAD. SCI. U.S.A. 57:901, 1967; B: FROM P. R. GROSS ET AL., PROC. NAT'L. ACAD. SCI. U.S.A. 51:409, 1964.)

protein (Figure 12.57*a*). However, if this same preparation of eggs is fertilized by addition of sperm, the incorporation of labeled amino acids becomes measurable, and the rate of incorporation slowly rises over the next few hours (Figure 12.57*b*) and then levels off as the embryo develops into a blastula of a thousand or so cells. The change from an inactive to active state occurs so rapidly that the synthe-

sis of proteins after fertilization cannot depend on synthesis of new mRNAs. Rather, the messenger RNAs that are translated immediately following fertilization are present in the egg before its contact with a sperm. If sea urchin eggs are fertilized and raised in the presence of actinomycin D (a potent inhibitor of mRNA synthesis), the activation of protein synthesis after fertilization occurs to the same degree as in control cultures (Figure 12.57*b*). Because new RNAs cannot be produced in the presence of the drug, the proteins must be synthesized on preexisting mRNA templates that become active following fertilization. The activation of these stored mRNAs is thought to involve two distinct events: release of bound inhibitory proteins and an increase in the length of the poly(A) tails by action of an enzyme residing in the egg cytoplasm.

A number of mechanisms have been discovered that regulate the rate of translation of mRNAs in response to changing cellular requirements. Some of these mechanisms can be considered to act *globally* because they affect translation of all messages. When a cell is subjected to certain stressful stimuli, such as viral infection, nutrient starvation, or heat shock, a kinase is activated that phosphorylates the initiation factor eIF2, which blocks further protein synthesis. As discussed on page 480, eIF2-GTP delivers the initiator tRNA to the small ribosomal subunit, after which it is converted to eIF2-GDP and released. The phosphorylated version of eIF2 cannot exchange its GDP for GTP, which is required for eIF2 to become engaged in another round of initiation of translation. Another example of global regulation of translation is seen during mitosis, when the rate of protein synthesis drops to a small fraction of the previous level. This drop occurs as the initiation factor eIF4E is inhibited from binding to the methylated cap at the 5′ end of the mRNA (page 481). Those mRNAs that continue to be translated during mitosis contain sites within their 5′ UTR that can recruit the small ribosomal subunit to initiate translation by a mechanism that does not involve the 5′ cap. Other mechanisms act to alter the rate of translation of *specific* mRNAs. One of the best-studied examples involves the mRNA that codes for the protein ferritin.

Ferritin is a protein that sequesters iron atoms in the cytoplasm of cells, thereby protecting the cells from the toxic effects of the free metal. The translation of ferritin mRNAs is regulated by a specific repressor, called the *iron regulatory protein (IRP)*, whose activity depends on the concentration of unbound iron in the cell. At low iron concentrations, IRP binds to a specific sequence in the 5′ UTR of the message called the *iron-response element (IRE)* (Figure 12.58). The bound IRP interferes physically with the binding of a ribosome to the 5′ end of the message, thereby inhibiting the initiation of translation. At high iron concentrations, the IRP is modified such that it loses its affinity for the IRE. Dissociation of the IRP from the ferritin mRNA gives the translational machinery access to the mRNA, and the encoded protein is synthesized.

The Control of mRNA Stability
The longer an mRNA is present in a cell, the more times it can serve as a template

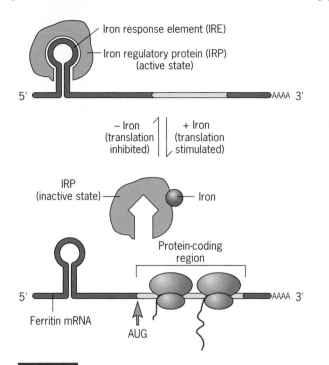

Figure 12.58 The control of ferritin mRNA translation. When iron concentrations are low, an iron-binding repressor protein, called the iron regulatory protein (IRP), binds to a specific sequence in the 5′ UTR of the ferritin mRNA, called the iron-response element (IRE), which is folded into a hairpin loop. When iron becomes available, it binds to the IRP, changing its conformation and causing it to dissociate from the IRE, allowing the translation of the mRNA to form ferritin.

for synthesis of a polypeptide. If a cell is to control gene expression, it is just as important to regulate the survival of an mRNA as it is to regulate the synthesis of that mRNA in the first place. Unlike prokaryotic mRNAs, which begin to be degraded at their 5′ end even before their 3′ end has been completed, most eukaryotic mRNAs are relatively long-lived. Even so, the half-life of eukaryotic mRNAs is quite variable. The c-*fos* mRNA, for example, which is involved in the control of cell division, is degraded rapidly in the cell (half-life of 10 to 30 minutes). As a result, Fos is produced for only a short period. In contrast, mRNAs that code for the production of the dominant proteins of a particular cell, such as hemoglobin in an erythrocyte precursor or ovalbumin in a cell of a hen's oviduct, typically have half-lives of more than 24 hours. Thus, as with mRNA localization or the rate of initiation of mRNA translation, specific mRNAs can be recognized by the cell's regulatory machinery and given differential treatment.

It was shown in an early experiment that mRNAs lacking poly(A) tails were rapidly degraded following injection into cells, whereas the same mRNAs possessing tails were relatively stable. This was the first piece of evidence that suggested that the longevity of an mRNA is related to the length of its poly(A) tail. When a typical mRNA leaves the nucleus, it contains a tail of approxi-

mately 200 adenosine residues. The poly(A) tail is not present as naked RNA but is bound by a specific protein, the *poly(A)-binding protein (PABP)*. Each PABP molecule is bound to approximately 30 adenosine residues. PABP is thought to have a dual function. On one hand, the protein protects the tail from general nuclease activity, but on the other hand it appears to increase the sensitivity of the tail to a specific poly(A) ribonuclease.

A model of a pathway for the degradation of an mRNA in a mammalian cell is shown in Figure 12.59. As an mRNA remains in the cytoplasm, its poly(A) tail tends to be gradually reduced in length as it is nibbled away by the poly(A) ribonuclease (step 2, Figure 12.59). No effect on the stability of the mRNA is observed until the tail becomes reduced to approximately 30 residues, which is probably too short a length to retain a bound PABP molecule (step 3). Once the tail is shortened to this length, the mRNA is usually rapidly degraded. Interestingly, degradation of the mRNA begins at its 5′ end following removal of the poly(A) at the 3′ end of the message. The fact that the poly(A) tail at the 3′ end of the message protects the cap at the 5′ end of the molecule suggests that the two ends of the mRNA are held in close proximity (Figure 11.48). Once the 3′ tail is removed (step 4, Figure 12.57), the message is decapped (step 5) and degraded from the 5′ end toward the 3′ end (step 6).

There must be more to mRNA longevity than simply the length of the poly(A) tail, as mRNAs having very different half-lives begin with a similar-size tail. Once again, differences in the nucleotide sequence of the 3′ UTR have been shown to play a role in the rate at which the poly(A) tail becomes shortened. The 3′ UTR of an α-globin mRNA, for example, contains a number of CCUCC repeats that are thought to serve as binding sites for specific proteins that stabilize the message. If these sequences are mutated, the mRNA is destabilized. In contrast, short-lived mRNAs often contain AU-rich sequences (e.g., AUUUA repeats) in their 3′ UTR that are thought to bind proteins that destabilize the message. If one of these destabilizing sequences is introduced into the 3′ UTR of a globin gene, the stability of the mRNA produced from the modified gene is reduced from a half-life of 10 hours to a half-life of 90 minutes. The importance of these destabilizing sequences (and the overall mRNA instability they produce) can be appreciated by considering the normally short-lived c-*fos* mRNA mentioned above. If the destabilizing sequence from the c-*fos* gene is lost through a deletion, the half-life of c-*fos* mRNA increases, and the cells often become malignant. Destabilizing sequences in the 3′ UTR are thought to bind proteins (e.g., AUF1) that promote shortening of the poly(A) tail.

Before leaving the subjects of translational-level controls and the control of gene expression in general, it can be noted that the past decade has seen the discovery of a number of startling new genetic phenomena. So far, these processes have been found in only a few systems, and thus their general applicability remains uncertain. Among these processes are

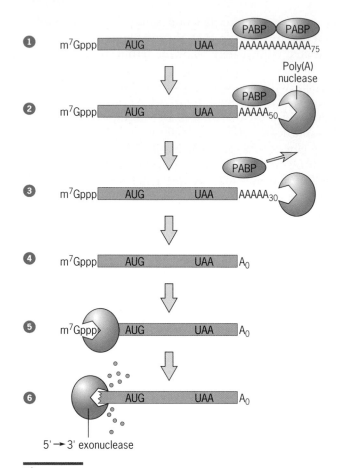

Figure 12.59 mRNA degradation in mammalian cells. The steps depicted in the drawing are described in the text.

- Translational frameshifting, in which the ribosome changes its reading frame at some point during its journey along the mRNA by moving either backward or forward one nucleotide. Frameshifting can be caused by slippage of a tRNA from one codon to an overlapping triplet, which then becomes the new codon. Because frameshifting can occur with an efficiency of less than 100 percent, two different polypeptides may be generated from the same mRNA.

- Termination codon read-through, in which the ribosome continues past a termination codon.

- Messenger RNA editing, in which specific nucleotides are converted to other nucleotides after the RNA has been transcribed (posttranscriptionally). Messenger RNA editing occurs primarily in mitochondria, but the best-studied case is the nuclear-encoded, mammalian protein apolipoprotein B, which is translated from an mRNA approximately 14,000 nucleotides long. In the intestine, the cytidine at nucleotide residue 6666 in the RNA is converted enzymatically to a uridine, which generates a stop codon (UAA) that terminates translation. The shortened version of the protein, called apolipoprotein B48, is produced only in cells of the small intestine.

- Alternative translation initiation, in which different AUG codons in the same mRNA are used as the initia-

tion codon in synthesizing polypeptides. In some cases, this produces a short and a long version of the same polypeptide. In other cases, when the two AUG codons are in different reading frames, two unrelated polypeptides can be generated.

- Translational bypassing, in which the ribosome skips over a sequence of nucleotides in a specific mRNA, leaving an internal portion of the message untranslated.

- Protein splicing, in which a segment from a specific polypeptide is excised and the two edges are covalently joined.

REVIEW ?

1. Describe the different levels at which gene expression is regulated to allow a β-globin gene with the following structure to direct the formation of a protein that accounts for over 95 percent of the protein of the cell.

 exon 1—intron—exon 2—intron—exon 3

2. How are the functions of a bacterial *lac* repressor and a mammalian glucocorticoid receptor similar? How are they different?

3. What types of regulatory sequences are found in the regulatory regions of the DNA upstream from a gene, such as that which codes for PEPCK? What is the role of these various sequences in controlling the expression of the nearby gene(s)?

4. What are some of the ways that nucleosome positioning may stimulate or inhibit the transcription of nearby genes?

5. Describe an example of alternate splicing? What is the value to a cell of this type of control? How might a cell regulate the sites on the pre-mRNA that are chosen for splicing?

6. Describe three different ways in which gene expression might be controlled at the translational level. Cite an example of each of these control mechanisms.

7. What is the role of poly(A) in mRNA stability? How might the cell regulate the stability of different mRNAs?

12.4 POSTTRANSLATIONAL CONTROL: DETERMINING PROTEIN STABILITY

We have seen how cells possess elaborate mechanisms to control the rates at which proteins are synthesized. It is not unexpected that cells would also possess mechanisms to control the length of time that proteins survive once they are fully functional. Although the subject of protein stability does not fall technically under the heading of control of gene expression, it is a logical extension of that topic and thus appears in this place in the text.

Degradation of cellular proteins is carried out by barrel-shaped, protein-degrading machines called **proteasomes** that are found in both the nucleus and cytosol of cells. Proteasomes consist of four rings of polypeptide subunits stacked one on top of the other with a cap attached at either end of the stack (Figure 12.60a). The two central rings consist of polypeptides (β subunits) that function as proteolytic enzymes. The active sites of these subunits face the enclosed central chamber, where proteolytic digestion can occur in a protected environment that is separated from proteins of the cytosol.

Proteasomes digest proteins that have been specially selected and marked for destruction as described below. Some proteins are selected because they are recognized as abnormal—either misfolded or incorrectly associated with other proteins. Included in this group are abnormal proteins that had been produced on membrane-bound ribosomes of the rough ER (page 298). The selection of "normal" proteins for proteasome destruction is based on the protein's biological stability. Every protein is thought to have a characteristic longevity, which is expressed as its **half-life**, or the period of time during which it has a 50 percent likelihood of being destroyed. Some proteins, such as the enzymes of glycolysis or the globin molecules of an erythrocyte, have a half-life that can be measured in days to weeks. Other proteins that are required for a specific, fleeting activity, such as regulatory proteins that initiate DNA replication or trigger cell division, may have a half-life of only a few minutes. All of these proteins, regardless of their half-life, are degraded by proteasomes, as can be demonstrated using drugs that specifically inhibit proteasomal digestion.

The factors that control a protein's lifetime are not well understood. One of the determinants is the specific amino acid that resides at the N-terminus of a polypeptide chain. Polypeptides that terminate in arginine or lysine, for example, are typically short-lived. A number of proteins that act at specific times within the cell cycle are marked for destruction when certain residues become phosphorylated. Still other proteins carry a specific internal sequence of amino acids that ensures they will not survive long within the cell. Failure to destroy certain of these proteins in a timely manner can cause a cell to become malignant.

The steps in degradation of a protein by a proteasome are depicted in Figure 12.60b. Proteins to be degraded by proteasomes are marked for destruction by covalent linkage to a small, highly-conserved protein named **ubiquitin** (step 1, Figure 12.60b). Ubiquitin is transferred enzymatically to a lysine residue on the target protein. Enzymes that transfer ubiquitin to target proteins comprise a large family in which different members recognize proteins bearing different degradation signals. These enzymes play a crucial role in determining the life or death of key proteins and are a focus of current research.

Most target proteins eventually contain a string of ubiquitin molecules joined together end-to-end to form a polyubiquitin chain. Polyubiquinated proteins are recognized by the cap of the proteasome (step 2), which removes the ubiquitin chain and unfolds the target protein. The un-

folded, linear polypeptide is then threaded through the narrow opening in the ring of α-subunits and passed into the central chamber of the proteasome (step 3), where in most cells it is digested into small peptides (steps 4 and 5). The peptide products are released back into the cytosol where they are degraded into their component amino acids.

(a) (b)

(c)

Figure 12.60 Proteasome structure and function. (*a*) High-resolution electron micrograph of an isolated *Drosophila* proteasome. (*b*) Model of a proteasome based on high-resolution electron microscopy and X-ray crystallography. Each proteasome consists of two large caps on either end of a barrel-shaped core that is formed by a stack of four rings. Each ring consists of seven subunits that are divided into two classes, α-type and β-type. The two inner rings are composed of β subunits, which surround a central chamber. The subunits are drawn in different shades of color because they are similar, but not identical, polypeptides. Three of the seven β subunits in each ring possess proteolytic activity, the other four are inactive in eukaryotic cells. (Prokaryotic cells also possess proteasomes, but they have a simpler construction, and all β subunits are active.) The two outer rings are composed of enzymatically inactive α subunits that form a narrow opening (approximately 13 Å) through which unfolded polypeptide substrates are threaded to reach the central chamber, where they are degraded. (*c*) Steps in the degradation of proteins by a proteasome. In step 1, the protein to be degraded is linked covalently to a string of ubiquitin molecules. This step requires the participation of three distinct enzymes (E1, E2, and E3) in a process that is not discussed in the text. In step 2, the ubiquinated target protein binds to the cap of the proteasome. The ubiquitin chain is then removed, and the unfolded polypeptide is threaded into the central chamber of the proteasome (step 3), where it is degraded by the catalytic activity of the β subunits (steps 4 and 5). (A: FROM H. HÖLZL ET AL., COURTESY OF WOLFGANG BAUMEISTER, J. CELL BIOL. 150:126, 2000; BY COPYRIGHT PERMISSION OF THE ROCKEFELLER UNIVERSITY PRESS.)

SYNOPSIS

The nucleus of a eukaryotic cell is a complex structure bounded by the nuclear envelope, which controls the exchange of materials between the nucleus and cytoplasm, maintaining the unique composition of the cell's two major compartments. The nuclear envelope consists of several distinct components, including an inner and outer nuclear membrane separated by a perinuclear space and a variable number of nuclear pores. Nuclear pores are sites where the inner and outer nuclear membranes are fused to form a circular opening that is filled with a complex structure called the nuclear pore complex (NPC). The NPC has a basketlike structure with octagonal symmetry composed of rings, spokes, particles, and filaments. The nuclear pores are the sites through which materials pass between the nucleus and cytoplasm. Studies indicate that low-molecular-weight substances diffuse through the NPC, probably passing between the spokes. RNA and proteins move through the NPC in both

directions by a poorly understood process of selective transport. Proteins that normally reside within the nucleus contain a stretch of amino acids called the nuclear localization signal (NLS) that allows them to bind to a receptor at the cytoplasmic surface of the NPC, which apparently triggers a conformational change in the central transporter that allows the bound molecules entry into the nucleus. The inner surface of the nuclear envelope is lined by a fibrillar meshwork called the nuclear lamina, which consists of proteins (lamins) that are members of the family of proteins that make up intermediate filaments. The fluid of the nucleus is called the nucleoplasm. (p. 495)

The chromosomes of the nucleus contain a defined complex of DNA and histone proteins that form characteristic nucleoprotein filaments, representing the first step in packaging the genetic material. Histones are small, basic proteins that

are divided into five distinct classes. Histones and DNA are organized into nucleosome core complexes, which consist of two molecules each of histones H2A, H2B, H3, and H4 encircled by almost two turns of DNA. Nucleosome core particles are connected to one another by stretches of linker DNA. Together, the core particles and DNA linkers generate a nucleosome filament that resembles a chain of beads on a string. The interaction between histones and DNA is primarily structural and is relatively independent of nucleotide sequence, though important exceptions are known in which nucleosomes are positioned at specific sites on the DNA. Packaging of DNA into nucleosomes reduces its length approximately one-seventh relative to the fully extended condition. (p. 501)

Chromatin is not present in cells in the highly extended nucleosome filament; it is compacted into higher levels of organization. Each nucleosome core particle contains a molecule of H1 histone bound to the DNA. Both core and H1 histones mediate an interaction between neighboring nucleosomes that generates a 30-nm fiber, which represents a higher level of chromatin organization. The 30-nm fibers are in turn organized into looped domains, which are most readily visualized when mitotic chromosomes are subjected to procedures that remove histones. Mitotic chromosomes represent the most compact state of chromatin, one in which the loops that compose a single chromosome are collected together into a highly compact structure. A certain fraction of the chromatin, called heterochromatin, remains in this highly compact state throughout interphase. Heterochromatin is divided into constitutive heterochromatin, which remains condensed in all cells at all times, and facultative heterochromatin, which is specifically inactivated during certain phases of an organism's life. One X chromosome in each cell of a female mammal is subjected to an inactivation process during embryonic development that converts it into transcriptionally inactive, facultative heterochromatin. Because X inactivation occurs randomly, the paternally derived X chromosome is inactivated in half the cells of the embryo, while the maternally derived X chromosome is inactivated in the other half. As a result, adult females are genetic mosaics with respect to genes present on the X chromosome. (p. 504)

Mitotic chromosomes possess several clearly recognizable features. Mitotic chromosomes can be visualized by lysing cells that have been arrested in mitosis and then staining them to generate identifiable chromosomes with predictable banding patterns. Each mitotic chromosome contains a marked indentation, called the centromere, which contains highly repeated DNA sequences and serves as a site for the attachment of microtubules during mitosis. The ends of the chromosome are the telomeres, which are maintained from one cell generation to the next by a special enzyme, called telomerase, that contains RNA as an integral component. Telomeres appear to play a number of important roles in regulating cellular activity. (p. 507)

The nucleus is an ordered cellular compartment. Observations that support this are the following: specific chromosomes are confined to particular regions of the nucleus; the telomeres may be associated with the nuclear envelope; RNPs involved in pre-mRNA splicing are restricted to particular sites. Evidence indicates that the network of protein filaments that makes up the nuclear matrix plays a key role in maintaining the ordered structure of the nucleus. (p. 515)

In prokaryotes, genes are organized into regulatory units called operons. Operons are clusters of structural genes that typically code for different enzymes in the same metabolic pathway. Because all of the structural genes are transcribed into a single mRNA, their expression can be regulated in a coordinated manner. The level of gene expression is controlled by a key metabolic compound, such as the inducer lactose, which attaches to a protein repressor and changes its shape. This event alters the ability of the repressor to bind to the operator site on the DNA, thereby blocking transcription. (p. 517)

The rate of synthesis of a particular polypeptide is determined by a complex series of regulatory events that operate primarily at three distinct levels. (1) Transcriptional control mechanisms determine whether or not a gene is transcribed and, if so, how often; (2) processing-level control mechanisms determine the path by which the primary RNA transcript is processed; and (3) translational control mechanisms determine whether a particular mRNA is actually translated and, if so, how often and for how long a period. (p. 521)

All of the differentiated cells of a eukaryotic organism retain all of the genetic information. Different genes are expressed by cells at different stages of development, by cells in different tissues, and by cells exposed to different stimuli. The transcription of a particular gene is controlled by transcription factors, which are proteins that bind to specific sequences located at sites outside of a gene's coding region. The closest upstream regulatory sequence is the TATA box, which is the major component of the gene's promoter and the site of assembly of the preinitiation complex. The activity of proteins at the TATA box is thought to depend on interactions with other proteins bound to other sites, including various types of response elements and enhancer elements. Enhancer elements are distinguished by the fact that they can be moved from place to place within the DNA or even reversed in orientation. Some enhancers may be located tens of thousands of base pairs upstream from the gene whose transcription they stimulate. Proteins bound to an enhancer and a promoter are thought to be brought into contact by loops in the DNA. (p. 521)

Determination of the three-dimensional structure of a number of complexes between transcription factors and DNA indicate that these proteins bind to the DNA through a limited number of structural motifs. Transcription factors typically contain at least two domains, one whose function is to recognize and bind to a specific sequence of base pairs in the DNA, and another whose function is to activate transcription by interacting with other proteins. Most transcription factors bind to DNA as dimers, either heterodimers or homodimers, that recognize sequences in the DNA that possess twofold symmetry. Most of the DNA-binding motifs contain a segment, often an α helix, that is inserted into the major groove of the DNA, where it recognizes the sequence of base pairs that line the groove. Among the most common motifs that occur in DNA-binding proteins are the zinc finger, the helix-loop-helix, the leucine zipper, and the HMG box. Each of these motifs provides a structurally stable framework on which the specific DNA-recognizing surfaces of the protein can interact with the DNA double helix. Though most transcription factors are probably stimulatory, some act by inhibiting transcription. (p. 529)

Eukaryotic genes are thought to be silenced when the cytosine bases of certain nucleotides residing in GC-rich regions are methylated. Methylation is a dynamic modification; enzymes are present to remove methyl groups as well as to add them. Addition of methyl groups in key regulatory regions upstream from genes is correlated with a decrease in transcription of the gene, whereas removal of such methyl groups is correlated with an increased level of transcription. Methylation is particularly evident in chromatin that has been rendered transcriptionally inactive by heterochromatization, such as the inactive X chromosome in the cells of female mammals. Another phenomenon associated with DNA methylation is genomic imprinting, which affects a small number of genes that are either transcriptionally active or inactive in the embryo depending on their parental source. (p. 534)

The incorporation of DNA into nucleosomes adds another dimension to the control of transcription by creating opportunities for transcriptional control. The precise positioning of nucleosomes within the regulatory region of a gene plays a key role in permitting DNA-binding proteins to stimulate transcription of that gene. The interaction of a transcription factor with the DNA can induce secondary changes in chromatin structure that make other sites more or less accessible to subsequent protein binding. In addition, transcriptionally active chromatin possesses certain distinctive properties that distinguish it from inactive chromatin, most notably sites that are hypersensitive to enzymatic DNA digestion. Some evidence indicates that nucleosomes undergo modifications that make the DNA more accessible to polymerase molecules. (p. 536)

Gene expression is regulated at the processing level by alternate splicing in which a single gene can encode two or more related proteins. Many primary transcripts can be processed by more than one pathway, producing mRNAs containing different combinations of exons. In the simplest case, a specific intron can be either spliced out of the transcript or retained as part of the final mRNA. The specific pathway taken during pre-mRNA processing is thought to depend on the presence of proteins that control the splice sites that are recognized for cleavage. (p. 540)

Gene expression is regulated at the translational level by various processes, including mRNA localization, control of translation of existing mRNAs, and mRNA longevity. Most of these regulatory activities are mediated by interactions with the 5′ and 3′ untranslated regions (UTRs) of the mRNA. For example, cytoplasmic localization of mRNAs in particular regions of the eggs of fruit flies is thought to be mediated by proteins that recognize localization sequences in the 3′ UTR. The presence of an mRNA in the cytoplasm does not ensure its translation. The protein synthetic machinery of a cell may be controlled globally so that translation of all mRNAs is affected, or the translation of specific mRNAs can be targeted, as occurs in the control of ferritin synthesis by iron levels acting on proteins that bind to the 5′ UTR of the ferritin mRNA. One of the primary factors controlling the longevity (stability) of mRNAs is the length of the poly(A) tail. This portion of the mRNA is normally protected by a poly(A)-binding protein. Specific nucleases in the cell gradually shorten the poly(A) tail until a point is reached where the protective protein can no longer stay bound to the tail. Once the protein is lost, the mRNA is decapped, and degradation occurs in a 5′ to 3′ direction. (p. 540)

ANALYTIC QUESTIONS

1. What are some of the consequences you might expect as the result of mutations in various proteins that make up the nuclear matrix?

2. How many of each type of core histone would it take to wrap the entire human genome into nucleosomes? How has evolution solved the problem of producing such a large number of proteins in a relatively short period of time?

3. Suppose you discovered a temperature-sensitive mutant whose nucleus failed to accumulate certain nuclear proteins at an elevated (restrictive) temperature but continued to accumulate other nuclear proteins. What conclusions might you draw about nuclear localization and the nature of this mutation?

4. Humans born with three X chromosomes but no Y chromosomes often develop into females of normal appearance. How many Barr bodies would you expect the cells of these women to have? Why?

5. Suppose that X inactivation were not a random process, but always led to the inactivation of the X chromosome derived from the father. What effect would you expect this to have on the phenotype of females?

6. The chromosomes shown in Figure 12.16b were labeled by incubating the preparation with DNA fragments known to be specific for each of the chromosomes. Suppose one of the chromosomes in the field contained regions having two different colors. What might you conclude about this chromosome?

7. What advantage might be gained by having transcripts synthesized and processed in certain regions of the nucleus rather than randomly throughout the nucleoplasm?

8. Compare and contrast the effect of a deletion in the operator gene of the lactose operon with one in the operator of the histidine operon.

9. If you were to find a mutant of *E. coli* that produced continuous polypeptide chains containing both β-galactosdase and galactoside permease (encoded by the *y* gene), how might you explain how this happened?

10. You suspect that a new hormone you are testing functions to stimulate myosin synthesis by acting at the transcriptional level. What type of experimental evidence would support this contention?

11. Suppose you had conducted a series of experiments in which you had transplanted nuclei from several different adult tissues into an activated, enucleated mouse egg and found that the egg did not develop past the blastocyst stage. Could you conclude that the transplanted nucleus had lost genes that were required for postblastula development? Why or why not? What does this type of experiment tell you more generally about interpreting negative results?

12. It was noted on page 526 that DNA footprinting allows isolation of DNA sequences that bind specific transcription factors. Describe an experimental protocol to identify transcription factors that bind to an isolated DNA sequence. (You might consider the techniques discussed in Section 18.7.)

13. How do you explain why enhancers can be moved around within the DNA without affecting their activity, whereas the TATA box can only operate at one specific site?

14. Suppose that you are working with a cell that exhibits a very low level of protein synthesis, and you suspect the cells are subject to a global translational-control inhibitor. What experiment might you perform to determine whether this is the case?

15. The signal sequences that direct the translocation of proteins into the endoplasmic reticulum are cleaved by a signal peptidase, whereas the NLSs and NESs required for movement of a protein into or out of the nucleus remain as part of that protein. Consider a protein such as hnRNPA1, which is involved in the export of mRNA to the cytoplasm. Why is it important that the transport signal sequences for this protein remain as part of the protein, whereas the signal sequence for ER proteins can be cleaved.

16. When methylated DNA is introduced into cultured mammalian cells, it is generally transcribed for a period before it becomes repressed. Why would you expect this type of delay before inhibition of transcription would occur?

17. Suppose you had isolated a new transcription factor and wanted to know which genes this protein might regulate. Is there any way that you could use a cDNA microarray of the type shown in Figure 12.30 to approach this question?

ADDITIONAL READING

Nuclear Structure and Function

BLOBEL, G. & WOZNIAK, R. W. 2000. Proteomics for the pore. *Nature* 403:835–836.

GÖRLICH, D. & KUTAY, U. 1999. Transport between the cell nucleus and the cytoplasm. *Annu. Rev. Cell Dev. Biol.* 15:607–660.

HEGELE, R. A. 2000. The envelope please: Nuclear lamins and disease. *Nature Med.* 6:136–137.

KAFFMAN, A. & O'SHEA, E. K. 1999. Regulation of nuclear localization. *Annu. Rev. Cell Dev. Biol.* 15:291–339.

LAMOND, A. I. & EARNSHAW, W. C. 1998. Structure and function in the nucleus. *Science* 280:547–553.

MATTAJ, I. W. & ENGLMEIER, L. 1998. Nucleocytoplasmic transport: The soluble phase. *Annu. Rev. Biochem.* 67:265–306.

MICHAEL, W. M. 2000. Nucleocytoplasmic shuttling signals: Two for the price of one. *Trends Cell Biol.* 10:46–50.

MISTELI, T. & SPECTOR, D. L. 1998. The cellular organization of gene expression. *Curr. Opin. Cell Biol.* 10:323–331.

SHOPLAND, L. S. & LAWRENCE, J. B. 2000. Seeking common ground in nuclear complexity. *J. Cell Biol.* 150:F1–F4.

SLEEMAN, J. E. & LAMOND, A. I. 1999. Nuclear organization of pre-mRNA splicing factors. *Curr. Opin. Cell Biol.* 11:372–377.

TALCOTT, B. & MOORE, M. S. 1999. Getting across the nuclear pore complex. *Trends Cell Biol.* 9:312–318.

WEIS, K. 1998. Importins and exportins: How to get in and out of the nucleus. *Trends Biochem. Sci.* 23:185–189.

WENTE, S. R. 2000. Gatekeepers of the nucleus. *Science* 288:1374–1376.

ZENK, D. & CREMER, T. 1998. Cell nucleus: Chromosomal dynamics in nuclei of living cells. *Curr. Biol.* 8:R321–R324.

Chromatin Structure, Chromosomes, and Transcription

BRYAN, T. M. & CECH, T. R. 1999. Telomerase and the maintenance of chromosome ends. *Curr. Opin. Cell Biol.* 11:318–324.

CHOO, K. H. A. 2000. Centromerization. *Trends Cell Biol.* 10:182–188.

COLLINS, K. 2000. Mammalian telomeres and telomerase. *Curr. Opin. Cell Biol.* 12:378–383.

DE LANGE, T. & DEPINHO, R. A. 1999. Unlimited mileage from telomerase. *Science* 283:947–949.

GREIDER, C. W. 1998. Telomeres and senescence: The history, the experiment, the future. *Curr. Biol.* 8:R178–R181.

GREIDER, C. W. 1999. Telomerase activation: One step on the road to cancer. *Trends Gen.* 15:109–112.

KORNBERG, R. D. & LORCH, Y. 1999. Twenty-five years of the nucleosome, fundamental particle of the eukaryotic chromosome. *Cell* 98:285–294.

LUGER, K. & RICHMOND, T. J. 1998. DNA binding within the nucleosome core. *Curr. Opin. Struct. Biol.* 8:33–40.

LYON, M. F. 1992. Some milestones in the history of X-chromosome inactivation. *Annu. Rev. Gen.* 26:17–28.

LYON, M. F. 1999. X-chromosome inactivation. *Curr. Biol.* 9:R235–R237.

MCKENZIE, K. E., ET AL. 1999. Applications of telomerase research in the fight against cancer. *Mol. Med. Today* 5:114–122.

MURPHY, T. D. & KARPEN, G. H. 1998. Centromeres take flight: Alpha satellite and the quest for the human centromere. *Cell* 93:317–320.

PANNING, B. & JAENISCH, R. 1998. RNA and the epigenetic regulation of X-chromosome inactivation. *Cell* 93:305–308.

PRESCOTT, J. C. & BLACKBURN, E. H. 1999. Telomerase: Dr. Jekyll or Mr. Hyde? *Curr. Opin. Genes Dev.* 9:368–373.

PRICE, C. M. 1999. Telomeres and telomerase: Effects on cell growth. *Curr. Opin Genes Dev.* 9:218–224.

SHERR, C. J. & DEPINHO, R. A. 2000. Cellular senescence: Mitotic clock or culture shock? (telomeres) *Cell* 102:407–410.

STRAHL, B. D. & ALLIS, D. 2000. The language of covalent histone modifications. *Nature* 403:41–45.

THOMAS, J. O. 1999. Histone H1: Location and role. *Curr. Opin. Cell Biol.* 11:312–317.

TRAVERS, A. 1999. The location of the linker histone on the nucleosome. *Trends Biochem. Sci.* 24:4–7.

WIDOM, J. 1998. Structure, dynamics, and function of chromatin in vitro. *Annu. Rev. Biophys. Biomol. Struct.* 27:285–327.

WILLARD, H. F. 1998. Centromeres: The missing link in the development of human artificial chromosomes. *Curr. Opin. Genes Dev.* 8:219–225.

WOLFFE, A. P. & KURUMIZAKA, H. 1998. The nucleosome: A powerful regulator of transcription. *Prog. Nuc. Acid Res. Mol. Biol.* 61:379–422.

ZUMSTEIN, L. A. & LUNDBLAD, V. 1999. Telomeres: Has cancer's Achilles heel been exposed? *Nature Med.* 5:1129–1130.

Transcription and Transcriptional-Level Control

BROWN, C. E., ET AL. 2000. The many HATs of transcription coactivators. *Trends Biochem. Sci.* 25:15–19.

BROWN, P. O. & BOTSTEIN, D. 1999. Exploring the world of the genome with DNA microarrays. *Nature Gen.* 21:33–37.

CHEN, L. 1999. Combinatorial gene regulation by eukaryotic transcription factors. *Curr. Opin. Struct. Biol.* 9:48–55.

CHO, R. J. & CAMPBELL, M. J. 2000. Transcription, genomes, and function. *Trends Gen.* 16:409–415.

FEREA, T. L. & BROWN, P. O. 1999. Observing the living genome. *Curr. Opin. Genes Dev.* 9:715–722.

FREEDMAN, L. P. 1999. Increasing the complexity of coactivation in nuclear receptor signaling. *Cell* 97:5–8.

GREEN, M. R. 2000. TBP-associated factors (TAF$_{II}$s): Multiple, selective transcription mediators in common complexes. *Trends Biochem. Sci.* 25:59–63.

HAMPSEY, M. & REINBERG, D. 1999. RNA polymerase II as a control panel for multiple coactivator complexes. *Curr. Opin. Genes Dev.* 9:132–139.

JOHNSTON, M. 1998. Gene chips: Array of hope for understanding gene regulation. *Curr. Biol.* 8:R171–R174.

KORNBERG, R. D. 1999. Eukaryotic transcriptional control. *Trends Biochem. Sci.* 24:M46–M49. (Dec.)

KUMAR, M. V. & TINDALL, D. J. 1998. Transcriptional regulation of the steroid receptor genes. *Prog. Nuc. Acid Res. Mol. Biol.* 59:289–306.

LANDER, E. S., ET AL. 1999. The chipping forecast. *Nature Gen.* 21, suppl. p. 1–60.

LOCKHART, D. J. & WINZELER, E. A. 2000. Genomics, gene expression, and DNA arrays. *Nature* 405:827–836.

MALIK, S. & ROEDER, R. G. 2000. Transcription regulation through Mediator-like coactivators in yeast and metazoan cells. *Trends Biochem. Sci.* 25:277–283.

MANNERVIK, M., ET AL. 1999. Transcriptional coregulators in development. *Science* 284:606–609.

ORPHANIDES, G. & REINBERG, D. 2000. RNA polymerase II elongation through chromatin. *Nature* 407:471–475.

PABO, C. O. & SAUER, R. T. 1992. Transcription factors: Structural families and principles of DNA recognition. *Annu. Rev. Biochem.* 61:1053–1095

PETERSON, C. L. & WORKMAN, J. L. 2000. Promoter targeting and chromatin remodeling by the SWI/SNF complex. *Curr. Opin. Genes Dev.* 10:187–192.

PTASHNE, M. & GANN, A. 1997. Transcriptional activation by recruitment. *Nature* 386:569–577.

STILLMAN, B., ET AL. 1998. *Mechanisms of Transcription.* Cold Spring Harbor Symp. Quant. Biol. vol. 63.

STRUHL, K. 1998. Histone acetylation and transcription regulatory mechanisms. *Genes Dev.* 12:599–606.

Struhl, K. 1999. Fundamentally different logic of gene regulation in eukaryotes and prokaryotes. *Cell* 98:1–4.

WOLFBERGER, C. 1999. Multiprotein-DNA complexes in transcription regulation. *Annu. Rev. Biophys. Biomol. Struct.* 28:29–56..

WORKMAN, J. L. & KINGSTON, R. E. 1998. Alterations of nucleosome structure as a mechanism of transcription regulation. *Annu. Rev. Biochem.* 67:545–579.

DNA Methylation and Genomic Imprinting

BARTOLOMEI, M. S. & TILGHMAN, S. M. 1997. Genomic imprinting in mammals. *Annu. Rev. Gen.* 31:493–525.

Bird, A. 1999. DNA methylation de novo. *Science* 286:2287–2288.

BIRD, A. P. & WOLFFE, A. P. 1999. Methylation-induced repression—Bells, braces, and chromatin. *Cell* 99:451–454.

BRANNAN, C. I. & BARTOLOMEI, M. S. 1999. Mechanisms of genomic imprinting. *Curr. Opin. Genes Dev.* 9:164–170.

ENG, C., ET AL. 2000. A bird's eye view of global methylation. *Nature Gen.* 24:101–102.

FINNEGAN, E. J., ET AL. 2000. Reviews on DNA methylation. *Curr. Opin. Genes Dev.* 10:217–233.

HENDRICH, B. 2000. Methylation moves into medicine. *Curr. Biol.* 10:R60–R63.

JONES, P. A. 1999. The DNA methylation paradox. *Trends Gen.* 15:34–37.

LI, E. 1999. The mojo of methylation. *Nature Gen.* 23:5–6.

MORISON, I. M. & REEVE, A. E. 1998. Insulin-like growth factor 2 and overgrowth: Molecular biology and clinical implications. *Mol. Med. Today* 4:110–113.

WOLFFE, A. P. 2000. Imprinting insulation. *Curr. Biol.* 10:R463–R465.

Processing-Level Control

BLACK, D. L. 2000. Protein diversity from alternate splicing. *Cell* 103:367–370.

BLENCOWE, B. J. 2000. Exonic splicing enhancers. *Trends Biochem. Sci.* 25:106–110.

LOPEZ, A. J. 1998. Alternate splicing and pre-mRNA: Developmental consequences and mechanisms of regulation. *Annu. Rev. Gen.* 32:279–305.

MISTELI, T. 2000. Different site, different splice. *Nature cell Biol.* 2:E98–E100.

Translational-Level Control

BASHIRULLAH, A., ET AL. 1998. RNA localization in development. *Annu. Rev. Biochem.* 67:335–394.

CONNE, B., ET AL. 2000. The 3′ untranslated region of messenger RNA: A molecular "hotspot" for pathology? *Nature Med.* 6:637–641.

GRAY, N. K. & WICKENS, M. 1998. Control and translation initiation in animals. *Annu. Rev. Cell Dev. Biol.* 14:399–458.

HAYS, T. & KARESS, R. 2000. Swallowing dynein: A missing link in RNA localization? *Nature Cell Biol.* 2:E60–E62.

HERSHEY, J. W. B. 1991. Translational control in mammalian cells. *Annu. Rev. Biochem.* 60:717–755.

MITCHELL, P. & TOLLERVEY, D. 2000. mRNA stability in eukaryotes. *Curr. Opin. Genes Dev.* 10:193–198.

OLEYNIKOV, Y. & SINGER, R. H. 1998. RNA localization: Different zipcodes, same postman. *Trends Cell Biol.* 8:381–383.

SCHNAPP, B. J. 1999. RNA localization: A glimpse of the machinery. *Curr. Biol.* 9:R725–R727.

WILSON, G. & BREWER, G. 1999. The search for trans-acting factors controlling mRNA decay. *Prog. Nuc. Acid Res. Mol. Biol.* 62:257–291.

13

R

DNA Replication and Repair

eproduction is a fundamental property of all living systems. The process of reproduction can be observed at several levels: organisms duplicate by asexual or sexual reproduction; cells duplicate by cellular division; and the genetic material duplicates by **DNA replication**. The machinery that replicates DNA is also called into action in another capacity: to repair the genetic material after it has sustained damage. These two processes—DNA replication and DNA repair—are the subjects addressed in this chapter.

The capacity for self-duplication is presumed to have been one of the first critical properties to have appeared in the evolution of the first primitive life forms. Without the ability to propagate, any primitive assemblage of biological molecules would be destined for oblivion. The first carriers of genetic information were probably RNA molecules that were able to self-replicate (page 470). As evolution progressed and RNA molecules were replaced by DNA molecules as the primary storehouse of genetic information, the process of replication became considerably more complex, requiring a large number of auxiliary components. Thus, although a DNA molecule contains the information for its own duplication, it lacks the ability to perform the activity itself. As Richard Lewontin expressed it, "the com-

Three-dimensional model of a DNA helicase encoded by the bacteriophage T7. The protein consists of two rings, each composed of six subunits. In this model, the central hole encircles only one of the two DNA strands. Driven by ATP hydrolysis, the protein moves in a 5' → 3' direction along the strand to which it is bound, displacing the complementary strand and unwinding the duplex. DNA helicase activity is required for DNA replication. (COURTESY OF EDWARD H. EGELMAN, UNIVERSITY OF VIRGINIA.)

mon image of DNA as a self-replicating molecule is about as true as describing a letter as a self-replicating document. The letter needs a photocopier; the DNA needs a cell." Let us see then how the cell carries out this activity.

13.1 DNA REPLICATION

The proposal for the structure of DNA by Watson and Crick in 1953 was accompanied by a proposal for its "self-duplication." The two strands of the double helix are held together by hydrogen bonds between the bases. Individually, these hydrogen bonds are weak and readily broken. Watson and Crick envisioned that replication occurred by gradual separation of the strands of the double helix (Figure 13.1), much like the separation of two halves of a zipper. Because the two strands are complementary to each other, each strand contains the information required for construction of the other strand. Thus once the strands are separated, each can act as a template to direct the synthesis of the complementary strand and restore the double-stranded state.

Semiconservative Replication

The Watson-Crick proposal made certain predictions concerning the behavior of DNA during replication. Ac-

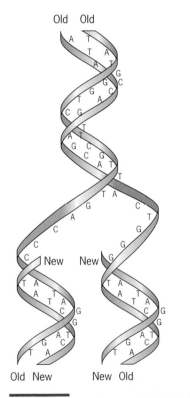

Figure 13.1 The original Watson-Crick proposal for the replication of a double-helical molecule of DNA. During replication, the double helix unwinds, and each of the parental strands serves as a template for the synthesis of a new complementary strand. As discussed at length in this chapter, the basic tenets of this proposal have been borne out.

Figure 13.2 Three alternate schemes of replication. Semiconservative replication is depicted in scheme 1, conservative replication in scheme 2, and dispersive replication in scheme 3. A description of the three alternate modes of replication is given in the text.

cording to the proposal, each of the daughter duplexes should be composed of one complete strand inherited from the parental duplex and one complete strand that has been newly synthesized. Replication of this type (Figure 13.2, scheme 1) is said to be **semiconservative** because each daughter duplex contains one strand from the parent structure. In the absence of information on the mechanism responsible for replication, two other types of replication had to be considered. In *conservative* replication (Figure 13.2, scheme 2), the two original strands would remain together (after serving as templates), as would the two newly synthesized strands. As a result, one of the daughter duplexes would contain only parental DNA, while the other daughter duplex would contain only newly synthesized DNA. In *dispersive* replication (Figure 13.2, scheme 3), the integrity of each of the parental strands would be disrupted. As a result, the daughter duplexes would contain strands that were composites of old and new DNA; that is, neither the parental strands nor the parental duplex is conserved.

To decide among these three possibilities, it was necessary to distinguish newly synthesized DNA strands from the original DNA strands that served as templates. This was first accomplished in studies on bacteria in 1958 by Matthew Meselson and Franklin Stahl of the California Institute of Technology who used heavy (^{15}N) and light (^{14}N) isotopes of nitrogen to distinguish between parental and newly synthesized DNA strands (Figure 13.3). These researchers grew bacteria for many generations in medium containing ^{15}N-ammonium chloride as the sole nitrogen source. Consequently, the nitrogen-

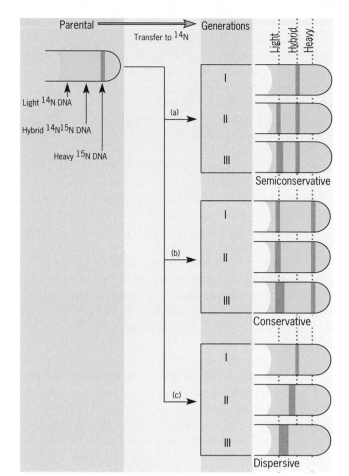

(a)

Figure 13.3 Experiment demonstrating that DNA replication in bacteria is semiconservative. DNA was extracted from bacteria at different stages in the experiment, mixed with a concentrated solution of the salt cesium chloride (CsCl), placed into a centrifuge tube, and centrifuged to equilibrium at high speed in an ultracentrifuge. Cesium ions have sufficient atomic mass to be affected by the centrifugal force, and they form a density gradient during the centrifugation period with the lowest concentration (lowest density) of Cs at the top of the tube and the greatest concentration (highest density) at the bottom of the tube. During centrifugation, DNA fragments within the tube become localized at a position having a density equal to their own density, which in turn depends on the ratio of ^{15}N/^{14}N that is present in their nucleotides. The greater the ^{14}N content, the higher in the tube the DNA fragment is found at equilibrium. (*a*) The results expected in this type of experiment for each of the three possible schemes of replication. The single tube on the left indicates the position of the parental DNA and the positions at which totally light or hybrid DNA fragments would band. (*b*) Experimental results obtained by Meselson and Stahl. The appearance of a hybrid band and the disappearance of the heavy band after one generation eliminates conservative replication. The subsequent appearance of two bands, one light and one hybrid, eliminates the dispersive scheme. (B: FROM M. MESELSON AND F. STAHL, PROC. NAT'L. ACAD. SCI. U.S.A. 44:671, 1958.)

(b)

containing bases of the DNA of these cells contained only the heavy nitrogen isotope. Cultures of "heavy" bacteria were washed free of the old medium and incubated in fresh medium with light, ^{14}N-containing compounds, and samples were removed at increasing intervals over a period of several generations. DNA was extracted from the samples of bacteria and subjected to equilibrium density-gradient centrifugation (Section 18.11). In this procedure, the DNA is mixed with a concentrated solution of cesium chloride and centrifuged until the double-stranded DNA molecules reach equilibrium according to their density.

In the Meselson-Stahl experiment, the density of a DNA molecule is directly related to the percentage of ^{15}N and ^{14}N atoms it contains. If replication is semiconservative, one would expect that the density of DNA molecules would gradually decrease during culture in the ^{14}N-containing medium (upper set of centrifuge tubes of Figure 13.3a). The density would decrease as light strands were synthesized in association with heavy strands. After one generation, all DNA molecules would be ^{15}N-^{14}N hybrids, and their buoyant density would be halfway between that expected for totally heavy and totally light DNA (Figure 13.3a). As replication continued beyond the first generation, the newly synthesized strands would continue to contain only light isotopes, and two types of duplexes would appear in the gradients: those containing ^{15}N-^{14}N hybrids and those containing only ^{14}N. As the time of growth in the light medium continued, a greater and greater percentage of the DNA molecules present would be light. However, as long as replication continued semiconservatively, the original heavy parental strands would remain intact and present in hybrid DNA molecules that occupy a smaller and smaller percentage of the total DNA (Figure 13.3a). The results of the density-gradient experiments obtained by Meselson and Stahl are shown in Figure 13.13b, and they indicate that replication occurs semiconservatively. The results that would have been obtained if replication occurred by conservative or dispersive mechanisms are indicated in the lower two sets of centrifuge tubes of Figure 13.3a.

Replication was soon demonstrated to occur semiconservatively in eukaryotes as well. The drawing and photograph of Figure 13.4 show the results of an experiment in which cultured mammalian cells were allowed to undergo replication in bromodeoxyuridine (BrdU), a compound that is incorporated into DNA in place of thymidine. Following replication, a chromosome is made up of two chromatids. After one round of replication in BrdU, both chromatids of each chromosome contained BrdU (Figure 13.4a). After two rounds of replication in BrdU, one chromatid of each chromosome was composed of two BrdU-containing strands, whereas the other chromatid was a hybrid consisting of a BrdU-containing strand and a thymidine-containing strand (Figure 13.4a,b). The thymidine-containing strand had been part of the original parental DNA molecule prior to addition of BrdU to the culture.

Replication in Bacterial Cells

We will focus, in this section of the chapter, on bacterial cells, because prokaryotic replication is better understood than the corresponding process in eukaryotes. The progress in bacterial research has been driven by certain genetic and biochemical approaches that, until recently, could not be used in studying eukaryotic systems. These include

- The availability of mutants that cannot synthesize one or another protein required for the replication process. The isolation of mutants unable to replicate their chromosome may seem paradoxical: how can cells with this defect be cultured? Mutant cells that are unable to synthesize a particular metabolite can be grown by supplying the required compound, but there is nothing that can be supplied to mutants deficient in a protein required for replication that enables them to overcome their deficiency. The solution to this type of experimental difficulty is solved by the isolation of **temperature-sensitive (ts)** mutants. The deficiency only reveals itself in the mutant at an elevated temperature, termed the *nonpermissive* (or *restrictive*) temperature. When grown at the lower (*permissive*) temperature, the mutant protein can function sufficiently well to carry out its required activity, and the cells can continue to grow and divide. Temperature-sensitive mutants have been isolated that affect virtually every type of physiological activity (see also page 283), and they have been particularly important in the study of DNA synthesis as it occurs in replication, DNA repair, and genetic recombination.

- The development of in vitro systems in which replication can be studied using purified cellular components. In some studies, the DNA molecule to be replicated is incubated with cellular extracts from which specific proteins suspected of being essential have been removed. In other studies, the DNA is incubated with a variety of purified proteins whose activity is to be tested.

Taken together, these approaches have revealed the activity of at least 30 different proteins that are required to replicate the chromosome of *E. coli*. In the following pages, we will discuss the activities of several of these proteins whose functions have been clearly defined. Replication in prokaryotes and eukaryotes occurs by very similar mechanisms, and thus most of the information presented in the discussion of bacterial replication applies to eukaryotic cells as well.

Replication Forks and Bidirectional Replication Replication begins at a specific site on the bacterial chromosome called the **origin**. The origin of replication on the *E. coli* chromosome is a sequence of approximately 245 base pairs called *oriC*, to which a number of proteins bind to **initiate** the process of replication. Thus, in many ways, a replication origin is analogous to a promoter for transcription in that both types of regulatory sequences act as binding sites for sequence-specific DNA-binding proteins

Chromosome — DNA strand

Chromosome contains
only thymidine

Replicates
in BrdU

Chromatid

Both chromatids contain one strand with BrdU and
one strand with thymidine

Continued replication
in BrdU-containing
medium

One chromatid of each chromosome
contains thymidine

(a)

(b)

Figure 13.4 **Experiment demonstrating that DNA replication in eukaryotic cells is semiconservative.** (*a*) Schematic diagram of the results of an experiment in which cells were transferred from a medium containing thymidine to one containing bromodeoxyuridine (BrdU) and allowed to complete two successive rounds of replication. DNA strands containing BrdU are shown in red. (*b*) The results of an experiment similar to that shown in *a*. In this experiment, cultured mammalian cells were grown in BrdU for *two* rounds of replication before mitotic chromosomes were prepared and stained by a procedure using fluores-

cent dyes and the Giemsa stain. Using this procedure, chromatids containing thymidine within one or both strands stain darkly, whereas chromatids containing only BrdU stain lightly. The photograph indicates that, after two rounds of replication in BrdU, one chromatid of each duplicated chromosome contains only BrdU, while the other chromatid contains a strand of thymidine-labeled DNA. (Some of the chromosomes are seen to have exchanged homologous portions between sister chromatids. This process of sister chromatid exchange is common during mitosis, but is not discussed in the text.) (B: COURTESY OF SHELDON WOLFF.)

that initiate DNA or RNA synthesis at a specific site along the template.

Once initiated, replication proceeds outward from the origin in both directions, that is, **bidirectionally** (Figure 13.5). The points at which the pair of replicated segments come together and join the nonreplicated segments are termed **replication forks**. Each replication fork corresponds to a site where (1) the parental double helix is undergoing strand separation, and (2) nucleotides are being incorporated into the newly synthesized complementary strands. The two replication forks move in opposite directions until they meet at a point across the circle from the origin, where replication is terminated (Figure 13.5). The two newly replicated duplexes detach from one another and are ultimately directed into two different cells.

Unwinding the Duplex and Separating the Strands Separation of the strands of a circular, helical DNA duplex (or a giant, linear DNA duplex within a eukaryotic chromosome) poses major topological problems. To visualize the difficulties, we can briefly consider an analogy between a DNA duplex and a two-stranded helical rope. Consider what would happen if you placed a linear piece of this rope on the ground, took hold of the two strands at one end, and began to pull the strands apart just as DNA is pulled apart during replication. It is apparent that separation of the strands of a double helix is also a process of *unwinding* the structure. In the case of a rope, which is free to rotate around its axis, separation of the strands at one end would be accompanied by rotation of the entire fiber as it resisted the development of tension. Now, consider what would

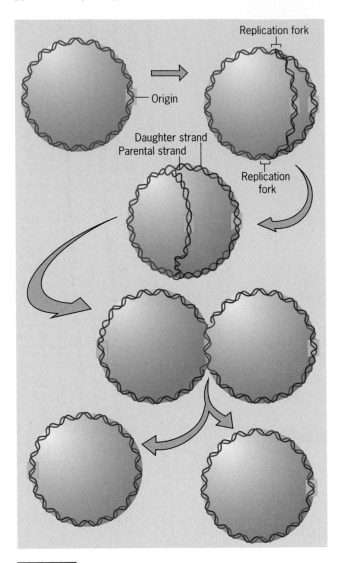

Figure 13.5 Model of a circular chromosome undergoing bidirectional, semiconservative replication. Two replication forks move in opposite directions from a single origin. When the replication forks meet at the opposite point on the circle, replication is terminated, and the two replicated duplexes detach from one another. New DNA strands are shown in red.

Figure 13.6 The effect of unwinding a two-stranded rope that has one end attached to a hook. The same problems arise when attempting to unwind a double-stranded, circular DNA molecule.

cular chromosome of *E. coli* contains approximately 400,000 turns and is replicated by two forks within 40 minutes, the magnitude of the problem becomes apparent.

It was noted on page 410 that cells contain enzymes, called topoisomerases, that can change the state of supercoiling in a DNA molecule. One enzyme of this type, called **DNA gyrase**, a type II topoisomerase, relieves the mechanical strain that builds up during replication. DNA gyrase molecules travel along the DNA ahead of the replication fork, removing positive supercoils. DNA gyrase accomplishes this remarkable feat by cleaving both strands of the DNA duplex, passing a segment of DNA through the double-stranded break to the other side, and then sealing the cuts, a process that is driven by the energy released during ATP hydrolysis (see Figure 10.14*b*). Eukaryotic cells possess similar enzymes that carry out this required function.

The Properties of DNA Polymerases We begin our discussion of the mechanism of DNA replication by describing some of the properties of **DNA polymerases**, the enzymes that synthesize new DNA strands. Study of these enzymes was begun in the 1950s by Arthur Kornberg at Washington University. In their initial experiments, Kornberg and his colleagues purified an enzyme from bacterial extracts that incorporated radioactively labeled DNA precursors into an acid-insoluble polymer identified as DNA. The enzyme was named *DNA polymerase* (and later, after the discovery of additional DNA-polymerizing enzymes, it was named *DNA polymerase I*). For the reaction to proceed, the enzyme required the presence of DNA and all four deoxyribonucleoside triphosphates (dTTP, dATP, dCTP, and dGTP). The newly synthesized, radioactively labeled DNA had the same base composition as the original unlabeled DNA, which indicated that the original DNA strands had served as **templates** for the polymerization reaction.

In principle, DNA polymerase catalyzes a reaction similar to that catalyzed by an RNA polymerase; both types of polymerase move along a template DNA strand and incorporate a complementary nucleotide onto the end of the strand being assembled. Thus the reaction in which a new DNA strand is synthesized is essentially the same as

happen if the other end of the rope were attached to a hook on a wall (Figure 13.6). Under these circumstances, separation of the two strands at the free end would generate increasing torsional stress in the rope and cause the unseparated portion to become more tightly wound. Separation of the two strands of a circular DNA molecule (or a linear molecule that is not free to rotate, as is the case in a large eukaryotic chromosome) is analogous to having one end of a linear molecule attached to a wall; in all of these cases, tension that develops in the molecule cannot be relieved by rotation of the entire molecule. Unlike a rope, which can become tightly overwound (as in Figure 13.6), an overwound DNA molecule becomes positively supercoiled (page 409). Consequently, movement of the replication fork generates positive supercoils in the unreplicated portion of the DNA ahead of the fork. When one considers that a complete cir-

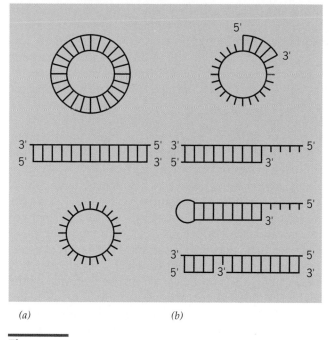

(a) *(b)*

Figure 13.7 Templates and nontemplates for DNA polymerase activity. (*a*) Examples of DNA structures that do not stimulate the synthesis of DNA in vitro by DNA polymerase isolated from *E. coli*. (*b*) Examples of DNA structures that stimulate the synthesis of DNA in vitro. In all cases, the molecules in *b* contain a template strand to copy and a primer strand with a 3′ OH on which to add nucleotides.

that shown in Figure 11.4 for RNA synthesis, except that the precursors for RNA polymerases are ribonucleoside triphosphates, whereas those for DNA polymerases are deoxyribonucleoside triphosphates.

As additional properties of the DNA polymerase were uncovered, it became apparent that replication was more complex than previously thought. When various types of template DNAs were tested, it was found that the template DNA had to meet certain structural requirements if it was to promote the incorporation of labeled precursors (Figure 13.7). An intact, double-stranded DNA molecule, for example, did not stimulate incorporation. This was not surprising considering the requirement that the strands of the helix must be separated for replication to occur. It was less obvious why a single-stranded, circular molecule was also devoid of activity; one might expect this structure to be an ideal template to direct the manufacture of a complementary strand. In contrast, addition of a partially double-stranded DNA molecule to the reaction mixture produced an immediate incorporation of nucleotides.

It was soon discovered that a single-stranded DNA circle cannot serve as a template for DNA polymerase because the enzyme cannot *initiate* the formation of a DNA strand. Rather, it can only add nucleotides to the 3′ hydroxyl terminus of an existing strand. The strand that provides the necessary 3′ OH terminus is called a **primer**. All DNA polymerases—both prokaryotic and eukaryotic—have these same two basic requirements (Figure 13.8*a*): a template

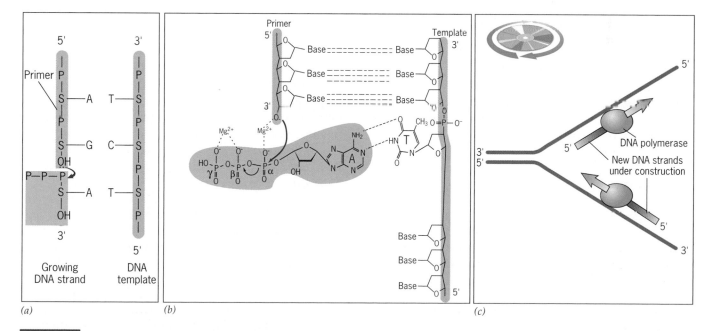

(a) *(b)* *(c)*

Figure 13.8 The incorporation of nucleotides onto the 3′ end of a growing strand by a DNA polymerase. (*a*) The polymerization of a nucleotide onto the 3′ end of the primer strand. The selection of the nucleotide is determined by the base of the nucleotide of the template strand. (*b*) A simplified model of the two-metal-ion mechanism for the reaction in which nucleotides are incorporated into a growing DNA strand by a DNA polymerase. In this model, one of the magnesium ions draws the

proton away from the 3′ hydroxyl group of the terminal nucleotide of the primer, facilitating the nucleophilic attack of the negatively charged 3′ oxygen atom on the α phosphate of the incoming nucleoside triphosphate. The second magnesium ion promotes the release of the pyrophosphate. The two metal ions are bound to the enzyme by highly conserved aspartic acid residues. (*c*) Schematic diagram showing the direction of movement of the polymerase along the two template strands.

DNA strand to copy and a primer strand to which nucleotides can be added. These requirements explain why certain DNA structures fail to promote DNA synthesis (Figure 13.7a). An intact, linear double helix provides the 3′ hydroxyl terminus but lacks a template. A circular single strand, on the other hand, provides a template but lacks a primer. The partially double-stranded molecule (Figure 13.7b) satisfies both requirements and thus promotes nucleotide incorporation. The finding that DNA polymerase cannot initiate the synthesis of a DNA strand raises a critical question: how is the synthesis of a new strand initiated in the cell? We will return to this question shortly.

The DNA polymerase purified by Kornberg had another property that was difficult to understand in terms of its presumed role as a replicating enzyme: it only synthesized DNA in a 5′-to-3′ (written 5′ → 3′) direction. The diagram first presented by Watson and Crick (see Figure 13.1) depicted events as they would be *expected* to occur at the replication fork. The diagram suggested that one of the newly synthesized strands is polymerized in a 5′ → 3′ direction, while the other strand is polymerized in a 3′ → 5′ direction. Is there some other enzyme responsible for the construction of the 3′ → 5′ strand? Does the enzyme work differently in the cell than under in vitro conditions? We will return to this question as well.

During the 1960s, there were hints that the "Kornberg enzyme" was not the only DNA polymerase in a bacterial cell. Then in 1969, a mutant strain of *E. coli* was isolated that had less than 1 percent of the normal activity of the enzyme, yet was able to multiply at the normal rate. Further studies revealed that the Kornberg enzyme, or **DNA polymerase I**, was only one of three distinct DNA polymerases present in bacterial cells; the others were named **DNA polymerase II** and **DNA polymerase III**. A typical bacterial cell contains 300 to 400 molecules of DNA polymerase I but only about 40 copies of DNA polymerase II and 10 copies of DNA polymerase III. The presence of DNA polymerases II and III had been masked by the much greater amounts of DNA polymerase I in the cell. But the discovery of other DNA polymerases did not answer the two basic questions posed above; none of the three enzymes can initiate DNA chains, nor can any of them construct strands in a 3′ → 5′ direction.

Semidiscontinuous Replication The lack of polymerization activity in the 3′ → 5′ direction has a straightforward explanation: DNA strands cannot be synthesized in that direction. Rather, both newly synthesized strands are assembled in a 5′ → 3′ direction. During the polymerization reaction, the OH group at the 3′ end of the primer carries out a nucleophilic attack on the 5′ α-phosphate of the incoming nucleoside triphosphate, as shown in Figure 13.8b. The polymerase molecules responsible for construction of the two new strands of DNA both move in a 3′-to-5′ direction *along the template*, and both construct a chain that grows from its 5′-P terminus (Figure 13.8c). Consequently, one of the newly synthesized strands grows

toward the replication fork where the parental DNA strands are being separated, while the other strand grows away from the fork.

Although this solves the problem concerning an enzyme that synthesizes a strand in only one direction, it creates an even more complicated dilemma. It is apparent that the strand that grows toward the fork in Figure 13.8c can be constructed by the continuous addition of nucleotides to its 3′ end. But how is the other strand synthesized? Evidence was soon gathered to indicate that the strand that grows away from the replication fork is synthesized *discontinuously*, that is, as fragments (Figure 13.9). Before the synthesis of a fragment can be initiated, a suitable stretch of template must be exposed by movement of the replication fork. Once initiated, each fragment grows away from the replication fork toward the 5′ end of a previously synthesized fragment to which it is subsequently linked. Thus, the two newly synthesized strands of the daughter duplexes are synthesized by very different processes. The strand that is synthesized continuously is called the **leading strand** because its synthesis continues as the replication fork advances. The strand that is synthesized discontinuously is called the **lagging strand** because initiation of each fragment must wait for the parental strands to separate and expose additional template (Figure 13.9). As discussed on page 561, both strands are probably synthesized simultaneously, so that the terms leading and lagging may not be as appropriate as was thought when they were first coined. Because one strand is synthesized continuously and the other discontinuously, replication is said to be *semidiscontinuous*.

The discovery that one strand was synthesized as small fragments was made by Reiji Okazaki of Nagoya University, Japan, following various types of labeling experiments. Okazaki found that if bacteria were incubated in [³H]thymidine for a few seconds and immediately killed, most of the radioactivity could be found as part of small DNA fragments 1,000 to 2,000 nucleotides in length. In contrast, if cells were incubated in the labeled DNA precursor for a minute or two, most of the incorporated radioactivity became part of much larger DNA molecules (Figure 13.10). These results indicated that a portion of the DNA was constructed in small segments (later called **Okazaki fragments**) that were rapidly linked to longer pieces that had been synthesized previously. The enzyme that joins the Okazaki fragments into a continuous strand is called **DNA ligase**.

The discovery that the lagging strand is synthesized in pieces raised a new set of perplexing questions about the initiation of DNA synthesis. How does the synthesis of each of these fragments begin when none of the DNA polymerases are capable of strand initiation? Further studies revealed that initiation is not accomplished by a DNA polymerase but, rather, by a distinct type of RNA polymerase, called **primase**, that constructs a short primer composed of RNA, not DNA. The leading strand, whose synthesis begins at the origin of replication, is also initi-

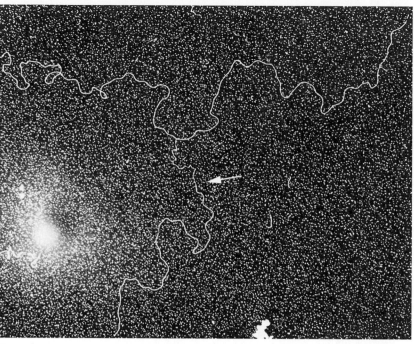

(a) *(b)*

Figure 13.9 The two strands of a double helix are synthesized by a different sequence of events. DNA polymerase molecules move along a template only in a 3′ → 5′ direction. As a result, the two newly assembled strands grow in opposite directions, one growing toward the replication fork and the other growing away from it. One strand is assembled in continuous fashion, the other as fragments that are joined together enzymatically. (*a*) Schematic diagram depicting the differences in synthesis of the two strands. (*b*) Electron micrograph of a repli-

cating bacteriophage DNA molecule. The left two limbs are the replicated duplexes, and the right end is the unreplicated duplex. The lagging strand of the newly replicated DNA is seen to contain an exposed, single-stranded (thinner) portion, which runs from the replication fork to the arrow. (B: From J. Wolfson and David Dressler, reprinted with permission from the Annual Review of Microbiology, volume 29, © 1975, by Annual Reviews, Inc.)

ated by a primase molecule. The short RNAs synthesized by the primase at the 5′ end of the leading strand and the 5′ end of each Okazaki fragment serve as the required primer for the synthesis of DNA by a DNA polymerase. The RNA primers are subsequently removed, and the resulting gaps in the strand are filled with DNA and then sealed by DNA ligase. These events are illustrated schematically in Figure 13.11. The formation of transient RNA primers during the process of DNA replication is a curious activity. It is thought that the likelihood of mistakes is greater during initiation than during elongation, and the use of a short removable segment of RNA avoids the inclusion of mismatched bases.

The Machinery Operating at the Replication Fork
Replication involves more than incorporating nucleotides. We saw on page 556 that bacterial cells possess an enzyme called DNA gyrase that converts positively supercoiled DNA that develops ahead of the replication fork into negatively supercoiled DNA. Negatively supercoiled DNA is under mechanical stress and is already primed to unwind and separate. Even with this motive force behind it, unwinding of the duplex and separation of the strands re-

quire the aid of two types of proteins that bind to the DNA, a **helicase** (or DNA unwinding enzyme) and **single-stranded DNA-binding (SSB) proteins**. Helicases unwind the DNA duplex in a reaction that uses energy released by ATP hydrolysis to break the hydrogen bonds that hold the two strands together, exposing the single-stranded DNA templates. *E. coli* has at least 12 different helicases for use in various aspects of DNA (and RNA) metabolism. One of these helicases—the product of the *dnaB* gene—serves as the major unwinding machine during replication. The DnaB helicase consists of six subunits arranged to form a ring-shaped protein that encircles a single DNA strand (Figure 13.12*a*). The DnaB helicase is first loaded onto the DNA at the origin of replication (with the help of the protein DnaC) and moves in a 5′ → 3′ direction along the lagging-strand template, unwinding the helix as it proceeds (Figure 13.12). A model of a similar shaped bacteriophage helicase engaged in strand separation during replication is depicted on page 551. DNA unwinding by the helicase is aided by the attachment of SSB proteins to the separated DNA strands (Figure 13.12). These proteins bind selectively to single-stranded DNA, keeping it in an extended state and

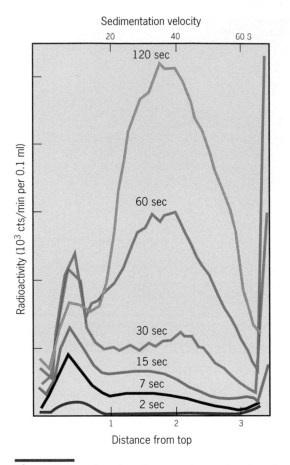

Figure 13.10 **Results of an experiment showing that part of the DNA is synthesized as small fragments.** Sucrose density gradient profiles of DNA from a culture of phage-infected *E. coli* cells. The cells were labeled for increasing amounts of time, and the sedimentation velocity of the labeled DNA was determined. When DNA was prepared after very short pulses, a significant percentage of the radioactivity appeared in very short pieces of DNA (represented by the peak near the top of the tube on the left). After longer labeling periods, the labeled DNA fragments had become joined to high-molecular-weight molecules. (FROM R. OKAZAKI ET AL., COLD SPRING HARBOR SYMP. QUANT. BIOL. 33:130, 1968.)

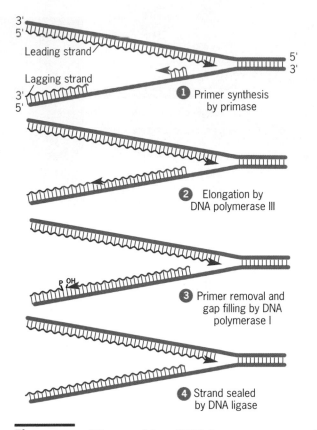

Figure 13.11 **The use of short RNA fragments as removable primers in initiating synthesis of each Okazaki fragment of the lagging strand.** The major steps are indicated in the drawing and discussed in the text. The role of various accessory proteins in these activities is indicated in the following figures.

preventing it from becoming rewound. A visual portrait of the combined action of a DNA helicase and SSB proteins on the structure of the DNA double helix is illustrated in the electron micrographs of Figure 13.12*b*.

Recall that an enzyme called a primase initiates the synthesis of each Okazaki fragment. In bacteria, the primase and the helicase associate transiently to form what is called a "primosome." Of the two members of the primosome, the helicase moves along the lagging-strand template processively (i.e., without being released from the template strand during the lifetime of the replication fork). As the helicase "motors" along the lagging-strand template, opening the strands of the duplex, the primase periodically binds to the helicase and synthesizes the short RNA primers that begin the formation of each Okazaki

fragment. As noted above, the RNA primers are subsequently extended as DNA by a DNA polymerase.

Given the fact that the initiation, elongation, and completion of Okazaki fragments occur very rapidly, the question arises as to whether the same DNA polymerase III molecule synthesizes successive fragments of the lagging strand or new polymerase molecules are recruited from the cytoplasm for each fragment. This question has been approached by in vitro experiments in which a reaction mixture carrying out DNA replication is suddenly diluted. If the polymerase remains attached to the DNA (either directly or indirectly through interactions with other proteins) as it synthesizes successive Okazaki fragments, then dilution of the medium should not affect its activity. In contrast, if new polymerases have to be recruited from the medium, dilution should greatly slow the synthesis of the lagging strand. Results from these experiments support the former proposal: a single polymerase III molecule is recycled from the site where it has just completed one Okazaki fragment to the next site along the lagging strand template. Once at the new site, the polymerase attaches to the 3′ OH of the RNA primer that has just been laid down by a primase and begins to incorporate deoxyribonucleotides onto the end of the short RNA.

(a)

(b)

Figure 13.12 **The role of the DNA helicase, single-stranded DNA-binding proteins, and primase at the replication fork.** (*a*) The helicase moves along the DNA, catalyzing the ATP-driven unwinding of the duplex. As the DNA is unwound, the strands are prevented from reforming the duplex by single-stranded DNA-binding proteins (SSBs). The primase associated with the helicase synthesizes the RNA primers that begin each Okazaki fragment. The RNA primers, which are about 10 nucleotides long, are subsequently removed. (*b*) A series of electron micrographs showing DNA molecules incubated with a viral DNA helicase (T antigen, page 566) and *E. coli* SSB proteins. The DNA molecules are progressively unwound from left to right. The helicase appears as the circular particle at the fork and the SSB proteins are bound to the single-stranded ends, giving them a thickened appearance. (B: FROM RAINER WESSEL, JOHANNES SCHWEIZER, AND HANS STAHL, J. VIROL. 66:807, 1992. COPYRIGHT © 1992, AMERICAN SOCIETY FOR MICROBIOLOGY.)

How does a polymerase III molecule move from one site on the lagging strand template to another site that is closer to the replication fork? The enzyme does this by "hitching a ride" with the DNA polymerase that is moving in that direction along the leading strand template. Thus even though the two polymerases are moving in opposite directions with respect to each of their templates, they are, in fact, part of a single protein complex (Figure 13.13). The two tethered polymerases can replicate both strands by looping the DNA of the lagging strand template back on itself, causing this template to have the same orientation as the leading strand template. Both polymerases then can move together as part of a single replicative complex, or *replisome*, without violating the "5′ → 3′ rule" for synthesis of a DNA strand (Figure 13.13). Once the polymerase assembling the lagging strand reaches the 5′ end of the Okazaki fragment synthesized during the previous round, the lagging-strand template is released and the polymerase begins work at the 3′ end of the next RNA primer toward the fork. The model depicted in Figure 13.13 is often referred to as the "trombone model" because the looping DNA repeatedly grows and shortens during the replication of the lagging strand, reminiscent of the movement of the brass "loop" of a trombone as it is played.

The Structure and Functions of DNA Polymerases

We are still left with the question as to why a bacterial cell needs more than one DNA polymerase. All of the DNA polymerases (in both prokaryotes and eukaryotes) have the same basic catalytic activity; they add deoxyribonucleotides

onto the growing end of a single-stranded primer (Figure 13.8). DNA polymerases differ, however, in their various roles within the cell. Based primarily on the analysis of mutants, the following assignments have been made. The enzyme that synthesizes DNA strands during replication in *E. coli* is DNA polymerase III, which is part of a large complex called the *DNA polymerase III holoenzyme* (Figure 13.14) The holoenzyme contains 10 different subunits having various functions in the replication process. One of the noncatalytic components of the holoenzyme, called the β clamp, keeps the polymerase associated with the DNA template. DNA polymerases (like RNA polymerases) possess two rather contrasting properties: (1) they must remain associated with the template over long stretches if they are to synthesize a continuous complementary strand, and (2) they must be attached loosely enough to the template to move from one nucleotide to the next. These contrasting properties are provided by the doughnut-shaped β clamp that encircles the DNA (Figure 13.15*a*).

The ring-shaped β "sliding clamp," allows the enzyme to move processively from one nucleotide to the next without diffusing away from the template. When the polymerase completes the synthesis of an Okazaki fragment, it disengages from the β clamp and is cycled to a new β clamp that has been assembled at an RNA primer-DNA template junction located closer to the replication fork (Figure 13.15*b*). The assembly of the β clamp around the DNA and its disassembly from the DNA require a multisubunit clamp loader that is also part of the DNA polymerase III holoenzyme (Figure 13.14).

The function of DNA polymerase II is as yet uncertain. Bacterial mutants lacking this polymerase have been

Figure 13.13 Replication of the leading and lagging strands in *E. coli* is accomplished by two DNA polymerases working together as part of a single complex. (*a*) The two DNA polymerase III molecules travel together, even though they are moving toward the opposite ends of their respective templates. This is accomplished by causing the lagging-strand template to form a loop. (*b*) The polymerase releases the lagging strand template when it encounters the previously synthesized Okazaki frag-

ment. (*c*) The polymerase that was involved in the assembly of the previous Okazaki fragment has now rebound the lagging-strand template farther along its length and is synthesizing DNA onto the end of RNA primer 3 that has just been constructed by the primase. (AFTER D. VOET AND J. G. VOET, BIO-CHEMISTRY, 2D ED., COPYRIGHT © 1995 JOHN WILEY AND SONS, INC. REPRINTED BY PERMISSION OF JOHN WILEY AND SONS, INC.)

isolated and have no evident deficiency. DNA polymerase I, which consists of only a single subunit, is involved primarily in DNA repair, a process by which damaged sections of DNA are corrected (page 572). DNA polymerase I also removes the RNA primers at the 5′ end of each Okazaki fragment and replaces them with DNA. The enzyme's ability to accomplish this feat is discussed in the following section.

Exonuclease Activities of DNA Polymerases Now that we have explained several of the puzzling properties of

DNA polymerase I that were originally discovered by Arthur Kornberg, such as the enzyme's inability to initiate strand synthesis, we can consider another curious observation. Kornberg found that DNA polymerase I preparations always contained exonuclease activities; that is, they were able to degrade DNA polymers by removing one or more nucleotides from the end of the molecule. At first, Kornberg assumed this activity was due to a contaminating enzyme because the action of exonucleases is so dramatically opposed to that of DNA synthesis. Nonetheless, the exonuclease activity could not be removed from the

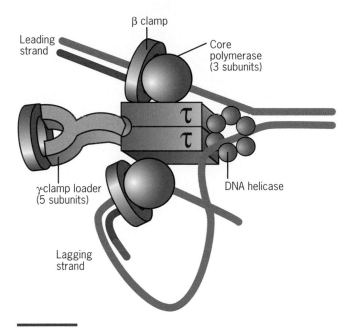

Figure 13.14 Schematic representation of DNA polymerase III holoenzyme. Included as part of the holoenzyme are (1) two core polymerases, each consisting of three subunits: an α subunit, which contains the polymerization active site; an ε subunit, which contains the 3′ → 5′ exonuclease, and a θ subunit, (2) two τ subunits, which hold the core polymerases in the complex, (3) one or more β clamps, each consisting of two identical subunits, and (4) a clamp loading (γ) complex of five different subunits that loads each sliding clamp onto the DNA. (BASED ON DRAWINGS BY M. O'DONNELL.)

polymerase preparation and was, in fact, a true activity of the polymerase molecule. It was subsequently shown that all of the bacterial polymerases possess exonuclease activity. Exonucleases can be divided into 5′ → 3′ and 3′ → 5′ exonucleases, depending on the direction in which the strand is degraded. DNA polymerase I has both 3′ → 5′ and 5′ → 3′ exonuclease activities, in addition to its polymerizing activity (Figure 13.16). These three activities are found in different domains of the single polypeptide. Thus, remarkably, DNA polymerase I is three different enzymes in one. The two exonuclease activities have entirely different roles in replication. We will consider the 5′ → 3′ exonuclease activity first.

Most nucleases are specific for either DNA or RNA, but the 5′ → 3′ exonuclease of DNA polymerase I can degrade either type of nucleic acid. Initiation of Okazaki fragments by the primase leaves a stretch of RNA at the 5′ end of each fragment, which is removed by the 5′ → 3′ exonuclease activity of DNA polymerase I (Figure 13.16*a*). As the enzyme removes ribonucleotides of the primer, its polymerase activity simultaneously fills the resulting gap with deoxyribonucleotides. The last deoxyribonucleotide incorporated is subsequently joined covalently to the 5′ end of the previously synthesized DNA fragment by DNA ligase. The role of the 3′ → 5′ exonuclease activity will be apparent in the following section.

(*a*)

(*b*)

Figure 13.15 The β sliding clamp of DNA polymerase III. (*a*) Space-filling model showing the two subunits that make up the doughnut-shaped β sliding clamp of the DNA polymerase III holoenzyme. The DNA is shown in blue within the β clamp. (*b*) Schematic diagram of polymerase cycling on the lagging strand. The polymerase is held to the DNA by the β sliding clamp as it moves along the template strand and synthesizes the complementary strand. Following completion of the Okazaki fragment, the enzyme disengages from its β clamp and

cycles to a recently assembled clamp "waiting" on an upstream RNA primer-DNA template junction. The assembly of a β clamp at the end of each RNA primer displaces the primase (not shown) that had synthesized the short RNA segment. The original β clamp is left behind for a period on the finished Okazaki fragment, but it is eventually disassembled and reutilized. (A: FROM JOHN KURIYAN, CELL, 69:427, 1992; © CELL PRESS. B: AFTER P. T. STUKENBERG, J. TURNER, AND M. O'DONNELL, CELL 78:878, 1994, BY PERMISSION OF CELL PRESS.)

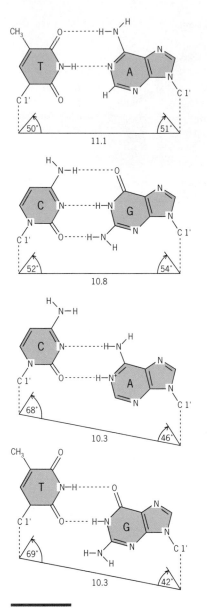

Figure 13.16 **The exonuclease activities of DNA polymerase I.** (*a*) The 5′ → 3′ exonuclease function removes several nucleotides at a time from the 5′ end of a single-strand nick. This activity plays a key role in removing the RNA primers. (*b*) The 3′ → 5′ exonuclease function removes mispaired nucleotides from the 3′ end of the growing DNA strand. This activity plays a key role in maintaining the accuracy of DNA synthesis. (FROM D. VOET AND J. G. VOET, BIOCHEMISTRY, 2D ED., COPYRIGHT © 1995 JOHN WILEY AND SONS, INC. REPRINTED BY PERMISSION OF JOHN WILEY AND SONS, INC.)

Figure 13.17 **Geometry of proper and mismatched base pairs.** (FROM H. ECHOLS AND M. F. GOODMAN, REPRODUCED WITH PERMISSION FROM THE ANNUAL REVIEW OF BIOCHEMISTRY, VOLUME 60, © 1991, BY ANNUAL REVIEWS INC.)

Ensuring High Fidelity During DNA Replication The survival of an organism depends on the accurate duplication of the genome. A mistake made in the synthesis of a messenger RNA molecule by an RNA polymerase results in the synthesis of defective proteins, but an mRNA molecule is only one short-lived template among a large population of such molecules; therefore, little lasting damage results from the mistake. In contrast, a mistake made during DNA replication results in a permanent mutation and the possible elimination of that cell's progeny. It is estimated that the chance that an incorrect nucleotide will be inserted during replication and will remain through the process in *E. coli* is about 10^{-9}. In other words, only 1 out of 1 billion nucleotides incorporated is not complementary to the corresponding nucleotide on the template strand. Because the genome of *E. coli* contains approximately 4×10^6 nucleotide pairs, this error rate corresponds to fewer than 1 nucleotide alteration for every 100 replication cycles. This represents the *spontaneous mutation rate* in this bacterium.

Incorporation of a particular nucleotide onto the end of a growing strand depends on the incoming nucleoside triphosphate being able to form an acceptable base pair with the nucleotide of the template strand (see Figure 13.8*b*). Analysis of the distances between atoms and bond angles indicates that A-T and G-C base pairs have nearly identical geometry. Any deviation from those pairings results in a different geometry, as shown in Figure 13.17. At each site along the template, DNA polymerase must discriminate among four different potential precursors as they move in and out of the active site. Among the four possible incoming nucleoside triphosphates, only one forms a proper geometric fit with the template, producing either an A-T or a G-C base pair that can fit into the active site of the enzyme. When an acceptable pairing oc-

curs, the enzyme incorporates the nucleotide into the growing strand.

On occasion, the polymerase incorporates an incorrect nucleotide, resulting in a mismatched base pair, that is, a base pair other than A-T or G-C. It is estimated that an incorrect pairing of this sort occurs once for every 10^5–10^6 nucleotides incorporated, a frequency that is 10^3–10^4 times greater than the spontaneous mutation rate of approximately 10^{-9}. How is the mutation rate kept so low? Part of the answer lies in the second of the two exonuclease activities mentioned above, the $3' \rightarrow 5'$ activity (Figure 13.16b), which is found in DNA polymerases I, II, and III. When an incorrect nucleotide is incorporated by the polymerase, the new end of the strand has an increased tendency to separate from the template and form a single-stranded $3'$ terminus, which enters the exonuclease site (Figure 13.18). When this occurs, the polymerase stalls at that site, providing time for the relatively slow-acting $3' \rightarrow 5'$ exonuclease to remove the mismatched nucleotide. This job of "proofreading" is one of the most remarkable of all enzymatic activities and illustrates the sophistication to which biological molecular machinery has evolved. The $3' \rightarrow 5'$ exonuclease activity removes about 99 out of every 100 mismatched bases, raising the fidelity to about 10^{-7}–10^{-8}. In addition, bacteria possess a mechanism called mismatch repair that operates after replication (page 574) and corrects nearly all of the mismatches that escape the proofreading step. Together these processes reduce the overall observed error rate to about 10^{-9}. Thus the fidelity of DNA replication can be traced to three distinct activities: (1) accurate selection of nucleotides, (2) immediate proofreading, and (3) postreplicative mismatch repair.

Another remarkable feature of bacterial replication is its rate. The replication of an entire bacterial chromosome in approximately 40 minutes at 37°C requires that each replication fork move about 1,000 nucleotides per second, which is equivalent to the length of an entire Okazaki fragment. Thus the entire process of Okazaki fragment synthesis, including formation of an RNA primer, DNA elongation and simultaneous proofreading by the DNA polymerase, excision of the RNA, its replacement with DNA, and strand ligation, occurs within about 1 second. Although it takes *E. coli* approximately 40 minutes to replicate its DNA, a new round of replication can begin before the previous round has been completed. Consequently, when these bacteria are growing at their maximal rate, they double their numbers in about 20 minutes.

Replication in Eukaryotic Cells

As might be expected, our understanding of replication in eukaryotic cells has lagged behind that in prokaryotic cells. This imbalance is rapidly changing with the development of experimental eukaryotic systems that parallel those used for decades to study prokaryotic replication.

(b)

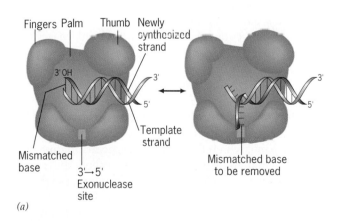

(a)

Figure 13.18 **Activation of the $3' \rightarrow 5'$ exonuclease of DNA polymerase I.** (a) A schematic model of a portion of DNA polymerase I known as the Klenow fragment, which contains the polymerase and $3' \rightarrow 5'$ exonuclease active sites. The $5' \rightarrow 3'$ exonuclease activity is located in a different portion of the polypeptide, which is not shown here. The regions of the Klenow fragment are often likened to the shape of a partially opened right hand, hence the portions labeled as "fingers," "palm," and "thumb." The catalytic site for polymerization is located in the central "palm" region. The $3'$ terminus of the grow-

ing strand can be shuttled between the polymerase and exonuclease active sites. Addition of a mismatched base to the end of the growing strand produces a frayed (single-stranded) $3'$ end that enters the exonuclease site, where it is removed. (b) A molecular model of the Klenow fragment complexed to DNA. The template DNA strand being copied is shown in blue, and the primer strand to which the next nucleotides would be added is shown in red. (A: AFTER T. A. BAKER AND S. P. BELL, CELL 92:296, 1998, AFTER A DRAWING BY C. M. JOYCE AND T. A. STEITZ, BY PERMISSION OF CELL PRESS; B: COURTESY OF THOMAS A. STEITZ.)

In figure (a): Fingers, Palm, Thumb, Newly synthesized strand, 3'OH, 3', 5', Template strand, Mismatched base, $3' \rightarrow 5'$ Exonuclease site, Mismatched base to be removed, 3', 5'

These include

■ The isolation of mutant yeast cells unable to produce specific gene products required for various aspects of replication. Because the replication proteins employed by yeast cells are remarkably similar in structure to those of higher eukaryotic cells, information obtained from studying yeast has proven to be largely applicable to mammalian cells.

■ The development of in vitro systems in which purified DNA is replicated using cellular extracts or purified components. Most in vitro studies of mammalian replication employ templates prepared from the simpler genomes of viruses that normally replicate in these cells. DNA from the monkey virus SV40 has proven particularly useful. SV40 DNA contains a single origin of replication, which is the site for binding of a viral encoded protein, called the *large T antigen*. The large T antigen serves two major functions: it binds at the SV40 origin to initiate replication, and it acts as the helicase that unwinds DNA at the replication fork. Other than the T antigen, the virus relies on the host cell's replication machinery to duplicate its genome, thus allowing researchers to study the roles of specific cellular replication proteins in the test tube.

Initiation of Replication in Eukaryotic Cells Replication in *E. coli* begins at only one site along the single, circular chromosome (Figure 13.5). Cells of higher organisms have much more DNA than bacteria, and their polymerases incorporate nucleotides into DNA at much slower rates. To accommodate these differences, eukaryotic cells replicate their genome in small portions, termed *replicons*. Replicons in eukaryotic chromosomes are generally between 15 and 100 μm in length (50 to 300 kilobases), and each has its own origin from which replication forks proceed outward in both directions. The existence of replicons was first demonstrated in autoradiographic experiments in which single DNA molecules were shown to be replicated simultaneously at several sites along their length (Figure 13.19).

The initiation of DNA synthesis in a given replicon is subject to regulation. Replicons located close together in a given chromosome tend to undergo replication simultaneously (as evident in Figure 13.19). Moreover, those replicons active at a particular time during one round of DNA synthesis (in the S phase of the cell cycle, Figure 14.1) tend to be active at a comparable stage in succeeding rounds. The timing of replication of a chromosomal region may be determined primarily by the activity of the genes in the region and/or its state of compaction: the less active, more tightly compacted the DNA, the later the stage at which it is replicated. The most highly compacted regions of the chromosome are packaged in the form of heterochromatin (page 505), and they are the last regions to be replicated. This difference in timing of replication is not due to DNA sequences because the inactive, hete-

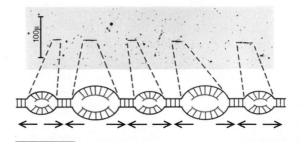

Figure 13.19 Results of an experiment demonstrating that replication in eukaryotic chromosomes begins at many sites along the DNA. Cells were incubated in [³H] thymidine for a brief period before preparation of DNA fibers for autoradiography. The lines of black silver grains indicate sites that were incorporating the radioactive DNA precursor during the labeling period. It is evident that synthesis is occurring at separated sites along the same DNA molecule. As indicated in the accompanying line drawing, initiation begins in the center of each site of thymidine incorporation, forming two replication forks that travel away from each other until they meet a neighboring fork. (MICROGRAPH COURTESY OF JOEL HUBERMAN.)

rochromatized X chromosome in the cells of female mammals (page 507) is replicated very late in S phase, whereas the active, euchromatic X chromosome is replicated at an earlier stage.

The mechanism by which replication is initiated in eukaryotes has been the subject of considerable research over the past decade. The greatest progress in this area has been made with yeast cells because the origins of replication can be removed from the yeast chromosome and inserted into other DNA molecules, conferring on them the ability to replicate either within a yeast cell or in cellular extracts containing the required replication proteins. Because these sequences promote replication of the DNA in which they are contained, they are referred to as *autonomous replicating sequences (ARSs)*.

The chromosomes of a yeast cell contain approximately 400 origins of replication (or ARSs) scattered throughout their length. Those ARSs that have been isolated and analyzed share several distinct elements. The core element of an ARS consists of a conserved sequence of 11 base pairs, which functions as a specific binding site for an essential multiprotein complex called the *origin recognition complex (ORC)*. If the ARS is mutated so that it is unable to bind the ORC, initiation of replication cannot occur. Studies indicate that an ORC remains bound to each origin throughout the cell cycle and that initiation of replication is triggered by the binding of other proteins to the ORC-origin complexes (discussed in the following section and illustrated in Figure 13.21).

Replication origins have proven more difficult to study in vertebrate cells than in yeast. Part of the problem stems from the fact that virtually any type of purified, naked DNA is suitable for replication using cellular extracts from frog eggs,. These studies suggested that, un-

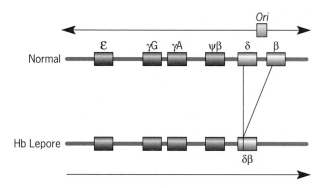

Figure 13.20 Location of an origin of replication within the human β-globin gene locus. Replication begins at the site indicated at the top and progresses outward in both directions. If this origin of replication happens to be deleted, as in the case of persons with a particular type of β-thalassemia who produce an abnormal hemoglobin (Hb Lepore), the β-globin genes are replicated by forks originating outside of the locus. (FROM J. F. X. DIFFLEY, CURR. OPIN. CELL BIOL. 6:370, 1994.)

like yeast, vertebrate DNA might not possess specific ARSs at which replication is initiated. However, studies of replication of intact mammalian chromosomes in vivo suggest that replication does begin at specific sites along the DNA, rather than by random selection as occurs in the frog egg extract. It would appear that a DNA molecule contains many sites where DNA replication *can* be initiated, but the effects of nucleosome positioning and higher order chromatin structure suppress initiation at most of these sites while promoting initiation at specific sites that serve as replication origins.

A site where bidirectional replication is initiated has been pinpointed within the human β-globin gene cluster. If this specific DNA segment is transplanted experimentally to a new chromosomal location in a monkey cell, it functions as an origin of replication in its new DNA environment. This experiment demonstrates that a mammalian origin of replication can depend on the presence of a specific DNA sequence. Conversely, if this specific DNA segment is deleted, as occurs in patients suffering from a specific hemoglobin deficiency known as the hemoglobin Lepore syndrome, a new origin of replication does not appear at the site. Instead, the DNA in this part of the chromosome is replicated by a replication fork that moves into the region from an unidentified origin residing outside of the β-globin gene cluster (Figure 13.20). Thus, the removal of an origin does not stimulate initiation at a new origin sequence. Instead, the cell takes advantage of the presence of a distant, existing origin.

Restricting Replication to Once Per Cell Cycle It is essential that each portion of the genome is replicated once, and only once, during each cell cycle. Consequently, some mechanism must exist to prevent the reinitiation of replication at a site that has already been duplicated. It is cur-

rently thought that the initiation of replication at a particular origin requires passage of the origin through several distinct states. The steps that are thought to occur at an origin of replication in a yeast cell are illustrated in Figure 13.21.

1. In step 1 (Figure 13.21), the origin of replication is bound by an ORC protein complex, which remains associated with the origin throughout the cell cycle.

2. Proteins referred to as "licensing factors" bind to the origins of replication (step 2, Figure 13.21) to assemble a protein-DNA complex, called the **prereplication complex**, that is "licensed" (competent) to initiate replication. Studies of the molecular nature of the licensing factors have focused on a set of six related Mcm proteins (Mcm2–Mcm7). The Mcm proteins are loaded onto the replication origin, with the aid of another protein called Cdc6, at a late stage of mitosis or soon after mitosis has been completed. The assembly of the prereplication complex can be followed by treating isolated chromatin with DNA-digesting enzymes. Prior to the assembly of prereplication complexes, only a small region of DNA at each origin of replication is protected from nuclease digestion by the bound ORC proteins. The protected region of DNA, which is called a *DNA footprint* (page 526), becomes greatly expanded following the association of Cdc6 and Mcm proteins with the previously bound ORC. Homologues of all of these proteins are found in frogs and mammals, suggesting that the basic mechanism of initiation of replication is conserved among eukaryotes.

3. Just before the beginning of the S phase of the cell cycle, an activation process begins (step 3, Figure 13.21) that leads to the initiation of replication. (The nature of this activation process is discussed in Section 14.1.) The Mcm proteins are thought to move with the replication fork (step 4) and are essential for completion of replication of a replicon. The fate of the Mcm proteins after replication depends upon the species studied. In yeast, the Mcm proteins are displaced from the chromatin and exported from the nucleus (step 5). In contrast, the Mcm proteins in mammalian cells are displaced from the DNA but remain in the nucleus. Studies indicate that the Mcm2–Mcm7 proteins are capable of associating into a ring-shaped complex that possesses helicase activity (as in step 4, Figure 13.21). Researchers studying DNA replication in eukaryotes have had great trouble identifying the major replicative helicase; that is, the helicase responsible for unwinding DNA at the replication fork. As suggested in Figure 13.21, the Mcm protein complex is a strong candidate for the long sought eukaryotic replicative helicase (analogous to DnaB in *E. coli*). Regardless of whether this is the case, Mcm proteins cannot reassociate with an origin of replication that has already "fired." Consequently, each origin can only be activated once per cell cycle.

Prereplication complex

The Eukaryotic Replication Fork Overall, the activities that occur at replication forks are quite similar, regardless of the type of genome being replicated—whether viral, prokaryotic, or eukaryotic. The various proteins in the replication "tool kit" of eukaryotic cells are listed in Table 13.1 and depicted in Figure 13.22. All replication systems

Figure 13.21 Steps leading to the replication of a yeast replicon. Yeast origins of replication contain a conserved sequence (ARS) that binds the multisubunit origin recognition complex (ORC) (step 1). The presence of the bound ORC is required for initiation of replication. The ORC is bound to the origin throughout the cell cycle. In step 2, licensing factors (identified as Mcm proteins) bind to the origin during or following mitosis, establishing a prereplication complex that is competent to initiate replication, given the proper stimulus. In step 3, the binding of an activation factor (identified as the protein kinase Cdc7/Dbf4) to the prereplication complex prior to S phase initiates replication. Step 4 shows a stage where replication has proceeded a short distance in both directions from the origin. In this model, the Mcm proteins form a replicative DNA helicase that unwinds DNA of the oppositely directed replication forks. The other proteins required for replication are not shown in this illustration but are indicated in the next figure. In step 5, the two strands of the original duplex have been replicated, an ORC is present at both origins, and the replication proteins, including the Mcm helicases, have been displaced from the DNA. In yeast, the Mcm proteins are exported from the nucleus and reinitiation of replication cannot occur until the cell has passed through mitosis.

require helicases, single-stranded DNA-binding proteins, topoisomerases, primase, DNA polymerase, and DNA ligase. As discussed in the previous section, the helicase that carries out DNA unwinding during replication has not been identified with certainty. In vitro studies utilize the SV40 large T antigen, which initiates replication and also possesses helicase activity (as in Figure 13.22). As in prokaryotes, the DNA of eukaryotic cells is synthesized in a semidiscontinuous manner, although the Okazaki fragments of the lagging strand are considerably smaller than in bacteria, averaging about 150 nucleotides in length. Like DNA polymerase III of *E. coli*, the eukaryotic DNA polymerase δ has been shown to exist as a dimer, suggesting that the leading and lagging strands are synthesized in a coordinate manner by a single replicative complex, or replisome.

To date, five "classic" DNA polymerases have been isolated from eukaryotic cells, and they are designated α, β, γ, δ, and ε. Two of these DNA polymerases, γ and β, are not involved in nuclear DNA replication. Polymerase γ replicates mitochondrial DNA, and polymerase β functions in DNA repair. The other three polymerases have replicative functions. Polymerase α is tightly associated with the primase, and together they initiate the synthesis of each Okazaki fragment. After the primase synthesizes a short RNA primer, a number of deoxyribonucleotides are added to it by polymerase α. Polymerase δ is thought to be the primary DNA-synthesizing enzyme during replication. Like the major replicating enzyme of *E. coli*, polymerase δ requires a "sliding clamp" that tethers the enzyme to the DNA, allowing it to move processively along a template. The sliding clamp of eukaryotic cells is very similar in structure and function to the β subunit of *E. coli*

Table 13.1	Some of the Proteins Required for Replication	

E. coli protein	Eukaryotic protein	Function
DnaA	ORC proteins	Recognition of origin of replication
Gyrase	Topoisomerase I/II	Relieves positive supercoils ahead of replication fork
DnaB	Mcm	DNA helicase that unwinds parental duplex
DnaC	?	Loads helicase onto DNA
SSB	RPA	Maintains DNA in single-stranded state
γ-complex	RFC	Subunits of the DNA polymerase holoenzyme that load the clamp onto the DNA
pol III core	pol δ/ε	Primary replicating enzyme; synthesizes entire leading strand and Okazaki fragments; has proofreading capability
β subunit	PCNA	Ring-shaped subunit of DNA polymerase holoenzyme that clamps replicating polymerase to DNA; works with pol III in *E. coli* and pol δ or ε in eukaryotes
Primase	Primase	Synthesizes RNA primers
—	pol α	Synthesizes short DNA oligonucleotides as part of RNA-DNA primer
DNA ligase	DNA ligase	Seals Okazaki fragments into continuous strand
pol I	FEN-1	Removes RNA primers; pol I of *E. coli* also fills gap with DNA
RNase H1	RNase H1	Secondary mechanism for RNA primer removal

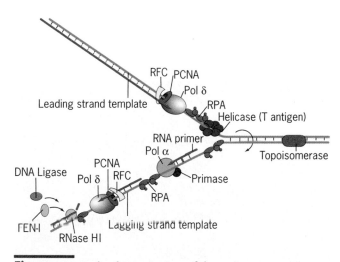

Figure 13.22 A schematic view of the major components at the eukaryotic replication fork. The viral T antigen is drawn as the replicative helicase in this figure because it is prominently employed in in vitro studies of DNA replication, and the actual helicase remains uncertain. DNA polymerase δ is thought to be the primary DNA synthesizing enzyme, but polymerase ε may also participate in this activity. PCNA acts as a sliding clamp for both polymerases δ and ε. The sliding clamp is loaded onto the DNA by a protein called RFC (replication factor C), which is similar in structure and function to the γ-clamp loader of *E.coli*. RPA is a trimeric single-stranded DNA-binding protein comparable in function to that of SSB utilized in *E. coli* replication. The RNA-DNA primers synthesized by the polymerase α-primase complex are removed by the combined action of two nucleases, FEN-1 and RNase H1. The gap is filled by polymerase δ and sealed by a DNA ligase. As in *E. coli*, a topoisomerase is required to remove the positive supercoils that develop ahead of the replication fork. (FROM G. S. BRUSH AND T. J. KELLY, IN M. L. DePAMPHILIS, ED., DNA REPLICATION IN EUKARYOTES, P. 28, COLD SPRING HARBOR LABORATORY PRESS, 1996.)

polymerase III illustrated in Figure 13.14. In eukaryotes, the sliding clamp is called PCNA.[1] The clamp loader that loads PCNA onto the DNA is called RFC and is analogous to the *E. coli* polymerase III clamp loader complex. After synthesizing an RNA primer and a short DNA segment, the polymerase α-primase complex is replaced at the template-primer junction by the PCNA-polymerase δ complex, which completes the synthesis of the Okazaki fragment. When the newly synthesized strand reaches the 5′ end of the previously synthesized Okazaki fragment, the RNA primer of that next fragment is removed and the gap is filled by the continued movement of polymerase δ. Recall that the RNA primer in bacterial cells is removed by the 5′ → 3′ exonuclease activity of DNA polymerase I (page 563). Eukaryotic DNA polymerases do not have this type of exonuclease activity, and the primers are removed by a pair of nucleases, RNase H1 and FEN-1 (Table 13.1). Adjoining fragments are ultimately sealed by a DNA ligase. Determining the role of polymerase ε has proven problematic; it appears to be involved in nuclear DNA replication because that process cannot be completed in cells that lack this polymerase. On the other hand, polymerase ε is not required for in vitro replication of SV40 DNA. Several other DNA polymerases (η, σ, and ι) have a specialized function that allows cells to replicate damaged DNA as described on page 575.

Like prokaryotic polymerases, all of the eukaryotic polymerases elongate DNA strands in the 5′ → 3′ direction by the addition of nucleotides to a 3′ hydroxyl group, and none of them is able to initiate the synthesis of a DNA chain without a primer. Polymerases γ, δ, and ε possess a

[1]PCNA is an acronym for *proliferating cell nuclear antigen*. This protein was first discovered as an antigen that reacted to autoantibodies in the serum of patients with lupus erythematosus, and it was subsequently localized to the nuclei of rapidly proliferating cells. Only later was its role in replication discovered.

$3' \rightarrow 5'$ exonuclease, whose proofreading activity ensures that replication occurs with very high accuracy.

Replication and Nuclear Structure Up to this point in the chapter, the illustrations of replication depict a replicating polymerase moving like a locomotive along a stationary DNA track. But the replication apparatus consists of a huge complex of proteins that operates within the confines of a structured nucleus. Increasing evidence suggests that the replication machinery is present in tight association with the nuclear matrix (page 515). When cells are given very short pulses of radioactive DNA precursors, over 80 percent of the incorporated label is associated with the nuclear matrix. If, instead of fixing the cells immediately after a pulse, the cells are allowed to incorporate *unlabeled* DNA precursors for an hour or so before fixation, most of the radioactivity is chased from the matrix into the surrounding DNA loops. This latter finding suggests that rather than remaining stationary, replicating DNA moves like a conveyer belt through an immobilized replication apparatus (Figure 13.23).

Further studies suggest that replication forks that are active at a given time are not distributed randomly throughout the cell nucleus, but instead are localized within 50 to 250 sites, called **replication foci** (Figure 13.24) It is estimated that each of the bright red dots indi-

cated in Figure 13.24 contains approximately 40 replication forks incorporating nucleotides into DNA strands simultaneously. The clustering of replication forks may provide a mechanism for coordinating the replication of adjacent replicons on individual chromosomes (as in Figure 13.19). When cells are treated to remove most of the chromatin, the replication foci remain attached to the nuclear matrix.

Chromatin Structure and Replication The chromosomes of eukaryotic cells consist of DNA tightly complexed to regular arrays of histone proteins. Examination of a replicating DNA molecule in the electron microscope reveals nucleosomes on both daughter duplexes very near the replication fork (Figure 13.25a), indicating that the assembly of DNA into nucleosomes is a very rapid event. Recall from page 502 that the core histone octamer of a nucleosome consists of an (H3H4)$_2$ tetramer together with a pair of H2A/H2B dimers. In vitro studies suggest that the (H3H4)$_2$ tetramers present prior to replication remain intact and are distributed randomly between the two daughter duplexes. As a result, old and new (H3H4)$_2$ tetramers are intermixed on each daughter DNA molecule as indicated in Figure 13.25b. In contrast, the two H2A/H2B dimers of each parental nucleosome do not remain together as the replication fork moves through the

Figure 13.23 **The involvement of the nuclear matrix in DNA replication.** The origins of replication are indicated by the dots, and the arrows indicate the direction of elongation of growing strands. According to this schematic model, it isn't the replication machinery that moves along stationary DNA tracks, but the DNA that is spooled through the replication apparatus, which is firmly attached to the nuclear matrix.

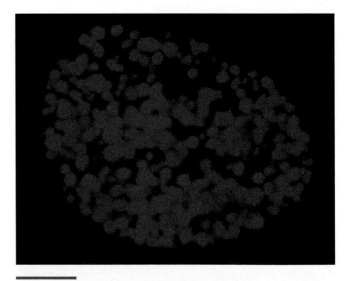

Figure 13.24 **Demonstration that replication activities do not occur randomly throughout the nucleus but are confined to distinct sites.** Prior to the onset of DNA synthesis at the start of S phase, various factors required for the initiation of replication are assembled at discrete sites within the nucleus, forming prereplication centers. These sites appear as discrete red objects in the micrograph, which has been stained with a fluorescent antibody against a protein called replication factor A (RPA), which is a single-stranded DNA-binding protein required for the initiation of replication. Other replication factors, such as PCNA and the polymerase-primase complex, are also localized to these foci. (FROM YASUHISA ADACHI AND ULRICH K. LAEMMLI, EMBO J. VOL. 13, COVER NO. 17, 1994.)

(a)

(b)

Figure 13.25 **The distribution of histone core complexes to daughter strands following replication.** (*a*) Electron micrograph of chromatin isolated from the nucleus of a rapidly cleaving *Drosophila* embryo showing a pair of replication forks (arrows) moving away from each other in opposite directions. Between the two forks one sees regions of newly replicated DNA that are already covered by nucleosomal core particles to the same approximate density as the parental strands that have not yet undergone replication. (*b*) Schematic diagram of the arrangement of nucleosomes after DNA replication. Each nucleosome core particle is shown schematically to be composed of a central (H3H4)$_2$ tetramer flanked by two H2A/H2B dimers. Histones that were present in parental nucleosomes prior to replication are indicated in blue; newly synthesized histones are indicated in red. The (H3H4)$_2$ tetramers remain intact and are distributed randomly to both daughter duplexes. The pairs of H2A/H2B dimers present in parental nucleosomes are thought to separate and recombine randomly with the (H3H4)$_2$ tetramers on the daughter duplexes. (A: Courtesy of Steven L. McKnight and Oscar L. Miller, Jr.)

chromatin. Instead, the dimers of a nucleosome separate from one another and appear to bind randomly to the new and old (H3H4)$_2$ tetramers already present on the daughter duplexes (Figure 13.25*b*).

REVIEW **?**

1. The original Watson-Crick proposal for DNA replication envisioned the continuous synthesis of DNA strands. How and why has this concept been modified over the intervening years?

2. What does it mean that replication is semiconservative? How was this feature of replication demonstrated in bacterial cells? In eukaryotic cells?

3. Why are there no heavy bands in the top three centrifuge tubes of Figure 13.3*a*?

4. How is it possible to obtain mutants whose defects lie in genes that are required for an essential activity such as DNA replication?

5. Describe the events that occur at an origin of replication during the initiation of replication in yeast cells. What is meant by replication being bidirectional?

6. Why do the DNA molecules depicted in Figure 13.7*a* fail to stimulate the polymerization of nucleotides by DNA polymerase I? What are the properties of a DNA molecule that allow it to serve as a template for nucleotide incorporation by DNA polymerase I?

7. Describe the mechanism of action of DNA polymerases operating on the two template strands and the effect this has on the synthesis of the lagging versus the leading strand.

8. Contrast the role of DNA polymerases I and III in bacterial replication.

9. Describe the role of the DNA helicase, the SSBs, the β clamp, the DNA gyrase, and the DNA ligase during replication in bacteria.

10. What is the advantage of having the DNA of the lagging strand template looped back on itself as in Figure 13.13*a*?

11. How do the two exonuclease activities of DNA polymerase I differ from one another? What are their respective roles in the overall process of replication?

12. Describe the factors that contribute to the high fidelity of DNA replication.

13. What is the major difference between bacterial and eukaryotic replication that allows a eukaryotic cell to replicate its DNA in a reasonable amount of time?

13.2 DNA REPAIR

Life on Earth is subject to a variety of destructive forces that originate in both the internal and external environments of an organism. Of all the molecules in a cell, DNA is placed in the most precarious position. On one hand, it is essential that the genetic information remain essentially unchanged as it is passed from cell to cell and individual to individual. On the other hand, DNA is one of the molecules in a cell that is most susceptible to environmental damage. When struck by ionizing radiation, the backbone of a DNA molecule is often broken; when subjected to a variety of common chemicals, some of which are produced by a cell's own metabolism, the bases of a DNA molecule may be altered structurally; when exposed to ultraviolet radiation, adjacent pyrimidines on a DNA strand have a tendency to interact with one another to form a covalent complex, that is, a dimer (Figure 13.26). Even the absorption of thermal energy by the DNA of a warm-blooded bird or mammal is sufficient to split adenine and guanine bases from their attachment to the sugars of the backbone. The magnitude of these spontaneous alterations, or *lesions*, can be appreciated from the estimate that each cell of a warm-blooded mammal loses approximately 10,000 bases per day! Failure to repair such lesions produces permanent alterations, or mutations, in the DNA. If the mutation occurs in a cell destined to become a gamete, the genetic alteration may be passed on to the next generation. Mutations also have effects in somatic cells (i.e., cells that are not in the germline): they can interfere with transcription and replication, lead to the malignant transformation of a cell, or speed the process by which an organism ages.

Considering the potentially drastic consequences of alterations in DNA molecules and the high frequency at which they occur, it is essential that cells possess mechanisms for repairing DNA damage. In fact, cells have a bewildering variety of repair systems that correct virtually any type of damage to which a DNA molecule is vulnerable. It is estimated that less than one base change in a thousand escapes a cell's repair systems. The existence of these systems provides an excellent example of the molecular mechanisms that maintain cellular homeostasis.

Both prokaryotic and eukaryotic cells possess a variety of proteins that patrol the DNA, searching for alterations and distortions. In some cases, damage can be repaired directly; there is an enzyme, for example, that uses the energy of sunlight to break the bond holding two pyrimidines together in a pyrimidine dimer, thereby returning the molecule to its original condition. Most repair systems, however, require that a damaged section of the DNA be *excised*, that is, selectively removed. One of the great virtues of the DNA duplex is that each strand contains the information required for constructing its partner. Consequently, if one or more nucleotides is removed from one strand, the complementary strand can serve as a template for reconstruction of the duplex. Because DNA repair usually involves DNA synthesis, its occurrence can be

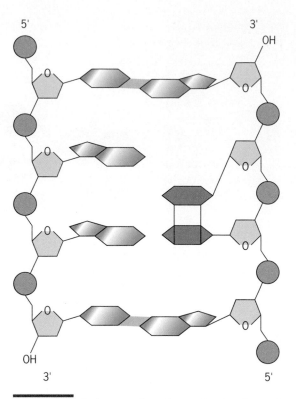

Figure 13.26 A pyrimidine dimer that has formed within a DNA duplex following UV irradiation.

revealed by the incorporation of [^3H]thymidine at times when the cell is not engaged in DNA replication (see Figure 3 in the Chapter 13 Experimental Pathways, which can be accessed on the Web at www.wiley.com/college/karp). As will be apparent in the following discussion, DNA replication and DNA repair have many features in common and, not surprisingly, share many of the same "parts and services."

Nucleotide Excision Repair

Nucleotide excision repair (NER) operates by a cut-and-patch mechanism that removes a variety of bulky lesions, including pyrimidine dimers and nucleotides to which various chemical groups have become attached. Two distinct NER pathways can be distinguished:

1. A *transcription-coupled pathway* in which the template strands of genes that are being actively transcribed are preferentially repaired. Repair of a template strand is thought to occur as the DNA is being transcribed, and the presence of the lesion may be signalled by a stalled RNA polymerase. This preferential repair pathway ensures that those genes of greatest importance to the cell, which are the genes the cell is actively transcribing, receive the highest priority on the "repair list."

2. A slower, less efficient *global pathway* that corrects DNA strands in the remainder of the genome.

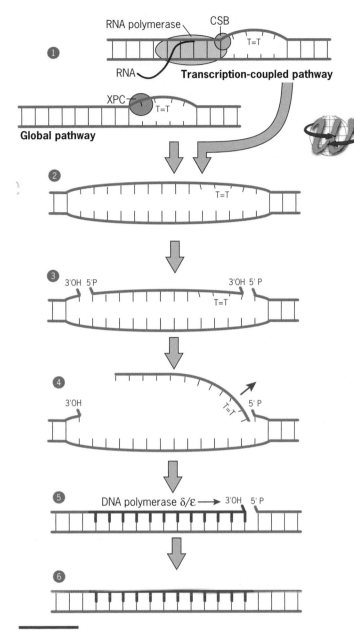

Figure 13.27 Nucleotide excision repair. The following steps are depicted in the drawing and discussed in the text: (1) damage recognition in the global pathway is mediated by an XPC protein, whereas damage recognition in the transcription-coupled pathway is thought to be mediated by a stalled RNA polymerase in conjunction with a CSB protein; (2) DNA strand separation (by XPB and XPD proteins, two helicase subunits of TFIIH); (3) incision (by XPG on the 3′ side and the XPF-ERCC1 complex on the 5′ side); (4) excision, (5) DNA repair synthesis (by DNA polymerase δ and/or ε); and (6) ligation (by DNA ligase I).

Although the recognition of the lesion is probably accomplished by different proteins in the two NER pathways (step 1, Figure 13.27), the steps that occur during repair of the lesion are thought to be very similar, as indicated in

steps 2–6 of Figure 13.27. One of the key components of the NER repair machinery is TFIIH, a huge protein that also participates in the initiation of transcription. The discovery of the involvement of TFIIH established a crucial link between transcription and DNA repair, two processes that were previously assumed to be independent of one another (discussed in the Experimental Pathways, which can be accessed on the Web at www.wiley.com/college/karp). Included among the various subunits of TFIIH are two subunits (XPB and XPD) that possess helicase activity; these enzymes separate the two strands of the duplex (step 2, Figure 13.27) in preparation for removal of the lesion. The damaged strand is then cut on both sides of the lesion by a pair of endonucleases (step 3), and the segment of DNA between the incisions is released (step 4). Once excised, the gap is filled by a DNA polymerase (step 5), and the strand is sealed by DNA ligase (step 6).

Base Excision Repair

A separate excision repair system operates to remove altered nucleotides generated by reactive chemicals present in the diet or produced by metabolism. The steps in this repair pathway in eukaryotes, which is called **base excision repair (BER)**, are shown in Figure 13.28. BER is initiated by a *DNA glycosylase* that recognizes the alteration (step 1, Figure 13.28) and removes the base by cleavage of the glycosidic bond holding the base to the deoxyribose sugar (step 2). A number of different DNA glycosylases have been identified, each more-or-less specific for a particular type of altered base, including uracil (formed by the hydrolytic removal of the amino group of cytosine), 8-hydroxyguanine (caused by damage from oxygen free radicals, page 35), and 3-methyladenine (produced by transfer of a methyl group from a methyl donor, page 451). Once the altered purine or pyrimidine is removed, the "beheaded" deoxyribose phosphate remaining in the site is removed by the combined action of an endonuclease and a phosphodiesterase (steps 3 and 4). The gap is filled by DNA polymerase β, and the strand is sealed by DNA ligase III (steps 5 and 6).

The fact that cytosine can be converted to uracil may explain why natural selection favored the use of thymine, rather than uracil, as a base in DNA, even though uracil was presumably present in RNA when it served as genetic material during the early evolution of life. If uracil had been retained as a DNA base, it would have caused difficulty for repair systems in distinguishing between a uracil that "belonged" at a particular site and one that resulted from an alteration of cytosine. Interestingly, the enzyme that removes uracil from DNA (uracil-DNA glycosylase) is one of the most conserved proteins between *E. coli* and humans, sharing 56 percent identity in amino acid sequence. Structural studies have shown that the enzyme binds to the DNA and causes the uracil to flip out of the helix and into the active site of the enzyme, where it is removed.

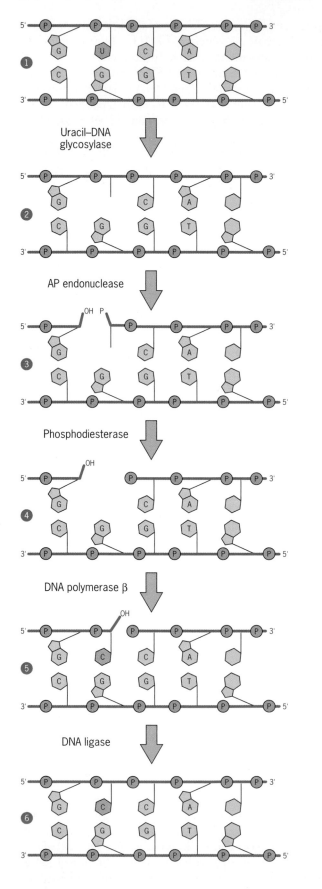

Figure 13.28 **Base excision repair.** The steps are described in the text. Other pathways for BER are also known.

Mismatch Repair

It was noted in an earlier section (page 565) that cells can remove mismatched bases that are incorporated by the DNA polymerase and escape the enzyme's proofreading exonuclease. This process is called mismatch repair. A mismatched base pair causes a distortion in the geometry of the double helix that can be recognized by a repair enzyme. But how does the enzyme "recognize" which member of the mismatched pair is the incorrect nucleotide? If it were to remove one of the nucleotides at random, it would make the wrong choice 50 percent of the time, creating a permanent mutation at that site. Thus, for a mismatch to be repaired after the DNA polymerase has moved past a site, it is important that the repair system distinguish the newly synthesized strand, which contains the incorrect nucleotide, from the parental strand, which contains the correct nucleotide. In *E. coli*, the two strands are distinguished by the presence or absence of methyl groups. The parental strand has methyl groups linked to certain adenosine residues, whereas the newly synthesized strand remains unmethylated for a period following replication. The repair system patrols the DNA prior to this methylation step; when a mismatch is recognized, enzymes always remove and replace nucleotides from the nonmethylated strand, which guarantees that it restores the original base pairs. DNA methylation does not appear to be utilized by the mismatch repair system in eukaryotes, and the mechanism of identification of the newly synthesized strand remains unclear.

Double-Strand Breakage Repair

X rays, gamma rays, and particles released by radioactive atoms are all described as *ionizing radiation* because they generate ions as they pass through matter. When these forms of radiation collide with a fragile DNA molecule, they often break both strands of the double helix. Double-strand breaks (DSBs) can also be caused by certain chemicals, including several (e.g., bleomycin) used in cancer chemotherapy, and free radicals produced by normal cellular metabolism (page 35). A single double-strand break can cause serious chromosome abnormalities and, ultimately, prove lethal for a cell. DSBs can be repaired by several alternate pathways. In mammalian cells, the simplest and most commonly employed pathway is called *nonhomologous end joining (NHEJ)*, in which a complex of proteins binds to the broken ends of the DNA duplex and catalyzes a series of reactions that rejoin the broken strands. The steps that occur during NHEJ are shown in Figure 13.29 and described in the accompanying legend. Cells that lack one of the proteins required for NHEJ are very sensitive to ionizing radiation. Another DSB repair pathway includes genetic recombination and is considerably more complex.

The importance of DNA repair can be appreciated by examining the effects on humans that result from DNA repair deficiencies, a subject discussed in the accompanying Human Perspective. The study of these human diseases has led to major advances in our understanding of

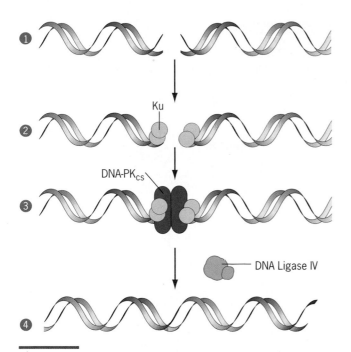

Figure 13.29 Repairing double-strand breaks (DSBs) by nonhomologous end joining. In this simplified model of double-strand break repair, the lesion (step 1) is detected by a heterodimeric protein called Ku, that binds to the broken ends of the DNA (step 2). The DNA-bound Ku recruits another protein, called DNA-PKcs, which is the catalytic subunit of a DNA-dependent protein kinase (step 3). The substrates phosphorylated by this protein kinase are unknown. The presence of these 3 subunits (two of Ku and the DNA-PKcs) brings the ends together in such a way that they can be joined by DNA ligase IV to regenerate an intact DNA duplex (step 4).

DNA repair mechanisms. These studies, which are described in the Experimental Pathway (www.wiley.com/college/karp on the Web), represent one of the most exciting areas of current study in molecular biology.

REVIEW

1. Contrast the events of nucleotide excision repair and basc excision repair.
2. Why is it important in mismatch repair that the cell distinguish the parental strands from the newly synthesized strands? How is this recognition accomplished?

13.3 BETWEEN REPLICATION AND REPAIR

The human perspective describes an inherited disease—xeroderma pigmentosum (XP)—that leaves patients with an inability to repair certain lesions caused by exposure to ultraviolet radiation. Patients described as having the "clas-

sical" form of XP have a defect in one of seven different genes involved in **nucleotide excision repair (NER)**. These genes are designated *XPA, XPB, XPC, XPD, XPE, XPF*, and *XPG*, and some of their roles in NER are indicated in the legend of Figure 13.27. Another group of patients were identified that, like those with XP, were highly susceptible to developing skin cancer as the result of sun exposure. However, unlike the cells from XP sufferers, cells from these patients were capable of nucleotide excision repair and were only slightly more sensitive to UV light than normal cells. This heightened UV sensitivity revealed itself during replication, as these cells often produced fragmented daughter strands following UV irradiation. Patients in this group were classified as having a variant form of XP, which was designated XP-V. We will return to the basis of the XP-V defect in a moment.

We have seen in the previous section that cells can repair a great variety of DNA lesions. On occasion, however, a DNA lesion is not repaired by the time that segment of DNA is scheduled to undergo replication. On these rare occasions, the replication machinery arrives at the site of damage on the template strand and becomes stalled there. When this happens, some type of signal is emitted that leads to the recruitment of a specialized polymerase that is able to bypass the lesion. Suppose the lesion in question is a thymidine dimer (Figure 13.26) in a skin cell that was caused by exposure to ultraviolet radiation. When the replicative polymerase (pol δ or ϵ) reaches the obstacle, the enzyme is temporarily replaced by a DNA polymerase designated pol η, which is able to insert two A residues into the newly synthesized strand across from the two T residues that are covalently linked as part of the dimer. Once this "damage bypass" is accomplished, the cell switches back to the normal replicative polymerase and DNA synthesis continues without leaving any trace that a serious problem had been resolved. As you might have guessed from the juxtaposition of topics, patients afflicted with XP-V have a mutation in the gene encoding pol η and thus have difficulty replicating past thymidine dimers.

Polymerase η is a member of a superfamily of recently discovered DNA polymerases whose function in replication is to bypass various types of DNA lesions or misalignments between the template and primer strands. Unlike the replicative polymerases whose active sites have very strict geometric requirements (page 564), the bypass polymerases must allow considerably more flexibility in choice of an incoming nucleotide. The polymerases of this superfamily are said to engage in *translesion synthesis (TLS)*. The TLS polymerases have a different basic structure than the classic DNA polymerases described earlier in the chapter. In addition, they are only capable of incorporating one to a few nucleotides into a DNA strand (they lack processivity); they have no proofreading capability; and they are much more likely to incorporate an incorrect (i.e., noncomplementary) nucleotide than the classic polymerases. Analysis of DNA sequence data suggests that the human genome has at least 30 TLS polymerases, several of which have received preliminary study.

THE HUMAN PERSPECTIVE

The Consequences of DNA Repair Deficiencies

We owe our lives to the light from the sun, which provides the energy captured during photosynthesis. But the sun also emits a constant stream of ultraviolet rays that ages and mutates the cells of our skin. The hazardous effects of the sun are most dramatically illustrated by the rare recessive genetic disorder, *xeroderma pigmentosum (XP)*. Patients with XP possess a deficient repair system that cannot remove segments of DNA damaged by ultraviolet radiation. As a result, persons with XP are extremely sensitive to sunlight; even very limited exposure to the direct rays of the sun can produce large numbers of dark-pigmented spots on exposed areas of the body (Figure 1) and a greatly elevated risk of developing disfiguring and fatal skin cancers.

XP is not the only genetic disorder characterized by nucleotide excision repair deficiency. Cockayne syndrome (CS) is an inherited disorder characterized by acute sensitivity to light, neurological dysfunction due to demyelination of neurons, and dwarfism, but no evident increase in the frequency of skin cancer. Cells from persons with CS are deficient in the preferential pathway by which transcriptionally active DNA is repaired (page 572). The remainder of the genome is repaired at the normal rate, presumably accounting for the normal levels of skin cancer. But why are persons with a defective repair mechanism subject to such widespread abnormalities? Most cases of CS can be traced to a mutation in one of two genes, either *CSA* or *CSB*, which are thought to be involved in coupling transcription to DNA repair. Mutations in these genes, in addition to disturbing DNA repair, may also affect the transcription of certain genes, perhaps leading to growth retardation and abnormal development of the nervous system that characterize this disease. This latter possibility is strengthened by the finding that, in rare cases, the symptoms of CS can also occur in persons with XP caused by mutations in the *XPB* or *XPD* genes. As noted on page 573, the *XPB* and *XPD* genes encode subunits of the transcription factor TFIIH, which has a role in both transcription and DNA repair. Persons who suffer from the combined CS/XP syndromes are thought to produce a TFIIH whose defects render it unable to carry out either transcription or DNA repair in a normal fashion.

Persons with xeroderma pigmentosum are not the only individuals who should worry about exposure to the sun. Even in a skin cell whose repair enzymes are functioning at optimal levels, a certain fraction of the lesions fail to be excised and replaced. Alterations in DNA lead to mutations that can cause a cell to become malignant (Chapter 16). Thus, one of the consequences of the failure to correct UV-induced damage is the risk of skin cancer. Consider the following statistics: over 600,000 persons develop one of three forms of skin cancer every year in the United States, and most of these cases are attributed to overexposure to the sun's ultraviolet rays. Fortunately, the two most common forms of skin cancer—basal cell carcinoma and squamous cell carcinoma—rarely spread to other parts of the body and can usually be ex-

Figure 1 Darkly pigmented regions of the skin are evident in this boy with xeroderma pigmentosum. The area of skin below the chin, which is protected from the sun, is relatively devoid of the lesions. (KEN GREER/VISUALS UNLIMITED.)

cised in a doctor's office. Both of these types of cancer originate from the skin's epithelial cells.

However, malignant melanoma, the third type of skin cancer, is a potential killer. Unlike the others, melanomas develop from pigment cells in the skin. The number of cases of melanoma diagnosed in the United States is climbing at the alarming rate of 4 percent per year due to the increasing amount of time people have spent in the sun over the past few decades. Studies suggest that one of the greatest risk factors to developing melanoma as an adult is the occurrence of a severe, blistering sunburn as a child or adolescent. It is obviously important, therefore, to prevent such burns in children.

XP is a very rare condition, but colon cancer is relatively common. It is estimated that up to 15 percent of colon cancer cases can be attributed to mutations in the genes that encode the proteins required for mismatch repair. Mutations that cripple the mismatch repair system inevitably lead to a higher mutation rate in other genes because mistakes made during replication are not corrected. The highest frequency of replication errors are made when DNA polymerases copy short, repeated sequences (microsatellites, page 415), which are sites where the enzyme tends to slip as it moves along the template. One gene, which codes for the receptor of a growth factor called TGF-β, is particularly susceptible to mutation

because its sequence contains a string of A residues where the replication enzymes tend to make mistakes. If both copies of this gene become mutated, the cell can no longer respond to the growth factor and has a greatly increased likelihood of developing into a tumor.

Cancer is also one of the consequences of double-strand DNA breaks that have either gone unrepaired or been repaired incorrectly. Breaks in DNA can be caused by a variety of environmental agents to which we are commonly exposed, including X rays, gamma rays, and radioactive emissions. The most serious environmental hazard in this regard is probably radon (specifically ^{222}Rn), a radioactive isotope formed during the disintegration of uranium. Some areas of the planet contain relatively high levels of uranium in the soil, and houses built in these regions can contain dangerous levels of radon gas. It is estimated that approximately one percent of houses in the United States have radon levels that produce more than 10 picocuries per liter of radiation. When the gas is breathed into the lungs it can lead to double-strand DNA breaks that increase the risk of lung cancer. A significant fraction of lung-cancer deaths in nonsmokers is probably due to radon exposure.

SYNOPSIS

DNA replication is semiconservative, which indicates that one half of the parent duplex is transmitted to each of the daughter cells during cell division. This mechanism of replication was first suggested by Watson and Crick as part of their model of DNA structure. They suggested that replication occurred by gradual separation of the strands by means of hydrogen bond breakage, so that each strand could serve as a template for the synthesis of a complementary strand. This model was soon confirmed in both bacterial and eukaryotic cells by showing that cells transferred to labeled media for one generation produce daughter cells whose DNA has one labeled strand and one unlabeled strand. (p. 552)

The mechanism of replication is best understood in bacterial cells. Replication begins at a single origin on the circular bacterial chromosome and proceeds outward in both directions as a pair of replication forks. Replication forks are sites where the double helix is unwound and nucleotides are incorporated into both newly synthesized strands. (p. 554)

DNA synthesis is catalyzed by a family of DNA polymerases. The first of these enzymes to be characterized was DNA polymerase I of *E. coli*. To catalyze the polymerization reaction, the enzyme requires all four deoxyribonucleoside triphosphates, a template strand to copy, and a primer containing a free 3' OH to which nucleotides can be added. The primer is required because the enzyme cannot initiate the synthesis of a DNA strand. Rather, it can only add nucleotides to the 3' hydroxyl terminus of an existing strand. Another unexpected characteristic of DNA polymerase I is that it only polymerizes a strand in a 5' → 3' direction. It had been presumed that the two new strands would be synthesized in opposite directions by polymerases moving in opposite directions along the two parental template strands. This finding was explained when it was shown that the two strands were synthesized quite differently. (p. 556)

One of the newly synthesized strands (the leading strand) grows toward the replication fork and is synthesized continuously. The other newly synthesized strand (the lagging strand) grows away from the fork and is synthesized discontinuously. In bacterial cells, the lagging strand is synthesized as fragments approximately 1,000 nucleotides long, called Okazaki fragments, that are covalently joined to one another by a DNA ligase. In contrast, the leading strand is synthesized as a single continuous strand. Neither the continuous strand nor any of the Okazaki fragments can be initiated by the DNA polymerase but instead begin as a short RNA primer that is synthesized by a type of RNA polymerase called primase. After the RNA primer is assembled, the DNA polymerase continues to synthesize the strand or fragment as DNA. The RNA is subsequently degraded, and the gap is filled in as DNA. (p. 558)

Events at the replication fork require a variety of different types of proteins having specialized functions. These include a DNA gyrase, which is a type II topoisomerase required to relieve the tension that builds up ahead of the fork as a result of DNA unwinding; a DNA helicase that unwinds the DNA by separating the strands; single-stranded DNA-binding proteins that bind selectively to single-stranded DNA and prevent reassociation; a primase that synthesizes the RNA primers; and a DNA ligase that seals the fragments of the lagging strand into a continuous polynucleotide. DNA polymerase III is the primary DNA-synthesizing enzyme that adds nucleotides to each RNA primer, whereas DNA polymerase I is responsible for removing the RNA primers and replacing them with DNA. Two molecules of DNA polymerase III are thought to move together as a complex along their respective template strands. This is accomplished as the lagging strand template loops back on itself. (p. 559)

DNA polymerases possess separate catalytic sites for polymerization and degradation of nucleic acid strands. Most DNA polymerases possess both 5' → 3' and 3' → 5' exonuclease activities. The first acts to degrade the RNA primers that begin each Okazaki fragment, and the second removes inappropriate nucleotides following their mistaken incorporation, thus contributing to the fidelity of replication. It is estimated that approximately one in 10^9 nucleotides is incorporated incorrectly during replication in *E. coli*. (p. 562)

Replication in eukaryotic cells follows a similar mechanism and employs similar proteins to those of prokaryotes. All of the DNA polymerases involved in replication elongate DNA strands in the 5' → 3' direction. None of them initiates the synthesis of a chain without a primer. Most possess a 3' → 5' exonuclease activity, ensuring that replication occurs with high fidelity. Unlike prokaryotes, replication in eukaryotes is initiated simultaneously at many sites along a chromosome, with replication forks proceeding outward in both directions from each site of initiation. (p. 565)

Replication in eukaryotic cells is intimately associated with nuclear structures. Evidence indicates that much of the machinery required for replication is associated with the nuclear matrix. In addition, replication forks that are active at any given time are localized within about 50 to 250 sites called replication foci. Newly synthesized DNA is rapidly associated with nucleosomes. (H3H4)$_2$ tetramers present prior to replication remain intact and are passed on to the daughter duplexes, whereas H2A/H2B dimers separate from one another and bind randomly to new and old (H3H4)$_2$ tetramers on the daughter duplexes. (p. 570)

DNA is subject to damage by many environmental influences, including ionizing radiation, common chemicals, and ultraviolet radiation. Cells possess a variety of systems to recognize and repair the resulting damage. It is estimated that less than one base change in a thousand escapes a cell's repair systems. Four major types of DNA repair systems are discussed. Nucleotide excision repair (NER) systems operate by removing a small section of a DNA strand containing a bulky lesion, such as a pyrimidine dimer. During NER, the strands of DNA containing the lesion are separated by a heli-case; paired incisions are made by endonucleases; the gap is filled by a DNA polymerase; and the strand is sealed by a DNA ligase. Base excision repair removes a variety of altered nucleotides that produce minor distortions in the DNA helix. Cells possess a variety of glycosylases that recognize and remove various types of altered bases. Once the base is removed, the remaining portion of the nucleotide is removed by an endonuclease, the gap is enlarged by a phosphodiesterase, and the gap is filled and sealed by a polymerase and ligase. Mismatch repair is responsible for removing incorrect nucleotides incorporated during replication that escape the proofreading activity of the polymerase. Double-strand breaks are repaired as proteins bind to the broken strands and join the ends together. (p. 572)

In addition to the classic DNA polymerases involved in DNA replication and repair, cells also possess an array of DNA polymerases that facilitate replication at sites of DNA lesions or misalignments. These polymerases, which engage in translesion synthesis, lack processivity and proofreading capability and are more error-prone than classic polymerases. (p. 575)

ANALYTIC QUESTIONS

1. Suppose that Meselson and Stahl had carried out their experiment by growing cells in medium with ^{14}N and then transferring the cells to medium containing ^{15}N. How would the bands within the centrifuge tubes have appeared if replication were semiconservative? If replication were conservative? If replication were dispersive?

2. Replication in bacteria occurs at a remarkably rapid rate (1,000 nucleotides per second). Suppose a DNA molecule were 1 meter in diameter rather than 2 nm. How fast (in km/hour) would the polymerase have to move to replicate one of these giant-sized DNA molecules in the same period of time? (Each nucleotide of a 2-nm DNA molecule is 3.5 Å in length.)

3. How would the chromosomes from the experiment on eukaryotic cells depicted in Figure 13.4 have appeared if replication occurred by a conservative or a dispersive mechanism?

4. We have seen that cells possess a special enzyme to remove uracil from DNA. What do you suppose would happen if the uracil groups were not removed? (You might consider the information presented in Figure 11.44 on the pairing properties of uracil.)

5. Draw a partially double-stranded DNA molecule that would not serve as a template for DNA synthesis by DNA polymerase I.

6. Some temperature-sensitive bacterial mutants stop replication immediately following elevation of temperature, whereas others continue to replicate their DNA for a period of time before they cease this activity, and still others continue until a round of replication is completed. How might these three types of mutants differ?

7. Suppose the error rate during replication in human cells were the same as that of bacteria (about 10^{-9}). How would this impact the two cells differently?

8. Figure 13.19 shows the results from an experiment in which cells were incubated with [^3H] thymidine for less than 30 minutes prior to fixation. How would you expect this photograph to appear after a one-hour labeling period? Can you conclude that the entire genome is replicated within an hour? If not, why not?

9. How is it that strand separation during initiation at a bacterial origin can be accomplished without the aid of a DNA helicase, but strand separation at other places requires this type of enzyme?

10. What advantages might you expect for DNA replication to occur in conjunction with the nuclear matrix as opposed to the nucleoplasm? What are the advantages of replication occurring in a small number of replication foci?

11. What are some of the reasons you might expect human cells to have more efficient repair systems than those of a frog?

12. Suppose you were to compare autoradiographs of two cells that had been exposed to [^3H]thymidine, one that was engaged in DNA replication (S phase) and another that was not. How would you expect autoradiographs of these cells to differ?

13. Construct a model that would explain how transcriptionally active DNA is repaired preferentially over transcriptionally silent DNA.

ADDITIONAL READING

Replication in Bacteria and General Topics

BAKER, T. A. & BELL, S. P. 1998. Polymerases and the replisome: Machines within machines. *Cell* 92:295–305.

BIRD, L. E., ET AL. 1998. Helicases: A unifying structural theme? *Curr. Opin. Struct. Biol.* 8:49–53.

DOUBLIÉ, S., ET AL. 1999. An open and closed case for all polymerases. *Struct.* 7:R31–R35.

ECHOLS, H. & GOODMAN, M. F. 1991. Fidelity mechanisms in DNA replication. *Annu. Rev. Biochem.* 60:477–511.

FRIEDBERG, E. C., ET AL. 2000. The many faces of DNA polymerases: Strategies for mutagenesis and for mutational avoidance. *Proc. Nat'l Acad. Sci. U.S.A.* 97:5681–5683.

FRIEDBERG, E. C. & GERLACH, V. L. 1999. Novel DNA polymerases offer clues to molecular basis of mutagenesis. *Cell* 98:413.

GOODMAN, M. F. 1998. Purposeful mutations. *Nature* 395:221–222.

HINGORANI, M. M. & O'DONNELL, M. 2000. Sliding clamps: A (tail)ored fit. *Curr. Biol.* 10:R25–R29.

HÜBSCHER, U., ET AL. 2000. Eukaryotic DNA polymerases, a growing family. *Trends Biochem. Sci.* 25:143–147.

JOHNSON, R. E., ET AL. 1999. Bridging the gap: A family of novel DNA polymerases that replicate faulty DNA. *Proc. Nat'l. Acad. Sci. U.S.A.* 96:12224–12226.

KORNBERG, A. 1989. Never a dull enzyme. *Annu. Rev. Biochem.* 58:1–30.

KORNBERG, A. & BAKER, T. A. 1988. *DNA replication.* 2d ed., W. H. Freeman.

MARIANS, K. J. 1992. Prokaryotic DNA replication. *Annu. Rev. Biochem.* 61:673–719.

MESELSON, M. & STAHL, F. W. 1958. The replication of DNA in *Escherichia coli. Proc. Nat'l. Acad. Sci. U.S.A.* 44:671–682.

POSTOW, L., ET AL. 1999. Knot what we thought before: The twisted story of replication. *BioEss.* 21:805–808.

RADMAN, M. 1999. Enzymes of evolutionary change. *Nature* 401:866–868.

STEITZ, T. A. 1999. DNA and RNA polymerases: Structure, diversity, and common mechanisms. *Harvey Lects.* 93:75–94

STILLMAN, B. 1994. Smart machines at the DNA replication fork. *Cell* 78:725–728.

TOONE, W.M., ET AL. 1997. Getting started: Regulating the initiation of replication in yeast. *Ann. Rev. Microbiol.* 51:125–149.

Replication in Eukaryotic Cells

ADAMS, C. R. & KAMAKAKA, R. T. 1999. Chromatin assembly: Biochemical identities and genetic redundancy. *Curr. Opin. Genes Dev.* 9:185–190.

COOK, P. R. 1999. The organization of replication and transcription. *Science* 284:1790–1795.

DEPAMPHILIS, M. L., ED. 1996. *DNA Replication in Eukaryotic Cells.* Cold Spring Harbor Laboratory Press.

DEPAMPHILIS, M. L. 1999. Replication origins in metazoan chromosomes: Fact or fiction? *BioEss.* 21:5–16.

DIFFLEY, J. F. X. 1998. Replication control: Choreographing replication origins. *Curr. Biol.* 8:R771–R773.

DONALDSON, A. D. & BLOW, J. J. 1999. The regulation of replication origin activation. *Curr. Opin. Genes Dev.* 9:62–68.

HUBERMAN, J. A. 1998. Choosing a place to begin. *Science* 281:929–930.

KEARSEY, S. E. & LABIB, K. 1998. MCM proteins: Evolution, properties, and role in DNA replication. *Biochim. Biophys. Acta* 1398:113–116.

KELLY, T. J. & BROWN, G. W. 2000. Regulation of chromosome replication. *Annu. Rev. Biochem.* 69:829–880.

KRUDE, T. 1999. Chromatin assembly during DNA replication in somatic cells. *Eur. J. Biochem.* 263:1–5.

KUNKEL, T. A. & BEBENEK, K. 2000. DNA replication fidelity. *Annu. Rev. Biochem.* 69:497–530.

NEWLON, C. S. 1997. Putting it all together: Building a prereplicative complex. *Cell* 91:717.

TAYLOR, J. H. 1997. Tritium-labeled thymidine and early insights into DNA replication and chromosome structure. *Trends Biochem. Sci.* 22:447–450.

TYE, B. K. 1999. MCM proteins in DNA replication. *Annu. Rev. Biochem.* 68:649–686.

WAGA, S. & STILLMAN, B. 1998. The replication fork in eukaryotic cells. *Annu. Rev. Biochem.* 67:721–751.

DNA Repair and XP

BAYNTON, K. & FUCHS, R.P.P. 2000. Lesions in DNA: Hurdles for polymerases. *Trends Biochem. Sci.* 25:74–79.

BUERMEYER, A. B., ET AL. 1999. Mammalian DNA mismatch repair. *Annu. Rev. Gen.* 33:533–564.

CARREAU, M. & BUCHWALD, M. 1998. Fanconi's anemia: What have we heard from the genes so far? *Mol. Med. Today* 4:201–206.

CITTERIO, E., ET AL. 2000. Transcriptional healing. *Cell* 101:447–450.

FEATHERSTONE, C. & JACKSON, S. P. 1999. DNA double-strand break repair. *Curr. Biol.* 9:R759–R761.

GOODMAN, M. F. & TIPPEN, B. 2000. Sloppier copier DNA polymerases involved in DNA repair. *Curr. Opin. Genes Dev.* 10:162–168.

HABER, J. E. 2000. Partners and pathways: Repairing a double-strand break. *Trends Gen.* 16:259–264.

HANAWALT, P. C. 2000. The bases for Cockayne syndrome. *Nature* 405:415–416.

KARRAN, P. 2000. DNA double-strand break repair in mammalian cells. *Curr. Opin. Genes Dev.* 10:144–150.

LEHMANN, A. R. 1998. Dual functions of DNA repair genes: Molecular, cellular, and clinical implications. *BioEss.* 20:146–155.

MCCULLOUGH, A. K., ET AL. 1999. Initiation of base excision repair. *Annu. Rev. Biochem.* 68:255–285.

VAN STEEG, H. & KRAEMER, K. H. 1999. Xeroderma pigmentosa and the role of UV-induced DNA damage in skin cancer. *Mol. Med. Today* 5:86–94.

WOOD, R. D. 1999. DNA repair: Variants on a theme. *Nature* 399:639–640.

14

A Cellular Reproduction

ccording to the third tenet of the cell theory, new cells originate only from other living cells. The process by which this occurs is called **cell division**. For a multicellular organism, such as a human or an oak tree, countless divisions of a single-celled zygote produce an organism of astonishing cellular complexity and organization. Cell division does not stop with the formation of the mature organism but continues throughout its life. It is estimated that more than 25 million cells undergo division each second in an adult human. This enormous output of cells is needed to replace cells that have aged or died. Old, worn-out blood cells, for example, are removed and replaced at the rate of about 100 million per minute.

Each dividing cell is called a *mother cell*, and its descendants are called *daughter cells*. There is a reason for using these "familial" terms. The mother cell transmits copies of its genetic information to its daughter cells, which become the next cell generation. In turn, the daughter cells can become mother cells to their own daughter cells, passing along the same genes they inherited from their

Fluorescence micrograph of a cultured newt lung cell at the early prometaphase stage of mitosis, just after the nuclear envelope has broken down and the microtubules of the mitotic spindle can interact with the chromosomes. The mitotic spindle appears green after labeling with a monoclonal antibody against tubulin, whereas the chromosomes appear blue after labeling with a fluorescent dye. (FROM ALEXY KHODJAKOV, WADSWORTH CENTER, ALBANY, NEW YORK.)

mother to yet another cellular generation. For this reason, cell division is often referred to as *cellular reproduction*.

Cell division is more than just a means of reproducing cells; it is the basis for reproducing more organisms through the formation of cellular gametes. Cell division, therefore, links parents to their offspring, living species to their extinct ancestors, and humans to the earliest, most primitive cellular organisms.

14.1 THE CELL CYCLE

We are used to thinking of biological processes in terms of cycles, including life cycles, metabolic cycles, and physiological cycles. Life is a continuum from one generation to the next, and there must be a continual renewal or return to an initial state for the process to occur once again. Any given cell can be said to have a life, just as any given organism. The life of a cell begins with its formation by cell division of a mother cell and ends with either the formation of daughter cells or with its death. The stages through which a cell passes from one cell division to the next constitute the **cell cycle** (Figure 14.1).

The cell cycle can be divided into two major phases based on cellular activities readily visible with a light microscope: M phase and interphase. **M phase** includes (1) the process of **mitosis**, during which duplicated chromosomes are separated into two nuclei, and (2) **cytokinesis**,

during which the entire cell divides into two daughter cells. Only a small percentage of cells in a tissue or cell culture are observed to be in mitosis at any given time, which indicates that cells spend the majority of time in **interphase**. Whereas M phase usually lasts only 30 to 60 minutes, interphase may extend for hours, days, weeks, or longer, depending on the cell type and the conditions. As discussed below, M phase is a time when the cell is involved in activities necessary for cell division—a time when the synthesis of macromolecules is largely shut down.

Although M phase is the period when the contents of a cell are actually divided, numerous preparations for an upcoming mitosis occur during interphase, including replication of the cell's DNA. One might guess that a cell engages in replication throughout interphase. However, studies in the early 1950s on asynchronous cultures (i.e., cultures whose cells are randomly distributed throughout the cell cycle) showed that this is not the case. If [^3H]thymidine is given to a culture of cells for a short period (e.g., 30 minutes) and a sample of the cell population is fixed, dried onto a slide, and examined by autoradiography, only a certain percentage of the cells are found to have radioactive nuclei. Among cells that were engaged in mitosis at the time of fixation (as evidenced by their compacted chromosomes) none is found to have a radioactively labeled nucleus. These mitotic cells have unlabeled chromosomes because they were not engaged in DNA replication during the labeling period.

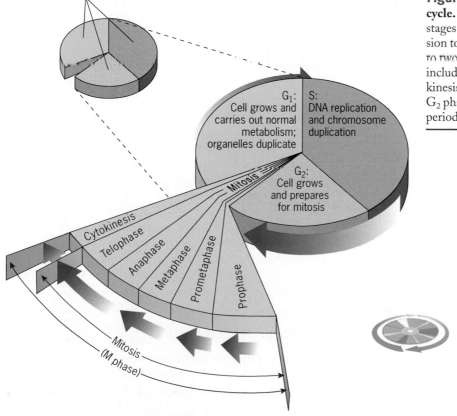

Figure 14.1 An overview of the eukaryotic cell cycle. This diagram of the cell cycle indicates the stages through which a cell passes from one division to the next. The cell cycle can be divided into two stages: M phase and interphase. M phase includes the successive stages of mitosis and cytokinesis. Interphase is divided into G_1, S, and G_2 phases, with S phase being equivalent to the period of DNA synthesis.

If labeling is allowed to continue for one to a few hours before the cells are sampled, there are still no cells with labeled mitotic chromosomes (Figure 14.2). We can conclude from these results that there is a definite period of time between the end of DNA synthesis and the beginning of M phase. This period is termed G_2 (for second gap). The duration of G_2 is revealed as one continues to take samples of cells from the culture until labeled mitotic chromosomes are observed. The first cells whose mitotic chromosomes are labeled must have been at the last stages of DNA synthesis at the start of the incubation with [³H]thymidine. The length of time between the start of the labeling period and the appearance of cells with labeled mitotic figures corresponds to the duration of G_2.

DNA replication occurs during a period of the cell cycle termed **S phase**. S phase is also the period when the cell synthesizes the additional histones that will be needed as the cell doubles the number of nucleosomes in its chromosomes (see Figure 13.25). The length of S phase can be determined directly. In an asynchronous culture, the percentage of cells engaged in a particular activity is an approximate measure of the percentage of time that this activity occupies in the lives of cells. Thus, if we know the length of the entire cell cycle, the length of S phase can be calculated directly from the percentage of the cells whose nuclei are radioactively labeled during a brief pulse with [³H]thymidine. Similarly, the length of M phase can be calculated from the percentage of cells in the population that are seen to be engaged in mitosis or cytokinesis. When one adds up the periods of $G_2 + S + M$, it is apparent that there is an additional period in the cell cycle yet to be accounted for. This other phase, termed G_1 (for first

Figure 14.2 Experimental results that demonstrate that replication occurs during a defined period of the cell cycle. HeLa cells were cultured for 30 minutes in medium containing [³H]thymidine and then incubated (chased) for various times in unlabeled medium before being fixed and prepared for autoradiography. The culture dish was scanned for cells that were in mitosis at the time they were fixed, and the percentage of those mitotic cells whose chromosomes were labeled was plotted as shown. (FROM A STUDY BY R. BASERGA AND F. WIEBEL.)

gap), is the period following mitosis and preceding DNA synthesis.

Cell Cycles in Vivo

One of the properties that distinguishes various types of cells within a multicellular plant or animal is their capacity to grow and divide. We can recognize three broad categories of cells:

1. *Cells, such as nerve cells, muscle cells, or red blood cells, that are highly specialized and have lost the ability to divide.* Once these cells have differentiated, they remain in that state until they die.

2. *Cells that normally do not divide but can be induced to begin DNA synthesis and divide when given an appropriate stimulus.* Included in this group are liver cells, which can be induced to proliferate by the surgical removal of part of the liver, and lymphocytes, which can be induced to proliferate by interaction with an appropriate antigen.

3. *Cells that normally possess a relatively high level of mitotic activity.* Certain tissues of the body are subject to continual renewal, with new cells continually formed by cell division. Included in this category are the spermatogonia that give rise to the male gametes, the hematopoietic stem cells that give rise to red and white blood cells, and the epithelial cells that line the body cavities and the body surface. The relatively unspecialized cells of the apical meristems located near the tips of plant roots and stems also exhibit rapid and continual cell division.

Analysis of a large variety of cells has revealed great variability in the lengths of their cell cycles. Cell cycles can range from less than 30 minutes in the rapidly dividing cells of cleaving embryos (other than mammals, which cleave very slowly) to cycles lasting several months in slowly growing tissues, such as the mammalian liver. Rapidly growing adult cells typically divide every 12 to 36 hours. Of the three stages of interphase, G_1 is the most variable, although major differences can also exist for S and G_2. With a few notable exceptions, cells that have stopped dividing, whether temporarily or permanently, whether in the body or in culture, are present in a stage preceding the initiation of DNA synthesis. Cells that are arrested in this state are usually said to be in the G_0 state to distinguish them from the typical G_1-phase cells that may soon enter S phase. As will be discussed below, a cell must generate an internal signal to proceed from G_1 into S phase. Once the signal to begin DNA replication has been generated, the cell invariably completes that round of DNA synthesis and continues through mitosis.

Control of the Cell Cycle

The study of the cell cycle is not only important in basic cell biology, but it has enormous practical implications in combating cancer, a disease that results from a breakdown in a

cell's ability to regulate its own division. In 1970, a series of cell fusion experiments carried out by Potu Rao and Robert Johnson of the University of Colorado helped open the door to understanding how the cell cycle is regulated.

Rao and Johnson wanted to know whether the cytoplasm of cells contains regulatory factors that affect cell-cycle activities. They approached the question by fusing cells that were in different stages of the cell cycle. In one experiment, they fused cells in G_1 with cells in S and asked the following question: Does the cytoplasm donated by the nonreplicating G_1 cell contain factors that block replication in the S-phase nucleus, or does the cytoplasm of the replicating S-phase cell contain factors that stimulate replication in the G_1-phase nucleus? They found that the nucleus in the hybrid cell that had been donated by the G_1-phase cell was activated by S-phase cytoplasm to begin DNA replication. These results suggested that the cytoplasm of a replicating cell contains factors that stimulate initiation of DNA synthesis in G_1-phase cells. In contrast, when G_2-phase and S-phase cells were fused, the G_2-phase nuclei did not initiate another round of DNA synthesis. This observation suggested that G_2-phase nuclei, which have already replicated their DNA, can no longer respond to the initiation factors present in the S-phase cell cytoplasm.

Results of other cell fusion experiments suggested that the transition from G_2 to M was also induced by cyto-plasmic factors. For example, when mitotic cells were fused with cells in other stages of the cell cycle, the mitotic cell always induced compaction of the chromatin in the nucleus of the nonmitotic cell (Figure 14.3), as if the mitotic phase was dominant over other phases of the cell cycle. If a G_1-phase and an M-phase cell were fused, the chromatin of the G_1-phase nucleus underwent *premature chromosomal compaction* to form a set of elongated compacted chromosomes (Figure 14.3a). If a G_2-phase and M-phase cell were fused, the G_2 chromosomes also underwent premature chromosome compaction, but unlike those of a G_1 nucleus, the compacted G_2 chromosomes were visibly doubled, reflecting the fact that replication had already occurred (Figure 14.3c). If a mitotic cell was fused with an S-phase cell, the S-phase chromatin also became compacted (Figure 14.3b). However, replicating DNA is particularly sensitive to damage, so that compaction in the S-phase nucleus led to the formation of "pulverized" chromosomal fragments rather than intact, compacted chromosomes. Taken together, the results of these experiments suggested that the transitions from G_1 to S and from G_2 to M were both under positive control; that is, both were induced by the presence of some stimulatory agent.

The Role of Protein Kinases While the cell fusion experiments revealed the existence of factors that regulated the

(a) *(b)* *(c)*

Figure 14.3 Experimental demonstration that cells contain factors that stimulate entry into mitosis. The photographs show the results of the fusion of an M-phase HeLa cell with a rat kangaroo PtK2 cell that had been in (*a*) G_1 phase, (*b*) S phase, or (*c*) G_2 phase at the time of cell fusion. As described in the text, the chromatin of the G_1-phase and G_2-phase PtK2 cells undergoes premature compaction, whereas that of the S-phase cell becomes pulverized. The elongated chromatids of the G_2-phase cell in *c* are doubled in comparison with those of the G_1 cell in *a*. (FROM KARL SPERLING AND POTU N. RAO, HUMANGENETIK 23:437, 1974.)

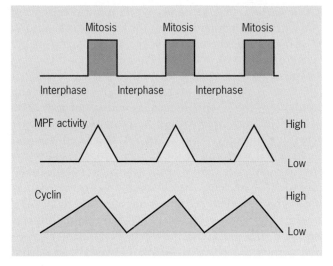

Figure 14.4 **Fluctuation of cyclin and MPF levels during the cell cycle.** This drawing depicts the cyclical changes that occur during early frog development when mitotic divisions occur synchronously in all cells of the embryo. The top tracing shows the alternation between periods of mitosis and interphase, the middle tracing shows the cyclical changes in MPF activity, and the lower tracing shows the cyclical changes in the concentrations of cyclins that control the relative activity of the MPF kinase. (REPRINTED WITH PERMISSION FROM A. W. MURRAY AND M. W. KIRSCHNER, SCIENCE 246:616, 1989; © COPYRIGHT 1989, AMERICAN ASSOCIATION FOR THE ADVANCEMENT OF SCIENCE.)

cell cycle, they provided no information about the biochemical properties of these factors. Insights into the nature of the agents that trigger DNA replication and promote entry of a cell into mitosis (or meiosis) were first gained in a series of experiments on the oocytes and early embryos of frogs and invertebrates. These experiments are described in the Experimental Pathways at the end of this chapter. To summarize here, it was shown that entry of a cell into M phase is initiated by a protein kinase called *maturation-promoting factor (MPF)*. MPF consists of two subunits: a catalytic subunit that transfers phosphate groups from ATP to specific serine and threonine residues of specific protein substrates, and a regulatory subunit called **cyclin**. The term "cyclin" was coined because the concentration of this regulatory protein rises and falls in a predictable pattern as the cell cycle progresses (Figure 14.4). When the cyclin concentration is low, the kinase lacks the cyclin subunit and, as a result, is inactive. When the cyclin concentration rises, the kinase is activated, causing the cell to enter M phase. These results suggested that (1) progression of cells into mitosis depends on an enzyme whose sole activity is to phosphorylate other proteins, and (2) the activity of this enzyme is controlled by a subunit whose concentration varies from one stage of the cell cycle to another.

Over the past decade, a large number of laboratories have focused on MPF-like enzymes, called **cyclin-dependent kinases (Cdks)**. It has been found that Cdks

are not only involved in M phase but are the key agents that orchestrate activities throughout the cell cycle. Yeast cells have been particularly useful in studies of the cell cycle because of the availability of temperature-sensitive mutants whose abnormal proteins affect various cell cycle processes. As discussed on page 554, temperature-sensitive mutants are used to study the role of genes whose encoded proteins perform indispensable functions. These mutants can be grown in a relatively normal manner at a lower (permissive) temperature and then shifted to a higher (restrictive) temperature to study the effect of the mutant gene product. The literature on the genetic control of the cell cycle is complicated by the fact that researchers have focused on two distantly related yeast species, the budding yeast *Saccharomyces cerevisiae*, which reproduces through the formation of buds at one end of the cell, and the fission yeast, *Schizosaccharomyces pombe*, which reproduces by elongating itself and then splitting into two equal-sized cells. The molecular basis of cell cycle regulation has been remarkably conserved throughout the evolution of eukaryotes. Once a gene involved in cell cycle control has been identified in one of the two yeast species, homologues are sought—and usually found—in the genomes of higher eukaryotes, including humans. By combining genetic, biochemical, and structural analyses, investigators have gained a comprehensive understanding of the major activities that allow a cell to grow and reproduce in a laboratory culture dish.

Research into the genetic control of the cell cycle in yeast began in the 1970s in two laboratories, that of Leland Hartwell at the University of Washington working on budding yeast and that of Paul Nurse at the University of Oxford working on fission yeast. Both laboratories found a gene that, when mutated, would cause the growth of cells to stop at one of two distinct points in the cell cycle, either just prior to DNA replication or just prior to mitosis. The product of this gene, which was called *cdc2* in fission yeast (and *CDC28* in budding yeast), was eventually found to be homologous to the catalytic subunit of MPF; in other words, it was a cyclin-dependent kinase. Subsequent research on yeast as well as many different mammalian cells has supported the concept that the progression of a eukaryotic cell through its cell cycle is regulated primarily at two distinct stages, one occurring near the end of G_1 and the other near the end of G_2. These stages represent points in the cell cycle where a cell becomes committed to beginning a crucial event—either initiating replication or entering mitosis.

In fission yeast cells, the same Cdk (cdc2) is responsible for passage through both points of commitment, though in partnership with different cyclins (Figure 14.5). The first transition point, which is called START, occurs just before the end of G_1. Passage through START requires the activation of cdc2 by one or more G_1 *cyclins*, whose levels peak just before the onset of DNA synthesis. Activation of cdc2 by the G_1 cyclins in fission yeast cells (Figure 14.5) leads to the initiation of replication by the

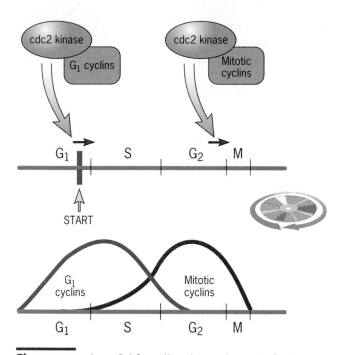

Figure 14.5 A model for cell cycle regulation in fission yeast. The cell cycle is controlled primarily at two points, START and the G_2-M transition. In fission yeast, cyclins can be divided into two groups, G_1 cyclins and mitotic cyclins. Passage of a cell through these two critical junctures requires the activation of the same cdc2 kinase by a different type of cyclin.

prereplication complexes that had assembled at the origins of replication earlier in G_1 (see Figure 13.21). Once a cell has passed START, it is irrevocably committed to replicating its DNA and, ultimately, completing the cell cycle.[1] Passage from G_2 to mitosis requires activation of cdc2 by a different group of cyclins—the *mitotic cyclins*. Cdks containing a mitotic cyclin (e.g., MPF described above) phosphorylate substrates that are required for the cell to enter mitosis. Included among the substrates are proteins required for the dynamic changes in organization of both the chromosomes and cytoskeleton that characterize the shift from interphase to mitosis. Cells make a third commitment during the middle of mitosis, which determines whether they will complete cell division and reenter G_1 of the next cycle. Exit from mitosis and entry into G_1 depends on a rapid decrease in Cdk activity that results from a plunge in concentration of the mitotic cyclins (Figure 14.5), an event that will be discussed on page 600 in conjunction with other mitotic activities.

Cyclin-dependent kinases are often described as the "engines" that drive the cell cycle through its various

[1]Mammalian cells pass through a comparable point during G_1, referred to as the *restriction point*, at which time they become committed to DNA replication and ultimately to completing mitosis. Prior to the restriction point, mammalian cells require the presence of growth factors in their culture medium if they are to progress in the cell cycle. After they have passed the restriction point, these same cells will continue through the remainder of the cell cycle without external stimulation.

stages. The activities of these enzymes are regulated by a number of factors that operate in combination. These include:

Cyclin concentration We have seen in the previous section that Cdks are activated by association with a regulatory subunit, or cyclin. The presence in the cell of a particular cyclin follows the activation of transcription of the gene encoding that cyclin. Different cyclin genes are transcribed at different stages during the cell cycle. When a cyclin is present in the cell, it binds to the catalytic subunit of the Cdk, causing a major change in the conformation of the catalytic subunit. X-ray crystallographic structures of various cyclin-Cdk complexes indicate that cyclin binding causes the movement of a flexible loop of the Cdk polypeptide chain away from the opening to the enzyme's active site, allowing the Cdk to phosphorylate its protein substrates.

Cdk phosphorylation state Even though the effect of cyclin binding is dramatic, it is not sufficient by itself to turn on the Cdk's kinase activity. For activation to occur, the Cdk subunit must also be phosphorylated at a critical threonine residue (Thr 161 in Figure 14.6), which is accomplished by another protein kinase called CAK (Cdk-activating kinase). The importance of Cdk phosphorylation is illustrated in Figure 14.6, which depicts the changes in the Cdk as a fission yeast cell moves from G_2 through M and into G_1. Whereas phosphorylation of Thr 161 activates the Cdk kinase, phosphorylation of a key tyrosine residue (Tyr 15 in Figure 14.6) inhibits the activity of the kinase. This inhibitory phosphate is added by another kinase called wee1. Thus, as shown in step 1 of Figure 14.6, phosphorylation by wee1 during G_2 inactivates the cdc2 enzyme. The enzyme remains in this inactivated state until the end of G_2 when the inhibitory phosphate at Tyr 15 is removed by a phosphatase named cdc25 (step 2, Figure 14.6). The removal of the phosphate activates the Cdk, driving the yeast cell into mitosis. The balance between wee1 kinase and cdc25 phosphatase activities is regulated in turn by other kinases and phosphatases. The involvement of a series of enzymes in the activation and inhibition of Cdks provides a number of targets at which information from both inside and outside the cell can alter the progress of the cell cycle. These types of "molecular communication pathways" will be explored in detail in the following chapter.

Cdk inhibitors Cdk activity can be blocked by a variety of inhibitors. In budding yeast, for example, a protein called Sic1 acts as a Cdk inhibitor during G_1. The degradation of Sic1 allows the cyclin-Cdk that is present in the cell to initiate DNA replication. The role of Cdk inhibitors in mammalian cells is discussed on page 589.

Controlled proteolysis. It is evident from Figures 14.4 and 14.5 that the cyclin concentrations oscillate during each

Figure 14.6 Progression through the cell cycle requires the phosphorylation and dephosphorylation of critical cdc2 residues. This figure describes basic events that occur in the cell cycle of both yeast and mammalian cells, though the precise positions of the phosphorylated residues can vary in the polypeptides of different species. During G_2, the cdc2 kinase interacts with a mitotic cyclin but remains inactive as the result of phosphorylation of a key tyrosine residue (Tyr 15 in fission yeast) by wee1 (step 1). (In mammalian cells, a second inhibitory phosphate is added to a threonine residue by a kinase called Myt1.) A separate kinase, called CAK, transfers a phosphate to another residue (Thr 161), which is required for cdc2 kinase activity later in the cell cycle. When the cell reaches a critical size, an enzyme called cdc25 phosphatase is activated, which removes the inhibitory phosphate(s) on the cdc2 subunit. The resulting activation of the cdc2 kinase then drives the cell into mitosis (step 2). By the end of mitosis (step 3), the cyclin dissociates from the cdc2 kinase and the stimulatory phosphate group is removed from Thr 161 by another phosphatase. The free cyclin is subsequently degraded, and the cell begins another cycle. (AFTER T. R. COLEMAN AND W. G. DUNPHY, CURR. OPIN. CELL BIOL. 6:877, 1994.)

cell cycle, which leads to changes in the activity of Cdks. Cells regulate the concentration of cyclins, and other key cell cycle proteins, by adjusting both the rate of synthesis and the rate of destruction of the molecule at different points in the cell cycle. Degradation is accomplished by means of the ubiquitin-proteasome pathway described on page 545. Regulation of the cell cycle requires two classes of multisubunit complexes (SCF and APC complexes) that recognize proteins marked for degradation and link these proteins to a polyubiquitin chain, which ensures their destruction in a proteasome. SCF complexes probably act throughout the cell cycle (see Figure 14.25*a*) and mediate the destruction of G_1 cyclins, Cdk inhibitors, and other cell cycle proteins. These proteins become targets for an SCF after they are phosphorylated by the protein kinases (i.e., Cdks) that regulate the cell cycle. Mutations that inhibit SCFs from mediating proteolysis of key proteins, such as G_1 cyclins or the Sic1 inhibitor mentioned above, can prevent cells from replicating their DNA. The other type of complex, called APC, acts in mitosis and degrades a number of key mitotic proteins, including the mitotic cyclins. Destruction of the mitotic cyclins allows a cell to exit mitosis and enter a new cell cycle (page 600).

Subcellular localization Cells contain a number of different compartments in which regulatory molecules can either be united with or separated from the proteins they interact with. Subcellular localization is a dynamic phenomenon characterized by the movement of cell cycle regulators into different compartments at different stages. For example, one of the major mitotic cyclins in animal cells (cyclin B1) shuttles between the nucleus and cytoplasm until G_2, when it accumulates in the nucleus just prior to the onset of mitosis (Figure 14.7). Nuclear accumulation of cyclin B1 is facilitated by phosphorylation of

an amino acid residue that resides in its nuclear export signal (NES, page 501). Phosphorylation at this site presumably blocks subsequent export of the cyclin back to the cytoplasm. If nuclear accumulation of the cyclin is blocked, cells fail to initiate cell division.

As noted above, the proteins and processes that control the cell cycle are remarkably conserved among eukaryotes from yeast to higher plants and animals. Although the same families of proteins are involved, a number of different cyclins and cyclin-dependent kinases have been implicated in the control of the mammalian cell cycle. Successive waves of synthesis and degradation of different cyclins play a key role in driving the cell from one stage to the next. The pairing between cyclins and Cdks is highly specific, and only certain combinations are found (Figure 14.8). For example, a complex formed between cyclin E and Cdk2 acts to drive the cell into S phase, whereas a complex between a B cyclin and Cdk1 acts to drive the cell into mitosis.

Studies aimed at identifying the roles of the various cyclins and Cdks in mammalian cells have utilized genetically engineered (knockout) mice that cannot synthesize that particular protein. The phenotypes of these mice depend on the gene that has been eliminated. Mice that are unable to synthesize Cdk1 or cyclin B1, for example, do not survive, indicating these genes are essential for cell proliferation. In contrast, mice that lack a gene encoding several other Cdks or cyclin subunits develop surprisingly well, which suggests that other members of the gene family can take over the functions of those that are lacking. Even with this "built-in genetic redundancy," the absence of any of these cell cycle regulatory proteins produces distinct abnormalities. Cdk4-deficient mice, for example, are much smaller than control animals, which stems from a

(a)　　　　　　　　　　　　　　　　*(b)*

Figure 14.7 **An example of the importance of subcellular localization in cell cycle regulation.** Micrographs of a living HeLa cell that has been injected with cyclin B1 linked to the green fluorescent protein (page 283). The cell shown in *a* is in the G_2 phase of its cell cycle, and the fluorescently labeled cyclin B1 is localized almost entirely in the cytoplasm. The micrograph in *b* shows the same cell in prophase of mitosis, and the labeled cyclin B1 is now seen to be concentrated in the cell nucleus. The basis for this change in localization is discussed in the text. (FROM PAUL CLUTE AND JONATHAN PINES, NATURE CELL BIOL. 1:83, 1999 © COPYRIGHT 1999 MACMILLAN MAGAZINES LIMITED.)

Figure 14.8 **Combinations between various cyclins and cyclin-dependent kinases at different stages in the mammalian cell cycle.** Cell cycle activities during mid G_1 are accomplished primarily by Cdk4 and Cdk6 associated with the D-type cyclins (D1, D2, and D3). Among the substrates for these Cdks is an important regulatory protein called Rb (Section 16.3). The phosphorylation of Rb leads to the transcription of a number of genes, including those that code for cyclins E and A, Cdk1, and proteins involved in replication. The G_1-S transition, which includes the initiation of replication, is driven by the activity of the cyclin E-Cdk2 and cyclin A-Cdk2 complexes. The transition from G_2 to M is driven by the activity of cyclin A-Cdk 1 and cyclin B-Cdk 1 complexes, which are thought to phosphorylate such diverse substrates as cytoskeletal proteins, histones, and proteins of the nuclear envelope. Destruction of cyclins A and B is a key factor in the inactivation of Cdk1 and the transition of the cell from mitosis back into G_1. (AFTER C. J. SHERR, CELL 73:1060, 1993; BY PERMISSION OF CELL PRESS.)

reduction in the level of cell division throughout the body. In addition, Cdk4-deficient mice are diabetic as the result of a specific failure in differentiation of the insulin-secreting cells of the pancreas. Mice lacking a gene for cyclin D1 also exhibit growth retardation, but display a particular lack of cell proliferation during development of the retina.

Another technique has been employed in recent years to examine the role of a particular protein in mouse development. Rather than simply removing a gene as is done in the formation of a knockout, the coding region of one gene can be replaced with that of a different gene, which leads to the formation of a *knockin*. Since the regulatory regions of the replaced gene remain intact, the substitute gene is transcribed whenever the replaced gene would normally have been expressed. As indicated in Figure 14.8, cyclin D1 is synthesized during mid-G_1 and functions in the activation of transcription of genes involved in the G_1-S transition, including the gene encoding cyclin E. When the cyclin D1 gene is replaced by the cyclin E gene, the knockin mouse that develops shows none of the abnormalities characteristic of the D1 knockout (e.g., small size and defective retina). It would appear from this experiment that cyclin E can substitute functionally for the missing D1 cyclin and suggests that the major role of cyclin D1 is to activate the transcription of cyclin E.

Checkpoints, Kinase Inhibitors, and Cellular Responses

Ataxia-telangiectasia (AT) is an inherited recessive disorder characterized by a host of diverse symptoms, including

a greatly increased risk for certain types of cancer.[2] During the late 1960s—following the deaths of several individuals undergoing radiation therapy—it was discovered that patients with AT are extremely sensitive to ionizing radiation. So too are the cells from these patients, which lack a crucial protective response found in normal cells. When normal cells are subjected to treatments that damage DNA, such as ionizing radiation or DNA-altering drugs, their progress through the cell cycle stops while the damage is repaired. If, for example, a normal cell is irradiated during the G_1 phase of the cell cycle, it delays progression into S phase. Similarly, cells irradiated in S phase delay further DNA synthesis, whereas cells irradiated in G_2 delay entry into mitosis. Studies of this type carried out in yeast gave rise to a concept, formulated by Leland Hartwell and Ted Weinert in 1988, that cells possess **checkpoints** as part of their cell cycle. Checkpoints are surveillance mechanisms that halt progress of the cell cycle if (1) any of the chromosomal DNA is damaged, or (2) certain critical processes, such as DNA replication during S phase or chromosome alignment during M phase, have not been properly completed. Checkpoint control is thought to require at least three distinct classes of proteins:

1. Sensors that detect abnormalities and emit an appropriate signal

2. Transmitters that send the signal along the proper pathways within the cell

3. Effectors that respond to the signal and inhibit the cell cycle machinery

If surveillance systems fail to detect abnormalities, the cell progresses to the next stage of the cycle. If, however, a sensor protein detects the presence of DNA damage or some other defect, such as an incompletely replicated chromosome or a misaligned mitotic chromosome, it triggers a response that temporarily arrests further cell cycle progress. The cell can then use the delay to repair the damage or finish replicating its DNA rather than progressing to the next stage of the cell cycle. This is especially important because cells that undergo division with genetic damage run a high risk of becoming transformed into a cancer cell. If the DNA is damaged beyond repair, the checkpoint mechanisms of higher eukaryotes can transmit a signal that leads to the death of the cell, thus removing any possibility that the cell might become cancerous.

Remarkably, the presence of a single break in one of the cell's DNA molecules is sufficient to cause cell cycle arrest. The gene responsible for ataxia-telangiectasia (the *ATM* gene) encodes a protein kinase that is thought to be part of a multisubunit complex that recognizes certain DNA lesions, including double-stranded breaks (page

574). Once bound to the damaged DNA, ATM emits signals, mediated by protein phosphorylation, that trigger cell cycle arrest. Mutations in the *ATM* gene produce checkpoint defects that allow cells with damaged DNA to progress through the cycle and into mitosis, which is the reason that these individuals have such a high risk of developing cancers.

How does a cell stop its progress from one stage of the cell cycle to the next? We will briefly examine two pathways available to cells to arrest their cell cycle. As indicated in Figure 14.9, and the following discussion, both pathways involve the action of a series of protein kinases. These types of signaling pathways are discussed in detail in the following chapter.

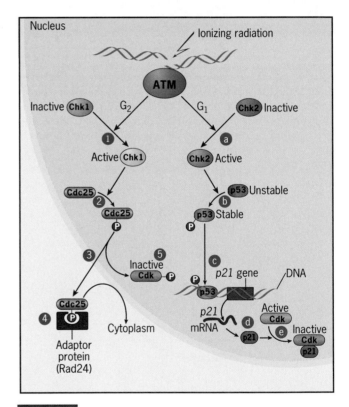

Figure 14.9 **A model for the mechanism of action of a DNA-damage checkpoint.** Recognition of double-strand breaks by the protein ATM leads to the arrest of the cell cycle in either G_1 or G_2 by activating signaling pathways, such as those described in this illustration. In the G_2 pathway, ATM phosphorylates and activates the checkpoint kinase Chk1 (step 1), which phosphorylates and inactivates the phosphatase Cdc25 (step 2), which normally shuttles between the nucleus and cytoplasm (step 3). Once phosphorylated, Cdc25 is bound by an adaptor protein in the cytoplasm (step 4) and cannot be reimported into the nucleus, which leaves the Cdk in its inactivated, phosphorylated state (step 5). In the G_1 pathway, ATM phosphorylates and activates the checkpoint kinase Chk2 (step a), which phosphorylates p53 (step b). p53 is normally very short-lived, but phosphorylation by Chk2 stabilizes the protein, enhancing its ability to activate *p21* transcription (step c). Once transcribed and translated (step d), p21 directly inhibits the Cdk (step e).

[2]Other symptoms of the disease include unsteady posture (ataxia) resulting from degeneration of nerve cells in the cerebellum, permanently dilated blood vessels (telangiectasia) in the face and elsewhere, immune dysfunction, and cells with an abnormally high number of chromosome aberrations. The basis for the first two symptoms has yet to be determined.

1. As illustrated in Figure 14.6, and described on page 585, the entry of a cell into mitosis is triggered by the removal of inhibitory phosphates from the Cdk by the phosphatase Cdc25. When the cell detects the presence of DNA damage, a signal is transmitted by ATM that leads to the phosphorylation and activation of a checkpoint kinase, called Chk1 (step 1, Figure 14.9). Once activated, Chk1 phosphorylates Cdc25 on a particular serine residue (Ser 216 in humans). Once this serine residue is phosphorylated (step 2), the Cdc25 molecule becomes a target for a special adaptor protein that binds to Cdc25 in the cytoplasm (step 4), inhibiting its phosphatase activity and preventing it from being reimported into the nucleus. The absence of Cdc25 from the nucleus leaves the Cdk in an inactive state (step 5) and the cell stuck in G_2.

2. Damage to DNA also leads to the synthesis of proteins that directly inhibit the cyclin-Cdk complex that drives the cell cycle. For example, cells irradiated in G_1 synthesize a protein called p21 (molecular mass of 21 kDa) that inhibits the kinase activity of the G_1 Cdk and thus prevents the cells from entering S phase. ATM is also involved in this checkpoint mechanism because ATM activates another checkpoint kinase called Chk2 (step a, Figure 14.9) that phosphorylates a transcription factor (p53) (step b), which leads to the transcription and translation of the *p21* gene (steps c, d) and subsequent inhibition of Cdk (step e). Approximately 50 percent of all human tumors show evidence of mutations in the *p53* gene, which reflects its importance in the control of cell growth. The role of p53 is discussed at length in Chapter 16.

p21 is only one of at least seven known Cdk inhibitors. The interaction between a different Cdk in-

hibitor (p27) and one of the cyclin-Cdk complexes is shown in Figure 14.10*a*. In this model, the p27 molecule drapes itself across both subunits of the Cdk complex, changing the conformation of the catalytic subunit and inhibiting its kinase activity.

Cdk inhibitors, such as p21 and p27, are also active in cell differentiation. Just before cells begin to differentiate—whether into muscle cells, liver cells, blood cells, or some other type—they typically withdraw from the cell cycle and stop dividing. Cdk inhibitors are thought to either allow or directly induce cell cycle withdrawal. Just as the functions of specific Cdks and cyclins have been studied in knockout mice, so too have their inhibitors. Knockout mice that lack the *p27* gene show a distinctive phenotype: they are larger than normal mice (Figure 14.10*b*), and certain organs, such as the thymus gland and spleen, contain a significantly greater number of cells than those of a normal animal (Figure 14.10*c*). In normal mice, the cells of these particular organs synthesize relatively high levels of p27, and it is presumed that the absence of this protein in the p27-deficient animals allows the cells to divide several more times before they differentiate.

In addition to their roles in checkpoint control and cell differentiation, certain Cdk inhibitors prevent uncontrolled growth that can lead to the development of cancer. This property becomes most apparent when the inhibitor is no longer synthesized so that checkpoint control is disturbed. For example, the gene encoding a Cdk inhibitor called p16 is often deleted in a variety of human tumors. As would be predicted from this observation, knockout mice that lack a gene for the p16 inhibitor exhibit a greatly increased incidence of cancer. The discovery of Cdk inhibitors has created an opportunity for the synthesis of new types of anticancer drugs that block uncontrolled cell growth by mimicking the effects of these inhibitory proteins.

(a) *(b)* *(c)*

Figure 14.10 p27: a Cdk inhibitor that arrests cell cycle progression. (*a*) Three-dimensional structure of a p27-cyclin-Cdk complex. Interaction of the p27 molecule with the cyclin-Cdk dimer (labeled CYCA/CDK2) alters the conformation of the Cdk catalytic subunit, inhibiting its protein kinase activity. (*b*) A pair of littermates at 12 weeks of age. In addition to possessing different genes for coat color, the mouse with dark fur has been genetically engineered to lack both copies of the *p27* gene (denoted as *p27*^{−/−}), which accounts for its larger size. (*c*) Com-

parison of the thymus glands from a normal (left) and a *p27*^{−/−} mouse (right). The gland from the *p27* knockout mouse is much larger owing to an increased number of cells. (A: REPRINTED WITH PERMISSION FROM ALICIA A. RUSSO ET AL., NATURE 382:327, 1995, COURTESY OF NIKOLA PAVLETICH, ASSOCIATE INVESTIGATOR WITH THE HOWARD HUGHES MEDICAL INSTITUTE; © COPYRIGHT 1995 MACMILLAN MAGAZINES LIMITED; B,C: FROM KEIKO NAKAYAMA ET AL., COURTESY OF KEI-ICHI NAKAYAMA, CELL 85:710, 711, 1996; BY PERMISSION OF CELL PRESS.)

1. What is the cell cycle? What are the stages of the cell cycle? How does the cell cycle vary among different types of cells?

2. Describe how the use of [³H]thymidine and autoradiography can be used to determine the length of the various periods of the cell cycle.

3. What is the effect of fusing a cell in G_1 with one in S; of fusing a G_1-phase cell with one in M; of fusing a G_2- or S-phase cell with one in M?

4. How does the activity of MPF vary throughout the cell cycle? How is this correlated with the concentration of cyclins? How does the cyclin concentration affect MPF activity?

5. What are the respective roles of CAK, wee1, and cdc25 in controlling Cdk activity in yeast cells?

6. What is meant by a cell cycle checkpoint? What is its importance? How does a cell stop its progress at one of these checkpoints?

14.2 M PHASE: MITOSIS AND CYTOKINESIS

The name "mitosis" comes from the Greek word *mitos*, meaning "thread." The name was first used in the 1870s to describe the threadlike chromosomes that mysteriously appeared in cells just before they divided in two. **Mitosis** is a process of nuclear division in which replicated DNA molecules of each chromosome are faithfully partitioned into two nuclei.

Mitosis is usually accompanied by **cytokinesis**, a process by which a dividing cell splits in two, partitioning the cytoplasm into two packages. The two daughter cells resulting from mitosis and cytokinesis possess a genetic content identical to each other and to the mother cell from which they arose. Mitosis, therefore, maintains the chromosome number and generates new cells for the growth and maintenance of an organism. Mitosis can take place in either haploid or diploid cells. Haploid mitotic cells are found in fungi, plant gametophytes, and a few animals (including male bees known as drones). Mitosis is a stage of the cell cycle when the cell devotes virtually all of its energy to a single activity—chromosome segregation. As a result, most metabolic activities of the cell, including transcription and translation, are curtailed during mitosis, and the cell becomes relatively unresponsive to external stimuli.

We have seen in previous chapters how much can be learned about the factors responsible for a particular process by carrying out that process outside of a living cell (page 286). Our understanding of the events that occur during mitosis has been greatly aided by the use of extracts prepared from frog eggs. These extracts contain stockpiles of all the materials (histones, tubulin, etc.) necessary to support mitosis, whether it be the original frog nucleus that is present in the egg or a foreign nucleus that is added to the egg extract. In many experiments, the role of a particular protein is studied by removing that protein from the extract by immunodepletion (addition of an antibody against it) and determining whether the process can continue in the absence of that substance (see Figure 14.20 for an example).

Mitosis is generally divided into five stages (Figure 14.11), *prophase*, *prometaphase*, *metaphase*, *anaphase*, and *telophase*, each characterized by a particular series of events. Keep in mind that each of these stages represents a segment of a continuous process; the division of mitosis into arbitrary phases is done only for the sake of discussion and experimentation.

Prophase

During the first stage of mitosis, that of **prophase**, the duplicated chromosomes are prepared for segregation, and the mitotic machinery is assembled.

Formation of the Mitotic Chromosome The nucleus of an interphase cell contains tremendous lengths of chromatin fibers. The extended state of interphase chromatin is ideally suited for the processes of transcription and replication but not for segregation into two daughter cells. Before segregating its chromosomes, a cell converts them into much shorter, thicker structures by a remarkable process of **chromosome compaction** (or *chromosome condensation*), which occurs during early prophase (Figures 14.11 and 14.12).

As described on page 504, the chromatin of an interphase cell is organized into fibers approximately 30 nm in diameter. Mitotic chromosomes are composed of similar types of fibers as seen by electron microscopic examination of whole chromosomes isolated from mitotic cells (Figure 14.13a). It appears that chromosome compaction does not alter the nature of the chromatin fiber, but rather the way that the chromatin fiber is packaged. The internal organization of mitotic chromosomes is revealed by treatment of the chromosomes with solutions that solubilize the histones and the majority of the nonhistone proteins. Electron micrographs of such chromosomes show them to contain a structural framework or scaffold that retains the basic shape of the mitotic chromosome (Figure 14.13b). Loops of DNA are attached at their base to the nonhistone proteins that make up the chromosome scaffold (shown at higher magnification in Figure 12.13a). During interphase, the proteins of the chromosome scaffold are dispersed within the nucleus, probably forming part of the nuclear matrix (page 516).

Given the importance of histones in chromatin structure, one would expect these small basic proteins to be intimately involved in chromosome compaction. It was thought for many years that phosphorylation of histone H1 played a key role in chromosome compaction—until it

Prophase

1. Chromosomal material condenses to form compact mitotic chromosomes. Chromosomes are seen to be composed of two chromatids attached together at the centromere.

2. Cytoskeleton is disassembled and mitotic spindle is assembled.

3. Golgi complex and ER fragment. Nuclear envelope disperses.

Prometaphase

1. Chromosomal microtubules attach to kinetochores of chromosomes.

2. Chromosomes are moved to spindle equator.

Metaphase

1. Chromosomes are aligned along metaphase plate, attached by chromosomal microtubules to both poles.

Anaphase

1. Centromeres split, and chromatids separate.

2. Chromosomes move to opposite spindle poles.

3. Spindle poles move farther apart.

Telophase

1. Chromosomes cluster at opposite spindle poles.

2. Chromosomes become dispersed.

3. Nuclear envelope assembles around chromosome clusters.

4. Golgi complex and ER reforms.

5. Daughter cells formed by cytokinesis.

Figure 14.11 The stages of mitosis. (MICROGRAPHS FROM ANDREW BAJER.)

was shown that compaction could occur in frog extracts even in the absence of this abundant linker histone. Phosphorylation of histone H3, one of the core histones, has been shown to be required for compaction, but its role in the process remains unclear. In recent years, the focus of attention in chromosome compaction has shifted from the

5 μm

Figure 14.12 **The chromosomes of this early prophase nucleus** have begun the process of compaction that converts them into short, rodlike mitotic chromosomes that separate at a later stage in mitosis. (FROM A. T. SUMNER, CHROMOSOMA 100:411, 1991.)

histones to a protein complex called *condensin*. The proteins of condensin were discovered by incubating nuclei in frog extracts (page 590) and identifying those proteins that associated with the chromosomes as they underwent compaction. Removal of any of the condensin proteins from the extracts prevented chromosome compaction. How is condensin able to induce such dramatic changes in chromatin architecture?

Supercoiled DNA occupies a much smaller volume than relaxed DNA (see Figure 10.12), and recent studies suggest that DNA supercoiling plays a key role in compacting a chromatin fiber into the tiny volume occupied by a mitotic chromosome. In the presence of a type II topoisomerase and ATP, condensin is able to bind to DNA in vitro and curl the DNA into positively supercoiled loops (Figure 14.14). This finding fits nicely with the observation that chromosome compaction at prophase in vivo requires topoisomerase II, which is present as one of the major nonhistone proteins of the scaffold of a mitotic chromosome. Condensin is activated at the onset of mitosis by phosphorylation of several of its subunits by the mitotic Cdk responsible for driving cells from G_2 into mitosis. Thus condensin is presumably one of the targets through which Cdks are able to trigger cell cycle activities.

As the result of compaction, the chromosomes of a mitotic cell appear as distinct, rodlike structures. Examination of mitotic chromosomes under the microscope reveals each of them to be composed of two distinct components (Figure 14.15a). Each of these rod-shaped members

(a)

(b)

3 μm

Figure 14.13 **Chromosome compaction during mitosis.**
(a) Electron micrograph of a whole-mount preparation of a human mitotic chromosome. The structure is seen to be composed of a knobby fiber 30 nm in diameter, which is similar to that found in interphase chromosomes. (b) Appearance of a mitotic chromosome after the histones and most of the nonhistone proteins have been removed. The residual proteins form a scaffold from which loops of DNA are seen to emerge (the DNA loops are shown more clearly in Figure 12.13a). (A: COURTESY OF GUNTHER F. BAHR; B: FROM JAMES R. PAULSON AND ULRICH K. LAEMMLI, CELL 12:820, 1977; BY PERMISSION OF CELL PRESS.)

is called a **chromatid** and is formed during replication in the previous interphase. The two "sister" chromatids of each mitotic chromosome are connected to each other se-

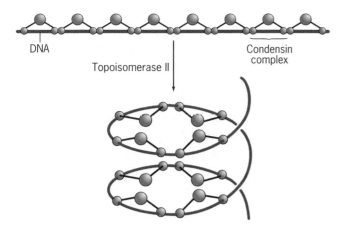

Figure 14.14 **The role of condensin in chromosome compaction.** Each multisubunit condensin complex binds to an extended DNA molecule and, in the presence of topoisomerase II, wraps the DNA into positively supercoiled loops that can be compacted into a mitotic chromosome. (FROM ILLUSTRATIONS BY N.R. COZZARELLI.)

curely at their centromeres and more loosely along their entire length (Figure 14.15a). The connections between chromatids consist of a protein "glue" called *cohesin* that contains a number of highly conserved subunits. In vertebrates, most of the cohesin is released from the chromosomes during prophase, but some of the "glue" remains to hold the sister chromatids together (Figure 14.15b). Mutations in genes encoding the cohesin subunits cause sister chromatids to separate from one another prematurely during mitosis. Sister chromatids are also linked by their DNA molecules, which fail to completely separate from one another during replication. As a result of these two types of connections, segregation of sister chromatids at anaphase requires a protease to cleave the cohesin complex and topoisomerase II to unlink the DNA molecules.

Centromeres and Kinetochores The most notable landmark on a mitotic chromosome is an indentation or *primary constriction*, which marks the position of the centromere (Figure 14.13a). The centromere is the residence of highly repeated DNA sequences (see Figure 10.22) that serve as the binding sites for specific proteins. Examination of sections through a mitotic chromosome reveals the presence of a proteinaceous, buttonlike structure, called the **kineto-chore**, at the outer surface of the centromere of each chromatid (Figure 14.16a,b). The kinetochore assembles on the centromere during prophase. As will be apparent shortly, the kinetochore functions (1) as the site of attachment of the chromosome to the dynamic microtubules of the mitotic spindle (as in Figure 14.16a,b), (2) as the site of localization of several microtubule-based motor proteins (Figure 14.16c), and (3) as a component of an important mitotic checkpoint (see Figure 14.27).

Formation of the Mitotic Spindle We discussed in Chapter 9 how microtubule assembly in animal cells is initiated by a special microtubule-organizing structure, the **centrosome** (page 347). As a cell progresses past G$_2$ and into mitosis, the microtubules of the cytoskeleton undergo sweeping disassembly in preparation for their reassembly as components of a microtubule-containing "machine" called the **mitotic spindle**. As discussed on page 353, the rapid disassembly of the interphase cytoskeleton is thought to be accomplished by the inactivation of proteins that stabilize microtubules (e.g., microtubule-associated proteins, or MAPs) and the activation of proteins that destabilize these polymers.

(a) 1 μm

(b)

Figure 14.15 **Each mitotic chromosome is comprised of a pair of sister chromatids connected to one another by the protein complex cohesin.** (*a*) Scanning electron micrograph of several human metaphase chromosomes showing the paired identical chromatids associated along their length and joined tightly at the centromere. The chromatids are not split apart from one another until anaphase. (*b*) Micrograph of a *Xenopus* metaphase chromosome that has undergone compaction in vitro and then was stained for cohesin (green), which is seen to be present between sister chromatids (red) and presumably holding them together. (A: FROM A. T. SUMNER, CHROMOSOMA 100:415, 1991; B: FROM ANA LOSADA AND TATSUYA HIRANO, CURR. BIOL. 10:R615, 2000.)

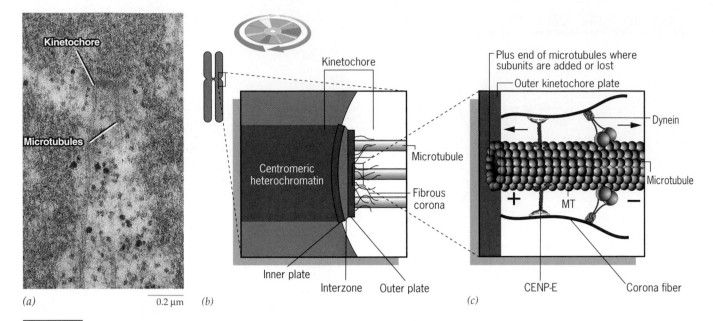

(a) 0.2 μm *(b)* *(c)*

Figure 14.16 **The kinetochore.** (*a*) Electron micrograph of a section through the kinetochore of a mammalian metaphase chromosome, showing its three-layered (trilaminar) structure. Microtubules of the mitotic spindle can be seen to terminate at the kinetochore. (*b*) Schematic representation of the kinetochore, which contains an electron-dense inner and outer plate separated by a lightly staining interzone. The inner plate is thought to consist of a specialized layer of chromatin that is attached to the centromeric heterochromatin of the chromosome, whereas the outer plate contains proteins that are thought to bind to the plus ends of microtubules. Associated with the outer plate is the fibrous corona, which binds motor proteins involved in chromosome movement. (*c*) A schematic model of the disposition of motor proteins at the outer surface of the kinetochore. Cytoplasmic dynein moves toward the minus end of a microtubule, whereas CENP-E, which is a member of the kinesin superfamily, moves toward the plus end. These motors may also play a role in tethering the microtubule to the kinetochore. (A: COURTESY OF PENTTI T. JOKELAINEN, J. ULTRASTRUC. RES. 19:19, 1967.)

To understand the formation of the mitotic spindle, we need to first examine the **centrosome cycle** as it progresses in concert with the cell cycle (Figure 14.17*a*). When an animal cell exits mitosis, it contains a single centrosome containing two centrioles situated at right angles to one another. It had been assumed that the two centrioles remained in close proximity, but recent observations of fluorescently labeled centrioles have revealed that one member of the pair wanders widely through the cytoplasm during G$_1$. The reason for this seemingly aimless migration is obscure. As DNA replication begins in the nucleus at the onset of S phase, the two centrioles of the centrosome initiate their "replication" in the cytoplasm. The formation of each new centriole begins with the appearance of a smaller *daughter* centriole next to each existing centriole and oriented at right angles to it (Figure 14.17*b*). Subsequent elongation of the microtubules of the daughter centriole converts it into a full-length centriole. The initiation of centrosome duplication at the G$_1$-S transition is triggered by cyclin E-Cdk2, the same agent responsible for the onset of DNA replication (Figure 14.8). Errors in centrosome duplication can lead to abnormal cell division and contribute to the development of cancer.

The first stage in the formation of the mitotic spindle in a typical animal cell is the appearance of microtubules in a "sunburst" arrangement, or **aster** (Figure 14.18), around each centrosome during early prophase. The increased microtubule-nucleating activity of the centrosome at mitosis is thought to be stimulated by phosphorylation of key proteins by the newly activated mitotic Cdk, which controls the progression of the cell from G$_2$ to M (page 585).

The process of aster formation is followed by separation of the centrosomes from one another and their subsequent movement around the nucleus toward opposite ends of the cell. Centrosome separation is driven by motor proteins associated with the adjacent microtubules. As the centrosomes separate, the microtubules stretching between them increase in number and elongate (Figure 14.18). Eventually, the two centrosomes reach points opposite one another, thus establishing the two poles of a *bipolar* mitotic spindle (as in Figure 14.17*a*). Following mitosis, one centrosome is distributed to each daughter cell.

Centrosomes are not essential components in the formation of bipolar mitotic spindles. A number of different types of animal cells (including those of the early mouse embryo) lack centrosomes, as do the cells of higher plants, yet all of these cells undergo a relatively typical mitosis. Mitotic spindles can even form in mutant *Drosophila* cells that lack centrosomes and in cells in which the centrosome has been experimentally removed. In all of these cases, the bipolar mitotic spindle forms by a very different pathway than it would in a cell that contained a centro-

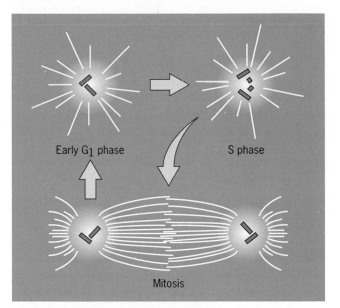

Early G₁ phase

S phase

Mitosis

(a)

(b) 0.3 μm

Figure 14.17 **The centrosome cycle of an animal cell.** (*a*) At the end of mitosis, the centrosome contains a single pair of centrioles oriented at right angles to one another. During G₁ one of the centrioles (the daughter centriole from the previous replication) wanders away from the other centriole. This movement is greatly restricted during S and G₂. During S phase, as the DNA is being replicated, daughter centrioles begin to form adjacent to parental centrioles so that two pairs of centrioles become visible within the centrosome (see *b*). The daughter centrioles continue to elongate during G₂ phase, and at the beginning of mitosis, the two pairs of centrioles move away from one another to opposite poles of the cell. Each pair of centrioles is now part of its own centrosome. As they separate, the centrosomes organize the microtubule fibers that make up the mitotic spindle. (*b*) The centrosome of this cultured cell is seen to contain two pairs of centrioles. The shorter daughter centrioles (or *procentrioles*) are indicated by the arrows. (A: AFTER D. R. KELLOGG, ET AL., REPRODUCED WITH PERMISSION FROM THE ANNUAL REVIEW OF BIOCHEMISTRY, VOL. 63, © 1994, BY ANNUAL REVIEWS INC.; B: FROM J. B. RATTNER AND STEPHANIE G. PHILLIPS, J. CELL BIOL. 57:363, 1973; BY COPYRIGHT PERMISSION OF THE ROCKEFELLER UNIVERSITY PRESS.)

20 μm

Figure 14.18 **Formation of the mitotic spindle.** During prophase, as the chromosomes are beginning to condense, the centrosomes move apart from one another as they organize the bundles of microtubules that form the mitotic spindle. This micrograph shows a cultured newt lung cell in early prophase that has been stained with fluorescent antibodies against tubulin, which reveals the distribution of the cell's microtubules (green). The microtubules of the developing mitotic spindle are seen to emanate as an aster from two sites within the cell. These sites correspond to the locations of the two centrosomes that are moving toward opposite poles at this stage of prophase. The centrosomes are situated above the cell nucleus, which appears as an unstained dark region. (FROM JENNIFER WATERS SHULER, RICHARD W. COLE, AND CONLY L. RIEDER, J. CELL BIOL. 122:364, 1993; BY COPYRIGHT PERMISSION OF THE ROCKEFELLER UNIVERSITY PRESS.)

some. Rather than being nucleated at the poles, the microtubules of the mitotic spindle are nucleated near the chromosomes. Then, once they are polymerized, the minus ends of the microtubules are brought together (i.e., focused) at each spindle pole through the activity of motor proteins, such as cytoplasmic dynein (Figure 14.19*a*). It would appear from these studies that cells possess fundamentally different mechanisms to achieve a similar end result. Remarkably, if one mechanism is rendered inactive, the alternate mechanism is able to finish the job.

Spindle assembly has also been studied in extracts from frog eggs, which are cells that lack centrosomes. Figure 14.19*b* shows a bipolar mitotic spindle (shown in green fluorescence) that had assembled outside a living cell in an extract prepared from *Xenopus* eggs. Not only did the spindle displayed in Figure 14.19*b* form in the absence of centrosomes, it also formed in the absence of kinetochores and chromosomes. To carry out the experiment, the researchers incubated extracts of *Xenopus* interphase eggs with bacterial DNA that had been linked to beads. In the presence of the extract, the bacterial DNA combined with histones and was converted into chromatin that remained associated with the beads. When these chromatin-covered beads were subsequently incubated in extracts from mitotic

(a)

(b)

Figure 14.19 Formation of a spindle pole in the absence of centrosomes. (*a*) In this model, each motor protein has multiple heads, which are bound to different microtubules. The movement of these minus-end-directed motor proteins causes the minus ends of the microtubules to converge to form a distinct spindle pole. This type of mechanism is thought to facilitate the formation of spindle poles in the absence of centrosomes. (*b*) As discussed in the text, this mitotic spindle assembled in extracts of *Xenopus* eggs, which are cells that lack a centrosome. The red spheres consist of chromatin-covered beads that became aligned at the center of the bipolar spindle. The microtubules that gave rise to the spindle shown in this photograph initiated their assembly near the chromatin-covered beads and grew outward, rather than initiating near the poles and growing inward toward the chromosomes, as would occur in a normal cell. Once formed, the microtubules were reorganized, with the help of motor proteins in the extract, into this bipolar spindle. (A: FROM A. A. HYMAN AND E. KARSENTI, CELL 84:406, 1996; BY PERMISSION OF CELL PRESS; B: REPRINTED WITH PERMISSION FROM R. HEALD, ET AL., NATURE VOL. 382, COVER OF 8/1/96; © COPYRIGHT 1996, MACMILLAN MAGAZINES LTD.)

eggs, they acted like artificial chromosomes and induced the nucleation of microtubules. Once assembled, the microtubules became organized into a bipolar spindle by the action of motor proteins in the extract.

The Dissolution of the Nuclear Envelope and the Partitioning of Cytoplasmic Organelles The mitotic spindle is assembled in the cytoplasm, and the chromosomes are compacted in the nucleoplasm. In most eukaryotic cells,

interaction between the spindle and chromosomes is made possible by the disappearance of the nuclear envelope at the end of prophase. The disassembly of the nuclear lamina, which constitutes the inner layer of the nuclear envelope, is promoted by phosphorylation of the lamin molecules by the mitotic Cdk kinase. Two views concerning the fate of the membranous portion of the nuclear envelope are widely held. According to one proposal, the nuclear membranes are progressively fragmented, initially into flattened, saclike vesicles (cisternae) that surround the condensing chromatin, and eventually into a population of small (70 nm) spherical vesicles that disperse throughout the mitotic cell. According to an alternate proposal, the integral proteins of the nuclear membranes become dispersed into the adjoining endoplasmic reticulum prior to dissolution and, consequently, the mitotic cell does not possess a population of vesicles that are derived exclusively from the nuclear membranes. In either case, the dissolution of the nuclear membranes is accompanied by the dismantling of the nuclear pore complexes.

Some of the membranous organelles of the cytoplasm remain relatively intact through mitosis; these include mitochondria, lysosomes, and peroxisomes, as well as the chloroplasts of a plant cell. Considerable controversy has been generated in recent years over the mechanism by which the Golgi complex and endoplasmic reticulum are partitioned during mitosis. According to one view, the forward (anterograde) movement of membrane and materials from the ER to the Golgi complex ceases during prophase, but the movement of vesicles in the opposite (retrograde) direction continues. As a result, the contents of the Golgi complex become incorporated into the ER and the Golgi complex ceases to exist briefly as a distinct organelle. According to an alternate view, the Golgi membranes become fragmented to form a distinct population of small vesicles that are divided between daughter cells. It is generally held that the endoplasmic reticulum undergoes fragmentation during prophase to give rise to large numbers of vesicles that become dispersed throughout the cytoplasm. Fragmentation and dispersal facilitate the partitioning of this membranous network into the two daughter cells.

Prometaphase

The dissolution of the nuclear envelope marks the start of the second phase of mitosis, **prometaphase**, during which mitotic spindle assembly is completed and the chromosomes are moved into position at the center of the cell. At the beginning of prometaphase, compacted chromosomes are scattered throughout the space that was the nuclear region. The chapter-opening photograph on page 580 shows a micrograph of a mitotic cell at this stage. As the microtubules of the spindle penetrate into the central region of the cell, the free ends of the microtubules are seen to grow and shrink in a highly dynamic fashion, as if they were "searching" for a chromosome. Those microtubules that happen to come into proximity of a centromere are "captured" by a kinetochore.

A kinetochore typically makes initial contact with the wall of a microtubule rather than its end. Once the initial contact is made, some chromosomes move actively along the wall of the microtubule, powered by motor proteins located in the kinetochore (Figure 14.16c). Soon, however, the kinetochore tends to become stably associated with the plus end of one or more spindle microtubules. Eventually, the unattached kinetochore on the sister chromatid captures its own microtubules from the opposite pole. As a result, the two sister chromatids of each mitotic chromosome ultimately become connected by their kinetochores to microtubules that extend from opposite poles.

Observations in living cells indicate that prometaphase chromosomes are not moved directly to the center of the spindle, but rather oscillate back and forth in both a poleward and anti-poleward direction. Ultimately, the chromosomes of a prometaphase cell are moved by a process called *congression* toward the center of the mitotic spindle, midway between the poles. The forces required for these movements are generated by motor proteins associated with both the kinetochores and arms of the chromosomes (see Figure 14.30a). Figure 14.20 shows the effects on congression of the absence of one of the kinesin-like chromosomal proteins.

As the chromosomes move toward the center of the mitotic spindle, the longer microtubules are shortened, while the shorter microtubules are elongated. Shortening and elongation of microtubules occur primarily by loss or gain of subunits at the plus end of the microtubule (Figure 14.21). Remarkably, this dynamic activity occurs while the plus end of each microtubule remains firmly attached to a kinetochore. The motor proteins located at the kinetochores (Figure 14.16c) may play a key role in tethering the microtubule to the chromosome while it loses and gains subunits.

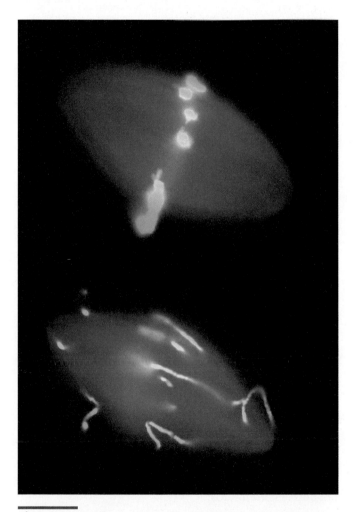

Figure 14.20 The consequence of a missing motor protein on chromosome alignment during prometaphase. The top micrograph shows a mitotic spindle that has assembled in a complete frog egg extract. The lower micrograph shows a mitotic spindle that has assembled in a frog egg extract that has been depleted of a particular kinesinlike protein called Xkid that is present along the arms of prometaphase chromosomes. In the absence of this motor protein, the chromosomes fail to align at the center of the spindle and instead are found stretched along spindle microtubules. Xkid is thought to be required for movements of the chromosomes away from the poles (see Figure 14.30a). (FROM CELIA ANTONIO ET AL, COURTESY OF ISABELLE VERNOS, CELL VOL. 102, COVER #4, 2000, BY PERMISSION OF CELL PRESS.)

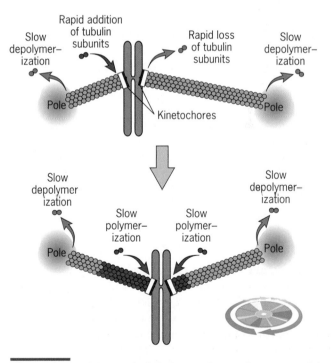

Figure 14.21 Microtubule behavior during formation of the metaphase plate. Initially, the chromosome is connected to microtubules from opposite poles that are very different in length. As prometaphase continues, this imbalance is corrected as the result of the shortening of microtubules from one pole, due to the rapid loss of tubulin subunits at the kinetochore, and the lengthening of microtubules from the opposite pole, due to the rapid addition of tubulin subunits at the kinetochore. These changes are superimposed over a much slower polymerization and depolymerization (lower drawing) that occur continually during prometaphase and metaphase, giving the microtubules of the mitotic spindle a dynamic character.

10 μm

Figure 14.22 **The engagement of a chromosome during prometaphase and its movement to the metaphase plate.** This series of photographs taken from a video recording shows the movements of the chromosomes of a newt lung cell over a period of 100 seconds during prometaphase. Although most of the cell's chromosomes were nearly aligned at the metaphase plate at the beginning of the sequence, one of the chromosomes had failed to become attached to spindle fibers (arrow). The wayward chromosome has attached to the spindle in B and then moves poleward with variable velocity until it reaches a stable position in F. The position of a pole is indicated by the arrowhead in A. (FROM STEPHEN P. ALEXANDER AND CONLY L. RIEDER, J. CELL BIOL. 113:807, 1991; BY COPYRIGHT PERMISSION OF THE ROCKEFELLER UNIVERSITY PRESS.)

Eventually, each chromosome moves into position along a plane at the center of the spindle, so that microtubules from each pole are equivalent in length. The movement of a wayward chromosome from a peripheral site to the center of the mitotic spindle during prometaphase is shown in the series of photos in Figure 14.22.

Metaphase

Once all of the chromosomes have become aligned at the spindle equator—with one chromatid of each chromosome connected by its kinetochore to microtubules from one pole and its sister chromatid connected by its kinetochore to microtubules from the opposite pole—the cell has reached the stage of **metaphase** (Figure 14.23). The plane of alignment of the chromosomes at metaphase is referred to as the *metaphase plate*. The mitotic spindle of the metaphase cell contains a highly organized array of microtubules that is ideally suited for the task of separating the duplicated chromatids positioned at the center of the cell. Functionally, the microtubules of the metaphase spindle of an animal cell can be divided into three groups of fibers (Figure 14.23):

1. *Astral microtubules* that radiate outward from the centrosome into the region outside the body of the spindle. They probably help position the spindle apparatus in the cell and determine the plane of cytokinesis.

2. *Chromosomal* (or *kinetochore*) *microtubules* that extend from the centrosome to the kinetochores of the chromosomes. During metaphase, the chromosomal microtubules exert a pulling force on the kinetochores. As a result, the chromosomes are maintained in the equatorial plane by a "tug-of-war" between balanced pulling forces exerted by chromosomal spindle fibers from opposite poles. During anaphase, chromosomal

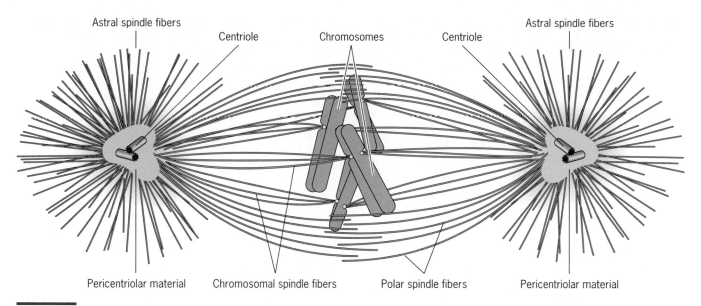

Astral spindle fibers Centriole Chromosomes Centriole Astral spindle fibers

Pericentriolar material Chromosomal spindle fibers Polar spindle fibers Pericentriolar material

Figure 14.23 The mitotic spindle of an animal cell. Each spindle pole contains a pair of centrioles surrounded by amorphous pericentriolar material at which the microtubules are nucleated. Three types of spindle fibers—astral, chromosomal, and polar spindle fibers—are evident, and their functions are described in the text. All of the spindle microtubules have their minus ends at the centrosome.

microtubules are required for the movement of the chromosomes toward the poles.

3. *Polar* (or *interpolar*) *microtubules* that extend from the centrosome past the chromosomes. The polar microtubules form a structural basket that maintains the integrity of the spindle. Individual microtubules from one centrosome do not extend across the entire gap to the other pole but interdigitate with their counterparts from the opposite centrosome.

As one watches films or videotapes of mitosis, metaphase appears as a stage during which the cell pauses for a brief period, as if all the mitotic activities suddenly come to a halt. However, experimental analysis reveals that metaphase is a time when important events occur.

Microtubule Flux in the Metaphase Spindle Even though there is no obvious change in length of the chromosomal microtubules as the chromosomes are aligned at the metaphase plate, studies using fluorescently labeled tubulin indicate that the microtubules exist in a highly dynamic state. Subunits are rapidly lost and added from the plus ends of the chromosomal microtubules, even though these ends are presumably attached to the kinetochore. Thus, the kinetochore does not act like a cap at the end of the microtubule, blocking the entry or exit of terminal subunits, but rather it is the site of dynamic activity. Because more subunits are added to the plus end than are lost, there is a net addition of subunits at the kinetochore. Meanwhile, the minus end of the microtubules experiences a net loss, and thus subunits are thought to move ("treadmill") along the chromosomal microtubules from the kinetochore toward the pole. The poleward flux of tubulin subunits in a mitotic spindle is indicated in the experiment illustrated in Figure 14.24.

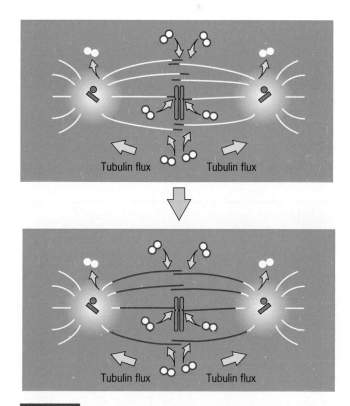

Tubulin flux Tubulin flux

Tubulin flux Tubulin flux

Figure 14.24 Tubulin flux through the microtubules of the mitotic spindle at metaphase. Even though the microtubules appear stationary at this stage, injection of fluorescently labeled tubulin subunits indicates that the components of the spindle are in a dynamic state of flux. Subunits are incorporated preferentially at the kinetochores of the chromosomal microtubules and the equatorial ends of the polar microtubules, and they are lost preferentially from the minus ends of the microtubules in the region of the poles. It would appear that tubulin subunits move through the microtubules of a metaphase spindle at a rate of about 1 μm/min.

Anaphase

Anaphase begins when the sister chromatids of each chromosome split apart and begin their movement toward opposite poles.

The Control of Anaphase Genetic and biochemical approaches over the past decade have produced a large body of information on the mechanism responsible for initiation of anaphase. It was pointed out on page 586 that two distinct multiprotein complexes, SCF and APC, act at different stages of the cell cycle to ubiquinate proteins and target them for destruction by a proteasome. The periods during the cell cycle in which the SCF and APC complexes are active is shown in Figure 14.25a. As discussed on page 586 and illustrated in Figure 14.25a, SCF acts primarily during interphase. The second of these complexes, the **anaphase promoting complex**, or **APC**, could be described as the "mitotic terminator" because it plays a key role in regulating the events that occur toward the end of mitosis. APC contains about a dozen subunits, one of which plays a key role in determining which proteins will serve as the APC substrate. This "substrate-targeting" subunit is represented by members of a family of proteins. Two members of this family in budding yeast cells—Cdc20 and Cdh1—play an important role in substrate selection during mitosis. APC complexes containing one or the other of these subunits are designated as APC^{Cdc20} or APC^{Cdh1} (Figure 14.25a).

APC^{Cdc20} becomes activated at the metaphase/anaphase transition (Figure 14.25a). When present as a subunit of the APC, Cdc20 directs the enzyme complex to ubiquinate a major anaphase inhibitor called *securin* be-

cause it is involved in securing the attachment between sister chromatids. The destruction of securin by proteasomes at the end of metaphase starts a chain reaction that leads to the cleavage of the cohesin complex that holds sister chromatids together (Figure 14.25b). Destruction of cohesin triggers the separation of the chromatids and their migration to separate spindle poles during anaphase.

Near the end of mitosis in budding yeast cells, the Cdc20 subunit of the APC is replaced by the other substrate-targeting subunit, Cdh1 (Figure 14.25a). When Cdh1 is associated with the APC, the enzyme shows a preference for ubiquinating certain mitotic cyclins. Destruction of these mitotic cyclins leads to a precipitous drop in activity of the mitotic Cdk and the progression of the cell out of mitosis and into the G_1 phase of the next cell cycle (Figure 14.25b).

The Events of Anaphase All the chromosomes of the metaphase plate are split in synchrony at the onset of anaphase, and the chromatids (now referred to as chromosomes, because they are no longer attached to their duplicates) begin their poleward migration (see Figure 14.11). The movement of each chromosome toward a pole is accompanied by the shortening of the microtubules attached to its kinetochore. Shortening results primarily from the net loss of subunits at the kinetochore during anaphase. This represents a reversal of the tendency for tubulin subunits to be added to this end of the microtubule during prometaphase and metaphase (Figures 14.21 and 14.24). As the chromosome moves during anaphase, its centromere is seen at its leading edge with the arms of the chromosome trailing behind (Figure 14.26).

Figure 14.25 SCF and APC activities during the cell cycle. SCF and APC are multisubunit complexes that ubiquinate substrates, leading to their destruction by proteasomes. (*a*) SCF is active primarily during interphase, whereas APC (anaphase promoting complex) is active during mitosis. Two different versions of APC are indicated. These two APCs differ in containing either a Cdc20 subunit or a Cdh1 subunit, which alters the substrates recognized by the APC. APC^{Cdc20} is active earlier in mitosis than is APC^{Cdh1}. (*b*) APC^{Cdc20} is responsible for destroying proteins, such as securin, that inhibit anaphase. Destruction of these substrates promotes the metaphase–anaphase transition. APC^{Cdh1} is responsible for ubiquinating proteins, such as mitotic cyclins, that inhibit exit from mitosis. Destruction of these substrates promotes the mitosis-G_1 transition. (A: REPRINTED FROM J.-M. PETERS, CURR. OPIN. CELL BIOL. 10:762, 1998, COPYRIGHT 1998, WITH PERMISSION FROM ELSEVIER SCIENCE.)

The movement of chromosomes toward opposite poles is very slow relative to other types of cellular movements, proceeding at approximately 1 μm per minute, a value calculated by one mitosis researcher to be equivalent to a trip from North Carolina to Italy that would take approximately 14 million years. Because distances in the cell are small, chromosome movement is completed in anywhere from 2 to 60 minutes. The slow rate of chromosome movement ensures that the chromosomes segregate accurately and without entanglement. The forces thought to power chromosome movement during anaphase are discussed in a following section.

The movement of the chromosomes toward the poles is referred to as **anaphase A** to distinguish it from a separate but simultaneous movement, called **anaphase B**, in which the two spindle poles move farther apart. Recall that the polar microtubules from opposite poles overlap in the region of the spindle equator (Figures 14.23 and 14.26). Consequently, separation of the poles during anaphase B could be accomplished by the sliding of overlapping microtubules from opposite poles over one another in opposite directions. The elongation of the mitotic spindle during anaphase B is accompanied by the net addition of tubulin subunits to the plus ends of the polar microtubules. Thus, subunits can be preferentially added to polar microtubules and removed from chromosomal microtubules at the same time in different regions of the same mitotic spindle.

The Spindle Checkpoint As discussed on page 588, cells possess **checkpoint mechanisms** that monitor the status of events during the cell cycle. One of these checkpoints operates at the transition between metaphase and anaphase. The **spindle checkpoint**, as it is often called, is best revealed when one or more chromosomes fails to become aligned properly at the metaphase plate. When this happens, the checkpoint mechanism delays the onset of anaphase until the misplaced chromosome has assumed its proper position along the spindle equator, as is evident in Figure 14.22. If a cell were not able to postpone chromosome segregation, it would greatly elevate the risk of the daughter cells receiving an abnormal number of chromosomes.

How does the cell determine whether or not all of the chromosomes are properly aligned at the metaphase plate? Studies suggest that cells monitor at least two distinct properties of the mitotic spindle that provide this information.

1. Cells that possess mutant copies of a gene called *MAD2* fail to arrest at metaphase when their chromosomes are misaligned. Moreover, the protein encoded by this gene is normally localized at the kinetochores of prometaphase chromosomes and misaligned metaphase chromosomes, but it is not present on chromosomes that have become properly aligned at the metaphase plate (Figure 14.27). The presence of Mad2 at the kinetochores provides a "wait" signal that delays a cell's progression into anaphase. Studies on both yeast and human cells indicate that the "wait"

Figure 14.26 The mitotic spindle and chromosomes at late anaphase. The arms of the chromosomes are seen to trail behind as the kinetochores attached to the chromosomal spindle fibers lead the way toward the respective poles. The chromosomal spindle fibers at this late anaphase stage are extremely short and are no longer evident between the forward edges of the chromosomes and the poles. The polar spindle fibers, however, are clearly visible in the interzone between the separating chromosomes. Activities of these polar fibers are thought to be responsible for causing the separation of the poles that occurs during anaphase B. (FROM J. RICHARD MCINTOSH, IN MICRO-TUBULES, J. HYAMS AND C. L. LLOYD, EDS. COPYRIGHT © 1994. REPRINTED BY PERMISSION OF WILEY-LISS, INC., A SUBSIDIARY OF JOHN WILEY & SONS, INC.)

Figure 14.27 The spindle checkpoint. Fluorescence micrograph of a mammalian cell in late prometaphase labeled with antibodies against the spindle checkpoint protein Mad2 (pink) and tubulin of the microtubules (green). The chromosomes appear blue. Only one of the chromosomes of this cell is seen to contain Mad2, and this chromosome has not yet become aligned at the metaphase plate. The presence of Mad2 on the kinetochore of this chromosome is sufficient to arrest the initiation of anaphase. (FROM JENNIFER WATERS SHULER, REY-HUEI CHEN, ANDREW W. MURRAY, AND E. D. SALMON, J. CELL BIOL. VOL. 141, COVER #5, 1998, BY COPYRIGHT PERMISSION OF THE ROCKEFELLER UNIVERSITY PRESS.)

signal is transmitted by Mad2 molecules that interfere with the proteolytic events that drive the onset of anaphase. For example, Mad2 has been shown to bind to Cdc20, inhibiting its activation of the APC, an event that is required for the metaphase-to-anaphase transition (Figure 14.25). Cells that possess mutant forms of Cdc20 that are unable to bind Mad2 fail to arrest mitosis in the presence of spindle damage. As each chromosome becomes aligned at the metaphase plate, its kinetochores lose all of their Mad2 molecules. It is only after the protein is absent from all of the chromosomes that APC activation can occur and anaphase can begin.

2. If a metaphase cell contains a chromosome that is connected by microtubules to only one spindle pole (a "monoattached" chromosome), the cell delays the onset of anaphase until the chromosome becomes connected to the other spindle pole (to become a "biattached" chromosome) and aligned at the equator. One of the properties that distinguishes a monoattached from a biattached chromosome is the type of mechanical tension to which it is subjected. A biattached chromosome is subjected to tension from both poles, whereas a monoattached chromosome is subjected to tension from only one pole (Figure 14.28a). Mechanical tension can be applied artificially to a chromosome of a living cell by hooking the chromosome with a glass microneedle and applying a tiny pulling force. Pulling on a monoattached metaphase chromosome with a microneedle mimics the tension normally applied from the opposite spindle pole and turns off the inhibitory signal from the chromosomes (Figure 14.28b). As a result, the cell proceeds into anaphase. This finding suggests that anaphase normally begins only after all of the chromosomes are subjected to equal mechanical tension exerted from both spindle poles.

These two signals may operate independently, or they may operate at different points along a common pathway. For example, the establishment of tension on a specific kinetochore may provide the stimulus that releases the Mad2 protein from that particular chromatid. The spindle checkpoint is lacking in many tumor cells, so they continue to divide despite the presence of marked chromosome abnormalities (see Figure 16.4).

Telophase

As the chromosomes near their respective poles, they tend to collect in a mass, which marks the beginning of the final stage of mitosis, or **telophase** (Figures 14.11 and 14.29). During telophase, daughter cells return to the interphase condition. The nuclear envelope reforms, and the chromosomes become more and more dispersed until they disappear from view under the microscope. Meanwhile, the vesicles that had been part of the endoplasmic reticulum fuse to begin the reformation of the cell's membranous cy-

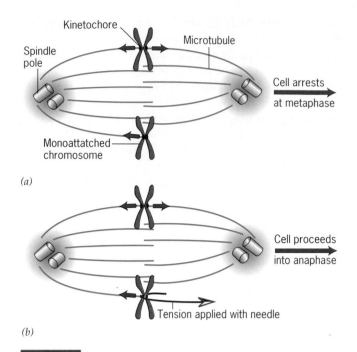

(a)

(b)

Figure 14.28 Experimental demonstration of the importance of mechanical tension in metaphase checkpoint control. (*a*) The biattached chromosome at the top of the drawing is subjected to equivalent tension from both poles (indicated by the red arrows), whereas the monoattached chromosome at the bottom is subjected to tension from only one pole. As a result of this imbalance, the cell is arrested at metaphase. (*b*) When the monoattached chromosome is pulled toward the unattached pole by a glass needle, the artificially applied tension mimics that present in a biattached chromosome, and the cell overrides the arrest and proceeds into anaphase. (The actual experiment performed by X. Li and Bruce Nicklas was carried out on meiotic spermatocytes, but the results are thought to apply just as well to mitotic cells.)

toplasmic network. The actual partitioning of the cytoplasm into two daughter cells occurs by a process to be discussed shortly. First, however, let's consider the forces required for chromosome movements during mitosis.

Forces Required for Mitotic Movements

Mitosis is characterized by extensive movements of cellular structures. Prophase is accompanied by movement of the spindle poles to opposite ends of the cell, prometaphase by movement of the chromosomes to the spindle equator, anaphase A by movement of the chromosomes from the spindle equator to its poles, and anaphase B by the elongation of the spindle. Over the past decade, a variety of different molecular motors have been identified in different locations in mitotic cells of widely diverse species. All of the motors thought to be involved in mitotic movements are microtubule motors, including a number of different kinesin-related proteins (page 344) and cytoplasmic dynein. Some of the motors move toward

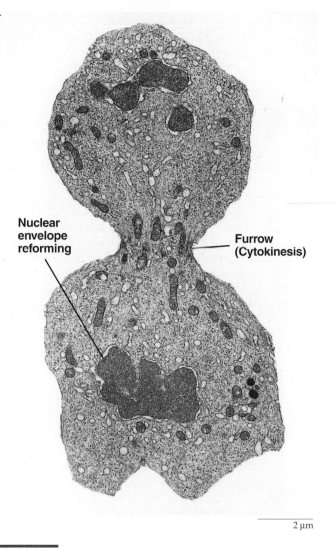

Nuclear envelope reforming

Furrow (Cytokinesis)

2 μm

Figure 14.29 Telophase. Electron micrograph of a section through an ovarian granulosa cell in telophase. (FROM J. A. RHODIN, HISTOLOGY, OXFORD, 1974.)

the plus end of the microtubule, others toward the minus end. Motor proteins have been localized at the spindle poles, along the spindle fibers, and within both the kinetochores and arms of the chromosomes.

The roles of specific motor proteins have been gleaned primarily from three types of studies: (1) analysis of the phenotype of cells that lack the motor because they carry a mutation in a gene that encodes part of the motor protein, (2) injection of antibodies against the motor into cells at various stages of mitosis, and (3) depletion of the motor from cell extracts in which mitotic spindles are formed. Although firm conclusions about the functions of various motor proteins cannot yet be drawn, a number of tentative interpretations are possible (Figure 14.30):

- Motor proteins located along the polar microtubules probably contribute by keeping the poles apart (Figure 14.30a,b).

- Motor proteins residing on the chromosomes are probably important in the movements of the chromosomes during prometaphase (Figure 14.30a), in maintaining the chromosomes at the metaphase plate (Figure 14.30b), and in separating the chromosomes during anaphase (Figure 14.30c), as discussed in the next section.

- Motor proteins situated along the interdigitating (polar) microtubules in the region of the spindle equator are probably responsible for elongating the spindle during anaphase B (Figure 14.30c).

Forces Required for Chromosome Movements at Anaphase Observations of time-lapse movies of mitosis and examination of sections of dividing cells in the electron microscope led early researchers to conclude that the chromosomes were pulled toward the poles during anaphase A in the same manner that a fish might be pulled into a boat with a line and fishing reel. According to this view, which is known as the *traction fiber hypothesis*, chromosomal microtubules act as the line pulling on the chromosomes. The observed shortening of the chromosomal microtubules would occur as they were disassembled by loss of subunits at the poles. This seemingly logical and uncomplicated view of chromosome movement was largely abandoned when studies on cultured cells indicated that depolymerization of chromosomal microtubules during anaphase occurs at the kinetochores, not at the poles.

Cytoplasmic dynein and at least two kinesin-related proteins have been identified at the kinetochores of mitotic chromosomes, indicating that chromosomes are endowed with all of the motor equipment needed to move themselves from one place to another along a microtubule. Fluorescently labeled antibodies against dynein indicate that, during interphase, the protein is localized in the cytoplasm. However, as the cell enters mitosis, the motor protein becomes localized to both the spindle poles and the kinetochores. Cytoplasmic dynein moves toward the minus end of a microtubule and, consequently, would tend to move an attached chromosome toward the poles. If cytoplasmic dynein is involved in chromosome movements at anaphase, this activity is accompanied by the simultaneous depolymerization of chromosomal microtubules at the kinetochore.

An alternate hypothesis to account for the forces required to move chromosomes toward the poles during anaphase was initially proposed by Shinya Inoué of the Marine Biological Laboratory at Woods Hole. Inoué proposed that the disassembly of chromosomal microtubules during anaphase was not simply a consequence of chromosome movement but the cause of it. Inoué suggested that disassembly of the microtubules that compose a spindle fiber could generate sufficient force to pull a chromosome forward. One would not expect the dissolution of a fiber at one end to provide much pulling force; however,

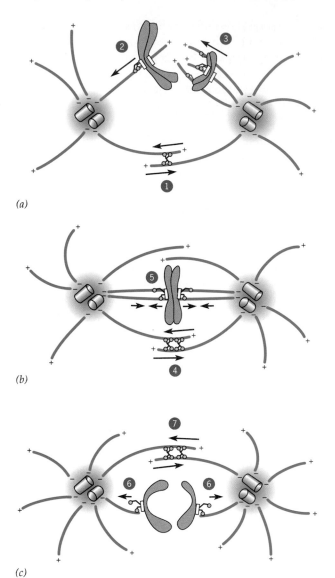

(a)

(b)

(c)

Figure 14.30 Proposed activity of motor proteins during mitosis. (*a*) Prometaphase. The two halves of the mitotic spindle are moving apart from one another to opposite poles, which is thought to result from the action of plus-end-directed motors that are associated with the polar microtubules (step 1). Meanwhile, the chromosomes have become attached to the chromosomal microtubules and can be seen oscillating back and forth along the microtubules. Ultimately, the chromosomes are moved to the center of the spindle, midway between the poles. Prometaphase movements are thought to result from activity of minus-end-directed motors (i.e., cytoplasmic dynein) residing at the kinetochore (step 2) and plus-end-directed motors (i.e., kinesin-like proteins) residing at the kinetochore and especially along the chromosome arms (step 3)(see Figure 14.20). (*b*) Metaphase. The two halves of the spindle maintain their separation as the result of plus-end-directed motor activity associated with the polar microtubules (step 4). The chromosomes are thought to be maintained at the equatorial plane by the balanced activity of motor proteins residing at the kinetochore (step 5). (*c*) Anaphase. The movement of the chromosomes toward the poles is thought to require the activity of kinetochore motors (step 6) that either move along the chromosomal microtubules or anchor the chromosomes to the microtubules as they depolymerize. The separation of the poles (anaphase B) is thought to result from the continuing activity of the plus-end-directed motors of the polar microtubules (step 7). (AFTER K. E. SAWIN AND J. M. SCHOLEY, TRENDS CELL BIOL. 1:122, 1991.)

calculations reveal that very little force is required to move an object as small as a chromosome such a short distance—probably as little as 10^{-8} dyne (equivalent to hydrolyzing 20–30 molecules of ATP).

Experimental support for the disassembly model has come from studies in which chromosomes that are attached to microtubules undergo movement as the result of the depolymerization of the microtubules. An example of one of these experiments is shown in Figure 14.31. In this case, the movement of a microtubule-bound chromosome (arrow) occurs in vitro following dilution of the medium. Dilution reduces the concentration of soluble tubulin, which in turn promotes the depolymerization of the microtubules. Chromosome movement depicted in Figure 14.31 occurs in the absence of ATP, which would be required if the force were generated by a motor protein. It would appear from this and other experiments that microtubule disassembly alone can generate sufficient force to pull chromosomes considerable distances, but whether or not this mechanism is utilized by the cell remains open to debate.

Even if microtubule disassembly is not responsible for *generating* the forces involved in chromosome movement, shortening of the chromosomal fibers is still expected to be the slowest step in the overall process and thus control the *rate* of chromosome movement. Conversely, even if microtubule motors do not serve as force-generating agents, they could still play a critical role in maintaining the dynamic attachment of kinetochores to the plus ends of chromosomal microtubules as they lose subunits.

Cytokinesis

Mitosis accomplishes the segregation of duplicated chromosomes into daughter nuclei, but the cell is divided into two daughter cells by a separate process called **cytokinesis**. The splitting of a cell in two has intrigued cell biologists for many years. The first hint of cytokinesis in animal cells appears during late anaphase as an indentation of the cell surface in a narrow band around the cell. As time progresses, the indentation deepens to form a furrow completely encircling the cell. The plane of the furrow lies in the same plane previously occupied by the chromosomes of the metaphase plate, so that the two sets of chromosomes are ultimately partitioned into different cells (as in Figure 14.29). The furrow continues to deepen until opposing surfaces make contact with one another and fuse in the center of the cell, thus splitting the cell in two (Figure 14.32).

Our present concept of the mechanism responsible for cytokinesis stems from a proposal made by Douglas

Figure 14.31 Experimental demonstration that microtubule disassembly can move attached chromosomes in vitro. The structure at the lower left is the remnant of a lysed protozoan. In the presence of tubulin, the basal bodies at the surface of the protozoan were used as sites for the initiation of microtubules, which grew outward into the medium. Once the microtubules had formed, condensed mitotic chromosomes were introduced into the chamber and allowed to bind to the ends of the microtubules. The arrow shows a chromosome attached to the end of a bundle of microtubules. The concentration of soluble tubulin within the chamber was then decreased by dilution, causing the depolymerization of the microtubules. As shown in this video sequence, the shrinkage of the microtubules was accompanied by the movement of the attached chromosome. Bar equals 5 mm (FROM MARTINE COUE, VIVIAN A. LOMBILLO, AND J. RICHARD MCINTOSH, J. CELL BIOL. 112:1169, 1991, BY COPYRIGHT PERMISSION OF THE ROCKEFELLER UNIVERSITY PRESS.)

Marsland in the 1950s known as the **contractile ring theory** (Figure 14.33a). Marsland proposed that the force required to cleave a cell is generated in a thin band of contractile cytoplasm located in the *cortex*, just beneath the plasma membrane of the furrow. Microscopic examination of the cortex beneath the furrow of a cleaving cell reveals the presence of large numbers of actin filaments (Figure 14.33b, 14.34a) aligned in parallel array.

Interspersed among the actin filaments are a smaller number of short, bipolar myosin filaments. These filaments are composed of myosin II, as evidenced by their binding of antimyosin II antibodies (Figure 14.34b). The importance of myosin II in cytokinesis is evident from the fact that (1) anti-myosin II antibodies cause the rapid cessation of cytokinesis when injected into a dividing cell (Figure 14.34c), and (2) cells lacking a functional myosin II gene carry out nuclear division by mitosis but cannot divide normally into daughter cells. The force-generating mechanism operating during cytokinesis is thought to be similar to the actinomyosin-based contraction of muscle cells. Whereas the sliding of actin filaments of a muscle cell brings about the shortening of the muscle fiber, sliding of the filaments of the contractile ring pulls the cortex and attached plasma membrane toward the center of the cell. As a result, the contractile ring constricts the equatorial region of the cell, much like pulling on a purse string narrows the opening of a purse.

One of the most remarkable aspects of the cortical ring is the rapidity with which it is assembled just prior to cytokinesis and then dismantled. Actin filaments that are assembled in the contractile ring are composed of the same subunits that had previously been present as part of the cytoskeleton of the interphase cell. What determines the site of actin filament assembly? Early studies on sea urchin eggs demonstrated that the contractile ring forms in a plane

Figure 14.32 Division of a sea urchin egg into two cells by cytokinesis. (Courtesy of Tryggve Gustafson.)

(a)

(b)

Figure 14.33 **The formation and operation of the contractile ring during cytokinesis.** (a) Actin filaments become assembled in a ring at the cell equator. Contraction of the ring, which requires the action of myosin, causes the formation of a furrow that splits the cell in two. (b) Electron micrograph of a section through the cleavage furrow of a dividing sea urchin egg, showing the actin filaments aligned beneath the surface of the plasma membrane. (B: COURTESY OF THOMAS E. SCHROEDER.)

(a) (b) 10 μm

(c) 30 μm

Figure 14.34 **Experimental demonstration of the importance of myosin in cytokinesis.** (a, b) Localization of actin and myosin II in a *Dictyostelium* amoeba during cytokinesis as demonstrated by double-stain immunofluorescence. (a) Actin filaments (orange) are located at both the cleavage furrow and the cell periphery where they play a key role in cell movement (Section 9.7). (b) Myosin II (green) is localized at the cleavage furrow, part of a contractile ring that encompasses the equator. (c) A starfish egg that had been microinjected with an antibody against starfish myosin, as observed under polarized light (which causes the mitotic spindles to appear either brighter or darker than the background due to the presence of oriented microtubules; see Section 18.1). While cytokinesis has been completely suppressed by the antibodies, mitosis (as revealed by the mitotic spindles) continues unaffected. (A: COURTESY OF YOSHIO FUKUI; B: FROM DANIEL P. KIEHART, ISSEI MABUCHI, AND SHINYA INOUÉ, J. CELL BIOL. 94:167, 1982; BY COPYRIGHT PERMISSION OF THE ROCKEFELLER UNIVERSITY PRESS.)

midway between the spindle poles, even if one of the poles is displaced by a microneedle inserted into the cell. An example of the relationship between the position of the spindle poles and cleavage plane is shown in the micrographs of Figure 14.35. These studies suggest that the site of actin assembly, and thus the plane of cytokinesis, is determined

(a)

(b)

(c)

Figure 14.35 The site of formation of the cleavage plane and the time at which cleavage occurs depends on the position of the mitotic spindle. (*a*) This echinoderm egg was allowed to divide once to form a two-cell embryo. Then, once the mitotic spindle appeared in each of the two cells, one of the cells was drawn into a micropipette, causing it to assume a cylindrical shape. The two dark spots in each cell are the spindle poles that have formed prior to the second division of each cell. (*b*) Nine minutes later the cylindrical cell has completed cleavage, while the spherical cell has not yet begun to cleave. These photos indicate that (1) the cleavage plane forms between the spindle poles, regardless of their position, and (2) cleavage occurs more rapidly in the cylindrical cell. Bar = 80 μm. (*c*) These results can be explained by assuming that (1) the cleavage plane (red bar) forms where the astral microtubules overlap, and (2) cleavage occurs earlier in the cylindrical cell because the distance from the poles (green spheres) to the site of cleavage is reduced, thereby shortening the time it takes the cleavage signal to reach the surface. (A,B: FROM CHARLES B. SHUSTER AND DAVID R. BURGESS, J.CELL BIOL. 146:987, 1999, BY COPYRIGHT PERMISSION OF THE ROCKEFELLER UNIVERSITY PRESS.)

by a signal emanating from the spindle poles. The signal is thought to travel from the spindle poles to the cell cortex along the astral microtubules. When the distance between the poles and the cortex is modified experimentally, the timing of cytokinesis can be dramatically altered (Figure 14.35). Studies on cultured mammalian cells support the concept of a cytokinesis signal but suggest that the source of the stimulus in mammalian cells emanates from the central part of the mitotic spindle rather than its poles.

We have seen in previous chapters that a rise in cytosolic Ca^{2+} concentration often triggers cellular responses, including muscle contraction (page 378). The signal that stimulates cytokinesis may also act by increasing cytosolic Ca^{2+} levels through the release of these ions from storage sites in the cytoplasm. Studies using calcium-sensitive fluorescent dyes indicate that the concentration of free Ca^{2+} in the cell cortex rises dramatically during furrow formation (Figure 14.36). It is speculated that the elevated Ca^{2+} concentration activates an enzyme called myosin light-chain kinase, which phosphorylates a particular serine residue of one of the light chains of the myosin molecule leading to contraction of the actin-containing cortical ring.

Cytokinesis in Plant Cells: Formation of the Cell Plate
Plant cells, which are enclosed by a relatively inextensible cell wall, undergo cytokinesis by a very different mechanism. Unlike animal cells, which are constricted by a furrow that advances inward from the outer cell surface, plant cells must construct an extracellular wall inside a living cell. Wall formation starts in the center of the cell and grows outward to meet the existing lateral walls. The formation of a new cell wall begins with the construction of a simpler precursor, which is called the **cell plate**.

The first sign of cell plate formation is seen in late anaphase–early telophase with the appearance of the phragmoplast in the center of the dividing cell. The **phragmoplast** consists of clusters of interdigitating microtubules ori-

Figure 14.36 Cytokinesis is associated with a rise in Ca^{2+} concentration. These amphibian eggs were injected with Calcium Green, a compound that fluoresces when complexed with calcium ions. First cleavage began at 0′, and the subsequent changes in Ca^{2+} concentration were monitored. The scale indicates the relative concentration of Ca^{2+} with red representing 2.5 times that of the resting egg. Arrowheads indicate site of cleavage furrow. (FROM AKIRA MUTO ET AL., J. CELL BIOL. 135:184, 1996, BY COPYRIGHT PERMISSION OF THE ROCKEFELLER UNIVERSITY PRESS.)

(a)

Figure 14.37 The formation of a cell plate between two daughter plant nuclei during cytokinesis. (*a*) A low-magnification electron micrograph showing the formation of the cell plate between the future daughter cells. Secretory vesicles derived from nearby Golgi complexes have become aligned along the equatorial plane (arrow) and are beginning to fuse with one another. The membrane of the vesicles forms the plasma membranes of the two daughter cells, and the contents of the vesicles will provide the material that forms the cell plate separating the cells. (*b*) Steps in the formation of the cell plate as described in the text. (A: DAVID PHILLIPS/PHOTO RESEARCHERS; B: AFTER A. L. SAMUELS, T. H. GIDDINGS, JR., AND L. A. STAEHELIN, J. CELL BIOL. 130:1354, 1995; BY COPYRIGHT PERMISSION OF THE ROCKEFELLER UNIVERSITY PRESS.)

(b)

ented perpendicular to the future plate (see Figure 9.23), together with membranous vesicles and electron-dense material. The microtubules of the phragmoplast arise largely from remnants of the mitotic spindle. After formation of the phragmoplast, small Golgi-derived secretory vesicles move into the region, probably transported along the microtubules, and become aligned along a plane between the daughter nuclei (Figure 14.37*a*). Electron micrographs of rapidly frozen tobacco cells have revealed the steps by which the Golgi-derived vesicles become reorganized into the cell plate (Figure 14.37*b*). To begin the process (step 1, Figure 14.37*b*), the vesicles send out fingerlike tubules that contact and fuse with neighboring vesicles to form an interwoven tubular network in the center of the cell (step 2). Additional vesicles are then directed along microtubules to the lateral edges of the network. The newly arrived vesicles continue the process of tubule formation and fusion, which extends the network in an outward direction (step 2). Eventually, the leading edge of the growing network contacts the parent plasma membrane at the boundary of the cell (step 3). Ultimately, the tubular network loses its cytoplasmic gaps and matures into a continuous, flattened partition. The membranes of the tubular network become the plasma

membranes of the two adjacent daughter cells, whereas the secretory products that had been carried within the vesicles contribute to the intervening cell plate. Once the cell plate is completed, additional materials are added over time to produce the mature cell wall.

REVIEW ?

1. How do the events of mitotic prophase prepare the chromatids for later separation at anaphase?
2. What are some of the activities of the kinetochore during mitosis?
3. Describe the events that occur in a cell during prometaphase and during anaphase.
4. What types of force-generating mechanisms might be responsible for chromosome movement during anaphase?
5. Contrast the events that occur during cytokinesis in typical plant and animal cells.

14.3 MEIOSIS

The production of offspring by sexual reproduction includes the union of two cells, each with a complete haploid set of chromosomes. As discussed in Chapter 10, the doubling of the chromosomes at fertilization is compensated by an equivalent reduction in chromosome number at a stage prior to formation of the gametes. Reduction in chromosome number is accomplished by **meiosis**. Meiosis ensures production of a haploid phase in the life cycle, and fertilization ensures a diploid phase. Without meiosis, the chromosome number would double with each generation, and sexual reproduction would not be possible.

To compare the events of mitosis and meiosis, we need to consider the fate of the chromatids. Prior to both mitosis and meiosis, diploid G_2 cells contain pairs of homologous chromosomes, with each chromosome consisting of two chromatids. During mitosis, the chromatids of each chromosome are split apart and separate into two daughter nuclei in a *single* division. As a result, cells produced by mitosis contain pairs of homologous chromosomes like their parents, but the chromosomes of the G_1 cells no longer contain two chromatids. The events of mitosis contrast with those of meiosis, in which the four chromatids of a pair of replicated homologous chromosomes are distributed among four daughter nuclei. Meiosis accomplishes this with two sequential divisions (Figure 14.38). In the first meiotic division, each chromosome (consisting of two chromatids) is separated from its homologue. As a result, each daughter cell contains only one member of each pair of homologous chromosomes. For this to occur, homologous chromosomes are paired during prophase of the first meiotic division (prophase I, Figure 14.38) by an elaborate process that has no counterpart in mitosis. As they are paired, homologous chromosomes engage in a process of genetic recombination that produces chromosomes with new combinations of maternal and paternal alleles (see metaphase I, Figure 14.38). In the second meiotic division, the two chromatids of each chromosome are separated from one another (anaphase II, Figure 14.38).

A survey of various eukaryotes reveals marked differences with respect to the stage within the life cycle at which meiosis occurs and the duration of the haploid phase. The following three groups (Figure 14.39) can be identified on these bases:

1. *Gametic or terminal meiosis.* In this group, which includes all multicellular animals and many protists, the meiotic divisions are closely linked to the formation of the gametes (Figure 14.39, left). In male vertebrates (Figure 14.40a), for example, meiosis occurs just prior to the differentiation of the spermatozoa. *Spermatogonia* that are committed to undergo meiosis become *primary spermatocytes*, which then undergo the two divisions of meiosis to produce four relatively undifferentiated *spermatids*. Each spermatid undergoes a complex differentiation to become the highly specialized

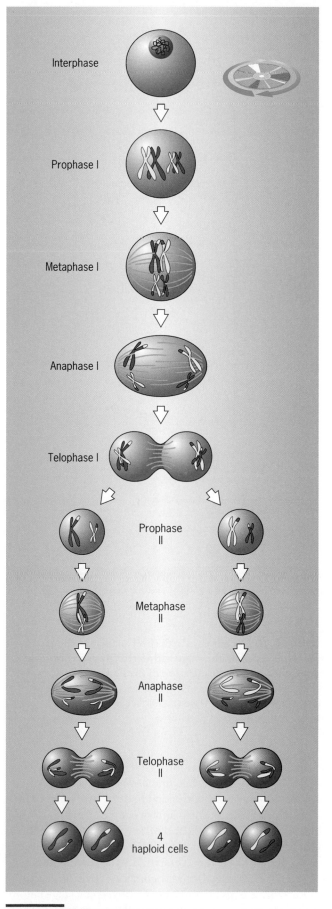

Figure 14.38 The stages of meiosis.

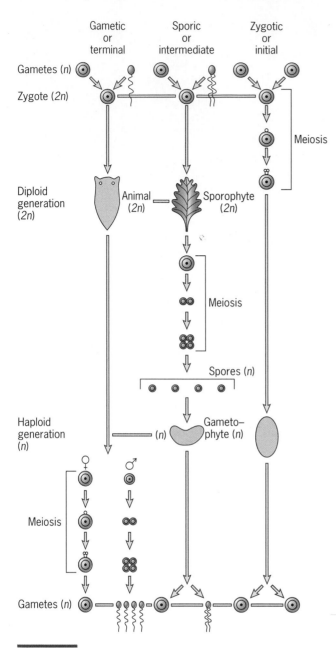

Figure 14.39 **A comparison of three major groups of organisms based on the stage within the life cycle at which meiosis occurs and the duration of the haploid phase.** (ADAPTED FROM E. B. WILSON, THE CELL IN DEVELOPMENT AND HEREDITY, 3D ED., 1925. REPRINTED WITH PERMISSION OF MACMILLAN PUBLISHING CO., INC.)

Meiosis is completed after fertilization, while the sperm resides in the egg cytoplasm.

2. *Zygotic or initial meiosis.* In this group, which includes only protists and fungi, the meiotic divisions occur just after fertilization (Figure 14.39, right) to produce haploid spores. The spores divide by mitosis to produce a haploid adult generation. Consequently, the diploid stage of the life cycle is restricted to a brief period after fertilization when the individual is still a zygote.

3. *Sporic or intermediate meiosis.* In this group, which includes all plants and some algae, the meiotic divisions take place at a stage unrelated to either gamete formation or fertilization (Figure 14.39, center). If we begin the life cycle with the union of a male gamete (the pollen grain) and a female gamete (the egg), the diploid zygote that results undergoes mitosis and develops into a **diploid sporophyte**. At some stage in the development of the sporophyte, *sporogenesis* (which includes meiosis) occurs, producing spores that germinate directly into a haploid **gametophyte**. The gametophyte can be either an independent stage or, as in the case of seed plants, a tiny structure retained within the ovules. In either case, the gametes are produced from the haploid gametophyte by *mitosis*.

The Stages of Meiosis

As is the case in any cell division, the prelude to meiosis includes the replication of the DNA. The premeiotic S phase generally takes several times longer than a premitotic S phase. Prophase of the first meiotic division (i.e., prophase I) is typically lengthened in extraordinary fashion when compared to prophase of a mitotic division. In the human female, for example, every oocyte in the ovary enters meiotic prophase I about the time of birth. Many of these oocytes remain arrested in the same approximate stage of prophase for several decades. The first meiotic prophase is very long and complex and is customarily divided into several stages that are similar in all sexually reproducing eukaryotes (Figure 14.41).

The first stage of prophase I is *leptotene*, during which the chromosomes become gradually visible in the light microscope. Although the chromosomes have replicated at an earlier stage, there is no indication that each chromosome is actually composed of a pair of identical chromatids. In the electron microscope, however, the chromosomes are revealed to be composed of paired chromatids.

The compaction of the chromosomes continues through leptotene until a point is reached when the homologues can be seen to associate with one another. This process of association is called **synapsis** and occurs during *zygotene*, the second stage of prophase I. Synapsis is an intriguing event about which there remain important unanswered questions: On what basis do the homologues recognize one another? How does the pair become so perfectly aligned? When does recognition between homo-

sperm cell (spermatozoon). In female vertebrates (Figure 14.40*b*), *oogonia* become *primary oocytes*, which then enter a greatly extended meiotic prophase. During this prophase, the primary oocyte grows and becomes filled with yolk and other materials. It is only after the differentiation of the oocyte is complete (i.e., the oocyte has reached essentially the same state as when it is fertilized) that the meiotic divisions occur. Vertebrate eggs are typically fertilized at a stage before the completion of meiosis (usually at metaphase II).

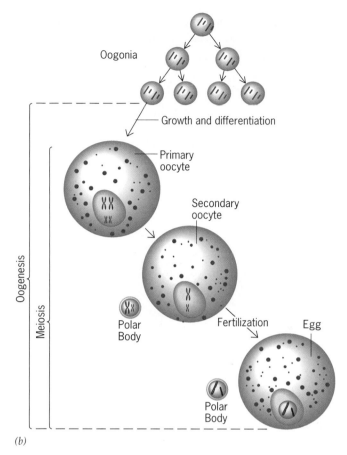

(a)

(b)

Figure 14.40 The stages of gametogenesis in vertebrates: a comparison between the formation of sperm and eggs. In both sexes, a relatively small population of primordial germ cells present in the embryo proliferates by mitosis to form a population of gonial cells (spermatogonia or oogonia) from which the gametes differentiate. In the male (*a*), meiosis occurs before differentiation, whereas in the female (*b*), both meiotic divisions occur after differentiation. Each primary spermatocyte generally gives rise to four viable gametes, whereas each primary oocyte forms only one fertilizable egg and two or three polar bodies.

logues first occur? Recent studies have shed considerable light on these questions. It had been assumed for years that interaction between homologous chromosomes first begins as chromosomes initiate synapsis. However, studies on yeast cells by Nancy Kleckner and her colleagues at Harvard University demonstrated that homologous regions of DNA from homologous chromosomes are already in contact with one another during leptotene. Chromosome compaction and synapsis during zygotene simply make this arrangement visible under the microscope. As will be discussed on page 618, the first step in genetic recombination is the occurrence of double-stranded breaks in aligned DNA molecules. Studies on yeast indicate that such breaks occur during leptotene, prior to synapsis, suggesting that genetic recombination begins before the chromosomes are visibly paired.

These findings are supported by recent studies aimed at locating particular DNA sequences within the nuclei of premeiotic and meiotic cells. We saw on page 515 that individual chromosomes are thought to occupy discrete regions within nuclei rather than being randomly dispersed throughout the nuclear space. When yeast cells about to enter meiotic prophase are examined, each pair of homologous chromosomes is found to share a joint territory distinct from the territories shared by other pairs of homologues. This finding suggests that homologous chromosomes are paired to some extent before meiotic prophase begins. Studies on maize cells also suggest that homologous chromosomes become paired very early in meiotic prophase, but not as early as in yeast. The telomeres (terminal segments) of maize leptotene chromosomes are distributed throughout the nucleus. Then, near the end of leptotene, there is a dramatic reorganization so that the telomeres become localized at the inner surface of the nuclear envelope at one side of the nucleus. Clustering of telomeres at one end of the nuclear envelope occurs in a wide variety of eukaryotic cells and causes the chromosomes to resemble a bouquet of flowers (Figure 14.42). The clustering of telomeres at the nuclear envelope is thought to facilitate the alignment of homologues in preparation for synapsis. In most organisms, synapsis begins at one end of each pair of homologues and progresses along the length of the chromosomes during zygotene, as illustrated in Figure 14.41.

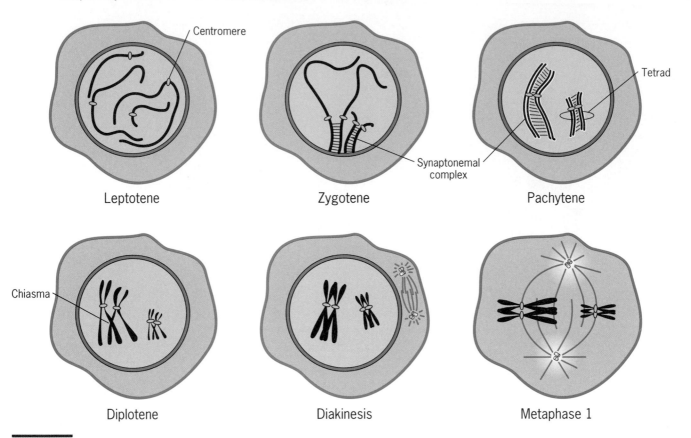

Figure 14.41 **The stages of prophase I.** The events at each stage are described in the text.

Figure 14.42 **Association of the telomeres of meiotic chromosomes with the nuclear envelope.** The chromosomes of a stage of meiotic prophase of the male grasshopper in which homologous chromosomes are physically associated as part of a bivalent. The bivalents are arranged in a well-defined "bouquet" with their terminal regions clustered near the inner surface of the nuclear envelope, at the base of the photo. (COURTESY OF BERNARD JOHN, FROM B. JOHN, MEIOSIS. REPRINTED WITH THE PERMISSION OF CAMBRIDGE UNIVERSITY PRESS, 1990.)

Electron micrographs indicate that chromosome synapsis is accompanied by the formation of a complex structure called the synaptonemal complex. The **synaptonemal complex (SC)** is a ladderlike structure (Figure 14.43) composed of three parallel bars with many transverse filaments connecting the central element with the two lateral elements. The chromatin of each homologue is organized into loops that extend from one of the lateral elements of the SC (Figure 14.43b). For many years, the SC was thought to hold each pair of homologous chromosomes in the proper position to initiate genetic recombination between strands of DNA that projected into the center of the structure. It is now evident that the SC is not required for genetic recombination. Not only does the SC form after genetic recombination has been initiated, but mutant yeast cells unable to assemble an SC can still engage in the exchange of genetic information between homologues. It is currently thought that the SC functions primarily as a scaffold to allow interacting chromatids to complete their crossover activities, as described below.

The complex formed by a pair of synapsed homologous chromosomes is called a **bivalent** or a **tetrad**. The former term reflects the fact that the complex contains two homologues, whereas the latter term calls attention to the presence of four chromatids. The end of synapsis marks the end of zygotene and the beginning of the next stage of prophase I, called *pachytene*, which is characterized by a fully formed synaptonemal complex. Whereas leptotene and zygotene stages generally last for a few

hours, pachytene is often extended for a period of days or weeks. During this period, the homologues are held closely together along their length by the SC (Figure 14.43). Under the electron microscope, a number of electron-dense bodies about 100 nm in diameter are seen at irregular intervals within the center of the SC. These structures have been named *recombination nodules* because they correspond to the sites where crossing over is taking place, as evidenced by the associated synthesis of DNA that occurs during intermediate steps of recombination. Recombination nodules are presumed to contain some of the enzymatic machinery that facilitates genetic recombination, which is completed by the end of pachytene.

The beginning of *diplotene*, the next stage of meiotic prophase I (Figure 14.41), is recognized by the dissolution of the SC and the tendency of the homologous chromosomes of the bivalents to pull away somewhat from each other. As the homologous chromosomes move apart, they are seen to remain attached to one another at specific points by X-shaped structures, termed **chiasmata** (singular **chiasma**) (Figure 14.44). Chiasmata are formed from junctions between two of the four chromatids of a bivalent, one from each homologue. Chiasmata are located at sites on the chromosomes where crossing-over between DNA molecules from the two chromosomes had previously occurred. These points of attachment provide a striking visual portrayal of the extent of genetic recombination.

Diplotene is an extremely extended phase of oogenesis during which the bulk of oocyte growth occurs. Thus the diplotene stage, occurring as it does in the middle of the meiotic prophase, is a period of intense metabolic activity. This activity can be followed by observing the behavior of the chromosomes. In many spermatocytes and oocytes, the diplotene chromosomes become dispersed into a particular configuration not found at any other time during the life cycle of the organism. These chromosomes, called *lampbrush chromosomes* (see Figure 12.13*b*), have an axial backbone from which pairs of loops extend out in opposite directions. Loops arise in pairs because each chromosome consists of a pair of duplicated chromatids, and each loop is a projection from a single chromatid. The two homologous chromosomes still remain attached to each other at the chiasmata. The DNA situated between the

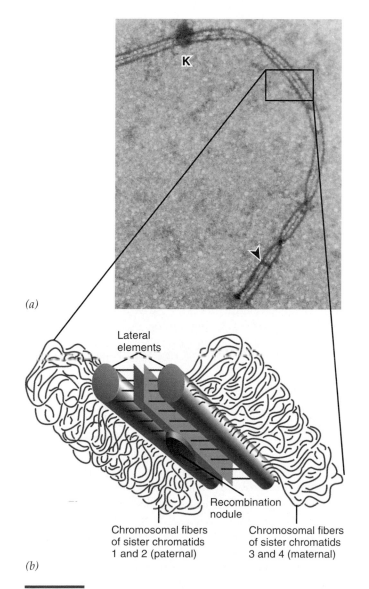

(*a*)

Lateral
elements

Recombination
nodule

Chromosomal fibers
of sister chromatids
1 and 2 (paternal)

Chromosomal fibers
of sister chromatids
3 and 4 (maternal)

(*b*)

(*c*) 0.2 µm

Figure 14.43 The synaptonemal complex. (*a*) Electron micrograph of a human pachytene bivalent showing the pair of chromosomes held in a tightly ordered parallel array. K, kinetochore. (*b*) Schematic diagram of the synaptonemal complex and its associated chromosomal fibers. The dense granules (recombination nodules) seen in the center of the SC (indicated by the arrowhead in part *a*) are thought to contain the enzymatic machinery required to complete genetic recombination, which is thought to begin at a much earlier stage in prophase I. (*c*) Electron micrograph of the synaptonemal complex after treatment with DNase to remove chromosomal fibers. The structure remaining is the proteinaceous, ladderlike SC itself. (A: Courtesy of Alberto J. Solari, Chromosoma 81:330, 1980; c: from D. Comings and T. Okada, Exp. Cell Res. 65:104, 1971.)

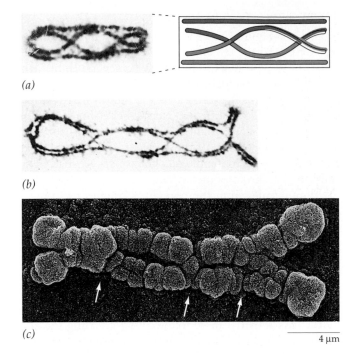

(a)

(b)

(c)

4 µm

Figure 14.44 **Visible evidence of crossing over.** (*a*, *b*) Pairs of diplotene bivalents from the grasshopper showing the chiasmata formed between chromatids of each homologous chromosome. The accompanying inset indicates the crossover that has presumably occurred within the bivalent in *a*. The chromatids of each diplotene chromosome are closely apposed except at the chiasmata. (*c*) Scanning electron micrograph of a bivalent from the desert locust with three chiasmata (arrows). (A, B: FROM BERNARD JOHN, MEIOSIS, CAMBRIDGE, 1990. REPRINTED WITH THE PERMISSION OF CAMBRIDGE UNIVERSITY PRESS. C: FROM KLAUS WERNER WOLF, BIOESS. 16:108, 1994.)

loops is tightly compacted and transcriptionally inactive. In contrast, the DNA of the loops is highly extended and the site of intense transcriptional activity. Transcription of lampbrush chromosomes in the oocyte provides the RNA utilized for protein synthesis during both oogenesis and early embryonic development following fertilization.

During the final stage of meiotic prophase I, called *diakinesis*, the meiotic spindle is assembled, and the chromosomes are prepared for separation. In those species in which the chromosomes become highly dispersed during diplotene, the chromosomes become recompacted during diakinesis. Diakinesis ends with the disappearance of the nucleolus, the breakdown of the nuclear envelope, and the movement of the tetrads to the metaphase plate. In vertebrate oocytes, these events are triggered by an increase in the level of the protein kinase activity of MPF (maturation-promoting factor). As discussed in the Experimental Pathways at the end of the chapter, MPF was first identified by its ability to initiate these events, which represent the *maturation* of the oocyte (page 619).

In most eukaryotic species, chiasmata can still be seen in homologous chromosomes aligned at the metaphase plate of meiosis I. In fact, chiasmata hold the homologues together as a bivalent during this stage. In humans and

other vertebrates, every pair of homologues typically contains at least one chiasma, and the longer chromosomes tend to have two or three of them. It is thought that some mechanism exists to ensure that even the smallest chromosomes form a chiasma. If a chiasma does not occur between a pair of homologous chromosomes, the chromosomes of that bivalent tend to separate from one another after dissolution of the SC. This premature separation of homologues often results in nondisjunction of homologues and formation of nuclei with an abnormal number of chromosomes. The consequence of such an event is discussed in the accompanying Human Perspective. Just as there is a mechanism to ensure that every pair of homologues develops a chiasma, there is also a mechanism to ensure that none of the homologous pairs is connected by more than two or three chiasmata, regardless of the length of the chromosomes. The inhibition of excess chiasmata is referred to as *crossover interference* and is thought to be mediated by the synaptonemal complex.

At metaphase I, the two homologous chromosomes of each bivalent are connected to the chromosomal fibers from opposite poles. Abnormalities in the formation of the metaphase I spindle trigger the arrest of meiosis by a checkpoint mechanism similar to that operating during mitosis (page 601). During anaphase I, cohesion between chromosome arms is lost, which allows homologous chromosomes of each bivalent to separate from one another. At the same time, cohesion between the joined centromeres of sister chromatids remains strong, so that sister chromatids remain firmly attached to one another as they move to one of the spindle poles.

The orientation of the maternal and paternal chromosomes of each bivalent on the metaphase I plate is random; the maternal member of a particular bivalent has an equal likelihood of facing either pole. Consequently, when chromosomes separate during anaphase I, each pole receives a random assortment of maternal and paternal chromosomes (see Figure 14.38). Thus, anaphase I is the cytological event that corresponds to Mendel's law of independent assortment (page 399). Each anaphase I chromosome is composed of two associated chromatids still connected to one another at their centromeres.

Telophase I of meiosis I produces less dramatic changes than telophase of mitosis. Although in many cases the chromosomes do undergo some dispersion, they do not reach the extremely extended state of the interphase nucleus. The nuclear envelope may or may not reform during telophase I. The stage between the two meiotic divisions is called *interkinesis* and is generally short-lived. In animals, cells in this stage are referred to as *secondary spermatocytes* or *secondary oocytes*. These cells are characterized as being haploid because they contain only one member of each pair of homologous chromosomes. Even though they are haploid, they have twice as much DNA as a haploid gamete because each chromosome is still represented by a pair of attached chromatids. Secondary spermatocytes are said to have a 2C amount of DNA, half as much as a primary spermatocyte, which has

(a) 10 μm

(b) 5 μm

Figure 14.45 An unfertilized amphibian egg arrested at metaphase II. As in other vertebrates, meiosis proceeds to the second meiotic metaphase where it awaits fertilization before completing meiosis. (*a*) A lateral view showing the chromo- somes and meiotic spindle. (*b*) A polar view of the same stage showing the distribution of chromosomes around the cell periphery. (FROM DAVID L. GARD,. DEV. BIOL. 151:523, 1992.)

a 4C DNA content, and twice as much as a sperm cell, which has a 1C DNA content.

Interkinesis is followed by prophase II, a much simpler prophase than its predecessor. If the nuclear envelope had reformed in telophase I, it is broken down again. The chromosomes become recompacted and line up at the metaphase plate. The kinetochores of sister chromatids of metaphase II face opposite poles and become attached to opposing sets of chromosomal spindle fibers. The progression of meiosis in vertebrate oocytes stops at metaphase II (Figure 14.45). The arrest of meiosis at metaphase II is thought to be caused by a protein kinase, called Mos, that inhibits the degradation of cyclin B. As long as cyclin B levels remain high within the oocyte, Cdk activity is maintained, and the cells cannot progress to the next meiotic stage. Metaphase II arrest is released only when the oocyte (now called an egg) is fertilized, which leads to the destruction of Mos and the subsequent inactivation of the Cdk. The fertilized egg responds to these changes by completing the second meiotic division. Anaphase II begins with the synchronous splitting of the centromeres, which hold the sister chromatids together, allowing them to move toward opposite poles of the cell. Meiosis II ends with telophase II, in which the chromosomes are once again enclosed by a nuclear envelope. The products of meiosis are haploid cells with a 1C amount of nuclear DNA.

Genetic Recombination During Meiosis

In addition to reducing the chromosome number as required by sexual reproduction, meiosis increases the genetic variability in a population of organisms from one generation to the next. Independent assortment allows maternal and paternal chromosomes to become shuffled during formation of the gametes, and genetic recombination (crossing over) allows maternal and paternal alleles on a given chromosome to become shuffled as well (see Figure 14.38). Without genetic recombination, the alleles along a particular chromosome would remain tied together, generation after generation. By mixing maternal and paternal alleles between homologous chromosomes, meiosis generates organisms with novel genotypes and phenotypes on which natural selection can act. An example of recombination between alleles on one of the chromosomes of the fruit fly was illustrated in Figure 10.7. It was also noted in Chapter 10 that the frequency of recombination between two alleles on a chromosome is proportional to the distance between them, a feature that has allowed geneticists to map the relative positions of genes along chromosomes in organisms ranging from bacteria to humans.

The first insights into the process of recombination were made by classical geneticists during the first decade of the century. In the early view of T. H. Morgan and others, the tension generated as the chromosomes twisted around each other during meiotic prophase caused the chromosomes to become fractured in such a way that pieces of chromatids were exchanged between homologous chromosomes. According to this concept, crossing-over involves an actual physical *breakage and reunion* of chromosomal material. The proposal that recombination occurs by breakage and reunion was first confirmed autoradiographically using meiotic chromosomes of the grasshopper that had been labeled with radioactive thymidine. In these experiments, which were conducted by J. Herbert Taylor of Columbia University, the germ cells were exposed to the radioactive DNA precursor during

THE HUMAN PERSPECTIVE

Meiotic Nondisjunction and Its Consequences

Meiosis is a complex process, and meiotic mistakes in humans appear to be surprisingly common. Homologous chromosomes may fail to separate from each other during meiosis I, or sister chromatids may fail to come apart during meiosis II. When either of these situations occurs, gametes are formed that contain an abnormal number of chromosomes—either an extra chromosome or a missing chromosome (Figure 1). If one of these gametes happens to fuse with a normal gamete, a zygote with an abnormal number of chromosomes forms, and serious consequences arise. In most cases, the zygote develops into an abnormal embryo that dies at some stage between conception and birth. In a few cases, however, the zygote develops into an infant whose cells have an abnormal chromosome number, a condition known as *aneuploidy*. The effect of aneuploidy depends on which chromosome or chromosomes are affected.

The normal human chromosome complement is 46: 22 pairs of autosomes and one pair of sex chromosomes. An extra chromosome (producing a total of 47 chromosomes) creates a condition referred to as a *trisomy* (Figure 2). A person whose cells contain an extra chromosome 21, for example, has trisomy 21. A missing chromosome (producing a total of 45 chromosomes) produces a *monosomy*. We will begin by considering the effects of an abnormal number of autosomes.

The absence of one autosomal chromosome, regardless of which chromosome is affected, invariably proves to be lethal at some stage during embryonic or fetal development. Consequently, a zygote containing an autosomal monosomy does not give rise to a fetus that is carried to term. Although one might not expect that possession of an extra chromosome would create a life-threatening condition, the fact is that trisomies do not fare much better than monosomic zygotes. Of the 22 different autosomes in the human chromosome complement, only persons with trisomy 21 can survive beyond the first few weeks or months of life. Most of the other possible trisomies are lethal during development, whereas trisomies for chromosomes 13 and 18 are often born alive but have such severe abnormalities that they succumb soon after birth. Approximately 25 percent of fetuses that spontaneously abort carry a chromosomal trisomy. It is thought that many more zygotes carrying abnormal chromosome numbers produce embryos that die at an early stage of development before the pregnancy is recognized. For example, for every trisomic zygote formed at fertilization, there is presumably an equal number of monosomic zygotes that fare even less well. Results of one study suggest that as many as 33 percent of pregnant women abort spontaneously prior to the stage where the pregnancy is recognizable. If these numbers are correct, then meiotic nondisjunction occurs at a remarkably high rate in humans. Recent studies suggest that the high failure rate during meiosis in females is a result of a lack of certain checkpoints that operate during oogenesis.

Even though chromosome 21 is the smallest human chromosome, the presence of an extra copy of this genetic

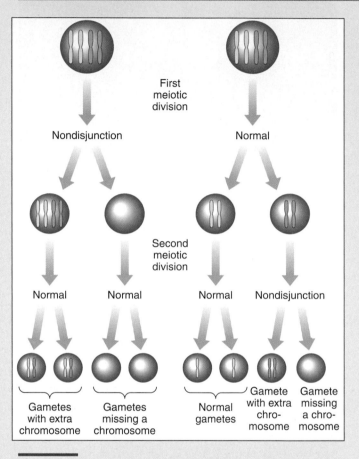

Figure 1 Meiotic nondisjunction occurs when chromosomes fail to separate from each other during meiosis. If the failure to separate occurs during the first meiotic division, which is called *primary nondisjunction*, all of the haploid cells have an abnormal number of chromosomes. If nondisjunction occurs during the second meiotic division, which is called *secondary nondisjunction*, only two of the four haploid cells are affected.

material has serious consequences, producing a condition called *Down syndrome*. Persons with Down syndrome exhibit varying degrees of mental impairment, alteration in some body features, circulatory problems, increased susceptibility to infectious diseases, a greatly increased risk of developing leukemia, and the early onset of Alzheimer's disease. All of these medical problems are thought to result from an abnormal level of expression of genes located on chromosome 21.

The presence of an abnormal number of sex chromosomes is much less disruptive to human development. A zygote with only one X chromosome and no second sex chromosome (denoted as XO) develops into a female with *Turner syndrome*, in which genital development is arrested in the juvenile state, the ovaries fail to develop, and body structure is slightly abnormal. Because a Y chromosome is male deter-

Figure 2 The karyotype of a person with Down syndrome. The karyotype shows an extra chromosome 21 (trisomy 21).

mining, persons with at least one Y chromosome develop as males. A male with an extra X chromosome (XXY) develops *Klinefelter syndrome*, which is characterized by mental retardation, underdevelopment of genitalia, and the presence of feminine physical characteristics (such as breast enlargement). Alternatively, a zygote with an extra Y (XYY) develops into a man who appears normal except that he will probably be taller than average. Considerable controversy developed surrounding claims that XYY males tend to exhibit more aggressive, antisocial, and criminal behavior than do XY males, but this hypothesis has never been substantiated.

The likelihood of having a child with Down syndrome rises dramatically with the age of the mother—from 0.05 percent for mothers 19 years of age to nearly 3 percent for women over the age of 45. Most studies show that no such correlation is found between the age of the father and the likelihood of bearing a child with trisomy 21. The most recent estimates, based on comparisons of DNA sequences between the offspring and the parents, indicate that approximately 95 percent of trisomies 21 can be traced to nondisjunction having occurred in the mother. Why is an older mother more likely to give birth to a child with Down syndrome? Fetuses with chromosomal abnormalities are usually aborted spontaneously by the mother, so one possibility is that older women are less likely to abort a fetus with trisomy 21 than a younger woman. This proposal is contradicted by studies of aborted fetuses from mothers of varying ages that suggest that younger women are as likely to carry a fetus with trisomy 21 to term as an older woman. Rather, the evidence supports the most straightforward explanation: older women are more prone to meiotic nondisjunction than younger women. Most geneticists believe that the risk of nondisjunction increases as the oocyte remains for longer periods in the ovary. It was noted in the chapter that all of the oocytes in a human ovary enter meiosis by about the time of birth. Those ova that are fertilized in a woman's latter reproductive years have remained suspended in prophase I for several decades. To understand why a delay in the completion of oogenesis should lead to an increase in the likelihood of nondisjunction, it is necessary to know more about the stage of meiosis that seems to be affected.

It was noted above that an abnormal chromosome number can result from nondisjunction at either of the two meiotic divisions (Figure 1). Although these different nondisjunction events produce the same effect in terms of chromosome numbers in the zygote, they can be distinguished by genetic analysis. Primary nondisjunction transmits two homologous chromosomes to the zygote, whereas secondary nondisjunction transmits two sister chromatids (most likely altered by crossing over) to the zygote. Studies indicate that most of the mistakes occur during meiosis I. For example, in one study of 433 cases of trisomy 21 that resulted from maternal nondisjunction, 373 were the result of errors that had occurred during meiosis I and 60 were the result of errors during meiosis II.

Why should meiosis I be more susceptible to nondisjunction than meiosis II? In one study, in which trisomies for chromosome 21 were genetically mapped, it was found that the chromosomes inherited from the mother exhibited a greatly decreased likelihood of having undergone genetic recombination. It was noted on page 614, that chiasmata—which are visual indicators of genetic recombination—play an important role in holding a tetrad together. It follows that a reduction in genetic recombination, which leads to a reduction in the likelihood of the formation of chiasmata, would lead to an increased risk of nondisjunction at meiosis I.

the S phase of the previous *mitotic* division so that, by the time they had reached the first meiotic prophase, only one chromatid of each chromosome was labeled. Autoradiographs were then made after crossing-over had occurred, and it was found that certain of the chromatids contained silver grains covering only a part of their length (Figure 14.46). Moreover, the labeled and unlabeled segments of one chromatid showed a reciprocal pattern to that of another chromatid of the same tetrad. This indicated that physical exchange had occurred between labeled and unlabeled chromatids.

When the concept of breakage and reunion was first proposed, it was presumed that each homologous chromosome was split entirely across into two pieces, and a piece from one chromosome was exchanged with the corresponding piece from the homologue. But meiotic chromosomes contain large numbers of DNA fibers lying side by side. Breakage of an entire compacted chromatid would sever the DNA of each chromatid many times, splitting hundreds of genes that could not possibly be reunited with their counterparts on another chromosome. As will be discussed shortly, recombination involves the

2 µm

Figure 14.46 Experimental demonstration of chromosomal breakage and reunion during crossing over. Autoradiograph of a pair of sister chromatids that were isolated from a grasshopper testis at metaphase II using colchicine to arrest events at that stage. The cell from which this chromosome was prepared had incorporated [³H]thymidine one cycle prior to the premeiotic interphase, so that in the absence of crossing over, one chromatid of each duplicated chromosome should be completely labeled and one should be completely unlabeled. The fact that chromatids contain both labeled and unlabeled segments indicates that exchanges of chromosomal material between chromatids had occurred in the preceding prophase I. The arrow indicates the position of the terminal centromere, which is located where the two sister chromatids come together (as indicated in the inset). (FROM J. HERBERT TAYLOR, J. CELL BIOL. 25:63, 1965; BY COPYRIGHT PERMISSION OF THE ROCKEFELLER UNIVERSITY PRESS.)

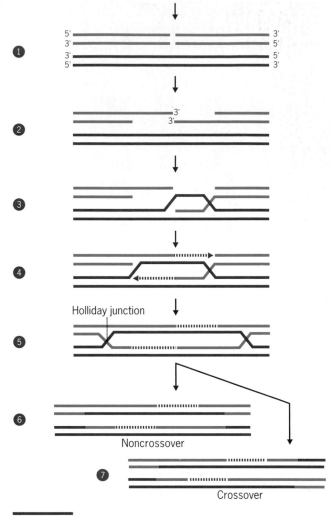

Holliday junction

Noncrossover

Crossover

Figure 14.47 A proposed mechanism for genetic recombination initiated by double-strand breaks. The steps are described in the text.

physical breakage of individual DNA molecules and the ligation of the split ends from one DNA duplex with the split ends of the duplex from the homologous chromosome. Recombination is a remarkably precise process that occurs between corresponding sites on homologous DNA molecules without the addition or loss of a single base pair. The precision of recombination is ensured by the involvement of DNA repair enzymes that fill gaps that develop during the exchange process.

A simplified model for the major steps during recombination in eukaryotic cells is shown in Figure 14.47. In this model, two DNA duplexes that are about to recombine are aligned next to one another as the result of some type of *homology search* in which homologous DNA molecules associate with one another in preparation for recombination. Once they are aligned, a double-stranded break is introduced into one duplex (step 1, Figure 14.47), and the gap is subsequently widened by the action of a 5′ → 3′

exonuclease (step 2). As a result of exonucleolytic digestion, the broken strands possess exposed single-stranded tails, each bearing a 3′-OH terminus. In the model shown in Figure 14.47, one of the single-stranded tails leaves its own duplex and invades the unbroken DNA molecule, hydrogen bonding with the complementary strand in the neighboring duplex (step 3). In *E. coli*, this process in which a single strand invades an homologous duplex and displaces the corresponding strand in that duplex is catalyzed by a multifunctional protein, called the *RecA protein*. The RecA protein polymerizes to form a filament that binds along a length of single-stranded DNA and facilitates that strand's invasion of an uncut, homologous double helix. Eukaryotic cells have homologues of RecA (e.g., Rad51) that are thought to catalyze strand invasion. Strand invasion activates a DNA repair activity (Section 13.3) that fills the gaps as shown in step 4.

As a result of the reciprocal exchange of DNA strands, the two duplexes are covalently linked to one an-

other to form a joint molecule (or *heteroduplex*) that contains a pair of DNA crossovers, or *Holliday junctions*, that flank the region of strand exchange (step 4, 5, Figure 14.47). These junctions are named after Robin Holliday, the researcher who proposed their existence in 1964. This type of recombination intermediate need not be a static structure because the point of linkage may move in one direction or another (an event known as *branch migration*) by breaking the hydrogen bonds holding the original pairs of strands and reforming hydrogen bonds between strands of the newly joined duplexes (step 5).

The recombination intermediate shown in Figure 14.47, step 5, is formed by the breakage of one of the duplexes and subsequent strand invasion. To resolve the interconnected Holliday junctions and restore the DNA to two discrete duplexes, another round of DNA cleavage must occur (indicated by the arrows). Depending upon the particular DNA strands that are cleaved and ligated, two alternate products can be generated. In one case, the two duplexes contain only short stretches of genetic exchange, which represents a noncrossover (step 6). In the alternate pathway of breakage and ligation, the duplex of

one DNA molecule is joined to the duplex of the homologous molecule, creating a fusion of the maternal and paternal chromosomes (step 7), that is, a crossover. It is this latter type of event that is detected in studies that monitor the frequency of recombination between paternal and maternal alleles (page 402).

REVIEW ?

1. Contrast the overall roles of mitosis and meiosis in the lives of a plant or animal. How do the nuclei formed by these two processes differ from one another?
2. Contrast the events that take place during prophase I and prophase II of meiosis.
3. Contrast the timing of meiosis in spermatogenesis versus oogenesis.
4. How was it demonstrated that genetic recombination took place by a breakage-and-reunion mechanism?

EXPERIMENTAL PATHWAYS

The Discovery and Characterization of MPF

As an amphibian oocyte nears the end of oogenesis, the large nucleus (called a germinal vesicle) moves toward the periphery of the cell. In subsequent steps, the nuclear envelope disassembles, the compacted chromosomes become aligned along a metaphase plate near one end (the animal pole) of the oocyte, and the cell undergoes the first meiotic division to produce a large secondary oocyte and a small polar body. The processes of germinal vesicle breakdown and first meiotic division are referred to as *maturation* and can be induced in fully grown oocytes by treatment with the steroid hormone progesterone. The first sign of maturation in the hormone-treated oocyte is seen 13–18 hours following progesterone treatment as the germinal vesicle moves near the oocyte surface. Germinal vesicle breakdown soon follows, and the oocyte reaches metaphase of the second meiotic division by about 36 hours after hormone treatment. Progesterone induces maturation only if it is applied to the external medium surrounding the oocyte; if the hormone is injected into the oocyte, the oocyte shows no response.[1] It appears that the hormone acts at the cell surface to trigger secondary changes in the cytoplasm of the oocyte that lead to germinal vesicle breakdown and the other changes associated with maturation.

To learn more about the nature of the cytoplasmic change that was responsible for triggering maturation, Yoshio Masui of the University of Toronto and Clement Markert of Yale University began a series of experiments in which they removed cytoplasm from isolated frog oocytes at various stages following progesterone treatment and injected 40–60 nanoliters (nl) of the donor cytoplasm into fully grown, immature oocytes that had not been treated with the hormone.[2] They found that cytoplasm taken from oocytes during the first 12 hours following progesterone treatment had little or no effect on recipient oocytes. After this period, however, the cytoplasm gained the ability to induce maturation in the recipient oocyte. The cytoplasm from the donor oocyte was maximally effective about 20 hours after progesterone treatment, and its effectiveness declined by 40 hours (Figure 1). However, cytoplasm taken from early embryos continued to show some ability to induce oocyte maturation. Masui and Markert referred to the cytoplasmic substance(s) that induce maturation in recipient oocytes as "maturation promoting factor," which became known as MPF.

Because it was assumed that MPF was involved specifically in triggering oocyte maturation, relatively little interest was paid at first to the substance or its possible mechanism of

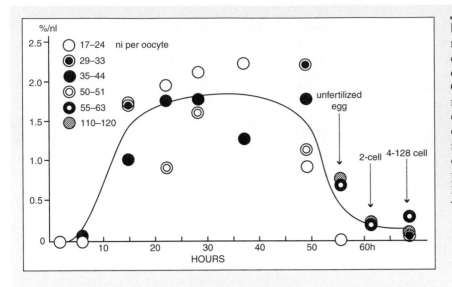

Figure 1 Change of activity of the maturation-promoting factor in the oocyte cytoplasm of *Rana pipiens* during the course of maturation and early development. Ordinate: the ratio of frequency of induced maturation to volume of injected cytoplasm. The higher the ratio, the more effective the cytoplasm. nl, nanoliters of injected cytoplasm. Abscissa: age of the donors (hours after administration of progesterone). (FROM Y. MASUI AND C. L. MARKERT, J. EXP. ZOOL. 177:142, 1971.)

action. Then in 1978, William Wasserman and Dennis Smith of Purdue University published a report on the behavior of MPF during early amphibian development.[3] It had been assumed that MPF activity present in early embryos was simply a residue of activity that had been present in the oocyte. But Wasserman and Smith discovered that MPF activity undergoes dramatic fluctuations in cleaving eggs that correlate with changes in the cell cycle. It was found, for example, that cytoplasm taken from cleaving frog eggs within 30–60 minutes after fertilization contains little or no detectable MPF activity, as assayed by injection into immature oocytes (Figure 2). However, if cytoplasm is taken from an egg at 90 minutes after fertilization, MPF activity can again be demonstrated. MPF activity reaches a peak at 120 minutes after fertilization and starts to decline again at 150 minutes (Figure 2). At the time the eggs undergo their first cytokinesis at 180 minutes, no activity is detected in the eggs. Then, as the second cleavage cycle gets underway, MPF activity once again reappears, reaching a peak at 225 minutes postfertilization, and then declines again to a very low level. Similar results were found in *Xenopus* eggs, except that the fluctuations in MPF activity occur more rapidly than in *Rana* and correlate with the more rapid rate of cleavage divisions in the early *Xenopus* embryo. Thus, MPF activity disappears and reappears in both amphibian species on a time scale that correlates with the length of the cell cycle. In both species, the peak of MPF activity corresponds to the time of nuclear membrane breakdown and the entry of the cells into mitosis. These findings suggested that MPF does more than simply control the time of oocyte maturation and, in fact, may play a key role in regulating the cell cycle of dividing cells.

It became apparent about this same time that MPF activity is not limited to amphibian eggs and oocytes but is present in a wide variety of organisms. It was found, for example, that mammalian cells growing in culture also possess MPF activity as assayed by the ability of mammalian cell extracts to induce germinal vesicle breakdown when injected into amphibian oocytes.[4] MPF activity of

Figure 2 Cycling of MPF activity in fertilized *R. pipiens* eggs. Ordinate: percent recipient oocytes undergoing germinal vesicle breakdown in response to 80 nanoliters of cytoplasm from fertilized eggs. Abscissa: time after fertilization when the *Rana* egg cytoplasm was assayed for MPF activity. Arrows indicate time of cleavage divisions. (FROM W. J. WASSERMAN AND L. D. SMITH, J. CELL BIOL. 78:R17, 1978; BY COPYRIGHT PERMISSION OF THE ROCKEFELLER UNIVERSITY PRESS.)

mammalian cells fluctuates with the cell cycle, as it does in dividing amphibian eggs. Extracts from cultured HeLa cells prepared from early G_1-, late G_1-, or S-phase cells lack MPF activity (Figure 3). MPF appears in early G_2, rises dramatically in late G_2, and reaches a peak in mitosis.

Another element of the machinery that regulates the cell cycle was discovered in studies on sea urchin embryos. Sea urchin eggs are favorite subjects for studies of cell divi-

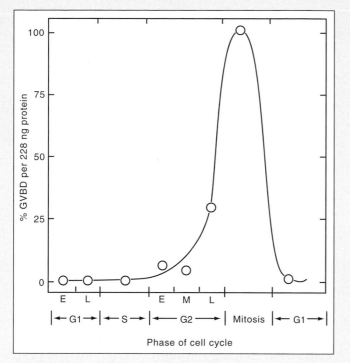

Figure 3 Maturation-promoting activity of HeLa cell extracts during different stages of the cell cycle. Because 228 ng of mitotic protein induced germinal vesicle breakdown (GVBD) in 100 percent of the cases, the percent activity for other phases of the cell cycle was normalized to that amount of protein. E, early; M, mid-, and L, late. (FROM P. S. SUNKARA, D. A. WRIGHT, AND P. N. RAO, PROC. NAT'L ACAD. SCI. U.S.A. 76:2801, 1979.)

Figure 4 Correlation of the level of cyclin with the cell division cycle. A suspension of eggs was fertilized, and after 6 minutes, [35S] methionine was added. Samples were taken for analysis by gel electrophoresis at 10-minute intervals, starting at 16 minutes after fertilization. The autoradiograph of the electrophoretic gel was scanned for label density, and the data were plotted as shown. Protein A, which varies according to the cell cycle and is named cyclin, is shown as solid circles. Protein B (not to be confused with cyclin B), which shows no cell cycle fluctuation, is plotted as solid triangles. The percentage of cells undergoing division at any given time period is given as the cleavage index (open squares). (FROM T. EVANS ET AL., CELL 33:391, 1983; BY PERMISSION OF CELL PRESS.)

sion because the mitotic divisions following fertilization occur rapidly and are separated by highly predictable time intervals. If sea urchin eggs are fertilized in sea water containing an inhibitor of protein synthesis, the eggs fail to undergo the first mitotic division, arresting at a stage prior to chromosome compaction and breakdown of the nuclear envelope. Similarly, each of the subsequent mitotic divisions can also be blocked if an inhibitor of protein synthesis is added to the medium at a time well before the division would normally occur. This finding had suggested that one or more proteins must be synthesized during each of the early cell cycles if the ensuing mitotic division is to occur. But early studies on cleaving sea urchin eggs failed to reveal the appearance of new species of proteins during this period.

In 1983, Tim Hunt and his colleagues at the Marine Biological Laboratory at Woods Hole reported on several proteins that are synthesized in fertilized sea urchin eggs but not unfertilized eggs.[5] To study these proteins further, they incubated fertilized eggs in sea water containing [35S]methionine and withdrew samples at 10-minute intervals beginning at 16 minutes after fertilization. Crude protein extracts were prepared from the samples and subjected to polyacrylamide gel electrophoresis, and the labeled proteins were located autoradiographically. Several prominent bands were labeled in gels from fertilized egg extracts that were not evi-

dent in comparable extracts made from unfertilized eggs. One of the bands that appeared strongly labeled at early stages after fertilization virtually disappeared from the gel by 85 minutes after fertilization, suggesting that the protein had been selectively degraded. This same band then reappeared in gels from eggs sampled at later times and disappeared once again in a sample taken at 127 minutes after fertilization. The fluctuations in the amount of this protein are plotted in Figure 4 (protein band A) together with the cleavage index, which indicates the time course of the first two cell divisions. The degradation of the protein occurs at about the same time that the cells undergo the first and second division. A similar protein was found in the eggs of the surf clam, another invertebrate whose eggs are widely studied. Hunt and colleagues named the protein "cyclin" and noted the striking parallel in behavior between the fluctuations in cyclin levels in their investigation and MPF activity in the earlier studies. Subsequent studies showed that there were two distinct cyclins, A and B, which are degraded at different times during the cell cycle. Cyclin A is degraded during a 5–6 minute period beginning just before the

metaphase–anaphase transition, and cyclin B is degraded a few minutes after this transition.

The first clear link between cyclin and MPF was demonstrated by Joan Ruderman and her colleagues at the Woods Hole Marine Biological Laboratory.[6] In these studies, an mRNA encoding cyclin A was transcribed in vitro from a cloned DNA fragment that contained the entire cyclin A coding sequence. The identity of this mRNA was verified by translating it in vitro and finding that it encoded authentic clam cyclin A. When the synthetic cyclin mRNA was injected into *Xenopus* oocytes, the cells underwent germinal vesicle breakdown and chromosome compaction over a time course not unlike that induced by progesterone treatment (Figure 5). These results suggested that the rise in cyclin A, which occurs normally during meiosis and mitosis, has a direct role in promoting entry into M phase. The amount of cyclin A normally drops rapidly and must be resynthesized prior to the next division or the cells cannot reenter M phase.

But what is the relationship between cyclins and MPF? One of the difficulties in answering this question was the use of different organisms. MPF had been studied primarily in amphibians, and cyclins in sea urchins and clams. Evidence indicated that frog oocytes contain a pool of inactive pre-MPF molecules, which are converted to active MPFs during meiosis I. Cyclin, on the other hand, is totally absent from clam oocytes but appears soon after fertilization. Ruderman considered the possibility that cyclin A is an activator of MPF. We will return to this shortly.

Meanwhile, another line of research was initiated to purify and characterize the substance responsible for MPF activity. In 1980, Michael Wu and John Gerhart of the University of California, Berkeley, accomplished a 20- to 30-fold purification of MPF by precipitating the protein in ammonium sulfate and subjecting the redissolved material to column chromatography. In addition to stimulating oocyte maturation, injections of the partially purified MPF stimulated the incorporation of ^{32}P into proteins of the amphibian oocyte.[7] When partially purified MPF preparations were incubated with [^{32}P]ATP in vitro, proteins present within the sample became phosphorylated, suggesting that MPF induced maturation by acting as a protein kinase.

MPF was finally purified in 1988 by a series of six successive chromatographic steps.[8] MPF activity in these purified preparations was consistently associated with two polypeptides, one having a molecular mass of 32 kDa and the other of 45 kDa. The purified MPF preparation possessed a high level of protein kinase activity, as determined by the incorporation of radioactivity from [^{32}P]ATP into proteins. When the purified preparation was incubated in the presence of [^{32}P]ATP, the 45 kDa polypeptide became labeled.

By the end of the 1980s, the efforts to uncover the role of cyclins and MPF had begun to merge with another line of research that had been conducted on fission yeast by

Figure 5 Kinetics of *Xenopus* oocyte activation by progesterone and cyclin A mRNA. Large, immature oocytes were isolated from fragments of ovaries and were incubated with progesterone or microinjected with varying amounts of cyclin A mRNA. At 3 to 4 hours after injection, obviously damaged oocytes were removed (about 2–4 per starting group of 20), and the remaining ones (which represent the 100% value) were allowed to develop. Germinal vesicle breakdown (GVBD) and oocyte activation were indicated by the formation of a white spot in the region of the animal pole and were confirmed by dissection of the oocytes. (FROM K. I. SWENSON, K. M. FARRELL, AND J. V. RUDERMAN, CELL 47:865, 1986; BY PERMISSION OF CELL PRESS.)

Paul Nurse and his colleagues at the University of Oxford.[9] It had been shown that yeast produced a protein kinase with a molecular weight of 34 kDa whose activity was required for these cells to enter M phase (discussed on page 584). The yeast protein was called p34^{cdc2} or simply cdc2. The first evidence of a link between cdc2 and MPF came as the result of a collaboration between yeast and amphibian research groups.[10,11] Recall from the previous study that MPF was found to contain a 32 and 45 kDa protein. Antibodies formed against cdc2 from fission yeast were shown to react specifically with the 32 kDa component of MPF isolated from *Xenopus* eggs. These findings indicate that this component of MPF is a homologue of the 34 kDa yeast kinase and, therefore, that the machinery controlling the cell cycle in yeast and vertebrates contains evolutionarily conserved components.

A similar study using antibodies against yeast cdc2 showed that the homologous protein in vertebrates does not fluctuate during the cell cycle.[12] This supports the proposal that the 32 kDa protein kinase in vertebrate cells depends on another protein. The modulator was predicted to be cyclin, which rises in concentration during each cell

cycle and is then destroyed as cells enter anaphase. This proposal was subsequently verified in a number of studies in which the MPF was purified from amphibians, clams, and starfish, and its polypeptide composition analyzed.[13-15] In all of these cases, it was shown that the active MPF present in M-phase animal cells is a complex consisting of two types of subunits: (1) a 34 kDa subunit that contains the protein kinase active site and is homologous to the yeast cdc2 protein kinase, and (2) a larger subunit (45 kDa) identified as a cyclin whose presence is required for kinase activity. The studies described in this Experimental Pathway provided a unified view of the regulation of the cell cycle in all eukaryotic organisms. In addition, they set the stage for the analysis of the numerous factors that control the activity of MPF (cdc2) at various points during yeast and mammalian cell cycles, which have become the focus of attention in the past few years. The most important findings of these recent studies are discussed in the first section of this chapter.

References

1. SMITH, L. D. & ECKER, R. E. 1971. The interaction of steroids with *R. pipiens* oocytes in the induction of maturation. *Dev. Biol.* 25:233–247.

2. MASUI, Y. & MARKERT, C. L. 1971. Cytoplasmic control of nuclear behavior during meiotic maturation of frog oocytes. *J. Exp. Zool.* 177:129–146.

3. WASSERMAN, W. J. & SMITH, L. D. 1978. The cyclic behavior of a cytoplasmic factor controlling nuclear membrane breakdown. *J. Cell Biol.* 78:R15–R22.

4. SUNKARA, P. S., WRIGHT, D. A., & RAO, P. N. 1979. Mitotic factors from mammalian cells induce germinal vesicle breakdown and chromosome condensation in amphibian oocytes. *Proc. Nat'l Acad. Sci. U.S.A.* 76:2799–2802.

5. EVANS, T., ET AL. 1983. Cyclin: A protein specified by maternal mRNA in sea urchin eggs that is destroyed at each cleavage division. *Cell* 33:389–396.

6. SWENSON, K. I., FARRELL, K. M., & Ruderman, J. V. 1986. The clam embryo protein cyclin A induces entry into M phase and the resumption of meiosis in *Xenopus* oocytes. *Cell* 47:861–870.

7. WU, M. & GERHART, J. C. 1980. Partial purification and characterization of the maturation-promoting factor from eggs of *Xenopus laevis*. *Dev. Biol.* 79:465–477.

8. LOHKA, M. J., HAYES, M. K., & MALLER, J. L. 1988. Purification of maturation-promoting factor, an intracellular regulator of early mitotic events. *Proc. Nat'l Acad. Sci. U.S.A.* 85:3009–3013.

9. NURSE, P. 1990. Universal control mechanism regulating onset of M-phase. *Nature* 344:503–507.

10. GAUTIER, J., ET AL. 1988. Purified maturation-promoting factor contains the product of a *Xenopus* homolog of the fission yeast cell cycle control gene *cdc2+*. *Cell* 54:433–439.

11. DUNPHY, W. G., ET AL. 1988. The *Xenopus* cdc2 protein is a component of MPF, a cytoplasmic regulator of mitosis. *Cell* 54:423–431.

12. LABBE, J. C., ET AL. 1988. Activation at M-phase of a protein kinase encoded by a starfish homologue of the cell cycle control gene *cdc2+*. *Nature* 335:251–254.

13. LABBE, J. C., ET AL. 1989. MPF from starfish oocytes at first meiotic metaphase is a heterodimer containing one molecule of cdc2 and one molecule of cyclin B. *EMBO J.* 8:3053–3058.

14. DRAETTA, G., ET AL. 1989. cdc2 protein kinase is complexed with both cyclin A and B: Evidence for proteolytic inactivation of MPF. *Cell* 56:829–838.

15. GAUTIER, J., ET AL. 1990. Cyclin is a component of maturation-promoting factor from *Xenopus*. *Cell* 60:487–494.

SYNOPSIS

The stages through which a cell passes from one cell division to the next constitute the cell cycle. The cell cycle is divided into two major phases: M phase, which includes the process of mitosis, in which duplicated chromosomes are separated into two nuclei, and cytokinesis, in which the entire cell is physically divided into two daughter cells; and interphase. Interphase is typically much longer than M phase and is subdivided into three distinct phases based on the time of replication, which is confined to a definite period within the cell cycle. G_1 is the period following mitosis and preceding replication; S is the period during which DNA synthesis (and histone synthesis) occurs; and G_2 is the period following replication and preceding the onset of mitosis. The length of the cell cycle, and the phases of which it is made up, vary greatly from one type of cell to another. Cell cycles can range from as short as approximately 30 minutes in rapidly dividing embryos to a period of months in very slowly growing tissues. Certain types of terminally differentiated cells, such as vertebrate skeletal muscle and nerve cells, have lost the ability to divide. Most synthetic activities, other than that of DNA and histones, continue throughout interphase, but they either drop markedly or cease entirely during mitosis. (p. 581)

The passage of a cell through the cycle is controlled by a variety of factors. The first indication that the cytoplasm contained factors that control the cell cycle came from studies of cell fusion using cells in different stages of the cycle. These studies indicated that a replicating cell contains one or more factors that stimulate the initiation of DNA synthesis, and that a mitotic cell contains one or more factors that induce the compaction of the chromatin. Subsequent experiments showed that entry of a cell into M phase is triggered by the activation of a protein kinase called MPF (maturation promoting factor). MPF consists of two subunits: a catalytic subunit that transfers phosphate groups to specific serine and threonine residues of specific protein substrates, and a regulatory subunit consisting of a member of a family of proteins called cyclins. The catalytic subunit is called a cyclin-dependent ki-

nase (Cdk). When the cyclin concentration is low, the kinase lacks the cyclin subunit and is inactive. When the cyclin concentration reaches a sufficient level, the kinase is activated, triggering the entry of the cell into M phase. (p. 582)

The activities that control the cell cycle are focused primarily at two points: the transition between G_1 and S and the transition between G_2 and the entry into mitosis. Passage through each of these points requires the transient activation of a Cdk by a specific cyclin. In yeast, the same Cdk is active at both G_1-S and G_2-M, but it is stimulated by a different cyclin. The concentrations of the various cyclins rise and fall during the cell cycle as the result of changes in the rate of synthesis and destruction of the protein molecules. In addition to regulation by cyclins, Cdk activity is also controlled by the state of phosphorylation of the catalytic subunit, which in turn is controlled in yeast by at least two kinases (CAK and wee1) and a phosphatase (cdc25). In mammalian cells, at least eight different cyclins and a half-dozen different Cdks play a role in cell cycle regulation. (p. 584)

Cells possess feedback controls that monitor the status of cell cycle events, such as replication and chromosome compaction, and determine whether or not the cycle continues. The G_1-S and G_2-M transitions are points at which primary control is exercised. Exogenous growth factors, for example, influence the activity of cyclin-dependent kinases, which determine whether a cell enters S phase. If a cell is subjected to treatments that damage DNA, passage between G_1 and S is delayed until the damage is repaired. Arrest of a cell at one of the checkpoints of the cell cycle is accomplished by inhibitors whose synthesis is stimulated by events such as DNA damage. Once the cell has been stimulated by external agents to traverse the G_1-S checkpoint and initiate replication, the cell generally continues without further external stimulation through the subsequent mitosis. (p. 587)

Mitosis ensures that two daughter nuclei receive a complete and equivalent complement of genetic material. Mitosis is divided into prophase, prometaphase, metaphase, anaphase, and telophase. Prophase is characterized by the preparation of chromosomes for segregation and the assembly of the machinery required for chromosome movement. Whereas interphase chromosomes exist in a highly extended configuration, those of mitosis are highly compacted, rod-shaped structures. Each mitotic chromosome can be seen to be split longitudinally into two chromatids, which are duplicates of one another formed by replication during the previous S phase. The primary constriction on the mitotic chromosome marks the centromere, which houses a platelike structure, the kinetochore, to which the microtubules of the spindle attach. The first stage in spindle formation in a typical animal cell is the appearance of microtubules in a "sunburst" arrangement, or aster, around each of the two centrosomes during early prophase. Aster formation is followed by the movement of the centrosomes away from one another toward the poles. As this happens, the microtubules that stretch between them increase in number and elongate. Eventually, the two centrosomes reach opposite points in the cell, establishing the two poles. A number of types of cells, including those of plants, assemble a mitotic spindle in the absence of centrosomes. The end of prophase is marked by the fragmentation of the nuclear envelope and the dispersal of its membranes in the form of small vesicles. (p. 590)

During prometaphase and anaphase, individual chromosomes are first attached to spindle microtubules emanating from both poles and then moved to a plane at the center of the spindle. At the beginning of prometaphase, microtubules from the forming spindle penetrate into the region that had been the nucleus and make attachments to the kinetochores of the condensed chromosomes. Initially, chromosomes can be seen to move actively along the lateral surfaces of microtubules, powered by motor proteins located in the kinetochore. Soon, however, the kinetochores become stably associated with the plus ends of chromosomal microtubules from both poles of the spindle. Eventually, each chromosome is moved into position along a plane at the center of the spindle, a process that is accompanied by the shortening of some microtubules due to loss of tubulin subunits and the lengthening of others due to addition of subunits. Once the chromosomes are stably aligned, the cell has reached metaphase. The mitotic spindle of a typical animal cell at metaphase consists of astral microtubules, which radiate out from the centrosome; chromosomal microtubules, which are attached to the kinetochores; and polar microtubules that extend from the centrosome past the chromosomes and form a structural basket that maintains the integrity of the spindle. The microtubules of the metaphase spindle exhibit dynamic activity as demonstrated by the directed movement of fluorescently labeled subunits. (p. 596)

During anaphase and telophase, sister chromatids are separated from one another into separate regions of the dividing cell, and the chromosomes are returned to their interphase condition. Anaphase begins as the sister chromatids suddenly split away from one another, a process that is triggered by the ubiquitin-mediated destruction of cohesin, a protein complex that holds the sisters together. The separated chromosomes then move toward their respective poles accompanied by the shortening of the attached chromosomal microtubules, which results primarily from the net loss of subunits at the kinetochore. The movement of the chromosomes, which is called anaphase A, is typically accompanied by the elongation of the mitotic spindle and consequent separation of the poles, which is referred to as anaphase B. Telophase is characterized by the reformation of the nuclear envelope, the dispersal of the chromosomes, and the reformation of membranous cytoplasmic networks. Anaphase movements may be brought about by microtubular motors, such as kinesin and dynein, and/or microtubule depolymerization. (p. 600)

Cytokinesis, which is the division of the cytoplasm into two daughter cells, occurs by constriction in animal cells and by construction in plant cells. Animal cells are constricted in two by an indentation or furrow that forms at the surface of the cell and moves inward. The advancing furrow contains a band of actin filaments, which are thought to slide over one another, driven by small, force-generating myosin II filaments. The site of cytokinesis is thought to be selected by a signal diffusing from the mitotic spindle. Plant cells undergo cytokinesis by building a cell membrane and cell wall in a plane lying between the two poles. The first sign of cell plate formation is the appearance of clusters of interdigitating microtubules and interspersed, electron-dense material. Small vesicles then move into the region and become aligned into a plane. The vesicles fuse with one another to form a membranous network that develops into a cell plate. (p. 604)

Meiosis is a process that includes two sequential nuclear divisions, producing haploid daughter nuclei that contain only one member of each pair of homologous chromosomes, thus reducing the number of chromosomes in half. Meiosis can occur at different stages in the life cycle, depending on the type of organism. To ensure that each daughter nucleus has only one set of homologues, an elaborate process of chromosome pairing occurs during prophase I that has no counterpart in mitosis. The pairing of chromosomes is accompanied by the formation of a proteinaceous, ladderlike structure called the synaptonemal complex (SC). The chromatin of each homologue is intimately associated with one of the lateral bars of the SC. During prophase I, homologous chromosomes engage in genetic recombination that produces chromosomes with new combinations of maternal and paternal alleles. Following recombination, the chromosomes become further condensed, and the homologues remain attached to one another at specific points called chiasmata, which represent sites where recombination had occurred at a previous stage. The paired homologues (called bivalents or tetrads) become oriented at the metaphase plate so that both chromatids of one chromosome face the same pole. During anaphase I, the homologous chromosomes separate from one another. Because there is no interaction between one tetrad and another, the maternal and paternal chromosomes of each tetrad segregate independently of one another. The cells, which are now haploid in chromosome content, progress into the second meiotic division during which the sister chromatids of each chromosome are separated into different daughter nuclei. (p. 609)

Genetic recombination during meiosis occurs as the result of the breakage and reunion of DNA strands from different homologues of a tetrad. Recombination is a poorly understood process in which homologous regions on different DNA strands are exchanged without the addition or loss of a single base pair. In an initial step, two duplexes that are about to recombine become aligned next to one another. Once they are aligned, breaks occur in either one or both strands of one of the duplexes. In following steps, DNA strands from one duplex invade the other duplex, forming an interconnected structure. Subsequent steps may include the activity of nucleases and polymerases to create and fill gaps in the various strands similar to the way in which DNA repair occurs. (p. 615)

ANALYTIC QUESTIONS

1. In what way is cell division a link between humans and the earliest eukaryotic cells?

2. What types of synthetic events would you expect to occur in G_1 that do not occur in G_2?

3. Suppose you are labeling a population of cells growing asynchronously with [^3H]thymidine. G_1 is 6 hours, S is 6 hours, G_2 is 5 hours, and M is 1 hour. What percentage of the cells would be labeled after a 15-minute pulse? How long would you have to chase these cells before you saw any labeled mitotic chromosomes? What percentage of the cells would have labeled mitotic chromosomes if you chased the cells for 18 hours?

4. Suppose you take a culture of the same cells used in the previous question, but instead of pulse labeling them with [^3H]thymidine, you labeled them continuously for 20 hours. Plot a graph showing the amount of radioactive DNA that would be present in the culture over the 20-hour period. What would be the minimum amount of time needed to ensure that all cells had some incorporated label in the above experiment? How could you determine the length of the cell cycle in this culture without use of a radioactive label?

5. Fusing G_1- and S-phase cells produces different results from fusing a G_2-phase cell with one in S. What would you expect this difference to be and how can it be explained?

6. Of the genes that code for CAK, wee1, and cdc25, which would you expect, when mutated, to stop a cell from continuing to divide? Which would you expect, when mutated, to cause the cell to proliferate in an uncontrolled manner?

7. Give four distinct mechanisms by which a Cdk can be inactivated.

8. A syncytium is a "cell" that contains more than one nucleus; examples are a skeletal muscle fiber and a blastula of a fly embryo. These two types of syncytia arise by very different pathways. What two mechanisms can you envision that could result in the formation of such syncytia? What does this tell you about the relationship between mitosis and cytokinesis?

9. How could you determine experimentally if the microtubules of the polar microtubules were in a state of dynamic flux during anaphase? Knowing the events that occur at this stage, what would you expect to see?

10. If you were to add [^3H]thymidine to a cell as it underwent replication (S phase) prior to beginning meiosis, what percentage of the chromosomes of the gametes produced would be labeled? If one of these gametes (a sperm) were to fertilize an unlabeled egg, what percentage of the chromosomes of the two-cell stage would be labeled?

11. If the haploid number of chromosomes in humans is 23 and the amount of nuclear DNA in a sperm is 1C, state how many chromosomes a human cell has in the following stages: metaphase of mitosis, prophase I of meiosis, anaphase I of meiosis I, prophase II of meiosis, anaphase II of meiosis. How many chromatids does the cell have at each of these stages? How much DNA (in terms of numbers of C) does the cell have at each of these stages?

12. Plot the amount of DNA in the nucleus of a spermatogonia from the G_1 stage prior to the first meiotic division through the completion of meiosis. Label each of the major stages of the cell cycle and of meiosis on the graph.

13. How many centrioles does a cell have at metaphase of mitosis?

14. Suppose you were told that most cases of trisomy result from aging of an egg in the oviduct as it awaits fertilization. What type of evidence could you obtain from examining spontaneously aborted fetuses that would confirm

this suggestion? How would this fit with data that has already been collected?

15. Suppose you incubate a meiotic cell in [³H]thymidine between the leptotene and zygotene stages, then you fix the cell during pachytene and prepare an autoradiograph. You find that the chiasmata are sites of concentrations of silver grains. What does this tell you about the mechanism of recombination?

16. It was discussed on page 601 that two types of signals have been implicated in metaphase checkpoint arrest. Consider the results of a recent experiment in which the unattached kinetochore of a monoattached chromosome was destroyed by a laser beam. It was found in this study that, following destruction of the kinetochore, the cell entered anaphase even though this chromosome was not properly aligned at the metaphase plate. How do you interpret this experiment in terms of the signals discussed in the chapter?

17. Assume for a moment that crossing over did not occur. Would you agree that you received half of your chromosomes from each parent? Would you agree that you received one-quarter of your chromosomes from each grandparent? Would the answer to these questions change if you allowed for crossing over to have occurred?

18. It was noted on page 583 that G_2-phase cells cannot be stimulated to undergo replication when fused with S-phase cells. Can you explain this observation based on information presented in Figure 13.21?

19. What type of phenotype would you expect of a cell whose Cdk subunit was lacking each of the following residues as the result of mutation: Tyr 15, Thr 161? Whose Cdc25 was lacking Ser 216?

20. It was noted on page 594 that centrosome duplication and DNA synthesis are both initiated by cyclin E-Cdk2, which becomes active at the end of G_1. A recent study found that if cyclin E-Cdk2 is activated at an earlier stage, such as the beginning of G_1, that centrosome duplication begins at that point in the cell cycle, but that DNA replication is not initiated until S phase would normally begin. Provide a hypothesis to explain why DNA synthesis does not begin as well. You might look back at Figure 13.21 for further information.

21. What type of phenotype would you expect of a cell whose Cdc20 polypeptide was mutated so that (1) it was unable to bind to Mad2, or (2) it was unable to bind to the other subunits of the APC, or (3) it failed to dissociate from the APC at the end of anaphase?

ADDITIONAL READING

The Cell Cycle

CASPARI, T. 2000. Checkpoints: How to activate p53. *Curr. Biol.* 10:R315–R317.

DEN BOER, B.G.W. & MURRAY, J.A.H. 2000. Triggering the cell cycle in plants. *Trends Cell Biol.* 10:245–250.

HALL, J. & ANGILE, S. 1999. Radiation, DNA damage, and cancer. *Mol. Med. Today* 5:157–164.

JACKS, T. & WEINBERG, R. A. 1998. The expanding role of cell cycle regulators. *Science* 280:1035–1036.

JOHNSON, D. G. & WALKER, C. L. 1999. Cyclins and cell cycle checkpoints. *Annu. Rev. Pharmacol. Toxicol.* 39:295–312.

KOEPP, D. M., ET AL. 1999. How the cyclin became a cyclin: regulated proteolysis in the cell cycle. *Cell* 97:431–434.

NURSE, P. 2000. A long twentieth century of the cell cycle and beyond. *Cell* 100:71–78.

NURSE, P, MASUI, Y., & HARTWELL, L. 1998. Understanding the cell cycle. *Nature Med.* 4:1103–1106.

O'CONNELL, M. J., ET AL. 2000. The G2-phase DNA damage checkpoint. *Trends Cell Biol.* 10:296–303.

OHI, R. & GOULD, K. L. 1999. Regulating the onset of mitosis. *Curr. Opin. Cell Biol.* 11:267–273.

PINES, J. 1999. Four-dimensional control of the cell cycle. *Nature Cell Biol.* 1:E73–E79.

PIWNICA-WORMS, H. 1999. Cell cycle: Fools rush in. *Nature* 401:535–536.

PRINZ, S. & AMON, A. 1999. Dual control of mitotic exit. *Nature* 402:133–134.

ROBERTS, J. M. 1999. Evolving ideas about cyclins. *Cell* 98:129–132.

SHERR, C. J. & ROBERTS, J. M. 1999. CDK inhibitors: Positive and negative regulators. *Genes Dev.* 13:1501–1512.

TYERS, M. & JORGENSEN, P. 2000. Proteolysis and the cell cycle: With this RING I do thee destroy. *Curr. Opin. Genes Dev.* 10:54–64.

WEINERT, T. 1998. DNA damage and checkpoint pathways: Molecular anatomy and interactions with repair. *Cell* 94:555–558.

YANG, J. & KORNBLUTH, S. 1999. All aboard the cyclin train: Subcellular trafficking of cyclins and their CDK partners. *Trends Cell Biol.* 9:207–213.

Mitosis and Cytokinesis

AMON, A. 1999. The spindle checkpoint. *Curr. Opin. Genes Dev.* 9:69–75.

BURKE, D. J. 2000. Complexity in the spindle checkpoint. *Curr. Opin. Genes Dev.* 10:26–31.

CLARKE, D. J. & GIMÉNEZ-ABIAN, J. F. 2000. Checkpoints controlling mitosis. *BioEss.* 22:351–363.

COLLAS, P. & COURVALIN, J.-C. 2000. Sorting nuclear membrane proteins at mitosis. *Trends Cell Biol.* 10:5–8.

COMPTON, D. A. 2000. Spindle assembly in animal cells. *Annu. Rev. Biochem.* 69:95–114.

ENDOW, S. A. & GLOVER, D. M., EDS. 1998. *Dynamics of Cell Division.* Oxford.

FIELD, C., ET AL. 1999. Cytokinesis in eukaryotes: A mechanistic comparison. *Curr. Opin. Cell Biol.* 11:68–80.

GARDNER, R. D. & BURKE, D. J. 2000. The spindle checkpoint: Two transitions, two pathways. *Trends Cell Biol.* 10:154–158.

GERSICH, G. & WEBER, I. 2000. Cytokinesis without myosin II. *Curr. Opin. Cell Biol.* 12:126–132.

GORBSKY, G. J., ET AL. 1999. Protein dynamics at the kinetochore: Cell cycle regulation and the metaphase to anaphase transition. *FASEB J.* 13:S231–S234.

GRANCELL, A. & SORGER, P. K. 1998. Chromosome movement: Kinetochores motor along. *Curr. Biol.* 8:R382–R385.

HEALD, R. 2000. Motor function in the mitotic spindle. *Cell* 102:399–402.

HIRANO, T. 1999. SMC-mediated chromosome mechanics: A conserved scheme from bacteria to vertebrates. *Genes Dev.* 13:11–19.

HIRANO, T. 2000. Chromosomal cohesion, condensation, and separation. *Annu. Rev. Biochem.* 69:115–144.

HOLMES, V. F. & COZZARELLI, N. R. 2000. Closing the ring: Links between SMC proteins and chromosome partitioning. *Proc. Nat'l. Acad. Sci. U.S.A.* 97:1322–1324.

KOSHLAND, D. E. & GUACCI, V. 2000. Sister chromatid cohesion: The beginning of a long and beautiful relationship. *Curr. Opin. Cell Biol.* 12:297–301.

MANEY, T., ET AL. 1999. The kinetochore of higher eukaryotes: A molecular view. *Int. Revs. Cytol.* 194:67–131.

MCINTOSH, J. R. & HERING, G. E. 1991. Spindle fiber action and chromosome movement. *Annu. Rev. Cell Biol.* 7:403–426.

MORGAN, D. O. 1999. Regulation of the APC and the exit from mitosis. *Nature Cell Biol.* 1:E47–E53.

MURRAY, A. W. 1998. How to compact DNA. *Science* 282:425–426.

MURRAY, A. W. & HUNT, T. 1994. *The Cell Cycle: An Introduction.* Oxford.

NASMYTH, K. 1999. Separating sister chromatids. *Trends Biochem. Sci.* 24:98–103.

NASMYTH, K., ET AL. 2000. Splitting the chromosome: Cutting the ties that bind sister chromatids. *Science* 288:1379–1384.

NELSON, W. J. 2000. W(h)ither the Golgi during mitosis? *J. Cell Biol.* 149:243–248.

ORR-WEAVER, T.L. 1999. The difficulty in separating sisters. *Science* 285:344–345.

PAGE, A. M. & HIETER, P. 1999. The anaphase-promoting complex: New subunits and regulators. *Annu. Rev. Biochem.* 68:583–609.

PETERS, J.-M. 1998. SCF and APC: The Yin and Yang of cell cycle regulated proteolysis. *Curr. Opin. Cell Biol.* 10:759–768.

RAPPAPORT, R. 1996. *Cytokinesis in Animal Cells.* Cambridge.

RIEDER, C. L. & SALMON, E. D. 1998. The vertebrate cell kinetochore and its roles during mitosis. *Trends Cell Biol.* 8:310–318.

ROBINSON, D. N. & SPUDICH, J. A. 2000. Towards a molecular understanding of cytokinesis. *Trends Cell Biol.* 10:228–237.

SHARP, D. J., ET AL. 2000. Microtubule motors in mitosis. *Nature* 407:41–47.

SKIBBENS, R. V. & HIETER, P. 1998. Kinetochores and the checkpoint mechanism that monitors for defects in the chromosome segregation machinery. *Annu. Rev. Gen.* 32:307–337.

ZACHARIAE, W. 1999. Progression into and out of mitosis. *Curr. Opin. Cell Biol.* 11:708–716.

Meiosis, Recombination, and Nondisjunction

DEJ, K. J. & ORR-WEAVER, T. L. 2000. Separation anxiety at the centromere. *Trends Cell Biol.* 10:392–399.

HABER, J. E. 1998. Meiosis: Avoiding inappropriate relationships. *Curr. Biol.* 8:R832–R835.

HABER, J. E. 1998. Searching for a partner. *Science* 279:823–824.

HANDEL, M. A., ED. 1997. Meiosis and Gametogenesis. *Curr. Topics Develop. Biol.*, vol. 37.

HIOM, K. 2000. Homologous recombination. *Curr. Biol.* 10:R359–R361.

JOHN, B. 1990. *Meiosis.* Cambridge.

OSMAN, F. & SUBRAMANI, S. 1998. Double-strand break-induced recombination in eukaryotes. *Prog. Nuc. Acid Res. Mol. Biol.* 58:263–299.

ROEDER, G. S. 1997. Meiotic chromosomes: It takes two to tango. *Genes Dev.* 11:2600–2621.

ROEDER, G. S. & BAILIS, J. M. 2000. The pachytene checkpoint. *Trends Gen.* 16:395–403.

SLUDER, G. & MCCOLLUM, D. 2000. The Mad ways of meiosis. *Science* 289:254–255.

STAHL, F. 1996. Meiotic recombination in yeast: Coronation of the double-strand break repair model. *Cell* 87:965–968.

ZICKLER, D. & KLECKNER, N. 1999. Meiotic chromosomes: Integrating structure and function. *Annu. Rev. Gen.* 33:603–754.

ZICKLER, D. & KLECKNER, N. 1998. The leptotene-zygotene transition in meiosis. *Annu. Rev. Gen.* 32:619–697.

15

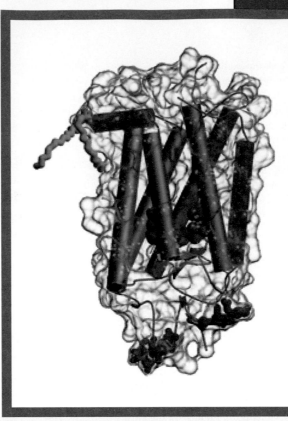

Cell Signaling: Communication Between Cells and Their Environment

The English poet John Donne expressed his belief in the interdependence of humans in the line "No man is an island." The same can be said of cells. To survive, cells must communicate with their neighbors, monitor the conditions in their environment, and respond appropriately to a host of different stimuli that impinge on their surface. Cells carry out these interactions by a phenomenon known as **cell signaling**, in which information is relayed across the plasma membrane to the cell interior and often transmitted to the cell nucleus. An overview of the process of cell signaling is shown in Figure 15.1. In most systems, cell signaling includes:

- Recognition of the stimulus at the outer surface of the plasma membrane by a specific receptor embedded in the membrane.

- Transfer of a signal across the plasma membrane to its cytoplasmic surface.

Three-dimensional structure of rhodopsin, the light-sensitive protein found in the rods of the retina. Rhodopsin is a member of a huge family of receptors characterized by seven membrane-spanning α helices (shown as cylinders) that respond to extracellular stimuli and transmit the signal into the cytoplasm. Rhodopsin is activated when its light-absorbing retinal group (shown in red) absorbs a photon of light, which leads to a conformational change in the protein that is transmitted to a G protein at the inner surface of the membrane. (COURTESY OF CRAIG A. BEHNKE, EMERALD BIOSTRUCTURES.)

Site of cell-cell contact

Extracellular ligand

Cytoskeleton

Activation of enzyme activity

Change in cytoskeletal organization

Change in ion permeability

Activation of DNA synthesis

Activation of RNA synthesis

Extracellular matrix

Figure 15.1 An overview of some of the ways that intracellular signals may be initiated and the types of responses that can be generated. The plasma membrane of a cell contains receptors that bind specifically to soluble ligands, extracellular matrices, and components of the surfaces of other cells. The stimulation arising from these contacts may be converted (transduced) by the cell membrane into intracellular signals that are transmitted along diverse pathways. Ultimately, these signals can lead to activation of specific enzymes, changes in cytoskeletal organization, changes in ion permeability, initiation of DNA synthesis, or the activation or repression of gene expression.

■ Transmission of the signal to specific effector molecules on the inner surface of the membrane or within the cytoplasm that trigger the cell's response. Depending on the type of cell and stimulus, the response may involve a change in gene expression, an alteration of the activity of metabolic enzymes, a reconfiguration of the cytoskeleton, a change in ion permeability, activation of DNA synthesis, or even the death of the cell.

■ Cessation of the response as a result of the destruction or inactivation of the signaling molecules, combined with a decrease in the level of the extracellular stimulus.

Before we examine these steps in more detail, it should be noted that not all information is transmitted from the extracellular space to the cell interior by way of a cell-surface receptor. Steroid hormones, for example, act on target cells by diffusing through the plasma membrane and binding to a cytoplasmic receptor (page 527). We will discuss another example in the present chapter in which an intercellular messenger called nitric oxide (NO) passes through the plasma membrane, rather than acting at the cell surface. Conversely, a few extracellular agents can evoke their response directly without having to initiate a signal inside the cell. The neurotransmitter acetylcholine

acts by such a mechanism when it binds to a receptor on a skeletal muscle cell, opening an ion channel that lies within the receptor itself (page 175).

Cell signaling can affect virtually every aspect of cell structure and function, which is one of the primary reasons this chapter appears near the end of the book. On one hand, an understanding of cell signaling requires knowledge about other types of cellular activity. On the other hand, insights into cell signaling can tie together a variety of seemingly independent cellular activities. Cell signaling is also intimately involved in the regulation of cell growth and division. This makes the study of cell signaling crucially important in understanding how a cell can lose growth control and develop into a malignant tumor. ■

15.1 BASIC CHARACTERISTICS OF CELL-SIGNALING SYSTEMS

The term *cell signaling* conveys the concept that a cell responds to a stimulus from its environment by relaying information to its internal compartment. In most cases, the stimulus is a molecule that has been secreted into the extracellular space by another cell, but a stimulus may also originate from contact with another cell or a noncellular substratum (Figure 15.1). Another term that is commonly used to describe this phenomenon is **signal transduction**, which indicates that the stimulus received by the cell-surface receptor is different from the signal released to the cell interior.

As will be evident shortly, signal transduction is a complex process in which information is passed along **signaling pathways** that consist of a series of distinct proteins. Each protein in the pathway typically acts by altering the conformation of the subsequent (or *downstream*) protein in the series, an event that activates or inhibits that protein (Figure 15.2). It should come as no surprise, after reading about other topics in cell biology, that alterations in the conformation of signaling proteins are usually accomplished by protein kinases and phosphatases that add or remove phosphate groups from other proteins (Figure 15.2). It is estimated that the human genome encodes as many as 2000 different protein kinases and more than 300 different protein phosphatases. Some of these kinases and phosphatases are soluble cytoplasmic proteins; others are integral membrane proteins. Some of these enzymes have numerous proteins as their substrates, whereas others phosphorylate or dephosphorylate only a single protein substrate. Many of the protein substrates of these enzymes are other enzymes—most often other kinases and phosphatases—but the substrates also include ion channels, transcription factors, and various types of regulatory subunits. Approximately one-third of the proteins of a cell are thought to be subject to phosphorylation.

Another feature of cell-signaling pathways that is shared by other cellular processes is the involvement of **GTP-binding proteins** (or **G proteins**) as switches that turn activities on or off. G proteins can act as switches be-

Figure 15.2 Signal transduction pathways consist primarily of protein kinases and protein phosphatases whose catalytic actions change the conformations, and thus the activities, of the proteins they modify. In the example depicted here, protein kinase 1 is activated by protein kinase 2. Once activated, protein kinase 1 phosphorylates protein kinase 3, activating the enzyme. Protein kinase 3 then phosphorylates a transcription factor, increasing its affinity for a site on the DNA, which initiates the response. Each of these activation steps is reversed by a phosphatase.

cause they exist in two interchangeable states: an active state that contains a bound GTP and an inactive state that contains a bound GDP. The difference in conformation of a G protein in the GTP-and GDP-bound states is shown in Figure 15.3*a*. Although the difference in conformation may appear subtle, the active, GTP-bound protein is able to bind specific proteins (i.e., downstream targets) with which the GDP-bound protein cannot interact. In most instances, binding of the GTP-bound protein activates the downstream target, allowing it to carry out a particular function. Hydrolysis of the bound GTP causes the G protein to revert to the inactive state, and it dissociates

(a)

(b)

Figure 15.3 The structure of a G protein and the G protein cycle. *(a)* Comparison of the tertiary structure of the active GTP-bound state (red) and inactive GDP-bound state (green) of the small G protein Ras. A bound guanine nucleotide is depicted in the ball-and-stick form. The differences in conformation occur in two regions of the molecule known as switch I and switch II. The difference in conformation shown here affects the molecule's ability to bind to other proteins. *(b)* The G protein cycle. G proteins are in their inactive state when they are bound by a molecule of GDP. If the inactive G protein interacts with a guanine nucleotide dissociation inhibitor (GDI), release of the GDP is inhibited and the protein remains in the inactive state (step 1a). If the inactive G protein interacts with a guanine nucleotide exchange factor (GEF; step 1b), the G protein exchanges its GDP

for a GTP (step 2), which activates the G protein so that it can bind to a downstream target protein (step 3). Binding to the GTP-bound G protein activates the target protein, which is typically an enzyme such as a protein kinase or a protein phosphatase. This has the effect of transmitting the signal farther downstream along the signaling pathway. G proteins have a weak intrinsic GTPase activity that is stimulated by interaction with a GTPase-activating protein (GAP) (step 4). The degree of GTPase stimulation by a GAP determines the length of time that the G protein is active. Consequently, the GAP serves as a type of clock that regulates the duration of the response (step 5). Once the GTP has been hydrolyzed, the complex dissociates, and the inactive G protein is ready to begin a new cycle (step 6). (A: FROM STEVEN J. GAMBLIN AND STEPHEN J. SMERDON, STRUCT. 7:R200, 1999.)

from the downstream target. GTP-binding proteins were discussed in connection with vesicle budding and fusion in Chapter 8, microtubule dynamics in Chapter 9, protein synthesis in Chapters 8 and 11, and nucleocytoplasmic transport in Chapter 12. In the present chapter, we will explore their role in transmitting messages along "cellular information circuits."

The cycling of G proteins between active and inactive states is aided by accessory proteins that bind to the G protein and regulate its activity (Figure 15.3*b*). These accessory proteins include

1. *GTPase-activating proteins (GAPs)*. Most G proteins possess some capability to hydrolyze a bound GTP, but this capability is greatly accelerated by interaction with specific GAPs. Because they stimulate hydrolysis of the bound GTP, which inactivates the G protein, GAPs dramatically shorten the duration of a G protein-mediated response.

2. *Guanine nucleotide-exchange factors (GEFs)*. An inactive G protein is converted to the active form when the bound GDP is replaced with a GTP. GEFs are proteins that bind to an inactive G protein and stimulate dissociation of the bound GDP. Once the GDP is released, the G protein rapidly binds a GTP, which is present at relatively high concentration in the cell, thereby activating the G protein.

3. *Guanine nucleotide-dissociation inhibitors (GDIs)*. GDIs are proteins that inhibit the release of a bound GDP from a G protein, thus maintaining the protein in the inactive, GDP-bound state.

The discussion of cell signaling in this chapter will focus on two contrasting types of signal transduction pathways (Figure 15.4). In one type of pathway, the attachment of a ligand to the cell-surface receptor is signaled through activation of a GTP-binding protein, whereas in the other type of response, ligand binding is signaled through direct activation of an enzymatic activity associated with the receptor.

REVIEW ?

1. Contrast the roles of GAPs, GEFs, and GDIs in the G protein cycle.
2. What is meant by the term *signal transduction*?

15.2 SECOND MESSENGERS AND G PROTEIN-COUPLED RECEPTORS

The oxidation of glucose by glycolysis and the TCA cycle provides cells with their primary source of energy. Because glucose utilization is so important in energy metabolism, biochemists and physiologists have been studying the mechanisms that control it for more than a hundred years.

As discussed in Chapter 3, glucose is stored in animal cells as glycogen, an insoluble glucose polymer. The breakdown of glycogen into its energy-rich subunits is controlled in vertebrates by a number of hormones, most notably glucagon, which is secreted by the pancreas, and epinephrine, which is secreted by the adrenal gland. Each of these hormones binds to its specific receptor, which protrudes from the plasma membrane of a target cell. Binding of either hormone initiates a series of reactions that leads to the activation of the enzyme phosphorylase, which catalyzes the breakdown of glycogen into glucose 1-phosphate, the first step in the catabolism of the sugar (Figure 15.5). In addition, the binding of either of these hormones leads to the inhibition of the enzyme glycogen synthase, which catalyzes the opposing reaction in which glucose is polymerized to form glycogen. Thus, two different stimuli (glucagon and epinephrine), recognized by different receptors, can induce the same response in a single target cell. In the following section we will see how this is possible.

The Discovery of a Second Messenger: Cyclic AMP

How does the binding of a hormone to the plasma membrane change the activity of cytoplasmic enzymes, such as phosphorylase and glycogen synthase? The answer to this

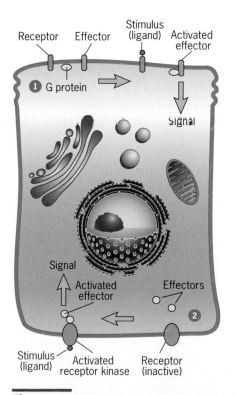

Figure 15.4 Two alternate types of signal transduction pathways. In pathway 1, the bound ligand activates a G protein that activates an effector that emits a signal. In pathway 2, the bound ligand activates an enzymatic (kinase) activity of the receptor, which activates an effector that emits a signal.

Figure 15.5 **The reactions that lead to glucose storage or mobilization**. The activities of two of the key enzymes in these reactions, phosphorylase and glycogen synthase, are controlled by hormones that act through signal transduction pathways.

question was provided by studies that began in the mid-1950s in the laboratories of Earl Sutherland and his colleagues at Case Western Reserve University and Edwin Krebs and Edmond Fischer at the University of Washington. Sutherland's goal was to develop an in vitro system to study the physiological responses to hormones. After considerable effort, he was able to activate phosphorylase in a preparation of *broken* cells that had been incubated with glucagon or epinephrine. This broken-cell preparation could be divided by centrifugation into a particulate fraction consisting primarily of cell membranes and a soluble supernatant fraction. Even though phosphorylase was present only in the supernatant fraction, the particulate material was required to obtain the hormone response. Subsequent experiments indicated that the response occurred in at least two distinct steps. If the particulate fraction of a liver homogenate was isolated and incubated with the hormone, some substance was released that, when added to the supernatant fraction, activated the soluble phosphorylase molecules. Sutherland identified the substance released by the membranes of the particulate fraction as cyclic adenosine monophosphate (**cyclic AMP**, or simply **cAMP**). As will be discussed below, cAMP stimulates glucose mobilization by activating a protein kinase that adds a phosphate to a specific serine residue of the phosphorylase polypeptide.

Cyclic AMP is a **second messenger**, a substance that is released into the interior of the cell following the binding of a first messenger—a hormone or other ligand—to a receptor at the outer surface of the cell. Cyclic AMP is only one of a number of substances that act as second messengers in eukaryotic cells. Whereas the first messenger binds exclusively to a single receptor species, the second messenger generated within the cytoplasm often stimulates a variety of cellular activities. As a result, second messengers enable cells to mount a large-scale, coordinated response following stimulation by a single extracellular ligand. We will return in a moment to the mechanism by which the signal from the bound ligand is transmitted

across the membrane, which is one of the major steps in the process of cell signaling, but first we will examine the mechanism of second messenger action by looking at how cAMP is formed and how it triggers the mobilization of glucose.

Glucose Mobilization: An Example of a Response Induced by cAMP

Cyclic AMP is synthesized by an integral membrane protein called *adenylyl cyclase*, whose catalytic domain resides at the inner surface of the plasma membrane (Figure 15.6). Because it brings about (effects) the cellular response by synthesizing cAMP, adenylyl cyclase is an **effector**. Cyclic AMP evokes a response that leads to glucose mobilization by initiating a chain of reactions, as illustrated in Figure 15.7. The first step in this *reaction cascade* occurs as the hormone binds to its receptor, activating its adenylyl cyclase effector, which catalyzes the formation of cAMP (steps 1 and 2, Figure 15.7). The steps leading to the activation of adenylyl cyclase are discussed in detail in the following section.

Once formed, cyclic cAMP molecules diffuse into the cytoplasm where they bind to an allosteric site on a regulatory subunit of a specific cAMP-dependent protein kinase, called *protein kinase A (PKA)* (step 3). The regulatory subunits normally inhibit the catalytic activity of the enzyme. Cyclic AMP binding causes the dissociation of the inhibitory subunits, releasing the catalytic subunits of PKA in their active form. The target substrates of PKA in a liver cell include two enzymes that play a pivotal role in glucose metabolism, glycogen synthase and phosphorylase kinase (steps 4 and 5). Phosphorylation of glycogen synthase inhibits its catalytic activity and thus prevents the conversion of glucose to glycogen. In contrast, phosphorylation of phophorylase kinase activates the enzyme to catalyze the transfer of phosphate groups to phosphorylase molecules. As discovered by Krebs and Fischer, the addition of a single phosphate group to a specific serine

Figure 15.6 Formation of cyclic AMP from ATP is catalyzed by adenylyl cyclase, an integral membrane protein that consists of two parts, each containing six transmembrane helices (shown here in two dimensions). The enzyme's active site is located on the inner surface of the membrane in a cleft situated between two similar cytoplasmic domains. The activity of the enzyme is regulated by GTP-binding proteins, as discussed later in the text and illustrated in Figure 15.10. The breakdown of cAMP (not shown) is accomplished by a phosphodiesterase, which converts the cyclic nucleotide to a 5′ monophosphate.

Active site

ATP cAMP

residue in the phosphorylase molecule activates this enzyme (step 6), stimulating the breakdown of glycogen (step 7). The glucose 1-phosphate formed in the phosphorylase reaction is converted to glucose, which diffuses into the blood and reaches the tissues of the body (step 8).

The binding of a single hormone molecule at the cell surface can activate a number of adenylyl cyclase molecules, each of which can produce a large number of cAMP messengers in a short period of time. Thus, the production of a second messenger provides a mechanism to greatly amplify the signal generated from the original message. Many of the steps in the reaction cascade illustrated in Figure 15.7 result in amplification of the signal

Figure 15.7 The response by a liver cell to glucagon or epinephrine. The steps in the response to hormonal stimulation that lead to glucose mobilization are described in the text. Many of the steps in the reaction cascade are accompanied by a dramatic amplification of the signal. Steps leading to amplification are indicated by the blue arrows.

(these steps are indicated by the blue arrows). Each cAMP molecule activates a single PKA molecule, which phosphorylates a large number phosphorylase kinase molecules, which in turn phosphorylate an even larger number of phosphorylase molecules, which in turn can catalyze the formation of a much larger number of glucose phosphates. Thus, what begins as a barely perceptible stimulus at the cell surface is rapidly transformed into a major mobilization of glucose within the cell.

Other Aspects of cAMP Signal Transduction Pathways

Although the most rapid and best-studied effects of cAMP are produced in the cytoplasm, the nucleus and its genes also participate in the response. A fraction of the activated PKA molecules translocate into the nucleus where they phosphorylate key nuclear proteins (step 9, Figure 15.7), most notably a transcription factor called *CREB* (*cAMP response element-binding protein*). The phosphorylated version of CREB binds as a dimer to sites on the DNA (step 10) containing a particular nucleotide sequence (TGACGTCA), known as the *cAMP response element (CRE)*. Recall from page 527 that response elements are sites in the DNA where transcription factors bind and increase the rate of initiation of transcription. CREs are located in the regulatory regions of genes that play a role in the response to cAMP. In liver cells, for example, several of the enzymes involved in gluconeogenesis, a pathway by which glucose is formed from the intermediates of glycolysis (see Figure 3.30), are encoded by genes that contain nearby CREs. Thus, epinephrine and glucagon not only activate catabolic enzymes involved in glycogen breakdown, they promote the synthesis of anabolic enzymes required to synthesize glucose from smaller precursors.

As one might expect, a mechanism must exist to reverse the steps discussed above; otherwise the cell would remain in the activated state indefinitely. Liver cells contain phosphatases that remove the phosphate groups added by the kinases. The most important of these enzymes, phosphatase-1, can remove phosphates from all of the phosphorylated enzymes of Figure 15.7: phophorylase kinase, glycogen synthase, and phosphorylase. The destruction of cAMP molecules present in the cell is accomplished by an enzyme, cAMP phosphodiesterase, that helps terminate the response.

Cyclic AMP is produced in many different cells in response to a wide variety of different ligands (i.e., first messengers). Several of the hormonal responses mediated by cAMP in mammalian cells are listed in Table 15.1. Cyclic AMP pathways have also been implicated in processes occurring in the nervous system, including learning, memory, and drug addiction. Chronic use of opiates, for example, leads to elevated levels of adenylyl cyclase and PKA, which may be partially responsible for the physiological responses that occur during drug withdrawal. Another cyclic nucleotide, cyclic GMP, also acts as a second messenger in certain cells as illustrated by the induced relaxation of smooth muscle cells discussed on page 661. Cyclic GMP also plays a key role in the signaling pathway involved in vision (page 638).

Because cAMP exerts most of its effects by activating PKA, the response of a given cell to cAMP is typically determined by the specific proteins phosphorylated by this kinase (Figure 15.8 and Table 15.2). Although activation of PKA in a liver cell in response to epinephrine leads to the breakdown of glycogen, activation of the same enzyme in a kidney tubule cell in response to vasopressin causes an increase in the permeability of the membrane to water, and activation of the enzyme in a thyroid cell in response to TSH leads to the secretion of thyroid hormone. The phosphorylation of specific substrates by PKA is facilitated by a family of proteins, called *AKAPs*, that anchor the protein kinase to particular locations within a cell. For example, one member of the AKAP family, a protein called yotiao, is present at the plasma membrane of certain nerve cells, where it binds PKA molecules in close proximity to their substrate, which is one of the ion channels of the membrane.

Table 15.1	Examples of Hormone-Induced Responses Mediated by cAMP

Tissue	Hormone	Response
Liver	Epinephrine and glucagon	Glycogen hydrolysis, glucose synthesis (gluconeogenesis), inhibition of glycogen synthesis
Skeletal muscle	Epinephrine	Glycogen hydrolysis, inhibition of glycogen synthesis
Cardiac muscle	Epinephrine	Increased contractility
Adipose	Epinephrine, ACTH, and glucagon	Triacylglycerol catabolism
Kidney	Vasopressin (ADH)	Increased permeability of epithelial cells to water
Thyroid	TSH	Secretion of thyroid hormones
Bone	Parathyroid hormone	Increased calcium resorption
Ovary	LH	Increased secretion of steroid hormones
Adrenal cortex	ACTH	Increased secretion of glucocorticoids

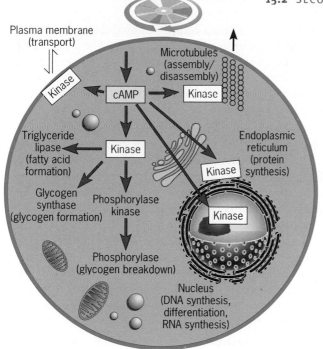

Figure 15.8 **Schematic illustration of the variety of processes that can be affected by changes in cAMP concentration.** All of these effects are thought to be mediated by activation of the same basic enzyme, protein kinase A. In fact, the same hormone can elicit very different responses in different cells, even when it binds to the same receptor. Epinephrine, for example, binds to a similar β-adrenergic receptor in liver cells, fat cells, and smooth muscle cells of the intestine, causing the production of cAMP in all three cell types. The responses, however, are quite different: glycogen is broken down in the liver cell, triacylglycerols are broken down in the fat cell, and the smooth muscle cells undergo relaxation.

Table 15.2	A Sample of Known PKA Substrates

Muscle glycogen synthase (la)
Phosphorylase kinase α
Protein phosphatase-1
Pyruvate kinase
CREB
Liver tryosine hydroxylase
Acetylcholine receptor δ
Protein phosphatase inhibitor-1
S6 ribosomal protein
Rabbit heart troponin
Hormone sensitive lipase
Phosphofructokinase
Myosin light-chain kinase
Fructose bisphosphatase
Phosphorylase kinase β
Muscle glycogen synthase
Acetyl CoA carboxylase

SOURCE: D. A. Walsh and S. M. van Patten, *FASEB J.* 8:1234, 1994.

The Structure and Function of G Protein-Coupled Receptors

Glucagon is a small protein of 29 amino acids, whereas epinephrine is a small hydrophilic amine that is synthesized from the amino acid tyrosine. These two molecules have virtually nothing in common structurally, yet both of them activate the same effector (adenylyl cyclase), leading to the synthesis of cyclic AMP and the subsequent phosphorylation of phosphorylase. The receptors for these two very different ligands are members of the same superfamily of integral membrane proteins, which are characterized by seven membrane-spanning α helices (Figure 15.9). The transmembrane segments are connected to one another by short extracellular and intracellular loops. The two receptors differ from one another primarily in the structure of the ligand-binding pocket on the extracellular surface of the receptor, which is specific for one or the other hormone.

Not only are the two receptors similar in structure, so too are the proteins that transmit the signal from the ligand-bound receptor to the adenylyl cyclase effector. Transmission of the signal from receptor to effector is accomplished by a third component of the transduction system, a **heterotrimeric G protein**, which is held at the plasma membrane by covalently linked lipid chains (Figure 15.9). Heterotrimeric G proteins were discovered and characterized by Martin Rodbell and his colleagues at the National Institutes of Health in the early 1970s; these studies are discussed in the Experimental Pathways, which can be found on the Web at www.wiley.com/

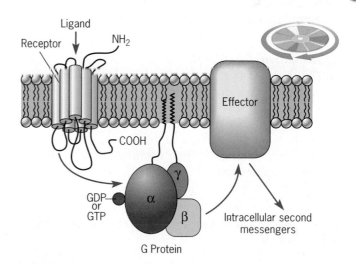

Figure 15.9 **The membrane-bound machinery for transducing signals by means of a seven-helix transmembrane receptor and a heterotrimeric G protein.** Receptors of this type, including those that bind epinephrine and glucagon, contain seven membrane-spanning helices. When bound with their ligand, the receptor interacts with a trimeric G protein, which activates an effector, such as adenylyl cyclase. As indicated in the figure, the α and γ subunits of the G protein are linked to the membrane by lipid groups that are embedded in the lipid bilayer.

college/karp. These proteins are referred to as G proteins because they bind guanine nucleotides, either GDP or GTP, as described on page 630. They are described as heterotrimeric because all of them consist of three different polypeptide subunits, called α, β, and γ. This property distinguishes them from another type of G protein to be discussed later in the chapter that consists of only a single subunit.

Receptors containing seven transmembrane helices comprise a superfamily of proteins that are usually referred to as **G protein-coupled receptors (GPCRs)** because they always work in concert with heterotrimeric G proteins. Hundreds of different GPCRs have been identified in organisms ranging from yeast to flowering plants and mammals that together mediate an extraordinary spectrum of cellular processes. In fact, GPCRs constitute the single largest superfamily of proteins encoded by animal genomes. For example, the nematode *Caenorhabditis elegans*, whose genome contains approximately 19,000 genes, encodes approximately 1000 different GPCRs. Included among the natural ligands that bind to GPCRs are a diverse array of hormones, neurotransmitters, opium derivatives, chemoattractants (molecules that attract phagocytic cells of the immune system), odorants and tastants (molecules detected by olfactory and gustatory receptors eliciting the senses of smell and taste), and photons. A list of some of the ligands that operate by means of this pathway and the effectors through which they act are listed in Table 15.3.

The pathway from receptor activation to G protein activation to activation of a membrane-bound effector follows a similar sequence of events regardless of the particular stimulus. This proposed sequence is indicated in Figure 15.10 and described as follows:

1. *Activation of the G protein by the receptor.* When a ligand binds to a G protein-coupled receptor, it induces a change in the conformation of the receptor that increases its affinity for the corresponding G protein. This is presumably the sole function of the extracellular ligand. The ligand-bound receptor binds to the G protein on the inner surface of the membrane, forming a receptor-G protein complex (step 1, Figure 15.10). Interaction with

the receptor induces a conformational change in the α subunit that causes the subunit to release its bound GDP and bind a GTP replacement (step 2). While in the activated state, a single receptor can activate a succession of G protein molecules, providing the first amplification step in the pathway.

2. *Relay of the signal from G protein to effector.* The exchange of GDP by GTP alters the conformation of the G_α subunit, causing it to dissociate from the receptor and from the other two subunits of the G protein, which remain together as a $G_{\beta\gamma}$ complex (step 3, Figure 15.10). Each dissociated G_α subunit (with GTP attached) is free to activate an effector molecule, such as adenylyl cyclase, which puts the second messenger system into operation (step 4). In this case, activation of the effector leads to the production of cAMP molecules (step 5).[1] The $G_{\beta\gamma}$ subunit complex may bind its own downstream effectors, most often an ion channel, providing an additional pathway for transmission of signals in a target cell.

3. *Ending the response.* At this point in the discussion, binding of a ligand to a GPCR has generated two activated signaling proteins: a) a ligand-bound GPCR, which acts on G proteins, and b) freed GTP-bound G_α subunits, which act on effectors. We will consider inactivation of each of these activated signaling proteins in turn.

a) Until recently, it was presumed that a GPCR stopped sending signals when the activating ligand was no longer present in the medium and bound to the receptor. Although a drop in concentration of ligand may be a factor in ending some responses, it has been found that many GPCRs, including β-adrenergic receptors and rhodopsin, can be "turned off" even while they are still interacting with ligand. This process of receptor inactivation is called *desensitization* and takes place in two steps. In the first step, the cytoplasmic domain of the activated GPCR is phosphorylated by a specific type of kinase called a *G protein-coupled receptor kinase (or GRK)* (Figure 5.10, step 6a). Phosphorylation of the GPCR sets the stage for the second step,

[1]Examples are known in which the α subunit of the G protein inhibits the activity of the effector rather than stimulating it. Binding of ligand to certain dopamine receptors, for example, leads to inhibition of adenylyl cyclase.

| Table 15.3 | **Physiological Processes Mediated by Heterotrimeric G Proteins** |

Stimulus	*Receptor*	*Effector*	*Physiological Response*
Epinephrine	β-Adrenergic receptor	Adenylyl cyclase	Glycogen breakdown
Serotonin	Serotonin receptor	Adenylyl cylase	Behavioral sensitization and learning in *Aplysia*
Light	Rhodopsin	cGMP phosphodiesterase	Visual excitation
IgE-antigen complexes	Mast cell IgE receptor	Phospholipase C	Secretion
f-Met Peptide	Chemotactic receptor	Phospholipase C	Chemotaxis
Acetylcholine	Muscarinic receptor	Potassium channel	Slowing of pacemaker activity

Adapted from L. Stryer and H. R. Bourne, reproduced with permission from the *Annual Review of Cell Biology*, vol. 2, © 1986, by Annual Reviews Inc.

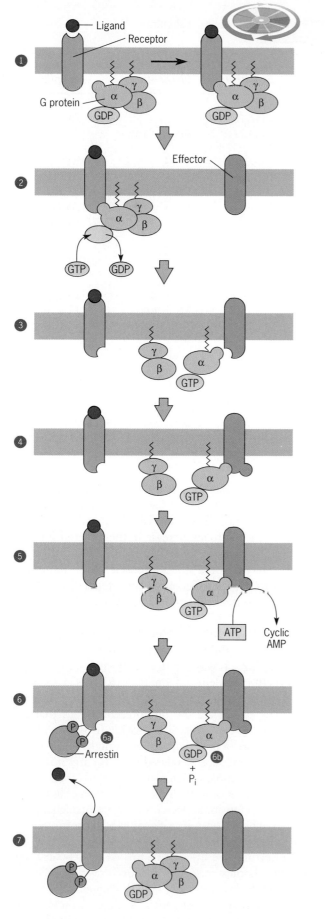

Figure 15.10 The mechanism of receptor-mediated activation (or inhibition) of effectors by means of heterotrimeric G proteins. In step 1, the ligand binds to the receptor, altering its conformation and increasing its affinity for the G protein to which it binds. In step 2, the G_α subunit loses its GDP, which is replaced by GTP. In step 3, the G_α subunit dissociates from the $G_{\beta\gamma}$ complex. In step 4, the G_α subunit binds to an effector (in this case adenylyl cyclase), activating the effector. The $G_{\beta\gamma}$ may also bind to an effector (not shown), such as an ion channel or an enzyme. In step 5, activated adenylyl cyclase produces cAMP. In step 6a, the receptor has been phosphorylated by a GRK and the phosphorylated receptor has been bound by an arrestin molecule, which inhibits the ligand-bound receptor from activating additional G proteins. In step 6b, the GTPase activity of G_α hydrolyzes the bound GTP, deactivating G_α. Hydrolysis of GTP is greatly accelerated by GTPase-activating proteins (GAPs), which are not shown (see Figure 15.3). In step 7, G_α reassociates with $G_{\beta\gamma}$, reforming the trimeric G protein, and the effector ceases its activity. The receptor bound to arrestin is likely to be taken up by endocytosis.

which is the binding of a protein, called *arrestin* (step 6a), that inhibits the receptor's ability to activate additional heterotrimeric G proteins. This action is termed desensitization because the cell stops responding to the stimulus, even while that stimulus is still acting on the outer surface of the cell. Desensitization is one of the mechanisms that allows a cell to respond to a change in its environment, rather than continuing to "fire" endlessly in the presence of an unchanging environment. Mutations that interfere with phosphorylation of the rhodopsin receptor by a GRK lead to the death of the photoreceptor cells in the retina and are thought to be one of the causes of blindness resulting from the disease retinitis pigmentosa.

While they are bound to phosphorylated GPCRs, arrestin molecules are also capable of binding to clathrin molecules that are situated in clathrin-coated pits (page 319). Interaction between bound arrestin and clathrin is thought to promote the uptake of phosphorylated GPCRs by endocytosis. Depending upon the circumstances, receptors that have been removed from the surface by endocytosis may be dephosphorylated and returned to the plasma membrane, or they may be destroyed by the endocytic machinery of the cell.

b) Signaling by the activated G_α subunit is terminated by a very different mechanism: the bound GTP molecule is simply hydrolyzed to GDP (step 6b). Thus, the strength and duration of the signal are determined in part by the rate of GTP hydrolysis by the G_α subunit. G_α subunits possess a weak GTPase activity, which allows them to slowly hydrolyze the bound GTP and inactivate themselves. Termination of the response is accelerated by interaction with an accessory protein (a GAP, page 631) that increases the efficiency of GTPase catalysis by the G_α subunit. Once the GTP is hydrolyzed, the G_α-GDP can reassociate with the $G_{\beta\gamma}$ subunits to reform the inactive

trimeric complex and return the system to the resting state (step 7). The cholera toxin (produced by the bacterium *Vibrio cholerae*) exerts its effect by modifying the G_α subunit and inhibiting its GTPase activity in the cells of the intestinal epithelium. As a result, adenylyl cyclase molecules remain in an activated mode, churning out cAMP, which causes large volumes of intestinal fluid to be secreted into the intestinal lumen. The loss of water associated with this inappropriate response often leads to death due to dehydration.

These mechanisms for transmitting signals across the plasma membrane are of ancient evolutionary origin and highly conserved. This is illustrated by an experiment in which yeast cells were genetically engineered to express a receptor for the mammalian hormone somatostatin. When these yeast cells were treated with somatostatin, the mammalian receptors at the cell surface interacted with the yeast heterotrimeric G proteins at the inner surface of the membrane and triggered a response leading to the proliferation of the yeast cells.

The effects of certain mutations in the function of G protein-coupled receptors are discussed in the accompanying Human Perspective.

The Specificity of G Protein-Coupled Responses It is evident that a wide variety of different agents, including hormones, neurotransmitters, and sensory stimuli, use a similar mechanism for transmitting information across the plasma membrane, triggering a wide variety of cellular responses. This does not mean that all of the various parts of the signal transduction machinery are identical in every cell type. The receptor for a given ligand can exist in several different versions (isoforms) that have different affinities for the ligand and for a particular G protein. For example, researchers have identified 9 different isoforms of the adrenergic receptor, which binds epinephrine, and 15 different isoforms of the receptor for serotonin, a powerful neurotransmitter released by nerve cells in parts of the brain governing emotions. Different isoforms of a receptor may coexist in the same plasma membrane, or they may occur in the membranes of different types of target cells. The heterotrimeric G proteins that transmit signals from receptor to effector can also exist in multiple forms, as can many of the effectors. At least 20 different G_α subunits, 5 different G_β subunits, and 12 different G_γ subunits have been identified, along with 9 isoforms of the effector adenylyl cyclase. Different combinations of specific subunits construct G proteins having different capabilities of reacting with specific isoforms of both receptors and effectors.

As mentioned on page 636, some G proteins act by inhibiting their effectors. Whether a G protein is stimulatory or inhibitory depends on the nature of the α subunit. Gα subunits are divided into four classes—$G_{\alpha s}$, $G_{\alpha i}$, $G_{\alpha q}$, and $G_{\alpha 12-13}$, whose tissue distribution and effectors are listed in Table 15.4. The same stimulus can activate a stimulatory G protein (one with a $G_{\alpha s}$ subunit) in one cell and an inhibitory G protein (one with a $G_{\alpha i}$ subunit) in a

| Table 15.4 | G protein α-subunits |

	Toxin Sensitivity	Tissue Distribution	Effector Coupling
G_s			
$G\alpha_s$	Cholera	Wide	AC
$G\alpha_{olf}$	Cholera	Brain/olfactory	AC
G_i			
$G\alpha_{i\text{-}1}$	Pertussis	Wide	AC/K⁺
$G\alpha_{i\text{-}2}$	Pertussis	Wide	AC/K⁺
$G\alpha_{i\text{-}3}$	Pertussis	Wide	AC/K⁺
$G\alpha_{oA}$	Pertussis	Brain	AC/K⁺
$G\alpha_{oB}$	Pertussis	Brain	AC/K⁺
$G\alpha_{t1}$	Pertussis	Retina	AC/PDE
$G\alpha_{t2}$	Pertussis	Retina	AC/PDE
$G\alpha_z$			
G_q			
$G\alpha_{15}$		Myeloid	PLC-β
$G\alpha_{16}$		Myeloid	PLC-β
$G\alpha_{14}$		Stromal/epithelial	PLC-β
$G\alpha_{11}$		Wide	PLC-β
$G\alpha_q$		Wide	PLC-β
$G_{12\text{-}13}$			
$G\alpha_{12}$		Wide	Cl⁻ channel
$G\alpha_{13}$		Wide	

SOURCE: A. J. Morris and C. C. Malbon, *Phys. Revs.* 79:1380, 1999. AC, adenylyl cylase; PDE, phosphodiesterase; PLC, phospholipase C; K⁺, K⁺ channel.

different cell. For example, when epinephrine binds to a β-adrenergic receptor on a cardiac muscle cell, a G protein with a $G_{\alpha s}$ subunit is activated, which stimulates cAMP production, leading to an increase in the rate and force of contraction. In contrast, when epinephrine binds to an α-adrenergic receptor on a smooth muscle cell in the intestine, a G protein with a $G_{\alpha i}$ subunit is activated, which inhibits cAMP production, producing muscle relaxation.

The Role of G Protein-Coupled Receptors in Sensory Perception Our ability to see, taste, and smell depends largely on GPCRs. It was mentioned above that rhodopsin, whose structure is depicted on page 628, is a GPCR. Rhodopsin is the light-sensitive protein present in the rods of our retina, which are the photoreceptor cells that respond to low light intensity and provide us with a black-and-white picture of our environment at night or in a darkened room. Several closely related GPCRs are present in the cones of the retina, which provide us with color vision under conditions of brighter light. Absorption of light by a molecule of rhodopsin induces a conformational change in the protein, which transmits a signal to a heterotrimeric G protein (called *transducin*) and leads to the activation of a coupled effector. The effector in this case is the enzyme cGMP phosphodiesterase, which hydrolyzes the cyclic nucleotide cGMP. Hydrolysis of cGMP triggers the closure of certain cation-specific channels, leading to the generation of membrane potentials that may be transmitted as action potentials along the optic nerve.

THE HUMAN PERSPECTIVE

Disorders Associated with G Protein-Coupled Receptors

The human genome may encode as many as 2000 different GPCRs. Their importance in human biology is reflected by the fact that more than one-quarter of all prescription drugs act as ligands that bind to this huge superfamily of receptors. A number of inherited disorders have been traced to defects in both GPCRs (Figure 1) and heterotrimeric G proteins (Table 1). Congenital nephrogenic diabetes insipidus (CNDI) is a rare inherited disease in which infants suffer serious dehydration as the result of an inability of their kidneys to produce a concentrated urine. If not diagnosed promptly, the chronic dehydration can produce mental retardation, inadequate growth, and even death. The disorder results from the inability of the cells of the kidneys to respond to the hormone vasopressin (antidiuretic hormone). As noted on page 153, some cases of this disease result from mutations in aquaporins, the water channels of the plasma membrane. In most cases, however, the fault lies in the vasopressin receptor, which is typically shortened as the result of a mutation that introduces a stop codon into the mRNA, causing premature termination of polypeptide synthesis (page 484). A different type of debilitat-

ing mutation in this same GPCR leads to an amino acid substitution at the junction between the third transmembrane segment and the second intracellular loop (site 4, Figure 1). Even though this receptor can still bind vasopressin at its external surface, the receptor cannot activate the G protein and thus cannot pass the signal downstream to the effector.

CNDI results from a mutation that leads to a loss of function of the encoded receptor. Many mutations that alter the structure of signaling proteins can have an opposite effect, leading to what is described as a "gain of function." In one such case, mutations have been found to cause a type of benign thy-

Table 1	Human Diseases Linked to the G Protein Pathway

Disease	Defective G Protein*
Albright's hereditary osteodystrophy and pseudohypoparathyroidisms	$G_{s\alpha}$
McCune-Albright syndrome	$G_{s\alpha}$
Pituitary, thyroid tumors (*gsp* oncogene)	$G_{s\alpha}$
Adrenocortical, ovarian tumors (*gip* oncogene)	$G_{i\alpha}$
Combined precocious puberty and pseudo-hypoparathyroidism	$G_{s\alpha}$

Disease	Defective G Protein-Coupled Receptor
Familial hypocalciuric hypercalcemia	Human analogue of BoPCAR1 receptor
Neonatal severe hyperparathyroidism	Human analogue of BoPCAR1 receptor (homozygous)
Hyperthyroidism (thyroid adenomas)	Thyrotropin receptor
Familial male precocious puberty	Luteinizing hormone receptor
X-linked nephrogenic diabetes insipidus	V2 vasopressin receptor
Retinitis pigmentosa	Rhodopsin receptor
Color blindness, spectral sensitivity variations	Cone opsin receptor
Familial glucocorticoid deficiency and isolated glucocorticoid deficiency	Adrenocorticoptropic hormone (ACTH) receptor

*As described in the text, a G protein with a $G_{s\alpha}$ acts to stimulate the effector, whereas a G protein with a $G_{i\alpha}$ inhibits the effector.
Source: D. E. Clapham, reprinted with permission from *Nature*, vol. 371, p. 109, 1994. © Copyright 1994, by Macmillan Magazines Ltd.

Figure 1 **Two-dimensional representation of a "composite" transmembrane receptor showing the approximate sites of a number of mutations responsible for causing human diseases.** Most of the mutations (numbers 1, 2, 5, 6, 7, and 8) result in constitutive stimulation of the effector, but others (3 and 4) result in blockage of the receptor's ability to stimulate the effector. Mutations at sites 1 and 2 are found in the MSH (melanocyte-stimulating hormone) receptor; 3 in the ACTH (adrenocorticotrophic hormone) receptor; 4 in the vasopressin receptor; 5 and 6 in the TSH (thyroid-stimulating hormone) receptor; 7 in the LH (luteinizing hormone) receptor; and 8 in rhodopsin, the light-sensitive pigment of the retina.

roid tumor, called an adenoma. Unlike normal thyroid cells that secrete thyroid hormone only in response to stimulation by the pituitary hormone TSH, the cells of these thyroid adenomas secrete large quantities of thyroid hormone without having to be stimulated by TSH (the receptor is said to act *constitutively*). The TSH receptor in these cells contains an amino acid substitution that affects the structure of the third intracellular loop of the protein (Figure 1, mutations at sites 5 or 6). As a result of the mutation, the TSH receptor constitutively activates a G protein on its inner surface, sending a continual signal through the pathway that leads not only to excessive thyroid hormone secretion but to the excessive cell proliferation that causes the tumor. This conclusion was verified by introducing the mutant gene into cultured cells that normally lack this receptor and demonstrating that the synthesis of the mutant protein and its incorporation into the plasma membrane led to the continuous production of cAMP in the genetically engineered cells.

The mutation that causes thyroid adenomas is not found in the normal portion of a patient's thyroid but only in the tumor tissue, indicating that the mutation was not inherited but arose in one of the cells of the thyroid, which then proliferated to give rise to the tumor. A mutation in a cell of the body, such as a thyroid cell, is called a *somatic mutation* to distinguish it from an inherited mutation that would be present in all of the individual's cells. As will be evident in the following chapter, somatic mutations are a primary cause of human cancer. At least one cancer-causing virus has been shown to encode a protein that acts as a constitutively active GPCR. The virus is a type of herpes virus that is responsible for Kaposi's sarcoma, which causes purplish skin lesions and is prevalent in AIDS patients. The virus genome encodes a constitutively active receptor for interleukin-8, which stimulates signaling pathways that control cell proliferation.

As noted in Table 1, mutations in genes that code for the subunits of heterotrimeric G proteins can also lead to inherited disorders. This is illustrated by a report on two male patients suffering from a rare combination of endocrine disorders: precocious puberty and hypoparathyroidism. Both patients were found to contain a single amino acid substitution in one of the G_α isoforms. The alteration in amino acid sequence caused two effects on the mutant G protein. At temperatures below normal body temperature, the mutant G protein remained in the active state, even in the absence of a bound ligand. In contrast, at normal body temperatures, the mutant G protein was inactive, both in the presence and absence of bound ligand. The testes, which are housed outside of the body's core, have a lower temperature than the body's visceral organs (33°C versus 37°C). Normally, the endocrine cells of the testes initiate testosterone production at the time of puberty in response to the pituitary hormone LH, which begins to be produced at that time. The circulating LH binds to LH receptors on the surface of the testicular cells, inducing the synthesis of cAMP and subsequent production of the male sex hormone. The testicular cells of the patients bearing the G protein mutation were stimulated to synthesize cAMP in the absence of the LH ligand, leading to premature synthesis of testosterone and precocious puberty. In contrast, the mutation in this same G_α subunit in the cells of the parathyroid glands, which function at a temperature of 37°C, caused the G protein to remain inactive. As a result, the cells of the parathyroid gland could not respond to stimuli that would normally cause them to secrete parathyroid hormone, leading to the condition of hypoparathyroidism. The fact that most of the bodily organs functioned in a normal manner in these patients suggests that this particular G_α isoform is not essential in the activities of most other cells.

Our sense of smell depends on nerve impulses transmitted along olfactory neurons that extend from the epithelia that line our nasal cavity to the olfactory bulb that is located in our brainstem. The distal tips of these neurons, which are located in the nasal epithelium, contain GPCRs, or *olfactory receptors*, that are able to bind various chemicals that enter our nose. It is estimated that we have more than 500 different olfactory receptors that, taken together, are able to combine with a large variety of different chemical structures. Each olfactory neuron probably contains only one of the hundreds of different olfactory receptors and, consequently, is only capable of responding to one or a few related chemicals. As a result, activation of different neurons containing different olfactory receptors provides us with the perception of different aromas. Mutations in a specific gene encoding a particular olfactory receptor can leave a person with the inability to detect a particular chemical in their environment that most other members of the population can perceive. When activated by bound ligands, olfactory receptors are thought to signal through heterotrimeric G proteins to adenylyl cyclase, resulting in the synthesis of cAMP and the opening of a cAMP-gated ion channel. This response ultimately leads to the generation of action potentials that are transmitted to the brain.

Our perception of taste operates at a much cruder level than our perception of smell. The taste buds in our tongue contain receptor cells that can only transmit a sense of one of four basic tastes: namely, salty, sour, sweet, or bitter. (A fifth taste is transmitted by a specialized taste receptor cell that responds to monosodium glutamate or disodium guanylate, which are commonly added to processed foods to enhance flavor.) The perception that a food or beverage is salty or sour is elicited directly by sodium ions or protons in the food that enter cation channels in the plasma membrane of the taste receptor cell, leading to a depolarization of the membrane (page 167). In contrast, the perception that a food is sweet or bitter depends on an ingredient interacting with a GPCR at the surface of the receptor cell. Studies reported in 2000 have provided considerable information about our perception of bitter-tasting substances. Humans were found to contain 50 to 80 genes that encode a family of taste receptors called T2Rs, all of which are thought to be coupled to the same heterotrimeric G protein. As a group, these taste receptors bind a diverse array of different compounds, including plant alkaloids or cyanides, that evoke a bitter taste in our mouths. For the most part, substances that evoke this perception are toxic compounds that elicit a distasteful, protective response that causes us to expel the

food matter from our mouth. Unlike the olfactory cells that contain a single receptor protein, a single taste-bud cell that evokes a bitter sensation contains a wide variety of different receptors that respond to many unrelated noxious substances. As a result, all of these diverse substances evoke the same basic taste, which is simply that the food we have eaten is bitter and disagreeable. Fortunately, food that is chewed releases odorants that travel via the throat to olfactory receptor cells in our nasal mucosa, allowing the brain to learn much more about the food we have eaten than the simple messages provided by the taste receptors. It is this merged input from both olfactory and taste (gustatory) neurons that provides us with our rich sense of taste. The importance of olfactory neurons in our perception of taste becomes more evident when we have a cold that causes us to lose some of our appreciation for the taste of food.

Lipid-Derived Second Messengers

It wasn't very long ago that the phospholipids of cell membranes were considered strictly as structural molecules that made membranes cohesive and impermeable to aqueous

solutes. Our appreciation of phospholipids has increased with the realization that these molecules form the precursors for a variety of second messengers. The phospholipids of cell membranes are converted into second messengers by activated phospholipases, which are hydrolytic enzymes that split phospholipids into their component parts. Figure 15.11a shows the cleavage sites within a generalized phospholipid that are attacked by the main classes of phospholipases. All four of the enzyme classes depicted in Figure 15.11a are involved in signaling pathways. In this section, we will focus on the best-studied lipid second messengers, which are derived from phosphatidylinositol, and are generated following the transmission of signals by G protein-coupled receptors. Another group of lipid second messengers derived from sphingomyelin are not discussed.

Phosphatidylinositol-Derived Second Messengers
When the neurotransmitter acetylcholine binds to the surface of a smooth muscle cell within the wall of the stomach, the muscle cell is stimulated to contract. When a foreign antigen binds to the surface of a mast cell, the cell is stimulated to secrete histamine, a substance that can trigger the

(a)

(b)

Figure 15.11 Phospholipid-based second messengers. *(a)* The structure of a generalized phospholipid (see Figure 2.22). Phospholipids are subject to attack by four types of phospholipases that cleave the molecule at the indicated sites. Of these enzymes, we will focus on PLC, which splits the phosphorylated head group from the diacylglycerol. *(b)* A model showing the interaction between a portion of a PLC molecule containing a PH domain that binds to the phosphorylated inositol ring of a phosphoinositide. This interaction holds the enzyme to the inner surface of the plasma membrane. *(c)* Fluorescence micrograph of a cell that has been stimulated to move toward a chemoattractant (i.e., a chemical to which the cell is attracted). This cell has been stained with an antibody that binds specifically to PI 3,4,5-trisphosphate (PIP_3), which is seen to be localized at the leading edge of the migrating cell (arrows). Bar = 15 μm. (b: FROM JAMES H. HURLEY AND JAY A. GROBLER, CURR. OPIN. STRUCT. BIOL. 7:559, 1997; c: FROM PAULA RICKERT ET AL, COURTESY OF HENRY R. BOURNE, TRENDS CELL BIOL. 10:470, 2000.)

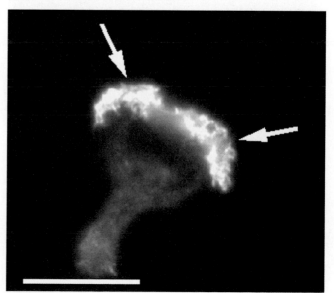

(c)

symptoms of an allergy attack. Both of these responses, one leading to contraction and the other to secretion, are triggered by the same cellular messenger, a substance derived from the compound phosphatidylinositol, a minor component of most cellular membranes (see Figure 4.25).

The first indication that phospholipids might be involved in cellular responses emerged from studies carried out in the early 1950s by Lowell and Mabel Hokin of Montreal General Hospital and McGill University. These investigators had set out to study the effects of acetylcholine on RNA synthesis in the pancreas. To carry out these studies, they incubated slices of pigeon pancreas in $[^{32}P]O_4$, which was the usual procedure for obtaining radiolabeled nucleic acids. They found, however, that treatment of the tissue with acetylcholine led to the incorporation of radioactivity into the phospholipid fraction of the cell. Further analysis revealed that the isotope was incorporated primarily into phosphatidylinositol (PI), which was rapidly converted to other phosphorylated derivatives *(phosphoinositides)*. These researchers went on to describe many of the steps in a PI-mediated response.

Several of the reactions of phosphoinositide metabolism are shown in Figure 15.12. As indicated on the left side of this figure, the inositol ring, which resides at the inner polar surface of the bilayer, has six carbon atoms. These carbons can be phosphorylated by various phosphoinositide kinases present in cells. The most prevalent of these enzymes are PI 3-kinases, PI 4-kinases, and PI 5-kinases, which transfer phosphates from ATP to the 3, 4, or 5 position on the inositol ring, respectively. For example, transfer of a single phosphate group to the 4-position of

the inositol sugar of PI by PI 4-kinase (PI4K) generates PI 4-phosphate (PIP), which can be phosphorylated again by PIP 5-kinase (PIP5K) to form PI 4,5-bisphosphate (PIP$_2$; steps 1 and 2, Figure 15.12), which can be phosphorylated again by PI 3-kinase to form PI 3,4,5-trisphosphate (PIP$_3$)(not shown in figure). All of these phospholipid species remain in the inner leaflet of the plasma membrane. Phosphorylated inositol rings of phosphoinositides bind to complementary modules, called **PH domains**, on specific proteins with which they interact (Figure 15.11*b*). PH domains have been identified in over 150 different proteins. Binding of a protein to a PIP$_2$ or PIP$_3$ messenger, recruits the protein to the cytosolic face of the membrane where it can interact with other membrane-bound proteins.

Figure 15.11*c* shows an example where PIP$_3$ is specifically localized to a particular portion of the plasma membrane of a cell. This particular cell is engaged in chemotaxis, which is to say that it is moving toward an increasing concentration of a particular chemical in the medium that serves as the chemoattractant. This is the mechanism that causes phagocytic cells such as macrophages to move toward bacteria or other targets that they engulf. Recent studies indicate that chemotaxis depends on the localized production of phosphoinositide messengers that acts like a compass to inform the cell of the target location (as illustrated in Figure 15.11*c*).

Not all inositol-containing messengers remain in the lipid bilayer of a membrane as described above. When acetylcholine binds to a smooth muscle cell, or an antigen binds to a mast cell, the bound receptor activates a heterotrimeric G protein (Figure 15.12, step 3), which in turn

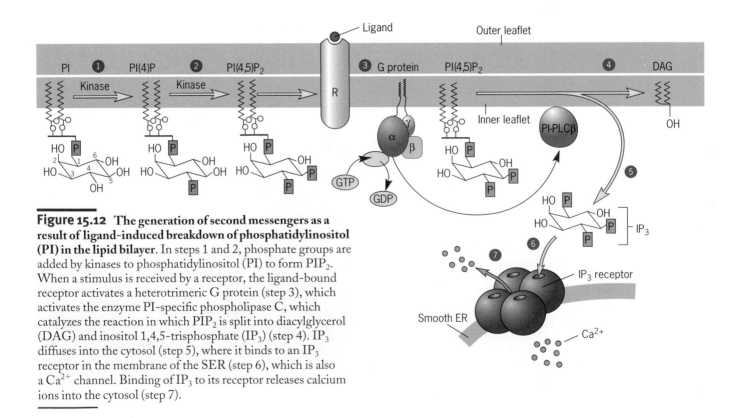

Figure 15.12 The generation of second messengers as a result of ligand-induced breakdown of phosphatidylinositol (PI) in the lipid bilayer. In steps 1 and 2, phosphate groups are added by kinases to phosphatidylinositol (PI) to form PIP$_2$. When a stimulus is received by a receptor, the ligand-bound receptor activates a heterotrimeric G protein (step 3), which activates the enzyme PI-specific phospholipase C, which catalyzes the reaction in which PIP$_2$ is split into diacylglycerol (DAG) and inositol 1,4,5-trisphosphate (IP$_3$) (step 4). IP$_3$ diffuses into the cytosol (step 5), where it binds to an IP$_3$ receptor in the membrane of the SER (step 6), which is also a Ca^{2+} channel. Binding of IP$_3$ to its receptor releases calcium ions into the cytosol (step 7).

activates the effector **phosphatidylinositol-specific phospholipase C-β (PI-PLCβ)**[2] Like the protein depicted in Figure 15.11*b*, PI-PLCβ is situated at the inner surface of the membrane (Figure 15.12), bound there by interaction between its PH domain and a PIP$_2$ molecule embedded in the bilayer. PI-PLCβ catalyzes a reaction that splits PIP$_2$ into two molecules, **inositol 1,4,5-trisphosphate (IP$_3$)** and **diacylglycerol (DAG;** step 4), both of which play important roles as second messengers in cell signaling. We will examine each of these second messengers separately.

Diacylglycerol (DAG) Diacylglycerol (Figure 15.12) is a lipid molecule that remains in the plasma membrane following its formation by PI-PLCβ. There it recruits and activates an effector, called **protein kinase C,** that phosphorylates serine and threonine residues on target proteins. Like protein kinase A (the enzyme activated by cAMP), protein kinase C is a multifunctional serine and threonine kinase that phosphorylates a wide variety of proteins. Protein kinase C has a number of important roles in cellular growth and differentiation, cellular metabolism, and transcriptional activation (Table 15.5).

The apparent importance of protein kinase C in growth control is seen in studies with a group of powerful plant compounds, called *phorbol esters*, that resemble DAG. These compounds activate protein kinase C in a variety of cultured cells, causing them to lose growth control and behave temporarily as malignant cells. When the phorbol ester is removed from the medium, the cells recover their normal growth properties. In contrast, cells that have been genetically engineered to constitutively express protein kinase C exhibit a permanent malignant phenotype in cell culture and can cause tumors in susceptible mice.

Inositol 1,4,5-trisphosphate (IP$_3$) Inositol 1,4,5-trisphosphate (IP$_3$) is a sugar phosphate—a small, water-soluble molecule capable of rapid diffusion. IP$_3$ molecules formed at the membrane diffuse into the cytosol (step 5, Figure 15.12) and bind to a specific IP$_3$ receptor located at the surface of the smooth endoplasmic reticulum (step 6, Figure 15.12). It was noted on page 288, that the SER is a site of calcium storage in a variety of cells. The IP$_3$ receptor does more than bind a ligand; it is also a tetrameric Ca^{2+} channel. Binding of IP$_3$ opens the channel, allowing Ca^{2+} ions to diffuse into the cytoplasm (step 7, Figure 15.12). Calcium ions can also be considered as intracellular messengers because they bind to various target molecules, triggering specific responses. In the two examples used above, contraction of a smooth muscle cell and exocytosis of histamine-containing secretory granules in a mast cell, both are triggered by elevated calcium levels. So, too, is

Table 15.5	Examples of Responses Mediated by Protein Kinase C
Tissue	**Response**
Blood platelets	Serotonin release
Mast cells	Histamine release
Adrenal medulla	Secretion of epinephrine
Pancreas	Secretion of insulin
Pituitary cells	Secretion of GH and LH
Thyroid	Secretion of calcitonin
Testes	Testosterone synthesis
Neurons	Dopamine release
Smooth muscle	Increased contractility
Liver	Glycogen hydrolysis
Adipose tissue	Fat synthesis

SOURCE: U. Kikkawa and Y. Nishizuka, reproduced with permsision from the *Annual Review of Cell Biology*, vol. 2, © 1986, by Annual Reviews Inc.

the response of a liver cell to the hormone vasopressin (the same hormone that has antidiuretic activity in the kidney). Vasopressin binds to its receptor at the liver cell surface and causes a series of IP$_3$-mediated bursts of Ca^{2+} release, which appear as oscillations of free calcium concentration in the recording shown in Figure 15.13. The effect of IP$_3$ is usually transient because it is rapidly inactivated enzymatically. A list of some of the responses mediated by IP$_3$

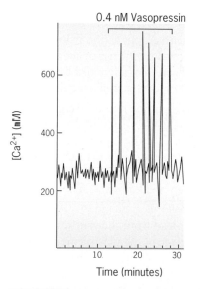

Figure 15.13 **Experimental demonstration of changes in free calcium concentrations in response to hormone stimulation.** A single liver cell was injected with aequorin, a protein extracted from certain jellyfish that luminesces when it binds calcium ions. The intensity of the luminescence is proportional to the concentration of free calcium ions. Exposure of the cell to vasopressin leads to several waves of free calcium at periodic intervals. Higher concentrations of hormone do not increase the height (amplitude) of the spikes, but they do increase their frequency. (REPRINTED WITH PERMISSION FROM N. M. WOODS, K. S. CUTHBERTSON, AND P. H. COBBOLD, NATURE 319:601, 1986. COPYRIGHT © 1986 BY MACMILLAN JOURNALS LIMITED.)

[2]The phospholipase C activated by G proteins is identified as PLCβ to distinguish it from an isoform called PLCγ that is activated by receptor tyrosine kinases (see Figure 15.23). Both forms catalyze the same reaction but have different properties.

Table 15.6	Summary of Cellular Responses Elicited by Adding IP₃ to Either Permeabilized or Intact Cells

Cell type	Response
Vascular smooth muscle	Contraction
Stomach smooth muscle	Contraction
Skeletal muscle	Contraction
Slime mold	Cylic GMP formation, actin polymerization
Blood platelets	Shape change, aggregation
Salamander rods	Modulation of light response
Xenopus oocytes	Calcium mobilization, membrane depolarization
Sea urchin eggs	Membrane depolarization, cortical reaction
Lacrimal gland	Increased potassium current

Adapted from M. J. Berridge, reproduced with permission from the *Annual Review of Biochemistry*, vol. 56, © 1987, by Annual Reviews Inc.

is indicated in Table 15.6. We will have more to discuss about Ca^{2+} ions in the following section.

REVIEW ?

1. What is the role of G proteins in a signaling pathway?

2. Describe Sutherland's experiment that led to the concept of the second messenger.

3. What is meant by the term amplification in regard to signal transduction? How does the use of a reaction cascade result in amplification of a signal? How does it increase the possibilities for metabolic regulation?

4. How is it possible that the same first messenger, such as epinephrine, can evoke different responses in different target cells? That the same second messenger, such as cAMP, can also evoke different responses in different target cells? That the same response, such as glycogen breakdown, can be initiated by different stimuli?

5. Describe the steps that lead from the synthesis of cAMP at the inner surface of the plasma membrane of a liver cell to the release of glucose into the bloodstream. How is this process controlled by GRKs and arrestin? By protein phosphatases? By cAMP phosphodiesterase?

6. Describe the steps between the binding of a ligand such as glucagon to a seven helix receptor and the activation of an effector, such as adenylyl cyclase. How is the response normally attenuated?

7. Name two extracellular stimuli that lead to the formation of IP_3. What is the mechanism of formation of this second messenger? What is the relationship between the formation of IP_3 and an elevation of intracellular $[Ca^{2+}]$?

8. Describe the relationship between phosphatidylinositol, diacylglycerol, calcium ions, and protein kinase C. How do phorbol esters interfere with signal pathways involving DAG?

15.3 THE ROLE OF CALCIUM AS AN INTRACELLULAR MESSENGER

Calcium ions play a significant role in a remarkable variety of cellular activities, including muscle contraction, cell division, secretion, endocytosis, fertilization, synaptic transmission, metabolism, and cell movement. Our understanding of the role of Ca^{2+} ions in cellular responses has been greatly furthered by the development of indicator molecules that emit light in the presence of free calcium. The first of these indicator molecules, which was purified in the early 1960s, was aequorin, a protein produced by certain jellyfish that becomes luminescent when it binds free Ca^{2+} ions. The protein could be microinjected into living cells, where changes in luminescence revealed changes in the concentration of Ca^{2+} ions during a particular cellular response (as in Figure 15.13). In the early 1990s, new types of highly sensitive, fluorescent, calcium-binding compounds (e.g., *fura-2*) were developed. These compounds are synthesized in a form that can enter a cell by diffusing across its plasma membrane. Once inside a cell, the compound is modified to a form that is unable to leave the cell. Using these probes, the concentration of free calcium ions in different parts of a living cell can be determined by monitoring the light emitted using a fluorescence microscope and computerized imaging techniques. Use of calcium-sensitive, light-emitting molecules has provided dramatic portraits of the complex spatial and temporal changes in free cytosolic calcium concentration that occur in a single cell in response to various types of stimuli. This is one of the advantages of studying calcium-mediated responses compared to responses mediated by other types of messengers whose location in a cell cannot be readily visualized.

Depending on the type of responding cell, a particular stimulus may induce sharp oscillations in the concentration of free calcium ions, as shown in Figure 15.13; cause a wave of Ca^{2+} release that spreads from one end of the cell to the other (see Figure 15.16); or trigger a localized and transient release of Ca^{2+} in one part of the cell. Figure 15.14 shows a Purkinje cell, a type of neuron in the mammalian cerebellum that maintains synaptic contact with thousands of other cells through an elaborate network of postsynaptic dendrites. The micrograph in Figure 15.14 shows the release of free calcium in a localized region of the dendritic "tree" of the cell following synaptic activation. The burst of calcium release remains restricted to this region of the cell.

A calcium ion is very different in structure from a cyclic nucleotide or an inositol phosphate and is not a substance that is synthesized or degraded enzymatically. In-

transport systems of the plasma and ER membranes normally pump calcium out of the cytosol.

IP$_3$ receptors described above are one of two main types of Ca^{2+} ion channels present in the ER membrane; the other type has been named *ryanodine receptors (RyRs)* because they bind the plant alkaloid ryanodine. Ryanodine receptors are found primarily in excitable cells and are best studied in skeletal muscle cells, where they mediate the rise in Ca^{2+} levels following the arrival of an action potential (page 378). Depending on the type of cell in which they are found, ryanodine receptors can be opened by a variety of agents, but apparently not by IP$_3$. Among the agents that can open ryanodine Ca^{2+} channels is calcium itself. The influx of a limited amount of calcium through open channels in the plasma membrane induces the opening of ryanodine receptors in the ER, causing the release of Ca^{2+} into the cytosol (Figure 15.15). This phe-

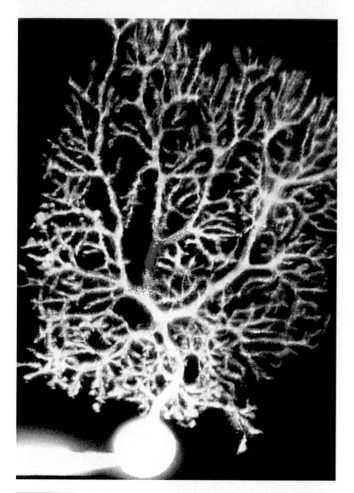

Figure 15.14 **Experimental demonstration of localized release of intracellular Ca^{2+} within a single dendrite of a neuron.** The mechanism of IP$_3$-mediated release of Ca^{2+} from intracellular stores was described in Figure 15.12. In the micrograph shown here, which pictures a portion of an enormously complex Purkinje cell (neuron) of the cerebellum, calcium ions have been released locally within one of the dendrites (red). Calcium release was induced in the dendrite following the local production of IP$_3$, which followed repetitive activation of a nearby synapse. The sites of release of cytosolic Ca^{2+} are revealed by fluorescence from a fluorescent calcium indicator that was loaded into the cell prior to stimulation of the cell. (FROM ELIZABETH A. FINCH AND GEORGE J. AUGUSTINE, NATURE, VOL. 396, COVER OF 12/24/98. COPYRIGHT © 1998 BY MACMILLAN MAGAZINES LIMITED.)

stead, the concentration of calcium ions in a particular cellular compartment is controlled by the regulated activity of Ca^{2+} transporters and Ca^{2+} ion channels located within the membranes that surround the compartment. The concentration of Ca^{2+} in the cytosol of a resting cell is maintained at very low levels, typically about 10^{-7} M. In contrast, the concentration of this ion in the extracellular space or within the lumen of the ER or a plant cell vacuole may be more than 10,000 times higher than the cytosol. The cytosolic calcium level is kept very low because (1) the Ca^{2+} ion channels of the plasma and ER membranes are normally kept closed, making these membranes highly impermeable to this ion, and (2) the ATP-driven Ca^{2+}

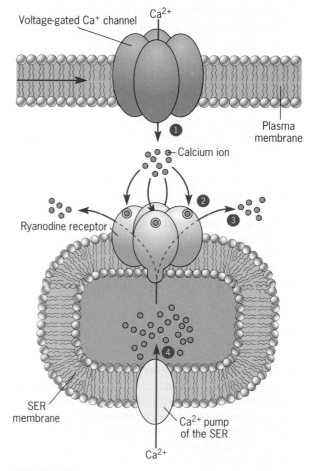

Figure 15.15 **Calcium-induced calcium release**, as it occurs in excitable cells, such as cardiac muscle. A change in membrane voltage causes the opening of voltage-gated calcium channels in the plasma membrane, allowing entry of a small amount of Ca^{2+} (step 1). The calcium ions bind to the ryanodine receptor in the SER membrane (step 2), leading to the release of stored Ca^{2+} (step 3). The calcium ions are subsequently removed from the cytosol by the action of the Ca^{2+} pump located in the membrane of the SER (step 4). (REPRINTED WITH PERMISSION FROM M. J. BERRIDGE, NATURE 361:317, 1993. COPYRIGHT © 1993 BY MACMILLAN MAGAZINES LIMITED.)

Figure 15.16 **Calcium wave in a starfish egg induced by a fertilizing sperm.** The unfertilized egg was injected with a calcium-sensitive fluorescent dye, fertilized, and photographed at five-second intervals. The release of Ca^{2+} is seen to spread from the point of sperm entry (arrow) throughout the entire egg. The blue color indicates low free $[Ca^{2+}]$, whereas the red color indicates high free $[Ca^{2+}]$. (COURTESY OF STEPHEN A. STRICKER.)

15 μm

nomenon is called *calcium-induced calcium release (CICR)*. One of the effects of caffeine is to lower the sensitivity of ryanodine receptors to Ca^{2+}, thereby increasing the likelihood that these ion channels will be opened by CICR, making cells more excitable.

Extracellular signals that are transmitted by Ca^{2+} ions typically act by opening a small number of Ca^{2+} ion channels at the cell surface at the site of the stimulus. As Ca^{2+} ions pass through these channels and enter the cytosol, they act on nearby Ca^{2+} ion channels in the plasma membrane and ER, causing these channels to open and release additional calcium into adjacent regions of the cytosol. In some responses, the elevation of Ca levels remains localized to a small region of the cytosol (as in Figure 15.14), whereas in other cases, a propagated wave of calcium release spreads through the entire cytoplasmic compartment. One of the most dramatic Ca^{2+} waves occurs within the first minute or so following fertilization and is induced by the sperm's contact with the plasma membrane of the egg (Figure 15.16). The sudden rise in cytoplasmic calcium concentration following fertilization triggers a number of events, including activation of cyclin-dependent kinases (page 584) that drive the zygote toward its first mitotic division. Calcium waves are transient because the ions are rapidly pumped out of the cytosol and back into the ER and/or the extracellular space.

Unlike cAMP, whose action is invariably mediated by stimulation of a protein kinase, calcium can affect a number of different types of cellular effectors, including protein kinases (Table 15.7). Depending on the cell type, calcium ions can activate or inhibit various enzyme and transport systems, change the ionic permeability of mem-

branes, induce membrane fusion, or alter cytoskeletal structure and function. Calcium does not bring about these responses by itself but acts in conjunction with a number of **calcium-binding proteins**. The best-studied calcium-binding protein is **calmodulin**.

Each molecule of calmodulin (Figure 15.17a) contains four binding sites for calcium. Calmodulin does not have sufficient affinity for Ca^{2+} to bind the ion in a nonstimulated cell. If, however, the Ca^{2+} concentration rises in response to a stimulus, the ions bind to calmodulin, changing the conformation of the protein and increasing its affinity for a variety of effectors. Depending on the cell type, the calcium-calmodulin complex may bind to a protein kinase, a cyclic nucleotide phosphodiesterase, ion channels, or even to the calcium-transport system of the plasma membrane. In the latter instance, rising levels of calcium activate the system responsible for ridding the cell of excess quantities of the ion, thus constituting a self-regulatory mechanism for maintaining low intracellular calcium concentrations. Calmodulin is found universally in plants, animals, and eukaryotic microorganisms, and it has virtually the same amino acid sequence from one end of the eukaryotic spectrum to the other. The distribution of calmodulin in the compound eye of a fruit fly is shown in Figure 15.17b.

Regulating Calcium Concentrations in Plant Cells

Calcium ions (acting in conjunction with calmodulin) are important intracellular messengers in plant cells. The levels of cytosolic calcium change dramatically within certain

Table 15.7	Examples of Mammalian Proteins Activated by Ca²⁺

Protein	Protein Function
Troponin C	Modulator of muscle contraction
Calmodulin	Ubiquitous modulator of protein kinases and other enzymes (MLCK, CaM kinase II, adenylyl cyclase I)
Calretinin, retinin	Activator of guanylyl cyclase
Calcineurin B	Phosphatase
Calpain	Protease
PI-specific PLC	Generator of IP₃ and diacylglycerol
α-Actinin	Actin-bundling protein
Annexin	Implicated in endo- and exocytosis, inhibition of PLA₂
Phospholipase A₂	Producer of arachidonic acid
Protein kinase C	Ubiquitous protein kinase
Gelsolin	Actin-severing protein
IP₃ receptor	Effector of intracellular Ca²⁺ release
Ryanodine receptor	Effector of intracellular Ca²⁺ release
Na⁺-Ca²⁺ exchanger	Efector of the exchange of Ca²⁺ for Na⁺ across the plasma membrane
Ca²⁺-ATPase	Pumps Ca²⁺ across membranes
Ca²⁺ antiporters	Exchanger of Ca²⁺ for monovalent ions
Caldesmon	Regulator of muscle contraction
Villin	Actin organizer
Arrestin	Terminator of photoreceptor response
Calsequestrin	Ca²⁺ buffer

Adapted from D. E. Clapham, *Cell* 80:260, 1995, by copyright permission of Cell Press.

(a)

(b)

Figure 15.17 Calmodulin. *(a)* A ribbon diagram of calmodulin with four bound calcium ions. *(b)* The compound eye of *Drosophila* consists of hundreds of individual photo units called ommatidia. Each ommatidium contains eight photoreceptor cells. Six of the eight photoreceptor cells extend the full depth of the ommatidia, whereas the other two occupy either the upper or lower portions of an ommatidium. Consequently, only seven photoreceptor cells are present in any given cross section (left). The spatial distribution of the calcium-binding protein calmodulin is revealed by immunofluorescence (right) and is seen to occupy a portion of each of the photoreceptors. (B: REPRINTED WITH PERMISSION FROM JEFFREY A. PORTER ET AL., COURTESY OF CRAIG MONTELL, SCIENCE 262:1039, 1993; © COPYRIGHT 1993 AMERICAN ASSOCIATION FOR THE ADVANCEMENT OF SCIENCE.)

plant cells in response to a variety of stimuli, including changes in light, pressure, gravity, and the concentration of plant hormones such as abscisic acid. The concentration of Ca²⁺ in the cytosol of a resting plant cell is kept very low by the action of transport proteins situated in the plasma membrane and vacuolar membrane (tonoplast).

The role of Ca²⁺ in plant cell signaling is seen most dramatically in guard cells that regulate the diameter of the microscopic pores (stomata) of a leaf (Figure 15.18*a*). Stomata are a major site of water loss in plants (page 237), and the diameter of their aperture is tightly controlled, which prevents desiccation. The diameter of the stomatal pore decreases as the fluid (turgor) pressure in the guard cell decreases. The drop in turgor pressure is caused in turn by a decrease in the ionic concentration (osmolarity) of the guard cell. Adverse conditions, such as high temperatures and low humidity, stimulate the release of abscisic acid, which opens the calcium channels in the guard cell plasma membrane (Figure 15.18*b*). The resulting influx of Ca²⁺ into the cytosol triggers the release of additional Ca²⁺ from intracellular stores. The elevated cytosolic Ca²⁺ concentration closes K⁺ influx channels in the plasma membrane, while opening K⁺ efflux channels. These changes produce a net outflow of potassium ions (and accompanying anions) and a decrease in turgor pressure.

REVIEW [?]

1. How is the [Ca²⁺] of the cytosol maintained at such a low level? How does the concentration change in response to stimuli?

2. What is the role of calcium-binding proteins such as calmodulin in eliciting a response?

3. Describe the role of calcium in mediating the diameter of stomata in guard cells.

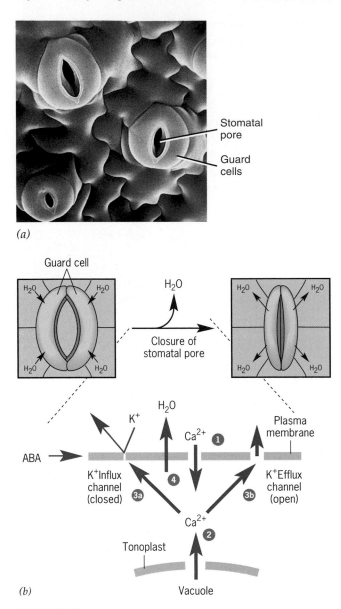

(a)

(b)

Figure 15.18 The role of calcium as an intracellular messenger in guard cell closure. *(a)* Photograph of stomatal pores, each flanked by a pair of guard cells. The stomata are kept open as turgor pressure is kept high within the guard cells, causing them to bulge outward as seen here. *(b)* One of the factors controlling stomatal pore size is the hormone abscisic acid (ABA). When ABA levels rise, calcium ion channels in the plasma membrane are opened, allowing the influx of Ca^{2+} (step 1), which triggers the release of Ca^{2+} from internal stores (step 2). The subsequent elevation of intracellular $[Ca^{2+}]$ closes K^+ influx channels (step 3a) and opens K^+ efflux channels (step 3b), leading to a drop in internal solute concentration and the osmotic loss of water (step 4). (A: JEREMY BURGESS/PHOTO RESEARCHERS.)

15.4 RECEPTOR TYROSINE KINASES: A SECOND MAJOR TYPE OF SIGNALING PATHWAY

We began the discussion of signal transduction with glucagon and epinephrine, two hormones that act on liver cells to promote the release of glucose into the blood-stream. Insulin acts on liver and muscle cells to initiate an opposite series of reactions in which glucose is removed from the bloodstream and polymerized as glycogen. In addition to its effects on glucose uptake and polymerization, insulin is a powerful stimulant of lipid synthesis (in fat cells), protein synthesis, as well as cell growth and proliferation. Insulin elicits these responses by means of a signaling pathway that is different in many respects from that taken by glucagon and epinephrine.

The Mechanism of Action of Insulin: Signaling by an RTK

Cells that respond to insulin possess an insulin receptor at their surface. The insulin receptor is more than just a protein that binds a ligand; it is also an enzyme—a **protein tyrosine kinase**—that adds phosphate groups to specific tyrosine residues of other proteins. Tyrosine kinases constitute a superfamily distinct from the serine and threonine kinases that operate in the reaction cascade discussed earlier in connection with glucose mobilization (see Figure 15.7).[3] Tyrosine kinases are involved primarily in the control of cell growth and differentiation, rather than the control of intermediary metabolism. Because the insulin receptor has this enzymatic activity, it is referred to as a **receptor tyrosine kinase** (or **RTK**). Over 50 different RTKs have been identified. Unlike the G protein-coupled receptors, which have seven transmembrane segments, each RTK monomer traverses the membrane only once.

The insulin receptor is a tetrameric protein composed of two α and two β polypeptide chains linked by disulfide bonds. The α chains reside on the extracellular surface of the membrane and contain the insulin-binding sites, whereas the β chains span the lipid bilayer and transmit the signal across the membrane to its inner surface (Figure 15.19). In the absence of bound insulin, the tyrosine kinase function of the receptor is inactive. Binding of insulin to the receptor's α subunit changes the conformation of the β subunit, activating its tyrosine kinase. The activated tyrosine kinase transfers phosphate groups to (1) specific tyrosine residues of the other β subunit of the complex, a reaction termed *autophosphorylation*, and (2) a dozen or more tyrosine residues of a small family of soluble protein substrates, called **insulin receptor substrates** (**IRSs**; Figure 15.19). Phosphorylated IRSs appear to have only one function, which is to bind and activate a variety of downstream effectors. To understand how this occurs, we need to digress for just a moment.

RTKs do not phosphorylate every tyrosine in a substrate protein; they phosphorylate only those that are present within certain amino acid sequences that we will refer to as *phosphotyrosine motifs*. A variety of proteins involved in cell signaling contain modules, called **SH2 domains**, that possess high-affinity binding sites for these phospho-

[3]A small number of tyrosine kinases have dual specificity and can also phosphorylate serine and threonine residues. The MAPKK of Figure 15.25 is an example of a dual specificity tyrosine kinase.

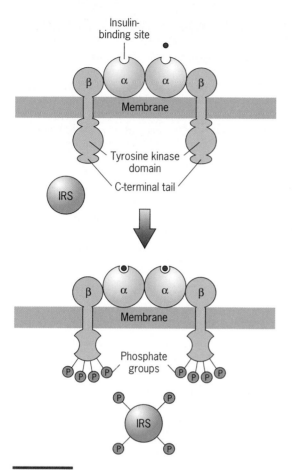

Figure 15.19 The response of the insulin receptor to ligand binding. The insulin receptor is a tetramer consisting of two α and two β subunits. Binding of insulin to the α subunits causes a conformational change in the β subunits, which activates the tyrosine kinase activity of the β subunits. The activated β subunits phosphorylate tyrosine residues located in the cytoplasmic domain of the receptor as well as several insulin receptor substrates (IRSs).

Figure 15.20 The interaction between an SH2 domain of a protein and a peptide containing a phosphotyrosine residue. The SH2 domain of the protein is shown in a cutaway view with the accessible surface area represented by red dots and the polypeptide backbone as a purple ribbon. The phosphotyrosine-containing heptapeptide (Pro-Asn-pTyr-Glu-Glu-Ile-Pro) is shown as a space-filling model whose side chains are colored green and backbone is colored yellow. The phosphate group is shown in light blue. The phosphorylated tyrosine residue and the isoleucine residue (+3) are seen to project into pockets on the surface of the SH2 domain, creating a tightly fitting interaction, but only when the key tyrosine residue is phosphorylated. (FROM GABRIEL WAKSMAN ET AL., COURTESY OF JOHN KURIYAN, CELL 72:783, 1993 BY PERMISSION OF CELL PRESS.)

tyrosine motifs (Figure 15.20).[4] SH2 domains have little or no affinity for proteins whose tyrosine residues are not phosphorylated. It is only after an IRS has been phosphorylated by the insulin receptor that the IRS acts as a "docking station" that binds proteins possessing SH2 domains. The consequence of this interaction depends on the particular proteins involved. The interaction may turn on an enzymatic activity of the SH2 protein, cause the SH2 protein to bind other proteins, or cause the SH2 protein to be translocated to a different part of the cell. The SH2 domain is only one of a dozen or more common modules used by signaling proteins to mediate interactions with other proteins. The presence of these "protein interaction modules" allows proteins to be strung together into multi-

protein assemblies that facilitate transmission of signals along a chain of participants (see Figure 15.23).

Even though they operate by different mechanisms, numerous parallels can be drawn between the heterotrimeric G protein and RTK signaling pathways. In heterotrimeric G-protein pathways, a change in conformation of the G protein is the molecular switch that transmits the signal into the cell. In an RTK-mediated pathway, tyrosine phosphorylation performs a similar function. Proteins with SH2 domains can be considered as the effectors of RTKs, just as adenylyl cyclase or phospholipase C are effectors for G protein-coupled receptors. Because different cells may contain different effectors possessing similar SH2 domains, the same ligand can trigger different responses in different target cells.

Let us return to the insulin response. Once the insulin receptor has been activated, the phosphorylated IRS proteins serve as docking sites for a number of different SH2-containing proteins, each of which may activate a separate signaling pathway (Figure 15.21a). As a result, the message that insulin has bound to the cell surface may radiate through the cell interior along several independent pathways.

[4]SH2 domain is an acronym for *src-homology* domain, denoting that this phosphotyrosine-binding domain was first identified in the protein encoded by the *src* oncogene. A number of proteins have a different phosphotyrosine-binding motif called a *PTB domain*, which is not discussed.

(a)

(b)

Figure 15.21 **The role of tyrosine-phosphorylated IRS in activating a variety of signaling pathways.** *(a)* Phosphorylation of IRSs by the activated insulin receptor is known to activate PI3K, PI-PLCγ, and Ras pathways, all of which are discussed in the chapter. Other pathways that are less well defined are also activated by IRSs. *(b)* Activation of PI3K leads to the formation of membrane-bound phosphoinositides PIP$_2$ and PIP$_3$. One of the key kinases in numerous signaling pathways is PKB, which interacts with PIP$_2$ by means of a PH domain on the protein. This interaction changes the conformation of the PKB molecule, making it a substrate for another kinase (PDK1), which phosphorylates and fully activates PKB. Once activated, PKB is a major component of a number of separate signaling pathways that mediate the insulin response. These pathways lead to translocation of glucose transporters to the plasma membrane, synthesis of glycogen, and the synthesis of new proteins in the cell. PKB is inactivated by dephosphorylation by the protein phosphatase PP2A.

nase, or **PI3K**. PI3K catalyzes a reaction in which a phosphate group is added to the 3′ position of the sugar ring of phosphatidylinositol (PI). The products of the enzyme, which include PI 3,4-bisphosphate (PIP$_2$) and PI 3,4,5-trisphosphate (PIP$_3$), remain in the cytosolic leaflet of the lipid bilayer where they serve as membrane-bound lipid messengers that have diverse functions in different cells.

Just as phosphorylated tyrosine motifs bind to complementary SH2 domains on target proteins, phosphorylated inositol rings of phospholipids bind to complementary PH domains on the proteins with which they interact (page 642). In cells responding to insulin, the lipid products of PI3K bind to the PH domain of a key enzyme, **protein kinase B**. Binding by PIP$_2$ recruits a PKB molecule to the plasma membrane and alters its conformation, making it a suitable substrate for activation by another protein kinase (Figure 15.21*b*). Activation of PKB has important consequences in an insulin-target cell, leading to (1) the transfer of glucose transporters to the plasma membrane, where they function in glucose uptake (see Figure 4.43); (2) increased protein synthesis, which is a hallmark of insulin action; and (3) stimulation of glycogen synthase, leading to conversion of glucose to glycogen. Another signaling pathway activated by IRS phosphorylation leads to a protein called Ras, which has a key role in stimulating cell growth and proliferation. We will explore this growth-stimulating pathway further in the following section.

Before leaving the subject of the insulin response it should be mentioned that one of the most prevalent diseases in Western countries—type 2 (or adult-onset) diabetes—is thought to result, at least in part, from a resistance of cells to insulin. Insulin levels typically remain high, but target cells don't respond properly to the hormone. It was thought for years that the disease resulted from alterations in insulin receptor function, though this belief is no longer widely held. However, recent studies in mice have rekindled hope that the disease might be

One of the best-studied effectors that contains an SH2 domain and is known to bind to a phosphorylated IRS is the enzyme **phosphatidylinositol 3-hydroxy ki-**

treated by therapies aimed at the insulin receptor. It was found that mice genetically engineered to lack a particular phosphatase (PTP-1B) exhibited a heightened sensitivity to insulin. In fact, an injection of insulin into these mice led to the removal of so much glucose from the bloodstream that many of the mice lost consciousness as the result of hypoglycemia (low blood sugar levels). Normally, mice that are fed high-fat diets develop insulin resistance characteristic of type 2 diabetes, and they also become obese. In contrast, PTP-1B knockout mice fed a high-fat diet displayed enhanced insulin sensitivity and maintained a normal blood-glucose level and body weight. PTP-1B is a phosphatase that is thought to remove phosphate groups from tyrosine residues on the insulin receptor, thereby inactivating the receptor and stopping the hormone response. The absence of the phosphatase in these knockout mice is presumed to interfere with receptor inactivation, thereby increasing the animals' sensitivity to insulin. These studies have suggested a treatment for diabetes through the use of drugs that specifically inhibit the human version of PTP1B. One such drug is being currently tested in clinical trials on diabetics and may also be tested as an anti-obesity medication.

The Role of RTKs in Other Cellular Activities

A wide variety of extracellular agents interact with RTKs on the surfaces of target cells to regulate such diverse processes as cell growth and proliferation, cell differentiation, phagocytosis, cell motility, and cell survival. Among the best studied of these agents are hormones, such as insulin and growth hormone; and growth factors, such as epidermal growth factor (EGF), platelet-derived growth factor (PDGF), and fibroblast growth factor (FGF).

Unlike the insulin receptor, most RTKs are present as monomers in an unstimulated cell. It is only after a ligand binds to the RTK monomers that they interact with one another to form dimers. The simplest case is shown in Figure 15.22, in which receptor dimerization occurs following the binding of either a dimeric ligand, such as PDGF, or a monomeric ligand that has two distinct binding sites for its receptor, such as EGF. Dimerization of the RTK polypeptides activates their tyrosine kinase, causing one subunit of the dimer to phosphorylate a number of tyrosine residues on the cytoplasmic domain of the adjacent subunit of the dimer. Phosphorylation of the RTK creates docking sites that can recruit a variety of specific SH2-containing effectors (Figure 15.23), which become activated as a result of their association with the receptor. In humans, a defect in an RTK, or in a key component of one of the signaling pathways leading from it, can lead to serious disorders, including cancer, diabetes, and atherosclerosis.

A key component of many RTK reaction cascades is the protein Ras.

The Importance of Ras
The *ras* gene was originally discovered as a viral oncogene, that is, a gene carried by certain tumor viruses that enables the virus to transform nor-

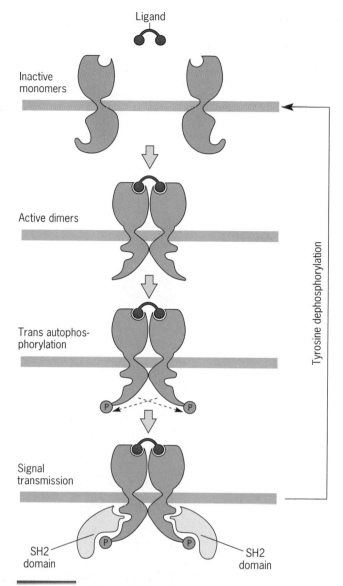

Figure 15.22 Steps in the activation of a receptor tyrosine kinase (RTK), such as the epidermal growth factor (EGF) receptor. In the nonactivated state, the receptors are present in the membrane as monomers. Binding of the ligand leads to dimerization of the receptor and activation of its kinase activity, causing it to add phosphate groups to the cytoplasmic domain of the receptor. The newly formed phosphotyrosine residues of the receptor serve as binding sites for target proteins containing SH2 domains, which become activated as a result of their interaction with the receptor. (This figure illustrates the simplest type of RTK activation. Different RTK monomers, bound by different ligands, can also come together and form RTK heterodimers, making this step in signal transduction potentially complex.) (AFTER J. SCHLESSINGER AND A. ULLRICH, NEURON 9:384, 1992; BY PERMISSION OF CELL PRESS.)

mal cells into malignant ones. Subsequent research showed that, like other oncogenes, *ras* was also present as part of the normal genome of animals, including humans. During the early 1980s, several human tumor cells were found to contain a mutant version of *ras*. Subsequent stud-

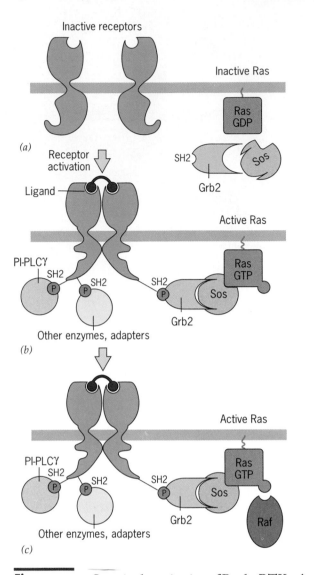

Figure 15.23 Steps in the activation of Ras by RTKs. A number of different receptors, including those that bind epidermal growth factor (EGF) and platelet-derived growth factor (PDGF), act as indicated in this figure. *(a)* In the inactive state, the receptors are present in the membrane as monomers, and Ras is combined with GDP and thus is inactive. *(b)* Binding of ligand triggers receptor dimerization, leading to autophosphorylation of the cytoplasmic domains of the receptor dimer. The newly formed phosphotyrosine residues act as sites for binding Grb2-Sos, which induces the exchange of GTP for GDP on Ras. *(c)* The GTP-activated Ras acts as a binding site for Raf, which causes this protein kinase to become localized at the plasma membrane, where it triggers the MAP kinase cascade (detailed in Figure 15.25).

ies have found that an altered version of the *ras* gene is found in about 30 percent of all human tumors. Given its importance in the development of human cancers, the product of the *ras* gene, a protein called Ras, became a focus of research.

Ras is a small G protein that is held at the inner surface of the plasma membrane by a lipid group that is embedded in the inner leaflet of the bilayer. Unlike het-

erotrimeric G proteins, Ras consists of a single small (21 kDa) subunit. Like other G proteins, Ras cycles between an inactive GDP-bound form and an active GTP-bound form. In its active form, Ras stimulates effectors that lie downstream in the signaling pathway. Ras has a very weak GTPase activity and, left by itself, would remain in the active, GTP-bound state for 30 minutes or so. In the cell, Ras activity is regulated by a number of GTPase-activating proteins (GAPs, page 631) that stimulate the Ras GTPase about 10^5-fold and return the protein to the inactive state. Ras-GAPs have a unique way of stimulating the enzyme activity of their Ras target: they provide an amino acid residue (an arginine) that becomes a critical part of the active site of the Ras GTPase. Mutations in the *ras* gene that lead to tumor formation prevent the protein from hydrolyzing the bound GTP back to the GDP form, even in the presence of the GAP. As a result, the mutant version of Ras remains in the "on" position, sending a continual message downstream along the signaling pathway, keeping the cell in the proliferative mode. Diseases involving abnormal cell proliferation also result from alterations in Ras-GAPs. Mutations in one of the Ras-GAP genes *(NF1)* cause neurofibromatosis 1, a disease in which patients develop large numbers of benign tumors (neurofibromas) along the sheaths that line the nerve trunks. Drugs have been developed that prevent attachment of Ras to the lipid group that holds the protein to the membrane. These drugs inhibit the enzyme farnesyltransferase and are currently in clinical trials on cancer patients.

In its best-studied role, Ras is a key component of a signaling pathway that leads all the way from the outer surface of the plasma membrane to the DNA of the nucleus. The pathway is activated when a growth factor, such as EGF or PDGF, binds to the extracellular domain of its receptor (as in Figure 15.22).

The MAP Kinase Cascade The steps in the Ras signaling pathway that follow binding of a growth factor to an RTK are depicted in Figure 15.23. Briefly, the phosphotyrosines that are generated in the cytoplasmic domain of the RTK act as binding sites for a specific SH2 protein called Grb2. Grb2 is not a protein with catalytic activity, but one that functions solely as an adaptor molecule that links other proteins into a complex. The structure of Grb2 is shown in Figure 15.24, where it can be seen to consist of three distinct domains, each of which binds to another protein. As indicated in Figure 15.23, one domain of the Grb2 molecule binds to the phosphorylated RTK at the inner surface of the membrane, while another domain binds to a protein called Sos (Figure 15.23a).[5] It is here that Ras comes into the picture.

Sos is a guanine nucleotide exchange factor for Ras, that is, a Ras-GEF. Recall from page 631 that GEFs activate G proteins by stimulating the exchange of GDP with

[5]Grb2 is an acronym for *g*rowth factor *r*eceptor *b*inding protein; and sos is an acronym for *s*on *of* *s*evenless, where *sevenless* is a gene identified in *Drosophila* that is involved in photoreceptor differentiation.

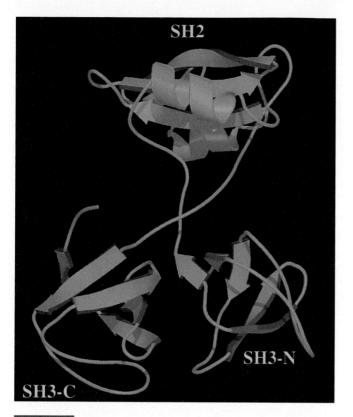

Figure 15.24 Tertiary structure of an adaptor protein, Grb2. Cells have a number of proteins, such as Grb2, whose only apparent function is to form a link between other proteins (as in Figure 15.23). Grb2 consists of three parts: two SH3 and one SH2 domains. SH2 domains bind to a protein (e.g., the activated EGF receptor) containing a particular motif that includes a phosphotyrosine residue. SH3 domains bind to a protein (e.g., Sos) that contains a particular motif that is rich in proline residues. Dozens of proteins that bear these domains have been identified. Interactions involving SH3 and SH2 domains are shown in Figures 2.42*a* and 15.20, respectively. Other adaptor proteins include Nck, Shc, and Crk. (REPRINTED WITH PERMISSION FROM SÉBASTIEN MAIGNAN ET AL. SCIENCE 268:291, 1995. © COPYRIGHT 1995 AMERICAN ASSOCIATION FOR THE ADVANCEMENT OF SCIENCE. COURTESY OF ARNAUD DUCRUIX.)

GTP. In the unstimulated cell, Ras remains bound to GDP. When a ligand binds to the RTK and recruits the Grb2-Sos to the inner surface of the membrane, the Sos protein binds to Ras, causing it to lose its GDP, which is replaced by GTP, thus activating Ras (Figure 15.23*b*). The primary, and possibly sole, function of Ras-GTP is to recruit another protein, called Raf, to the plasma membrane (Figure 15.23*c*). Once it is localized at the plasma membrane, Raf becomes activated as a protein kinase that initiates an orderly chain of phosphorylation reactions called the **MAP kinase cascade,**[6] which is detailed in Figure 15.25. The MAP kinase cascade is similar in many

ways to the cascade of reactions triggered by cAMP during glucose mobilization (see Figure 15.7), but even more complex (Figure 15.26). Both the cAMP and MAP kinase cascades provide pathways by which extracellular signals can influence gene expression. In the cAMP cascade, PKA translocates into the nucleus, whereas in the MAP kinase cascade, it is MAPK that carries out this essential step. Once in the nucleus, MAPK phosphorylates and activates specific transcription factors, such as Elk-1. Elk-1 binds to the promoter regions of a number of genes, including c-*fos* and c-*jun*. The products of these genes, Fos and Jun, interact to form a heterodimeric transcription factor called AP-1 (see Figure 12.31) that activates genes involved in cell proliferation, including cyclin D1, which plays a key role in driving a cell from G_1 into S (page 587).

As discussed in the following chapter, oncogenes are identified by their ability to cause cells to become cancerous. Oncogenes are derived from normal cellular genes that have either become mutated or are overexpressed. Many of the proteins that constitute the Ras signaling pathway were discovered because they were encoded by cancer-causing oncogenes. This includes the genes for Ras, Raf, and a number of the transcriptional factors activated at the end of the pathway (e.g., Fos and Jun). Genes for several of the RTKs situated at the beginning of the pathway, including the receptors for both EGF and PDGF, have also been identified among the several dozen known oncogenes. The fact that so many proteins in this pathway are encoded by genes that can cause cancer when mutated emphasizes the importance of the pathway in the control of cell growth and proliferation.

Adapting the MAP Kinase Cascade to Transmit Different Types of Information The same basic pathway from an RTK through Ras to the activation of transcription factors, as illustrated in Figure 15.25, is found in all eukaryotes investigated, from yeast through flies and nematodes to mammals. Evolution has adapted the pathway to meet many different ends. In yeast, for example, the MAP kinase cascade is required for cells to respond to mating pheromones; in fruit flies, the pathway is utilized during the differentiation of the photoreceptors in the compound eye; and in flowering plants, the pathway transmits signals that initiate a defense against pathogens. Each of the components indicated in Figure 15.25, such as MAPKKK, MAPKK, and MAPK, is represented by a small family of proteins. To date, 14 different MAPKKKs, 7 different MAPKKs, and 13 different MAPKs have been identified in mammals. By utilizing different members of these protein families, mammals are able to assemble a number of different MAP kinase pathways that transmit different types of extracellular signals. We have already described how mitogenic stimuli are transmitted along one type of MAP kinase pathway that leads to cell proliferation. In contrast, when cells are exposed to stressful stimuli, such as X rays or damaging chemicals, signals are transmitted along different MAP kinase pathways that cause the cell to withdraw from the cell cycle, rather than

[6]MAP kinase is an acronym for *m*itogen-*a*ctivated *p*rotein kinase because it is a kinase that is activated by mitosis-stimulating growth factors (i.e., mitogens).

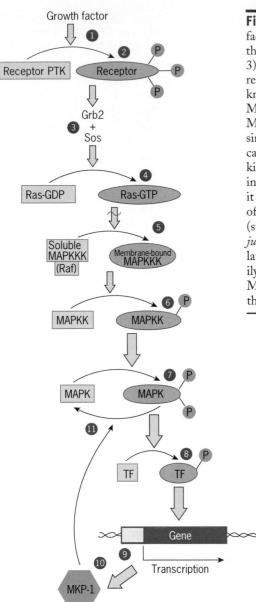

Figure 15.25 **The steps of a generalized MAP kinase cascade**. Binding of growth factor to its receptor (step 1) leads to the autophosphorylation of tyrosine residues of the receptor (step 2) and the subsequent recruitment of the Grb-Sos proteins (step 3). This activated complex causes the GTP-GDP exchange of Ras (step 4), which recruits the protein Raf to the membrane, where it is activated (step 5). Raf is also known as a MAPKKK (MAP kinase kinase kinase) because it phosphorylates MAPKK (MAP kinase kinase), thereby activating its kinase function (step 6). MAPKK is a dual-specificity kinase, a term denoting that it can phosphorylate tyrosine as well as serine and threonine residues. All MAPKs have a tripeptide near their catalytic site with the sequence Thr-X-Tyr. MAPKK phosphorylates MAPK (MAP kinase) on both the threonine and tyrosine residue of this sequence, thereby activating the enzyme (step 7). Once activated, MAPK translocates into the nucleus where it phosphorylates transcription factors (TF, step 8), such as Elk-1. Phosphorylation of the transcription factors increases their affinity for regulatory sites on the DNA (step 9), leading to an increase in the transcription of specific genes (e.g., c-*fos* and c-*jun*) involved in the growth response. One of the genes whose expression is stimulated encodes a MAPK phosphatase (MKP-1; step 10). Members of the MKP family can remove phosphate groups from both tyrosine and threonine residues of MAPK (step 11), which inactivates MAPK and stops further signaling activity along the pathway. (AFTER H. SUN AND N. K. TONKS, TRENDS BIOCHEM. SCI. 19:484, 1994.)

teins. Spatial localization is accomplished by structural (i.e., nonenzymatic) proteins referred to as **scaffolding proteins** whose apparent function is to tether the appropriate members of a signaling pathway in a specific spatial orientation that enhances their mutual interactions. In addition to facilitating a particular series of reactions, scaffolding proteins prevent proteins involved in one signaling pathway from participating in other pathways.

REVIEW ?

1. Describe the steps between the binding of an insulin molecule at the surface of a target cell and the activation of the effector PI3K. How does the action of insulin differ from other ligands that act by means of receptor tyrosine kinases?

2. What is the role of Ras in signaling pathways? How is this affected by the activity of a Ras-GAP? How does Ras differ from a heterotrimeric G protein?

3. What is an SH2 domain, and what role does it play in signaling pathways?

4. How does the MAP kinase cascade alter the transcriptional activity of a cell?

progressing through it as indicated in Figure 15.25. Withdrawal from the cell cycle gives the cell time to repair the damage resulting from the adverse conditions.

Recent studies have focused on the specificity of MAP kinase cascades in an attempt to understand how cells are able to utilize similar proteins as components of pathways that elicit different cellular responses. Studies of amino acid sequences and protein structure suggest that part of the answer lies in selective interactions between enzymes and substrates. For example, certain members of each MAPKKK family (e.g., MLK3) phosphorylate specific members of each MAPKK family (e.g., MKK7), which in turn phosphorylate specific members of each MAPK family (e.g., JNK1). But many members of these families can participate in more than one MAPK signaling pathway. Specificity in MAP kinase pathways is also achieved by spatial localization of the component pro-

15.5 SIGNALS THAT ORIGINATE FROM CONTACTS BETWEEN THE CELL SURFACE AND THE SUBSTRATUM

One of the clearest distinctions between normal and malignant cells is the inability of normal cells to proliferate or to even survive in culture under conditions in which they

Figure 15.26 (ART BY TONY BRAMLEY, FROM TRENDS BIOCHEM. SCI. 19:469, 1994.)

cannot attach to the surface of a culture dish. Unlike malignant cells, which grow and divide while suspended in a culture medium, normal cells require contact between their surface and an extracellular substratum. As discussed in Chapter 7, contacts between a cell and its substratum are mediated by integrins, which are transmembrane proteins that possess binding sites for a variety of extracellular materials, including fibronectin, collagen, and proteoglycans. Integrins do more than attach a cell to its substratum; they provide a pathway by which the cell's extracellular environment can regulate activities within the cell.

Interaction between the extracellular domain of an integrin and an extracellular ligand, such as fibronectin, can generate a variety of signals, including the release of Ca^{2+} into the cytosol, synthesis of phosphoinositide second messengers, and tyrosine phosphorylation of cellular proteins. Such signals generated by contact with the ECM can have profound effects on a cell, influencing such basic processes as growth, migration, differentiation, and survival. This is the reason that normal cells do not grow in suspension; the interior of the cell fails to receive the necessary growth-promoting signals that are normally trans-

mitted from ligand-bound integrins situated at the cell surface. Let us look more closely at how integrin-mediated signaling occurs.

When cells settle onto a culture dish, they contact the surface of the dish at specialized sites called focal adhesions (see Figure 7.17). Focal adhesions contain clustered integrins, cytoplasmic proteins, and bundles of actin filaments known as stress fibers (see Figure 9.68). Focal adhesions first came to the attention of investigators interested in the regulation of cell growth when it was discovered that a tyrosine kinase called Src becomes localized to these structures. Src is encoded by the gene *src*, which was the first oncogene to be discovered. Cells that contain mutant versions of *src* may lose their normal growth control and behave as malignant cells.

The interaction of integrins with extracellular materials leads to clustering of the integrins and activation of cytoplasmic protein kinases, including Src. One of the proteins phosphorylated by Src following integrin engagement is another tyrosine kinase called *focal adhesion kinase*, or *FAK*. The effects of the catalytic activity of Src and FAK (and other tyrosine kinases as well) can be seen

Figure 15.27 Experimental demonstration of phosphotyrosine residues at the sites of focal adhesions in rat fibroblasts. Actin filaments, which radiate from the areas of contact at focal adhesions, are stained with fluorescein-conjugated phalloidin (green). Phosphotyrosine residues are localized by a red fluorescent antibody. The sites of overlap of the actin filaments and phosphotyrosine residues in the regions of the focal adhesions appear orange. (COURTESY OF KEITH BURRIDGE.)

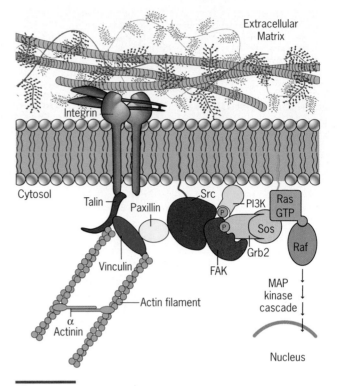

Figure 15.28 Schematic model of the protein-protein interactions of a focal adhesion complex. Focal adhesions perform (1) a mechanical-structural function, which is carried out by actin filaments and associated proteins, and (2) a signaling function, which is carried out by tyrosine kinases, such as Src and FAK. In the model depicted here, Src phosphorylates FAK, which becomes linked to Grb2-Sos, which activates Ras, which sends signals along the MAP kinase pathway. Signals are presumably transmitted into the nucleus, where they stimulate the transcription of genes involved in cell growth and proliferation.

in the micrograph of Figure 15.27, which shows the focal adhesions present where the underlying surface of a rat fibroblast contacts the culture dish. Actin filaments appear green in the micrograph, whereas the phosphotyrosine residues that are generated by the activated tyrosine kinases are stained red. Orange areas are sites where the green microfilaments and the red phosphotyrosine residues coincide in the focal adhesions.

Some of the interactions and signaling events thought to occur at a focal adhesion are depicted in the model of Figure 15.28. Many of the key signaling molecules of a focal adhesion are the same proteins discussed in the previous section, which illustrates how similar signaling pathways can become activated by diverse stimuli. When Src phosphorylates FAK, it creates a phosphotyrosine residue that can bind the SH2 domain of the adaptor protein Grb2 (page 652). Binding of Grb2 (and the associated Sos) to FAK is thought to activate Ras, which can transmit growth-promoting signals to the cell nucleus via the MAP kinase pathway. This may be the primary route by which cell-surface contacts lead to cell proliferation. In the absence of these contacts, the Ras-Raf-MAPK pathway appears to be inhibited, blocking progression of nonadherent cells through the cell cycle.

The nucleus is not the only target of signals that originate from the extracellular matrix. It was noted above that attachment of cells to a substratum may be necessary for cell survival, a phenomenon that is particularly evident in cultured epithelial cells. Within the body, epithelial cells are held in tight contact with one another, and it is inappropriate for one of these cells to detach from its substratum or its neighbors. A detached epithelial represents a

potential cancer cell, so it is not surprising that organisms have evolved mechanisms to destroy such cells.

As long as epithelial cells adhere to a suitable substratum, whether in the body or in cell culture, they are protected from self-destruction. Under these conditions, signals are apparently transmitted from ligand-bound integrins through FAK to the lipid phosphorylating enzyme PI3K that lead to cell survival. As illustrated in Figure 15.21b, lipids produced by PI3K, namely PIP$_2$ and PIP$_3$ bind to protein kinase B (PKB), which becomes activated. These same steps involving PI3K and PKB appear to play a key role in cell survival. For example, cells that express constitutively active forms of PI3K or PKB, that is, forms of these enzymes that are always active regardless of conditions, will survive in culture even when they are not in contact with a substratum. Conversely, epithelial cells that express inactive, mutant versions of PI3K or PKB fail to survive, even if they are adhering to a substratum. Studies indicate that PKB mediates cell survival by phosphorylating, and thus inactivating proteins (e.g., Bad and caspase-9, page 665), that are required to bring about cell death.

(a)

(b)

(c)

(d)

20 µm

Figure 15.29 The assembly of focal adhesions in response to stimuli that activate Rho. The cells in *a* and *c* have been stained with fluorescently labeled phalloidin to reveal the location of actin filaments, whereas the cells in *b* and *d* have been stained with fluorescently labeled antibodies (against vinculin) to reveal the location of focal adhesions. The cells shown in *a* and *b* are quiescent fibroblasts that had been growing in the absence of serum. These cells contain very few organized actin filaments *(a)* or vinculin-containing focal adhesions *(b)*. The cells shown in *c* and *d* have been stimulated by addition of a growth factor (lysophosphatidic acid) to the medium. Lysophosphatidic acid, which is present in serum, binds to a receptor at the cell surface, sending signals into the cell that activate Rho, which leads to the assembly of actin-containing stress fibers *(c)* and focal adhesions *(d)*. (REPRINTED WITH PERMISSION FROM ALAN HALL, SCIENCE 279:509, 1998, COURTESY OF CATHERINE D. NOBES; © COPYRIGHT 1998 AMERICAN ASSOCIATION FOR THE ADVANCEMENT OF SCIENCE.)

Controlling the Assembly of Focal Adhesions

Focal adhesions are relatively complex macromolecular structures involving extracellular components, plasma membrane and cytosolic proteins, and actin filaments of the cytoskeleton. The study of focal adhesion assembly provides insights into mechanisms by which cells regulate shape and internal structure through directed changes in the organization of the cytoskeleton. The assembly of focal adhesions is initiated when integrins of the plasma membrane become bound by extracellular ligands. The signal for the assembly of a focal adhesion is routed from the bound integrin to a key GTP binding protein called Rho. Rho is a small G protein that is similar in size and structure to Ras; in fact, both are members of the same superfamily of regulatory proteins.[7] Like Ras, Rho is a molecular switch that determines whether or not signals are transmitted along certain pathways. In this case, the signaling pathways regulate the organization of the cell's actin cytoskeleton, as illustrated in Figure 15.29. The cells shown in Figure 15.29*a* and *b* are quiescent (nonstimulated) cells that lack both focal adhesions and organized actin filaments. When these cells are exposed to certain growth factors, signals are transmitted through Rho, leading to the assembly of focal adhesions (Figure 15.29*d*) that contain well-defined stress fibers (Figure 15.29*c*).

Figure 15.30 shows two pathways activated by integrin binding through which Rho can influence cytoskeletal organization. In one of these pathways, Rho activates a lipid kinase (phosphatidylinositol 5-kinase) that transfers a phosphate group onto the inositol ring of phosphatidyl 4-phosphate (PIP of Figure 15.12) to produce phosphatidyl 4,5-bisphosphate (PIP$_2$ of Figure 15.12). PIP$_2$ acts as a second messenger by binding to a number of target proteins. Included among the PIP$_2$ targets are several actin-binding proteins (e.g., profilin and gelsolin) that control the state of polymerization of actin molecules (page 382). Binding of PIP$_2$ to these proteins stimulates the polymerization of actin monomers into actin filaments. In the other pathway depicted in Figure 15.30, Rho activates a kinase, called Rho kinase, which phosphorylates and *inactivates* the enzyme myosin light-chain phosphatase. Myosin activity is stimulated by light-chain phosphorylation and inhibited by dephosphorylation. Consequently, inactivation of myosin phosphatase leads to activation of myosin, which plays a key role in bundling actin filaments into organized stress fibers (page 385).

[7]The proteins of the Ras superfamily are small (about 21 kDa), monomeric GTP-binding proteins with very weak GTPase activity that is accelerated by association with a GAP. The Ras superfamily contains numerous subfamilies of proteins, including Ras, Rab, Arf, Ran, and Rho subfamilies, all of which are conserved from yeast to mammals. Some of the roles of these proteins have been discussed in previous chapters. In mammals, the Rho subfamily includes over a dozen proteins and is subdivided into several groups. We will restrict the present discussion to one of these groups, which contains at least three proteins (RhoA, RhoB, and RhoC) that are involved in the assembly of stress fibers and focal adhesions. For the present purposes, these proteins will be referred to collectively as Rho.

REVIEW ?

1. What is the role of FAK and Src in transmitting signals from extracellular substrates? How are such signals transmitted to the nucleus?

2. What is the role of PIP$_2$ and Rho in altering the organization of the actin cytoskeleton?

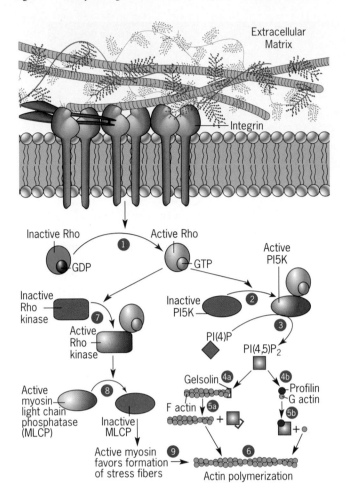

Figure 15.30 Signal pathways that lead to the assembly of the stress fibers of a focal adhesion. Binding of ligands to integrins leads by an undefined pathway to the activation of Rho (i.e., the conversion of Rho-GDP to Rho-GTP, step 1). Rho-GTP activates two distinct signaling pathways. In one pathway, shown on the right, Rho-GTP binds and activates a phosphatidylinositol 5-kinase denoted as PI5K (step 2), which transfers phosphate groups to the 5′ position of PI(4)P to generate PI(4,5)P_2 (step 3). PI(4,5)P_2 binds to a number of target proteins. As shown in the figure, PI(4,5)P_2 binds to gelsolin and profilin molecules (steps 4a and 4b), causing these proteins to dissociate from their complexes with actin filaments and monomeric actin, respectively (steps 5a and 5b). The release of actin monomers leads to their polymerization onto the newly exposed end of the actin filament (step 6). In the other pathway, shown on the left, Rho-GTP binds to the enzyme Rho-kinase, activating the kinase activity of the protein (step 7), which phosphorylates and inactivates the enzyme myosin light chain phosphatase (step 8). Inhibition of the phosphatase leaves myosin molecules in an active, phosphorylated state and favors the formation of stress fibers (step 9).

15.6 CONVERGENCE, DIVERGENCE, AND CROSSTALK AMONG DIFFERENT SIGNALING PATHWAYS

The signaling pathways described above and illustrated schematically in Figure 15.25 and Figure 15.30 depict linear pathways leading directly from a receptor at the cell surface to an end target. In actual fact, signaling pathways in the cell are much more complex. For example:

- Signals from a variety of unrelated receptors, each binding to its own ligand, can *converge* to activate a common effector, such as Ras or Raf.

- Signals from the same ligand, such as EGF or insulin, can *diverge* to activate a variety of different effectors, leading to diverse cellular responses.

- Signals can be passed back and forth between different pathways, a phenomenon known as *crosstalk*.

These characteristics of cell-signaling pathways are illustrated schematically in Figure 15.31.

Signaling pathways provide a mechanism for routing information through a cell, not unlike the way the central nervous system routes information to and from the various organs of the body. Just as the central nervous system collects information about the environment from various sense organs, the cell receives information about its environment through the activation of various surface receptors, which act like sensors to detect extracellular stimuli. Like sense organs that are sensitive to specific forms of stimuli (e.g., light, pressure, or sound waves), cell-surface receptors can bind only to specific ligands and are unaffected by the presence of a large variety of unrelated molecules. A single cell may have dozens of different receptors sending signals to the cell interior simultaneously. Once they have been transmitted into the cell, signals from these receptors can be selectively routed along a number of different signaling pathways that may cause a cell to divide, change shape, activate a particular metabolic pathway, or even commit suicide (discussed in a following section). In this way, the cell integrates information arriving from different sources and mounts an appropriate and comprehensive response.

Examples of Convergence, Divergence, and Crosstalk Among Signaling Pathways

1. Convergence We have discussed three distinct types of cell-surface receptors in this chapter: G protein-coupled receptors, receptor tyrosine kinases, and integrins. Although these three types of receptors may bind to very different ligands, all of them can lead to the formation of phosphotyrosine docking sites for the SH2 domain of the adaptor protein Grb2 (Figure 15.32). The recruitment of Grb2 results in the activation of Ras and transmission of signals down the MAP kinase pathway. As a result of this convergence, signals from diverse receptors can lead to

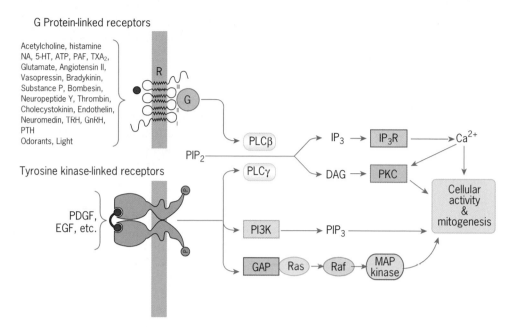

Figure 15.31 **Examples of convergence, divergence, and crosstalk among various signal-transduction pathways.** This drawing shows the outlines of signal-transduction pathways initiated by receptors that act by means of both heterotrimeric G proteins and receptor tyrosine kinases. The two are seen to converge by the activation of different phospholipase C isoforms, both of which lead to the production of the same second messengers (IP_3 and DAG). Activation of the receptor tyrosine kinase by either PDGF or EGF leads to the transmission of signals along three different pathways, an example of divergence. Crosstalk between the two types of pathways is illustrated by calcium ions, which are released from the SER by action of IP_3 and can then act on various proteins, including protein kinase C (PKC), whose activity is also stimulated by DAG. (After M. J. Berridge, reprinted with permission from Nature vol. 361, p. 315, 1993. © Copyright 1993 by Macmillan Journals Limited.)

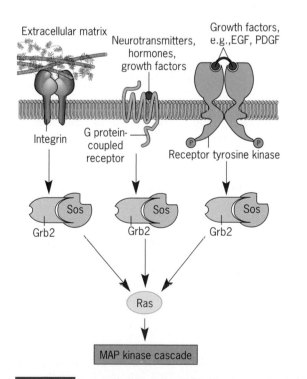

Figure 15.32 **Signals transmitted from a G protein-coupled receptor, an integrin, and a receptor tyrosine kinase all converge on Ras and are then transmitted along the MAP kinase cascade.**

the transcription and translation of a similar set of growth-promoting genes in each target cell.

2. Divergence Evidence of signal divergence has been evident in virtually all of the examples of signal transduction that have been described in this chapter. A quick look at Figures 15.21, 15.23, and 15.31 illustrates how a single stimulus—a ligand binding to an insulin receptor, a growth factor receptor, or an integrin—sends signals out along a variety of different pathways. The interaction of insulin with its receptor, for example, sends signals along separate pathways headed by Ras, PI3K, and PLCγ (Figure 15.21), each of which initiates distinct cellular responses.

3. Crosstalk In previous sections, we examined a number of signaling pathways as if each operated as an independent, linear chain of events. In fact, the information circuits that operate in cells are more likely to resemble an interconnected web in which components produced in one pathway can participate in events occurring in other pathways. The more that is learned about information signaling in cells, the more crosstalk between signaling pathways that is discovered. Rather than attempting to catalog the ways that information can be passed back and forth within a cell, we will look at an example involving cAMP that illustrates the importance of this type of crosstalk.

Figure 15.33 An example of crosstalk between two major signaling pathways. Cyclic AMP acts in some cells, by means of the cAMP-dependent kinase PKA, to block the transmission of signals from Ras to Raf, which inhibits the activation of the MAP kinase cascade. In addition, both PKA and the kinases of the MAP kinase cascade phosphorylate the transcription factor CREB on the same serine residue, activating the transcription factor and allowing it to bind to specific sites on the the DNA.

Cyclic AMP was depicted on page 632 as an initiator of a reaction cascade leading to glucose mobilization. However, cAMP can also inhibit the growth of a variety of cells, including fibroblasts and fat cells, by blocking signals transmitted through the MAP kinase cascade. Cyclic AMP is thought to accomplish this by activating PKA, the cAMP-dependent kinase, which can phosphorylate and inhibit Raf, the protein that heads the MAP kinase cascade (Figure 15.33). These two pathways also intersect at another important signaling effector, the transcription factor CREB. CREB was described on page 634 as a terminal effector of cAMP-mediated pathways. It was assumed for years that CREB could only be phosphorylated by the cAMP-specific kinase, PKA. Over the past few years it has become apparent that CREB is a substrate of a much wider range of kinases. For example, one of the kinases that phosphorylates CREB is Rsk-2, which is a substrate of MAPK of the MAP kinase pathway (Figure 15.33). In fact, both PKA and RSK-2 phosphorylate CREB on precisely the same amino acid residue, Ser233,

which should endow the transcription factor with the same potential in both pathways.

A major unanswered question is raised by these examples of convergence, divergence, and crosstalk: How are different stimuli able to evoke distinct responses, even though they utilize similar pathways? PI3K, for example, is an enzyme that is activated by a remarkable variety of stimuli, including cell adhesion to the ECM, insulin, and EGF. How is it that activation of PI3K in an insulin-stimulated liver cell promotes protein synthesis, whereas activation of PI3K in an adherent epithelial cell promotes cell survival? Ultimately, these contrasting cellular responses must be due to differences in the protein composition of different cell types. Part of the answer probably lies in the fact that different cells have different versions (isoforms) of these various proteins, including PI3K. Some of these isoforms are encoded by different, but related genes, whereas others are generated by alternate splicing (page 540), or other mechanisms. Different isoforms of PI3K, PKB, or PLC, for example, may bind to different sets of upstream and downstream components, which could allow similar pathways to evoke distinct responses. But it isn't likely that isoform variation can explain the extraordinary diversity of cellular responses any more than differences in the structures of neurons can explain the range of responses evoked by the nervous system. Hopefully, as the signaling pathways of more and more cells are described, we will gain a better understanding of the specificity that can be achieved through the use of similar signaling molecules.

15.7 OTHER SIGNALING PATHWAYS

The best-studied signaling pathways are initiated by contact between extracellular ligands and G protein-coupled receptors, receptor tyrosine kinases, or integrins, but numerous other avenues leading from the extracellular medium to the cell interior have also been identified. A few of the most important of these pathways will be discussed in this section.

The Role of NO as an Intercellular Messenger

During the 1980s it was learned that the bacteria-killing activity of macrophages in cell culture depends on the presence of arginine in the medium; in the absence of this amino acid, these phagocytes cannot destroy invading pathogens. It was soon discovered that arginine is a substrate for a small family of enzymes called *nitric oxide synthases* that produce nitric oxide (NO) as one of the products of arginine oxidation. This was the first indication that NO, a substance considered solely as a toxic pollutant in the atmosphere, is an important biological molecule.[8]

[8]Nitric oxide (NO) should not be confused with nitrous oxide (N_2O) or "laughing gas."

Nitric oxide synthase is induced in macrophages and other phagocytic white blood cells as part of the normal inflammatory response. Nitric oxide is a free radical (i.e., it contains an unpaired electron, $\cdot N{=}O$) and reacts with other free radicals, particularly superoxide, $O_2 \cdot^-$. Reaction between these two free radicals in the macrophage cytoplasm produces the potent oxidizing agent peroxynitrite, $ONOO^-$, which can react with lipids, DNA, and proteins. Peroxynitrite is thought to be the primary agent produced in macrophages that kills invading pathogens.

Though relatively high concentrations of NO can have a cell-killing role, much lower concentrations of the substance act as a rapidly diffusible messenger, which is the reason this subject is being considered in the present chapter. As with many other biological phenomena, the discovery that NO functions as a second messenger began with an accidental observation. It had been known for many years that acetylcholine acts in the body to relax the smooth muscle cells of blood vessels, but the response could not be duplicated in vitro. When portions of a major blood vessel such as the aorta were incubated in physiological concentrations of acetylcholine in vitro, the preparation usually showed little or no response. In the late 1970s, Robert Furchgott, a pharmacologist at a New York State medical center, was studying the in vitro response of pieces of rabbit aorta to various agents. In his earlier studies, Furchgott used strips of aorta that had been dissected from the organ. For technical reasons, Furchgott switched from strips of aortic tissue to aortic rings and discovered that the new preparations responded to acetylcholine by undergoing relaxation. Further investigation revealed that the strips had failed to display the relaxation response because the delicate endothelial layer that lines the aorta had been rubbed away during the dissection. This surprising finding suggested that the endothelial cells were somehow involved in the response by the adjacent muscle cells. In subsequent studies, it was found that acetylcholine binds to receptors on the surface of endothelial cells, leading to the production and release of a diffusible agent that causes the muscle cells to relax. The diffusible agent was identified in 1986 as nitric oxide by Louis Ignarro at UCLA and Salvador Moncada at the Wellcome Research Labs in England. The steps in the acetylcholine-induced relaxation response are illustrated in Figure 15.34.

The binding of acetylcholine to the outer surface of an endothelial cell (step 1, Figure 15.34) signals a rise in cytosolic Ca^{2+} concentration (step 2) that activates nitric oxide synthase (step 3). The NO formed in the endothelial cell diffuses across the plasma membrane and into the adjacent smooth muscle cells (step 4), where it binds and stimulates guanylyl cyclase (step 5), the enzyme that synthesizes cyclic GMP (cGMP), which is an important second messenger similar in structure to cAMP. Cyclic GMP leads to a decrease in cytosolic Ca^{2+} concentration, which causes relaxation of the muscle cell (step 6) and dilation of the blood vessel.

The discovery that NO acted as an activator of guanylyl cyclase was made in the late 1970s by Ferid Murad and

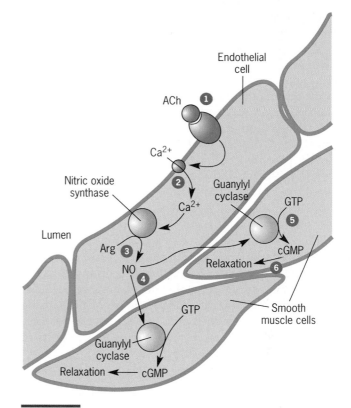

Figure 15.34 A signal transduction pathway that operates by means of NO and cyclic GMP that leads to the dilation of blood vessels. The steps illustrated in the figure are described in the text. (FROM R. G. KNOWLES AND S. MONCADA, TRENDS BIOCHEM. SCI. 17:401, 1992.)

colleagues at the University of Virginia. Murad was working with azide (N_3), a potent inhibitor of electron transport, and chanced to discover that the molecule stimulated cGMP production in cellular extracts. Murad and colleagues ultimately demonstrated that azide was being converted enzymatically into nitric oxide, which served as the actual guanylyl cyclase activator. These studies also explained the action of nitroglycerine, which had been used since the 1800s to treat the pain of angina that results from an inadequate flow of blood to the heart. Nitroglycerine is metabolized to nitric oxide, which stimulates the relaxation of the smooth muscles lining the blood vessels of the heart, increasing blood flow to the organ.

The importance of NO in maintaining blood flow is also illustrated by the often fatal condition of septic shock, in which blood vessels become dilated throughout the body, leading to a dramatic drop in blood pressure. Septic shock is triggered by cell wall molecules released during infections by certain gram-negative bacteria. The cell wall material triggers the systemic release of NO by macrophages, which in turn triggers the widespread dilation of blood vessels. Inhibitors of nitric oxide synthase have shown promise as an effective treatment for septic shock.

The discovery of NO as a second messenger has had a practical impact on millions of lives because it led to the

development of Viagra (sildenafil). During sexual arousal, nerve impulses to the penis trigger the release of NO, which causes relaxation of smooth muscle cells in the lining of penile blood vessels and engorgement of the organ with blood. As described above, NO mediates this response in smooth muscle cells by activation of the enzyme guanylyl cyclase and subsequent production of cGMP. Viagra doesn't influence the release of NO or the activation of guanylyl cyclase, but instead acts as an inhibitor of cGMP phosphodiesterase, the enzyme that destroys cGMP. Inhibition of this enzyme leads to maintained, elevated levels of cGMP, which promotes the development and maintenance of an erection. Viagra is quite specific for one particular isoform of cGMP phosphodiesterase, PDE5, which is the version that acts in the penis. Another isoform of the enzyme, PDE3, plays a key role in the regulation of heart muscle contraction, but fortunately is not inhibited by the drug. As a last word on nitric oxide, it can be noted that the highest concentrations of nitric oxide synthase are found in the brain, suggesting that this messenger is likely involved in many unidentified aspects of central nervous system function.

The Role of Phosphatases in Cell Signaling

The role of protein kinases in stimulating (and inhibiting) the activities of other proteins has been stressed in this chapter. The phosphorylation of proteins is a reversible modification, so that every phosphate that is added by a protein kinase can be removed by a protein phosphatase. It is evident that kinases and phosphatases have opposite effects on their substrate: when phosphorylation activates the substrate, dephosphorylation inactivates it, and vice versa. Thus, even though they have not been discussed, phosphatases are as important as kinases in regulating cellular activities. Consider the drugs cyclosporin and Fk506, which are given to recipients of organ transplants to suppress their immune system and prevent organ rejection. Both of these drugs act by inhibiting calcineurin, a calcium-activated protein phosphatase. Calcineurin-inhibiting drugs may also hold promise in the treatment of congestive heart failure, which can result from the collective enlargement of heart muscle cells. Experiments with cultured heart muscle cells indicate that hormones that trigger cell enlargement, such as angiotensin II, function by activating calcineurin. Inhibitors of calcineurin, such as cyclosporin, can prevent heart enlargement in strains of mice that are susceptible to the disease.

Like protein kinases, some phosphatases are multifunctional and can remove phosphates from a variety of different proteins, whereas others are quite specific, removing phosphates from only one or two substrates. One protein phosphatase, PP2A, can associate with more than 15 different regulatory subunits, and it is these subunits that determine the substrate specificity and subcellular localization of the phosphatase. Most phosphatases, like the kinases they reverse, remove phosphates from either serine and threonine residues or tyrosine residues, but not both. However, a few dual-specificity phosphatases are known that can remove phosphates from all three types of amino acid residues.

Pathways That Lead to Cell Death

Consider the following events, all of which occur during embryonic development: many more neurons grow out of the central nervous system toward a target organ than are required to innervate that organ; T lymphocytes (immune cells responsible for killing abnormal or infected cells) are produced that can attack the body's own cells; the earliest form of the human hand resembles a paddle without any space between the tissues that become the fingers. In each of these situations, cells that are superfluous (such as the excess neurons or the cells between the fingers) or that could harm the individual (such as the T lymphocytes) undergo a type of "cell suicide," called **apoptosis**. Apoptosis does not stop with the end of embryonic development. In adults, apoptosis leads to the death of cells that are no longer required (such as activated T cells that have responded to an infectious agent that has been eliminated) or cells that have sustained irreparable genetic damage and have the potential to progress into cancer cells. In fact, apoptosis is one of the body's main weapons against the development of cancer; cells that lose the ability to undergo apoptosis have taken a major step on the pathway to becoming cancerous.

Apoptosis can be divided into two phases: an activation phase, when the cell responds to "death signals" that commit it to undergoing self-destruction, and an execution phase, when the death sentence is carried out. We will begin with the execution phase, which is best understood. Death by apoptosis is a neat, orderly process (Figure 15.35) characterized by the overall shrinkage in volume of the cell and its nucleus, the loss of adhesion to neighboring cells, the formation of blebs at the cell surface, the dissection of the chromatin into small fragments, and the rapid engulfment of the dying cell by phagocytosis. Apoptotic cells are recognized by phagocytes because they carry exposed markers, called "eat me" signals, that are not present on healthy cells. The best studied "eat me" signal is the presence of phosphatidylserine molecules in the outer leaflet of the plasma membrane of apoptotic cells. These phospholipids are restricted to the inner leaflet of normal cells.

Insight into the molecular basis of apoptosis was first revealed in studies on the nematode worm *C. elegans*, whose cells can be followed with absolute precision during embryonic development. Of the 1090 cells produced during the development of this worm, 131 cells are normally destined to die by apoptosis. In 1986, Robert Horvitz and his colleagues at the Massachusetts Institute of Technology discovered that worms carrying a mutation in the *CED-3* gene proceed through development without losing any of their cells to apoptosis. This finding suggested

(a)

(b)

(c) 2 μm

Figure 15.35 A comparison of normal and apoptotic cells.
(a, b) Scanning electron micrographs of a normal (a) and apoptotic (b) T-cell hybridoma. The apoptotic cell exhibits many surface blebs that are budded off in the cell. Bar equals 4 mm. (c) Transmission electron micrograph of an apoptotic cell treated with an inhibitor that arrests apoptosis at the membrane blebbing stage. (A, B: FROM Y. SHI AND D. R. GREEN, IN S. J. MARTIN, ET AL., TRENDS BIOCHEM. SCI. 19:28, 1994; C: COURTESY OF NICOLA J. MCCARTHY.)

that the product of the *CED*-3 gene played a crucial role in the process of apoptosis in this organism. Once a gene has been identified in one organism, such as a nematode, researchers can search for homologous genes in other organisms, such as humans or other mammals. The identification of the *CED*-3 gene in nematodes led to the discovery of a homologous family of proteins in mammals, which are now called **caspases**. Caspases are a distinctive group of proteolytic enzymes that are activated at an early stage of apoptosis and are responsible for triggering most, if not all, of the changes observed during the execution phase. Caspases accomplish this feat by cleaving a highly select group of essential proteins. Among the targets of caspases are the following:

■ *More than a dozen protein kinases, including focal adhesion kinase (FAK), PKB, PKC, and Raf1.* Inactivation of FAK, for example, is presumed to disrupt cell adhesion, leading to detachment of the apoptotic cell from its neighbors.

■ *Lamins,* which make up the inner lining of the nuclear envelope. Cleavage of lamins leads to the disassembly of the nuclear lamina and shrinkage of the nucleus.

■ *Proteins required for cell structure,* such as those of intermediate filaments, actin, and gelsolin. Cleavage and consequent inactivation of these proteins lead to changes in cell shape.

■ *An endonuclease called caspase activated DNase (CAD),* which is activated following caspase cleavage of an inhibitory protein. Once activated, CAD translocates from the cytoplasm to the nucleus where it attacks DNA, severing it into fragments.

■ *Enzymes involved in DNA repair,* which are inactivated by caspase cleavage. DNA repair is a homeostatic activity that is inappropriate in an apoptotic cell in which the DNA is being dismantled.

Recent studies have focused on the events that lead to the activation of a cell's suicide program. Apoptosis can be triggered by both internal stimuli, such as abnormalities in the DNA, and external stimuli, such as the removal of growth factors from the medium. For example, epithelial cells of the prostate become apoptotic when deprived of the male sex hormone testosterone. This is the reason why prostate cancer that has spread to other tissues is often treated with drugs that interfere with testosterone production. Studies indicate that external stimuli activate apoptosis by a signaling pathway that is distinct from that utilized by internal stimuli. We will begin with activation by external stimuli.

The best understood apoptotic pathway activated by stimuli that originate in the extracellular environment is illustrated in Figure 15.36. In this case, the stimulus for apoptosis is carried by a protein called tumor necrosis factor (TNF), which was named for its ability to kill tumor cells. TNF is produced by certain cells of the immune system in response to adverse conditions, such as exposure to

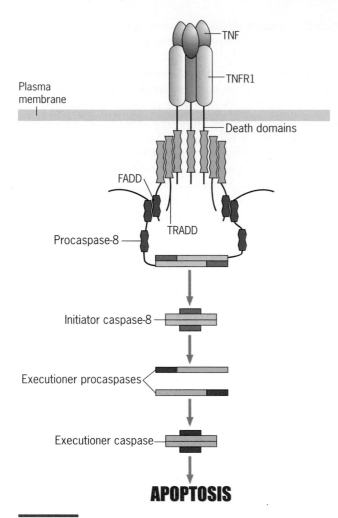

Figure 15.36 **The receptor-mediated pathway of apoptosis.**
When TNF binds to a TNF receptor (TNFR1), the activated
receptor binds two different cytoplasmic adaptor proteins
(TRADD and FADD) and procaspase-8 to form a multipro-
tein complex at the inner surface of the plasma membrane. The
cytoplasmic domains of the TNF receptor, FADD, and
TRADD interact with one another by homologous regions
called death domains that are present in each protein (indicated
as green boxes). Procaspase-8 and FADD interact by means of
homologous regions called death effector domains (indicated as
brown boxes). Once assembled in the complex, the two procas-
pase molecules cleave one another to generate an active caspase-
8 molecule containing four polypeptide segments. Caspase-8 is
an initiator complex that cleaves downstream (executioner) cas-
pases that carry out the death sentence. It can be noted that the
interaction between TNF and TNFR1 also activates other sig-
naling pathways, one of which leads to cell survival rather than
self-destruction.

The cytoplasmic domain of each TNF receptor subunit
contains a segment of about 70 amino acids called a "death
domain" (each green segment in Figure 15.36) that medi-
ates protein-protein interactions. Binding of TNF to the
trimeric receptor produces a change in conformation of the
receptor's death domain, which leads to the recruitment of
a number of proteins, as indicated in Figure 15.36.

The last proteins to join the complex that assembles at
the inner surface of the plasma membrane are two procas-
pase-8 molecules (Figure 15.36). These proteins are called
"procaspases" because each is a precursor of a caspase; it
contains an extra portion that must be removed by prote-
olytic processing. Unlike most proenzymes, procaspases
exhibit a low level of proteolytic activity. Consequently,
when two or more procaspases are held in close association
with one another, as they are in Figure 15.36, they are ca-
pable of cleaving one another's polypeptide chain and
converting the other molecule to the fully active caspase.
The final mature enzyme (i.e., caspase-8) contains four
polypeptide chains, derived from two procaspase precur-
sors as illustrated in the figure.

Activation of caspase-8 is similar in principle to the
activation of effectors by a hormone (page 632) or growth
factor (page 651). In all of these signaling pathways, the
binding of an extracellular ligand causes a change in con-
formation of a receptor that leads to the binding and acti-
vation of proteins situated downstream in the pathway.
Caspase-8 is described as an *initiator* caspase because it
initiates apoptosis by cleaving and activating downstream,
or *executioner*, caspases, that carry out the controlled self-
destruction of the cell as described above.

Internal stimuli, such as irreparable genetic damage,
extremely high concentrations of cytosolic Ca^{2+}, or severe
oxidative stress (i.e., the production of large numbers of
destructive free radicals, page 35) lead to the activation of
proapoptotic cytoplasmic factors. The best-studied of
these factors are members of the Bcl–2 family of proteins.
The Bcl–2 family comprises an unusual collection of regu-
latory proteins, some of which promote apoptosis (e.g.,
Bad, Bid, and Bax) and others of which (e.g., Bcl-X, Bcl-
w, and Bcl–2, itself) prevent apoptosis.

Irreparable genetic damage and other catastrophic
events occurring within a cell trigger apoptosis by a path-
way illustrated in Figure 15.37. The death stimulus acti-
vates proapoptotic members of the Bcl-2 family, such as
Bad, which translocate from the cytosol to the outer mito-
chondrial membrane into which they become inserted. If
you look back at Figure 5.16, you will notice that cy-
tochrome c, while loosely associated with the outer surface
of the inner mitochondrial membrane, actually resides in
the intermembrane space. The attachment of Bad, or
other proapoptotic proteins, to the outer mitochondrial
membrane alters the permeability of that membrane and
promotes the release of cytochrome c from the intermem-
brane space into the cytosol. Virtually all of the cy-
tochrome c molecules present in all of a cell's mitochon-
dria may be released from an apoptotic cell in a period as
short as five minutes. Release of cytochrome c, along with

ionizing radiation, elevated temperature, viral infection, or
toxic chemical agents such as those used in cancer
chemotherapy. Like other types of first messengers dis-
cussed in this chapter, TNF evokes its response by binding
to a transmembrane receptor, TNFR1. TNFR1 is a mem-
ber of a family of related "death receptors" that mediate
apoptosis. Recent studies suggest that the TNF receptor is
present in the plasma membrane as a preassembled trimer.

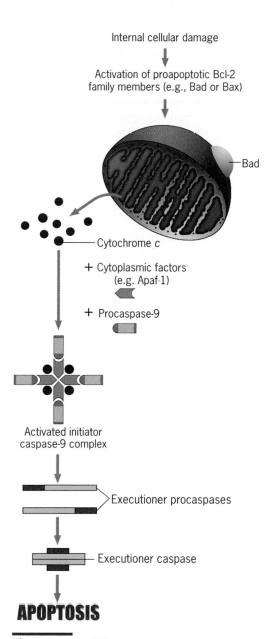

Internal cellular damage

Activation of proapoptotic Bcl-2
family members (e.g., Bad or Bax)

Bad

Cytochrome *c*

+ Cytoplasmic factors
(e.g. Apaf-1)

+ Procaspase-9

Activated initiator
caspase-9 complex

Executioner procaspases

Executioner caspase

APOPTOSIS

Figure 15.37. The mitochondria-mediated pathway of apoptosis. Various types of cellular stress cause proapoptotic members of the Bcl-2 family of proteins, such as Bad or Bax, to become inserted into the outer mitochondrial membrane. Insertion of these proteins leads, by an unidentified mechanism, to the release of cytochrome *c* molecules from the intermembrane space of the mitochondria. Once in the cytosol, the cytochrome *c* molecules form a multisubunit complex with a cytosolic protein called Apaf-1 and procaspase-9 molecules. Procaspase-9 molecules are apparently activated to their full proteolytic capacity as the result of a conformational change induced by association with Apaf-1. According to this model, each active caspase-9 retains an Apaf-1 molecule as a type of cofactor. Regardless of the mechanism of activation, caspase-9 molecules cleave and activate executioner caspases, which carry out the apoptotic response.

protein complex that also includes several molecules of procaspase-9. Recent studies suggest that procaspase-9 becomes activated by simply joining the multiprotein complex and doesn't require proteolytic cleavage (Figure 15.37). Like caspase-8, which is activated by the receptor-mediated pathway described above, caspase-9 is an initiator caspase that activates downstream executioner caspases, which bring about apoptosis. The external (receptor-mediated) and internal (mitochondria-mediated) pathways ultimately converge by activating the same executioner caspases, which cleave the same cellular targets.

You might be wondering why cytochrome *c*, a component of the electron transport chain, and the mitochondrion, an organelle that functions primarily in aerobic respiration, would be involved in initiating apoptosis. There is no obvious answer to this question at the present time. The key role of mitochondria in apoptosis is even more perplexing when one considers that these organelles have evolved from prokaryotic endosymbionts and that prokaryotes do not undergo apoptosis and do not possess caspases. Caspases are also lacking in yeast and plants, and the evolutionary origin of these unique proteases remains to be clarified.

Just as there are signals that commit a cell to self-destruction, there are opposing signals that maintain cell survival. In fact, interaction of TNF with a TNF receptor often transmits two distinct and opposing signals into the cell interior: one stimulating apoptosis and another stimulating cell survival. As a result, most cells that possess TNF receptors do not undergo apoptosis when treated with TNF. This was a disappointing finding because it was originally hoped that TNF could be used as an agent to kill tumor cells. Cell survival is typically mediated by the antiapoptotic members of the Bcl-2 family, such as Bcl-2 and Bcl-X, that were mentioned above. These proteins also bind to the outer mitochondrial membrane where they prevent the cell-killing activities of the proapoptotic proteins, such as Bad, and block the release of cytochrome *c*.

You might recall that the subject of cell survival was also brought up on page 656 in connection with integrin signaling. It was noted there that FAK can activate PI3K, which activates PKB, which phosphorylates Bad and caspase-9, inactivating these effectors. Now that we have discussed the roles of Bad and caspase-9 in apoptosis, it becomes apparent why this integrin-based pathway leads to cell survival. Another important survival pathway that is signaled by TNFR, and many other factors, leads to the activation of a key transcription factor called NF-κB, which mediates the expression of genes encoding cell-survival proteins. We will return to the subject of apoptosis when we consider the development of malignant tumors in the following chapter.

Signaling Pathways in Plants

Although research into signaling pathways in plant cells has lagged well behind that on animal cells, an increasing number of studies is being carried out on plants. Evidence

a number of other essential intermembrane proteins, may be the crucial event that commits the cell to apoptosis. Once in the cytosol, cytochrome *c* forms part of a multi-

from these studies indicates that plants and animals share certain basic signaling mechanisms, including the use of Ca^{2+} and phosphoinositide messengers, but that certain pathways are unique to each major kingdom. For example, cyclic nucleotides, which may be the most ubiquitous animal cell messengers, appear to play little, if any, role in plant cell signaling. Receptor tyrosine kinases are also lacking in plant cells. On the other hand, as discussed below, plants contain a type of protein kinase that is absent from animal cells.

It has long been known that bacterial cells have a protein kinase that phosphorylates histidine residues and mediates the cell's response to a variety of environmental signals. Until 1993, these enzymes were thought to be restricted to bacterial cells but were then discovered in both yeast and flowering plants. In both types of eukaryotes, the enzymes are transmembrane proteins with an extracellular domain that acts as a receptor for external stimuli and a cytoplasmic, histidine kinase domain that transmits the signal to the cytoplasm. One of the best studied of these plant proteins is encoded by the *Etr1* gene. The product of the *Etr1* gene encodes a receptor for the gas ethylene (C_2H_4), a plant hormone that regulates a diverse array of developmental processes, including seed germination, flowering, and fruit ripening. Binding of ethylene to its receptor leads to transmission of signals along a pathway that is very similar to the MAP kinase

cascade found in yeast and animal cells. As in other eukaryotes, the downstream targets of the MAP kinase pathway in plants are transcription factors that activate expression of specific genes encoding proteins required for the hormone response. As researchers analyze the massive amount of data obtained from sequencing *Arabidopsis* and other plant genomes, the similarities and differences between plant and animal signaling pathways should become more apparent.

REVIEW ?

1. Describe the steps in the signaling pathway by which nitric oxide mediates dilation of blood vessels. How does the role of NO in macrophage function differ from its role in vasodilation?

2. What are some of the functions of apoptosis in vertebrate biology? Describe the steps that occur between a) the time that a TNF molecule binds to its receptor and the eventual death of the cell and 2) the time a proapoptotic Bcl–2 member binds to the outer mitochondrial membrane and the death of the cell.

3. What is the role of the formation of caspase-containing complexes in the process of apoptosis?

SYNOPSIS

Cell signaling is a phenomenon in which information is relayed across the plasma membrane to the cell interior and often to the cell nucleus. Cell signaling typically includes recognition of the stimulus at the outer surface of the plasma membrane; transfer of the signal across the plasma membrane; and transmission of the signal to the cell interior, triggering a response. Responses may include a change in gene expression, an alteration of the activity of metabolic enzymes, a reconfiguration of the cytoskeleton, a change in ion permeability, the activation of DNA synthesis, or the death of the cell. This process is often called signal transduction because the stimulus received at the cell surface is entirely different from the signal released to the cell interior. Within the cell, information passes along signaling pathways, which often include protein kinases and protein phosphatases that activate or inhibit their substrates through changes in conformation. Another prominent feature of signaling pathways is the involvement of GTP-binding proteins that serve as switches that turn the pathway on or off. (p. 629)

Many extracellular stimuli (first messengers) initiate responses by interacting with a G protein-coupled receptor on the outer cell surface and stimulating the release of a second messenger within the cell. The uptake and utilization of glucose is controlled by this type of signaling pathway. The breakdown of glycogen into glucose is stimulated by the hormones epinephrine and glucagon, which act as first messengers by binding to their respective receptors on the outer

surface of target cells. Binding of the hormones activates an effector, adenylyl cyclase, on the membrane's inner surface, leading to the production of the diffusible second messenger cyclic AMP (cAMP). Cyclic AMP evokes its response through a reaction cascade in which a series of enzymes are covalently modified. Cyclic AMP molecules bind to the regulatory site of a cAMP-dependent protein kinase called PKA, which phosphorylates phosphorylase kinase and glycogen synthase, leading to the activation of the former enzyme and inhibition of the latter one. The activated phosphorylase kinase molecules add phosphates to glycogen phosphorylase, activating the latter enzyme, leading to the breakdown of glycogen to glucose 1-phosphate, which is converted to glucose. As a result of this reaction cascade, the original message—as delivered by the binding of the hormone at the cell surface—is greatly amplified, and the response time is greatly reduced. Reaction cascades of this type also provide varied sites of regulation. The addition of phosphate groups by kinases is reversed by phosphatases that remove the phosphates. Cyclic AMP is produced in many different cells in response to a wide variety of first messengers. The course of events that occurs in the target cell depends on the specific proteins phosphorylated by the cAMP-dependent kinase. (p. 631)

Both glucagon and epinephrine, as well as many other first messengers, act by binding to receptors that are integral membrane proteins containing seven membrane-spanning α helices. The signal is transmitted from the receptor to the ef-

fector by a heterotrimeric G protein. These proteins are referred to as heterotrimeric because they have three subunits (α, β, and γ) and as G proteins because they bind guanine nucleotides, either GDP or GTP. Each G protein can exist in two states: an active state with a bound GTP or an inactive state with a bound GDP. Hundreds of different G protein-coupled receptors have been identified that respond to a wide variety of stimuli. All of these receptors act through a similar mechanism. The binding of the ligand to its specific receptor causes a change in the conformation of the receptor that increases its affinity for the G protein. As a result, the ligand-bound receptor binds to the G protein, causing the G protein to release its bound GDP and bind a GTP replacement, which switches the G protein into the active state. Exchange of guanine nucleotides changes the conformation of the G_α subunit, causing it to dissociate from the other two subunits, which remain together as a $G_{\beta\gamma}$ complex. Each dissociated G_α subunit with its attached GTP can activate specific effector molecules, such as adenylyl cyclase. The dissociated G_α subunit is also a GTPase and, with the help of an accessory protein (a GAP), hydrolyzes the bound GTP to form bound GDP, which shuts off the subunit's ability to activate additional effector molecules. The G_α-GDP then reassociates with the $G_{\beta\gamma}$ subunits to reform the trimeric complex and return the system to the resting state. Each of the three subunits that make up a heterotrimeric G protein can exist in different isoforms. Different combinations of specific subunits compose G proteins having different properties in their interactions with both receptors and effectors. (p. 635)

Phospholipase C is another important effector on the inner surface of the plasma membrane that can be activated by heterotrimeric G proteins. PI-phospholipase C splits phosphatidylinositol 4,5-bisphosphate (PIP$_2$) into two different second messengers, inositol 1,4,5,-trisphosphate (IP$_3$) and 1,2-diacylglycerol (DAG). DAG remains in the plasma membrane where it activates the enzyme protein kinase C, which phosphorylates serine and threonine residues on a variety of target proteins, including enzymes of intermediate metabolism, such as glycogen synthase, and transcription factors, such as myogenin. Constitutive activation of protein kinase C leads to loss of growth control. IP$_3$ is a small, water-soluble molecule that can diffuse into the cytoplasm where it binds to IP$_3$ receptors located on the surface of the smooth ER. IP$_3$ receptors are tetrameric calcium ion channels; binding of IP$_3$ leads to an opening of the ion channels and diffusion of Ca^{2+} into the cytosol. (p. 641)

The rapid elevation of cytosolic Ca^{2+} whether brought about by the opening of ion channels either in cytoplasmic membranes or in the plasma membrane, triggers a wide variety of cellular responses. The concentration of Ca^{2+} is normally maintained at about 10^{-7} M within the cytosol by the action of Ca^{2+} pumps located in the plasma membrane and the SER membrane. Many different stimuli—ranging from a fertilizing sperm to a nerve impulse arriving at a muscle cell—cause a sudden rise in cytosolic [Ca^{2+}], which may follow the opening of the plasma membrane Ca^{2+} channels, IP$_3$ receptors, or ryanodine receptors, which are a different type of calcium channel located in the SER membrane. Depending on the type of cell, ryanodine channels may be opened by an action potential arriving at the cell, or by the influx of a small amount of Ca^{2+} through the plasma membrane. Among the responses, elevated [Ca^{2+}] can lead to the activation or inhibition of various enzymes and transport systems, membrane fusion, or alterations in cy-

toskeletal or contractile functions. Calcium does not act on these various targets in the free ionic state, but instead it binds to one of a small number of calcium-binding proteins, which in turn elicit the response. The most widespread of these proteins is calmodulin, which contains four calcium-binding sites. The calcium ion is also an important intracellular messenger in plant cells, where it mediates responses to a variety of stimuli, including changes in light, pressure, gravity, and the concentration of plant hormones such as abscisic acid. (p. 644)

Many extracellular stimuli initiate a cellular response by binding to the extracellular domain of a receptor tyrosine kinase (RTK), which activates the tyrosine kinase domain located at the inner surface of the plasma membrane. Insulin mediates many of its actions on target cells through interaction with the insulin receptor, which is an RTK. The activated kinase adds phosphate groups to tyrosine residues located on both the receptor and on select substrates (IRSs). The phosphorylated IRSs can then bind and activate a variety of downstream effectors. The phosphorylated tyrosine residues in an IRS serve as docking sites for proteins that have SH2 domains, which become activated upon binding the IRS. A number of different proteins that contain SH2 domains bind to phosphorylated IRSs, and thus several separate signaling pathways may become activated. One pathway may stimulate DNA synthesis and cell division, another may stimulate the movement of glucose transporters to the cell membrane, and yet another may activate transcription factors that turn on the expression of a set of insulin-specific genes. (p. 648)

Various extracellular agents interact with RTKs on the surfaces of specific target cells and regulate diverse functions, including cell growth and proliferation, the course of cell differentiation, the uptake of foreign particles, and cell survival. The best-studied growth-stimulating ligands, such as PDGF, EGF, and FGF, activate a signaling pathway called the MAP kinase cascade that includes a small monomeric GTP-binding protein called Ras. Like other G proteins, Ras cycles between an inactive GDP-bound form and an active GTP-bound form. In its active form, Ras stimulates effectors that lie downstream in the signaling pathway. Like other G proteins, Ras has a GTPase activity (stimulated by a GAP) that hydrolyzes the bound GTP to form bound GDP, thus switching itself off. When a ligand binds to the RTK, autophosphorylation of the cytoplasmic domain of the receptor leads to the recruitment of Ras to the inner surface of the membrane and exchange of its bound GDP with GTP, thus activating the G protein. Activated Ras has an increased affinity for another protein called Raf, which is recruited to the plasma membrane where it becomes an active protein kinase that initiates an orderly chain of phosphorylation reactions that are outlined in Figure 15.25. The ultimate targets of the MAP kinase cascade are transcription factors that stimulate the expression of genes whose products play a key role in activating the cell cycle, leading to the initiation of DNA synthesis and cell division. The MAP kinase cascade is found in all eukaryotes from yeast to mammals, though it has adapted through evolution to evoke different responses in different types of cells. (p. 651)

Different signaling pathways are often interconnected. As a result, signals from a variety of unrelated ligands can converge to activate a common effector, such as Ras; signals from the same ligand can diverge to activate a variety of different effectors; and signals can be passed back and forth between different pathways (crosstalk). (p. 658)

Nitric oxide (NO) acts as an intercellular messenger that diffuses directly through the plasma membrane of the target cell. Included among the activities stimulated by NO is the relaxation of the smooth muscle cells that line blood vessels. NO is produced by the enzyme nitric oxide synthase, which uses arginine as a substrate. NO functions by activating guanylyl cyclase to produce the second messenger cGMP. (p. 660)

Signaling pathways may end in apoptosis—the programmed death of the cell. Examples of apoptosis include the death of excess nerve cells, the death of unnecessary embryonic tissue, the death of T lymphocytes that react with the body's own tissues,

and the death of potential cancer cells. Death by apoptosis is characterized by the overall compaction of the cell and its nucleus and the orderly dissection of the chromatin by special endonucleases. Apoptosis is mediated by proteolytic enzymes called caspases that activate or deactivate key protein substrates by removing a portion of their polypeptide chain. Two distinct pathways of apoptosis have been identified, one triggered by extracellular stimuli acting through death receptors, such as TNFR1, and the other triggered by internal cellular stress acting through the release of cytochrome c from the intermembrane space of mitochondria and the activation of proapoptotic members of the Bcl-2 protein family (p. 662)

ANALYTIC QUESTIONS

1. The subject of cell signaling was placed near the end of the book because it ties together so many different topics in cell biology. Now that you have read the chapter, would you agree or disagree with this statement? Support your conclusions by example.

2. Suppose the signaling pathway in Figure 15.2 were to lead to the activation of a gene that inhibits a cyclin-dependent kinase. How would a debilitating mutation in phosphatase 2 affect the growth of the cell? How would a similar mutation in protein kinase 3 affect the cell's growth?

3. What might be the effect on liver function of a mutation in a gene that encodes a cAMP phosphodiesterase? Of a mutation in a gene encoding a glucagon receptor? Of a mutation in a gene encoding phosphorylase kinase? Of a mutation that altered the active site of the GTPase of a G_α subunit? (Assume in all cases that the mutation causes a loss of function of the gene product.)

4. Ca^{2+}, IP_3, and cAMP have all been described as second messengers. In what ways are their mechanisms of action similar? In what ways are they different?

5. In the reaction cascade illustrated in Figure 15.25, which steps lead to amplification and which do not?

6. Suppose that epinephrine and norepinephrine could initiate a similar response in a particular target cell. How could you show that the two compounds act by binding to the same cell-surface receptor?

7. One of the key experiments to show that gap junctions (page 270) allowed the passage of small molecules was carried out by allowing cardiac muscle cells (which respond to norepinephrine by contraction) to form gap junctions with ovarian granulosa cells (which respond to FSH by undergoing various metabolic changes). The researchers then added FSH to the mixed cell culture and observed the contraction of the muscle cells. How could muscle cells respond to FSH, and what does this tell you about the structure and function of gap junctions?

8. How would you expect a nonhydrolyzable GTP analogue to affect signaling events that take place during the stimulation of a liver cell by glucagon? What would be the effect of the same analogue on signal transduction of an epithelial cell after exposure to epidermal growth factor (EGF)? How would this compare to the effects of the cholera toxin (page 638) on these same cells?

9. You suspect that phosphatidylcholine might be serving as a precursor for a second messenger that triggers secretion of a hormone in a type of cultured endocrine cell that you are studying. Furthermore, you suspect that the second messenger released by the plasma membrane in response to a stimulus is choline phosphate. What type of experiment might you perform to verify your hypothesis?

10. Figure 15.14 shows the localized changes in $[Ca^{2+}]$ within the dendrites of a Purkinje cell. Calcium ions are small, rapidly diffusible agents. How is it possible for a cell to maintain different concentrations of this free ion in different regions of its cytosol? What do you suspect would happen if you injected a small volume of a calcium chloride solution into a cell that had been injected previously with a fluorescent calcium probe?

11. Formulate a hypothesis that might explain how contact of the outer surface of an egg by a fertilizing sperm at one site causes a wave of Ca^{2+} release that spreads through the entire egg, as shown in Figure 15.16.

12. Because calmodulin activates many different effectors (e.g., protein kinases, phosphodiesterases, calcium transport proteins), a calmodulin molecule must have many different sites on its surface. Would you agree with this statement? Why or why not?

13. Diabetes is a disease that can result from a number of different defects, all of which involve insulin function. Describe three different molecular abnormalities in a liver cell that could cause different patients to exhibit a similar clinical picture, including, for example, high concentrations of glucose in the blood and urine.

14. Would you expect a cell's response to EGF to be more sensitive to the fluidity of the plasma membrane than its response to insulin? Why or why not?

15. Would you expect a mutation in Ras to act dominantly or recessively as a cause of cancer? Why? (A dominant mutation causes its effect when only one of the homologous alleles is mutated, whereas a recessive mutation requires that both alleles of the gene are mutated.)

16. Speculate on a mechanism by which apoptosis might play a crucial role in combatting the development of cancer, a topic discussed in the following chapter.

17. You are working with a type of fibroblast that normally responds to epidermal growth factor by increasing its rate of growth and division, and to epinephrine by lower-

ing its rate of growth and division. You have determined that both of these responses require the MAP kinase pathway, and that EGF acts by means of an RTK and epinephrine by means of a G protein-coupled receptor. Suppose you identified a mutant strain of these cells that can still respond to EGF but is no longer inhibited by epinephrine. You suspect that the mutation is affecting the crosstalk between two pathways (shown in Figure 15.33). Which component in this figure might be affected by such a mutation?

18. Gas samples taken from the lumen of the large intestine of persons suffering from the disease ulcerative colitis were found to contain 100 times the level of nitric oxide as compared to those taken from controls. What does this tell you about the probable nature of this disease?

19. It was noted in the Human Perspective that some receptor mutations result in a loss of function, whereas others result in a gain of function. Which of these would you expect to be inherited as recessive mutations? Which as dominant mutations (see question 15 for definitions)? Why?

20. One of the genes of the cowpox virus encodes a protein called CrmA that is a potent inhibitor of caspases. What effect would you expect this inhibitor to have on an infected cell? Why is this advantageous to the infecting virus?

21. Most RTKs act directly on downstream effectors, whereas the insulin RTK acts through an intermediate, the insulin receptor substrates (IRSs). Are there any advantages in signaling that might accrue from the use of these IRS intermediates.

22. Researchers have reported that 1) virtually all of the physiological effects of insulin on target cells can be blocked by incubation of cells with wortmannin, a compound that specifically inhibits the enzyme PI3K and 2) that causing cells to overexpress a constitutively active form of PKB (i.e., a form of the enzyme that is continually active regardless of circumstances) induces a response in cells that is virtually identical to addition of insulin to these cells. Looking at Figure 15.21, are these observations ones that you might have predicted? Why or why not?

23. Knockout mice that are unable to produce caspase-9 die as a result of a number of defects, most notably a greatly enlarged brain. Why would these mice have this phenotype? How would you expect the phenotype of a cytochrome *c* knockout mice to compare with that of the caspase-9 knockout?

24. Why do you suppose that some people find a compound called PROP to have a bitter taste, whereas others do not report this perception?

ADDITIONAL READING

General Signaling and Heterotrimeric G Protein-Mediated Responses

BOCKAERT, J. & PIN, J. P. 1999. Molecular tinkering of G protein-coupled receptors: an evolutionary success. *EMBO J.* 18:1723–1729.

CLAPHAM, D. E. & NEER, E. J. 1997. G-protein βγ-subunits. *Annu. Rev. Pharmacol. Toxicol.* 37:167–203.

COHEN, P. 1998. Protein phosphorylation and hormone action. *Proc. Roy. Soc. London* B234:115–144.

DECESARE, D. & SASSONE-CORSI, P. 2000. Transcription regulation by cyclic AMP-responsible factors. *Prog. Nuc. Acid Res. Mol. Biol.* 64:343–369.

DEVRIES, L., ET AL. 2000. The regulator of G protein signaling family. *Annu. Rev. Pharmacol. Toxicol.* 40:235–271.

DEKKER, L. V. & SEGAL, A. W. 2000. Signals to move cells. *Science* 287:982–984.

LEFKOWITZ, R. J. 2000. The superfamily of heptahelical receptors. *Nature Cell Biol.* 2:E133–E136.

MONTELL, C. 2000. PLC fills a GAP in G-protein coupled signalling. *Nature Cell Biol.* 2:E82–E83.

ROSS, E. M. & WILKIE, T. M. 2000. GTPase-activating proteins for the heterotrimeric G proteins. *Annu. Rev. Biochem.* 69:795–827.

SHAYWITZ, A. J. & GREENBERG, M. E. 1999. CREB: A stimulus-induced transcription factor activated by a diverse array of extracellular signals. *Annu. Rev. Biochem.* 68:821–861.

The Human Perspective

CLAPHAM, D. E. 1994. Why testicles are cool. *Nature* 371:109–110.

MORRIS, A. J. & MALBON, C. C. 1999. Physiological regulation of G protein-linked signaling. *Physiol. Revs.* 79:1373–1430.

POST S. R., ET AL. 1999. β-adrenergic receptors and receptor signaling in heart failure. *Annu. Rev. Pharmacol. Toxicol.* 39:343–360.

SPIEGEL, A. M. 1996. Defects in G protein-coupled receptors in human disease. *Annu. Rev. Physiol.* 58:143–170.

SPIEGEL, A. M., ED. 1998. *G Proteins, Receptors, and Disease.* Humana.

RTK-Mediated Responses

ALPER, J. 2000. New insights into Type 2 diabetes. *Science* 289:37–39.

BOURNE, H. R. 1994. The importance of being GTP. *Nature* 369:611–612.

COBB, M. H. & GOLDSMITH, E. J. 2000. Dimerization in MAP-kinase signaling. *Trends Biochem. Sci.* 25:7–9.

ELION, E. A. 1998. Routing MAP kinase cascades. *Science* 281:1625–1626.

HUBBARD, S. R. & TILL, J. H. 2000. Protein tyrosine kinase structure and function. *Annu. Rev. Biochem.* 69:373–398.

PAWSON, T. & SAXTON, T. M. 1999. Signaling networks—do all roads lead to the same genes. *Cell* 97:675–678.

ROOVERS, K. & ASSOIAN, R. K. 2000. Integrating the MAP kinase signal into the G1 phase cell cycle machinery. *BioEss.* 22:818–825.

SCHAEFFER, H. J. & WEBER, M. J. 1999. Mitogen-activated protein kinases: specific messages from ubiquitous messengers. *Mol. Cell Biol.* 19:2435–2444.

SHARROCKS, A. D, ET AL. 2000. Docking domains and substrate-specificity determination for MAP kinases. *Trends Biochem. Sci.* 25:448–453.

SCHLESSINGER, J. 2000. Cell signaling by receptor tyrosine kinases. *Cell* 103:211–235.

SCOTT, J. D. & PAWSON, T. 2000. Cell communication: the inside story. *Sci. Am.* 282:72–79. (June)

Figure 16.4 **Karyotype of a cell from a breast cancer line showing a highly abnormal chromosome complement.** A normal diploid cell would have 22 pairs of autosomes and two sex chromosomes. The two members of a pair would be identical, and each chromosome would be a single continuous color (as in the karyotype of a normal cell in Figure 12.16*b* that uses this same spectral visualization technique). The chromosomes of this cell are highly deranged as evidenced by the presence of extra and missing chromosomes and chromosomes of more than one color. These multicolored chromosomes reflect the large numbers of translocations that have occurred in previous cell generations. A cell with normal cell cycle checkpoints and apoptotic pathways could never have attained a chromosome complement approaching that seen here. (COURTESY OF J. DAVIDSON AND PAUL A. W. EDWARDS).

(a)

(b) ————— 10 μm

Figure 16.5 **A comparison of the microtubular cytoskeleton in control and transformed cell cultures.** The arrangement of microtubules in these cells was revealed by immunofluorescence using anti-tubulin antibodies. (*a*) These fibroblasts had been infected with a temperature-sensitive tumor virus and cultured at the higher, restrictive temperature at which the protein that transforms the cells is not functional. The growth properties and arrangement of microtubules are normal. (*b*) Cells from the same culture shown in *a* that were grown at the lower, permissive temperature at which the transformed phenotype is expressed. The cytoskeletal apparatus is highly disorganized. (FROM RONALD L. BROWN ET AL., PROC. NAT'L ACAD. SCI. U. S. A. 78:5595, 1981.)

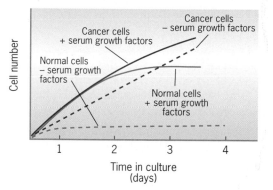

Figure 16.6 **The effects of serum deprivation on the growth of normal and transformed cells.** Whereas the growth of cancer cells continues regardless of the presence or absence of exogenous growth factors, normal cells require these substances in their medium for growth to continue. The growth of normal cells levels off as the growth factors in the medium are depleted.

dergo an aging process that renders them unfit to continue to grow and divide. Cancer cells, on the other hand, are seemingly immortal because they continue to divide indefinitely. This difference in growth potential is often attributed to the presence of telomerase in cancer cells and its absence in normal cells. Recall from page 512 that telomerase is the enzyme that maintains the telomeres at the ends of the chromosomes, thus allowing cells to continue to divide.

REVIEW ?

1. Describe some of the properties that distinguish cancer cells from normal cells.

2. How do the properties of cancer cells manifest themselves in culture?

16.2 THE CAUSES OF CANCER

In 1775, Percival Pott, a British surgeon, made the first known correlation between an environmental agent and the development of cancer. Pott concluded that the high incidence of cancer of the nasal cavity and the skin of the scrotum in chimney sweeps was due to their chronic exposure to soot. Within the past several decades, the carcinogenic chemicals in soot have been isolated, along with hundreds of other compounds shown to cause cancer in laboratory animals. In addition to a diverse array of chemicals, a number of other types of agents are also carcinogenic, including ionizing radiation and a variety of DNA- and RNA-containing viruses. All of these agents have one property in common: they alter the genome. Carcinogenic chemicals, such as those present in soot or cigarette smoke, can almost always be shown either to be directly mutagenic or to be converted to mutagenic compounds by cellular enzymes. Similarly, ultraviolet radiation, which is the leading cause of skin cancer, is also strongly mutagenic.

A number of viruses can infect mammalian cells growing in cell culture, transforming them into cancer cells. These viruses are broadly divided into two large groups: **DNA tumor viruses** and **RNA tumor viruses**, depending on the type of nucleic acid found within the mature virus particle. Among the DNA viruses capable of transforming cells are polyoma virus, simian virus 40 (SV40), adenovirus, and herpeslike viruses. RNA tumor viruses, or retroviruses, are similar in structure to HIV (see Figure 1.23*b*) and are the subject of the Experimental Pathways, which can be found at the end of the chapter. Tumor viruses can transform cells because they carry genes whose products interfere with the cell's normal growth-regulating activities. Although tumor viruses were an invaluable tool for researchers in identifying numerous genes involved in tumorigenesis, viruses are associated with only a small fraction of human cancers. In most cases, these viruses greatly increase a person's risk of developing the cancer, rather than being the sole determinant responsible for the disease. This relationship between viral infection and cancer is illustrated by human papilloma virus (HPV), which can be transmitted through sexual intercourse and is increasing in frequency in the population. Although the virus is present in about 90 percent of cervical cancers, indicating its importance in development of the disease, the vast majority of women who have been infected with the virus will never develop this malignancy. Other viruses linked to human cancers include hepatitis B virus, which is associated with liver cancer; Epstein-Barr virus, which is associated with Burkitt's lymphoma; a herpes virus (HHV-8), which is associated with Kaposi's sarcoma; and the retrovirus HTLV-1, which is associated with adult T-cell leukemia. Certain gastric lymphomas are associated with chronic infection with the stomach-dwelling bacterium *Helicobacter pylori*, which is also responsible for ulcers. Unlike any other known cancer, treatment of the patient with a bacteria-killing antibiotic will usually cure the patient of the lymphoma.

Determining the causes of different types of cancer is an endeavor carried out by *epidemiologists*. The causes of certain cancers are obvious: smoking causes lung cancer and exposure to ultraviolet radiation causes skin cancer. But despite a large number of studies, we are still uncertain as to the causes of most types of human cancer. Humans live in complex environments and are exposed to many potential carcinogens in a changing pattern over a period of decades. Attempting to determine the causes of cancer from a mountain of statistical data obtained from the answers to questionnaires about individual lifestyles has proven very difficult. However, there have been several important correlations. For example, part of the increased incidence of breast cancer in Western countries in recent decades is attributed to improved diet, which has led to an earlier age of onset of menstruation, and to a later age of first pregnancy. Both are risk factors for breast cancer.

There is a general consensus among epidemiologists that some ingredients in the diet, such as animal fat and alcohol, can increase the risk of developing cancer, whereas certain compounds found in fruits and vegetables can reduce that risk. Continued use of nonsteroidal anti-inflammatory drugs (NSAIDs) such as aspirin, sulindac, and indomethacin has been shown to markedly decrease the risk of colon cancer. They are thought to have this effect by inhibiting cyclooxygenases (page 70), enzymes that catalyze the synthesis of hormonelike prostaglandins, which promote the growth of intestinal polyps. Reservatrol, a compound found in grapes (and wine), also inhibits cyclooxygenases and exhibits anticancer activity.

More specific information as to the cause of a few cancers has been obtained by analyzing the types of mutations caused by specific carcinogens. Examples of several types of nucleotide changes induced by carcinogens are shown in Figure 16.7. For instance, aflatoxin B, produced by certain molds, is a major contributor to the high incidence of liver cancer in Asia, where nuts and grains are stored under conditions that favor growth of the mold. Aflatoxin B causes a characteristic G→T substitution in a base pair of codon 249 within the tumor-suppressor gene *p53* (Figure 16.7), thus allowing epidemiologists to make some determination as to the effect of this carcinogen in the population. Although mutagens in the diet and environment are undoubtedly a factor in carcinogenesis, most cancer-causing mutations are thought to result from

DNA damage caused by normal metabolic reactions (page 572).

REVIEW ?

1. Describe some of the ways in which normal cells can be transformed into cancer cells.

2. How can a researcher determine whether a particular case of liver cancer is likely due to the ingestion of the mold product aflatoxin?

16.3 THE GENETICS OF CANCER

Cancer is one of the two leading causes of death in Western countries, afflicting approximately one in every three individuals. Viewed in this way, cancer is a very common disease. But at the cellular level, the development of a cancer is a remarkably rare event. Whenever the cells of a cancerous tumor are genetically scrutinized, they are invariably found to have arisen from a single cell. Thus, unlike other diseases that require modification of a large number of cells, cancer results from the uncontrolled proliferation of a single wayward cell (cancer is said to be *monoclonal*). Consider for a moment that the human body contains trillions of cells, billions of which undergo cell division on any given day. Though almost any one of these dividing cells has the potential to change in genetic composition and grow into a malignant tumor, this only occurs in about one-third of the human population during its entire lifetime.

One of the primary reasons why a greater number of cells do not give rise to cancerous tumors is that malignant transformation requires more than a single genetic alteration. The development of a malignant tumor (*tumorigenesis*) is a multistep process characterized by a **progression** of genetic alterations in a single line of cells that makes the cells increasingly less responsive to the body's normal regulatory machinery and better able to invade normal tissues. Even after they have become malignant, cancer cells continue to accumulate mutations that confer new properties on the cells, making them even more dangerous. The genetic instability of cancer cells also makes them difficult to treat by conventional chemotherapy because cells often arise within the tumor mass that are resistant to the drug.

One of the first steps in the development of some malignant tumors is the formation of a **benign tumor**, which is a tumor composed of cells that are no longer responsive to normal growth controls, but that lack the capability to invade normal tissues or metastasize to distant sites. Some benign tumors pose little threat of becoming malignant, but others, such as the polyps that can develop in the wall of the colon (see Figure 16.18), are very likely to produce cells that cross the boundary from the benign to the malignant state. In some cases, premalignant cells can be identi-

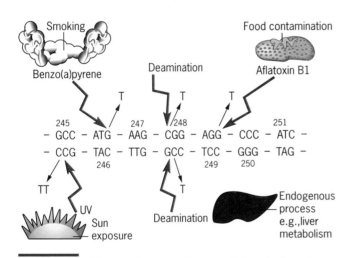

Figure 16.7 The mechanism of action of chemical carcinogens. The figure depicts the nucleotide sequence of seven tandem codons in the *p53* gene, which is deleted or mutated in approximately 50 percent of human cancers. Several different types of chemicals have been found to cause alterations in this part of the gene, including benzo[a] pyrene, which is found in cigarette smoke; aflatoxin B1, which is produced by mold cells that grow on nuts or grain; and ultraviolet light in sunlight. Each of these carcinogens produces a distinctive nucleotide substitution. The loss of amino groups (deamination) from cytosine residues by endogenous metabolism also produces a distinctive C→T mutation. (FROM T. SOUSSI, MOL. MED. TODAY 2:36, 1996. COPYRIGHT 1996, WITH PERMISSION OF ELSEVIER SCIENCE.)

(a)

(b)

Figure 16.8 **Detection of abnormal (premalignant) cells in a Pap smear.** (*a*) Normal squamous epithelial cells of the cervix. The cells have a uniform shape with a small centrally located nucleus. (*b*) Abnormal cells from a case of carcinoma in situ, which is a preinvasive cancer of the cervix. The cells have heterogeneous shapes and large nuclei. (MICROGRAPHS FROM © VU/CABISCO/VISUALS UNLIMITED.)

fied by their morphology. The *Pap smear* is a test for detecting precancerous cells in the epithelial lining of the cervix. The development of cervical cancer typically progresses over a period of more than 10 years and is characterized by cells that appear increasingly abnormal (less well differentiated than normal cells, with larger nuclei, as in Figure 16.8). When cells having an abnormal appearance are detected, the precancerous site in the cervix can be located and destroyed by laser treatment, freezing, or surgery.

As will be evident in the following discussion, the genes involved in carcinogenesis constitute a specific subset of the genome whose products are involved in such activities as the progression of a cell through the cell cycle, adhesion of a cell to its neighbors, apoptosis, and the repair of DNA damage. A model of genetic progression in colon cancer is illustrated in Figure 16.9. Different genes tend to be mutated at different stages in the development of this cancer. Mutations in the *APC* gene are found in over 60 percent of the smallest benign adenomas of the colon, suggesting that mutations in this gene represent a first step in the formation of colon cancers (Figure 16.9). In contrast, the *p53* gene tends to mutate only at later stages along the path. The functions of the *APC* and *p53* genes are discussed later in the chapter.

In the past few years, a new technology for analyzing gene expression has been developed that may, one day, have considerable impact on the way cancer is diagnosed and treated. This technology, which uses DNA microarrays (or DNA chips), was described in detail on page 523. Briefly a glass slide is prepared that contains thousands of spots of DNA, each spot containing the DNA corresponding to a single, known gene. Any particular set of genes, such as those thought to be involved in growth and division, or those thought to be involved in the development and differentiation of lymphocytes or some other cell type, can be included within the microarray. Once prepared, the slide is incubated with fluorescently labeled cDNAs synthesized from the mRNAs of a particular population of cells, such as those from a tumor mass that has been removed during surgery, or from the cancerous blood cells of a patient with leukemia. The fluorescently labeled cDNAs hybridize to spots of complementary DNA immobilized on the slide, and subsequent analysis of the fluorescence pattern tells the investigators which mRNAs are present within the tumor cells and their relative abundance within the mRNA population.

Studies using DNA microarrays that have been published to date have focused on a number of questions, including the following: Is the development of a particular type of tumor correlated with the expression of particular types of genes? Can different types of cancers be distinguished on the basis of their gene-expression profiles? Does the gene-expression profile of a secondary tumor change notably from the profile of the primary tumor from which it arose following metastasis? Results from these

Figure 16.9 **One of a variety of possible sequences of genetic changes in a cell lineage that can lead to the development of colon cancer.** The functions of most of these genes are discussed later in the chapter.

studies indicate that different types of tumors have characteristic gene-expression profiles. This is readily seen in the results shown in Figure 16.10, which depicts the levels at which 50 different genes were found to be transcribed in two different types of leukemia, acute lymphoblastic leukemia (ALL) and acute myeloid leukemia (AML). The genes used in this figure, which are named on the right side, represent those whose transcription showed the

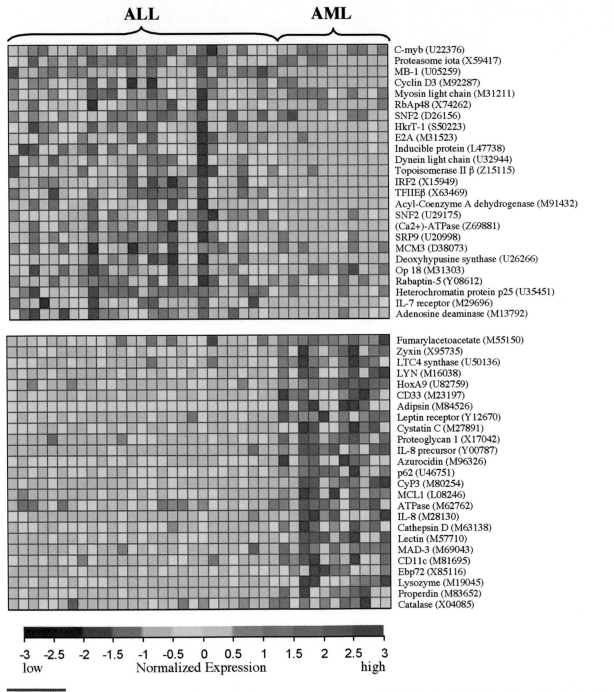

Figure 16.10 Gene-expression profiling that distinguishes two types of leukemia. Each row depicts the expression level of a single gene that is named to the right of the row. Altogether the levels of expression of 50 different genes are indicated. The color key is shown at the bottom of the figure, indicating the lowest level of expression is in dark blue and highest in dark red. Each column represents data from a different sample (patient). The columns on the left show the expression profiles from patients with acute lymphoblastic leukemia (ALL), whereas the columns on the right show the profiles from patients with acute myeloid leukemia (AML) (indicated by the brackets at the top). It is evident that the genes in the upper box are expressed at a much greater level in patients with ALL, whereas the genes in the lower box are expressed at a much greater level in patients with AML. (The genes selected for inclusion in this figure were chosen for the illustration because of these differences in expression between the two diseases.) (FROM T. R. GOLUB ET AL., SCIENCE 286:534, 1999; © COPYRIGHT 1999 AMERICAN ASSOCIATION FOR THE ADVANCEMENT OF SCIENCE.)

greatest difference between these two types of blood cell cancers. Each column represents the results from a single patient with ALL or AML, so that different columns allow one to compare the similarities in expression of the gene from one patient to the next. Gene-expression levels are indicated from dark blue (lowest level) to deep red (highest level). The top half of the figure identifies genes that are transcribed at a much higher level in ALL cells, whereas the bottom half identifies genes that are transcribed to a much higher level in AML cells. These studies make it evident that there are many differences in gene expression between different types of tumors. Some of these differences can be correlated with biological differences between the tumors; one, for example, is derived from a myeloid progenitor and the other from a lymphoid progenitor (see Figure 17.4). Most differences, however, cannot be explained. Why, for example, is the gene encoding catalase (the last gene on the list) expressed at a low level in ALL and a high level in AML? Even if these studies can't answer this question, they can provide cancer researchers with a list of genes to look at more closely as potential targets for therapeutic drugs.

It is also hoped that gene-expression profiles can be used eventually to improve the diagnosis and optimize treatment for individual patients. To illustrate how this might happen, we can examine one study that focused on a particular type of non-Hodgkin's lymphoma called DLBCL. All of the patients in this study were thought to have the same type of cancer, but cDNA microarray analysis revealed that members of the study group showed two distinct gene-expression profiles. In other words, even though all patients appeared to have the same disease based on current diagnostic criteria, microarray analysis indicated they actually suffered from at least two distinct diseases. This finding suggests that DNA microarrays will provide criteria for identifying previously hidden types of cancers. More importantly, differences in gene-expression profiles in the DLBCL study correlated strongly with the outcome of drug treatment. All of the patients had been given the same type of chemotherapy, but the patients who exhibited one type of gene-expression profile had a much more positive outcome from the treatment than those with the alternate gene-expression profile. This type of results suggests that DNA microarrays might be used to predict what type of regimen is most likely to be effective in treating a particular patient based on the gene-expression profile of their particular tumor. These types of studies have just begun, and it will be a long time before we know how useful they will be. For the present, they represent, at the least, a new way to study the genetics of cancer.

Before launching into a detailed examination of some of the genes that figure most prominently in the development of human cancers, it should be noted that genetic changes are not the only factors of importance. A cell may possess a number of genetic changes that would be expected to cause it to grow into a fully malignant tumor, yet it may remain in a dormant state for the remainder of the life of the individual. Among the external influences that affect tumorigenesis are those that promote tumor cell growth and proliferation. This is illustrated by the hormone estrogen.

Estrogen may act as a tumor promoter in the development of breast cancer. Epidemiological evidence indicates a correlation between the length of time a woman is exposed to circulating estrogen and her risk of developing breast cancer. For example, women who have had their ovaries removed early in life and are not given estrogen replacement therapy rarely develop breast cancer. Because estrogen is not mutagenic, it is thought to elevate the risk of tumorigenesis by promoting the growth and division of target cells. Conversely, compounds that block estrogen action decrease the risk of breast cancer. Tamoxifen is a nonsteroidal compound that binds to estrogen receptors and blocks estrogen binding, thus inhibiting the growth of breast cancer cells, which often depend on estrogen stimulation. Tamoxifen has been used for more than a decade to treat breast cancer patients, but it has more recently been approved for use by healthy women whose family history indicates they are at high risk of developing the disease. The decision to take tamoxifen is not a simple one, because tamoxifen treatment increases a woman's risk of developing uterine cancer and blood clots.

The common carcinomas—such as those of the breast, colon, prostate, and lung—arise in epithelial tissues that are normally engaged in a relatively high level of cell division. The same is true of leukemias, which develop in rapidly dividing blood cell precursors. These tissues contain a population of stem cells that divide mitotically, maintaining a relatively constant number of stem cells while also providing the cells that differentiate into the short-lived, specialized cells of the epithelium or blood. As they divide, stem cells may accumulate the mutations required for malignant transformation.

Tumor-Suppressor Genes and Oncogenes: Brakes and Accelerators

The genes that have been implicated in carcinogenesis are divided into two broad categories: tumor-suppressor genes and oncogenes. **Tumor-suppressor genes** act as a cell's brakes; they encode proteins that restrain cell growth and prevent cells from becoming malignant (Figure 16.11a). The existence of such genes originally came to light from studies in the late 1960s in which normal and malignant rodent cells were fused to one another. Some of the cell hybrids formed from this type of fusion lost their malignant characteristics, suggesting that a normal cell possesses factors that can suppress the uncontrolled growth of a cancer cell. Further evidence for the existence of tumor-suppressor genes was gathered from observations that specific regions of particular chromosomes are consistently deleted in cells of certain types of cancer. If the absence of such genes is correlated with the development of a tumor, then it follows that the presence of these genes normally suppresses the formation of the tumor.

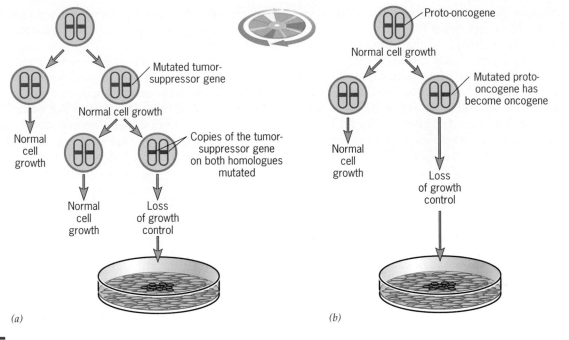

Figure 16.11 Contrasting effects of mutations in tumor-suppressor genes (a) and in oncogenes (b). Whereas a mutation in one of the two copies (alleles) of an oncogene may be sufficient to cause the cell to lose growth control, both copies of a tumor-suppressor gene must be knocked out to induce the same effect. As discussed shortly, oncogenes arise from proto-oncogenes as the result of gain-of-function mutations, that is, mutations that cause the gene product to exhibit new functions that lead to malignancy. Tumor-suppressor genes, in contrast, suffer loss-of-function mutations that render them unable to restrain cell growth.

Oncogenes, on the other hand, encode proteins that promote the loss of growth control and the conversion of a cell to a malignant state (Figure 16.11*b*). As a result, oncogenes act as accelerators of cell proliferation and tumorigenesis. The existence of oncogenes was discovered through a series of investigations on RNA tumor viruses that is documented in the Experimental Pathways for this chapter, which can be found at the end of the chapter. These viruses transform a normal cell into a malignant cell because they carry a gene that encodes a protein that interferes with the cell's normal activities. The turning point in these studies came in 1976, when it was discovered that an oncogene called *src*, carried by an RNA tumor virus called avian sarcoma virus (ASV), was actually present in the genome of uninfected cells. The oncogene, in fact, was not a viral gene, but a cellular gene that had become incorporated into the viral genome during a previous infection. It soon became evident that cells possess a variety of genes, now referred to as **proto-oncogenes**, that have the potential to subvert the cell's own activities and push the cell toward the malignant state (Figure 16.12).

As discussed below, proto-oncogenes encode proteins that have various functions in a cell's *normal* activities. Proto-oncogenes can be converted into oncogenes (i.e., *activated*) by several mechanisms (Figure 16.12):

1. The gene can be mutated in a way that alters the properties of the gene product so that it can no longer carry out its normal activity (Figure 16.12, path a).

2. A mutation in a nearby regulatory sequence can alter the expression of the gene, so that an excessive quantity of the gene product is produced (Figure 16.12, path b).

3. A chromosome rearrangement can occur that brings a DNA sequence from a distant site in the genome into close proximity of the gene, which can either alter the expression of the gene or the nature of the gene product (Figure 16.12, path c).

Any of these genetic alterations can cause a cell to become less responsive to normal growth controls, causing it to behave as a malignant cell. Oncogenes act *dominantly*, which is to say that a single copy of an oncogene can cause the cell to express the altered phenotype, regardless of whether or not there is a normal, unactivated copy of the gene on the homologous chromosome (Figure 16.11*b*). Researchers have taken advantage of this property to identify oncogenes by introducing the DNA suspected of containing the gene into cultured cells and monitoring the cells for evidence of altered growth properties.

We saw earlier that the development of a human malignancy requires more than a single genetic alteration. The reason becomes more apparent with the understanding that there are two types of genes responsible for tumor formation. As long as a cell has its full complement of tumor-suppressor genes, it is thought to be protected against the effects of an oncogene for reasons that will be evident when the functions of these genes are discussed next. Most tumors contain alterations in both tumor-

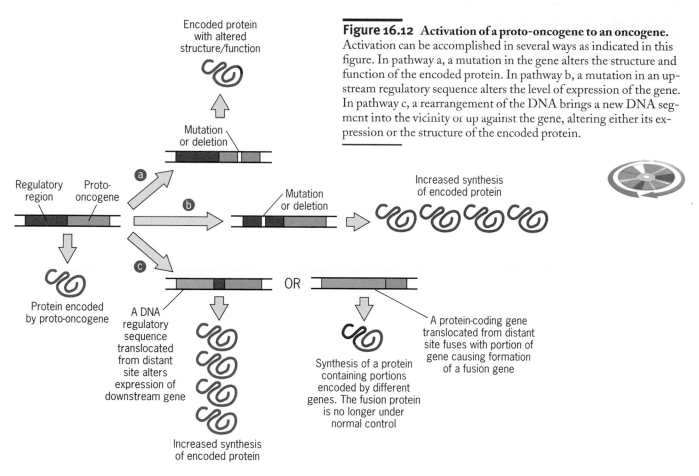

Figure 16.12 Activation of a proto-oncogene to an oncogene. Activation can be accomplished in several ways as indicated in this figure. In pathway a, a mutation in the gene alters the structure and function of the encoded protein. In pathway b, a mutation in an upstream regulatory sequence alters the level of expression of the gene. In pathway c, a rearrangement of the DNA brings a new DNA segment into the vicinity of up against the gene, altering either its expression or the structure of the encoded protein.

suppressor genes and oncogenes, suggesting that the loss of a tumor-suppressor function within a cell must be accompanied by the conversion of a proto-oncogene into an oncogene before the cell becomes fully malignant. Even then, the cell may not exhibit all of the properties required to invade surrounding tissues or to form secondary colonies by metastasis. Mutations in additional genes, such as those encoding cell adhesion molecules or extracellular proteases (discussed on page 265), may be required before these cells acquire the full life-threatening phenotype. In general, the number of altered genes increases along with a progression in the malignant virulence of the tumor. Studies of colorectal carcinomas indicate, for example, that mutations in as many as seven different genes may be necessary for the development of a fully malignant tumor (see Figure 16.9).

We can now turn to the functions of the products encoded by both tumor-suppressor genes and oncogenes and examine how mutations in these genes can cause a cell to become malignant.

Tumor-Suppressor Genes The transformation of a normal cell to a cancer cell is accompanied by the loss of function of one or more tumor-suppressor genes. At the present time, approximately two dozen genes have been implicated as tumor suppressors in humans, some of which are listed in Table 16.1. Included among the genes on the list are those that encode transcription factors (e.g.,

$p53$ and $WT1$), cell-cycle regulators (e.g., RB and $p16$), components that regulate signaling pathways ($NF1$), a phosphoinositide phosphatase ($PTEN$), and a protein that regulates RNA polymerase II elongation (VHL). In one way or another, most of the proteins encoded by tumor-suppressor genes act as negative regulators of cell proliferation, which is why their elimination promotes uncontrolled cell growth. Some of these genes are involved in the development of a wide variety of different cancers, whereas others play a role in the formation of one or a few cancer types.

It is common knowledge that members of some families are at high risk of developing certain types of cancers. Although these inherited cancer syndromes are rare, they provide an unprecedented opportunity to identify tumor-suppressor genes that, when missing, contribute to the development of both inherited and sporadic (i.e., noninherited) forms of cancer. The first tumor-suppressor gene to be studied and eventually cloned—and one of the most important—is associated with a rare childhood cancer of the retina of the eye, called *retinoblastoma*. The gene responsible for this disorder is named *RB*. The incidence of retinoblastoma follows two distinct patterns: it occurs at high frequency and at young age in members of certain families, and it occurs sporadically at an older age among members of the population at large. The fact that retinoblastoma runs in certain families suggested that the cancer can be inherited. Examination of cells from chil-

Table 16.1	Tumor-Suppressor Genes		
Gene	**Primary Tumor**	**Proposed Function**	**Inherited Syndrome**
APC	Colorectal	Binds β-catenin acting as transcription factor	Familial adenomatous polyposis
BRCA1	Breast	Transcription factor, DNA repair	Familial breast cancer
MSH2, MLH1	Colorectal	Mismatch repair	HNPCC
E-Cadherin	Breast, colon, etc.	Cell adhesion molecule	Familial gastric cancer
INK4a	Melanoma, pancreatic	p16: Cdk inhibitor p19ARF: stabilizes p53	Familial melanoma
NF1	Neurofibromas	Activates GTPase of Ras	Neurofibromatosis type 1
NF2	Meningiomas	Links membrane to cytoskeleton	Neurofibromatosis type 2
p16 (MTS1)	Melanoma	Cdk inhibitor	Familial melanoma
p53	Sarcomas, lymphomas, etc.	Transcription factor (cell cycle and apoptosis)	Li-Fraumeni syndrome
PTEN	Breast, thyroid	PIP$_3$ phosphatase	Cowden disease
RB	Retinal	Binds E2F (cell cycle transcrip. regulation)	Retinoblastoma
VHL	Kidney	Regulates RNA pol II elongation	von Hippel-Lindau syndrome
WT1	Wilms tumor of kidney	Transcription factor	Wilms tumor

dren suffering from retinoblastoma revealed that one member of the thirteenth pair of homologous chromosomes was missing a small piece from the interior portion of the chromosome. The deletion was present in all of the children's cells—both the cells of the retinal cancer and cells elsewhere in the body—indicating that the chromosomal aberration had been inherited from one of the parents.

Retinoblastoma is inherited as a dominant genetic trait because members of high-risk families that develop the disease inherit one normal allele and one abnormal allele. But unlike most dominantly inherited conditions, such as Huntington's disease, where an individual who inherits a missing or altered gene invariably develops the disorder, children who inherit a chromosome missing the retinoblastoma gene inherit a strong disposition toward developing retinoblastoma, rather than the disorder itself. In fact, approximately 10 percent of individuals who inherit a chromosome with an *RB* deletion never develop the retinal cancer. How is it that a small percentage of these predisposed individuals escape the disease?

The genetic basis of retinoblastoma was explained in 1971 by Alfred Knudson of the University of Texas. Knudson proposed that the development of retinoblastoma requires that both copies of the *RB* gene of a retinal cell be either eliminated or mutated before the cell can give rise to a retinoblastoma. In other words, the cancer arises as the result of two independent "hits" in a single cell. In cases of sporadic retinoblastoma, the tumor develops from a retinal cell in which both copies of the *RB* gene have undergone successive spontaneous mutation (Figure 16.13*a*). Because the chance that both alleles of the same gene will be the target of debilitating mutations in the same cell is extremely unlikely, the incidence of the cancer in the general population is extremely low. In contrast, the

cells of a person who inherits a chromosome with an *RB* deletion are already halfway along the path to becoming malignant. A mutation in the remaining *RB* allele in any of the cells of the retina produces a cell that lacks a normal *RB* gene and thus cannot produce a functional *RB* gene product (Figure 16.11*b*). This explains why individuals who inherit an abnormal *RB* gene are so highly predisposed to developing the cancer. The second "hit" fails to occur in approximately 10 percent of these individuals, who do not develop the disease. Knudson's hypothesis was subsequently confirmed by examining cells from patients with an inherited disposition to retinoblastoma and finding that, as predicted, both alleles of the gene were mutated in the cancer cells. Individuals with sporadic retinoblastomas had normal cells that lacked *RB* mutations and tumor cells in which both alleles of the gene were mutated.

Although deficiencies in the RB gene are first manifested in the development of retinal cancers, this is not the end of the story. People who suffer from the inherited form of retinoblastoma are also at high risk of developing other types of tumors later in life, particularly soft-tissue sarcomas (tumors of mesenchymal origin). The consequences of *RB* mutations are not confined to persons who inherit a mutant allele. Mutations in *RB* alleles are a common occurrence in sporadic breast, prostate, and lung cancers among individuals who have inherited two normal *RB* alleles. When cells from these tumors are cultured in vitro, the reintroduction of a wild-type *RB* gene back into the cells is sufficient to suppress their cancerous phenotype, indicating that the loss of this gene function contributes significantly to tumorigenesis. Given the importance of the *RB* gene in regulating the growth of a wide variety of tissues, a number of laboratories have directed their attention to determining the role of the product of the RB gene.

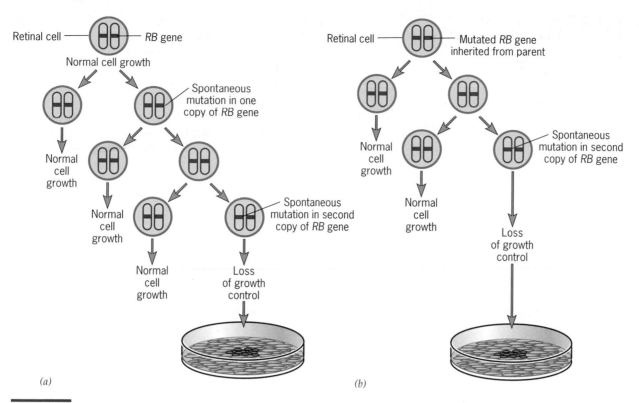

Figure 16.13 Mutations in the *RB* gene that can lead to retinoblastoma. (*a*) In sporadic (nonfamilial) cases of the disease, an individual begins life with two normal copies of the *RB* gene in the zygote, and retinoblastoma occurs only in those rare individuals in whom a given retinal cell accumulates independent mutations in both alleles of the gene. (*b*) In familial (inherited) cases of the disease, an individual begins life with one abnormal allele of the *RB* gene, usually present as a deletion. Thus all the cells of the retina have at least one of their two RB genes nonfunctional. If the other *RB* allele in a retinal cell becomes inactivated, usually as the result of a point mutation, that cell gives rise to a retinal tumor. The chance of the second *RB* allele incurring such a mutation is approximately 90 percent.

The Role of pRb in Regulating the Cell Cycle The importance of the cell cycle in cell growth and proliferation was discussed in Chapters 14 and 15, where it was noted that factors that control the cell cycle can play a pivotal role in the development of cancer. The protein encoded by the *RB* gene, **pRb**, helps regulate the passage of cells from the G_1 stage of the cell cycle into S phase, during which DNA synthesis occurs. As discussed on page 584, the transition from G_1 to S is a time of commitment for the cell; once a cell enters S phase, it invariably proceeds through the remainder of the cell cycle and into mitosis. The transition from G_1 to S is accompanied by the activation of many different genes that encode proteins ranging from DNA polymerases to cyclins and histones. Among the transcription factors involved in activating genes required for S-phase activities are members of the E2F family, which are key targets of pRb. The role of pRb in controlling E2F activity is illustrated in Figure 16.14. During G_1, the E2F transcription factors of the cell are normally bound to pRb, which prevents them from activating a number of genes encoding proteins required for S-phase activities (e.g., cyclin E and DNA polymerase α). Studies suggest (as indicated in step 1 of Figure 16.14) that the E2F-pRB complex is associated with the DNA but acts as a gene repressor rather than a gene activator. As the end of G_1 approaches, the pRb subunit of the pRB-E2F complex is phosphorylated by the cyclin-dependent kinases that regulate the G_1-S transition. Once phosphorylated, pRB releases its bound E2F, allowing the transcription factor to activate gene expression, which marks the cell's irreversible commitment to enter S phase. A cell that loses pRb activity as the result of *RB* mutation would be expected to lose its ability to inactivate E2F, thereby removing certain restraints over the entry to S phase. E2F is only one of more than 40 proteins capable of binding to pRb, suggesting that pRb has numerous other functions. The complexity of pRB interactions is also suggested by the fact that the protein contains at least 16 different serine and threonine residues that can be phosphorylated by cyclin-dependent kinases. It is likely that phosphorylation of different combinations of amino acid residues allow the protein to interact with different downstream targets.

The importance of pRb in regulating the cell cycle has been demonstrated in numerous experiments. For example, injection of excess quantities of the nonphosphorylated pRb protein into cells during G_1 blocks the progression of the cells into S. Even more telling is the fact that several DNA tumor viruses (including adenoviruses, human papilloma virus, and SV40) encode a protein that binds to pRb, blocking its ability to bind to E2F. The abil-

Figure 16.14 The role of pRb in controlling transcription of genes required for progression of the cell cycle. During most of G_1, the unphosphorylated pRB is bound to the E2F protein. The E2F-pRB complex binds to regulatory sites in the promoter regions of numerous genes involved in cell-cycle progression, acting as a repressor that blocks gene expression. Repression probably involves a histone deacetylase (not shown) that modulates chromatin architecture (page 539). Activation of the cyclin-dependent kinase (Cdk) leads to the phosphorylation of pRB, which can no longer bind the E2F protein (step 2). Loss of the bound pRB converts the DNA-bound E2F into a transcription activator, leading to transcription of the genes being regulated (step 3). The mRNA is translated into proteins (step 4) that are required for the progression of cells from G_1 into S phase of the cell cycle (step 5).

ity of these viruses to induce cancer in infected cells depends on their ability to block the negative influence that pRb has on progression of a cell through the cell cycle. By using these pRb-blocking proteins, these viruses accomplish the same result as when the *RB* gene is deleted, leading to the development of human tumors.

Some of the most interesting data on the retinoblastoma gene have come from studies on knockout mice, that is, mice that have been genetically engineered so that they lack both alleles of the *RB* gene (designated $RB^{-/-}$). Given the apparent role of pRb in restraining the cell cycle, the cells of $RB^{-/-}$ mice might be expected to divide in an uncontrolled manner and produce totally aberrant

embryos. But in fact, the very early stages of development proceed rather normally, without any obvious cell cycle defects. $RB^{-/-}$ embryos die on day 12 or 13 as the result of widespread apoptosis in cells of the central nervous system and liver. The role of the *RB* gene in the formation of tumors is more evident in mice that are heterozygous for the gene ($RB^{+/-}$). Animals with this genotype proceed through embryonic development and then exhibit a greatly elevated risk of developing cancer, just as humans that have an $RB^{+/-}$ genotype. Unlike humans, however, these mice are afflicted primarily with tumors of the pituitary, rather than retinoblastomas (which have never been observed to occur in mice). Although these results confirm the importance of the *RB* gene in suppressing tumor formation, they add to a large body of evidence indicating that important differences exist between rodent and human tumorigenesis. Consequently, analysis of risk assessments of carcinogens conducted on mice and rats is often difficult to extrapolate to humans.

The Role of p53: Guardian of the Genome Despite its unassuming name, the *p53* gene may have more to do with the development of human cancer than any other component of the genome. The gene gets its name from the product it encodes, p53, which is a polypeptide having a molecular mass of 53,000 daltons. In 1990, *p53* was recognized as the tumor-suppressor gene that, when absent, is responsible for a rare inherited disorder called Li-Fraumeni syndrome. Victims of this disease are afflicted with a very high incidence of certain cancers, including breast and brain cancer and leukemia. Like individuals with the inherited form of retinoblastoma, persons with Li-Fraumeni syndrome inherit one normal and one abnormal (or deleted) allele of the *p53* tumor-suppressor gene and are thus highly susceptible to cancers that result from random mutations in the normal allele.

The importance of p53 as an anti-tumor weapon is most evident from the finding that more than 50 percent of all human cancers contain cells with point mutations or deletions in both alleles of the *p53* gene. Furthermore, tumors composed of cells bearing *p53* mutations are correlated with a poorer survival rate than those containing a wild-type *p53* gene. Clearly, the elimination of *p53* function is an important step in the progression of many cancer cells toward the fully malignant state. More than a thousand different *p53* mutations have been identified among human tumor samples, indicating that the proper functioning of the protein is very sensitive to even slight changes in amino acid sequence (Figure 16.15).

Why is the presence of p53 so important in preventing a cell from becoming malignant? p53 is a transcription factor that activates the expression of other genes. One of the genes activated by p53 encodes a protein called p21 that inhibits the cyclin-dependent kinase that normally drives a cell through the G_1 checkpoint (page 589). As the level of p53 rises in the damaged G_1 cell, expression of the *p21* gene is activated, and progression through the cell

(b)

Figure 16.15 **The amino acid residues most frequently mutated in p53 from human tumors.** (*a*) A mutation frequency map of the p53 core domain. The amino acid sequence is indicated in single-letter nomenclature (see Figure 2.26). The underlined residues are most highly conserved in the protein from different species. The number of tumor-derived missense (single base substitution) mutations at each residue is indicated by the height of the bars. The six most frequently mutated residues are identified. The arrows and cylinders represent β strands and α helices, respectively. (*b*) A ribbon drawing showing the three-dimensional structure of the portion of the p53 polypeptide whose sequence was indicated in part *a* and its interaction with the DNA helix. The residues most frequently mutated in human cancers occur at or near the protein-DNA interface. (REPRINTED WITH PERMISSION FROM Y. CHO, S. GORINA, P. D. JEFFREY, N. P. PAVLETICH, SCIENCE 265:352, 1994; © COPYRIGHT 1994 AMERICAN ASSOCIATION FOR THE ADVANCEMENT OF SCIENCE.)

cycle is arrested. This gives the cell time to repair the genetic damage before it initiates DNA replication. When both copies of the *p53* gene in a cell have been mutated so that their product is no longer functional, the cell can no longer produce the p21 inhibitor or exercise the feedback control that prevents it from entering S phase when it is not prepared to do so. Failure to repair DNA damage leads to the production of abnormal cells that have the potential to become malignant.

Cell cycle arrest is not the only way that p53 protects an organism from developing cancer. Alternatively, p53 can direct a genetically damaged cell along a pathway that leads to death by apoptosis, thereby ridding the body of cells with a malignant potential. The p53 protein is thought to accomplish this by activating the expression of the *bax* gene, whose encoded product (Bax) initiates apoptosis (664). If both alleles of *p53* should become inac-

tivated, a cell that is carrying damaged DNA fails to be destroyed, even though it lacks the genetic integrity required for controlled growth (Figure 16.16). When a normal *p53* gene is reintroduced into a cancer cell that has lost both of its own alleles of *p53*, the genetically engineered cell often undergoes apoptosis.

The level of p53 in a healthy G_1 cell is very low. However, if a G_1 cell sustains genetic damage, as occurs if the cell is subjected to ultraviolet light or chemical carcinogens, the concentration of p53 rises rapidly. A similar response can be elicited simply by injecting a cell with DNA containing broken strands. The increase in p53 levels is not due to increased expression of the gene but to a decrease in the protein's degradation. p53 degradation is facilitated by a protein called MDM2, which binds to p53 and escorts it out of the nucleus and into the cytosol where the p53 molecule is ubiquinated and destroyed (page 545). How does

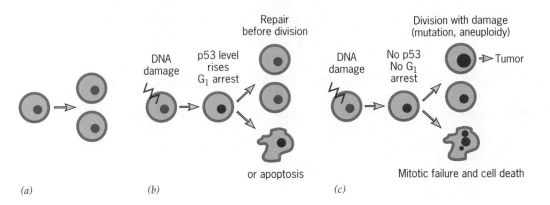

Figure 16.16 A model for the function of p53. (*a*) Cell division does not normally require the involvement of p53. (*b*) If, however, the DNA of a cell becomes damaged as the result of exposure to mutagens, the level of p53 rises and acts either to arrest the progression of the cell through G_1 or to direct the cell toward apoptosis. (*c*) If both copies of the *p53* gene are inactivated, the cell loses the ability to arrest the cell cycle or commit the cell to apoptosis following DNA damage. As a result, the cell either dies from mitotic failure or continues to proliferate with genetic abnormalities that may lead to the formation of a malignant growth. (AFTER D. P. LANE, REPRINTED WITH PERMISSION FROM NATURE 358:15, 1992. © COPYRIGHT 1992, MACMILLAN MAGAZINES LTD.)

DNA damage lead to stabilization of p53? We saw on page 588 that persons suffering from ataxia telangiectasia lack a protein kinase called ATM and are unable to respond properly to DNA-damaging radiation. ATM is normally activated following DNA damage and p53 is one of the proteins it is thought to phosphorylate. The phosphorylated version of the p53 molecule is no longer able to interact with MDM2, which stabilizes existing p53 molecules in the nucleus and allows them to activate the expression of genes such as *p21* and *bax* (see Figure 16.19).

Some tumor cells have been found that contain a wild-type *p53* gene but extra copies of *MDM2*. Such cells are thought to produce excessive amounts of MDM2, which prevents p53 levels from building to required levels to stop the cell cycle or induce apoptosis following DNA damage. The relationship between MDM2 and p53 has also been demonstrated using gene knockouts. Mice that lack a gene encoding MDM2 die at an early stage of development, presumably because their cells undergo p53-dependent apoptosis. This interpretation is supported by the finding that mice lacking genes that encode *both* MDM2 and p53 (double knockouts) survive to term. Because these embryos cannot produce p53, they don't require a protein such as MDM2 that facilitates p53 destruction. This observation illustrates an important principle in cancer genetics: even if a "crucial" gene such as *RB* or *p53* is not mutated or deleted, the *function* of that gene can be affected as the result of alterations in other genes whose products are part of the same pathway as the "crucial" gene. In this case, overexpression of MDM2 can have the same effect as the absence of p53. Similarly, the presence of pRb can be negated by the simultaneous presence of a mutant E2F protein that is not subject to pRb inhibition. As long as the tumor-suppressor pathway is blocked, the tumor-suppressor gene itself need not be mutated. Some cancer biologists believe that both the p53 and pRb pathways have to be inactivated, one way or another, to allow the progression of virtually all tumor cells.

Because of its ability to trigger apoptosis, p53 may play a pivotal role in treatment of cancer by radiation and chemotherapy. It was assumed for many years that cancer cells are more susceptible than normal cells to drugs and radiation because cancer cells divide more rapidly. But some cancer cells divide more slowly than their normal counterparts, yet they are still more sensitive to drugs and radiation. An alternate theory suggests that normal cells are more resistant to drugs or radiation because, once they sustain genetic damage, they either arrest their cell cycle until the damage is repaired or they undergo apoptosis. In contrast, cancer cells that have sustained DNA damage are more likely to become apoptotic—as long as they possess a functioning *p53* gene. If cancer cells lose p53 function, they often cannot be directed into apoptosis and they become highly resistant to further treatment (Figure 16.17). This may be the primary reason why tumors that typically lack a functional *p53* gene (e.g., melanoma, colon cancer, prostate cancer, and pancreatic cancer) respond much more poorly to radiation and chemotherapy than tumors that possess a wild-type copy of the gene (e.g., testicular cancer and childhood acute lymphoblastic leukemias).

Other Tumor-Suppressor Genes Although mutations in *RB* and *p53* are associated with a wide variety of human malignancies, mutations in a number of other tumor-suppressor genes are detected in only a few types of cancer. A number of tumor-suppressor genes, for example, are commonly inactivated in cells taken from cancers of the colon (Table 16.2 and Figure 16.9). As with the *RB* gene, the tumor-suppressor genes associated with colon cancer were originally identified by studying the genomes of members of families who are at high risk of developing the disease.

Familial adenomatous polyposis coli (FAP) is an inherited disorder in which individuals develop hundreds or even thousands of premalignant polyps (adenomas) from epithelial cells that line the colon wall (Figure 16.18). If

| | Untreated | 5-Fluorouracil | Etoposide | Adriamycin |

(+/+)				
(+/–)				
(–/–)				

Figure 16.17 Experimental demonstration of the role of p53 in the survival of cells treated with chemotherapeutic agents. Cells were cultured from mice that had two functional alleles of a *p53* gene (top row), one functional allele of a *p53* gene (middle row), or were lacking a functional allele of the gene (bottom row). Cultures of each of these cells were grown either in the absence of a chemotherapeutic agent (first column) or in the presence of one of the three compounds indicated at the top of the other three columns. It is evident that the compounds had a dramatic effect on arresting growth and inducing cell death (apoptosis) in normal cells, whereas the $p53^{-/-}$ cells continued to proliferate in the presence of these compounds. (FROM SCOTT W. LOWE, H. E. RULEY, T. JACKS, AND D. E. HOUSMAN, CELL 74:959, 1993; BY PERMISSION OF CELL PRESS.)

not removed, cells within some of these polyps are very likely to progress to a fully malignant stage. The cells of patients with this condition were found to contain a deletion of a small portion of chromosome 5, which was subsequently identified as the site of a tumor-suppressor gene called *APC*. A person inheriting an *APC* deletion is in a similar position to one who inherits an *RB* deletion: if the second allele of the gene is knocked out in a given cell, the protective value of the gene function is lost. The loss of the second allele of *APC* causes the cell to lose growth control and proliferate to form a polyp rather than differentiating into a normal epithelial cell of the intestinal wall. The conversion of cells in a polyp to the more malignant state, characterized by the ability to metastasize and invade other tissues, is presumably gained by the accumulation of additional mutations, including those in *p53*. Mutated *APC* genes are found not only in inherited forms of colon cancers, but also in up to 80 percent of sporadic colon tumors, suggesting that the gene plays a major role in the development of this disease. The protein encoded by the *APC* gene binds a number of different proteins, and its mechanism of action is complex. In its best-studied role,

Table 16.2 Genes Altered in Colon Cancers

Gene	Chromosome	Tumors with Mutations	Class	Action
HNPCC	2	~15%	Onocogene	Maintains DNA replication accuracy
K-Ras	12	~50%	Oncogene	Intracellular signaling molecule
Cyclins	Various	4%	Oncogene	Regulates cell cycle
neu/HER2	17	2%	Oncogene	Growth factor receptor
myc	8	2%	Oncogene	Regulates gene activity
APC	5	>70%	Tumor suppressor	Regulates gene activity
p53	17	>70%	Tumor suppressor	Regulates gene activity

Source: J. Marx, *Science* 260:752, 1993. © Copyright 1993 American Association for the Advancement of Science.

Figure 16.18 Premalignant polyps in the epithelium of the human colon. Photograph of a portion of a colon removed from a patient with Gardner's syndrome that demonstrates the pattern of polyp formation. Similar patterns are found in the colon of persons with the inherited condition of familial adenomatous polyposis coli (FAP). (COURTESY OF RANDALL W. BURT.)

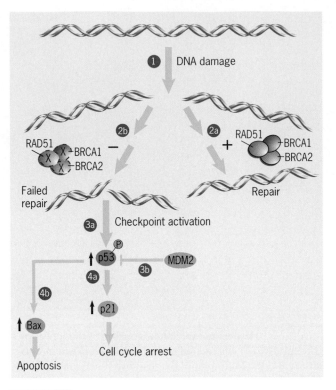

Figure 16.19 DNA damage initiates activity of a number of proteins encoded by both tumor-suppressor genes and proto-oncogenes. In this figure, DNA damage is seen to cause double-strand breaks in the DNA (step 1) that are repaired by a complex that includes BRCA1, BRCA2, and Rad51 (step 2a). Mutations in any of the three genes that encode these proteins can block the repair process (step 2b). If DNA damage is not repaired, a checkpoint is activated that leads to a rise in the level of p53 activity (step 3a). The p53 protein is normally inhibited by interaction with the protein MDM2 (step 3b). p53 is a transcription factor that activates expression of either (1) the *p21* gene (step 4a), whose product (p21) causes cell cycle arrest, or (2) the *bax* gene (step 4b), whose product (Bax) causes apoptosis. (REPRINTED WITH PERMISSION AFTER J. BRUGAROLAS AND T. JACKS, NATURE MED. 3:721, 1997. COPYRIGHT 1997, MACMILLAN MAGAZINES LTD.)

APC restrains cell growth by interfering with the transcription of genes (e.g., *myc*) that stimulate cell proliferation.

It is estimated that breast cancer strikes approximately one in eight women living in the United States, Canada, and Europe. Of these cases, approximately 5 percent can be traced to the inheritance of a gene that predisposes the individual to the development of the disease. After an intensive effort by several laboratories, two genes named *BRCA1* and *BRCA2* were identified in 1994 as being responsible for the majority of the inherited cases of breast cancer. *BRCA1* mutations also predispose a woman to the development of ovarian cancer, which has an especially high mortality rate.

BRCA1 and BRCA2 are thought to have at least two distinct functions in a cell: as a transcription factor and as a part of the cell's DNA-repair machinery. In this latter capacity, BRCA1 and BRCA2 bind to each other and to a third protein called Rad51 to form a complex that repairs DNA strand breaks that occur following exposure to certain types of radiation. Failure to repair the damage leads to the activation of p53, which causes the cell to either arrest cell cycle progress or undergo apoptosis, as illustrated in Figure 16.19.

We have seen in this chapter that apoptosis is one of the body's primary mechanisms of ridding itself of potential tumor cells. The mechanism of apoptosis was discussed in the last chapter, as were pathways that promoted cell survival rather than cell destruction. The best-studied cell-survival pathway was described on page 656 and involves

the activation of a kinase called PKB by the phosphoinositide PIP$_3$. Whether a cell lives or dies following a particular event depends to a large degree on the balance between proapoptotic and antiapoptotic signals. Mutations that affect this balance, such as those that lead to the alteration of *p53* genes can shift this balance in the favor of cell survival, which can provide a potential cancer cell with a tremendous advantage. Another protein that can affect the balance between life and death of a cell is the lipid phosphatase, PTEN, which removes the phosphate group from the 3-position of PIP$_3$, converting the molecule into PI(4,5)P$_2$, which cannot activate PKB. Cells in which one or both copies of the *PTEN* gene are inactivated tend to have an excessively high level of PIP$_3$, which leads to an overactive population of PKB molecules, and an increased likelihood that a cell will survive a stimulus that normally would lead to its destruction. Like the other tumor-suppressor genes listed in Table 16.1, mutations in *PTEN*

cause a rare hereditary disease characterized by an increased risk of cancer, and such mutations are also found in a variety of sporadic cancers. When a normal *PTEN* gene is introduced into tumor cells that lack a functional copy of the gene, the cells typically undergo apoptosis, as would be expected.

Oncogenes As described above, oncogenes encode proteins that promote the loss of growth control and the conversion of a cell to a malignant state. Oncogenes are derived from proto-oncogenes (page 680), which are genes that encode proteins having a function in the normal cell. Most known proto-oncogenes play a role in the control of cell growth, including growth stimulation by external ligands, signal transduction within the cell, or progression through the cell cycle. Approximately 100 different oncogenes have been identified, most of them included as part of the genomes of RNA tumor viruses. Even though the viral versions of each of these genes are derived from cellular versions present in the mammalian genome, only about a dozen of them have been shown to play a role in human carcinogenesis (Table 16.3). The oncogene mutated most frequently in human tumors is *ras*, which encodes a GTP-binding protein (Ras) that functions as an on-off switch for a key signaling pathway controlling cell proliferation (page 651). Oncogenic ras mutants typically encode a protein whose GTPase activity cannot be stimulated, which leaves the molecule in an active GTP-bound form, sending continuous proliferation signals along the pathway. The functions of a number of oncogenes are summarized in Figure 16.20 and discussed below.

Oncogenes That Encode Growth Factors or Their Receptors
The first connection between oncogenes and growth factors was made in 1983, when it was discovered that the cancer-causing simian sarcoma virus (SIS) contained an oncogene (*sis*) derived from the cellular gene for platelet-derived growth factor (PDGF), a protein present in human blood. Cultured cells that are infected with this virus become cancerous because they secrete large amounts of PDGF into the medium, which causes the cells to proliferate in an uncontrolled fashion. Overexpression of PDGF has been implicated in the development of brain tumors (gliomas).

Another oncogenic virus, avian erythroblastosis virus, has been found to carry an oncogene (*erbB*) that directs the formation of an EGF receptor that is missing part of the extracellular domain of the protein that binds the growth

Table 16.3	Proto-Oncogenes and Human Tumors: Some Consistent Incriminations	
Proto-oncogene	**Neoplasm(s)**	**Lesion**
abl	Chronic myelogenous leukemia	Translocation
bcl–2	B-cell lymphoma	Translocation
CYCD1	Carcinoma of breast	Translocation
Cdk4	Sarcomas	Amplification
erbB	Squamous cell carcinoma; astrocytoma	Amplification
neu/HER2	Adenocarcinoma of breast, ovary, and stomach	Amplification
gip	Carcinoma of ovary and adrenal gland	Point mutations
gsp	Adenoma of pituitary gland; carcinoma of thyroid	Point mutations
myc	Burkitt's lymphoma	Translocation
	Carcinoma of lung, breast, and cervix	Amplification
L-*myc*	Carcinoma of lung	Amplification
N-*myc*	Neuroblastoma, small cell carcinoma of lung	Amplification
H-*ras*	Carcinoma of colon, lung, and pancreas; melanoma	Point mutations
K-*ras*	Acute myeloid and lymphoblastic leukemia; carcinoma of thyroid; melanoma	Point mutations
N-*ras*	Carcinoma of genitourinary tract and thyroid; melanoma	Point mutations
ret	Carcinoma of thyroid	Rearrangement
trk	Carcinoma of thyroid	Rearrangement

Source: After J. M. Bishop, *Cell* 64:236, 1991, by permission of Cell Press.

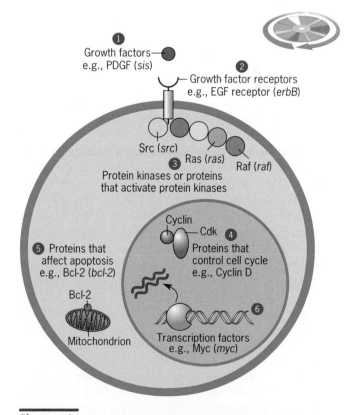

Figure 16.20 A schematic diagram summarizing the types of proteins encoded by proto-oncogenes. These include growth factors (1), receptors for growth factors (2), protein kinases and the proteins that activate them (3), proteins that regulate the cell cycle (4), proteins that either activate or inhibit apoptosis (5), and DNA-binding proteins (6).

factor. One might expect that the altered receptor would be unable to signal a cell to divide, but just the reverse is true. This altered version of the receptor stimulates the cell constitutively, that is, regardless of whether or not the growth factor is present in the medium. This is the reason why cultured cells that carry the altered gene proliferate in an uncontrolled manner. A number of spontaneous human cancers have been found to contain cells with genetic alterations that affect growth factor receptors, including erbB. Most commonly, the malignant cells contain a much larger number of the receptors in their plasma membranes than do normal cells. The presence of excess receptors makes the cells sensitive to much lower concentrations of the growth factor, and thus, they are stimulated to divide under conditions that would not affect normal cells. Proto-oncogenes that encode growth factor receptors can also be activated by mutations or translocations that cause the receptor monomers to undergo dimerization in the absence of exogenous ligand, which leads to the activation of their protein kinase activity (see Figure 15.22).

Oncogenes That Encode Cytoplasmic Protein Kinases A number of cytoplasmic protein kinases, including both serine/threonine kinases and tyrosine kinases, are included among the list of oncogenes. Raf, for example, is a serine-threonine protein kinase that heads the MAP kinase cascade, the primary growth-controlling signaling pathway in cells (page 653). It is evident that Raf is well positioned to wreak havoc within a cell should its enzymatic activity become altered as the result of a mutation in the *raf* gene. As with the growth factor receptors and Ras, mutations that turn Raf into an enzyme that remains in the "on" position are most likely to convert the proto-oncogene into an oncogene and contribute to the cell's loss of growth control.

The first oncogene to be discovered, *src*, is also a protein kinase, but one that phosphorylates tyrosine residues on protein substrates rather than serine and threonine residues. Transformation of a cell by a *src*-containing tumor virus is accompanied by the phosphorylation of a wide variety of proteins. Included among the apparent substrates of Src are proteins involved in signal transduction, control of the cytoskeleton, and cell adhesion. For an unknown reason, *src* mutations appear only rarely among the repertoire of genetic changes in human tumor cells.

Oncogenes That Encode Nuclear Transcription Factors A number of oncogenes encode proteins that act as transcription factors. The progression of cells through the cell cycle requires the timely activation (and repression) of a large variety of genes whose products contribute in various ways to cell growth and division. It is not surprising, therefore, that alterations in the proteins that control the expression of these genes could seriously disturb a cell's normal growth patterns. Probably the best-studied oncogene whose product acts as a transcription factor is *myc*.

It was noted on page 582 that cells that are not actively growing and dividing tend to withdraw from the cell cycle and enter a stage referred to as G_0, from which they can be retrieved. The Myc protein is normally one of the first proteins to appear when a cell in this quiescent stage has been stimulated by growth factors to reenter the cell cycle and divide. When *myc* expression is selectively blocked using antisense oligonucleotides, the progression of the cell through G_1 is blocked.

The *myc* gene is one of the oncogenes found to be altered in human cancers, often being amplified within the genome or rearranged as the result of a chromosome inversion or translocation. These chromosomal changes are thought to remove the *myc* gene from its normal regulatory influences and increase its level of expression in the cell, producing an excess of the Myc protein. One of the most common types of cancer among populations in Africa, called Burkitt's lymphoma, results from the translocation of a *myc* gene to a position adjacent to an antibody gene. The disease occurs primarily in persons who have also been infected with the herpeslike Epstein-Barr virus. For some unknown reason, this same virus causes only minor infections (e.g., mononucleosis) in people living in Western countries and is not associated with tumorigenesis.

Oncogenes That Encode Products That Affect Apoptosis Apoptosis is one of the body's key mechanisms to rid itself of tumor cells at an early stage in their progression toward malignancy. Consequently, any alteration that diminishes a cell's ability to self-destruct would be expected to increase the likelihood of that cell giving rise to a tumor. The oncogene most closely linked to apoptosis is *bcl-2*, which encodes a membrane-bound protein that inhibits apoptosis (page 664).

The role of *bcl-2* in apoptosis is most clearly revealed in the phenotypes of knockout mice that are lacking a *bcl-2* gene. Once formed, the lymphoid tissues of these mice undergo dramatic regression as the result of widespread apoptosis. Like *myc*, the product of the *bcl-2* gene becomes oncogenic when it is expressed at higher-than-normal levels, as can occur when the gene is translocated to an abnormal site on the chromosome. Certain human lymphoid cancers (called follicular B-cell lymphomas) are correlated with the translocation of the *bcl-2* gene next to a gene that codes for the heavy chain of antibody molecules. It is suggested that overexpression of the *bcl-2* gene leads to the suppression of apoptosis in lymphoid tissues, allowing abnormal cells to proliferate to form lymphoid tumors. The *bcl-2* gene may also play a role in reducing the effectiveness of chemotherapy by keeping tumor cells that have been damaged by the drugs alive and proliferating.

Genes That Encode Proteins Involved in DNA Mismatch Repair If one considers cancer as a disease that results from alterations in the DNA of somatic cells, then it follows that any activity that increases the frequency of genetic mutations is likely to increase the risk of developing cancer. It was described in Chapter 13 how nucleotides that are incorporated incorrectly during replication are selectively removed from the newly synthesized strand by a process

called mismatch repair (page 574). Mismatch repair requires the cooperative efforts of a number of proteins, including proteins that recognize the lesion, remove a portion of the strand containing the lesion, and replace the missing segment with complementary nucleotides. If any of these proteins are defective, the affected cell can be expected to display an abnormally high mutation rate, which in turn increases its risk of becoming malignant.

The first hard evidence that a defect in mismatch repair might be a factor in the development of cancer came in 1993 in studies on the cells of patients with the most common hereditary form of colon cancer, called hereditary nonpolyposis colon cancer (HNPCC) to distinguish it from the polyp-forming type (FAP) described earlier. A defective gene responsible for HNPCC is carried by approximately 0.5 percent of the population and accounts for about 5 percent of all colon cancer cases. It was noted on page 415 that the genome contains large numbers of very short, repetitive DNA sequences called microsatellites. Analysis of the DNA of persons with HNPCC revealed that microsatellites of the tumor cells are often of different length than the corresponding sequence in the DNA of normal cells. One would expect such variations in the DNA from different individuals, but not from the DNA of different cells of the same person.

Changes in the length of a microsatellite sequence arise as an error during replication when the template strand being copied and the new strand being synthesized slip in their position relative to one another. This type of slippage occurs most frequently in regions containing short, highly repeated sequences. Depending on the direction of slippage, the newly synthesized strand contains either additional nucleotides or missing nucleotides compared to the template strand. The difference in the lengths of the nucleotide sequences in the two strands generates a small loop of unpaired DNA in the daughter duplex, which is normally recognized by the mismatch repair enzymes and repaired. The discovery of microsatellite sequence variation in these hereditary cancers—and in sporadic colon cancers as well—suggested that a deficiency in the mismatch repair system is likely at fault. Support for this proposal has come from studies of persons with HNPCC. Whereas extracts from normal cells carry out mismatch repair in vitro, extracts from HNPCC tumor cells display repair deficiencies. Analysis of the DNA from a large number of people with HNPCC has revealed deletions or debilitating mutations in any one of a number of genes that code for proteins that form the DNA mismatch repair pathway. Cells with mismatch repair deficiencies accumulate secondary mutations throughout the genome. Genes containing long, simple, repetitive sequences in their coding region are particularly susceptible to this type of mutation. When these mutations occur in tumor-suppressor genes or oncogenes, the cells are at greatly increased risk of becoming malignant. In fact, one of the genes most closely linked to colon cancer, the *APC* gene, contains a string of adenosines that appear to become mutated in certain familial colon cancers.

Concluding Remarks on Cancer Genetics Now that we have covered some of the recent findings on the genetic basis of tumor formation, you can appreciate why the war against cancer may ultimately require that the genetic composition of the cells of a tumor be permanently altered. The finding that many different types of cancers share the same genetic defects, such as an alteration in *p53*, *RB*, and/or *ras*, raises the hope that many different cancers can be treated by a common approach. If, for example, a drug could be developed that would mimic the effects of p53 or inhibit Ras, then a large number of different cancers might be treatable with the same agent.

The identification of the specific genes involved in cancer has also led to the development of screening procedures that can identify persons who are at elevated risk of developing particular types of cancers. The earlier a cancer is discovered, the greater the chance of survival. Consequently, these screening procedures could have a significant impact in lowering the mortality from cancer.

REVIEW ?

1. Contrast benign tumor and malignant tumor; tumor-suppressor gene and oncogenes; dominantly acting and recessively acting mutations; proto-oncogenes and oncogenes.
2. What is meant by the statement that cancer arises as the result of a genetic progression?
3. Why is p53 described as the "guardian of the genome"?
4. Give two mechanisms by which p53 acts to prevent a cell from becoming malignant.
5. How can DNA microarrays be used to determine the type of cancer a patient suffers from? How might they be used to optimize cancer treatment?
6. What types of proteins are encoded by proto-oncogenes, and how do mutations in each type of proto-oncogene cause a cell to become malignant?

16.4 | NEW STRATEGIES FOR COMBATING CANCER

It is painfully evident that conventional approaches to combating cancer, namely surgery, chemotherapy, and radiation, often cannot rid a patient of all cancer cells. At the present time, a large number of clinical trials are being conducted to test a variety of new anti-cancer strategies. Before a particular drug or procedure is tested on humans, it has to be shown effective in preventing tumor growth in laboratory animals. Most animal studies employ a strain of mutant mice (shown in Figure 1.20c) whose compromised immune system makes it impossible for them to reject foreign cells (called *xenografts*). In these studies, a piece of human tumor tissue is grafted beneath the skin of the mouse, and subsequent growth of the tumor tissue is

monitored after various types of treatments. It has become increasingly clear in recent years that this widely used model system is not as predictive of a treatment's success as originally assumed. Numerous procedures that rid laboratory animals of their human tumors have proven much less effective when subsequently tested in human patients. In fact, most recently published reports of clinical trials of new innovative therapies that have worked well in laboratory animals have proven disappointing in humans. Yet despite this harsh reality, there are so many new drugs and therapeutic approaches being developed and tested, there is reason for considerable optimism that new treatments will become available in the next few years. Furthermore, as discussed below, there have been a few notable successes in small clinical trials, which provide further reason to believe that our considerable insight into the genetics of cancer will form the basis for new therapeutic strategies.

The anticancer strategies to be discussed in the following sections can be divided into four groups: (1) those that depend on antibodies or immune cells to attack tumor cells, (2) those that introduce a gene that either kills tumor cells or causes them to regain normal properties, (3) those that inhibit the activity of cancer-promoting proteins, and (4) those that prevent the growth of blood vessels that nourish the tumor. Another innovative strategy based on the use of antisense oligonucleotides was discussed on page 473.

Immunotherapy

We have all heard or read about persons with metastatic cancer who were told they had only months to live, yet they defy the prognosis and remain alive and cancer-free years later. The best-studied cases of such "spontaneous remissions" were recorded in the late 1800s by a New York physician named William Coley. Coley's interest in the subject began in 1891 when he came across the hospital records of a patient with an inoperable tumor of the neck, who had gone into remission after contracting a streptococcal infection beneath the skin. Coley located the patient and found him to be free of the cancer that had once threatened his life. Coley spent his remaining years trying to develop a bacterial extract that, when injected under the skin, would stimulate a patient's immune system to attack and destroy the malignancy. The work was not without its successes, particularly against certain uncommon soft-tissue sarcomas. Although the use of Coley's toxin, as it was later called, was never widely embraced, Coley's results confirmed anecdotal observations that the body has the capability to destroy a tumor, even after it has become well established.

Immunotherapy is an approach that tries to get the immune system more involved in the fight against malignant cells. The immune system has evolved to recognize and destroy foreign materials, but cancers are derived from a person's own cells. Although many tumors do contain proteins that may not be normally expressed by their normal counterparts, or mutated proteins that are different

from those present in normal cells, they are still basically host proteins present in host cells. As a result, the immune system typically fails to recognize these proteins as "inappropriate." Many different strategies have been formulated to overcome this hurdle and stimulate the immune system to mount a specific response against tumor cells. Most have not been very successful to date, and will not be discussed, but one experimental protocol has shown particular promise.

As discussed in the following chapter, cells of the immune system normally respond to fragments of proteins that are presented on the surface of *antigen-presenting cells (APCs)*. Recent attention has focused on one particular type of APC called a *dendritic cell*. Dendritic cells (DCs) are particularly effective in stimulating the immune system to respond to antigen presentation by producing both antibodies and reactive T lymphocytes. Although they are normally present as a very small fraction of cells in the blood, immature DCs can be separated from other blood cells and placed into culture where they will proliferate. Once in culture, the DCs can be exposed to tumor proteins in several ways. In some experiments, the DCs are simply incubated with either purified proteins or crude tumor lysates. In either case, proteins are taken up from the medium and processed by the cells. Alternatively, DCs can be induced to fuse directly with tumor cells producing hybrid cells (as in the example on page 143). In a third approach, DCs can be genetically modified to express any particular protein that is also preferentially expressed by the tumor cells. This is most easily accomplished by infecting the cells with a virus containing a gene for the protein in question. Under these circumstances, DCs produce the protein in large quantities within their cytoplasm and present its fragments on their surface. Regardless of the technique used, once they have been exposed to tumor proteins, the DCs are injected into the patient to initiate an immune response. Studies suggest that DCs are such effective antigen-presenters that they can stimulate the immune system to mount a response against the tumor cells, despite the fact that the proteins inducing the response are actually host proteins. One of the consequences of this type of treatment is the risk that the injected cells will induce an immune response against normal tissues within the body (an autoimmune response), but so far that does not seem to be a major problem. Although early clinical trials with DCs have shown considerable promise, the value of this approach remains to be determined.

Gene Therapy

Gene therapy is a treatment in which a patient's genotype is modified by the addition, deletion, or alteration of a specific gene. When gene therapy was first considered, inherited diseases such as cystic fibrosis and muscular dystrophy were presumed to be the likely targets because these diseases could be corrected by a normal gene introduced into the cells of affected tissues and organs (page

164). Although several inherited diseases are the subjects of promising clinical trials, the focus of gene therapy has shifted to more common diseases with a more complex genetic basis, most notably cancer. Of 350 gene therapy clinical trials being conducted in the United States in March 2000, 67 percent focused on cancer treatment.

We have seen in this chapter that tumor cells lack functional tumor-suppressor genes whose products restrain the growth of normal cells. Just as a cystic fibrosis cell should be returned to normal by acquisition of a wild-type CF gene, a tumor cell should be converted to the nonmalignant state by acquisition of a wild-type tumor-suppressor gene, such as p53. The wild-type p53 can be delivered to tumor cells through infection with a virus that has been engineered to carry the gene. p53-bearing viruses have been shown to restore normal growth to malignant cells both in culture and in laboratory animals. Currently, p53-bearing viruses are being tested in clinical trials, but no major successes have been reported. The major obstacle to this type of gene therapy lies with the inability of current vectors to deliver replacement genes to the tumor cells at a high enough efficiency to bring about a significant clinical response.

A genetically modified adenovirus called ONYX-015 is being used in another gene-therapy protocol. Like that of other viruses, the adenovirus genome encodes proteins that interfere with a host's defense mechanisms. E1B is a protein encoded by adenovirus that binds to p53, preventing p53 from either stopping the cell cycle or triggering apoptosis. Consider what could happen if an adenovirus that lacked the E1B gene were to infect a cell. As long as that cell produced a normal p53 protein, it would be expected to either arrest its progress through the cell cycle or undergo apoptosis, killing both itself and the virus it harbored. Now consider what might happen to a cell that lacked p53, as do many types of cancer cells. The genetically altered adenovirus would be expected to replicate inside the p53-deficient cell, leading to the lysis of the host tumor cell and the release of infective virus particles. This is exactly what happens. Unlike viruses used simply as gene vectors, these lytic virus particles can infect additional tumor cells, leading to their lysis and, ultimately, to the destruction of the tumor. Preclinical tests showed the engineered adenovirus to be 100 times more toxic to cultured tumor cells lacking p53 than to normal cells and to destroy human tumor tissue grafted to mice. A major clinical trial of ONYX-015 on patients with squamous cell carcinoma of the head and neck has recently shown considerable success. In this trial, the use of ONYX-015 was combined with standard chemotherapy, which by itself is not very effective in treating these types of tumors. The combined treatment led to a significant response (at least 50 percent reduction in tumor mass) in most patients, including a complete response in 8 patients. Several other tumor-specific lytic (oncolytic) viruses are being tested in other clinical trials that include patients with inoperable prostate cancer and brain tumors.

Inhibiting the Activity of Cancer-Promoting Proteins

Cancer cells behave as they do because they contain proteins that either are present at abnormal concentration or display abnormal activity. A number of these proteins are illustrated in Figure 16.20. If the activity of these proteins can be selectively blocked, it should be possible to stop the uncontrolled growth and invasive properties of malignant cells. With this goal in mind, researchers have synthesized a virtual arsenal of small-molecular-weight compounds that inhibit the activity of cancer-promoting proteins. Some of these drugs were custom-designed to inhibit a particular protein (page 70), whereas others were identified by randomly screening large numbers of compounds that have been synthesized by pharmaceutical companies. Although a number of such compounds have shown some promise in halting the growth of various types of tumors, one compound has had unparalleled success in clinical trials on patients with chronic myelogenous leukemia (CML). It was noted on page 513 that certain types of cancers are caused by specific chromosome translocations. CML is caused by a translocation that brings a proto-oncogene (*abl*) in contact with another gene (*bcr*) to form a chimeric gene (*bcr-abl*). Blood-forming cells that carry this translocation produce a protein that expresses a high level of Abl tyrosine kinase activity, which causes the cells to proliferate uncontrollably, producing the disease. A compound called STI-571 has been identified that selectively inhibits Abl kinase by binding to the inactive form of the protein and preventing its phosphorylation by another kinase, which is required for Abl activation. Initial clinical trials on STI-571 demonstrated that nearly all CML patients who received the drug at sufficiently high dose went into remission. Moreover, the drug had only minor adverse side effects. This study confirms the belief that oncogenes represent valid targets for therapeutic drugs. It also demonstrates that "old-fashioned" enzyme inhibitors may play an important role in the treatment of cancer.

Proteins can also be inhibited when they interact with antibodies. HER2 is a receptor that binds a growth factor that stimulates the proliferation of breast cancer cells. Approximately 30 percent of breast cancers are composed of cells that overexpress the HER2 gene, which causes these cells to be especially sensitive to growth factor stimulation. Clinical trials demonstrated that an antibody (called Herceptin) directed against the HER2 protein is effective in combination with Taxol in stopping the growth of breast cancers in a significant fraction of patients. Herceptin was approved by the FDA in 1998 and is currently in use.

Inhibiting the Formation of New Blood Vessels (Angiogenesis)

As a tumor grows in size, it stimulates the formation of new blood vessels, a process termed *angiogenesis* (Figure 16.21). Blood vessels are required to deliver nutrients and

Figure 16.21 Angiogenesis and tumor growth. Steps in the vascularization of a primary tumor. In step 1, the tumor proliferates to form a small mass of cells. As long as it is avascular (without blood vessels), the tumor remains very small (1–2 mm). In step 2, the tumor mass has produced angiogenic factors that stim-ulate the endothelial cells of nearby vessels to grow out toward the tumor cells. In step 3, the tumor has become vascularized and is now capable of unlimited growth. (AFTER B. R. ZETTER, REPRODUCED WITH PERMISSION FROM THE ANNUAL REVIEW OF MEDICINE, VOLUME 49, © 1998, BY ANNUAL REVIEWS WWW.ANNUALREVIEWS.ORG)

Figure 16.22 The response of a solid tumor grown in mice to treatment with angiogenesis inhibitors. (*a*) When mice bearing tumors are treated with endostatin, the tumor shrinks, and it then regrows if treatment is suspended. The tumor continues to respond in successive rounds of endostatin treatment. Not only does the tumor fail to develop resistance to endostatin, it eventually reaches a point where it remains very small and dormant, even after treatment is suspended. (*b*) Treatment of a tumor with conventional chemotherapeutic agents typically results in tumor regression followed by tumor regrowth from resistant cells. (*c*) Results of endostatin treatment. The photo on the left shows a mouse bearing a tumor that has regrown after the first round of endostatin treatment. The photo on the right shows the mouse 12 days after the second round of endostatin treatment. The tumor has shrunk to a size where it is no longer evident. (A, B: REPRINTED WITH PERMISSION FROM R. S. KERBEL, NA-TURE 390:335, 1997; C: REPRINTED WITH PERMISSION FROM THOMAS

BOEHM, JUDAH FOLKMAN, TIMOTHY BROWDER, AND MICHAEL S. O'REILLY, NATURE 390:405, 1997. ALL © COPYRIGHT 1997, MACMILLAN MAGAZINES LTD.)

oxygen to the rapidly growing tumor cells and to remove their waste products. Blood vessels also provide the conduits for cancer cells to spread to other sites in the body. In 1971, Judah Folkman of Harvard University suggested that solid tumors might be destroyed by inhibiting their ability to form new blood vessels. After a quarter of a century of relative obscurity, this idea has blossomed into a promising anticancer strategy.

Cancer cells promote angiogenesis by secreting growth factors that act on the endothelial cells of surrounding blood vessels, stimulating them to proliferate and develop into new vessels. Just as there are stimulants of angiogenesis, there are also inhibitors. The growing list of angiogenic inhibitors includes antibodies and synthetic compounds directed against endothelial growth factors and their receptors (VEGF and VEGFRs), the drug thalidomide (infamous for causing birth defects in the 1950s), and a number of naturally occurring proteins, including endostatin. Endostatin is a small protein derived by proteolytic cleavage from collagen molecules present in the basement membrane lining the endothelium of blood vessels (page 245. The results of one major preclinical study with endostatin is shown in Figure 16.21. In this study, Folkman and his colleagues administered endostatin in repeated cycles to mice bearing one of three different types of tumors. When mice with well-developed tumors were treated with endostatin, the tumors shrank dramatically (Figure 16.22*a,c*). Not only did endostatin halt the development of new vessels, but it also caused the

destruction of blood vessels that had been formed earlier by the tumor. Blood vessels present in normal tissue were not affected, nor were any serious side effects evident. When endostatin treatment was suspended, the blood supply to the tumor redeveloped and the tumor regrew. If endostatin treatment was resumed, the tumor once again shrank dramatically, but it then regrew following withdrawal of the inhibitor. This cycle of tumor shrinkage and expansion was repeated several times, demonstrating that, unlike other cancer treatments (as shown in Figure 16.22*b*), the tumor did not become resistant to renewed drug application. Tumor cells become resistant to the usual chemotherapeutic agents because the cells are genetically unstable and can evolve into resistant forms. Inhibitors of angiogenesis, however, target normal endothelial cells, not abnormal tumor cells, so it is not surprising that the animals fail to develop resistance to these natural agents. After a number of cycles of endostatin treatment, a striking change in the behavior of the tumors was observed. The tumors stopped growing and remained as dormant microscopic cell clusters, even in the absence of endostatin treatment. Antiangiogenic agents, including endostatin, have entered clinical trials, but at the present time it isn't evident whether or not they will prove to be useful therapeutic weapons. If endostatin and other antiangiogenic agents ultimately fail to demonstrate antitumor activity, then Figure 16.22 provides an excellent example of the ease with which cancer can be cured in mice and the difficulty in transferring these therapies to humans.

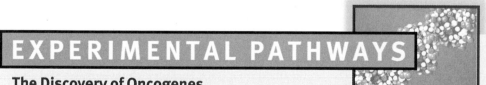

EXPERIMENTAL PATHWAYS

The Discovery of Oncogenes

In 1911, Peyton Rous of the Rockefeller Institute for Medical Research published a paper that was less than one page in length (it shared the page with a note on the treatment of syphilis) and had virtually no impact on the scientific community. Yet this paper reported one of the most farsighted observations in the field of cell and molecular biology.[1] Rous had been working with a chicken sarcoma that could be propagated from one hen to another by inoculating a host of the same strain with pieces of tumor tissue. In this paper, Rous described a series of experiments that strongly suggested that the tumor could be transmitted from one animal to another by a "filterable virus," which is a term that had been coined a decade or so earlier to describe pathogenic agents that were small enough to pass through filters that were impermeable to bacteria.

In his experiments, Rous removed the tumors from the breasts of hens, ground the cells in a mortar with sterile sand, centrifuged the particulate material into a pellet, removed the supernatant, and forced the supernatant fluid through filters of varying porosity including those small enough to prevent the passage of bacteria. He then injected the filtrate into the breast muscle of a recipient hen and found that a significant percentage of the injected animals developed the tumor.

The virus discovered by Rous in 1911 is an RNA-containing virus. By the end of the 1960s, similar viruses were found to be associated with mammary tumors and leukemias in rodents and cats. Certain strains of mice had been bred that developed specific tumors with very high frequency. RNA-containing viral particles could be seen within the tumor cells and also budding from the cell surface, as shown in the mi-

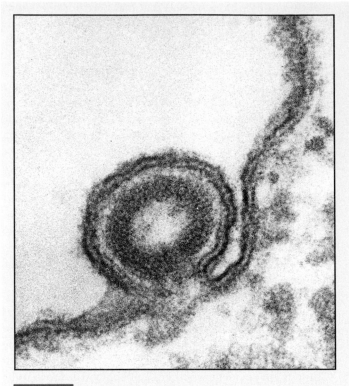

Figure 1 Electron micrograph of a Friend mouse leukemia virus budding from the surface of a cultured leukemic cell. (COURTESY OF E. DE HARVEN.)

Table 1		Characterization of the Polymerase Product	
Expt.	Treatment	Acid-Insoluble Radioactivity	Percentage Undigested Product
1	Untreated	1,425	(100)
	20 μg deoxyribo- nuclease	125	9
	20 μg micrococcal nuclease	69	5
	20 μg ribonuclease	1,361	96
2	Untreated	1,644	(100)
	NaOH hydrolyzed	1,684	100

Source: D. Baltimore, reprinted with permission from *Nature* 226:1210, 1970. © Copyright 1970, Macmillan Magazines Ltd.

crograph of Figure 1. It was apparent that the gene(s) causing tumors in these inbred strains are transmitted vertically, that is, through the fertilized egg from mother to offspring, so that the adults of each generation invariably develop the tumor. These studies provided evidence that the viral genome can be inherited through the gametes and subsequently transmitted from cell to cell by means of mitosis without having any obvious effect on the behavior of the cells. The presence of inherited viral genomes is not a peculiarity of inbred laboratory strains, because it was shown that wild (feral) mice treated with chemical carcinogens develop tumors that often contain the same antigens characteristic of RNA tumor viruses and that exhibit virus particles under the electron microscope.

One of the major questions concerning the vertical transmission of RNA tumor viruses was whether the viral genome is passed from parents to progeny as free RNA molecules or is somehow integrated into the DNA of the host cell. Evidence indicated that infection and transformation by these viruses required the synthesis of DNA. Howard Temin of the University of Wisconsin suggested that the replication of RNA tumor viruses occurs by means of a DNA intermediate—a provirus—which then serves as a template for the synthesis of viral RNA. But this model requires a unique enzyme—an RNA-dependent DNA polymerase—which had never been found in any type of cell. Then in 1970, an enzyme having this activity was discovered independently by David Baltimore of the Massachusetts Institute of Technology and by Temin and Satoshi Mizutani.[2,3]

Baltimore examined the virions (the mature viral particles) from two RNA tumor viruses, Rauscher's mouse

leukemia virus (R-MLV) and Rous sarcoma virus (RSV). A preparation of purified virus was incubated under conditions that would promote the activity of a DNA polymerase, including Mg^{2+} (or Mn^{2+}), NaCl, dithiothreitol (which prevents the —SH groups of the enzyme from becoming oxidized), and all four deoxyribonucleoside triphosphates, one of which (TTP) was labeled with 3H. Under these conditions, the preparation incorporated the labeled DNA precursor into an acid-insoluble product that exhibited the properties of DNA (Table 1). As is characteristic of DNA, the reaction product was rendered acid soluble (indicating that it had been converted to low-molecular-weight products) by treatment with pancreatic deoxyribonuclease or micrococcal nuclease, but it was unaffected by pancreatic ribonuclease or by alkaline hydrolysis (to which RNA is sensitive; Table 1). The DNA-polymerizing enzyme was found to cosediment with the mature virus particles, suggesting that it was part of the virion itself and not an enzyme donated by the host cell. Although the product was insensitive to treatment with pancreatic ribonuclease, the template was very sensitive to this enzyme (Figure 2), particularly if the virions were pretreated with the ribonuclease prior to addition of the other components of the reaction mixture (Figure 2, curve 4). These results strengthened the suggestion that the viral RNA was providing the template for synthesis of a DNA copy, which presumably served as a template for the synthesis of viral mRNAs required for infection and transformation. Not only did these experiments suggest that cellular transformation by RNA tumor viruses proceeded through a DNA intermediate, they also overturned the long-standing concept originally proposed by Francis Crick and known as the Central Dogma, which stated that information in a cell always flowed from DNA to RNA to protein. The RNA-dependent DNA polymerase became known as reverse transcriptase.

During the 1970s, attention turned to the identification of the genes carried by tumor viruses that were responsible for transformation and the mechanism of action of the gene products. Evidence from genetic analyses indicated that mutant strains of viruses could be isolated that retained the ability to grow in host cells, but were unable to transform the cell into one exhibiting malignant properties.[4] Thus, the capacity to transform a cell resided in a restricted portion of the viral genome.

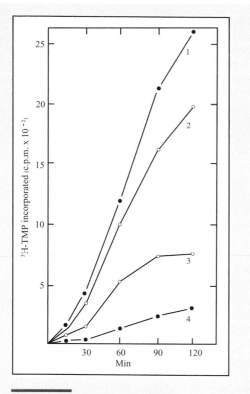

Figure 2 Incorporation of radioactivity from [³H]TTP into an acid-insoluble precipitate by the Rauscher murine leukemia virus DNA polymerase in the presence and absence of ribonuclease. (Note: the labeled TTP precursor is converted to TMP as it is incorporated into DNA.) Curve 1, no added ribonuclease; curve 2, preincubated without added ribonuclease for 20 minutes before addition of [³H]TTP; curve 3, ribonuclease added to the reaction mixture; curve 4, preincubated with ribonuclease before addition of [³H]TTP. (REPRINTED WITH PERMISSION FROM D. BALTIMORE, NATURE 226:1210, 1970. © COPYRIGHT 1970, MACMILLAN MAGAZINES LTD.)

These findings set the stage for a series of papers by Harold Varmus, J. Michael Bishop, Dominique Stehelin, and their coworkers at the University of California, San Francisco. These researchers began by isolating mutant strains of the avian sarcoma virus (ASV) carrying deletions of 10 to 20 percent of the genome that render the virus unable to induce sarcomas in chickens or to transform fibroblasts in culture. The gene responsible for transformation, which is missing in these mutants, was referred to as *src* (for sarcoma). To isolate the DNA corresponding to the deleted regions of these mutants, which presumably carry the genes required for transformation, the following experimental strategy was adopted.[5] RNA from the genomes of complete (oncogenic) virions was used as a template for the formation of a radioactively labeled, single-stranded, complementary DNA (cDNA) using reverse transcriptase. The labeled cDNA (which is present as fragments) was then hybridized to RNA obtained from one of the deletion mutants. Those DNA fragments that failed to hybridize to the RNA represent portions of the genome that had been deleted from the transformation-defective mutant and thus were presumed to contain the

gene required by the virus to cause transformation. DNA fragments that did not hybridize to the RNA were separated from those that were part of DNA-RNA hybrids by column chromatography. Using this basic strategy, a DNA sequence referred to as cDNA$_{sarc}$ was isolated, which corresponded to approximately 16 percent of the viral genome (1600 nucleotides out of a total genomic length of 10,000 nucleotides).

Once isolated, cDNA$_{sarc}$ proved to be a very useful probe. It was first shown that this labeled cDNA hybridizes to DNA extracted from cells of a variety of avian species (chicken, turkey, quail, duck, and emu) indicating that the cellular genomes of these birds contain a DNA sequence that is closely related to *src*.[6]

These findings provided the first strong evidence that a gene carried by a tumor virus that causes cell transformation is actually present in the DNA of normal (uninfected) cells and thus is presumably a part of the cells' normal genome. These results indicated that the transforming genes of the viral genome (the oncogenes) are not true viral genes, but rather are cellular genes that were picked up by RNA tumor viruses during a previous infection. Possession of this cell-derived gene apparently endows the virus with the power to transform the very same cells in which this gene is normally found. The fact that the *src* sequence is present in all of the avian species tested suggests that the sequence has been conserved during avian evolution and, thus, is presumed to govern a basic activity of normal cells. In a subsequent study, it was found that cDNA$_{sarc}$ binds to DNA from all vertebrate classes, including mammals, but not to the DNA from sea urchins, fruit flies, or bacteria. Based on these results it was possible to conclude that the *src* gene is not only present in the RNA of the ASV genome and the genome of the chicken cells it can infect, but a homologous gene is also present in the DNA of distantly related vertebrates, suggesting that it plays some critical function in the cells of all vertebrates.[7]

These findings raised numerous questions; foremost among these were (1) what is the function of the *src* gene product, and (2) how does the presence of the viral *src* gene (referred to as v-*src*) alter the behavior of a normal cell that already possesses a copy of the cellular gene (referred to as c-*src*)?

The product of the *src* gene was initially identified by Ray Erikson and coworkers at the University of Colorado by two independent procedures: (1) precipitation of the protein from extracts of transformed cells by antibodies prepared from RSV-infected animals, and (2) synthesis of the protein in a cell-free protein-synthesizing system using the isolated viral gene as a template. Using these procedures, the *src* gene product was found to be a protein of 60,000 daltons, which they named pp60src.[8] When pp60src was incubated with [³²P]ATP, radioactive phosphate groups were transferred to the heavy chains of the associated antibody (IgG) molecules used in the immunoprecipitation. This finding suggested that the *src* gene codes for an enzyme that possesses protein kinase activity.[9] When cells infected with ASV were fixed, sectioned, and incubated with ferritin-labeled antibodies against pp60src, the antibodies were found to be localized on the inner surface of the plasma membrane, suggesting a concentration of the *src* gene product in this part of the cell (Figure 3).[10]

These were the first studies to elucidate the function of an oncogene. A protein kinase is the type of gene product that might be expected to have potential transforming activity, be-

0.1 μm

Figure 3 Electron micrograph of a section through a pair of adjacent fibroblasts that had been treated with ferritin-labeled antibodies against the pp60^{src} protein. The protein is localized (as revealed by the dense ferritin granules) at the plasma membrane of the cell and is particularly concentrated at the sites of gap junctions. (FROM MARK C. WILLINGHAM, GILBERT JAY, AND IRA PASTAN, CELL 18:128, 1979, BY PERMISSION OF CELL PRESS.)

cause it can regulate the activities of numerous other proteins, each of which might serve a critical function in one or another activity related to cell growth. Further analysis of the role of the *src* gene product turned up an unexpected finding. Unlike all the other protein kinases whose function had been studied, pp60*src* transferred phosphate groups to tyrosine residues on the substrate protein rather than to serine or threonine residues.[11] The existence of phosphorylated tyrosine residues had escaped previous detection because phosphory-

lated serine and threonine residues are approximately 3000 times more abundant in cells than phosphotyrosine, and because phosphothreonine and phosphotyrosine residues are difficult to separate from one another by traditional electrophoretic procedures. Not only did the product of the viral *src* gene (v-*src*) code for a tyrosine protein kinase, so too did c-*src*, the cellular version of the gene. However, the number of phosphorylated tyrosine residues in proteins of RSV-transformed cells was approximately eight times higher than that of control cells. This finding suggested that the viral version of the gene may induce transformation because it functions at a higher level of activity than the cellular version.

The results from the study of RSV provided preliminary evidence that an increased activity of an oncogene product could be a key to converting a normal cell into a malignant cell. Evidence soon became available that the malignant phenotype could also be induced by an oncogene that contained an altered nucleotide sequence. A key initial study was conducted by Robert Weinberg and his colleagues at the Massachusetts Institute of Technology using the technique of DNA transfection.[12]

Weinberg began the studies by obtaining 15 different malignant cell lines that were derived from mouse cells that had been treated with a carcinogenic chemical. Thus these cells had been made malignant without exposing them to viruses. The DNA from each of these cell lines was extracted and used to transfect a type of nonmalignant mouse fibroblast called an NIH3T3 cell. NIH3T3 cells were selected for these experiments because they take up exogenous DNA with high efficiency and they are readily transformed into malignant cells in culture. After transfection with DNA from the tumor cells, the fibroblasts were grown in vitro, and the cultures were screened for the formation of clumps (foci) that contained cells that had been transformed by the added DNA. Of the 15 cell lines tested, five of them yielded DNA that could transform the recipient NIH3T3 cells. DNA from normal cells lacked this capability. These results demonstrated that carcinogenic chemicals produced alterations in the nucleotide sequences of genes that gave the altered genes the ability to transform other cells. Thus cellular genes could be converted into oncogenes in two different ways: as the result of becoming incorporated into the genome of a virus or by becoming altered by carcinogenic chemicals.

Up to this point, virtually all of the studies on cancer-causing genes had been conducted in mice, chickens, or other organisms whose cells were highly susceptible to transformation. In 1981, attention turned to human cancer when it was shown that DNA isolated from human tumor cells can transform mouse NIH3T3 cells following transfection.[13] Of 26 different human tumors that were tested in this study, two provided DNA that was capable of transforming mouse fibroblasts. In both cases, the DNA had been extracted from cell lines taken from a bladder carcinoma (identified as EJ and J82). Extensive efforts were undertaken to determine if the genes had been derived from a tumor virus, but no evidence of viral DNA was detected in these cells. These results provided the first evidence that some human cancer cells contain an activated oncogene that can be transmitted to other cells, causing their transformation.

The finding that cancer can be transmitted from one cell to another by DNA fragments provided a basis for determin-

ing which genes in a cell, when activated by mutation or some other mechanism, are responsible for causing the cell to become malignant. To make this determination, it was necessary to isolate the DNA that was taken up by cells, causing their transformation. Once the foreign DNA responsible for transformation had been isolated, it could be analyzed for the presence of cancer-causing alleles. Within two months of one another in 1982, three different laboratories reported the isolation and cloning of an unidentified gene from human bladder carcinoma cells that can transform mouse NIH3T3 fibroblasts.[14–16]

Once the transforming gene from human bladder cancer cells had been isolated and cloned, the next step was to determine if this gene bears any relationship to the oncogenes carried by RNA tumor viruses. Once again, within two months of one another, three papers appeared from different laboratories reporting similar results.[17–19] All three showed that the oncogene from human bladder carcinomas that transforms NIH3T3 cells is the same oncogene (named *ras*) that is carried by the Harvey sarcoma virus, which is a rat RNA tumor virus. Preliminary comparisons of the two versions of *ras*–the viral version and its cellular homologue–failed to show any differences, indicating that the two genes are either very similar or identical. These findings suggested that cancers that develop spontaneously in the human population are caused by a genetic alteration that is similar to the changes in cells that have been virally transformed in the laboratory. It is important to note that the types of cancers induced by the Harvey sarcoma virus (sarcomas and erythroleukemias) are quite different from the bladder tumors, which have an epithelial origin. This was the first indication that alterations in the same gene—*ras*—can cause a wide range of different tumors.

By the end of 1982, three more papers from different laboratories reported on the precise changes in the *ras* gene that leads to its activation as an oncogene.[20–22] Once the section of the large DNA fragment that is responsible for causing transformation was pinned down, nucleotide sequence analysis indicated that the DNA from the malignant bladder cells is activated as a result of a single base substitution within the coding region of the gene. Remarkably, cells of both human bladder carcinomas studied (identified as EJ and T24) contain DNA with precisely the same alteration: a guanine-containing nucleotide at a specific site in the DNA of the proto-oncogene had been converted to a thymidine in the activated oncogene. This base substitution results in the replacement of a valine for a glycine as the twelfth amino acid residue of the polypeptide.

Determination of the nucleotide sequence of the v-*ras* gene carried by the Harvey sarcoma virus revealed an alteration in base sequence that affected precisely the same codon found to be modified in the DNA of the human bladder carcinomas. The change in the viral gene substitutes an arginine for the normal glycine. It was apparent that this particular glycine residue plays a critical role in the structure and function of this protein. It is interesting to note that the *ras* gene is a proto-oncogene that, like *src*, can be activated by linkage to a viral promoter. Thus *ras* can be activated to induce transformation by two totally different pathways: either by increasing its expression or by altering the amino acid sequence of its encoded polypeptide.

The research described in this Experimental Pathway provided a great leap forward in our understanding of the genetic basis of malignant transformation. Much of the initial research on RNA tumor viruses stemmed from the belief that these agents may be an important causal agent in the development of human cancer. The search for viruses as a cause of cancer led to the discovery of the oncogene, which led to the realization that the oncogene is a cellular sequence that is acquired by the virus, which ultimately led to the discovery that an oncogene can cause cancer without the involvement of a viral genome. Thus tumor viruses, which are not themselves directly involved in most human cancers, have provided the necessary window through which we can view our own genetic inheritance for the presence of information that can lead to our own undoing.

References

1. ROUS, P. 1911. Transmission of a malignant new growth by means of a cell-free filtrate. *J. Am. Med. Assoc.* 56:198.

2. BALTIMORE, D. 1970. RNA-dependent DNA polymerase in virions of RNA tumour viruses. *Nature* 226:1209–1211.

3. TEMIN, H. & MIZUTANI, S. 1970. RNA-dependent DNA polymerase in virions of Rous sarcoma virus. *Nature* 226:1211–1213.

4. MARTIN, G. S. 1970. Rous sarcoma virus: A function required for the maintenance of the transformed state. *Nature* 227:1021–1023.

5. STEHELIN, D., ET AL. 1976. Purification of DNA complementary to nucleotide sequences required for neoplastic transformation of fibroblasts by avian sarcoma viruses. *J. Mol. Biol.* 101:349–365.

6. STEHELIN, D., ET AL. 1976. DNA related to the transforming gene(s) of avian sarcoma viruses is present in normal avian DNA. *Nature* 260:170–173.

7. SPECTOR, D. H., VARMUS, H. E., & BISHOP, J. M. 1978. Nucleotide sequences related to the transforming gene of avian sarcoma virus are present in DNA of uninfected vertebrates. *Proc. Nat'l Acad. Sci. U. S. A.* 75:4102–4106.

8. PURCHIO, A. F., ET AL. 1978. Identification of a polypeptide encoded by the avian sarcoma virus src gene. *Proc. Nat'l Acad. Sci. U. S. A.* 75:1567–1671.

9. COLLETT, M. S. & ERIKSON, R. L. 1978. Protein kinase activity associated with the avian sarcoma virus src gene product. *Proc. Nat'l Acad. Sci. U. S. A.* 75:2021–2924.

10. M. C. WILLINGHAM, G. JAY, & I. PASTAN. 1979. Localization of the ASV src gene product to the plasma membrane of transformed cells by electron microscopic immunocytochemistry. *Cell* 18:125–134.

11. HUNTER, T. & SEFTON, B. M. 1980. Transforming gene product of Rous sarcoma virus phosphorylates tyrosine. *Proc. Nat'l Acad. Sci. U. S. A.* 77:1311–1315.

12. SHIH, C., ET AL. 1979. Passage of phenotypes of chemically transformed cells via transfection of DNA and chromatin. *Proc. Nat'l Acad. Sci. U. S. A.* 76:5714–5718.

13. KRONTIRIS, T. G. & COOPER, G. M. 1981. Transforming activity of human tumor DNAs. *Proc. Nat'l Acad. Sci. U. S. A.* 78:1181–1184.

14. GOLDFARB, M., ET AL. 1982. Isolation and preliminary characterization of a human transforming gene from T24 bladder carcinoma cells. *Nature* 296:404–409.

15. SHIH, C. & WEINBERG, R. A. 1982. Isolation of a transforming sequence from a human bladder carcinoma cell line. *Cell* 29:161–169.

16. PULCIANI, S., ET AL. 1982. Oncogenes in human tumor cell lines: Molecular cloning of a transforming gene from human bladder carcinoma cells. *Proc. Nat'l Acad. Sci. U. S. A.* 79:2845–2849.

17. PARADA, L. F., ET AL. 1982. Human EJ bladder carcinoma oncogene is a homologue of Harvey sarcoma virus ras gene. *Nature* 297:474–478.

18. DER, C. J., ET AL. 1982. Transforming genes of human bladder and lung carcinoma cell lines are homologous to the ras genes of Harvey and Kirsten sarcoma viruses. *Proc. Nat'l Acad. Sci. U.S.A.* 79:3637–3640.

19. SANTOS, E., ET AL. 1982. T24 human bladder carcinoma oncogene is an activated form of the normal human homologue of BALB- and Harvey-MSV transforming genes. *Nature* 298:343–347.

20. TABIN, C. J., ET AL. 1982. Mechanism of activation of a human oncogene. *Nature* 300:143–149.

21. REDDY, E. P., ET AL. 1982. A point mutation is responsible for the acquisition of transforming properties by the T24 human bladder carcinoma oncogene. *Nature* 300:149–152.

22. TAPAROWSKY, E., ET AL. 1982. Activation of the T24 bladder carcinoma transforming gene is linked to a single amino acid change. *Nature* 300:762–765.

SYNOPSIS

Cancer is a disease involving heritable defects in cellular control mechanisms that result in the formation of invasive tumors capable of releasing cells that can spread the disease to distant sites in the body. Many of the characteristics of tumor cells can be observed in culture. Whereas normal cells proliferate until they form a single layer (monolayer) over the bottom of the dish, cancer cells continue to grow in culture, piling on top of one another to form clumps. Other characteristics often revealed by cancer cells are an ability to grow while suspended in soft agar; a tendency to exhibit an abnormal number of chromosomes; an ability to continue to divide indefinitely; a disorganized cytoskeleton; and a lack of responsiveness to neighboring cells. (p. 672)

Normal cells can be converted to cancer cells by treatment with a wide variety of chemicals, ionizing radiation, and a variety of DNA- and RNA-containing viruses; all of these agents act by causing changes in the genome of the transformed cell. Analysis of the cells of a cancerous tumor almost always shows that the cells have arisen by growth of a single cell (the tumor is said to be monoclonal). The development of a malignant tumor is a multistep process characterized by a progression of genetic alterations that make the cells increasingly less responsive to the body's normal regulatory machinery and better able to invade normal tissues. The genes involved in carcinogenesis constitute a specific subset of the genome whose products are involved in such activities as control of the cell cycle, intercellular adhesion, and DNA repair. The level of expression of particular genes in different types of cancers can be determined using DNA microarrays. In addition to genetic alterations, growth of tumor cells is also influenced by nongenetic or epigenetic influences that allow the cell to express its malignant phenotype. Estrogen, for example, promotes the development of breast tumors. (p. 675)

Genes that have been implicated in carcinogenesis are divided into two broad categories: tumor-suppressor genes and oncogenes. Tumor-suppressor genes encode proteins that restrain cell growth and prevent cells from becoming malignant. Tumor-suppressor genes act recessively since both copies must be deleted or mutated before their protective function is lost. Oncogenes, in contrast, encode proteins that promote the loss of growth control and malignancy. Oncogenes arise from proto-oncogenes—genes that encode proteins having a role in a cell's normal activities. Mutations that alter either the protein or its expression cause the proto-oncogene to act abnormally and promote the formation of a tumor. Oncogenes act dominantly; that is, a single copy will cause the cell to express the altered phenotype. Most tumors contain alterations in both tumor-suppressor genes and oncogenes. As long as a cell retains at least one copy of all of its tumor-suppressor genes, it should be protected against the consequences of oncogene formation. Conversely, the loss of a tumor-suppressor function should not be sufficient, by itself, to cause the cell to become malignant. (p. 679)

The first tumor-suppressor gene to be identified was *RB*, which is found to be responsible for a rare retinal tumor called retinoblastoma, which occurs with high frequency in certain families but can also occur sporadically. Children with the familial form of the disease inherit one mutated copy of the gene. These individuals develop the cancer only after sporadic damage to the second allele in one of the retinal cells. *RB* encodes a protein called pRB, which is involved in regulating passage of a cell from G_1 to S in the cell cycle. The unphosphorylated form of pRb interacts with certain transcription factors, preventing them from binding to the DNA and activating the genes required for certain S phase activities. Once a pRB has been phosphorylated, the protein releases its bound transcription factor, which can then activate gene expression, leading to the initiation of S phase. (p. 681)

The tumor-suppressor gene most often implicated in human cancer is *p53*, whose product (p53) may be able to suppress cancer formation by several different mechanisms. In one of its actions, p53 acts as a transcription factor that activates the expression of a protein (p21) that inhibits the cyclin-dependent kinase that moves a cell through the cell cycle. Damage to DNA triggers the phosphorylation and stabilization of p53, leading to the arrest of the cell cycle until the

damage can be repaired. p53 can also redirect cells that are on the path toward malignancy onto an alternate path leading to programmed death, or apoptosis. *p53* knockout mice begin to develop tumors several weeks after birth. Other tumor-suppressor genes include *APC* that, when mutated, predisposes an individual to developing colon cancer, and *BRCA1* and *BRCA2* that, when mutated, predispose an individual to developing breast cancer. (p. 684)

Most of the known oncogenes are derived from proto-oncogenes that play a role in the pathways that transmit growth signals from the extracellular environment to the cell interior, particularly the cell nucleus. A number of oncogenes have been identified that encode growth factor receptors, including *sis*, the gene that encodes the platelet-derived growth factor (PDGF) and *erbB*, the gene that encodes the epidermal growth factor (EGF) receptor. Malignant cells may contain a much larger number of one of these growth factor receptors in their plasma membranes when compared to normal cells. The excess receptors make the cells sensitive to lower concentrations of the growth factor and, thus, are stimulated to divide under conditions that would not affect normal cells. A number of cytoplasmic protein kinases, including both serine/threonine and tyrosine kinases, are included on the list of oncogenes. These include *raf*, which encodes a protein kinase in the MAP kinase cascade. Mutations in *ras* are among the most common oncogenes found in human cancers. As discussed in Chapter 15, Ras activates the protein kinase activity of Raf. If Raf remains in the activated state, it sends continual signals along the MAP kinase pathway, leading to the continual stimulation of cell proliferation. A number of oncogenes, such as *myc*, encode proteins that act as transcription factors. Myc is normally one of the first proteins to appear when a cell is stimulated to reenter the cell cycle from the quiescent G_0 phase. Overexpression of *myc* may cause cells to proliferate by overriding the inhibitory influences exerted by the products of the tumor-suppressor genes. Another group of oncogenes, such as *bcl-2*, encode proteins involved in apoptosis.

Overexpression of the *bcl-2* gene leads to the suppression of apoptosis in lymphoid tissues, allowing abnormal cells to proliferate to form lymphoid tumors. (p. 689)

Genes that encode proteins involved in DNA repair have also been implicated in carcinogenesis. The genome of patients with the most common hereditary form of colon cancer, called hereditary nonpolyposis colon cancer (HNPCC), contains microsatellite sequences of abnormal nucleotide number. Changes in the length of a microsatellite sequence arise as an error during replication, which is normally recognized by mismatch repair enzymes, suggesting that defects in such systems may be responsible for the cancers. This conclusion is supported by findings that extracts from HNPCC tumor cells display DNA repair deficiencies. Cells that contain such deficiencies would be expected to display a greatly elevated level of mutation in tumor-suppressor genes and oncogenes that would lead to a greatly increased risk of malignancy. (p. 690)

Cancer is currently treated by surgery, chemotherapy, and radiation. Several other strategies are being tested; these include immunotherapy, gene therapy, and inhibition of angiogenesis. In one promising immunotherapy strategy, immature dendritic cells are purified from the blood, exposed to tumor proteins, and injected into the patient in an attempt to mount an immune response against the tumor cells. In one type of gene therapy, patients are infected with a lytic adenovirus lacking a particular gene that makes it unable to divide in normal cells, but capable of dividing in cells lacking the *p53* gene. In another strategy, compounds are being developed that inhibit key cancer-promoting proteins, such as the Abl kinase of patients with CML. In antiangiogenic strategies, an attempt is made to prevent a solid tumor from inducing the formation of new blood vessels (angiogenesis) that are needed to supply the tumor cells with nutrients and other materials. A number of agents have been identified that block angiogenesis in mice. (p. 691)

ADDITIONAL READING

AMES, B. N. 1999. Cancer prevention and diet: help from single nucleotide polymorphisms. *Proc. Nat'l Acad. Sci. U. S. A.* 96:12216–12218.

ANDERSON, W. F. 2000. Gene therapy scores against cancer. *Nature Med.* 6:862–863.

ARTANDI, S. E. & DEPINHO, R. A. 2000. A critical role for telomeres in suppressing and facilitating carcinogenesis. *Curr. Opin. Genes Dev.* 10:39–46.

BARON, J. A. & SANDLER, R. S. 2000 Nonsteroidal anti-inflammatory drugs and cancer prevention. *Annu. Rev. Med.* 51:511–523.

BERNS, A. 2000. Gene expression in diagnosis. *Nature* 403:491–492.

BIRGYN, A. & KWAK, L. W. 2000. Designer cancer vaccines are still in fashion. *Nature Med.* 6:966–968.

BISHOP, J. M. & WEINBERG, R. A., EDS. 1996. *Molecular Oncology.* Sci. Amer. Press.

BOEHM-VISWANATHAN, T. 2000. Is angiogenesis inhibition the Holy Grail of cancer therapy? *Curr. Opin. Oncol.* 12:89–94.

BRINCKERHOFF, L. H., ET AL. 2000. Melanoma vaccines. *Curr. Opin. Oncol.* 12:163–173.

BROWN, M. A. & SOLOMON, E. 1997. Studies on inherited cancers: Outcomes and challenges of 25 years. *Trends Gen.* 13:202–206.

CAHILL, D. P., ET AL. 1999. Genetic instability and darwinian selection in tumours. *Trends Biochem. Sci.* 24:M57–M60. (Dec.)

CANTLEY, L. C. & NEEL, B. G. 1999. New insights into tumor suppression: PTEN suppresses tumor formation by restraining the phosphoinositide 3-kinase/AKT pathway. *Proc. Nat'l Acad. Sci. U. S. A.* 96:4240–4245.

CARMELIET, P. & JAIN, R. K. 2000. Angiogenesis in cancer and other diseases. *Nature* 407:248–257.

CHANG, E. H., ET AL. 2000. *TP53* gene therapy: A key to mdulating resistance to anticancer therapies? *Mol. Med. Today* 6:358–364.

CHRISTOFORI, G. & SEMB, H. 1999. The role of the cell adhesion molecule E-cadherin as a known tumour-suppressor gene. *Trends Biochem. Sci.* 24:73–76.

CHUNG-FAYE, G. A., ET AL. 2000. Gene therapy strategies for colon cancer. *Mol. Med. Today* 6:82–87.

DENG, C.-X. & BRODIE, S. G. 2000. Roles of BRCA1 and its interacting proteins. *BioEss.* 22:728–737.

FEARON, E. R. & DANG, C. V. 1999. Cancer genetics: Tumor suppressor meet oncogene. *Curr. Biol.* 9:R62–R65.

FINKEL, E. 1999. Does cancer therapy trigger cell suicide? *Science* 286:2256–2258.

GIBBS, J. B. 2000. Mechanism-based target identification and drug discovery in cancer research. *Science* 287:1969–1973.

HARBOUR, J. W. & DEAN, D. C. 2000. Rb functions in cell-cycle regulation and apoptosis. *Nature Cell Biol.* 2:E65–E67.

HUNTER, T. 1997. Oncoprotein networks. *Cell* 88:333–346.

JONES, P. A. 2001. Death and methylation. (cancer cells and apoptosis). *Nature* 409:141–144.

KINZLER, K. W. & VOGELSTEIN, B. 1996. Lessons from hereditary colorectal cancer. *Cell* 87:159–170.

KLOOG, Y. & COX, A. D. 2000. RAS inhibitors: Potential for cancer therapeutics. *Mol. Med. Today* 6:398–402.

KUFE, D.W. 2000. Smallpox, polio, and now a cancer vaccine. *Nature Med.* 6:252–253.

LEES, J. A. & WEINBERG, R. A. 1999. Tossing monkey wrenches into the clock: New ways to treat cancer. *Proc. Nat'l Acad. Sci. U. S. A.* 96:4221–4223.

LLOYD, A. C. 2000. p53: Only ARF the story. *Nature Cell Biol.* 2:E48–E50.

MACLEOD, K. 2000. Tumor suppressor genes. *Curr. Opin. Genes Dev.* 10:81–93.

MCCARTNEY, B. M. & PEIFER, M. 2000. Teaching tumor suppressors new tricks. *Nature Cell Biol.* 2:E58–E60.

MCCAWLEY, L. J. & Matrisian, L. M. 2001. Tumor progression. *Curr. Biol.* 11:R25–R-27.

MCCORMICK, F. 1999. Signaling networks that cause cancer. *Trends Biochem. Sci.* 24:M53–M56. (Dec.)

MAEHAMA, T. & DIXON, J. E. 1999. PTEN: A tumor suppressor that functions as a phospholipid phosphatase. *Trends Cell Biol.* 9:125–128.

MARX, J. 2000. DNA arrays reveal cancer in its many forms. *Science* 289:1670–1672.

MOLL, J., ET AL. 2000. Antiangiogenesis therapy: Preclinical premise and promise. *Mol. Med. Today* 6:188–189.

MÜLLER, H. & HELIN, K. 2000. The E2F transcription factors: Key regulators of cell proliferation. *Biochim. Biophys. Act* 1470:M1–M12.

NICHOLSON, D. W. 2000. From bench to clinic with apoptosis-based therapeutic agents. *Nature* 407:810–816.

PARDOLL, D. M. 1999. Inducing autoimmune disease to treat cancer. *Proc. Nat'l Acad. Sci. U. S. A.* 96:5340–5342.

PERERA, F. P. 1996. Uncovering new clues to cancer risk. *Sci. Am.* 274:54–62. (May)

PEROU, C. M., ET AL. 2000. Molecular portraits of human breast tumors. *Nature* 406:747–752.

RABBITTS, T. H. 1994. Chromosomal translocations in human cancer. *Nature* 372:143–149.

RICH, T., ET AL. 2000. Defying death after DNA damage. *Nature* 407:777–783.

ROWLEY, J. D. 1998. The critical role of chromosome translocations in human leukemias. *Annu. Rev. Gen.* 32:495–519.

SHERR, C. J. & WEBER, J.D. 2000. The ARF/p53 pathway. *Curr. Opin. Gen. Dev.* 10:94–99.

SIEBER, O., ET AL. 2000. The adenomatous polyposis coli (APC) tumour suppressor-genetics, function, and disease. *Mol. Med. Today* 6:462–469.

SOUSSI, T. 1996. The p53 tumor suppressor gene: A model for molecular epidemiology of human cancer. *Mol. Med. Today* 2:32–37.

TIMMERMAN, J. M. & LEVY, R. 1999. Dendritic cell vaccines for cancer immunotherapy. *Annu. Rev. Med.* 50:507–529.

VENKITARAMAN, A. R. 1999. Breast cancer genes and DNA repair. *Science* 286:1100–1102.

VILE, R. G., ET AL. 2000. Cancer gene therapy: hard lessons and new courses. *Gene Ther.* 7:2–8.

WEINBERG, R. A., ET AL. 1996. Special issue on cancer. *Sci. Am.* 275:#3. (Sept.)

WOOSTER, R. 2000. Cancer classification with DNA microarrays. *Trends Gen.* 16:327–329.

ZETTER, B. R. 1998. Angiogenesis and tumor metastasis. *Annu. Rev. Med.* 49:407–424.

The Immune Response

iving organisms provide ideal habitats in which other organisms can grow. It is not surprising, therefore, that animals are subject to infection by viruses, bacteria, protists, fungi, and animal parasites. Vertebrates have evolved several mechanisms that allow them to recognize and destroy these infectious agents. As a result, vertebrates are able to develop an **immunity** against invading pathogens. Immunity results from the combined activities of many different cells, some of which patrol the body, whereas others are concentrated in lymphoid organs, such as the bone marrow, thymus, spleen, and lymph nodes (Figure 17.1). Together, these dispersed cells and discrete organs form the body's **immune system**. The cells of the immune system engage in a type of molecular screening by which they recognize "foreign" macromolecules, that is, ones whose structure is different from those of the body's normal macromolecules. If foreign material is encountered, the immune system mounts a specific and concerted attack against it. The weapons of the immune system include (1) cells

T Lymphocytes are cells of the immune system that become activated when they contact cells displaying foreign peptides on their surface. This computer-generated image shows portions of the molecules that participate in this cell-cell interaction. The T-cell receptor that projects from the T lymphocyte is shown in yellow, the MHC protein that projects from the surface of an infected cell is shown in blue, and a peptide derived from a human retrovirus is shown in white. The viral peptide is held in the "grip" of the MHC protein while it is presented to the T lymphocyte. (FROM DAVID N. GARBOCZI, ET AL., COURTESY OF DON C. WILEY, HARVARD UNIVERSITY. REPRINTED BY PERMISSION FROM NATURE 384:137, 1996. COPYRIGHT 1996 BY MACMILLAN MAGAZINES LTD.)*

ACOC segment type="table_of_contents">
17.1 An Overview of the Immune Response
17.2 The Clonal Selection Theory as It Applies to B Cells
17.3 T Lymphocytes: Activation and Mechanism of Action
17.4 Selected Topics on the Cellular and Molecular Basis of Immunity

THE HUMAN PERSPECTIVE: Autoimmune Diseases
EXPERIMENTAL PATHWAYS: The Role of the Major Histocompatibility Complex in Antigen Presentation

703

Figure 17.1 **The human immune system** includes various lymphoid organs, such as the thymus, bone marrow, spleen, lymph nodes, and scattered cells located as patches within the small intestine, adenoids, and tonsils.

that kill or ingest infected or altered cells and (2) soluble proteins that can neutralize, immobilize, agglutinate, or kill pathogens. The immune system is also implicated in the body's fight against cancer, but the degree to which the system can recognize and kill cancer cells remains controversial. In some cases, the immune system may mount an inappropriate response that attacks the body's own tissues. As discussed in the Human Perspective on page 728, these incidents can lead to serious disease.

It is impossible to cover the entire subject of immunity in a single chapter. Instead, we will focus on a number of selected aspects that illustrate principles of cellular and molecular biology that were discussed in previous chapters. First, however, it is necessary to examine the basic events in the body's response to the presence of an intruding microbe. ■

17.1 AN OVERVIEW OF THE IMMUNE RESPONSE

The outer surface of the body and the linings of its internal tracts provide an excellent barrier to penetration by viruses, bacteria, and parasites. If these surface barriers are breached, a series of **immune responses** are initiated that contain the invasion. Immune responses can be divided into two general categories: innate responses and acquired (or adaptive) responses. Both types of responses depend on the ability of the body to distinguish between materials that are supposed to be there (i.e., "self") and those that are not (i.e., foreign, or "nonself"). We can also distinguish two categories of pathogens: those that occur primarily inside a host cell (all viruses, some bacteria, and certain protozoan parasites) and those that occur primarily in the extracellular compartments of the host (most bacteria and other cellular pathogens). Different types of immune mechanisms have evolved to combat these two types of infections. An overview of some of these mechanisms is shown in Figure 17.2.

Innate Immune Responses

Innate immune responses are those the body mounts immediately without requiring previous contact with the microbe. Thus, they provide the body with a first line of defense. Innate responses are characterized by their lack of specificity; the same defensive activities are effective regardless of the pathogen presenting the threat. Innate responses to invading bacteria are typically accompanied by a process of **inflammation** at the site of infection where

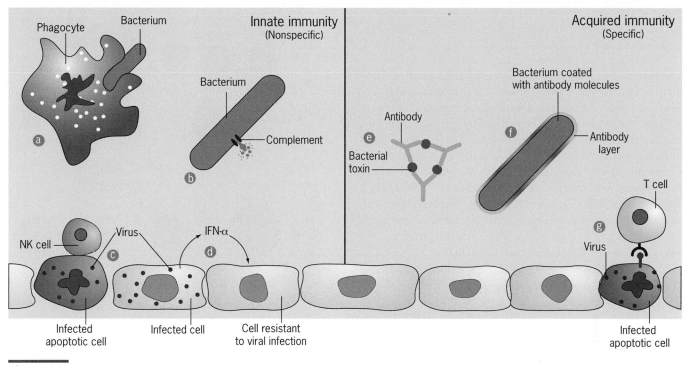

Figure 17.2 An overview of some of the mechanisms by which the immune system rids the body of invading pathogens. The left panel depicts several types of innate immunity: (*a*), phagocytosis of a bacterial cell; (*b*), bacterial cell killing by complement; (*c*), apoptosis induced in an infected cell by a natural killer (NK) cell; and (*d*), induction of viral resistance by interferon α (IFN-α). The right panel depicts several types of acquired immunity: (*e*), neutralization of a bacterial toxin by

antibody; (*f*), bacterial cell coated with antibody (opsonized), which makes it susceptible to phagocytosis or complement-induced death; and (*g*), apoptosis induced in an infected cell by an activated T lymphocyte (T cell). Innate and acquired immune responses are linked to one another because cells, such as macrophages, that phagocytize pathogens, can use the foreign proteins to stimulate the production of specific antibodies and T cells directed against the pathogen.

fluid, cells, and dissolved substances leak out of the blood and into the affected tissues (page 264). These events are accompanied by local redness, swelling, and fever. Inflammation provides a means for concentrating the body's defensive agents at the site where they are needed. During inflammation, phagocytic cells, such as neutrophils and macrophages, leave the bloodstream and migrate toward a site of infection in response to chemicals (chemoattractants) released at the site. Once there, these cells recognize, engulf, and destroy the pathogen (Figure 17.2*a*). Blood also contains a group of soluble proteins called **complement** that bind to extracellular pathogens, triggering their destruction. In one of the complement pathways, an activated assembly of these proteins perforates the plasma membrane of a bacterial cell, leading to cell lysis and death (Figure 17.2*b*). Innate immune responses are particularly effective against certain bacteria whose cell walls contain unique polysaccharides that are immediately identified as foreign by host cells.

Innate responses against intracellular pathogens, such as viruses, are targeted primarily against cells that are already infected. Cells infected with certain viruses are recognized by a type of nonspecific lymphocyte called a **natural killer (NK) cell**. As its name implies, NK cells act by causing the death of the infected cell (Figure 17.2*c*). The

NK cell induces the infected cell to undergo apoptosis (page 662). NK cells can also kill certain types of cancer cells in vitro (Figure 17.3) and may provide a mechanism for destroying such cells before they develop into a tumor.

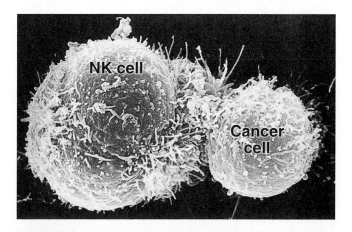

Figure 17.3 Innate immunity. Scanning electron micrograph of a natural killer cell bound to a target cell, in this case, a malignant erythroleukemia cell. (COURTESY OF GIUSEPPE ARANCIA, DEPT. OF ULTRASTRUCTURES, INSTITUTO SUPERIORE DI SANITÁ, ROME, FROM BLOOD CELLS 17:165, 1991.)

Normal (i.e., noninfected and nonmalignant) cells possess surface molecules that protect them from attack by NK cells (see Figure 17.23).

Another type of antiviral response is initiated within the infected cell itself. Virus-infected cells produce substances called type 1 **interferons** (interferon α and interferon β) that are secreted into the extracellular space, where they bind to the surface of noninfected cells rendering them resistant to subsequent infection (Figure 17.2*d*). Interferons accomplish this by activating a signal transduction pathway that results in the phosphorylation and consequent inactivation of the translation factor eIF2 (page 480). Cells that have undergone this response cannot synthesize viral proteins that are required for virus replication.

Acquired Immune Responses

Unlike innate responses, **acquired immune responses** require a lag period during which the immune system gears up for an attack against a foreign agent. Unlike innate responses, acquired responses are highly specific; that is, they are directed only against specific molecules present in the foreign material. For example, the blood of a person who has just recovered from measles contains antibodies that react with the virus that causes measles, but not with a related virus, such as that which causes mumps.

Whereas all animals possess some type of innate immunity against microbes and parasites, only vertebrates are known to mount an acquired response. There are two broad categories of acquired immunity:

■ **Humoral immunity**, which is carried out by **antibodies** (Figure 17.2*e, f*). Antibodies are globular, blood-borne proteins of the **immunoglobulin (Ig)** superfamily.

■ **Cell-mediated immunity**, which is carried out by cells (Figure 17.2*g*).

Both types of acquired immunity are mediated by **lymphocytes**, which are nucleated leukocytes (white blood cells) that circulate between the blood and lymphoid organs. Humoral immunity is mediated by **B lymphocytes** (or **B cells**) that, when activated, differentiate into cells that secrete antibodies. Antibodies are directed primarily against foreign materials that are situated outside the body's cells. Such materials include the protein and polysaccharide components of bacterial cell walls, bacterial toxins, and viral coat proteins. In some cases, antibodies can bind to a bacterial toxin or virus particle and directly prevent the agent from entering a host cell (Figure 17.2*e*). In other cases, antibodies function as "molecular tags" that bind to an invading pathogen and mark it for destruction. Bacterial cells coated with antibody molecules (Figure 17.2*f*) are rapidly ingested by wandering phagocytes or destroyed by complement molecules carried in the blood. Antibodies are not effective against pathogens that are present inside cells, hence the need for a second type of

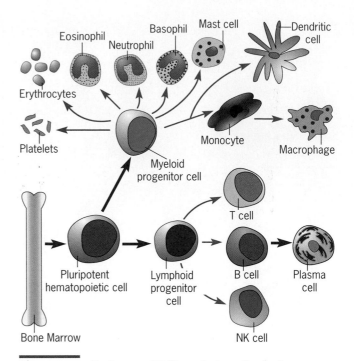

Figure 17.4 **Pathways of differentiation of a pluripotent hematopoietic stem cell of the bone marrow.** A hematopoietic stem cell can give rise to two different progenitor cells: a myeloid progenitor cell that can differentiate into most of the various blood cells (e.g., erythrocytes, basophils, and neutrophils), macrophages, or dendritic cells; or a lymphoid progenitor cell that can differentiate into any of the various types of lymphocytes (NK cells, T cells, or B cells). T-cell precursors migrate to the thymus where they undergo differentiation into T cells. In contrast, B cells undergo differentiation in the bone marrow.

weapon system. Cell-mediated immunity is carried out by **T lymphocytes** (or **T cells**) that, when activated, can specifically recognize and kill an infected (or foreign) cell (Figure 17.2*g*).

B and T cells arise from the same type of precursor cell (a *pluripotent hematopoietic stem cell*), but they differentiate along different pathways in different lymphoid organs. A summary of the various pathways of differentiation of the pluripotent hematopoietic stem cell is shown in Figure 17.4. B lymphocytes differentiate in the fetal liver or adult bone marrow, whereas T lymphocytes differentiate in the thymus gland, an organ located in the chest that reaches its peak size during childhood. Because of these differences, cell-mediated and humoral immunity can be dissociated to a large extent. For instance, humans may suffer from a rare disease called congenital agammaglobulinemia in which humoral antibody is deficient and cell-mediated immunity is normal.

REVIEW ?

1. Contrast the general properties of innate and acquired immune responses.

2. List four types of innate immune responses. Which would be most effective against a pathogen inside an infected cell?

3. What is meant by the terms "humoral" and "cell-mediated" immunity?

17.2 THE CLONAL SELECTION THEORY AS IT APPLIES TO B CELLS

If a person is infected with a virus or exposed to foreign material, their blood soon contains a high concentration of antibodies capable of reacting with the foreign substance, which is known as an **antigen**. Most antigens consist of proteins or polysaccharides, but lipids and nucleic acids can also serve in this capacity. How can the body produce antibodies that react *specifically* with an antigen to which the body is exposed? In other words, how does an antigen induce an acquired immune response? For many years it was thought that antigens somehow *instructed* lymphocytes to produce complementary antibodies. It was suggested that an antigen wraps itself around an antibody molecule, molding the antibody into a shape capable of combining with that particular antigen. In this "instructive" model, the lymphocyte only gains its ability to produce a specific antibody after its initial contact with antigen. In 1955, Niels Jerne, a Danish immunologist, proposed a radically different mechanism. Jerne suggested

that the body produced small amounts of randomly structured antibodies in the absence of any antigen. As a group, these antibodies could combine with any type of antigen to which a person might someday be exposed. According to Jerne's model, when a person is exposed to an antigen, the antigen combines with a specific antibody, which leads to the subsequent production of that particular antibody molecule. Thus in Jerne's model, the antigen *selects* those preexisting antibodies capable of binding to it. In 1957, the concept of antigen selection of antibodies was expanded into a comprehensive model of antibody formation by the Australian immunologist F. MacFarlane Burnet. Burnet's **clonal selection theory** has gained virtually complete acceptance and remains compatible with all the experimental evidence accumulated since its original proposal. An overview of the steps that occur during the clonal selection of B cells is shown in Figure 17.5. More detailed discussion of these events is provided later in the chapter. The clonal selection of T cells is described in the following section. The main features of B-cell clonal selection are the following:

1. *Each B cell becomes committed to produce one species of antibody.* B cells arise from a population of undifferentiated and indistinguishable stem cells. As a stem cell becomes a B cell (Figure 17.4), it becomes committed as the result of DNA rearrangements (see Figure 17.15) to producing only one species of antibody molecule (Figure 17.5, step 1). Thousands of different DNA rearrangements are possible, so that different B cells produce different antibody molecules. Thus, even

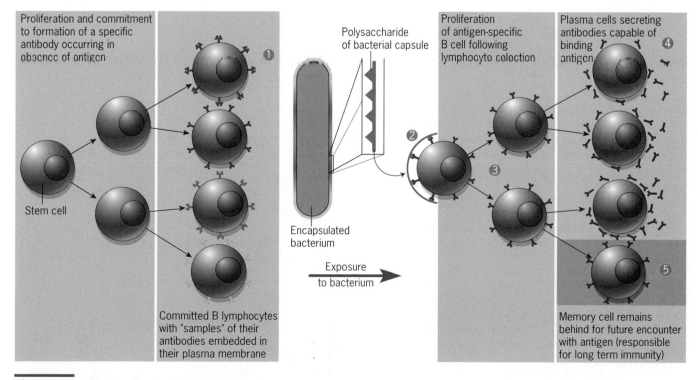

Proliferation and commitment to formation of a specific antibody occurring in absence of antigen ❶

Stem cell

Committed B lymphocytes with "samples" of their antibodies embedded in their plasma membrane

Polysaccharide of bacterial capsule

Encapsulated bacterium

Exposure to bacterium ❷

Proliferation of antigen-specific B cell following lymphocyte selection ❸

Plasma cells secreting antibodies capable of binding antigen ❹

Memory cell remains behind for future encounter with antigen (responsible for long term immunity) ❺

Figure 17.5 **The clonal selection of B cells by a thymus-independent antigen.** The steps are described in the figure and also in the text.

though mature B cells appear identical under the microscope, they can be distinguished by the antibodies they produce.

2. *B cells become committed to antibody formation in the absence of antigen.* The entire repertoire of antibody-producing cells a person will ever possess is already present within the lymphoid tissues *prior to stimulation by an antigen* and is independent of the presence of foreign materials. Each B cell displays its particular antibody on its surface with the antigen-reactive portion facing outward. As a result, the cell is coated with antigen receptors that can bind specifically with antigens having a complementary structure. Although most lymphoid cells are never required during a person's lifetime, the immune system is primed to respond immediately to any antigen that a person may be exposed to. The presence of cells with different membrane-bound antibodies can be demonstrated experimentally, as shown in Figure 17.6.

3. *Antibody production follows selection of B cells by antigen.* In most cases, the activation of a B cell by antigen requires the involvement of T cells (discussed on pages 712 and 727). A few antigens, however, such as the polysaccharides present in bacterial cell walls, activate B cells by themselves; antigens of this type are described as *thymus-independent antigens.* For simplicity, we will restrict the discussion at this point in the chapter to a thymus-independent antigen. Suppose a person were to be exposed to *Haemophilus influenzae* type B, an encapsulated bacterium that can cause fatal meningitis. The capsule of these bacteria contains a polysaccharide that can bind to a small fraction of the body's B cells (Figure 17.5, step 2). The B cells that bind the polysaccharide contain membrane-bound antibodies whose combining site allows them to interact

specifically with that antigen. In this way, an antigen *selects* those lymphocytes that produce antibodies capable of interacting with that antigen. Antigen binding activates the B cell, causing it to proliferate (Figure 17.5, step 3) and form a population (or *clone*) of lymphocytes, all of which make the same antibody. Some of these activated cells differentiate into short-lived **plasma cells** that secrete large amounts of antibody molecules (Figure 17.5, step 4). Unlike their B-cell precursors (Figure 17.7*a*), plasma cells possess an extensive rough ER characteristic of cells that are specialized for protein synthesis and secretion (Figure 17.7*b*).

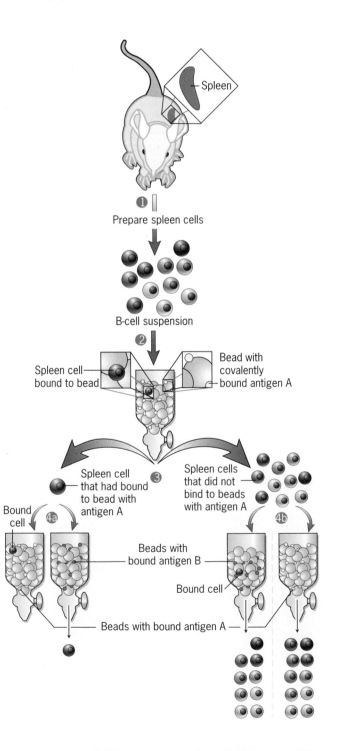

Figure 17.6 **Experimental demonstration that different B cells each contain a different membrane-bound antibody and that these antibodies are produced in the absence of antigen.** In this experiment, B cells are prepared from a mouse spleen (step 1). In step 2, the spleen cells are passed through a column containing beads coated with an antigen (antigen A) to which the mouse had never been exposed. A small fraction of the spleen cells are found to bind to the beads, while the vast majority of spleen cells pass directly through the column (shown in step 3). In step 4, spleen cells from the previous experiment are passed through one of two different columns: a column whose beads are coated with antigen A or a column whose beads are coated with an unrelated antigen (antigen B) to which the mouse had never been exposed. In step 4a, the spleen cells tested are those that had bound to the beads in the previous step. These cells are found to rebind to beads coated with antigen A, but do not bind to beads coated with antigen B. In step 4b, the spleen cells tested are those that did not bind to bind to the beads in the previous step. None of these cells bind to beads coated with antigen A, but a small fraction binds to beads coated with antigen B.

4. ***Immunologic memory provides long-term immunity.*** Not all B lymphocytes that are activated by antigen differentiate into antibody-secreting plasma cells. Some remain in lymphoid tissues as **memory B cells** (Figure 17.5, step 5) that can respond rapidly at a later date if the antigen reappears in the body. Although plasma cells die off following removal of the antigenic stimulus, memory B cells may persist for a person's lifetime. When stimulated by the same antigen, some of the memory B cells rapidly proliferate into plasma cells, generating a *secondary immune response* in a matter of hours rather than the days required for the original response (see Figure 17.10). The presence of memory cells ensures that we only suffer once in our lifetime from the effects of a particular strain of virus or bacteria.

5. ***Immunologic tolerance prevents the production of antibodies against self.*** As discussed below, genes encoding antibodies are generated by a process in which DNA segments are randomly combined. As a result, genes are invariably formed that encode antibodies that can react with the body's own tissues, which could produce widespread organ destruction and subsequent disease. It is obviously in the best interest of the body to prevent the production of such proteins, which are called **autoantibodies**. During the development of the immune system, cells producing autoantibodies either die by apoptosis (a process called *clonal deletion*) or become inactivated and unable to respond to antigen (a

process called *clonal anergy*). As a result, the body develops an immunologic tolerance toward itself. At the same time, it is evident that many cells capable of producing autoantibodies normally remain in the body. As discussed in the Human Perspective, a breakdown of the tolerant state can lead to the development of debilitating autoimmune diseases, such as lupus erythematosus and rheumatoid arthritis.

Several principles of the clonal selection theory can be illustrated by briefly considering the subject of vaccination.

Vaccination

Edward Jenner practiced medicine in the English countryside at a time when smallpox was one of the most prevalent and dreaded diseases. Over the years, he noticed that the maids who tended the cows were typically spared the ravages of the disease. Jenner concluded that milkmaids were somehow "immune" to smallpox because they were infected at an early age with cowpox, a harmless disease they contracted from their cows. Cowpox produces blisters that resemble the pus-filled blisters of smallpox, but the cowpox blisters are localized and disappear, causing nothing more serious than a scar at the site of infection.

In 1796, Jenner performed one of the most famous (and risky) medical experiments of all time. First, he infected an eight-year-old boy with cowpox and gave the boy time to recover. Six weeks later, Jenner intentionally infected the boy with smallpox by injecting pus from a smallpox lesion directly under the boy's skin. The boy showed no signs of the deadly disease. Within a few years, thousands of people had become immune to smallpox by intentionally

(a)

(b)

Figure 17.7 Comparison of the structure of a B cell (*a*) and a plasma cell (*b*). The plasma cell has a much larger cytoplasmic compartment than the B cell, with more mitochondria and an extensively developed rough endoplasmic reticulum. These characteristics reflect the synthesis of large numbers of antibody molecules by the plasma cell. (A: © VU/DAVID PHILLIPS/VISUALS UNLIMITED; B: © VU/K. G. MURTI/VISUALS UNLIMITED.)

infecting themselves with cowpox. This procedure was termed **vaccination**, after "vacca," the Latin word for cow.

Jenner's smallpox vaccination produces immunity by stimulating T cells, which is the subject of the next section. Most vaccines currently in use are B-cell vaccines, such as that employed to fight tetanus. Tetanus results from infection by the anaerobic soil bacterium *Clostridium tetani*, which can enter the body through a puncture wound. As they grow, the bacteria produce a powerful neurotoxin that blocks transmission across inhibitory synapses on motor neurons, leading to sustained muscle contraction and asphyxiation. At two months of age, most infants are **immunized** against tetanus by inoculation with a modified and harmless version of the tetanus toxin (called a *toxoid*). The tetanus toxoid binds to the surfaces of B cells whose membrane-bound antibody molecules have a complementary binding site. These B cells proliferate to form a clone of cells that produce antibodies capable of binding to the actual tetanus toxin. This initial response soon wanes, but the person is left with memory cells that respond rapidly if the person should happen to develop a *C. tetani* infection at a later date. Unlike most immunizations, immunity to the tetanus toxin does not always last a lifetime, which is the reason that people are given a booster shot every 10 years or so. The booster shot contains the toxoid protein and stimulates the production of additional memory cells. What if a person receives a wound that has the potential to cause tetanus, and they cannot remember ever receiving a booster shot? In these cases, the person is likely to be given a *passive immunization*, consisting of antibodies that can bind the tetanus toxin. Passive immunization is effective for only a short period of time and does not protect the recipient against a subsequent infection.

REVIEW **?**

1. Contrast instructive and selective mechanisms of antibody production.
2. What are the basic tenets of the clonal selection theory?
3. What does it mean for a B cell to become committed to antibody formation? How is this process influenced by the presence of antigen? What role does antigen play in antibody production?
4. What is meant by the terms "immunologic memory" and "immunologic tolerance"?

17.3 T LYMPHOCYTES: ACTIVATION AND MECHANISM OF ACTION

Like B cells, T cells are also activated by a process of clonal selection. T cells possess a cell-surface protein, called a **T-cell receptor**, that allows them to interact specifically with a particular antigen. Like the antibody molecules that serve as B-cell receptors, the proteins that serve as T-cell receptors exist as a large population of molecules that have differently shaped combining sites. Just as each B cell produces only one species of antibody, each T cell has only a single species of T-cell receptor. It is estimated that adult humans possess approximately 10^{12} T cells that, collectively, exhibit more than 10^7 different antigen receptors.

Unlike B cells, which are activated by soluble, intact antigens, T cells are activated by fragments of antigens that are displayed on the surfaces of other cells, called **antigen-presenting cells (APCs)**. Consider what would happen if a liver or kidney cell were infected with a virus. The infected cell would display portions of the viral proteins on its surface (see Figure 17.22), enabling the infected cell to bind to a T cell with the appropriate T-cell receptor. As a result of this presentation, the immune system becomes alerted to the entry of a specific pathogen. The process of antigen presentation is discussed at length later in the chapter (page 721) and is also the subject of the Experimental Pathways.

Whereas any infected cell can serve as an APC in activating T cells, certain types of "professional" APCs are specialized for this function. Professional APCs include dendritic cells and macrophages, both of which play a key role in initiating an immune response (Figure 17.8). We will focus on dendritic cells (DCs), which are often described as the "sentinels" of the immune system. Dendritic cells have earned this title because they "stand guard" in the body's peripheral tissues where pathogens are likely to enter (such as the skin and airways). They also accumulate at sites of infection and inflammation. As immature cells in the body's tissues, DCs engage primarily in the uptake of materials by endocytosis and phagocytosis. Once an antigen is taken into a dendritic cell, it has to be processed before it can be presented to another cell. Antigen processing requires that the ingested material is fragmented enzymatically in the cytoplasm and the fragments moved to the cell surface (see Figure 17.21). DCs that have processed antigen migrate to nearby lymph nodes where they differentiate into mature antigen-presenting cells. Once in a lymph node, DCs come into contact with a large pool of resident T cells, including a minute percentage whose T-cell receptors can bind specifically with the processed foreign antigen, which activates the T cell. Once activated, a T cell proliferates to form a clone of T cells having the same T-cell receptor. It is estimated that a single activated T cell can divide as many as 15 times in less than a week, generating a tremendous population of T cells capable of interacting with the foreign antigen. The massive proliferation of specific T lymphocytes in response to an infection is often reflected in the enlargement of local lymph nodes. Once the foreign antigen has been cleared, the vast majority of the expanded T-cell population dies off, leaving behind a relatively small population of memory T cells capable of responding rapidly in the event of future contact with the same pathogen.

T cells function at close range by interacting with other cells, which include APCs, B cells, other T cells, or target cells located anywhere in the body. In addition to

(a)

(b)

Figure 17.8 Professional antigen-presenting cells (APCs).
Colorized scanning electron micrographs of a macrophage *(a)*
and a dendritic cell *(b)*. Dendritic cells are irregular-shaped cells
with long cytoplasmic processes that resemble the dendrites of a
nerve cell, hence their name. (A: FROM A. POLLIACK/PHOTO RE-
SEARCHERS, INC.; B: COURTESY OF DENDREON CORPORATION.)

Table 17.1 Selected Cytokines

Cytokine	Source	Major Functions
IL-1	Diverse	Induces inflammation, stimulates T_H-cell proliferation
IL-2	T_H cells	Stimulates T-cell and B-cell proliferation
IL-4	T cells	Induces IgM to IgG class switching in B cells, suppresses inflammatory cytokine action
IL-5	T_H cells	Stimulates B-cell differentiation
IL-10	T cells, macrophages	Inhibits macrophage function, suppresses inflammatory cytokine action
IFN-γ	T_H cells, CTLs	Induces MHC expression in APCs, activates NK cells
TNF-a	Diverse	Induces inflammation, activates NO production in macrophages (page 000)
GM-CSF	T_H cells, CTLs	Stimulates growth and proliferation of granulocytes and macrophages

Two major subclasses of T cells can be distinguished by
the proteins on their surfaces and their biological functions.

1. **Cytotoxic T lymphocytes** (CTLs or T_C cells) screen
 the cells of the body for abnormalities. Under normal
 circumstances, healthy cells are not harmed by CTLs,
 but aged or infected cells, and possibly malignant cells,
 are attacked and killed. CTLs kill target cells by in-
 ducing them to undergo apoptosis. Two distinct cell-
 killing pathways have been discovered. In one path-
 way, the CTL releases perforins and granzymes into
 the space between the cells. *Perforins* are proteins that
 assemble within the membrane of the target cell to
 form transmembrane channels. *Granzymes* are prote-
 olytic enzymes that enter the perforin channels and
 activate caspases, which are the proteolytic enzymes
 that initiate the apoptotic response (page 663). In the
 alternate pathway, the CTL binds to a receptor on the
 target cell surface, activating a suicide pathway in the
 target cell similar to that of Figure 15.36. By killing
 infected cells, CTLs eliminate viruses, bacteria, yeast,
 protozoa, and parasites after they have found their way
 into host cells and are no longer accessible to circulat-
 ing antibodies. CTLs possess a surface protein called
 CD8 (cluster designation 8) and are referred to as
 $CD8^+$ cells.

2. **Helper T lymphocytes** (T_H cells) are regulatory cells,
 not killers. They are distinguished from CTLs by the

direct cell contact, many of these interactions are medi-
ated by highly active chemical messengers, called **cy-
tokines**, that work at very low concentrations. Cytokines
are small proteins produced by a wide variety of cells and
include interferons (IFNs), interleukins (ILs), and tumor
necrosis factors (TNFs). Cytokines bind to specific recep-
tors on the surface of a responding cell, generating an in-
ternal signal that alters the activity of the cell. In respond-
ing to a cytokine, a cell may prepare to divide, undergo
differentiation, or secrete its own cytokines. One family of
small cytokines, called *chemokines*, acts as chemoattrac-
tants by stimulating the migration of leukocytes. Different
types of lymphocytes and phagocytes possess receptors for
different chemokines, so that their migration patterns can
be separately controlled. A list of some of the best-studied
cytokines is given in Table 17.1.

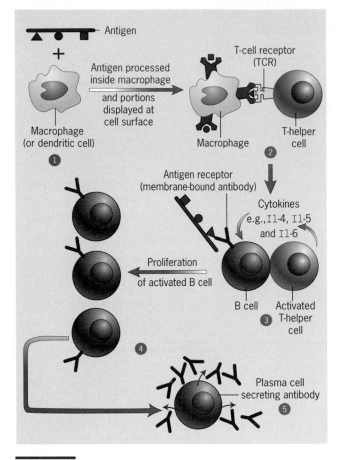

Figure 17.9 Highly simplified, schematic drawing showing the role of T_H cells in antibody formation. In step 1, the macrophage interacts with the complex antigen. The antigen is taken into the macrophage and cleaved into fragments, which are displayed at the cell surface. In step 2, the macrophage interacts with a T_H cell whose TCR is bound to one of the displayed antigen fragments (the green membrane protein is an MHC molecule, page 719). This interaction activates the T cell. In step 3, the activated T_H cell interacts with a B cell whose antigen receptor is bound to an intact, soluble antigen. B-cell activation is stimulated by cytokines (e.g., IL-4, IL-5, and IL-6) released by the T_H cell into the space separating it from the adjacent B cell. Interaction with the T_H cell activates the B cell, causing it to proliferate (step 4). The progeny of the activated B cell differentiate into plasma cells that synthesize antibodies that can bind the antigen (step 5).

presence of the protein CD4 at their surface rather than CD8.[1] T_H cells are activated by professional antigen-presenting cells, such as dendritic cells and macrophages (Figure 17.9). This is one of the first and most important steps in initiating an acquired immune response. Once activated, T_H cells regulate sub-

sequent immune responses by recognizing and activating other lymphocytes that are specific for the same antigen. Nearly all B cells require the help of T_H cells before they can mature and differentiate into antibody-secreting plasma cells.[2] B cells are activated by direct interaction with a T_H cell, as shown in Figure 17.9 (and in more detail in Figure 17.25). Thus, the formation of antibodies requires the activation of both T cells and B cells capable of interacting specifically with the same antigen. The importance of T_H cells becomes apparent when one considers the devastating effects wrought by HIV, the virus that causes AIDS. T_H cells are the primary targets of HIV, which enters a host cell by binding to its CD4 surface protein. Most HIV-infected people remain free of symptoms as long as their T_H-cell count remains relatively high—above 500 cells per mm³ (the normal count is over 1,000 cells per mm³). Once the count drops below approximately 200 cells per mm³, a person develops full-blown AIDS and becomes prone to attack by viral and cellular pathogens.

REVIEW ?

1. How does an infected cell in the body reveal its condition to a T cell? What is the T cell's response?

2. What is an APC? What types of cells can act as APCs?

3. Compare and contrast the properties and functions of a T_H and CTL.

17.4 SELECTED TOPICS ON THE CELLULAR AND MOLECULAR BASIS OF IMMUNITY

The Modular Structure of Antibodies

Antibodies are proteins produced by B cells and their descendants (plasma cells). B cells incorporate antibody molecules into their plasma membrane, where they serve as antigen receptors, whereas plasma cells secrete these proteins into the blood or other bodily fluids, where they serve as a molecular arsenal in the body's war against invading pathogens. Interaction between soluble antibodies and antigens on the surface of a virus or bacterial cell can neutralize the pathogen's ability to infect a host cell and facilitate the pathogen's ingestion and destruction by wandering phagocytes. The immune system produces millions of different antibody molecules that, taken together, can bind any type of foreign substance to which

[1]There are two major classes of helper T cells, T_H1 and T_H2 cells, which can be distinguished by the cytokines they secrete and their basic function. T_H1 cells produce IFN-γ and IL-2, and they protect the body against intercellular pathogens by activating macrophages to kill pathogens they might harbor. T_H2 cells produce other interleukins, and they protect against extracellular pathogens by activating B cells to produce antibodies. The two types of T_H cells differentiate from a common precursor following stimulation by different cytokines.

[2]As noted on page 708, a small number of antigens are able to elicit antibody production by B cells without the involvement of T cells. Included among these so-called "thymus-independent" antigens are large polymeric molecules with repeating substructures such as lipopolysaccharide, a component of bacterial cell walls.

Table 17.2			Classes of Human Immunoglobulins	

Class	Heavy Chain	Light Chain	Molecular Mass (kDa)	Properties
IgA	α	κ or λ	360–720	Present in tears, nasal mucus, breast milk, intestinal secretions
IgD	δ	κ or λ	160	Present in B-cell plasma membranes; function unknown
IgE	ε	κ or λ	190	Binds to mast cells, releasing histamine responsible for allergic reactions
IgG	γ	κ or λ	150	Primary blood-borne soluble antibodies; crosses placenta
IgM	μ	κ or λ	950	Present in B-cell plasma membranes; mediates initial immune response; activates bacteria-killing complement

the body may be exposed. Though the immune system exhibits great diversity through the antibodies it produces, a single antibody molecule can interact with only one or a few closely related antigenic structures.

Antibodies are globular proteins called **immunoglobulins**. Immunoglobulins are built of two types of polypeptide chains, larger **heavy chains** (molecular mass of 50,000 to 70,000 daltons) and smaller **light chains** (molecular mass of 23,000 daltons). The two types of chains are linked to one another in pairs by disulfide bonds. Five different classes of immunoglobulin (**IgA, IgD, IgE, IgG,** and **IgM**) have been identified. The different immunoglobulins appear at different times after exposure to a foreign substance and have different biological functions (Table 17.2). IgM molecules are the first antibodies secreted by B cells following stimulation by an antigen, appearing in the blood after a lag of a few days (Figure 17.10). IgM molecules have a relatively short half-life (about 5 days), and their appearance is followed by secretion of longer-lived IgG and/or IgE molecules. IgG molecules are the predominant antibodies found in the blood and lymph during a secondary response to most antigens (Figure 17.10). IgE molecules are produced at high levels in response to many parasitic infections. IgE molecules are also bound with high affinity to the surface of mast cells, triggering histamine release, which causes inflammation and symptoms of allergy. IgA is the predominant antibody in secretions of the respiratory, digestive, and urogenital tracts. The function of IgD is unclear.

There are two types of light chains, kappa (κ) chains and lambda (λ) chains, both of which are present in the immunoglobulins of all five classes. In contrast, each immunoglobulin class has a unique heavy chain that defines that class (Table 17.2).[3] We will focus primarily on the structure of IgGs. An IgG molecule is composed of two identical light chains and two identical heavy chains arranged to form a Y-shaped molecule, as shown in Figure 17.11*a* and described below.

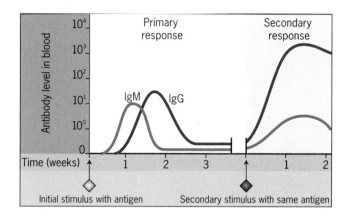

Figure 17.10 Primary and secondary antibody responses. A primary response, which is elicited by an initial exposure to an antigen, leads first to the production of soluble IgM antibody molecules, followed by production of soluble IgG antibody molecules. Appearance of blood-borne antibodies follows a delay of several days. When the antigen is reintroduced at a later time, a secondary response is initiated. In contrast with the primary response, the secondary response begins with production of IgG (as well as IgM) molecules, leads to a much higher antibody level in the blood, and occurs with almost no delay.

To determine the basis of antibody specificity, it was first necessary to determine the amino acid sequence of a number of specific antibodies. Normally, the first step in amino acid sequencing is to purify the particular protein to be studied. Under normal conditions, however, it is impossible to obtain a purified preparation of a specific antibody from the blood, because each person produces a large number of different antibody molecules that are too similar in structure to be separated from one another. The problem was solved when it was discovered that the blood of patients suffering from a type of lymphoid cancer called multiple myeloma contained large quantities of a single antibody molecule.

As described in Chapter 16, cancer is a monoclonal disease; that is, the cells of a tumor arise from the proliferation of a single wayward cell. Because a single lymphocyte *normally* produces only a single species of antibody, a

[3]Humans actually produce four related heavy chains as part of their IgG molecules (forming IgG1, IgG2, IgG3, and IgG4) and two related heavy chains as part of their IgA molecules (forming IgA1 and IgA2) (see Figure 17.17). These differences will not be mentioned in the following discussion.

(a)

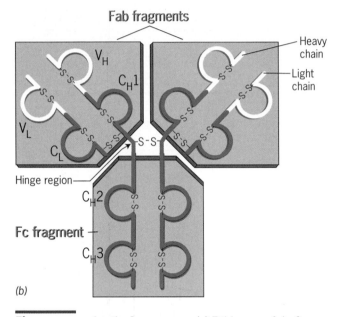

(b)

Figure 17.11 **Antibody structure.** (*a*) Ribbon model of an IgG molecule. The molecule contains four polypeptide chains: two identical light chains and two identical heavy chains. One of the heavy chains is shown in blue, the other in yellow, while both light chains are shown in red. The domains of each chain (two per light chain and four per heavy chain) are evident. (*b*) Schematic model showing the domain structure of an IgG molecule. The tertiary structure of each Ig domain is maintained by a disulfide bond. Domains comprising a constant region of the polypeptide chain are indicated by the letter C; domains comprising a variable region are indicated by the letter V. Each heavy chain contains three C_H regions (C_H1, C_H2, C_H3) and one V_H region at the N-terminus of the polypeptide. Each light chain contains one C_L and one V_L region at its N-terminus. The variable regions of each light and heavy chain form an antigen-combining site. Each Y-shaped IgG molecule contains two antigen-combining sites. Each IgG molecule can be fragmented by mild proteolytic treatment into two Fab fragments that contain the antigen-combining sites and an Fc fragment, as indicated. (A: COURTESY OF ALEXANDER MCPHERSON.)

patient with multiple myeloma produces large amounts of the specific antibody that is synthesized by the particular cell that became malignant. One patient produces one

highly abundant antibody species, and other patients produce different antibodies. As a result, investigators were able to purify substantial quantities of several antibodies from a number of patients and compare their amino acid sequences. An important pattern was soon revealed. It was found that half of each kappa light chain (110 amino acids at the amino end of the polypeptide) has a constant amino acid sequence among all kappa chains, whereas the other half varies from patient to patient. A similar comparison of amino acid sequences of several lambda chains from different patients revealed that they too consist of a section of constant sequence and a section whose sequence varies from one immunoglobulin to the next. The heavy chains of the purified IgGs also contain a variable (V) and a constant (C) portion. A schematic structure of one of these IgG molecules is shown in Figure 17.11*b*.

It was further found that, whereas approximately half of each light chain consists of a variable region (V_L), only one-quarter of each heavy chain is variable (V_H) among different patients; the remaining three-quarters of the heavy chain (C_H) is constant for all IgGs. The constant portion of the heavy chain can be divided into three sections of approximately equal length that are clearly homologous to one another. These homologous Ig units are designated C_H1, C_H2, and C_H3 in Figure 17.11*b*. It appears that the three sections of the C part of the IgG heavy chain (as well as those of heavy chains of the other Ig classes and the C portions of both kappa and lambda light chains) arose during evolution by the duplication of an ancestral gene that coded for an Ig unit of approximately 110 amino acids. The variable regions (V_H or V_L) are also thought to have arisen by evolution from the same ancestral Ig unit. Structural analysis indicates that each of the homologous Ig units of a light or heavy chain folds independently to form a compact domain that is held together by a disulfide bond (Figure 17.12). In an intact IgG molecule, each light chain domain associates with a heavy chain domain as shown in Figure 17.11*a,b*. Genetic analysis indicates that each domain is encoded by its own exon.

The specificity of an antibody is determined by the amino acids of the antigen-combining sites at the ends of each arm of the Y-shaped antibody molecule (Figure 17.11). The two combining sites of a single IgG molecule are identical, and each is formed by the association of the variable portion of a light chain with the variable portion of a heavy chain (Figure 17.11). The assembly of antibodies from different combinations of light and heavy chains allows a person to produce a tremendous variety of antibodies from a modest number of different polypeptides (page 718).

A closer look at the polypeptides of immunoglobulins reveals that the variable portions of both heavy and light chains contain subregions that are especially variable, or *hypervariable*, from one antibody to another (labeled Hv in Figure 17.12). Light and heavy chains both contain three hypervariable stretches that are clustered at the ends of each arm on the antibody molecule. As might be expected, the hypervariable regions play a prominent role in

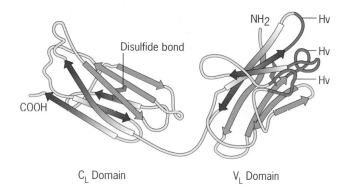

Figure 17.12 Antibody domains. Schematic drawing of a human lambda light chain synthesized by cells from a patient with multiple myeloma. The polypeptide undergoes folding so that the constant and variable portions are present in separate domains. Thick arrows represent β strands, which are assembled into β sheets. Each domain has two β sheets, which are distinguished by the red and orange colors. The three hypervariable (Hv) segments of the chain are clustered at one end of the variable domain, which forms part of the antigen-combining site of the antibody. (AFTER M. SCHIFFER ET AL., BIOCHEMISTRY 12:4628, 1973. © COPYRIGHT 1973, AMERICAN CHEMICAL SOCIETY.)

forming the structure of the antigen-combining site, which can range from a deep cleft to a narrow groove or a relatively flat pocket. Variations in the amino acid sequence of hypervariable regions account for the great diversity of antibody specificity, allowing these molecules to bind to antigens of every conceivable shape.

The combining site of an antibody has a complementary stereochemical structure to a particular portion of the antigen, which is called the **epitope** (or **antigenic determinant**). Because of their close fit, antibodies and antigens form stable complexes, even though they are joined only by noncovalent forces that individually are quite weak. The precise interaction between a particular antigen and antibody, as determined by X-ray crystallography, is shown in Figure 17.13.

An antigen typically contains a number of different epitopes and, thus, stimulates the production of a variety of different antibodies. Even if an antigen contained only a single epitope, its introduction into the body would still generate numerous antibody species, all capable of binding to the antigen, but in different configurations having different affinities. Because individual antigens can interact with different B-cell clones containing different membrane-bound immunoglobulins, the response is said to be *polyclonal*. Preparations of a single species of antibody, called a **monoclonal antibody**, directed against a specific epitope can be prepared in vitro by a technique described in Section 18.14. Monoclonal antibodies have proven invaluable tools in basic and applied biomedical research. One of their many uses is illustrated in the fluorescence micrographs seen throughout this book. In most of these micrographs, the location of a particular antigen within a cell is determined by incubation of a tissue section with a monoclonal antibody that has been covalently linked to a fluorescent dye, such as fluorescein.

Whereas the hypervariable portions of light and heavy chains determine an antibody's combining site specificity, the remaining portions of the variable domains provide a scaffold that maintains the overall structure of the combining site. The constant portions of antibody molecules are also important. Different classes of antibodies (IgA, IgD, IgE, IgG, and IgM) have different heavy chains whose constant regions differ considerably in length and sequence. These differences enable antibodies of various classes to carry out different biological functions. For example, heavy chains of an IgM molecule bind and activate one of the proteins of the complement system, which leads to the lysis of bacterial cells to which the IgM molecules are bound. Heavy chains of IgE molecules play an important role in allergic reactions by binding to specific receptors on the surfaces of mast cells, triggering the release of histamine. In contrast, heavy chains of an IgG molecule bind specifically to the surface receptors of macrophages and neutrophils, inducing these phagocytic cells to ingest the particle to which the antibodies are

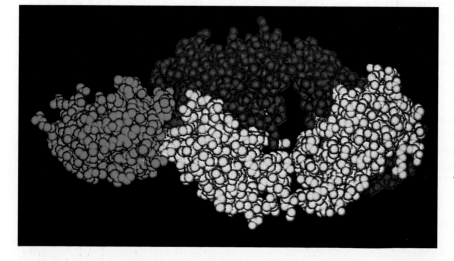

Figure 17.13 Antigen-antibody interaction. Space-filling model based on X-ray crystallography of a complex between lysozyme (green) and the Fab portion of an antibody molecule (see Figure 17.11). The heavy chain of the antibody is blue; the light chain is yellow. A glutamine residue of lysozyme is red. (REPRINTED WITH PERMISSION FROM A. G. AMIT, R. A. MARIUZZA, S. E. V. PHILLIPS, AND R. J. POLJAK, SCIENCE 233:749, 1986. © COPYRIGHT 1986, AMERICAN ASSOCIATION FOR THE ADVANCEMENT OF SCIENCE.)

bound. The heavy chains of IgG molecules are also important in allowing this class of antibodies to pass from the blood vessels of a mother to those of her fetus during pregnancy. While this provides the fetus and newborn with passive immunity to infectious organisms, it may cause a life-threatening condition called erythroblastosis fetalis. For this condition to occur, an Rh^- mother must have given birth to a child with an Rh^+ phenotype (Rh^+/Rh^- genotype) during a previous pregnancy. The mother is usually exposed to the Rh^+ fetal antigen during delivery of the first child, who is not affected. If, however, the mother should have a second Rh^+ pregnancy, antibodies present in her bloodstream can enter the fetal circulation and destroy the red blood cells of the fetus. Babies born with this condition are given a blood transfusion, which cleanses the blood of maternal antibodies.

DNA Rearrangement of Genes Encoding B- and T-Cell Antigen Receptors

As discussed above, each IgG molecule is composed of two light (L) chains and two heavy (H) chains. Both types of polypeptides consist of two recognizable parts—a variable (V) portion, whose amino acid sequence varies from one antibody species to another, and a constant (C) portion, whose amino acid sequence is identical among all H or L chains of the same class. What is the genetic basis for the synthesis of polypeptides having a combination of shared and unique amino acid sequences?

In 1965, William Dreyer of the California Institute of Technology and J. Claude Bennett of the University of Alabama put forward the "two gene–one polypeptide" hypothesis to account for antibody structure. In essence, Dreyer and Bennett proposed that each antibody chain is encoded by two separate genes—a C gene and a V gene—that somehow combine to form one continuous "gene" that codes for a single light or heavy chain. In 1976, Susumu Tonegawa, working at a research institute in Basel, Switzerland, provided clear evidence in favor of the

DNA rearrangement hypothesis. The basic outline of the experiment is shown in Figure 17.14. In this experiment, Tonegawa and his colleagues compared the length of DNA between the nucleotide sequences coding for the C and V portions of a specific antibody chain in two different types of mouse cells: early embryonic cells and malignant, antibody-producing myeloma cells. The DNA segments encoding the C and V portions of the antibody were widely separated in embryonic DNA but were very close to each other in DNA obtained from antibody-producing myeloma cells (Figure 17.14). These findings strongly suggested that DNA segments encoding the parts of antibody molecules became rearranged during the formation of antibody-producing cells.

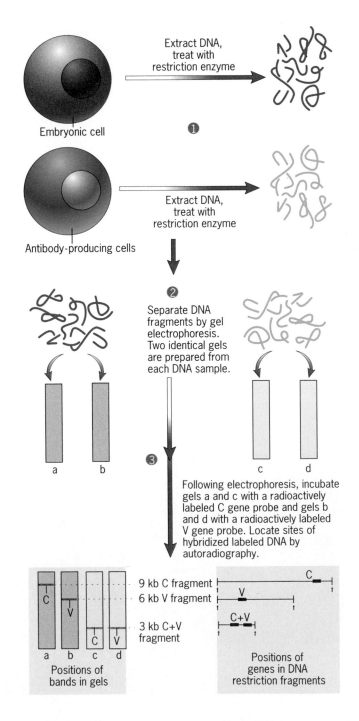

Figure 17.14 Experimental demonstration that genes encoding antibody light chains are formed by DNA rearrangement. DNA is first extracted from either embryonic cells or antibody-producing cancer cells and fragmented by a restriction endonuclease (step 1). The DNA fragments from the two preparations are fractionated separately by gel electrophoresis; two identical gels are prepared from each DNA sample (step 2). Following electrophoresis, each of the gels is incubated with a labeled probe containing either the variable (V) or constant (C) gene sequence (step 3). The location of the bound, labeled DNA in the gel is revealed by autoradiography and shown at the bottom of the figure. Whereas the V and C gene sequences are located on separate fragments in DNA from embryonic cells, the two sequences are located on the same small fragment in DNA from the antibody-producing cells. The V and C gene sequences are brought together during B-cell development by a process of DNA rearrangement.

Subsequent research revealed the precise arrangement of the DNA sequences that give rise to antibody genes. To simplify the discussion, we will consider only those DNA sequences involved in the formation of human kappa light chains, which are located on chromosome 2. The organization of sequences in germ-line DNA (i.e., the DNA of a sperm or egg) that are involved in the formation of human kappa light chains is shown in the top line (step 1) of Figure 17.15. In this case, a variety of different V_κ genes are located in a linear array and separated from a single C_κ gene by some distance. Nucleotide-sequence analysis of these V genes indicated that they are shorter than required to encode the V region of the kappa light chain. The reason became clear when other segments in the region were sequenced. The stretch of nucleotides that encodes the 13

amino acids at the carboxyl end of a V region is located at some distance from the remainder of the V_κ gene sequence. This small portion that encodes the carboxyl end of the V region is termed the *J segment*. As shown in Figure 17.15 there are five distinct J_κ segments of related nucleotide sequence arranged in tandem. The cluster of J_κ segments is separated from the C_κ gene by an additional stretch of over 2,000 nucleotides. A complete kappa V gene is formed, as shown in Figure 17.15 (steps 2–3), as a specific V_κ gene is joined to one of the J_κ segments with the intervening DNA deleted as described below. This process is catalyzed by a protein complex called *V(D)J recombinase*. As indicated in Figure 17.15, the V_κ gene sequence generated by this DNA rearrangement is still separated from the C_κ gene by 2,000 to 4,000 nucleotides. No further DNA rearrangement occurs in kappa gene assembly prior to transcription; the entire genetic region is transcribed into a large primary transcript (step 4) from which introns are excised by RNA splicing (step 5).

The process of DNA rearrangement (or DNA joining) is facilitated by highly conserved **signal sequences** located adjacent to the amino acid coding sequences in the DNA of the germ line (shown in Figure 17.16*a*). These signal sequences are located at the border of each V segment and each J segment. The signal sequences located

Figure 17.15 DNA rearrangements that lead to the formation of a functional gene that encodes an immunoglobulin κ chain. The organization of the variable (V), joining (J), and constant (C) DNA sequences within the genome is shown in step 1. Steps leading to the synthesis of the mature mRNA that codes for the κ chain polypeptide are described in the text. The random union of a V and J segment (steps 2 and 3) determines the amino acid sequence of the polypeptide. The space between the "chosen" J segment and the C segment (which may contain one or more J segments, as shown in the figure) remains as an intron in the gene. The portion of the primary transcript (step 4) that corresponds to this intron is removed during RNA processing (step 5).

Figure 17.16 Signal sequences involved in V_κ–J_κ joining. (*a*) The V_κ gene is flanked by a conserved heptamer and nonamer separated by a 12-base-pair spacer, whereas the J_κ gene is flanked by the same conserved heptamer and nonamer separated by a 23-base-pair spacer. (*b*) Hypothetical base-paired structure formed by complementary base sequences in the signal sequences flanking the V_κ and J_κ genes. This structure brings the ends of the V and J segments into close proximity to facilitate DNA joining.

adjacent to V and J segments are similar. They consist of a conserved heptamer (a 7-base-pair sequence) and a conserved nonamer (a 9-base-pair sequence) separated by a nonconserved spacer of *either* 12 or 23 base pairs in length. V_κ segments have a signal sequence with a 12-base-pair spacer, whereas J_κ segments have a signal sequence with a 23-base-pair spacer (Figure 17.16*a*). DNA joining always occurs between a signal sequence containing a 12-base-pair spacer and one containing a 23-base-pair spacer. The two segments of DNA are thought to base pair to one another, forming a stem-and-loop structure, as shown in Figure 17.16*b*. The difference between these two spacers—11 base pairs—is approximately one turn of the DNA double helix (see Figure 10.10*a*). This feature ensures that the same faces of the heptamer and nonamer would be aligned in both cases, allowing the same proteins (called RAG1 and RAG2) to recognize and cleave at both types of signals. The joining of two unlike signals, known as the 12/23 rule, ensures that a V segment is never joined to another V segment, and a J segment is never joined to another J segment.

To begin rearrangement, double-stranded cuts are made in the DNA between a V gene and its signal sequence and a J gene and its signal sequence. The four free ends that are generated are then joined in such a way that the V and J coding segments are linked to form an exon that encodes the variable region of the polypeptide chain, while the two signal sequences are linked to form a small circular piece of DNA that is displaced from the chromosome (Figure 17.15, step 3).

The rearrangement of Ig DNA sequences has important consequences for a lymphocyte. Once a specific V_κ sequence is joined to a J_κ sequence, no other species of kappa chain can be synthesized by that cell. It is estimated that the DNA of human germ cells contains approximately 40 functional V_κ genes. Thus, if we assume that any V sequence may join to any J sequence, we expect that a person can synthesize approximately 200 different kappa chains (5 J_κ segments × 40 V_κ genes). But this is not the only source of diversity among these polypeptides. The site at which a J sequence is joined to a V sequence can vary somewhat from one rearrangement to another so that the same V_κ and J_κ genes can be joined in two different cells to produce kappa light chains having different amino acid sequences. Additional variability is achieved by the enzyme deoxynucleotidyl transferase, which inserts nucleotides at the sites of strand breakage. These sources of additional variability increase the diversity of kappa chains an additional tenfold, bringing the number to about 2,000 species. The site at which V and J sequences are joined is part of one of the hypervariable regions of each antibody polypeptide (Figure 17.12). Thus, slight differences at a joining site can have important effects on antibody–antigen interaction.

We have restricted our discussion to kappa light chains for the sake of simplicity. Similar types of DNA rearrangement occur during the commitment of a cell to the synthesis of a particular lambda light chain and to a particular heavy chain. Whereas the variable regions of light chains are formed from two distinct segments (V and J segments), the variable regions of heavy chains are formed from three distinct segments (V, D, and J segments) by similar types of rearrangement. The human genome contains approximately 50 different functional V_H segments, approximately 30 different D_H segments, and 6 different J_H segments. Given the additional diversity stemming from the variability in V_H–D_H and D_H–J_H joining, a person is able to synthesize approximately 100,000 different heavy chains. The antigen receptors of T cells (TCRs) also consist of a type of heavy and light chain, whose variable regions are formed by a similar process of DNA rearrangement.

The formation of antibody genes by DNA rearrangement illustrates the potential of the genome to engage in dynamic activities. Because of this rearrangement mechanism, a handful of DNA sequences that are present in the germ line can give rise to a remarkable diversity of gene products. As discussed above, a person synthesizes roughly 2,000 different species of kappa light chains and 100,000 different species of heavy chains. If any kappa light chain can combine with any heavy chain, a person can theoretically produce more than 200 million different antibody species from a few hundred genetic elements present in the germ line.[4]

We have seen how antibody diversity arises from (1) the presence of multiple V exons, J exons and D exons in the DNA of the germline, (2) variability in V–J and V–D–J joining, and (3) the enzymatic insertion of nucleotides. An additional mechanism for generating antibody diversity, referred to as **somatic hypermutation**, occurs long after DNA rearrangement is complete. When a specific antigen is reintroduced into an animal following a period of time, antibodies produced during the secondary response have a much greater affinity for the antigen than those produced during the primary response. Increased affinity is due to small changes in amino acid sequence of variable regions of the heavy and light antibody chains. These sequence changes result from mutations in the genes that encode these polypeptides. It is estimated that rearranged DNA elements encoding antibody V regions have a mutation rate 10^5 times greater than that of other genetic loci in the same cell. The mechanism responsible for this increased level of V-region mutation is not understood, but it is thought to involve some form of error-prone DNA repair. Regardless, those B cells that carry mutations producing Ig molecules with greater antigen affinity are preferentially selected following antigen reexposure. Selected cells proliferate to form clones that undergo additional rounds of somatic mutation and selection, whereas nonselected cells that express low affinity Igs undergo apoptosis. In this way, the antibody response to recurrent or chronic infections improves over time.

[4]A roughly comparable number of antibodies containing lambda light chains can also be generated.

Figure 17.17 Arrangement of the C genes for the various human heavy chains. In humans, the heavy chains of IgM, IgD, and IgE are encoded by a single gene, whereas those of IgG are encoded by four different genes and IgA by two different genes (see footnote, page 713).

Once a cell is committed to form a specific antibody, it can switch the class of Ig it produces (e.g., from IgM to IgG) by changing the heavy chain produced in the cell. This process, known as **class switching**, occurs without changing the combining site of the antibodies synthesized. Recall that there are five different types of heavy chains distinguished by their constant regions. The genes that encode the constant regions of heavy chains (C_H portions) are clustered together in a complex, as shown in Figure 17.17. Class switching is accomplished by moving a different C_H gene next to the VDJ gene that was formed previously by DNA rearrangement. Class switching is under the direction of cytokines secreted by helper T cells during their interaction with the B cell producing the antibody molecule. For example, a helper T cell that secretes IFN-γ, induces a switch in the adjacent B cell from synthesis of IgM to synthesis of one of the IgG classes. Class switching allows a lineage of B cells to continue to produce antibodies having the same specificity but different effector functions.

Membrane-Bound Antigen Receptor Complexes

The recognition of antigen by both B and T lymphocytes occurs at the cell surface. An antigen receptor on a B cell (a B-cell receptor, or BCR) consists of a membrane-bound immunoglobulin that binds selectively to a portion of an intact antigen (i.e., the epitope) (Figure 17.18a). In contrast, the antigen receptor on a T cell (a T-cell receptor, or TCR, Figure 17.18b) recognizes and binds to a small fragment of an antigen, typically a peptide about 7 to 25 amino acids in length, that is held at the surface of another cell (described below). Both types of antigen receptors are part of large membrane-bound protein complexes that include invariant proteins as depicted in Figure 17.18. The invariant polypeptides associated with BCRs and TCRs play a key role in signal transduction.

Each subunit of a TCR contains two Ig-like domains, indicating that they share a common ancestry with BCRs. Like the heavy and light chains of the immunoglobulins, one of the Ig-like domains of each subunit of a TCR has a variable amino acid sequence; the other domain has a constant amino acid sequence (Figure 17.18). Recent X-ray crystallographic studies have shown that the two types of antigen receptors also share a similar three-dimensional shape.

Figure 17.18 Structure of the antigen-receptors of a B cell and a T cell. (*a*) The BCR of a B cell is a membrane-bound version of an immunoglobulin associated with a pair of invariant α chains and a pair of invariant β chains. The α and β chains are also members of the Ig superfamily. (*b*) The TCR of a T cell consists of an α and β polypeptide chain linked to one another by a disulfide bridge. Each polypeptide contains a variable domain that forms the antigen-binding site and a constant domain. The TCR is associated with six other invariant polypeptides as indicated in the illustration. (A small fraction of T cells contain a different type of TCR consisting of a γ and δ subunit. These cells are not restricted to recognition of MHC-peptide complexes, and their function is not well understood.)

The Major Histocompatibility Complex

During the first part of the twentieth century, clinical researchers discovered that blood could be transfused from one person to another, as long as the two individuals were compatible for the ABO blood group system. The success of blood transfusion led to the proposal that skin might also be grafted between individuals. This idea was tested during World War II when skin grafts were attempted on pilots and other military personnel who had received serious burns. The grafts were rapidly and completely rejected. After the war, researchers set out to determine the basis for tissue rejection. It was discovered that skin could be grafted successfully between mice of the same inbred strain, but that grafts between mice of different strains were rapidly rejected. Mice of the same inbred strain are like identical twins; they are genetically identical. Subsequent studies revealed that the genes that governed tissue graft rejection were clustered in a region of the genome that was named the **major histocompatibility complex (MHC)**. The MHC consists of approximately 15 different genes, most of which are highly polymorphic: over 500 different alleles of MHC genes have been identified (Table 17.3), far more than any other loci in the human genome. It is very unlikely, therefore, that two individuals

Table 17.3	MHC class II alleles
Locus	*Number of alleles*
HLA-DRA	2
HLA-DRB	179
HLA-DQA	18
HLA-DQB	29
HLA-DPA	8
HLA-DPB	69
HLA-DMA	4
HLA-DMB	5
Total	314

MHC class I alleles	
Locus	*Number of alleles*
HLA-A	67
HLA-B	149
HLA-C	39
HLA-E	5
HLA-F	1
HLA-G	6
Total	266

Source: P. Parham and T. Ohta, *Science* 272:67, 1996. © Copyright 1996, American Association for the Advancement of Science.

(a)

(b)

Figure 17.19 Human APCs can present large numbers of peptides. (*a*) Schematic model of the variety of class I MHC molecules an individual might possess. As indicated in Table 17.3, this class of MHC protein is encoded by six alleles, three of which are represented by a large number of alleles. This particular individual is heterozygous at the HLA-A, -B, and -C loci and homozygous at the HLA-E, -F, and -G loci, which gives them a total of 9 different MHC class I molecules. (The difference between class I and II MHC is discussed on page 721). (*b*) Schematic model illustrating the variety of peptides that can be presented by the protein encoded by a single MHC allele.

in a population have the same combination of MHC alleles. This is the reason that transplanted organs are so likely to be rejected and why transplant patients are given drugs, such as cyclosporin A, to suppress the immune system following surgery. Cyclosporin A is a cyclic peptide produced by a soil fungus. Cyclosporin A inhibits a particular phosphatase in the signaling pathway leading to the production of cytokines required for T-cell activation. Although these drugs help prevent graft rejection, they make patients susceptible to opportunistic infections similar to those that strike people with immunodeficiency diseases such as AIDS.

It is obvious that proteins encoded by the MHC did not evolve to prevent indiscriminate organ transplantation, which raises the question of their normal role. Long after their discovery as transplantation antigens, MHC proteins were shown to be involved in antigen presentation. Some of the key experiments that led to our current understanding of antigen presentation are discussed in the Experimental Pathways at the end of the chapter.

It was noted earlier that T cells are activated by an antigen that has been dissected into small peptides and displayed on the surface of an antigen-presenting cell (an APC). These small fragments of antigen are held at the surface of the APC in the grip of MHC proteins. Each species of MHC molecule can bind a large number of different peptides that share structural features that allow them to fit into its binding site (see Figure 17.22). For example, all of the peptides that are capable of binding to a protein encoded by a particular MHC allele, such as

HLA-B8, may contain a specific amino acid at a certain position, which allows it to fit into the MHC groove.

Given the fact that each individual expresses a number of different MHC proteins (as in Figure 17.19*a*), and each MHC variant may be able to bind large numbers of different peptides (as in Figure 17.19*b*), a dendritic cell or macrophage should be able to display a vast array of peptides. At the same time, not every person is capable of presenting every possible peptide in an effective manner, which is thought to be a major factory in determining differences in susceptibility in a population to different infectious diseases, including AIDS. For example, the HLA-DRB1*1302 allele is correlated with resistance to a certain type of malaria and hepatitis B infection. The MHC alleles present in a given population have been shaped by natural selection. Those persons who possess MHC alleles that are best able to present the peptides of a particular infectious agent will be the most likely to survive an infection by that agent. Conversely, persons who lack these alleles are more likely to die without passing on their

alleles to offspring. As a result, populations tend to be more resistant to diseases to which their ancestors have been routinely exposed. This could explain why Native American populations have been devastated by certain diseases, such as measles, that produce only mild symptoms in persons of European ancestry.

The entire process of T-cell-mediated immunity rests on the basis that small peptides derived from the proteins of a pathogen differ in structure from those derived from the proteins of the host. Consequently, one or more peptides held at the surface of an APC serve as a small representation of the pathogen, providing the cells of the immune system with a "glimpse" of the type of pathogen that is hidden within the cytoplasm of the infected cell. As described on page 723, nearly any cell in the body can function as an APC. Most cells present antigen as an incidental activity that alerts the immune system to the presence of a pathogen, but some professional APCs (e.g., dendritic cells, macrophages, and B cells) are specialized for this function, as discussed later in the chapter.

When a T cell interacts with an APC, it does so by having its TCRs dock onto the MHC molecules projecting from the APC's surface (Figure 17.20). This interaction brings a TCR of a T cell into an orientation that allows it to recognize the specific peptide displayed within a groove of an MHC molecule. The interaction between MHC proteins and TCRs is strengthened by additional contacts that form between cell-surface components, such as occur between CD4 or CD8 molecules on a T cell and MHC proteins on an APC (Figure 17.20).

MHC proteins can be subdivided into two major groups, **MHC class I** and **MHC class II** molecules. MHC class I molecules consist of one polypeptide chain encoded by an MHC allele (known as the heavy chain) associated noncovalently with a non-MHC polypeptide known as β_2-microglobulin (see Figure 2, page 733). Differences in the amino acid sequence of the heavy chain are responsible for dramatic changes in the shape of the molecule's peptide-binding groove. MHC class II molecules also consist of a heterodimer, but both subunits are encoded by MHC alleles. Both classes of MHC molecule as well as β_2-microglobulin contain Ig-like domains and are thus members of the immunoglobulin superfamily. Whereas most cells of the body express MHC class I molecules on their surface, MHC class II molecules are expressed primarily by professional APCs.

The two classes of MHC molecules display antigens that originate from different sites within a cell, though some overlap has been reported. MHC class I molecules are predominantly responsible for displaying antigens that originate within the cytosol of a cell, that is, *endogenous* proteins. In contrast, MHC class II molecules primarily display fragments of exogenous antigens that are taken into a cell by phagocytosis. The proposed pathways by which these two classes of MHC molecules pick up their antigen fragments and display them on the plasma membrane are described here and shown in Figure 17.21.

(a)

(b)

Figure 17.20 Interaction between a macrophage and a T cell during antigen presentation. (*a*) Electron micrograph of the two types of cells as they interact. (*b*) Schematic model showing some of the proteins involved in an interaction between an APC and an immature cytotoxic T lymphocyte (CTL) or helper T (T_H) cell. Antigen recognition occurs as the TCR of the T cell recognizes a peptide fragment of the antigen bound to a groove in an MHC molecule of the APC. As discussed in the text, CTLs recognize antigen in combination with an MHC class I molecule, whereas T_H cells recognize antigen in combination with an MHC class II molecule. (A: COURTESY OF ALAN S. ROSENTHAL; B: AFTER L. CHATENOUD, MOL. MED. TODAY 4:26, 1998, WITH PERMISSION FROM ELSEVIER SCIENCE.)

■ **Processing of class I MHC-peptide complexes** (Figure 17.21*a*). Antigens located in the cytosol of an APC are degraded into short peptides by proteases that are part of multimeric complexes called proteasomes (page 545). These proteases cleave the antigen into fragments approximately 8 to 10 residues long that are suitable for binding within a groove of an MHC class I molecule (Figure 17.22*a*). The peptides are then transported across the membrane of the rough endoplasmic reticulum and into its lumen by a dimeric protein called TAP

Figure 17.21 **Processing pathways for antigens that become associated with MHC class I and class II molecules.** (*a*) A proposed pathway for the assembly of an MHC class I-peptide complex. In step 1, the heavy chain of the MHC protein is synthesized by a membrane-bound ribosome and translocated into the ER membrane. The MHC heavy chain becomes associated with calnexin (step 2), a chaperone in the ER membrane, and the dimeric complex binds the invariant β_2m chain (step 3). The MHC complex then becomes associated with another ER membrane protein, TAP (step 4). Meanwhile, cytosolic antigens are taken into proteasomes (step A) and degraded into small peptides (step B). The peptides are transported into the ER lumen by the TAP protein, where they become bound within the groove of the MHC molecule (step 5). Calnexin and TAP dissociate from the MHC complex (step 6), which is transported along the biosynthetic/secretory pathway through the Golgi complex (step 7) to the plasma membrane, where it is ready to interact with the TCR of a CTL. (*b*) A proposed path-

way for the assembly of an MHC class II-peptide complex. In step 1, the MHC protein is synthesized by a membrane-bound ribosome and translocated into the ER membrane, where it becomes associated with Ii (step 2), a protein that blocks the MHC-peptide-binding site. The MHC-Ii complex passes through the Golgi complex (step 3) and into a transport vesicle (step 4). Meanwhile, an extracellular protein antigen is taken into the APC by endocytosis (step A) and delivered to a lysosome (step B), where the antigen is fragmented into peptides. The lysosome containing antigenic fragments fuses with the transport vesicle containing the MHC-Ii complex (step 5), leading to the degradation of the Ii protein and association between the antigenic peptide fragment and the MHC class II molecule (step 6). The MHC-peptide complex is transported to the plasma membrane (step 7) where it is ready to interact with the TCR of a T_H cell. (A: FROM D. B. WILLIAMS ET AL., TRENDS CELL BIOL. 6:271, 1996, WITH PERMISSION FROM ELSEVIER SCIENCE.)

(a)

(b)

Figure 17.22 Peptides produced by antigen processing bind within a groove of the MHC protein molecule. These models illustrate the binding of peptides to an MHC class I (*a*) and MHC class II (*b*) molecule. The molecular surfaces of the MHC molecules are shown in white and the peptide in the peptide-binding site in color. The peptide in *a* is derived from the influenza virus matrix protein, and the peptide in *b* is derived from influenza virus hemagglutinin protein. The N-terminus of each peptide is on the left. (A, B: COURTESY OF T. JARDETZKY; FROM TRENDS BIOCHEM SCI. 22:378, 1997, WITH PERMISSION FROM ELSEVIER SCIENCE.)

(an acronym for *t*ransporter *a*ssociated with antigen *p*rocessing) (Figure 17.21*a*). Once in the ER lumen, the peptide can bind to a newly synthesized MHC class I molecule, which is an integral protein of the ER membrane. The MHC-peptide complex moves through the biosynthetic pathway (page 280) until it reaches the plasma membrane where the peptide is displayed.

■ **Processing of class II MHC-peptide complexes** (Figure 17.21*b*). MHC class II molecules are also synthesized as membrane proteins of the RER, but they are joined noncovalently to a protein called Ii, which blocks the peptide-binding site of the MHC molecule (Figure 17.21*b*). Following its synthesis, the MHC class II-Ii complex moves out of the ER along the biosynthetic pathway, directed by targeting sequences located within the cytoplasmic domain of Ii. MHC class I and II molecules are thought to separate from one another in the *trans* Golgi network (TGN), which is the primary sorting compartment along the biosynthetic pathway (page 307). An MHC class I-peptide complex is directed toward the cell surface, whereas an MHC class II-Ii complex is directed into an endosome or lysosome, where the Ii protein is digested by acid proteases. An MHC class II molecule is then free to bind to peptides digested from antigens that were taken into the cell and directed along the endocytic pathway (Figure 17.21*b*).[5] The MHC class II-peptide complex is then moved to the plasma membrane where it is displayed, as shown in Figure 17.22*b*.

Once on the surface of an APC, MHC molecules direct the cell's interaction with different types of T cells (Figure 17.20). Cytotoxic T lymphocytes (CTLs) recognize their antigen in association with MHC class I molecules; they are said to be *MHC class I restricted*. Under normal circumstances, cells of the body that come into contact with CTLs display fragments of their own normal proteins in association with their MHC class I molecules. Normal cells displaying normal protein fragments are ignored by the body's T cells, because T cells capable of binding with high affinity to peptides derived from normal cellular proteins are eliminated as they develop in the thymus. In contrast, when a cell is infected, it displays fragments of viral proteins in association with its MHC class I molecules. These cells are recognized by CTLs bearing TCRs whose binding sites are complementary to the viral peptides, and the infected cell is destroyed. The presentation of a single foreign peptide on the surface of a cell is probably sufficient to invite an attack by a CTL. Because virtually all cells of the body express MHC class I molecules on their surface, CTLs can combat an infection regardless of the type of cell affected. CTLs may also recognize and destroy cells that display abnormal (mutated) proteins on their surfaces, which could play a role in the elimination of potentially life-threatening tumor cells.

Another role for MHC class I molecules has come to light. One of the strategies employed by some viruses to evade a host's immune system is to suppress the expression

[5]Peptides generated in lysosomes and attached to MHC class II molecules tend to be longer (10 to 25 residues) than those generated in proteasomes and attached to MHC class I molecules (8 to 10 residues).

of MHC class I molecules. This effectively makes the infected host cell "invisible" to cytotoxic T cells. Some metastatic tumor cells also lose MHC expression. Though such cells may become resistant to CTL attack, they become sensitive to attack by another arm of the body's defense system, namely, the NK cells of the innate immune system (page 705). NK cells bear receptors on their surface that recognize self MHC class I proteins on the surfaces of other cells in the body. When these receptors bind to MHC class I proteins on a normal body cell, the cytotoxic activity of the NK cell is inhibited. In contrast, when an infected cell loses one or more of its MHC class I proteins, it becomes an NK cell target (Figure 17.23).

In contrast to CTLs, helper T cells recognize their antigen in association with MHC class II molecules; they are said to be *MHC class II restricted*. As a result, helper T cells are activated primarily by exogenous (i.e., extracellular) antigens (Figure 17.21*b*), such as those that occur as part of bacterial cell walls or bacterial toxins. MHC class II molecules are found predominantly on B cells, dendritic cells, and macrophages. These are the lymphoid cells that ingest foreign, extracellular materials and present the fragments to helper T cells, as discussed later. Helper T cells that are activated in this way can then stimulate B cells to produce soluble antibodies that bind to the exogenous antigen wherever it is located in the body.

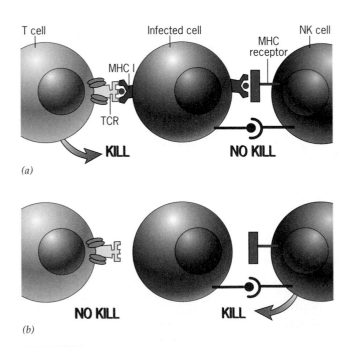

(a)

(b)

Figure 17.23 MHC proteins and viral infections. (*a*) Cytotoxic T lymphocytes (CTLs) recognize infected cells by the viral peptides presented in combination with an MHC molecule. Natural killer (NK) cells, on the other hand, do not kill a cell displaying MHC molecules, even if it is infected. (*b*) Some viruses evade CTL destruction by causing the host cell to stop displaying MHC molecules. The absence of MHC proteins on a cell's surface makes it a target for an NK cell. (REPRINTED WITH PERMISSION FROM K. KÄRRE AND R. M. WELSH, NATURE 386:446, 1997. © COPYRIGHT 1997, BY MACMILLAN MAGAZINES LIMITED.)

MHC molecules are obviously important in antigen presentation. But if this were the only function of the MHC, one wouldn't expect the genetic loci encoding these proteins to be so highly polymorphic (Table 17.3). It has long been speculated that MHC polymorphism gives members of a population an individuality by which they can be distinguished from other members. Evidence favoring this idea has recently been obtained from a series of fascinating behavioral studies in both mice and humans. These studies indicate that characteristic differences in body odor from one person to another (or between mice of different inbred strains) can be attributed to differences in specific MHC alleles. These differences may arise from soluble MHC proteins excreted in sweat and the effects of these proteins on the growth of bacterial flora. Not only does the MHC contribute to body odor, it also contributes to olfactory perception, particularly in female mammals. Studies indicate that mating preferences in mice are strongly influenced by MHC genotypes, and the same may be true of humans. This is supported by a preliminary study in which women were given sweaty clothing that had been worn by different men and asked to select items with a pleasurable aroma. The selected items were generally from men whose MHC loci are the most different from their own. Mating between individuals having different MHC alleles would provide offspring with the greatest variety of different MHC molecules, which in turn would give them the ability to display the widest possible range of peptides.

Distinguishing Self from Nonself

T cells gain their identity in the thymus. When an immature T cell migrates to the thymus from the bone marrow, it lacks the cell surface proteins that mediate T-cell function, most notably its TCRs. Once a T cell undergoes the DNA rearrangements that enable it to produce a specific TCR, the cell is subjected to a complex screening process in the thymus that selects for cells having potentially useful T-cell receptors (Figure 17.24). T cells whose TCRs have a high affinity for peptides derived from the body's own proteins are destroyed (Figure 17.24*a*). This process of *negative selection* greatly reduces the likelihood that the immune system will attack its own tissues. Both types of T cells (CTLs and T$_H$ cells) are subjected to negative selection, and thus the body is protected from both humoral and cell-mediated autoimmune responses.

The generation of T cells requires more than negative selection. When a TCR interacts with a foreign peptide on the surface of an APC, it must recognize both the peptide and the MHC molecule holding that peptide (discussed on page 731). Consequently, T cells whose TCRs do not recognize self-MHC molecules are of little value. The immune system screens out such cells by requiring that T cells recognize self-MHC—self-peptide complexes with low affinity. T cells whose TCRs are unable to recognize self-MHC complexes undergo apoptosis in the thymus, a process referred to as "death by neglect" (Figure 17.24*b*).

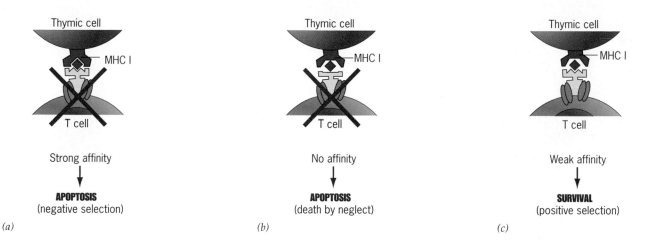

(a) (b) (c)

Figure 17.24 Determining the fate of a newly formed T cell in the thymus. A screening process takes place in the thymus that selects for T cells with appropriate TCRs. (*a*) Those T cells whose TCR exhibits strong affinity for MHC molecules bearing self-peptides are eliminated by apoptosis (negative selection). (*b*) Those T cells whose TCR fails to recognize MHC molecules bearing self-peptides also die by apoptosis (death by neglect). (*c*) In contrast, those T cells whose TCR exhibits weak affinity for MHC molecules bearing self-peptides survive (positive selection) and ultimately leave the thymus to constitute the peripheral T-cell population of the body.

In contrast, T cells whose TCRs exhibit weak (low affinity) recognition toward self-MHC—self-peptide complexes are stimulated to remain alive but are not activated (Figure 17.24*c*). This process of selective survival is referred to as *positive selection*.

Recognition of class I self-MHC complexes commits an immature T cell to differentiate into a CD8$^+$ killer T cell, whereas recognition of class II self-MHC complexes commits an immature T cell to differentiate into a CD4$^+$ helper T cell. It is estimated that less than five percent of T cells that mature in the thymus survive these negative and positive screening events. T cells that do survive leave the thymus and migrate through the peripheral lymphoid tissues. Once in the periphery, these T cells react strongly to the presence of cells bearing self-MHC molecules bound to foreign peptides. These same T cells also respond strongly to cells bearing nonself-MHC molecules, such as the cells of a transplanted organ from a mismatched donor. In this latter case, they initiate a widespread attack against the graft cells that can result in organ rejection.

Several different types of cancer are currently treated by bone marrow transplantation, which creates special immune problems. In this procedure, the patient is first irradiated to destroy his or her immune cells and then injected with donor bone marrow that contains lymphoid stem cells that can repopulate the patient's lymphoid tissues. Because the patient's own immune system is destroyed by the radiation, there is no threat that the graft will be rejected. However, there is a threat that the grafted cells will produce T cells that recognize the patient's cells as foreign and mount an attack against them. This response is called a **graft-versus-host (GVH)** reaction. The threat of a GVH reaction is minimized by matching donor and recipient tissues, but small differences in MHC molecules can still have serious repercussions. To minimize the risk, T cells that react with host cells can be removed from the donated bone marrow by treatment of the graft with antibodies that react specifically with this population of cells. Attention has recently become focused on an alternative source of lymphoid stem cells: blood from umbilical cords. The stem cells of cord blood are much less likely to generate a GVH reaction than those of bone marrow, even in recipients who are immunologically dissimilar to the donor. In addition, unlike bone marrow, which must be surgically extracted from a donor, cord blood, which is normally discarded at birth, can be kept in blood banks and is available immediately when needed.

Lymphocytes Are Activated by Cell-Surface Signals

Lymphocytes communicate with other cells through an array of cell-surface proteins. As discussed earlier, T-cell activation requires an interaction between the TCR of the T cell and an MHC-peptide complex on the surface of another cell. This interaction provides specificity, which ensures that only T cells that bind the antigen are activated. T-cell activation also requires a second signal, called the **costimulatory signal**, which is delivered through a second type of receptor on the surface of a T cell. This receptor is distinct and spatially separate from the TCR. Unlike the TCR, the receptor that delivers the costimulatory signal is not specific for a particular antigen and does not require an MHC molecule to bind. The best-studied of these interactions is between helper T cells and professional antigen-presenting cells (e.g., dendritic cells and macrophages).

Activation of Helper T Cells by Professional APCs T$_H$ cells recognize antigen fragments on the surface of dendritic cells and macrophages that are lodged in the binding cleft of MHC class II molecules. A costimulatory signal is delivered to a T$_H$ cell as the result of an interaction be-

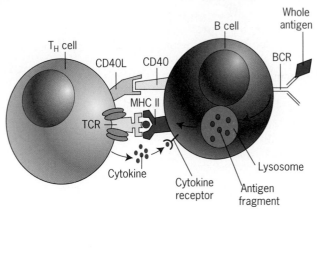

(a) (b)

Figure 17.25 Lymphocyte activation. The left portion of this illustration, designated part *a*, depicts the interaction between a professional APC, in this case a dendritic cell, and a T$_H$ cell. The right portion, designated part *b*, depicts the interaction between the activated T$_H$ cell and a B cell. (*a*) Specificity in this cell-cell interaction derives from recognition by the TCR on the T$_H$ cell of the MHC class II-peptide complex displayed on the surface of the dendritic cell. Interaction between CD28 of the T cell and a B7 protein of the dendritic cell provides a nonspecific costimulatory signal that is required for T-cell activation.

(*b*) Specificity in this cell-cell interaction derives from recognition by the TCR on the T$_H$ cell of the MHC class II-peptide complex displayed on the surface of the B cell. The peptide displayed by the B cell is derived from protein molecules that had initially bound to the BCRs at the cell surface. These bound antigens are taken up by endocytosis, fragmented in lysosomes, and bound to MHC class II molecules, as in Figure 17.21*b*. (AFTER E. LINDHOUT ET AL., IMMUNOL. TODAY 18:574, 1997, WITH PERMISSION FROM ELSEVIER SCIENCE.)

tween a protein known as CD28 on the surface of the T$_H$ cell and a member of the B7 family of proteins on the surface of the APC (Figure 17.25*a*). The B7 protein appears on the surface of the APC after the phagocyte ingests foreign antigen. If the T$_H$ cell does not receive this second signal from the APC, rather than becoming activated, the T$_H$ cell either becomes nonresponsive (anergized) or is stimulated to undergo apoptosis (deleted). Because professional APCs are the only cells capable of delivering the costimulatory signal, they are the only cells that can initiate a T$_H$ response. As a result, normal cells of the body that bear proteins capable of combining with the TCRs of a T cell cannot activate T$_H$ cells. Thus, the requirement by a T$_H$ cell for two activation signals protects normal body cells from autoimmune attack involving T$_H$ cells.

Prior to its interaction with an APC, a T$_H$ cell can be described as a resting cell, that is, one that has withdrawn from the cell cycle (a G$_0$ cell, page 582). Once it receives the dual activation signals, a T$_H$ cell is stimulated to reenter the G$_1$ phase of the cell cycle and eventually to progress through S phase into mitosis. Thus, interaction of T cells with a specific antigen leads to the proliferation (*clonal expansion*) of those cells capable of responding to that antigen. In addition to triggering cell division, activation of a T$_H$ cell causes it to synthesize and secrete cytokines (most notably IL-2). Cytokines produced by activated T$_H$ cells

act on other cells of the immune system (including B cells and macrophages) and also back on the T$_H$ cells that secreted the molecules. The source and function of various cytokines were indicated in Table 17.1.

We have seen in this chapter how immune responses are stimulated by ligands that activate receptor signaling pathways. But many of these events are also influenced by inhibitory stimuli, so that the ultimate response by the cell is determined by a balance of positive and negative influences. For example, interaction between CD28 and a B7 protein delivers a positive signal to the T cell leading to its activation. Once the T cell has been activated, it produces another cell-surface protein called CTLA4 that is similar to CD28 in structure and also interacts with B7 proteins of the APC. Unlike the CD28-B7 interaction, however, contact between CTLA4 and B7 leads to inhibition of the T cell's response, rather than activation. The need for balance between activation and inhibition is most evident in mice that have been genetically engineered to lack the gene encoding CTLA4. These mice die as a result of massive overproliferation of T cells. Findings with CTLA4 have recently been put to clinical use. Soluble forms of CTLA4 are capable of interacting with B7 receptors on the surface of APCs, leading to inactivation (anergization) of T cells. It was noted above that one of the major risks of bone marrow transplantation is an immunological attack

by transplanted T cells against host tissues. The risk of this problem is particularly acute when the donor and recipients are not very well matched. Results of one preliminary clinical study indicate that treatment of poorly matched human bone marrow cells with CTLA4 can dramatically lower the risk that the donor cells will attack the recipient's tissue.

Activation of B Cells by T$_H$ Cells T$_H$ cells bind to B cells whose receptors recognize the same antigen. The antigen initially binds to the immunoglobulin (BCR) at the B-cell surface. The bound antigen is then taken into the B cell, where it is processed enzymatically, and its fragments are displayed in combination with MHC class II molecules (Figure 17.25*b*). Activation of a B cell follows transmission of several signals from the T$_H$ cell to the B cell. Some of these signals are transmitted directly from one cell surface to the other through an interaction between complementary proteins, such as CD40 and CD40 ligand (CD40L) (Figure 17.25*b*). Binding between CD40 and CD40L generates signals that help move the B cell from the G$_0$ resting state back into the cell cycle. Other signals are transmitted by cytokines released by the T cell into the space separating it from the nearby B cell. This process is not unlike the way neurotransmitters act across a neural synapse (page 171). Cytokines released by the T cells into the "immunological synapse" include IL-4, IL-5, IL-6, and IL-10. Interleukin 4 is thought to stimulate the B cell to switch from producing the IgM class to producing the IgG or IgE class. Other cytokines induce the proliferation, differentiation, and secretory activities of B cells.

Signal Transduction Pathways Used in Lymphocyte Activation

We saw in Chapter 15 how hormones, growth factors, and other chemical messengers bind to receptors on the outer surfaces of target cells to initiate a process of signal transduction, which transmits information into a cell's internal compartments. We saw in that chapter how a large variety of extracellular messengers transmit information along a small number of shared signal transduction pathways. The stimulation of lymphocytes occurs by a similar mechanism and utilizes many of the same components employed by hormones and growth factors that act on other cell types.

When a T cell is activated by contact with an MHC-peptide complex, for example, a signal is transmitted from plasma membrane to cytoplasm by tyrosine kinases, similar to the signals described in Chapter 15 for insulin and growth factors. Unlike the receptors for insulin and growth factors (page 648), lymphocyte antigen receptors lack an inherent tyrosine kinase activity. Instead, ligand binding to antigen receptors leads to the recruitment of cytoplasmic tyrosine kinase molecules to the inner surface of the plasma membrane. Several different tyrosine kinases have been implicated in signal transduction during lymphocyte activation, including members of the Src fam-

ily (e.g., Lck and Fyn). Src was the first tyrosine kinase to be identified and is the product of the first cancer-causing oncogene to be discovered (Section 16.3).

Activation of various tyrosine kinases leads to a cascade of protein phosphorylation and the subsequent activation of at least three distinct signal transduction pathways:

1. Activation of phospholipase C, which leads to the formation of IP$_3$ and DAG. As discussed on page 643, IP$_3$ causes a marked elevation in the levels of cytosolic Ca^{2+}, whereas DAG stimulates protein kinase C activity.
2. Activation of Ras, which leads to the activation of the MAP kinase cascade (page 651).
3. Activation of phosphatidylinositol 3-hydroxy kinase or PI3K, which catalyzes the formation of membrane-bound lipid messengers having diverse functions in cells (page 641).

Transmission of signals along these various pathways, and others, leads to the transcriptional activation of dozens of genes that are not expressed in the resting T cell.

As noted above, one of the most important responses by an activated lymphocyte is the production and secretion of cytokines, some of which may act back on the cell that released them. Like other extracellular signals, cytokines bind to receptors on the outer surface of target cells, generating cytoplasmic signals that act on various intracellular targets. Cytokines utilize a novel signal transduction pathway referred to as the **JAK-STAT pathway**. The "JAK" portion of the name is an acronym for Janus kinases, a family of cytoplasmic tyrosine kinases whose members become activated following the binding of a cytokine to a cell-surface receptor. (Janus is a two-faced Roman god who protected entrances and doorways.) "STAT" is an acronym for "signal transducers and activators of transcription," a family of transcription factors that become activated when one of their tyrosine residues is phosphorylated by a JAK. Once phosphorylated, STAT molecules interact to form dimers that translocate from the cytoplasm to the nucleus where they bind to specific DNA sequences, such as an interferon-stimulated response element (ISRE). ISREs are found in the regulatory regions of a dozen or so genes that are activated when a cell is exposed to the cytokine interferon α (IFN-α).

As with the hormones and growth factors discussed in Chapter 15, the specific response of a cell depends on the particular cytokine receptor engaged and the particular JAKs and STATs that are present in that cell. For example, it was noted earlier that IL-4 acts to induce Ig class switching in B cells. This response follows the IL-4-induced phosphorylation of the transcription factor STAT6, which is present in the cytoplasm of activated B cells. Resistance to viral infection induced by interferons (page 706) is mediated through the phosphorylation of STAT1. Phosphorylation of other STATs may lead to the progression of the target cell through the cell cycle.

REVIEW ?

1. Draw the basic structure of an IgG molecule bound to an epitope of an antigen. Label the heavy and light chains; the variable and constant regions of each chain; the regions that would contain hypervariable sequences.

2. Draw the basic arrangement of the genes of the germ line that are involved in encoding the light and heavy chains of an IgG molecule. How does this differ from their arrangement in the genome of an antibody-producing cell? What steps occur to bring about this DNA rearrangement?

3. Give three different mechanisms that contribute to variability in the V regions of antibody chains.

4. Compare and contrast the structure of antigen receptors on B and T cells.

5. Describe the steps in processing of a cytosolic antigen in an APC. What is the role of MHC proteins in this process?

6. Compare and contrast the roles of an MHC class I and class II protein molecule. What types of APCs utilize each class of MHC molecule, and what types of cells recognize them?

7. What is a graft-versus-host reaction? Under what circumstances would such a reaction be expected to occur?

8. Describe the steps between the stage when a bacterium is ingested by a macrophage and the stage when plasma cells are producing antibodies that bind to the bacterium and neutralize its infectivity.

THE HUMAN PERSPECTIVE

Autoimmune Diseases

The immune system requires complex and highly specific interactions between many different types of cells and molecules. Numerous events must take place before a humoral or cell-mediated immune response can be initiated, which makes these processes vulnerable to disruption at various stages by numerous factors. Included among the various types of immune dysfunction are **autoimmune diseases**, which result when the body mounts an immune response against part of itself.

Because the specificity of the antigen receptors of both T and B cells is determined by a process of random gene rearrangement (page 717), it is inevitable that some members of these cell populations possess receptors directed against the body's own proteins (*self-antigens*). Lymphocytes that bind self-antigens with high affinity tend to be removed from the lymphocyte population, making the immune system **tolerant** toward self. However, some of the self-reactive lymphocytes generated in the thymus and bone marrow escape the body's negative selection processes, giving them the *potential* to attack normal body tissues. The presence of B and T lymphocytes capable of reacting against the body's tissues is readily demonstrated in healthy individuals. For example, when T cells are isolated from the blood and treated in vitro with a normal self-protein, together with the cytokine IL-2, a small number of cells in the population are likely to proliferate to form a clone of cells that can react to that self-antigen. Similarly, if laboratory animals are injected with a purified self-protein together with an *adjuvant*, which is a nonspecific substance that enhances the response to the injected antigen, they mount an immune response against the tissues in which that protein is normally found. Under normal physiological conditions, self-reactive lymphocytes are prevented from becoming activated by a number of poorly understood mechanisms. A breakdown in these mechanisms leads to the production of autoantibodies and autoreactive T cells that can cause chronic tissue damage. It is estimated that about 5 percent of the adult population suffers from some type of autoimmune disease, including the following:

1. **Multiple sclerosis (MS)** is a disease that typically strikes young adults, causing severe and progressive neurological damage. MS results from an attack by both T cells and antibodies against components of the myelin sheath that surround the axons of nerve cells (page 127). These sheaths form the white matter of the central nervous system. The demyelination of nerves that results from this immunologic attack interferes with the conduction of nerve impulses along axons, leaving the person with diminished eyesight, problems with motor coordination, and disturbances in sensory perception. The serum of MS patients typically contains antibodies against myelin basic protein and proteolipid protein, which are two major proteins of myelin cell membranes. A disease similar to MS, called experimental allergic encephalomyelitis, can be induced in laboratory animals by injection of myelin basic protein and adjuvant.

2. **Insulin-dependent diabetes mellitus (IDDM)**, which is often called juvenile-onset or type 1 diabetes because it tends to arise in children, results from the autoimmune destruction of the insulin-secreting β cells of the pancreas. The serum of patients with IDDM typically contains antibodies directed against the hormone insulin, the enzyme

glutamic acid decarboxylase, and the chaperonin Hsp60 (page 76), which are all major proteins of pancreatic β cells. At present, patients with IDDM are administered daily doses of insulin. While the hormone allows them to survive, these patients are still subject to degenerative kidney, vascular, and retinal disease.

3. **Grave's disease and thyroiditis** are autoimmune diseases of the thyroid that produce very different symptoms. In Grave's disease, the target of the immune attack is the TSH receptor on the surface of thyroid cells that normally binds the pituitary hormone thyroid-stimulating hormone (TSH). In patients with this disease, autoantibodies bind to the TSH receptor, causing the prolonged stimulation of thyroid cells, leading to hyperthyroidism (i.e., elevated blood levels of thyroid hormone). Thyroiditis (or Hashimoto's thyroiditis) develops from an immune attack against one or more common proteins of thyroid cells, including thyroglobulin. The resulting destruction of the thyroid gland leads to hypothyroidism (i.e., decreased blood levels of thyroid hormone).

4. **Rheumatoid arthritis** affects approximately 1 percent of the population and is characterized by the progressive destruction of the body's joints due to a cascade of inflammatory responses. In a normal joint, the synovial membrane that lines the synovial cavity is only a single cell layer thick. In persons with rheumatoid arthritis, this membrane becomes inflamed and thickened due to the infiltration of T and B lymphocytes, plasma cells, macrophages, neutrophils, and dendritic cells into the joint. These various cells release a mix of inflammatory cytokines, hydrolytic enzymes, and other agents leading to destruction of cartilage and bone. Over time, the cartilage is replaced by fibrous tissue, which causes immobilization of the joint.

5. **Systemic lupus erythematosus (SLE)** gets its name "red wolf" from a reddish rash that develops on the cheeks during the early stages of the disease. Unlike the other autoimmune diseases discussed above, SLE is not confined to a particular organ but often attacks tissues throughout the body, including the central nervous system, kidneys, and heart. The serum of patients with SLE contains antibodies directed against a number of components that are found in the nuclei of cells, including small nuclear ribonucleoproteins (snRNPs); proteins of the centromeres of chromosomes; and, most notably, double-stranded DNA. The incidence of SLE is particularly high in women of child-bearing age, suggesting a role for female hormones in triggering the disease.

Not everyone in the population is equally susceptible to developing one of these autoimmune diseases. Most of these disorders appear much more frequently in certain families than in the general population, indicating a strong genetic component to their development. Though many different genes have been shown to increase susceptibility to autoimmune diseases, those that encode MHC class II polypeptides are most strongly linked. For example, people who inherit certain alleles of the MHC locus known as HLA-DQB are particularly susceptible to developing type 1 diabetes (IDDM). Analysis of MHC polypeptides encoded by various alleles of the HLA-DQB locus has become focused on one particular residue (number 57) of the polypeptide chain, which is situated at one end of the peptide-binding groove. Resistance to development of IDDM is correlated with the presence of an aspartic acid residue at this position, whereas susceptibility is correlated with a serine, alanine, or valine at this position. These amino acid differences can influence the type of peptide that the MHC molecule can bind and present to other cells. MHC molecules *lacking* aspartate at position 57 can bind peptides containing negatively charged glutamic or aspartic acid residues at their C terminus, whereas MHC molecules containing aspartic acid at this position cannot bind these peptides. It is thought that cells bearing MHC molecules encoded by the susceptible allele can bind some particular peptide that stimulates the formation of autoantibodies against insulin-secreting β cells of the pancreas.

Possession of high-risk alleles may be necessary for an individual to develop certain autoimmune diseases, but it is not the only contributing factor. Studies of monozygotic (genetically identical) twins indicate that if one twin develops an autoimmune disease, the likelihood that the other twin will also develop the disease ranges from about 25 to 75 percent, not 100 percent as expected if genetics were the only contributing factor. Studies of this type demonstrate that environmental factors also play a role. The importance of pathogens in the development of autoimmune disorders was first demonstrated in studies of rheumatic fever, a condition that can develop in children several weeks after a streptococcal infection of the throat ("strep throat"). Rheumatic fever develops when heart tissue is attacked by antibodies that were produced in response to the streptococcal bacteria. Heart tissue becomes a target of these antibodies because of a phenomenon called "molecular mimicry." In this case, one of the components of the bacterial cell wall is similar to that of a glycoprotein on the surface of cells that line the heart valves. As a result, antibodies produced in response to the bacterial infection can cross-react with heart tissue. In cases of multiple sclerosis, it is found that respiratory or intestinal viral infections can activate autoreactive T cells and trigger relapses in patients who have gone into remission. In fact, some researchers believe that viral infection is an underlying cause of MS. Studies on animals have also indicated the importance of pathogens. For example, mice that lack a gene for the cytokine IL-2 (IL-2 knockout mice) develop an inflammatory bowel disorder resembling human ulcerative colitis. This condition only develops in these laboratory animals if they are exposed to infectious agents of the normal environment; if they are raised under germ-free conditions, they remain disease-free.

The great progress that has been made over the past two decades in our understanding of the cellular and molecular basis of immunity has led to promising new treatments for a number of autoimmune diseases. These treatments have been tested in animal models (i.e., animals that can be made to develop diseases similar to those of humans), and clinical trials are underway. Several distinct approaches are being tested:

■ Treatment with immunosuppressive drugs, such as cyclosporin A that block the autoimmune response. Because these drugs are nonspecific, they inhibit all types of immune responses and, therefore, render the patient highly susceptible to dangerous infections.

- Induction of a return to a tolerant state so that the body stops producing autoantibodies and autoreactive T cells. One way to induce tolerance to specific antigens is to administer altered peptide ligands (APLs) that are thought to bind to TCRs in a suboptimal manner, blocking T-cell activation and reducing the secretion of inflammatory cytokines (e.g., TNF-α and IFN-γ). One drug of this type (Copaxone) has been shown to reduce the frequency of relapses in patients with multiple sclerosis. Copaxone is a synthetic peptide whose structure resembles that of myelin basic protein. Other peptide-based drugs are in clinical trials.

- Interfering with the autoimmune response by administering antibodies. Among the types of antibodies that have shown promise in animal tests are ones directed against the CD28 costimulatory molecule on the surface of T cells (page 725). Blocking costimulation blocks activation of effector T cells, possibly leading either to anergization or death by apoptosis. Antibodies against helper T cells, cytokines, and offending antigens have all shown some success in animal models. For example, antibodies against the cytokine TNF-α have shown promise in treatment of rheumatoid arthritis.

- Blocking the effect of inflammatory cytokines by administration of suppressive cytokines. For example, IL-4 has shown promise in the treatment of IDDM, and IFN-β1a (Avonex) has been approved for the treatment of MS.

- Blocking the body's ability to mount an immune response by immunization. Trials have been carried out on patients with MS who have been immunized with a component of the TCR that recognizes the myelin basic protein self-antigen. The body responds to the immunization by producing antibodies that attack those specific T cells that are responsible for the disease.

EXPERIMENTAL PATHWAYS

The Role of the Major Histocompatibility Complex in Antigen Presentation

In 1973, Hugh McDevitt and his colleagues at the Scripps Foundation in La Jolla, California, and at Stanford University, demonstrated that the susceptibility of mice to a particular pathogen depends on the allele present at one of the MHC loci.[1] They found that lymphocytic choriomeningitis virus (LCMV) causes a lethal brain infection in mice that are either homozygous or heterozygous for the $H\text{-}2^q$ allele but does not cause infections in mice that are homozygous for the $H\text{-}2^k$ allele at this locus.

These findings led Rolf Zinkernagel and Peter Doherty of the Australian National University to examine the role of cytotoxic T lymphocytes (CTLs) in the development of this disease. Zinkernagel and Doherty planned experiments to correlate the level of CTL activity with the severity of the disease in mice having different MHC genotypes (or haplotypes, as they are called). Cytotoxic T lymphocyte activity was monitored using the following experimental protocol. Monolayers of fibroblasts (L cells) from one mouse were grown in culture and subsequently infected with the LCM virus. The infected fibroblasts were then overlaid by a preparation of spleen cells from a mouse that had been infected with the LCM virus seven days earlier. Waiting for seven days gives the animal's immune system time to generate CTLs against virus-infected cells. The CTLs become concentrated in the infected animal's spleen. To monitor the effectiveness of the attack by the CTLs on the cultured L cells, the L cells were first labeled with a radioisotope of chromium (^{51}Cr). ^{51}Cr is used as a marker for cell viability: as long as a cell remains alive, the radioisotope remains inside the cell. If a cell should be lysed during the experiment by a CTL, the ^{51}Cr is released into the medium.

Zinkernagel and Doherty found that the level of CTL activity against the cultured fibroblasts—as measured by the release of ^{51}Cr—depended on the relative genotypes of the fibroblasts and spleen cells (Table 1).[2] All of the fibroblasts used to obtain the data shown in Table 1 were taken from an

Table 1 Cytotoxic Activity of Spleen Cells from Various Strains of Mice Injected 7 Days Previously with LCM Virus for Monolayers of LCM-Infected or Normal C3H (H-2k) Mouse L Cells

Exp.	*Mouse* Strain	H-2 Type	% ^{51}CR Release Infected	% ^{51}CR Release Normal
1	CBA/H	k	65.1 ± 3.3	17.2 ± 0.7
	Balb/C	d	17.9 ± 0.9	17.2 ± 0.6
	CB57Bl	b	22.7 ± 1.4	19.8 ± 0.9
	CBA/H × C57Bl	k/b	56.1 ± 0.5	16.7 ± 0.3
	CB57Bl × Balb/C	b/d	24.8 ± 2.4	19.8 ± 0.9
	nu/+ or +/+		42.8 ± 2.0	21.9 ± 0.7
	nu/nu		23.3 ± 0.6	20.0 ± 1.4
2	CBA/H	k	85.5 ± 3.1	20.9 ± 1.2
	AKR	k	71.2 ± 1.6	18.6 ± 1.2
	DBA/2	d	24.5 ± 1.2	21.7 ± 1.7
3	CBA/H	k	77.9 ± 2.7	25.7 ± 1.3
	C3H/HeJ	k	77.8 ± 0.8	24.5 ± 1.5

Reprinted with permission from R. M. Zinkernagel and P. C. Doherty, *Nature* 248:701, 1974. © Copyright 1974, Macmillan Magazine Ltd.

inbred strain of mice that were homozygous for the allele H-2^k at the H-2 locus. When spleen cells were prepared from mice having an H-2^k allele (e.g., CBA/H, AKR, and C3H/HeJ strains of mice), the L cells were effectively lysed. However, spleen cells taken from mice bearing H-2^b or H-2^d alleles at this locus were unable to lyse the infected fibroblasts. (The ^{51}Cr released is approximately the same when noninfected fibroblasts are used in the assay, as shown in the right column of Table 1.)

It was essential to show that the results were not peculiar to H-2^{k-} bearing mice. To make this determination, Zinkernagel and Doherty tested LCMV-activated spleen cells from H-2^b mice against various types of infected cells. Once again, the CTLs would only lyse infected cells having the same H-2 genotype, in this case H-2^b. These studies provided the first evidence that the MHC molecules on the surface of an infected cell restricts its interactions with T cells. T-cell function is said to be *MHC restricted*.

These and other experiments during the 1970s raised questions about the role of MHC proteins in immune cell function. Meanwhile, another line of investigation was focusing on the mechanism by which T cells were stimulated by particular antigens. Studies had indicated that T cells respond to antigen that is bound to the surface of other cells. It was presumed that the antigen being displayed had simply bound to the surface of the antigen-presenting cell (APC) from the extracellular medium. During the mid-1970s and early 1980s, studies by Alan Rosenthal at NIH, Emil Unanue at Harvard University, and others demonstrated that antigen had to be internalized by the APC and subjected to some type of processing before it could stimulate T-cell proliferation. Most of these studies were carried out in cell culture utilizing T cells activated by macrophages that had been previously exposed to bacteria, viruses, or other foreign material.

One of the ways to distinguish between antigen that is simply bound to the surface of an APC and antigen that has been processed by metabolic activities is to compare events that can occur at low temperatures (e.g., 4°C) where metabolic processes are blocked with those that occur at normal body temperatures. In one of these earlier experiments shown in Figure 1, macrophages were incubated with antigen for one hour at either 4°C or 37°C and then tested for their ability to stimulate T cells prepared from lymph nodes.[3] T-cell stimulation is measured by the incorporation of [^3H]thymidine into DNA that accompanies cell proliferation. At lower antigen concentrations, macrophages were nearly 10 times more effective at stimulating T cells at 37°C than at 4°C, suggesting that antigen processing requires active metabolic events. Treatment of cells with sodium azide, a cytochrome oxidase inhibitor, also inhibits the appearance of antigen on the surface of T cells, indicating that antigen presentation requires metabolic energy.[4]

Subsequent experiments by Kirk Ziegler and Emil Unanue provided evidence that sequestration occurred as extracellular antigens were taken into the macrophage by endocytosis and delivered to the cell's lysosomal compartment.[5] One approach to determining whether lysosomes are involved in a particular process is to treat the cells with substances, such as ammonium chloride or chloroquine, that disrupt lysosomal enzyme activity. Both of these agents raise the pH of the lysosomal compartment, which inactivates the acid

Figure 1 Temperature dependence of antigen processing. Macrophages were incubated with antigen (PPD, purified protein derivative) for one hour at either 4°C (open circles) or 37°C (closed circles). After the incubation period, the macrophages were washed free of the protein-containing medium and mixed with lymphocytes capable of responding to PPD antigen. Activation of the lymphocytes by the macrophages was measured by the incorporation of [^3H]thymidine into DNA. (FROM J. A. WALDRON, JR., R. G. HORN, AND A. S. ROSENTHAL, J. IMMUNOL. 112:749, 1974. COPYRIGHT 1974 THE AMERICAN ASSOCIATION OF IMMUNOLOGISTS.)

hydrolases (page 311). Table 2 shows the effects of these treatments on processing and presentation of antigen derived from the bacteria *Listeria monocytogenes*. It can be seen from the table that neither of these substances affected the uptake (endocytosis) of antigen, but both substances markedly inhibited the processing of the antigen and its ability to stimulate binding of T cells to the macrophage. These data were among the first to suggest that fragmentation of extracellular antigens by lysosomal proteases may be an essential step in preparation of extracellular antigens prior to presentation.

Other studies continued to implicate MHC molecules in the interaction between APCs and T cells. In one series of experiments, Ziegler and Unanue treated macrophages with antibodies directed against MHC proteins encoded by the H-2 locus. They found that these antibodies had no effect on the uptake or catabolism of antigen,[6] but markedly inhibited the macrophages from interacting with T cells.[7] Inhibition of binding of T cells to macrophages was obtained even when macrophages were exposed to the antibodies before addition of antigen.

Evidence from these and many other studies indicated that the interaction between a T cell and a macrophage depended on the recognition of two components on the surface of the antigen-presenting cell: the antigen fragment being displayed and an MHC molecule. But there was no clear-cut picture as to how the antigen fragment and MHC molecule were related. Two models of antigen recognition were considered likely possibilities. According to one model, T cells possess

| | Table 2 | Inhibition of Antigen Presentation with NH₄Cl and Chloroquine |

		10 mM NH4Cl		0.1 mM *Chloroquine*	
	Control	Observed	Inhibition	Observed	Inhibition
Assay	(%)	(%)	(%)	(%)	(%)
Antigen uptake	15 ± 1	13 ± 2	13	15 ± 2	0
Antigen ingestion	66 ± 2	63 ± 2	5	67 ± 6	−2
Antigen catabolism	29 ± 4	13 ± 3	55	14 ± 6	52
T cell-macrophage binding					
before antigen handling	70 ± 7	26 ± 8	63	30 ± 8	57
after antigen handling	84 ± 8	70 ± 11	17	60 ± 10	24

Source: H. K. Ziegler and E. R. Unanue, *Pro. Nat'l Acad. Sci. U. S. A.* 79:176, 1982

two distinct receptors, one for the antigen and another for the MHC protein. According to the other model, a single T-cell receptor recognizes both the MHC protein and the antigen peptide on the APC surface simultaneously. The balance of opinion began to shift in favor of the one-receptor model as evidence began to point to a physical association between MHC proteins and displayed antigens. In one study, for example, it was shown that antigen that had been processed by T cells could be isolated as a complex with MHC proteins.[8] In this experiment, cultured T cells taken from H-2k mice were incubated with radioactively labeled antigen for 40 minutes. After the incubation period, processed antigen was prepared from the cells and passed through a column containing beads coated with antibodies directed against MHC proteins. When the beads were coated with antibodies against H-2k protein, an MHC molecule present in the T cells, large amounts of radioactive antigen adhered to the beads, indicating the association of the processed antigen with the MHC protein. If the beads were instead coated with antibodies against H-2b protein, an MHC protein that was not present in the T cells, relatively little radioactive antigen remained in the column.

Over the past decade, investigators have focused their attention on the atomic structure of the molecules involved in T-cell interactions. Rather than utilizing MHC class II molecules on the surfaces of macrophages, structural studies have examined MHC class I molecules of the type found on the surfaces of virally infected cells. The first three-dimensional portrait of an MHC molecule was published in 1987 based on X-ray crystallographic studies by Don Wiley and colleagues at Harvard University.[9] MHC class I molecules consist of (1) a heavy chain containing three extracellular domains (α_1, α_2, and α_3) and a single membrane-spanning segment and (2) an invariant β_2m polypeptide (see Figure 17.21). Wiley and colleagues examined the structure of the extracellular (soluble) portion of the MHC molecule (α_1, α_2, α_3, and β_2m) after removing the transmembrane anchor. A ribbon model of the observed structure is shown in Figure 2a, with the outer (antigen-bearing) portion of the protein constructed from the α_1 and α_2 domains. It can be seen from this model that the inner surfaces of these domains form the walls of a deep groove approximately 25 Å long and 10 Å wide. It is this groove that acts as the binding site for peptides produced by antigen processing in the cytoplasm. As shown in Figure 2b, the sides of

the antigen-binding pocket are lined by α helices from the α_1 and α_2 domains, and the bottom of the pocket is lined by β sheet that extends from these same domains across the midline. The helices are thought to form relatively flexible side walls enabling peptides of different sequence to bind within the groove.

Subsequent X-ray crystallographic studies described the manner in which peptides are positioned within the MHC antigen-binding pocket. In one of these studies, the spatial arrangement of several naturally processed peptides situated within the antigen-binding pocket of a single MHC class I molecule (HLA-B27) was determined.[10] The backbones of all the peptides bound to HLA-B27 share a single, extended conformation running the length of the binding cleft. The N- and C-termini of the peptides are precisely positioned by numerous hydrogen bonds at both ends of the cleft. The hydrogen bonds link the peptide to a number of conserved residues in the MHC molecule that are part of both the sides and bottom of the binding groove.

In another study, Ian Wilson and colleagues at the Scripps Research Institute in La Jolla, California, reported on the X-ray crystallographic structure of a mouse MHC class I protein complexed with two peptides of different length.[11,12] The overall structure of the mouse MHC protein is similar to that of the human MHC protein shown in Figure 2a. In both cases, the peptides are bound in an extended conformation deep within the binding groove of the MHC molecule (Figure 3). This extended conformation allows for numerous interactions between the side chains of the MHC molecule and the backbone of the bound peptide. Because the MHC interacts primarily with the backbone of the peptide rather than its side chains, there are very few restrictions on the particular amino acid residues that can be present at various sites of the binding pocket. As a result, each MHC molecule can bind a diverse array of antigenic peptides.

An MHC-peptide complex projecting from the surface of an infected cell is only half the story of immunologic recognition; the other half is represented by the T-cell receptor (TCR) projecting from the surface of the cytotoxic T cell. It has been evident for more than a decade that a TCR can somehow recognize both an MHC and its contained peptide, but the manner by which this occurs eluded researchers because of difficulties in preparing protein crystals of the TCR that were

Figure 2 (*a*) Schematic representation of a MHC class I molecule, in this case the human protein HLA-A2. The molecule consists of two subunits: a heavy chain made up of three domains (α_1, α_2, and α_3) and a β_2m chain. The membrane-spanning portion of the heavy chain would connect with the polypeptide at the site marked C (for C-terminus of the remaining portion). Disulfide bonds are indicated as two connected spheres. The peptide-binding groove is shown at the top of the drawing, between the α helical segments of the α_1 and α_2 domains of the heavy chain. (*b*) Schematic representation of the peptide-binding pocket of the MHC protein as viewed from the top of the molecule. The bottom of the pocket is lined by β sheet (red-purple arrows) and the walls by α helices (green). The α_1 domain is shown in red and light green; the α_2 domain is shown in purple and dark green. (REPRINTED WITH PERMISSION FROM P. J. BJORKMAN, ET AL., NATURE 329:508, 509, 1987. © COPYRIGHT 1987, BY MACMILLAN MAGAZINES LTD.)

suitable for X-ray crystallography. These difficulties were eventually overcome, and in 1996, reports were published by both the Wiley and Wilson laboratories that provided a three-dimensional portrait of the interaction between an MHC-peptide and TCR.[13,14] The overall structure of the complex formed between the two proteins is shown in Figure 4, where the backbones of the polypeptides are portrayed as tubes. A space-filling model of a similar complex is shown in the chapter-opening photograph on page 703.

Figure 3 Three-dimensional models of a peptide (shown in ball-and-stick structure) bound within the antigen-binding pocket of a mouse MHC class I protein (H-2Kb). The peptide in *a* consists of eight amino acid residues; the peptide in *b* consists of nine residues. The peptides are seen to be buried deep within the MHC binding groove. (REPRINTED WITH PERMISSION FROM MASAZUMI MATSUMURA, DAVED H. FREMONT, PER A. PETERSON, AND IAN A. WILSON, SCIENCE 257:932, 1992. © COPYRIGHT 1992, AMERICAN ASSOCIATION FOR THE ADVANCEMENT OF SCIENCE.)

(*a*) (*b*)

The structure shown in Figure 4 depicts the portions of the proteins that project between a CTL and virally infected host cell. The lower half of the image shows the structure and orientation of the MHC class I molecule with the extended peptide antigen (green) embedded within the protein's binding pocket. The upper half of the image shows the structure and orientation of the TCR. As indicated in Figure 17.18b, a TCR consists of α and β polypeptide chains, each chain consisting of a variable (V) and constant (C) portion. Like immunoglobulins (Figure 17.12), the variable portion of each TCR subunit contains regions that are exceptionally variable (i.e., hypervariable). The hypervariable regions form protruding loops (shown as the colored segments of the two TCR polypeptides in Figure 4) that fit snugly over the outer end of the MHC-peptide complex. The hypervariable regions are referred to as *complementarity-determining regions*, or *CDRs*, because they determine the binding properties of the TCR. The CDRs of the TCR interact with the α helices of the α_1 and α_2 domains of the MHC, as well as the exposed residues of the bound peptide. The central CDRs of the TCR, which exhibit the greatest sequence variablity, interact primarily with the centrally situated bound peptide, whereas the outer CDRs, which have a less variable sequence, interact most closely with the α helices of the MHC.[15] Because of these interactions, the TCR meets both of its recognition "responsibilities": it recognizes the bound peptide as a foreign antigen and the MHC as a self-protein.

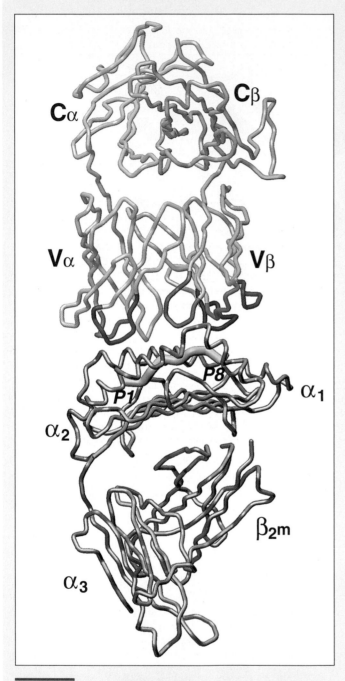

Figure 4 Representation of the interaction between an MHC-peptide complex (at the bottom) and a TCR (at the top). The hypervariable regions (CDRs) of the TCR are shown as colored loops that form the interface between the two proteins. The bound peptide is shown in green as it is situated within the binding groove of the MHC class I molecule. The peptide backbones are represented as tubes. (REPRINTED WITH PERMISSION FROM K. CHRISTOPHER GARCIA ET AL., COURTESY OF IAN WILSON, SCIENCE 274:217, 1996. © COPYRIGHT 1996, AMERICAN ASSOCIATION FOR THE ADVANCEMENT OF SCIENCE.)

References

1. OLDSTONE, M. B. A., ET AL. 1973. Histocompatibility-linked genetic control of disease susceptibility. *J. Exp. Med.* 137:1201–1212.

2. ZINKERNAGEL, R. M. & DOHERTY, P. C. 1974. Restriction of in vitro T cell-mediated cytotoxicity in lymphocytic choriomeningitis within a syngeneic or semiallogeneic system. *Nature* 248:701–702.

3. WALDRON, J. A., ET AL. 1974. Antigen-induced proliferation of guinea pig lymphocytes in vitro: Functional aspects of antigen handling by macrophages. *J. Immunol.* 112:746–755.

4. WEKERLE, H., ET AL. 1972. Fractionation of antigen reactive cells on a cellular immunoadsorbant. *Proc. Nat'l Acad. Sci. U. S. A.* 69:1620–1624.

5. ZIEGLER, K. & UNANUE, E. R. 1982. Decrease in macrophage antigen catabolism caused by ammonia and chloroquine is associated with inhibition of antigen presentation to T cells. *Proc. Nat'l Acad. Sci. U. S. A.* 79:175–178.

6. ZIEGLER, K. & UNANUE, E. R. 1981. Identification of a macrophage antigen-processing event required for I-region-restricted antigen presentation to T lymphocytes. *J. Immunol.* 127:1869–1875.

7. ZIEGLER, K. & UNANUE, E. R. 1979. The specific binding of Listeria monocytogenes-immune T lymphocytes to macrophages. I. Quantitation and role of H-2 gene products. *J. Exp. Med.* 150: 1142–1160.

8. PURI, J. & LONAI, P. 1980. Mechanism of antigen binding by T cells H-2 (I-A)-restricted binding of antigen plus Ia by helper cells. *Eur. J. Immunol.* 10:273–281.

9. BJORKMAN, P. J., ET AL. 1987. Structure of the human class I histocompatibility antigen, HLA-A2. *Nature* 329:506–512.

10. MADDEN, D. R., ET AL. 1992. The three-dimensional structure of HLA-B27 at 2.1 Å resolution suggests a general

11. FREMONT, D. H., ET AL. 1992. Crystal structures of two viral peptides in complex with murine MHC class I H-2Kb. *Science* 257:919–927.

12. MATSUMURA, M., ET AL. 1992. Emerging principles for the recognition of peptide antigens by MHC class I molecules. *Science* 257:927–934.

mechanism for tight peptide binding to MHC. *Cell* 70:1035–1048.

13. K. C. GARCIA, ET AL. 1996. An αβ T cell receptor structure at 2.5 Å and its orientation in the TCR-MHC complex. *Science* 274:209–219.

14. GARBOCZI, D. N., ET AL. 1996. Structure of the complex between human T-cell receptor, viral peptide and HLA-A2. *Nature* 384:134–140.

15. WILSON, I.A. 1999. Class-conscious TCR? *Science* 286:1867–1868.

SYNOPSIS

Vertebrates are protected from infection by immune responses carried out by cells that can distinguish between materials that are "supposed" to be there (i.e., "self") and those that are not (i.e., foreign, or "nonself"). Innate responses occur rapidly but lack specificity, whereas acquired responses are highly specific but require a lag period of several days. Innate responses are carried out by patrolling phagocytic cells, blood-borne molecules (complement) that can lyse bacterial cells, antiviral proteins (interferons), and natural killer cells that can cause infected cells to undergo apoptosis. Two broad categories of acquired immunity can be distinguished: (1) humoral (blood-borne) immunity mediated by antibodies that are produced by cells derived from B lymphocytes (B cells) and (2) cell-mediated immunity carried out by T lymphocytes (T cells). B and T cells both arise from the same stem cell, which also gives rise to other types of blood cells. (p. 704)

The cells of the immune system develop by clonal selection. During its development, each B cell becomes committed to producing only one species of antibody molecule, which is initially displayed as an antigen receptor in the cell's plasma membrane. B-cell commitment occurs in the absence of antigen, so that the entire repertoire of antibody-producing cells is already present prior to stimulation by antigen. When a foreign substance appears in the body, it functions as an antigen by interacting with B cells that contain membrane-bound antibodies capable of binding that substance. Antigen binding activates the B cell, causing it to proliferate and form a clone of cells that differentiate into antibody-producing plasma cells. In this way, antigen selects those B cells that produce antibodies capable of interacting with it. Some of the antigen-selected B cells remain as undifferentiated memory cells that can respond rapidly if the antigen is reintroduced. B cells whose antigen-receptors are capable of reacting to the body's own tissues are either inactivated or eliminated, causing the body to develop an immunologic tolerance toward itself. (p. 707)

T lymphocytes recognize antigens by their T-cell receptors (TCRs). T cells can be divided into two distinct subclasses: helper T cells (T$_H$ cells) that activate B cells and cytotoxic T lymphocytes (CTLs) that kill infected cells. Unlike B cells, which are activated by soluble, intact antigens, T cells are activated by fragments of antigens that are displayed on the surfaces of other cells, called antigen-presenting cells (APCs). Whereas any infected cell can activate a CTL, only professional APCs, such as macrophages and dendritic cells, that function in antigen ingestion and processing can activate T$_H$ cells. Cytotoxic T cells kill target cells by secreting proteins that permeabilize the membrane of the target cell and induce it to undergo apoptosis. (p. 710)

Antibodies are globular proteins called immunoglobulins (Igs) that are built of two types of polypeptide chains: heavy and light chains. Antibodies are divided into several classes (IgA, IgG, IgD, IgE, and IgM) depending on their heavy chain. Heavy and light chains both consist of (1) constant (C) regions, that is, regions whose amino acid sequence is virtually identical among all heavy or light chains of a given class, and (2) variable (V) regions, that is, regions whose sequence varies from one antibody species to another. Different classes of antibody appear at different times after exposure to an antigen and have different biological functions. IgG is the predominant form of blood-borne antibody. Each Y-shaped IgG molecule contains two identical heavy chains and two identical light chains. The antigen-combining sites are formed by association of the V region of a light chain with the V region of a heavy chain. Included within the V regions are hypervariable regions that form the walls of the antigen-combining site. (p. 712)

Antigen receptors on both B and T cells are encoded by genes produced by DNA rearrangement. Each variable region of a light Ig chain is composed of two DNA segments (a V and J segment) that are joined together, whereas each variable region of a heavy Ig chain is composed of three joined segments (a V, J, and D segment). DNA rearrangement is facilitated by highly conserved signal sequences located adjacent to the coding sequences of the germ-line DNA. Variability arises as the result of the joining of different V genes with different J genes in different antibody-producing cells. Additional variability is introduced by the enzymatic insertion of nucleotides and variation in the V-J joining site. It is estimated that an individual can synthesize more than 2,000 different species of kappa light chains and 100,000 species of heavy chains, which can combine randomly to form more than 200 million antibodies. Additional diversity arises in antibody-producing cells by somatic hypermutation in which the rearranged V genes experience a mutation rate much greater than the remainder of the genome. (p. 716)

Antigens are taken up by APCs, fragmented into short peptides, and combined with proteins encoded by the major histocompatibility complex (MHC) for presentation to T cells. MHC proteins can be subdivided into two classes: class I molecules and class II molecules. MHC class I molecules primarily display peptides derived from endogenous cytosolic proteins, including those derived from an infecting virus. Nearly any

cell of the body can display peptides through MHC class I molecules to CTLs. If the TCR of the CTL recognizes a foreign peptide displayed by a cell, the CTL becomes activated to kill the target cell. Professional APCs, including macrophages and dendritic cells, display peptides in combination with MHC class II peptides. The MHC-peptide complexes on the APC are recognized by TCRs on the surface of T$_H$ cells. Helper T cells that become activated by antigens displayed on an APC subsequently associate with and activate B cells whose antigen receptors (BCRs consisting of Igs) recognize the same antigen. Helper T cells stimulate B cells by direct interaction and by secretion of cytokines. Activation of a T cell by a TCR leads to the stimulation of protein tyrosine kinases within the T cell, which in turn leads to the activation of a number of signaling pathways, including the activation of ras and the MAP kinase cascade and the activation of phospholipase C. (p. 719)

ADDITIONAL READING

The following consist entirely of reviews in Immunology:

Advances in Immunology

Annual Review of Immunology

Critical Reviews in Immunology

Current Opinion in Immunology

Immunological Reviews

Immunology Today

Other Readings

ABBAS, A. K. & JANEWAY, C. A., Jr. 2000. Immunology: Improving on Nature in the twenty-first century. *Cell* 100:129–138.

ABBAS. A. K. & SHARPE, A. H. 1999. T-cell stimulation: An abundance of B7s. *Nature Med.* 5:1345–1346.

AKASHI, K., ET AL. 1999. Signaling and gene expression in the immune system. *Cold Spring Harbor Symp. Quant. Biol.* vol. 64.

ARSTILA, T. P., ET AL. 1999. A direct estimate of the human αβ T cell receptor diversity. *Science* 286:958–961.

BAGGIOLINI, M. 1998. Chemokines and leukocyte traffic. *Nature* 392:565–568.

BANCHEREAU, J. & STEINMAN, R. M. Dendritic cells and the control of immunity. *Nature* 392:245–252.

BENJAMINI, E., 1996. *Immunology: A Short Course*, 3d ed. Wiley-Liss.

BEVAN, M. J. & GOLDRATH, A. W. 2000. T-cell memory: You must remember this. *Curr. Biol.* 10:R338–R340.

BRUNNER, T. & MUELLER, C. 1999. Is autoimmunity coming to a Fas(t) end? *Nature Med.* 5:19–20.

CYSTER, J. G. 1999. Chemokines: Chemokines and cell migration in secondary lymphoid organs. *Science* 286:2098–2102.

DARNELL, J. E., Jr. 1997. STATs and gene regulation. *Science* 277:1630–1635.

DATTA, S. K. 2000. Positive selection for autoimmunity. *Nature Med.* 6:259–261.

DAVOUST, J. & BANCHEREAU, J. 2000. Naked antigen-presenting molecules on dendritic cells. *Nature Cell Biol.* 2:E46–E48.

DUSTIN, M. L. & CHAN, A. C. 2000. Signaling takes shape in the immune system. *Cell* 103:283–294.

GENAIN, C. P. & ZAMVIL, S. S. 2000. Specific immunotherapy: One size does not fit all. *Nature Med.* 6:1098–1100.

GOLDRATH, A. W. & BEVAN, M. J. 1999. Selecting and maintaining a diverse T-cell repertoire. *Nature* 402:255–262.

GREENBERG, P. D. & RIDDELL, S. R. 1999. Deficient cellular immunity-finding and fixing the defects. *Science* 285:546–551.

HAGMANN, M., ET AL. 2000. Reviews on cellular immunology. *Science* 290:80–100.

HAGMANN, M 1999. Keeping bone marrow grafts in check. *Science* 285:310–311.

HILL, A. V. 1999. Defence by diversity. *Nature* 398:668–669.

JANEWAY, C. A., Jr. & TRAVERS, P. 1996. *Immunobiology*, 2d ed. Garland.

LAABI, Y. & STRASSER , A. 2000. Lymphocyte survival—ignorance is BLys. (on control of B cell apoptosis). *Science* 289:883–884.

MAIZELS, N. 1995. Somatic hypermutation: How many mechanisms diversify V region sequences? *Cell* 83:9–12.

MCDEVITT, H. O. 2000. Discovering the role of the major histocompatibility complex in the immune response. *Ann. Rev. Immunol.* 18:1–17.

MUELLER, D. L. 2000. T cells: A proliferation of costimulatory molecules. *Curr. Biol.* 19:R227–R230.

NEUBERGER, M.S. & SCOTT, J. 2000. RNA editing AIDs antibody diversification. (on somatic hypermutation and Ig class switching). *Science* 289:1705–1706.

NOSEWORTHY, J. H. 1999. Progress in determining the cause and treatment of multiple sclerosis. *Nature* 399, Suppl. A40–A47.

O'GARRA, A. 2000. Count ye helpers. *Nature* 404:719–720.

PALUCKA, K. & BANCHEREAU, J. 1999. Linking innate and adaptive immunity. *Nature Med.* 5:868–870.

PARHAM, P. 1999. Soaring costs in defence. *Nature* 401:870–871.

PODULSKY, S. H. & TAUBER, A. I. 1997. *The Generation of Diversity: Clonal Selection Theory and the Rise of Molecular Immunology.* Harvard Press.

RIDGWAY, W. M., ET AL. 1999. A new look at MHC and autoimmune disease. *Science* 284:749–751.

RODEWALD, H.-R. 1998. The thymus in the age of retirement. *Nature* 396:630–631.

ROITT, I., ET AL. 1996. *Immunology*, 4th ed. Mosby.

SCHWARTZ, R. H. 2001. It takes more than two to tango. (on costimulatory signals). *Nature* 409:31–32.

SCHATZ, D. G. 1999. Developing B-cell theories. *Nature* 400:614–618.

STARKEBAUM, G. 1998. Role of cytokines in rheumatoid arthritis. *Sci. & Med.* 5:6–15. (March/April)

STAVNEZER, J. 2000. A touch of antibody class. (on Ig class switching) *Science* 288:984–985.

STEINMAN, L. 2000. Multiple approaches to multiple sclerosis. *Nature Med.* 6:15–16.

TISCH, R., MCDEVITT, H. ET AL., 1996. Reviews on autoimmunity. *Cell* 85:291–318.

TIAN, M. & ALT, F. W. 2000. RNA editing meets DNA shuffling. *Nature* 407:31–33.

WHITACRE, C. C., ET AL. 1999. A gender gap in autoimmunity. *Science* 283:1277–1278.

ZAVAZAVA, N. & EGGERT, F. 1997. MHC and behavior. *Immunol. Today* 18:8–10.

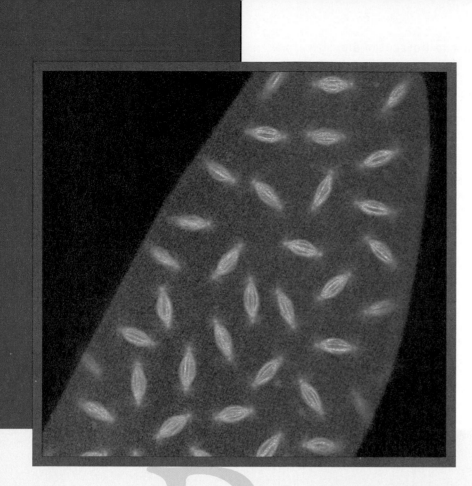

Techniques in Cell and Molecular Biology

Because of the very small size of the subject matter, cell and molecular biology is more dependent on the development of new instruments and technologies than any other branch of biology. Consequently, it is difficult to learn about cell and molecular biology without also learning about the technology that is required to collect data. We will survey the methods used most commonly in the field without becoming immersed in all of the details and variations that are employed. These are the goals of the present chapter: to describe the ways that various techniques are used and to provide examples of the types of information that can be learned by using these techniques. We will begin with the instrument that enabled biologists to discover the very existence of cells, providing the starting point for all of the information that has been presented in this text.

Fluorescence micrograph of a live Drosophila *embryo expressing a fusion protein composed of the green fluorescent protein (GFP) and one of the kinesin-related motor proteins. Localization of GFP fluorescence indicates the motor protein is present as part of the mitotic spindle and centrosomes of the embryo.* (COURTESY OF SHARYN ENDOW, DUKE UNIVERSITY.)

18.1 THE LIGHT MICROSCOPE

Microscopes are instruments that produce an enlarged image of an object. Figure 18.1 shows the most important components of a compound light microscope. A light source, which may be external to the microscope or built into its base, illuminates the specimen. The substage **condenser lens** gathers the diffuse rays from the light source and illuminates the specimen with a small cone of bright light that allows very small parts of the specimen to be seen after magnification. The light rays focused on the specimen by the condenser lens are then collected by the microscope's **objective lens**. From this point, we need to consider two sets of light rays that enter the objective lens: those that the specimen has altered and those that it hasn't (Figure 18.2). The latter group consists of light from the condenser that passes directly into the objective lens, forming the background light of the visual field. The former group of light rays emanates from the many parts of the specimen. These light rays are brought to focus by the objective lens to form a real, enlarged image of the object within the column of the microscope (Figure 18.1). The image formed by the objective lens is used as an object by a second lens system, the **ocular lens**, to form an enlarged and virtual image. A third lens system located in the front part of the eye uses the virtual image produced by the ocular lens as an object to produce a real image on the retina. When the focusing knob of the light microscope is turned, the relative distance between the specimen and the objective lens changes, allowing the final image to become focused precisely on the plane of the retina. The total magnification attained by the microscope is the product of the magnifications produced by the objective lens and the ocular lens.

Resolution

Up to this point we have considered only the magnification of an object without paying any attention to the quality of the image produced, that is, the extent to which the detail of the specimen is retained in the image. Suppose you are looking at a structure in the microscope using a relatively high-power objective lens (for example, 63X) and an ocular lens that magnifies the image of the objective lens another fivefold (a 5X ocular). Suppose the field is composed of chromosomes and it is important to determine the number present, but some of them are very close together and cannot be distinguished as separate structures (Figure 18.3a). One solution to the problem might be to change oculars to increase the size of the object being viewed. If you were to switch from the 5X to a 10X ocular, you would most likely increase your ability to determine the number of chromosomes present (Figure 18.3b) because you have now spread the image of the chromosomes produced by the objective lens over a greater part of your retina. The more photoreceptors that provide information about the image, the more detail that can be seen (Figure 18.4). If, however, you switch to a 20X ocular, you are not

Figure 18.1 **Sectional diagram through a compound light microscope**; that is, a microscope that has both an objective and an ocular lens.

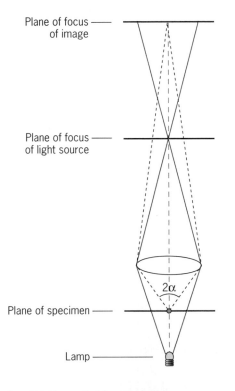

(•••) Light rays that form the image

(—) Background light of the field

Figure 18.2 **The paths taken by light rays that form the image of the specimen and those that form the background light of the field.** Light rays from the specimen are brought to focus on the retina, whereas background rays are out of focus, producing a diffuse bright field. As discussed later in the text, the resolving power of an objective lens is proportional to the sine of the angle α. Lenses with greater resolving power have shorter focal lengths, which means that the specimen is situated closer to the objective lens when it is brought into focus.

(a) (b) (c)

Figure 18.3 Magnification versus resolution. The transition from (a) to (b) provides the observer with increased magnification and resolution, whereas the transition from (b) to (c) provides only increased magnification (empty magnification).

Figure 18.4 The resolving power of the eye. A highly schematic illustration of the relationship between the stimulation of individual photoreceptors (left) and the resulting scene one would perceive (right). The diagram illustrates the value of having the image fall over a sufficiently large area of the retina.

likely to see additional detail, although the image is larger (Figure 18.3c), that is, occupies more retinal surface. The change of ocular fails to provide additional information because the image produced by the objective lens does not possess any further detail to be enhanced by the increased power of the ocular lens. The second switch in oculars provides only empty magnification (as in Figure 18.3c).

The optical quality of an objective lens is measured by the extent to which the fine detail present in a specimen can be discriminated, or resolved. The **resolution** attained by a microscope is limited by diffraction. Because of diffraction, light emanating from a point in the specimen can never be seen as a point in the image, but only as a small disk. If the disks produced by two nearby points overlap, the points cannot be distinguished in the image (as in Figure 18.4). Thus the resolving power of a microscope can be defined in terms of the ability to see two neighboring points in the visual field as distinct entities. The resolving power of a microscope is limited by the wavelength of the illumination according to the equation

$$d = \frac{0.61\lambda}{n \sin \alpha}$$

where d is the minimum distance that two points in the specimen must be separated to be resolved, λ is the wavelength of light (527 nm is used for white light), and n is the refractive index of the medium present between the specimen and objective lens. Alpha (α) is equal to half the angle of the cone of light entering the objective lens as shown in Figure 18.2. Alpha is a measure of the light-gathering ability of the lens and is directly related to its aperture.

The denominator of the equation written above is called the *numerical aperture (N.A.)*. The numerical aperture is a constant for each lens, a measure of its light-gathering qualities. For an objective that is designed for use in air, the maximum possible N.A. is 1.0 because the sine of the maximum possible angle of α, 90°, is 1, and the refractive index of air is 1.0. For an objective designed to be immersed in oil, the maximum possible N.A. is approximately 1.5. As a common rule of thumb, the maximum useful magnification of a microscope ranges between 500 and 1000 times the numerical aperture of the objective lens being used. Attempts to enlarge the image beyond this point result in empty magnification, and the quality of

the image deteriorates. High numerical aperture is achieved by using lenses with a short focal length, which allows the lens to be placed very close to the specimen.

If we substitute the minimum possible wavelength of illumination and the greatest possible numerical aperture in the preceding equation, we can determine the **limit of resolution** of the light microscope. When these substitutions are made, a value of slightly less than 0.2 mm (or 200 nm) is obtained, which is sufficient to resolve larger cellular organelles, such as nuclei and mitochondria. In contrast, the limit of resolution of the naked eye, which has a numerical aperture of about 0.004, is approximately 0.1 mm (0.0036 inches).

In addition to these theoretical factors, resolving power is also affected by optical flaws, or *aberrations*. There are seven important aberrations, and they are the handicaps that lensmakers must overcome to produce objective lenses whose actual resolving power approaches that of the theoretical limits. Objective lenses are made of a complex series of lenses, rather than a single convergent lens, to eliminate these aberrations. One lens unit typically affords the required magnification, while the others compensate for errors in the first lens to provide a corrected overall image.

Visibility

On the more practical side of microscopy from that of limits of resolution is the subject of *visibility*, which is concerned with factors that allow an object actually to be observed. This may seem like a trivial matter; if an object is there, it should be seen. Consider the case of a glass bead. Under most conditions, against most backgrounds, the bead is clearly visible. If, however, a glass bead is dropped into a beaker of immersion oil having the same refractive index as glass, the bead disappears from view because it no longer affects light in any obvious manner that is different from the background fluid. Anyone who has spent any time searching for an amoeba can appreciate the problem of visibility when using the light microscope.

What we see through a window or through a microscope are those objects that affect the light differently from their background. Another term for visibility in this sense of the word is **contrast**, or the difference in appear-

ance between adjacent parts of an object or an object and its background. The need for contrast can be appreciated by considering the stars. Whereas a clear night sky can be filled with stars, the same sky during the day appears devoid of celestial bodies. The stars have disappeared from view, but they haven't disappeared from the sky. They are no longer visible against the much brighter background.

In the macroscopic world, we examine objects by having the light fall on them and then observing the light that is reflected back to our eyes. In contrast, when we use a microscope, we place the specimen between the light source and our eyes and view the light that is transmitted through the object (or more properly, diffracted by the object). If you take an object and go into a room with one light source and hold the object between the light source and your eye, you can appreciate part of the difficulty in such illumination; it requires that the object being examined be nearly transparent, that is, translucent. Therein lies another aspect of the problem: objects that are "nearly transparent" can be difficult to see.

One of the best ways to make a thin, translucent specimen visible under the microscope is to stain it with a dye, which absorbs only certain wavelengths within the visible spectrum. Those wavelengths that are not absorbed are transmitted to the eye, causing the stained object to appear colored. Different dyes bind to different types of biological molecules, and therefore, not only do these procedures heighten visibility of the specimen, they can also indicate where in the cells or tissues different types of substances are found. A good example is the Feulgen stain, which is specific for DNA, causing the chromosomes to appear colored under the microscope (Figure 18.5). A problem with stains is that they generally cannot be used with living cells; they are usually toxic, or the staining conditions are toxic, or the stains do not penetrate the plasma membrane. The Feulgen stain, for example, requires that the tissue be hydrolyzed in acid before the stain is applied. A few stains called *vital dyes* are suitable for use with living cells (as exemplified by the mitochondria seen in Figure 5.20), but their use is limited.

Different types of light microscopes use different types of illumination. In a **bright-field microscope**, the cone of light that illuminates the specimen is seen as a bright background against which the image of the specimen must be contrasted. Bright-field microscopy is ideally suited for specimens of high contrast, such as stained sections of tissues, but it may not provide optimal visibility for other specimens. In the following sections, we will consider various alternate means of making specimens more visible in a light microscope.

Phase-Contrast Microscopy

Small, unstained specimens such as a living cell can be very difficult to see in a bright-field microscope (Figure 18.6a). The **phase-contrast microscope** makes highly transparent objects more visible (Figure 18.6b). We can distinguish different parts of an object because they affect light differently

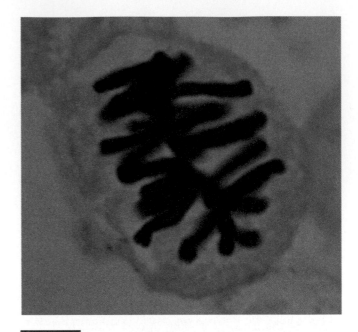

Figure 18.5 **The Feulgen stain.** This staining procedure is specific for DNA as indicated by the localization of the dye to the chromosomes of this onion root tip cell that was in metaphase of mitosis at the time it was fixed. (BY ED RESCHKE/PETER ARNOLD, INC.)

from one another. One basis for such differences is refractive index. Cell organelles are made up of different proportions of various molecules: DNA, RNA, protein, lipid, carbohydrate, salts, and water. Regions of different composition are likely to have different refractive indexes. Normally such differences cannot be detected by our eyes. However, the phase-contrast microscope converts differences in refractive index into differences in intensity (relative brightness and darkness), which are visible to the eye. Phase contrast microscopes accomplish this result by (1) separating the direct light that enters the objective lens from the diffracted light emanating from the specimen and (2) causing light rays from these two sources to *interfere* with one another. The relative brightness or darkness of each part of the image reflects the way in which the light from that part of the specimen interferes with the direct light.

Phase-contrast microscopes are most useful for examining intracellular components of living cells at relatively high resolution. For example, the dynamic motility of mitochondria, mitotic chromosomes, and vacuoles can be followed and filmed with these optics. Simply watching the way tiny particles and vacuoles of cells are bumped around in a random manner in a living cell conveys an excitement about life that is unattainable by observing stained, dead cells. The greatest benefit derived from the invention of the phase-contrast microscope has not been in the discovery of new structures, but in its everyday use in research and teaching labs for observing cells in a more revealing way.

The phase-contrast microscope is only suitable for observing single cells or thin cell layers. In addition, it has optical handicaps that result in loss of resolution, and the

(a)

(b)

(c)

Figure 18.6 A comparison of cells seen with different types of light microscopes. Light micrographs of yeast cells (*Candida albicans*) that were growing on the vaginal epithelium as observed under bright-field (*a*), phase-contrast (*b*), and differential interference contrast (DIC) (or Nomarski) optics (*c*). Note that the yeast cells are virtually invisible under bright-field illumination but clearly seen under phase-contrast and DIC microscopy. (MICROGRAPHS BY M. I. WALKER/ PHOTO RESEARCHERS, INC.)

image suffers from interfering halos and shading produced where sharp changes in refractive index occur. The phase-contrast microscope is a type of *interference microscope*. Other types of interference microscopes minimize these optical artifacts by achieving a complete separation of direct and diffracted beams using complex light paths and prisms. Another type of interference system, termed **differential interference contrast (DIC)**, or sometimes Nomarski interference after its developer, delivers an image that has an apparent three-dimensional quality (Figure 18.6*c*). Contrast in DIC microscopy depends on the rate of change of refractive index across a specimen. As a consequence, the edges of structures, where the refractive index varies markedly over a relatively small distance, are seen with especially good contrast.

Fluorescence Microscopy

Certain compounds (called *fluorochromes or fluorophores*) absorb invisible, ultraviolet radiation and release a portion of the energy in the longer, visible wavelengths, a phenomenon called *fluorescence*. The presence of fluorochromes within cells is observed using a **fluorescence microscope**. The light source in a fluorescence microscope produces a beam of ultraviolet light that travels through a filter, which blocks all wavelengths except that which is capable of exciting the fluorochrome. The beam of monochromatic light is focused on the specimen containing the fluorochrome, which becomes excited and emits light of a visible wavelength that is focused by the objective lens into an image that can be seen by the viewer. Because the light source produces only ultraviolet (black) light, objects stained with a fluorochrome appear brightly colored against a black background, providing very high contrast.

In many studies, the fluorochrome is injected into a living cell, and its fate is followed as the cell carries out its dynamic activities. An example of the use of a fluorochrome is shown in Figure 18.7, which shows a neuron from the sea hare *Aplysia* that had been microinjected with a fluorescently labeled protein kinase (PKA) that is activated by cyclic AMP (page 632). Changes in the concentration of cAMP produce changes in the intensity of fluorescence, which are converted to differences in color as revealed in the micrographs. The series of micrographs shown in this figure was taken over a 25-minute period during which the neuron was stimulated with the neurotransmitter 5-hydroxytryptamine (serotonin), which produces a sharp but transient rise in intracellular cAMP concentrations. In other examples, fluorochromes have been used to study the sizes of molecules that can pass between cells (see Figure 7.33), as indicators of transmembrane potentials (see Figure 5.20), or as probes to determine the free Ca^{2+} concentration in the cytosol (see Figure 15.14). The use of calcium-sensitive fluorochromes is discussed on page 644.

In its most common application, a fluorochrome (such as rhodamine or fluorescein) is covalently linked (*conjugated*) to an antibody to produce a fluorescent antibody that can be used to determine the location of a spe-

Figure 18.7 Use of a fluorochrome to follow the formation of cAMP in a living cell in response to specific stimulation. This series of photographs shows a sensory nerve cell from the sea hare *Aplysia*. The concentration of free cAMP (as indicated in μM by the color bar on the right side of the illustration) was measured indirectly by microinjection of a fluorescently labeled cAMP-dependent protein kinase labeled with both fluorescein and rhodamine on different subunits. Energy transfer between the subunits provides a measure of cAMP concentration. The upper left image shows the intracellular cAMP concentrations in the unstimulated neuron, and the next four images show the effects of stimulation by the neurotransmitter serotonin (5-hydroxytryptamine) at the times indicated. When the cells are washed free of serotonin, the cAMP levels return to their unstimulated values. (REPRINTED WITH PERMISSION FROM BRIAN J. BACKSAI ET AL., SCIENCE 260:223, 1993; © COPYRIGHT 1993, AMERICAN ASSOCIATION FOR THE ADVANCEMENT OF SCIENCE.)

cific protein within the cell. This technique is called *immunofluorescence* and is discussed further on page 785. Suppose, for example, you were interested in determining the changes in location of a particular motor protein, such as cytoplasmic dynein, before and during mitosis. To make this determination, you would make preparations of either whole mounts or sections of dividing and nondividing cells, and then treat the preparations with a fluorescently labeled anti-dynein antibody. Examination of the preparation in a fluorescence microscope would reveal the location of dynein in the two stages of the cell cycle.

Fluorescently labeled proteins can also be used to study dynamic processes as they occur in a living cell. For example, a specific fluorochrome can be linked to a cellular protein, such as actin or tubulin, and the fluorescently labeled protein injected into a living cell. In recent years, a noninvasive approach has been employed that utilizes a fluorescent protein (called green fluorescent protein, or GFP) from the jellyfish *Aequorea victoria* as illustrated in the chapter opening photo on page 737. In most of these studies, a recombinant DNA is constructed in which the coding region of the GFP is joined to the coding region of the protein under study. This recombinant DNA is used to transfect cells, which then synthesize a chimeric protein containing fluorescent GFP fused to the protein under study. Use of GFP to study membrane dynamics is discussed on page 283. In all of these strategies, the labeled proteins participate in the normal activities of the cell, and their location can be followed microscopically to reveal the dynamic activities in which the protein participates (see Figures 9.2 and 9.4). Fluorochromes can also be used to locate DNA or RNA molecules that contain specific nucleotide sequences as described on page 769 and illustrated in Figure 10.22.

Polarized Light Microscopy

Light (and other forms of electromagnetic radiation) normally vibrates in all directions. A **polarizing microscope** utilizes polarizing filters, which are capable of selecting light that vibrates only in a single plane. By using these filters, the polarizing microscope can provide information about cellular structures that are composed of well-aligned elements, such as microtubules or microfilaments, that are normally below the limit of resolution of a light microscope. This is possible because such structures possess an optical property called *birefringence*. An object manifests birefringence by appearing bright when placed between two polarizing filters oriented with their respective plane of transmission perpendicular to one another (Figure 18.8). Light passing through the first filter becomes plane polarized; that is, the light waves vibrate in only one direction. In the absence of a birefringent specimen, none of the plane polar-

(a)

(b)

Figure 18.8 The visualization of subcellular structures composed of aligned elements in the polarizing microscope. (*a*) The metaphase spindle of an oocyte displays striking birefringence due to the large number of oriented microtubules. (*b*) The A bands of the sarcomeres of a single isolated myofibril from a skeletal muscle appear birefringent due to the aligned myosin filaments. (A: FROM SHINYA INOUÉ AND HIDEMI SATO, J. GEN. PHYSIOL. 50:262, 1967; B: FROM RICHARD H. COLBY, J. CELL BIOL. 51:765, 1971; BY COPYRIGHT PERMISSION OF THE ROCKEFELLER UNIVERSITY PRESS.)

ized light rays can pass through the second filter, causing the field to appear black. However, if the specimen contains oriented elements that rotate plane polarized light, the object appears bright against a black background. The birefringence of the spindle seen in Figure 18.8*a* results from the many aligned microtubules of which it is composed. Other cell structures that are birefringent include myosin filaments in striated muscle (Figure 18.8*b*), chloroplasts (owing to their stacks of internal membranes), plant cell walls, and various types of crystalline inclusions.

Video Microscopy and Image Processing

Just as a microscopic field can be seen with the eyes or filmed by a camera, it can also be viewed electronically and taped using a video camera. Video cameras offer a number of advantages for viewing specimens. Special types of video cameras are constructed to be very sensitive to light, which allows them to image specimens at very low illumination. This is particularly useful when observing live specimens,

which are easily damaged by the heat from a light source, and fluorescently stained specimens, which fade rapidly on exposure to light. In addition, video cameras can greatly increase the contrast of an image so that very small objects become visible. The photographs in Figure 9.6, for example, show an image of an individual microtubule (diameter of 0.025 μm) that is far below the limit of resolution of a standard light microscope (0.2 μm). As an added advantage, images produced by video cameras can be converted to digital electronic images and subjected to computer processing to greatly increase their information content.[1] In one technique, the distracting out-of-focus background in a visual field is stored by the computer and then subtracted from the image containing the specimen. This greatly increases the clarity of the image. Similarly, differences in brightness in an image can be converted to differences in color, which makes them much more apparent to the eye. Figure 18.7 provides an example of this technique in which differences in color are used to display spatial differences in cAMP concentration within the cytoplasm.

Confocal Scanning Light Microscopy

The use of video cameras, electronic images, and computer processing has led to a renaissance in light microscopy over the past couple of decades. So too has the development of a new type of light microscope. Normally, when a whole cell or a section of an organ is examined under the light microscope, the observer views the specimen at different depths by changing the position of the objective lens by rotating the focusing knob. As this is done, different parts of the specimen go in and out of focus. But the fact that the specimen contains different levels of focus reduces the ability to form a crisp image because those parts of the specimen above and below the plane of focus interfere with the light rays from that part that is in the plane of focus. In the late 1950s, Marvin Minsky of the Massachusetts Institute of Technology invented a revolutionary new instrument called a **confocal scanning light microscope** that produces an image of a thin plane situated within a much thicker specimen. A schematic diagram of the optical components and light paths within a modern version of a fluorescence confocal scanning light microscope is shown in Figure 18.9. In this type of microscope, the specimen is illuminated by a finely focused laser beam that rapidly scans across the specimen at a single depth, thus illuminating only a thin plane (or "optical section") within the specimen. As described above, short-wavelength incident light is absorbed by the specimen and reemitted at longer wavelength. Light emitted from the specimen is brought to focus at a site within the microscope that contains a pinhole aperture. Thus, the aperture and the illuminated plane in the specimen are confocal. Light rays emitted from the

[1]The conversion of an analog electronic image, such as that obtained by a video camera, to a digital electronic image, such as that obtained by a digital camera and stored on a floppy disc, is called digitization. Digital images consist of a discrete number of picture elements (pixels) each of which has an assigned color and brightness value corresponding to that site in the original image.

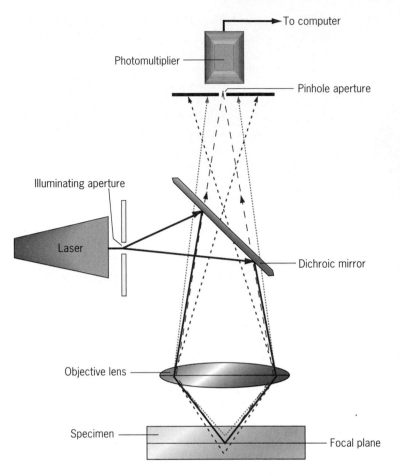

Figure 18.9 The light paths in a fluorescence confocal scanning light microscope. Light of short (blue) wavelength is emitted by a laser light source, passes through a tiny aperture, and is reflected by a dichroic mirror (a type of mirror that reflects certain wavelengths and transmits others) into an objective lens and focused onto a spot in the plane of the specimen. Fluorochromes in the specimen absorb the incident light and emit light of longer wavelength, which is able to pass through the dichroic mirror and come to focus in a plane that contains a pinhole aperture. The light then passes into a photomultiplier tube that amplifies the signal and is transmitted to a computer which forms a processed, digitized image. Any light rays that are emitted from above or below the optical plane in the specimen are prevented from passing through the aperture and thus do not contribute to formation of the image. This diagram shows the illumination of a single spot in the specimen. Different sites within this specimen plane are illuminated by means of a laser scanning process.

illuminated plane of the specimen can pass through the aperture, whereas any light rays that might emanate from above or below this plane are prevented from participating in image formation. As a result, out-of-focus points in the specimen become invisible.

Figure 18.10 shows micrographs of a single cell nucleus taken at three different planes within the specimen. It is evident that objects out of the plane of focus have little effect on the quality of the image of each section. If desired, the images from separate optical sections can be stored in a computer and used to reconstruct a three-dimensional model of the entire object that can be viewed from any desired angle.

Preparation of Specimens for Light Microscopy

Specimens to be observed with the light microscope are broadly divided into two categories: whole mounts and sections. A **whole mount** is an intact object, either living or dead, and can consist of an entire microscopic organism such as a protozoan or a small part of a larger organism. As long as the object is sufficiently transparent, it can be viewed with transmitted light. Often, opaque objects can be made translucent by substituting the water with alcohol and immersing the object in solvents, such as toluene or xylene, in which they become clear.

Most tissues of plants and animals are much too opaque for microscopic analysis unless examined as a very

thin slice, or **section**. To prepare a section, the cells are first killed by immersing the tissue in a chemical solution, called a **fixative**. A good fixative rapidly penetrates the cell membrane and immobilizes all of its macromolecular material so that the structure of the cell is maintained as close as possible to that of the living state. The most common fixatives for the light microscope are solutions of formaldehyde, alcohol, or acetic acid.

After fixation, the tissue is dehydrated by transfer through a series of alcohols and then embedded in paraffin (wax), which provides mechanical support during sectioning. Paraffin is used as an embedding medium because it is readily dissolved by organic solvents. Slides containing adherent paraffin sections are immersed in toluene, which dissolves the wax, leaving the thin slice of tissue attached to the slide, where it can be stained or treated with enzymes, antibodies, or other agents. After staining, a coverslip is mounted over the tissue using a mounting medium that has the same refractive index as the glass slide and coverslip.

18.2 TRANSMISSION ELECTRON MICROSCOPY

The electron micrographs shown in this text were taken by two different types of electron microscopes. **Transmission electron microscopes (TEMs)** form images using electrons that are transmitted through a specimen,

(a)

(b)

Figure 18.11 **A comparison between the information contained in images taken by a light and electron microscope at a comparable magnification of 4500 times actual size.** (*a*) A photo of skeletal muscle tissue that had been embedded in plastic, sectioned at 1 μm, and photographed with a light microscope under an oil immersion objective lens. (*b*) An adjacent section to that used for part *a* that was cut at 0.025 μm and examined under the electron microscope at comparable magnification to that in *a*. The resulting image displays a one- to two-hundred-fold increase in resolution. Note the difference in the details of the muscle myofibrils, mitochondria, and the red blood cell contained in the capillary. Whereas the light microscope cannot provide any additional information to that of *a*, the electron microscope can provide much more information, producing images, for example, of the structure of the individual membranes within a small portion of one of the mitochondria (as in Figure 5.21). (COURTESY OF DOUGLAS E. KELLY AND M. A. CAHILL.)

Figure 18.10 **Confocal scanning micrographs** of three optical sections 0.3 μm thick of a yeast nucleus stained with two different fluorescently labeled antibodies. The red fluorescent antibody has stained the DNA within the nucleus, and the green fluorescent antibody has stained a telomere-binding protein that is localized at the periphery of the nucleus. (FROM THIERRY LAROCHE AND SUSAN M. GASSER, CELL 75:543, 1993, BY PERMISSION OF CELL PRESS.)

whereas **scanning electron microscopes (SEMs)** utilize electrons that have bounced off the surface of the specimen. All of the comments in this section of the chapter concern the use of the TEM; the SEM is discussed separately on page 751.

The transmission electron microscope can provide vastly greater resolution than the light microscope. This is readily illustrated by comparing the two photos in Figure 18.11, which show images of adjacent sections of the same tissue at the same magnification using a light or an electron microscope. Although the photo in Figure 18.11*a* is near the limit of resolution of the light microscope, the photo in Figure 18.11*b* is a very low power electron micrograph. The great resolving power of the electron microscope derives from the wave properties of electrons. As indicated in the equation on page 739, the limit of resolution of a microscope is directly proportional to the wavelength of the illuminating light: the longer the wavelength, the poorer the resolution. Unlike a photon of light, which has a constant wavelength, the wavelength of an electron varies with the speed at which the particle is traveling, which in turn depends on the accelerating voltage applied in the microscope. This relationship is defined by the equation

$$\lambda = \sqrt{150/V}$$

where λ is the wavelength in angstroms and *V* is the accelerating voltage in volts. Standard TEMs operate with a voltage range from 10,000 to 100,000 V. At 60,000 V, the

wavelength of an electron is approximately 0.05 Å. If the numerical aperture attainable with the light microscope were substituted into the equation on page 739, the limit of resolution would be about 0.03 Å. If one considers that the typical distance between the centers of atoms within a molecule is approximately 1 Å, the prospects are astonishing. In actual fact, the resolution attainable with a standard transmission electron microscope is about two orders of magnitude less than its theoretical limit. This is due to serious spherical aberration of electron-focusing lenses, which requires that the numerical aperture of the lens be made very small (generally between 0.01 and 0.001). The practical limit of resolution of standard TEMs is in the range of 3 to 5 Å. The actual limit when observing cellular structure is more typically in the range of 10 to 15 Å.

Electron microscopes consist largely of a tall, hollow cylindrical column through which the electron beam passes and a console containing a panel of dials that electronically control the operation in the column. The top of the column contains the cathode, a tungsten wire filament that is heated to provide a source of electrons. Electrons are drawn from the hot filament and accelerated as a fine beam by the high voltage applied between the cathode and anode. Air is pumped out of the column prior to operation, producing a vacuum through which the electrons travel. If the air were not removed, electrons would be prematurely scattered by collision with gas molecules.

Just as a beam of light rays can be focused by a ground glass lens, a beam of negatively charged electrons can be focused by electromagnetic lenses, which are located in the wall of the column. The strength of the magnets is controlled by the current provided them, which is determined by the positions of the various dials of the console. A comparison of the lens systems of a light and electron microscope is shown in Figure 18.12. The condenser lenses of an electron microscope are placed between the electron source and the specimen, and they focus the electron beam on the specimen. The specimen is supported on a small, thin metal grid (3-mm diameter) that is inserted with tweezers into a grid holder, which in turn is inserted into the column of the microscope.

Because the focal lengths of the lenses of an electron microscope vary depending on the current supplied to them, one objective lens provides the entire range of magnification delivered by the instrument. As in the light microscope, the image from the objective lens serves as the object for an additional lens system. The image provided by the objective lens of an electron microscope is only magnified about 100 times, but unlike the light microscope, there is sufficient detail present in this image to magnify it an additional 10,000 times. By altering the current applied to the various lenses of the microscope, magnifications can vary from about 1000 times to 250,000 times. Electrons that have passed through the specimen are brought to focus on a screen situated at the bottom of the column. Electrons striking the screen excite a coating of fluorescent crystals, which emit their own visible light that is perceived by the eye as an image of the specimen.

Figure 18.12 **A comparison of the lens systems of a light and electron microscope.** (FROM W. AGAR, PRINCIPLES AND PRACTICE OF ELECTRON MICROSCOPE OPERATION, ELSEVIER/ NORTH-HOLLAND, 1974.)

Image formation in the electron microscope depends on differential scattering of electrons by parts of the specimen. Consider a beam of electrons emitted by the filament and focused on the screen. If no specimen were present in the column, the viewing screen would be evenly illuminated by the beam of electrons, producing an image that is uniformly bright. By contrast, if a specimen is placed in the path of the beam, some of the electrons strike atoms in the specimen and are scattered away. Electrons that bounce off the specimen cannot pass through the very small aperture at the back focal plane of the objective lens and are, therefore, lost as participants in the formation of the image.

The scattering of electrons by a part of the specimen is proportional to the size of the nuclei of the atoms that make up the specimen. Because the insoluble material of cells consists of atoms of relatively low atomic number—carbon, oxygen, nitrogen, and hydrogen—biological material possesses very little intrinsic capability to scatter electrons. To increase electron scattering and obtain required contrast, tissues are fixed and stained with solutions of heavy metals (described next). These metals penetrate into the structure of the cells and become selectively complexed with different parts of the organelles. Those parts of a cell that bind the greatest number of metal atoms allow passage of the least number of electrons. The fewer electrons that are focused on the screen at a given

spot, the darker that spot. Photographs of the image are made by lifting the viewing screen out of the way and allowing the electrons to strike a photographic plate in position beneath the screen. Because photographic emulsions are directly sensitive to electrons, much as they are to light, an image of the specimen can be recorded directly on film.

Specimen Preparation for Electron Microscopy

As with the light microscope, tissues to be examined in the electron microscope must be fixed, embedded, and sectioned. Fixation of tissue for electron microscopy (Figure 18.13) is much more critical than for light microscopy because the sections are subjected to much greater

scrutiny. A fixative must stop the life of the cell without significantly altering the structure of that cell. At the level of resolution of the electron microscope, relatively minor damage, such as swollen mitochondria or ruptured endoplasmic reticulum, becomes very apparent. To obtain the most rapid fixation and the least cellular damage, very small pieces of tissue (less than 0.5 mm) are fixed and embedded. Fixatives are chemicals that denature and precipitate cellular macromolecules. Chemicals having such action may cause the coagulation or precipitation of materials that had no structure in the living cell, leading to the formation of an **artifact**. The best argument that a particular structure is not an artifact is the demonstration of its existence in cells fixed in a variety of different ways

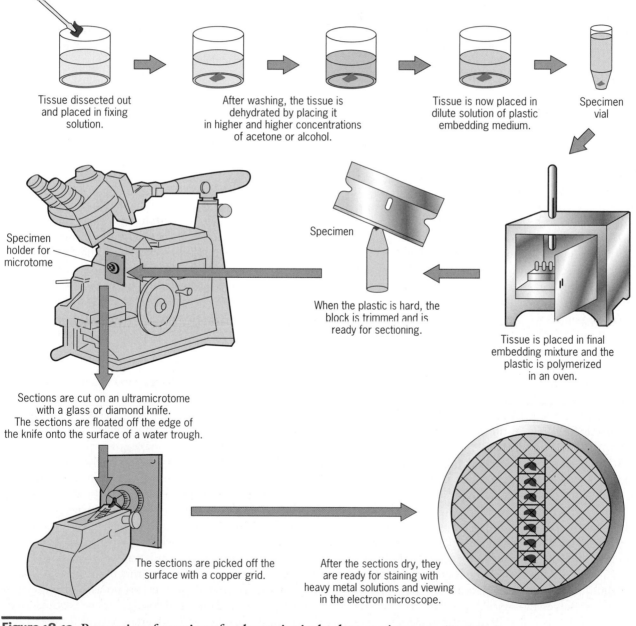

Tissue dissected out and placed in fixing solution.

After washing, the tissue is dehydrated by placing it in higher and higher concentrations of acetone or alcohol.

Tissue is now placed in dilute solution of plastic embedding medium.

Specimen vial

Specimen holder for microtome

Specimen

When the plastic is hard, the block is trimmed and is ready for sectioning.

Tissue is placed in final embedding mixture and the plastic is polymerized in an oven.

Sections are cut on an ultramicrotome with a glass or diamond knife. The sections are floated off the edge of the knife onto the surface of a water trough.

The sections are picked off the surface with a copper grid.

After the sections dry, they are ready for staining with heavy metal solutions and viewing in the electron microscope.

Figure 18.13 **Preparation of a specimen for observation in the electron microscope.** (FROM W. A. JENSEN AND R. B. PARK, CELL ULTRASTRUCTURE, 1ST EDITION © 1967. REPRINTED WITH PERMISSION OF WADSWORTH, AN IMPRINT OF THE WADSWORTH GROUP, A DIVISION OF THOMSON LEARNING. FAX 800/730–2215.)

or, even better, not fixed at all. To view cells that have not been fixed, the tissue is rapidly frozen, and special techniques are utilized to reveal its structure (see cryofixation and freeze-fracture replication, described later). The most common fixatives for electron microscopy are glutaraldehyde and osmium tetroxide. Glutaraldehyde is a 5-carbon compound with an aldehyde group at each end of the molecule. The aldehyde groups react with amino groups in proteins and cross-link the proteins into an insoluble network. Osmium is a heavy metal that reacts primarily with fatty acids leading to the preservation of cellular membranes. In addition, bound osmium atoms are capable of scattering electrons and thus help make membranes visible under the microscope. Glutaraldehyde and osmium tetroxide are often used sequentially to ensure optimal tissue fixation.

Once the tissue has been fixed, the water is removed by dehydration in alcohol, and the tissue spaces are filled with a material that supports tissue sectioning. The demands of electron microscopy require the sections to be very thin. The wax sections cut for light microscopy are rarely thinner than about 5 μm, whereas sections for conventional electron microscopy are best when cut at less than 0.1 μm (equivalent in thickness to about four ribosomes). These sections are so small and thin that if all the sections cut by an electron microscopist during a lifetime were stacked together, they would probably be equivalent to less than a 1-cm cube.

Tissues to be sectioned for electron microscopy are usually embedded in epoxy resins, such as Epon or Araldite. Sections are cut by slowly bringing the plastic block down across an extremely sharp cutting edge (Figure 18.13) made of cut glass or a finely polished diamond face. The sections coming off the knife edge float onto the surface of a trough of water that is contained just behind the knife edge. The sections are then picked up with the metal specimen grid and dried onto its surface. The tissue is stained by floating the grids on drops of heavy metal solutions, primarily uranyl acetate and lead citrate. These heavy metal atoms bind to macromolecules and provide the atomic density required to scatter the electron beam. In addition to the standard stains, tissue sections can be treated with metal-tagged antibodies or other materials that react with specific molecules in the tissue section. Studies with antibodies are usually carried out on tissues embedded in acrylic resins (e.g., Lowicryl), which are more permeable to large molecules than epoxy resins.

Cryofixation and the Use of Frozen Specimens Cells and tissues do not have to be fixed with chemicals and embedded in plastic resins in order to be observed with the electron microscope. Alternatively, cells and tissues can be rapidly frozen. Just as a chemical fixative stops metabolic processes and preserves biological structure, so too does rapid freezing, which is called **cryofixation**. Because cryofixation accomplishes these goals without altering the cell's macromolecules, it is less likely to lead to the formation of fixation artifacts. The major difficulty with cryofix-

ation is the formation of ice crystals, which grow outward from sites where nucleation occurs. As an ice crystal grows, it destroys the fragile contents of the cell in which it develops. One way to avoid ice crystal formation is to immerse the tissue in *cryoprotectants*, such as glycerol, but this creates other types of artifacts and limits the usefulness of the preparation. Alternatively, a specimen can be frozen so rapidly that crystals don't have time to grow large enough to visibly disrupt cellular structure. There are several techniques employed to achieve such ultra-rapid freezing rates. For example, smaller specimens can be plunged into a liquid of very low temperature (such as liquid propane, boiling point of $-42°C$) or placed against a metal block that has been cooled by liquid helium (boiling point of $-273°C$). Larger specimens (e.g., 100 μm diameter) are best treated by high-pressure freezing. In this technique, the specimen is subjected to high hydrostatic pressure and sprayed with jets of liquid nitrogen. High pressure lowers the freezing point of water, which reduces the rate of ice crystal growth.

You might not think that a frozen block of tissue would be of much use to a microscopist, but a surprising number of approaches can be taken to visualize frozen cellular structure in the light or electron microscope. For example, a frozen block of tissue can be sectioned with a special microtome (a cryomicrotome) in a manner similar to that of a paraffin or plastic block of tissue. Frozen sections (cryosections) are particularly useful for studies on enzymes, whose activities tend to be denatured by chemical fixatives. Because frozen sections can be prepared much more rapidly than paraffin or plastic sections, they are often employed by pathologists to examine the light microscopic structure of tissues removed during surgery. As a result, the determination as to whether or not a tumor is malignant can be made while the patient is still on the operating table. A technique to view the ultrastructure of frozen tissue that does not require sectioning is discussed on page 749.

Negative Staining The electron microscope is also well suited for examining very small particulate materials, including high-molecular-weight aggregates such as viruses, ribosomes, multisubunit enzymes, cytoskeletal elements, and protein complexes. The shapes of individual proteins and nucleic acids can also be resolved by the microscope as long as they are made to have sufficient contrast from their surroundings. One of the best ways to make such substances visible is to employ **negative staining** procedures in which heavy-metal deposits are collected everywhere on a specimen grid except where the particles are present. As a result, the specimen stands out by its relative brightness on the viewing screen. For this procedure, a drop of stain solution (uranyl acetate or potassium phosphotungstate) is placed on a grid containing the particles to be examined. Excess liquid is then removed, and the residual stain solution is allowed to dry onto the grid. As a result of surface tension, the stain tends to accumulate around the edge of the particle and penetrates into any holes and

crevices at the particle's surface. The remainder of the particle collects little stain. The method is particularly well suited for revealing subunit organization within the particle. An example of a negatively stained specimen is shown in Figure 18.14*a*.

(a)

(b)

Figure 18.14 Examples of negatively stained and metal-shadowed specimens. Electron micrographs of a tobacco rattle virus after negative staining with potassium phosphotungstate (*a*) or shadow casting with chromium (*b*). (COURTESY OF M. K. CORBETT.)

Shadow Casting Another technique to visualize isolated particles is to have the objects cast shadows. The technique is described in Figure 18.15. The grids containing the specimens are placed in a sealed chamber, which is then evacuated by vacuum pump. The chamber contains a filament composed of a heavy metal (usually platinum) together with carbon. The filament is heated to high temperature, causing it to evaporate and deposit a metallic coat over accessible surfaces within the chamber. As a result, the metal is deposited on those surfaces facing the filament, while the opposite surfaces of the particles and the grid space in their shadow remain uncoated. When the grid is viewed in the electron microscope, the areas in shadow appear bright on the viewing screen, whereas the metal-coated regions appear dark. This relationship is reversed on the photographic plate, which is a negative of the image. The convention for illustrating shadowed specimens is to print a negative image in which the particle appears illuminated by a bright, white light (corresponding to the coated surface) with a dark shadow cast by the particle (Figure 18.14*b*). The technique provides excellent contrast for isolated materials and produces a three-dimensional effect.

Freeze-Fracture Replication and Freeze-Etching As noted above, a number of electron microscopic techniques have been adapted to work with frozen tissues. The ultrastructure of frozen cells is often viewed using the technique of **freeze-fracture replication**, which is illustrated in Figure 18.16. Small pieces of tissue are placed on a small metal disk and rapidly frozen. The disk is then mounted on a cooled stage within a vacuum chamber, and

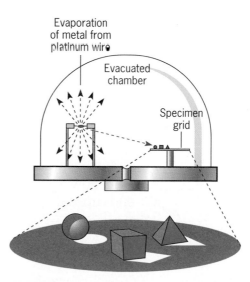

Figure 18.15 The procedure used for shadow casting as a means to provide contrast in the electron microscope. This procedure is often used to visualize small particles, such as the virus shown in the previous figure. DNA and RNA molecules are often made visible by a modification of this procedure known as rotary shadowing in which the metal is evaporated at a very low angle while the specimen is rotated.

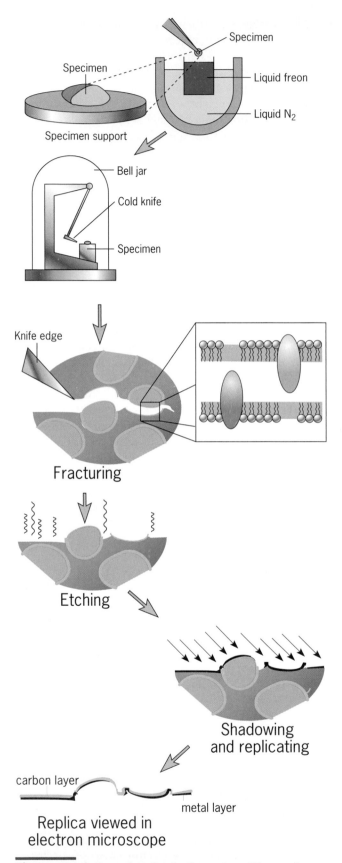

Figure 18.16 Procedure for the formation of freeze-fracture replicas as described in the text. Freeze-etching is an optional step in which a thin layer of covering ice is evaporated to reveal additional information about the structure of the surface of the fractured specimen.

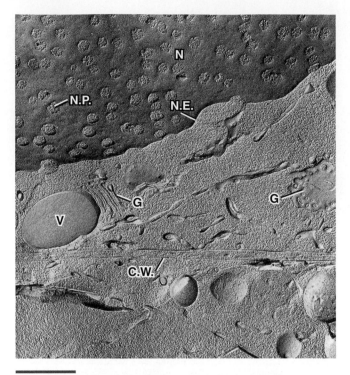

Figure 18.17 Replica of a freeze-fractured onion root cell showing the nuclear envelope (NE) with its pores (NP), the Golgi complex (G), a cytoplasmic vacuole (V), and the cell wall (CW). (COURTESY OF DANIEL BRANTON.)

the frozen tissue block is struck by a knife edge. The resulting fracture plane spreads out from the point of contact, splitting the tissue into two pieces, not unlike the way that an axe blade splits a piece of wood in two.

Consider what might happen as a fracture plane spreads through a cell containing a variety of organelles of different composition. These structures tend to cause deviations in the fracture plane, either upward or downward, giving the fracture face elevations, depressions, and ridges that reflect the contours of the protoplasm traversed. Consequently, the surfaces exposed by the fracture contain information about the contents of the cell. The goal is to make this information visible. The *replication* process accomplishes this by using the fractured surface as a template on which a heavy-metal layer is deposited. The heavy metal is deposited on the newly exposed surface of the frozen tissue in the same chamber where the tissue was fractured. The metal is deposited at an angle to provide shadows that accentuate local topography (Figure 18.17), as described in the previous section on shadow casting.

A carbon layer is then deposited on top of the metal layer from directly overhead, rather than at an angle, so that a uniform layer of carbon is formed that cements the patches of metal into a solid layer. Once a cast of the surface has been made, the tissue that provided the template can be thawed, removed, and discarded; it is the metal-carbon **replica** that is placed on the specimen grid and viewed in the electron microscope. Variations in thickness of the metal in different parts of the replica cause variations in the

numbers of penetrating electrons to reach the viewing screen, producing the necessary contrast in the image. As discussed in Chapter 4, fracture planes take the path of least resistance through the frozen block, which often carries them through the center of cellular membranes. As a result, this technique is particularly well suited for examining the distribution of integral membrane proteins as they span the lipid bilayer (Figure 4.13*b*). Such studies carried out by Daniel Branton and others played an important role in the formulation of the fluid mosaic structure of cellular membranes in the early 1970s (page 126).

Freeze-fracture replication by itself is an extremely valuable technique, but it can be made even more informative by including a step called **freeze etching** (Figure 18.16). In this step, the frozen, fractured specimen, while still in place within the cold chamber, is exposed to a vacuum at an elevated temperature for one to a few minutes, during which a layer of ice can evaporate (sublime) from the exposed surface. Once some of the ice has been removed, the surface of the structure can be coated with heavy metal and carbon to create a metallic replica that reveals both the external surface and internal structure of cellular membranes. The development of *deep-etching* techniques, in which greater amounts of surface ice are re-

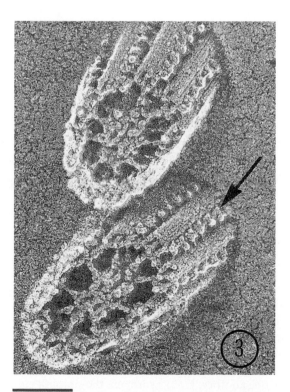

Figure 18.18 Deep etching. Electron micrograph of ciliary axonemes from the protozoan *Tetrahymena*. The axonemes were fixed, frozen, and fractured, and the frozen water at the surface of the fractured block was evaporated away, leaving a portion of the axonemes standing out in relief, as visualized in this metal replica. The arrow indicates a distinct row of outer dynein arms. (FROM URSULA W. GOODENOUGH AND JOHN E. HEUSER, J. CELL BIOL. 95:800, 1982, BY COPYRIGHT PERMISSION OF THE ROCKEFELLER UNIVERSITY PRESS.)

moved, has led to a fascinating new look at cellular organelles. Examples of specimens prepared by this technique are shown in Figures 18.18, 8.41, and 9.45, where it can be seen that the individual parts of the cell stand out in deep relief against the background. The technique delivers very high resolution and can be used to reveal the structure and the distribution of macromolecular complexes, such as those of the cytoskeleton, as they are presumed to exist within the living cell.

18.3 SCANNING ELECTRON MICROSCOPY

The TEM has been exploited most widely in the examination of the internal structure of cells. In contrast, the scanning electron microscope (SEM) is utilized primarily to examine the surfaces of objects ranging in size from a virus to an animal head (Figure 18.19). The construction and operation of the SEM are very different from that of the TEM. The goal of specimen preparation for the SEM is to produce an object that has the same shape and surface properties as the living state, but is devoid of fluid, as required for observing the specimen under vacuum. Because water constitutes such a high percentage of the weight of living cells and is present in association with virtually every macromolecule, its removal can have a very destructive effect on cell structure. When cells are simply air dried, destruction results largely from surface tension at air-water interfaces. Specimens to be examined in the SEM are fixed, passed through a series of alcohols, and then dried by a process of *critical-point drying*. Critical-point drying takes advantage of the fact that a critical temperature and pressure exist for each solvent at which the density of the vapor is equal to the density of the liquid. At this point, there is no surface tension between the gas and the liquid. The solvent of the cells is replaced with a liquid transitional fluid (generally carbon dioxide), which is vaporized under pressure so that the cells are not exposed to any surface tension that might distort their three-dimensional configuration.

Once the specimen is dried, it is coated with a thin layer of metal (usually gold or gold-palladium), which makes it suitable as a target for an electron beam. In the TEM, the electron beam is focused by the condenser lenses to simultaneously illuminate the entire viewing field. In the SEM, electrons are accelerated as a fine beam that scans the specimen. In the TEM, electrons pass through the specimen to form the image. In the SEM, the image is formed by electrons that are reflected back from the specimen (back-scattered) or by secondary electrons given off by the specimen after being struck by the primary electron beam. These electrons strike a detector that is located near the surface of the specimen.

Image formation in the SEM is indirect. In addition to the beam that scans the surface of the specimen, another electron beam synchronously scans the face of a cathode-ray tube, producing an image similar to that seen

(a)

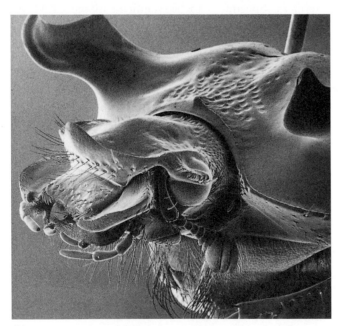

(b)

Figure 18.19 Scanning electron microscopy. Scanning electron micrographs of (*a*) a T4 bacteriophage (X275,000) and (*b*) the head of an insect (X40). (A: REPRINTED WITH PERMISSION FROM A. N. BROERS, B. J. PANESSA, AND J. F. GENNARO, SCIENCE 189:635, 1975; © COPYRIGHT 1975 AMERICAN ASSOCIATION FOR THE ADVANCEMENT OF SCIENCE; B: COURTESY OF H. F. HOWDEN AND L. E. C. LING.)

on a television screen. The electrons that bounce off the specimen and reach the detector control the strength of the beam in the cathode-ray tube. The more electrons collected from the specimen at a given spot, the stronger the signal to the tube and the greater the intensity of the beam on the screen at the corresponding spot. The result is an image on the screen that reflects the surface topology of the specimen because it is this topology (the crevices, hills, and pits) that determines the number of electrons collected from the various parts of the surface.

As evident in the micrographs of Figure 18.19, an SEM can provide a great range of magnification (from about 15 to 150,000 times for a standard instrument). Resolving power of an SEM is related to the diameter of the electron beam. Newer models are capable of delivering resolutions of less than 5 nm, which can be used to localize gold-labeled antibodies bound to a cell's surface. The SEM also provides remarkable depth of focus, approximately 500 times that of the light microscope at a corresponding magnification. This property gives SEM images their three-dimensional quality. At the cellular level, the SEM allows the observer to appreciate the structure of the outer cell surface and all of the various processes, extensions, and extracellular materials that interact with the environment.

18.4 THE USE OF RADIOISOTOPES

A tracer is a substance that reveals its presence in one way or another and thus can be localized or monitored during the course of an experiment. Depending on the substance and the type of experiment, a tracer might be fluorescently labeled; spin labeled; density labeled; or most often, radioactively labeled. In each case, a labeled group enables a molecule to be detected without affecting the specificity of its interactions. Radioactive molecules, for example, participate in the same reactions as nonradioactive species, but their location can be followed and the amount present can be measured.

The identity of an atom (whether it be an iron atom, a chlorine atom, or some other type), and thus its chemical properties, is determined by the number of positively charged protons in its nucleus. All hydrogen atoms have a single proton, all helium atoms have two protons, all lithium atoms have three protons, and so forth. Not all hydrogen, helium, or lithium atoms, however, have the same number of neutrons. Atoms having the same number of protons and different numbers of neutrons are said to be **isotopes** of one another. Even hydrogen, the simplest element, can exist as three different isotopes, depending on whether the atom has 0, 1, or 2 neutrons in its nucleus. Of these three isotopes of hydrogen, only the one containing two neutrons is radioactive; it is tritium (3H).

Isotopes are radioactive when they contain an unstable combination of protons and neutrons. Atoms that are unstable tend to break apart, or disintegrate, thus achieving a more stable configuration. When an atom disintegrates, it releases particles or electromagnetic radiation

that can be monitored by appropriate instruments. Radioactive isotopes occur throughout the periodic table of elements, and they can be produced from nonradioactive elements in the laboratory. All types of biological molecules are available in a radioactive state, that is, containing one or more radioactive atoms as part of their structure. Other than the identity of the element, the most important properties of a radioactive isotope are (1) the type of radiation that is emitted, (2) the energy of the radiation, and (3) the half-life of the isotope.

Three main forms of radiation can be released by atoms during their disintegration. The atom may release an **alpha particle**, which consists of two protons and two neutrons and is equivalent to the nucleus of a helium atom; a **beta particle**, which is equivalent to an electron; and/or **gamma radiation**, which consists of electromagnetic radiation or photons.

Of the three types of radiation, alpha particles are the least energetic, and gamma "rays" are the most energetic. Most biologically important radioisotopes emit beta particles, which are the easiest type of radiation to detect and quantitate and the only type that will be considered in the following discussion. Beta particles can range in energy from very weak, such as those emitted by a tritium atom (0.018 million eV), to highly energetic, such as those emitted by a ^{32}P atom (1.71 million eV). This difference in energy content is manifested in the distance that the beta particle travels, a value known as its *path length*, and in its penetrating power. Whereas the beta particle emitted by a ^3H atom can travel only a few micrometers and cannot penetrate the walls of a glass container, the beta particle from a ^{32}P atom can travel across a room. Because of its

high energy, concentrated solutions of ^{32}P are kept behind a lead shield to protect workers.

The **half-life ($t_{\frac{1}{2}}$)** of a radioisotope is a measure of its instability. The more unstable a particular isotope, the greater the likelihood that a given atom will disintegrate in a given amount of time. If one starts with one Curie[2] of tritium, half that amount of radioactive material will be left after approximately 12 years (which is the half-life of this radioisotope). The half-life of a substance is not usually very important in biological investigations as long as it is sufficiently long to allow it to be used in experiments. In the early years of research into photosynthesis and other metabolic pathways, the only available radioisotope of carbon was ^{11}C, which has a half-life of approximately 20 minutes. Experiments with ^{11}C were literally carried out on the run so that the amount of incorporated isotope could be measured before the substance disappeared. The availability in the 1950s of ^{14}C, a radioisotope having a half-life of 5700 years, was greeted with great celebration. The radioisotopes of greatest importance in cell biological research are listed in Table 18.1, along with information on their half-lives and the nature of their radiation.

Radioactive isotopes are widely used because they are easily detected and accurately quantified. The most commonly used isotopes are beta emitters, which are monitored by either of two different methodologies: liquid scintillation spectrometry or autoradiography. A liquid scintillation counter is used to determine the amount of radioactivity in a given sample, such as the contents of a fraction taken from

[2]A Curie is that amount of radioactivity required to yield 3.7×10^{10} disintegrations per second.

Table 18.1 **Properties of a Variety of Radioisotopes Commonly Used in Biological Research**

Atomic Number	Symbol and Atomic Weight	Half-life	Type of Particle(s) Emitted	Energy of Particle (million eV)
1	^3H	12.3 yr	Beta	0.018
6	^{11}C	20 min	Beta	0.981
	^{14}C	5700 yr	Beta	0.155
11	^{24}Na	15.1 hr	Beta	1.39
			Gamma	2.75, 1.37
15	^{32}P	14.3 d	Beta	1.71
16	^{35}S	87.1 d	Beta	0.167
19	^{42}K	12.4 hr	Beta	3.58, 2.04
			Gamma	1.395
20	^{45}Ca	152 d	Beta	0.260
26	^{59}Fe	45 d	Beta	0.460, 0.26
			Gamma	1.30, 1.10
27	^{60}Co	5.3 yr	Beta	0.308
			Gamma	1.317, 1.115
29	^{64}Cu	12.8 hr	Beta	0.657, 0.571
			Gamma	1.35
30	^{65}Zn	250 d	Beta	0.32
			Gamma	1.11
53	^{131}I	8.0 d	Beta	0.605, 0.250
			Gamma	0.164, etc.

the sucrose gradient depicted in Figure 18.34c. In contrast, autoradiography is used when one wants to know where a particular isotope is located, whether in a cell, in a polyacrylamide gel, or on a nitrocellulose filter. Autoradiography will be discussed in the following section.

Liquid scintillation spectrometry is based on the property of certain molecules, termed *phosphors* or *scintillants*, to absorb some of the energy of an emitted particle and release that energy in the form of light. In preparing a sample for liquid scintillation counting, one mixes the sample with a solution of the phosphor in a glass or plastic scintillation vial. This brings the phosphor and radioactive isotope into very close contact so that radiation from even the weakest beta emitters can be efficiently measured. Once mixed, the vial is placed into the counting instrument, where it is lowered into a well whose walls contain an extremely sensitive photodetector. As radioactive atoms within the vial disintegrate, the emitted particles activate the scintillants, which emit flashes of light. The light is detected by a photocell, and the signal is amplified by a photomultiplier tube within the counter. After correcting for background noise, the amount of radioactivity present in each vial is displayed on a printout as counts per minute.

Autoradiography

Autoradiography is a broad-based technique used to localize sites containing radioisotopes. The importance of autoradiography in providing information about the synthetic activities of cells was described on page 282. In these experiments, cells were incubated with radioactively labeled amino acids, and the fate of the incorporated radioactivity was followed from the site of protein synthesis in the rough endoplasmic reticulum to its discharge into the extracellular space (see Figure 8.3).

Autoradiography takes advantage of the ability of a particle emitted from a radioactive atom to activate a photographic emulsion, much like light or X rays activate the emulsion that coats a piece of film. If the photographic emulsion is brought into close contact with a radioactive source, the particles emitted by the source leave tiny, black silver grains in the emulsion after photographic development. Autoradiography is used to localize radioisotopes within tissue sections that have been immobilized on a slide or TEM grid. The steps involved in the preparation of a light microscopic autoradiograph are shown in Figure 18.20. The emulsion is applied to the sections on the slide or grid as a very thin overlying layer, and the specimen is put into a lightproof container to allow the emulsion to be exposed by the emissions. The longer the specimen is left before development, the greater the number of silver grains that are formed. When the developed slide or grid is examined in the microscope, the location of the silver grains in the layer of emulsion just above the tissue indicates the location of the radioactivity in the underlying cells (Figure 18.21).

The most serious limitation of electron microscopic autoradiography in the localization of radioisotopes

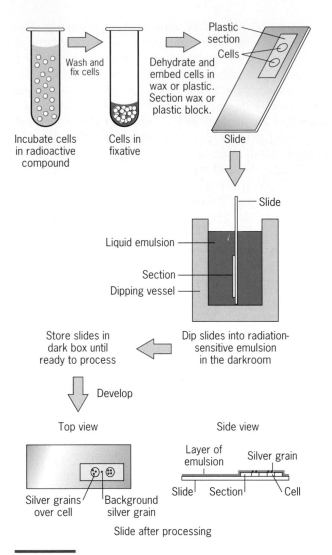

Figure 18.20 Step-by-step procedure for the preparation of an autoradiograph.

within cells is its resolution. The localization of radioactivity within a cell is revealed by the positions of the overlying silver grains. But a silver grain may be larger than the organelle from which the radioactivity is emerging, which makes it impossible to pinpoint the precise site of origin. Moreover, the particle emitted by the radioactive source can travel a certain distance from its point of origin before it strikes the emulsion. As a result, the site where the particle interacts with the emulsion may be some distance from the source of emission within the tissue. For this reason, only those radioisotopes that emit particles with short path lengths, such as 3H and ^{14}C, are suitable for widespread use in microscopic autoradiography.

18.5 CELL CULTURE

Throughout this book we have stressed the approach in cell biology that attempts to understand particular processes by their analysis in a simplified, controlled, in

(a) (b)

Figure 18.21 Examples of a light microscopic and electron microscopic autoradiograph. (*a*) Light microscopic autoradiograph of a polytene chromosome from the insect *Chironomus*, showing extensive incorporation of [³H]uridine into RNA in the puffed regions of the chromosome. Autoradiographs of this type confirmed that the chromosome puffs are sites of transcription. This micrograph can be compared with that of Figure 10.8*b*, which shows the same chromosome as seen in the scanning electron microscope. (*b*) Electron microscopic autoradiograph of a bone marrow cell incubated in [³⁵S]O₄ for 5 minutes and immediately fixed. Sulfate incorporation—as revealed by the black silver grains—is localized in the Golgi complex (arrow). (A: FROM C. PELLING, CHROMOSOMA 15:98, 1964; B: FROM R. W. YOUNG, J. CELL BIOL. 57:177, 1973; BY COPYRIGHT PERMISSION OF THE ROCKEFELLER UNIVERSITY PRESS.)

vitro system. The same approach can be applied to the study of cells themselves because they too can be removed from the influences they are normally subject to within a complex multicellular organism. Learning to grow cells outside the organism, that is, in **cell culture**, has proved to be one of the most valuable technical achievements in the entire study of biology. A quick glance through any journal in cell biology reveals that the majority of the articles describe research carried out on cultured cells. The reasons for this are numerous: cultured cells can be obtained in large quantity; most cultures contain only a single type of cell; many different cellular activities, including endocytosis, cell movement, cell division, membrane trafficking, and macromolecular synthesis, can be studied in cell culture; cells can differentiate in culture; and cultured cells respond to treatment with drugs, hormones, growth factors, and other active substances. Cell culture is gaining new importance in medicine as strategies are being developed to implant cultured cells into patients that suffer from specific cellular deficiencies.

As described on page 390, the first successful attempt to culture living vertebrate cells outside the body was made in 1907. Over the next few decades, a number of researchers determined the optimal conditions for growing animal cells outside the body and keeping the cell cultures free from contamination by microorganisms. The early tissue culturists employed media containing a great variety of unknown substances. Cell growth was accomplished by adding fluids obtained from living systems, such as lymph, blood serum, or embryo homogenates. It was found that cells required a considerable variety of nutrients, hormones, growth factors, and cofactors to remain healthy and proliferate. Even today, most culture media contain large amounts of serum.

One of the primary goals of cell culturists has been to develop a defined, serum-free medium that supports the growth of cells. Using a pragmatic approach in which combinations of various ingredients are tested for their ability to support cell growth and proliferation, a growing number of cell types have been successfully cultured in "artificial" media that lack serum or other natural fluids. As would be expected, the composition of these chemically defined media is relatively complex; they consist of a mixture of nutrients and vitamins, together with a variety of purified proteins, including insulin, epidermal growth factor, and transferrin (which provides the cells with iron). Table 18.2 lists the supplements most commonly employed in serum-free media.

In addition to a complex medium, most cells need an appropriate surface on which to grow. Unlike most cancer cells, which can grow in suspension, normal cells require a solid substratum, such as glass or plastic. Once the cells settle to the bottom of the culture dish, they secrete extracellular materials, such as collagen, fibronectin, and laminin, which help attach them to their substratum. Addition of these extracellular proteins to a culture dish helps most cells adhere and grow.

Table 18.2	Serum-Free Medium Supplements	
		Concentration
Hormones and growth factors		
Insulin		0.1–10 g/ml
Glucagon		0.05–5 g/ml
Epidermal growth factor		1–100 ng/ml
Nerve growth factor		1–10 ng/ml
Gimmel factor		0.5–10 g/ml*
Fibroblast growth factor		1–100 ng/ml
Follicle-stimulating hormone		50–500 ng/ml
Growth hormone		50–500 ng/ml
Luteinizing hormone[†]		0.5–2 g/ml
Thyrotropin-releasing hormone		1–10 ng/ml
Luteinizing hormone-releasing hormone		1–10 ng/ml
Prostaglandin E_1		1–100 ng/ml
Prostaglandin F_2		1–100 ng/ml
Triiodothyronine		1–100 pM
Parathyroid hormone		1 ng/ml
Somatomedin C		1 ng/ml
Hydrocortisone		10–100 nM
Progesterone		1–100 nM
Estradiol		1–10 nM
Testosterone		1–10 nM
Binding proteins		
Transferrin		0.5–100 g/ml
Fatty acid–free BSA[‡]		1 mg/ml
Attachment and spreading factors		
Cold-insoluble globulin		2–10 g/ml
Serum spreading factor		0.5–5 g/ml
Fetuin[†]		0.5 mg/ml
Collagen gel		Substrate coat
Polylysine coat		Substrate coat
Low-molecular-weight nutrient factors		
H_2SeO_3		10–100 nM
$CdSO_4$		0.5 μM
Putrescine		100 μM
Ascorbic acid		10 μg/ml
α-Tocopherol		10 μg/ml
Retinol		10 ng/ml
Linoleic acid[§]		3–5 μg/ml

*Concentration depends on purity of the preparation.
[†]Growth-promoting activity is due to impurities.
[‡]Not stimulatory unless added with *cis*-unsaturated fatty acid.
[§]Not stimulatory unless added with BSA.
Source: D. Barnes and G. Sato, *Anal. Biochem.* 102:258, 1980.

The first step in cell culture is to obtain the cells. In most cases, one need only remove a vial of frozen, previously cultured cells from a tank of liquid nitrogen, thaw the vial, and transfer the cells to the waiting medium. A culture of this type is referred to as a **secondary culture** because the cells are derived from a previous culture. In a **primary culture**, on the other hand, the cells are obtained directly from the organism. Most primary cultures of animal cells are obtained from embryos, whose tissues are more readily dissociated into single cells than those of adults. Dissociation is accomplished with the aid of a pro-

teolytic enzyme, such as trypsin, which digests the extracellular domains of proteins that mediate cell adhesion (Chapter 7). The tissue is then washed free of the enzyme and usually suspended in a saline solution that lacks Ca^{2+} ions and contains a substance, such as ethylenediamine tetraacetate (EDTA), that binds (chelates) calcium ions. As discussed in Chapter 7, calcium ions play a key role in cell-cell adhesion, and their removal from tissues greatly facilitates the separation of cells.

Once the tissue has been dissociated into a single-cell suspension, the cells can be cultured directly, or they can be separated according to cell type and then placed in culture. Cells can be separated by a number of techniques, including differential centrifugation or the use of a *fluorescence-activated cell sorter*. In this latter approach, the cell suspension is treated with a fluorescent antibody that binds specifically to the surface of the cell type to be cultured, and the suspension is passed through an electronic instrument that deflects the fluorescently labeled cells away from nonlabeled cells.

Once the cells have been prepared, there are basically two types of primary cultures that can be started. In a **mass culture**, a relatively large number of cells are added to a culture dish; they settle and attach to the bottom and form a relatively uniform layer of cells. Those cells that survive will grow and divide and, after a number of generations, form a monolayer of cells that covers the bottom of the dish (Figure 18.22*a*). In a **clonal culture**, a relatively small number of cells is added to a dish, so that each cell resides at some distance from its neighbors. Under these conditions, each surviving cell proliferates to form a separate colony or clone (Figure 18.22*b*) whose members are all derived from the same original cell.

Normal (nonmalignant) cells can divide a limited number of times (typically 50 to 100) before they undergo senescence and death. Because of this, many of the cells that are commonly used in tissue culture studies have undergone genetic modifications that allow them to be grown indefinitely. Cells of this type are referred to as a **cell line**, and they typically grow into malignant tumors when injected into susceptible laboratory animals. The frequency with which a normal cell growing in culture spontaneously transforms into a cell line depends on the organism from which it was derived. Mouse cells, for example, transform at a relatively high frequency; human cells transform only rarely, if ever. Human cell lines (e.g., HeLa cells) are typically derived from human tumors or from cells treated with cancer-causing viruses or chemicals.

Many different types of plant cells can also grow in culture. To begin a culture, plant cells are treated with the enzyme cellulase, which digests away the surrounding cell wall, releasing the naked cell, or **protoplast**. Protoplasts can then be grown in a chemically defined medium that promotes their growth and division. Under suitable conditions, the cells can grow into an undifferentiated clump of cells, called a *callus*, which can be induced to develop shoots from which a plant can regenerate.

(a)

(b)

Figure 18.22 Two types of cell cultures. (*a*) Light micrograph showing a small portion of a mass culture of mouse L cells growing on the surface of a culture dish in a chemically defined medium. (*b*) Low-power micrograph of colonies scattered over the surface of a culture dish. In this clonal culture, each colony contains cells that are descendants of a single original cell. This culture was begun by the addition of only about 100 cells to the dish. (COURTESY OF CHARITY WAYMOUTH.)

18.6 THE FRACTIONATION OF A CELL'S CONTENTS BY DIFFERENTIAL CENTRIFUGATION

Most cells contain a variety of different organelles. If one sets out to study a particular function of mitochondria or to isolate a particular enzyme from the Golgi complex, it is useful to first isolate the relevant organelle in a purified state. Isolation of a particular organelle in bulk quantity is generally accomplished by the technique of **differential centrifugation**, which depends on the principle that, as long as they are more dense than the surrounding medium, particles of different size and shape travel toward the bottom of a centrifuge tube at different rates when placed in a centrifugal field.

To carry out this technique, cells are first broken open by mechanical disruption using a mechanical *homogenizer*. Cells are homogenized in an isotonic buffered solution (often containing sucrose), which prevents the rupture of membrane vesicles due to osmosis. The homogenate is then subjected to a series of sequential centrifugations at increasing centrifugal forces (as illustrated in Figure 18.23*a*). Initially, the homogenate is subjected to low centrifugal forces for a short period of time so that only the largest cellular organelles, such as nuclei (and any remaining whole cells), are sedimented into a pellet. At greater centrifugal forces, larger cytoplasmic organelles (mitochondria, chloroplasts, lysosomes, and peroxisomes) can

be spun out of suspension. In subsequent steps, microsomes (the fragments of vacuolar and reticular membranes of the cytosol) and ribosomes are removed from suspension. This last step requires the ultracentrifuge, which can generate speeds of 75,000 revolutions per minute, producing forces equivalent to 500,000 times that of gravity. Once ribosomes have been removed, the supernatant consists of the cell's soluble phase and those particles too small to be removed by sedimentation.

The initial steps of differential centrifugation do not yield pure preparations of a particular organelle, so that further steps are usually required. In many cases, further purification is accomplished by centrifugation of the crude preparation through a density gradient, as shown in Figure 18.23*b*, which distributes the contents of the sample into various layers according to their density. The composition of various fractions can be determined by microscopic examination, or by measuring the amounts of particular proteins known to be specific for particular organelles.

Cellular organelles isolated by differential centrifugation retain a remarkably high level of normal activity, as long as they are not exposed to denaturing conditions during their isolation. Organelles isolated by this procedure can be used in *cell-free systems* to study a wide variety of cellular activities, including the synthesis of membrane-bound proteins (page 286), the formation of coated vesicles (see Figure 8.6), and the transport of solutes and development of ionic gradients.

Remove tissue.
Mince with blade into small pieces.
Suspend in cold isotonic homogenization medium (e.g., 0.25M sucrose).
Homogenize the tissue fragment.

Pestle of homogenizer

Close-fitting glass tube

Ice

Homogenate (suspension of material from broken cells)

Isolate subcellular component from homogenate by differential centrifugation.

Homogenate → 800 g/10 min → Postnuclear supernatant / Pellet containing nuclei and unbroken cells

Postnuclear supernatant → 12,000 g/20 min → Pellet containing large particles (mitochondria, lysosomes, peroxisomes)

Postmitochondrial supernatant → 50,000 g/2 hr → Pellet containing microsomes (small fragments of membrane from ER and Golgi)

Postmicrosomal supernatant → 300,000 g/3 hr → Postribosomal supernatant / Pellet containing free ribosomes

(a)

Resuspended material from 12,000g pellet

1 molar sucrose → 65,000 g/2 hr → Lysosomes filled with Triton WR1339 (1.12 gram/ml) / Mitochondria (1.18 gram/ml) / Peroxisomes (1.23 gram/ml)

Sucrose density gradient

2 molar sucrose

(b)

Figure 18.23 Cell fractionation. (*a*) Step-by-step procedure for the purification of organelles by differential centrifugation. (See Figure 8.5*b* for an example.) (*b*) Subsequent purification steps by density-gradient equilibrium centrifugation. In this particular example, the medium is composed of a continuous density gradient, and the particles sediment until they reach a place in the tube equal to their own density, where they form bands.

18.7 ISOLATION, PURIFICATION, AND FRACTIONATION OF PROTEINS

During the course of this book, we have considered the properties of many different proteins. Before information about the structure or function of a particular protein can be obtained, that protein must be isolated in a relatively pure state. Because most cells contain thousands of different proteins, the purification of a single species can be a challenging mission, particularly if the protein is present in the cell at low concentration. In this section, we will briefly survey a few of the techniques used to purify proteins.

Purification of a protein is generally performed by the stepwise removal of contaminants. Two proteins may be very similar in one property, such as overall charge, and very different in another property, such as molecular mass. Consequently, the complete purification of a given protein usually requires the use of successive techniques that take advantage of different properties of the proteins being separated.

Purification is measured as an increase in **specific activity**, which is the ratio of the amount of that protein to the total amount of protein present in the sample. Some identifiable feature of the specific protein must be utilized as an **assay** to determine the relative amount of that protein in the sample. If the protein is an enzyme, its catalytic activity may be used as an assay to monitor purification. Alternatively, assays can be based on immunologic, electrophoretic, electron microscopic, or other criteria. Measurements of total protein in a sample can be made using various properties, including total nitrogen, which can be very accurately measured and is quite constant at about 16 percent of the dry weight for all proteins.

Selective Precipitation

The first step in purification should be one that can be carried out on a highly impure preparation and can yield a large increase in specific activity. The first step usually takes advantage of solubility differences among proteins by selectively precipitating the desired protein. The solubility properties of a protein is determined largely by the distribution of hydrophilic and hydrophobic side chains on its surface. A protein's solubility in a given solution de-

pends on the relative balance between protein-solvent interactions, which keep it in solution, and protein-protein interactions, which cause it to aggregate and precipitate from solution. The ionic strength of the solution is particularly important in determining which type of interactions predominates. At low ionic strength, interactions with the solvent are enhanced, and proteins tend to remain in solution; they are said to be *salted in*. At higher ionic strengths, protein solubility can rapidly decrease. The salt most commonly employed for selective protein precipitation is ammonium sulfate, which is extremely soluble in water and has high ionic strength. Purification is achieved by gradually adding a solution of saturated ammonium sulfate to the crude protein extract. As addition of salt continues, precipitation of contaminating proteins increases, and the precipitate can be discarded. Ultimately, a point is reached at which the protein being sought comes out of solution. This point is recognized by the loss of activity in the soluble fraction when tested by the particular assay being used. Once the desired protein is precipitated, contaminating proteins are left behind in solution, while the protein being sought can be redissolved.

Liquid Column Chromatography

Chromatography is a term for a variety of techniques in which a mixture of dissolved components is fractionated as it moves through some type of porous matrix. In liquid chromatographic techniques, components in a mixture can become associated with one of two alternative phases: a mobile phase, consisting of a moving solvent, and an immobile phase, consisting of the matrix through which the solvent is moving.[3] In the chromatographic procedures described below, the immobile phase consists of materials that are packed into a column. The proteins to be fractionated are dissolved in the solvent and then passed through the column. The materials that make up the immobile phase contain sites to which the proteins in solution can bind. As individual protein molecules interact with the materials of the matrix, their progress through the column is retarded. Thus the greater the affinity of a particular protein for the matrix material, the slower its passage through the column. Because different proteins in the mixture have different affinity for the matrix, they are retarded to different degrees. As the solvent passes through the column and drips out the bottom, it is collected as **fractions** in a series of tubes. Those proteins in the mixture with the least affinity for the column appear in the first fractions to emerge from the column. The resolution of many chromatographic procedures has been improved in recent years with the development of *high performance liquid chromatography (HPLC)*, in which long, narrow columns are used, and the mobile phase is forced through a tightly packed noncompressible matrix under high pressure.

Ion-Exchange Chromatography Proteins are large, polyvalent electrolytes, and it is unlikely that many proteins in a partially purified preparation have the same overall charge. Ionic charge is used as a basis for purification in a variety of techniques, including **ion-exchange chromatography**. The overall charge of a protein is a summation of all the individual charges of its component amino acids. Because the charge of each amino acid depends on the pH of the medium (see Figure 2.27), the charge of each protein also depends on the pH. As the pH is lowered, negatively charged groups become neutralized and positively charged groups become more numerous. The opposite occurs as the pH is increased. A pH exists for each protein at which the total number of negative charges equals the total number of positive charges. This pH is the **isoelectric point**, at which the protein is neutral. The isoelectric point of most proteins is below pH 7.

Ion-exchange chromatography depends on the ionic bonding of proteins to an inert matrix material, such as cellulose, containing covalently linked charged groups. Two of the most commonly employed ion-exchange resins are diethylaminoethyl (DEAE) cellulose and carboxymethyl (CM) cellulose. DEAE-cellulose is positively charged and therefore binds negatively charged molecules; it is an *anion exchanger*. CM-cellulose is negatively charged and acts as a *cation exchanger*. The resin is packed into a column, and the protein solution is allowed to percolate through the column in a buffer whose composition promotes the binding of some or all of the proteins to the resin. Proteins are bound to the resin reversibly and can be displaced by increasing the ionic strength of the buffer (which adds small ions to compete with the charged groups of the macromolecules for sites on the resin) and/or changing its pH. Displacement of the bound proteins can be done stepwise by the sequential addition of a series of different buffers or continuously by adding a solution of continually changing ionic strength or pH (a gradient). In either case, proteins are eluted from the column in order from the least strongly bound to the most strongly bound. Figure 18.24 shows a schematic representation of the separation of two protein species by stepwise elution from an ion-exchange column.

Gel Filtration Chromatography **Gel filtration** separates proteins (or nucleic acids) primarily by molecular mass. Like ion-exchange chromatography, the separation material consists of tiny beads that are packed into a column through which the protein solution slowly passes. The materials used in gel filtration are composed of cross-linked polysaccharides (dextrans or agarose) of different porosity, which allow proteins to diffuse in and out of the beads. The technique is best illustrated by example (Figure 18.25).

Suppose the protein being purified has a molecular mass of 125,000 daltons and is present in solution with two contaminating proteins of 250,000 daltons and 75,000 daltons. One way the protein might be purified is

[3]Liquid chromatography is distinguished from gas chromatography in which the mobile phase is represented by an inert gas.

Figure 18.24 Ion-exchange chromatography. The separation of two species of protein by DEAE-cellulose. In this case, a positively charged ion-exchange resin is used to bind the more negatively charged protein.

to pass the mixture through a column of Sephadex G-150 beads, which allows entry to molecules less than 200 kDa. When the protein mixture passes through the column bed, the 250 kDa protein is unable to enter the beads and remains dissolved in the moving solvent phase. As a result, the 250 kDa protein is eluted as soon as the preexisting solvent in the column (the bed volume) has dripped out. In contrast, the other two proteins can diffuse into the interstices within the beads and are retarded in their passage through the column. As more and more solvent moves through the column, these proteins move down its length and out the bottom, but they do so at different rates. Among those proteins that enter the beads, smaller

Figure 18.25 Gel filtration chromatography. The separation of three proteins on the basis of molecular mass, as described in the text. Larger proteins are eluted before smaller ones.

species are retarded to a greater extent than larger ones. Consequently, the 125 kDa protein is eluted in a purified state, while the 75 kDa protein remains in the column.

Affinity Chromatography The techniques described up to this point utilize the bulk properties of a protein to effect purification or fractionation. Another purification technique called **affinity chromatography** takes advantage of the unique structural properties of a protein, allowing one protein species to be specifically withdrawn from solution while all others remain behind in solution (Figure 18.26). Proteins interact with specific substances: enzymes with substrates, receptors with ligands, antigens with antibodies, and so forth. Each of these types of proteins can be removed from solution by passing a mixture of proteins through a column in which the specific interacting molecule (substrate, ligand, antibody, etc.) is covalently linked to an inert, immobilized material (the matrix). If, for example, an impure preparation of an insulin receptor is passed through a column containing agarose beads to which the protein insulin is attached, the receptor binds specifically to the beads as long as the conditions in the column are suitable to promote the interaction. Once all the contaminating proteins have passed through the column and out the bottom, the insulin receptor molecules can be displaced from their binding sites on the matrix by changing the ionic composition and/or pH of the solvent in the column. Thus, unlike the other chromatographic procedures that separate proteins on the basis of

Figure 18.26 Affinity chromatography. (*a*) Schematic representation of the coated agarose beads to which only a specific protein can combine. (*b*) Steps in the chromatographic procedure.

size or charge, affinity chromatography can achieve a near-total purification of the desired molecule in a single step. Examples of the use of these various chromatographic procedures are illustrated in the Experimental Pathways of Chapters 4 and 15.

Determining Protein-Protein Interactions One of the ways to learn about the function of a protein is to identify the proteins with which it interacts. Several techniques are available for determining which proteins in a cell might be capable of interacting with a given protein that has already been identified. One of these techniques was just discussed: affinity chromatography. Another technique uses antibodies. Consider, for example, that protein A, which has already been identified and purified, is part of a complex with two other proteins in the cytoplasm, proteins B and C. Once protein A has been purified, an antibody against this protein can be obtained and used as a probe to bind and remove protein A from solution. If a cell extract is prepared that contains the protein A-B-C complex, and the extract is incubated with the anti-A antibody, then binding of the antibody to the A protein will usually result in the *coprecipitation* of other proteins that are bound to A, in this case proteins B and C, which can then be identified.

The technique most widely used to search for protein-protein interactions is the **yeast two-hybrid system**, which was invented in 1989 by Stanley Fields and Ok-kyu Song at the State University of New York in Stony Brook. The technique is illustrated in Figure 18.27 and depends on the expression of a reporter gene, such as β-galactosidase (*lacZ*), whose activity is readily monitored by a test that detects a color change when the enzyme is present in a population of yeast cells. Expression of the *lacZ* gene in this system is activated by a particular protein—a transcription factor—that contains two domains, a DNA-binding domain and an activation domain (Figure 18.27*a*). The DNA-binding domain mediates binding to the promoter, and the activation domain mediates interaction with other proteins involved in the activation of gene expression. Both domains must be present for transcription to occur. To employ the technique, two different types of recombinant DNA molecules are prepared. One DNA molecule contains a segment of DNA encoding the DNA-binding domain of the transcription factor linked to a segment of DNA encoding the "bait" protein (X). The bait protein is the protein that has been characterized and the one for which potential binding partners are being sought. When this recombinant DNA is expressed in a yeast cell, a hybrid protein such as that depicted in Figure 18.27*b* is produced in the cell. The other DNA molecule contains a portion of the transcription factor encoding the activation domain linked to a DNA encoding an unknown protein (Y). Such DNAs (or cDNAs as they are called) are prepared from mRNAs by reverse transcriptase as described on page 775. Let's assume that Y is a protein capable of binding to the bait protein. When a recombinant DNA encoding Y is expressed in a yeast cell, a hybrid protein such as that depicted in Figure 18.27*c* is produced

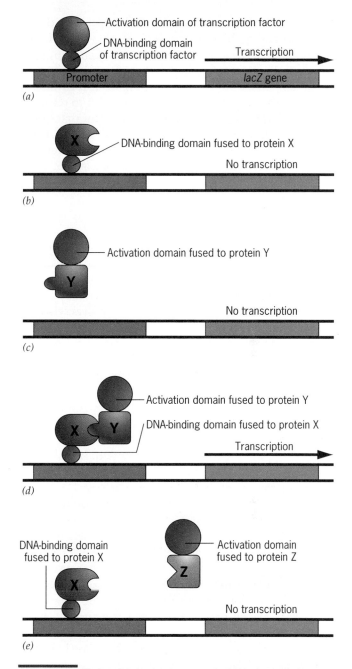

Figure 18.27 Use of the yeast two-hybrid system. This test for protein-protein interaction depends on a cell being able to put together two parts of a transcription factor. (*a*) The two parts of the transcription factor—the DNA-binding domain and the activation domain—are seen here as the transcription factor binds to the promoter of a gene encoding β-galactosidase (*lacZ*). (*b*) In this case, a yeast cell has synthesized the DNA-binding domain of the transcription factor linked to a known "bait" protein X. This complex cannot activate transcription. (*c*) In this case, a yeast cell has synthesized the activation domain of the transcription factor linked to an unknown "fish" protein Y. This complex cannot activate transcription. (*d*) In this case, a yeast cell has synthesized both X and Y protein constructs, which reconstitutes a complete transcription factor, allowing *lacZ* expression, which is readily detected. (*e*) If the second DNA had encoded a protein, e.g., Z, that could not bind to X, no expression of the reporter gene would have been detected.

in the cell. When produced in a cell alone, neither the X- nor Y-containing hybrid protein is capable of activating transcription of the *lacZ* gene (Figure 18.27*b,c*). However, if a yeast cell is transfected with both of these particular recombinant DNA molecules (as in Figure 18.27*d*), the X and Y proteins can interact with one another to reconstitute a functional transcription factor, an event that can be detected by the cell's ability to produce β-galactosidase. Using this technique, researchers are able to "fish" for proteins encoded by unknown genes that are capable of interacting with the "bait" protein.

Polyacrylamide Gel Electrophoresis

Another powerful technique that is widely used to fractionate proteins is **electrophoresis**. Electrophoresis depends on the ability of charged molecules to migrate when placed in an electric field. The electrophoretic separation of proteins is usually accomplished using **polyacrylamide gel electrophoresis (PAGE)**, in which the proteins are driven by an applied current through a gelated matrix. The matrix is composed of polymers of a small organic molecule (acrylamide) that are cross-linked to form a molecular sieve. A polyacrylamide gel may be formed as a thin slab between two glass plates or as a cylinder within a glass tube. Once the gel has polymerized, the slab (or tube) is suspended between two compartments containing buffer in which opposing electrodes are immersed. In a slab gel, the concentrated, protein-containing sample is layered in slots along the top of the gel, as shown in step 1 of Figure 18.28. The protein sample is prepared in a solution containing sucrose or glycerol, whose density prevents the sample from mixing with the buffer in the upper compartment. A voltage is then applied between the buffer compartments, and current flows across the slab, causing the proteins to move toward the oppositely charged electrode (step 2). Separations are typically carried out using alkaline buffers, which make the proteins negatively charged and cause them to migrate toward the positively charged anode at the opposite end of the gel. Following electrophoresis, the slab is removed from the glass plates and stained (step 3).

The relative movement of proteins through a polyacrylamide gel depends on the *charge density* (charge per unit of mass) of the molecules. The greater the charge density, the more forcefully the protein is driven through the gel, and thus the more rapid its rate of migration. But charge density is only one important factor in PAGE fractionation; size and shape also play a role. Polyacrylamide forms a cross-linked molecular sieve that entangles proteins passing through the gel. The larger the protein, the more it becomes entangled, and the slower it migrates. Shape is also a factor because compact globular proteins move more rapidly than elongated fibrous proteins of comparable molecular mass. The concentration of acrylamide (and cross-linking agent) used in making the gel is also an important factor. The lower the concentration of acrylamide, the less the gel becomes cross-linked, and the

more rapidly a given protein molecule migrates. A gel containing 5 percent acrylamide might be useful for sepa-

Figure 18.28 Polyacrylamide gel electrophoresis. The protein samples are typically dissolved in a sucrose solution whose density prevents the sample from mixing with the buffer and then loaded into the wells with a fine pipette as shown in step 1. In step 2, a direct current is applied across the gel, which causes the proteins to move into the polyacrylamide along parallel lanes. When carried out in the detergent SDS, which is usually the case, the proteins move as bands at rates that are inversely proportional to their molecular mass. Once electrophoresis is completed, the gel is removed from the glass frame and stained in a tray (step 3).

rating proteins of 60 to 250 kDa, whereas a gel of 15 percent acrylamide might be useful for separating proteins of 10 to 50 kDa.

The progress of electrophoresis is followed by watching the migration of a charged *tracking dye* that moves just ahead of the fastest proteins (step 2, Figure 18.28). After the tracking dye has moved to the desired location, the current is turned off, and the gel is removed from its container. Gels are typically stained with Coomassie Blue or silver stain to reveal the location of the proteins. If the proteins are radioactively labeled, their location can be determined by pressing the gel against a piece of X-ray film to produce an autoradiograph, or the gel can be sliced into fractions and individual proteins isolated. Alternatively, the proteins in the gel can be transferred by a second electrophoretic procedure to a nitrocellulose membrane to form a blot (page 770). Proteins become absorbed onto the surface of the membrane in the same relative positions they occupied in the gel. In a *Western blot*, the proteins on the membrane are identified by their interaction with specific antibodies.

SDS-PAGE PAGE is usually carried out in the presence of the negatively charged detergent sodium dodecyl sulfate (SDS), which binds in large numbers to all types of protein molecules. The electrostatic repulsion between the bound SDS molecules causes the proteins to unfold into a similar rodlike shape, thus eliminating differences in shape as a factor in separation. The number of SDS molecules that bind to a protein is roughly proportional to the protein's molecular mass (about 1.4 g SDS/g protein). Consequently, each protein species, regardless of its size, has an equivalent charge density and is driven through the gel with the same force. However, because the polyacrylamide is highly cross-linked, larger proteins are held up to a greater degree than smaller proteins. As a result, proteins become separated by SDS-PAGE on the basis of a single property—their molecular mass. In addition to separating the proteins in a mixture, SDS-PAGE can be used to determine the molecular mass of the various proteins by comparing the positions of the bands to those produced by proteins of known size. Examples of SDS-PAGE are shown on pages 149 and 175.

Isoelectric Focusing **Isoelectric focusing** is a type of electrophoresis in which the supporting matrix consists of a gel containing a mixture of *ampholytes*, low-molecular-weight polymers having varying ratios of positively charged amino groups and negatively charged carboxyl groups. When a voltage is applied across the gel, the ampholyte molecules become redistributed, with the most negative ones moving closest to the anode and the most positive ones moving closest to the cathode. This movement sets up a stable pH gradient within the gel from one end to the other. As proteins migrate through the gel in response to the electric field, they are exposed to a continually changing pH, which produces a continuous change in their ionic charge. At some point along the gel, each protein encounters a pH that is equal to its isoelectric point, which converts the protein into a neutral molecule and causes its migration to stop. As a result, each species of protein becomes focused in a very sharp band at a specific position along the length of the gel.

Two-Dimensional Gel Electrophoresis In 1975, a technique called **two-dimensional gel electrophoresis** was developed by Patrick O'Farrell at the University of California, San Francisco, to fractionate complex mixtures of proteins using two different properties of the molecules. Proteins are first separated according to their isoelectric point by isoelectric focusing within a tubular gel. After separation, the gel is removed and placed on top of a slab of SDS-saturated polyacrylamide and subjected to SDS-PAGE. The proteins move into the slab gel and become separated according to their molecular mass (Figure 18.29). Once separated, individual proteins can be removed from the gel and digested into peptide fragments that can be analyzed by mass spectrometry. The resolution of the technique is so great that the majority of proteins present in a cell can be distinguished. Because of its great resolving power, two-dimensional gel electrophoresis is ideally suited to detect changes in the proteins present in a cell under different conditions or at different stages in development or the cell cycle. The technique, however, is not suitable for distinguishing among proteins that have high molecular mass, that are highly hydrophobic, or that are present at very low copy numbers per cell.

Figure 18.29 **Two-dimensional gel electrophoresis.** A two-dimensional polyacrylamide gel of HeLa cell nonhistone chromosomal proteins labeled with [^{35}S]methionine. Over a thousand different proteins can be resolved by this technique. (FROM J. L. PETERSON AND E. H. McCONKEY, J. BIOL. CHEM. 251:550, 1976.)

18.8 DETERMINATION OF PROTEIN STRUCTURE BY X-RAY DIFFRACTION ANALYSIS

The first step in the analysis of a particular protein by any technique is to obtain the most purified preparation of that protein possible. **X-ray diffraction** (or **X-ray crystallography**) utilizes protein crystals, the preparation of which is one of the best guarantees of molecular purity. A crystal is composed of a regularly repeating arrangement of a unit cell, in this case an individual protein molecule. For X-ray diffraction analysis, a crystal is bombarded with a fine beam of X rays (Figure 18.30). The radiation that is scattered (diffracted) by the electrons of the protein's atoms strikes an electron-sensitive plate (e.g., a piece of photographic film) placed behind the crystal. The diffraction pattern produced by a crystal is determined by the structure within the protein; the large number of molecules in the crystal reinforces the reflections, causing it to behave as if it were one giant molecule.

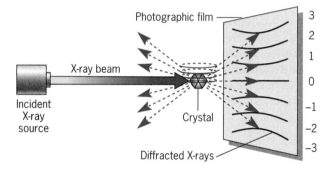

Figure 18.30 X-ray diffraction analysis. Schematic diagram of the diffraction of X rays by atoms of one plane of a crystal onto a photographic plate. The ordered array of the atoms within the crystal produces a repeating series of overlapping circular waves that spread out and intersect the sheet of film. As with diffraction of visible light, the waves form an interference pattern, reinforcing each other at some points on the film and canceling one another at other points.

The positions and intensities of the reflections (which appear as spots on the plates) can be related mathematically to the electron densities within the protein because it is the electrons of the atoms that produced them. Because of the way the crystal is bombarded, each plate represents a slice of the molecule, and many plates must be analyzed before the total diffraction pattern is obtained. A single photographic plate used in the determination of the three-dimensional structure of myoglobin is shown in Figure 2.33. Those spots closer to the center of the pattern result from X rays scattered at smaller angles from the crystal and provide information about the grosser aspects of the protein, that is, the long spacings within the molecule. The spots closer to the periphery of the photograph contain information about the closely spaced aspects of the molecules within the crystal. The resolution obtained by X-ray diffraction depends on the number of spots that are analyzed.

Myoglobin was the first protein whose structure was determined by X-ray diffraction. The protein was analyzed successively at 6, 2, and 1.4 Å, with years elapsing between each completed determination. Considering that covalent bonds are between 1 and 1.5 Å in length and noncovalent bonds between 2.8 and 4 Å in length, the information gathered for a protein depends on the resolution achieved. This is illustrated by a comparison of the electron density of a small organic molecule at four levels of resolution (Figure 18.31). In myoglobin, a resolution of 6 Å was sufficient to show the manner in which the polypeptide chain is folded and the location of the heme moiety, but it was not sufficient to show structure within the chain. At a resolution of 2 Å, groups of atoms could be separated from one another, whereas at 1.4 Å, the positions of individual atoms were determined. Over the years, X-ray diffraction technology has improved greatly (see p. 98) and the technique has been been applied to the analysis of larger and larger protein structures. In one recent study of the blue tongue virus particle, the positions of nearly 3,000,000 atoms were determined from the analysis of approximately 21 million reflections.

X-ray diffraction analysis has also been used in the study of DNA, and it served as one of the most important

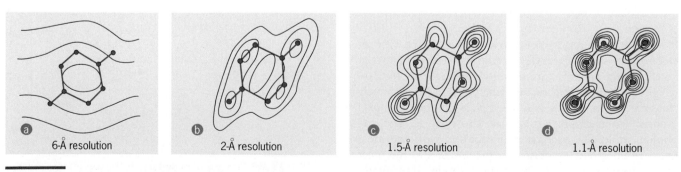

Figure 18.31 Electron density distribution of a small organic molecule (diketopiperazine) calculated at several levels of resolution. At the lowest resolution (*a*), only the ring nature of the molecule can be distinguished, whereas at the highest resolution (*d*), the electron density around each atom (indicated by the circular contour lines) is revealed. (AFTER D. HODGKIN. REPRINTED WITH PERMISSION FROM NATURE 188:445, 1960; © COPYRIGHT 1960, MACMILLAN MAGAZINES LTD.)

sources of data that led Watson and Crick to their proposal for the structure of the double helix in 1953. The analysis of DNA is made using fibers containing oriented DNA molecules rather than crystals and does not provide as high a level of resolution as that obtained with proteins

X-ray diffraction is ideally suited for determining the structure of soluble proteins, but is less appropriate for study of complex molecular structures, such as ribosomes or proteasomes, or for membrane proteins, where it is difficult to obtain the three-dimensional crystals required for analysis. Structural analysis of these types of specimens is often conducted using an alternate technique known as *electron cryomicroscopy* that takes advantage of the tremendous resolving power of the electron microscope and computer-based image processing techniques. Unfixed specimens are typically frozen in liquid nitrogen and maintained on the electron microscope grid in a hydrated, frozen state. A model of a eukaryotic ribosome based on electron cryomicroscopy is illustrated in Figure 2.51. Electron cryomicroscopy is particularly well suited for the study of membrane proteins, such as gap-junction connexons (shown in Figure 7.32c) or nicotinic acetylcholine receptors (see the Experimental Pathway for Chapter 4) that can be closely packed at very low temperatures (e.g., −195°C) into two-dimensional crystalline arrays within the plane of the membrane. The structures illustrated in these figures were determined from combined, high-resolution, electron-microscopic images of many different specimens tilted at various angles.

18.9 PURIFICATION AND FRACTIONATION OF NUCLEIC ACIDS

The steps taken in the purification of nucleic acids are very different from those used in the purification of proteins, which reflects the basic difference in structure of these two types of macromolecules. The first step in DNA purification is generally to homogenize the cells and isolate the nuclei from which the DNA is extracted. The nuclei are then extracted with a buffered salt solution containing a detergent, such as SDS, that serves to lyse the nuclei and release the DNA. As the DNA is released, the viscosity of the solution rises markedly. The detergent also inhibits any nuclease activity present in the preparation.

The major goal of subsequent purification steps is to separate the DNA from contaminating materials, such as RNA and protein. Deproteinization is usually accomplished by shaking the mixture with a volume of phenol. Phenol (or alternatively phenol/chloroform) is an active protein denaturant that causes proteins in the preparation to lose their solubility and precipitate from solution. Because phenol and buffered saline solutions are immiscible, the suspension is simply centrifuged to separate the phases, which leaves the DNA (and RNA) in solution in the upper aqueous phase and the protein present as a precipitate at the boundary between the two phases. The

aqueous phase is removed from the tube and subjected to repeated cycles of shaking with phenol and centrifugation until no further protein is removed from solution. The nucleic acids are then precipitated from solution by addition of cold ethanol. The cold ethanol is often layered on top of the aqueous DNA solution, and the DNA is wound on a glass rod as it comes out of solution at the interface between the ethanol and saline. In contrast, RNA comes out of solution as a flocculent precipitate that settles at the bottom of the vessel. Following this initial purification procedure, the DNA is redissolved and treated with ribonuclease to remove contaminating RNA. The ribonuclease is then destroyed by treatment with a protease, the protease is subsequently removed by deproteinization with phenol, and the DNA is reprecipitated with ethanol.

RNA can be purified in a similar manner using DNase in the final purification steps rather than ribonuclease. An alternative procedure for isolating RNA in a single step was published in 1987. In this technique, tissues are homogenized in a solution containing 4 M guanidine thiocyanate, and the RNA extract is mixed with phenol and shaken with chloroform (or bromochloropropane). The suspension is then centrifuged, leaving RNA in the upper aqueous phase and both DNA and protein at the interface between the two phases.

Separation of DNAs by Gel Electrophoresis

Of the various techniques used in the fractionation of proteins discussed earlier, one of them, gel electrophoresis, is also widely used to separate nucleic acids of different molecular mass (i.e., nucleotide length). Small RNA or DNA molecules of a few hundred nucleotides or less are generally separated by polyacrylamide gel electrophoresis. Larger molecules have trouble making their way through the cross-linked polyacrylamide and are generally fractionated on agarose gels, which have greater porosity. Agarose is a polysaccharide extracted from sea weed; it is dissolved in hot buffer, poured into a mold, and caused to gelate simply by lowering the temperature. Gels of lower agarose concentration (as low as 0.3 percent) are used to separate larger DNA fragments. Separation of DNA molecules greater than about 25 kb is generally accomplished using the technique of pulsed-field electrophoresis in which the direction of the electric field in the gel is periodically changed, which causes the DNA molecules to reorient themselves during their migration.

The separation of nucleic acids by gel electrophoresis is based on similar principles to those that operate during the fractionation of proteins by SDS-PAGE. Unlike proteins, all nucleic acid molecules, regardless of their length, have a similar charge density (the number of negative charges per unit of mass), and therefore, all have an equivalent potential for migration in an electric field. The gelated polyacrylamide or agarose provides the necessary resistance to migration, so that the greater the length of an RNA or DNA molecule, the more slowly it makes its way through the gel. The sensitivity of gel electrophoresis is so

Figure 18.32 Separation of DNA restriction fragments by gel electrophoresis. (*a*) Identifying the restriction fragments that contain the nucleotide sequence of the insulin gene. Once the DNA fragments have been separated by electrophoresis through the agarose gel, the DNA is blotted onto a sheet of nitrocellulose filter paper (not shown in drawing) and then incubated with the labeled insulin DNA probe. After a period of incubation, the excess labeled probe is washed away, and the location of the bound radioactivity is determined auto-radiographically. (*b*) All of the DNA fragments that are present in a gel can be revealed by immersing the gel in a solution of ethidium bromide and then viewing the gel under an ultra-violet light. (B: PHOTOGRAPH BY PHILLIPE PLAILLY/SCIENCE PHOTO LIBRARY/PHOTO RESEARCHERS.)

great that DNA or RNA molecules that differ by only a single nucleotide can be separated from one another by this technique, a feature that has given rise to an invalu-able method for DNA sequencing (page 781). An exam-ple of how gel electrophoresis might be used to identify the DNA fragments that contain a particular gene, such as the gene that codes for human insulin, is illustrated in Fig-ure 18.32*a*. In this example, the DNA fragments being sought are identified by their binding a labeled DNA probe that contains the insulin sequence. Alternatively, all of the DNA fragments in the gel can be localized by im-mersing the gel in a solution of ethidium bromide, which intercalates into the double helix and causes the DNA bands to appear fluorescent when viewed under ultraviolet light (Figure 18.32*b*).

18.10 MEASUREMENTS OF PROTEIN AND NUCLEIC ACID CONCENTRATION BY SPECTROPHOTOMETRY

One of the simplest and most widely used methods to de-termine the amount of protein or nucleic acid present in a given solution is to measure the amount of light of a spe-cific wavelength that is absorbed by that solution. The in-strument used for this measurement is a **spectropho-tometer**. To make this type of measurement, the solution is placed in a special, flat-sided, quartz container (quartz is used because, unlike glass, it does not absorb ultraviolet light), termed a *cuvette*, which is then placed in the light beam of the spectrophotometer. The amount of light that passes through the solution unabsorbed (i.e., the trans-mitted light) is measured by photocells on the other side of the cuvette.

Of the 20 amino acids incorporated into proteins, two of them, tyrosine and phenylalanine, absorb light in the ultraviolet range with an absorbance maximum at about 280 nm. If the proteins being studied have a typical per-centage of these amino acids, then the absorbance of the solution at this wavelength provides a measure of protein concentration. Alternatively, one can use a variety of

chemical assays, such as the Lowry or Biuret technique, in which the protein in solution is engaged in a reaction that produces a colored product whose concentration is pro-portional to the concentration of protein. Nucleic acids have their maximal absorption at 260 nm (see Figure 10.15), which is thus used as the wavelength of choice for measurement of DNA or RNA concentrations.

18.11 ULTRACENTRIFUGATION

Common experience indicates that the stability of a solution (or suspension) depends on the components. Cream floats to the top of raw milk, a fine precipitate gradually settles to the bottom of a container, and a solution of sodium chloride remains stable indefinitely. Numerous factors determine whether or not a given component will settle through a liquid medium, including the size, shape, and density of the substance and the density and viscosity of the medium. If a component in a solution or suspension is denser than its medium, then centrifugal force causes it to become concentrated toward the bottom of a centrifuge tube. Larger particles sediment more rapidly than smaller particles of similar shape and density. The tendency for molecules to become concentrated during centrifugation is counteracted by the effects of diffusion, which causes the molecules to become redistributed in a more uniform (random) arrangement. With the development of ultracentrifuges, it has become possible to generate centrifugal forces as high as 500,000 times the force of gravity, which is great enough to overcome the effects of diffusion and cause macromolecules to sediment toward the bottom of a centrifuge tube.

The velocity of a given particle at any given time during centrifugation in a liquid medium is directly proportional to the square of the angular velocity (v) and the distance of the particle from the center of the rotor (r):

$$v = s(\omega^2 r)$$

In this equation, $\omega^2 r$ is termed the *radial acceleration* and is given in centimeters per second squared. The term s, which is given in seconds, is the *sedimentation coefficient*, which is equivalent to the average velocity per unit acceleration. Centrifugal accelerations are generally expressed relative to the Earth's gravitational acceleration, which has a value of 980 cm/sec². For example, a value for radial acceleration of 4.9×10^7 cm/sec² is equivalent to 50,000 times the force of gravity.

Throughout this book we have referred to various macromolecules and their complexes as having a particular S value. The unit S (or Svedberg, after the inventor of the ultracentrifuge) is equivalent to a sedimentation coefficient of 10^{-13} sec. Because the velocity at which a particle moves through a liquid column depends on a number of factors, including shape, the determination of sedimentation coefficients does not by itself provide the molecular mass. However, as long as one is dealing with the same type of molecule, the S value provides a good measure of relative size. For example, the three ribosomal RNAs of *E. coli*, the 5S, 16S, and 23S molecules, have lengths of 120, 1600, and 3200 nucleotides, respectively.

There are basically two types of ultracentrifuges, analytic models and preparative models. Preparative ultracentrifuges simply generate large centrifugal forces for a set period of time. Centrifugation proceeds in a near vacuum to minimize frictional resistance. Some rotors (centrifuge

Figure 18.33 A variety of rotors used in a preparative ultracentrifuge. Each of these rotors has a slot at the bottom of the central cylindrical portion that fits securely onto the top of a shaft at the bottom of the centrifuge chamber. After placing the rotor onto the shaft, the lid is closed and the air in the chamber is evacuated before the centrifuge motor is started. The rotors in the rear can be seen to contain hanging buckets. As the rotor gains speed, the buckets swing outward, as indicated in the following figure. The tubes in the other rotors fit into slots that remain at a fixed angle. (COURTESY OF BECKMAN INSTRUMENTS, INC.)

heads) are constructed to allow the tubes to swing out, causing the particles to move in a direction parallel to the walls of the tube. This type of rotor is said to contain *swinging buckets* (Figure 18.33). Swinging buckets are useful for separating different sized molecules as in Figure 18.34. In *fixed-angle* rotors, the tubes fit into fixed slots and are maintained at a specific angle (between 14 and 40°) during centrifugation. Fixed angle rotors are useful for sedimenting particles to the bottom of the tube. In preparative ultracentrifugation, all determinations are made after the tube is removed from the centrifuge. The term *preparative* refers to the use of this type of centrifuge to purify components for further use. The analytic ultracentrifuge contains instrumentation to follow the progress of the substances in the tubes (called *cells*) during the centrifugation (as illustrated in Figure 13.3*b*). This allows researchers to obtain estimates of the sedimentation velocity of various components and to make measurements of molecular mass, purity, macromolecular interactions, and so forth.

Sedimentation Behavior of Nucleic Acids

DNA (and **RNA**) molecules are extensively analyzed by techniques utilizing the ultracentrifuge. For the present purpose, we will consider two of the most commonly employed centrifugation techniques used in the study of nucleic acids, which are illustrated in Figure 18.34. In *velocity (or rate-zonal) sedimentation*, nucleic acid molecules are separated according to nucleotide length. The sample containing the mixture of nucleic acid molecules is care-

Figure 18.34 Techniques of nucleic acid sedimentation.
(*a*) Separation of different-sized DNA molecules by velocity sedimentation. The sucrose density gradient is formed within the tube (step 1) by allowing a sucrose solution of increasing concentration to drain along the wall of the tube. Once the gradient is formed, the sample is carefully layered over the top of the gradient (steps 2 and 3), and the tube is subjected to centrifugation (e.g., 50,000 rpm for 5 hours) as illustrated in step 4. The DNA molecules are separated on the basis of their size (step 5). (*b*) Separation of DNA molecules by equilibrium sedimentation on the basis of differences in base composition. The DNA sample is mixed with the CsCl solution (step 1) and subjected to extended centrifugation (e.g., 50,000 rpm for 72 hours). The CsCl gradient forms during the centrifugation (step 2), and the DNA molecules then band in regions of equivalent density (step 3). (*c*) The tube from the experiment of *b* is punctured and the contents are allowed to drip into successive tubes, thereby fractionating the tube's contents. The absorbance of the solution in each fraction is measured and plotted as shown.

(*a*)

(*b*)

(*c*)

fully layered over a solution containing an increasing concentration of sucrose (or other suitable substance). This preformed gradient increases in density (and viscosity) from the top to the bottom. When subjected to high centrifugal forces, the molecules move through the gradient at a rate determined by their sedimentation coefficient. The greater the sedimentation coefficient, the farther a molecule moves in a given period of centrifugation. Because the density of the medium is less than that of the nucleic acid molecules, even at the bottom of the tube (approximately 1.2 g/ml for the sucrose solution and 1.7 g/ml for the nucleic acid), these molecules continue to sediment as long as the tube is centrifuged. In other words, centrifugation never reaches equilibrium. After a prescribed period, the tube is removed from the centrifuge, its contents are fractionated (as shown in Figure 18.34*c*), and the relative positions of the various molecules are determined. The presence of the viscous sucrose prevents the contents of the tube from becoming mixed due to either convection or handling, allowing molecules of identical S value to remain in place in the form of a band. If marker molecules of known sedimentation coefficient are present, the S values of unknown components can be determined. Experimental results obtained by sucrose-density gradient centrifugation are shown in Figures 11.13 and 11.17.

In the other type of centrifugation technique, *equilibrium* (or *isopycnic*) *centrifugation* (Figure 18.34*b*), nucleic acid molecules are separated on the basis of their buoyant density. In this procedure, one generally employs a highly concentrated solution of the salt of the heavy metal cesium. The analysis is begun by mixing the DNA with the solution of cesium chloride or cesium sulfate in the centrifuge tube and then subjecting the tube to extended centrifugation (e.g., 2 to 3 days at high forces). During the centrifugation, the heavy cesium ions are slowly driven toward the bottom of the tube, forming a continuous density gradient through the liquid column. After a time, the tendency for cesium ions to be concentrated toward the bottom of the tube is counterbalanced by the opposing tendency for them to become redistributed by diffusion, and the gradient becomes stabilized. As the cesium gradient is forming, individual DNA molecules are driven downward or move buoyantly upward in the tube until they reach a

position that has a buoyant density equivalent to their own, at which point they are no longer subject to further movement. Molecules of equivalent density form narrow bands within the tube. This technique is sensitive enough to separate DNA molecules having different base composition (as illustrated in Figure 18.34*b*) or ones containing different isotopes of nitrogen (^{15}N versus ^{14}N, as shown in Figure 13.3*b*).

18.12 NUCLEIC ACID HYBRIDIZATION

Nucleic acid hybridization includes a variety of related techniques that are based on the observation that two single-stranded nucleic acid molecules of complementary base sequence can form a double-stranded hybrid. Consider a situation in which one has a mixture of hundreds of fragments of DNA of identical length and overall base composition that differ from one another solely in their base sequence. Assume, for example, that one of the DNA fragments constitutes a portion of the β-globin gene and all the other fragments contain unrelated genes. The only way to distinguish between the fragment encoding the β-globin polypeptide and all the others is to carry out a molecular hybridization experiment using complementary molecules as probes.

In the present example, incubation of the mixture of denatured DNA fragments with an excess number of β-globin mRNAs would drive the globin fragments to form double-stranded DNA-RNA hybrids, leaving the other DNA fragments single-stranded. There are a number of ways one could separate the DNA-RNA hybrids from the single-stranded fragments. For example, the mixture could be passed through a column of hydroxylapatite under ionic conditions in which the hybrids would bind to the calcium phosphate salts in the column, while the nonhybridized DNA molecules would pass through unbound. The hybrids could then be released from the column by increasing the concentration of the elution buffer.

Experiments using nucleic acid hybridization require the incubation of two populations of complementary single-stranded nucleic acids under conditions (ionic strength, temperature, etc.) that promote formation of double-stranded molecules. Depending on the type of experiment being conducted, the two populations of reacting molecules may both be present in solution, or one of the populations may be immobilized, for example, by localization within a chromosome (as in Figure 10.22) or by adsorption to a nylon or nitrocellulose membrane (as in Figure 18.35).

One of the populations of single-stranded nucleic acids in a hybridization experiment is often immobilized within an electrophoretic gel (Figure 18.35). To carry out

| Electrophoretic gel | Blotting procedure for transfer of DNA from gel to nitrocellulose membrane. | Nitrocellulose membrane | Incubate membrane with labeled DNA or RNA probes to allow hybridization, then wash and prepare autoradiogram. | Autoradiogram |

Electrophoretic gel containing fractionated DNA fragments. The DNA is made single-stranded (denatured) by alkali treatment.

Nitrocellulose membrane with adsorbed DNA fragments following heat treatment that fixes the DNA to the membrane.

Autoradiogram showing location of DNA fragments complementary to labeled probe.

Figure 18.35 Determining the location of specific DNA fragments in a gel by a Southern blot. As described in the figure, the fractionated DNA fragments are denatured, transferred to a nitrocellulose membrane, which is incubated with radioactively labeled DNA (or RNA) probes, and the location of the hybridized fragments determined autoradiographically. During the blotting procedure, capillary action draws the buffer upwards into the paper towels. As the buffer moves through the electrophoretic gel, it dissolves the DNA fragments and transfers them to the surface of the adjacent membrane.

the hybridization, the single-stranded DNA is transferred from the gel to a nitrocellulose membrane, and fixed onto the membrane by heating it to 80°C in a vacuum. The procedure by which DNA is transferred to a membrane is termed *blotting*. Once the DNA is attached, the membrane is incubated with a radioactively labeled, single-stranded DNA (or RNA) probe capable of hybridizing to a complementary group of fragments. Unbound radioactivity is then washed away, and the location of the bound probe is determined autoradiographically as shown in Figure 18.35. This procedure in which specific DNA fragments that have been fractionated by gel electrophoresis are identified with a labeled DNA probe, is called a **Southern blot** (named after Edwin Southern, its developer). One or a few restriction fragments that contain a particular nucleotide sequence can be identified in a Southern blot, even if there are thousands of unrelated fragments present in the gel. An example of a Southern blot is shown in Figure 10.21. RNA molecules separated by electrophoresis can also be identified with a labeled DNA probe after being blotted onto a membrane. An example of this procedure, which is called a Northern blot, is shown in Figure 11.38.

DNA probes are typically made radioactive using the following procedure. A purified preparation of DNA is treated with trace amounts of the enzyme DNase I, which introduces single-stranded nicks in the double-stranded molecules. The preparation is then incubated with DNA polymerase I, an enzyme that possesses both polymerase and exonuclease activities (page 562), in the presence of radioactively labeled DNA precursors. The polymerase molecules attach to the nicks in the duplex and move down the nicked strand toward its 3' end, removing the existing nucleotides and replacing them with radioactively labeled versions. This technique is called *nick translation* because the nick moves down the DNA molecule as nucleotides are removed and replaced.

Nucleic acid hybridization can also provide a measure of the similarity in nucleotide sequence between two samples of DNA, as might be obtained, for example, from two different organisms. The more distant the evolutionary relationship between the two species, the greater the divergence of their DNA sequences. If purified DNA from species A and B is mixed together, denatured, and allowed to reanneal, a percentage of the DNA duplexes are formed by DNA strands from the two species. Because they contain mismatched bases, such duplexes are less stable than those formed by DNA strands of the same species, and this instability is reflected in the lower temperature at which they melt. When DNAs from different species are allowed to reanneal in different combinations, the melting temperature (T_m, page 412) of the hybrid duplexes provides a measure of the evolutionary distance between the organisms. Two other types of nucleic acid hybridization protocols are discussed in detail in the text: *in situ hybridization* on page 416 and hybridization to *cDNA microarrays* on page 522.

18.13 RECOMBINANT DNA TECHNOLOGY

Over the past twenty-five years, tremendous advances have been made in the analysis of eukaryotic genomes. This progress began as molecular biologists learned to construct **recombinant DNA** molecules, which are molecules containing DNA sequences derived from more than one source. Recombinant DNAs can be used in myriad ways. We will begin by considering one of the most important applications: the isolation from the genome of a particular segment of DNA that codes for a particular polypeptide.

Suppose you were interested in isolating the gene that codes for human insulin so that you could synthesize this important protein in vitro. The human genome contains approximately three billion base pairs, whereas the insulin gene makes up only a few hundred of these nucleotides. One of the first steps in the isolation of the insulin gene is to fragment the large DNA molecules extracted from human cells into a defined population of smaller fragments by incubation with a restriction enzyme. As discussed on page 427, restriction enzymes recognize specific sequences (**restriction sites**) of four to eight nucleotides and make cuts at specific sites on both strands of the duplex. Because particular nucleotide sequences of this length occur quite frequently simply by chance, they appear in all types of DNA—viral, bacterial, plant, and animal—and are thus susceptible to fragmentation by these enzymes.

Recombinant DNAs can be formed in a variety of ways. In the method shown in Figure 18.36, DNA molecules from two different sources are treated with a restriction enzyme that makes staggered cuts in the DNA duplex. Staggered cuts leave short, single-stranded tails that act as "sticky ends" that can bind to a complementary single-stranded tail on another DNA molecule to restore a double-stranded molecule.[4] In the case depicted in Figure 18.36, one of the preparations of DNA fragments is obtained by adding a restriction enzyme to a preparation of purified bacterial plasmids, which are small, circular, double-stranded DNA molecules that are separate from the main bacterial chromosome. The other preparation of DNA fragments shown in Figure 18.36 is obtained by treatment of DNA extracted from human cells with the same restriction enzyme.

Let's assume that one of the fragments contains the insulin gene that you are seeking to purify. This particular fragment represents only a minuscule portion of the total number of fragments that are formed. The goal is to isolate the fragment being sought from this large, heterogeneous population. The first step toward this goal is to in-

[4]Recombinant DNAs can also be formed from DNA fragments that have been generated by restriction enzymes that produce blunt ends. The ends of the plasmid and the restriction fragment are subsequently modified to allow them to "stick" to one another.

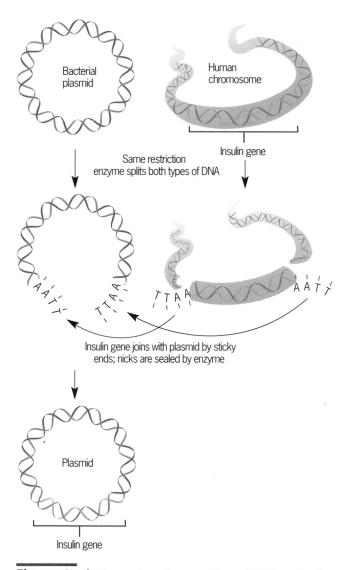

Figure 18.36 Formation of a recombinant DNA molecule. In this example, both plasmid DNA and human DNA containing the insulin gene are treated with the same restriction enzyme so that the DNA from the two sources have complementary "sticky ends." As a result, the two DNA molecules become noncovalently joined to each other and are then covalently sealed by DNA ligase, forming a recombinant DNA molecule.

cubate the plasmid and human DNA digests together in the presence of DNA ligase. During the incubation, the two types of DNAs become hydrogen bonded to one another by their sticky ends and then ligated to form circular DNA recombinants. The first recombinant DNA molecules were formed by this method in 1973 by Paul Berg, Herbert Boyer, Annie Chang, and Stanley Cohen of Stanford University and the University of California, San Francisco, marking the birth of modern genetic engineering.

By following the procedure just described, you will have produced a large number of different recombinant molecules, each of which contains a bacterial plasmid with a segment of human DNA incorporated into its circular

structure. Because your goal is to obtain a purified preparation of the one type of recombinant DNA that contains the insulin-coding fragment, you must separate this one fragment from all of the others. This is done by a process called DNA cloning. We will return to the search for the insulin gene after describing the basic methodology behind DNA cloning.

DNA Cloning

DNA cloning is a technique to produce large quantities of a specific DNA segment. The DNA segment to be cloned is first linked to a vector DNA, which is a vehicle for carrying foreign DNA into a suitable host cell, such as the bacterium *E. coli*. The vector contains sequences that allow it to be replicated within the host cell. Two types of vectors are commonly used to clone DNAs within bacterial hosts. In one approach, the DNA segment to be cloned is introduced into the bacterial cell by joining it to a plasmid, as described previously, and then causing the bacterial cells to take up the plasmid from the medium. In an alternate approach, the DNA segment is joined to a portion of the genome of the bacterial virus lambda (λ), which is then allowed to infect a culture of bacterial cells, producing large numbers of viral progeny, each of which contains the foreign DNA segment. Either way, once the DNA segment is inside a bacterium, it is replicated along with the bacterial (or viral) DNA and partitioned to the daughter cells (or progeny viral particles). Thus, the number of recombinant DNA molecules increases in proportion to the number of bacterial cells (or viral progeny) that are formed. From a single recombinant plasmid or viral genome inside a single bacterial cell, millions of copies of the DNA can be formed within a short period of time. Once the amount of DNA has been sufficiently amplified, the recombinant DNA can be purified and used in other procedures. In addition to providing a means to amplify the amount of a particular DNA sequence, cloning can also be used as a technique to isolate a pure form of any specific DNA fragment among a large, heterogeneous population of DNA molecules. We will begin by discussing DNA cloning using bacterial plasmids.

Cloning Eukaryotic DNAs in Bacterial Plasmids The foreign DNA to be cloned is first inserted into the plasmid to form the recombinant DNA molecule. Plasmids used for DNA cloning are modified versions of those found in bacterial cells. Like the natural counterparts from which they are derived, these plasmids contain an origin of replication and one or more genes that make the recipient cell resistant to one or more antibiotics. Antibiotic resistance allows investigators to select for those cells that contain the recombinant plasmid.

As first demonstrated by Avery, Macleod, and McCarty (page 433), bacterial cells can take up DNA from their medium. This phenomenon, known as transformation, forms the basis for cloning plasmids in bacterial cells

E. coli
chromosome

Plasmid

Antibiotic
resistance gene

Purify
human DNA

Purify
plasmid DNA

Treat with *Eco*R1 to
cleave both human and
bacterial DNA into
fragments

Join fragments
into recombinant
DNAs with DNA
ligase

Population of plasmids
containing different segments
of human DNA

Incubate *E. coli*
cells under conditions
in which they will take
up plasmids from the
medium. Grow cells in
medium that selects
for those containing a
recombinant plasmid

Plasmid-free *E. coli*

Insulin gene ? ? ? Ribosomal ?
 RNA gene

Figure 18.37 An example of DNA cloning using bacterial plasmids. DNA is extracted from human cells, fragmented with *Eco*R1, and the fragments are inserted into a population of bacterial plasmids. Techniques are available to prevent the formation of plasmids that lack a foreign DNA insert. Once a bacterial cell has picked up a recombinant plasmid from the medium, the cell gives rise to a colony of cells containing the recombinant DNA molecule. In this example, most of the bacteria contain eukaryotic DNAs of unknown function (labeled as ?), while one contains a portion of the DNA coding for ribosomal RNA, and another contains DNA coding for insulin.

(Figure 18.37). In the most commonly employed technique, recombinant plasmids are added to a bacterial culture that has been pretreated with calcium ions. When subjected to a brief heat shock, such bacteria are stimulated to take up DNA from their surrounding medium. Usually only a small percentage of cells are competent to pick up and retain one of the recombinant plasmid molecules. Once it is taken up, the plasmid replicates autonomously within the recipient and is passed on to the progeny during cell division. Those bacteria containing a recombinant plasmid can be selected from the rest by growing the cells in the presence of the antibiotic to which resistance is conferred by the gene on the plasmid.

We began this discussion with the goal of isolating a small DNA fragment that contains the sequence for the insulin gene. To do this, we have formed a population of bacteria containing different recombinant plasmids, very few of which contain the gene being sought (as in Figure 18.37). One of the great benefits of DNA cloning is that, in addition to producing large quantities of particular DNAs, it allows one to separate different DNAs from a mixture. It was noted above that plasmid-containing bacteria can be selected from those without plasmids by treatment with antibiotics. Once this has been done, the plasmid-bearing cells can be grown at low density on petri dishes so that all of the progeny of each cell (a clone of cells) remain physically separate from the progeny of other cells. Because a large number of different recombinant plasmids were present initially in the medium, the different cells plated on the dish contain different foreign DNA fragments. Once the cells containing the various plasmids have grown into separate colonies, the investigator can search through the colonies for the few that contain the gene being sought—in this case, the insulin gene.

Culture dishes that contain bacterial colonies (or phage plaques) are screened for the presence of a particular DNA sequence using the combined techniques of *replica plating* and *in situ hybridization*. As illustrated in Figure 18.38*a*, replica plating allows one to prepare numerous dishes containing representatives of the same bacterial colonies in precisely the same position on each dish. One of the replica plates is then used to localize the DNA sequence being sought (Figure 18.38*b*), a procedure that requires that the cells be lysed and the DNA fixed onto the surface of a nylon or nitrocellulose membrane. Once the DNA is fixed in place, the DNA is denatured to prepare it for in situ hybridization, during which the membrane is incubated with a labeled, single-stranded DNA probe containing the complementary sequence to that being sought. Following incubation, the unhybridized probe is washed from the membrane, and the location of the labeled hybrids is determined by autoradiography. Live representatives of the identified clones can be found at corresponding sites on the original plates, and these cells can be grown into large colonies, which serves to amplify the recombinant DNA plasmid. After the desired amount of amplification, the cells are harvested, the DNA is ex-

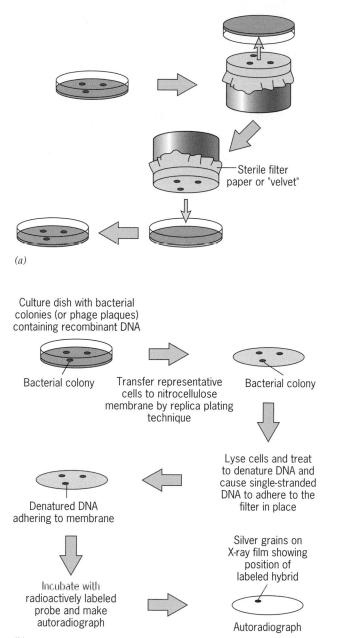

(a)

Culture dish with bacterial colonies (or phage plaques) containing recombinant DNA

Bacterial colony Transfer representative cells to nitrocellulose membrane by replica plating technique Bacterial colony

Denatured DNA adhering to membrane Lyse cells and treat to denature DNA and cause single-stranded DNA to adhere to the filter in place

Incubate with radioactively labeled probe and make autoradiograph Silver grains on X-ray film showing position of labeled hybrid

Autoradiograph

(b)

Sterile filter paper or "velvet"

Figure 18.38 Locating a bacterial colony containing a desired DNA sequence by replica plating and in situ hybridization. (*a*) Once the bacterial cells that were plated on the culture dish have grown into colonies, the dish is inverted over a piece of filter paper, allowing some of the cells from each colony to become adsorbed to the paper. Empty culture dishes are then inoculated by pressing them against the filter paper to produce replica plates. (*b*) Procedure for screening the cells of a culture dish for those colonies that contain the recombinant DNA of interest. Once the relevant colonies are identified, cells can be removed from the original dish and grown separately to yield large quantities of the DNA fragments being sought.

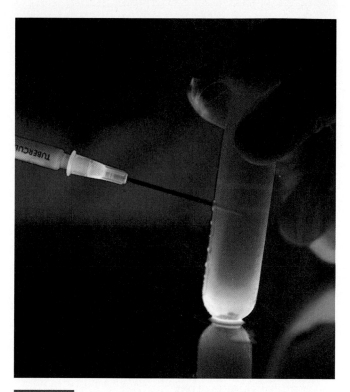

Figure 18.39 Separation of plasmid DNA from that of the main bacterial chromosome by CsCl equilibrium centrifugation. This centrifugation tube can be seen to contain two bands, one of plasmid DNA carrying a foreign DNA segment that has been cloned within the bacteria, and the other containing chromosomal DNA from these same bacteria. The two types of DNA have been separated during centrifugation. The researcher is removing the DNA from the tube with a needle and syringe. The DNA in the tube is made visible using the DNA-binding compound ethidium bromide, which fluoresces under ultraviolet light. (PHOTOGRAPH BY TED SPEIGEL/BLACK STAR.)

tracted, and the recombinant plasmid DNA is readily separated from the much larger chromosome by various techniques, including equilibrium centrifugation (Figure 18.39). The isolated recombinant plasmids can then be treated with the same restriction enzyme used in their formation, which releases the cloned DNA segments from the remainder of the DNA that served as the vector. The cloned DNA can then be separated from the plasmid by centrifugation.

Cloning Eukaryotic DNAs in Phage Genomes Another major cloning vector is the bacteriophage lambda, which is shown in Figure 18.40. The genome of λ is a linear, double-stranded DNA molecule approximately 50 kb in length. The modified strain used in most cloning experiments contains two cleavage sites for the enzyme *Eco*R1, which fragments the genome into three large segments. Conveniently, all the information essential for infection and cell lysis is contained in the two outer segments, so that the dispensable middle fragment can be replaced by a piece of DNA up to approximately 25 kb. Recombinant DNA molecules can be packaged into phage heads in vitro, and these genetically engineered phage particles can be used to infect host bacteria. (Phage DNA molecules

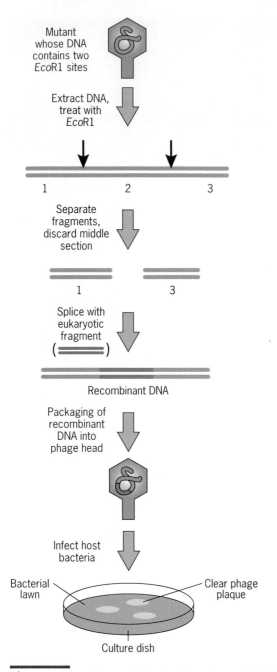

Figure 18.40 Protocol for cloning eukaryotic DNA fragments in lambda phage. The steps are described in the text.

(2) virtually every phage containing a recombinant DNA is capable of infecting a bacterial cell; and (3) a single petri dish can accommodate more than 100,000 different plaques. Once the phage plaques have been formed, the particular DNA fragment being sought is identified by a similar process of replica plating and in situ hybridization described in Figure 18.38 for cloning of recombinant plasmids.

Formation of a DNA Library DNA cloning is often used to produce **DNA libraries**, which are collections of cloned DNA fragments. Two basic types of DNA libraries can be created: genomic libraries and cDNA libraries. **Genomic libraries** are produced from total DNA extracted from nuclei and contain all of the DNA sequences of the species. Once a genomic library of a species is available, researchers can use the collection to isolate specific DNA sequences, such as those containing the human insulin gene. **cDNA libraries**, on the other hand, are derived from DNA copies of an RNA population. cDNA libraries are typically produced from messenger RNAs present in a particular cell type and thus correspond to the genes that are active in that type of cell. We will first examine the production of a genomic library.

In one approach, genomic DNA is treated with one or two restriction enzymes that recognize very short nucleotide sequences under conditions of low enzyme concentration, such that only a small percentage of susceptible sites are actually cleaved. Two commonly used enzymes that recognize tetranucleotide sequences are *Hae*III (recognizes GGCC) and *Sau*3A (recognizes GATC). A given tetranucleotide would be expected to occur by chance with such a high frequency that any sizable segment of DNA is sensitive to fragmentation. Once the DNA is partially digested, the digest is fractionated by gel electrophoresis or density gradient centrifugation, and those fragments of suitable size (e.g., 20 kb in length) are incorporated into lambda phage particles. These phage are used to generate the million or so plaques needed to ensure that every single segment of a mammalian genome is represented.

Because the DNA is treated with enzymes under conditions in which most susceptible sites are not being cleaved, for all practical purposes, the DNA is randomly fragmented. Once the phage recombinants are produced, they can be stored for later use and, as such, constitute a permanent collection of all the DNA sequences present in the genome of the species. Whenever an investigator wants to isolate a particular sequence from the library, the phage particles can be grown in bacteria and the various plaques (each originating from the infection of a single recombinant phage) screened for the presence of that sequence by in situ hybridization.

The use of randomly cleaved DNA has an advantage in the construction of a library because it generates overlapping fragments that can be used in the analysis of regions of the chromosome extending out in both directions

lacking the insert are too short to be packaged.) Once in the bacteria, the eukaryotic DNA segment is amplified along with the viral DNA and then packaged into a new generation of virus particles, which are released when the cell is lysed. The released particles infect new cells, and soon a clear spot (or *plaque*) in the bacterial "lawn" is visible at the site of infection. Each plaque contains millions of phage particles, each carrying a single copy of the same eukaryotic DNA fragment.

Lambda phage is appealing as a cloning vector for a number of reasons: (1) the DNA is nicely packaged in a form in which it can be readily stored and easily extracted;

from a particular sequence, a technique known as *chromosome walking*. For example, if one isolates a fragment containing the coding region of a globin gene, that particular fragment can be labeled and used as a probe to screen the genomic library for fragments with which it overlaps. The process is then repeated using the new fragments as labeled probes in successive screening steps, as one gradually isolates a longer and longer part of the original DNA molecule. Using this approach, one can study the organization of linked sequences in an extended region of a chromosome.

Cloning Larger DNA Fragments in Specialized Cloning Vectors

Neither plasmid nor lambda phage vectors are suitable for cloning DNAs larger than about 20 to 25 kb in length. Several vectors have been developed that allow investigators to clone much larger pieces of DNA. One of the most important of these vectors is the **yeast artificial chromosome (YAC)**, which can accept segments of foreign DNA as large as 1000 kb (one million base pairs). As the name implies, YACs are artificial versions of a normal yeast chromosome. They contain all of the elements of a yeast chromosome that are necessary for the structure to be replicated during S phase and segregated to daughter cells during mitosis, including one or more origins of replication, telomeres at the ends of the chromosomes, and a centromere to which the spindle fibers can attach during chromosome separation. In addition to these elements, YACs are constructed to contain (1) a gene whose encoded product allows those cells containing the YAC to be selected from those that lack the element and (2) the DNA fragment to be cloned. Like other cells, yeast cells can take up DNA from their medium, which provides the means by which YACs are introduced into the cells. Over the past few years, laboratories involved in sequencing genomes have relied heavily on an alternate cloning vector, called a **bacterial artificial chromosome (BAC)**, which is also able to accept large foreign DNA fragments (up to about 300 kb). BACs are specialized bacterial plasmids (F factors) that contain a bacterial origin of replication and the genes required to regulate their own replication. BACs have the advantage over YACs in high-speed sequencing projects because they can be cloned in *E. coli*, which readily picks up exogenous DNA, has an extremely short generation time, can be grown at high density in simple media, and does not "corrupt" the cloned DNA through recombination.

DNA fragments cloned in YACs and BACs are typically greater than 100 kb in length. Fragments of such large size are usually produced by treatment of DNAs with restriction enzymes that recognize particularly long nucleotide sequences (7–8 nucleotides) containing CG dinucleotides. As noted on page 534, CG dinucleotides have special functions in the mammalian genome, and presumably because of this they do not appear nearly as often as would be predicted by chance. The restriction enzyme *Not*1, for example, which recognizes the 8-nucleotide sequence GCGGCCGC, typically cleaves mammalian

DNA into fragments several hundred thousand base pairs long. Fragments of this length can then be incorporated into YACs or BACs and cloned within host yeast or bacterial cells.

cDNA Libraries Up to this point, the discussion has been restricted to cloning DNA fragments isolated from extracted DNA, that is, genomic fragments. When working with genomic DNA, one is generally seeking to isolate a particular gene or family of genes from among hundreds of thousands of unrelated sequences. In addition, the isolation of genomic fragments allows one to study a variety of topics, including (1) regulatory sequences flanking the coding portion of a gene; (2) noncoding intervening sequences; (3) various members of a multigene family, which often lie close together in the genome; (4) evolution of DNA sequences, including their duplication and rearrangement as seen in comparisons of the DNA of different species; and (5) interspersion of transposable genetic elements.

The cloning of cDNAs, as opposed to genomic DNAs, has also been important in the analysis of gene structure and gene expression. To produce a cDNA library, a population of mRNAs is isolated and used as a template by reverse transcriptase to form a population of DNA-RNA hybrids, as shown in Figure 18.41. The DNA-RNA hybrids are converted to a double-stranded cDNA population by nicking the RNA with RNase H and replacing it with DNA by DNA polymerase I. The double-stranded cDNA is then combined with the desired vector (in this case a plasmid) and cloned as shown in Figure 18.41*b*. Messenger RNA populations typically contain thousands of different messages, but individual species may be present in markedly different numbers (abundance). As a result, a cDNA library has to contain a million or so different cDNA clones to be certain that all of the rarer mRNAs will be represented. As with experiments using genomic DNA fragments, clones must be screened to isolate one particular sequence from a heterogeneous population of recombinant molecules.

The analysis of cloned cDNAs serves several functions. It is generally easier to study a diverse population of cDNAs than the corresponding population of mRNAs, so one can use the cDNAs to learn about the variety of mRNAs present in a cell, the percentage of mRNAs shared by two different types of cells, or the number of copies of different mRNAs present in a cell. A single cloned and amplified cDNA molecule is also very useful. The cDNA contains only that information present in the mRNA, so comparisons between a cDNA and its corresponding genomic locus can provide information on the precise locations of the noncoding intervening sequences within the DNA. In more practical endeavors, the purified cDNA can be readily sequenced to provide a shortcut in determining the amino acid sequence of a polypeptide; labeled cDNAs are used as probes to screen for complementary sequences among recombinant clones; and be-

Figure 18.41 Synthesizing cDNAs for cloning in a plasmid.
(*a*) In this method of cDNA formation, a short oligo(dT) primer is bound to the poly(A) of each mRNA, and the mRNA is transcribed by reverse transcriptase (which requires the primer to initiate DNA synthesis). Once the DNA-RNA hybrid is formed, the RNA is nicked by treatment with RNase H, and DNA polymerase is added to digest the RNA, and replace it with DNA, just as it does during DNA replication in a bacterial cell (page 563). (*b*) To prepare the blunt-ended cDNA for cloning, a short stretch of poly(G) is added to the 3′ ends of the cDNA and a complementary stretch of poly(C) is added to the 3′ ends of the plasmid DNA. The two DNAs are mixed and allowed to form recombinants, which are sealed and used to transform bacterial cells in which they are cloned.

cause they lack introns, cDNAs have an advantage over genomic fragments when one is attempting to synthesize eukaryotic proteins in bacterial cell cultures.

Chemical Synthesis and Site-Directed Mutagenesis

The development of chemical techniques to synthesize polynucleotides having a specific base sequence was begun by H. Gobind Khorana in the early 1960s as part of an attempt to decipher the genetic code. Khorana and coworkers continued to refine their techniques, and a decade after their initial work on the code, they succeeded in synthesizing a complete bacterial tyrosine tRNA gene, including the nontranscribed promoter region. The gene totaling 126 base pairs was put together from over 20 segments, each of which was individually synthesized and later joined enzymatically. This artificial gene was then introduced into bacterial cells carrying mutations for this tRNA, and the synthetic DNA was able to replace the previously deficient function.

A second landmark in the field of gene synthesis came in 1977 when a gene encoding the small hypothalmic hormone somatostatin (14 amino acid residues) was synthesized by Keiichi Itakura and coworkers of the City of Hope Medical Center. The gene was inserted into a specially constructed plasmid downstream from bacterial regulatory sequences and introduced into *E. coli*, where it was transcribed and translated. A gene for the first "average-sized" protein, human interferon, was synthesized in 1981, an effort that required the synthesis and assembly of 67 different fragments to produce a single duplex of 514 base pairs containing initiation and termination signals recognized by the bacterial RNA polymerase.

In the past decade or so, the development of chemical procedures required to assemble nucleotides has led to the construction of automated DNA-synthesizing machines that can synthesize polynucleotides of any desired sequence (or a randomly generated sequence) approximately 75–100 nucleotides in length. Each fragment is assembled one nucleotide at a time from the 3′ to the 5′ end of the

segment. Once such fragments are synthesized, they can be covalently joined to one another to generate synthetic DNA molecules of considerable length.

As is evident throughout this book, the isolation of naturally occurring mutants has played an enormously important role in determining the function of genes and their products. But natural mutations are rare events, and it isn't feasible to use such mutations to study the role of particular amino acid residues in the function of a particular protein. Rather then waiting for an organism to appear with an unusual phenotype and identifying the responsible mutation, researchers can now mutate the gene (or its associated regulatory regions) in a desired way and observe the resulting phenotypic change. Using in vitro techniques, small sections of a DNA sequence can be deleted, additional nucleotides inserted, or one specific base can be changed to another one. This procedure was developed by Michael Smith of the University of British Columbia and is termed **site-directed mutagenesis (SDM)**. SDM is usually accomplished by first synthesizing a DNA oligonucleotide containing the desired change, allowing this oligonucleotide to hybridize to a single-stranded preparation of the normal DNA, and then using the oligonucleotide as a primer for DNA polymerase. The polymerase elongates the primer by adding nucleotides that are complementary to the normal DNA. The modified DNA can then be cloned and the effect of the genetic alteration determined by introducing the DNA into an appropriate host cell. If the site under investigation is part of a regulatory region, then the effect of the alteration on the rate of gene expression can be monitored. If the site is located within the coding region of a gene, the resulting alteration in amino acid sequence provides insight into the role of that site in the overall structure and function of the protein.

Gene Transfer into Eukaryotic Cells and Mammalian Embryos

We have discussed in previous sections how eukaryotic genes can be isolated, modified, and amplified. In this section we will consider some of the ways in which genes can be introduced into eukaryotic cells, where they are generally transcribed and translated. One of the most widely used strategies to accomplish this goal is to incorporate the DNA into the genome of a nonreplicating virus and allow the virus to infect the cell. Viral-mediated gene transfer is called **transduction**. Depending upon the type of virus used, the gene of interest can be expressed transiently for a period of hours to days, or it can be stably integrated into the genome of the host cell. Stable integration is usually accomplished using modified retroviruses, which contain an RNA genome that is reverse transcribed into DNA inside the cell. The DNA copy is then inserted into the DNA of the host chromosomes. Retroviruses have been used in many of the recent attempts at gene therapy to transfer a gene into the cells of a patient that is

lacking that gene. Most of these clinical trials have not had much success due to the low efficiency of infection of current viral vectors.

A number of procedures are available to introduce naked DNA into cultured cells, a process called **transfection**. Most often the cells are treated with either calcium phosphate or DEAE-dextran, both of which form a complex with the added DNA that promotes its adherence to the cell surface. It is estimated that only about 1 in 10^5 cells takes up the DNA and incorporates it stably into the chromosomes. It is not known why this small percentage of cells in the population are competent to be transfected, but those that are transfected generally pick up several fragments. One way to select for those cells that have taken up the foreign DNA is to include a gene that allows the transfected cells to grow in a particular medium in which nontransfected cells cannot survive. Because transfected cells typically take up more than one fragment, the gene used for selection need not be located on the same DNA fragment as the gene whose role is being investigated (called the **transgene**).

Two other procedures to transfect cells are electroporation and lipofection. In *electroporation*, the cells are incubated with DNA in special vials that contain electrodes that deliver a brief electric pulse. Application of the current causes the plasma membrane to become transiently permeable to DNA molecules, some of which find their way into the nucleus and become integrated into the chromosomes. In *lipofection*, cells are treated with DNA that is bound to positively charged lipids (cationic liposomes) that are capable of fusing with the lipid bilayer of the cell membrane and delivering the DNA to the cytoplasm.

One of the most direct ways to introduce foreign genes into a cell is to microinject DNA directly into the cell's nucleus. The nuclei of oocytes and eggs are particularly well suited for this approach. *Xenopus* oocytes, for example, have long been used to study the expression of foreign genes. The nucleus of the oocyte contains all the machinery necessary for RNA synthesis; when foreign DNA is injected into the nucleus, it is readily transcribed. In addition, the RNAs synthesized from the injected templates are processed normally and transported to the cytoplasm, where they are translated into proteins that can be detected immunologically or by virtue of their specific activity.

Another favorite target for injected DNA is the nucleus of a mouse embryo (Figure 18.42). The goal in such experiments is not to monitor the expression of the gene in the injected cell, but to have the foreign DNA become integrated into the egg's chromosomes to be passed on to all the cells of the embryo and subsequent adult. Animals that have been genetically engineered so that their chromosomes contain foreign genes are called **transgenic animals**, and they provide a means to monitor when and where in the embryo particular genes are expressed (as in Figure 12.41), and to determine the impact of extra copies of particular genes on the development and life of the animal.

Figure 18.42 **Microinjection of DNA into the nucleus of a recently fertilized mouse egg.** The egg is held in place by a suction pipette shown at the bottom, while the injection pipette is shown penetrating the egg at the top. (FROM THOMAS E. WAGNER ET AL., PROC. NAT'L ACAD. SCI. U. S. A. 78:6377, 1981.)

Figure 18.43 **Transgenic mice.** This photograph shows a pair of littermates at age 10 weeks. The larger mouse developed from an egg that had been injected with DNA containing the rat growth hormone gene placed downstream from a metallothionein promoter. The larger mouse weighs 44 g; the smaller, uninjected control weighs 29 g. The rat GH gene was transmitted to offspring that also grew larger than the controls. (COURTESY OF RALPH L. BRINSTER, FROM WORK REPORTED IN R. D. PALMITER ET AL., NATURE 300:611, 1982.)

Transgenic Animals and Plants In 1981, Ralph Brinster of the University of Pennsylvania and Richard Palmiter of the University of Washington succeeded in introducing a gene for rat growth hormone (GH) into the fertilized eggs of mice. The injected DNA was constructed so as to contain the coding portion of the rat GH gene in a position just downstream from the promoter region of the mouse metallothionein gene. Under normal conditions, the synthesis of metallothionein is greatly enhanced following the administration of metals, such as cadmium or zinc, or glucocorticoid hormones. The metallothionein gene carries a strong promoter, and it was hoped that placing the GH gene downstream from it might allow the GH gene to be expressed following treatment of the transgenic mice with metals or glucocorticoids. As illustrated in Figure 18.43, this expectation was fully warranted.

The formation of transgenic animals provides a useful means to determine the effects of overexpression of a particular DNA sequence. An example of this type of experiment is illustrated in Figure 9.43, which shows a section of

the spinal cord from a mouse that had overexpressed the gene for one of the polypeptides that make up neurofilaments. Overexpression of the gene leads to a neurodegenerative disorder resembling that of the human disease amyotrophic lateral sclerosis (ALS).

This latter example illustrates how genetic engineering can produce **animal models**, which are laboratory animals that exhibit a particular human disease they would normally not be subject to. Transgenic animals are also being developed as part of agricultural biotechnology. For example, pigs born with foreign growth hormone genes incorporated into their chromosomes grow much leaner than control animals lacking the genes. The meat of the transgenic animals is leaner because the excess growth hormone stimulates the conversion of nutrients into protein rather than fat.

Plants are also prime candidates for genetic engineering. It was indicated on page 521 that whole plants can be grown from individual cultured plant cells. This provides the opportunity to alter the genetic composition of plants by introducing DNA into the chromosomes of cultured cells that can subsequently be grown into mature plants. One way to introduce foreign genes is to incorporate them into the *Ti plasmid* of the bacterium *Agrobacterium tumefaciens*. Outside of the laboratory, this bacterium lives in symbiotic association with dicotyledonous plants, where it causes the formation of tumorous lumps called crown galls. During an infection, the Ti plasmid is passed from the bacterium into the plant cell, where it becomes incor-

Figure 18.44 **Formation of transgenic plants using the Ti plasmid.** The transgene is spliced into the DNA of the Ti plasmid, which is reintroduced into host bacteria. Bacteria containing the recombinant plasmid are then used to transform plant cells, in this case cells of the meristem at the tip of a dissected shoot apex. The transformed shoots are transferred to a selection medium where they develop roots. The rooted plants can then be transferred to potting soil.

porated into the plant cell's chromosomes and induces the cell to proliferate and provide nutrients for the bacterium. The Ti plasmid can be isolated from bacteria and linked with foreign genes to produce a recombinant plasmid that is taken up in culture by undifferentiated dicotyledonous plant cells, including those of carrots and tobacco. Figure 18.44 shows the use of recombinant Ti plasmids to introduce foreign DNA into meristem cells at the tip of newly formed shoots. The shoots can then be rooted and grown into mature plants. This technique, which is called *T-DNA transformation*, has been used to transform plant cells with genes derived from bacteria that code for insect-killing toxins, thus protecting the plants from insect predators.

Other procedures are currently being developed to introduce foreign genes into cells of monocotyledonous plants. In one of the more exotic approaches, plant cells can serve as targets for microscopic, DNA-coated tungsten pellets fired by a "gene gun." This technique has been used to genetically alter a number of different types of plant cells.

The two most important activities that plant geneticists might improve by genetic engineering are photosynthesis and nitrogen fixation, both crucial bioenergetic functions. Any significant improvement in photosynthetic efficiency would mean great increases in crop production. It is hoped that a modified version of the CO_2-fixing enzyme might be engineered that is less susceptible to photorespiration (page 236). Nitrogen fixation is an activity carried out by certain genera of bacteria (e.g., *Rhizobium*) that live in a symbiotic relationship with certain plants (e.g., soybean, peanut, clover, alfalfa, and pea). The bacteria reside in swellings, or *leguminous nodules*, located on the roots, where they remove N_2 from the atmosphere, reduce it to ammonia, and deliver the product to the cells of the plant. Geneticists are looking for a way to isolate genes involved in this activity from the bacterial genome and introduce them into the chromosomes of nonleguminous plants that presently depend heavily on added fertilizer for their reduced nitrogenous compounds. Alternatively, it might be possible to alter the genome of either plant or bacteria so that new types of symbiotic relationships can be developed.

Knockout Mice We have described in a number of places in this text the phenotypes of mice that lack a functional copy of a particular gene. For example, it was noted that mice lacking a functional copy of the *p53* gene invariably

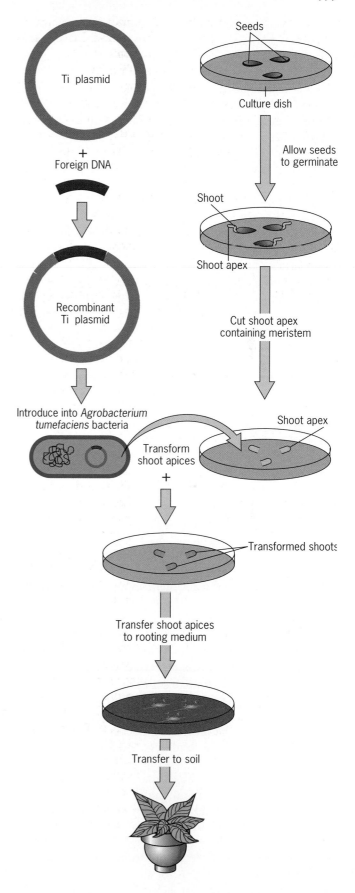

develop malignant tumors (page 684). These animals, which are called **knockout mice**, can provide a unique insight into the genetic basis of a human disease, as well as a

mechanism for studying the various cellular activities in which the product of a particular gene might be engaged. Knockout mice are born as the result of a series of experimental procedures depicted in Figure 18.45.

The technique used to generate knockout mice was developed in the late 1980s by Mario Capecchi at the University of Utah. The first step is to isolate an unusual type of cell that has virtually unlimited powers of differentiation. These cells, called **embryonic stem cells**, are found in the mammalian blastocyst, which is an early stage of embryonic development comparable to the blastula stage of other animals. A mammalian blastocyst (Figure 18.45) is composed of two distinct parts. The outer layer makes up the *trophoectoderm*, which gives rise to most of the extraembryonic membranes characteristic of a mammalian embryo. The inner surface of the trophectoderm contacts a cluster of cells called the *inner cell mass* that projects into the spacious cavity (the blastocoel). The inner cell mass gives rise to the cells that make up the embryo. The inner cell mass contains the embryonic stem (ES) cells, which differentiate into all of the various tissues of which a mammal is composed.

ES cells can be isolated from blastocysts and cultured in vitro under conditions where the cells grow and proliferate. The ES cells are then transfected with a DNA fragment containing a nonfunctional, mutant allele of the gene to be knocked out, as well as antibiotic-resistance genes that can be used to select for cells that have incorporated the altered DNA into their genome. Of those cells that take up the DNA, approximately one in 10^4 undergo a process of *homologous recombination* in which the transfecting DNA replaces the homologous DNA sequence. Through use of this procedure, ES cells that are heterozygous for the gene in question are produced and then selected for on the basis of their antibiotic resistance. In the next step, a number of these donor ES cells are injected into the blastocoel of a recipient mouse embryo. In the protocol depicted in Figure 18.45, the recipient embryo is obtained from an albino (nonpigmented) strain. The recipient embryo is then implanted into the oviduct of a female mouse that has been hormonally prepared to carry the embryo to term. As the embryo develops in its surrogate mother, the injected ES cells join the embryo's own inner cell mass and contribute to formation of embryonic tissues, including the germ cells of the gonads. These chimeric mice can be recognized because their coat contains characteristics of both the donor and recipient strains. To determine whether the germ cells also contain the knockout mutation, the chimeric mice are mated to a member of an inbred albino strain. If the germ cells do contain the knockout mutation, the offspring will be heterozygous for the gene in all of their cells. Heterozygotes can be distinguished by their coat coloration. These heterozygotes are then mated to one another to produce offspring that are homozygous for the mutant allele. These are the knockout mice that lack a functional copy of the gene. In some cases, the deletion of a particular gene can lead to the absence of a particular process, which provides convincing evidence that the gene is essential for that

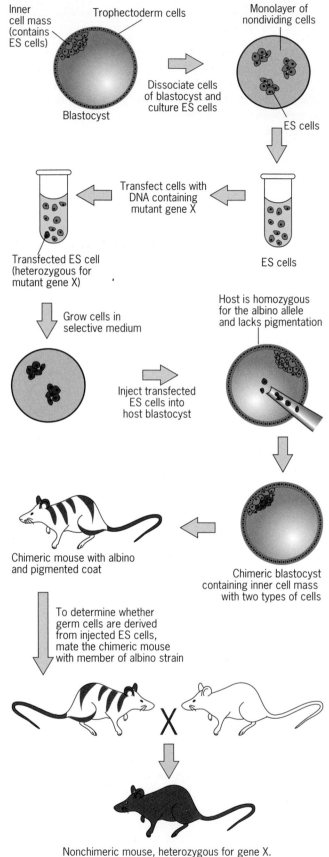

Figure 18.45 Formation of knockout mice. The steps are described in the text.

process. Often, however, the deletion of a gene that is thought to participate in an essential process causes little or no alteration in the animal's phenotype. Such results can be difficult to interpret. It is possible, for example, that the gene is not involved in the process being studied or, as is usually the case, the absence of the gene product is compensated by the product of an entirely different gene. Compensation by one gene for another can be verified by producing mice that lack both of the genes in question (i.e., a double knockout).

Enzymatic Amplification of DNA by PCR

In 1983, a new technique was conceived by Kary Mullis of Cetus Corporation that has become widely used to amplify specific DNA fragments without the need for bacterial cells. This technique is known as **polymerase chain reaction (PCR)**. There are many different PCR protocols used for a multitude of different applications in which anywhere from one to a large population of DNAs can be amplified. PCR amplification is readily adapted to RNA templates by first converting them to complementary DNAs using reverse transcriptase.

The basic procedure used in PCR is depicted in Figure 18.46. The technique employs a heat-stable DNA polymerase, called *Taq polymerase*, that was originally isolated from *Thermus aquaticus*, a bacterium that lives in hot springs at temperatures above 90°C. In the simplest protocol, a sample of DNA is mixed with an aliquot of Taq polymerase and all four deoxyribonucleotides, along with a large excess of two short, synthetic DNA fragments (oligonucleotides) that are complementary to DNA sequences at the 3′ ends of the region of the DNA to be amplified. The oligonucleotides serve as primers (page 557) to which nucleotides are added during the following replication steps. The mixture is then heated to about 93°, which is hot enough to cause the DNA molecules in the sample to separate into their two component strands. The mixture is then cooled to about 60°C to allow the primers to bind to the strands of the target DNA, and the temperature is raised to about 72°C, to allow the thermophilic polymerase to add nucleotides to the 3′ end of the primers. As the polymerase extends the primer, it selectively copies the target DNA, forming new complementary DNA strands. The temperature is raised once again, causing the newly formed and the original DNA strands to separate from each other. The sample is then cooled to allow the synthetic primers in the mixture to bind once again to the target DNA, which is now present at twice the original amount. This cycle is repeated over and over again, each time doubling the amount of the specific region of DNA that is flanked by the bound primers. Billions of copies of this one specific region can be generated in just a few hours using a thermal cycler that automatically changes the temperature of the reaction mixture to allow each step in the cycle to take place.

In addition to its use in the amplification of specific DNA fragments, PCR can generate large amounts of DNA from minuscule starting samples. For example, PCR has been used in criminal investigations to generate quantities of DNA from a spot of dried blood left on a crime suspect's clothing or even from the DNA present in part of a single hair follicle left at the scene of a crime. For this purpose, one selects regions of the genome for amplification that are highly polymorphic (i.e., vary at high frequency within the population), so that no two individuals will have the same-sized DNA fragments (as in Figure 10.21). This same procedure can be used to study DNA fragments from well-preserved fossil remains that may be millions of years old.

DNA Sequencing

By 1970, the amino acid sequence of a long list of proteins had been determined, yet virtually no progress had been made toward the sequencing of nucleotides in DNA. There were several reasons for this state of affairs. Unlike DNA molecules, polypeptides come in defined and manageable lengths; a given polypeptide species could be readily purified; a variety of techniques were available to cleave the polypeptide at various sites to produce overlapping fragments; and the presence of 20 different amino acids having widely varying properties made separation and sequencing of small peptides a straightforward task. Then in the mid-1970s, a revolution in DNA sequence technology took place. By 1977, the complete nucleotide sequence of an entire viral genome was reported, that of φX174, some 5375 nucleotides in length. This milestone in molecular biology was accomplished in the laboratory of Frederick Sanger, who had determined the first amino acid sequence of a polypeptide (insulin) some 25 years earlier.

In 1970 it might have taken over a year to sequence a DNA fragment a few dozen nucleotides in length; 15 years later the same task could be accomplished in a single day. Several developments were responsible for this changing situation. Most important, the availability of restriction enzymes provided the means necessary to prepare a population of small, defined DNA fragments. Then with the advent of DNA cloning technology, any given fragment could be amplified to produce enough material to carry out the necessary biochemical procedures. But, beyond these advances, a new sequencing methodology was needed that did not depend on the generation of overlapping fragments as required in polypeptide sequencing. Almost simultaneously, two different techniques were developed, an enzymatic approach taken by Sanger and A. R. Coulson of the Medical Research Council in Cambridge, England, and a chemical approach taken by Allan Maxam and Walter Gilbert of Harvard University. In this section, we will briefly consider the Sanger-Coulson (or chain termination) method, which is most widely used.

In this procedure, one begins with a population of identical DNA fragments up to about 500 base pairs in length that are obtained by treatment of DNA with a restriction enzyme. The preparation is then divided into four samples, each of which is treated in a slightly differ-

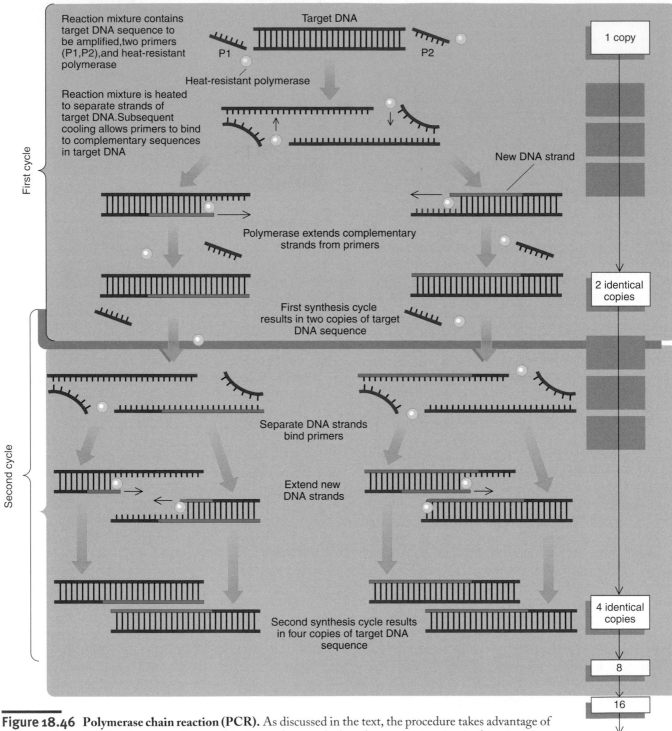

Figure 18.46 Polymerase chain reaction (PCR). As discussed in the text, the procedure takes advantage of a heat-resistant DNA polymerase whose activity is not destroyed when the temperature is raised to separate the two strands of the double helix. With each cycle of duplication, the strands are separated, flanking segments (primers) bind to the ends of the selected region, and the polymerase copies the intervening segment.

ent fashion. The DNA of each sample is denatured into its single strands (step 1, Figure 18.47) and then incubated with a short, radioactively labeled oligonucleotide that is complementary to the 3′ end of one of the single-stranded fragments (step 2). The reaction mixture also contains a DNA polymerase (most often bacterial DNA polymerase I), all four deoxyribonucleoside triphosphate precursors (dNTPs), and a low concentration of a modified precursor called a **dideoxyribonucleoside triphosphate**, or **ddNTP**. A different ddNTP (ddATP, ddCTP, ddTTP, or ddGTP) is added to each of the four samples.

When the reaction mixtures are incubated under the appropriate conditions, a labeled oligonucleotide binds to the 3′ end of each single-stranded fragment (step 3, Figure

Figure 18.47 DNA sequencing. Sequencing of a small hypothetical fragment by the Sanger-Coulson (dideoxy) technique, as described in the text. For major projects, such as sequencing of genomes, this process is carried out by sequencing machines using fluorescently labeled DNA fragments.

High-resolution gel electrophoresis can separate fragments that differ by only one nucleotide in length. If, for example, the initial fragment contained 100 nucleotides, then each of the 100 labeled fragments of different length would migrate to a specific location in the gels. Among the four lanes, one should find labeled bands present at all the 100 possible locations. Because successive labeled bands correspond to fragments with a common 5′ end but an increasing length of one nucleotide, and because the 3′ terminal nucleotide of each fragment is known by virtue of the lane in which it lies, the sequence of the entire molecule can be read directly from the positions of the bands in the gel.

Once the nucleotide sequence of a segment of DNA has been determined, it is a relatively simple task to deduce the amino acid sequence of an encoded polypeptide using the codon chart shown in Figure 11.41. To accomplish this goal, it is necessary (1) to determine the site in the sequence at which the initiation of translation begins and terminates, and (2) to identify each of the introns in the DNA that interrupts coding. Because each intron-exon boundary begins and ends with a precise nucleotide sequence, introns can usually be identified without much difficulty. Once the amino acid sequence has been deduced, it can be compared to other known sequences to provide information about the polypeptide's possible function. The amino acid sequence also provides clues as to the tertiary structure of the protein, particularly those parts of the polypeptide that may act as membrane-spanning segments of integral membrane proteins. As discussed in the Human Perspective in Chapter 10, DNA sequencing technology has been used as part of a worldwide cooperative attempt to determine the nucleotide sequence of the entire human genome.

18.14 THE USE OF ANTIBODIES

As discussed in Chapter 17, antibodies (or immunoglobins) are proteins produced by lymphoid tissues in response to the presence of foreign materials, or antigens. One of the most striking properties of antibodies, and one that makes them so useful to biological researchers, is their remarkable specificity. A cell may contain thousands of different proteins, yet a preparation of antibodies binds only to those select molecules within the cell having a small part that fits into the antigen-binding sites of the antibody molecules. Antibodies can often be obtained that distinguish between two polypeptides that differ by only a single amino acid.

Biologists have long taken advantage of antibodies and have developed a wide variety of techniques involving their use. There are basically two different approaches to the preparation of antibody molecules that interact with a given antigen. In the traditional approach, an animal (typically a rabbit or goat) is repeatedly injected with the antigen, and after a period of several weeks, blood is drawn that contains the desired antibodies. Whole blood is treated to remove cells and clotting factors so as to produce an **antiserum**, which can be tested for its antibody titer and from which the immunoglobulins can be puri-

18.47) and serves as a primer for the addition of nucleotides into a growing complementary chain. Dideoxyribonucleotides lack a hydroxyl group at both their 2′ and 3′ positions. Once one of these nucleotides has been incorporated onto the end of a growing chain, the lack of the 3′ OH makes it impossible for the polymerase to add another nucleotide, thus causing chain termination (step 4). Because the ddNTP is present at much lower concentration in the reaction mixture than the corresponding dNTP, the incorporation of the dideoxyribonucleotide is infrequent and random; it may be incorporated near the beginning of one chain, near the middle of another chain, or not until the end of a third chain. Regardless, when the ddNTP is incorporated, growth of the chain ceases at that point.

Each of the four samples contains a population of incomplete, radioactively labeled DNA fragments of different lengths, all of which end at the same nucleotide. The four different samples have fragments that end at the four different nucleotides. The products of the four reaction mixtures are loaded into separate wells at the top of a high-resolution electrophoretic gel and fractionated into their components along parallel lanes (step 5, Figure 18.47).

fied. Although this method of antibody production remains in use, it suffers from certain inherent disadvantages. Because of the mechanism of antibody synthesis, an animal invariably produces a variety of different species of immunoglobulin, that is, immunoglobulins with different V regions in their polypeptide chains, even when challenged with a highly purified preparation of antigen. An antiserum that contains a variety of immunoglobulins that bind to the same antigen is said to be *polyclonal*. Because immunoglobulins are too similar in structure to be fractionated, it becomes impossible to obtain a preparation of a single purified species of antibody by this technique.

In 1975, Cesar Milstein and Georges Köhler of the Medical Research Council in Cambridge, England, carried out a far-reaching set of experiments that led to the development of univalent preparations of antibodies. To understand their work, it is necessary to digress briefly. A given clone of antibody-producing cells (which is derived from a single B lymphocyte) synthesizes antibodies having identical antigen-combining sites. The heterogeneity of the antibodies produced when a single purified antigen is injected into an animal results from the fact that many different B lymphocytes are activated, each having membrane-bound antibodies with an affinity for a different part of the antigen. An important question arose: was it possible to circumvent this problem and obtain a single species of antibody molecule? Consider for a moment the results of a procedure in which one injects an animal with a purified antigen, waits a period of weeks for antibodies to be produced, removes the spleen or other lymphoid organs, prepares a suspension of single cells, isolates those cells producing the desired antibody, and grows these particular cells as separate colonies so as to obtain large quantities of this particular immunoglobulin. If this procedure were followed, one should obtain a preparation of antibody molecules produced by a single colony (or clone) of cells, which is referred to as a **monoclonal antibody**. But antibody-producing cells do not grow and divide in culture, so an additional manipulation had to be introduced to obtain monoclonal antibodies.

Malignant myeloma cells are a type of cancer cell that grows rapidly in culture and produces large quantities of antibody. Although the antibody molecules produced by myeloma cells were of great value in the study of antibody structure (page 713), they are of little use as analytic tools because they are not formed in response to a particular antigen. Instead, myeloma cells develop from the random conversion of a normal lymphocyte to the malignant state, and therefore, they produce the antibody that was being synthesized by the particular lymphocyte before it became malignant.

Milstein and Köhler combined the properties of these two types of cells, the normal antibody-producing lymphocyte and the malignant myeloma cell. They accomplished this feat by fusing the two types of cells to produce hybrid cells, called **hybridomas**, that grow and proliferate, producing large amounts of a single (monoclonal) antibody. The antibody produced is one that was being syn-

thesized by the normal lymphocyte prior to fusion with the myeloma cell.

The procedure for the production of monoclonal antibodies is illustrated in Figure 18.48. The antigen (whether present in a soluble form or as part of a cell) is in-

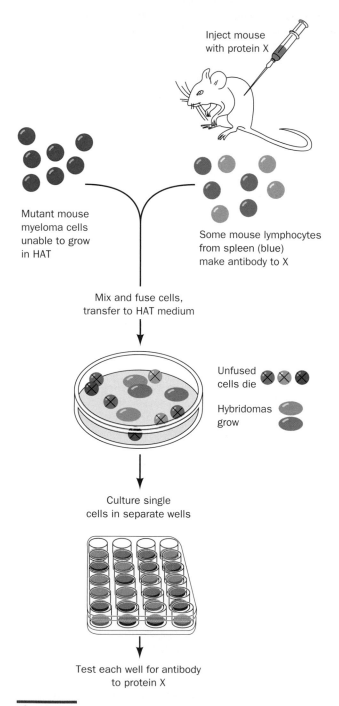

Figure 18.48 The procedure used in the formation of monoclonal antibodies. The HAT medium is so-named because it contains hypoxanthine, aminopterin, and thymidine. This medium allows cells with a functional hypoxanthine-guanine phosphoribosyl transferase (HGPRT) to grow, but it does not support the growth of cells lacking this enzyme, such as the unfused myeloma cells used in this procedure. (D. VOET & J. G. VOET, BIOCHEMISTRY, 2D ED. COPYRIGHT © 1995, JOHN WILEY & SONS, INC. REPRINTED BY PERMISSION OF JOHN WILEY & SONS, INC.)

jected into the mouse to cause the proliferation of specific antibody-forming cells. After a period of several weeks, the spleen is removed, and the antibody-producing lymphocytes are fused with a population of malignant myeloma cells, which makes the hybrid cells immortal, that is, capable of unlimited cell division. Hybrid cells (hybridomas) are selected from nonfused cells by placing the cell mixture in a medium in which only the hybrids can survive. Hybridomas are then grown clonally in separate wells and individually screened for production of antibody against the antigen being studied. Hybrid cells containing the appropriate antibody can then be cloned in vitro or in vivo (by growing as tumor cells in a recipient animal), and unlimited amounts of the monoclonal antibody can be prepared. Once the hybridomas have been produced, they can be stored indefinitely in a frozen state and aliquots can be made available to researchers around the world.

One of the most important features of this methodology is that one does not have to begin with a purified antigen to obtain an antibody. In fact, the antigen to which the monoclonal antibody is sought may even be a minor component of the entire mixture. In addition to their use in research, monoclonal antibodies are playing a valuable role in diagnostic medicine in tests to determine the concentration of specific proteins in the blood or urine. For example, an unusually high blood level of a prostate-specific antigen (PSA), which is measured by its interaction with a monoclonal antibody, provides an early warning that a man may have developed prostate cancer.

Once a preparation of antibody molecules is in hand, whether obtained by conventional immunologic techniques or by use of hybridomas, these molecules can be used as highly specific probes in a variety of analytic techniques. For example, antibodies can be used in protein purification. When a purified antibody is added to a crude mixture of proteins, the specific protein being sought selectively combines with the antibody and precipitates from solution. Antibodies can also be used in conjunction with various types of fractionation procedures to identify a particular protein (antigen) among a mixture of proteins. In a Western blot, for example, a mixture of proteins is first fractionated by two-dimensional gel electrophoresis (as in Figure 18.29). The fractionated proteins are then transferred to a sheet of nitrocellulose filter, which is then incubated with a preparation of antibodies that have been labeled either radioactively or fluorescently. The location on the filter of the specific protein bound by the antibody can be determined from the location of the bound radioactivity or fluorescence.

A quick scan through this text will reveal numerous micrographs depicting the immunolocalization of a particular protein within a cell as visualized in the light or electron microscope. Immunolocalization of proteins within a cell depends on the use of antibodies that have been specifically prepared against that particular protein. Once prepared, the antibody molecules are linked (conjugated) to a substance that makes them visible under the microscope, but does not interfere with the specificity of their interactions. For use with the light microscope, antibodies are generally complexed with small fluorescent molecules, such as fluorescein or rhodamine, to form derivatives that are then incubated with the cells or sections of cells. The binding sites are then visualized with the fluorescence microscope. This technique is called **direct immunofluorescence**. It is often preferable to carry out antigen localization by a variation of this technique called **indirect immunofluorescence**. In indirect immunofluorescence, the cells are incubated with *unlabeled* antibody, which is allowed to form a complex with the corresponding antigen. The location of the antigen-antibody couple is then revealed in a second step using a preparation of fluorescently labeled antibodies whose combining sites are directed against the *antibody* molecules used in the first step. Indirect immunofluorescence results in a brighter image because numerous secondary antibody molecules can bind to a single primary antibody. Indirect immunofluorescence also has a practical advantage; the conjugated (fluorescent) antibody is readily purchased from vendors. Immunofluorescence provides remarkable clarity because only the proteins bound by the antibody are revealed to the eye; all of the unlabeled materials remain invisible. Localization of antigens in the electron microscope is accomplished using antibodies that have been tagged with electron-dense materials, such as the iron-containing protein ferritin or gold particles. An example of this technique is shown in Figure 4.31*b*.

Update 2003

At the time of this writing, a little more than one year has elapsed since the completion of the manuscript for the Third edition of *Cell and Molecular Biology*. Some areas of interest have seen important recent advances, and I have tried in these supplementary pages to update the discussion of some of these topics. I have focused on the Human Perspectives because they hold particular interest to many readers and they tend to be subjects of ongoing research. Unlike a revision, where new findings are integrated with older material, this update will focus on recent studies, which will be given more individual coverage than would be the case in the text itself.

THE HUMAN PERSPECTIVE

Replacing Damaged Cells
and Organs

The prospect of treating chronic diseases by cell transplantation has continued to be a focus of research and speculation. Possibly the most controversial recent study concerned the treatment of Parkinson's disease. Up until 2001, all of the studies in which fetal cells were implanted into the brains of Parkinson's patients were *uncontrolled*, which is to say that all of the patients in the study group received identical treatment. In 2001, preliminary results from the first of two planned double-blind cell-transplant studies on Parkinson's patients were published. A *double-blind study* is one in which:

1. The patients under study are divided into two groups that are treated similarly except that one group (the experimental group) is given the curative factor (cells, drugs, etc.) being investigated and the other group (the control group) has that factor withheld, and

2. The patients are randomly assigned to one or the other group and identified in a way that doesn't allow the researchers or patients to know whether or not they are receiving the experimental treatment.

In this particular study reported in the *New England Journal of Medicine* (vol. 344, p. 710, 2001), the experimental and control groups each consisted of 20 long-term Parkinson's patients. Four small holes were drilled through the skulls of patients in both groups, but only those in the experimental group had replacement cells injected into their brain. The injected cells were originally obtained from brain tissue of aborted fetuses and then allowed to proliferate in cell culture. One year after the surgery, patients were asked to evaluate the severity of their symptoms as compared to their condition before the procedure. Patients in the control group reported no change. Younger patients (under 60) in the experimental group generally reported a modest improvement in the state of the disease, whereas older patients in this group reported little or no improvement. Furthermore, a significant fraction of the experimental group reported serious side effects from the treatment that left them with severe jerky, uncontrolled movements. Fortunately, patients who developed these side effects have been successfully treated with drugs or deep-brain electrical stimulation.

The results of the study were interpreted in the media as being very disappointing and led some to question the value of fetal cell transplants for the treatment of Parkinson's disease. Most researchers in the field, however, did not share in this evaluation. Follow-up studies on the above groups of patients using standardized objective tests of motor function (rather than subjective evaluation by the patients) indicated that their motor skills had generally improved, considerably. Moreover, autopsies on three patients in the study who had died revealed that transplanted cells remained alive and capable of developing into nerve cells up to eight years after the transplant operation (see *Nature*, vol. 416, p. 666, 2002).

The particular surgical procedures used by these researchers have also been questioned as has their decision to use cells that had been subjected to cell culture. The underlying causes of the side effects in some of the treated patients has also been the subject of controversy, because these problems had generally not arisen in previous uncontrolled studies where fetal cell transplantation had been performed. This study and the controversy it stirred illustrate the difficult path that is often taken in the development of complex clinical procedures and the need for a number of studies to be carried out in different ways by different groups of researchers.

Regardless of the success or failure of the type of study just described, the future of cell transplantation therapy does not lie in the use of fetal cells, which will never be readily available, but lies instead in the use of stem cells and their differentiated end-products. Toward this end, considerable advances have been made in the development of strategies to direct stem cells into differentiating into particular cell types. If one considers that the body contains as many as 300 distinct cell types, the major challenge is to develop various types of culture media that restrict differentiation along undesirable pathways while at the same time inducing differentiation along one or another desired pathway. Consider the beta cells of the pancreas. These are one of several different types of pancreatic cells that secrete important hormones. The beta cells secrete insulin, and they are the cells that are attacked and destroyed in childhood (type I) diabetes (see p. 728 of the text). If stem cells can be made to differentiate into pancreatic beta cells in culture, they could be injected into a diabetic's own pancreas where they might restore the patient's ability to produce insulin when needed. Researchers at the National Institutes of Health have reported (*Science*, vol. 292, p. 1389, 2001) on a protocol that leads to the conversion of mouse embryonic stem (ES) cells into clusters of differentiated cells that resemble the clusters of beta cells scattered throughout the pancreas. When these cultured cells are exposed to elevated glucose levels, they respond by secreting insulin, just as normal beta cells would within the body. Moreover, when these "pancreatic beta" cells are injected into a diabetic mouse, they secrete small amounts of insulin and significantly prolong the survival of the animal. To get the mouse ES cells to follow this differentiation pathway the researchers (1) grew them in the presence of a protein called LIF in which they proliferated, (2) removed the LIF, which caused the cells to form clumps

called embryoid bodies, (3) grew the clumps in a simple medium that lacked serum, which killed off 90 percent of the cells, (4) exposed the surviving cells to a protein called bFGF, which caused them to proliferate and to develop into the undifferentiated precursors of pancreatic beta cells, and (5) removed the bFGF and added nicotinamide, which led to the differentiation of some of the precursor cells into clusters of beta cells. This example illustrates the way in which researchers use a trial-and-error approach to screen large numbers of proteins to determine which combination of ingredients happens to induce cells to differentiate along a particular pathway and produce a pure population of a particular cell type. Even though there are established clues to follow when testing various culture additives, determination of these complex experimental protocols can be very time consuming.

Another study reported at a recent meeting has raised questions as to whether or not it is even necessary to have stem cells undergo differentiation in culture before they are implanted into the body. In this case, mice were first made to develop symptoms resembling those of Parkinson's disease and then given implants of undifferentiated ES cells. The cells were implanted into the area of the brain that was undergoing degeneration. Once in place, the cells differentiated into dopamine-producing neurons, which are the type of neurons that are destroyed in patients with Parkinson's disease. These findings suggest that the transplanted, undifferentiated cells were programmed to develop specifically into dopaminergic neurons by influences from the environment in which they found themselves. It may turn out that different areas of the body provide specific influences that cause undifferentiated stem cells to develop into the same types of cells that are found in that part of the body in which they are implanted. As pointed out in the text, there is always the spectre that undifferentiated ES cells might differentiate into malignant teratomas following implantation, but these types of tumors were not observed in those mice that received ES cell brain implants.

The past year has also seen a continuation of reports that describe the remarkable capabilities of *adult* stem cells to differentiate into tissues that are dramatically different from those in which the cells were initially isolated. A striking example was reported in the Journal of Cell Biology in November of 2001 (*JCB*, vol. 155, p. 699 & 703, 2001) where researchers injected fluorescently labeled adult bone marrow cells into adult mice and found that some of the injected cells eventually differentiated into Purkinje cells, which could be identified by their fluorescent label. Purkinje cells, which are key neurons of the cerebellum, the motor control center of the brain, are arguably the most complex of all neurons. Purkinje cells, as pictured on page 645 of the text, contain a multitude of processes and make synaptic connections with thousands of other neurons. That a relatively "simple" blood-forming cell has the capability to differentiate into a nerve cell of such immense complexity would have been beyond the wildest imagination of most researchers only a few years ago.

In another study reported in *Nature* (vol. 410, p. 701, 2001), stem cells present in the bone marrow of adult mice were found to be capable of differentiating into both heart muscle cells (cardiomyocytes) and cells that line the blood vessels of the heart. In this study, one of the coronary arteries of the mice was tied off, stopping blood to the heart through that artery. Interruption of blood flow through the artery led to the death of an area of heart muscle and a resulting decreased heart function in the animal. The same consequence is seen in humans who suffer heart attacks as the result of blood clots that form in a coronary artery. In the study on mice, stem cells were first isolated from bone marrow and then injected near the damaged heart tissue a few hours after the experimentally induced heart attack. Within nine days, transplanted cells (which could be identified by a fluorescent protein they carried) were seen to fill a major portion of the area of damaged heart muscle. The transplanted cells were found to have differentiated into new, replacement heart muscle tissue and to have formed new blood capillaries in the area. Even more importantly, the pumping capabilities of the damaged hearts improved significantly following cell transplantation. Studies of this type hold out great promise that one day humans who have suffered serious heart attacks may be treated by transplantation of stem cells isolated from their own bone marrow or other bodily tissue (even skin). Development of treatments that utilize an individual's own cells pose no ethical dilemmas that are inherent in the use of fetal tissue and, moreover, ensure that the transplanted cells will not be subject to immunologic rejection.

The most heated discussions over the past year or so—both in the media and in government circles—have focused on the use of embryonic versus adult stem cells as material for cell transplantation. As was well publicized in 2001, President Bush decided to restrict government-funded research on embryonic stem (ES) cells in the United States to those ES cell lines that had already been established by August 9, 2001. As a result, the National Institutes of Health was no longer allowed to fund studies that utilized stem cells removed from living embryos after that date. The ban covered the use of all embryos, including those that might be donated by fertility clinics as well as embryos that might be produced in the laboratory simply to provide a source of ES cells. This ruling allowed government-funded researchers to work with only about 65 existing cell lines, most of which had not been very well characterized. Research activities being carried out by private biotechnology companies in the United States was not affected by the ruling.

The studies described in this Human Perspective have focused on the use of adult stem cells versus embryonic stem cells as the basis of cell transplantation therapies. As discussed in the text, each type of stem cell has its own advantages and disadvantages, and it is much too early to predict the direction this area of experimental medicine will take. It can be noted that some of the excitement over the prospect of using adult stem cells to generate a diverse array

of differentiated cells has waned over the past months. Several reports have suggested that adult stem cells are not as malleable as previous publications have indicated. Two studies have generated particular interest (see *Nature*, vol. 416, p. 485, 2002). In both of these studies, it was found that adult stem cells obtained their ability to differentiate into diverse cell types by fusing with ES cells that were present in the same culture dish. Whether or not these findings have any relevance to studies in which adult stem cells have been injected into humans or animals remains to be seen. Most researchers don't believe that fusion of transplanted cells with host cells plays any role in the differentiation of the donor stem cells. Regardless, investigators have begun to avoid culturing stem cells for long periods prior to transplantation. In addition, these negative reports have caused stem cell researchers to carry out more comprehensive analyses of the stem cells they are transplanting and the differentiated cells derived from them.

THE HUMAN PERSPECTIVE

Protein Misfolding Can Have
Deadly Consequences

No major breakthroughs in our understanding or treatment of either prion diseases or Alzheimer's disease (AD) have been reported recently, but continuing advances have been made. We will begin with a discussion of AD.

The greatest excitement—and disappointment—in the field of AD research over the past few years has centered on the possibility of a vaccine. It was hoped that immunization of patients with the Aβ42 peptide—the cellular product thought to be responsible for causing neurodegeneration—could elicit an immune response capable of preventing or even reversing the devastating effects of the disease. As discussed in the text (page 69), immunization with Aβ42 was shown to reduce amyloid deposits in the brains of mice that have been genetically engineered to carry a mutated gene responsible for AD in humans. Reduction in plaque deposits was accompanied by an improvement in the ability of these mice to perform tasks that require learning and memory skills. These studies on mice led to a clinical trial in which 360 Alzheimer's patients were immunized with a synthetic version of the Aβ42 peptide. In March 2002, the trial was terminated after at least 15 members of the group had developed serious brain inflammation. It is suspected that the amyloid peptide injected into the patients had induced the formation of antibodies that reacted with more than the abnormal amyloid deposits they were designed to attack. This type of *autoimmune* response is not unexpected because the substance used to induce antibody formation is a naturally occurring human peptide. Interactions between antibodies and naturally occurring substances can cause a number of serious diseases (see The Human Perspective, Chapter 17, page 728).

Immunization therapy is not the only strategy being considered in the treatment of Alzheimer's disease. Two other approaches—use of compounds that inhibit either the β-secretase or γ-secretase enzymes and β-sheet breakers—were discussed in the text. Studies in the past year or so have also led researchers to consider other possible options. These include:

1. *Nonsteroidal anti-inflammatory drugs (NSAIDs).* Epidemiological studies are ones that try to make correlations between various aspects of a population's lifestyle and human health. Millions of people, most notably chronic arthritis sufferers, take aspirin or other NSAIDs on a regular basis. According to numerous epidemiological studies, persons who have taken NSAIDs for long periods of time are at a somewhat lower risk of developing Alzheimer's disease when compared to those who do not take such drugs. Although it was not mentioned in the text, the brains of patients who have died of AD show evidence of marked inflammation. Inflammation is characterized by the presence in a tissue of cells from the immune system that mediate various types of immune responses, some of which can provoke cellular damage. Just as anti-inflammatory drugs relieve inflammatory conditions such as arthritis, they may also prevent or reduce the rate of progression of AD. Studies on laboratory animals have supported this conclusion, and clinical trials of NSAID treatment for Alzheimer's patients are currently in progress. Before you conclude that an aspirin a day will prevent the future development of Alzheimer's disease, you should be aware that the inflammatory response can have beneficial as well as injurious consequences. For example, certain immune cells of the brain—the microglia—may remove preexisting amyloid deposits and help prevent neurodegeneration. It is evident that much more needs to be learned about these activities before we will understand the consequences of interfering with immune responses in the brain.

2. *Serum cholesterol-lowering drugs.* During the examination of brain tissue from persons who had died of various causes, it was noted that individuals who had died of heart disease had a much greater likelihood of having amyloid deposits in their brain than did persons who had died of other diseases. One of the major risks for cardiovascular disease is high serum cholesterol, which has led millions of people with high cholesterol levels to take a type of drug called a statin. Statins block the synthesis of cholesterol and lower blood cholesterol levels. Recent epidemiological studies have suggested that long-term administration of statins significantly lowers the risk of a person developing Alzheimer's disease. Studies on animal models have shown that administration of cholesterol-lowering drugs reduces the production of Aβ peptides and the formation of amyloid plaques. Although the basis for a relationship between cholesterol levels, plaque development, and Alzheimer's disease has not been fully explained, there are some intriguing clues.

In 1993, a series of reports from researchers at Duke University caused great excitement among investigators studying AD. The reports centered on the gene that codes for the protein apolipoprotein E (apoE), which is normally involved in transporting cholesterol through the blood. Three different versions (alleles) of the *apoE* gene are found at high frequency in the human population. These alleles code for the proteins apoE2, apoE3, and apoE4, which vary from one another by one or two amino acids. Persons with late-onset AD, which accounts for approximately 95 percent of all cases, were found to have a greatly increased likelihood of possessing at least one copy

of the *apoE4* allele. Thus, while the *apoE4* allele does not *invariably* cause a person to develop AD (as do the mutations in the *APP* or *PS1* genes discussed in the text), it is a very strong risk factor. How does the possession of a copy of the *apoE4* gene predispose a person to develop AD? This question has not yet been satisfactorily answered. Suffice it to say that the involvement of apoE4 in the development of AD may explain why cholesterol-lowering statins tend to decrease the risk of the disease. Given the fact that most cholesterol-lowering drugs have proven safe when taken for long periods, these drugs may ultimately be prescribed as a preventive treatment for individuals who are thought to be at a higher risk for the development of AD, especially those individuals who possess one or two copies of the *apoE4* allele. Clinical trials are currently underway to determine if statins have value in slowing progression of Alzheimer's in persons suffering from early-stage disease.

3. *Drugs that bind metal ions, particularly copper and zinc ions (Cu²⁺ and Zn²⁺).* Considerable attention has been paid over the years by both the media and some researchers to a possible link between the intake of particular metals, especially aluminum, and the development of Alzheimer's disease. While the case for a link between aluminum ingestion and AD has not been substantiated, the amyloid plaques characteristic of AD do bind copper and zinc ions. Studies on slices of brain tissue taken from individuals who have died of AD indicate that removal (chelation) of these metal ions can lead to the solubilization of the β-amyloid peptides. Given this observation, it has been speculated that removal of metal ions from amyloid deposits in the brain of a living person might greatly alleviate the symptoms of the disease. To achieve this end, researchers have screened chemical compounds for their ability to remove copper and zinc ions from amyloid deposits. A number of compounds having this ability have been identified; one such compound, known by the trade name Clioquinol, has received particular attention. Clioquinol was prescribed widely as an antibiotic during the early 1970s for the treatment of diarrhea. For the most part, the drug had an excellent safety profile, but it was found to be responsible for causing a severe neurodegenerative disease in a small number of users, particularly in Japan. A strong causal link between the disease and Clioquinol use was never definitively established but, regardless, the drug was withdrawn from the market. Given its ability to dissolve amyloid deposits, Clioquinol has been retrieved from the "pharmaceutical trash heap" and given a new life as a possible AD medication. When transgenic mice that are prone to developing Alzheimer's symptoms are administered Clioquinol, they exhibit a marked reduction in amyloid deposits and increased cognitive abilities. Based on these types of studies, the drug is currently being tested in clinical trials of patients with mild to moderate Alzheimer's disease. Preliminary results released in April 2002 suggest the drug is slowing the rate of mental deterioration in these patients.

Controversy has increased recently as to the identity of the substance responsible for the death of neurons in persons with this disease. The text suggests that the amyloid plaques, which are composed largely of insoluble Aβ clumps and fibrils, are responsible for killing nerve cells and causing AD. The greatest stumbling block with this hypothesis is that a greater degree of dementia does not always correlate with a greater concentration of amyloid plaques. This is particularly true in certain mouse models of AD, where the animals display significant memory impairment despite the fact that their brains appear to lack visible plaque formation. Increasing evidence shows that smaller, soluble complexes of the Aβ peptide, which are precursors to the large insoluble plaques, are themselves toxic and may, in fact, be the primary cause of neuronal death. For example, cultured nerve cells are more likely to be damaged by a medium containing Aβ peptide oligomers (complexes composed of a small number of peptides) than they are by large, insoluble amyloid deposits.

The other major topic of the Human Perspective on page 67 of the text is vCJD, a fatal neurodegenerative disease that has been traced to the ingestion of contaminated beef. At the present time, just over 100 residents of Britain and a few other European countries have contracted the disease. A major question concerns the scope of the epidemic. Is the relatively small number of reported cases the tiny tip of a huge iceberg, or is the worst of the outbreak largely over? The answer depends primarily on (1) the length of the incubation period for the disease (i.e., the time between exposure to the infectious agent and the onset of symptoms); and (2) the degree of variation in that incubation period. Since the incubation period in cases of CJD that have developed from surgical contamination tends to average about 15 years, it is likely that the average incubation period in cases of vCJD from contaminated beef will be at least this long.

A number of epidemiological reports published in the past year make predictions concerning the eventual size of the epidemic. Depending on the assumptions made by the researchers, these predictions range from a few hundred cases at the low end of the spectrum to hundreds of thousands of cases at the high end. One fact has emerged that has added to the uncertainty. To date, all individuals diagnosed with vCJD have the amino acid methionine at position #129 in both copies of the PrP protein that they produce. In other words, they are homozygous for a gene that encodes methionine at that position. Persons with this particular genotype, however, represent only about 38 percent of the total British population. The remainder are either homozygous for valine at that position or are heterozygotes that produce both types of PrP proteins. Are methionine homozygotes the only individuals to have contracted vCJD because those with other genotypes are resistant to the disease? Or do persons with other genotypes simply display a longer incubation period before the onset of symptoms? This question has yet to be answered, although it is known that individuals with valine at position #129 of the PrP protein are

subject to contracting kuru through the ingestion of infected brain tissue (page 67). If data concerning kuru has any bearing on contraction of vCJD, then another wave of the epidemic could begin at a future date when members of the population who produce the valine-containing PrP protein begin to show evidence of infection.

Another question that has been raised over the past year concerns the safety of the blood supply. Unlike classical CJD, in which the abnormal PrP molecules appear to be restricted to brain tissue, persons who have been diagnosed with vCJD show evidence of the infective prion particles in other parts of the body, particularly in tissues of the immune system, such as the tonsils and spleen. The presence of prion particles in these tissues raises the spec-

tre that they may also be present in white blood cells, or even the fluid phase of the blood, and thus represent a risk factor in blood transfusions. Although infective prion particles have never been demonstrated in human blood, the United States and several other countries are screening donors on the basis of their past residence. Persons who have lived in England for a period of several months between 1980 and 1996 have been prevented from donating blood in the United States for fear of transmission of vCJD. A number of companies are working to develop highly sensitive tests for the detection of blood-borne prions. Use of such tests would allow blood banks to screen blood samples for the presence of these potentially infective proteins.

THE HUMAN PERSPECTIVE

**The Growing Problem
of Antibiotic Resistance**

Antibiotic treatment made headlines in 2001 after a substantial number of people in the United States were exposed to potentially lethal spores of the Gram-positive bacterium *Bacillus anthracis* that had been transported through the mail. *B. anthracis* is the agent that causes anthrax, a disease once common in domestic animals, and occasionally contracted by humans who worked with sheep or wool. Until 2001, there hadn't been a confirmed U.S. case of inhalation anthrax—an anthrax infection of the respiratory tract—since 1978. Bacterial spores, such as those that were sent through the mail, are a dormant, inactive stage in the organism's life cycle. Such spores can germinate in a nutrient-rich environment, such as the respiratory tract or skin. Once they germinate, the *B. anthracis* cells grow and divide, and it is during this stage that they can be killed by a wide variety of antibiotics. As the infection progresses, the bacteria produce lethal toxins, which are the agents that can cause the death of the human host. That is why it is crucial for an infected person to receive antibiotic treatment as early as possible, because once the toxins have built up to a high enough concentration within the body, extermination of the bacteria can no longer save the individual.

People that were exposed to the anthrax spores in the autumn of 2001 were given the antibiotic ciprofloxacin, made by Bayer under the brand name Cipro. Ciprofloxacin is a quinolone, a class of synthetic antibiotics mentioned on page 104 as acting as an inhibitor of the enzyme DNA gyrase, which is required for the replication of bacterial DNA. As it happens, the strain of anthrax that was used in the 2001 attacks was susceptible to a variety of agents, including penicillin. When the outbreak occurred, however, it wasn't immediately known if the bacteria involved had been genetically modified to be resistant to penicillin or other widely used antibiotics. Ciprofloxacin was an antibiotic that had been developed fairly recently, so it was much less likely that the person(s) responsible for the attack would have had possession of bacterial spores that had been made resistant to Cipro. In fact, a panel of microbiologists had met in the late 1990s and recommended the use of ciprofloxacin to treat an outbreak of anthrax resulting from a bioterrorist attack (published in the *J. American Medical Assoc.*, vol. 281, p. 1735, 1999). Fortunately, Cipro proved highly effective against the bacterium and might have saved the lives of many individuals. Given Cipro's success, laboratories involved in biological warfare will most likely develop strains of the anthrax bacterium that are resistant to this antibiotic. Therefore, we must not be dependent wholly on this antibiotic in the event of future attacks.

Ciprofloxacin is also one of the major antibiotics considered for the treatment of vancomycin-resistant infections. In other words, ciprofloxacin is one of the "last resort drugs" for treatment of infections that are resistant to an antibiotic (vancomycin) that had previously been described as the "drug of last resort." As in the case of vancomycin-like drugs, quinolones are also being fed in large quantities to livestock, and many scientists are concerned that this alternate use will have an impact on resistance in human infections (see the debate in *Science*, vol. 291, p. 2550, 2001).

THE HUMAN PERSPECTIVE

Defects in Ion Channels as a Cause of Human Disease

It has been known for some time that CFTR, the protein whose deficiency is responsible for cystic fibrosis (CF), is an ion channel, or more specifically a chloride ion channel. Although there has been evidence for years that CFTR is also capable of conducting bicarbonate (HCO_3^-) ions, only over the past year or so has this additional function of CFTR been appreciated. It was noted on page 162 that CF often affects the pancreas, an organ responsible for producing and secreting hydrolytic enzymes that are shipped to the large intestine where they digest food matter. These enzymes are secreted into the pancreatic ducts in an alkaline solution whose pH is maintained at an elevated level by the presence of a high concentration of bicarbonate ions (140 mM). Bicarbonate ions are key buffering agents in extracellular fluids (see page 41). Bicarbonate ions move across the plasma membrane of the pancreatic epithelial cells through the CFTR protein. Thus, unlike the respiratory tract and sweat glands where chloride conductance appears to be the major CFTR activity, it is HCO_3^- conductance that is of primary importance in the pancreas.

In some CF patients, the secretion of pancreatic digestive enzymes is severely limited, a condition described as pancreatic insufficiency. Other CF patients show very little pancreatic dysfunction. As shown in Figure 2 on page 163, many different *CFTR* mutations have been identified as causing CF. It has been found that chloride and bicarbonate conductance can be affected quite differently by some of these mutations. In fact, a number of mutant proteins exhibit a marked reduction in bicarbonate conductance, even though they are capable of normal or even elevated conductance of chloride ions. As might be predicted, those patients whose mutant CFTR protein is unable to conduct bicarbonate ions are the ones who suffer the greatest degree of pancreatic insufficiency (*Nature*, vol. 410, p. 94, 2001). These findings have raised speculation that decreased bicarbonate ion secretion could play a more important role in the respiratory-tract complications experienced by CF patients than previously thought. Decreased bicarbonate conductance could lower the pH of the fluid that lines the respiratory airway, which in turn could increase the viscosity of respiratory mucus and the ability of infectious bacteria to adhere to the epithelial cells that line the tract.

Another subject dealt with in the Human Perspective of Chapter 4 is gene therapy, a strategy to treat human disease through the transfer of new genetic material into the body. Rather than restricting this discussion to the treatment of cystic fibrosis, we will consider a broader examination of the subject of gene therapy itself. The first attempts to treat humans by gene therapy took place at the National Institutes of Health in 1990 and 1991. The patients were two young girls who suffered from a disease called ADA⁻ SCID. These patients had a severely crippled immune system due to the hereditary deficiency of a single enzyme, adenosine deaminase (ADA). The researchers took samples of blood from the two girls, isolated a particular type of immune cell called a T lymphocyte, infected these T lymphocytes with a retrovirus that carried a normal copy of the ADA gene, grew the genetically modified (transduced) cells in culture, and then infused the cells back into the bloodstream of the children. This approach is referred to as *ex vivo* gene therapy because the cells are genetically altered while they are outside the body and then reintroduced into the bloodstream. The procedure was repeated a number of times to maximize the number of T lymphocytes capable of producing ADA. Tests on the children over the next few years revealed that the normal copy of the gene had indeed integrated itself into the chromosomes of a significant proportion of the patients' T lymphocytes and that these cells were producing the normal version of the ADA enzyme. At last report, both children are doing well, but they are still dependent upon periodic injections of purified ADA, which indicates that these patients are not producing sufficient quantities of the normal enzyme to endow them with a functional immune system.

These initial trials were hailed as a great success and generated considerable optimism both among physicians and the informed public that gene therapy would soon lead to the treatment of many different disorders. Unfortunately, this has not proven to be the case. The major stumbling block has been the development of an agent (or vector) that is capable of genetically altering a sufficient percentage of target cells to bring about the required effect. To date, most gene therapy studies have utilized one of several different types of viruses (retroviruses, adenoviruses, or adeno-associated viruses) to deliver DNA into the target cells (*Nature Med.*, vol. 7, p. 33, 2001). Viruses are well-suited for this task because this is how they normally operate; they bind to receptors on a cell's surface, introduce their genetic material into that cell, which then manufactures proteins encoded by the virus. DNA has also been commonly transferred into cells using various types of lipid-DNA complexes. Unfortunately, none of these vectors has exhibited a gene-transfer efficiency high enough to cure an inherited disease. This is illustrated by gene therapy trials with patients afflicted by cystic fibrosis.

Ever since the discovery of the *CFTR* gene, cystic fibrosis has been considered as a model human disease for the development of treatments based on the delivery

of a normal gene to mutant cells. This is based primarily on the fact that (1) the cells that are most affected by the disease can be reached by inhaled aerosols and (2) the transfer of a normal gene to as few as five percent of the cells of the respiratory epithelium might be expected to alleviate symptoms of the disease. Despite these apparent advantages, treatment of CF by gene therapy has not been successful. As noted on page 164, attempts to deliver a normal *CFTR* gene have utilized either adenoviruses or liposomes as vectors. In the latest report on a clinical trial using an adenoviral vector, researchers found that no more than 2.4 percent of respiratory epithelial cells contained evidence of the virally transferred *CFTR* DNA (*Human Gene Ther.*, vol. 12, p. 1383, 2001). A trial in which patients inhaled an aerosol containing a lipid-DNA conjugate also proved unsuccessful *(Human Gene Ther.*, vol. 12, p. 751, 2001). Taken together with earlier reports, major improvements in vector technology are required before it will be feasible to correct CF by gene therapy. This is likely to take quite a few years. Perhaps other inherited diseases, most notably hemophilia, will prove to be more suited to current gene-therapy technology (see *Science*, vol. 291, p. 1692, 2001).

THE HUMAN PERSPECTIVE

Diseases That Result from Abnormal Mitochondrial or Peroxisomal Function

One of the best ways to learn about the underlying causes of a human disease, and to test possible treatments for that disease, is to study laboratory animals that develop a similar condition. These *animal models*, as they are called, may occur naturally in animal populations as they do in human populations, or they may have to be produced in the laboratory by some type of genetic engineering. Most human diseases that can be caused by identifiable genetic mutations, such as Alzheimer's disease or muscular dystrophy, can be simulated in laboratory animals, particularly mice. In most cases, this is accomplished by causing the animals to carry the mutated human gene (as on pages 69, 777) or by "knocking out" the corresponding gene in the animal's own genome (pages 684, 779). These procedures work well for nuclear genes that are present in only one copy per haploid set of chromosomes. But they don't work for genes located in mitochondria that are present in hundreds or even thousands of copies per cell. During the past couple of years, techniques have been developed that have led to the production of animal models for human diseases caused by mutant mtDNA (*PNAS*, vol. 97, p. 14461, 2000; *Nat. Gen.*, vol. 26, p. 177, 2000). The following paragraph describes one of these experimental protocols.

The first step in formation of these animal models is to prepare cells that lack mitochondria entirely. This is accomplished by allowing cells to grow and divide in the presence of a mitochondrial toxin. Over time, cells emerge that lack mitochondria and depend upon glycolysis and fermentation for ATP production. These mitochondria-deficient cells are then provided with a new population of mitochondria containing mtDNA molecules bearing a particular mutation or deletion to be studied. Mitochondria can be introduced into amitochondrial cells by microinjection or by other techniques. Cells formed by this experimental procedure that contain nuclear DNA from one source and mitochondrial DNA from another source are called *cybrids*. The nucleus of the cybrid can then be removed mechanically producing an anucleate cell, called a *cytoplast*. Cytoplasts containing mutant mitochondrial DNA can then be fused with a fertilized egg so that the mutant mitochondria are introduced into the egg's cytoplasm. The mouse that develops from the fertilized egg will be composed of cells whose mitochondria contain the mutant mtDNA.

Study of these models should shed light on several key questions regarding mitochondrial diseases. Most importantly, what is the relationship between a particular mitochondrial mutation and the development of a particular disease? Why, for example, does a mutation in a gene for one of the mitochondrial ribosomal RNAs cause humans to develop a particular type of deafness? This type of mutation is assumed to stop mitochondrial ribosomes from manufacturing mitochondrial proteins. As discussed in Chapter 5, all of the proteins produced in the mitochondria are part of either the electron transport chain or the ATP synthase, so that one can presume that the major defect that results from this type of mutation is a deficiency in ATP production. But why should such a deficiency cause a person to be deaf, but otherwise essentially normal? According to one view, certain cells involved in hearing depend upon a Ca^{2+} pump that requires high ATP concentrations. If ATP concentrations remain low because of a particular mitochondrial deficiency, then the pump cannot work properly, and hearing loss results. But if all mutations in mitochondrial DNA have the same basic end result, namely the loss of ATP synthesis, then why do some mutations cause deafness and other mutations cause diabetes, heart failure, or some other condition? Early results from some mouse models also raise questions. In one study, mice that developed from fertilized eggs whose mtDNA contained a large deletion covering numerous genes died at an early age of kidney failure, a condition that is not seen in humans with comparable mtDNA deletions. This is the major problem with all types of animal models; they attempt to use mice or other laboratory animals as substitute humans. Although the genetics and physiology of humans and mice are remarkably similar, no one has ever mistaken a member of one species for that of the other. This is the reason that drugs proven safe and effective in studies on animal models must still pass through years of carefully controlled human trials before they are approved for public use.

THE HUMAN PERSPECTIVE

Diseases That Result from the Expansion of Trinucleotide Repeats

Although there have been no breakthroughs in the prevention or treatment of Huntington's disease, or any of the other disorders caused by trinucleotide expansion, there has been considerable recent progress in our understanding of the mechanism by which these genetic changes may cause degeneration in parts of the brain. As noted on page 418, the mutant version of huntingtin, the protein encoded by the *HD* locus, acquires toxic properties not exhibited by the wild-type version of the protein. Because of their expanded polyglutamine tract, mutant huntingtin molecules are thought to undergo abnormal folding to produce a misfolded protein molecule that (1) binds to other huntingtin molecules to form aggregates and (2) binds to a number of unrelated proteins that do not interact with normal huntingtin molecules. Among the proteins bound by mutant huntingtin are several transcription factors, which are proteins involved in the regulation of gene expression. Two of the most important transcription factors present in cells are TBP (discussed on page 455) and CBP (discussed on page 537), both of which are bound by mutant protein molecules containing expanded polyglutamine tracts. Protein aggregates found in degenerating neurons of HD patients contain both of these key transcription factors. This finding suggests that mutant huntingtin sequesters these transcription factors, thereby removing them from the remainder of the nucleus and disrupting the transcription of genes that are required for the health and survival of the affected neurons. Other proteins are also bound and sequestered by mutant huntingtin, including proteins that are normally involved in the destruction of misfolded protein molecules (page 545). As a result, the production of abnormal huntingtin molecules may interfere with the very system used by the cell to rid itself of unwanted proteins. The ensuing buildup of aberrant proteins within affected cells would likely contribute to neurodegeneration.

Recent investigations have also led to a revision in our view of the mechanism that causes trinucleotide expansion. As stated in the text, "expansion is thought to occur during DNA replication." DNA replication is a process that occurs prior to either mitosis or meiosis, but recent studies on mice have suggested that expansion of the CAG trinucleotide that causes HD occurs in the latter stages of sperm formation, after meiosis has been completed (see Figure 14.40), and in adult neurons, which had completed their last mitosis much earlier. Consequently, rather than occurring during DNA replication, CAG expansion may occur during the process of DNA repair, which normally serves to mend damaged DNA (page 572). The presence of extra CAG nucleotides appears to cause the DNA molecule to form abnormal structures (e.g., base-paired loops), which in turn attract DNA-repair proteins to these sites. Once bound to the DNA, the repair machinery introduces additional CAG trinucleotides into the DNA, resulting in polyglutamine expansion. This conclusion is supported by studies on mice that lack certain DNA-repair proteins. When such mice are also made to carry a mutant *HD* gene, the CAG tract in the mutant *HD* gene is not expanded as occurs in mice that have a full complement of repair proteins.

THE HUMAN PERSPECTIVE

Sequencing the Human Genome

The years 2000 and 2001 marked a turning point in the science of biology and witnessed a great leap forward in the level of understanding that humans have of themselves. In 2000 the rough draft of the nucleotide sequence of the human genome was published. The sequence was described as "rough" because each segment was sequenced an average of about four times, which is not enough to ensure complete accuracy. This rough draft covered approximately 90 percent of the genome, excluding certain regions that proved difficult to sequence. The following year saw the first attempts to *annotate* the human genomic sequence, that is, to interpret the sequence in terms of the numbers and types of genes it encoded. The single-most striking "fact" to emerge from these initial analyses concerned gene number: researchers concluded that the human genome probably contained roughly 30,000 protein-coding genes.

Despite continued sequencing efforts, some uncertainty remains about the actual number of protein-coding genes in the human genome. The true number probably lies somewhere between 30,000 and 40,000 genes, but that could still change. Why is it so hard to come up with a firm number? Part of the problem lies in the fact that the sequencing of most chromosomes still remains "unfinished" (the exceptions at the time of this writing are chromosomes 20, 21, and 22). In other words, large numbers of gaps still exist in the sequences of the other chromosomes. These gaps contain segments of DNA that are difficult to analyze. But these parts of the genome consist largely of repeated sequences or inactive (heterochromatic) regions and are not likely to be a source of very many additional genes. The major "sticking point" in determining whether or not a particular stretch of nucleotides in a DNA molecule contains a gene centers on certain difficulties that are inherent in the problem of gene identification. As was touched on in Chapter 10 and as discussed in detail in Chapter 11, genes consist of alternating coding regions (exons) and noncoding regions (introns). The coding regions are typically small (approximately 150 nucleotides in length), whereas the noncoding regions tend to be much larger, sometimes several thousand nucleotides. Although there are certain clues in a sequence as to where an exon is likely to begin and end, these clues are not always obvious enough for a gene-hunting computer program to pick out the exons and introns of a human gene simply by analyzing its nucleotide sequence.

The surest way to locate a gene within a given stretch of DNA is to know something about the product it encodes. Researchers know the amino acid sequence of thousands of different proteins, and the genes that encode these proteins are readily identified. Just as importantly, most genes are members of gene families that have related (homologous) nucleotide sequences. Once one member of a family has been identified, other members can usually be detected using computer software programs that have been developed to detect sequence homologies. Even if the protein encoded by a gene has not been isolated and studied, it is likely that something is known about the messenger RNA transcribed from that gene. As discussed on page 775 of the text, researchers are able to extract the entire mRNA population from a given tissue or organ and to copy all of the contained mRNA species into DNA, which produces a cDNA library. Since the various cDNAs that make up the library are all derived from mRNAs, and mRNAs are all derived from protein-coding genes, it is a relatively straightforward process to use the nucleotide sequences of cDNAs to identify the genes from which these mRNAs were initially transcribed. For this reason, a major effort is underway at the present time to obtain as many different full-length cDNAs as possible.

Those genes for which researchers have a corresponding protein or cDNA are the easiest targets in the human genome to identify. But what about those genes that encode proteins that have yet to be characterized, or those genes that are only transcribed during a brief period of embryonic development for which cDNA libraries have not yet been obtained? Because we don't know how many genes fall into these categories, current estimates of the total number of human genes are probably best described as "educated guesses" or "predictions." Ultimately, the discovery of hard-to-find genes will probably depend heavily on data obtained from sequencing the genomes of other species, most notably that of the mouse. How can information about a mouse help us with the search for human genes?

Consider the following facts: (1) The great majority of the genome consists of DNA that resides between the genes, and thus represents intergenic DNA, (2) each of the 30,000–40,000 protein-coding genes consists largely of noncoding portions (intronic DNA). Taken together, these facts indicate that the protein-coding portion of the genome represents a remarkably small percentage of the total DNA. Most of the intergenic and intronic DNA of the genome does not contribute to the reproductive fitness of an individual organism and thus is not subject to any significant degree of natural selection. As a result, intergenic and intronic sequences tend to change dramatically as organisms evolve. In other words, these sequences tend not to be conserved. In contrast, those small portions of

the genome that code for protein sequences are subject to great selective pressure and tend to be highly conserved among related species.

Just as Celera Genomics and the public Human Genome Project were the two key groups responsible for sequencing the human genome, these same groups of researchers have been actively sequencing the genome of several strains of mice. A draft sequence covering approximately 96 percent of the mouse genome was made available over the internet in May 2002. Like the human genome, the genome of the mouse also contains approximately 3 billion base pairs of DNA and includes at least 30,000 protein-coding genes. Despite the fact that human and mouse species have not shared a common ancestor for tens of millions of years, the two species share similar genes, which tend to be clustered in a similar pattern (see *Science*, vol. 296, p. 1617, 2002). For example, the distribution of human globin genes depicted in Figure 10.26 of the text is basically similar to that found on a comparable segment of DNA in the mouse genome. Consequently, when corresponding segments of mouse and human DNA are aligned with one another, those sequences that are conserved between the two DNAs are expected to represent functionally important segments of the genome, either coding portions of genes or important regulatory regions (see page 526). In contrast, those sequences that are not conserved between the two DNAs likely represent noncoding regions. It is hoped that this comparative genomic approach will identify genes that would otherwise be difficult to reveal.

The primary goal of geneticists continues to center on gene function. Once a human gene is identified by its sequence similarity to a mouse gene, some idea of the function of that gene may be ascertained by studying the phenotypes of mice in which that gene has been incapacitated (using the technique described on page 780). If, for example, the elimination of a particular mouse gene leads to the birth of animals that are deaf, then it is likely that the human counterpart to this gene is involved in some way in the process of sound perception.

The mouse is not the only mammalian species that is likely to play a key role in genomic research. Plans have been made to sequence the chimpanzee genome, which is thought to be approximately 98.7 percent similar in overall nucleotide sequence to that of the human genome. Chimpanzees are thought to be our closest living relative, having shared a common ancestor as recently as 4.6–6.2 million years ago. A detailed analysis of the relatively small differences in DNA sequence and gene organization between chimpanzees and humans may tell us a great deal about the genetic basis for recently evolved features that make us uniquely human. The human brain, for example, has a volume of about 1300 cm^3, which is nearly four times that of a chimpanzee.

Gene identification is only a beginning in genomic research. As pointed out on page 430, a single gene can give rise to a number of different proteins by means of a process called alternate splicing (p. 540). As a result, the information content of the genome is probably two- or three-fold greater than revealed through simple DNA sequence analysis. To identify all of the different polypeptides that humans can produce will be a difficult task. The eventual goal of genomic researchers is to understand the role of each gene in the development and maintenance of the human organism, a task that will include (1) identification of each protein encoded by the genome, (2) determination of the three-dimensional structure of each protein, (3) elucidation of the specific function of each protein, (4) identification of other proteins with which each protein interacts, and (5) recognition of the specific mechanisms by which the expression of each gene is regulated. The Human Genome Project has given biologists enough work to keep them busy for a long time.

Let us turn to another subject discussed in the Human Perspective: gene patents. Arguments continue to rage between those at one extreme who believe that genes should not be patented and those at the other extreme who believe that gene-patent holders possess a monopoly on all commercial activities involving that particular gene. The United States Patent Office continues to hold to the position that gene sequences are inventions (rather than discoveries) and therefore subject to patent (see *Trends Genetics*, vol. 17, p. 670, 2001). The patent office has, however, recently tightened the requirements for gene patenting, requiring that applicants provide a useful commercial application for the gene in question.

The most recent newsworthy event in the controversy over gene patenting has come from a challenge by a coalition of European institutes over a patent for the gene *BRCA1* held by Myriad Genetics, a biotechnology company in Salt Lake City, Utah. In 1994, researchers at Myriad Genetics isolated the *BRCA1* gene, whose mutated form had been found to predispose women to the development of breast and ovarian cancer. Myriad Genetics received a U.S. patent for *BRCA1* in 1999 that gave them exclusive rights to all commercial ventures involving the gene. As part of their work with *BRCA1*, Myriad Genetics developed an automated diagnostic test to identify mutant versions of the gene in DNA samples from women whose family history suggested that they might be carrying a gene that made them susceptible to the development of breast cancer. The patent was later extended to Europe by the European Patent Office in January 2001, which meant that DNA samples from European women were supposed to be sent to Myriad Genetics' laboratories in Salt Lake City where the diagnostic test would be run for a fee. This directive was largely ignored as European medical facilities continued to send DNA samples to regional labs where tests for *BRCA1* mutations were performed. The European medical community, in essence, does not believe that a private company should have the exclusive right to make determinations regarding the presence or absence of mutations in a human gene. They view such a patent as an unwarranted restriction on the provision of medical services. The issue came to the forefront following the publication in 2001 by a team of French and American scientists

that the Myriad Genetics test failed to detect a large deletion in the *BRCA1* gene from a DNA sample taken from a woman who, along with other members of her family, had been diagnosed with breast and ovarian cancer. Myriad Genetics responded to this finding by assuring clinicians that they would modify the test to detect the presence of such deletions. Regardless, in the wake of this finding, several European institutes banded together to mount a court challenge to the concept that all diagnostic tests involving the *BRCA1* gene had to be conducted by Myriad Genetics and a fee paid to that company for the service. The challengers argue that the initial sequencing of the *BRCA1* gene by Myriad Genetics utilized a technology that was developed by previous research efforts and was in the public domain and, moreover, that the sequence described in the original patent application contained inaccuracies that should invalidate the patent. Ultimately, the courts will likely determine any restrictions to be placed on gene patents.

Another issue that is likely to make it to the courts concerns the fact, as discussed above, that a significant fraction of human genes encode more than one protein. In many cases, proteins encoded by the same gene can have different tissue-specific functions, or at least different roles to play in the development and maintenance of the organism. Even if a gene only encodes a single protein, it is often possible to produce laboratory variants of that protein that prove to be more effective for a particular clinical procedure, or may even have uses for which the natural protein is unsuited. To what degree are protein variants distinct property from the gene from which they are all derived? This question has not yet been satisfactorily answered. As the criteria for patenting a gene becomes increasingly dependent on the potential use of the gene product, the overall value of gene patents has come into some question. To avoid potential legal battles with companies that specialize in patenting proteins, several of the large genomics companies are establishing centers to study the structures and functions of the proteins that are encoded by the genes they already hold patents on. By doing this, these companies hope to be the first to learn about the potential uses of the various proteins that might be encoded by their patented genes.

With recent advances in genome sequencing, a new term—**proteome**—has come into widespread use. A proteome is the complete collection of proteins produced by a particular organism. The term is also used to describe the entire set of proteins found in a particular organ, tissue, cell type, or organelle. The proteins manufactured by a cell are encoded by the genome, which is stored in that cell's nucleus. Consequently, the study of proteomes is actually a study of genome function.

Researchers engaged in the analysis of proteomes are attempting to separate and identify the tens of thousands of proteins produced by complex organisms and answer a host of questions about each protein molecule. These questions include: What are the properties and functions of the protein? When does the protein appear in the development of the organism? Where in the cell is the protein is the protein localized? How much of the protein is present and how long does it survive before being degraded? Which other proteins in the cell does it interact with?

Just as the human genome was sequenced through a coordinated effort of many independent laboratories (under the leadership of the Human Genome Organization, or HUGO), a similar consortium is being organized under the heading of the Human Proteome Organization, or HUPO. It is hoped that this concerted effort by public institutions will prevent this valuable information from being gathered solely by private companies and retained under the veil of trade secrecy.

Proteome studies have already begun to have an impact in medicine, particularly in the area of disease diagnosis. It has been known for years, for example, that men with prostate cancer have elevated levels in their blood of a protein called prostate specific antigen (PSA). Elevated PSA levels are an early warning sign of prostate cancer, and a test for this blood protein is widely used to screen for this disease. Recent studies suggest that persons with other types of cancer also contain abnormal levels of one or more proteins in their blood. Patients with ovarian cancer, for example, have a telltale pattern of blood proteins that distinguishes them from normal individuals. It is hoped, based on this finding, that a diagnostic test can be developed to detect ovarian cancer at an early stage when it is still curable. Many clinical researchers believe that advances in early cancer detection will have more impact on cancer survival in the coming years than advances in cancer treatment.

THE HUMAN PERSPECTIVE

Clinical Application of Ribozymes and Antisense Oligonucleotides

Over the past year or so, a variety of antisense oligodeoxyribonucleotides (ODNs) have entered clinical trials, and a number of preliminary reports of these studies have been published. Several of the most promising clinical trials depend on an antisense ODN that targets mRNA molecules encoding the protein Bcl-2. As discussed on page 664, Bcl-2 acts to protect cells from self-destruction (apoptosis). While this might sound at first like a beneficial protein, its synthesis in a cancer cell can prevent that cell from being destroyed by the body's own screening mechanisms. Bcl-2 can also protect cancer cells against many of the drugs used during cancer chemotherapy, thereby preventing the chemicals from ridding the body of the cancer cells. Therefore, antisense ODNs directed against Bcl-2 mRNAs should enhance the cell-killing activity of cancer therapeutics. Antisense ODNs targeting the Bcl-2 mRNA have been tested in patients with lymphomas, malignant melanomas, hormone-insensitive prostate cancers, breast cancers, and other malignancies. In one study, for example, 14 patients with advanced malignant melanoma were treated with this antisense ODN in combination with chemotherapy. Six of the patients showed antitumor responses and two others exhibited a stabilization of their disease. Overall, the estimated survival time of the 14 patients at the time of the publication exceeded 12 months, compared to an expected survival time of 4–5 months in patients that did not receive the treatment. Possibly the best results to date have been obtained on patients with a type of lung cancer (NSCLC) treated with chemotherapy in conjunction with an antisense ODN directed against synthesis of a protein called protein kinase C, which has been implicated in tumor formation (page 643). Of 48 patients treated with this antisense oligonucleotide, 20 exhibited an antitumor response, 20 others showed stabilization of their disease, while the disease progressed unabated in the remainder of the group. It was estimated that treatment with the oligonucleotide extended the survival time of these patients from about 8 months with chemotherapy alone to at least 18 months when treatment included the antisense ODN. Large-scale phase III trials—the last and most demanding steps leading to drug approval—have begun on a number of different antisense ODNs in the treatment of various types of cancer. So far, none of the earlier phase I/II clinical trials have indicated that these drugs are likely to result in an actual cure of the disease.

Although not discussed in the text, the field of antisense biotechnology has been wracked with controversy and skepticism. Whereas antisense oligodeoxynucleotides were originally expected to be highly specific therapeutic agents, attacking only those RNAs having complementary nucleotide sequences, antisense ODNs have exhibited a range of nonspecific effects on cells at times. For example, ODNs are often able to stimulate B lymphocytes, the cells that produce the body's antibodies, and that their therapeutic effect in some instances can be explained by this mechanism. It is important in any given study to be certain that the effect of the ODN is being achieved by its antisense properties rather than by some other means. Appropriate controls include:

○ testing the effects of ODNs having similar nucleotide sequence to the one being tested, but are not able to form tight hybrids with the mRNA being targeted. If these types of mismatched ODNs also prove beneficial, the response can be attributed to a nonspecific effect.

○ demonstrating that the expression of the protein encoded by the target mRNA is selectively inhibited.

While studies using these "classic" antisense agents are continuing, researchers have become excited in the past year over the prospects of a different type of technology that also uses nucleic acids to recognize and inactivate specific mRNAs. During the 1990s, studies on a wide variety of organisms including plants, nematodes, and fruit flies revealed a phenomenon that became known as *RNA interference (RNAi)*. It was found, in a nutshell, that double-stranded RNAs (dsRNAs) could be taken up by cells where they induced a response in which mRNAs having the same sequence as the added dsRNA were subject to selective degradation. Let's say, for example, one sought to stop the production of the enzyme phosphorylase within the cells of a nematode worm so that the effect of this enzyme deficiency on the phenotype of the worm could be determined. Remarkably, this result could be accomplished by simply injecting the worm with a solution of dsRNA in which one of the RNA strands of the duplex contained the sense sequence of the mRNA and the other strand contained the complementary antisense sequence. Even more remarkable, the effect on enzyme production is not just restricted to the cells that take up the dsRNA around the site of injection but is propagated throughout the worm, in this case blocking production of the phosphorylase enzyme in the entire organism. This phenomenon is similar in effect, although entirely different in mechanism, to the formation of knockout mice that are lacking a particular gene that encodes a particular protein (page 780).

The phenomenon of dsRNA-mediated interference (RNAi), which can be demonstrated in virtually every type of eukaryotic organism, is thought to have evolved as

a means to protect organisms from the presence of unwanted genetic material. To be more specific, RNAi probably evolved as a mechanism to block the replication of viruses and/or to suppress the movements of transposons within the genome, because both of these potentially dangerous processes typically involve the formation of double-stranded RNA intermediates. Cells can recognize dsRNAs as "undesirable" because such molecules are not produced by the cell's normal genetic activities.

How can the presence of dsRNA within a cell block the formation of a particular protein, and how can this effect spread from one part of the organism to another? The apparent answers to these questions have come from studies on the mechanism of RNAi in the test tube. There RNA interference can be made to occur in cell-free extracts, most notably those prepared from the cells of fruit flies. Studies on these extracts indicate that the double-stranded RNA that initiates the response is first cleaved into small (21–23 nucleotide) fragments by a particular type of ribonuclease. The small fragments of dsRNA, along with associated protein, then bind to the target mRNA, which is an mRNA containing a region that has the same nucleotide sequence as the dsRNA fragment. Once the dsRNA complex has bound to the target mRNA, it initiates a reaction that causes the bound mRNA to be cleaved by an associated ribonuclease. In other words, the dsRNA fragment acts like a specific guide to direct the activity of an associated enzyme toward a target RNA, not unlike the way that antisense snoRNAs act as guides to direct the activity of methylation enzymes at specific sites on ribosomal RNAs (page 452). In the case of RNAi, the dsRNA has the effect of destroying the mRNA that encodes the protein encoded by the sense strand of the dsRNA. The fact that the phenomenon of RNAi can spread from one part of an organism to other parts suggests that the process incorporates an amplification step in which the small dsRNA fragments are amplified and exported from the cell (see *Science*, vol. 296, p. 1263, 2002). Studies have revealed the existence of an RNA-dependent RNA polymerase that may be responsible for this amplification process. Since the dsRNA employed in these experiments does not have a direct effect on the transcription of the gene in question, but rather leads to the destruction of preexisting mRNAs, the process is also described as *posttranscriptional gene silencing (PTGS)*.

Up until 2001, RNA interference could only be reliably demonstrated to occur in nonmammalian cells, which is the primary reason it was not discussed in the third edition of the text. When a normal-sized dsRNA (e.g., 50–500 nucleotides) is added to mammalian cells growing in culture or injected directly into the body of a mammal,

rather than stopping the translation of a *specific* protein it initiates a *global* response that inhibits protein synthesis in general. This global shutdown of protein synthesis (discussed on pages 543 and 706 of the text) is thought to have evolved in early mammals as a means to protect cells from infection by viruses. As noted above, many viruses replicate through the formation of a double-stranded RNA, so that the natural presence of such dsRNA within a cell is highly indicative of the existence of a potentially dangerous virus. Consequently, even if sequence-specific RNAi does occur in mammals, as expected, it is difficult to demonstrate experimentally because of the much larger nonspecific global response.

What if it were possible to induce the specific RNAi response without triggering the nonspecific global response? Could this be accomplished by treating mammalian cells with small dsRNAs equivalent in size to those produced as intermediates during the process of RNA interference in other types of organisms (i.e., dsRNAs of 21 nucleotides in length)? This experiment was carried out by scientists at the Max Plank Institute in Göttingen, Germany, and published in 2001. These researchers found that dsRNAs of 21 base-pair length not only failed to induce the global inhibition of protein synthesis but were capable of bringing about a marked inhibition in synthesis of the specific protein encoded by an mRNA having a sequence matching the dsRNA with which the cells were treated (*Nature*, vol. 311, p. 494, 2001). Proteins encoded by other mRNAs were not affected. Thus, by using dsRNAs of the size produced by the ribonuclease involved in RNAi, this team of researchers was able to demonstrate that specific posttranscriptional gene silencing could be accomplished in mammalian cells.

These studies have led to speculation that small dsRNAs might be used clinically in the same way that ribozymes or antisense ODNs are currently being tested, that is, to destroy specific mRNAs that encode proteins responsible for human infection or disease. Even if RNAi is not found to have clinical value, it is likely to be of great importance in helping researchers determine the function of many of the genes whose existence and identity have been uncovered during genome sequencing studies. Once the nucleotide sequence of a particular gene has been determined, cells can be treated with a dsRNA expected to silence that gene, allowing researchers to determine the phenotypic effect of the absence of the corresponding protein. Unfortunately, the technology being used to study gene silencing has been caught up in a landslide of competing patent claims, which has clouded the future of this promising field of research (see *Nature*, vol. 417, p. 779, 2002).

Glossary

Many key terms and concepts are defined here. The parenthetical numbers following most definitions identify the chapter and section in which the term is first defined. For example, a term followed by (3.2) is first defined in Chapter 3, Section 2: "Enzymes." Terms defined in *The Human Perspective* or *Experimental Pathways* essays are identified parenthetically with HP or EP. For example, a term followed by (EP1) is first defined in the *Experimental Pathways* essay of Chapter 1.

A (aminoacyl) site Site into which aminoacyl-tRNAs enter the ribosome-mRNA complex. (11.5)

Absorption spectrum A plot of the intensity of light absorbed relative to its wavelength. (6.3)

Acetyl CoA A metabolic intermediate produced through catabolism of many compounds, including fatty acids, and used as the initial substrate for the central respiratory pathway, the TCA cycle. (5.2)

Acid hydrolases Hydrolytic enzymes with optimal activity at an acid pH. (8.6)

Acquired immune response A specific response to a pathogen that requires previous exposure to that agent. Includes responses mediated by anitbodies and T lymphocytes. (17)

Actin-binding proteins Any of nearly 100 different proteins belonging to numerous families that affect the assembly of the actin filaments, their physical properties, and their interactions with one another and with other cellular organelles. (9.7)

Action potential The collective changes in membrane potential, beginning with depolarization to threshold and ending with return to resting potential, that occur with stimulation of an excitable cell and act as the basis for neural communication. (4.8)

Activation energy The minimal kinetic energy needed for a reactant to undergo a chemical reaction. (3.2)

Active site The part of an enzyme molecule that is directly involved in binding substrate. (3.2)

Active transport The energy-requiring process in which a substance binds to a specific transmembranous protein, changing its conformation to allow passage of the substance through the membrane against the electrochemical gradient for that substance. (4.7)

Adaptors Protein complexes that cover the cytosol-facing surface of clathrin-coated vesicle membranes of the plasma membrane and *trans* Golgi network. (8.4)

Adenosine triphosphate (ATP) Nucleotide consisting of adenosine bonded to three phosphate groups; it is the principal immediate-energy source for prokaryotic and eukaryotic cells. (3)

Adherens junctions (zonulae adherens) Adherens junctions are a type of specialized adhesive junction particularly common in epithelia. The plasma membranes in this region are separated by 20 to 35 nm and are sites where cadherin molecules are concentrated. The cells are held together by linkages between the extracellular domains of cadherin molecules that bridge the gap between neighboring cells. (7.3)

Aerobes Organisms dependent on the presence of oxygen to utilize energy-rich compounds. (5.1)

Affinity chromatography Protein purification technique that utilizes the unique structural properties of a protein that allow the molecule to be specifically withdrawn from solution while all other molecules remain in solution. The solution is passed through a column in which a specific interacting molecule (such as a substrate, ligand, or antigen) is immobilized by linkage to an inert material (the matrix). (17.7)

Alleles Alternate forms of the same gene. (10.1)

Allosteric modulation Modification of the activity of an enzyme through interaction with a compound that binds to a site (i.e., allosteric site) other than the active site. (3.3)

Alpha (α) helix One possible secondary structure of polypeptides, in which the backbone of the chain forms a spiral (i.e., helical) conformation. (2.5)

Alpha particle A particle released by atoms during their disintegration, consisting of two protons and two neutrons that is equivalent to the nucleus of a helium atom. (18.4)

Alternate splicing Widespread mechanism by which a single gene can encode two or more related proteins. (12.3)

Amide bond The chemical bond that forms between carboxylic acids and amines (or acidic and amino functional groups) while producing a molecule of water. (2.4)

Amino acids The monomeric units of proteins, each is composed of three functional groups attached to a central, α carbon: an amino group, a defining side chain, and a carboxyl group. (2.5)

Aminoacyl-tRNA synthetase An enzyme that covalently links amino acids to the 3′ ends of their cognate tRNA(s). Each amino acid is recognized by a specific aminoacyl-tRNA synthetase. (11.4)

Amphipathic The biologically important property of a molecule having both hydrophobic and hydrophilic regions. (2.5)

Amphoteric Structural property allowing the same molecule to act as an acid or as a base. (2.3)

Anabolic pathway A metabolic pathway resulting in the synthesis of relatively complex products. (3.3)

Anaerobes Organisms that utilize energy-rich compounds through oxygen-independent metabolic pathways such as glycolysis and fermentation. (5.1)

Anaphase The stage of mitosis during which sister chromatids separate from one another. (14.2)

Anaphase A The movement of the chromosomes toward the poles during mitosis. (14.2)

Anaphase B Elongation of the mitotic spindle, which causes the two spindle poles to move farther apart. (14.2)

Angstrom The nonmetric unit, equivalent to 0.1 nm, that is still in use to describe atomic and molecular dimensions. (1.3)

Animal models Laboratory animals that exhibit characteristics of a particular human disease. (HP17)

Anion An ionized atom or molecule with a net negative charge. (2.1)

Antenna The light harvesting molecules of a photosynthetic unit that trap photons of varying wavelengths and transfer the energy of excitation to the pigment molecule at the reaction center. (6.4)

Antibody A globular protein produced by plasma cells derived from B lymphocytes that interacts with the surface of a pathogen or foreign substance to facilitate its destruction. (17.2)

Antigen Any substance recognized by an immune system as foreign to the organism. (17.2)

Antigen-presenting cells (APCs) Cells that display portions of protein antigens at their surface. The portions are derived by proteolysis of larger antigens. The antigenic peptides are displayed in conjunction with MHC molecules. Virtually any cell in the body can function as an APC by displaying peptides in combination with MHC class I molecules, which provides a mechanism for the destruction of infected cells. In contrast, macrophages, dendritic cells, and B cells are referred to as "professional" APCs because they function to ingest materials, process them, and display them to T_H cells in combination with MHC class II molecules. (17.4)

Antiserum Fluid containing desired antibodies that remains after the removal of the cells and clotting factors from whole blood that has been exposed to given antigen. (18.14)

Apoptosis A type of orderly, or programmed, cell death in which the cell responds to certain signals by initiating a normal response that leads to the death of the cell. Death by apoptosis is characterized by the overall compaction of the cell and its nucleus, the orderly dissection of the chromatin into pieces at the hands of a special DNA-splitting endonuclease, and the rapid engulfment of the dying cell by phagocytosis. (15.7)

Artifact An image seen in a microscope that results from the coagulation or precipitation of materials that had no structure in the living cell. (18.2)

Assay Some identifiable feature of a specific protein, such as the catalytic activity of an enzyme, used to determine the relative amount of that protein in a sample. (18.7)

Aster "Sunburst" arrangement of microtubules around each centrosome during prophase. (14.2)

ATP synthase The ATP-synthesizing enzyme of the inner mitochondrial membrane, which is composed of two chief components: the F_1 headpiece and the F_0 basal piece, the latter of which is embedded in the membrane itself. (5.5)

Autoantibodies Antibodies capable of reacting with the body's own tissues. (HP17)

Autoimmune disorders Diseases characterized by an attack of the immune system against the body's own tissues. Includes multiple sclerosis, insulin dependent diabetes mellitus, and rheumatoid arthritis. (HP17)

Autophagy The destruction of organelles and their replacement during which an organelle is surrounded by a membrane donated by the endoplasmic reticulum. The membrane surrounding the organelle then fuses with a lysosome. (8.6)

Autoradiography A technique for visualizing biochemical processes by allowing an investigator to determine the location of radioactively labeled materials within a cell. Tissue sections containing radioactive isotopes are covered with a thin layer of photographic emulsion, which is exposed by radiation emanating from the tissue. Sites in the cells containing radioactivity are revealed under the microscope by silver grains after development of the overlying emulsion. (8.2)

Autotroph Organism capable of surviving on CO_2 as its principal carbon source. (6.1)

B lymphocytes (B cells) Lymphocytes that respond to antigen by proliferating and differentiating into plasma cells that secrete blood-borne antibodies. These cells attain their differentiated state in the bone marrow. (17.2)

Bacterial artificial chromosome (BAC) Cloning vector capable of accepting large foreign DNA fragments that can be cloned in bacteria. Consists of an F plasmid with an origin of replication and the genes required to regulate replication. BACs have played key role in genome sequencing. (18.13)

Bacteriophages A group of viruses that require bacteria as host cells. (1.4)

Base composition analysis Determination of the relative amounts of each base in samples of DNA. (10.2)

Basement membrane (basal lamina) Thickened layer of approximately 50 to 200 nm of extracellular matrix that surrounds muscle and fat cells and underlies the basal surface of epithelial tissues such as the skin, the inner lining of the digestive and respiratory tracts, and the inner lining of blood vessels. (7.1)

Benign tumor A tumor composed of cells that are no longer responsive to normal growth controls but lack the capability to invade normal tissues or metastasize to distant sites. (16.3)

Beta particle A particle released by atoms during their disintegration, equivalent to an electron. (18.4)

Beta (β) pleated sheet One possible secondary structure of a polypeptide, in which several β-strands lie parallel to each other, creating the conformation of a sheet. (2.5)

Beta (β) strand One possible secondary structure of a polypeptide, in which the backbone of the chain assumes a folded (or pleated) conformation. (2.5)

Biochemicals Compounds synthesized by living organisms. (2.4)

Biosynthetic pathway (secretory pathway) Route through the cytoplasm by which materials are synthesized in the endoplasmic reticulum or Golgi complex, modified during passage through the Golgi complex, and transported within the cytoplasm to various destinations such as the plasma membrane, a lysosome, or a large vacuole of a plant cell. The alternative term secretory pathway has been used because many of the materials synthesized in the pathway are destined to be discharged outside the cell.

Bivalent (tetrad) The complex formed during meiosis by a pair of synapsed homologous chromosomes. (14.3)

Bright-field microscope A microscope in which light from the illuminating source is caused to converge on the specimen by the substage condenser, thereby forming a cone of bright light that can enter the objective lens. (18.1)

Buffers Compounds that can interact with either free hydrogen or hydroxyl ions, minimizing a change in pH. (2.3)

C₃ pathway (Calvin cycle) The metabolic pathway by which carbon dioxide is assimilated into the organic molecules of the cell during photosynthesis. Ribulose 1,5-bisphosphate (RuBP) is used as the initial CO_2 acceptor, which then fragments into two three-carbon PGA molecules. (6.6)

C4 (Hatch-Slack) pathway Alternate pathway for carbon fixation utilizing phosphoenolpyruvate combined with CO_2 to produce four-carbon compounds (predominantly malate and oxaloacetate). (6.6)

C₄ plants Plants, primarily tropical grasses, that use the C_4 carbon fixation pathway. (6.6)

Cadherins A family of related glycoproteins that mediate Ca^{2+}-dependent cell adhesion. (7.3)

Calcium-binding proteins Proteins, such as calmodulin, that bind calcium and allow the calcium to elicit a variety of cellular responses. (15.2)

Calmodulin A small, calcium-binding protein that is widely distributed. Each molecule of calmodulin contains four binding sites for calcium. (15.3)

CAM plants Plants that utilize PEP carboxylase in fixing CO_2 just as C_4 plants do, but conduct the light-dependent reactions and carbon fixation at different times of the day, so that stomata can be closed during the peak water loss hours of the day. (6.6)

Carbohydrates Organic molecules including simple sugars (saccharides) and multisaccharide polymers, which largely serve as energy-storage and structural compounds in cells. (2.5)

Catabolic pathway A metabolic pathway in which relatively complex molecules are broken down into simpler products. (3.3)

Cation An ionized atom or molecule with a net positive charge. (2.1)

Cell culture The technique used to grow cells outside the organism. (18.5)

Cell cycle The stages through which a cell passes from one cell division to the next. (14.1)

Cell division The process by which new cells originate from other living cells. (14)

Cell fractionation Bulk separation of the various cell organelles by differential centrifugation. (8.2)

Cell fusion Technique whereby two different types of cells (from one organism of from different species) are joined to produce one cell with one, continuous plasma membrane. (4.5)

Cell plate Structure between the cytoplasm of two newly formed daughter cells that gives rise to a new cell wall in plant cells. (7.6)

Cell signaling Communication in which information is relayed across the plasma membrane to the cell interior and often to the cell nucleus by means of a series of molecular interactions. (15)

Cell theory Theory of biological organization, which has three tenets: all organisms are made up of one or more cells; the cell is the structural basis of life; cells only arise from the division of preexisting cells. (1.1)

Cell wall A rigid, nonliving structure that provides support and protection for the cell it surrounds. (7.6)

Cellulose Unbranched glucose polymer with β(1–4) linkages that tends to aggregate in a cablelike formation and serves as a principal structural element of plant cell walls. (2.5)

Centrioles Cylindrical structures, about 0.2 μm across and typically about twice as long, that contain nine evenly spaced fibrils, each of which appears in cross section as a band of three microtubules. Centrioles are nearly always found in pairs, with each of the members situated at right angles to one another. (9.3)

Centromere Marked indentation on a mitotic chromosome that serves as the site of kinetochore formation. (14.2)

Centrosome A complex structure that contains two barrel-shaped centrioles surrounded by amorphous, electron-dense pericentriolar material where microtubules are nucleated. (9.3)

Chaperones Proteins that bind to other polypeptides, preventing their aggregation and promoting their folding and/or assembly into multimeric proteins. (EP2)

Chaperonins Members of the Hsp60 class of chaperones, e.g., GroEL, that form a cylindrical complex of 14 subunits within which the polypeptide folding reaction takes place. (EP2)

Chemiosmosis The mechanism for ATP synthesis whereby the movement of electrons through the electron-transport chain results in establishment of a proton gradient across the inner mitochondrial membrane, with the gradient acting as a high-energy intermediate, linking oxidation of substrates to the phosphorylation of ADP. (EP5)

Chemoautotroph An autotroph that utilizes the energy stored in inorganic molecules (such as ammonia, hydrogen sulfide, or nitrites) to convert CO_2 into organic compounds. (6.1)

Chiasmata Specific points of attachment between homologous chromosomes of bivalents, observed as the homologous chromosomes move apart at the beginning of the diplotene stage of meiotic prophase 1. The chiasmata are generally located at the site on the chromosomes at which genetic exchange during crossing over had previously occurred. (14.3)

Chlorophyll The most important light-absorbing photosynthetic pigment. (6.3)

Chloroplast A specialized membrane-bound cytoplasmic organelle that is the principal site of photosynthesis in eukaryotic cells. (6.1)

Cholesterol Sterol found in animal cells that can constitute up to half of the lipid in a plasma membrane, with the relative proportion in any membrane affecting its fluid behavior. (4.3)

Chromatid Paired, rod-shaped members of mitotic chromosomes that together represent the duplicated chromosomes formed during replication in the previous interphase. (14.2)

Chromatin Fibers of DNA and associated protein that constitute chromosomes. (12.1)

Chromatography A term used for a wide variety of techniques in which a mixture of dissolved components is fractionated as it moves through some type of immobile matrix. (18.7)

Chromosomes Threadlike strands that are composed of the nuclear DNA of eukaryotic cells and are the carriers of genetic information. (10.1)

Ciliary dynein A huge protein (up to 2 million daltons) responsible for conversion of the chemical energy of ATP into the mechanical energy of ciliary locomotion. (9.3)

Cis **cisternae** The cisternae of the Golgi complex closest to the endoplasmic reticulum. (8.4)

Cisternal (luminal) space The region of the cytoplasm enclosed by the membranes of the endoplasmic reticulum. (8.3)

Clonal culture A type of primary culture in which relatively few cells are added to a dish, each of which, after settling and attaching to the surface, is at some distance from its neighbors. The proliferation of the cells generates individual colonies, or clones, of cells whose members are all derived from the same original cell. (18.5)

Clonal selection theory Well supported theory that B and T lymphocytes develop their ability to produce specific antibodies or T cell receptors prior to exposure to antigen. Should an antigen then enter the body, it can interact specifically with those B and T lymphocytes bearing complementary receptors. Interaction between the antigen and the B or T lymphocytes leads to proliferation of the lymphocyte to form a clone of cells capable of responding to that specific antigen. (17.2)

Codons Sequences of three nucleotides (nucleotide triplets) that specify amino acids. (11.4)

Coenzyme An organic, nonprotein component of an enzyme. (3.2)

Cofactor The nonprotein component of an enzyme, it can be either inorganic or organic. (3.2)

Collagens A family of fibrous glycoproteins known for their high tensile strength that function exclusively as part of the extracellular matrix. (7.1)

Competitive inhibitor An enzyme inhibitor that competes with substrate molecules for access to the active site. (3.2)

Complement A system of blood-plasma proteins that act as part of the innate immune system to destroy invading microorganisms either directly (by making their plasma membrane porous) or indirectly (by making them susceptible to phagocytosis). (17.1)

Complementarity The relationship between the sequence of bases in the two strands of the double helix of DNA. Structural restrictions on the configurations of the bases limits bonding to the two pairs: adenine-thymine and guanine-cytosine. (10.2)

Conductance The movement of small ions across membranes. (4.7)

Confocal scanning microscope A microscope in which the specimen is illuminated by a finely focused laser beam that rapidly scans across the specimen at a single depth, thus illuminating only a thin plane (or "optical section") within the object. The microscope is typically used with fluorescently stained specimens, and the light emitted from the illuminated optical section is used to form an image of the section on a video screen. (17.1)

Conformation The three-dimensional arrangement of the atoms within a molecule, often important in understanding the biological activity of proteins and other molecules in a living cell. (2.5)

Conformational change A predictable movement within a protein that is associated with biological activity. (2.5)

Conjugate acid Paired form created when a base accepts a proton in an acid-base reaction. (2.3)

Conjugate base Paired form created when an acid loses a proton in an acid-base reaction. (2.3)

Conjugated protein Protein linked covalently or noncovalently to substances other than amino acids, such as metals, nucleic acids, lipids, and carbohydrates. (2.5)

Connexon Multisubunit complex of a gap junction formed from the clustering within plasma membrane of integral membrane protein called connexin. Each connexon is composed of six connexin subunits arranged around a central opening (or annulus) about 1.5 nm in diameter. (7.5)

Conserved sequences Refers to the amino acid sequences of particular polypeptides or the nucleotide sequences of particular nucleic acids. If two sequences are similar to one another, i.e., homologous, they are said to be conserved, which indicates that they have not diverged very much from a common ancestral sequence over long periods of evolutionary time. Conserved sequences are ones that are rendered nonfunctional by nucleotide or amino acid substitutions. (2.4)

Constant regions The portions of light and heavy antibody polypeptide chains that have the same amino acid sequence. (17.4)

Constitutive heterochromatin Chromatin that remains in the condensed state in all cells at all times and, thus, represents DNA that is permanently silenced. It consists primarily of highly repeated sequences. (12.1)

Constitutive secretion Discharge of materials synthesized in the cell into the extracellular space in a continual, unregulated manner. (8.1)

Contrast The difference in appearance between adjacent parts of an object or an object and its background. (18.1)

Cotransport A process that couples the movement of two solutes across a membrane, termed symport if the two solutes are moved in the same direction and antiport if the two move in opposite directions. (4.7)

Covalent bond The type of chemical bond in which electron pairs are shared between two atoms. (2.1)

Cristae The many deep folds that are characteristic of the inner mitochondrial membrane, which contain the molecular machinery of oxidative phosphorylation. (5.1)

Crossing over (genetic recombination) Reshuffling of the genes on chromosomes (thereby disrupting linkage groups) that occurs during meiosis as a result of breakage and reunion of segments of homologous chromosomes. (10.1)

Cyanobacteria Evolutionarily important and structurally complex prokaryotes that possess photosynthetic cytoplasmic membranes. (1.3)

Cyclic photophosphorylation The formation of ATP in chloroplasts that is carried out by PSI independent of PSII. (6.5)

Cyclin-dependent kinases (Cdks) Enzymes that control progression of cells through the cell cycle. (14.1)

Cytochromes A type of electron carrier in which a protein is bound to a heme group. (5.3)

Cytokine Proteins secreted by cells of the immune system that alter the behavior of other immune cells. (17.3)

Cytokinesis The part of the cell cycle during which physical division of the

cell into two daughter cells occurs. (14.2)

Cytoplasmic dynein A huge protein (molecular mass over 1 million daltons) composed of numerous polypeptide chains. The molecule contains two large globular heads that act as force-generating engines. Evidence suggests that cytoplasmic dynein functions in the movement of chromosomes during mitosis and also as a minus end-directed microtubular motor for the movement of vesicles and membranous organelles through the cytoplasm. (9.3)

Cytoskeleton An elaborate interactive network composed of three well-defined filamentous structures: microtubules, microfilaments, and intermediate filaments. These elements function to provide structural support; an internal framework responsible for positioning the various organelles within the interior of the cell; as part of the machinery required for movement of materials and organelles within cells; as force-generating elements responsible for the movement of cells from one place to another; as sites for anchoring messenger RNAs and facilitating their translation into polypeptides; and as a signal transducer, transmitting information from the cell membrane to the cell interior. (9)

Cytosolic space The region of fluid content of the cytoplasm outside of the membranous organelles. (8.3)

Dalton A measure of molecular mass, with one dalton equivalent to one unit of atomic mass (e.g., the mass of a ^1H atom). (1.4)

Dehydrogenase An enzyme that catalyzes a redox reaction by removing a hydrogen atom from one reactant. (3.3)

Deletion A chromosomal aberration that occurs when a portion of a chromosome is missing. (HP12)

Deletion Loss of a segment of DNA caused by the misalignment of homologous chromosomes during meiosis. (10.4)

Denaturation Separation of the DNA double helix into its two component strands. (10.3)

Depolarization A decrease in the electrical potential difference across a membrane. (4.8)

Desmosomes (maculae adherens) Disc-shaped adhesive junction containing cadherins found in a variety of tissues, but most notably epithelia, where they are located basal to the ad-

herens junction. Dense cytoplasmic plaques on the inner surface of the plasma membranes in this region serve as sites of anchorage for looping intermediate filaments that extend into the cytoplasm. (7.3)

Differential centrifugation A technique used to isolate a particular organelle in bulk quantity, which depends on the principle that, as long as they are more dense than the surrounding medium, particles of different size and shape travel toward the bottom of a centrifuge tube at different rates when placed in a centrifugal field. (18.6)

Differentiation The process through which unspecialized cells become more complex and specialized in structure and function (1.3)

Diffusion The spontaneous process in which a substance moves from an area of higher concentration to one of lower concentration, eventually reaching the same concentration in all areas. (4.7)

Direct immunofluorescence Technique using antibodies as intracellular localization probes for an antigen, in which antibodies are conjugated with small fluorescent molecules to form derivatives that are then incubated with the cells or sections of cells. The binding sites are then visualized with the fluorescence microscope. (18.13)

DNA cloning A broad-based technique for producing large quantities of a specific DNA segment. (18.12)

DNA gyrase A type of topoisomerase that is able to change the state of supercoiling in a DNA molecule by relieving the tension that builds up during replication. It does this by traveling along the DNA and acting like a "swivel," changing the positively supercoiled DNA into negatively supercoiled DNA. (13.1)

DNA library Population of DNA fragments representing the entire genome of an organism, or some other collection. (18.13)

DNA ligase The enzyme responsible for joining DNA fragments into a continuous strand. (13.1)

DNA microarrays These are prepared by spotting DNAs from different genes in a known, ordered location on a glass slide. The slide is then incubated with fluorescently labeled cDNAs whose hybridization level provides a measure of the level of expression of each gene in the array. (12.3)

DNA polymerases The enzymes responsible for constructing new DNA strands during replication or DNA repair. (13.1)

DNA-dependent RNA polymerases (RNA polymerases) The enzymes responsible for transcription in both prokaryotic and eukaryotic cells. (11.2)

DNA tumor viruses Viruses capable of infecting vertebrate cells, transforming them into cancer cells. DNA viruses have DNA in the mature virus particle. (16.2)

Dolichol phosphate Hydrophobic molecule built from more than 20 isoprene units that assembles the basal, or core, segment of carbohydrate chains for manufacture of glycoproteins. (8.3)

Domain A region within a protein that folds and functions in a semi-independent manner. (2.5)

Duplication An extra segment of DNA caused by the misalignment of homologous chromosomes during meiosis so that one chromosome receives both copies of a gene. (10.4)

Effector A substance that brings about a cellular response to a signal. (15.2)

Electrochemical gradient The overall difference in electrical charge and in solute concentration that determines the ability of an electrolyte to diffuse between two compartments. (4.7)

Electrogenic Any process that contributes directly to a separation of charge across a membrane. (4.7)

Electron-transfer potential The relative affinity for electrons, such that a compound with a low affinity has a high potential to transfer one or more electrons in a redox reaction (and thus act as a reducing agent). (5.3)

Electron-transport chain Membrane-embedded electron carriers that accept high-energy electrons and sequentially lower the energy state of the electrons as they pass through the chain, with the net results of capturing energy for use in synthesizing ATP or other energy-storage molecules. (5.3)

Electronegativity The relative tendency of an atom, when forming bonds, to attract electrons belonging to other atoms. (2.1)

Electrophoresis Protein fractionation techniques that depend on the ability of charged molecules to migrate when placed in an electric field. (18.7)

Embryonic stem cells A type of cell that has virtually unlimited powers of differentiation, found in the mammalian blastocyst, which is an early stage of embryonic development comparable to the blastula of other animals. (HP1)

Endergonic reactions Reactions that are thermodynamically unfavorable and cannot occur spontaneously, possessing a $+\Delta G$ value. (3.1)

Endocytic pathway Route for moving materials from outside the cell (and from the membrane surface of the cell) to compartments, such as endosomes and lysosomes, located within the cell interior. (8.8)

Endomembrane system Functionally and structurally interrelated group of membranous cytoplasmic organelles including the endoplasmic reticulum, Golgi complex, endosomes, lysosomes, and vacuoles. (8)

Endoplasmic reticulum (ER) A system of tubules, cisternae, and vesicles that divides the fluid content of the cytoplasm into a luminal space within the ER membrane and a cytosolic space outside the membranes. (8.3)

Endosymbiont theory Proposal, which is based on considerable evidence, that mitochondria and chloroplasts arose from symbiotic prokaryotes that took up residence within a primitive host cell. (EP1)

Endothermic reactions Those gaining heat under conditions of constant pressure and volume. (3.1)

Energy The capacity to do work, it exists in two forms: potential and kinetic. (3.1)

Enthalpy change (ΔH) The change during a process in the total energy content of the system. (3.1)

Entropy (S) A measure of the relative disorder of the system or universe associated with random movements of matter; because all movements cease at absolute zero ($0°K$), entropy is zero only at that temperature. (3.1)

Enzyme inhibitor Any molecule that can bind to an enzyme and decrease its activity, classified as noncompetitive or competitive based on the nature of the interaction with the enzyme. (3.2)

Enzyme-substrate complex The physical association between an enzyme and its substrate(s), during which catalysis of the reaction takes place. (3.2)

Enzymes The vitally important protein catalysts of cellular reactions. (3.2)

Epitope (or antigenic determinant) Portion of an antigen that binds to the antigen-combining site of a specific antibody. (17.4)

Equilibrium constant of a reaction (K_{eq}) The ratio of product concentrations to reactant concentrations when a reaction is at equilibrium. (3.1)

Ester bond The chemical bond that forms between carboxylic acids and alcohols (or acidic and alcoholic functional groups) while producing a molecule of water. (2.4)

Euchromatin Chromatin that returns to its dispersed state during interphase. (12.1)

Eukaryotic cells Cells (e.g., plant, animal, protist, fungal) characterized by an internal structure based on organelles such as the nucleus, derived from *eu-karyon*, or true nucleus. (1.3)

Excitation-contraction coupling The steps that link the arrival of a nerve impulse at the muscle plasma membrane to the shortening of the sarcomeres deep within the muscle fiber. (9.6)

Excited state Electron configuration of a molecule after absorption of a photon has energized an electron to shift from an inner to an outer orbital. (6.3)

Exergonic reactions Reactions that are thermodynamically favorable, possessing a $-\Delta G$ value. (3.1)

Exocytosis The process of membrane fusion and content discharge during which the membrane of a secretory granule comes into contact with the overlying plasma membrane with which it fuses, thereby forming an opening through which the contents of the granule can be released. (8.5)

Exon shuffling The movement of genetic "modules" among different genes. (11.3)

Exons Those parts of a split gene that contribute to a mature RNA product. (11.3)

Exothermic reactions Those releasing heat under conditions of constant pressure and volume. (3.1)

Extracellular matrix An organized network of extracellular materials that is present beyond the immediate vicinity of the plasma membrane. It may play an integral role in determining the shape and activities of the cell. (7.1)

Facilitated diffusion Process by which the normal diffusion rate of a substance is increased through interaction with a substance-specific membrane protein. (4.7)

Facilitative transporter A transmembrane protein that binds a specific substance and, in so doing, changes conformation so as to facilitate diffusion of the substance down its concentration gradient. (4.7)

Facultative heterochromatin Chromatin that has been specifically inactivated during certain phases of an organism's life. (12.1)

Fats Polymers consisting of a glycerol backbone linked by ester bonds to three fatty acids, also termed triacylglycerols. (2.5)

Fatty acid Long, unbranched hydrocarbon chain with a carboxylic acid group at one end. (2.5)

Feedback inhibition A mechanism to control metabolic pathways where the end product interacts with the first enzyme in the pathway, resulting in inactivation of the enzyme. (3.3)

Fermentation An anaerobic metabolic pathway in which pyruvate is converted to a simpler molecule (often lactate or ethanol, depending on the organism) and NAD^+ is regenerated for use in the earlier metabolic pathway for glucose catabolism, glycolysis. (3.3)

Fibrous protein One with a tertiary structure that is greatly elongated, resembling a fiber. (2.5)

Fixative A chemical solution that kills cells by rapidly penetrating the cell membrane and immobilizing all of its macromolecular material in such a way that the structure of the cell is maintained as close as possible to that of the living state. (18.1)

Flavoproteins A type of electron carrier in which a polypeptide is bound to one of two related prosthetic groups, either FAD or FMN. (5.3)

Fluid-mosaic model Model presenting membranes as dynamic structures in which both lipids and associated proteins are mobile and capable of moving within the membrane to engage in interactions with other membrane molecules. (4.2)

Fluorescence recovery after photobleaching (FRAP) Technique to study movement of membrane components that consists of three steps: (1) linking membrane components to a fluorescent dye, (2) irreversibly bleaching (removing the visible fluorescence of) a portion of the membrane, (3) monitoring the reappearance of fluorescence (due to random movement of fluorescent-dyed components from outside of the bleached

area) in the bleached portion of the membrane. (4.5)

Focal adhesions (focal contacts) Adhesive structures characteristic of cultured cells adhering to the surface of a culture dish. The plasma membrane in the region of the focal adhesion contains clusters of integrins that connect the extracellular material that coats the culture dish to the actin-containing microfilament system of the cytoskeleton. (7.2)

Fractionated Disassembly of a preparation into its component ingredients so that the properties of individual species of molecules can be examined. (18.7)

Frameshift mutations Mutations in which a single base pair is either added to or deleted from the DNA, resulting in an incorrect reading frame from the point of mutation through the remainder of the coding sequence. (11.4)

Free energy change (ΔG) The change during a process in the amount of energy available to do work. (3.1)

Free radical Highly reactive atom or molecule that contains a single unpaired electron. (HP2)

Freeze-etch technique Analytic technique in which tissue is freeze fractured, then exposed briefly to a vacuum so a thin layer of ice can evaporate from above and below the fractured surfaces, exposing features for identification by electron microscopy. (18.2)

Freeze-fracture replication Analytic technique in which a tissue sample is first frozen and then struck with a blade that fractures the tissue block along the lines of least resistance, often resulting in a fracture line between the two leaflets of the lipid bilayer; metals are then deposited on the exposed surfaces to create a shadowed replica that is analyzed by electron microscopy. (4.4)

Functional groups Particular groupings of atoms that tend to act as a unit, often affecting the chemical and physical behavior of the larger organic molecules to which they belong. (2.4)

G protein A GTP-binding protein that acts as a switch to turn activities on and off. (15.1)

G protein-coupled receptors A group of over 100 related receptors that span the plasma membrane seven times. The binding of the ligand to its specific receptor causes a change in the conformation of the receptor that increases its affinity for the G protein initiating a response within the cell. (15.2)

G_1 Period within the cell cycle following mitosis and preceding the initiation of DNA synthesis. (14.1)

G_2 Period within the cell cycle between the end of DNA synthesis and the beginning of M phase. (14.1)

Gametophyte The haploid stage of the life cycle of plants that begins with spores generated during the sporophyte stage. During the gametophyte stage, gametes form through the process of mitosis. (14.3)

Gap junctions Sites between animal cells that are specialized for intercellular communication. Plasma membranes of adjacent cells come within about 3 nm of each other, and the gap is spanned by very fine "strands," or molecular "pipelines," that allow the passage of small molecules. (7.5)

Gated channel An ion channel that can change conformation between a form open to its solute ion and one closed to the ion; such channels can be voltage gated or chemical gated depending on the nature of the process that triggers the change in conformation. (4.7)

Gel filtration Purification technique in which separation of proteins (or nucleic acids) is based primarily on molecular mass. The separation material consists of tiny porous beads that are packed into a column through which the solution of protein slowly passes. (18.7)

Gene regulatory protein A protein that is capable of recognizing a specific sequence of base pairs within the DNA and binding to that sequence with high affinity thereby altering gene expression. (12.2)

Gene therapy The process by which a patient is cured by altering his or her genotype. (HP4)

General transcription factors Auxiliary proteins necessary for RNA polymerase to initiate transcription. These factors are termed "general" because the same ones are required for transcription of a diverse array of genes by the polymerase. (11.3)

Genes In nonmolecular terms, a unit of inheritance that governs the character of a particular trait. In molecular terms, a segment of DNA containing the information for a single polypeptide or RNA molecule, including transcribed but non-coding regions. (10.1)

Genetic code Manner in which the nucleotide sequences of DNA encode the information for making protein products. (11.4)

Genetic map Assignment of genetic markers to relative positions on a chromosome based on crossover frequency. (10.5)

Genetic recombination (crossing over) Reshuffling of the genes on chromosomes (thereby disrupting linkage groups) that occurs as a result of breakage and reunion of segments of homologous chromosomes. (10.1)

Genome The complement of genetic information unique to each species of organism. Equivalent to the DNA of a haploid set of chromosomes from that species. (10.3)

Globular protein One with a tertiary structure that is compact, resembling a globe. (2.5)

Glycocalyx A layer closely applied to the outer surface of the plasma membrane. It contains membrane carbohydrates along with extracellular materials that have been secreted by the cell into the external space, where they remain closely associated with the cell surface. (7.1)

Glycogen Highly branched glucose polymer that serves as readily available chemical energy in most animal cells and consists primarily of $\alpha(1-4)$ glycosidic bonds. (2.5)

Glycolipids Lipid molecules linked to carbohydrates, often active components of plasma membranes. (4.3)

Glycolysis The first pathway in the catabolism of glucose, it does not require oxygen and results in the formation of pyruvate. (3.3)

Glycosidic bond The chemical bond that forms between sugar molecules. (2.5)

Glyoxysomes Organelles found in plant cells that serve as sites for enzymatic reactions including the conversion of stored fatty acids to carbohydrate. (5.6)

Golgi complex Network of smooth membranes organized into a characteristic morphology, consisting of flattened, disk-like cisternae with dilated rims and associated vesicles and tubules. The Golgi complex functions primarily as a processing plant where proteins newly synthesized in the endoplasmic reticulum are modified in specific ways. (8.4)

Grana Orderly, stacked arrangement of thylakoids. (6.1)

GTPase-activating proteins (GAPs) Proteins that bind to G proteins, activating their GTPase activity. As a result, GAPs shorten the duration of a G protein-mediated response. (15.1)

Guanine nucleotide-dissociation inhibitors (GDIs) Proteins that bind to G proteins, inhibiting the dissociation of the bound GDP, thus maintaining the G protein in the inactive state. (15.1)

Guanine nucleotide-exchange factors (GEFs) Proteins that bind to a G protein, stimulating the exchange of a bound GDP with a GTP, thereby activating the G protein. (15.1)

Half-life A measure of the instability of a radioisotope, or, equivalently, the amount of time required for one-half of the radioactive material to disintegrate. (18.4)

Head group The polar, water-soluble region of a phospholipid that consists of a phosphate group linked to one of several small, hydrophilic molecules. (4.3)

Heavy chain One of the two types of polypeptide chains in an antibody, usually with a molecular mass of 50 to 70 kD. (17.4)

Helicase A protein that unwinds the DNA (or RNA) duplex in a reaction in which energy released by ATP hydrolysis is used to break the hydrogen bonds that hold the two strands together. (13.1)

Hemidesmosome Specialized adhesive structure at the basal surface of epithelial cells that functions to attach the cells to the underlying basement membrane. The hemidesmosome contains a dense plaque on the inner surface of the plasma membrane, with keratin-containing filaments coursing out into the cytoplasm. (7.2)

Hemolysis The permeabilization of red blood cell membranes, experimentally done through placement of cells in hypotonic solution, where they swell before bursting, releasing cellular contents and leaving membrane ghosts. (4.6)

Heterochromatin Chromatin that remains condensed during interphase. (12.1)

Heterogeneous nuclear RNAs (hnRNAs) A large group of RNA molecules that share the following properties: (1) they heave large molecular weights (up to about 80S, or 50,000 nucleotides); (2) they represent many different nucleotide sequences; and (3) they are found only in the nucleus. Includes pre-mRNAs. (11.3)

Heterotrimeric G protein A component of certain signal transduction systems, referred to as G proteins because they bind guanine nucleotides (either GDP or GTP), and described as heterotrimeric because all of them consist of three different polypeptide subunits. (15.2)

Heterotroph Organism that depends on an external source of organic compounds. (6.1)

High-performance liquid chromatography (HPLC) A high-resolution type of chromatography in which long, narrow columns are used, and the mobile phase is forced through a tightly packed matrix under high pressure. (18.7)

Highly repeated fraction Typically short (a few hundred nucleotides at their longest) DNA sequences that are present in at least 10^5 copies per genome. The highly repeated sequences typically account for about 10 percent of the DNA of vertebrates. (10.3)

Histones A collection of small, well-defined, basic proteins of chromatin. (12.1)

Homogenize To experimentally rupture cells. (8.2)

Homologous chromosomes Paired chromosomes of diploid cells, each carrying one of the two copies of the genetic material carried by that chromosome. (10.1)

Homologous sequences When the amino acid sequences of two or more polypeptides, or the nucleotide sequences of two or more genes, are similar to one another, they are presumed to have evolved from the same ancestral sequence. Such sequences are said to be *homologous*, a term reflecting an evolutionary relatedness. (2.4)

Humoral immunity Immunity mediated by blood-borne antibodies. (17.2)

Hybridomas Hybrid cells produced by fusion of a normal antibody-producing lymphocyte and a malignant myeloma cell. Hybridomas proliferate, producing large amounts of a single (monoclonal) antibody that was being synthesized by the normal cell prior to fusion with the myeloma cell. (18.13)

Hydrocarbons The simplest group of organic molecules, consisting solely of carbon and hydrogen. (2.4)

Hydrogen bond The weak, attractive interaction between a hydrogen atom covalently bonded to an electronegative atom (thus, with a partial positive charge) and a second electronegative atom. (2.2)

Hydrophilic The tendency of polar molecules to interact with surrounding water molecules, which are also polar; derived from "water loving." (2.2)

Hydrophobic interaction The tendency of nonpolar molecules to aggregate so as to minimize their collective interaction with surrounding polar water molecules; derived from "water fearing." (2.2)

Hypertonic Property of one compartment having a higher solute concentration compared with that in a given compartment. (4.7)

Hypervariable regions The subsections of antibody variable regions that vary more significantly in sequence from one molecule to another and are associated with antigen specificity. (17.4)

Hypotonic Property of one compartment having a lower solute concentration compared with that in a given compartment. (4.7)

Immune response Response elicited by cells of the immune system upon contact with foreign materials, including invading pathogens. Includes innate and acquired responses. Acquired immune responses can be divided into primary responses that follow initial exposure to an antigen and secondary responses that follow reexposure to that antigen. (17)

Immune system Physiological system consisting of organs, scattered tissues, and independent cells that protect the body from invading pathogens and foreign materials. (17)

Immunity State in which the body is not susceptible to infection by a particular pathogen. (17)

Immunoglobulin superfamily (IgSF) A wide variety of proteins that contain domains composed of 70 to 110 amino acids that are similar to domains that compose the polypeptide chains of blood-borne antibodies. (7.3)

Immunologic tolerance A state in which the body does not react against a particular substance, such as the body's own proteins, because cells that could engage in such a response have either been inactivated or destroyed. (17.1, HP 17)

Imprinted Differential expression of genes depending solely on whether they were contributed to the zygote by the sperm or the egg. (12.3)

In situ hybridization Technique to localize a particular DNA or RNA sequence in a cell or on a culture plate or electrophoretic gel. (10.3)

Indirect immunofluorescence A variation of direct immunofluorescence in which the cells are treated with unlabeled antibody, which is allowed to form a complex with the corresponding antigen. The location of the antigen-antibody couple is then revealed in a second step using a preparation of fluorescently labeled antibodies whose combining sites are directed against the antibody molecules used in the first step. (18.13)

Induced fit The conformational change in an enzyme after the substrate has been bound that allows the chemical reaction to proceed. (3.2)

Inducible operon An operon in which the presence of the key metabolic substance induces transcription of the structural genes. (12.2)

Inflammation The localized accumulation of fluid and leukocytes in response to injury or infection, leading to redness, swelling, and soreness. (17.1)

Initiation codon The triplet AUG, the site to which the ribosome attaches to the mRNA to assure that the ribosome is in the proper reading frame to correctly read the entire message. (11.4)

Initiation The first step in polypeptide chain synthesis in which the ribosome attaches to the mRNA at a fixed point so that the message is read in the correct reading frame. (11.6)

Innate immune response A nonspecific response to a pathogen that does not require previous exposure to that agent, includes responses mediated by NK cells, complement, phagocytes, and interferon. (17.1)

Insulin receptor substrates (IRSs) Protein substrates that, when phosphorylated in response to insulin, bind and activate a variety of "downstream" effectors. (15.3)

Integral protein A membrane-associated protein that penetrates or spans the lipid bilayer. (4.4)

Integrins A superfamily of integral membrane proteins that bind specifically to extracellular molecules. (7.2)

Intermediate filaments (IFs) Tough, ropelike cytoskeletal fibers approximately 10 nm in diameter that, depending on the cell type, may be composed of a variety of protein subunits capable of assembling into similar types of filaments. IFs are thought to provide mechanical stability to cells and provide specialized, tissue-specific functions. (9.4)

Interphase The portion of the cell cycle between periods of cell division. (14.1)

Intervening sequences (introns) Regions of DNA that are between coding sequences of a gene and that are therefore missing from corresponding mRNA. (11.3)

Introns Those parts of a split gene that correspond to the intervening sequences. (11.3)

Inversion A chromosomal aberration that results after a chromosome is broken in two places and the resulting center segment is reincorporated into the chromosome in reverse order. (HP12)

In vitro Outside the organism. Cells grown in culture are said to be grown in vitro, and studies on cultured cells are described as in vitro studies. (18.5)

Ion An atom or molecule with a net positive or negative charge because it has lost or gained one or more electrons during a chemical reaction. (2.1)

Ion channel A transmembranous structure (e.g., an integral protein with an aqueous pore) permeable to a specific ion or ions. (4.7)

Ion-exchange chromatography A technique for protein purification in which ionic charge is used to separate different proteins. (18.7)

Ionic bond A chemical bond occurring between oppositely charged ions, also called a salt bridge. (2.2)

Iron-sulfur proteins A group of protein electron carriers with an inorganic, iron-sulfur center. (5.3)

Irreversible inhibitor An enzyme inhibitor that binds tightly, often covalently, thus inactivating the enzyme molecule permanently. (3.2)

Isoelectric focusing A type of electrophoresis in which proteins are separated according to isoelectric point. (18.7)

Isoelectric point The pH at which the negative charges of the component amino acids of a protein equal the positive charges of the component amino acids, so the protein is neutral. (18.7)

Karyotype Image in which paired homologous chromosomes are ordered in decreasing size. (12.1)

Kinesin A large plus end-directed motor protein that moves membranous vesicles and other organelles along microtubules through the cytoplasm. (9.3)

Kinetic energy Energy released from a substance through atomic or molecular movements. (3.1)

Kinetochore A platelike structure situated at the outer surface of the centromere to which the microtubules of the spindle attach. (14.2)

Knockout mice Mice, born as the result of a series of experimental procedures, that are lacking a functional gene that they would normally contain. (18.13)

Lagging strand The newly synthesized daughter DNA strand that is synthesized discontinuously, so called because the initiation of each fragment must wait for the parental strands to separate and expose additional template. (13.1)

Lamellipodium The leading edge of a moving fibroblast, which is extended out from the cell as a broad, flattened, veil-like projection that glides over the substratum. (9.7)

Leading strand The newly synthesized daughter DNA strand that is synthesized continuously, so called because its synthesis continues as the replication fork advances. (13.1)

Ligand Any molecule that can bind to a receptor because it has a complementary structure. (15.1)

Light chain Smaller of the two types of polypeptide chains in an antibody, with a molecular mass of 23 kDa. (17.4)

Light-harvesting complex II (LHCII) Pigment-protein complex, located outside the photosystem itself, that contains most of the antenna pigments that collect light for PSII. It can also be associated with PSI. (6.4)

Light-dependent reactions First of two series of reactions that compose photosynthesis. In these reactions, energy from sunlight is absorbed and converted to chemical energy that is stored in ATP and NADPH. (6.2)

Light-independent reactions Second of two series of reactions that compose photosynthesis. In these reactions, carbohydrates are synthesized from carbon dioxide using the energy stored in the ATP and NADPH molecules formed in the light reactions. (6.2)

Limit of resolution The resolution attainable by a microscope is limited by

the wavelength of the illumination according to the equation $D = 0.61 \lambda/n\sin\alpha$, where D is the minimum distance that two points in the specimen must be separated to be resolved, λ is the wavelength of light, and n is the refractive index of the medium. Alpha is a measure of the light-gathering ability of the lens and is directly related to its aperture. For a light microscope, the limit of resolution is slightly less than 200 nm. (18.1)

Linkage groups Groups of genes that reside on the same chromosome casing nonindependent segregation of traits controlled by these genes. (10.1)

Lipid-anchored protein A membrane-associated protein that is located outside the bilayer but is covalently linked to a lipid molecule within the bilayer. (4.4)

Lipid bilayer Phospholipids self-assembled into a bimolecular structure based on hydrophobic and hydrophilic interactions; biologically important as the core organization of cellular membranes. (4.2)

Lipids Nonpolar organic molecules, including fats, steroids, and phospholipids, whose common property of not dissolving in water contributes to much of their biological activity. (2.5)

Liposome An artificial lipid bilayer that aggregates into a spherical vesicle or vesicles when in an aqueous environment. (4.3)

Liquid scintillation spectrometry A technique for measuring the amount of radioactivity present in a sample, based on the property of certain molecules, termed phosphors or scintillants, to absorb some of the energy of an emitted particle and release that energy in the form of light. (18.4)

Locus The position of a gene on a chromosome. (10.1)

Luminal (cisternal) space The region of fluid content of the cytoplasm enclosed by the membranes of the endoplasmic reticulum or Golgi complex. (8.3)

Lymphocytes Nucleated leukocytes (white blood cells) that circulate between the blood and lymph organs that mediate acquired immunity. Includes both B cells and T cells. (17)

Lysosomal storage disorders Diseases characterized by the deficiency of a lysosomal enzyme and the corresponding accumulation of undegraded substrate. (HP8)

M phase The part of the cell cycle that includes the processes of mitosis, during which duplicated chromosomes are separated into two nuclei, and cytokinesis, during which the entire cell is physically divided into two daughter cells. (14.1)

Macromolecules Large, highly organized molecules crucial to the structure and function of cells; divided into polysaccharides, lipids, proteins, and nucleic acids. (2.4)

Major histocompatibility complex (MHC) A region of the genome that encodes MHC proteins. The genes that encode these proteins tend to be highly polymorphic, being represented by a large number of different alleles. These genetic differences between humans account for the tendency of a person to reject a transplant from another person other than an identical twin. (17.4)

Mass culture A type of primary culture in which large numbers of cells are added to a culture dish, where they settle, attach to the bottom, and form a relatively uniform layer of cells. (18.5)

Maximal velocity (V_{max}) The highest rate achieved for a given enzymatically catalyzed reaction, it occurs when the enzyme is saturated with substrate. (3.2)

Medial cisternae The cisternae of the Golgi complex between the *cis* and *trans* cisternae. (8.4)

Meiosis The process during which the chromosome number is reduced so that cells are formed that contain only one member of each pair of homologous chromosomes. (14.3)

Membrane potential The electrical potential difference across a membrane. (4.8)

Messenger RNA (mRNA) The intermediate molecule between a gene and the polypeptide for which it codes. Messenger RNA is assembled as a complementary copy of one of the two DNA strands that encodes the gene. (11.1).

Metabolic intermediate A compound produced during one step of a metabolic pathway. (2.4)

Metabolic pathway A series of chemical reactions that results in the synthesis of an end product important to cellular function. (2.4)

Metabolism The total of the chemical reactions occurring within a cell or multicellular organism. (3.3)

Metaphase The stage of mitosis during which all of the chromosomes have become aligned at the spindle equa-

tor, with one chromatid of each chromosome connected to one pole and its sister chromatid connected to the opposite pole. (14.2)

Methylguanosine cap Modification of the 5′ end of an mRNA precursor molecule, so that the terminal "inverted" guanosine is methylated at the 7 position on its guanine base, while the nucleotide on the internal side of the triphosphate bridge is methylated at the 2′ position of the ribose. This cap prevents the 5′ end of the mRNA from being digested by nucleases, aids in transport of the mRNA out of the nucleus, and plays a role in the initiation of mRNA translation. (11.3)

MHC molecules Proteins encoded by the MHC region of the genome that bind processed antigens (antigenic peptides) and display them on the surface of the cell. Divided into two major classes, MHC class I molecules produced by virtually all cells of the body, and MHC class II molecules produced by "professional" APCs, such as macrophages and dendritic cells. (17.4)

Michaelis constant (K_M) In enzyme kinetics, the value equal to the substrate concentration present when reaction rate is one-half of the maximal velocity. (3.2)

Microfilaments Solid, 8-nm thick, cytoskeletal structures composed of a double-helical polymer of the protein actin. They play a key role in virtually all types of contractility and motility within cells. (9.5)

Microsomes A heterogeneous collection of vesicles formed from the endomembrane system (primarily the endoplasmic reticulum and Golgi complex) after homogenization. (8.2)

Microtubule-organizing centers (MTOCs) A variety of specialized structures that, because of their role in initiating microtubule formation, are called microtubule-organizing centers. (9.3)

Microtubule-associated proteins (MAPs) Proteins other than tubulin contained in microtubules obtained from cells. MAPs may interconnect microtubules to form bundles and can be seen as cross-bridges connecting microtubules to each other. Other MAPs may increase the stability of microtubules, alter their rigidity, or influence the rate of their assembly. (9.3)

Microtubules Hollow, cylindrical cytoskeletal structures, 25 nm in diameter, whose wall is composed of the

protein tubulin. Microtubules are polymers assembled from αβ-tubulin heterodimers that are arranged in rows, or protofilaments. Because of their rigidity, microtubules often act in a supportive capacity. (9.3)

Mitochondrial matrix The aqueous compartment within the interior of a mitochondrion. (5.1)

Mitochondrial membranes The outer membrane serves as a boundary with the cytoplasm and is relatively permeable, and the inner membrane houses respiratory machinery in its many invaginations and is highly impermeable. (5.1)

Mitochondrion The cellular organelle in which aerobic energy transduction takes place, oxidizing metabolic intermediates such as pyruvate to produce ATP. (5.1)

Mitosis Process of nuclear division in which duplicated chromosomes are faithfully separated from one another, producing two nuclei, each with a complete copy of all the chromosomes present in the original cell. (14.2)

Mitotic spindle Microtubule-containing "machine" that functions in the organization and sorting of duplicated chromosomes during mitotic cell division. (14.2)

Moderately repeated fraction DNA sequences that are repeated from a few to several hundred thousand times within a eukaryotic genome. The moderately repeated fraction of the DNA can vary from about 20 to about 80 percent of the total DNA. These sequences may be identical to each other or nonidentical but related. (10.3)

Motif A substructure found among many different proteins, such as the αβ barrel, which consists of β strands connected by an α-helical region. (2.5)

Monoclonal antibody A preparation of antibody molecules produced from a single colony (or clone) of cells. (18.14)

Multiprotein complex The interaction of more than one complete protein to form a larger, functional complex. (2.5)

Muscle fiber A skeletal muscle cell, referred to as a muscle fiber because of its highly ordered, multinucleated, cablelike structure composed of hundreds of thinner, cylindrical strands. (9.6)

Mutant An individual having an inheritable characteristic that distinguishes it from the wild type. (10.1)

Mutation A spontaneous change in a gene that alters it in a permanent fashion so that it causes heritable change. (10.1)

Myofibrils The thin, cylindrical strands that form muscle fibers. Each myofibril is composed of repeating linear arrays of contractile units, called sarcomeres, that give skeletal muscle cells their striated appearance. (9.6)

Natural killer (NK) cell A type of lymphocyte that engages in a nonspecific attack on an infected host cell, leading to apoptosis. (17.1)

Negative staining Procedures in which heavy-metal deposits are collected everywhere on the specimen grid of an electron microscope except in the locations of very small particulate materials, including high-molecular-weight aggregates such as viruses, ribosomes, multisubunit enzymes, cytoskeletal elements, and protein complexes. (18.1)

Nerve impulse The process through which an action potential is propagated along the membrane of a neuron by sequentially triggering action potentials in adjacent stretches of membrane. (4.8)

Neuromuscular junction The point of contact of a terminus of an axon with a muscle fiber, the neuromuscular junction is the site of transmission of nerve impulses from the axon across the synaptic cleft to the muscle fiber. (9.6)

Neurotransmitter A chemical that is released from a presynaptic terminal and binds to the postsynaptic, target cell, altering the membrane potential of the target cell. (4.8)

Nitrogen fixation The process through which nitrogen gas is chemically reduced and converted into a component of organic compounds. (1.3)

Noncompetitive inhibitor An enzyme inhibitor that does not bind at the same site as the substrate, and so the level of inhibition depends only on the concentration of inhibitor. (3.2)

Noncovalent bond A relatively weak chemical bond based on attractive forces between oppositely charged regions within a molecule or between two nearby molecules. (2.2)

Noncyclic photophosphorylation The formation of ATP during the process of oxygen-releasing photosynthesis in which electrons move in a linear path from H_2O to $NADP^+$. (6.5)

Nonpolar molecules Molecules with a symmetric distribution of charge because the component atoms have approximately the same electronegativities. (2.1)

Nonrepeated fraction Those sequences in the genome that are present in only one copy per haploid set of chromosomes. These sequences contain the greatest amount of genetic information including the codes for virtually all proteins other than histones. (10.3)

Nonsense mutations Mutations that produce stop codons within genes, thereby causing premature termination of the coded polypeptide chain. (11.5)

Nuclear envelope The complex, double-membrane structure that divides the eukaryotic nucleus from its cytoplasm. (12.1)

Nuclear envelope breakdown The disassembly of the nuclear envelope at the end of prophase. (14.2)

Nuclear lamina A dense, fibrillar meshwork composed of intermediate filaments that lines the inner surface of the nuclear envelope. (12.1)

Nuclear matrix Insoluble fibrillar protein network that crisscrosses through nuclear space. (12.1)

Nuclear pore complex (NPC) Complex, basketlike apparatus that fills the nuclear pore like a stopper, projecting outward into both the ctyoplasm and the nucleoplasm. (12.1)

Nucleic acid Polymers composed of nucleotides, which in living organisms are based on one of two sugars, ribose or deoxyribose, yielding the terms ribonucleic acid (RNA) and deoxyribonucleic acid (DNA). (2.5)

Nucleic acid hybridization A variety of related techniques that are based on the fact that two single-stranded nucleic acid molecules of complementary base sequence will form a double-stranded hybrid. (18.12)

Nucleoid The poorly defined region of a prokaryotic cell that contains its genetic material. (1.3)

Nucleoli Irregular-shaped nuclear structures that function as ribosome-producing organelles. (11.3)

Nucleosomes Repeating subunits of DNA and histones. Each nucleosome contains a nucleosome core particle, consisting of 146 base pairs of supercoiled DNA wrapped almost twice around a disc-shaped complex of eight histone molecules. The nucleosome core particles are connected to one another by a stretch of linker

DNA that is typically about 60 base pairs in length. (12.1)

Nucleotide The monomer of nucleic acids, each consists of three parts: a sugar (either ribose or deoxyribose), a phosphate group, and a nitrogenous base, with the phosphate linked to the sugar at the $5'$ carbon and the base at the $1'$ carbon. (2.5)

Nucleus The organelle that contains a eukaryotic cell's genetic material. (12.1)

Objective lens The lens of a light microscope that focuses light rays from the specimen to form a real, enlarged image of the object within the column of the microscope. (18.1)

Okazaki fragments Small segments of DNA that are rapidly linked to longer pieces that have been synthesized previously, to form the lagging strand. (13.1)

Oncogenes Genes that encode proteins that promote the loss of growth control and the conversion of the cell to a malignant state. These genes have the ability to transform cells. (16.3)

Operator Binding site for bacterial repressors that is situated between the polymerase binding site and the first structural gene. (12.2)

Operon A functional complex on a bacterial chromosome comprising a cluster of genes including structural genes, a promoter region, an operator region, and a regulatory gene. (12.2)

Organelles The organizationally and functionally diverse, membranous or membrane-bounded, intracellular structures that are the defining feature of eukaryotic cells. (1.3)

Osmosis The property of water passing through a semipermeable membrane from a region of lower solute concentration to one of higher solute concentration, with the tendency of eventually equalizing solute concentration in the two compartments. (4.7)

Oxidation The process through which an atom loses one or more electrons to another atom, in which the atom gaining electrons is considered to be reduced. (3.3)

Oxidation-reduction (redox) potential The separation of charge, measured in voltage, for any given pair of oxidizing-reducing agents, such as NAD^+ and NADH, relative to a standard (e.g., H^+ and H_2). (5.3)

Oxidation-reduction (redox) reaction One in which a change in the electronic state of the reactants occurs. (3.3)

Oxidative phosphorylation ATP formation driven by energy derived from high-energy electrons removed during substrate oxidation in pathways such as the TCA cycle, with the energy released for ATP formation by passage of the electrons through the electron-transport chain in the mitochondrion. (5.2)

Oxidizing agent The substance in a redox reaction that becomes reduced, causing the other substance to become oxidized. (3.3)

P (peptidyl) site Site in the ribosome from which the tRNA donates amino acids to the growing polypeptide chain. (11.5)

P680 The reaction center of photosystem II. "P" stands for pigment and "680" is the wavelength of light that this molecule absorbs most strongly. (6.5)

P700 The reaction center of photosystem I. "P" stands for pigment, and "700" is the wavelength of light that this molecule absorbs most strongly. (6.4)

Partition coefficient The ratio of a solute's solubility in oil to that in water, it is a measure of the relative polarity of a biological substance. (4.5)

Patch clamping A technique to study ion movement through ion channels that is accomplished by clamping, or maintaining voltage, across a patch of membrane by sealing a micropipette electrode to the surface and then measuring the current across that portion of membrane. (4.5)

Pathogen Any agent capable of causing infection or disease in a cell or organism. (2.5)

Peroxisomes Simple, membrane-bounded vesicles that participate in diverse enzymatic activities within eukaryotic cells, derived from *peroxisome*, or peroxide body, because they from hydrogen peroxide. (5.6)

Peptide bond The chemical bond linking amino acids in a protein, which forms when the carboxyl group of one amino acid reacts with the amino group of a second amino acid. (2.5)

Pericentriolar material Amorphous, electron-dense material that surrounds the centrioles in an animal cell. (9.3)

Peripheral protein A membrane-associated protein that is located entirely outside of the lipid bilayer and interacts with it through noncovalent bonds. (4.4)

pH The standard measure of relative acidity, it mathematically equals $-\log[H^+]$. (2.3)

Phase-contrast microscope A microscope that converts differences in refractive index into differences in intensity (relative brightness and darkness), which are then visible to the eye, making highly transparent objects more visible. (18.1)

Phosphatidylinositol 3-hydroxy kinase [or PI3K] One of the best-studied effectors containing an SH2 domain. The products of the enzyme serve as precursors for a number of inositol-containing cellular messengers that have diverse functions in cells. (15.3)

Phospholipase C An enzyme that catalyzes a reaction that splits PIP_2 into two molecules: inositol 1,4,5,-trisphosphate (IP_3) and diacylglycerol (DAG), both of which play important roles as second messengers in cell signaling. (15.2)

Phospholipid transfer proteins Proteins whose function is to transport specific phospholipids through the aqueous cytosol from one type of membrane compartment to another. (8.3)

Photoautotroph An autotroph that utilizes the radiant energy of the sun to convert CO_2 into organic compounds. (6.1)

Photon Packet of light energy. The shorter the wavelength, the greater the energy of the photons. (6.3)

Photorespiration A series of reactions in which O_2 is attached to RuBP, and eventually resulting in the release of recently fixed CO_2 from the plant. (6.6)

Photosynthesis The pathway converting the energy of sunlight into chemical energy that is usable by living organisms. (3.1)

Photosynthetic unit A group of about 300 chlorophyll molecules acting together to trap photons and transfer energy to the pigment molecule at the reaction center. (6.4)

Photosystem I (PSI) One of two spatially separated pigment complexes, which are necessary to boost the energy of a pair of electrons sufficiently to remove them from a molecule of water and transfer them to $NADP^+$. Photosystem I boosts electrons from an energy level at about the midway point to an energy level above $NADP^+$. (6.4)

Photosystem II (PSII) One of two spatially separated pigment complexes, which are necessary to boost the

energy of a pair of electrons sufficiently to remove them from a molecule of water and transfer them to $NADP^+$. Photosystem II boosts electrons from an energy level below that of water at the bottom of the energy trough to about the midway point. (6.4)

Phragmoplast Dense material roughly aligned in the equatorial plane of the previous metaphase plate in plant cells, consisting of clusters of interdigitating microtubules oriented perpendicular to the future cell plate, together with some type of associated electron-dense material. (14.2)

Physical map A map of chromosomes that utilizes cloned and sequenced DNA fragments rather than genetic markers. (10.5)

Pigments Molecules that contain a chromophore, a chemical group capable of absorbing light of particular wavelength(s) within the visible spectrum. (6.3)

Plasma cells Terminally differentiated cells that develop from B lymphocytes that synthesize and secrete large amounts of blood-borne antibodies. (17.2)

Plasma membrane The membrane serving as a boundary between the interior of a cell and its extracellular environment. (4.1)

Plasmodesmata Cylindrical cytoplasmic channels, 30 to 60 nm in diameter, that connect most plant cells and extend between adjacent cells directly through the cell wall. Plasmodesmata are lined with plasma membrane and usually contain a dense central rod, the desmotubule, derived from the endoplasmic reticulum of the two cells. (7.5)

Polar molecules Molecules with an uneven distribution of charge because the component atoms have different electronegativities. (2.1)

Poly(A) tail A string of adenosine residues at the 3′ end of an mRNA added posttranscriptionally. (11.3)

Polyacrylamide gel electrophoresis (PAGE) Protein fractionation technique in which the proteins are driven by an applied current through a gel composed of a small organic molecule (acrylamide) that is cross-linked to form a molecular sieve. (18.7)

Polymerase chain reaction (PCR) A technique in which a single region of DNA which may be present in vanishingly small amounts, can be amplified cheaply and rapidly. (18.13)

Polymorphic genes Genes that are known to exist in several allelic forms, such as those that determine blood type. (10.5)

Polytene chromosomes Giant chromosomes of insects that contain perfectly aligned, duplicated DNA strands, with as many as 1,024 times the number of DNA strands of normal chromosomes. (10.1)

Porins Integral proteins found in mitochondrial and chloroplast outer membranes that act as large, nonselective channels. (5.1)

Potential difference The difference in charge between two compartments, often measured as voltage across the separating membrane. (4.7)

Potential energy Stored energy that can be used to perform work. (3.1)

Preinitiation complex The assembled association of general transcription factors and RNA polymerase, required before transcription of the gene can be initiated. (11.3)

Primary culture Culturing of cells obtained directly from the organism. (18.5)

Primary electron acceptor Molecule that receives the photoexcited electron from reaction-center pigments in both photosystems. (6.4)

Primary structure The linear sequence of amino acids within a polypeptide chain. (2.5)

Primary transcript (pre-RNA) The initial RNA molecule synthesized from DNA, which is equivalent in length to the DNA from which it was transcribed. Primary transcripts typically have a fleeting existence, being processed into smaller, functional RNAs by a series of processing reactions. (11.3)

Primary walls The walls of a growing plant cell. They allow for extensibility. (7.6)

Primase Type of RNA polymerase that assembles the short RNA primers that begin the synthesis of each Okazaki fragment of the lagging strand. (13.1)

Primer The DNA strand that provides DNA polymerase with the necessary 3′ OH terminus. (13.1)

Prion An infectious agent associated with certain mammalian degenerative diseases that is composed solely of protein; derived from *pro*-teinacious *in*-fectious particle. (HP2)

Processing-level control Determination of the path by which a primary RNA transcript is processed into a messenger RNA that can be translated into a polypeptide. (12.3)

Progression Multistep process of malignant tumor generation involving a series of genetic alterations that makes the cells increasingly less responsive to the body's normal regulatory machinery and better able to invade normal tissues, which makes the tumor more life threatening. (16.3)

Prokaryotic cells Structurally simple cells of bacteria that do not have membrane-bounded organelles; derived from *pro-karyon*, or "before the nucleus." (1.3)

Prometaphase The phase of mitosis during which the definitive mitotic spindle is formed and the chromosomes are moved into position at the center of the cell. (14.2)

Promoter The site on the DNA to which an RNA polymerase molecule binds prior to initiating transcription. The promoter contains information that determines which of the two DNA strands is transcribed and the site at which transcription begins. (11.2)

Prophase The first stage of mitosis during which the duplicated chromosomes are prepared for segregation and the mitotic machinery is assembled. (14.2)

Prosthetic group A portion of a protein that is not composed of amino acids, such as the heme group within hemoglobin and myoglobin. (2.5)

Proteasome Barrel-shaped, multiprotein complex in which cytoplasmic proteins are degraded. Proteins selected for destruction are linked to ubiquitin molecules and threaded into the central chamber of the proteasome. (12.4)

Protein kinase An enzyme that transfers phosphate groups to other proteins, often having the effect of regulating the activity of the other proteins. (3.3)

Protein kinase C A family of related enzymes that are multifunctional, serine and threonine kinases, which are able to phosphorylate a wide variety of different proteins. (15.2)

Protein tyrosine kinase Enzyme that phosphorylates specific tyrosine residues of other proteins. (15.3)

Proteins Structurally and functionally diverse group of polymers built of amino acid monomers. (2.5)

Proteoglycan A protein-polysaccharide complex consisting of a core protein molecule to which chains of glycosaminoglycans are attached. Due to

the acidic nature of the glycosamino-glycans, proteoglycans are capable of binding huge numbers of cations, which in turn draw huge numbers of water molecules. As a result, proteoglycans form a porous, hydrated gel that acts like a "packing" material to resist compression. (7.1)

Proto-oncogenes A variety of genes that have the potential to subvert the cell's own activities and push the cell toward the malignant state. Proto-oncogenes encode proteins that have various functions in a cell's normal activities. Proto-oncogenes can be converted to oncogenes. (16.3)

Protofilaments Longitudinally arranged rows of globular subunits of a microtubule that are aligned parallel to the long axis of the tubule. (9.3)

Protoplast A naked plant cell whose cell wall has been digested away by the enzyme cellulase. (18.5)

Provirus The term for viral DNA when it has been integrated into its host cell's chromosome(s). (1.4)

Pseudogenes Sequences that are clearly homologous to functional genes, but have accumulated mutations that render them nonfunctional. (10.4)

Pseudopodia Broad, rounded protrusions formed during amoeboid movement as portions of the cell surface are pushed outward by a column of cytoplasm that flows through the interior of the cell toward the periphery. (9.7)

Purine A class of nitrogenous base found in nucleotides that has a double-ring structure, including adenine and guanine, which are found in both DNA and RNA. (2.5)

Pyrimidine A class of nitrogenous base found in nucleotides that has a single-ring structure, including cytosine and thymine, which are found in DNA, and cytosine and uracil, which are found in RNA. (2.5)

Quaternary structure The three-dimensional organization of a protein that consists of more than one polypeptide chain, or subunit. (2.5)

rDNA The DNA sequences encoding rRNA that are normally repeated hundreds of times and are typically clustered in one or a few regions of the genome. (11.3)

Reaction-center chlorophyll The single chlorophyll molecule of the 300 or so in the photosynthetic unit that is capable of transferring electrons to an electron acceptor. (6.4)

Reading frame The consecutive blocks of three nucleotides decoded, beginning from a particular site on the mRNA. (11.4)

Reannealing (renaturation) Reassociation of complementary single strands of a DNA double helix that had been previously denatured. (10.3)

Receptor Any substance that can bind to a specific molecule (ligand), often leading to uptake or signal transduction. (15.1)

Receptor tyrosine kinase (or RTK) Cell-surface receptor that can phosphorylate tyrosine residues on itself and/or on cytoplasmic substrates. It is involved primarily in the control of cell growth and differentiation. (15.3)

Recombinant DNA Molecules containing DNA sequences derived from more than one source. (18.13)

Reducing agent The substance in a redox reaction that becomes oxidized, causing the other substance to become reduced. (3.3)

Reducing power The potential in a cell to reduce metabolic intermediates into products, usually measured through the size of the NADPH pool. (3.3)

Reduction The process through which an atom gains one or more electrons from another atom, in which the atom losing electrons is considered to be oxidized. (3.3)

Refractory period The brief period of time following the end of an action potential during which an excitable cell cannot be restimulated to threshold. (4.8)

Regulated secretion Discharge of materials synthesized in the cell that have been stored in membrane-bound secretory granules in the peripheral regions of the cytoplasm, occurring in response to an appropriate stimulus. (8.1)

Regulatory gene Gene that codes for a repressor protein. (12.2)

Renaturation (reannealing) Reassociation of complementary single strands of DNA double helix that had been previously denatured. (10.3)

Replica Metal-carbon cast of a tissue surface used in electron microscopy. Variations in the thickness of the metal in different parts of the replica cause variations in the number of penetrating electrons to reach the viewing screen. (18.2)

Replication Duplication of the genetic material. (13)

Replication foci Localization of active replication forks in the cell nucleus. There are about 50 to 250 sites, each of which contains approximately 40 replication forks incorporating nucleotides into DNA strands simultaneously. (13.1)

Replication forks The points at which the pair of replicated segment of DNA come together and join the nonreplicated segments. Each replication fork corresponds to a site where (1) the parental double helix is undergoing strand separation, and (2) nucleotides are being incorporated into the newly synthesized complementary strands. (13.1)

Repressor A gene regulatory protein that binds to DNA and inhibits transcription. (12.2)

Resolution The ability to see two neighboring points in the visual field as distinct entities. (18.1)

Resting potential The electrical potential difference measured for an excitable cell when it is not subject to external stimulation. (4.8)

Restriction endonucleases (restriction enzymes) Nucleases contained in bacteria that recognize short nucleotide sequences within duplex DNA and cleave the backbone at highly specific sites on both strands of the duplex. (10.5)

Restriction map A type of physical map of the chromosome based on the identification and ordering of sets of fragments generated by restriction enzymes. (10.5)

Retrotransposons Transposable elements that require reverse transcriptase for their movements. (10.4)

Ribozyme An RNA molecule that functions as a catalyst in cellular reactions. (2.5)

RNA polymerase I The transcribing enzyme found in eukaryotic cells that synthesizes the large (28S, 18S, and 5.8S) ribosomal RNAs. (11.3)

RNA polymerase II The transcribing enzyme found in eukaryotic cells that synthesizes messenger RNAs and most small nuclear RNAs. (11.3)

RNA polymerase III The transcribing enzyme found in eukaryotic cells that synthesizes the various transfer RNAs and the 5S ribosomal RNA. (11.3)

RNA splicing The process of removing the intervening DNA sequences (introns) from a primary transcript. (11.3)

RNA tumor viruses Viruses capable of infecting vertebrate cells, transforming

them into cancer cells. RNA viruses have RNA in the mature virus particle. (16.2)

Rough endoplasmic reticulum (RER) That part of the endoplasmic reticulum that has ribosomes attached. The RER appears as an extensive membranous organelle composed primarily of flattened sacs (cisternae) separated by a cytosolic space. RER functions include synthesis of secretory proteins, lysosomal proteins, integral membrane proteins, and membrane lipids. (8.3)

S phase The phase of the cell cycle in which replication occurs. (14.1)

Saltatory conduction Propagation of a nerve impulse when one action potential triggers another at the adjacent stretch of unwrapped membrane (i.e., propagating by causing the action potentials to jump form one node of Ranvier to the next). (4.8)

Sarcomeres Contractile units of myofibrils that are endowed with a characteristic pattern of bands and stripes that give skeletal muscle cells their striated appearance. (9.6)

Sarcoplasmic reticulum A system of cytoplasmic SER membranes in muscle cells that forms a membranous sleeve around the myofibril. (9.6)

Saturated fatty acids Those having only single bonds between carbons. (2.5)

Scanning electron microscopes (SEMs) Microscopes that utilize electrons that have bounced off the surface of the specimen to create an image. (18.2)

Second messenger A substance that is released into the interior of the cell as the result of the binding of a first messenger—a hormone or other ligand—to a receptor at the outer surface of the cell. (15.2)

Secondary culture Transfer of previously cultured cells to a culture medium. (18.5)

Secondary structure The three-dimensional arrangement of portions of a polypeptide chain. (2.5)

Secondary walls Thicker cell walls found in most mature plant cells. (7.6)

Secreted Discharged outside the cell. (8.1)

Secretory pathway (biosynthetic pathway) Route through the cytoplasm by which materials are synthesized in the endoplasmic reticulum or Golgi complex, modified during passage through the Golgi complex, and transported within the ctyoplasm to various destinations such as the plasma membrane, a lysosome, or a large vacuole of a plant cell. Many of the materials synthesized in the endoplasmic reticulum or Golgi complex are destined to be discharged outside the cell; hence the term secretory pathway has been used. (8.1)

Section A very thin slice of tissue of a plant or animal. (18.1)

Selectins A family of integral membrane glycoproteins that recognize and bind to specific arrangements of carbohydrate groups projecting from the surface of the other cells. (7.3)

Selectively permeable barrier Any structure, such as a plasma membrane, that allows some substances to freely pass through while denying passage to others. (4.1)

Self-assembly The property of proteins (or other structures) to assume the correct conformation based on the chemical behavior dictated by the amino acid sequence. (EP2)

Semiconservative Replication in which each daughter cell receives one strand of the parent DNA helix. (13.1)

Semipermeable The membrane property of being freely permeable to water while allowing much slower passage to small ions and polar solutes. (4.7)

Serial sections A series of successive sections cut from a block of tissue. (18.1)

SH2 domains Domains with high-affinity binding sites for "phosphotyrosine motifs." Found in a variety of proteins involved in cell signaling. (15.3)

side chain The defining functional group of an amino acid, which can range from a single hydrogen to complex polar or non-polar units in the 20 amino acids most commonly found in cells. (2.5)

Signal peptidase Proteolytic enzyme that removes the N-terminal portion including the signal peptide of a nascent polypeptide synthesized in the RER. (8.3)

Signal recognition particle (SRP) A particle consisting of six distinct polypeptides and a small RNA molecule, called the 7S RNA, that recognizes the signal sequence as it emerges from the ribosome. SRP binds to the signal sequence and then to an ER membrane. (8.3)

Signal sequence Special series of amino acids located at the N-terminal portion of newly forming proteins that triggers the attachment of the protein-forming ribosome to an ER membrane and the movement of the nascent polypeptide into the cisternal space of the ER. (8.3)

Signal transduction The ability of a cell to convert an external stimulus to an appropriate cellular response. (15)

Signal transduction pathways The steps, involving a series of distinct proteins, along which information is passed from the point of stimulation on the cell surface to the interior of the cell. (15.1)

Single nucleotide polymorphism (SNP) Sites in the genome where alternate bases are found with high frequency in the population. SNPs are excellent genetic markers for genome mapping studies. (10.6)

Single particle tracking A technique for studying movement of membrane proteins that consists of two steps: (1) linking the protein molecules to visible substances such as colloidal gold particles and (2) monitoring the movements of the individual tagged particles under the microscope. (4.6)

Single-stranded DNA-binding (or SSB) proteins Proteins that facilitate the separation of the DNA strands by their attachment to bare, single DNA strands, keeping them in an extended state and preventing them from becoming rewound. (13.1)

Site-directed mutagenesis A research technique to modify a gene in a predetermined way so as to produce a protein with a specifically altered amino acid sequence. (18.13)

Small nuclear RNAs (snRNAs) RNAs required for RNA processing that are small (90 to 300 nucleotides long) and that function in the nucleus. (11.3)

Smooth endoplasmic reticulum (SER) That part of the endoplasmic reticulum that is without attached ribosomes. The membranous elements of the SER are typically tubular and form an interconnecting system of pipelines curving through the cytoplasm in which they occur. The SER functions vary from cell to cell and include the synthesis of steroid hormones, detoxification of a wide variety of organic compounds, mobilization of glucose from glucose 6-phosphate, and sequestration of calcium ions. (8.3)

snRNPs Distinct ribonucleoprotein particles contained in spliceosomes, so called because they are composed of snRNAs bound to specific proteins. (11.3)

Sodium-potassium pump (Na$^+$/K$^+$-ATPase) The active-transport protein that uses ATP as the energy source for transporting sodium and potassium ions, with the result that each conformational change transports three sodium ions out of the cell and two potassium ions into the cell. (4.7)

Specific activity The ratio of the amount of a protein of interest to the total amount of protein present in a sample, which is used as a measure of purification. (18.7)

Spectrophotometer The instrument used to measure the amount of light of a specific wavelength that is absorbed by a solution. If one knows the absorbance characteristics of a particular type of molecule, then the amount of light of the appropriate wavelength absorbed by a solution of that molecule provides a sensitive measure of its concentration. (18.10)

Splice sites The 5′ and 3′ edges of each intron. (11.3)

Spliceosome A macromolecular complex containing a variety of proteins and a number of distinct ribonucleoprotein particles that functions in removal of introns from a primary transcript. (11.3)

Split genes Genes with intervening sequences. (11.3)

Spontaneous reactions Reactions that are thermodynamically favorable, occurring without any input of external energy. (3.1)

Sporophyte A diploid stage of the life cycle of plants that begins with the union of two gametes to form a zygote. During the sporophyte stage, meiosis occurs, producing spores that germinate directly into a haploid gametophyte. (14.3)

Standard free-energy difference ($\Delta G°′$) The difference in free energy when one mole of each reactant is converted to one mole of each product under defined standard conditions: temperature of 298°K and pressure of 1 atm. (3.1)

Starch Mixture of two glucose polymers, amylose and amylopectin, that serves as readily available chemical energy in most plant cells. (2.5)

Steady state Metabolic condition in which concentrations of reactants and products are essentially constant, although individual reactions may not be at equilibrium. (3.1)

Stereoisomers Two molecules that structurally are mirror images of each other and may have vastly different biological activity. (2.5)

Steroid Lipid group based on a characteristic four-ring hydrocarbon skeleton, including cholesterol and hormones such as testosterone and progesterone. (2.5)

Stomata Openings at the surface of leaves through which gas and water are exchanged between the plant and the air. (6.6)

Stop codons Three of the 64 possible trinucleotide codons whose function is to terminate polypeptide assembly. (11.5)

Stress fibers Closely packed bundles of actin filaments located just beneath the plasma membrane at tightly adherent focal adhesions between cultured cells and the substratum. (9.7)

Stroma Space outside the thylakoid but within the relatively impermeable inner membrane of the chloroplast envelope. (6.1)

Stroma thylakoids Also called stroma lamellae, these are flattened membranous cisternae that connect some of the thylakoids of one granum with those of another. (6.1)

Structural genes Genes that code for protein molecules. (12.2)

Structural isomers Molecules having the same formula but different structures. (2.4)

Substrate The reactant bound by an enzyme. (3.2)

Substrate-level phosphorylation Direct synthesis of ATP through the transfer of a phosphate group from a substrate to ADP. (3.3)

Subunit A polypeptide chain that associates with one or more separate chains (subunits) to form a complete protein. (2.5)

Supercoiled A molecule of DNA that is twisted upon itself. (10.3)

Surface area/volume ratio The proportion between cellular dimensions that indicates how effectively (or whether) a cell can sustainably exchange substances with its environment. (1.3)

Synapse The specialized junction of a neuron with its target cell. (4.8)

Synapsis The process by which homologues become joined to one another during meiosis. (14.3)

Synaptic vesicles The storage sites for neurotransmitter within the terminal knobs of a neuronal axon. (4.8)

Synaptonemal complex (SC) A ladder-like structure composed of three parallel bars with many cross fibers. The SC holds each pair of homologous chromosomes in the proper position to allow the continuation of genetic recombination between strands of DNA. (14.3)

T-cell receptor (TCR) Proteins present on the surface of T lymphocytes that mediate interaction with specific cell-bound antigens. Like the immunoglobulin of B cells, these proteins are formed by a process of DNA rearrangement that generates a specific antigen-combining site. TCRs consist of two subunits, each containing both a variable and a constant domain. (17.4)

T lymphocytes (T cells) Lymphocytes that respond to antigen by proliferating and differentiating into either CTLs (cytotoxic lymphocytes) that attack and kill infected cells or T_H cells that are required for antibody production by B cells. These cells attain their differentiated state in the thymus. (17.2)

Tandem repeats A cluster in which a sequence repeats itself over and over again without interruption. (10.3)

Telomere An unusual stretch of repeated sequences, which forms a "cap" at each end of a chromosome. (12.1)

Tertiary structure The three dimensional shape of an entire polypeptide. (2.5)

Tetrad (bivalent) The complex formed during meiosis by a pair of synapsed homologous chromosomes that includes four chromatids. (14.3)

Thermodynamics Study of the changes in energy accompanying events in the physical universe. (3.1)

Thermodynamics, first law of Principle of conservation of energy that states energy can neither be created nor destroyed, merely converted from one form to another. (3.1)

Thermodynamics, second law of Principle that all events move from a higher energy state to a lower energy state, and thus are spontaneous. (3.1)

Thick filaments One of two distinct types of filaments that give sarcomeres their characteristic appearance. Thick filaments consist primarily of myosin and are surrounded by a hexagonal array of thin filaments. (9.6)

Thin filaments One of two distinct types of filaments that give sarcomeres their characteristic appearance. Thin filaments consist primarily of actin and are arranged in a hexagonal array around each thick filament, with

each thin filament situated between two thick filaments. (9.6)

Threshold The point during depolarization of an excitable cell where voltage-gated sodium channels open, with the resulting Na$^+$ influx causing a brief reversal in membrane potential. (4.6)

Thylakoids Flattened, membranous sacs formed by the chloroplast's internal membrane, which contain the energy-transducing machinery for photosynthesis. (6.1)

Tight junctions Specialized contacts that occur at the very apical end of the junctional complex that forms between adjacent epithelial cells. The adjoining membranes make contact at intermittent points, where integral proteins of the two adjacent membranes meet. (7.4)

Tonoplast The membrane that bounds the vacuole of a plant cell. (8.7)

Topoisomerases Enzymes found in both prokaryotic and eukaryotic cells that are able to change the supercoiled state of the DNA duplex. They are essential in processes, such as DNA replication and transcription, that require the DNA duplex to unwind. (10.2)

Trans **cisternae** The cisternae of the Golgi complex farthest from the endoplasmic reticulum. (8.4)

Trans **Golgi network (TGN)** A network of interconnected tubular elements at the *trans* end of the Golgi complex that sorts and targets proteins for delivery to their ultimate cellular or extracellular destination. (8.4)

Transcription The formation of an RNA from a DNA template. (11.1)

Transcription factors Auxiliary protein (beyond the 8 to 14 distinct polypeptides that make up the polymerases) that bind to specific sties in the DNA and alter the transcription of nearby genes. (11.3)

Transcription unit The corresponding segment of DNA on which a primary transcript is transcribed. (11.3)

Transcriptional-level control Determination whether a particular gene will be transcribed and, if so, how often. (12.3)

Transduction The conversion of energy from one form to another, as in the conversion of chemical energy into heat energy when fuel is burned. (3.1)

Transfection A process that introduces DNA into cultured cells. (18.13)

Transfer potential A measure of the ability of a molecule to transfer any group to another molecule, with molecules having a higher affinity for the group being the better acceptors and molecules having a lower affinity better donors. (3.3)

Transfer RNAs (tRNAs) A family of small RNAs that translate the information encoded in the nucleotide "alphabet" of an mRNA into the amino acid "alphabet" of a polypeptide. (11.1)

Transformed Normal cells that have been converted to cancer cells by treatment with a carcinogenic chemical, radiation, or an infective tumor virus. (16.1)

Transgenic animals Animals that have been genetically engineered so that their chromosomes contain foreign genes. (18.13)

Transition state The point during a chemical reaction at which bonds are being broken and reformed to yield products. (3.2)

Transition temperature The temperature at which a membrane is converted from a fluid state to a crystalline gel in which lipid-molecule movement is greatly reduced. (4.5)

Translation Synthesis of proteins in the cytoplasm using the information coded on an mRNA. (11.1)

Translational-level control Determination whether a particular mRNA is actually translated and, if so, how often and for how long a period. (12.3)

Translocation A chromosomal aberration that results when all or part of one chromosome becomes attached to another chromosome. (HP12)

Translocation The third step in the translation elongation cycle that involves (1) ejection of the uncharged tRNA from the P site and (2) the movement of the ribosome three nucleotides (one codon) along the mRNA in the 3′ direction. (11.6)

Transmembrane segment The portion of a membrane protein that passes through the lipid bilayer, often composed of non-polar amino acids in an α-helical conformation. (4.4)

Transmembrane signaling Transfer of information across the plasma membrane. (7.3, 15)

Transmission electron microscopes (TEMs) Microscopes that form images from electrons that are transmitted through a specimen. (18.2)

Transport vesicles The shuttles, formed by budding from a membrane compartment, that carry materials between organelles. (8.1)

Transposable elements DNA segments that move from one place on a chromosome to a completely different site, often affecting gene expression. (10.4)

Transposition Movement of DNA segments from one place on a chromosome to an entirely different site, often affecting gene expression. (10.4)

Transposons DNA segments capable of moving from one place in the genome to another. (10.4)

Transverse (T) tubules Membranous folds along which the impulse generated in a skeletal muscle cell is propagated into the interior of the cell. (9.6)

Triacylglycerols Polymers consisting of a glycerol backbone linked by ester bonds to three fatty acids, commonly called fats. (2.5)

Tricarboxylic acid cycle (TCA cycle) The circular metabolic pathway that oxidizes acetyl CoA, conserving its energy; the cycle is also known as the Krebs cycle or the citric acid cycle. (5.2)

Tubulin The protein that forms the walls of microtubules. (9.3)

Tumor-suppressor genes Genes that encode proteins that restrain cell growth and prevent cells from becoming malignant. (16.3)

Turnover The destruction of cellular materials and their replacement. (12.4)

Ubiquitin A small, highly conserved protein that is linked to proteins targeted for degradation in proteasomes. (12.4)

Unsaturated fatty acids Those having one or more double bonds between carbon atoms. (2.5)

Untranslated regions (UTRs) Noncoding segments contained at both 5′ and 3′ ends of mRNAs. (12.3)

Van der Waals force A weak attractive force due to transient asymmetries of charge within adjacent atoms or molecules. (2.2)

Variable regions The portions of light and heavy antibody polypeptide chains that differ in amino acid sequence from one specific antibody to another. (17.4)

V(D)J joining The DNA rearrangements that occur during the development of B cells that limit the cells to the production of a specific antibody species. (17.4)

Vector DNA A vehicle for carrying foreign DNA into a suitable host cell, such as the bacterium *E. coli*. The

vector contains sequences that allow it to be replicated inside the host cell. Most often, the vector is a plasmid or the bacterial virus lambda (λ). Once the DNA is inside the bacterium, it is replicated and partitioned to the daughter cells. (18.13)

Virion The form a virus assumes outside of a cell, which consists of a core of genetic material surrounded by a protein capsule. (1.4)

Viroids Small, obligatory intracellular pathogens, that, unlike viruses, consist only of an uncoated circle of genetic material, RNA. (1.4)

Viruses Small, obligatory intracellular pathogens that are not considered to be alive because they cannot divide directly, which is required by the cell theory of life. (1.4)

Whole mount A specimen to be observed with a light microscope that is an intact object, either living or dead, and can be an entire intact organism or a small part of a large organism. (18.1)

X-ray diffraction (or X-ray crystallography) A technique that bombards protein crystals with a thin beam of X-rays of a single (monochromatic) wavelength. The radiation that is scattered (diffracted) by the electrons of the protein atoms strikes a photographic plate placed behind the crystal. The diffraction pattern produced by the crystal is determined by the structure within the protein. (18.8)

Yeast artificial chromosomes (YACs) Cloning elements that are artificial versions of a normal yeast chromosome. They contain all of the elements of a yeast chromosome that are necessary for the structure to be replicated during S phase and segregated to daughter cells during mitosis, plus a gene whose encoded product allows those cells containing the YAC to be selected from those that lack the element and the DNA fragment to be cloned. (18.13)

Answers to Odd-Numbered Analytic Questions

CHAPTER 1

1. The specific answer would depend on the question asked. Advantages of working with a cell culture include the ability to study a single cell type, ease of obtaining large numbers of cells, ability to minimize number of unknown variables by using carefully controlled in vitro conditions. The advantage of using a whole organism is that information obtained is more meaningful in understanding the role of the process in the overall activity of the organism. For example, one might want to study glucose transport across the plasma membrane of cultured liver cells in response to insulin, but the results would not tell you much about the role of this process in maintaining proper blood glucose levels.

3. Cancer cells grow in a much less controlled manner than normal cells, which is why they continue to proliferate in the body. Cancer cells tend to have fewer requirements for growth and are therefore more readily cultured in diverse media than normal cells. Culturing such cells was a logical first step in the development of culture conditions.

5. These receptors mark specific cells as targets for particular hormones. Without such receptors, cells would not be able to bind a specific hormone and thus would not be able to respond to it. If all cells had the same receptors, hormones would not be able to selectively activate specific target cells. All cells would become potential targets.

7. Activities in cells are not directed by external agents. Activities that occur in a cell must include a mechanism for self-direction and self-control.

9. Such as, the capture and uptake of other organisms, the sensing of conditions in the external environment and making the appropriate responses, the ability for locomotion. Most activities of this cell are shared by a muscle or nerve cell.

11. You could examine the filtrate under the electron microscope and determine whether the infectious agent was cellular, i.e., bacterial, or noncellular, i.e., viral. You could try to culture the infective agent. If it was a bacterium, you should be able to culture the agent in the absence of host cells, but not so if it were a virus. You could determine the size of the genome (i.e., the RNA or DNA that constituted its genetic material). If it was cellular, it would be expected to have a much larger genome than if it was viral.

CHAPTER 2

1. Since leucine is similar in structure to valine, one would expect the protein to have similar abnormal properties. Since aspartic acid is similar to glutamic acid, which is the proper amino acid, relatively little effect might be expected.

3. Four (one straight chain and three branched). Three (two straight chain and one branched), ignoring cis-trans isomerism.

5. The fatty acids would tend to contain greater numbers of double bonds (more unsaturated), which would lower their melting point, thus keeping them liquid at the lower temperature.

7. 2 and 4. A rise in pH, would increase the loss of a proton from a $-COOH$ group and from a $-NH_3^+$ group.

9. It would be expected to slow the rates of the reaction because the products of the first two reactions would diffuse into the surrounding medium rather than be passed directly into the active sites of the second and third enzymes.

11. 8,000 (20^3). 4 carboxyl terminals, one per chain.

13. Glutamic acid has a pK of 4.3. Therefore, above pH 4.3, the side chain is predominantly negatively charged and below pH 4.3 uncharged. Arginine has a pK of 12.5. Therefore, below pH 12.5, the side chain is predominantly positively charged and above pH 12.5 uncharged. For an ionic bond to form, both side chains must be charged. At pH 4, arginine is charged, but glutamic acid is uncharged; at pH 13, glutamic acid is charged, but arginine is uncharged. However, at pH 7 and 12, both groups will be charged and an ionic bond can form.

15. There are two possibilities. 1)Even though the normal polypeptide normally folds into PrP^C, it is possible that on rare occasions it might fold abnormally to form PrP^{Sc}. If this were to occur, the formation of one or a few of the abnormal molecules could lead to a chain reaction that converted normal PrP molecules to the PrP^{Sc} form. 2) A mutation in a gene could arise in a somatic cell that led to the formation of the abnormal PrP^{Sc} in that cell. If that protein was released into the extracellular environment, it could be picked up by other cells whose PrP^C could then be converted to the abnormal form, thus causing the disease.

17. Convergent evolution. These examples indicate that a similar grouping of reactive side chains within the active site of a protein can evolve from ancestral proteins that are not related.

CHAPTER 3

1. In the case of chymotrypsin, a drop in pH might suppress the loss of a proton from the serine, whereas the hydroxyl ions resulting from a rise in pH might interfere with histidine's ability to influence the charge on the serine.

3. #a, b, e are true.

5. 0.5

7. That the ratio of ATP/ADP in the cell is much greater than that at standard conditions, which would be 1.0.

9. -4.2 kcal/mol; $+4.2$ kcal/mol; $10^{2.85}$.

11. Yes, because the ΔG for ATP hydrolysis in the cell is approximately -12 kcal/mol

13. $X + ATP \rightarrow XP + ADP; XP + Y \rightarrow Z + P$

15. -12.9 kcal/mol; the ratio of ADP/ATP would be approximately 10^6; -7.3 kcal/mol

17. $$\frac{[g6P]}{[g][Pi]} = \frac{83 \times 10^{-6}}{(5 \times 10^{-3})(1 \times 10^{-3})} = \frac{83 \times 10^{-6}}{5 \times 10^{-6}} = 16.6$$

 $Keq = 6.5 \times 10^{-3}$ (from problem 16)

 $16.6 >> Keq$, thus reaction will *not* go spontaneously from left to right. To determine the required glucose concentration.

 At equilibrium

 $$Keq = \frac{[g6P]}{[g][Pi]} - \frac{83 \times 10^{-6}}{[g](1 \times 10^{-3})} = \frac{83 \times 10^{-3}}{[g]} = 6.5 \times 10^{-3}$$

 $= 6.5 \times 10^{-3}$

 Thus glucose concentration = 12.8 M at equilibrium

 For the reaction to proceed from left to right

 [g] must be > 12.8M

19. $glu + NH_3 \rightleftharpoons gln + H_2O$ $\quad \Delta G° = 3.4$ kcal/mol

 $ATP + H_2O \rightleftharpoons ADP + Pi$ $\quad \Delta G° = -7.3$ kcal/mol

 net $= glu + ATP + NH_3$ $\quad \Delta G° = -3.9$ kcal/mol
 $\rightleftharpoons gln + ADP + Pi$

 $$\Delta G = \Delta G° + RT\ln \frac{[gln][ADP][Pi]}{[glu][ATP][NH_3]}$$

 $$= -3.9 \text{ kcal/mol} + 1.4 \text{ kcal/mol} \cdot \log \frac{[gln][ADP][Pi]}{[glu][ATP][NH_3]}$$

 $= -3.9$ kcal/mol $+ 1.4$ kcal/mol $\cdot \log$

 $$\left(\frac{10^{-2}M \cdot 10^{-2}M \cdot 10^{-2}M}{10^{-2}M \cdot 10^{-2}M \cdot [NH_3]} \right)$$

 $= -3.9$ kcal/mol $= 1.4$ kcal/mol $\cdot \log \frac{10^{-2}}{[NH_3]}$

 at equil, $\Delta G = 0$,

 $$3.9 \text{ kcal/mol} = 1.4 \text{ kcal/mol} \cdot \log \frac{10^{-2}M}{[NH_3]}$$

 $$2.9 = \log \frac{10^{-2}}{[NH_3]}$$

 $$[NH_3] = 12 \ \mu m$$

 so when $[NH_3] > 12 \ \mu M$, the reaction will go from left to right

CHAPTER 4

1. For example, integral proteins involved in cell-cell adhesion and cell-cell communication. Epithelial cells form a tightly cohering sheet that requires such proteins, whereas erythrocytes occur in the body as single cells. Epithelia would also be expected to have different types of transport proteins than erythrocytes whose primary transport activity is the exchange of bicarbonate and chloride ions.

3. It was presumed that the outer electron dense layers of the trilaminar image corresponded to proteins that had bound the heavy metal stains. Since they were seen as continuous layers, it was presumed that they reflected the presence of continuous layers of proteins situated outside the lipid bilayer.

5. Polysaccharides, such as starch and glycogen, are monotonous polymers of a single subunit, whereas the oligosaccharides at the cell surface contain a variety of different sugar subunits whose arrangement within the complex is highly defined. This is illustrated by the fact that an antiserum against the type A antigen will precipitate blood cells from a person with type A blood, but will not precipitate those from a person with type B or type O blood, and vice versa.

7. Some other type. The S4 segment has both nonpolar and polar residues, but the polar residues spiral around the helix rather than lining one side as would occur in a amphipathic helix.

9. Those with the smallest or most hydrophobic head groups. A flippase is actively moving PC from one leaflet to the other. The integral protein should show no detectable flip-flop since movement of the hydrophilic segments of the protein across the membrane would be extremely unfavorable.

11. Both ions would be expected to move into the medium when the axon is at rest; the K^+ would be diffusing through leak channels, while the Na^+ would be pumped out by active transport. The firing of an action potential would be accompanied primarily by the movement of labeled K^+ out of the membrane during the repolarization phase.

13. Lipids are less likely to form large complexes or to be linked to cytoskeletal or extracellular materials of the type that inhibit the movement of integral proteins.

15. The internal fluids of marine invertebrates have a much higher osmotic pressure (solute concentration) than those of marine vertebrates. It is thought that marine fishes evolved from species that arose in fresh water and had salt concentrations considerably less than that of sea water.

17. Action potentials might be able to move in both directions from sites along the axon.

19. Would be based on ΔG of 3.1 rather than 6.2. Substituting in the equation on p. 165, the value for X becomes -2.21 and $10^{-2.21} = 1/162$. The transporter could only work against a gradient about 160 fold, rather than 23,000 fold if two ions were transported per glucose.

21. The equation on p. 166 can be used to determine the membrane potential (V_x) that would be generated by each of the five ions whose concentration is given in the table. The equation can be rewritten

 $$V_x = \frac{59mV}{z} \log \frac{[x_o]}{[x_i]}$$

 $V_{K^+} = -85$ mV
 $V_{Na^+} = +61$ mV
 $V_{Cl^-} = -44$ mV
 $V_{Ca^{2+}} = +139$ mV
 $V_{H^+} = -23.6$ mV

 None of these ions are at equilibrium when the membrane potential (V_m) is -70 mV. The distance each is away from

equilibrium is equal to $V_x - V_m$. Under these conditions, K^+ and Cl^- would move out of the axon through an open channel for that ion, whereas Na^+, Ca^{2+}, and H^+ would move into the axon through an open channel.

23. An alpha helix is able to from internal hydrogen bonds, whereas a beta strand cannot; it must form H bonds with the residues on an adjacent backbone.

25. A channel is much more rapid, because it doesn't depend on a catalytic reaction. Each open channel can conduct millions of ions per second at near diffusion limiting rate, whereas each Na^+/K^+–ATPase, for example, can only move 2 or 3 ions per catalytic cycle (a fixed stoichiometry).

CHAPTER 5

1. 2 and 3 are correct. $2B + 2H^+ + 2e^- \rightarrow 2BH$

3. a) flavoprotein + $2H^+$ + $2e^- \rightarrow$ flavoprotein: H_2
 $2cytochrome_{ox} + 2e^- \rightarrow 2cytochrome_{red}$

 b) $AH_2 + 2B_{ox} \rightleftharpoons 2B_{red} + A + 2H^+$

 c) Reduced B and oxidized A would be present at highest concentrations.

5. It could abolish the voltage component of the proton-motive force.

7. Yes, because they occur at the expense of either the potential difference across the membrane (in the case of the ATP/ADP exchange) or of the proton gradient (in the case of the uptake of P_i).

9. One by substrate-level phosphorylation. Eleven by oxidative phosphorylation (assuming three from each NADH and two from each $FADH_2$). Two CO_2 molecules released. One molecule of FAD reduced. Five pairs of electrons stripped.

11. ATP/ADP levels in the cell are much higher than that at standard conditions, which requires a greater proton-motive force and thus the expenditure of more energy for ATP production than would be true at standard conditions.

13. If a potential difference of 59 mV is equivalent to a ΔG of 1.37 kcal/mol, then a proton-motive force of 220 mV is equivalent to 5.1 kcal/mol. The passage of 3 mol of protons across the membrane at this ΔG is sufficient to release 15.3 kcal/mol. It was estimated in Chapter 3 that the ΔG for ATP synthesis in the cell was about 11.5 kcal/mol (p. 90), which suggests that three protons release enough energy to generate a molecule of ATP.

15. 23 kcal/mol

17. The proton-motive force (Δp) = 59 mV \bullet ΔpH $- V_m$. If the ΔpH = 1 and the V_m = +59 mV, then Δp = 59 mV \bullet 1 $- 59$ mV = 0. There is no proton-motive force and thus no ATP synthesis.

19. The stator consists of the alpha, beta, and delta subunits of F_1, and the a and b subunits of F_0. The other subunits are part of the rotor. The b subunit of F_0 connects with the delta subunit of F_1, which holds the alpha and beta subunits in a fixed position.

CHAPTER 6

1. PSI operates at the most negative potential. PSI generates the strongest reducing agent. Both photosystems must absorb four photons during each round.

3. Reduced ferredoxin, Tyr_z^+, Tyr_z^+, reduced ferredoxin.

5. 236 mV

7. Isolated chloroplasts raise the pH of the medium, whereas isolated mitochondria lower it.

9. PSII is directly responsible for the translocation of protons and electrons, whereas PSI is responsible only for the movement of electrons, which are transferred to NAD^+ and do not contribute to the electrochemical gradient.

11. P680 and P700; P700 and NADPH

13. Tyr_z is the primary donor of PSII, and plastocyanin is the primary donor of PSI.

15. It would have a greater effect on C_3 plants, which are more sensitive to CO_2 levels because their only CO_2 fixing enzyme has relatively low affinity for CO_2.

17. Greater CO_2 levels would be expected to lead to an increased ratio of photosynthesis over photorespiration and, consequently, to increased plant life, which would increase O_2 levels. In fact, it is estimated that at current CO_2 levels of 380 ppm, O_2 levels might rise over a long period of time to up to 28% of atmospheric gases.

19. The products of photorespiration, e.g., glycolate, glycine, serine.

CHAPTER 7

1. Selectin-mediated cell adhesion would be inhibited by trypsin, neuraminidase, and EGTA. L1-mediated cell adhesion would be inhibited by trypsin and, possibly by RGD peptides if the adhesion was based on L1-integrin interactions, the integrin had an RGD-binding site, and this site was close enough to the L1-binding site to interfere.

3. Migration of neural crest cells. Any process that required tight linkage of cells to their extracellular matrix might be disrupted because fibronectin is an important linker between the cell surface and other materials of the ECM.

5. Both contain fibrous materials to resist pulling forces and an amorphous matrix to resist compression.

7. Cadherins. They are widely distributed among tissues, and they act by binding to the same cadherin on the surface of other cells, thus mediating adhesion of similar types of cells. Aggregation and sorting out could be carried out in the presence of an anti-cadherin antibody, which should block cadherin-cadherin interactions.

9. Nearly all animals possess some type of skeleton that maintains support, thereby eliminating the structural requirements for a cell wall. In addition, the cells of animals are bathed in a fluid that is maintained isotonic to the cells, thereby eliminating the need for a cell wall to resist the build up of osmotic pressure.

11. In a tight junction, the encircling nature of the cell-contact forms a seal, like a gasket, that prevents passage of solutes between the cells. If the junction were not continuous, it would leave openings between the cells. Desomosomes and gap junctions can mediate adhesion and communication without having to form such a seal.

13. Erythrocytes remain as free circulating cells; they are characterized by a lack of interaction with either other cells or an extracellular matrix.

CHAPTER 8

1. hemoglobin (the globin polypeptides that make up the protein) and ribosomal proteins.

3. Endoplasmic reticulum; *trans* golgi cisternae; TGN

5. The RER would have the SRP receptor and protein-translocating channel proteins. The SER would have a calcium pump (Ca^{2+}-ATPase), glucose 6-phosphatase, and cytochrome $P450$s.

7. They are mutant cells that lack the glycosyltransferases that add sugars onto the mannose residues.

9. [^3H]leucine in the cytosol and the RER; [^3H]sialic acid in the *trans* end of the Golgi complex; [^3H]mannose in the rough ER; [^3H]choline in the rough ER; [^3H]glucuronic acid in the Golgi complex; [^3H]pregnenolone in the SER; [^3H]rhamnose in the Golgi complex.

11. To the extracellular side. The cisternal (luminal) side of the Golgi becomes the inner (luminal) side of secretory vesicles and, following exocytosis, becomes the external side of the plasma membrane.

13. Throughout the entire biosynthetic pathway and the extracellular medium.

15. Binding to a ligand would change the conformation of the cytosolic domain of the receptor causing it to bind to adaptors on the inner surface of the plasma membrane.

17. If the bead remained in the cisterna and it traveled toward the trans end of the stack, it would provide evidence for the maturation or combined model. If it didn't move, yet other types of cargo continued to pass through the cell and were secreted, it would provide evidence for the vesicular transport model.

CHAPTER 9

1. It would inhibit contractility because the thin filaments would be pulled beyond the point where myosin heads could engage them. It would increase the H and I bands without affecting the A band.

3. The acrosomal reaction and bacterial movement in a phagocyte. Both depend on actin polymerization as a force-generating mechanism.

5. They would not be processive. Instead, they would fall off the microtubule after each cycle.

7. No, since you are providing the axoneme with substances whose concentration may be controlled by the plasma membrane. In fact, the plasma membrane controls cAMP and Ca^{2+} concentrations, both of which influence the assembly-disassembly equilibrium.

9. No. The centrosome is located at the opposite end from that where these events are occurring.

11. The doublets on one side of the axoneme or the other would be missing, depending on the direction the cilium is bent.

13. Changes in fluorescence can be followed during microscopic observation of a living cell, whereas radioactive localization requires that the cell be killed. Radioactivity is better suited for quantitative measurements, such as rates of tubulin synthesis or of tubulin incorporation into microtubules in a brief period of time.

15. Brain; muscle; and skin; keratin; MAPs, tropomyosin, troponin, or other actin-binding proteins.

17. Free radicals may act by damaging neurofilaments, which then can cause the disease.

19. Microtubules are oriented with their plus ends toward the nerve terminal so that movement of these structures is accomplished by a motor that would move toward the minus end of the transported microtubule. If the motor is stationary, the microtubule attached to the head of the dynein would move toward the end of the axon.

CHAPTER 10

1. Those two traits would not have assorted independently. Dihybrid crosses would not have produced $9:3:3:1$ ratios.

3. X and Z are at opposite ends with Y between them approximately $\frac{2}{3}$ of the way toward Z.

5. The DNA that codes for rRNA would reanneal in the second part of the curve with the moderately repeated fraction. The pure ribosomal DNA would reanneal rapidly, in the region where the satellite DNA fraction of the total DNA normally reanneals.

7. (2) Incubate the fragments for a sufficient period of time to allow fragments containing repeated sequences to reanneal, then separate the reannealed and nonreannealed DNA fragments as can be done by passing the mixture through a column of hydroxylapatite.

9. Labeled myoglobin mRNA would produce four spots, one on each half of each replicated chromosome of a homologous pair. The pattern with labeled histone mRNA would depend on whether the genes encoding histones were clustered or not. The number of sites would reflect the distribution among the cell's chromosomes. If they were all clustered, hybridization would produce four sites that were much more heavily labeled than those seen with myoglobin mRNA.

11. False. Any percentage from 0 to 70 is possible. However, 30 percent of the bases on the complementary strand will consist of Ts.

13. No. The genes from rodents and mammals both have two introns, which indicates it was present in the gene prior to the evolution of primates. The second intron must have been lost in the primate species.

CHAPTER 11

1. UGG for tryptophan, AUA for isoleucine, and AUG for methionine. Because the third letter of the codon can be replaced by a different letter (A or G, U or C) without changing the amino acid specified. As a result of the wobble at the third position, one tRNA can bind to at least two different codons that specify the same amino acid.

3. The curve for absorbance would show peaks at 18 and 28S, whereas the curve for radioactivity would show a single peak at 45S.

5. A polypeptide with alternating amino acids arginine and glutamic acid. Only one type of polypeptide would be produced, because it wouldn't matter where the ribosome first attached to begin reading. There would likely be differences as to the specific amino acid (Arg or Glu) at the two ends of the polypeptides.

7. It would be similar to that of Figure 11.21 except it would contain two intervening sequences within the coding region.

9. The reannealing curve for the oocyte DNA would have a much greater percentage of its DNA reanneal in the highly repeated fraction due to the amplification of the rDNA.

11. They would comprise a population of highly variable length. 16 of them (GAA, CAA, AAA, GAG, CAG,

AAG, GGA, CGA, AGA, UCA, UUA, UAC, UAU, UGC, UGU, UGG).

13. 2 and 3 are correct.

15. 5'-GCU-3'. This same tRNA could be used to incorporate serine when the codon was AGU instead of AGC.

17. Yes, because frameshift mutations are very likely to generate premature termination codons in their new reading frame.

CHAPTER 12

1. Possible collapse of the nucleus, disruptions in the transport of mRNAs to the cytoplasm, disruptions in both transcription and replication, and others.

3. Cells possess more than one NLS receptor that recognize different localization sequences, and that the mutation was occurring in a gene that encoded one of these NLS receptor species.

5. It would have serious effects, because any recessive alleles carried on the maternal X chromosome would be expressed. If this were to happen, X-linked inheritance in men and women would be similar, and diseases such as hemophilia would be as high in women as in men.

7. This would allow the cell to concentrate the factors required for transcription and RNA processing, which would greatly increase the rate of interaction with their substrates and thus increase the efficiency of the processes.

9. A mutation had occurred that changed the stop codon normally present at the end of the β-galactosidase gene.

11. No. It is just as likely that the transplanted nucleus could not adapt to the egg cytoplasm, which has a very different composition from that of the cytoplasm from which it had been removed. Negative results cannot provide definitive answers to an experimental question because one cannot be sure of the reason(s) for the results.

13. Enhancers are far enough away from the transcription initiation site to allow the DNA between them to form a loop, which allows the proteins bound at the enhancer to interact with the basal transcription machinery regardless of their orientation at the enhancer.

15. Many nuclear proteins move back and forth between the nucleus and cytoplasm which requires that they retain these localization signals. Even those nuclear proteins that remain in the nucleus during interphase must be reimported into the nucleus after mitosis following reformation of the nuclear envelope.

17. You could see which mRNAs are synthesized in a particular cell type under normal conditions and then determine which mRNAs are produced in a cell in which the transcription factor has been mutated, as in a knockout mouse lacking the factor entirely.

CHAPTER 13

3. If replication were conservative, one of the two chromatids would be labeled after one round and one of the four chromatids would be labeled after two rounds. If replication were dispersive, all chromatids would be labeled after both one or two rounds.

5. The molecule would have a free 5' phosphate, but would lack a free 3' OH group.

7. Because of their smaller genome, the mutation rate in bacteria is less than one per 100 replication cycles. In contrast,

human cells contain about 6×10^9 nucleotides, which would mean an average of 6 alterations per cell division. Since most of the human genome does not code for meaningful products, the impact is not as great as this, but one would still expect to see a much greater mutation rate in the human cells (assuming a comparable rate of DNA repair).

9. Origins of replication contain A-T-rich regions which are more readily unwound than other regions of the chromosome.

11. Human cells are typically found at a much higher temperature and would be expected to experience greater damage, as a result. In addition, humans have a much greater lifespan and, consequently, the DNA in their cells would be expected to accumulate less damage in a given period of time to allow them to continue to function normally over an extended period.

13. The model would likely propose that many of the key parts of the DNA repair process, such as the helicases, are also part of the transcription machinery, or at least accompany it.

CHAPTER 14

1. Both major types of reproduction—asexual and sexual—require cell division as an obligatory step. Every DNA strand of an offspring organism can be traced through either mitosis or meiosis to the DNA strands of its parents.

3. One-third of the cells, because one-third of the cells were in S phase during the 15-minute pulse. Nearly 6 hours for the cells that were at the end of S to pass through G_2. None of the cells, because 18 hours is one generation time; the cells in mitosis after an 18-hour chase would not have been in S phase during the 15-minute pulse.

5. A G_2-phase cell is not activated to initiate DNA synthesis because it has already duplicated its DNA and will not do so again. Each replicon can only undergo one round of replication during a single cell cycle.

7. Removal of activating phosphate, addition of inhibitory phosphates, degradation of cyclin, and binding by CKI

9. Inject fluorescently labeled tublin into the cell. Because the polar microtubules are presumed to be responsible for spindle elongation during anaphase B, you would expect to see tubulin subunits being added at their plus ends, which would be located away from the poles, such as at the cell equator.

11. 46; 46; 23; 23; 23 number of chromosomes. 92; 92; 46; 46; 0 number of chromatids. 4C; 4C; 2C; 2C; 1C amount of DNA.

13. Four.

15. That it required the synthesis of DNA.

17. You did receive half of your chromosomes from each parent but because of independent assortment, you could theoretically have received anywhere from zero to 23 chromosomes from a particular grandparent. Crossing over doesn't really change the answer, but means that the chromosomes from your grandparents are no longer strictly maternal or paternal, but have portions from each grandparent.

19. The Cdk would be 1) constitutively active, 2) an inactive mutant, 3) constitutively active.

21. 1) would fail to arrest progress of mitosis in the presence of an unattached chromosome, 2) would make the APC unable to ubiquinate the anaphase inhibitor so the cell

would not be able to exit metaphase, 3) the alternate substrate-targeting subunit (Cdh1) would not be able to bind to the APC so the cell would not be able to exit mitosis and enter G_1.

CHAPTER 15

3. Inactivation of cAMP phosphodiesterase would be expected to cause prolonged responses to hormonal stimulation and excessive glucose mobilization. An inactive glucagon receptor would prevent the cell from responding to this hormone, which would be expected to lead to a problem maintaining adequate blood glucose levels. Cells with an inactive phosphorylase kinase would be unable to mobilize glucose from glycogen because the phosphorylase molecules could not be activated. A G protein unable to hydrolyze GTP would be expected to continue to send stimulatory signals downstream, leading to a constitutively active adenylyl cyclase.

5. Amplification occurs when activated MAPKKK (which is Raf) phosphorylates MAPKK molecules; when activated MAPKK phosphorylates MAPK molecules; when activated MAPK phosphorylates transcription factors; and when activation of a single DNA sequence leads to the production of large numbers of gene products.

7. The cardiac muscle cells responded to FSH by contracting, which they would not normally do. That the second messenger (in this case cAMP) was diffusing through the gap junctions between the cells.

9. For example, incubate the cells with radioactively labeled choline to label the PC of the membrane, then chase the cells for a period to remove labeled choline from the cell, then stimulate the cells, homogenize the cells, discard the membrane fraction, and determine whether there is an increase in soluble label over that found in unstimulated controls. Subsequent experiments could analyze the nature of the labeled compound(s) that appear after stimulation to determine if the label is in choline phosphate.

11. The wave is the result of calcium-induced calcium release in which free calcium released at one site binds to calcium channels at nearby sites, causing them to open and release additional calcium. As a result, a wave of calcium release sweeps across the cell.

13. A deficiency in insulin production, a deficiency or alteration in the insulin receptor, a deficiency or alteration in an IRS or a component downstream in the pathway.

15. Dominantly

17. Protein kinase A may be mutated so that it no longer recognizes Raf, which it normally phosphorylates; or Raf may be altered so that it is no longer a substrate for protein kinase A.

19. Those resulting in loss of function would be recessive because both copies of the gene would have to be altered before the cell would lack the activity. Gain of function mutations are inherited as dominant mutations because only one gene need be altered to produce the mutant phenotype.

21. Use of the intermediate provides a major, early amplification step because each activated receptor can phosphorylate numerous IRS substrates. In addition, the IRSs are soluble proteins that can diffuse to other parts of the cell to activated downstream effectors.

23. They have an enlarged brain due to lack of death of excess neurons normally generated in the embryonic nervous system. Cytochrome c knockouts would be expected to die at a very early stage of development due to lack of ability to carry out electron transport.

Index